Mathematica® Navigator

D0219091

Mathematical Navigator

Mathematica® Navigator

Mathematics, Statistics, and Graphics

THIRD EDITION

Heikki Ruskeepää

Department of Mathematics

University of Turku, Finland

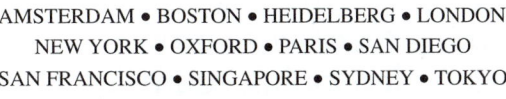

AMSTERDAM • BOSTON • HEIDELBERG • LONDON
NEW YORK • OXFORD • PARIS • SAN DIEGO
SAN FRANCISCO • SINGAPORE • SYDNEY • TOKYO

Academic Press is an imprint of Elsevier

ELSEVIER

The book is produced from PDF files prepared by the author with *Mathematica*®.

Academic Press is an imprint of Elsevier
30 Corporate Drive, Suite 400, Burlington, MA 01803, USA
525 B Street, Suite 1900, San Diego, California 92101-4495, USA
84 Theobald's Road, London WC1X 8RR, UK

Copyright © 2009, Elsevier Inc. All rights reserved.

No part of this publication may be reproduced or transmitted in any form or by any
means, electronic or mechanical, including photocopy, recording, or any information
storage and retrieval system, without permission in writing from the publisher.

Permissions may be sought directly from Elsevier's Science & Technology Rights
Department in Oxford, UK: phone: (+44) 1865 843830, fax: (+44) 1865 853333,
E-mail: permissions@elsevier.com. You may also complete your request online
via the Elsevier homepage (http://elsevier.com), by selecting "Support & Contact"
then "Copyright and Permission" and then "Obtaining Permissions."

Library of Congress Cataloging-in-Publication Data
Ruskeepää, Heikki.
  Mathematic navigator : mathematics, statistics, and graphics / Heikki Ruskeepää. – 3rd ed.
    p. cm.
  Includes bibliographical references and index.
  ISBN 978-0-12-374164-6 (pbk. : alk. paper) 1. Mathematics–Data processing. 2. Mathematica
(Computer file) I. Title.
  QA76.95.R87 2009
  510.285'5–dc22
                              2008044637

British Library Cataloguing-in-Publication Data
A catalogue record for this book is available from the British Library.

ISBN: 978-0-12-374164-6

For information on all Academic Press publications
visit our Web site at www.elsevierdirect.com

Printed in the United States of America
09 10 11 9 8 7 6 5 4 3 2 1

Working together to grow
libraries in developing countries

www.elsevier.com | www.bookaid.org | www.sabre.org

ELSEVIER   BOOK AID   Sabre Foundation
            International

To Marjatta

Contents

Preface

What is the difference between an applied mathematician and a pure mathematician?
An applied mathematician has a solution for every problem,
while a pure mathematician has a problem for every solution.

Welcome

The goals of this book, the third edition of *Mathematica Navigator: Mathematics, Statistics, Graphics, and Programming,* are as follows:

- to introduce the reader to *Mathematica;* and
- to emphasize mathematics (especially methods of applied mathematics), statistics, graphics, programming, and writing mathematical documents.

Accordingly, we navigate the reader through *Mathematica* and give an overall introduction. Often we slow down somewhat when an important or interesting topic of mathematics or statistics is encountered to investigate it in more detail. We then often use both graphics and symbolic and numerical methods.

Here and there we write small programs to make the use of some procedures easier. One chapter is devoted to *Mathematica* as an advanced environment of writing mathematical documents.

The online version of the book, which can be installed from the enclosed CD-ROM, makes the material easily available when working with *Mathematica.*

Changes in this third edition are numerous and are explained later in the Preface. The current edition is based on *Mathematica* 6. On the CD-ROM, there is material that describes the new properties of *Mathematica* 7.

■ Readership

The book may be useful in the following situations:

- for courses teaching *Mathematica;*
- for several mathematical and statistical courses (given in, for example, mathematics, engineering, physics, and statistics); and
- for self-study.

Indeed, the book may serve as a tutorial and as a reference or handbook of *Mathematica*, and it may also be useful as a companion in many mathematical and statistical courses, including the following:

differential and integral calculus • linear algebra • optimization • differential, partial differential, and difference equations • engineering mathematics • mathematical methods of physics • mathematical modeling • numerical methods • probability • stochastic processes • statistics • regression analysis • Bayesian statistics

■ **Previous Knowledge**

No previous knowledge of *Mathematica* is assumed. On the other hand, we assume some knowledge of various topics in pure and applied mathematics. We study, for example, partial differential equations and statistics without giving detailed introductions to these topics. If you are not acquainted with a topic, you can simply skip the chapter or section of the book considering that topic.

Also, to understand the numerical algorithms, it is useful if the reader has some knowledge about the simplest numerical methods. Often we introduce briefly the basic ideas of a method (or they may become clear from the examples or other material presented), but usually we do not derive the methods. If a topic is unfamiliar to you, consult a textbook about numerical analysis, such as Skeel and Keiper (2001).

■ **Recommendations**

If you are a newcomer to *Mathematica*, then Chapter 1, Starting, is mandatory, and Chapter 2, Sightseeing, is strongly recommended. You can also browse Chapter 3, Notebooks, and perhaps also Chapter 4, Files, so that you know where to go when you encounter the topics of these chapters. After that you can proceed more freely. However, read Section 13.1, "Basic Techniques," because it contains some very common concepts used constantly for expressions.

If you have some previous knowledge of *Mathematica*, you can probably go directly to the chapter or section you are interested in, with the risk, however, of having to go back to study some background material. Again, be sure to read Section 13.1.

Contents

The 30 chapters of the book can be divided into nine main parts:

Dependencies between the chapters are generally quite low. If you read Chapter 2, Sightseeing, you will get a background that may serve you well when reading most other chapters; in some chapters, you will also find references to previous chapters, where you will find the needed background.

The following bar chart shows the numbers of pages of the 30 chapters:

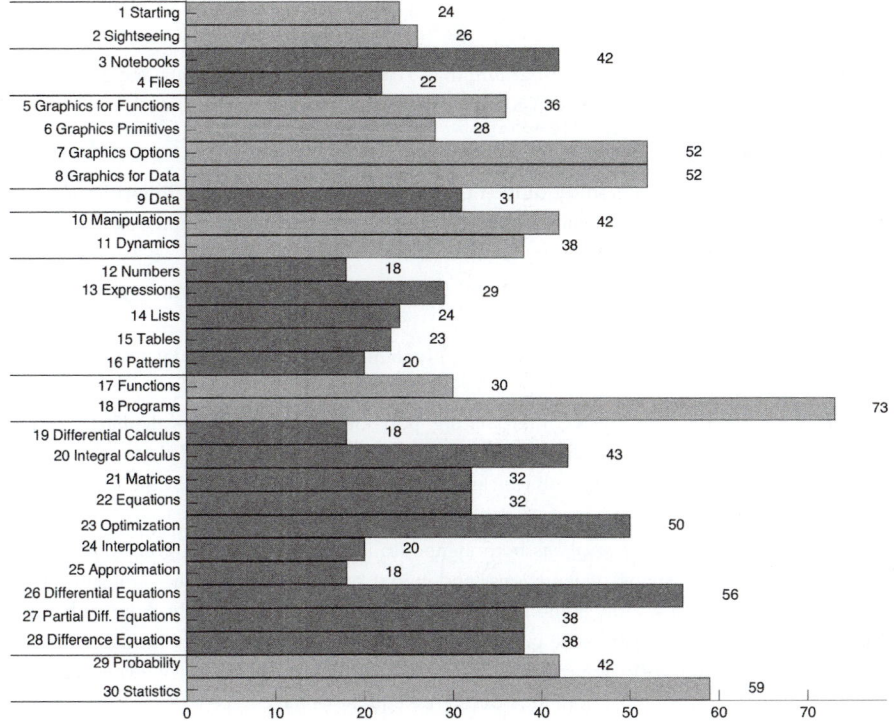

The six longest chapters are 7, Graphics Options; 8, Graphics for Data; 18, Programs; 23, Optimization; 26, Differential Equations; and 30, Statistics.

Next we describe the main parts of the book.

■ Introduction, Files, Graphics, Data, Dynamics, Expressions, and Programs

The first two chapters introduce *Mathematica* and give a short overview.

The next two chapters consider files, particularly files created by *Mathematica*, which are called *notebooks*. We show how *Mathematica* can be used to write mathematical documents. We also explain how to load packages, how to export and import data and graphics into and from *Mathematica*, and how to manage memory and computing time. You may skip these two chapters until you need them.

Then we go on to graphics. One of the finest aspects of *Mathematica* is its high-quality graphics, and one of the strongest motivations for studying *Mathematica* is to learn to illustrate mathematics with figures. We consider separately graphics for functions and graphics for data. In addition, we have chapters about graphics primitives and graphics options.

New in *Mathematica* 6 are the built-in data sources, covering topics such as chemistry, astronomy, particles, countries, cities, finance, polyhedrons, graphs, words, and colors.

The main new topic in *Mathematica* 6 is dynamics. This allows us to easily build interactive interfaces. The user of such an interface can choose some parameters or other options and the output will be changed dynamically, in real time. This helps in studying various models and phenomena.

Then we study various types of expressions, from numbers to strings, mathematical expressions, lists, tables, and patterns.

We have two chapters relating to programming. The first studies functions and the next various styles of programming. Four styles are considered: procedural, functional, rule-based, and recursive.

■ Mathematics and Statistics

In the remaining 12 chapters, we study different areas of pure and applied mathematics and statistics. The mathematical chapters can be divided into four classes, with each class containing chapters of more or less related topics. Descriptions of these classes follow.

Topics of traditional *differential* and *integral calculus* include derivatives, Taylor series, limits, integrals, sums, and transforms.

Then we consider vectors and *matrices*; linear, polynomial, and transcendental *equations*; and global, local, and classical *optimization*.

In *interpolation* we have the usual interpolating polynomial, a piecewise-calculated interpolating polynomial, and splines. In *approximation* we distinguish the approximation of data and functions. For the former, we can use the linear or nonlinear least-squares method, whereas for the latter we have, for example, minimax approximation.

Mathematica solves *differential equations* both symbolically and numerically. We can solve first- and higher-order equations, systems of equations, and initial and boundary value problems. For *partial differential equations,* we show how some equations can be solved symbolically, how to handle series solutions, and how to numerically solve problems with the method of lines or with the finite difference method. Then we consider *difference equations*. For linear difference equations, we can possibly find a solution in a closed form, but most nonlinear difference equations have to be investigated in other ways, such as studying trajectories and forming bifurcation diagrams.

Lastly, we study *probability* and *statistics*. *Mathematica* contains information about most of the well-known probability distributions. Simulation of various random phenomena (e.g., stochastic processes) is done well with random numbers. Statistical topics include descriptive statistics, frequencies, confidence intervals, hypothesis testing, regression, smoothing, and Bayesian statistics.

Special Aspects

The book explains a substantial portion of the topics of *Mathematica*. However, some topics are emphasized, some are given less emphasis, and some are even excluded. We describe these special aspects of the book here.

■ Breadth

We have had the goal of studying important topics in some breadth and depth. This may mean detailed explanations, clarifying examples, programs, and applications. It may also mean introducing topics for which there is little or no built-in material.

The headings of the chapters give a list of topics that are emphasized in this book and that are explained in some breadth. However, some emphasized topics cannot be identified from the chapter headings. One of them is numerical methods; they are used in every mathematical chapter. Another is methods relating to data. Indeed, we use several real-life and artificial data sets in chapters about data, graphics for data, approximation, differential and difference equations, probability, and statistics.

■ Depth

To give an impression of the depth of various topics, we next describe some special topics in various chapters of the book.

- Chapter 3, Notebooks: An introduction to *Mathematica* as an environment for preparing technical documents; writing mathematical formulas
- Chapter 5, Graphics for Functions: Stereographic figures; graphics for four-dimensional functions
- Chapter 8, Graphics for Data: Visualizations of several real-life data; dot plots; statistical plots
- Chapter 18, Programs: Four styles of programming (procedural, functional, rule-based, and recursive); emphasis on functional programming; many examples of programs
- Chapter 22, Equations: Iterative methods of solving linear equations; programs for nonlinear equations
- Chapter 23, Optimization: A program for numerical minimization; a program for classical optimization with equality and inequality constraints; dynamic programming
- Chapter 25, Approximation: Graphical diagnostics of least-squares fits
- Chapter 26, Differential Equations: Analyzing and visualizing solutions of systems of nonlinear differential equations; study of a predator–prey model, a competing species model, and the Lorenz model; numerical solution of linear and nonlinear boundary value problems; estimation of nonlinear differential equations from data; solving integral equations
- Chapter 27, Partial Differential Equations: Series solutions for partial differential equations; solving parabolic and hyperbolic problems by the method of lines; solving elliptic problems by the finite difference method
- Chapter 28, Difference Equations: The logistic model as an example of nonlinear difference equations; bifurcation diagrams, periodic points, Lyapunov exponents; a discrete-time predator–prey model as an example of a system of nonlinear difference equations; estimation of nonlinear difference equations from data; fractal images; Lindenmayer systems
- Chapter 29, Probability: Simulation of several stochastic processes
- Chapter 30, Statistics: Visualizing confidence intervals and types of errors in statistical tests; confidence intervals and tests for probabilities; local regression; Bayesian statistics; Gibbs sampling; Markov chain Monte Carlo

■ Programs

Mathematica has a large number of ready-to-use commands for symbolic and numerical calculations and for graphics. Nevertheless, in this book we also present approximately 130 of our own programs. Indeed, programming is one of the strongest points of *Mathematica*. It is often amazing how concisely and efficiently we can write a program even for a somewhat complex problem. We think that our own programs can be of some value, despite the fact that they are not so fine and powerful as *Mathematica*'s built-in commands. We have included our own programs for the following reasons:

1. A self-made implementation shows clearly how the algorithm works. You know (or should know) exactly what you are doing when you use your own implementation. The ready-made commands are often like black (or gray) boxes because we do not know much about the methods.

2. Writing our own implementations teaches us programming. We present short programs throughout the book (especially in the mathematical chapters). In this way, we hope that you will become steadily more familiar with programming and that you are encouraged to practice program writing.

3. A self-made implementation can be pedagogically worthwhile. For example, we implement Euler's method for differential equations. It has almost no practical value, but as the simplest numerical method for initial value problems, it has a certain pedagogical value. Also, programming a simple method first may help us to tackle a more demanding method later.

■ Other Special Aspects

We have integrated the so-called packages tightly into the material covered in this book. Instead of presenting a separate chapter about packages, each package is explained in its proper context.

We have tried to make the structure of the book such that finding a topic is easy. Usually a topic is considered in one and only one chapter or section so that you need not search in several places to find the whole story. Each numerical routine is also presented in the proper context after the corresponding symbolic methods. This helps you to find material for solving a given problem: It is usually best to try a symbolic method first and, if this fails, to then resort to a numerical method.

Some topics of a "pure" nature, such as finite fields, quaternions, combinatorics, computational geometry, and graph theory, are not considered in this book; *Mathematica* has packages for these topics. Commands for box and notebook manipulation are treated only briefly. We do not consider *MathLink* (a part of *Mathematica* that enables interaction between *Mathematica* and external programs), *J/Link* (a product that integrates *Mathematica* and Java), *XML* (a metamarkup language for the World Wide Web), or *MathML* (an XML-based markup language for representing mathematics). Also, we do not consider any of the many other *Mathematica*-related products, such as *webMathematica*, *gridMathematica*, *CalculationCenter*, or the *Applications Library* packages.

Mathematica 6

■ Introduction

Mathematica 6 contains a huge amount of new functionality. The following is a part of an on-line document:

> *Mathematica* 6.0 fundamentally redefines *Mathematica* and introduces a major new paradigm for computation. Building on *Mathematica*'s time-tested core symbolic architecture, version 6.0 adds nearly a thousand new functions—almost doubling the total number of functions in the system—dramatically increasing both the breadth and depth of *Mathematica*'s capabilities, as well as introducing hundreds of major original algorithms, and perhaps a thousand new ideas, large and small.

To study the new features, see the following on-line documentation (the use of the Documentation Center is explained in Section 1.4.2, p. 17):

- **Help** ▷ **Startup Palette**, the *What's New in 6* link to Wolfram's website
- **Help** ▷ **Documentation Center**, the *New in 6* links in the home page
- **Help** ▷ **Documentation Center**, the guide/SummaryOfNewFeaturesIn60 document
- **Help** ▷ **Documentation Center**, the guide/NewIn60AlphabeticalListing document
- **Help** ▷ **Function Navigator**, the *New In 6* item

If you are a new user of *Mathematica* and would like to study the basics of *Mathematica* 6, see the following documents:

- **Help** ▷ **Startup Palette**: the *First Five Minutes with Mathematica* button
- **Help** ▷ **Virtual Book**: the *Introduction* item

■ New Properties of Version 6

Because the new features are numerous, we do not list them all here. However, we mention some of the most remarkable new commands and features, classified according to the chapters of the book:

- Chapter 1, Starting: documentation is on-line in the form of Documentation Center, Function Navigator and Virtual Book (we do not have a printed manual); documentation is automatically updated via the Internet; writing *Mathematica* inputs is helped by syntax coloring
- Chapter 3, Notebooks: `Style`, `Text`, `Hyperlink`
- Chapter 4, Files: commands of many packages are now built-in; the remaining packages are rebuilt; look at Compatibility/guide/StandardPackageCompatibilityGuide in the Documentation Center to obtain information about how to replace the functionality of the old packages
- Chapter 5, Graphics for Functions: `GraphicsRow`, `GraphicsGrid`, `Tooltip`; graphics is handled like other expressions; the default font in graphics is Times instead of Courier; 3D graphics is adaptive; contours in contour plots have tooltips; density plots, by default, do not have meshes; 2D graphics can be interactively drawn and edited; 3D graphics can be interactively manipulated (e.g., rotated); for animation, use `Manipulate` or `Animate`
- Chapter 6, Graphics Primitives: `Arrow`, `Opacity`, `Inset`
- Chapter 7, Graphics Options: `Directive`, `BaseStyle`, `Filling`; the default value of `AspectRatio` in `Graphics` and `ParametricPlot` is `Automatic` instead of `1/GoldenRatio`
- Chapter 8, Graphics for Data: `ListLinePlot`, `GraphPlot`; plotting of several data sets
- Chapter 9, Data: `ElementData`, `CountryData`, `PolyhedronData`, etc.
- Chapter 10, Manipulations: `Manipulate` (for creating interactive dynamic interfaces)
- Chapter 11, Dynamics: `Dynamic` (for advanced dynamic interfaces), `MenuView`, `TabView`, etc.
- Chapter 15, Tables: `Grid`, `Row`, `Column`
- Chapter 16, Patterns: `DictionaryLookup`
- Chapter 21, Matrices: `Accumulate`, `PositiveDefiniteMatrixQ`
- Chapter 23, Optimization: `FindShortestTour`
- Chapter 29, Probability: `RandomReal`, `RandomInteger`, `RandomChoice`, `RandomSample`
- Chapter 30, Statistics: `Tally`, `BinCounts`, `FindClusters`

In my opinion, the most impressive new commands in version 6 are `Manipulate`, `Dynamic`, `GraphPlot`, and `Grid`.

Note that many familiar commands, such as `NIntegrate` or `NDSolve`, have also been enhanced in version 6.

In the forthcoming chapters, we mark with (❆6) the properties and commands of *Mathematica* available for the first time in version 6.

■ Obsolete Properties in Version 6

Version 6 makes obsolete some old commands and features, especially in graphics. First, here are some changes that relate to the display and arrangement of graphics:

- To prevent the display of graphics, end the plotting command with `;` instead of using the `DisplayFunction` option.
- In programs, enclose a plotting command with `Print` if that command is not the last command of the program and you would like the program to show that plot.
- `GraphicsArray` is obsolete. To show, for example, two plots `p1` and `p2` side by side, use one of the following ways: `{p1, p2}`, `Row[{p1, p2}]`, or `GraphicsRow[{p1, p2}]`. Use `GraphicsGrid` for arrays of plots.

- To show two plots side by side, you can also simply give a list of plotting commands {Plot[…], Plot[…]}.
- To show two plots on top of each other, simply write Show[Plot[…], Plot[…]]; the DisplayFunction option is no longer needed.
- Graphics and Graphics3D no longer need Show to display the graphics. Thus, write Graphics[{…}] instead of Show[Graphics[{…}]].
- Use Inset[gr, pos] instead of Rectangle[{x1, y1}, {x2, y2}, gr].

Some changes that relate to plotting of data are as follows:

- To plot data by connecting the points with lines, use ListLinePlot[data] instead of ListPlot[data, PlotJoined → True].
- To plot data by points and connecting lines, use ListLinePlot[data, Mesh → All] instead of ListPlot[data, PlotJoined → True, Epilog → {PointSize[s], Map[point, data]}].
- To plot data by points and vertical lines, use ListPlot[data, Filling → Axis] instead of resorting to Prolog or Epilog.
- To plot several data sets, use ListPlot[{data1, data2, … }] or ListLinePlot[{data1, data2, … }] instead of resorting to MultipleListPlot in a package.
- To plot several points, simply write Point[points] instead of Map[Point, points].

Here are some changes that relate to styles and options of graphics:

- Use Style instead of StyleForm.
- Use the BaseStyle option instead of the TextStyle option or the $TextStyle global constant.
- Use the MaxRecursion option instead of the PlotDivision option.
- Use the DataRange option instead of the MeshRange option.
- Use the Filling option instead of the FilledPlot command.

Some other changes are as follows:

- Use RandomReal[…], RandomInteger[…], and RandomComplex[…] instead of Random[Real, …], etc.
- For random numbers from probability distributions, use RandomReal[contDist, n] or RandomInteger[discrDist, n] instead of resorting to Random or RandomArray.
- Use Tally instead of Frequencies in a package.

The Third Edition

■ Main Changes

The text has been revised throughout. Indeed, *Mathematica* 6 brings up so much new and changed features that almost every topic has undergone a revision and new topics are included. Recall that the second edition of this book was based on *Mathematica* 5.

The main change in the structure of the book is that we have six new chapters: Chapter 6, Graphics Primitives; Chapter 9, Data; Chapter 10, Manipulations; Chapter 11, Dynamics; Chapter 15, Tables; and Chapter 16, Patterns. On the other hand, some chapters have been merged and the result is that the current edition has but one chapter about the following topics: graphics for functions, graphics for data, and graphics options (the second edition had two chapters for each of these topics, one for two-dimensional and one for three-dimensional graphics).

The main change in the contents of the book is the transition from version 5 to version 6. In addition, we have some other enhancements. The chapter on programming is much enhanced and enlarged and contains much more examples. The chapter about matrix calculus is also enhanced. The chapter about optimization now includes the method of dynamic programming. Chapters about graphics for data and optimization have undergone a restructuring.

Note that this book fully utilizes the new features of *Mathematica* 6. Because version 6 differs so much from earlier versions, this book cannot practically be used with older versions of *Mathematica*. If you have *Mathematica* 5.2 or an earlier version, please use the second edition of *Mathematica Navigator*.

The CD-ROM contains Help Browser material that describes the new properties of *Mathematica* 7.

■ Some Notes

New Features

Some of the new features of version 6 would have warranted a broader and deeper treatment and more examples of use throughout the book. These features include the creation of dynamic interfaces and the use of the built-in data sources. However, to keep the book at a reasonable size, we had to limit the treatment and the number of examples. We suggest that the reader consults the built-in documentation. The website http://demonstrations.wolfram.com contains thousands of examples of dynamic interfaces.

Environment

During the writing of this book, I used a Macintosh with MacOS X. *Mathematica* works in much the same way in various environments, but the keyboard shortcuts of menu commands vary among different environments. To some extent, we mention the shortcuts for the Microsoft Windows and Macintosh environments.

Options

Many commands of *Mathematica* have options for modifying them. All options have a default value, but we can input other values. When listing the options, we give either all possible values of them or some examples of possible values, but we do not explicitly mention the default values, to save space. In the context of this book, *the default value of an option is always the first value mentioned.* After that are other possible values or examples of other values.

Simulations

In several places in the book, we simulate various random phenomena. Usually, each time a simulation is run, a slightly different result is obtained. However, in experimenting with the examples of the book, the reader may want to get exactly the same result as printed in the book. This can be achieved by using a seed to the random number generator with **SeedRandom[n]** for a given integer **n**. With the same seed, the result of a simulation remains the same in repeated executions. We use **SeedRandom** quite often in this book. If you want to get other results of simulation than those of this book, give different seeds or do not execute **SeedRandom[n]** at all (in the latter case, the default seed is used).

CD-ROM

The entire book is contained on the CD-ROM that comes with it. With a few easy steps you can install the book into the Help Browser of *Mathematica* (the CD-ROM contains installation instructions). With the Help Browser you can easily find and read sections of the book, experiment with the commands, and copy material from the book to your document. You can see all of the figures of the book in color and interactively study the manipulations and animations. The material about the new properties of *Mathematica* 7 can also be installed into the Help Browser. In addition, the CD-ROM contains some data files that are used in the book.

Notation

Throughout the book, the adjectives one-, two-, three-, and four-dimensional are abbreviated 1D, 2D, 3D, and 4D, respectively. The symbol ⟍ is used as a hyphen for *Mathematica* commands. In addition, we use extensively the following handy short notation:

π Means the same as `Pi`. The symbol π can be written as [ESC]`p`[ESC].

∞ Means the same as `Infinity`. The symbol ∞ can be written as [ESC]`inf`[ESC].

⟦…⟧ Means the same as `[[…]]`. For example, `x[[3]]` can also be written as `x⟦3⟧`. The symbols ⟦ and ⟧ can be written as [ESC]`[[`[ESC] and [ESC]`]]`[ESC].

ᵀ Means the same as `Transpose`. For example, `Transpose[x]` can also be written as `xᵀ`. The symbol ᵀ can be written as [ESC]`tr`[ESC].

/@ Means the same as `Map`. For example, `Map[f[#]&, {a, b, c}]` can also be written as `f[#]& /@ {a, b, c}`. A third way is to write `Table[f[x], {x, {a, b, c}}]`.

The symbols π and ∞ can also be found from the **BasicMathInput** palette. For example, instead of

```
Map[#^2 &, Transpose[{{Pi, Infinity}}]]
```

$$\{\{\pi^2\}, \{\infty\}\}$$

we can write

```
#^2 & /@ {{π, ∞}}ᵀ
```

$$\{\{\pi^2\}, \{\infty\}\}$$

Questions

If you have questions about the use of *Mathematica*, do not hesitate to contact me. I try to answer when I have the time. Also, please send comments and corrections.

Acknowledgments

In preparing this book, the main source has been the excellent on-line Documentation Center of *Mathematica* 6. The technical support staff at Wolfram Research, Inc., helped me a lot; I especially thank Eric Bynum, Roberto Cavaliere, Huihua Huang, Yong Huang, Vivec Joshi, and Bruce Miller.

The entire book was written and produced with *Mathematica*; each chapter is a *Mathematica* notebook. The notebooks were connected into a single project by the **AuthorTools** package of *Mathematica*. The package then automatically generated the index (after we had attached the index entries with the cells of the book, also with the package), and the package also prepared the on-line Help Browser version of the book.

I have been lucky enough to enjoy excellent working conditions at the Department of Mathematics of the University of Turku. For this my sincere thanks are due to Professor Marko Mäkelä. I also thank Professor Juhani Karhumäki and Professor Matti Vuorinen for their support and encouragement.

The third edition is also published in India. I am deeply indepted to Professor Ponnusamy Saminathan, Indian Institute of Technology, Madras, Chennai, for suggesting and supporting the Indian edition.

For their review of the manuscript of the second edition, I am very thankful to Donald Balenovich, Indiana University of Pennsylvania; Joaquin Carbonara, Buffalo State University; William Emerson, Metropolitan State University; Jim Guyker, Buffalo State University; Mike Mesterton-Gibbons, Florida State University; and Fred Szabo, Concordia University. Their valuable comments and suggestions greatly improved the second edition.

I also thank the following people for taking the task of writing a review of the second edition of the book in some journals: Robert M. Lurie (*Mathematica in Education and Research*), Matti Vuorinen (*Zentralblatt MATH*), K. Waldhör (*Computing Reviews*), and John A. Wass (*Scientific Computing*).

Many readers of the second edition have sent me e-mail, giving feedback and asking questions. Thank you all! Your support has encouraged me in writing the third edition.

The anecdotes at the beginning of the chapters are from the wonderful book by MacHale (1993) (the anecdotes are reproduced or adapted with the permission of the publishers, Boole Press, 26 Temple Lane, Dublin 2, Ireland).

The editorial staff at Elsevier has done a fine work with the production of the book. Especially I would like to thank Phil Bugeau for efficient project management. I am also very grateful to Dan Hays and Kristen Cassereau Ng for copy-editing and proofreading the manuscript with great care.

For financial support I express my deep gratitude to Elsevier Academic Press and Suomen Tietokirjailijat (The Association of Finnish Non-Fiction Writers).

Lastly, I thank my wife, Marjatta, for her encouragement and support during the work.

Heikki Ruskeepää

Department of Mathematics
University of Turku
FIN-20014 Turku
Finland
ruskeepa@utu.fi

Starting

Introduction

> *In 1903 at a meeting of the American Mathematical Society, F. N. Cole read a paper entitled "On the Factorization of Large Numbers." When called upon to speak, Cole walked to the board and, saying nothing, raised two to its sixty-seventh power and subtracted one from the answer. Then he multiplied, longhand, 193,707,721 by 761,838,257,287 and the answers agreed. Without having said a word, Cole sat down to a standing ovation. Afterwards he announced that it had taken him twenty years of Sunday afternoons to factorize the Mersenne number $2^{67} - 1$.*

This chapter is intended to give you an impression of *Mathematica* and to teach you some of its basic techniques and commands. A more complete insight is given in the next chapter, in which we briefly present a selection of the most important commands of *Mathematica*.

Although this book puts some emphasis on the methods of applied mathematics, this chapter begins, in Section 1.1, with a "pure" example: factoring integers. We consider the problem mentioned in the anecdote above and show what we can do nowadays with such powerful systems as *Mathematica*. This example will enlighten you regarding some of the major aspects of the program. We emphasize that it is not intended that you do the calculations of this example, nor that you should understand the commands we use.

In Section 1.2, we give a brief overview of some of *Mathematica*'s basic techniques and commands, beginning with the classical starting example of calculating 1 + 2 and ending with calculus and graphics. Section 1.3 presents and explains the important conventions of *Mathematica*, which often cause trouble for beginners.

In Sections 1.4 and 1.5, we discuss how you can get help within *Mathematica* and how you can correct and edit what you have written. These two sections may give more information than you need now, but you can read the basic points and return to these sections later, when getting help and editing become more relevant concerns.

Parts of this chapter depend on the computer you use. We explain only the Windows and Macintosh environments, although some comments may be found about the basics of *Mathematica* in a Unix system.

1.1 What Is *Mathematica*?

1.1.1 An Example

■ **Verifying the Work of Cole**

(Note: It is not intended that you do the calculations of Section 1.1.1. The example is only intended to be read and to demonstrate certain aspects of *Mathematica*. Your actual lessons begin in Section 1.2.)

Did you read the anecdote about F. N. Cole at the beginning of this chapter? Cole sacrificed every Sunday afternoon for 20 years to study the Mersenne number $M_{67} = 2^{67} - 1$:

```
2^67 - 1
```
```
147 573 952 589 676 412 927
```

The first line is the command entered to *Mathematica,* and the second line is the answer given by *Mathematica*. At last he found that the number is the product of 193,707,721 and 761,838,257,287:

```
193 707 721 * 761 838 257 287
```
```
147 573 952 589 676 412 927
```

M_{67} is thus not a prime. Cole's feat was admirable. Now, after 100 years, we have *Mathematica,* and the situation is totally different. It now takes only a fraction of a second to do the factorization:

```
FactorInteger[2^67 - 1] // Timing
```
```
{0.016942, {{193 707 721, 1}, {761 838 257 287, 1}}}
```

Mathematica found that 193,707,721 and 761,838,257,287 are factors of multiplicity 1.

■ **Difficult Factors**

However, even today some problems can be surprisingly difficult. When *Mathematica*, in my computer (which is not very fast), factorizes M_{254}, it needs approximately 10,000 seconds or 3 hours and approximately 200 megabytes of RAM:

```
FactorInteger[2^254 - 1] // Timing
```
```
{10 521.7, {{3, 1}, {56 713 727 820 156 410 577 229 101 238 628 035 243, 1},
   {170 141 183 460 469 231 731 687 303 715 884 105 727, 1}}}
```
```
MaxMemoryUsed[]
```
```
207 758 616
```

However, note that M_{254} is very big:

```
2^254 - 1
```
```
28 948 022 309 329 048 855 892 746 252 171 976 963 317 496 166 410 141 009 864 396 001 978 282
   409 983
```

and two of the three factors are also big. Thus, factoring the number is obviously a difficult task. By the way, the difficulty of factoring large numbers is a key to some cryptographic methods.

However, for M_{254}, *Mathematica* can immediately tell that it is not a prime:

```
PrimeQ[2^254 - 1] // Timing
{0.000223, False}
```

Indeed, we can easily investigate Mersenne numbers for primality up to, for example, index 607:

```
(mp = Table[{i, PrimeQ[2^i - 1]}, {i, 2, 607}];) // Timing
{0.417603, Null}
```

The indices for which the corresponding Mersenne number is a prime are as follows:

```
Select[mp, #[[2]] == True &][[All, 1]]
{2, 3, 5, 7, 13, 17, 19, 31, 61, 89, 107, 127, 521, 607}
```

■ A Demanding Computation

To further illustrate the use of *Mathematica*, we now factor the Mersenne numbers M_2 to M_{250}. (Note that you are not supposed to do the calculations in this example. Just cast an admiring glance at the commands. Later in this book you will learn such commands as **Total**, **/@**, and **Range**.) We do not show the factors themselves; we only count the number of factors:

```
(t = Total[FactorInteger[2^# - 1][[All, 2]]] & /@ Range[2, 250]) // Timing
{5430.6, {1, 1, 2, 1, 3, 1, 3, 2, 3, 2, 5, 1, 3, 3, 4, 1, 6, 1, 6, 4, 4, 2, 7, 3, 3, 3, 6,
  3, 7, 1, 5, 4, 3, 4, 10, 2, 3, 4, 8, 2, 8, 3, 7, 6, 4, 3, 10, 2, 7, 5, 7, 3, 9, 6,
  8, 4, 6, 2, 13, 1, 3, 7, 7, 3, 9, 2, 7, 4, 9, 3, 14, 3, 5, 7, 7, 4, 8, 3, 10, 6,
  5, 2, 14, 3, 5, 6, 10, 1, 13, 5, 9, 3, 6, 5, 13, 2, 5, 8, 14, 2, 11, 2, 10, 11,
  6, 1, 15, 2, 12, 6, 11, 5, 9, 6, 9, 9, 6, 6, 17, 4, 3, 5, 8, 5, 14, 1, 9, 5, 9, 2,
  15, 3, 5, 10, 11, 2, 9, 2, 16, 6, 6, 6, 19, 5, 6, 7, 10, 2, 14, 5, 11, 8, 10, 8,
  18, 4, 5, 8, 13, 7, 16, 5, 10, 10, 8, 2, 19, 4, 7, 7, 10, 4, 11, 9, 14, 6, 5, 3,
  24, 4, 11, 5, 11, 5, 8, 5, 10, 10, 5, 16, 3, 7, 8, 11, 2, 17, 2, 20, 6, 4, 7,
  20, 6, 5, 9, 12, 6, 22, 3, 10, 7, 4, 8, 21, 6, 5, 7, 19, 4, 13, 6, 16, 14, 10, 2,
  17, 4, 12, 10, 12, 4, 16, 7, 13, 8, 11, 6, 23, 2, 8, 10, 9, 7, 12, 6, 12, 5, 11}}
MaxMemoryUsed[]
83 409 352
```

Thus, for example, M_4 has two factors, and M_{250} has 11 factors. The computations took approximately 1.5 hours and approximately 80 megabytes of RAM.

■ A Graphic Illustration

We continue studying Mersenne numbers and now form pairs from the indices and the numbers of factors:

```
s = {Range[2, 250], t}ᵀ;
```

(Here, ᵀ means a transpose.) Then we find a logarithmic least-squares fit for the number of factors as a function of the index of the Mersenne numbers:

```
lsq = Fit[s, {1, Log[i]}, i]
-3.05399 + 2.25251 Log[i]
```

We then plot the fit and the numbers of factors:

```
p1 = Plot[lsq, {i, 2, 250}, PlotStyle → Black];

p2 = ListLinePlot[s, PlotStyle → Black,
    Mesh → All, MeshStyle → {Black, PointSize[Small]}];
```

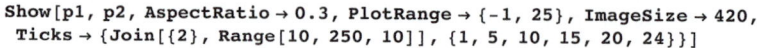

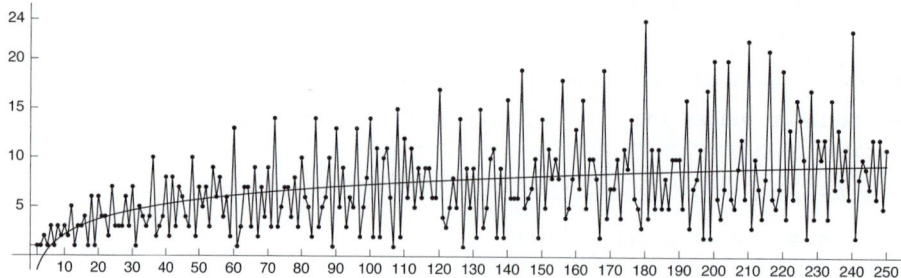

So, here are the numbers of factors of M_2 to M_{250}, together with the logarithmic fit. In the figure we see the prime Mersenne numbers with indices 2, 3, 5, 7, 13, 17, 19, 31, 61, 89, 107, and 127. These Mersenne primes had all been found by 1913. After the index 127, primes are not very common among Mersenne numbers. The next known primes occur with indices 521 (found in 1952); 607; 1279; 2203; 2281; 3217; 4253; 4423; 9689; 9941; 11,213; 19,937; 21,701; 23,209; 44,497; 86,243; 110,503; 132,049; 216,091; 756,839; 859,433; 1,257,787; 1,398,269; 2,976,221; 3,021,377; 6,972,593; 13,466,917; 20,996,011; 24,036,583; 25,964,951; 30,402,457 (found in 2005); 32,582,657 (found in 2006).

■ Lessons Learned

The preceding examples show how easy it now is to do long and complicated calculations and to visualize the results. *Mathematica* is one of the more popular systems for doing such calculations. However, even today, with powerful mathematical systems and machines, some problems remain very time-consuming.

The example also illustrates some aspects of *Mathematica*, such as working with exact and approximate quantities, using graphics, and making programs. In general, *Mathematica* integrates symbolic calculation, numerical calculation, graphics, and programming into one system.

Mathematica contains still another aspect: a document-making environment (in versions of *Mathematica* that support the notebook interface). In this environment, you can do symbolic and numerical calculations, produce graphics, and add text to explain what you have done. The result is a complete document of your work; in fact, this book was written with *Mathematica*. In addition, the document is interactive. You can change parameters and functions, redo calculations, show animations, create interactive graphical interfaces, and continuously develop the document. The notebook interface is included in both Windows and Macintosh versions of *Mathematica*. In plain Unix, the interface is text-based and thus all features of *Mathematica* are not supported, but the X Window System supports notebooks.

1.1.2 The Structure of *Mathematica*

Mathematica is a large system: The total installation of *Mathematica* takes up approximately 1.1 gigabytes (Gb) of space on the hard drive. The following are the essential parts of *Mathematica*:

- the *kernel*, with the name **MathKernel**;
- the *front end*, with the name **Mathematica**; and
- the *packages*, with the name **Packages**.

Next we consider each of these three components individually.

■ The Kernel

The main component of *Mathematica* is the kernel; it does all of the computations. It is written mainly with the C programming language and is one of the largest mathematical systems ever written. The total number of lines of code in the kernel is approximately 2.5 million. An important thing to understand is that the kernel is the same in all environments; this means that you get the same results in all environments (except possibly for the precision of floating-point calculations).

Mathematica commands are easy to use and quite versatile. However, behind the commands there is a huge amount of mathematical knowledge and a vast amount of work. For example, behind the single command **Integrate**, there are approximately 600 pages of C code and 500 pages of *Mathematica* code.

■ The Front End

The front end is an environment for communicating with the kernel. When you open *Mathematica*, you open the front end. Commands are entered into the front end, and they are sent automatically to the kernel (the communication between the front end and the kernel is done with *MathLink*, a system that handles communication between parts of *Mathematica* and between *Mathematica* and other programs). The result of a calculation is then displayed by the front end. Figures come from the kernel in the form of PostScript code, and the front end then creates a screen image from this code. The front end contains approximately 700,000 lines of system-independent C++ source code.

There are two types of front ends: *notebook* and *text-based*.

A notebook front end is, for example, in the Windows, Macintosh, and X Window versions of *Mathematica*. A notebook is an interactive document. It contains the commands you have entered and their results, graphics (including animations), dynamic graphical interfaces, and comments you have added. You can do any kind of correcting, editing, and formatting in a notebook. In fact, you can make a notebook into a whole document of your work, even a book. (The chapters of this book were made as separate notebooks.)

The text-based front end is not nearly so handy and versatile as the notebook front end. This type of front end is found in the plain Unix version of *Mathematica*. With this kind of front end you can enter commands and see the results, but editing is very limited. Another disadvantage is that pictures are displayed in separate windows.

■ The Packages

The packages supplement the kernel. They are not normally loaded when *Mathematica* is loaded; you have to manually load the packages you want to use. Packages contain commands considered not to be as central and as common as the commands in the kernel. Packages are written using the programming capabilities of *Mathematica*.

To the packages that come with *Mathematica* you can also add the vast collection of packages and other material found at http://library.wolfram.com/. In addition, you can write your own packages and programs.

Furthermore, there are collections of Wolfram application packages and third-party application packages; see www.wolfram.com/products.

1.2 First Calculations

1.2.1 Opening, Calculating, and Quitting

■ Opening Mathematica

To open *Mathematica*, do the following:

- In Windows and Macintosh environments: open *Mathematica* as other programs
- In Unix with X Window: type `mathematica`
- In plain Unix: type `math`

When you open *Mathematica* in a notebook environment, you get an empty notebook. A cursor appears once you have pressed the first key.

■ Calculating

Now, are you ready to begin? Let us start modestly and try to calculate 1 + 2. Press these three keys:

`1 + 2`

The command is executed (i.e., sent to the kernel for processing) by pressing a special key or key combination.

To execute a command, press ⌷ENTER⌷ or ⌷SHIFT⌷⌷RET⌷

In Windows and Macintosh interfaces, both ways work (⌷ENTER⌷ is the key at the bottom right corner of the keyboard; ⌷SHIFT⌷⌷RET⌷ means that you are holding down the Shift key while at the same time pressing the Return key ↵). In Unix, it may be that ⌷SHIFT⌷⌷RET⌷ is the only possibility.

Note that although we have spoken about "executing a command," this is not in keeping with the official terminology. We should speak about "*evaluating* an input." You should know the official term *evaluation*, but we ourselves feel free to use the more concrete term *execution*.

So, after typing the command or input `1 + 2`, press any of the executing or evaluating keys or key combinations. Now the kernel begins to evaluate the input. In most microcomputers, the kernel is not loaded until you ask it to do a calculation. This means that the first calculation takes some time, even if it is as simple as `1 + 2`. After you have entered the input, the label `In[1]:=` appears before the input, and the result has the label `Out[1]=`. You get

`1 + 2`

`3`

Note that here we do not show the `In` and `Out` labels. Indeed, by adjusting the preferences of *Mathematica*, we can turn the labels off. In a Macintosh, the preferences can be found from the **Mathematica ▷ Preferences...** menu. Go then into the *Evaluation* preferences and click *Show In/Out names* to be off. The input and the output can be distinguished by the font: The input is in boldface Courier, whereas the output is in plain Courier. Try the next command (note that although you again cannot see a cursor in the notebook, just start typing; the new command appears where you see a horizontal line):

`18 / 4`

$$\frac{9}{2}$$

Fractions are automatically simplified, and they are written in a 2D form (instead of 9/2). If the expression contains a decimal number, then the result is also a decimal number:

```
10 / 3.
3.33333
```

■ Correcting

You may observe an error in a command when you write it, or after executing the command *Mathematica* may give you an error message telling you that there is something wrong with the command.

In a notebook environment (e.g., Windows, Macintosh, or X Window), you can do standard editing as with a word processor. For example, to delete an incorrect character, just move the cursor to the desired location by pressing the arrow keys or with the mouse and press the backspace key or the delete key to remove the character to the left or right, respectively. You can also highlight a portion of an input, cut it (CTRL x; on a Macintosh, use ⌘ in place of CTRL) or copy it (CTRL c), and then paste it (CTRL v) to a new location.

Once you have a corrected input, execute it. Note that after correcting a command you can leave the cursor where it is and then execute the command; the cursor need not be at the end of the command when you execute it. We consider editing in more detail in Section 1.5, p. 22. (In plain Unix, possibly only the backspace key can be used to edit the input; arrow keys may not function. Using special editing commands such as **Edit** together with an editor such as Emacs can help a lot in plain Unix.)

■ Quitting *Mathematica*

> To quit *Mathematica*, do the following:
>
> • In Windows, Macintosh, and X Window: quit *Mathematica* as other programs
> • In plain Unix: execute the command **Quit**

Do not quit right now. Instead, continue reading and experimenting.

■ Aborting a Calculation

You may sometimes observe that a calculation is useless (perhaps because there was an error in the input or you do not have time to wait for the answer). You can then abort the calculation.

> To abort a calculation, do the following:
>
> • In Windows, Macintosh, and X Window: choose **Abort Evaluation** from the **Evaluation** menu, or press ALT . (⌘ . on a Macintosh)
> • In plain Unix: press CTRL c RET a RET

It may happen that when you try to abort a calculation it seems that the calculation just goes on and on. You can then quit *Mathematica* (after possibly saving the notebook) and start a new session. In notebook environments, you can also only quit the kernel by choosing **Evaluation ▷ Quit Kernel ▷ Local**. Start a new session by executing an input or by choosing **Evaluation ▷ Start Kernel ▷ Local**.

1.2.2 Names and Decimals

■ Before Continuing

Now we continue exploring *Mathematica*. Please note the following very important point:

• When you try the examples of the book with your machine, write the commands exactly as they are printed here.

Mathematica will not forgive even the smallest error in syntax. Be especially careful with small and capital letters: All *Mathematica* names such as `Sin` or `Integrate` begin with a capital letter. Also, you have to write all arguments in functions and commands in square brackets `[ ]`, for example, `Sin[x]` and `Integrate[a + b x, x]`; parentheses `( )` are not allowed. Parentheses are only used for grouping terms in expressions. Note that *Mathematica* automatically adds spaces in some places in your input (e.g., around `+` or `=`).

You probably will occasionally get error messages and wrong results because the syntax was not correct. Do not worry. This is normal. Getting used to *Mathematica* takes time, and only by working with the program can you learn to use it efficiently.

During a session, do not just enter one input after another in a hurry. Think carefully about each input and each result—how the input is written and what the result is. In this way, you will learn more effectively.

Also, once you have tried some of the examples in this book, you can try other examples. This is recommended because by trying similar examples you strengthen your skills and get a clearer impression of each command and technique.

■ Referring to Earlier Results

`%`	Refers to the last result
`%%`	Refers to the next-to-last result
`%%...%`	Refers to the *k*th previous result if there are *k* `%` marks
`Out[n]` or `%n`	Refers to the result in the output line `Out[n]`

Often you want to refer to earlier results. The percent mark `%` and the output names can be used for this. Try the following commands:

```
353^4
15 527 402 881

30^4 + 120^4 + 272^4 + 315^4
15 527 402 881

%% - %
0
```

Try also the output names by executing, for example, `Out[4] + Out[5]` or `%4 + %5`.

Using `%` can become a problem if the command you execute contains errors to be corrected. Suppose you first calculate

```
16^2
256
```

Then you want to calculate $16^2 - 15^2$, but you write

```
% + 15^2
481
```

If you now correct the command to read `% - 15^2`, you do not get what you want because `%` now refers to the result 481 of the last (wrong) command. So, you have to correct the command to read `%% - 15^2`.

In general, if you correct several times a command that originally contained a **%**, you have to add one **%** each time. This may become awkward. It may be clearer to assign names to expressions, as explained later.

A way to refer to the last command and its result is to choose **Insert ▷ Input from Above** or **Insert ▷ Output from Above**.

■ Giving Names

a = value Assign **value** for **a**
a Show the value of **a**
a =. Clear the value of **a**

Another technique for referring to earlier results is to give names to results. Later, you can use the names as needed. For example, what is the probability of getting two 6's when tossing a die six times?

```
a = Binomial[6, 2]
```

15

```
b = (1 / 6) ^ 2 * (5 / 6) ^ 4
```

$$\frac{625}{46\,656}$$

```
c = a * b
```

$$\frac{3125}{15\,552}$$

You can always ask the value of a variable simply by entering the name of the variable:

```
a
```

15

When a symbol is no longer used, it is useful to clear the value of the variable so that it does not cause trouble later:

```
a =.
```

Now **a** has no value, as we see if we ask the value of **a**:

```
a
```

a

Mathematica printed only the name of the variable.

Note that we can see from the color of a symbol whether it has a value: A blue symbol does not have a value, whereas a black symbol does have a value.

■ Decimal Values

expr//N or **N[expr]** Calculate a decimal value for **expr**

You have perhaps noted that all calculations with integers and fractions are kept in an exact form; a decimal value is not automatically computed. A decimal value can be asked for with **N**. It can be used in two equivalent forms. The form **expr//N** may be easier to write than the form **N[expr]**. For example,

```
b
  625
 ──────
 46 656
% // N
0.0133959

c // N
0.200939

N[c]
0.200939
```

We can also ask directly for the decimal value (without first asking for a simplified fraction):

```
(1 / 6) ^ 2 * (5 / 6) ^ 4 // N
0.0133959
```

Note that, from now on, we here and there show the results of commands next to each command to save space.

1.2.3 Basic Calculations and Plotting

■ Basic Arithmetic

```
a + b    a - b    a b or a*b    a/b    a^b
```

The basic arithmetic operations plus, minus, division, and power are expressed in the usual manner, but multiplication is different. With *Mathematica*, multiplication is usually expressed by a space (press the space bar once). If you are more comfortable with the asterisk *, you can use it. For example,

```
a = 5    5

b = 3    3

a b      15
```

Note that if you write **ab** without a space, *Mathematica* treats this as a single variable with the name **ab**:

```
ab       ab
```

We have not defined a value for the variable **ab**, so *Mathematica* just writes the name of the variable. This is a common error when using *Mathematica*. You have to write **a b** with a space or **a*b** with an asterisk if you want multiplication. Multiplication is explained further in Section 1.3, p. 13.

■ Basic Constants

```
Pi, E, I, Infinity
```

These well-known constants are usually denoted by π, e, i, and ∞ in mathematical texts. For example,

```
Pi // N    3.14159

E // N     2.71828

I ^ 2      -1
```

```
1 / Infinity    0
```

Negative infinity is **-Infinity**. Note that *Mathematica* writes **Pi** as π, **E** as e, **I** as i, and **Infinity** as ∞:

```
{Pi, E, I, Infinity}
```

$\{\pi, e, i, \infty\}$

We can also write **Pi** as π, **Infinity** as ∞, and so on by using the Escape key, as follows:

> To write π, type ⎋p⎋
> To write ∞, type ⎋inf⎋

Just press the ⎋, p, and ⎋ keys in turn. From now on in this book, we write **Pi** as π and **Infinity** as ∞.

■ Basic Functions

> **Sqrt[z]** or **z^(1/2)**   (square root)
>
> **Exp[z]** or **E^z**   (exponential function)
>
> **Log[z]**, **Log[b, z]**   (natural logarithm and logarithm to base **b**)
>
> **Abs[z]**   (absolute value)
>
> **Sin[z]**, **Cos[z]**, **Tan[z]**, **Cot[z]**, **Sec[z]**, **Csc[z]**
>
> **ArcSin[z]**, **ArcCos[z]**, **ArcTan[z]**, **ArcCot[z]**, **ArcSec[z]**, **ArcCsc[z]**
>
> **n!**, **Binomial[n, m]**
>
> **Max[{x, y, ...}]**, **Min[{x, y, ...}]**

Note that the natural logarithm is **Log[z]**. [In many mathematical texts, $\log(z)$ means a logarithm to base 10, whereas the natural logarithm is denoted by $\ln(z)$.] The arguments in trigonometric functions are in radians, and the values of the inverse trigonometric functions are in radians. **Binomial** gives the binomial coefficient. **Max** and **Min** give the maximum and the minimum of the arguments. For example,

```
Exp[Log[Sqrt[x]]]    √x
```

```
Cos[π / 4]          1
                   ———
                   √2
```

```
ArcCos[1 / Sqrt[2]]    π
                      ——
                      4
```

We throw a die 10 times and find the maximum of the results:

```
a = RandomInteger[{1, 6}, 10]
```

$\{3, 5, 2, 4, 3, 4, 3, 1, 5, 4\}$

```
Max[a]    5
```

The value of the variable **a** is a *list*—an ordered collection of elements enclosed in curly braces {}.

■ Basic Calculus

> **D[expr, x]**   Derivative of **expr** with respect to **x**
>
> **Integrate[expr, x]**   Indefinite integral of **expr** with respect to **x**
>
> **Integrate[expr, {x, a, b}]**   Definite integral of **expr** with respect to **x** from **a** to **b**
>
> **Simplify[expr]** or **expr//Simplify**   Simplify the expression

Again a note about terminology. We have spoken about commands such as **D** or **Integrate**, but the official term is a *function*. However, we feel free to speak about commands and use the term *function* mainly for such expressions as **Sin[x]**, which are official mathematical functions. For example,

D[x Sin[x], x]   x Cos[x] + Sin[x]

Integrate[x^2 Exp[x], {x, 0, 1}]   −2 + e

Integrate[p x / (q + r x), x] $p \left(\dfrac{x}{r} - \dfrac{q \, \text{Log}[q + r x]}{r^2} \right)$

We check the last integral by calculating the derivative of the result:

D[%, x] $p \left(\dfrac{1}{r} - \dfrac{q}{r \, (q + r x)} \right)$

After simplification, we get the desired result:

% // Simplify $\dfrac{p x}{q + r x}$

■ **Basic Plotting**

Plot[expr, {x, a, b}] Plot **expr** when **x** takes on values from **a** to **b**

Mathematica has many plotting commands, but **Plot** is the basic one. An example:

Plot[Exp[-x] Sin[2 x], {x, 0, 2 π}]

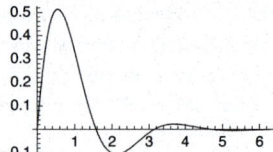

In notebooks, plotting is this easy. You can also change the size of a figure: Click on it and then drag one of the handles. In plain Unix, things may be not so simple. Ask for more information from a person who knows your environment.

Congratulations! Now you have used *Mathematica* for some simple calculations and you have an impression of how *Mathematica* works. We will give you a better overview of *Mathematica* in Chapter 2. However, first, in Section 1.3, we summarize the basic conventions of *Mathematica*. Then we explain, in Section 1.4, how you can get information about the commands of *Mathematica*. Section 1.5 considers writing, correcting, and editing in *Mathematica*.

1.3 Important Conventions

You have observed that all of the built-in *Mathematica* names we have presented have begun with a capital letter and that all arguments have been given in square brackets **[ ]**. These are two of the most important conventions in *Mathematica*. Here are the six most important ones:

> • All built-in *Mathematica* names begin with a capital letter.
> • Multiplication must be expressed by a space or an asterisk (∗). (For numerical multipliers or complete expressions, nothing is needed.)
> • All arguments are given in square brackets **[ ]**.
> • Parentheses **( )** are used only for grouping terms.
> • Curly braces **{ }** are used for lists.
> • Double square brackets **[[ ]]** are used to extract elements from lists.

It takes some time to get used to these conventions, and at the beginning you will often get error messages and wrong results because you have not remembered these rules. Later, you may see that the conventions have advantages. Let us consider the conventions in more detail.

■ Names

Mathematica is case sensitive. If a name is **Sin**, you cannot write **sin** or **SIN**; you must write **Sin** exactly. It is recommended that all names you introduce (like **a**, **b**, and **c** previously) begin with a small letter. If this convention is followed, then it is always clear which names are built-in and which are defined by the user. Such a distinction makes reading the *Mathematica* code easier; you need not remember whether a name is your own. Also, you cannot mistakenly define a symbol with the same name as a built-in command, thereby avoiding any confusion.

Many built-in names consist of several words run together, such as **FindMinimum**, and in these cases each individual word begins with a capital letter. If you define a name consisting of several words, you can use capital letters in the middle of the name, as in **randomWalk**; this makes reading the name easier.

Another convention is that all built-in names and words are written completely; abbreviations are not used. This can make some names long (e.g., **InverseLaplaceTransform** or **NoncentralChiSquare-Distribution**), but the advantage is that such complete names are often easier to remember than abbreviated names. Some abbreviations exist, however, such as **D** (derivative), **Det** (determinant), and **Tr** (trace). Names may be as long as you want. Names cannot begin with a number. User-defined names are also often written in full without abbreviations [the longest I have seen is in Shaw & Tigg (1994, p. 104): **NapoleonicMarchOnMoscowAndBackAgainPlot**].

Let us try out an example with the capital first letter. Instead of the correct form **Sin[π/2]**, write

$$\texttt{sin[}\pi\,\texttt{/ 2]} \qquad \sin\left[\frac{\pi}{2}\right]$$

Note that, in the command, **sin** remained blue, and this means that *Mathematica* does not know about **sin**. We did not get the expected result 1. We correct the command:

$$\texttt{Sin[}\pi\,\texttt{/ 2]} \qquad 1$$

■ Multiplication

Multiplication was already considered in Section 1.2.3, p. 10, but let us still try some examples:

```
a = 3     3

b = 4     4

{a b, a ∗ b, ab}
{12, 12, ab}
```

Recall that you cannot write **ab** if you want **a** times **b**. If you write **ab**, *Mathematica* understands it as a variable with the name **ab**. Some more examples:

```
{5 a, a5, d (e + f), (d + e) (f + g), Sin[x] Cos[y]}
{15, a5, d (e + f), (d + e) (f + g), Cos[y] Sin[x]}
```

Note that with a numeric multiplier, we do not need to write a multiplication indicator such as a space or an asterisk. We can write **5a**; *Mathematica* automatically adds a space between the terms. However, **a5** is interpreted as a name. No space or asterisk is needed with parentheses either: We can write **c(d+e)** and **(c+d)(e+f)**, and *Mathematica* adds the space. A multiplication indicator is generally not needed between complete expressions. For example, we can write **Sin[x]Cos[y]**, and, again, *Mathematica* adds the space.

If a multiplication occurs at the end of a line, then it is safe to use the asterisk (*). Place the asterisk either at the end of the first line or at the beginning of the next line. If you do not use the asterisk, *Mathematica* understands the two rows as separate commands, if they can be interpreted as complete commands.

■ Arguments

In traditional mathematical notation, parentheses are used for two purposes: for arguments and for grouping terms. *Mathematica* avoids this ambiguity by using different notation for these two purposes: square brackets for arguments and parentheses for grouping. For example, if we write, instead of the correct form **Sin[π/3]**, what you see below, we get a wrong result:

$$\text{Sin } (\pi \, / \, 3) \qquad \frac{\pi \, \text{Sin}}{3}$$

Mathematica interprets the expression according to its standard rules: **Sin** is a variable by which we want to multiply **Pi/3**. Note that the parentheses are red to remind that the syntax is incorrect. Here is the correct command:

$$\text{Sin}[\pi \, / \, 3] \qquad \frac{\sqrt{3}}{2}$$

■ Grouping

Be careful in entering expressions. Parentheses are sometimes easily forgotten, and the result will be incorrect. Special care is necessary with quotients and rational powers. Here are some examples of quotients:

```
{1 / 4 Sqrt[x] Log[x], 1 / (4 Sqrt[x]) Log[x], 1 / (4 Sqrt[x] Log[x])}
```

$$\left\{\frac{1}{4} \, \sqrt{x} \, \text{Log}[x], \, \frac{\text{Log}[x]}{4 \, \sqrt{x}}, \, \frac{1}{4 \, \sqrt{x} \, \text{Log}[x]}\right\}$$

Thus, **a/b*c** in interpreted as **(a/b)*c** and not as **a/(b*c)**. Here are some examples of powers:

```
{E^-1, E^-1 / 2, E^ (-1 / 2)}
```

$$\left\{\frac{1}{e}, \, \frac{1}{2 \, e}, \, \frac{1}{\sqrt{e}}\right\}$$

Thus, **a^b/c** is interpreted as **(a^b)/c** and not as **a^(b/c)**. Remember to write the necessary parentheses, and if you are uncertain whether you should use parentheses or not, go ahead and use them because unnecessary parentheses are harmless. Note that if you want to square **Sin[x]**, you can simply write **Sin[x]^2**. If you want to calculate the value of **Sin** at **x^2**, write **Sin[x^2]**.

■ Lists

Lists are like vectors: A list is, mathematically, an ordered set of elements. Lists are used to store data and expressions. Here is an example of a list with three elements:

c = {6, 2 E, Sin[1.2 π]}   {6, 2 e, −0.587785}

Curly braces are reserved for lists. Another example:

d = {Cosh[3], Pi, 2}   {Cosh[3], π, 2}

Calculations with lists are simple because all operations are automatically done element by element:

d^2   {Cosh[3]², π², 4}

c + d   {6 + Cosh[3], 2 e + π, 1.41221}

Double square brackets are used to extract elements from lists. For example:

c[[2]]   2 e

In place of [[...]] we can also use [[...]]; here, [[and]] can be written as ⎡ESC⎤ [[⎡ESC⎤ and ⎡ESC⎤]] ⎡ESC⎤.

1.4 Getting Help

1.4.1 Palettes

Some palettes can help you when you are entering input for *Mathematica* in notebook environments. Palettes can be accessed from the **Palettes** menu. Below we show four palettes: **AlgebraicManipulation**, **BasicMathInput**, **BasicTypesetting**, and **SpecialCharacters**.

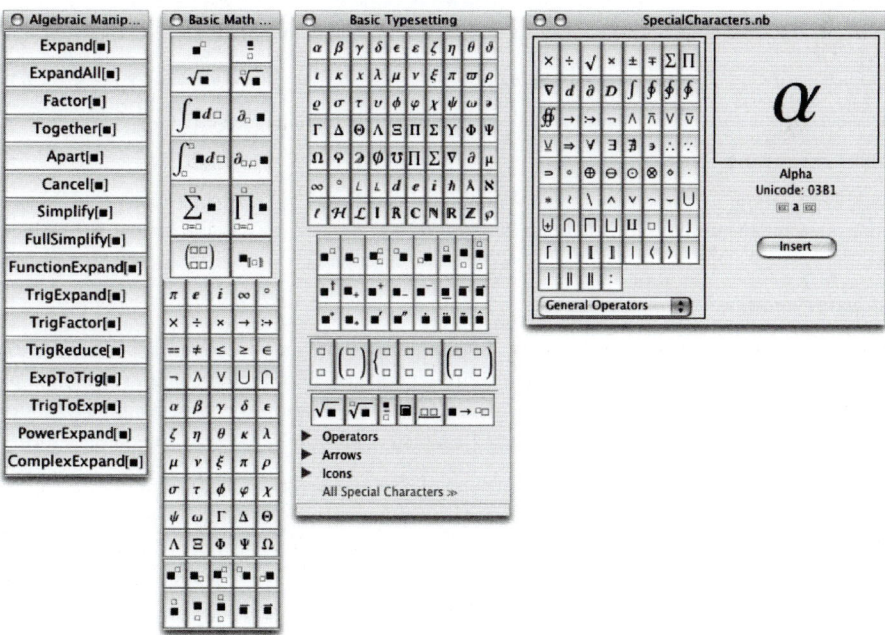

■ **AlgebraicManipulation**

The **AlgebraicManipulation** palette contains such commands as `Expand`, `Factor`, and `Simplify`. As an example, type the following:

   (f + g) ^ 6

Then select the whole expression with the mouse and click `Expand` in the palette. The expression is expanded. Then click `Factor` in the palette. The expression is now factored. In this way, whatever is currently selected in your notebook will be inserted into the position of the *selection placeholder*, ■, that can be seen in the commands of the palette.

■ **BasicMathInput**

The **BasicMathInput** palette contains buttons to perform some basic calculations and to input some basic symbols. Suppose you want to calculate the derivative of $x \sin(x) + \cos(x)$. First click the derivative button $\partial_\square \blacksquare$, then write **x**, press ⎇TAB⎇, write **(x Sin[x] + Cos[x])**, and execute the resulting command:

   ∂_x **(x Sin[x] + Cos[x])**   x Cos[x]

You can also do the following: Write **(x Sin[x] + Cos[x])**, select the whole expression, click the derivative button, press **x**, and execute.

For another example, suppose you want to calculate the definite integral of $x \sin(x) + \cos(x)$ on $(0, 2\pi)$. First click the integral button $\int_\square^\square \blacksquare \, d\square$, then write **0**, press ⎇TAB⎇, write **2**, click π on the palette, press ⎇TAB⎇, write **(x Sin[x] + Cos[x])**, press ⎇TAB⎇, write **x**, and execute:

   $\int_0^{2\pi}$ **(x Sin[x] + Cos[x]) dx** -2π

You can also do this the other way: Write first **(x Sin[x] + Cos[x])**, select the whole expression, click the integral button, and fill the limits and the integration variable with the help of ⎇TAB⎇.

This palette also contains buttons for powers, fractions, roots, sums, products, 2×2 matrices, and part extraction. Also included are the four basic symbols π (= 3.14159...), e (= 2.71828...), i (= $\sqrt{-1}$), and ∞ (= infinity).

■ **BasicTypesetting**

The **BasicTypesetting** palette contains many mathematical characters and constructs, useful especially in writing mathematical text with *Mathematica*. *Mathematica* as a writing environment is considered in Chapter 3.

■ **SpecialCharacters**

The **SpecialCharacters** palette contains all of the characters that can be entered into *Mathematica*. The characters are in groups such as Greek letters, script letters, general operators, and arrows. Just put the cursor in the place in your notebook where you want to add a character, choose a suitable group from the palette, and click on a character. The selected character can now be seen in an enlarged form. Then click *Insert*.

1.4.2 On-line Documentation

■ Documentation Center

Mathematica 6 used in notebook environments incorporates an excellent help system called the Documentation Center. Go to the **Help** menu and choose **Documentation Center** (or press [SHIFT][F1] in Windows or the Help key in Macintosh). The home page of the center appears:

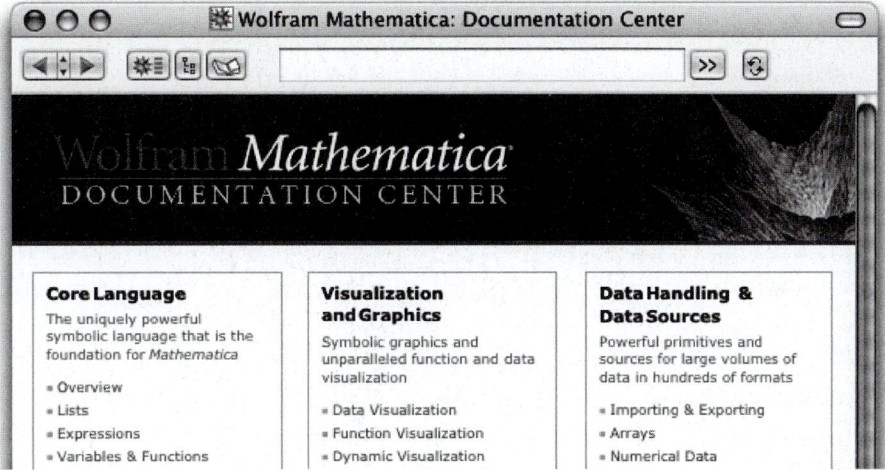

In the home page, the material about *Mathematica* is classified into seven topics such as Core Language and Visualization and Graphics. Inside each topic, we have a list of narrower topics; they are hyperlinks to the corresponding documents. In these documents, we have classified lists of suitable commands associated with the topic in question; such documents are called *guide documents*. Each command is again a hyperlink to a document in which the command is explained in detail; such documents are called *reference documents*. The guide and reference documents also have hyperlinks to documents explaining the use of the commands; such documents are called *tutorial documents*. Each document also has an input field for searching a topic.

The Documentation Center can also be used as follows. In your notebook (not in the Documentation Center), type a command such as **Solve**, leave the cursor at the end of the word, and press the F1 key (in Macintosh the Help key can also be pressed) or choose **Help ▷ Find Selected Function**. The page explaining the command appears.

Note that you can have multiple help windows open. Indeed, every time you choose **Help ▷ Documentation Center** a new help window opens.

The Documentation Center includes the equivalent of 50,000 pages of material, with more than 100,000 examples and more than 150,000 links. The center contains 345 guide documents, 655 tutorial documents, and thousands of reference documents (as of *Mathematica* 6.0.2).

The examples in the reference documents contain commands that have already been executed. You can also execute them anew and you can add new calculations. Note that all these calculations do not have any effect in your own notebook. For example, a package may have been loaded in a reference document. If you want to use the same package in your own document, you have to load the package in your document.

Note that the packages of *Mathematica* also have guide, tutorial, and reference documents; they can be accessed from the guide documents of the packages. Hyperlinks to these guide documents are given in Section 4.1.1, p. 94.

■ **Function Navigator and Virtual Book**

In addition to Documentation Center, *Mathematica* also contains two useful views of information: Function Navigator and Virtual Book (new in *Mathematica* 6.0.2). Below we show the Documentation Center, the Function Navigator, and the Virtual Book, all opened to show functions or topics in dynamic interactivity.

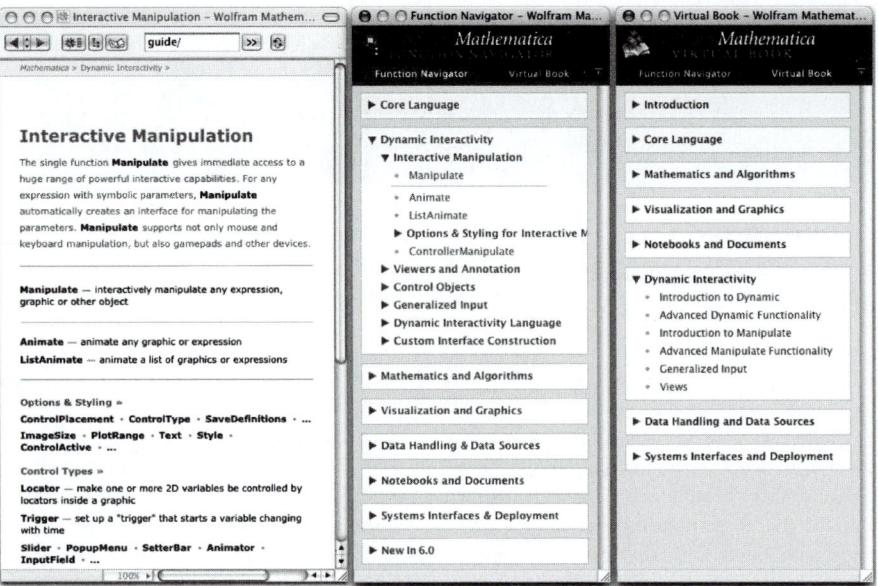

The Function Navigator and Virtual Book can be accessed from **Help ▷ Function Navigator** and **Help ▷ Virtual Book**. Actually, we only have one window that shows either the Function Navigator or the Virtual Book; the choice can be made at the top of the window. The main topics accessible in these windows are classified similarly as the topics in the Documentation Center.

A guide document shown by Documentation Center gives a short introduction to the topic and lists of commands with short descriptions for main commands; the page also has links to other guide pages (e.g., Options & Styling) that contain more complete lists of commands. The commands are links to the corresponding reference documents.

The Function Navigator centers to give lists of commands classified to several topics; each command is a link to the corresponding reference document. So, the Function Navigator can be used to quickly show lists of commands and to show reference documents.

The Virtual Book contains links to tutorial documents. The tutorials are comprehensive enough to constitute a virtual book about *Mathematica*. Users of previous versions of *Mathematica* have read *The Mathematica Book*, the complete manual of *Mathematica*, but in version 6 we no longer have the manual as a printed book. Instead, all documentation is now on-line. This has the advantages that there is no limit on the amount of material and the material can be continuously updated.

In addition to the Virtual Book, we mention the following guide and tutorial documents:

- guide/NewIn60AlphabeticalListing   • guide/SummaryOfNewFeaturesIn60
- guide/ListingOfNamedCharacters   • guide/ListingOfAllFormats
- tutorial/TraditionalFormReferenceInformation   • tutorial/IncompatibleChanges

See also the following web documents: Overview of *Mathematica*, Wolfram *Mathematica* 6, Wolfram *Mathematica* 6 Comparative Analyses, and The Wolfram Technology Guide.

■ Finding Information

With the Documentation Center, Function Navigator, and Virtual Book, finding suitable information is efficient and easy. The input field in the Documentation Center can also be used to search information about a given topic. The suitable document may open directly or, if there are several suitable documents, we get a list of links. From this list, we often easily find a suitable document.

If we would like to get directly to a document in the Documentation Center, we have to write its name. Generally, the names of the documents begin with *ref, guide,* or *tutorial*. Examples of names of documents are as follows:

Kernel:
ref/**Integrate**: detailed information about `Integrate`
guide/**Calculus**: lists of commands related to calculus
tutorial/**IndefiniteIntegrals**: a tutorial of indefinite integration

Packages:
LinearRegression/ref/**Regress**: detailed information about `Regress` from the **LinearRegression** package
LinearRegression/guide/**LinearRegressionPackage**: lists of commands from the package
LinearRegression/tutorial/**LinearRegression**: a tutorial of the package

Notebooks:
ref/character/**Rule**: detailed information about the → character
guide/**MathematicalTypesetting**: lists of commands related to mathematical typesetting
tutorial/**TwoDimensionalExpressionInputOverview**: a tutorial of entering 2D inputs

Formats:
ref/format/**EPS**: detailed information about the EPS format
guide/**ListingOfAllFormats**: a list of all formats
tutorial/**ImportingAndExportingData**: a tutorial of importing and exporting data

Messages:
ref/message/**General/plln**: detailed information about the General::plln message
guide/**Messages**: a list of commands related to messages
tutorial/**Messages**: a tutorial of messages

Menus:
ref/menuitem/**New**: explanation of the **File** ▷ **New** menu command
guide/**FileMenu**: a list of the **File** menu items

In the way, the documents of the Documentation Center are located at **Mathematica 6** ▷ **Documentation** ▷ **English** ▷ **System**. There, we have folders **ReferencePages**, **Guides**, and **Tutorials**.

■ Automatic Updates

Mathematica has an automated way to update the information in the Documentation Center. Information in the center is in the form of so-called *paclets*. Often they are documentation notebooks, but they may also be data. Wolfram Research has a paclet server containing the latest information. *Mathematica* in your computer weekly contacts this server and checks which paclets have updated documentation (the updates are not loaded at this stage, however). In this way, *Mathematica* maintains local indices of paclets in your computer. Now, when you ask information about, for example, **Solve**, *Mathematica* checks from the indices whether new information is available. If yes, *Mathematica* contacts the paclet server, loads the update, and then shows this latest information to you. If no, then an update is not needed and *Mathematica* shows the current information available in your computer.

Mathematica's use of the Internet can be controlled from **Help ▷ Internet Connectivity…**. Note that some data collection functions may not work without an Internet connection.

■ Introductions

If you are a new user of *Mathematica* and would like to study the basics of *Mathematica* 6, look at the following documents:

* **Help ▷ Startup Palette**: the *First Five Minutes with Mathematica* button
* **Help ▷ Virtual Book**: the *Introduction* item

If you are an old user of *Mathematica* and would like to study the new features in *Mathematica* 6, look at the following documents:

* **Help ▷ Startup Palette**: the *What's New in 6* link to Wolfram's website
* **Help ▷ Documentation Center**: the *New in 6* links in the home page
* **Help ▷ Documentation Center**: the guide/SummaryOfNewFeaturesIn60 document
* **Help ▷ Documentation Center**: the guide/NewIn60AlphabeticalListing document
* **Help ▷ Function Navigator**: the *New In 6* item

The Preface also contains information about the new functionality of *Mathematica* 6 and about obsolete properties.

■ Help Browser

The home page of the Documentation Center has a link *Installed Add-Ons*. By clicking this link we get a list of add-ons that have been installed into the so-called Help Browser of *Mathematica*. Prior to version 6, the Help Browser was the main help system. In version 6, the main help system is the Documentation Center, whereas the Help Browser only contains help material of add-ons of *Mathematica*. For example, the material on the CD-ROM of this book can be installed into the Help Browser, so that the book can be easily browsed within *Mathematica*.

1.4.3 Other Help

■ The Question Mark

With the question mark (**?**) and asterisk (*****), we can get lists of names containing certain characters. Suppose you are interested in finding a minimum of a function. Perhaps such a command has **Mini** in its name, and so we ask *Mathematica* to give a list of all names containing **Mini**:

`? *Mini*`

▼ System`

| DigitBlockMinimum | MinimalPolynomial | NMinimize |
| FindMinimum | Minimize | |

Now we can ask for information about these commands by clicking the names of the commands. For example, if we click `Minimize`, we get the following short information:

Minimize[f, {x, y, ...}] minimizes f with respect to x, y, ...
Minimize[{f, *cons*}, {x, y, ...}] minimizes f subject to the constraints *cons*.
Minimize[{f, *cons*}, {x, y, ...}, *dom*] minimizes
    with variables over the domain *dom*, typically Reals or Integers. ≫

By clicking the ≫ link, we get Documentation Center to open, and the information about `Minimize` is displayed.

Another way to use the question mark is to directly ask information about a particular command:

`? Minimize`

Minimize[f, {x, y, ...}] minimizes f with respect to x, y, ...
Minimize[{f, *cons*}, {x, y, ...}] minimizes f subject to the constraints *cons*.
Minimize[{f, *cons*}, {x, y, ...}, *dom*] minimizes
    with variables over the domain *dom*, typically Reals or Integers. ≫

`?Abcd*`	Give a table of all names beginning with `Abcd`
`?*abcd`	Give a table of all names ending with `abcd`
`?*abcd*`	Give a table of all names containing `abcd` somewhere
`?Name`	Give information about `Name`
`??Name`	Give also attributes and options of `Name`

The double question mark gives the same information as the single question mark and, in addition, information about attributes and options.

■ Completing Names and Making Templates

Mathematica has quite long names for some commands, but in notebook environments you can let *Mathematica* do some of the typing work. Use palettes or the following technique.

Suppose we want to write the command `InterpolatingPolynomial`. We first write, for example, `Interpo`, and then press ⌈CTRL⌉⌊k⌋ (or ⌘⌊k⌋ on a Macintosh); this is the same as choosing **Complete Selection** in the **Edit** menu. We get the following list:

InterpolatingFunction
InterpolatingPolynomial
Interpolation
InterpolationOrder
InterpolationPoints
InterpolationPrecision

Here are all the commands beginning with `Interpo`. From the list, we can choose the one we want by clicking with the mouse or by highlighting the appropriate command with the arrow keys and then pressing ⌊RET⌋. `Interpo` is then automatically completed according to your choice.

Now that we have the complete name of the command `InterpolatingPolynomial`, we can press [SHIFT][CTRL] k (or [SHIFT]⌘ k on a Macintosh); this is the same as choosing **Make Template** from the **Edit** menu. We get a template for the command:

`InterpolatingPolynomial`[{f_1, f_2, ...}, x]

This is useful if we do not remember the syntax of a command. Now we can replace {f_1, f_2, ...} with our data and x with our variable and then execute the command.

If the name of the command is uniquely determined by the first letters we have written, then [CTRL] k will complete the name directly, and [SHIFT][CTRL] k will complete the name and make a template for the command. Try typing `InterpolatingP` and then pressing [CTRL] k or [SHIFT][CTRL] k.

■ Syntax Coloring

In writing commands, *Mathematica* uses colors to indicate various aspects of syntax. To see the colors explained, open **Preferences** from a menu (in a Macintosh, this menu command is in the **Mathematica** menu) and go to *Appearance ▷ Syntax Coloring*. For example, variables for which we have not given a value are blue, syntax errors purple, emphasized syntax errors red on a yellow background, unrecognized options red, and arguments of functions green. Missing arguments are represented by a red ∧ symbol. The colors help in getting the commands into a correct form. From **Help ▷ Why the Coloring...** we can see real-time explanations of the colors as we write an input.

■ Balancing Paired Characters

When writing commands and programs with *Mathematica*, you may use many parentheses ( ), brackets [ ], and curly braces { }. It may be difficult to see whether they are correctly balanced. *Mathematica* helps with this as follows: Unbalanced characters are shown in purple, and each time you write),], or }, *Mathematica* highlights the corresponding (, [, or { for a short time.

You can also use **Check Balance** from the **Edit** menu or press [SHIFT][CTRL] b (or [SHIFT]⌘ b on a Macintosh). Place the cursor somewhere in the command and then press the previously mentioned key combination. The smallest balanced part containing the cursor will be highlighted; if a correctly balanced part cannot be found, nothing happens.

In the **Edit** menu, we also have the **Extend Selection** command ([CTRL] .). Put the cursor somewhere in a command. By repeatedly pressing [CTRL] . , larger and larger subexpressions are selected. Triple-click on a function to select the function and its arguments.

■ Help from Wolfram Research

Help can also be found at the website http://support.wolfram.com of Wolfram Research. To display the version of *Mathematica* you are using (and thus find the appropriate information), execute the command `$Version`. Execute `SystemInformation[]` to get a detailed summary of your *Mathematica* installation.

1.5 Editing

■ Correcting

In notebooks, you can use all the usual editing methods familiar from word processors. Look at the **Edit** menu for editing commands.

Note especially that *you can edit all old inputs and execute them anew*. To recalculate an input, simply place the cursor anywhere in the command and then execute the command.

In particular, if an input resulted in errors, you do not need to retype the input; simply correct the old input by means of standard editing and then execute the input again. Note also that the input and output numbers are assigned in the order of execution and not according to the physical order of the commands in the notebook.

■ Using Cells

You have probably noted the brackets at the right side of the *Mathematica* window. They indicate *cells*. An input is in a cell, the result is in another cell, and these two cells together form a higher-level cell.

A new input cell is automatically created when you start typing after a command is executed. A notebook is a structured document that is organized into a sequence of cells.

You can insert new cells between old cells. Simply place the cursor between two cells so that the cursor becomes horizontal, and then click with the mouse and start typing. In this way, you can insert new calculations among old ones.

The cells are handy for moving a part of a notebook to another place: Just click on the cell bracket so that it becomes black, and then cut or copy the cell and paste it to a new location. You can select several cells by dragging with the mouse over the cell brackets. You can copy material from one notebook to another notebook with the usual copy-and-paste techniques. Copying is easy to do by selecting cells.

If you want to re-execute several cells, select the cell brackets and then execute them in the standard way.

Double-clicking a cell bracket closes the cells inside it so that only the first cell is visible. In this way, you get a short outline of your notebook. Double-clicking the cell bracket of a closed cell opens the cell again. Double-clicking an input cell hides the output cell, and double-clicking an output cell hides the input cell.

■ Each Command into Its Own Cell

Mathematica is an interactive calculator that is designed such that we can easily proceed step by step: We execute one command and then proceed to the next command. A common bad habit is to write several commands in one cell and then execute all the commands at the same time. When this happens, the connection between inputs and the corresponding outputs becomes obscured. For example,

```
f = x Sin[x]
Integrate[f, x]
D[f, x]
```
$x \sin[x]$

$-x \cos[x] + \sin[x]$

$x \cos[x] + \sin[x]$

Here we put three commands into one cell. The outputs are shown one after another. This is not clear. In addition, if one of the commands contains an error, we have to execute all the commands anew instead of only re-executing the corrected command.

So, *write each command in its own cell:*

```
f = x Sin[x]
```
$x \sin[x]$

```
Integrate[f, x]
```
$-x \cos[x] + \sin[x]$

```
D[f, x]
```
$x \cos[x] + \sin[x]$

As previously stated, a new input cell is automatically created when you start typing after a command is executed. If you want to write several commands before you execute any of them, create a new cell for each command as follows: Write a command, press the down arrow key (↓), and start writing the next command. (Instead of the arrow key, you can also place the cursor below the last command so that the cursor becomes horizontal, click with the mouse, and start typing the next command.)

Note that if a calculation takes some time, you need not be idle. You can write (but not execute) new commands so that they are ready when the present command has been executed. You can also edit the notebook in any way you like while you await the execution of a command.

■ Editing Inputs

Sometimes you want to slightly modify an old input and then execute it again. You have several options. First, you can directly edit the old input and then re-execute it, as previously explained.

Second, you can select an old input or a part of it with the mouse, copy the selection, paste it to a new location, edit, and execute. A whole input can be selected by clicking the cell bracket, and the cell can then be copied and pasted.

Third, if the input you want to modify is the last input, you can get a copy of it by pressing ⟨CTRL⟩[l] (or ⌘[l] on a Macintosh); this is the same as choosing **Input from Above** in the **Insert** menu.

Mathematica divides long inputs automatically into several lines. However, you can also press the Return key (⟨RET⟩) here and there to make long code easier to read.

You can also execute a part of an input. Select with the mouse the part you want to execute, and then press ⟨ALT⟩⟨RET⟩ or ⌘⟨RET⟩ (this corresponds to **Evaluation ▷ Evaluate in Place**). Only the selected part is then executed.

■ Editing Outputs

You also have access to the results *Mathematica* writes. For example, you can copy a result or a part of it, paste the copy to a new cell, edit the new command, and then execute it.

If you want to edit the result of the last command, you can get a copy of the result by pressing ⟨SHIFT⟩⟨CTRL⟩[l] (or ⟨SHIFT⟩⌘[l] on a Macintosh); this is the same as choosing **Output from Above** in the **Insert** menu.

■ Opening a Saved Notebook

Note that *if you open a saved notebook and continue calculations, you cannot directly use any results in the saved document*. You have to recalculate all the commands with the results you need in your new session. Suppose, for example, that the saved notebook contains the result of `int = Integrate[Sqrt[x]` `Sin[x], x]`. When you open the notebook, `int` has no value. If you need the value of `int` in your new session, you have to execute anew the integrating command.

■ Writing a Document with *Mathematica*

You can write a whole document with *Mathematica* by calculating, plotting, and adding text comments to the results. To modify the look of the document, you can use *styles* and *style sheets*. These are considered in detail in Chapter 3.

After you have written your document with *Mathematica*, check spelling by placing the cursor at the beginning of the document and then choosing **Edit ▷ Check Spelling....** English is built into *Mathematica*, but you can buy dictionaries of other languages from Wolfram Research.

Sightseeing

Introduction

> *The relationship between pure and applied mathematics is based on trust and understanding:*
> *The pure mathematicians don't trust the applied mathematicians,*
> *and the applied mathematicians don't understand the pure mathematicians.*

Welcome all mathematicians, pure and applied, and everyone else, too, to a quick sightseeing tour through the vast and wonderful *Mathematica* factory, which produces graphics, eigenvalues, integrals, and so much more. We will visit the three main divisions of the factory: Graphics, Expressions, and Mathematics. In each division, we will show you only the most important or most basic machines (a total of approximately 60, leaving more than 3000 that are not shown). We will introduce each machine only briefly, but we encourage you to spend more time investigating and experimenting (at your own risk, of course; use a helmet, because every erroneous input is thrown out of the machine).

Later, we will explain all this and much more in detail, but the knowledge you get during this sightseeing tour may suffice for awhile. In fact, this short tour gives you snapshots of Chapters 5 through 28. Please note that what you will see in this chapter is not anything spectacular but only some very basic facts. Later, we will show you some much more impressive results.

2.1 Graphics

2.1.1 Graphics for Functions

■ **Graphics for 2D Functions**

> `Plot[f, {x, a, b}]` Plot **f** when **x** takes on values from **a** to **b**
>
> `Plot[{f1, f2}, {x, a, b}]` Plot **f1** and **f2** in the same figure
>
> `Show[p1, p2]` Show figures **p1** and **p2** superimposed

We plot two functions in the same figure (recall that π can be written as `Pi` or as ESC p ESC):

`Plot[{Sin[x], Cos[x]}, {x, 0, 2 π}]`

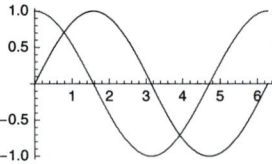

Note that the curves automatically have different colors. We can also first plot the two figures separately:

`p1 = Plot[Sin[x], {x, 0, 2 π}]`

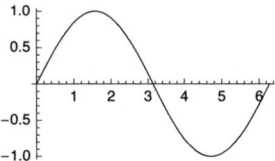

`p2 = Plot[Cos[x], {x, 0, 2 π}]`

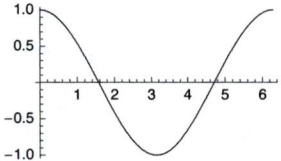

Then we can combine them (now the curves have the same color):

`Show[p1, p2]`

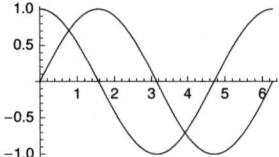

We could also do as follows:

Show[Plot[Sin[x], {x, 0, 2 π}], Plot[Cos[x], {x, 0, 2 π}]]

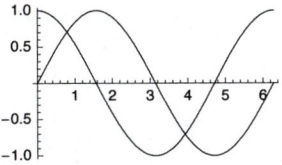

■ **Arranging Graphics**

{p1, p2} (⌘6) Show figures **p1** and **p2** side by side in a list form
Row[{p1, p2}] (⌘6) Show figures side by side in two graphics
GraphicsRow[{p1, p2}] (⌘6) Show figures side by side in one graphic
GraphicsGrid[{{p1, p2}, {p3, p4}}] (⌘6) Show figures as an array

The figures can be placed side by side:

{p1, p2}

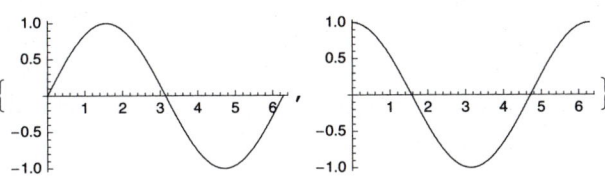

Here, note the list structure **{…, …}** of the output. In the following way we get the two figures side by side without the list structure:

Row[{p1, p2}]

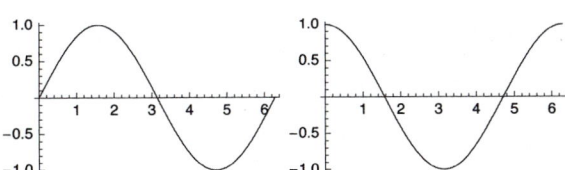

This output still contains two separate plots. If we like to get one figure containing the two plots, we can write as follows:

GraphicsRow[{p1, p2}, ImageSize → 260]

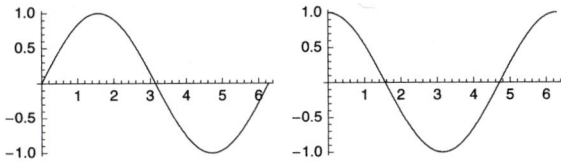

Here we used the **ImageSize** option to enlarge the figure (the arrow → can be written by pressing the hyphen and greater-than keys in turn; *Mathematica* will then replace them with a genuine arrow).

■ **Showing the Whole Function**

We can enter the expression to be plotted directly in the plotting command, as we have done thus far, but we can also first give a name to the expression:

```
f = x ^ 2 Exp[-x] Sin[x];
```

To save space, in this book we often suppress the display of outputs by ending the command with the semicolon. Then, plot the function:

```
Plot[f, {x, 0, 14}]
```

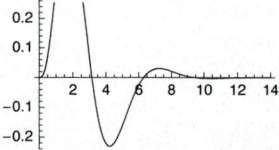

In this example, we do not see the whole function in the given interval. Indeed, sometimes *Mathematica* cuts a part of the figure out in order to give you a closer look at the more interesting parts of the curve. You can control the range of *y* values by using the `PlotRange` option. If you want to see the whole function, give the option `PlotRange → All`:

```
Plot[f, {x, 0, 14}, PlotRange → All]
```

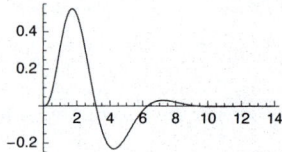

■ **Suppressing the Display**

With the semicolon, we can also suppress the display of graphics;

```
p1 = Plot[Sin[x], {x, 0, 2 π}];
p2 = Plot[Cos[x], {x, 0, 2 π}];
```

The plots were prepared, but they were not shown. The plots can then be superimposed or shown side by side:

```
Show[p1, p2]
```

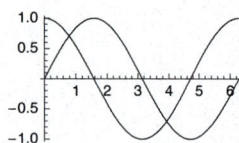

```
GraphicsRow[{p1, p2}, ImageSize → 230]
```

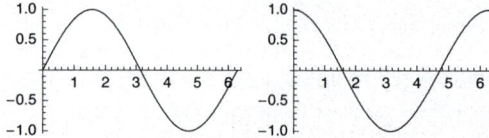

■ **Graphics for 3D Functions**

`Plot3D[f, {x, a, b}, {y, c, d}]`	Plot **f** as a surface
`ContourPlot[f, {x, a, b}, {y, c, d}]`	Plot **f** as contours
`DensityPlot[f, {x, a, b}, {y, c, d}]`	Plot **f** as a density plot

We plot a function of two variables as a surface, contour, and density plot:

```
f = Sin[x^2] Cos[Sqrt[y]];
```

```
{Plot3D[f, {x, 0, 3}, {y, 0, 4}],
 ContourPlot[f, {x, 0, 3}, {y, 0, 4}],
 DensityPlot[f, {x, 0, 3}, {y, 0, 4}]}
```

On the contours in a contour plot, the function takes on a constant value; dark areas are lower than light areas. By moving the mouse pointer over the contour plot (without pressing the mouse button), we can see the values of the contours. In the density plot, dark areas are also lower than light areas.

2.1.2 Graphics for Data

`ListPlot[data]`	Plot **data** as points
`ListPlot[data, Filling → Axis]` (❋6)	Plot **data** as points and vertical lines
`ListLinePlot[data]` (❋6)	Plot **data** as joining lines
`ListLinePlot[data, Mesh → All]` (❋6)	Plot **data** as joining lines and points

The data are given in either of the following forms:

```
{y1, y2, ..., yn}
```

```
{{x1, y1}, {x2, y2}, ..., {xn, yn}}
```

In the former case, the x values are automatically 1, 2, 3, …; in the latter case, we give explicit x values. Here are some data and three plots:

```
data = {{0, 5}, {1, 7}, {2, 8}, {3, 7}, {4, 9}, {5, 8},
    {6, 6}, {7, 5}, {8, 5}, {9, 4}, {10, 5}, {11, 3}, {12, 0}, {13, 1}};
```

```
GraphicsGrid[{{ListPlot[data], ListPlot[data, Filling → Axis]},
   {ListLinePlot[data], ListLinePlot[data, Mesh → All]}}, ImageSize → 300]
```

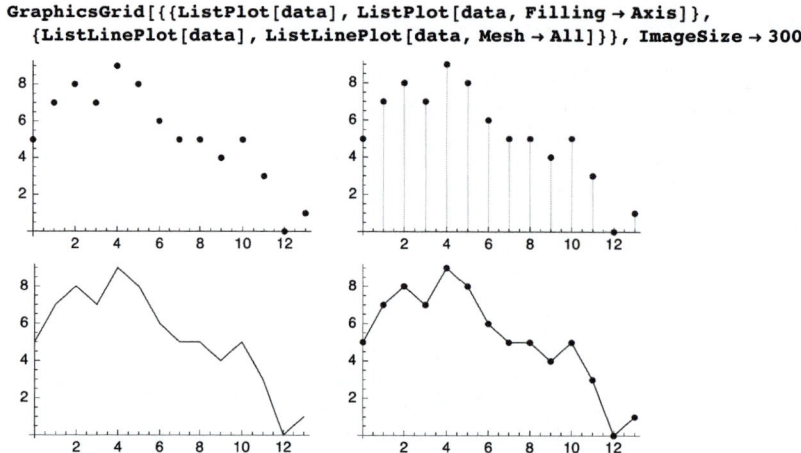

2.1.3 Manipulations

To create an interface enabling the interactive choice of the value of the parameter **u** with a slider and
showing the corresponding value of **expr**:

Manipulate[expr, {u, u$_{min}$, u$_{max}$}] (⌘6) u can have any value between u$_{min}$ and u$_{max}$

Manipulate[expr, {u, u$_{min}$, u$_{max}$, du}] u can have any value between u$_{min}$ and u$_{max}$ in steps of **du**

With a manipulation we can investigate how an expression changes when one or more parameters
change. The main use of manipulations is to study plots, but other expressions can also be manipulated.
Here are two examples. In the first manipulation, parameters **a** and **b** take on values from 0 to 2:

```
Manipulate[Plot[Sin[a x] Cos[b x], {x, 0, 2 π},
   PlotRange → {-1.05, 1.05}, ImageSize → 200], {a, 0, 2}, {b, 0, 2}]
```

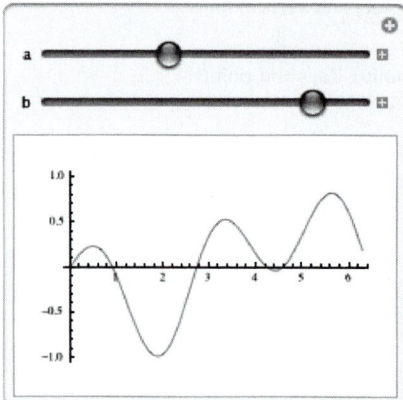

The sliders can be moved with the mouse so that we can see in real time how the plot changes. In the
second manipulation, the parameter **n** takes on values from 0 to 6 in steps of 1—that is, the discrete
values 0, 1, 2, ..., 6:

```
Manipulate[
 Plot[Sinc[n x], {x, 0, 2 π}, PlotRange → {-0.25, 1.05}, ImageSize → 200], {n, 0, 6, 1}]
```

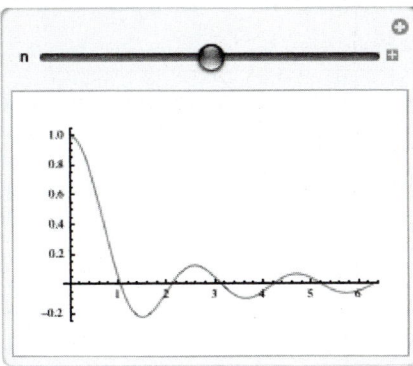

2.2 Expressions

2.2.1 Numbers and Expressions

■ **Numbers**

N[expr] or **expr//N** Calculate a decimal value of **expr**
N[expr, n] Calculate a decimal value of **expr** to **n**-digit precision

Here is a decimal value for π:

π // N   3.14159

Now we ask for a decimal value to 30-digit precision:

N[π, 30]   3.14159265358979323846264338328

RandomReal[{a, b}, n] (✿6) **n** random numbers from the interval (**a, b**)
RandomInteger[{k, l}, n] (✿6) **n** random numbers from the integers from **k** to **l**
SeedRandom[n] Reseed the random number generator with the integer n

Random numbers can be used in simulating various phenomena. Here are some uniform random numbers:

RandomReal[{0, 1}, 7]

{0.335378, 0.0610635, 0.892345, 0.0844273, 0.748569, 0.451667, 0.704084}

Then we simulate die tossing:

SeedRandom[1]; RandomInteger[{1, 6}, 24]

{5, 3, 5, 1, 2, 1, 1, 3, 1, 1, 4, 6, 3, 1, 4, 5, 5, 2, 4, 4, 5, 2, 5, 3}

Bad luck as usual: only one 6, although four were expected.

■ Calculating the Value of an Expression

> **x = a** Give **x** the value **a**
>
> **expr** Show the value of **expr** when **x** has the value **a**

If you want to calculate the value of an expression for a certain value of a variable, one possibility is to explicitly give the value for the variable and then ask the value of the expression:

```
expr = Sin[x] Cos[x];

x = π / 6;

expr
```
$$\frac{\sqrt{3}}{4}$$

This method has the drawback that from now on **x** has the value $\pi/6$ in all expressions, and this may give unintended results. For example, if you now try to calculate the derivative of the expression, you get an error message:

```
D[expr, x]
```

General::ivar : $\frac{\pi}{6}$ is not a valid variable. ≫

$$\partial_{\frac{\pi}{6}} \frac{\sqrt{3}}{4}$$

Mathematica could not calculate the derivative with respect to a constant $\pi/6$. So, if you give values for variables, remember to remove the values when you no longer need them:

```
x =.
```

> **expr /. x → a** Replace **x** by **a** in **expr**

The arrow can be written as **->** (*Mathematica* automatically replaces these characters by a genuine arrow →). This is the recommended method to calculate the value of an expression for a value of a variable. For example:

```
expr /. x → π / 6
```
$$\frac{\sqrt{3}}{4}$$

This is a very important technique. Here, **x → $\pi/6$** is a transformation rule. It can be applied to any expression by preceding the transformation rule with **/.** . Note that now **x** has no value:

```
x          x
```

Another example:

```
ArcSin[x] /. x → 1
```
$$\frac{\pi}{2}$$

■ Manipulating Expressions

> **Simplify[expr]** Simplify **expr**
>
> **FullSimplify[expr]** Simplify **expr** thoroughly
>
> **Factor[expr]** Factor **expr**
>
> **Expand[expr]** Expand **expr**
>
> **Apart[expr]** Give the partial fraction expansion of **expr**

Note that these commands can also be used in the following way: **expr // Simplify**. For example,

a = (1 + x)^2 + (1 + x) (2 + x) $(1 + x)^2 + (1 + x) (2 + x)$

a // Simplify $3 + 5 x + 2 x^2$

a // Factor $(1 + x) (3 + 2 x)$

a // Expand $3 + 5 x + 2 x^2$

FullSimplify is often good for simplifying special functions. In the following example, **Simplify** does not work, but **FullSimplify** does:

n! / (n – 1)! // Simplify $\dfrac{n!}{(-1 + n)!}$

n! / (n – 1)! // FullSimplify n

We calculate a partial fraction expansion:

(1 + x + x^2 – x^3) / (x + 2)^2 // Apart

$$5 - x + \frac{11}{(2 + x)^2} - \frac{15}{2 + x}$$

■ Function Application

f[expr]	Standard function application
expr // f	Postfix function application
f @ expr	Prefix function application

Thus far, we have seen that we can write **N[expr]** or **Simplify[expr]** but we can also write **expr // N** and **expr // Simplify**. The use of the square brackets is the standard way to apply functions. The use of **//** is called a postfix function application. We also have a prefix function application: **N @ expr**, **Simplify @ expr**. In the following example, we use all three function applications:

{N[π], π // N, N@π}

{3.14159, 3.14159, 3.14159}

The standard notation also applies for functions with several arguments: **f[x, y]**. The postfix and prefix notations can only be used for functions with one argument.

■ Some Display Techniques

Sometimes we need not see the result of a computation. For example, we already know the result for certain, the result is so large an expression that it is useless to see it, or it takes too much time to have it displayed on the screen. We can prohibit displaying the result by ending the command with a semicolon (**;**).

expr;	Calculate the value of **expr** but do not display the result

We do not want to see 100! (a number with 158 digits):

a = 100!;

a / 99! 100

We can execute several commands at the same time using the semicolon:

> `expr1; expr2; expr3`   Calculate the expressions; display the last result
> `expr1; expr2; expr3;`   Calculate the expressions; do not display anything

We calculate other factorials and display only the final result:

> `b = 97 !; c = 3 !; a / (b c)`   161 700

If you want to display the values of all expressions, you can place the expressions in a list with curly braces:

> `{expr1, expr2, expr3}`   Calculate the expressions; display all values

> `{Sin[π / 4], Sin[π / 5], Sin[π / 6], Sin[π / 7]}`

$$\left\{\frac{1}{\sqrt{2}}, \sqrt{\frac{5}{8} - \frac{\sqrt{5}}{8}}, \frac{1}{2}, \sin\left[\frac{\pi}{7}\right]\right\}$$

> `% // N`   `{0.707107, 0.587785, 0.5, 0.433884}`

For long expressions, it often suffices to see only some parts. This can be done with **Short**:

> `Short[expr]`   Give **expr** in a shortened form
> `Short[expr, c]`   Give **expr** in a shortened form having length **c**

We generate 50 uniform random numbers but show only a few of them:

> `t = RandomReal[{0, 1}, 50];`

> `Short[t]`
> `{0.214152, ≪48≫, 0.130847}`

> `Short[t, 4]`
> `{0.214152, 0.613783, 0.767945, ≪44≫, 0.011151, 0.44259, 0.130847}`

In the first case 48 values and in the latter case 44 values are not shown. (To find an appropriate length, such as 4, for the expression may take some experimenting.)

Before we continue, we clear the values of **a**, **b**, **c**, and **t**. We could write **a=.; b=.; c=.; t=.**, but a more convenient way is the following:

> `Clear[a, b, c, t]`

2.2.2 Lists and Tables

■ **Lists**

> `{a, b, c}` A 1D list
> `list[[i]]`   ith part of **list**
> `Length[list]` Number of elements in **list**
> `Sort[list]` Sort the elements of **list** into canonical order
>
> `{{a, b, c, d}, {e, f, g, h}}` A 2D list
> `list[[i, j]]` (i, j)th part of **list**

In place of `[[…]]` we can also use `⟦…⟧`; here, `⟦` and `⟧` can be written as ESC `[` ESC and ESC `]` ESC.

Lists are very basic objects in *Mathematica;* you will use them all the time. Vectors and matrices are in fact lists, and in many other computations you need lists. Lists can have as many elements as you want them to have (an empty list is `{}`). Lists with lists as elements are 2D, 3D, and higher-dimensional lists. We define two lists:

```
a = {x, y, z, v};
b = {{3, 2, 5, 4}, {4, 1, 6, 2}, {3, 1, 1, 6}};
```

Picking a part of a list is easy:

```
a[[3]]      z
b[[2]]      {4, 1, 6, 2}
b[[2, 3]]      6
```

Here are the lengths of the lists:

```
Length[a]      4
Length[b]      3
```

We try **Sort**:

```
Sort[{r, 4, P, 2, q, p, 3}]
{2, 3, 4, p, P, q, r}
```

Note that calculations with lists are automatically done element by element:

```
2 {3, 2, 5, 4}      {6, 4, 10, 8}
{3, 2, 5, 4}^2      {9, 4, 25, 16}
{3, 2, 5, 4} + {4, 1, 6, 2}      {7, 3, 11, 6}
```

■ Tables

`MatrixForm[m]`	Display matrix `m` in a 2D matrix form
`TableForm[m]`	Display `m` in a 2D tabular form
`Grid[m]` (⌘6)	Display `m` in a 2D tabular form
`Row[v]` (⌘6)	Form a row from a list `v`

Matrices can be displayed handily with **MatrixForm**:

```
b // MatrixForm
```

$$\begin{pmatrix} 3 & 2 & 5 & 4 \\ 4 & 1 & 6 & 2 \\ 3 & 1 & 1 & 6 \end{pmatrix}$$

TableForm prints a table:

```
b // TableForm
```

```
3  2  5  4
4  1  6  2
3  1  1  6
```

Grid also prints a table:

```
b // Grid
3  2  5  4
4  1  6  2
3  1  1  6
```

As we will see in Chapter 15, with **Grid** we can form advanced tables. With **Row** we can put elements side by side:

```
Row[{"1000th prime is ", Prime[1000]}]
1000th prime is 7919
```

■ **Statistics**

> **Total[list]**   The sum of the elements of **list**
> **Accumulate[list]** (⌘6)   The cumulative sums of the elements of **list**
> **Tally[list]** (⌘6)   The frequencies of the elements of **list**
> **Mean[list], Variance[list], StandardDeviation[list]**

We simulate the tossing of a die 20 times:

```
c = RandomInteger[{1, 6}, 20]
{4, 4, 3, 3, 2, 4, 6, 4, 1, 5, 3, 1, 6, 5, 2, 4, 1, 2, 3, 1}
```

Calculate the sum, mean, (unbiased) variance, and standard deviation:

```
{Total[c], Mean[c], Variance[c], StandardDeviation[c]} // N
{64., 3.2, 2.58947, 1.60918}
```

The cumulative sums are as follows:

```
Accumulate[c]
{4, 8, 11, 14, 16, 20, 26, 30, 31, 36, 39, 40, 46, 51, 53, 57, 58, 60, 63, 64}
```

Here are the frequencies:

```
Tally[c] // Sort
{{1, 4}, {2, 3}, {3, 4}, {4, 5}, {5, 2}, {6, 2}}
```

■ **Forming Lists**

> **Range[m]**   Form the list {1, 2, …, **m**}
> **Range[m, n]**   Form the list {**m**, **m** + 1, …, **n**}
> **Range[m, n, d]**   Form the list {**m**, **m** + **d**, **m** + 2 **d**, …, **n**}

With **Range**, we can easily form equally spaced numbers:

```
Range[6]     {1, 2, 3, 4, 5, 6}

Range[0, 6]     {0, 1, 2, 3, 4, 5, 6}

Range[0, 7, 2]     {0, 2, 4, 6}
```

> **Table[expr, {i, a, b}]**   Form a list of values of **expr** when **i** takes on values from **a** to **b** (in steps of 1)
> **Table[expr, {i, a, b}, {j, c, d}]**   Index **i** takes on values from **a** to **b** and, for each **i**, **j** takes on values from **c** to **d**

Table is one of the most useful commands in *Mathematica*. It forms a list from a general rule. Iteration specification of the form **{i, a, b}** is most common, but other forms can also be used:

{n}	Form a list from **n** values of **expr**
{i, b}	Index **i** has values from 1 to **b** (in steps of 1)
{i, a, b}	Index **i** has values from **a** to **b** (in steps of 1)
{i, a, b, d}	Index **i** has values from **a** to **b** in steps of **d**

As can be seen, if the starting value of **i** is 1, it can be left out (but it can also be written), and if the step size is 1, it, too, can be left out. For example,

```
Table[0, {10}]
```

$\{0, 0, 0, 0, 0, 0, 0, 0, 0, 0\}$

```
Table[n!, {n, 10}]
```

$\{1, 2, 6, 24, 120, 720, 5040, 40\,320, 362\,880, 3\,628\,800\}$

```
Table[Sin[n π / 6], {n, 0, 6}]
```

$$\left\{0, \frac{1}{2}, \frac{\sqrt{3}}{2}, 1, \frac{\sqrt{3}}{2}, \frac{1}{2}, 0\right\}$$

```
Table[Exp[x], {x, 0., 3., 0.5}]
```

$\{1., 1.64872, 2.71828, 4.48169, 7.38906, 12.1825, 20.0855\}$

```
Table[1 / (i + j - 1), {i, 4}, {j, 4}]
```

$$\left\{\left\{1, \frac{1}{2}, \frac{1}{3}, \frac{1}{4}\right\}, \left\{\frac{1}{2}, \frac{1}{3}, \frac{1}{4}, \frac{1}{5}\right\}, \left\{\frac{1}{3}, \frac{1}{4}, \frac{1}{5}, \frac{1}{6}\right\}, \left\{\frac{1}{4}, \frac{1}{5}, \frac{1}{6}, \frac{1}{7}\right\}\right\}$$

Often it is useful to make pairs of the value of the index and the corresponding value of the expression:

```
Table[{x, Exp[x]}, {x, 0., 3.}]
```

$\{\{0., 1.\}, \{1., 2.71828\}, \{2., 7.38906\}, \{3., 20.0855\}\}$

In the following example, we study the recursion formula $x_{i+1} = 3.7\,x_i(1 - x_i)$, $i = 0, 1, 2, \dots$ We start from $x_0 = 0.5$ and do 200 iterations:

```
x = 0.5; t = Table[x = 3.7 x (1 - x), {200}];
```

```
ListPlot[t, AspectRatio → 0.4, PlotRange → {-0.05, 1.05}, ImageSize → 300]
```

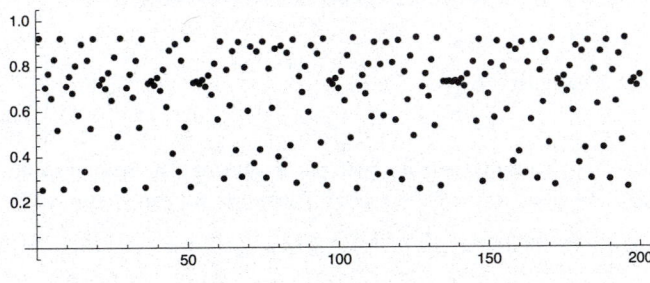

```
Clear[a, b, c, f, x, t]
```

■ Advanced List Manipulation

Sometimes we want to tabulate an expression for such irregular values of a variable that cannot be formed by an iteration specification; instead, we have the values of the variable as a list. **Table** can also be used in such cases, although **Map** is an alternative. **Map** can also be replaced with **/@**:

> **Table[f[x], {x, {a, b, c}}]** (❀6) Calculate {f[a], f[b], f[c]}
>
> **Map[f[#]&, {a, b, c}]** Calculate {f[a], f[b], f[c]}
>
> **f[#] & /@ {a, b, c}** Calculate {f[a], f[b], f[c]}

Examples:

> **Table[f[x], {x, {a, b, c}}]**   {f[a], f[b], f[c]}
>
> **Map[f[#] &, {a, b, c}]**   {f[a], f[b], f[c]}
>
> **f[#] & /@ {a, b, c}**   {f[a], f[b], f[c]}

Other examples:

> **Table[x^2, {x, {a, b, c}}]** $\{a^2, b^2, c^2\}$
>
> **Map[#^2 &, {a, b, c}]** $\{a^2, b^2, c^2\}$
>
> **#^2 & /@ {a, b, c}** $\{a^2, b^2, c^2\}$

Note that the following is the easiest way to form the squares of the elements of a list:

> **{a, b, c}^2** $\{a^2, b^2, c^2\}$

Map or **/@** is one of the most useful commands for manipulating lists. In this book, we use **Map** extensively. With **Map** we can map each element of a given list with a given function. The effect of **Map** is that each element of the list is substituted in turn for **#**, and a list is formed from the results. The function is given in a special form having the name *pure function*. In such a function, the argument is expressed as **#**, and at the end of the function we have **&**.

Next, we tabulate values of the sin function:

> **Sin[# π / 6] & /@ {1, 2, 4, 6}**
>
> $\left\{ \dfrac{1}{2}, \dfrac{\sqrt{3}}{2}, \dfrac{\sqrt{3}}{2}, 0 \right\}$

The result is the value of $\sin(n\pi/6)$ for $n = 1, 2, 4, 6$. Recall the **b** we defined previously:

> **b = {{3, 2, 5, 4}, {4, 1, 6, 2}, {3, 1, 1, 6}};**

Now we find the maximum element of each list of **b**:

> **Max[#] & /@ b**   {5, 6, 6}

If the function to be mapped is a single built-in command with one argument (e.g., **Max**), it suffices to write the name of the function (i.e., we need not write **#** and **&**). Therefore, we can simply write the following:

> **Max /@ b**   {5, 6, 6}
>
> **b =.**

2.2.3 Functions and Programs

■ **Functions**

`f[x_] := expr` Define a function

When defining a function, it is important to remember the underscore (_) after each argument (`x_` is actually a *pattern*). The underscore makes the function capable of calculating the value of the function for *any* value of the argument. The colon (`:`) before the equals sign results in the value of `expr` not being calculated in the definition; the value is calculated only when using the function.

In everyday use of *Mathematica*, we rarely need to define functions for expressions to be, for example, differentiated, integrated, or plotted. Mostly we can use the expressions as such or we can give a name to the expression and then use that name.

For example, consider integration. First, we directly enter the expression to be integrated into the command:

```
Integrate[x / (a + x), x]     x - a Log[a + x]
```

Then we give a name for the expression and use that name:

```
f = x / (a + x);

Integrate[f, x]     x - a Log[a + x]
```

We can also define a function and then use it:

```
g[x_] := x / (a + x)

Integrate[g[x], x]     x - a Log[a + x]
```

However, this is unnecessarily complicated. The first two methods are the most useful. Giving a name is especially handy if we do several calculations with the same expression.

Function definitions are mostly used to form more complicated functions and programs. For example, here is a function for calculating the characteristic polynomial of a matrix:

```
charPoly[m_, x_] := Det[m - x IdentityMatrix[Length[m]]]
```

The following is an example of using the function:

```
charPoly[{{2, 5}, {3, 1}}, x]     -13 - 3 x + x²
```

■ **Programs**

`f[x_] := Module[{local variables}, body]` Define a function as a module

More complicated programs are often written as modules. In a module, we can define local variables that are only used within the module and that have no values outside the module.

As an example, we develop a program to simulate random walk, in which the object starts at zero and moves one step up or down, each with a probability of 0.5. In the following way, we can generate the steps:

```
SeedRandom[2];
RandomInteger[{0, 1}, 20]
{1, 0, 1, 1, 1, 0, 0, 0, 1, 1, 1, 0, 0, 1, 0, 0, 1, 0, 1, 1}
```

```
2 % - 1
{1, -1, 1, 1, 1, -1, -1, -1, 1, 1, 1, -1, -1, 1, -1, -1, 1, -1, 1, 1}
```

Here, we first generated 20 random 0's and 1's. Each element of this list is multiplied with 2, and 1 is then subtracted from each element. The result is a list of random 1's and −1's. The random walk is the cumulative sum of the steps. Cumulative sums can easily be calculated with **Accumulate**:

```
Accumulate[%]
{1, 0, 1, 2, 3, 2, 1, 0, 1, 2, 3, 2, 1, 2, 1, 0, 1, 0, 1, 2}
```

Thus, a program for a random walk with n steps could be written as follows:

```
randomWalk[n_] := Module[{steps, walk},
  steps = 2 RandomInteger[{0, 1}, n] - 1;
  walk = Accumulate[steps];
  ListLinePlot[walk, ImageSize → 200]]
```

The variables **steps** and **walk** in the program are local and thus they have no value outside the module. We simulate 200 steps of the random walk:

```
SeedRandom[2];
randomWalk[200]
```

Commands in a program can often be nested, as in the following:

```
randomWalk2[n_] :=
  ListLinePlot[Accumulate[2 RandomInteger[{0, 1}, n] - 1]]
```

In this way, the code of the program becomes shorter, but the readability of the code may also become weaker.

2.3 Mathematics

2.3.1 Differential and Integral Calculus

■ **Differential Calculus**

D[f, x]	Derivative of **f** with respect to **x**
D[f, x, x]	Second-order derivative of **f** with respect to **x**
D[f, x, y]	Mixed second-order derivative of **f** with respect to **x** and **y**
Series[f, {x, a, n}]	nth-order Taylor series of **f** with respect to **x** at **a**
Limit[f, x → a]	Limit of **f** as **x** approaches **a**

Here are some examples:

```
a = x Sin[y];
```

`{D[a, x], D[a, y], D[a, x, x], D[a, x, y], D[a, y, y]}`

`{Sin[y], x Cos[y], 0, Cos[y], -x Sin[y]}`

`Series[Sin[Sqrt[x]], {x, 0, 4}]`

$$\sqrt{x} - \frac{x^{3/2}}{6} + \frac{x^{5/2}}{120} - \frac{x^{7/2}}{5040} + O[x]^{9/2}$$

`Limit[(1 + c / x) ^ x, x → ∞]` e^c

(Recall that ∞ can be written as **Infinity** or as ⎡ESC⎤inf⎡ESC⎤.)

■ **Integral Calculus**

`Integrate[f, x]`   Indefinite integral of **f** with respect to **x**

`Integrate[f, {x, a, b}]`   Definite integral of **f** when **x** varies from **a** to **b**

`NIntegrate[f, {x, a, b}]`   Calculate the definite integral by numerical methods

`Sum[f, {i, a, b}]`   Sum of the values of **f** when **i** varies from **a** to **b**

Prepare to see special functions when you integrate functions that are not easy:

`a = Integrate[Sin[Exp[x]], x]` $\mathrm{SinIntegral}\left[e^x\right]$

This is one of the many special functions in *Mathematica*. Do not worry! You can do the same with the special functions as you do with the more usual functions. For example, you can check the result by differentiation:

`D[a, x]` $\mathrm{Sin}\left[e^x\right]$

You can ask for a value:

`a /. x → 1.`   `1.82104`

You can ask for a plot:

`Plot[a, {x, 0, 3}]`

`a =.`

Sometimes even *Mathematica* does not know an integral:

`Integrate[Sin[Sin[x]], {x, 0, 1}]` $\int_0^1 \mathrm{Sin}[\mathrm{Sin}[x]]\, dx$

Mathematica just writes the command as such. You can then resort to numerical integration (Gaussian quadrature):

`NIntegrate[Sin[Sin[x]], {x, 0, 1}]`   `0.430606`

Sums are calculated like integrals:

`Sum[1 / 2 ^ n, {n, 1, 10}]` $\dfrac{1023}{1024}$

$$\texttt{Sum[1 / n \textasciicircum 2, \{n, 1, ∞\}]} \qquad \frac{\pi^2}{6}$$

$$\texttt{Sum[r\textasciicircum n, \{n, 1, m\}]} \qquad \frac{r\,(-1 + r^m)}{-1 + r}$$

2.3.2 Matrices

> **{a, b, c}** A vector
> **{{a, b, c, d}, {e, f, g, h}}** A matrix with two rows

A 1D list is also a vector; a 2D list is a matrix. Vectors and matrices can have as many elements as you want. *Mathematica* does not distinguish column and row vectors but, nevertheless, it does calculations with matrices and vectors so that the results are, almost always, what you intended. We define a vector:

    a = {2, 5};

A useful fact to know is that *Mathematica* automatically does all operations with vectors element by element:

    {a^2, Sqrt[a], a / {3, 6}}

$$\left\{\{4,\,25\},\,\left\{\sqrt{2}\,,\,\sqrt{5}\,\right\},\,\left\{\frac{2}{3},\,\frac{5}{6}\right\}\right\}$$

Here is a matrix:

    MatrixForm[m = {{2, 1}, {3, 2}}]

$$\begin{pmatrix} 2 & 1 \\ 3 & 2 \end{pmatrix}$$

Note that **MatrixForm** is used only in displaying matrices. You cannot do any calculations with such a form. If you would write **m = {{2, 1}, {3, 2}} // MatrixForm**, then the value of **m** would be the *matrix form* of the given matrix, and with such an **m** we cannot calculate. However, you could write **(m = {{2, 1}, {3, 2}}) // MatrixForm**.

> **a m** The product of a scalar **a** and a vector or matrix **m**
> **m + n** The sum of two vectors or matrices **m** and **n**
> **m^2** The squares of the elements of a vector or matrix **m**
> **m.n** The product of two vectors or matrices **m** and **n**
> **Transpose[m]** or **m$^\intercal$** The transpose of a matrix **m** (write $^\intercal$ as ⎋tr⎋)
> **Det[m]** The determinant of a square matrix **m**
> **Inverse[m]** The inverse of a square matrix **m**
> **Eigenvalues[m]** The eigenvalues of a square matrix **m**

Note that the point (.) has to be used when calculating products of vectors and matrices; the space and the asterisk do not work properly. You cannot use powers, either. Thus, to calculate the second matrix power of a matrix **m**, you have to write **m.m**; you cannot write **m^2** (this only calculates the squares of each element). Also, to calculate the inverse of a matrix **m**, you have to write **Inverse[m]**; you cannot write **m^-1**. The transpose command $^\intercal$ can be written as ⎋tr⎋.

In the following example, **a** is interpreted to be a row vector:

    a.m   {19, 12}

Here, **a** is a column vector:

m.a   {9, 16}

We calculate the square, transpose, determinant, inverse, and eigenvalues of **m**:

m.m   {{7, 4}, {12, 7}}

m$^\mathsf{T}$ {{2, 3}, {1, 2}}

Det[m]   1

Inverse[m]   {{2, -1}, {-3, 2}}

Eigenvalues[m] $\left\{2 + \sqrt{3}\,,\, 2 - \sqrt{3}\,\right\}$

Clear[a, m]

2.3.3 Equations

■ **Polynomial Equations: Exact Solutions**

> **expr1 == expr2** An equation (== can be written as ==)
> **Solve[eqn, x]** Solve a (polynomial) equation with respect to **x**
> **Solve[{eqn1, eqn2}, {x, y}]** Solve two (polynomial) equations with respect to **x** and **y**

Equations are formed with two equal signs (==), but *Mathematica* replaces them with the special symbol ==. Forgetting the second = is a common error; remember that = is used only to assign values for variables.

Here is a polynomial equation familiar to you (we give the name **eqn** to this equation):

eqn = a x^2 + b x + c == 0

$c + b\,x + a\,x^2 == 0$

sol = Solve[eqn, x]

$$\left\{\left\{x \to \frac{-b - \sqrt{b^2 - 4\,a\,c}}{2\,a}\right\}, \left\{x \to \frac{-b + \sqrt{b^2 - 4\,a\,c}}{2\,a}\right\}\right\}$$

The result is in the form of transformation rules. If you want only the values of **x**, apply the transformation rule to **x** (see Section 2.2.1, p. 32):

x /. sol

$$\left\{\frac{-b - \sqrt{b^2 - 4\,a\,c}}{2\,a},\, \frac{-b + \sqrt{b^2 - 4\,a\,c}}{2\,a}\right\}$$

We can also check that the solution is correct by inserting the solution into the equation:

eqn /. sol // Simplify

{True, True}

Then we solve two linear equations (larger systems are solved similarly). Enclose a system of equations and the variables within curly braces (**{ }**):

Solve[{2 x + 5 y == 4, x - 3 y == 3}, {x, y}]

$$\left\{\left\{x \to \frac{27}{11},\, y \to -\frac{2}{11}\right\}\right\}$$

■ **Polynomial Equations: Numerical Solutions**

`NSolve[eqn, x]`   Solve a (polynomial) equation with numerical methods

Polynomial equations of a degree higher than four can rarely be solved:

`eqn2 = x^5 - x^3 + x^2 - 2 == 0;`

`Solve[eqn2, x]`

$\left\{\left\{x \to \text{Root}\left[-2 + \#1^2 - \#1^3 + \#1^5 \&, 1\right]\right\},\right.$
$\left\{x \to \text{Root}\left[-2 + \#1^2 - \#1^3 + \#1^5 \&, 2\right]\right\}, \left\{x \to \text{Root}\left[-2 + \#1^2 - \#1^3 + \#1^5 \&, 3\right]\right\},$
$\left.\left\{x \to \text{Root}\left[-2 + \#1^2 - \#1^3 + \#1^5 \&, 4\right]\right\}, \left\{x \to \text{Root}\left[-2 + \#1^2 - \#1^3 + \#1^5 \&, 5\right]\right\}\right\}$

We did not obtain the solution in an explicit form (*Mathematica* only gives a symbolic list representing the five roots). We can now ask a decimal value:

`% // N`

$\{\{x \to 1.17525\}, \{x \to -1.09595 - 0.361002\, i\}, \{x \to -1.09595 + 0.361002\, i\},$
$\{x \to 0.508323 - 1.00984\, i\}, \{x \to 0.508323 + 1.00984\, i\}\}$

We can also directly resort to numerical methods:

`NSolve[eqn2, x]`

$\{\{x \to -1.09595 - 0.361002\, i\}, \{x \to -1.09595 + 0.361002\, i\},$
$\{x \to 0.508323 - 1.00984\, i\}, \{x \to 0.508323 + 1.00984\, i\}, \{x \to 1.17525\}\}$

■ **Transcendental Equations**

`Solve` can solve some transcendental equations:

`Solve[Exp[a x] == b, x]`

Solve::ifun : Inverse functions are being used by Solve, so some

solutions may not be found; use Reduce for complete solution information. ≫

$\left\{\left\{x \to \dfrac{\text{Log}[b]}{a}\right\}\right\}$

However, `FindRoot` is the general-purpose command for such equations. It calculates a zero iteratively by Newton's method and other methods.

`FindRoot[eqn, {x, x0}]`   Solve an equation with numerical methods, starting from **x0**

To find a zero for the following function, we first plot it:

`f = Exp[-x] - 0.5 x;`

`Plot[f, {x, 0, 3}]`

The zero seems to be near one, and so we start from this point:

`x0 = FindRoot[f == 0, {x, 1}]` $\{x \to 0.852606\}$

The value of the function at this point is zero, with a high degree of accuracy:

    f /. x0 -5.55112×10^{-17}

2.3.4 Optimization

■ **Global Optimization: Exact Solutions**

Minimize[f, vars]   Give the global minimum of **f** with respect to variables **vars**
Minimize[{f, cons}, vars]   Minimize subject to constraints **cons**

We also have **Maximize** that is used in the same way. Consider the following function:

    f = x^4 - 2 x^3 - 2 x^2 - 1;

    Plot[f, {x, -2, 3}]

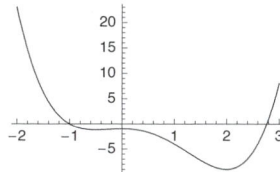

The function has a finite global minimum point but not a finite global maximum point:

    Minimize[f, x] $\{-9, \{x \to 2\}\}$

    Maximize[f, x]

    Maximize::natt : The maximum is not attained at any point satisfying the given constraints. ≫
    $\{\infty, \{x \to -\infty\}\}$

The function also has a *local* minimum point and a *local* maximum point. They can be found by constraining *x* suitably:

    Minimize[{f, x < 0}, x] $\left\{-\dfrac{19}{16}, \left\{x \to -\dfrac{1}{2}\right\}\right\}$

    Maximize[{f, -1 / 2 < x < 2}, x] $\{-1, \{x \to 0\}\}$

Next, we solve a linear programming problem:

    Minimize[{x + y, y ≥ x, y ≥ -2 x + 1}, {x, y}]

    $\left\{\dfrac{2}{3}, \left\{x \to \dfrac{1}{3}, y \to \dfrac{1}{3}\right\}\right\}$

■ **Global Optimization: Numerical Solutions**

NMinimize[f, vars]   Give the global minimum of **f** with respect to variables **vars**
NMinimize[{f, cons}, vars]   Minimize subject to constraints **cons**

We also have **NMaximize** that is used in the same way. For the following function, we do not get an explicit minimum point with **Minimize**:

    f = x^6 + x^4 + x + 1;

```
Minimize[f, x]
```

$$\left\{1 + \text{Root}\left[1 + 4\,\#1^3 + 6\,\#1^5\,\&,\,1\right] + \text{Root}\left[1 + 4\,\#1^3 + 6\,\#1^5\,\&,\,1\right]^4 + \text{Root}\left[1 + 4\,\#1^3 + 6\,\#1^5\,\&,\,1\right]^6,\right.$$
$$\left.\left\{x \to \text{Root}\left[1 + 4\,\#1^3 + 6\,\#1^5\,\&,\,1\right]\right\}\right\}$$

We can ask for the solution in a decimal form:

```
% // N        {0.569105, {x → -0.555036}}
```

We can also directly resort to numerical methods:

```
NMinimize[f, x]        {0.569105, {x → -0.555036}}
```

■ **Local Optimization**

`FindMinimum[f, {x, x0}]` Find a local minimum of **f** starting from **x0**
`FindMinimum[f, {{x, x0}, {y, y0}, … }]` Start from **x0, y0**, …

We also have `FindMaximum` that is used in the same way. These commands use iterative methods to find local minimum and maximum points. Consider the following function:

```
f = Cos[x] + Log[1 + x];
```

```
Plot[f, {x, 0, 2}]
```

The maximum seems to be near 0.5 and so we start from this point:

```
x0 = FindMaximum[f, {x, 0.5}]        {1.29686, {x → 0.650752}}
```

Thus, the local maximum is at the point 0.650752, and the maximum value is 1.29686. The derivative of the function at the given point is indeed zero with a high degree of accuracy:

```
D[f, x] /. x0[[2]]        6.83054 × 10^{-12}
```

2.3.5 Interpolation and Approximation

■ **Interpolation**

`Interpolation[data]` Find a piecewise third-degree interpolating polynomial for **data**

Let us first generate some data:

```
points = Table[{x, Cos[Exp[x]]}, {x, 0., 2., 0.2}]
{{0., 0.540302}, {0.2, 0.342328}, {0.4, 0.0788896},
 {0.6, -0.248685}, {0.8, -0.608957}, {1., -0.911734}, {1.2, -0.984107},
 {1.4, -0.610894}, {1.6, 0.238328}, {1.8, 0.972854}, {2., 0.448356}}
```

```
p1 = ListPlot[points]
```

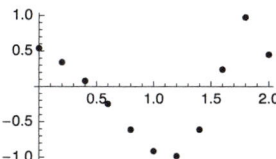

We now calculate a piecewise third-degree interpolating polynomial through the points:

```
int = Interpolation[points]
InterpolatingFunction[{{0., 2.}}, <>]
```

Mathematica calls the result an interpolating function. We do not see the actual function but, rather, only the interval where it is defined. However, we can calculate values of the interpolating function:

```
int[1.5]      -0.208865
```

We can plot it:

```
p2 = Plot[int[x], {x, 0, 2}]
```

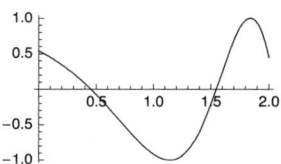

The interpolating function goes exactly through all the given points and is of degree 3 between each pair of points:

```
Show[p1, p2]
```

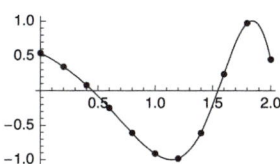

The result is a good representation of the data. In Section 2.3.6, when we solve differential equations numerically, we encounter these functions: The numerical solution is expressed as an interpolating function.

■ **Approximation**

`Fit[data, basis, var]`   Fit **data** by a linear combination of functions of **var** in **basis**

`Fit` calculates a least-squares function to smoothly represent data containing errors. Consider the following data:

```
data = {{0, 0.185}, {1, 0.935}, {2, 0.649}, {3, 1.231}, {4, 2.279},
    {5, 3.913}, {6, 4.670}, {7, 5.620}, {8, 6.767}, {9, 9.044}, {10, 11.045}};
```

`p1 = ListPlot[data]`

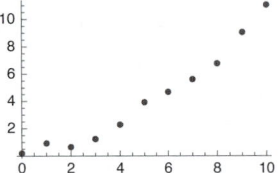

The points seem to follow roughly a quadratic pattern, and so we try a quadratic fit:

`lsq = Fit[data, {1, x, x^2}, x]`

$0.293972 + 0.146282\,x + 0.0910618\,x^2$

`p2 = Plot[lsq, {x, 0, 10}]`

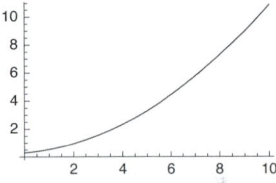

To see how close the fit is to the data, we can show both:

`Show[p1, p2]`

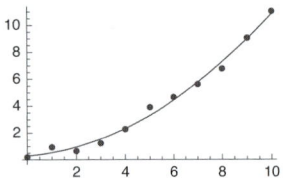

The fit seems to be good.

2.3.6 Differential Equations

■ **Symbolic Solutions**

`sol = y[t] /. DSolve[eqn, y[t], t]`   Give the general solution of a differential equation

`sol = y[t] /. DSolve[{eqn, y[a] == α}, y[t], t]`   Solve a first-order initial value problem

`sol = y[t] /. DSolve[{eqn, y[a] == α, y'[a] == β}, y[t], t]`   Solve a second-order initial value problem

`Plot[sol, {t, a, b}]`   Plot the solution

A differential equation is like a usual equation containing `==`, but now the equation contains an unknown function such as `y[t]` and its derivatives such as `y'[t]` and `y''[t]`. Note that initial conditions must also be written as equations (containing `==`).

`DSolve` can solve a large number of differential equations. We first ask for a general solution to the following linear equation with constant coefficients:

`eqn1 = y'[t] == a y[t] + b` $y'[t] == b + a\,y[t]$

DSolve[eqn1, y[t], t] $\left\{\left\{y[t] \rightarrow -\dfrac{b}{a} + e^{a\,t}\, C[1]\right\}\right\}$

The solution is in the familiar form of a transformation rule. The `C[1]` is an undetermined constant. Often, it is useful to directly ask the value of `y[t]`:

y[t] /. DSolve[eqn1, y[t], t] $\left\{-\dfrac{b}{a} + e^{a\,t}\, C[1]\right\}$

Then we solve a first-order initial value problem:

eqn2 = y'[t] + 2 t y[t] - t == 0;

sol2 = y[t] /. DSolve[{eqn2, y[0] == 0}, y[t], t]

$\left\{\dfrac{1}{2}\, e^{-t^2}\left(-1 + e^{t^2}\right)\right\}$

We plot the solution:

Plot[sol2, {t, 0, 3}]

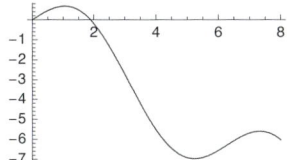

Next, we solve a second-order initial value problem and plot the solution:

eqn3 = y''[t] + y[t] == -t;

sol3 = y[t] /. DSolve[{eqn3, y[0] == 0, y'[0] == 1}, y[t], t]

$\{-t + 2 \sin[t]\}$

Plot[sol3, {t, 0, 8}]

■ **Numerical Solutions**

If *Mathematica* is not able to solve a differential equation symbolically with **DSolve**, then we can resort to numerical methods with **NDSolve**. We can simply replace **t** with **{t, a, b}** defining the interval (*a*, *b*) where the solution is calculated.

```
sol = y[t] /. NDSolve[{eqn, y[a] == α}, y[t], {t, a, b}]  Solve a first-order equation numeri-
  cally
sol = y[t] /. NDSolve[{eqn, y[a] == α, y'[a] == β}, y[t], {t, a, b}]  Solve a second-order
  equation numerically
Plot[sol, {t, a, b}]  Plot the solution
```

Mathematica cannot solve the following nonlinear initial value problem:

```
eqn4 = y''[t] == -y[t]^2 + t;
```

```
DSolve[{eqn4, y[0] == 0, y'[0] == 1}, y[t], t]
```

$$\text{DSolve}\Big[\big\{y''[t] == t - y[t]^2,\ y[0] == 0,\ y'[0] == 1\big\},\ y[t],\ t\Big]$$

However, we can use numerical methods and obtain an approximate solution:

```
sol4 = y[t] /. NDSolve[{eqn4, y[0] == 0, y'[0] == 1}, y[t], {t, 0, 7}]
```

$$\{\text{InterpolatingFunction}[\{\{0.,\ 7.\}\},\ <>][t]\}$$

The solution is expressed as an interpolating function—that is, as a piecewise third-degree interpolating polynomial (we considered these in Section 2.3.5, p. 46). We plot the solution:

```
Plot[sol4, {t, 0, 7}]
```

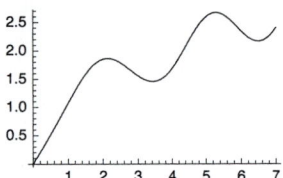

Notebooks

Introduction

> *During his stay in Berlin, Euler developed the habit of writing memoir after memoir,*
> *placing each when finished on top of a pile of manuscripts. Whenever material was needed*
> *to fill the journal of the Academy, the printers would help themselves to a few papers*
> *from the top of the pile. This meant that papers at the bottom remained there a long time*
> *and earlier papers often contained developments and improvements on later papers.*

Note that you may skip this chapter if you, right now, do not need to do the following:

- get more information about notebooks (saving, opening, printing, styles, style sheets, palettes, hyperlinks, slide shows);
- edit the outlook of notebooks;
- write special characters and 2D formulas with the keyboard; or
- write a mathematical document with display and inline formulas.

Regardless of your level of comfort with *Mathematica*, it may be helpful for you to read Sections 3.1.1, 3.1.2, and 3.3.1.

This chapter emphasizes *Mathematica* as a writing tool. Indeed, you can use *Mathematica* to write all kinds of mathematical and other documents. *Mathematica* is a strong alternative to traditional writing tools because it has the remarkable advantages that the whole document can be done with the same high-quality application (e.g., graphics, tables, or formulas need not be prepared with other applications) and that *Mathematica*'s computing power gives you excellent possibilities to do all kinds of calculations needed for the preparation of your document.

Adding notebook material into the *Help Browser* and other advanced matters about notebooks are explained in a separate document, **Using Author Tools.nb**, that can be found on the CD-ROM included with this book. Note also the GUIKit`, Notation`, and XML` packages.

3.1 Working with Notebooks

3.1.1 Saving, Opening, and Printing

■ **Saving**

Mathematica documents are called *notebooks*. To save a notebook for the first time, choose **Save As...** from the **File** menu; later, choose **Save** to save the modified notebook. It is customary to give the notebook a name ending with **.nb**. The handling of file names depends somewhat on the system used. For example, in Windows, the name of a notebook will automatically end with **.nb**, and without this extension the system does not automatically know that the file should be opened with *Mathematica*. On the other hand, in MacOS X, *Mathematica* does not suggest a name ending with **.nb**; indeed, the name can be without that extension.

You can arrange for the notebook to be automatically saved after each command (you must, however, do the first saving). Choose **Option Inspector...** from the **Format** menu and then choose *Scope* to be *Selected Notebook* and *View* to be *by category*. Then go to *Notebook Options ▷ File Options* and click the box for **NotebookAutoSave**.

The notebook can also be saved in certain special formats such as LaTeX, HTML, or PDF by choosing **Save As Special...** from the **File** menu.

■ **Opening**

A saved notebook can be opened by double-clicking the document or by choosing **Open...** from the **File** menu. Recently used notebooks are listed in the **File ▷ Open Recent** menu. When you open a notebook, you may observe that it no longer has the In- and Out-labels such as In[1] and Out[1].

After opening a notebook, you can continue working with it by adding new calculations, deleting old calculations, and doing other kinds of editing and modifying (this topic was considered in Section 1.5, p. 22).

One important thing to know is that when you open an old notebook, *you cannot directly use any of its results in the new session*. For example, suppose that in the old notebook you have defined **a = 5**. When you open the notebook, the variable **a** has no value. If you want **a** to have the value 5 in the new session, you have to execute the command **a = 5** anew (place the cursor anywhere in the relevant cell and execute). Similarly, all the results you want to use in the new session have to be recalculated. In other words, *you can use only the results that have been calculated in the current session*.

A straightforward way to continue working with an old notebook is to re-execute all its commands by choosing **Evaluation ▷ Evaluate Notebook**. However, some commands may require substantial time for execution. A better method in such a situation may be to use the old outputs directly.

For example, if you have calculated an integral, take a copy of the output cell, place the cursor at the beginning of the output, write `int =`, and execute. Now you have the value of the integral in `int` without having to recalculate the integral. You can also save results to files and then load them in later sessions (see Section 4.3.1, p. 109).

■ **Spell Checking and Hyphenating**

One of the last steps in preparing a document is the checking of the spelling. Choose **Edit** ▷ **Check Spelling…**. *Mathematica* comes with English spell checking, but you can buy spell checking for many other languages; contact Wolfram Research. A **Spelling Language** menu command then appears in either the **Edit** or the **Format** menu so that you may choose the language used. To find some options that control the spell checker, open the Option Inspector by choosing **Format** ▷ **Option Inspector…**, choose *Scope* to be *Selected Notebook*, and go into *Formatting Options* ▷ *Text Content Options* ▷ *SpellingOptions*.

Note that, if you want it to, *Mathematica* will automatically hyphenate words according to the rules of English. To turn hyphenation on or off, open Option Inspector, choose *Scope* to be *Selected Notebook*, go into *Formatting Options* ▷ *Text Layout Options*, and click the box for `Hyphenation`. The international spell-checking products also contain hyphenation rules for other languages.

■ **Adjusting Printing Settings**

Before printing the notebook, we can check, from **File** ▷ **Printing Settings**, the page setup, the margins, and the headers and footers. Headers and footers can be adjusted in the following window:

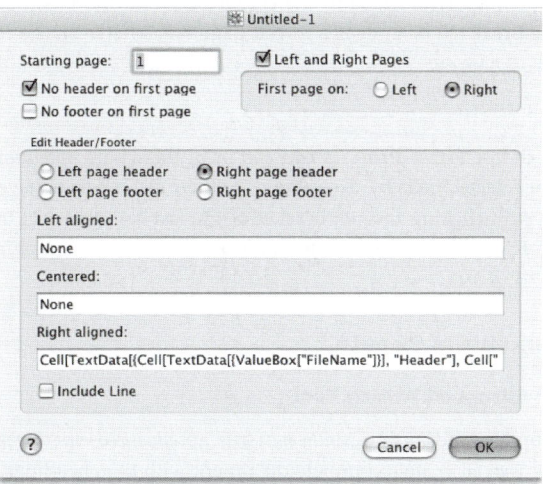

Here, we can adjust the starting page number; whether the notebook has separate left and right pages; and what to include in the left, center, and right of the headers and footers. For example, the top right header is, by default, as follows:

```
Cell[TextData[{
    Cell[TextData[{ValueBox["FileName"]}], "Header"],
    Cell["", "Header", CellFrame -> {{0, 0.5}, {0, 0}}, CellFrameMargins -> 4], "",
    Cell[TextData[{CounterBox["Page"]}], "PageNumber"]}],
  CellMargins -> {{Inherited, 0}, {Inherited, Inherited}}]]
```

This means that at the right we have the name of the file, a vertical line, and the page number. The code can be changed. For example, at the left of the right page header we can have the following code:

Cell[TextData["Chapter 3 • Notebooks"], "Header"]

At the right of the right page header we can have the following code:

Cell[TextData[CounterBox["Page"]], "PageNumber"]

If you want page numbers such as *i, ii, iii, iv*, etc., define them as follows:

CounterBox["Page", CounterFunction -> RomanNumeral]

The styles "Header" and "PageNumber" can be adjusted by the style sheet (see Section 3.2.2, p. 63) and the margin above the header by an option (see Section 3.2.3, p. 69).

■ Adjusting Page Breaks

Before printing, it may also be useful to see the page breaks. Choose **File ▷ Printing Settings ▷ Show Page Breaks**. The appearance of the notebook on the screen changes to reflect the printed output. A page break between two cells is shown by a gray, thick line, and a page break within a cell is shown by a short horizontal line at the right of the window. The current page number can be seen in the bottom left corner of the window.

In some special cases, you may want to manually adjust the automatically set page breaks. Suppose you want a page break between two cells. Put the cursor between the cells (a horizontal line appears) and choose **Insert ▷ Page Break**. Now there is a forced page break between the cells. Page break options are considered in Section 3.2.3, p. 69.

■ Printing

To print your notebook, choose **File ▷ Print...**. You can print the whole notebook or a selected range of pages. You can also print selected cells by dragging over their cell brackets (or clicking the cell bracket of the first cell and then shift-clicking the cell bracket of the last cell) to select them and then choosing **File ▷ Print Selection...**. If you have problems with fonts when printing graphics, you may find useful information in Section 7.2.4, p. 193.

3.1.2 Cell Styles and Style Sheets

■ *Mathematica* as an Advanced Writing Tool

An important point to understand is that *Mathematica is an advanced environment for technical writing.* Indeed, with *Mathematica* we can write a complete document with title, headings, texts, formulas, tables, graphics, inputs, and outputs. For example, each chapter of this book is a *Mathematica* notebook.

When using *Mathematica* as a writing environment, two things are important: cell styles and style sheets. These are considered next. *Mathematica* as a writing tool is considered in Section 3.4.

■ Cell Styles

In *Mathematica*, each cell has a *style*. The names of the styles can be seen by choosing **Format ▷ Style** (the styles can also be seen from a toolbar; choose **Window ▷ Show Toolbar**). For example, when we write a new command, the style is automatically **Input**, and the style of the result is **Output**. We can add text into the notebook with the **Text** style. Titles and headings of sections can be written with styles such as **Title**, **Section**, and **Subsection**.

The default style of a new cell is **Input**. However, we can change the style of a cell by highlighting its bracket and then choosing an appropriate style from **Format ▷ Style**. Or we can, before we write anything, choose the style of the next cell and then type the text. Within a cell of a certain type, we can apply other styles for smaller parts.

If we want to change the appearance of all cells of a certain type in our notebook, we can just change the style of that cell; see Section 3.2.2, p. 63.

One tip is that if you intend to write a new cell with the same style as the current cell (e.g., **Text**), choose **Insert ▷ Cell with Same Style** or press [ALT]+[ENTER] in Windows or [OPTION]+[RET] in Macintosh.

■ Style Sheets

The cell styles have default settings (font, face, size, color, etc.), but we can adjust these if we are not satisfied with them. One way to adjust the styles is to choose an appropriate *style sheet*. A style sheet is a special notebook that defines styles for a normal notebook. The style sheets can be seen by choosing **Format ▷ Stylesheet**. The default style sheet is **Default**, but there are several others, such as **JournalArticle**, **Textbook**, **StandardReport**, and **Correspondence**. In preparing a notebook, try various style sheets and choose the one you prefer.

Note that different style sheets define different sets of cell styles. For example, for the **Memo** style sheet, we have in **Format ▷ Style** only the styles **Section**, **Text**, and **Input**.

Note also that for a given style sheet, **Format ▷ Style** does not necessarily list all available styles. For example, for the **Default** style sheet, the following styles are not listed but are available: **Subsubsubsection**, **Subsubsubsubsection**, **SmallText**, **Subsubitem**, and various styles for headers, footers, and page numbers. These styles can be used by choosing **Format ▷ Style ▷ Other...**, typing the name of the style in a separate dialog, and clicking OK.

With more advanced style sheets, even more styles are not shown in the list of styles. For example, for the **Textbook** style sheet, a minority of available styles are listed. As an example, for equations, the style menu for this style sheet only lists the **Equation** and **EquationNumbered** styles, but we also have the following styles related with equations: **EquationGroup**, **EquationGroupAligned**, **EquationGroupNumbered**, **EquationGroupAlignedNumbered**, **EquationNumber**, **EquationGrid**, **SplitEquation**, **Piecewise**, and **Matrix**. The styles **BookChapterNumber** and **BookChapterTitle** (important with numbered equations) are also not listed.

After you have chosen an appropriate style sheet, it may be that you would still like to change the styles of some cells. This can easily be done, as is shown in Section 3.2.2, p. 63. We can even create new style sheets.

■ Screen Environments

In addition to style sheets, we have *screen environments* in **Format ▷ Screen Environment**. These allow you to modify the appearance of your notebook according to how you plan use it. When writing the notebook, you can use the **Working** environment (with large fonts); for presentations, the **Presentation** environment (with still larger fonts); for slide shows, the **SlideShow** environment (with slide show controllers); for small screens, the **Condensed** environment (with small fonts and a condensed style); and before printing, the **Printout** environment (to see how the notebook will look when printed).

A simple way to change the size of the text on the screen is to choose an appropriate magnification percentage from the bottom of the notebook window or from **Window ▷ Magnification**.

3.1.3 Palettes, Hyperlinks, and Slide Shows

■ Palettes

Palettes, which are available from the **Palettes** menu, help when you are writing inputs; see Section 1.4.1, p. 15. In addition to the built-in palettes, you may find other things you want to do with a palette. To create your own palette that pastes something when a button is pressed, use the following commands:

> `CreatePalette[buttons, WindowTitle → "title"]` (⌘6) Create a palette with the given buttons and given title
>
> `PasteButton[label, expr]` (⌘6) Create a button that pastes **expr** when the button is pressed; put **label** on the button
>
> `Button[label, action]` (⌘6) Create a button that does **action** when the button is pressed; put **label** on the button

As an example, we create a palette containing various plotting commands for data:

```
CreatePalette[Column[{
    PasteButton["Points", ListPlot[■]],
    PasteButton["Points and vertical lines", ListPlot[■, Filling → Axis]],
    PasteButton["Joining lines", ListLinePlot[■]],
    PasteButton["Joining lines and fill", ListLinePlot[■, Filling → Axis]],
    PasteButton["Joining lines and points", ListLinePlot[■, Mesh → All]],
    PasteButton["Joining lines, points, and fill",
      ListLinePlot[■, Mesh → All, Filling → Axis]]}],
  WindowTitle → "Plot Data as"] // Quiet
```

NotebookObject | 🔲 Plot Data as |

Here, we wrote the symbols ■ by writing \[SelectionPlaceholder] or [ESC]spl[ESC]; the symbol can also be found from the **SpecialCharacters** palette. With **Quiet**, we asked not to show some noninteresting messages. After executing the command, we can find the palette at the top right corner of the screen. The palette looks like the following:

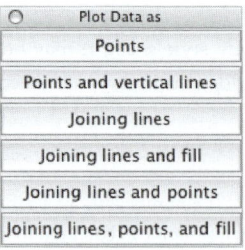

To use the palette, click, for example, **ListPlot[■]** and write, in place of the placeholder, a list of points to be plotted. Then execute the command. Or, write first the data to be plotted, select it, click a button in the palette, and execute.

To save the palette so that it appears in the **Palettes** menu, do as follows. Click the close button of the palette. You are asked to save the palette; save it to a place you prefer. Choose **Palettes ▷ Install Palette...**, select the file you saved, give the palette a suitable name, and click *Finish*. The palette now appears in the **Palettes** menu.

■ Hyperlinks

Hyperlinks are special buttons that consist of underlined words in blue type. When a hyperlink is clicked, *Mathematica* jumps to a cell of the current notebook, to a cell in another notebook, to another notebook, or to a URL. As an example, suppose that we want to create a hyperlink from HERE to the cell above containing the subsubsection heading Hyperlinks. Do the following:

> • Select the destination cell (the cell bracket containing Hyperlinks in our example).
> • Choose **Cell ▷ Cell Tags ▷ Add/Remove Cell Tags…**. A dialog box appears (see the figure on the left below). At the bottom of the dialog, write a word or phrase (e.g., "Hyperlinks") as the cell tag. It identifies the destination cell. Click **Add**; you can also close the dialog.
> • Select the word(s) (the word "HERE" in our example) in the notebook that will represent the hyperlink—that is, the words you want to be able to click in order to jump to the destination cell.
> • Choose **Insert ▷ Hyperlink…**. In the dialog box (see the figure on the right below), we can see all cell tags of the current notebook. Select the tag ("Hyperlinks" in our example) you created previously, and then click **OK**.

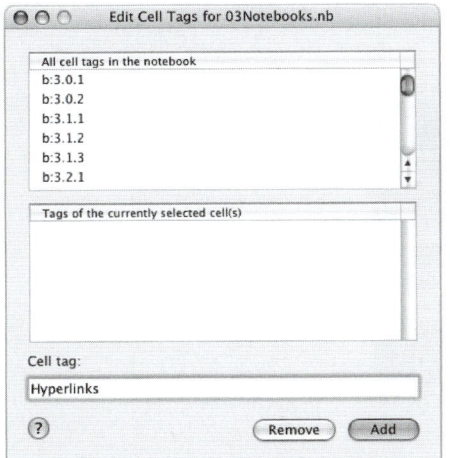 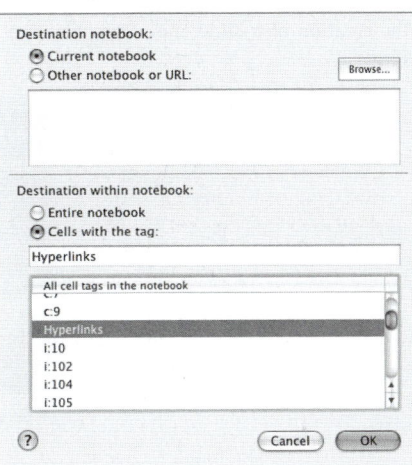

In the same way, we can create other kinds of hyperlinks. For example, to create a hyperlink to another notebook, simply select the word(s) in the notebook that will represent the hyperlink, choose **Insert ▷ Hyperlink…**, in the dialog box browse to the notebook, and click OK. To create a hyperlink to, for example, http://www.wolfram.com, write and select a word such as Wolfram, choose **Insert ▷ Hyperlink…**, in the dialog box write the address http://www.wolfram.com, and click OK. You will get a link such as Wolfram.

The text of the hyperlink button can be edited by clicking the text near the button and then moving the cursor with the arrow keys into the button.

Hyperlink["label", "URI"] (⌘6) Create a hyperlink that is displayed as **label**

Hyperlinks can also be created with the previous command. Here is an example:

```
Hyperlink["Wolfram", "http://www.wolfram.com"]

Wolfram
```

Similarly, we can create hyperlinks to the Documentation Center:

```
Hyperlink["Integrate", "paclet:ref/Integrate"]
Integrate

Hyperlink["calculus", "paclet:guide/Calculus"]
calculus

Hyperlink["indefinite integrals", "paclet:tutorial/IndefiniteIntegrals"]
indefinite integrals

Hyperlink["Fourier series package", "paclet:FourierSeries/tutorial/FourierSeries"]
Fourier series package
```

■ Slide Shows

To create a slide show, choose **File ▷ New ▷ Slide Show**. A template of a slide show opens. Modify the template as you want, and then choose **Format ▷ Screen Environment ▷ SlideShow**. To change the appearance of the slide show, try several style sheets.

In the creation of a slide show, we can also use the **Slide Show** palette available from **Palettes ▷ SlideShow**:

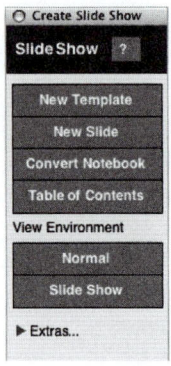

A new template for a slide show can be obtained by clicking *New Template*. A new slide can be added by clicking *New Slide*. The slide show environment can be chosen by clicking *Slide Show*; we can go back to normal view by clicking *Normal*. With *Table of Contents* we get the contents as a separate palette that has hyperlinks to the slides.

If you want to convert a usual notebook into a slide show, click *Convert Notebook*. A new window, *Notebook -> Slideshow,* appears:

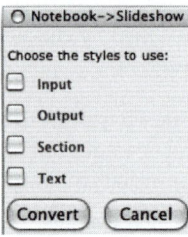

In this window, it reads "Choose the styles to use" and below it we have a list of the cell styles appearing in our notebook. This means that we have to select the cell styles according to which the notebook is divided into slides. For example, if our notebook contains several Section style cells and we want each section to become a slide, then we should choose the Section style by clicking the corresponding checkbox. Then click *Convert*. To see the new slide show, click *Slide Show* on the **SlideShow** palette.

With the palette, we can also manually convert a usual notebook into a slide show. Put the cursor at a location where you want to begin a new slide, click the triangle before *Extras…*, and choose *Paste… ▷ Navigation Bar* from the palette. If you want the Previous/Next buttons at the end of a slide, put the cursor at this location and choose *Paste… ▷ Previous/Next* from the palette. Do this for each slide. Then click *Slide Show*.

From the *Extras* we can also choose the style sheet, presentation size, and magnification of the slide show.

3.2 Editing Notebooks

3.2.1 Methods of Editing

■ Steps

When creating a document with *Mathematica,* you want to be aware of how the document looks on screen and when printed out. The look can be adjusted in many ways. Proceed with the following steps:

> • Step 1: *Select a style sheet.* Try several style sheets from **Format ▷ Stylesheet** (Section 3.1.2, p. 55), and choose the one you like most.
> • Step 2: *Modify styles.* If some styles do not satisfy you in all respects, modify these styles.
> • Step 3: *Adjust the notebook.* If you still find some small parts of the document needing adjustment, modify these parts directly in the document.

The key to modifying a notebook is the use of style sheets and styles. The notebook should be modified directly only in exceptional cases. Why? To save your work and to maintain the consistency of the look of the document. Suppose you want to change the font size of all text cells. You could manually change the style of each text cell, but if you then continue writing your document, you have to modify all new text cells as well. If you instead modify the styles, all text cells immediately change accordingly, and all new text cells also have the correct style.

Thus, resist direct modification of the notebook, and modify the styles instead. The editing of styles is described in Section 3.2.2, p. 63.

■ Methods

Whether you modify the styles or the document, you can use several methods in the modification:

> • Method 1: Use the **Format** menu.
> • Method 2: Use options directly.
> • Method 3: Use the Option Inspector.

In the following section, we consider these methods in turn. In each method, we set values of some *front-end options*. All options can be set with the third method and most options with the second method, whereas only a small number of options (although important ones) can be set with the **Format** menu.

The Option Inspector is a tool that is used to view and modify various options of cells and notebooks, among other things. The options can also be adjusted directly, without the Option Inspector. When we use the **Format** menu, we actually also adjust options. Many options are explained in the Documentation Center under tutorial/ManipulatingNotebooksOverview.

■ Method 1: Using the Format Menu

The editing tool used most often may be the **Format** menu. With it we can set various font options: **Font**, **Face**, **Size**, **Text Color**, and **Background Color**. We can also adjust **Cell Dingbat**, **Magnification** (magnifies a selection), **Text Alignment**, **Text Justification**, **Word Wrapping**, and **Spelling Language**. Furthermore, with **Format** we can choose the **Style** of each cell, use style sheets by **Stylesheet** and **Edit Stylesheet**, and choose a **Screen Environment**.

To use the **Format** menu, first select with the mouse the part of the document you want to modify. The part is typically a cell (click its cell bracket), but it can also be a part of a cell (drag over the part), several cells (drag over their cell brackets), or a group of cells (click its bracket). Then, choose a suitable command from the **Format** menu. As an example, write an **EmphasizedText** cell:

Here is some text.

Increase the size of the font to 14 so that the cell becomes

Here is some text.

To show that the **Format** menu actually uses options, we next show how to look at the internal code that *Mathematica* uses for cells.

■ Method 2: Using the Options Directly

Front-end options are normally not visible; we only see the effect of the options. However, with a special menu command, we can see the internal code behind a cell:

> • Put the cursor somewhere in the cell, or select the bracket of the cell.
> • Open the code of the cell by choosing **Cell ▷ Show Expression** or pressing ⇧SHIFT⌐CTRL⌐e⌐ (⇧SHIFT⌐⌘⌐e⌐ on a Macintosh).
> • Modify the code, if you so choose.
> • Close the code by choosing the same menu command again; the cell is then formatted according to the code.

As examples, here are the codes of the two text cells we considered previously:

```
Cell["Here is some text.", "EmphasizedText"]

Cell["Here is some text.", "EmphasizedText", FontSize->14]
```

In both cases, the text is in an **EmphasizedText** style cell and the latter cell has an option for the font size. So, we have shown that the **Format** menu actually uses options.

When the code is open, we also have the possibility to modify it. This is the second method of modification mentioned previously, "Use options directly." We can directly write new options, delete options, and change values of options. For example, we could have opened the code of the cell of "Here is some text.", written the option FontSize -> 14, and then closed the code. It may be that the more you use options and learn their names, the more you will like the direct modification of the code of the cells. In Section 3.2.3, p. 65, we list some useful options, many of which are not available with the **Format** menu.

■ The Structure of Expressions and Notebooks

It may be interesting to look at the code of a mathematical result. For example,

```
a = Sqrt[8] + Sin[Pi / 6]
```

$$\frac{1}{2} + 2\sqrt{2}$$

The code of the output cell is as follows (put the cursor in the result and press SHIFT-CTRL-(e)):

```
Cell[BoxData[
  RowBox[{
    FractionBox["1", "2"], "+",
    RowBox[{"2", " ",
      SqrtBox["2"]}]}]], "Output",
  CellChangeTimes->{3.3931306115955353`*^9}]
```

We see that the formula is built up from various box constructs.

A whole notebook consists of a list of cells, together with possible options. The general form of a notebook is thus `Notebook[{cell1, cell2, …}, options]`. In this way, *Mathematica* represents cells and notebooks as text expressions containing only 7-bit ASCII characters. From this, it follows that notebooks work independently of the platform they are opened with and thus can be used unchanged with any computer system. We shall not go into the details of cell and notebook expressions here. We only study the use of the front-end options.

■ Method 3: Using the Option Inspector

The Option Inspector is a special window where we can view and modify the options of the front end. The window can be opened by choosing **Format ▷ Option Inspector…**, and it looks like this (here we have opened some groups of options):

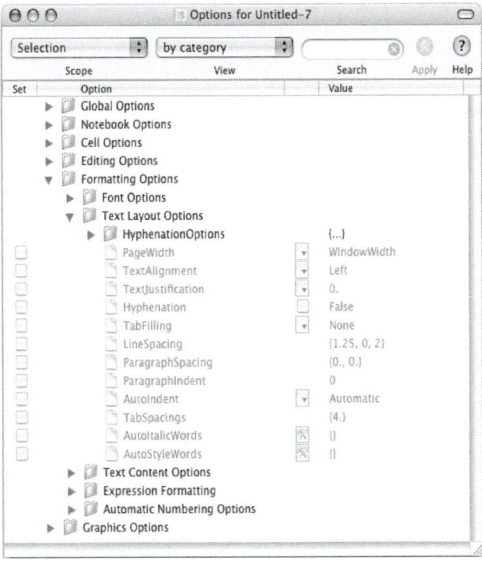

The options are grouped into six categories: *Global*, *Notebook*, *Cell*, *Editing*, *Formatting*, and *Graphics* options. In the figure above, we can see options such as **PageWidth** and **TextAlignment** together with their current values. There are several hundred options, but most of us will never need the majority of them. Clicking a triangle before a category opens the category into a list of subcategories or options. This categorical listing is the default way that the Option Inspector shows the options.

To use the Option Inspector to modify a part of your document, do the following:

- Select the part of your document that you want to modify with options (typically the part is a cell, but it can also be a part of a cell or several cells).
- Open the Option Inspector by choosing **Format ▷ Option Inspector…** or pressing ⌷SHIFT⌷⌷CTRL⌷⌷o⌷ (⌷SHIFT⌷⌷⌘⌷⌷o⌷ on a Macintosh).
- Set the values of the options you want. The values of many options can be set by check boxes, pop-up menus, or dialog boxes, but if you type the value, you have to *press the return key* ⌷RET⌷ after typing for your setting to go into effect.
- Go back to your document (you can leave the Option Inspector open, in case you need it again).

An option for which we have set a nondefault value has a check mark symbol before it. Clicking the check mark changes the value of the option to the default value and removes the check mark.

After experimenting with some options of a cell, we may find that we would like to return to the default option values for this cell. Select the cell bracket and choose **Format ▷ Clear Formatting**.

■ More about the Option Inspector

At the top of the window, we can select the scope of the option settings and the way the options are listed in the Option Inspector. We can also search suitable options.

The scope can be *Selection*, *Selected Notebook*, or *Global Preferences* (we can also choose from a list of open notebooks). Indeed, a key property of Option Inspector is that it allows us to specify the level at which we want to set the value of an option:

- If the scope is *Selection*, the option settings only affect the part of the current notebook that is selected with the mouse (the selected part may be a cell, several cells, or a part of a cell).
- If the scope is *Selected Notebook*, the option settings only affect the current notebook.
- If the scope is *Global Preferences*, the option settings affect the whole *Mathematica* application—the current session and all future sessions, all currently open notebooks, and all notebooks opened or created in the future.

When working with Option Inspector, it is important to choose the suitable level; otherwise, you can easily generate unwanted effects. Note also that some options cannot be set at all levels; options that cannot be set at the currently selected level are dimmed.

The options in the Option Inspector can be listed in three ways: *by category*, *alphabetically*, or *as text*. A listing by category is useful if you want to find out what options are available for a certain purpose. An alphabetical listing is useful if you already know the option you want to apply. The text view only shows the nondefault options as text, such as "FontSize→12"; this view cannot be used to set the value of an option.

At the top of the Option Inspector window is a field in which we can write a word to be searched from the names of the options. For example, write *font* in the field to get a list of options containing *font*.

In the next section, we show how we can edit styles, and the section following that gives lists of useful options (approximately 50 options).

3.2.2 Modifying Styles

■ Modifying Styles

As noted in Section 3.2.1, p. 59, notebooks should be edited mainly by modifying styles, and they should be directly edited only in exceptional cases. In this section, we show how to modify styles.

The starting point in modifying the styles of a notebook is the choice of a suitable style sheet from **Format ▷ Stylesheet**. Choose a style sheet that is as close to your needs as possible. If the styles do not fully satisfy you, you may want to make some modifications to the styles. Proceed as follows:

- Choose **Edit Stylesheet…** from the **Format** menu. A notebook appears with the title *Private Style Definitions for …*.
- From *Choose a Style to Modify*, choose the style(s) you want to modify. The corresponding *style definition cells* appear in the notebook. These cells show how cells having these styles look out.
- Modify the style definition cells by using the methods explained in Section 3.2.1, p. 59: Use the **Format** menu, use options directly, or use the Option Inspector. You can immediately see the corresponding changes in your notebook.
- If a style is not mentioned in *Choose a Style to Modify* or if you want to define a new style, use the *Enter a style name* input field to give a suitable name for the style; the corresponding style definition cell appears in the document. Edit this cell as you want. A new style appears in **Format ▷ Style**.
- The *Private Style Definitions* notebook can be left open (for later modifications) or it can be closed (without saving).

Note that in the previous procedure, the name of the menu command **Edit Stylesheet…** gives a somewhat wrong impression: We did not edit the style sheet. Rather, we created some private style definitions for our current notebook (and only for it). The private styles are based on the chosen style sheet; we only tell how we want to change the styles.

Below we show a *Private Style Definitions* notebook. The styles are based on the **Default** style sheet, or, as is said in the notebook, the base definitions of the styles are *inherited* from the **Default** style sheet. We have changed the style of **Section**, **Subsection**, and **Subsubsection** by adding color definitions.

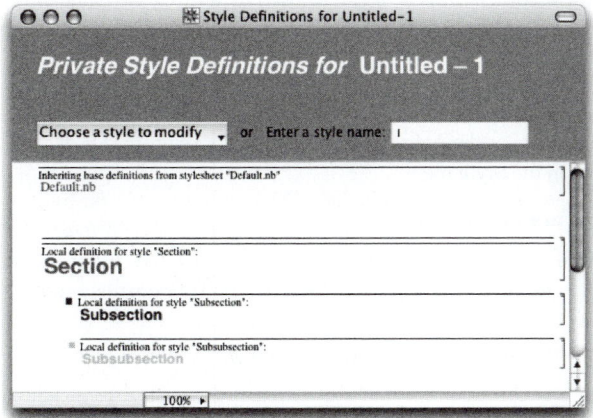

The modified and new styles work immediately. Whatever you have written in your notebook and whatever you write later will follow the modified styles. If you later edit the styles of the notebook in question, the same *Private Style Definitions* notebook opens and you can continue modifying the styles.

■ Creating New Style Sheets

As noted, the procedure we described previously only modifies the styles of the *current* notebook. If you want to use your own styles for *several* notebooks, you should create your own style sheet as follows:

> • Proceed as discussed previously: Choose **Format ▷ Edit Stylesheet…** and edit the styles and add the new styles you want.
> • Save the style sheet by choosing **Format ▷ Save As…**. You can save it with the name and to the place you want.
> • Install the new style sheet by choosing **File ▷ Install…**, select *Type of Item to Install* to be *Stylesheet*, select *Source* to be *From File…*, select the new style sheet, give the style sheet a suitable name, and click *Finish*. The new style sheet now appears in **Format ▷ Stylesheet**.

Later, if you want to continue editing the new style sheet, open a document using that style sheet and choose **Format ▷ Edit Stylesheet…**. The appearing *Private Style Definitions for …* notebook contains a hyperlink to the style sheet; click it. The style sheet opens and you can edit it; the changes are automatically saved.

■ Creating New Styles

If you find yourself occasionally manually modifying a new cell of your document, this may be an indication that you should define a new cell style to your style sheet. To create a new style, we can use the *Enter a style name* input field in the *Private Style Definitions for …* notebook as described previously. Another way is to copy a suitable style definition cell from a style sheet:

> • For a notebook, choose a style sheet that contains a style you want to copy. Choose **Format ▷ Edit Stylesheet**, and click the hyperlink to the style sheet.
> • Copy a style definition cell, paste the cell into your style sheet, select the cell bracket of the new cell, choose **Cell ▷ Show Expression**, write a new representative name into `StyleData["name"]`, and choose **Cell ▷ Show Expression** again.
> • Change the style of the cell according to your needs (e.g., using the **Format** menu). The new style then either appears in **Format ▷ Style** or can be used by writing the name of the style into the window opened by **Format ▷ Style ▷ Other…**.

The default is to place new styles at the end of the **Style** menu. With the `MenuPosition` option we can define a suitable place. Select the cell bracket of the new style, choose **Format ▷ Option Inspector…**, set *Scope* to be *Selection*, and set the value of the mentioned option to, for example, 10 if you want the style to be the 10th item in the **Style** menu.

■ Changing the Size of Printed Text

In some style sheets, the sizes of texts are quite small when printed. If you would like to increase the sizes of texts, this can be done by editing a style sheet; for more details, see the following FAQ document:

http://support.wolfram.com/mathematica/interface/print/increasingprintoutfont.html.

3.2.3 Useful Options

■ Formatting Options: Fonts

Most options concerning fonts can be set with the **Format** menu. These options can also be found from the Option Inspector, and they can be written directly into the code of the cells.

FontFamily Examples of values: **"Times"**, **"Arial"**, **"Helvetica"**, **"Courier"**

FontSize Examples of values: **12**, **10**, **9**

FontWeight Examples of values: **"Plain"**, **"Bold"**

FontSlant Examples of values: **"Plain"**, **"Italic"**, **"Oblique"**

FontTracking Examples of values: **"Plain"**, **"Condensed"**, **"Extended"**

FontColor Examples of values: **Automatic**, **RGBColor[1, 0, 0]**

FontOpacity Examples of values: **Automatic**, **0.5**

Background Examples of values: **None**, **RGBColor[1, 1, 0]**

Magnification Examples of values: **1**, **0.75**, **1.5**

FontVariations An example: **{"Underline" → True, "Outline" → True}**; possible variations:
   "Underline", **"Outline"**, **"Shadow"**, **"StrikeThrough"**, **"Masked"**, **"CompatibilityType"**, and
   "RotationAngle"; each can be set to **False** or **True**, except that **"CompatibilityType"** can be set
   to **"Normal"**, **"Superscript"**, or **"Subscript"**, and **"RotationAngle"** to an angle

---

These options can be found from the Option Inspector under *Formatting Options ▷ Font Options*. Note that the value of **FontColor** and **Background** for cells cannot be a special color directive such as **Red** or **Blue**; the value has to be given with **RGBColor**, **Hue**, **GrayLevel**, or **CMYKColor** (however, colors such as **Red** or **Blue** work in graphics).

For color definitions, see Section 6.2.8, p. 168. In the following example, we use several font options:

An example of font options

The code of this cell is as follows:

```
Cell[TextData[StyleBox["\tAn example of font options",
  Background->RGBColor[1, 1, 0]]],
  "Text",
  CellMargins->{{Inherited, Inherited}, {1, 2}},
  CellChangeTimes->{3.393143437924532*^9},
  FontFamily->"Times", FontSize->14,
  FontWeight->"Bold", FontSlant->"Italic",
  FontColor->RGBColor[1, 0, 0]]
```

■ Formatting Options: Text Layout

The look of a document is greatly affected by various margins and spacings; we present lists of options that can be used to adjust these. The values of many of these options are given in *printer's points*. One printer point is approximately 1/72 of an inch. The values of some options are given in *ems* or *x-heights*. An em is approximately the width of an "m," and x-height is the height of an "x."

LineSpacing Specifies the spacing between successive lines of text; value is of the form **{c, n}**,
   meaning that the height of each line is **c** times the height of the contents of the line plus **n** printer's
   points (**n** may be negative); value can also be of the form **{c, n, max}**, meaning that the extra space
   is limited to **max** times the height of a single line

ParagraphSpacing The extra space between two paragraphs (a new paragraph begins after each explicit RET character); value is of the form **{c, n}**, meaning that the extra space is **c** times the height of the font plus **n** printer's points

ParagraphIndent The indentation, in printer's points, of the first line of a paragraph (a new paragraph starts at the beginning of a cell and after each explicit RET); a negative value causes all but the first line to be indented

TabSpacings The number of spaces, in ems, that the cursor advances (at most) when TAB is pressed in a text cell; examples of values: **4**, **{10, 15, 12, 7}**

TabFilling Determines how a TAB is represented; examples of values: **None**, **"."**, **"Underline"**, **"GrayUnderline"**

These options can be found from the Option Inspector under *Formatting Options ▷ Text Layout Options.*

LineSpacing and **CellMargins**, together with fonts and their sizes, are important options that affect the overall look of the pages of your document. Usually, each paragraph is in its own cell, and then **CellMargins** determines the spacing between paragraphs. However, if you write several paragraphs into the same cell, then **ParagraphSpacing** determines the space between the paragraphs in such cells. With **LineSpacing**, any of **c**, **n**, or **max** can also be zero, and then the spacing is determined only with the height of the contents or only with the given printer's points. A typical value of this option is **{1, 3}**.

With a negative **ParagraphIndent**, we get paragraphs like this one. Usually, we need both indented and nonindented paragraphs. A normal paragraph is indented, but the first paragraph in a section is nonindented. If the same paragraph continues after a formula, we also need a nonindented paragraph. One way to use both indented and nonindented paragraphs is to let a text cell have a zero indentation and to press TAB if an indentation is needed; we can give **TabSpacings** a suitable value.

If one value, for example, m, is given for **TabSpacings**, then the width of the space between two tab stops is m; that is, tab stops are at positions m, $2m$, $3m$, etc. However, if the value is a list such as $\{m, n, k\}$, then the widths of the spaces between tab stops are m, n, k, k, k, etc. The default is that a tab is represented as white space, but we can use any character, such as a period **"."**, and also the values **Underline** and **GrayUnderline**.

TextAlignment Examples of values: **Left**, **Right**, **Center**

TextJustification Examples of values: **0** (natural spacing) **0.5**, **1** (full justification)

PageWidth Examples of values: **WindowWidth**, **PaperWidth**, **n** (in printer's points)

Hyphenation Possible values: **False**, **True** (see Section 3.1.1, p. 53)

HyphenationOptions → {"HyphenationMinLengths" → {m, n}} A minimum of **m** and **n** characters can be split off the start and end of a word, respectively; default value: **{3, 3}** (this option is most easily set with the Option Inspector)

These options can be found from the Option Inspector under *Formatting Options ▷ Text Layout Options.*

■ **Formatting Options: Formulas**

The following options are useful when writing mathematical formulas (the first option is explained in Section 3.4.3, p. 88).

LimitsPositioning How to display limits in constructs such as sum, product, union, intersection, lim, max, and min; possible values: **Automatic**, **True** (display the limits as in $\sum_{i=1}^{n}$ when the formula is inline), **False** (display the limits as in $\sum\limits_{i=1}^{n}$)

> **ScriptMinSize** Minimum font size used in subscripts, etc.; default values: **9** (on screen), **5** (on paper)
>
> **ScriptSizeMultipliers** How much smaller to render each successive level of subscripts, etc.; default value: **0.71**

These options can be found from the Option Inspector under *Formatting Options* ▷ *Expression Formatting*. The first is under *Specific Box Options* in *OverscriptBoxOptions*, *UnderscriptBoxOptions*, and *UnderoverscriptBoxOptions*. The last two options are under *Display Options*.

■ **Formatting Options: Frame Boxes**

If we want a frame around a smaller part than a cell, we select that part and choose **Insert** ▷ **Typesetting** ▷ **Add Frame**. An example is $\boxed{\sin(x)^2 + \cos(x)^2 = 1}$. As another example, we emphasize a display formula:

$$\boxed{\lim_{x \to \infty} \left(1 + \frac{a}{x}\right)^x = e}$$

Such frames can be controlled with **FrameBoxOptions**, presented later.

The values of some options we will present from now on may be given in the form **{{left, right}, {bottom, top}}**, meaning that numerical values are given that control the left, right, bottom, and top parts. This value is written below in the short form **{{l, r}, {b, t}}**.

> **FrameBoxOptions → {opts}** Options controlling the frame around a box (to get a frame, use **Insert** ▷ **Typesetting** ▷ **Add Frame**); the following options can be given:
>
> **BoxFrame** Whether to draw lines around a frame box. Value may be **True**, **False**, **f**, or **{{l, r}, {b, t}}**, where each number is the thickness of the frame or a part of it in printer's points (the value **True** implies the value 1).
>
> **FrameMargins** Margins between the contents of a grid box and the surrounding frame. Left and right margins are given in ems, and bottom and top margins are given in x-heights. Value may be **True**, **False**, **m**, or **{{l, r}, {b, t}}** (the value **True** implies the value {{1, 1}, {1, 1}}).
>
> **FrameStyle** Style of the frame; an example of value: **{RGBColor[0, 0, 1]}**
>
> **BaseStyle** Base style of the frame; an example of value: **{RGBColor[1, 0, 0]}**
>
> **Background** Background color; an example of value: **RGBColor[0, 1, 0]**

These options can be found from the Option Inspector under *Formatting Options* ▷ *Expression Formatting* ▷ *Specific Box Options* ▷ *FrameBoxOptions*.

■ **Cell Options: General**

> **CellMargins** Margins, in printer's points, around a cell; value is of the form **{{l, r}, {b, t}}**, where each number is the size of a part of the margin (note that left and right margins can be set with the ruler, available from **Window** ▷ **Show Ruler**; its unit can be set with **RulerUnits**)
>
> **CellDingbat** Dingbat to be used to emphasize a cell; examples of values: **None**, **"■"**, **"\ [FilledSmallSquare]"**
>
> **Background** Background color of a cell; examples of values: **Automatic**, **RGBColor[1,0,0]**
>
> **Magnification** Magnification factor for the cell; examples of values: **1**, **0.75**, **1.5**
>
> **ShowCellLabel** Whether to show cell labels; possible values: **True**, **False**

The first four options can be found from the Option Inspector under *Cell Options ▷ Display Options* and the last option under *Cell Options ▷ Cell Labels*.

`CellMargins` determines, in addition to the white space to the left and right of the content of the cell, the white space between cells. Usually, each paragraph of text is written into its own cell, and then the cell margins determine the spacing between paragraphs.

One application of `Magnification` may be as follows. When using magnification values of more than 100% from **Window ▷ Magnification**, all text remains in the window, but it may be that graphics can no longer be seen completely. Setting `Magnification → 1` for plots keeps the size of graphics the same, irrespective of the magnification percentage used for the notebook.

Normally, we have labels such as `In[1]:=` and `Out[1]=` before inputs and outputs. With the `ShowCellLabel` option, we can turn the labels off. This can also be done with **Mathematica ▷ Preferences ▷ Evaluation ▷ Show In/Out Names**.

■ Cell Options: Cell Frames

> `CellFrame` Whether a frame is drawn around a cell; value may be `False`, `True`, `f`, or `{{l, r}, {b, t}}`, where each number is the absolute thickness of the frame or a part of it in printer's points (the value `True` implies the value 0.25)
>
> `CellFrameMargins` Margins, in printer's points, inside a frame; value may be `m` or `{{l, r}, {b, t}}` (typical default value is 8)
>
> `CellFrameColor` Color of the frame; examples of values: `GrayLevel[0]`, `RGBColor[1,0,0]`
>
> `CellFrameLabels` Labels of the frame; value is of the form `{{left text, right text}, {bottom text, top text}}` (the style of frame labels can be adjusted by using the style sheet)
>
> `CellFrameLabelMargins` Margins, in printer's points, between a cell frame and the labels; value may be `m` or `{{l, r}, {b, t}}` (typical default value is 6)

These options can be found from the Option Inspector under *Cell Options ▷ Cell Frame Options*.

With `CellFrame`, we can form various frames, such as the following ones:

CellFrame → True

CellFrame → {{0, 0}, {0, 0.25}}

CellFrame → {{0, 0}, {0.25, 0}}

CellFrame → {{5, 0}, {0, 1}}

CellFrame → True, Background → GrayLevel[0.85]

In the following example, we have cell frame labels:

top

left | A text with a frame and frame labels. | *right*

bottom

■ **Cell Options: Page Breaks**

Page breaks can be seen by choosing **File ▷ Printing Settings ▷ Show Page Breaks**. Page breaks can be manually added with **Insert ▷ Page Break**. The following options give a more detailed control of the page breaks.

PageBreakAbove Whether a page break should be made above a cell; possible values: `Automatic`, `True`, `False`

PageBreakBelow Whether a page break should be made below a cell; possible values: `Automatic`, `True`, `False`

PageBreakWithin Whether a page break should be allowed within a cell; possible values: `Automatic`, `True`, `False`

GroupPageBreakWithin Whether a page break should be allowed within a group of cells; possible values: `Automatic`, `True`, `False`

PrintingOptions → {`"PageHeaderMargins"` → {`left`, `right`}} Vertical margins, in printer's points, above the header of the left and right facing pages, respectively (recall that the margins of the pages can be set with **File ▷ Printing Settings ▷ Printing Options…**)

The first four options can be found from the Option Inspector under *Cell Options ▷ Page Breaking*; the last option is in *Notebook Options ▷ Printing Options ▷ PrintingOptions*.

■ **Notebook Options**

Next, we list some options typically specified at the notebook or global level.

DragAndDrop Whether to allow us to drag a selection to a new location with the mouse; possible values: `False`, `True`

EvaluationCompletionAction What to do when an evaluation is completed; examples of values: {}, {`"ShowTiming"`} (see Section 4.4.1, p. 112)

`"GraphicsPrintingFormat"` Format in which PostScript graphics are to be printed; possible values: `"Automatic"`, `"RenderInFrontEnd"`, `"DownloadPostScript"`, `"Bitmap"` (see Section 7.2.4, p. 193)

ImageSize Size of an image; default value: {`350`, `350`} (see Section 5.1.1, p. 120)

InputAutoReplacements Sequences of characters that are automatically replaced with other characters; default value for input cells: {`"->"` → `"→"`, `":>"` → `":→"`, `"<="` → `"≤"`, `">="` → `"≥"`, `"!="` → `"≠"`, `"=="` → `"=="`, `ParentList`}

NotebookAutoSave Whether the notebook should automatically be saved after each command execution; possible values: `False`, `True` (see Section 3.1.1, p. 52)

RulerUnits Determines the units in the ruler toolbar; possible values: `"Inches"`, `"DecimalInches"`, `"Points"`, `"Picas"`, `"Centimeters"`, `"Millimeters"`

SpellingOptions Determines settings for spell checking (see Section 3.1.1, p. 53)

■ **Preferences**

The preferences of *Mathematica* can be found (in a Macintosh) from **Mathematica ▷ Preferences**. There, we can globally adjust many options. For example, the following figure shows how the display of numbers can be adjusted.

3.3 Inputs and Outputs

3.3.1 Forms of Input and Output

■ **Forms of Output**

Mathematica can show results in several forms. The forms are **InputForm**, **OutputForm**, **StandardForm**, and **TraditionalForm**. The default form is **StandardForm**. As an example, we calculate an integral and show the result with all four format types:

```
Integrate[Sqrt[x] / (a^2 + b^2 x), x] // InputForm
(2*Sqrt[x])/b^2 - (2*a*ArcTan[(b*Sqrt[x])/a])/b^3

Integrate[Sqrt[x] / (a^2 + b^2 x), x] // OutputForm
```

$$\frac{2 \text{ Sqrt}[x]}{b^2} - \frac{2 \text{ a ArcTan}[\frac{b \text{ Sqrt}[x]}{a}]}{b^3}$$

```
Integrate[Sqrt[x] / (a^2 + b^2 x), x] // StandardForm
```

$$\frac{2\sqrt{x}}{b^2} - \frac{2\,a\,\text{ArcTan}\left[\frac{b\sqrt{x}}{a}\right]}{b^3}$$

```
Integrate[Sqrt[x] / (a^2 + b^2 x), x] // TraditionalForm
```

$$\frac{2\sqrt{x}}{b^2} - \frac{2\,a\,\tan^{-1}\left(\frac{b\sqrt{x}}{a}\right)}{b^3}$$

InputForm is a linear or 1D form that uses only standard characters. **OutputForm** is a 2D form that also uses only standard characters. Both of these forms are only rarely used. One application of the input form is to ask for all of the 16 decimals that *Mathematica* uses in its internal calculations:

```
π // N     3.14159
```

```
% // InputForm     3.141592653589793
```

StandardForm is a 2D form and uses special characters such as $\sqrt{\ }$, special spacings, and special character sizes. This is the default form of output (so writing **//StandardForm** above was unnecessary).

TraditionalForm imitates all aspects of traditional mathematical notation. For example, variables are in italics, arguments of functions are in parentheses ( ) and not in square brackets [ ], and other traditional notations such as $\tan^{-1}$ (instead of ArcTan) are used. This form is mainly used in preparing mathematical documents.

In addition to the commands such as **InputForm**, we can also use the menu command **Cell ▷ Convert To**.

Note that *Mathematica* has many more formatting commands:

```
Names["*Form"]
```

```
{AccountingForm, BaseForm, BlankForm, BoxForm, CForm, ColonForm, ColumnForm,
 DisplayForm, EdgeForm, EngineeringForm, FaceForm, FontForm, FortranForm,
 FullForm, HoldForm, HorizontalForm, HornerForm, InputForm, LineForm, LongForm,
 MathMLForm, MatrixForm, NumberForm, OutputForm, PaddedForm, ParentForm,
 PointForm, PolynomialForm, PrecedenceForm, PrintForm, PromptForm, RealBlockForm,
 RecurringDigitsForm, RuleForm, ScientificForm, SequenceForm, ShowShortBoxForm,
 SpaceForm, StandardForm, StringForm, StyleForm, SyntaxForm, TableForm,
 TeXForm, TextForm, TraditionalForm, TreeForm, ValueForm, VerticalForm}
```

For dynamic ways to show outputs, see Chapters 12 and 13.

■ **Style and Text**

> **Style[expr, dirsOpts]** (⌘6) Show **expr** with the specified directives and options
> **Text[expr]** (⌘6) Show **expr** in plain text format

Style is mainly used to adjust font properties. Font directives and options are considered in Section 6.2.6, p. 163. We have such options as **FontSize**, **FontWeight**, **FontSlant**, **FontColor**, **FontFamily**, and **Background**. However, the size, weight, slant, and color can also be defined simply by giving the value of the option (i.e., the option itself need not be mentioned):

```
Style[Integrate[x Sin[x], x], 12, Bold, Italic,
 Blue, FontFamily → "Helvetica", Background → Yellow]
```
$-x\,Cos[x] + Sin[x]$

With **Text** we can easily change the usual Courier font in the outputs to the Times font:

```
Integrate[x Sin[x], x] // Text
```

$-x\,Cos[x] + Sin[x]$

One of the options of **Style** is **DefaultOptions**. This option may sometimes be useful to change the style of some objects without having to put the same set of options into each object:

```
Style[{Graphics[{Red, Circle[]}], Graphics[{Green, Rectangle[]}],
 Graphics[{Blue, Disk[]}]}, DefaultOptions → {Graphics → {Background → Yellow}}]
```

■ **Framed, Panel, Labeled, and Pane**

Framed[expr] (⌘6) Put a frame around **expr**

Panel[expr] (⌘6) Display **expr** as a panel
Panel[expr, title] Give the panel the title **title**
Panel[expr, title, pos] Put the title at position **pos**

Labeled[expr, label] (⌘6) Give **expr** the label **label**
Labeled[expr, label, pos] Put the label at position **pos**

Pane[expr, w] (⌘6) Show **expr** as a pane; set the width to **w** printer's points
Pane[expr, {w, h}] Set the width and height to **w** and **h** printer's points, shrinking the contents if
  necessary

With **Framed** we can use the options **Background**, **FrameMargins**, **FrameStyle**, and **ImageMargins**.

Integrate[x Sin[x], x] // Framed

$$-x\,Cos[x] + Sin[x]$$

Framed[Integrate[x Sin[x], x], FrameMargins → 7]

$$-x\,Cos[x] + Sin[x]$$

With a panel we get a gray background:

Panel[Integrate[x Sin[x], x]]

$$-x\,Cos[x] + Sin[x]$$

Panel[Integrate[x Sin[x], x], "Integral"]

Integral

$$-x\,Cos[x] + Sin[x]$$

Panel[Integrate[x Sin[x], x], "Integral", Bottom]

$$-x\,Cos[x] + Sin[x]$$

Integral

Examples of a label:

Labeled[Integrate[x Sin[x], x], "Integral" // Text]

$$-x\,Cos[x] + Sin[x]$$

     Integral

Labeled[Integrate[x Sin[x], x], "Integral" // Text, {{Top, Right}}]

       Integral

$$-x\,Cos[x] + Sin[x]$$

A pane is useful if we would like to put the output on an area of fixed width and/or fixed height:

Table[Prime[n], {n, 1, 30}]

{2, 3, 5, 7, 11, 13, 17, 19, 23, 29, 31, 37, 41, 43, 47,
 53, 59, 61, 67, 71, 73, 79, 83, 89, 97, 101, 103, 107, 109, 113}

```
Pane[Table[Prime[n], {n, 1, 30}], 200]
```

```
{2, 3, 5, 7, 11, 13, 17, 19, 23, 29, 31,
 37, 41, 43, 47, 53, 59, 61, 67, 71, 73,
 79, 83, 89, 97, 101, 103, 107, 109, 113}
```

Some commands have the **Frame** option and various options for labels. However, with **Framed** and **Labeled** we get somewhat different results. Here, we use **Plot** as an example:

```
{Plot[Sin[x], {x, 0, 2 π}, Frame → True],
 Plot[Sin[x], {x, 0, 2 π}] // Framed}
```

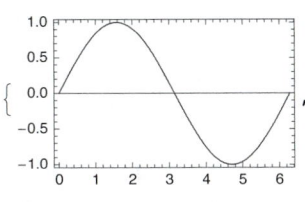 ,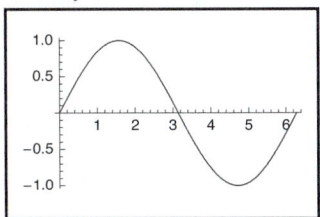

```
{Plot[Sin[x], {x, 0, 2 π}, PlotLabel → Sin[x]],
 Labeled[Plot[Sin[x], {x, 0, 2 π}], Sin[x] // Text // TraditionalForm]}
```

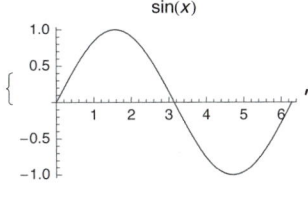 ,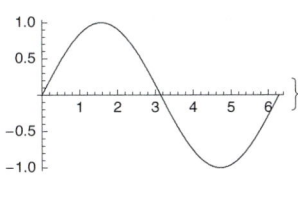

$\sin(x)$

■ Forms of Input

Inputs can be written both in a 1D and in a 2D form. For example:

```
Integrate[7 / 2 (x^2 + a Pi / x) Sqrt[x], {x, 0, 1}]      1 + 7 a π
```

$$\int_0^1 \frac{7}{2}\left(x^2 + \frac{a\,\pi}{x}\right)\sqrt{x}\;dx \qquad 1 + 7\,a\,\pi$$

In the latter form, we used the **BasicMathInput** palette; see Section 1.4.1, p. 15, regarding the use of palettes to write 2D inputs. In Section 3.3.3, p. 76, we learn how to write 2D inputs without palettes using special key combinations.

Choosing between the 1D and 2D input forms may mainly be a matter of habit. I personally use 1D inputs because they are straightforward to write with the keyboard and because I then do not need the mouse, the palettes, or special key combinations. However, 2D inputs are very clear and illustrative because they resemble traditional mathematical notation. From a 2D input, we can more easily check that the input is correct. Using palettes to form inputs may be handy at least in the early stages of learning the use of *Mathematica* because the user then does not need to remember the form of, for example, an integrating command. A compromise may be to write in 2D form only the parts of the input that can easily be written with the keyboard:

$$\text{Integrate}\left[\frac{7}{2}\left(x^2 + \frac{a\,\pi}{x}\right)\sqrt{x}\,,\,\{x,\,0,\,1\}\right] \qquad 1 + 7\,a\,\pi$$

Still another way is to write a 1D input and then convert it to 2D form with **Cell ▷ Convert To ▷ StandardForm**. In Sections 3.3.2 and 3.3.3, we learn how to write special characters and 2D inputs.

3.3.2 Writing Special Characters

■ **Introduction**

Mathematica has a large set of special characters such as π, ∞, $\in$, and $\sum$. To write these characters, we have several possibilities. As an example, we consider various ways to write ∞.

1. In an input, we can write **Infinity**.
2. From the keyboard, we can possibly find ∞ (on a Macintosh, press ⎇[5]).
3. All characters have a *full name*; for ∞, it is \[Infinity]. After we have typed], the sequence of characters transforms to ∞.
4. Many characters also have an *alias*. For ∞, it is [ESC]inf[ESC] ([ESC] means the escape key; it appears as ⦂ on the screen). After pressing the second [ESC], the sequence of characters transforms to ∞. This is a very handy way to type special characters, although we have to remember the aliases.
5. Many characters can be found in the **BasicMathInput** palette; more are found in the **BasicTypesetting** and **SpecialCharacters** palettes (the latter palette can also be opened from **Insert ▷ Special Character…**). By moving the mouse over the **BasicTypesetting** palette, we can see the aliases of the characters at the bottom of the palette. By clicking a character in the **SpecialCharacters** palette, we can see both the full name and the alias of the character. In this way, we can learn the aliases and full names so that we possibly more rarely need to use the palettes.

In addition, characters can be typed using TeX aliases such as [ESC]\infty[ESC] or HTML and SGML aliases.

During the early stages of learning to use *Mathematica,* it may be handy to use the palettes. In time, it may be more useful to learn the aliases of the characters you need often. When we next study in some detail the special characters of *Mathematica*, we will give the aliases of the characters. Characters are also explained in the Documentation Center under tutorial/MathematicalAndOtherNotationOverview.

■ **Greek and Other Letters**

char.	α	β	γ	δ	ϵ	ε	ζ	η	θ	ϑ	ι	κ	χ	λ	μ	ν	ξ	o	π	ϖ	ρ	ϱ	σ	τ	υ	ϕ	φ	χ	ψ	ω
alias	a	b	g	d	e	ce	z	h	q	cq	i	k	ck	l	m	n	x	om	p	cp	r	cr	s	t	u	f	j	c	y	o

char.	A	B	Γ	Δ	E		Z	H	Θ		I	K		Λ	M	N	Ξ	O	Π		P		Σ	T	Y	Υ	Φ		X	Ψ	Ω
alias	A	B	G	D	E		Z	H	Q		I	K		L	M	N	X	Om	P		R		S	T	U	cU	F		C	Y	O

The second and fourth rows tell how the alias of a Greek letter is formed: For example, to write γ, press [ESC]g[ESC], or to write Γ, press [ESC]G[ESC]. In general, these aliases are easy to remember: The alias is formed by the corresponding usual letter (exceptions are q for θ and y for ψ, among others). The "c" letter in some aliases comes from "curly." If you do not remember this kind of short alias, you can use a longer alias containing the whole name of the Greek letter, for example, [ESC]gamma[ESC] or [ESC]Gamma[ESC].

In addition to Greek letters, *Mathematica* has script letters such as a, b, c, d, e, $\mathcal{A}$, $\mathcal{B}$, $\mathcal{C}$, $\mathcal{D}$, $\mathcal{E}$, Gothic letters such as $\mathfrak{a}$, $\mathfrak{b}$, $\mathfrak{c}$, $\mathfrak{d}$, $\mathfrak{e}$, $\mathfrak{A}$, $\mathfrak{B}$, $\mathfrak{C}$, $\mathfrak{D}$, $\mathfrak{E}$, and double-struck letters such as $\mathbb{a}$, $\mathbb{b}$, $\mathbb{c}$, $\mathbb{d}$, $\mathbb{e}$, $\mathbb{A}$, $\mathbb{B}$, $\mathbb{C}$, $\mathbb{D}$, $\mathbb{E}$. Their aliases are formed by enclosing, for example, sca, scA, goa, goA, dsa, or dsA with [ESC] keystrokes.

In the following table, we have collected some special symbols:

char.	π	e	i	∞	$\mathbb{C}$	$\mathbb{R}$	$\mathbb{Q}$	$\mathbb{Z}$	$\mathbb{N}$
alias	p	ee	ii	inf	dsC	dsR	dsQ	dsZ	dsN

For example, to write π, type ESC p ESC. Note that among the Greek letters, only π has a special meaning in *Mathematica*. $\mathbb{C}$, $\mathbb{R}$, $\mathbb{Q}$, $\mathbb{Z}$, and $\mathbb{N}$ are often used to denote the set of complex, real, rational, integer, and natural numbers, respectively. In outputs, *Mathematica* uses the first four symbols given in the table for **Pi** ($= 3.141$), **E** ($= 2.718$), **I** ($= \sqrt{-1}$), and **Infinity**, respectively:

{Pi, E, I, Infinity} $\{\pi, e, i, \infty\}$

■ Mathematical Symbols

char.	±	∓	×	⨯	·	√	∂	∇	d	∫	∮	Σ	Π
alias	+−	−+	*	cross	.	sqrt	pd	del	dd	int	cint	sum	prod
char.	∈	∉	⊂	⊆	∪	∩	∅	≡	≈	≃	≅	≠	∝
alias	el	!el	sub	sub=	un	inter	es	===	~~	~=	~==	!=	prop
char.	⇒	⇔	→	⟶	↔	∀	∃	∄	∧	∨	¬	⟨	⟩
alias	=>	<=>	->	-->	<->	fa	ex	!ex	and	or	not	<	>
char.	*	ᵀ	†	ᴴ	⌊	⌋	⌈	⌉	⟦	⟧	{	°	∎
alias	co	tr	ct	hc	lf	rf	lc	rc	[[	]]	pw	deg	fssq

For example, to write ±, type ESC +− ESC. The symbol × means multiplication, ⨯ the cross-product of two vectors, ∂ partial differentiation, ∇ a gradient or a backward difference, d the differential in an integral, ∅ an empty set, * conjugate, ᵀ transpose, † conjugate transpose, ᴴ Hermitian conjugate, ⌊ and ⌋ left and right floor, ⌈ and ⌉ left and right ceiling, { a piecewise definition, and ° degree.

Among the symbols in the table, many have special mathematical meanings in *Mathematica*. Such symbols include ×, ⨯, √, ∂, ∫ Σ, Π, {, ∈, ∪, ∩, ⇒, ∀, ∃, ∧, ∨, ¬, *, ᵀ, †, ᴴ, ⌊ ⌋, ⌈ ⌉, {, and °. For example,

$$\Big\{2 \times 4,\ \{1, 2, 3\} \times \{a, b, c\},\ \sqrt[4]{\ },\ \partial_x \text{Sin}[x],\ \int \text{Sin}[x]\ dx,\ \sum_{i=1}^{4} i,$$

$$\{a, b, c\} \cap \{a, c, d\},\ \neg\,(\text{True} \wedge \text{False}),\ (2 - 3\,\text{I})^*,\ \lfloor 5.8 \rfloor,\ 180° \,//\, \text{N}\Big\}$$

$\{8,\ \{-3\,b + 2\,c,\ 3\,a - c,\ -2\,a + b\},\ 2,$
$\text{Cos}[x],\ -\text{Cos}[x],\ 10,\ \{a, c\},\ \text{True},\ 2 + 3\,i,\ 5,\ 3.14159\}$

■ Characters for Fine-Tuning

\[InvisibleSpace]	ESC is ESC	\[InvisibleComma]	ESC , ESC
\[VeryThinSpace]	ESC ␣ ESC	\[NonBreakingSpace]	ESC nbs ESC
\[ThinSpace]	ESC ␣␣ ESC	\[NoBreak]	ESC nb ESC
\[MediumSpace]	ESC ␣␣␣ ESC	\[IndentingNewLine]	ESC nl ESC
\[ThickSpace]	ESC ␣␣␣␣ ESC	\[AlignmentMarker]	ESC am ESC

For each character, the table shows the full name and the alias. *Mathematica* has some special characters that are useful for fine-tuning text and expressions. One of them is an invisible comma. An example is x_{ij}, where we used an invisible comma between i and j. If we write i and j one after another, the indices are not italicized: x_{ij}. If we write a space between i and j, the indices may be too far from each other: $x_{i\,j}$.

Another special character is an invisible space. An example is xy, where we used an invisible space between x and y. Such an xy may be used as the product of x and y if we think that a space between x

and y is not aesthetically what we want: $x\,y$. We also have an implicit plus to display mixed fractions such as $5\frac{3}{4}$; write the implicit plus as ESC+ESC.

We also have spaces of various widths; note that ⌴ means the space key. Not shown in the previous box are the corresponding negative spaces from \[NegativeVeryThinSpace] to \[NegativeThickSpace], which allow us to move characters nearer to each other. We also have a space that does not allow a break (i.e., the words on both sides of the nonbreaking space will always be on the same text line) and a character that does not allow a break. Note that the special spaces are of fixed width and thus do not stretch like a normal space.

An indenting character inserts a line break while always maintaining the correct indenting level [a usual new line character (↵) sets the indenting level at the time the new line is started]. With an alignment marker, we can tell how to align rows, for example, in a mathematical formula containing several rows (see Example 5 in Section 3.4.2, p. 83).

■ Keys on a Keyboard

char.	SPACE	⌴	TAB	↵	RET	ENTER	SHIFT	CTRL	ALT	:	ESC	CMD	⌘	OPTION	DEL	{	]	￢
alias	spc	space	tab	ret	‿ret	ent	sh	ctrl	alt	esc	‿esc	cmd	cl	opt	‿del	[	]	kb

These characters are useful in describing the use of the keyboard.

3.3.3 Writing 2D Expressions

■ 2D Constructions

Recall that we can write inputs in a 2D form by using palettes. Now we learn how to write such inputs with the keyboard. These techniques are also useful when writing formulas in a mathematical document (see Section 3.4).

2D expressions are written with the control key. Here is a list of mathematical constructions made in that manner (these can also be found by choosing **Typesetting** and **Table/Matrix** from the **Insert** menu).

CTRL /	Go to the numerator of a fraction to be written or to the denominator of a fraction whose numerator is already written
CTRL @ or CTRL 2	Go into a square root
CTRL ^ or CTRL 6	Go to the superscript position (e.g., of a power)
CTRL _ or CTRL −	Go to the subscript position (e.g., of a variable or derivative)
CTRL & or CTRL 7	Go to the overscript position (e.g., of a sum, product, or integral)
CTRL + or CTRL =	Go to the underscript position (e.g., of a sum, product, integral, or limit)
CTRL % or CTRL 5	Go from superscript position to subscript position or vice versa, or from overscript position to underscript position or vice versa, or from the root position to the exponent position or vice versa
CTRL ,	Add a column to a table or matrix
CTRL ↵	Add a row to a table or matrix
CTRL ⌴	Return from a special position
CTRL .	Select the next larger subexpression

The symbols [CTRL]/ mean that we hold down the [CTRL] key while pressing /. The symbol ↵ means the return key, and ␣ means the space key. Some constructions can be made in two ways. The first way mentioned should be possible with most keyboards, whereas the second should work with any keyboard. If both work with your keyboard, choose the way you feel more comfortable with or that is easier for you to remember.

In formulas, we can move the cursor with the arrow keys (←, →, ↑, and ↓). To go to the next selection placeholder (∎), press [TAB].

To form derivatives, integrals, sums, and products, we need the characters ∂, $\int$, d, Σ, and $\prod$ (d is a special differential used in integrals). They can be written by enclosing pd, int, dd, sum, and prod between two [ESC] keystrokes (see Section 3.3.2, p. 75).

Use [CTRL]␣ to get out of the denominator of a fraction, to get out of a square root, to get down from a power, to get to the baseline from an index, and so on.

∎ **Examples**

The following are simple examples of writing 2D formulas with the keyboard:

To write	in the form	type
a / b	$\frac{a}{b}$	a [CTRL]/ b [CTRL]␣
Sqrt[a]	$\sqrt{a}$	[CTRL]2 a [CTRL]␣
x^2	x^2	x [CTRL]^ 2 [CTRL]␣
x[n]	x_n	x [CTRL]␣ n [CTRL]␣
xbar	$\bar{x}$	x [CTRL]& ␣ [CTRL]␣

Next are more elaborate constructs:

To write	in the form	type
D[a + b x, x]	$\partial_x (a + b\,x)$	[ESC]pd[ESC] [CTRL]␣ x [CTRL]␣ (a + b x)
Integrate[a + b x, {x, c, d}]	$\int_c^d (a + b\,x)\,\mathrm{d}x$	[ESC]int[ESC] [CTRL]+ c [CTRL]% d [CTRL]␣ (a + b x) [ESC]dd[ESC] x
Sum[x^i, {i, 0, n}]	$\sum_{i=0}^n x^i$	[ESC]sum[ESC] [CTRL]+ i = 0 [CTRL]% n [CTRL]␣ x [CTRL]^ i [CTRL]␣
{{a, b, c}, {d, e, f}}	$\begin{pmatrix} a & b & c \\ d & e & f \end{pmatrix}$	(a [CTRL], b [CTRL], c [CTRL]↵ d [TAB] e [TAB] f [CTRL]␣)
Piecewise[{{0, x ≤ 0}, {1, x > 0}}]	$\begin{bmatrix} 0 & x \le 0 \\ 1 & x > 0 \end{bmatrix}$	[ESC]pw[ESC] 0 [CTRL], x ≤ 0 [CTRL]↵ 1 [TAB] x > 0

Note that parentheses are needed around the expression to be differentiated, integrated, or summed if the expression is in the form of a sum.

Note also that we have shown one way to write the expressions, but often we have several possibilities. For example, to write $\frac{a+b}{c+d}$, we can first press [CTRL]/ and then fill in the numerator and denominator. We can also first write $a + b$, then press [CTRL]. two times to select the numerator, then press [CTRL]/, and lastly type the denominator.

Remember that matrices of size 2×2 can be formed with the **BasicMathInput** palette. A handy way to create matrices is to use **Insert ▷ Table/Matrix**. In the appearing dialog box, ask for a matrix, and then enter the number of rows and columns. We get a blank matrix:

$$\begin{pmatrix} \square & \square & \square \\ \square & \square & \square \end{pmatrix}$$

By pressing TAB, we can then go through the matrix and enter the elements. If we ask for a table, the result does not have the parentheses.

Writing 2D formulas is easier than it seems to be when looking at the previous CTRL and ESC key sequences. Indeed, *Mathematica*'s way of writing formulas is one of the best and easiest in the market, if not *the* best and easiest. Try it.

3.4 Writing Mathematical Documents

3.4.1 Introduction

■ *Mathematica* as a Writing Tool

As noted in Section 3.1.2, p. 54, *Mathematica is an advanced environment for technical writing.* One of the advantages of using *Mathematica* as a writing tool is that the whole document can be prepared with just one application.

For example, graphics can be created with *Mathematica*, and *Mathematica* writes the results of mathematical computations in a form ready to be printed. Writing and computation take place in the same application so that they have a fruitful interaction: When writing the text, you may observe new needs for mathematical computation, and from the results, you may observe new things to be reported.

In addition, the properties of *Mathematica* needed for using *Mathematica* as a writing environment are of high quality. For example, with *Mathematica* we get first-rate

- *text layout* by using styles and style sheets (Section 3.1.2, p. 54);
- *formulas* by using the traditional form (Sections 3.4.2, p. 80, and 3.4.3, p. 86);
- *graphics* of PostScript form (Chapters 5 through 8);
- *tables* with many kinds of fine-tuning (Chapter 15);
- *indexes* with hyperlinks (see Author Tools, p. 79); and
- *Help Browser material* with hyperlinks (see Author Tools, p. 79).

Furthermore, *Mathematica*, as one of the most powerful mathematical systems available, helps you with all kinds of computations needed to prepare your document. *Mathematica* documents can also be converted into LATEX, HTML, PDF, RTF, and PS documents from **File ▷ Save As...**.

Whereas text layout, graphics, and tables are considered elsewhere, in Sections 3.4.2 through 3.4.4 we present some other material that is helpful to know when using *Mathematica* as a writing tool. In particular, we show how to write mathematical formulas and how to number formulas and sections automatically.

■ Selecting a Style Sheet

The key to writing with *Mathematica* is the use of *styles* (Section 3.1.2, p. 54). With styles, you can easily write titles, headings, text, and formulas. Styles help in getting a consistent look throughout the document.

Remember that the styles of cells vary according to the *style sheet* you use (see Section 3.1.2). You may start the writing work with the **Default** style sheet, but at some point you should consider the style sheet in more detail. Regarding equation numbers, with all style sheets we can get simple equation numbers such as (7) with the **DisplayFormulaNumbered** style. Structured equation numbers such as (2.7) can be obtained by using the **EquationNumbered** style with the **Textbook** style sheet (see Section 3.4.4, p. 90). After choosing a style sheet, you may want to modify it (see Section 3.2.2, p. 63).

■ Main and Working Documents

When writing a mathematical document with *Mathematica*, it may be useful to work simultaneously with two documents: a *main document* and a *working document*. The main document will grow into the final publication, whereas all computations are done in the working document. Mathematical results, tables, and graphics are copied from the working document into the main document. This division into two documents may be needed because the main document may not contain the *Mathematica* commands but only the results. The working document contains all used *Mathematica* commands so that all computations can easily be done again.

The working document should include the same sections as are in the main document so that you can easily find the computations of a certain section. Add into the working document comments about the computations, such as any difficulties that may arise; they may be valuable if you need to do similar computations at a later time.

When you have completed the writing project, you will then have the main document ready to be printed and the working document that will enable you to redo and modify computations as needed.

■ Author Tools

In preparing a long, possibly book-sized document, and adding documents into the Help Browser, the add-on package **AuthorTools** may be very valuable. With the author tools, we can do the following:

- create a table of contents with hyperlinks;
- add index entries into the document;
- create an index with hyperlinks;
- create a Browser Categories file (for adding information into the Help Browser); and
- create a Browser Index file (to add the index entries into the Master Index of the Help Browser).

With this package, we can also compare differences among notebooks; restore corrupted notebooks; create bilaterally formatted cells (for displaying examples of *Mathematica* calculations); extract all cells of a particular type and save them in a desired format; insert objects to display the current values of variables such as the date, time, and the file name; and set printing options such as headers and footers.

Each of these operations can be done either on a single notebook or on a set of notebooks. For example, we can generate a unified index or table of contents for a book consisting of several notebooks. We used the package in the writing of this book to add the index entries, to create an index, and to create the Browser Categories and Browser Index files.

Look at http://support.wolfram.com/mathematica/packages/authortools/authortoolsinv6.html for information about the **AuthorTools** package in *Mathematica* 6. This page says, "The **AuthorTools** app pack is included with *Mathematica* 6, but it is considered legacy code and may not work as well with version 6 as it did in previous versions of *Mathematica*. It is included to help you during the transition to newer functions and syntax."

3.4.2 Display Formulas

■ **Styles for Display Formulas**

We can distinguish two types of formulas: *display formulas* and *inline formulas*. Display formulas are displayed in their own lines of text, whereas inline formulas are written inside normal text. Before explaining how to write display formulas, we consider the styles used for them.

To write formulas, each style sheet (except **Demonstration**) has the styles **DisplayFormula** and **DisplayFormulaNumbered**. These styles can be used either directly with **Format ▷ Style** or by choosing **Format ▷ Style ▷ Other...** and then typing the name of the style; the latter method has to be used with the **Article** and **Book** type style sheets. The **Article** and **Book** type style sheets have the styles **Equation** and **EquationNumbered** in **Format ▷ Style** (other style sheets do not have these styles).

The styles **DisplayFormula** and **Equation** have some differences in the alignment of the formula. The styles **DisplayFormulaNumbered** and **EquationNumbered** have the difference that the former generates simple numbers such as (7) throughout the notebook, whereas the latter generates structured numbers such as (2.7), where the first number refers to the chapter number and the second number to the formula within that chapter [however, in the **JournalArticle** style sheet, the **EquationNumbered** style gives a simple number such as (7)].

Recommendations are as follows. If you want to use structured equation numbers, use the **Textbook** style sheet and the styles **Equation** and **EquationNumbered** (see more details in Section 3.4.4, p. 90). If unstructured equation numbers suffice, use any of the other style sheets and the styles **DisplayFormula** and **DisplayFormulaNumbered** (however, with the **JournalArticle** style sheet, it is easiest to use the styles **Equation** and **EquationNumbered** because these are in the style menu).

Numbered equations are considered in Section 3.4.4, p. 89.

■ **Writing Display Formulas**

To write a display formula (a mathematical formula in a separate line), complete the following simple steps:

- Write the formula in the usual **Input** style by using any of the various ways to write inputs (see Section 3.3.1, p. 73): Write usual 1D mathematica code (e.g., `Sum[Subscript[a, i], {i, n}]`) or write 2D formulas (e.g., $\sum_{i=1}^{n} a_i$) with the keyboard or by using the **BasicMathInput** and **SpecialCharacters** palettes.
- You can execute any part of the formula by selecting that part with the mouse and choosing **Evaluation ▷ Evaluate in Place** or by pressing ALT+RET (⌘+RET on a Macintosh).
- Transform the formula into the traditional mathematical form by letting the cursor be in the formula and choosing **Cell ▷ Convert To ▷ TraditionalForm** or by pressing SHIFT+CTRL+t (SHIFT+⌘+t on a Macintosh). Lastly, choose, from **Format ▷ Style**, the cell style to be one of **DisplayFormula**, **DisplayFormulaNumbered**, **Equation**, or **EquationNumbered**.

Sometimes you may want to nudge some parts of a formula. For example, to nudge to the left, press CTRL+← several times. Another way to nudge is the use of the menu commands, found in **Insert ▷ Typesetting**.

To write an inline formula (a mathematical formula among text), see Section 3.4.3, p. 86.

■ **Example 1**

Suppose we want to write the following formula:

$$f_n = \int_0^\pi \frac{\sin(x)}{n+x}\,dx,\ n = 1, 2, \ldots$$

First, we write the following input:

```
Subscript[f, n] = Integrate[Sin[x] / (n + x), {x, 0, π}]
```

Then we transform the cell into traditional form:

$$f_n = \int_0^\pi \frac{\sin(x)}{n+x}\,dx$$

Lastly, we transform the cell into the **Equation** style:

$$f_n = \int_0^\pi \frac{\sin(x)}{n+x}\,dx$$

We can write "$n = 1, 2, \ldots$" at the end. Note that here we showed the formula in three separate forms in three cells to show how the formula proceeds, but normally the conversions are done in only one cell.

■ **Comment**

The previous method of writing formulas works but has the restriction that the formula to be written has to follow correct *Mathematica* syntax; otherwise, the transformation into traditional form does not succeed. For example, we cannot write "$n = 1, 2, \ldots$" at the end of the input:

```
Subscript[f, n] = Integrate[Sin[x] / (n + x), {x, 0, π}], n = 1, 2, …
```

This input simply does not transform into the traditional form. Therefore, often we have to afterwards add some minor components such as "$n = 1, 2, \ldots$" into the formula.

Next, we present a modified way to write formulas; I have been comfortable with this way.

■ **Using a Template**

Now we present a way to write formulas in which we first prepare a template for a formula. Each time we would like to write a formula, we take a copy of this template and then write the formula. Thus, first prepare a template as follows:

- Write **template** in the **Input** style.
- Transform the cell into the traditional form by letting the cursor be in the input and choosing **Cell ▷ Convert To ▷ TraditionalForm** or by pressing SHIFT CTRL t (SHIFT ⌘ t on a Macintosh).
- Choose, from **Format ▷ Style**, the cell style to be one of **DisplayFormula**, **DisplayFormulaNumbered**, **Equation**, or **EquationNumbered**.

The result of these steps is a template like the following:

template

We can hold the template at the end of our document or in a separate notebook. Now, each time we want to write a formula, we take a copy of this cell, remove "template", and write a new formula. In writing, we can use palettes or write 2D formulas directly with the keyboard. We can also use *Mathematica* commands such as Integrate, but then we have to again transform the formula into the traditional form.

In this method, the formula need not be in the correct *Mathematica* syntax. On the other hand, if we now want to execute a part of the formula, we get the warning that the traditional form has some restrictions in mathematical interpretation.

■ Example 2

We want to write the same formula as in Example 1. We take a copy of the template and, with the aid of **BasicMathInput** palette or the keyboard, write the formula. It comes ready without any transformations:

$$f_n = \int_0^\pi \frac{\sin(x)}{n+x} \, dx, \, n = 1, 2, \ldots$$

In writing the formula, do not type any spaces because the traditional form automatically writes spaces at suitable places.

■ Example 3

Next, we want to write the following formula, which is a special case of the formula we considered in Examples 1 and 2:

$$f_1 = Ci(1) \sin(1) - Ci(1 + \pi) \sin(1) + \cos(1) (-Si(1) + Si(1 + \pi))$$

Copy the formula we wrote in Example 2, delete "$n = 1, 2, \ldots$", and replace n with 1:

$$f_1 = \int_0^\pi \frac{\sin(x)}{1+x} \, dx$$

Select the integral and choose **Evaluation ▷ Evaluate in Place**.

■ Example 4

Now we want to write this formula:

$$\int_0^\pi \frac{\sin(x)}{n+x} \, dx = Ci(n) \sin(n) - Ci(n + \pi) \sin(n) + \cos(n) (Si(n + \pi) - Si(n))$$

Because we now want to use an assumption, we write *Mathematica* code:

```
Integrate[Sin[x] / (n + x), {x, 0, π}] =
   Integrate[Sin[x] / (n + x), {x, 0, π}, Assumptions → n > 0]
```

Execute the right-hand side by pressing [ALT][RET]:

```
Integrate[Sin[x] / (n + x), {x, 0, π}] = CosIntegral[n] Sin[n] -
   CosIntegral[n + π] Sin[n] + Cos[n] (-SinIntegral[n] + SinIntegral[n + π])
```

Convert this into traditional form and equation:

$$\int_0^\pi \frac{\sin(x)}{n+x} \, dx = Ci(n) \sin(n) - Ci(n + \pi) \sin(n) + \cos(n) (Si(n + \pi) - Si(n))$$

■ **Example 5**

To write the formula,

$$d = \int_0^1 \frac{\log^7(x)}{\sqrt{1-x^2}}\,dx$$

$$= -\frac{1}{768}\,\pi\left\{275\,\pi^6 \log(4) + 133\,\pi^4\left[\log^3(4) + 12\,\zeta(3)\right] + 21\,\pi^2\left[\log^5(4) + 120\,\zeta(3)\log^2(4) + 720\,\zeta(5)\right] + \right.$$

$$\left. 3\left[\log^7(4) + 420\,\zeta(3)\log^4(4) + 15\,120\,\zeta(5)\log^2(4) + 10\,080\,\zeta(3)^2\log(4) + 90\,720\,\zeta(7)\right]\right\}$$

$$= -5040.39$$

so that the equal signs are aligned, first write as follows:

```
d = Integrate[Log[x]^7 / Sqrt[1 - x^2], {x, 0, 1}] =
  Integrate[Log[x]^7 / Sqrt[1 - x^2], {x, 0, 1}] =
  Integrate[Log[x]^7 / Sqrt[1 - x^2], {x, 0, 1}] // N
```

Then execute separately the second and the third integral:

```
d = Integrate[Log[x]^7 / Sqrt[1 - x^2], {x, 0, 1}] =
```

$$-\frac{1}{768}\,\pi\left(275\,\pi^6\,\text{Log}[4] + 133\,\pi^4\left(\text{Log}[4]^3 + 12\,\text{Zeta}[3]\right) + \right.$$

$$21\,\pi^2\left(\text{Log}[4]^5 + 120\,\text{Log}[4]^2\,\text{Zeta}[3] + 720\,\text{Zeta}[5]\right) +$$

$$3\left(\text{Log}[4]^7 + 420\,\text{Log}[4]^4\,\text{Zeta}[3] + 10\,080\,\text{Log}[4]\,\text{Zeta}[3]^2 + \right.$$

$$\left.\left. 15\,120\,\text{Log}[4]^2\,\text{Zeta}[5] + 90\,720\,\text{Zeta}[7]\right)\right) = -5040.389239250824\,\grave{}$$

Convert this to traditional form and equation:

$$d = \int_0^1 \frac{\log^7(x)}{\sqrt{1-x^2}}\,dx =$$

$$-\frac{1}{768}\,\pi\left(275\,\pi^6 \log(4) + 133\,\pi^4\left(\log^3(4) + 12\,\zeta(3)\right) + 21\,\pi^2\left(\log^5(4) + 120\,\zeta(3)\log^2(4) + 720\,\zeta(5)\right) + \right.$$

$$\left. 3\left(\log^7(4) + 420\,\zeta(3)\log^4(4) + 15\,120\,\zeta(5)\log^2(4) + 10\,080\,\zeta(3)^2\log(4) + 90\,720\,\zeta(7)\right)\right) = -5040.39$$

To fine-tune the formulas, add a return character before the second and third equal signs, write an alignment marker (ESC am ESC) before each equal sign, write a no break sign (ESC nb ESC) after the third equal sign, select the cell bracket, and choose **Format ▷ Text Alignment ▷ On AlignmentMarker**. Lastly, replace some usual parentheses with brackets ([ ]) and curly braces ({ }) to make the formula more readable.

■ **Example 6**

Next, we write the following formula:

$$\int x^a\,dx = \begin{cases} \dfrac{x^{a+1}}{a+1}, & a \neq -1, \\ \log(x), & a = -1. \end{cases}$$

To begin, write the integral with the template:

$$\int x^a \, dx =$$

Then write and execute the following command:

$$\texttt{Piecewise}\Big[\Big\{\Big\{\dfrac{\texttt{x}^{\texttt{a+1}}}{\texttt{a + 1}}, \ \texttt{a != -1}\Big\}, \ \{\texttt{Log[x]}, \ \texttt{a == -1}\}\Big\}\Big] \texttt{ // TraditionalForm}$$

$$\begin{cases} \frac{x^{a+1}}{a+1} & a \neq -1 \\ \log(x) & a = -1 \end{cases}$$

Copy this result and paste into the previous formula:

$$\int x^a \, dx = \begin{cases} \frac{x^{a+1}}{a+1} & a \neq -1 \\ \log(x) & a = -1 \end{cases}$$

Lastly, select the cell bracket of the formula and choose **Insert** ▷ **Typesetting** ▷ **Add Frame** or click the ▣ button in the **BasicTypesetting** palette. If the frame seems to be too thick, select the cell bracket of the formula and, with the Option Inspector, change the value of `BoxFrame` to, for example, 0.25:

$$\boxed{\int x^a \, dx = \begin{cases} \frac{x^{a+1}}{a+1} & a \neq -1 \\ \log(x) & a = -1 \end{cases}}$$

Previously, we wrote the piecewise definition as a *Mathematica* command. Another way to write the formula is to write directly a 2D formula. Take a copy of the template we created previously, paste it, and write, using the **BasicMathInput** palette or direct 2D input, as follows:

$$\int x^a \, dx = \begin{cases} \frac{x^{a+1}}{a+1} & a \neq -1 \\ \log(x) & a = -1 \end{cases}$$

Here, to begin the piecewise definition, type ⎋pw⎋. To create slots for the four parts of the definition, type ⎈⎵,⎵. Fill in the slots. To align the columns left, select the cell bracket of the formula, and using the Option Inspector, set the value of the `GridBoxAlignment` option to `{"Columns" -> {{Left}}}`.

To fine-tune the formula, type a comma or period at suitable places. If you want to adjust the frame (thickness, margins, background, etc.), select **Format** ▷ **Option Inspector**, go to *Formatting Options* ▷ *Expression Formatting* ▷ *Specific Box Options* ▷ *FrameBoxOptions*, and set the values of suitable options.

■ **Example 7**

To write the formula,

$$\lim_{x \to \infty} \left(\frac{a}{x} + 1\right)^x = e^a$$

use the **BasicMathInput** palette or direct 2D input. First, click the ■ button in the palette, and write lim into the upper box and $x \to \infty$ into the lower box. Then, write the rest of the formula. Another way is to write a *Mathematica* command:

$$\texttt{Limit[(1 + a / x)\^{}x, x} \to \infty] = \texttt{Limit[(1 + a / x)\^{}x, x} \to \infty]$$

Execute the right-hand side, and convert the cell to traditional form and equation.

■ **Example 8**

To write this one,

$$\underbrace{1\,1\,1\cdots 1}_{k}\underbrace{0\,0\,0\cdots 0}_{n}$$

use the template we created previously and click ■ two times from the **BasicMathInput** palette to get three placeholders. Write "1 1 1 ⋯ 1" in the first box; write ESC␣ESC between each character. The three dots ⋯ can be found from the **SpecialCharacters** palette under *Textual Forms*. Type ␣ from the same palette in the second box, and k in the last box. Copy this whole expression, paste it next to the expression, and replace the 1's with 0's and k with n.

■ **Example 9**

To write the matrix

$$\begin{pmatrix} a_{11} & a_{12} & \cdots & a_{1\,n} \\ a_{21} & a_{22} & \cdots & a_{2\,n} \\ \vdots & \vdots & & \vdots \\ a_{m\,1} & a_{m\,2} & \cdots & a_{mn} \end{pmatrix}$$

paste a copy of the template we created previously and insert a (4×4) empty matrix with **Insert ▷ Table/Matrix ▷ New…**. Then fill in the elements. The characters ⋯ (center ellipsis) and ⋮ (vertical ellipsis) can be found from the **SpecialCharacters** palette under *Textual Forms*. Note that for the elements in the last row, we have to type a space between the indices to get m and n italicized.

The space between the indices in the last column and last row may not satisfy you. To get the indices closer to each other, write an invisible comma between the indices with ESC,ESC. In this way, we get the following matrix:

$$\begin{pmatrix} a_{11} & a_{12} & \cdots & a_{1n} \\ a_{21} & a_{22} & \cdots & a_{2n} \\ \vdots & \vdots & & \vdots \\ a_{m1} & a_{m2} & \cdots & a_{mn} \end{pmatrix}$$

■ **Example 10**

With the **BasicMathInput** palette, we can write expressions such as $\overline{x}$ and $\vec{x}$. With the **BasicTypesetting** palette, many other similar expressions can be written: $x^\dagger$, x_+, x^+, x_-, x^-, $\underline{x}$, $\overline{x}$, $\vec{x}$, x^*, x_*, x', x'', $\dot{x}$, $\ddot{x}$, $\tilde{x}$, and $\hat{x}$.

Still more similar expressions can be written by clicking the buttons ■□, ■□, ■□, □■, □■, ■, ■, or ■ in the **BasicTypesetting** palette and filling the slots with suitable characters. Recall (Section 3.3.3, p. 76) that the constructions ■□, ■□, ■, and ■ can also be written with the keyboard by typing CTRL[^], CTRL[_], CTRL[&], or CTRL[+], respectively. In addition, we have the following characters, familiar from matrix calculus: * (ESC co ESC), ᵀ (ESC tr ESC), † (ESC ct ESC), and ᴴ (ESC hc ESC) (they are written without the ■□ construct).

3.4.3 Inline Formulas

■ **Writing Inline Formulas**

An inline formula is among the text of a **Text**-style cell. Such formulas are often small, such as i, a_i, x, $f(x)$, $f'(x)$, $\partial f / \partial x$, $\frac{\partial f}{\partial x}$, $\sqrt{x}$, $\sin(x)$, a/b, $\frac{a}{b}$, $\lim_{x \to \infty} \left(1 + \frac{a}{x}\right)^x = e^x$, $\int \sin(x)\, dx$, $\int_0^\pi \cos(x)\, dx$, or $\sum_{i=1}^\infty 1/i^2$. If the formula is larger, it is often more useful to put it in a separate line as a display formula.

Note that *Mathematica* formats inline formulas lower as display formulas (as is usual in traditional mathematical notation). For example, in a fraction, the font size is smaller; in a limit, the variable and limiting value are written as a subscript next to "lim"; and in a definite integral or sum, the lower and upper limits are written as sub- and superscripts (and not as under- and overscripts).

To write an inline formula, we begin and end the formula with the special key combinations CTRL((
and CTRL()). The simple steps are as follows:

- Begin the formula by typing CTRL((or CTRL(9.
- Write the formula.
- End the formula by typing CTRL() or CTRL(0.

The formula can be written in any way: linearly, in a 2D form with the keyboard (see Section 3.3.3, p. 76), or in a 2D form with the palettes. 2D formulas are automatically in the traditional form. Linear 1D formulas such as Integrate[$a/(1 + x)$, x] can be transformed into the traditional form by selecting the formula with the mouse and choosing **Cell ▷ Convert To ▷ TraditionalForm** or pressing SHIFT(CTRL(t)
(SHIFT(⌘(t on a Macintosh).

I often write the simplest formulas (e.g., variables, special characters, subscripts, powers, functions, square roots, and fractions) with the keyboard and less simple formulas (e.g., integrals and sums) with palettes.

Note that a part of an inline formula can also be executed by selecting that part and then choosing **Evaluation ▷ Evaluate in Place** or by pressing ALT(RET (⌘(RET on a Macintosh). However, this may seldom be needed because inline formulas are, as a rule, simple.

■ **Example 1**

We want to write a simple text cell:

Assume that x is positive. Then…

Just begin a **Text** cell by pressing ALT(7, write the text before x, then write CTRL((x CTRL()), and continue the text. Note that you need not worry about italicizing the variable; *Mathematica* does it for you. Another way to write x is to press CTRL(i x CTRL(i to italicize the variable. Choose the way you find easiest. I consistently use CTRL((… CTRL()) for all inline formulas, even formulas as simple as x.

■ **Example 2**

Next, we want to write the following text cell:

Consider the function $f(x) = a \sin(x + \pi)$. Let…

To write the formula, press [CTRL]([(], write f(x)=a_sin(x+[ESC]p[ESC]), press [CTRL]([)], and continue the text. Note that when writing the formula you should not press the space key in any other place besides in multiplications; *Mathematica* adds suitable spaces when needed (e.g., before and after = or +).

■ Example 3

Now, we write the following text cell:

Let $f(x) = \frac{a}{x}$ and assume that…

To write the formula, do one of the following:

- Use the keyboard: [CTRL]([(] f(x)=a [CTRL]([/] x [CTRL]([)] (see Section 3.3.3, p. 76).
- Use the **BasicMathInput** palette to write the fraction: [CTRL]([(] f(x)=$\frac{a}{x}$[CTRL]([)].
- Write the formula linearly as [CTRL]([(] f(x)=a/x [CTRL]([)], select a/x, and press [SHIFT][CTRL]([t].

■ Example 4

Let us now write the following formula in the text cell:

If the integral $\int_0^\infty f(x)\,dx$ converges, then…

Do one of the following:

- Use the keyboard to write the lower and upper bounds of the integral as sub- and superscripts: [CTRL]([(] [ESC]int[ESC] [CTRL]([_] 0 [CTRL]([%] [ESC]inf[ESC] [CTRL]([_] f(x) [ESC]dd[ESC] x [CTRL]([)]. (This is easier than it looks; try it.)
- Use the **BasicInput** palette to write the integral: [CTRL]([(]$\int_0^\infty f(x)\,dx$[CTRL]([)] (go with [TAB] from one selection placeholder to the next one).
- Write the formula linearly as [CTRL]([(]Integrate[f[x],{x,0,Infinity}][CTRL]([)], select the formula, and press [SHIFT][CTRL]([t].

■ Example 5

Now we want to write the following:

Because $\int \frac{1}{a+x^2}\,dx = \frac{1}{\sqrt{a}}\tan^{-1}\left(\frac{x}{\sqrt{a}}\right)$, then…

First, write the following with the **BasicMathInput** palette:

Because $\int \frac{1}{a+x^2}\,dx = \int \frac{1}{a+x^2}\,dx$, then…

Select the latter integral and execute it with [ALT][RET]. Here is the result:

Because $\int \frac{1}{a+x^2}\,dx = \dfrac{\tan^{-1}\left(\frac{x}{\sqrt{a}}\right)}{\sqrt{a}}$, then…

To make the formula lower, cut the denominator, insert a fraction after the equal sign, write 1 into the numerator, and paste the cut square root into the denominator:

Because $\int \frac{1}{a+x^2}\,dx = \frac{1}{\sqrt{a}}\tan^{-1}\left(\frac{x}{\sqrt{a}}\right)$, then…

In this way, you can also modify the results *Mathematica* gives.

■ Example 6

The following is another way to write the same formula:

Because $\int \frac{1}{a+x^2}\, dx =$, then…

Then execute the integral:

```
Integrate[1 / (a + x^2), x]
```

$$\frac{\text{ArcTan}\left[\frac{x}{\sqrt{a}}\right]}{\sqrt{a}}$$

Copy the result into the formula:

Because $\int \frac{1}{a+x^2}\, dx = \dfrac{\text{ArcTan}\left[\frac{x}{\sqrt{a}}\right]}{\sqrt{a}}$, then…

Select the value of the integral, convert it into traditional form, and then move the denominator as shown previously. In this way, results from calculations can be inserted into inline formulas.

■ Example 7

In display formulas, the default is that constructs such as a sum, product, union, intersection, lim, max, or min are written in a tall form:

$$\sum_{i=1}^{n} a_i + \bigcup_{i=1}^{n} c_i + \lim_{x \to 0} f(x) + \max_i \{d_i\}$$

However, in an inline formula the default is that these constructs are displayed in a low form:

$\sum_{i=1}^{n} a_i + \bigcup_{i=1}^{n} c_i + \lim_{x \to 0} f(x) + \max_i \{d_i\}$

This is a very useful property. However, we have an option, `LimitsPositioning`, that also enables us to get tall forms in inline formulas. This option can be found from the Option Inspector when the *View* is set to be *alphabetically*. In the Option Inspector, we can see that the option can be used in underscript-boxes ■, overscriptboxes ■, and underoverscriptboxes ■. If we select the cell bracket of the previous formula and set, with Option Inspector, `LimitsPositioning` (`UnderscriptBoxOptions`) to be `False`, the lim and max constructs become tall:

$\sum_{i=1}^{n} a_i + \bigcup_{i=1}^{n} c_i + \lim_{x \to 0} f(x) + \max_i \{d_i\}$

If we set `LimitsPositioning` (`UnderoverscriptBoxOptions`) to be `False`, the sum and union become tall:

$$\sum_{i=1}^{n} a_i + \bigcup_{i=1}^{n} c_i + \lim_{x \to 0} f(x) + \max_i \{d_i\}$$

If we set both `LimitsPositioning` (`UnderscriptBoxOptions`) and `LimitsPositioning` (`Underover` `scriptBoxOptions`) to be `False`, all constructs become tall:

$$\sum_{i=1}^{n} a_i + \bigcup_{i=1}^{n} c_i + \lim_{x \to 0} f(x) + \max_{i} \{d_i\}$$

3.4.4 Automatic Numbering

■ Introduction

Sections and some formulas in a mathematical document are often numbered so that we can easily refer to them. Often, we also want to refer to some page numbers of our document. With *Mathematica,* we can give the numbers manually and also write references to them manually, but we can also let *Mathematica* automatically choose the appropriate numbers.

In a small document, the numbers of sections and references to page numbers can easily be written manually, and to some extent the same is true for formulas. However, in a larger document the automatic numbering becomes more tempting because chances increase that sections and formulas are moved to other places and new sections and new formulas are written among old ones. After such a modification, automatic numbers and references to them are again correct without your having to manipulate them in any way, whereas manually given numbers and references to them have to be manually corrected. However, using the automatic numbering system of *Mathematica* often requires some work in the form of handling so-called cell tags. Next, we consider automatic numbering of pages, formulas, and sections.

■ Referring to Page Numbers

Suppose you want to write "According to Theorem 2.3 (see p. XXX) …", where XXX should be replaced with the correct page number. First, you have to assign a *cell tag* to the cell that contains Theorem 2.3. Select the cell bracket of the theorem, and do the following:

• Choose **Cell ▷ Cell Tags ▷ Add/Remove Cell Tags…**; the following window appears:

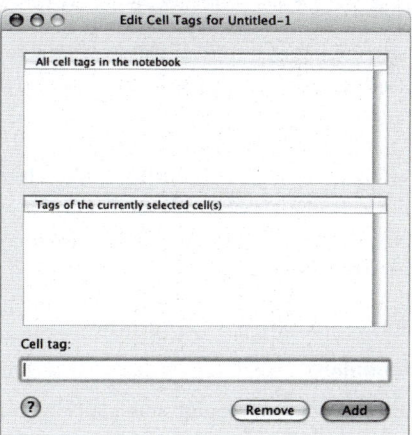

• Write a name (e.g., Theorem 2.3) for the cell into the *Cell tag* field and click *Add*.

To create a reference to the page number of Theorem 2.3, put the cursor in the place where you want to refer to the theorem (i.e., after "see p. ") and do the following:

• Choose **Insert ▷ Automatic Numbering...**; the following window appears:

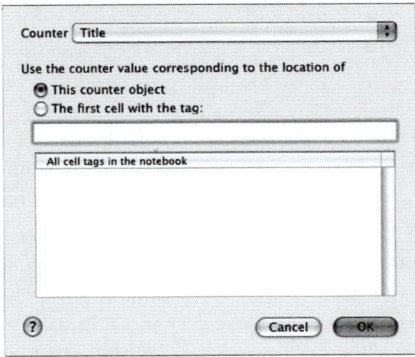

• Set *Counter* to be *Page*.
• From the list *All cell tags in the notebook*, click the cell tag of the theorem, and click *OK*.

The page number now appears, at the location of the cursor, in the form XXX. When the document is printed, XXX is automatically replaced with the correct page number. Before printing, you can see the correct page number by choosing **File ▷ Printing Settings ▷ Show Page Breaks**. The page number also has the useful property that when it is clicked on, the document is scrolled to the cell to which the page number refers.

By the way, a list of all cell tags of the current notebook can be seen by choosing **Cell ▷ Cell Tags ▷ Find Cell Tag**. By selecting a cell tag from the list, the notebook is scrolled to the corresponding cell. The cell tags can be seen in the notebook by choosing **Cell ▷ Cell Tags ▷ Show Cell Tags**.

■ Manual Numbering of Formulas

Before explaining the automatic numbering of formulas, we show how to manually add a number to a formula. First, write the formula (see Section 3.4.2, p. 80):

$$f = a + b\,x$$

Add the formula number as a cell frame label as follows. Select the cell bracket of the formula, choose **Format ▷ Option Inspector**, set *View* to be *alphabetically*, and find the option `CellFrameLabels`. The default value of this option is {{None, None}, {None, None}}. Replace the second None with "(3.1)" and press ⏎. The formula now has the given number:

$$f = a + b\,x \qquad\qquad\qquad\qquad\qquad\qquad\qquad\qquad\qquad\qquad\qquad (3.1)$$

■ Automatic Numbering of Formulas

All style sheets (except **Demonstration**) have styles for automatically numbered formulas; see Section 3.4.2, p. 80. To repeat, if you want to use structured equation numbers such as (2.7), use the **Textbook** style sheet and the styles **Equation** and **EquationNumbered**. If unstructured equation numbers such as (7) suffice, use any of the other style sheets and the styles **DisplayFormula** and **DisplayFormulaNumbered** (however, with the **JournalArticle** style sheet, it is easiest to use the styles **Equation** and **Equation-Numbered** because these are in the style menu). The automatic numbers of formulas are useful because they automatically change if we add or remove formulas.

To create a formula with an unstructured number, first write a formula, then select the cell bracket of the formula, and choose the style to be **DisplayFormulaNumbered**:

$$f = a + b\,x \qquad\qquad\qquad\qquad\qquad\qquad\qquad\qquad (1)$$

To create a formula with a structured number, use the **Textbook** style sheet. The first number, such as 2 in (2.7), refers to a *chapter*. Thus, you have to compose the document by chapters. Each chapter may contain sections, subsections, etc. For each chapter, you have to create, with the **BookChapterNumber** style, a cell containing the number of the chapter. Of course, it is also useful to create a chapter title; use the **BookChapterTitle** style. These two styles are not listed in the style menu, so choose **Format ▷ Style ▷ Other...** and type the name of the style.

If you would like to remove the number of an equation, just choose the style of the cell to be **DisplayFormula** or **Equation**.

■ **Referring to Numbered Formulas**

To be able to refer to a numbered formula, we have to first assign a cell tag to the formula. Select the cell bracket of the numbered formula, and do the following:

- Choose **Cell ▷ Cell Tags ▷ Add/Remove Cell Tags...**.
- Write a name for the formula into the *Cell tag* field and click *Add*.

To create a reference to a numbered formula, put the cursor in the place where you want to refer to the numbered formula, and do the following:

- Write the opening parenthesis (and choose **Insert ▷ Automatic Numbering...**.
- Set *Counter* to *DisplayFormulaNumbered (or EquationNumbered)*
- From the list *All cell tags in the notebook*, click the cell tag of the formula, click *OK*, and write the closing parenthesis).

The number of the formula now appears at the location of the cursor. This reference automatically changes if the number of the formula changes. By clicking the number, the notebook is scrolled to where the formula is located.

■ **Modifying Automatic Numbers of Formulas**

In the **Textbook** style sheet, the working of the numbering of formulas is based on chapters. However, you may want to create numbers where the first part is based on, for example, sections. Do as follows. Set the style sheet to be **Textbook** and choose **Format ▷ Edit Stylesheet...**.

From *Choose a style to modify* select *Section*, choose **Format ▷ Option Inspector...**, and find the option `CounterAssignments`. Its value is a list. Add the element `{"EquationNumbered", 0}` into the list. In this way, we tell that the counter for equations is reset to zero at the beginning of each section.

From *Choose a style to modify* select *EquationNumbered*, choose **Format ▷ Option Inspector...**, and find the option `CellFrameLabels`. In its value, replace `CounterBox["BookChapterNumber"]` with `CounterBox["Section"]`. In this way, we tell that the first number for an equation should be the number of the current section (and not the current chapter).

■ **Automatic Numbering of Sections**

Manual numbering of sections is, of course, very easy: Just write a suitable number at the beginning of the heading of the section. With automatic numbering, however, we have the advantage that the numbers of sections and references to them are correct after modifications in the order of the sections. To give an automatic number to a section, place the cursor at the beginning of the section heading and do the following:

• Choose **Insert** ▷ **Automatic Numbering...**.

• Set *Counter* to be *Section* and click *OK*.

If you want an automatic number for a subsection, such as 2.4, first create the number 2 of the section in the way we showed previously, type a period, and then create the number 4 of the subsection in the same way (just set *Counter* to *Subsection*). Similarly, we can create numbers for sub-subsections.

To make creating section numbers easier, you can create a notebook having one section, one subsection, and one sub-subsection with automatic numbers. Then copy a suitable heading from this notebook each time you would like to start a new section.

■ Referring to Numbered Sections

Suppose we want to write "In Section X we have shown...", where we want, in place of X, the correct section number to appear. In that section, we have a certain cell to which we want to refer. To write such a reference, first assign a cell tag to the corresponding cell: Select the cell bracket, choose **Cell** ▷ **Cell Tags** ▷ **Add/Remove Cell Tags...**, write a name for the cell, and then click *Add*.

A reference to the cell is then written in the same way a reference is written for a numbered equation: Choose **Insert** ▷ **Automatic Numbering...**, set *Counter* to *Section*, click the cell tag of the cell in question, and click *OK*.

To write a reference to a cell in a subsection, first assign a cell tag to that cell. To write a reference such as "In Subsection X.Y we will show that...", first create the section number X with **Insert** ▷ **Automatic Numbering...** as above, type a period, and then, in the same way, create the subsection number Y. Likewise, we can create a reference to a sub-subsection.

■ Automatic Numbering of Figures, Tables, etc.

To create automatic numbers for figures and tables, first choose a style sheet having the styles **Figure** and **Table**; such style sheets are among the **Article** and **Book** type style sheets.

After you have created a figure, select its cell bracket and set its style to be **Figure**. Create a caption below the figure by choosing the **FigureCaption** style and writing the caption. In the caption, create an automatic number for the figure by choosing **Insert** ▷ **Automatic Numbering...**, setting *Counter* to be *Figure*, and clicking *OK*.

After you have created a table, select its cell bracket and set its style to be **Table**. Create a title below the table by choosing the **TableTitle** style and writing the title. In the title, create an automatic number for the table by choosing **Insert** ▷ **Automatic Numbering...**, setting *Counter* to be *Table*, and clicking *OK*.

To refer to a figure or table, first create a cell tag to the corresponding cell. Then choose **Insert** ▷ **Automatic Numbering...**, set *Counter* to be, for example, *Figure*, click the cell tag, and click *OK*.

In the same way we can use automatic numbers for pictures and programs, for example.

Files

Introduction

> *The world's greatest and most powerful computer was constructed, so mathematicians decided to test it out by seeing if it could make any impression on some classical unsolved problems. They decided on Fermat's last theorem (this happened prior to the work of Andrew Wiles), namely that $x^n + y^n = z^n$ has no solutions over the natural numbers for $n \geq 3$. For days they fed it with every known piece of information, conjecture, and partial result, and at last they set it to work. After a few minutes it printed out: "I have a wonderful proof of this result, but my memory is too small to store it."*

Note that you may skip this chapter if you right now do not need (or already know how) to do the following:

- use the packages of *Mathematica*;
- read data or graphics from a file;
- write data or graphics into a file;
- save results for later use;
- convert results and notebooks for TeX, HTML, C, or Fortran;
- speed up calculations; or
- save memory consumption.

With packages we can give added functionality to *Mathematica*. Section 4.1 describes various types of packages, their loading, and what to do if we forget to load a package.

Read Section 4.2.1 if you have data in a file and you want to visualize or analyze it with *Mathematica* by, for example, using some plotting commands or some statistical methods. You then need to import the data in such a way that *Mathematica* understands it. This means that you must form lists that contain the elements of the data. We consider here some simple examples, but real-life data are considered, for example, in Chapter 10. Data can also be exported for use in other applications.

You may want to export a *Mathematica* plot to another application or to import into *Mathematica* a plot made by another application. Section 4.2.2 is devoted to these topics.

In Section 4.3, we consider saving and loading results. This may be useful if we want to continue calculations in later sessions. We also consider converting *Mathematica* notebooks and results to the forms required by TeX, HTML, C, and Fortran.

In Section 4.4, we show some ways to manage and save computing time and computer memory.

4.1 Loading Packages

4.1.1 Standard Packages

■ Types of Packages

Mathematica packages supplement the kernel by providing more commands. We have several types of packages. To see the types of packages, write as follows:

```
SetDirectory[$InstallationDirectory <> "/AddOns"]; FileNames["*Packages"]
```
```
{ExtraPackages, LegacyPackages, Packages}
```

The directory **Packages** contains the standard packages (or the standard extra packages, as they also are called) of *Mathematica* 6. In **ExtraPackages** we have a few extra packages. **LegacyPackages** contains the packages of *Mathematica* 5.2; note that they are now obsolete. Much of the functionality of the legacy packages has been included in the ordinary *Mathematica* 6, and for much of the functionality of the legacy packages that has not been included in the ordinary *Mathematica* 6, we have new packages. Let us look at the various packages in more detail.

■ Standard Packages

A list of the standard packages can be seen by clicking, in the home page of the Documentation Center, the *Standard Extra Packages* hyperlink. The resulting page has hyperlinks to the documentation of the packages. To see the names of all the packages, we could also write as follows:

```
SetDirectory[$InstallationDirectory <> "/AddOns/Packages"]; FileNames[]
```

However, we do not show the result here. Instead, we show the packages classified to several groups. Each package shown here is also a hyperlink to the guide page in the Documentation Center. For most packages, the guide page also contains a link to a tutorial page.

- **Advanced commands:** Developer`, Experimental`
- **Algebra:** FiniteFields`, Quaternions`
- **Calculus:** FourierSeries`, VariationalMethods`, VectorAnalysis`
- **Differential equations:** EquationTrekker`, NumericalDifferentialEquationAnalysis`
- **Discrete mathematics:** Combinatorica`, ComputationalGeometry`, GraphUtilities`
- **Geometry:** Polytopes`, PolyhedronOperations`

- **Graphics:** BarCharts`, ErrorBarPlots`, Histograms`, PieCharts`, PlotLegends`, VectorFieldPlots`, WorldPlot`
- **Miscellaneous:** Benchmarking`, Calendar`
- **Music:** Audio`, Music`
- **Notebooks:** GUIKit`, Notation`, XML`
- **Number theory:** PrimalityProving`
- **Numerical mathematics:** ComputerArithmetic`, FunctionApproximations`, NumericalCalculus`, Splines`
- **Physics:** BlackBodyRadiation`, Geodesy`, PhysicalConstants`, ResonanceAbsorptionLines`, StandardAtmosphere`, Units`
- **Statistics:** ANOVA`, HierarchicalClustering`, HypothesisTesting`, LinearRegression`, MultivariateStatistics`, NonlinearRegression`, RegressionCommon`, StatisticalPlots`

■ **Using Standard Packages**

The packages are not normally loaded when *Mathematica* is loaded. Instead, we have to take care of loading each package we intend to use. To load a package, do as in the following example:

```
<< NumericalCalculus`   Load the numerical calculus package
```

Instead of <<, we can also use the **Get** command. At the end of the name of a package, we have the backquote or grave accent character ` (this is actually a context mark in *Mathematica*). Note that after the ` you always have to press the space key so that the accent character appears. We now load the numerical calculus package:

```
<< NumericalCalculus`
```

We can then ask a table of the names defined in the package:

```
? NumericalCalculus`*
```

▼ NumericalCalculus`

EulerRatio	ExtraTerms	NLimit	NSeries	Terms
EulerSum	ND	NResidue	Radius	WynnDegr·. ee

NLimit[*expr*, $z \to z_0$] numerically finds the limiting value of *expr* as z approaches z_0. ≫

By clicking, in the table, a name such as **NLimit**, we get a short description of that name (as shown previously). To find more information about **NLimit**, click the ≫ button. At the bottom of the resulting page, we can click the *Numerical Calculus Package* hyperlink to get a description of the whole package.

After loading a package, we can use its commands:

```
NLimit[Sin[x] / x, x → 0]     1.
```

We can even open a package to look at the code. Just choose **Open** from the **File** menu and select a package from **$InstallationDirectory <> "/AddOns/Packages"**. Looking at the code of a package may be interesting if you are wondering how the package works. You can then perhaps even learn something about programming with *Mathematica*.

■ **Automating Loading**

In **$UserBaseDirectory**, we have a **Kernel** folder and there an initialization file called **init.m**. Each time the kernel is started, the commands in the initialization file are executed. So, if you frequently use, for example, the **NumericalCalculus`** package, write **<< NumericalCalculus`** in the initialization file. Then you can always use the commands of this package without loading it.

4.1.2 Forgetting to Load

■ **Forgetting to Load a Package: Shadowing**

Let us see what happens if we forget to load a package. We start a new session and try to use **NLimit**:

```
NLimit[Sin[x] / x, x → 0]
```
$$
\text{NLimit}\left[\frac{\text{Sin}[x]}{x}, \; x \to 0\right]
$$

It did not work because we forgot to load the appropriate package. Note that the **NLimit** command is blue, indicating that *Mathematica* does not know about it. So, we try to solve the problem by loading the package:

```
<< NumericalCalculus`
```

NLimit::shdw :
    Symbol NLimit appears in multiple contexts {NumericalCalculus`, Global`}; definitions in
        context NumericalCalculus` may shadow or be shadowed by other definitions. ≫

The package was loaded, but now the **NLimit** command we previously used became red to remind us that there is a problem with the command. As the message states, we now have two **NLimit** symbols: the one in the package and the one we (unintentionally) created when trying to use **NLimit**. One of the two symbols may "shadow" the other—that is, *Mathematica* has to use one of these symbols and disregard the other. The problem (shadowing) is explained in Section 17.3.2, p. 533. Let us now try **NLimit**:

```
NLimit[Sin[x] / x, x → 0]        1.
```

It worked. If we want to resolve the shadowing problem and get rid of the red color, we can remove our own **NLimit**:

```
Remove[Global`NLimit]
```

In summary, suppose that we forget to load a package—that is, we use a command of a package before loading the package. If we then load the package, it works but we get the warning about shadowing and the command we have used becomes red. (This is an improvement compared with earlier versions of *Mathematica*. Previously, if a package was loaded after trying to use one of its commands, the command still did not work.)

■ **Forgetting to Load a Package: Recommendations**

How to avoid the problem of shadowing? We simply remove, before loading the package, the command we have tried to use. Here is an example. We start a new session and try again to use **NLimit**:

```
NLimit[Sin[x] / x, x → 0]
```
$$
\text{NLimit}\left[\frac{\text{Sin}[x]}{x}, \; x \to 0\right]
$$

We observe that we forgot to load the package. Before loading the package, we remove the **NLimit** symbol we have (unintentionally) created:

```
Remove[NLimit]
```

Then we load the package:

```
<< NumericalCalculus`
```

Now **NLimit** becomes black so that we do not have any problems with it. Then we can use **NLimit**:

```
NLimit[Sin[x] / x, x → 0]     1.
```

Here is a summary:

If you forget to load a package before using one of its commands, do as follows:
- remove the name you have tried to use;
- load the package; and
- use the command of the package again.

However, note that even without removing the name, the package works, as we previously saw.

Another solution to the shadowing problem is to quit the kernel from **Evaluation ▷ Quit Kernel** and then restart the kernel from **Evaluation ▷ Start Kernel** (or simply by executing a command), but then we may need to do some calculations again.

4.1.3 Other Packages and Add-Ons

■ Extra Packages

Here are the extra packages:

```
SetDirectory[$InstallationDirectory <> "/AddOns/ExtraPackages"]; FileNames[]
{DifferentialEquations, Integration,
 LinearAlgebraExamples, Optimization, StatisticsExamples, Utilities}
```

Here, **DifferentialEquations`**, **Integration`**, **Optimization`**, and **Utilities`** actually contain several packages, as follows:

- **DifferentialEquations`**: InterpolatingFunctionAnatomy`, NDSolveProblems`, NDSolveUtilities`
- **Integration`**: NIntegrateUtilities`
- **Optimization`**: MPSData`, UnconstrainedProblems`
- **Utilities`**: CleanSlate`, URLTools`

Directories **Optimization/Data**, **LinearAlgebraExamples/Data**, and **StatisticsExamples/Data** contain data files:

- **Optimization/Data**: afiro.mps, ganges.mps, shell.mps, standmps.mps
- **LinearAlgebraExamples/Data**: can__229.psa, cavity01.rua, dwg961b.cua, dwg961b.mtx, dwt_1005.psa, gr_30_30.rsa, nos6.mtx, simplematrix.dat, west0381.mtx, wm1.rra
- **StatisticsExamples/Data**: iris.dat

■ Using Extra Packages

To load, for example, the **NDSolveProblems** package, write

```
<< DifferentialEquations`NDSolveUtilities`
```

The package defines the following names:

`? DifferentialEquations`NDSolveUtilities`*`

▼ DifferentialEquations`NDSolveUtilities`

CompareMethods	InvariantErrorPlot
FinalSolutions	InvariantErrorSampleRate
InvariantDimensions	RungeKuttaLinearStabilityFunction
InvariantErrorFunction	StepDataPlot

FinalSolutions[sys, sols] gives the end point solutions sols for the system
sys specified as an NDSolveProblem.

By clicking, in the previous table, a name such as **FinalSolutions**, we get a short description of that name (as shown previously). To get more information about the package, search, in **Help ▷ Documentation Center**, *Differential Equations package* and then click *Utility Packages for Numerical Differential Equation Solving*.

To use the data files in the extra packages, import them into *Mathematica*. Here is an example (we use the directory notation / of a Macintosh computer; we only show two rows of the file):

`(data = Import["StatisticsExamples/Data/iris.dat"]) // Short`

`{{5.1, 3.5, 1.4, 0.2}, <<148>>, {5.9, 3., 5.1, 1.8}}`

For **Import**, see Section 4.2.1, p. 100.

■ Legacy Packages

If you have used *Mathematica* 5.2 or earlier versions, note that the old packages, called the *legacy packages*, are now obsolete. They may work, but they are not documented in **Help ▷ Documentation Center**. Much of the functionality of the old packages has been included in the basic *Mathematica*. For example, **LogPlot**, earlier in the **Graphics`Graphics`** package, is now a built-in command. Other packages have been replaced with new packages. For example, the functionality of the old **Statistics`ConfidenceIntervals`** package is now in the new **HypothesisTesting`** package. For information about the legacy packages, look at Compatibility/guide/StandardPackageCompatibilityGuide in the Documentation Center.

If we load a legacy package, we get a message telling about the status of the package. For example,

`<< Graphics`Graphics``

General::obspkg :
  Graphics`Graphics` is now obsolete. The legacy version being loaded may conflict with current
    Mathematica functionality. See the Compatibility Guide for updating information. ≫

Thus, the package is obsolete. By clicking the ≫ button, we get more information about how to replace the package with new functionality of version 6. Here is another example:

`<< Statistics`ConfidenceIntervals``

General::newpkg :
  Statistics`ConfidenceIntervals` is now available as the Hypothesis Testing Package.
    See the Compatibility Guide for updating information. ≫

Thus, the package is replaced by a new package. Again, by clicking the ≫ button, we get more information about how to replace the package with the new package.

■ **Other Packages and Applications**

In addition to the packages that come with *Mathematica*, you may have additional packages and applications. If you want them to be available for all users, put them in the **Applications** folder that is located in the **$BaseDirectory**. For example, on my Macintosh with MacOS X, here is the location:

$BaseDirectory

/Library/Mathematica

On a Windows machine, the location may be **C:\Documents and Settings\All Users\Application Data\Mathematica**. On the other hand, if you want the packages to be available only for you, put them in the **Applications** folder that is located in the **$UserBaseDirectory**. For example, on my Macintosh, here is the location:

$UserBaseDirectory

/Users/heikki/Library/Mathematica

On a Windows machine, the location may be of the form **C:\Documents and Settings***username***\Application Data\Mathematica**.

Next, we describe a way to automatically install a palette, style sheet, or package to the correct location.

■ **Automatic Installation of Palettes, Style Sheets, and Packages**

If you have a new palette, style sheet, or package that you want to install into the correct location, use **File ▷ Install…**. The following dialog opens:

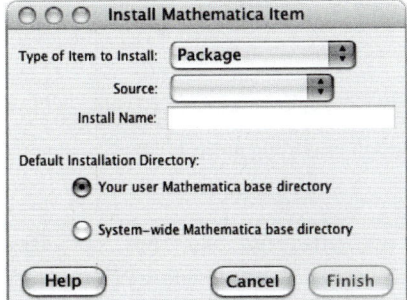

From *Type of Item to Install*, choose *Palette*, *Stylesheet*, or *Package*. From *Source*, select the file you want to install. In *Install Name*, type a suitable name for the palette, style sheet, or package. You can also choose the *Default Installation Directory*. Finally, click *Finish*.

Style sheets and palettes are put in the **SystemFiles** folder, which is located in the same directories as the **Application** folder we considered previously. Style sheets are put in **SystemFiles ▷ FrontEnd ▷ StyleSheets** and palettes in **SystemFiles ▷ FrontEnd ▷ Palettes**.

4.2 Exporting and Importing

4.2.1 Exporting and Importing Data

■ **Export and Import**

In Sections 4.2.1 and 4.2.2, we consider `Export` and `Import`. These are versatile commands to use for the writing and reading of data and graphics (and other material) in more than 100 formats. Next, we study the export and import formats—that is, file formats to which *Mathematica* is able to write *Mathematica* material and file formats from which *Mathematica* is able to read material into *Mathematica*. The formats that can be used both in exporting and in importing are (in a Macintosh) as follows:

```
ef = $ExportFormats;
if = $ImportFormats;

Intersection[ef, if]
```

```
{3DS, ACO, AIFF, AU, AVI, Base64, Binary, Bit, BMP, Byte, BYU, BZIP2, CDF,
 Character16, Character8, Complex128, Complex256, Complex64, CSV, DICOM,
 DIF, DXF, ExpressionML, FASTA, FITS, FLAC, GIF, Graph6, GZIP, HarwellBoeing,
 HDF, HDF5, HTML, Integer128, Integer16, Integer24, Integer32, Integer64,
 Integer8, JPEG, JPEG2000, JVX, List, LWO, MAT, MathML, MGF, MOL, MTX, MX, NB,
 NOFF, OBJ, OFF, Package, PBM, PCX, PDF, PGM, PLY, PNG, PNM, PPM, PXR, RawBitmap,
 Real128, Real32, Real64, RIB, RTF, SCT, SND, Sparse6, STL, String, Table, TAR,
 TerminatedString, Text, TGA, TIFF, TSV, UnsignedInteger128, UnsignedInteger16,
 UnsignedInteger24, UnsignedInteger32, UnsignedInteger64, UnsignedInteger8,
 UUE, WAV, Wave64, WDX, XBM, XHTML, XHTMLMathML, XLS, XML, XYZ, ZIP}
```

The formats that can only be used in exporting are as follows (in a Windows machine, this list also includes EMF and WMF):

```
Complement[ef, if]
```

```
{EPS, FLV, Maya, MIDI, PICT, POV, SVG, SWF, TeX, VRML, X3D, ZPR}
```

The formats that can only be used in importing are as follows:

```
Complement[if, ef]
```

```
{ApacheLog, CDED, CUR, DBF, Directory, EDF, GTOPO30, ICO, LaTeX, MBOX, MDB,
 MPS, MTP, NetCDF, ODS, PDB, QuickTime, RSS, SDTS, SXC, USGSDEM, VCF, XPORT}
```

A mathematician or statistician may mainly be interested in reading existing data files into *Mathematica* and writing *Mathematica* graphics into files that some other programs can use. Note that in place of **Export**, we can also use the menu command **File ▷ Save Selection As** to save material in PDF or HTML form and many graphics forms. Also, in place of **Import** we can use the menu command **Insert ▷ Picture ▷ From File** to import graphics.

The page guide/ListingOfAllFormats in Documentation Center lists all exporting and importing formats.

■ **Exporting and Importing Data**

`Export["file", data, "format"]` Write **data** into **file** in **format**
`data = Import["file", "format"]` Read **file** into **data** in **format**

`FilePrint["file"]` (⌘6) Look at **file**
`FileFormat["file"]` (⌘6) Try to determine the format of **file**

`$ExportFormats` Show all available export formats
`$ImportFormats` Show all available import formats

Examples of formats:

List A column. Exporting: put each item of a 1D list in its own line. Importing: form a 1D list.

Table A table with space- or tab-separated values; default file extension `.dat`. Exporting: put each sublist of a 2D list in its own line, separating items in a line with tabs. Importing: form a 2D list.

CSV A table with comma-separated values; default file extension `.csv`. Works similarly as **Table**.

Text A text file; default file extension `.txt`. Exporting: put the whole text into the file, using several lines if the text contains newline characters `\n`. Importing: form a single string from the text.

Lines Importing: form a string from each line of a text file to form a 1D list of strings.

Words Importing: form a string from each word of a text file to form a 1D list of strings.

Note that the **Table**, **CSV**, and **Text** formats can also be indicated by the extensions `.dat`, `.csv`, and `.txt` of the file. For example, instead of `Export["file", data, "Table"]`, we can write `Export["file.dat", data]`; instead of `Import["file.dat", "Table"]`, we can simply write `Import["file.dat"]`.

Export and **Import** accept various options; see Documentation Center for more details.

For easy reference, here are the two most important commands to use to read data files:

`data = Import["file", "List"]` Read a 1D table of **file** into **data**
`data = Import["file", "Table"]` Read a 2D table of **file** into **data**

Note that by default, **Export** writes the file into the current working directory; the command to view this directory is **Directory[]**. Also by default, **Import** searches for a file only from certain directories; these directories can be seen by asking the value of **$Path**. If you want to write a file into or read a file from a nondefault directory, you have to specify the full name of the file or modify the default directories (see Section 4.2.3, p. 107).

Next, we consider some special cases in more detail.

■ **A Column**

`Export["file", data, "List"]` Write 1D list **data** into a column file **file**
`data = Import["file", "List"]` Read a column file, forming a 1D list **data**

Consider the following example:

```
data1 = {23, 41.7, 39.5, 143, 8.4 × 10^-7};
```

Write the list into a file **columndata.dat**, putting each item in its own row:

```
Export["columndata.dat", data1, "List"]
columndata.dat
```

Look at the file:

```
FilePrint["columndata.dat"]
23
41.7
39.5
143
8.4e-7
```

Note that small and large numbers are written in a C- or Fortran-like e-form.

Suppose then that we have data in **columndata.dat**. The file in this example is written by *Mathematica*, but it could be done with a text editor by saving the file in a plain text format. Now we read the file, forming a 1D list:

```
data1a = Import["columndata.dat", "List"]
```

$\{23, 41.7, 39.5, 143, 8.4 \times 10^{-7}\}$

■ A Table

For 2D lists, we need not declare the type of the data if we use the extension **.dat**:

> **Export["file.dat", data]** Write 2D list **data** into a tab-separated file **file.dat**
> **data = Import["file.dat"]** Read a tab- or space-separated file, forming a 2D list **data**

If we do not use the standard extension, we have to declare the type as **Table**:

- **Export["file", data, "Table"]** Write 2D list **data** into a tab-separated file **file**
- **data = Import["file", "Table"]** Read a tab- or space-separated file, forming a 2D list **data**

Consider the following data:

```
data2 = {{23, 41.7, 39.5}, {143, 8, 56}, {28.8, 74, 13}};
```

Export it in a table form:

```
Export["tabledata.dat", data2];
```

Look at the resulting file:

```
FilePrint["tabledata.dat"]
23      41.7    39.5
143     8       56
28.8    74      13
```

Next, we read the file, forming sublists from the rows:

```
data2a = Import["tabledata.dat"]
{{23, 41.7, 39.5}, {143, 8, 56}, {28.8, 74, 13}}
```

■ A Table with Textual Items

Define the following data:

```
data3 = {{"Results from an experiment"}, {"Individual", "Measurement"},
    {"a", 23}, {"b", 41.7}, {"c", 39.5}, {"d", 143}, {"e", 8}};
```

Write it into a file:

```
Export["alphanumericdata.dat", data3];
```

When looking at the file, we see that the quotation marks have been dropped:

```
FilePrint["alphanumericdata.dat"]
```

```
Results from an experiment
Individual     Measurement
a       23
b       41.7
c       39.5
d       143
e       8
```

Then read the file:

```
data3a = Import["alphanumericdata.dat"]
```

```
{{Results, from, an, experiment}, {Individual, Measurement},
 {a, 23}, {b, 41.7}, {c, 39.5}, {d, 143}, {e, 8}}
```

All textual items were converted into strings, as we can see by asking the **InputForm**:

```
% // InputForm
```

```
{{"Results", "from", "an", "experiment"},
 {"Individual", "Measurement"}, {"a", 23},
 {"b", 41.7}, {"c", 39.5}, {"d", 143}, {"e", 8}}
```

We can drop the first two rows:

```
data3b = Drop[data3a, 2]
```

```
{{a, 23}, {b, 41.7}, {c, 39.5}, {d, 143}, {e, 8}}
```

If we begin to analyze the measurements, we can assign them to a variable **meas**:

```
{ind, meas} = data3b⊤
```

```
{{a, b, c, d, e}, {23, 41.7, 39.5, 143, 8}}
```

In general, a textual item can be a string such as `"Donkey"`, which has the quotation marks, or a word such as `Donkey`, which does not have quotation marks. When exporting a string, the text is written without quotation marks, but when exporting a textual item that is not a string, its *value* (if any) is written. When importing a textual item such as `Donkey`, the item is converted to a string (`"Donkey"`).

■ **Text**

Export["file.txt", data] Write *a single string* **data** into a text file **file.txt**; the file will contain several lines if **data** contains newline characters **\n**
Export["file.txt", data] Write *a list of strings* **data** into a text file **file.txt**; each string will be in its own line
data = Import["file.txt"] Read a text file; *the whole file* will be a single string; the string will contain newline characters **\n**, if the file contains several lines
data = Import["file.txt", "Lines"] Read a text file; *each line* of the file will be a string; the result is a 1D list of strings
data = Import["file.txt", "Words"] Read a text file; *each word* of the file will be a string; the result is a 1D list of strings

As shown previously, for text data, we need not declare the type of the data if we use the extension **.txt**. If we do not use this standard extension, we have to declare the type as **Text**—that is,

- instead of **Export["file.txt", data]** write **Export["file", data, "Text"]**;
- instead of **Import["file.txt"]** write **Import["file", "Text"]**.

As an example, consider the following text:

```
data5 = "A mnemonic for the digits of pi = 3.1415926535:\nMay
   I have a large container of coffee - sugar and cream?"
A mnemonic for the digits of pi = 3.1415926535:
May I have a large container of coffee - sugar and cream?
```

Here, \n is a newline character, which inserts a line break. Write the text into a text file (in **Text** form):

```
Export["textdata.txt", data5];
```

In the file, each \n has generated a new line:

```
FilePrint["textdata.txt"]
```

```
A mnemonic for the digits of pi = 3.1415926535:
May I have a large container of coffee - sugar and cream?
```

If we read the file in **Text** form, we get a single string with newline characters \n:

```
data4a = Import["textdata.txt"]
```

```
A mnemonic for the digits of pi = 3.1415926535:
May I have a large container of coffee - sugar and cream?
```

```
% // InputForm
```

```
"A mnemonic for the digits of pi = 3.1415926535:\nMay \
I have a large container of coffee - sugar and cream?"
```

As we can see from the **InputForm**, the whole file became a single string. Using the **Lines** format gives a list of two strings, one for each line:

```
data4b = Import["textdata.txt", "Lines"]
```

```
{A mnemonic for the digits of pi = 3.1415926535:,
 May I have a large container of coffee - sugar and cream?}
```

```
% // InputForm
```

```
{"A mnemonic for the digits of pi = 3.1415926535:", "May \
I have a large container of coffee - sugar and cream?"}
```

With **Words**, each word is transformed into a string:

```
data4c = Import["textdata.txt", "Words"]
```

```
{A, mnemonic, for, the, digits, of, pi, =, 3.1415926535:, May,
 I, have, a, large, container, of, coffee, -, sugar, and, cream?}
```

```
% // InputForm
```

```
{"A", "mnemonic", "for", "the", "digits", "of", "pi", "=",
 "3.1415926535:", "May", "I", "have", "a", "large",
 "container", "of", "coffee", "-", "sugar", "and", "cream?"}
```

■ Other Commands

For reading files, *Mathematica* also has **ReadList**, which allows, for example, detailed declaration of the types of items of the files. Files can also be read item by item with **OpenRead**, **Read**, **Skip**, and **Close**. For item-specific writing, we have **OpenWrite**, **OpenAppend**, and **Write**.

The contents of files can be searched with **FindList**:

FindList["file", "text"]   Get a list of all lines in **file** containing **text**

```
FindList["alphanumericdata.dat", "c"]
```

```
{c       39.5}
```

Mathematica has tools to work with databases; see guide/DatabaseConnectivity and *DatabaseLink*/tutorial/Overview in the Documentation Center.

4.2.2 Exporting and Importing Graphics

`Export["file", fig, "format"]` Write graphics `fig` into `file` in `format`

`fig = Import["file", "format"]` Read `file` into graphics `fig` in `format`

`$ExportFormats` Show all available export formats

`$ImportFormats` Show all available import formats

Examples of vector graphics formats:

`EPS` Encapsulated PostScript format (`.eps`) (only exporting)

`PDF` Adobe PDF format (`.pdf`)

`WMF, EMF` Windows metafile/enhanced metafile formats (`.wmf`, `.emf`) (only exporting)

Examples of raster image formats:

`GIF` GIF format (`.gif`)

`JPEG` JPEG format (`.jpeg`, `.jpg`)

`TIFF` TIFF format (`.tiff`, `.tif`)

`BMP` Microsoft bitmap format (`.bmp`)

`WMF, EMF` Windows metafile/enhanced metafile formats (`.wmf`, `.emf`) (only exporting)

`PICT` Macintosh PICT format (`.pict`) (only exporting)

Options of `Export` *include:*

`ImageSize` Absolute size of the image in printer's points (1/72 inch)

`ImageResolution` Resolution of the image in dpi (dots per inch)

`ImageRotated` Whether to rotate the image to get an image in the landscape form

---

In place of `Export`, we can also use the menu command **File ▷ Save Selection As** to save graphics in many different formats. This menu command supports, for example, the following formats: `EPS`, `PDF`, `GIF`, `JPEG`, `TIFF`, and `BMP`. Also, in place of `Import`, we can use the menu command **Insert ▷ Picture ▷ From File** to import graphics.

Most formats can be indicated by the extensions of the file names; the extensions are given in parentheses above. For example, instead of `Export["file", fig, "EPS"]`, we can write `Export["file.eps", fig]`; instead of `Import["file.gif", "GIF"]`, we can simply write `Import["file.gif"]`.

When exporting, the vector graphics formats are independent of the setting for `ImageResolution`, whereas the raster image formats depend on the value of this option. When exporting or importing, `Rasterize` (⌘6) can be used to get the file in raster format.

`Export` and `Import` accept various options; see Documentation Center for more details.

Note that by default, `Export` writes the file into the current working directory; the command to view this directory is `Directory[]`. Also by default, `Import` searches for a file only from certain directories; these directories can be seen by asking the value of `$Path`. If you want to write a file into or read a file from a nondefault directory, you have to specify the full name of the file or modify the default directories (see Section 4.2.3, p. 107).

■ **Example**

We first make a plot:

```
fig = Plot[Sin[x], {x, 0, 2 π}]
```

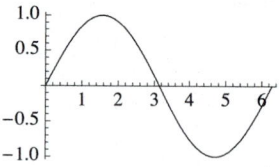

Then we export it into an EPS file:

```
Export["fig.eps", fig, ImageSize → 100]
fig.eps
```

This file can then be used, for example, in MS-Word. As an example of importing, we show one of the example figures of *Mathematica*:

```
Import["ExampleData/coneflower.jpg", ImageSize → 90]
```

■ **Exporting Graphics into Microsoft Word**

If you are preparing a mathematical document with Microsoft Word, you may want to produce some plots with *Mathematica* and insert them into your Word document. First, make the plot:

```
fig = Plot[Sin[x], {x, 0, 2 π}]
```

To export the plot into Word, the best method may vary from system to system, but the following method should work on a PC or Macintosh.

Adjust the size of the plot to be suitable for the Word document, click on the plot, choose **File ▷ Save Selection As**, choose the format of the file to be EPS, give the file a name ending with **.eps**, and save the plot in a suitable place. Then go to your Word document, place the cursor at a suitable point, choose **Insert ▷ Picture ▷ From File**, locate and click the file you saved from the dialog box, and click **Insert**.

The inserted figure will only be shown in Word as a box that contains some information about the figure; however, when printed, the figure should be OK. In Word, do not adjust the size of the plot because the size of all text in the plot then also changes, which should be avoided; any changes in size should be done with *Mathematica,* which does not change the size of text.

In exporting as EPS, we can use the **"PreviewFormat"** option to get the figure shown in Word:

```
Export["fig.eps", fig, "PreviewFormat" → "TIFF"]
```

Possible values of the option are **None** (the default), **"TIFF"**, **"Metafile"** (WMF), and **"Interchange"** (EPSI).

On some PC systems, **Save Selection As** may also yield a good result with the Metafile format. **Export** can also be tried:

```
Export["fig.wmf", fig, ImageSize → s]
```

■ **Exporting Graphics into LaTeX**

To export a plot into a LaTeX document, save the plot in EPS format, either with **File ▷ Save Selection As** or with `Export["fig.eps", fig]`. Save the file in the same folder where you have saved your LaTeX document.

In the starting rows of the document, add `\usepackage[dvips]{graphicx}`. At a suitable place in the document, add `\includegraphics{fig.eps}`. If you want to modify the plot, write `\includegraphics[key=value]{fig.eps}`, where key is **width**, **height**, **angle**, or **scale**. For example, write `\includegraphics[width=8cm]{fig.eps}`. This kind of command adds the plot in the place you have chosen.

A more recommended method is to use a **figure** environment, as follows:

```
\begin{figure}[htbp]
\includegraphics[width=8cm]{fig.eps}
\caption{caption to be added}
\label{name used to refer to the plot}
\end{figure}
```

Now LaTeX places the plot in the best position, a caption can be added, and we can refer to the plot with `\ref`. The placement of the plot in the example is directed with `[htbp]`. This asks to place the plot at the present point (**h** = here), if possible, but if not possible, then at the top of the page (**t** = top), then at the bottom of the page (**b** = bottom), and then in a separate page (**p** = page), in this order of preference.

4.2.3 Locating the File

■ **The Default Locations**

As noted previously, using short names of files such as `data1.dat` with **Export** and **Import** has some restrictions. With **Export**, *Mathematica* saves the file in a default location; the location can be seen by using the command `Directory[]`. With **Import**, *Mathematica* only searches the file from a list of default locations; this list can be seen by using the command `$Path`.

`Directory[]`	The default directory where **Export** writes a file
`$Path`	The default list of directories from which **Import** searches a file

More generally, `Directory[]` gives the *current working directory*, and `$Path` gives the *search path* (i.e., the default list of directories to search in attempting to find an external file). On my Macintosh with MacOS X, the directories are as follows:

`Directory[]`

`/Users/heikki`

`Short[$Path, 3]`

`{/Applications/Mathematica 6.0.app/SystemFiles/Links, ≪15≫,`
`  /Applications/Mathematica 6.0.app/Documentation/English/System}`

Thus, your files are, by default, saved in your home directory such as **"/Users/heikki"**. By investigating the output of `$Path`, we observe that when your files are read, they are searched, among others, from your home directory such as **"/Users/heikki"**, from the current working directory represented by **"."**, and from a directory such as **"Users/heikki/Library/Mathematica/Applications"**.

If it suits you that **Export** writes files into **Directory[]** and **Import** searches files only from **$Path**, then you can use short names for the files. Otherwise, we have two possibilities. Either use the full names of files or modify the values of **Directory[]** and **$Path**. First, we consider an example of using full names.

■ Using Full Names of Files

We have a folder **DataFiles** in our home directory **"/Users/heikki"**, and now we write a data file into this folder. Because this is not the default location, we have to give more information about the saving location. Because **DataFiles** is a subdirectory of our home directory, it suffices to specify the subdirectory:

```
data5 = {{46, 71}, {22, 38}};
```

```
Export["DataFiles/data5.dat", data5]
```
DataFiles/data5.dat

We could also use a full name of the file:

```
Export["/Users/heikki/DataFiles/data6.dat", data5]
```
/Users/heikki/DataFiles/data6.dat

Then we read these files. Again, because **DataFiles** is a subdirectory of a directory (**"/Users/heikki"**) in **$Path**, it suffices to specify the subdirectory:

```
data5a = Import["DataFiles/data5.dat"]
```
{{46, 71}, {22, 38}}

We could also use a full name of the file:

```
data6a = Import["/Users/heikki/DataFiles/data6.dat"]
```
{{46, 71}, {22, 38}}

Note that you need not write the full names by yourself; let *Mathematica* do it. Simply choose **Insert ▷ File Path** and then in the dialog box select the file in which you are interested. The full name of this file is now pasted at the current location of the cursor. Note also that the full names are different in form with various computer systems. **ToFileName** can be used to form file names from directories.

■ Modifying the Current Working Directory

Directory[]	Give the current working directory
SetDirectory["dir"]	Set the current working directory
ResetDirectory[]	Reset the current working directory to its previous value
FileNames[]	List all files in the current working directory

The current working directory can be changed by **SetDirectory**:

```
SetDirectory["/Users/heikki/DataFiles"]
```
/Users/heikki/DataFiles

This is now the current working directory:

```
Directory[]
```
/Users/heikki/DataFiles

Export now writes files into this default directory (without having to use full names of files). Therefore, exporting a file into **DataFile** is easy:

```
Export["data7.dat", data5]
```
```
data7.dat
```

Because **$Path** contains the current working directory ".", we can also import the file with a short name:

```
Import["data7.dat"]
```
```
{{46, 71}, {22, 38}}
```

We can go back to the original directory by **ResetDirectory[]**:

```
ResetDirectory[]
```
```
/Users/heikki
```

This is now the current working directory:

```
Directory[]
```
```
/Users/heikki
```

■ Modifying the Search Path

A nondefault folder can easily be added into the search path:

```
$Path = Append[$Path, "/Users/heikki/DataFiles"];
```

Now we can easily read files from this folder:

```
Import["data7.dat"]
```
```
{{46, 71}, {22, 38}}
```

If you will read data from a certain folder in several sessions, you may consider adding a command **$Path = Append[$Path, "…"]** in the **init.m** file, which can be found in a **Kernel** folder in **$UserBaseDirectory**. If you modify **$Path** in this way, then every time you open *Mathematica*, the **$Path** variable has the appropriate value and you can easily read the data.

4.3 Saving for Other Purposes

4.3.1 Continuing Work in Later Sessions

■ Continuing without Recalculation

As noted in Section 3.1.1, p. 52, when we open a notebook to continue calculations, we cannot directly use any results that are already in the notebook. Suppose a notebook contains the following calculation:

```
int = Integrate[2 x Sin[x] Exp[x], x] // Simplify
```
$$e^x (\text{Cos}[x] - x \text{Cos}[x] + x \text{Sin}[x])$$

Now we open this notebook and want to continue by differentiating the integral. Note that opening a notebook only shows the notebook on the screen; opening it does not execute any commands. Thus, after opening the notebook, we cannot write **D[int, x]** because *Mathematica* does not yet know the value of **int**. We have to tell *Mathematica* the value. We have at least three ways to do this:

- execute anew the command defining **int**;
- write **int =** at the beginning of the *value* of **int** and execute the resulting command; or
- save the value of **int** into a file; in a new session, load the file to get the value of **int**.

The first method is very straightforward if the execution of the command does not take much time.

The second method also is very handy and quick: We get the value of **int** without doing the calculation anew. Note that the output of the original command changes to an input. If you want to keep the original output untouched, take a copy of the output and write **int =** at the beginning of the copy:

    int = e^x (Cos[x] - x Cos[x] + x Sin[x])

    e^x (Cos[x] - x Cos[x] + x Sin[x])

The third method may be worth considering if you have a very large expression that is clumsy to keep and handle in the notebook but with which you want to continue calculations in later sessions. The expression may be, for example, a large symbolic expression, a complicated plot, or a large set of generated random numbers. Saving and loading an expression is considered in the following section.

The three methods also work for graphics. In the second method, take a copy of a plot of interest, paste the plot after **p1 =**, and execute. Now you have the plot in the variable **p1** and you can modify it with options by using **Show**. For a plot, to get its full code with primitives, directives, and options, take a copy of the plot, select the plot, and choose **Cell ▷ Convert To ▷ StandardForm**.

Remember also the mouse manipulations of plots mentioned in Section 5.1.1, p. 120, and interactive drawing (**Graphics ▷ Drawing Tools, Graphics ▷ Graphics Inspector**) considered in Section 5.1.3, p. 126.

■ **Saving and Loading Expressions**

> **a >> file**   Save the value of a variable **a** into **file** (clearing the file if it already exists)
> **FilePrint["file"]**   View **file**
> **a = << file**   Load **file** and assign the content as the value of **a**

In place of **>>** and **<<**, we can use **Put** and **Get**; remember that we already discussed **<<** in Section 4.1.1 as it pertains to loading packages. As an example, we save the value of **int** into a file:

    int >> intfile

We can check that everything is all right by viewing the file:

    FilePrint["intfile"]

    E^x*(Cos[x] - x*Cos[x] + x*Sin[x])

In a new session, we can then load the file:

    int = << intfile

    e^x (Cos[x] - x Cos[x] + x Sin[x])

Now **int** has the desired value.

> **Save["file", {a, b, …}]**   Save definitions of variables **a, b,** … into **file**
> **<< file**   Load **file** (variables **a, b,** … then have the saved values)

The value of a single variable can be saved with **>>**. With **Save**, we can save several values in the same file. Note that **Save** appends the values to the file if it already exists (remember that **>>** clears an existing file; **>>>** or **PutAppend** can also be used to append expressions to an existing file).

Another saving command is **DumpSave**. It saves expressions in a binary format and may be advantageous for very large and complicated expressions. These files can be read with **<<**.

Plots can also be saved and loaded. As an example, suppose we have a plot **p1**. We can save it with **p1 >> plot1file**. Later, we can load the plot with **p1 = << plot1file**. Now we can add options: **Show[p1, Frame → True]**.

4.3.2 Exporting to TeX, C, Fortran, and HTML

■ Exporting to TeX

> **File ▷ Save As**: Format: **LaTeX Document** Save the present notebook as an AMS-LaTeX file
> **Export["file.tex", expr]** Export an expression as an AMS-LaTeX file
> **TeXForm[expr]** Show the AMS-LaTeX form of an expression

For exporting graphics into LaTeX, see Section 4.2.2, p. 107. Here is an example of **TeXForm**:

```
i = Integrate[a^2 / (x^2 + 2 x + b), x]
```

$$\frac{a^2 \, \mathrm{ArcTan}\left[\frac{1+x}{\sqrt{-1+b}}\right]}{\sqrt{-1+b}}$$

```
TeXForm[i]
```

```
\frac{a^2 \tan ^{-1}\left(\frac{x+1}{\sqrt{b-1}}\right)}{\sqrt{b-1}}
```

■ Exporting to C and Fortran

> **CForm[expr]** Show the C form of the expression
> **FortranForm[expr]** Show the Fortran form of the expression

These commands help when you want to export *Mathematica* results into a C or Fortran program. For example,

```
CForm[i]
```

```
(Power(a,2)*ArcTan((1 + x)/Sqrt(-1 + b)))/Sqrt(-1 + b)
```

```
FortranForm[i]
```

```
(a**2*ArcTan((1 + x)/Sqrt(-1 + b)))/Sqrt(-1 + b)
```

If you write TeX, C, or Fortran code, you may also be interested in **Splice**. Suppose your C program needs the derivative of a function. In the C code, write **<*D[…,…]*>**, including a *Mathematica* command between **<*** and ***>**. Give the file a name ending with **.mc**. Then execute a **Splice** command, and you get a file in which the C code is as it was input, and the *Mathematica* command has been executed and written in a C form.

■ Exporting to HTML

> **File ▷ Save As**: Format: **Web Page** Save the present notebook as an HTML file
> **File ▷ Save Selection As**: Format: **HTML** Save the selection as an HTML file
> **Export["file.html", expr]** Export an expression as an HTML file
> **MathMLForm[expr]** Show the MathML form of the expression

4.4 Managing Time and Memory

4.4.1 Managing Time Consumption

■ **Information about Time**

We have various commands that relate to time:

> **Timing[expr]** Evaluate **expr**; give the CPU time in seconds the kernel has used, together with the result obtained
>
> **AbsoluteTiming[expr]** Evaluate **expr**; give the absolute elapsed time, together with the result obtained
>
> **TimeConstrained[expr, t]** Stop evaluating **expr** after **t** seconds
>
> **TimeUsed[]** Show the used CPU time in the current session
>
> **SessionTime[]** Show the elapsed time in the current session
>
> **DateList[]** (⌘6) Current time: {year, month, day, hour, minute, second}
>
> **DateString[]** (⌘6) Current time: "weekday day month year hour:minute:second"
>
> **DatePlus[date, n]** (⌘6) The date **n** days after **date**
>
> **DateDifference[date1, date2]** (⌘6) The number of days from **date1** to **date2**

The time is measured in steps of **$TimeUnit**, which, in many systems, has the default value of $1/100$ second. Note that doing the same calculation again or doing a similar calculation may take much less time because *Mathematica* may have already loaded some files or stored results. The following is an example of **Timing**:

```
FactorInteger[2^137 - 1] // Timing
{6.47204, {{32 032 215 596 496 435 569, 1}, {5 439 042 183 600 204 290 159, 1}}}
```

If we only want to see the time—not the result—we can use the semicolon. In place of the result, we then have **Null**:

```
(ran = Table[RandomReal[], {10^6}];) // Timing
{0.450717, Null}
```

The execution time can be seen in the lower left-hand corner of the window by choosing **Format ▷ Option Inspector...**, setting *Scope* to be *Selected Notebook*, going to **Notebook Options ▷ Evaluation Options**, and then selecting **ShowTiming** as the value of **EvaluationCompletionAction**. This time is somewhat longer than the time given by **Timing** because it includes the time to format and show the result.

■ **Time-Saving Tips**

• It is good to have plenty of random access memory (RAM). To some extent, the more memory *Mathematica* has available, the speedier it is. Also, remember that virtual memory on the hard disk is much slower than actual physical RAM.

• If you work with a heavy and time-consuming problem in several sessions, you perhaps need not start from scratch each time. Instead, try to continue the work from where you left off (see Section 4.3.1, p. 109).

• Avoid using so-called "arbitrary-precision" numbers. Calculations with these numbers are done by software in *Mathematica* and are much slower than calculations done with the hardware-implemented, machine-precision numbers. This point is explained in more detail in Sections 12.2.2, p. 404, and 12.3.1, p. 409. The precision used in calculations can often be set with the **WorkingPrecision** option. Resist changing the default value **MachinePrecision** of this option. Of course, if the problem is ill-conditioned and round-off errors have an effect, the option mentioned should be used by giving it a large enough value, such as 20. In this way, the problem may be solved without difficulties.

• In numerical routines such as **NIntegrate**, **FindRoot**, **FindMinimum**, and **NDSolve**, the default precision may sometimes be more than you actually need (e.g., if the result is used in plotting). Lowering the precision requirements **PrecisionGoal** and **AccuracyGoal** reduces the calculation time (for these options, see Section 12.3.1, p. 409).

• When plotting largely varying functions, you can sometimes get better results if you increase the value of the **PlotPoints** option. However, after a certain value you cannot see any difference in the plot, and you merely waste time because the larger the value of **PlotPoints**, the longer it takes to produce the plot.

• The execution time of a program may significantly depend on the design of the program; see Wagner (1996) for detailed experiments and recommendations. Some points are as follows. Use built-in commands if possible. If the result of the program consists of decimal numbers, use decimal numbers as early as possible (i.e., avoid calculating with exact numbers). Use the functional programming style (see Section 18.3, p. 568). Avoid using **Append**, **AppendTo**, **Prepend**, and **PrependTo**. Compile your functions (see Section 17.2.3, p. 528).

• While awaiting the result of a time-consuming command, you can edit the notebook by adding, for example, text to explain what you have done and what the results mean. You can also write new commands so that they are ready to be executed when the time-consuming command has printed its result.

• You can calculate with *Mathematica* while waiting for the result of a time-consuming command. This can be done by using a subsession. Once you have entered a subsession, the execution of the time-consuming command is interrupted, and you can do some shorter calculations during the subsession. After you exit the subsession, the execution of the time-consuming command continues. To enter a subsession, choose **Evaluation ▷ Interrupt Evaluation**. A dialog appears where you can ask to get a subsession; wait for the cell bracket to change to a special bracket. Now you can do some shorter calculations. To exit the subsession and continue the long calculation, execute **Return[ ]**.

• A good way to save time may be to use a remote kernel. If your machine is not powerful enough, you can save time by using the greater power of a remote machine. If your machine is powerful enough, you can still consider using a remote kernel or even several remote kernels to do parts of your calculations. In this way, you can do several calculations at the same time. To define a remote kernel, choose **Evaluation ▷ Kernel Configuration Options**.

4.4.2 Managing Memory Consumption

■ **Information about Memory**

In the following box, we list commands that relate to the consumption of memory. The kernel needs memory to store both the code of *Mathematica* and the results of computations. The latter material is called "data."

> `ByteCount[expr]` Bytes used by `expr` if sharing (see Memory-Saving Tips) is not used
>
> `MemoryConstrained[expr, b]` Stop evaluating `expr` if more than `b` bytes are needed for the evaluation
>
> `MemoryInUse[]` Memory in bytes currently being used to store data in the kernel
>
> `MemoryInUse[$FrontEnd]` (⌘6) Memory in bytes currently being used in the front end
>
> `MaxMemoryUsed[]` Maximum memory in bytes used to store data thus far

Note that the kernel of *Mathematica* keeps in RAM all the results of the current session, and this means that the memory consumption of the kernel steadily grows as your session continues. However, *Mathematica* uses kernel memory sparingly and deletes all intermediate results that are no longer needed.

It may be wise to divide long sessions into shorter ones by occasionally quitting *Mathematica* and starting a new session with a clean sheet, maximum memory, and maximum speed. In notebook environments, we can also quit the kernel only by choosing **Evaluation ▷ Quit Kernel**.

■ Memory-Saving Tips

• Ending a command with the semicolon (`;`) prevents the result from being displayed. This saves memory, especially for large expressions. It also saves your screen area by not cluttering it with uninteresting formulas or graphics and thus helps you to manage the flow of computation.

• You can delete all output cells by choosing **Cell ▷ Delete All Output**. This releases front-end memory and makes the notebook smaller. The memory saving may be very substantial if the notebook contains many plots. You can recalculate the commands later if needed.

• When you have finished a plot and no longer want to change or print it, you can greatly save front-end memory and the size of the notebook by converting the plot to bitmap form. Select the plot with the mouse and choose **Cell ▷ Convert To ▷ Bitmap**. The plot does not change in any way on the screen, but the PostScript code behind the plot is discarded.

• Use `Compress` to compress large expressions. They can be uncompressed with `Uncompress`.

• Execute `Share[]` to reduce the amount of kernel memory. This command shares the storage of common subexpressions between different parts of an expression or between different expressions. The output of the command is the memory saved.

• One method of saving memory is to run only the front end on your machine and to run the kernel in another machine. Then only the front end takes memory from your machine, and you can take advantage of the possibly larger RAM and speedier processor of the remote kernel. To define a remote kernel, choose **Evaluation ▷ Kernel Configuration Options**.

Graphics for Functions

Introduction

> *Straight line—the shortest way between two points.—Euclid*
> *Cycloid—the fastest way between two points.—Johann Bernoulli*
> *Curve—the loveliest way between two points.—Mae West*

The plotting capabilities of *Mathematica* are impressive. There are many ready-to-use commands such as **Plot, Plot3D, ParametricPlot, ParametricPlot3D, ListPlot, ListPlot3D**, and **ContourPlot**, which often give good results. In case we want to modify the plots, we have many options at our disposal that may help us to obtain just the result we want. If there is not a suitable plotting command, we can build the plot from so-called graphics primitives, or we can write a program.

This chapter explains graphics for functions. Graphics primitives and directives are addressed in Chapter 6 and options for graphics in Chapter 7. In Chapter 8, we consider graphics for data, and in Chapter 9, we discuss built-in data of *Mathematica*. Exporting and importing figures is explained in Section 4.2.2, p. 105.

Mathematica 6 contains an impressive new feature called *dynamics*. It enables us to create dynamic interfaces where we can adjust some parameters and look at how the result changes. This is very useful especially in studying how a plot depends on some parameters. A special case of dynamic interfaces is also animating. We consider dynamic interfaces and animations in Chapters 10 and 11.

Graphics is one of the central parts of *Mathematica* and contains a wealth of material. You may first want to read only the topics you are interested in now and go on to other topics later. More about *Mathematica* graphics can be found in Smith & Blachman (1995), Wickham-Jones (1994), and Trott (2004a).

The graphics functionality has significantly changed and been enhanced in *Mathematica* 6. If you would like to use version 5 graphics instead, execute **<<Version5`Graphics`**. To restore version 6 graphics capabilities, execute **<<Version6`Graphics`**.

5.1 Basic Plots for 2D Functions

5.1.1 Plotting One Curve

■ **The Basic Plotting Command**

> **Plot[f, {x, a, b}]**   Plot **f** when **x** takes on values from **a** to **b**

The **f** can be an explicit expression, the name of an expression, or the name of a function (for functions, see Section 2.2.3, p. 39). Therefore, to plot the density function of the standard normal distribution, we can write the following:

```
Plot[Exp[-x^2 / 2] / Sqrt[2 π], {x, -4, 4}]
```

or

```
f = Exp[-x^2 / 2] / Sqrt[2 π]
Plot[f, {x, -4, 4}]
```

or

```
g[x_] := Exp[-x^2 / 2] / Sqrt[2 π]
Plot[g[x], {x, -4, 4}]
```

The second method is often handy because from the output of **f = …** we can first check that the expression is correct and because the plotting command then becomes simpler and shorter. We try the second method:

```
f = Exp[-x^2 / 2] / Sqrt[2 π]
```

$$\frac{e^{-\frac{x^2}{2}}}{\sqrt{2\,\pi}}$$

```
p1 = Plot[f, {x, -4, 4}]
```

Plot works by first sampling the function to be plotted at 51 (almost) equally spaced points. If the function changes rapidly somewhere in the interval, then more points are automatically sampled in such regions. **Plot** is in this sense adaptive. The sampled points are then joined by straight lines. For more about the algorithm behind **Plot**, see Section 7.4.6, p. 207.

■ **Other Examples**

The expression to be plotted can consist of several definitions:

```
mu = 1; sigma = 0.7; c = 1 / (sigma Sqrt[2 π]);
g = Exp[-0.5 ((x - mu) / sigma) ^2];
p2 = Plot[c g, {x, -4, 4}]
```

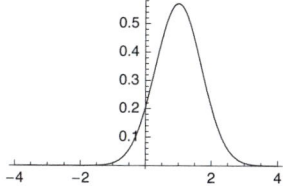

Next we try a function:

```
h[x_, mu_, sigma_] := 1 / (sigma Sqrt[2 π]) Exp[-1 / 2 ((x - mu) / sigma) ^2]
```

```
p3 = Plot[h[x, 0, 1.5], {x, -4.1, 4.1}]
```

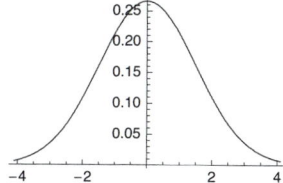

We can plot almost any kind of expression. Next we plot a function defined with an integral:

```
Plot[NIntegrate[Exp[-t^2], {t, 0, x}], {x, 0, 2}]
```

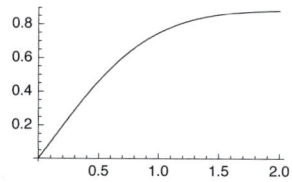

■ **Discontinuous Functions**

> **Plot[f, {x, a0, a1, …, ak}]**   Plot **f** when **x** takes on values from **a0** to **ak**, potentially breaking the curve at each of the **ai**
>
> **Plot[f, {x, a, b}, Exclusions → list]**   Plot **f** when **x** takes on values from **a** to **b**, excluding values of **x** given either explicitly or implicitly (in the form of equations) in the list **list**
>
> **Plot[f, {x, a, b}, Exclusions → None]**   Do not exclude any values of **x**

Here is a discontinuous function:

```
Plot[Tan[x], {x, 0, 2 π}, Ticks → {{π / 2, 3 π / 2}, Automatic}]
```

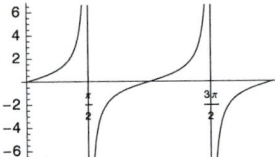

Strictly speaking, the vertical lines do not belong to the graph of the function. To get rid of the vertical lines, define intermediate points at the singularities. We can also use the **Exclusions** option to tell the points to be avoided either explicitly as a list or implicitly by an equation:

```
{Plot[Tan[x], {x, 0, π / 2, 3 π / 2, 2 π}],
 Plot[Tan[x], {x, 0, 2 π}, Exclusions → {π / 2, 3 π / 2}],
 Plot[Tan[x], {x, 0, 2 π}, Exclusions → {Cos[x] == 0}]}
```

Next, we plot another discontinuous function. Now *Mathematica* automatically excludes the point of discontinuity—that is, we do not have a vertical line at the discontinuity. If you want a vertical line, use the **Exclusions** option with value **None**:

```
{Plot[Piecewise[{{Sin[x], x < 3 π / 4}, {Cos[x], x ≥ 3 π / 4}}], {x, 0, π}],
 Plot[Piecewise[{{Sin[x], x < 3 π / 4}, {Cos[x], x ≥ 3 π / 4}}],
 {x, 0, π}, Exclusions → None]}
```

■ Using Options

Mathematica often draws very nice plots. Occasionally, however, we may want to make some adjustments. When this is the case, many options are available. Each option has a default value that is used if another value is not given. One of the options is **AspectRatio** (the ratio of height to width of the plot). The default value of this option is **1/GoldenRatio** = 0.618. With the default value, we get the following plot:

```
p4 = Plot[Sin[x], {x, 0, 2 π}]
```

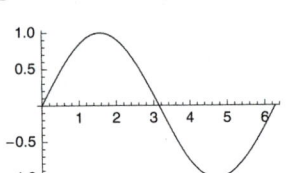

If we want to plot in such a way that one unit on the x axis has the same length as one unit on the y axis, then we can give **AspectRatio** the value **Automatic**:

  Plot[Sin[x], {x, 0, 2 π}, AspectRatio → Automatic]

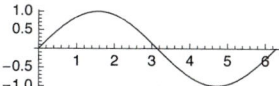

We could also use **Show** and define the option there:

  Show[p4, AspectRatio → Automatic]

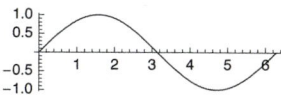

  The standard way when plotting is to first use the default values for the options (i.e., writing no options in the plotting command). If the result is not satisfactory, change the value of some of the options. To make changes, we have two approaches. First, we can write the options into the original plotting command and then execute the command anew. Second, we can write the options into a **Show** command. Note that **Show** does not redo the computations; only the way the figure is shown is changed. This means that computer time is saved, at least for complex figures.

■ Important Options

We consider in detail all the options of **Plot** in Chapter 7. However, here are some of the most important options. For each option, we mention several examples of values. Throughout this book, we use the convention that the default value of each option is always mentioned first.

AspectRatio Ratio of height to width of the plot; examples of values: **1/GoldenRatio** ($= 0.618$),
 Automatic (one unit on both axes has the same length), **0.4**
PlotRange Range of coordinates in the plot; examples of values: **{Full, Automatic}**, **All**, **{-1, 1}**
Ticks Ticks on the axes; examples of values: **Automatic**, **{{ π, 2 π, 3 π}, Automatic}**, **{{1, 2, 3},**
 {-2, -1, 1, 2}}

  Sometimes **Plot** cuts off low or high parts of the function in order to plot the remaining parts more accurately. If you want to see the whole function in the given interval, give **PlotRange** the value **All**. If the ticks of a plot do not satisfy you, define them with **Ticks**. You can specify the ticks on the x or y axis and let **Plot** choose the ticks on the other axis, or you can specify the ticks on both axes. Options can be written in any order, but they must be the last entries in the command—that is, they must be written after the expression and the plotting interval. For example,

  Plot[Sin[x], {x, 0, 2 π}, AspectRatio → Automatic,
  PlotLabel → Sin[x], Ticks → {{π, 2 π}, {-1, 1}}]

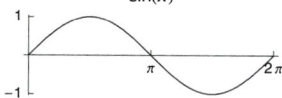

■ **Mouse Manipulations**

In notebook environments, we can change a plot in several ways using the mouse. First, click on the plot with the mouse; a selection rectangle appears.

• To *resize* the plot, drag with the mouse by any of the *handles* of the selection rectangle. When dragging, the width and height of the plot can be read from the bottom left-hand corner of the window (the size is expressed in printer's points; one printer's point is 1/72 of an inch). Note that the size of all text remains unchanged when resizing the plot. Also, the form of the plot remains the same. To change the form of the plot, hold down the ⇧ key and then drag (or use the **AspectRatio** option). Changing the default size of figures is considered soon.

• To *move* the plot on the screen (and thus adjust the margins around the plot), drag with the mouse by one of the *edges* of the selection rectangle (but not by any of its handles). The plot moves and a second selection rectangle appears outside the original seclection rectangle. The outer selection rectangle changes its size as you move the plot. The size of the outer selection rectangle can be changed by dragging by one of its handles. While dragging, the width and height of the outer selection rectangle can be read from the bottom left-hand corner of the window. By moving the plot on the screen, we can set the margins around the plot. To get equal margins on the top and bottom and on the left and right of the figure, shift-drag by one of the handles of the outer selection rectangle. To get rid of the margins, shift-drag by one of the handles of the outer section rectangle to the top left corner.

• To *crop* the plot, hold down the ⎈ key in Windows or the ⌘ key on Macintosh, and drag by one of the handles of the (inner) selection rectangle. The size of the plot can be read from the bottom left-hand corner of the window. By cropping, you can either make more room around the plot or cut off some parts of the plot (croppings are not permanent).

• To *go back* to the standard size and position, click on the plot and choose **Graphics ▷ Rendering ▷ Make Standard Size** (this does not affect a possible cropping).

• To *align* several plots according to the first plot, select the plots by dragging over the cell brackets so that all plots to be aligned become selected (do not worry if some nongraphics cells also become selected), and then choose **Graphics ▷ Rendering ▷ Align Selected Graphics**. Plots can be aligned with respect to any side of the first plot. Plots can also be made the same size as the first plot.

• To make a copied graphic *evaluatable*, select the graphic and choose **Cell ▷ ConvertTo ▷ StandardForm**.

To change the default size of all future plots in the current notebook or in all current and future notebooks, choose **Format ▷ Option Inspector**, select *Scope* to be *Selected Notebook* (if you want to change the size in the current notebook only) or *Global Preferences* (if you want to change the size in all notebooks), go to **Graphics Options ▷ Image Bounding Box**, and change the width and height values for **ImageSize**. The default value is 350 printer's points. Divide the value by 72 to get the width of figures in inches. For example, with the default value, the width of the figure is approximately 4.86 inches. Once you have typed a new value, remember to press ⏎. Note that the sizes of old plots do not change according to the new standard size unless you replot them.

Another way to change the default size of all future plots in the current session is to use **SetOptions**, as in the following example:

```
SetOptions[Plot, ImageSize → 150, BaseStyle → {FontSize → 7}];
```

Here, we also changed the size of the font in plots.

■ **Sound**

With the `Play` command, we can hear sound. For example, the following command plays a pure tone with a frequency of 400 hertz for 2 seconds:

```
Play[Sin[2 π 400 t], {t, 0, 2}]
```

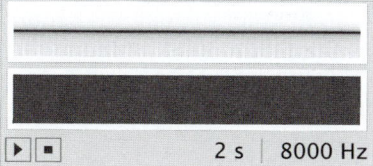

Note that we also have the Audio` and Music` packages.

5.1.2 Plotting Several Curves

■ **Suppressing the Display of a Plot**

Before discussing plots of several curves, we recall from Section 2.1.1 the use of the semicolon (`;`) to suppress the display of a plot:

> `Plot[f, {x, a, b}];` Plot `f` but do not show the plot

Often when we prepare a plot, we, naturally, would like to see the plot. However, when preparing several plots that will later be combined with commands such as `Show` or `GraphicsGrid`, we may want to suppress the display of the separate plots to make the presentation clearer and shorter both on the screen and on the paper. Suppressing the display is easy: Simply end the command with the semicolon `;` (in earlier versions of *Mathematica* we had to use the `DisplayFunction` option).

■ **Showing Several Curves in the Same Plot**

> `Plot[{f1, f2, … }, {x, a, b}]` Plot several expressions in the same figure
> `Show[p1, p2, … ]` Combine several plots in the same figure

To get several curves in the same plot, we have two methods. First, we can plot the curves in one command, giving the expressions as a list:

```
Plot[{Sin[x], Cos[x]}, {x, -π, π}]
```

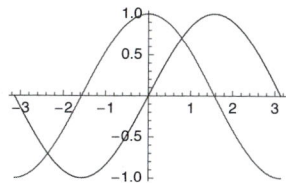

Note that the curves automatically have different colors. We can also plot each curve separately and then combine the plots with `Show`:

```
p1 = Plot[Sin[x], {x, -π, π}];
p2 = Plot[Cos[x], {x, -π, π}];
```

`Show[p1, p2]`

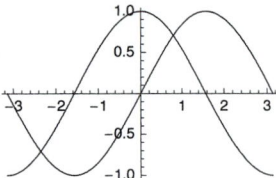

Now the curves have the same color. We can also write complete plotting commands inside **Show**:

`Show[Plot[Sin[x], {x, -π, π}], Plot[Cos[x], {x, -π, π}]]`

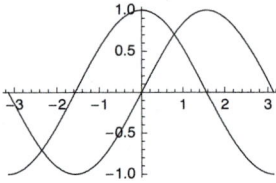

Here is another example. It illustrates the use of a precomputed list of expressions and the use of **Table** inside **Plot**:

`t = {Sin[x], Sin[2 x], Sin[3 x]};`

```
{Plot[{Sin[x], Sin[2 x], Sin[3 x]}, {x, 0, Pi}],
 Plot[t, {x, 0, Pi}],
 Plot[Table[Sin[n x], {n, 1, 3}], {x, 0, Pi}]}
```

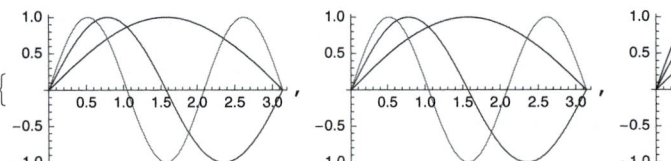

We can see that, with the exception of the last plot, the curves are given different colors. In the last plot, we get different colors if we add **//Evaluate** after the **Table** command.

■ **Tooltips**

> **Tooltip[f]** (⌘6) Show expression **f** as a tooltip when the mouse pointer is moved over the area
>   where **f** is plotted
> **Tooltip[f, label]** Show label **label** as a tooltip

With **Tooltip** we can see the expression plotted or another label when moving the mouse pointer over the plot:

```
{Plot[Table[Tooltip[Sin[n x]], {n, 3}] // Evaluate, {x, 0, π}],
 Plot[Table[Tooltip[Sin[n x], n], {n, 3}] // Evaluate, {x, 0, π}],
 Plot[Table[Tooltip[Sin[n x], Row[{"n = ", n}]], {n, 3}] // Evaluate, {x, 0, π}]}
```

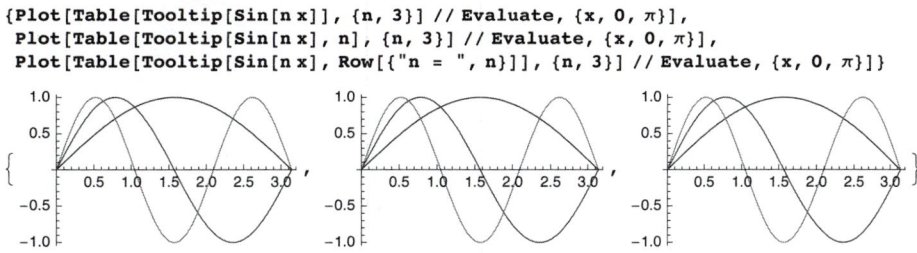

In the first plot, the tooltip is the expression plotted; in the second plot, it is the value of **n**; and in the third, it is a label such as $n = 2$. The label can also be a plot:

```
Table[Tooltip[Sin[n x], Plot[Sin[n x], {x, 0, π}]], {n, 3}]
```

```
{Sin[x], Sin[2 x], Sin[3 x]}
```

When we move the mouse pointer over the result, we can see the plots of the expressions.

■ **Showing Several Curves Separately**

{p1, p2, … }	Show a list of plots (with braces and commas)
Row[{p1, p2, … }]	Show a row of separate plots
GraphicsRow[{p1, p2, … }]	Show the plots in a row as a single graphic

We can show a list of plots:

```
t = Table[Plot[Sin[n x], {x, 0, Pi}, ImageSize → 95], {n, 4}]
```

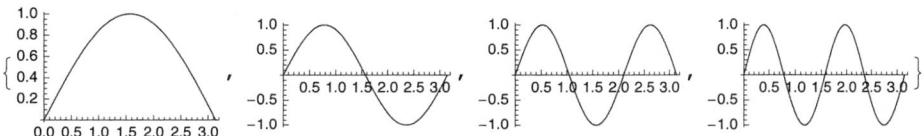

The list contains the braces at the ends and commas in between. We can also use **Row**, **Column**, or **Grid** to get arrays of plots:

```
Row[t, "    "]
```

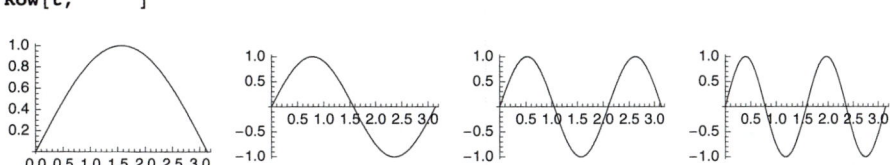

This is a row of four separate plots (**Row**, **Column**, and **Grid** are considered in Chapter 17). We also have **GraphicsRow**, **GraphicsColumn**, and **GraphicsGrid**. With these commands, we get as a result a single graphic:

`GraphicsRow[t]`

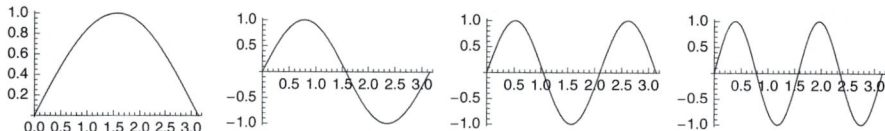

Here are the special commands to arrange plots.

`GraphicsRow[{p1, p2, … }]` (⌘6)   Show plots side by side

`GraphicsColumn[{p1, p2, … }]` (⌘6)   Show plots one below the other

`GraphicsGrid[{{p11, p12, … }, {p21, p22, … }, … }]` (⌘6)   Show plots as a 2D grid

`GraphicsGrid[Partition[t, 2]]`

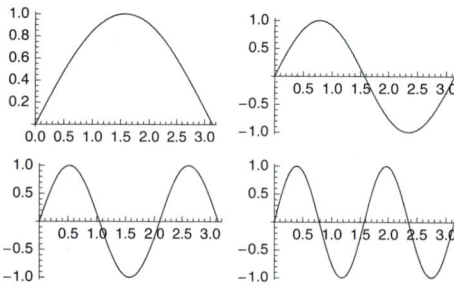

■ Options

`GraphicsRow`, `GraphicsColumn`, and `GraphicsGrid` have many options; execute, for example, `Options[GraphicsGrid]` to get a list of all of them. Here are some of the options.

Some options of `GraphicsRow`, `GraphicsColumn`, *and* `GraphicsGrid`:

Spacings   Horizontal and vertical space between plots; examples of values: `Scaled[0.1]`, `{Scaled[0.2], Scaled[0.1]}`

ImageSize   Size of the whole grid; examples of values: `Automatic, 300`

ImageMargins   Margins around the whole grid; examples of values: `0., 5`

AspectRatio   Ratio of height to width for the whole grid; examples of values: `Automatic, 0.5`

Background   Background color; examples of values: `None, LightGray`

Frame   Where to draw frames; examples of values: `None, True` (frame around the whole grid), `All` (all items become boxed), `{None, All}` (frame around each row)

FrameStyle   Style of frames; examples of values: `Automatic, Thick`

Dividers   Where to draw lines; examples of values: `None, All` (all items become boxed), `Center` (all interior dividers), `{None, All}` (no column lines, all row lines)

Many of the options have more advanced forms of values; see Section 15.2, p. 470, in which we consider the options of `Grid`. To give an example of `GraphicsGrid`, prepare four plots:

```
p1 = Plot[Sin[x], {x, 0, 2 π}, PlotLabel → Sin[x]];
p2 = Plot[Cos[x], {x, 0, 2 π}, PlotLabel → Cos[x]];
p3 = Plot[Tan[x], {x, 0, 2 π}, PlotLabel → Tan[x], PlotRange → {-11, 11}];
p4 = Plot[Cot[x], {x, 0, 2 π}, PlotLabel → Cot[x], PlotRange → {-11, 11}];
```

With the **PlotLabel** option we get the label above the grid:

```
GraphicsGrid[{{p1, p2}, {p3, p4}}, Spacings → Scaled[0.2],
  Frame → All, FrameStyle → Blue, Background → LightYellow, ImageSize → 260,
  PlotLabel → Style["Trigonometric functions", 10, Bold, FontFamily → "Times"]]
```

Trigonometric functions

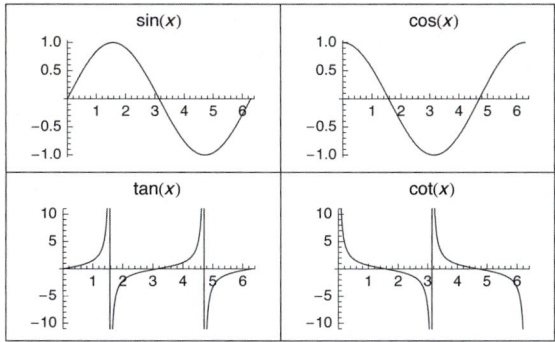

With the **Labeled** command we get the label below the grid:

```
Labeled[GraphicsGrid[{{p1, p2}, {p3, p4}}, Spacings → Scaled[0.2],
  Frame → All, FrameStyle → Blue, Background → LightYellow, ImageSize → 260],
  Style["Trigonometric functions", 10, Bold, FontFamily → "Times"]]
```

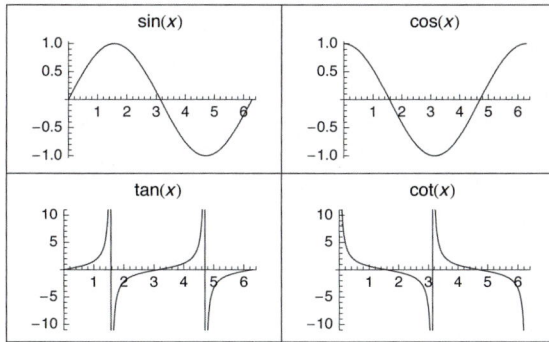

Trigonometric functions

■ How Show Handles Conflicting Settings

When **Show** is used to combine plots, *Mathematica* has to decide what to do if the plots have conflicting settings. Indeed, the plotting intervals may be different, and the values of some options may be different. *Mathematica* uses the following rules:

- use the plotting interval of the *first* plot;
- use the values of the options of the *first* plot; and
- options given in **Show** override those of the first plot.

The second rule, as inevitable it is for *Mathematica*, may cause trouble for us. It means that all option settings other than those in the first plot are simply discarded! We have perhaps carefully set various options in all plots, but when the plots are combined, only the options from the first plot are used. To correct the situation, in **Show**, we can give some of the options we have used in the plots. If you know in advance that you do not need the separate plots as such but will combine the plots into one figure, then fine-tune only the combined plot by giving suitable options in **Show**; do not fine-tune each plot.

As an example, consider the following plots:

```
p1 = Plot[Sin[x], {x, -π, π}];
p2 = Plot[Cos[x], {x, 0, 2π}];
p3 = Plot[Sin[x], {x, 0, 2π}, PlotRange → {0, 1}];
p4 = Plot[Cos[x], {x, 0, 2π}, PlotRange → {-1, 0}];
p5 = Plot[Cos[x], {x, -π, π}, Ticks → {{-π, π}, {1}}];
```

Try to combine some plots:

```
{Show[p1, p2], Show[p3, p4]}
```

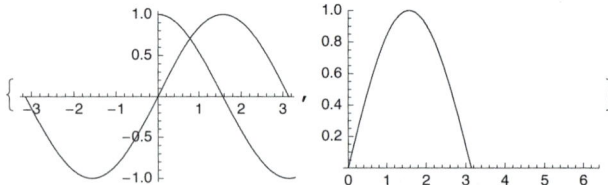

Plots **p1** and **p2** have different plotting intervals, so the combined plot uses the interval of the first plot. Plots **p3** and **p4** have conflicting plot ranges, so the plot range of the first plot is used.

```
{Show[p1, p5],
 Show[p1, p5, PlotRange → {0, 1}, Ticks → {{-π, π}, {1}}]}
```

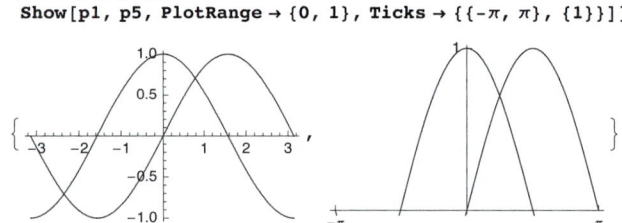

Plots **p1** and **p5** have conflicting ticks (the first plot uses the default ticks), so the ticks of the first plot are used. The last plot shows that we can adjust the options in **Show**.

5.1.3 Interactive Drawing

■ **Graphics Inspector and Drawing Tools**

In *Mathematica*, graphics are normally generated by commands such as **Plot** or **ListPlot**. The style of a plot can be changed in many ways with options such as **PlotStyle**; options are considered in Chapter 7. Sometimes we want to add into a plot some elements or primitives, as they are called in *Mathematica*, such as points, lines, or text. This can be done with the **Prolog** and **Epilog** options (see Section 7.3.6, p. 201). Sometimes we want to build up a graphic from scratch by using graphical primitives. This can be done with the **Graphics** (or **Graphics3D**) command (see Section 6.1.1, p. 152).

In addition to the use of the **PlotStyle**, **Prolog**, and **Epilog** options and the **Graphics** command, in *Mathematica* we can also draw interactively with the mouse. By choosing **Graphics ▷ Graphics Inspector** and **Graphics ▷ Drawing Tools**, we get the following windows:

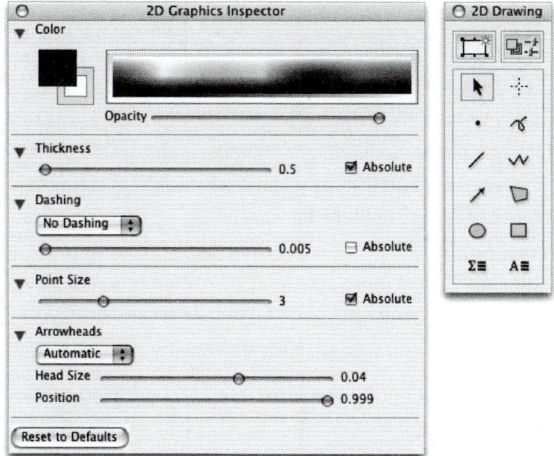

- With the **2D Graphics Inspector** window we can change the styles of components of a plot.
- With the **2D Drawing** window we can add primitives to a plot or build up a graphic from scratch.

The **2D Graphics Inspector** window can also be shown by clicking on the upper right button on the **2D Drawing** window.

■ Changing the Style of Plots

To change the style of a curve or point set,

- double-click on the curve or one of the points with the selection tool ▸ of the **2D Drawing** window; and
- choose, from the **2D Graphics Inspector**, the styles you want to apply for the curve.

In general,

- for a curve, we can change the color, thickness, and dashing; and
- for a set of points, we can change the color and point size.

As an example, here we have three plots:

```
{Plot[Sin[x], {x, 0, 2 π}],
 ListPlot[Table[{x, Sin[x]}, {x, 0, 2 π, π / 10}]],
 ListLinePlot[Table[{x, Sin[x]}, {x, 0, 2 π, π / 10}], Mesh → All]}
```

We have changed the plots as follows:

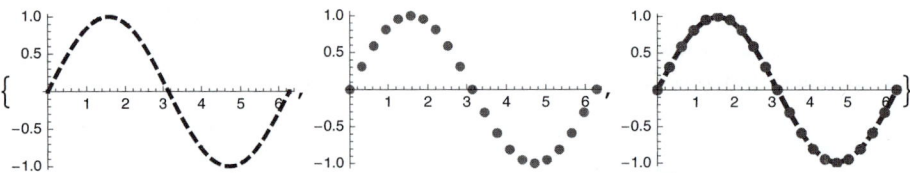

To get the first plot, we selected the curve by double-clicking on it. Be careful: The double-clicking has to be done *on the curve*, not on the white parts of the plot. Then we set, from the **2D Graphics Inspector**, the color to dark red, thickness to approximately 2, and dashing to **Dashed**.

To get the second plot, we first selected the point set by double-clicking on one of the points. We set the color to red and point size to approximately 4.

To get the third plot, we first selected the curve by double-clicking on it. We set the color to dark blue, thickness to approximately 2, and point size to approximately 4. To change the color of the points, we have to select the points by double-clicking one of them. Then we set the color to dark red.

With the **2D Graphics Inspector**, we can also select the opacity of a color.

By double-clicking the color panel on the inspector, we get a more advanced color selector (the same we get from **Insert ▷ Color**).

■ Using the 2D Drawing Tools

The 2D drawing tools can be used in two ways:

- Add some primitives to an existing plot
- Draw a graphic from scratch

To start the second method, simply click on the **New Graphic** button ⊡ on the **2D Drawing** window or choose **Graphics ▷ New Graphic**. We get an empty graphic. Then add some suitable primitives.

With the 2D drawing tools, we can draw

- point • ;
- line ╱ , arrow ➚ , line segments ╲╱╲ , and freehand line ⚡ ;
- text A≡ and traditional form text Σ≡ ; and
- filled or unfilled disk / circle ◯ , rectangle / square ▢ , and polygon ⬠ .

In addition, with the coordinate-picking tool ⊹ we can read and copy coordinates.

■ Selecting a Drawing Tool

On the **2D Drawing** window, we can select a drawing tool in two ways:

- To draw *a single* primitive, *click one time* on the corresponding tool on the **2D Drawing** window. Then click or drag on the plot.
- To draw *several* primitives of the same kind, *double-click* on the corresponding tool (to go back to the normal arrow mouse pointer, click on this tool on the **2D Drawing** window).

To select a drawing tool to draw a single primitive, we can also type a special key:

- point, **p**
- line, **l**; arrow, **a**; line segments, **s**; freehand line, **f**
- text, **t**; traditional form text, **m**
- disk/circle, **c**; rectangle, **q**; polygon, **g**
- mouse pointer, **o**

Thus, to draw, for example, an arrow, type **a** to get the arrow tool and then draw.

■ **Drawing Points, Lines, Arrows, and Text**

Points, lines, arrows, and text can be drawn as follows, after selecting the suitable tool:

- To draw a *point*, click on the plot at a suitable location.
- To draw a *line*, click on the starting point, keep the mouse button down, and drag to the end point.
- To draw an *arrow*, proceed as with a line.
- To draw *line segments*, click on the starting point, release the mouse button, move the mouse to the next point, click with the mouse, release the mouse button, …, move the mouse to the last point, and double click.
- To draw a *freehand line*, click on the starting point, keep the mouse button down, move the mouse as you want, and at the end point release the mouse button. The resulting curve consists of line segments with breaking points. The quality of the line depends on the speed of the mouse: The slower you move the mouse, the more line segments and breaking points the resulting curve will have.
- To add a *text*, click on the starting point of the text and type the text.
- To add a *traditional form text* such as a formula, click on the starting point of the formula and type the formula or build the formula with the aid of palettes.

In the following plot, we have added a point, an arrow, a text, and a formula.

`Plot[Sin[x], {x, 0, 2 π}]`

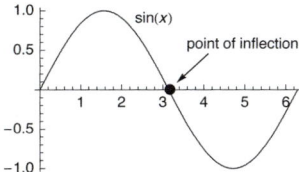

Note that once we have changed the style of a plot or added a primitive, the higher-level cell bracket at the right, grouping the command and the plot, disappears. This is a useful feature because if we execute the plotting command again, the modified plot remains on the notebook.

■ **Drawing Filled Primitives**

Filled ellipses, rectangles, and polygons can be drawn as follows:

- To draw an *elliptical disk* or *filled ellipse*, click on the center point, keep the mouse button down, and drag until the disk has a suitable size and form.
- To draw a *filled rectangle*, click on one of the vertices, keep the mouse button down, and drag to the opposite vertex.
- To draw a *filled polygon*, click on the starting point, release the mouse button, move the mouse to the next point, click with the mouse, release the mouse button, …, move the mouse to the last point, and double click.

■ Coloring Filled Primitives

A filled disk, rectangle, or polygon can be colored in two ways: Either the whole primitive has the same color, or the inside and the edge of the primitive have different colors. To set the colors of filled primitives, we need the color settings on the top left part of the **2D Graphics Inspector**. There, we have two squares:

- The upper left square shows the color of the *inside* (or face) of the filled primitive.
- The lower right square shows the color of the *edge* of the filled primitive.

To change the color of the inside or edge of a filled primitive, select the primitive, set (by clicking if necessary) one of the squares in the **2D Graphics Inspector** to be the topmost square, and choose a color from the panel in the inspector. Then set the other square to be the topmost square and choose a color.

■ Drawing Unfilled Primitives

To get an *unfilled primitive*, we can set the primitive to have nothing as the inside and black (or another color) as the edge. Indeed, by clicking repetitively on one of the two squares on the inspector window, the corresponding color setting toggles between a color and nothing. In this way, we can draw unfilled primitives:

- To get an *unfilled ellipse, rectangle, or polygon*, set, by clicking on the two squares on the inspector, the inside color to nothing and the edge color as black (the edge color can then be changed to another color). Then draw a disk, rectangle, or polygon.

■ Constraining Primitives

A primitive (other than a point or a text) can be constrained as follows, by *shift-dragging*:

- A *line* and arrow can be constrained to horizontal or vertical direction.
- A *line segment* can be constrained to have horizontal or vertical segments.
- A *disk* can be constrained to a circle form.
- A *rectangle* can be constrained to a square.
- A *polygon* can be constrained to have horizontal or vertical edges.

In the following plot, we started from an empty graphics, drew a disk (by constraining it to the form of a circle), selected it, copied it, and pasted it two times. Two of the copies were moved to other places. Suitable colors were given to the disks, and a suitable opacity was defined.

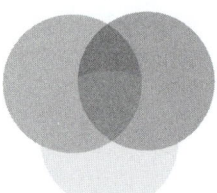

■ Selecting Primitives and Their Components

The primitives can be modified in various ways. In modifying, the depth at which the primitive is selected with the selection tool ▶ is essential. The depth of selection is controlled by clicking various times.

- At the lowest level, the selection rectangle appears around the primitive. At this level, the primitive as a whole can be modified. We say simply that *the primitive is selected*.
- At the next level (by continuing the clicking), the components of the primitive become visible; for example, for a line we can see the line and its end points. We say that the *components of the primitive are selected*.

■ Moving and Resizing Primitives

To change a primitive as a whole, first select the primitive. Then do the following:

- To *move* a primitive, drag either on an edge of the selection rectangle or on the primitive.
- To *change the size* of a primitive, drag on one of the handles of the selection rectangle. To change the size in such a way that the form of the primitive remains the same, hold down the shift key and then drag on a handle.
- To change the *color, point size, thickness,* or *dashing,* use the **2D Graphics Inspector**.

■ Modifying the Components of Primitives

To change the components of a primitive, first select the components of the primitive. Then do the following:

- Changing a *line*: To move the line, drag on the line. To change the starting or ending point of the line, drag on the corresponding point.
- Changing an *arrow*: Do as with a line. To change the form, size, or position of an arrowhead, either select the arrow or select the end point of the arrow. Then change the arrowhead from the **2D Graphics Inspector**.
- Changing a set of *line segments*: To move a line segment, drag on that segment. To move a breaking point, drag on that point.
- Changing a *freehand line*: Drag either on a line segment or on a breaking point.
- Changing a *text* or *traditional form text*: Set the cursor at a suitable location and modify or type.
- Changing a *disk*: To change the center point, drag on it. To change the size or form of the disk, drag on the point of one of the vertices around the disk.
- Changing a *rectangle*: To change the size or form, drag on one of the two points on opposite vertices.
- Changing a *polygon*: Drag either on a line segment or on a breaking point.

■ Grouping Primitives and Moving Them Front and Back

Primitives can be grouped together. A group can then be treated as one object, enabling, for example, the moving of the primitives at the same time. Primitives can also be moved to front or back; this may have an effect if the primitives are on top of each other.

- To group a set of primitives, select the first primitive and then shift-click to select the other primitives. Then select **Graphics ▷ Operations ▷ Group**.
- To move a primitive to the front, select the primitive and then choose **Graphics ▷ Operations ▷ Move To Front**.

■ Picking Coordinates

With the coordinate-picking tool ⊹ we can read and copy coordinates.

- To read coordinates, move the coordinate-picking tool on the plot (without pressing the mouse button).

- To mark a point, click, with the coordinate-picking tool, on the point. To mark several points on a path, drag with the coordinate-picking tool. To delete a marker, [CTRL]-click (⌘-click on a Macintosh) on the marker.
- To copy the coordinates of a point, mark the point and do standard copying; the coordinates can then be pasted.
- To mark a rectangle, [ALT]-drag ([OPTION]-drag on a Macintosh). To copy the coordinates of two opposite corners of the marked rectangle, do standard copying; the coordinates can then be pasted. If marked points are inside the rectangle, their coordinates are also copied and pasted.

5.2 Other Plots for 2D Functions

5.2.1 Parametric Plots

`ParametricPlot[{fx, fy}, {t, a, b}]` Make a parametric plot
`ParametricPlot[{{fx, fy}, {gx, gy}, … }, {t, a, b}]` Make several parametric plots

We can define a function parametrically by giving a list of two expressions `{fx, fy}`. The expressions define the x and y coordinates of the function and are functions of a parameter, often denoted by `t`. The usual plotting command `Plot[expr, {x, a, b}]` can be seen as the special case `ParametricPlot[{x, expr}, {x, a, b}]` of the parametric plotting command.

`ParametricPlot` has almost the same options and default values as `Plot` (the options are considered in detail in Chapter 7). One notable exception is the option `AspectRatio`. For `Plot`, the default value of this option is `1/GoldenRatio`, whereas for `ParametricPlot` the default value is `Automatic`. For parametric plots, this default value is useful because, for example, a circle looks like a circle (see `p1` below) and not like an ellipse. Here are some examples:

```
SetOptions[ParametricPlot, Ticks → None];
p1 = ParametricPlot[{Cos[t], Sin[t]}, {t, 0, 2 π}];
p2 = ParametricPlot[{2 Cos[t], Sin[t]}, {t, 0, 2 π}];
p3 = ParametricPlot[{(2 Cos[t] - 1) Cos[t], (2 Cos[t] - 1) Sin[t]}, {t, 0, 2 π}];
p4 = ParametricPlot[{t Cos[t], t Sin[t]}, {t, 0, 12 π}];
p5 = ParametricPlot[{t Cos[t] Sin[t], t Sin[t]^2}, {t, 0, 8 π}];
p6 = ParametricPlot[{Sin[2 t] + Sin[5 t], Cos[2 t] + Cos[5 t]}, {t, 0, 2 π}];
p7 = ParametricPlot[{Sin[2 t] Sin[5 t], Cos[2 t] Sin[5 t]}, {t, 0, 2 π}];
p8 = ParametricPlot[{Cos[t] + 1 / 2 Cos[7 t] + 1 / 3 Cos[-17 t + π / 2],
    Sin[t] + 1 / 2 Sin[7 t] + 1 / 3 Sin[-17 t + π / 2]}, {t, 0, 2 π}];
```

`GraphicsGrid[{{p1, p2, p3, p4}, {p5, p6, p7, p8}}, ImageSize → 340]`

With a rotation transform we can plot, for example, a rotated ellipse:

```
t =.; ParametricPlot[RotationTransform[Pi / 4][{2 Cos[t], Sin[t]}] // Evaluate,
   {t, 0, 2 π}, Ticks → {{-1, 1}, {-1, 1}}, ImageSize → 70]
```

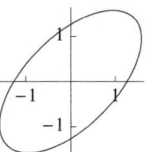

■ **Polar Plots**

```
PolarPlot[r, {θ, a, b}] (✤6)  Make a polar plot
```

In a polar plot, the radius **r** is a function of angle θ. A polar plotting command **PolarPlot[r, {θ, a, b}]** is the special case **ParametricPlot[{r Cos[θ], r Sin[θ]}, {θ, a, b}]** of the parametric plotting command. Examples:

```
SetOptions[PolarPlot, Ticks → None];

p1 = PolarPlot[1, {θ, 0, 2 π}];
p2 = PolarPlot[θ, {θ, 0, 30}];
p3 = PolarPlot[θ ^ (-1 / 2), {θ, 0.2, 50}, PlotRange → All];
p4 = PolarPlot[Sin[6 θ], {θ, 0, 2 π}];

GraphicsRow[{p1, p2, p3, p4}, ImageSize → 340]
```

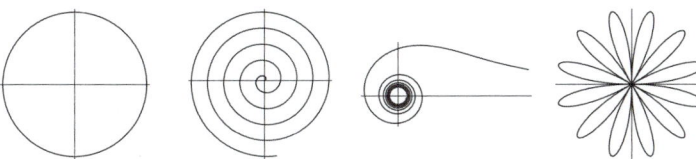

5.2.2 Logarithmic Plots

```
LogPlot[f, {x, a, b}] (✤6)  A plot of Log[f] as a function of x
LogLinearPlot[f, {x, a, b}] (✤6)  A plot of f as a function of Log[x]
LogLogPlot[f, {x, a, b}] (✤6)  A plot of Log[f] as a function of Log[x]
```

With **LogPlot**, exponential functions become straight lines. With **LogLinearPlot**, logarithmic functions become straight lines. With **LogLogPlot**, power functions become straight lines. Logarithmic plots have the same options as **Plot**.

```
{LogPlot[Exp[x], {x, 0, 4}, GridLines → Automatic],
 LogLinearPlot[Log[x], {x, 1, 100}, GridLines → Automatic],
 LogLogPlot[x ^ (1 / 2), {x, 1, 100}, GridLines → Automatic]}
```

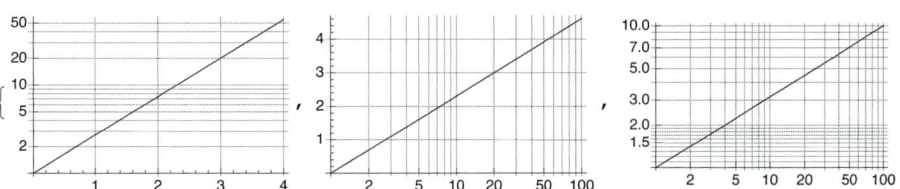

5.2.3 Implicit Plots

■ Plotting an Implicit Function

An implicit function is defined by an equation `expr1 == expr2` (note the two equals signs needed in equations) in which the expressions are functions of two variables, often denoted by **x** and **y**. From the equation we can, at least in principle, solve **y** for each **x**, and so we get a function **y** of **x**. With `ContourPlot`, we can plot implicit functions (`ContourPlot` is considered in more detail in Sections 5.3.1, p. 139, and 7.6.1, p. 226).

> `ContourPlot[expr1 == expr2, {x, a, b}, {y, c, d}]` (❀6)   Plot an implicit function defined by an equation

The default is that a contour plot has a frame, but we can ask not to draw a frame but, rather, the axes. Also, the default aspect ratio is 1 for a contour plot.

```
ContourPlot[x^3 - 2 x y + y^3 == 0, {x, -1.2, 1.1}, {y, -1, 1.1},
  PlotPoints → 40, Frame → False, Axes → True, AspectRatio → Automatic]
```

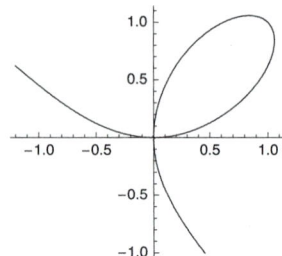

Note that if you give a name for the equation, you need `Evaluate` in the plotting command:

```
eqn = x^3 - 2 x y + y^3 == 0;
```

```
ContourPlot[eqn // Evaluate, {x, -1.2, 1.1}, {y, -1, 1.1}]
```

Whereas `ContourPlot` plots the points satisfying an equality, `RegionPlot` (see Section 5.2.5, p. 136) plots regions where one or more inequalities are satisfied.

■ Plotting a Function of an Implicit Function

We may have the situation in which **y** is defined implicitly as a function of **x** by an equation `expr1 == expr2` but we want to plot, by using **x** as the independent variable, not the plain **y** but an expression `expr3` containing **y** (and possibly also **x**). In this way, `expr3` is a function of the implicit function **y**.

As an example, consider the salmon model of Mesterton-Gibbons (1989, p. 62). In studying the stability of a period-2 equilibrium cycle, we have the following equation:

```
eqn = y Exp[r (1 - y)] == 2 - y;
```

This implicitly defines **y** for each **r**, and we want to plot the following expression as a function of **r**:

```
delta = Abs[(1 - r y) (1 - 2 r + r y)];
```

First, we plot the implicit function defined by **eqn**:

```
ContourPlot[eqn // Evaluate, {r, 1.9, 2.8}, {y, 0, 2}, Frame → False, Axes → True,
   AxesOrigin → {1.9, 0}, PlotPoints → 40, AspectRatio → 1 / GoldenRatio]
```

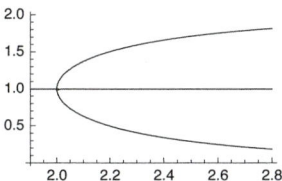

Thus, **y** as a function of **r** has multiple values. When plotting **delta** for **r** in (2, 2.8), we are interested in the smallest **y** value. Plot **delta** as follows (we also plot the line 1 with the **Epilog** option):

```
ListLinePlot[Table[{r, y = y /. FindRoot[y Exp[r (1 - y)] == 2 - y, {y, 0.1}]; delta},
   {r, 2, 2.8, 0.005}] // Quiet, Epilog → Line[{{0, 1}, {2.8, 1}}]]
y =.
```

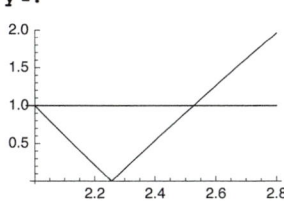

In **FindRoot**, we used the starting value 0.1, for which we get the smallest **y** value (as can be verified).

The stability of the cycle requires that **delta** be smaller than 1. From the previous plot, we see that **r** should be at most approximately 2.5. To get a more accurate value, we solve a system of two equations:

```
FindRoot[{delta == 1, eqn}, {r, 2}, {y, 0.1}]
```

$\{r \to 2.52647, y \to 0.277704\}$

5.2.4 Filled Plots

■ **Plotting One Curve with a Fill**

Filling (⌘6)   Type of filling to use; examples of values: **None**, **Axis**, **Bottom**, **Top**, 0.3

FillingStyle (⌘6)   Style of filling; examples of values: **Automatic**, Red, **{Blue, Red}** (different
    style for negative and positive values), **{{Opacity[0.3], Blue}}**

```
GraphicsRow[
   Plot[Cos[x], {x, 0, 2 π}, Filling → #] & /@ {Axis, Bottom, Top, -0.5}, ImageSize → 420]
```

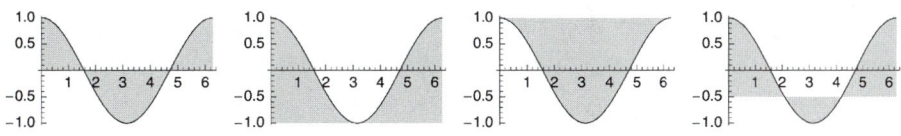

Next, we define custom styles:

```
GraphicsRow[Plot[Cos[x], {x, 0, 2 π}, Filling → Axis, FillingStyle → #] & /@
   {Red, {Blue, Red}, {{Opacity[0.3], Blue}}}, ImageSize → 310]
```

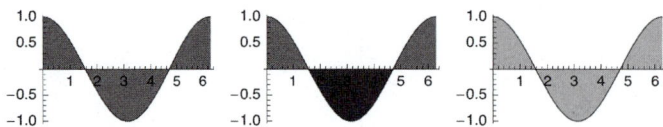

■ Plotting Two Curves with Fills

> **Filling** (⌘6)   Type of filling to use; examples of values: **Axis** (fills from the curves to the x axis),
>   **True** (a fill between the curves), **{1 → Axis}** (a fill between the first curve and x axis), **{2 → {1}}** (a
>   fill between the curves), **{1 → {{2}, Red}}** (define also the filling style), **{1 → {{2},**
>   **Directive[Opacity[0.3], Red]}}** (define style with several directives)
>
> **FillingStyle** (⌘6)   Style of filling; examples of values: **Red**, **{Blue, Red}** (different styles depend-
>   ing on which curve is the topmost), **{{Opacity[0.3], Blue}}**

In filling, several directives have to be collected together with **Directive**. Examples:

```
GraphicsRow[{Plot[{Cos[x], Sin[x]}, {x, 0, 2 π}, Filling → True],
   Plot[{Cos[x], Sin[x]}, {x, 0, 2 π}, Filling → Axis],
   Plot[{Sin[x], If[π / 4 < x < 3 π / 4, 0]}, {x, 0, π}, Filling → {1 → {2}}],
   Plot[{Cos[x], Sin[x]}, {x, 0, 2 π}, Filling →
      {1 → {Axis, Red}, 2 → {{1}, Directive[Opacity[0.3], Blue]}}}], ImageSize → 420]
```

5.2.5 Region Plots

> **RegionPlot[ineqs, {x, a, b}, {y, c, d}]** (⌘6)   Show by a filled plot the region where the given
>   logical combination of inequalities are satisfied (i.e., gives True)

RegionPlot has much the same options as **ContourPlot**. The style of the boundary of the region can be controlled with **BoundaryStyle** and the style of the fill with **PlotStyle**.

Inequalities can be combined with logical operations such as **&&** (AND), **||** (OR), **!** (NOT), or **Xor** (exclusive OR); these are explained in Section 13.3.5, p. 431. For example,

```
SetOptions[RegionPlot, Frame → False, Axes → True, AspectRatio → Automatic];

p1 = RegionPlot[x^2 < y < Sqrt[x], {x, 0, 1}, {y, 0, 1}];
p2 = RegionPlot[1 < x^2 - 1.5 x y + y^2 < 2, {x, -2.2, 2.2}, {y, -2.2, 2.2}];
p3 = RegionPlot[x^3 - 2 x y + y^3 < 0, {x, -1.2, 1.1}, {y, -1, 1.1}];
p4 = RegionPlot[
   Xor[(x + 1)^2 + y^2 < 2, (x - 1)^2 + y^2 < 3], {x, -2.5, 2.8}, {y, -1.8, 1.8}];
```

```
GraphicsRow[{p1, p2, p3, p4}, ImageSize → 420]
```

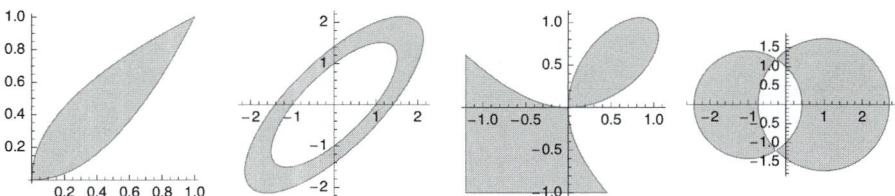

■ **Parametric Region Plots**

`ParametricPlot[{fx, fy}, {t, a, b}, {u, c, d}]` (⌘6)   Plot a parametrically defined region

```
{ParametricPlot[{r Cos[θ], r Sin[θ]}, {r, 1, 2}, {θ, 0, 2 π}],
 ParametricPlot[{r Cos[θ], Sin[θ]},
  {r, 1, 2}, {θ, 0, 2 π}, Mesh → False, Frame → False],
 ParametricPlot[RotationTransform[θ][{2 Cos[t], Sin[t]}] // Evaluate,
  {t, 0, 2 π}, {θ, 0, π / 4}, Ticks → {{-1, 1}, {-1, 1}}]}
```

5.2.6 Complex Plots

■ **Using Re, Im, and Abs**

If the function to be plotted has complex values for some values of the argument, nothing is plotted for these values of argument:

```
Plot[Sqrt[Cos[x]] - 0.5, {x, 0, 2 π}]
```

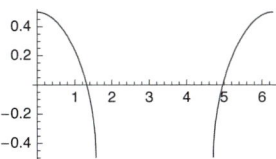

One possibility to plot functions that have complex values is to use **Re**, **Im**, or **Abs** to plot only the real or imaginary part or the absolute value:

```
{Plot[Re[Sqrt[Cos[x]] - 0.5], {x, 0, 2 π}],
 Plot[Im[Sqrt[Cos[x]] - 0.5], {x, 0, 2 π}],
 Plot[Abs[Sqrt[Cos[x]] - 0.5], {x, 0, 2 π}]}
```

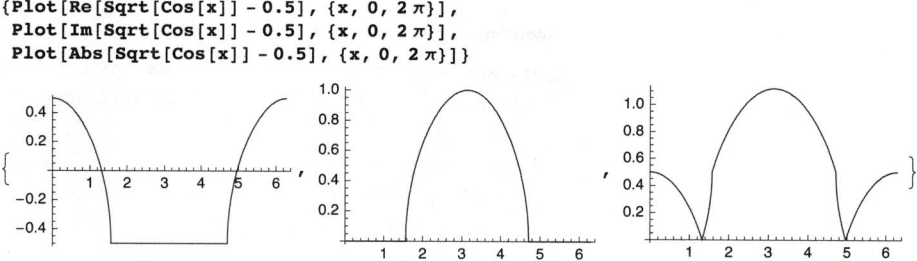

■ Complex Powers

Consider the function $x^{1/3}$. For a given x such as -2.0, it takes on three values, which can be obtained by solving the equation $y^3 = -2.0$:

```
Solve[y^3 == -2.0]
```
$$\{\{y \rightarrow -1.25992\}, \{y \rightarrow 0.629961 - 1.09112\,i\}, \{y \rightarrow 0.629961 + 1.09112\,i\}\}$$

Let us look at what value *Mathematica* gives us for $(-2)^{1/3}$:

```
(-2.)^(1/3)
```
$$0.629961 + 1.09112\,i$$

We have a complex value rather than the real value -1.25992. In general, *Mathematica* gives the value with the smallest positive argument. To plot a function such as $x^{1/3}$, one possibility is to plot the implicitly defined function $y^3 = x$:

```
ContourPlot[y^3 == x, {x, -2, 2}, {y, -1.2, 1.2},
 Frame → False, Axes → True, AspectRatio → 1/GoldenRatio]
```

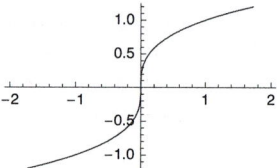

■ Region Plots

> `RegionPlot[ineqs, {x, a, b}, {y, c, d}]` (⌘6) Show by a filled plot the region where the given logical combination of inequalities is satisfied (i.e., gives True)

`RegionPlot` can also be used for inequalities with complex variables:

```
SetOptions[RegionPlot, Frame → False, Axes → True];

p1 = RegionPlot[Abs[x + I y] ≤ 1, {x, -1, 1}, {y, -1, 1}];
p2 = RegionPlot[Abs[1 - (x + I y)^3] ≤ 1, {x, -1, 1.3}, {y, -1.2, 1.2}]; p3 = RegionPlot[
   Abs[1 - (x + I y)^2] < Abs[1 - (x + I y) + (x + I y)^2], {x, -1.5, 2.1}, {y, -1, 1}];
p4 = RegionPlot[Sqrt[x^3 - y] ∈ Reals, {x, -1.5, 1.5}, {y, -1, 1}];
```

```
GraphicsRow[{p1, p2, p3, p4}, ImageSize → 420]
```

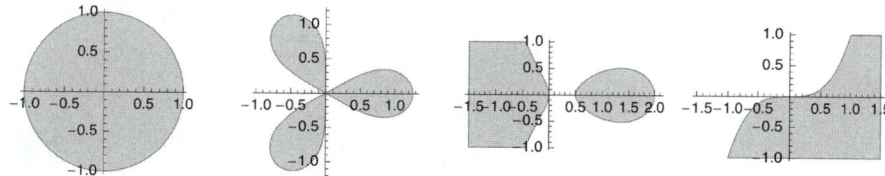

The last plot shows the region where the function $\sqrt{x^3 - y}$ is real.

■ Contour and Parametric Plots

```
ContourPlot[Abs[f], {x, a, b}, {y, c, d}]
ParametricPlot[{Re[f], Im[f]}, {x, a, b}, {y, c, d}] (❀6)
```

```
f = 1 / (x + I y); {ContourPlot[Abs[f], {x, -1, 1}, {y, -1, 1}],
  ParametricPlot[{Re[f], Im[f]}, {x, -0.1, 0.1}, {y, -0.1, 0.1}]}
```

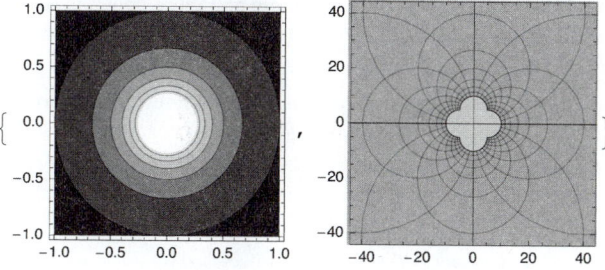

The package VectorFieldPlots` defines **PolyaFieldPlot**.

5.3 Plots for 3D Functions

5.3.1 Basic Plots

■ Surface, Contour, and Density Plots

```
Plot3D[f, {x, a, b}, {y, c, d}]    Plot f as a surface plot
ContourPlot[f, {x, a, b}, {y, c, d}]   Plot f as a contour plot
DensityPlot[f, {x, a, b}, {y, c, d}]   Plot f as a density plot
```

Surface, contour, and density plots are the three main plot types to illustrate 3D functions. For example,

```
f = Cos[x y] Cos[x];
```

```
{Plot3D[f, {x, 0, 3}, {y, 0, 4}, AxesLabel → {x, y, None}],
 ContourPlot[f, {x, 0, 3}, {y, 0, 4}, FrameLabel → {x, y}, RotateLabel → False],
 DensityPlot[f, {x, 0, 3}, {y, 0, 4}, FrameLabel → {x, y}, RotateLabel → False]}
```

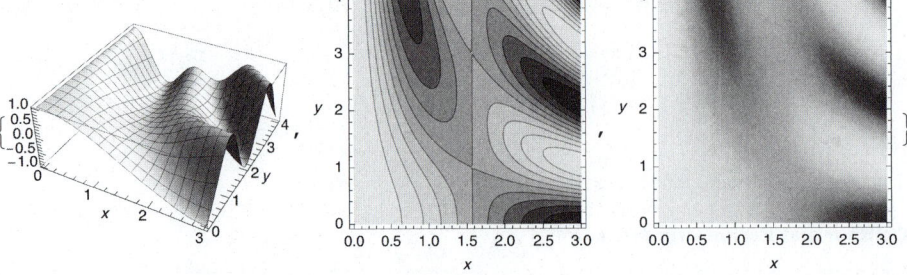

Plot3D shows a 3D surface. **ContourPlot** produces curves that are *contours of constant value*—that is, along each of the curves the value of the function is a certain constant. By moving the mouse pointer over a contour plot (without pressing the mouse button), we can see the constants corresponding to the contours. **DensityPlot** shows the highest parts of the function as light and the lowest parts as dark. All of these commands are adaptive (⌘6): They sample the function at more and more points until the plot is smooth enough. Note that only a surface plots is a true 3D plot; a contour or density plot is actually a 2D plot suitable to illustrate a 3D function.

We can show a surface with contours of constant value. We can also show several surfaces at the same time:

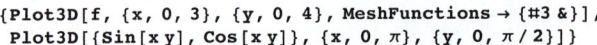

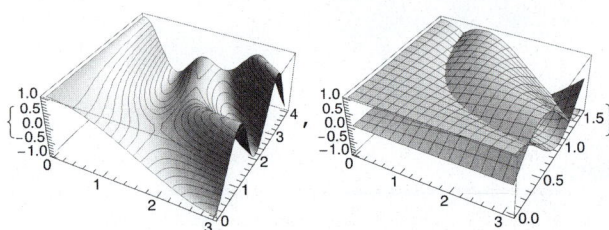

In Section 5.2.3, p. 134, we considered the use of **ContourPlot** to plot implicitly defined functions.

■ Mouse Manipulations

For 3D plots, we can perform similar manipulations with the mouse as we did for 2D plots (see Section 5.1.1, p. 120). First, click on the plot so that the selection rectangle appears. Then do the following:

- To *resize*, drag by a handle of the selection rectangle; ⇧SHIFT drag to change the aspect ratio.
- To *move*, drag by an edge of the selection rectangle.
- To *crop*, hold down the ⌃CTRL (Windows) or ⌘ (Macintosh) key and drag by a handle.
- To *go back* to the standard position and size, choose **Graphics ▷ Rendering ▷ Make Standard Size**.

In addition, we can rotate, zoom, and pan a plot produced by **Plot3D**:

- To *rotate*, move the pointer over the plot. The pointer changes to the rotate pointer ↺. Drag to rotate.

• To *rotate about the axis perpendicular to the screen,* move the pointer to a corner area of the plot. The pointer changes to the vertical rotate pointer ⊙. Drag clockwise or counterclockwise to rotate.

• To *zoom,* hold down the ⌜CTRL⌝ (Windows) or ⌘ (Macintosh) key and drag with the zoom pointer
⬀ upwards (to zoom in) or downwards (to zoom out).

• To *pan* the 3D object—that is, to move it inside the selection rectangle—shift-drag the object with the pan pointer ✛.

Note that interactive drawing (**Graphics ▷ Drawing Tools, Graphics ▷ Graphics Inspector**) does not work with plots produced by **Plot3D** or by other surface plotting commands.

Here is the Klein bottle:

```
a = 6 Cos[u] (1 + Sin[u]); b = 16 Sin[u]; c = 4 (1 - Cos[u] / 2);
fx = If[π < u ≤ 2 π, a + c Cos[v + π], a + c Cos[u] Cos[v]];
fy = If[π < u ≤ 2 π, b, b + c Sin[u] Cos[v]];
fz = c Sin[v];
ParametricPlot3D[{fx, fy, fz}, {u, 0, 2 π}, {v, 0, 2 π}, Boxed → False, Axes → False]
```

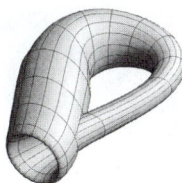

As an example of manipulating with the mouse, resize this plot larger and then rotate it to look inside the bottle. Try also zooming.

■ Some Options

We consider all of the options of the main 3D graphics commands in Chapter 9. Here, we only mention two options:

An option for **Plot3D**:
BoxRatios Ratios of side lengths of the bounding box; examples of values: **{1, 1, 0.4}, Automatic**

An option for **ContourPlot** *and* **DensityPlot**:
AspectRatio Ratio of height to width of the plotting rectangle; examples of values: **1, Automatic**

The default value **{1, 1, 0.4}** of **BoxRatios** for **Plot3D** means that the x and y axes have the same length and that the length of the z axis is 0.4 times the length of the other axes. If we set **BoxRatios →** **Automatic**, then one unit on each of the x, y, and z axes has the same length in the surface plot.

The default value **1** of **AspectRatio** for **ContourPlot** and **DensityPlot** gives a plot of a square form. If we set **AspectRatio → Automatic**, then one unit on both the x and y axes has the same length. Note that contour and density plots have a *frame* (and no axes), and so ticks on the frame can be set with **FrameTicks**.

5.3.2 Special Plots

■ **Parametric Plots**

`ParametricPlot3D[{fx, fy, fz}, {u, a, b}]`	Plot a parametric curve
`ParametricPlot3D[{fx, fy, fz}, {u, a, b}, {v, c, d}]`	Plot a parametric surface

The expressions **fx**, **fy**, and **fz** give the x, y, and z coordinates of the curve or surface.

```
SetOptions[ParametricPlot3D, Axes → False, Boxed → False];

p1 = ParametricPlot3D[{u, Cos[9 u], Sin[9 u]}, {u, 0, π}, PlotStyle → Black];
p2 = ParametricPlot3D[{v Cos[u], v Sin[u], 2 v}, {u, 0, 2 Pi}, {v, 0, 1}];
p3 = ParametricPlot3D[{Cos[u], Sin[u], 2 v}, {u, 0, 2 Pi}, {v, 0, 1}];
p4 = ParametricPlot3D[
    {Sin[u] Cos[v], Sin[u] Sin[v], 1 + Cos[u]}, {u, 0, π}, {v, 0, 2 π}];
p5 = ParametricPlot3D[2 {Sin[u] Cos[v], Sin[u] Sin[v], Cos[u]},
    {u, π / 2, π}, {v, -π, π}];

GraphicsRow[{p1, p2, p3, p4, p5}, Spacings → -80, ImageSize → 420]
```

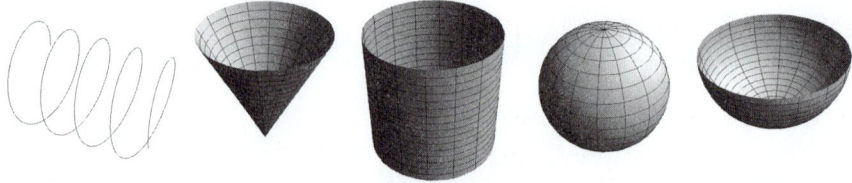

```
p6 = Show[{p4, p5}, PlotRange → All];
p7 = ParametricPlot3D[
    {Cos[u] (3 + Cos[v]), Sin[v], Sin[u] (3 + Cos[v])}, {u, 0, 2 π}, {v, 0, 2 π}];
p8 = ParametricPlot3D[{{Cos[u] (3 + Cos[v]), Sin[v], Sin[u] (3 + Cos[v])},
    {3 + Sin[u] (3 + Cos[v]), Cos[u] (3 + Cos[v]), Sin[v]}}, {u, 0, 2 π}, {v, 0, 2 π}];
p9 = ParametricPlot3D[{u Cos[v] Sin[u], u Cos[u] Cos[v], -u Sin[v]},
    {u, 0, 2 π}, {v, 0, π}];
p10 = ParametricPlot3D[1.2^v {Sin[u]^2 Sin[v], Sin[u]^2 Cos[v], Sin[u] Cos[u]},
    {u, 0, π}, {v, -π / 4, 5 π / 2}, PlotRange → All];

GraphicsRow[{p6, p7, p8, p9, p10}, Spacings → -80, ImageSize → 420]
```

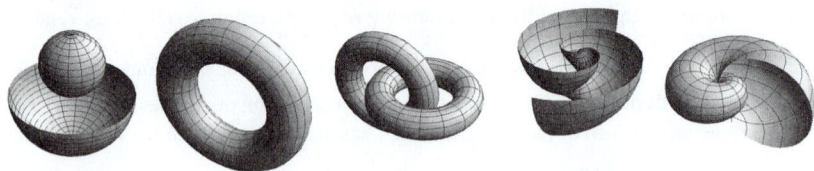

Like other plotting commands, **ParametricPlot3D** is adaptive. The default value of **BoxRatios** is **Automatic** (compared with **{1, 1, 0.4}** for **Plot3D**). This means that a parametric plot is shown in the natural scaling, where one unit on the x, y, and z axes has the same length.

■ **Spherical Plots**

> `SphericalPlot3D[r, {θ, a, b}, {φ, c, d}]` (⌘6)

`SphericalPlot3D` uses `ParametricPlot3D` with argument `r {Sin[θ] Cos[φ], Sin[θ] Sin[φ], Cos[θ]}`.

```
SetOptions[SphericalPlot3D, Axes → False, Boxed → False];
```

```
{SphericalPlot3D[1, {θ, 0, π}, {φ, 0, 2 π}],
 SphericalPlot3D[{1, 2}, {θ, 0, π}, {φ, 0, 3 π / 2}],
 SphericalPlot3D[θ - 1, {θ, 0, π}, {φ, 0, π}]}
```

{ 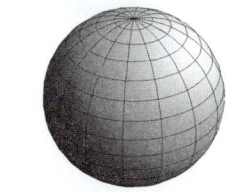 , 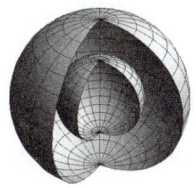 , 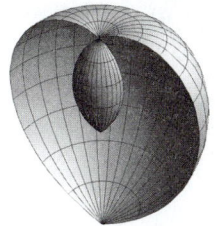 }

■ **Surfaces of Revolution**

> `RevolutionPlot3D[fz, {t, a, b}]` (⌘6)   Rotate around z axis the curve `fz` in the (t, z) plane
> `RevolutionPlot3D[{fx, fz}, {t, a, b}]`   Rotate around z axis the parametrically defined curve
> `RevolutionPlot3D[{fx, fy, fz}, {t, a, b}]`   Rotate around z axis the parametrically defined
>   space curve

We can add a third argument `{θ, c, d}` to these commands so that the curve is rotated only from angle **c** to angle **d**.

```
SetOptions[RevolutionPlot3D, Axes → False, Boxed → False];
```

```
{RevolutionPlot3D[Cos[t], {t, 0, 2 π}],
 RevolutionPlot3D[{t Cos[t], t Sin[t]}, {t, 0, π}, ViewPoint → {.7, -3.2, -.5}],
 RevolutionPlot3D[{t, Cos[2 t], Sin[2 t]}, {t, 0, π / 4}]}
```

{ 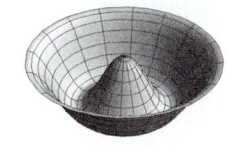 , 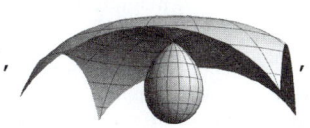 ,  }

■ **Region Plots**

> `RegionPlot3D[ineqs, {x, a, b}, {y, c, d}, {z, e, f}]` (⌘6)   Show the region that satisfies the
>   given logical combination of inequalities

```
{RegionPlot3D[ x ^ 2 + y ^ 2 + z ^ 2 ≤ 1, {x, -1, 1},
  {y, -1, 1}, {z, -1, 1}, Mesh → None, PlotPoints → 30],
 RegionPlot3D[ x ^ 2 + y ^ 2 + z ^ 2 ≤ 1 && -x + y + z ≥ 0.05, {x, -1, 1},
  {y, -1, 1}, {z, -1, 1}, Mesh → None, PlotPoints → 60],
 RegionPlot3D[Cos [x ^ 2 + y ^ 2 + z ^ 2] ≥ 0, {x, -1.5, 1.5}, {y, -1.5, 0},
  {z, -1.5, 1.5}, Mesh → None, BoxRatios → Automatic]}
```

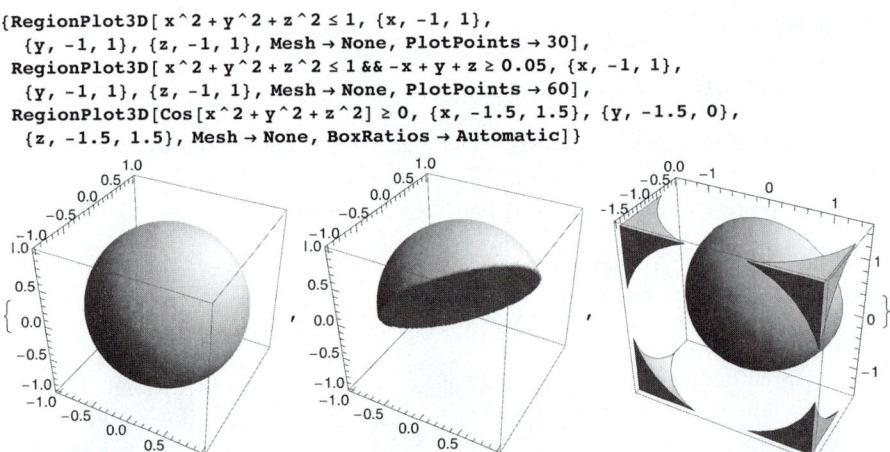

The default value of **BoxRatios** for **RegionPlot3D** is {1, 1, 1}.

Note that with **ContourPlot3D** we can plot surfaces of constant value (see Section 5.4.2, p. 149).

■ **Gradient Fields**

---

*In the **VectorFieldPlots`** package:*

VectorFieldPlot[{fx, fy}, {x, a, b}, {y, c, d}] (✿6) Plot the vector field of the vector-valued
   function **{fx, fy}**

GradientFieldPlot[f, {x, a, b}, {y, c, d}] (✿6) Plot the gradient vector field of the scalar-
   valued function **f**; that is, plot the vector field of $(\partial_x f, \partial_y f)$

HamiltonianFieldPlot[f, {x, a, b}, {y, c, d}] (✿6) Plot the Hamiltonian vector field of the
   scalar-valued function **f**; that is, plot the vector field of $(\partial_y f, -\partial_x f)$

Special options (in addition to the options of **Graphics**):

PlotPoints Number of evaluation points; examples of values: **15**, **{10, 15}**

ColorFunction Function to define the style of the vectors; examples of values: **None**, **(Hue[0] &)**,
   (RGBColor[#, 0, 1 - #] &)

ScaleFunction Function to use for rescaling the magnitude of the vectors; examples of values:
   None, **(0.2 # &)**

MaxArrowLength Eliminates vectors that are longer than the specified value (applied after
   ScaleFunction); default value: **None** (no vectors are removed)

ScaleFactor Lengths of vectors are linearly scaled so that the length of the longest vector is equal
   to the specified value (applied after **MaxArrowLength**); examples of values: **Automatic** (fits the
   vectors in the mesh), **None** (no rescaling is used; use this value with **ScaleFunction**)

Plotting a gradient field is a way to describe a 3D function. The gradient points to the direction of
maximum increase of the function.

```
<< VectorFieldPlots`

f = Cos [x y] Cos [x];
```

```
{Plot3D[f, {x, 0, 3 π / 4}, {y, 0, π / 2}, ImageSize → 190],
 GradientFieldPlot[f, {x, 0, 3 π / 4}, {y, 0, π / 2}, PlotPoints → 10, ImageSize → 210]}
```

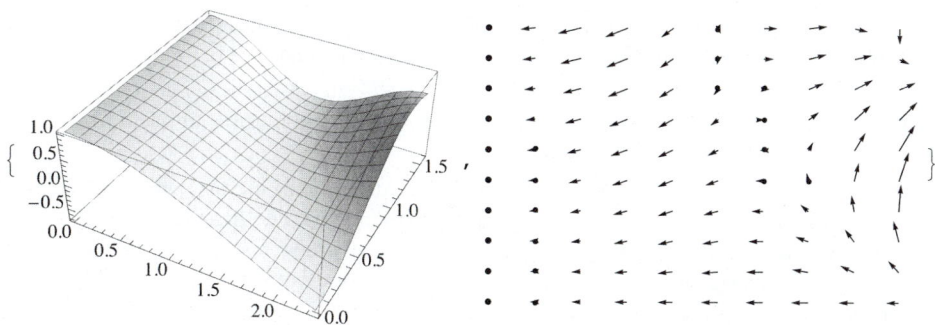

Plots of vector fields are handy in describing differential equations (see Sections 26.1.1, p. 832, 26.1.3, p. 839, and 26.3.2, p. 854) and difference equations (see Sections 28.1.1, p. 926, 28.1.2, p. 930, and 28.2.1, p. 937).

```
f =.
```

5.3.3 Stereograms

■ **Two-Image Stereograms**

An excellent 3D impression can be obtained by making two plots with slightly different viewpoints and placing the plots on top of each other using the eyes. Many of you are probably able to do this. For most people, it is easiest to try to focus beyond the paper surface. First, you get four plots, and the goal is to relax the eyes so much that two of the four plots are superimposed. The result is three plots, with the one in the middle giving the stereo view of the plot. For example,

```
p1 = ParametricPlot3D[{s Cos[t] Sin[s], s Cos[s] Cos[t], -s Sin[t]},
    {s, 0, 2 π}, {t, 0, π}, Axes → False, Boxed → False];
p2 = Show[p1, ViewPoint → {1.4, -2.3, 2.0}];

GraphicsRow[{p1, p2}, Spacings → -140,
  ImageSize → {400, 180}, PlotRegion → {{-0.08, 0.92}, {-0.4, 1.1}}]
```

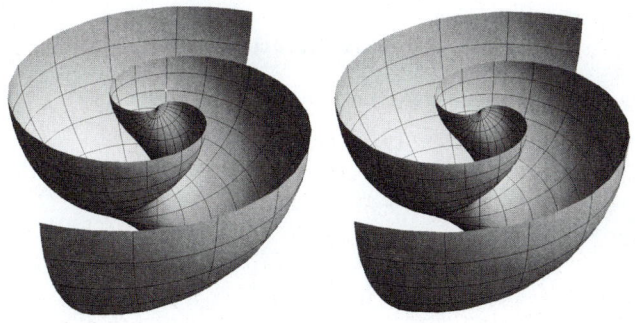

Here the first plot has the default viewpoint **{1.3, -2.4, 2.0}**. The other viewpoint **{1.4, -2.3, 2.0}** is at the same height (2.0) but the *x* and *y* coordinates are somewhat different. We showed the two figures very near to each other. For the stereographic method to be possible to the eyes, the distance of two corresponding points in the two plots should be at most approximately 4 or 5 cm.

The pair of figures we have considered gives the correct result if you focus your eyes beyond the paper. For some people, it is easier to focus on the front of the paper surface. In this case, the order of the figures has to be changed to get the correct result.

For surfaces, the improvement in the illusion of three dimensions is clear with two-image stereograms but not so remarkable as for curves, dots, and arrows in space. For these latter graphics objects, if plotted in the usual way, the eye has too few guides to obtain an adequate impression of the positions of the objects, but the stereo view makes a major improvement. For examples of arrows and space curves, see Sections 5.4.1, p. 149 (gradient field of a 4D function), and 26.3.3, p. 864 (solution of a system of three differential equations).

■ Single-Image Stereograms

Single-image (or random-dot) stereograms are explained in Maeder (1995a). The package of this article can be downloaded from the website www.mathematica-journal.com/backissues. From there, click "Vol. 5 (1)" and download the electronic supplement of this issue. See Section 4.1.3, p. 99, to learn about the correct place to put the **SIS.m** package. The **SIS-EX.MA** notebook contains some examples.

In the **SIS.m** *package:*

SIS[expr, {x, a, b}, {y, c, d}] Produce a single-image stereogram from the given function by looking from the positive *z* axis

Options:

PlotRange Expected range of the function; default value: **{0, 1}**

PlotPoints Number of random initial points; default value: **100**

PlaneDistance Distance from the back plane to the viewing plane; default value: **2**

EyeDistance Distance from the viewing plane to the eyes; default value: **2**

Guides Whether guide dots at the top of the plot are drawn; possible values: **True, False**

EyeSeparation Separation of the guide dots measured as a fraction of half of the horizontal width of the plot; default value: **1/4**

Object Objects used in the plot; default value: **Point**

PlotStyle Style of the objects; default value: **{PointSize[0.01]}**

As an example, we produce a single-image stereogram of the following function:

```
Plot3D[Cos[Sqrt[x^2 + y^2]], {x, -7, 7}, {y, -7, 7}]
```

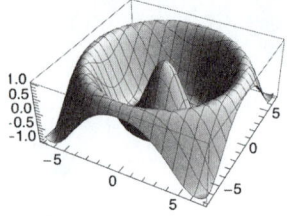

```
<< SIS.m
```

```
SIS[Cos[Sqrt[x^2 + y^2]], {x, -7, 7}, {y, -7, 7},
 PlotPoints → 150, PlotRange → {-1, 1}, ImageSize → 200]
```

The top of the plot contains two dots as guides for the eyes. You get the stereographic view if you can get from the two dots—by focusing beyond the paper surface—first four dots and then three dots (two of the four dots are then superimposed). We have presented the plot in a small size; enlarge the plot to have a more impressive experience.

5.4 Plots for 4D Functions

5.4.1 Simple Methods

2D, 3D, and 4D functions are of the forms $y = f(x)$, $z = f(x, y)$, and $v = f(x, y, z)$, respectively. For 2D and 3D functions, we can use 2D and 3D graphics; for example, use **Plot** or **Plot3D**. Graphical illustration of 4D functions involves difficulties, but something can be done with 3D and even 2D graphics.

■ **Plotting Values on Curves and Surfaces**

Consider the following function:

```
f[x_, y_, z_] := Cos[x^2 + y^2 + z^2]
```

With 2D graphics, we can show the values of the function along parametrically defined curves. For example, along the curves $x = t$, $y = z = 0$; $x = y = t$, $z = 0$; $x = y = z = t$; and $x = \cos(t)$, $y = \sin(t)$, $z = 0$, the value of the function is as follows:

```
p1 = Plot[f[t, 0, 0], {t, -1.5, 1.5}, PlotRange → 1, BaseStyle → 6];
p2 = Plot[f[t, t, 0], {t, -1.5, 1.5}, PlotRange → 1, BaseStyle → 6];
p3 = Plot[f[t, t, t], {t, -1.5, 1.5}, PlotRange → 1, BaseStyle → 6];
p4 = Plot[f[Cos[t], Sin[t], 0], {t, -1.5, 1.5}, PlotRange → 1, BaseStyle → 6];
```

```
GraphicsRow[{p1, p2, p3, p4}, ImageSize → 420, BaseStyle → 5]
```

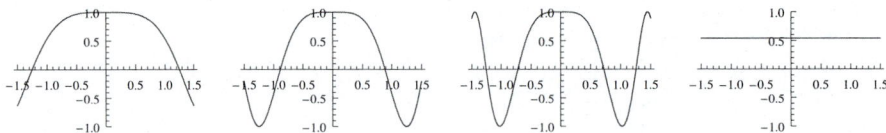

With 3D graphics, we can show the values of the function on parametrically defined surfaces. For example, on the surface $x = u$, $y = v$, $z = 0$ [i.e., on the (x, y) plane], the value of the function is as follows:

```
Plot3D[f[u, v, 0], {u, -1.5, 1.5}, {v, -1.5, 1.5}]
```

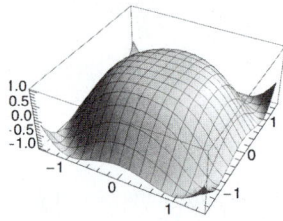

The properties of the first, second, and fourth 2D plots can also be seen (or at least guessed) from this single plot.

■ Gradient Fields

> *In the* **VectorFieldPlots`** *package:*
>
> **VectorFieldPlot3D[{gx, gy, gz}, {x, a, b}, {y, c, d}, {z, e, f}]** (❀6) Plot the vector field of the vector-valued function **{gx, gy, gz}**
>
> **GradientFieldPlot3D[g, {x, a, b}, {y, c, d}, {z, e, f}]** (❀6) Plot the gradient vector field of the scalar-valued function **g**; that is, plot the vector field of $(\partial_x g, \partial_y g, \partial_z g)$
>
> *Special options* (in addition to the options of **Graphics3D**):
> **PlotPoints, ColorFunction, ScaleFunction, MaxArrowLength, ScaleFactor, VectorHeads**

These commands have the same five options as the corresponding 2D commands (see Section 5.3.2, p. 144) plus the option **VectorHeads**. This new option can be used to tell whether the vectors should have a head (**True**) or not (**False**, the default). **PlotPoints** now has the default value 7; the value must be a single number.

One method for illustrating a 4D function is to plot the gradient field. The gradient points in the direction of maximum increase of the function. The longer the arrow, the stronger the growth. The heads of the arrows show the direction of growth. As an example, consider the same function as we studied previously and generate a two-image stereogram:

```
<< VectorFieldPlots`

p1 = GradientFieldPlot3D[Cos[x^2 + y^2 + z^2], {x, -1.5, 1.5},
    {y, -1.5, 1.5}, {z, -1.5, 1.5}, PlotPoints → 5, VectorHeads → True];
p2 = Show[p1, ViewPoint → {1.4, -2.3, 2.0}];
```

```
GraphicsRow[{p1, p2}, Spacings → -20, ImageSize → 320]
```

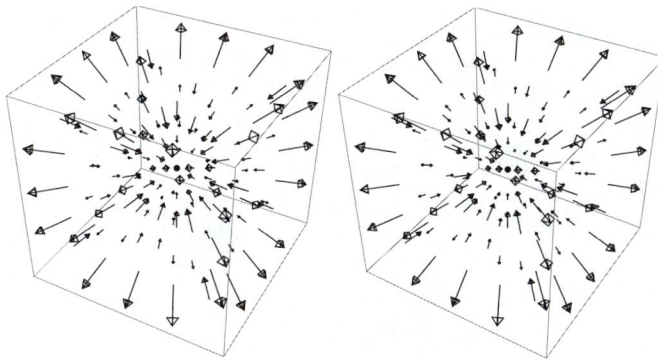

Near the boundary of the box, the function grows toward the outside; in the middle of the box, it grows toward the inside. This means that where the directions of the arrows change, the function has a surface of local minimum points.

5.4.2 Surfaces of Constant Value

With **ContourPlot**, we can plot *curves* of constant value for a 3D function $z = g(x, y)$. With **ContourPlot3D**, we can plot *surfaces* of constant value for a 4D function $v = g(x, y, z)$.

ContourPlot3D[g, {x, a, b}, {y, c, d}, {z, e, f}]   Plot some surfaces of constant value of **g**

ContourPlot3D[g == h, {x, a, b}, {y, c, d}, {z, e, f}]   Plot the surfaces where **g == h**

Special options (in addition to the options of **Plot3D**):

Contours   How many or what surfaces are drawn; examples of values: **Automatic**, 3, {0, 1, 2}

ContourStyle   Style of the surfaces; examples of values: **GrayLevel[1]**, **Opacity[0.8]**, {**Red**, **Green**}

The command has mostly the same options as **Plot3D**. Two new options are **Contours** and **ContourStyle** (**PlotStyle** does not exist). **BoundaryStyle** can be used to define the style of the boundaries of the surfaces. **BoxRatios** has the default value {1, 1, 1}.

By default, the command automatically chooses some constants for which the corresponding surfaces are plotted. The option **Contours** can be given either a constant value, telling how many surfaces are plotted, or a list of constants, telling the constants for which surfaces are plotted. Because the surfaces of constant value tend to give somewhat complex plots, it may be useful to plot separate figures for each value of the constant.

Plotting surfaces of constant value is a demanding task, so the computation time may be somewhat large. When experimenting with the command (so that the quality of the plot need not be high), the **MaxRecursion** option can be given a small enough value such as 0 to reduce the computation time. Also, **PlotPoints** can be given a small value such as 10. However, this may cause some surfaces to be missed. When producing the final plot, these options can be given higher values or the options can be dropped altogether to get a high-quality plot. **ContourStyle** can be useful to identify the surfaces if several are plotted at the same time; opacity may be useful to see surfaces behind other surfaces.

As an example, we consider the same function as in Section 5.4.1, p. 147. We ask for the surfaces where the function has the value −0.5, 0, and 0.5:

```
ContourPlot3D[Cos[x^2 + y^2 + z^2] == #, {x, -1.5, 1.5},
  {y, 0, 1.5}, {z, -1.5, 1.5}, ContourStyle → Opacity[0.8], Mesh → None,
  BoxRatios → Automatic, MaxRecursion → 0] & /@ {-0.5, 0, 0.5}
```

The function seems to grow when the argument approaches the origin but also when the argument moves near the corners of the bounding box. For another example of surfaces of constant value, see Section 27.2.6, p. 909, in which we plot the solution of a 3D elliptic partial differential equation.

Remember that with **RegionPlot3D**, we can plot 3D regions:

```
RegionPlot3D[Cos[x^2 + y^2 + z^2] ≥ 0, {x, -1.5, 1.5},
  {y, 0, 1.5}, {z, -1.5, 1.5}, Mesh → None, BoxRatios → Automatic]
```

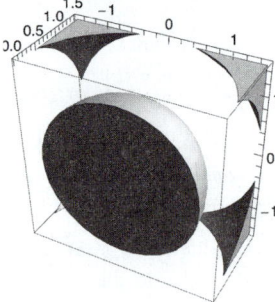

Graphics Primitives

Introduction

> *In the fourth century B.C., Alexander the Great asked his teacher, Menaechmus,*
> *for a shortcut to geometry and received the reply "Oh, King, for travelling over the*
> *country, there are royal roads for kings, but in geometry there is one road for all."*

For producing graphics, commands such as **Plot** or **ListPlot** are the ones we use most often. However, these commands do not cover all graphics. For example, for various geometric figures we need other techniques in which we can gather the plot from basic components such as points, lines, circles, and arcs.

Graphics is a command designed to build plots from so-called *graphics primitives*. We have such primitives as **Point, Line, Circle, Arrow, Text**, and **Rectangle**. The styles of the primitives can be adjusted with so-called *graphics directives*. We have such directives as **PointSize, Thickness, Dashing, GrayLevel, RGBColor**, and **Opacity**. Building plots from scratch with **Graphics** is one use of the primitives and directives. This may also be called *graphics programming*. We consider programming in Chapter 18, but graphics programming is covered in this chapter.

Another use of the directives is in enhancing usual plots with options such as **PlotStyle**: In defining a style, we use directives. Also, with the **Prolog** and **Epilog** options we can add primitives to usual plots. The importance of the primitives and directives is even more clear considering the fact that all plots in *Mathematica* are made up of these components (although we usually do not see them).

6.1 Introduction to Graphics Primitives

6.1.1 Introduction

■ Importance of Primitives and Directives

All plots in *Mathematica* are made up of a few basic components called *graphics primitives*. We have such primitives as **Point**, **Line**, **Circle**, **Rectangle**, and **Text**. The style of these primitives is controlled with *graphics directives* such as **PointSize**, **Thickness**, **Dashing**, **GrayLevel**, or **RGBColor**. The importance of the primitives and directives derives from three facts:

- *Mathematica* uses the primitives and directives in the construction of all plots.
- We can construct a plot directly from the primitives and directives (this can be called *graphics programming*).
- The primitives and directives can be used to modify plots with options.

In practice, the third fact is the most important. Indeed, all plotting commands have several options, such as **PlotStyle**, to modify the plot with directives (e.g., **PlotStyle → Thickness[0.02]**). In addition, with the **Prolog** and **Epilog** options, we can add primitives to plots (e.g., **Epilog → {Thickness[0.02], Line[{{0, 1}, {2, 1}}]}**). Options are considered in Chapter 7. Examples 1 and 2 illustrate the first and second facts.

■ Example 1

Here is a very simple plot:

```
p = Plot[1, {x, 0, 4}, Ticks → {{4}, None}, PlotPoints → 3, MaxRecursion → 0]
```

With **InputForm**, we can see how *Mathematica* made the plot:

```
Short[InputForm[p], 3.5]
Graphics[{{{}, {}, {Hue[0.67, 0.6, 0.6], Line[{{2.*^-6,
    1.}, {1.9236087977499787, 1.}, {3.999998, 1.}}]}}}, {<<7>>}]
```

(We showed only the basic part of **p**; seven options are not shown.) We see that the line is drawn with the **Line** primitive. It goes through three points. The color of the line is set with a **Hue** directive. **Graphics** then creates a *graphics object* from the directive and primitive. The front end automatically displays a graphics object as graphics.

With **FullGraphics**, we can see all the graphics primitives of a plot, including the primitives used for axes and ticks, among other things:

```
Style[InputForm[FullGraphics[p]], 8]
```

```
Graphics[{{{{}, {}, {Hue[0.67, 0.6, 0.6],
      Line[{{2.*^-6, 1.}, {1.9236087977499787, 1.}, {3.999998, 1.}}]}]}},
   {{GrayLevel[0.], AbsoluteThickness[0.25], Line[{{4., 0.}, {4., 0.02022542485937369}}]}},
    Text[4, {4., -0.04045084971874738}, {0., 1.}],
   {GrayLevel[0.], AbsoluteThickness[0.25], Line[{{0., 0.}, {4., 0.}}]},
   {GrayLevel[0.], AbsoluteThickness[0.25], Line[{{0., 0.}, {0., 2.}}]}}}]
```

Here we first see the line. Then we have the tick mark and the tick label 4 on the x axis, and lastly we can see the two lines forming the axes.

■ Example 2

Usually, it is most practical to use commands such as **Plot**, but if we so choose, we can collect a plot from graphics primitives. As an example, we make the same plot as we made previously, with slight simplification:

```
Graphics[{Line[{{0, 1}, {4, 1}}], Line[{{4, 0}, {4, 0.02}}],
   Text[4, {4, -0.04}, {0, 1}], Line[{{-0.1, 0}, {4.1, 0}}],
   Line[{{0, -0.05}, {0, 2.05}}]}, AspectRatio → 1 / GoldenRatio]
```

This is a small example of graphics programming: the building of graphics from primitives.

■ Building Plots from Primitives and Directives

The examples show that we get a graphics object from the primitives with **Graphics**.

Graphics[{directives and primitives**}, options]** Create and show a graphics object from directives and primitives, using some options

All primitives have their default styles. If these are used, we need not define directives; we simply write a list of primitives (as in Example 2). We can modify each primitive with one or more directives. They are written before the corresponding primitive and thus work like adjectives. However, note that a directive affects all primitives after the directive. Thus, if you want to apply a directive or several directives only to the next primitive, enclose the directives and the primitive in curly braces (**{ }**), as seen in the following example:

Graphics[{{dir$_{11}$, dir$_{12}$, …, prim$_{1}$**}, {**dir$_{21}$, dir$_{22}$, …, prim$_{2}$**}, … }]**

Graphics has many of the options of **Plot** and most default values are the same for these commands. Some exceptions are as follows. For **Graphics** we have

AspectRatio → Automatic, Axes → False, PlotRange → All, and **PlotRangeClipping → False**,

whereas for **Plot** we have

AspectRatio → 1/GoldenRatio, Axes → True, PlotRange → {Full, Automatic}, and **PlotRangeClipping → True**.

Graphics does not have certain options that **Plot** does, which control the sampling algorithm.

With the **Prolog** and **Epilog** options, we can add graphics primitives to plots produced by usual plotting commands such as **Plot** (see Section 7.3.6, p. 201):

Plot[f, {x, a, b}, Epilog → {directives and primitives}]   Add graphics primitives

---

6.1.2 Summary

■ Graphics Primitives

The following are all the built-in 2D graphics primitives and one primitive from a package. With **p**, **p1**, **p2**, and **pn** we denote a point, made up of the *x* and *y* coordinates, such as **{1, 4}**. For a mathematician, the primitives **Point**, **Line**, **Circle**, **Arrow**, and **Text** may be the most important.

Point[p]   Point at **p**

Line[{p1, …, pn}]   Line through points **p1**, …, **pn**

Circle[p, r]   Circle with center **p** and radius **r** (also ellipse)

Arrow[{p1, p2}] (※6)   Arrow from **p1** to **p2**

Spline[{p1, …, pn}, type, opts]   Spline through points **p1**, …, **pn**

Text[expr, p, opts]   Text **expr** centered at **p**

Rectangle[p1, p2]   Filled rectangle with two opposite corners at **p1** and **p2**

Polygon[{p1, …, pn}]   Filled polygon with vertices **p1**, …, **pn**

Disk[p, r]   Filled disk with center **p** and radius **r** (also filled ellipse)

Raster[colors]   Raster image

Spline is defined in the **Splines`** package. Note that **GraphicsComplex**, **Inset**, and **GraphicsGroup**, considered in Sections 6.2.9 and 6.2.10, can also be used like primitives. In addition, in Section 10 we encounter **Locator**, a graphics primitive useful in dynamic graphics.

In Section 6.2, we consider all the 2D primitives. 3D primitives are addressed in Section 6.2.11, p. 176. In Section 7.3.6, p. 201, and in numerous other sections, we present other examples of using graphics primitives.

■ Graphics Directives

Here are the most important graphics directives:

PointSize[d], AbsolutePointSize[d]

Thickness[d], AbsoluteThickness[d]

Dashing[{d1, d2, … }], AbsoluteDashing[{d1, d2 ,… }]

Arrowheads[s]

EdgeForm[dirs], FaceForm[dirs]

GrayLevel[g]   Gray **g**

Hue[h]   Hue **h**

RGBColor[r, g, b]   Red **r**, green **g**, and blue **b**

CMYKColor[c, m, y, k]   Cyan **c**, magenta **m**, yellow **y**, and black **k**

Opacity[o]   Opacity **o**

Each primitive can be modified with certain directives. A summary is provided here. With **Raster**, we can use no directives. The opacity and color of all other graphics primitives can be defined with **Opacity** and with one of **GrayLevel**, **Hue**, **RGBColor**, and **CMYKColor**, or with special colors (e.g., **Red**); colors are considered in Section 6.2.8, p. 168. Color and opacity are the only directives that can be used with **Text**. In addition to opacity and color, the style of **Rectangle**, **Polygon**, and **Disk** can be controlled with **EdgeForm** and **FaceForm**. In addition to opacity and color, the style of a point can be controlled with **PointSize** or **AbsolutePointSize** and the style of a line, circle, arrow, and spline with **Thickness** or **AbsoluteThickness** and **Dashing** or **AbsoluteDashing** or with special thickness or dashing specifications (e.g., **Thick** or **Dashed**). **Arrow** also has the special directive **Arrowheads**, and **Spline** and **Text** have some special options.

With **Directive**, we can collect together several directives and use the collection like a single directive:

Directive[dir1, dir2, …] (✻6) Represents a single graphics directive composed of the given directives

---

In various options for styles such as **PlotStyle**, **AxesStyle**, **FrameStyle**, **MeshStyle**, or **FillingStyle**, the use of **Directive** may be helpful in simplifying the use of braces.

6.2 Primitives and Directives

6.2.1 Point

Point[p] A point at **p**
Point[{p1, …, pn}] (✻6) Points at **p1, …, pn**

With **Point**, we can easily get the same result as with **ListPlot**:

```
t = Table[{x, Sin[x]}, {x, 0, 2 π, π / 5}];
```

```
{ListPlot[t],
 Graphics[Point[t], Axes → True, AspectRatio → 1 / GoldenRatio]}
```

Using **Graphics** is, indeed, a noteworthy method to use to plot data, particularly in some more complex situations, and we will use it again in Chapter 8.

■ **PointSize**

PointSize[d] The diameter of a point is a fraction **d** of the width of the graph
AbsolutePointSize[d] The diameter of a point is a multiple **d** of $1/72$ inch
Tiny, Small, Medium, Large (✻6) Special values of **d** (they correspond approximately to absolute point sizes 0, 2, 4.5, and 7)

The diameter of a point can be defined in two ways: either as a fraction of the width of the graph or as a multiple of the printer's point, which is approximately 1/72 inch. For those of us more familiar with millimeters, we note that 3 printer points is approximately 1 millimeter. The default point size is **AbsolutePointSize[3]**. In addition, we can use such point sizes as **PointSize[Small]** (it is approximately the same as **AbsolutePointSize[2]**). Note that when a notebook is printed, the size of graphics is reduced to 80% so that the absolute sizes do not exactly hold on paper.

The size of a point defined by **PointSize** depends on the size of the plot: The larger the plot, the larger the points (however, this does not hold for point sizes defined by **Tiny**, etc.). On the other hand, the size of a point defined with **AbsolutePointSize** does not depend on the size of the plot. In the following points, the absolute diameter varies from 1 to 7:

```
Graphics[Table[{AbsolutePointSize[i], Point[{i, 0}]}, {i, 7}], AspectRatio → 0.2]
```

·   ·   ·   •   •   ●   ●

6.2.2 Line

> **Line[{p1, p2}]**   A straight line between points **p1** and **p2**
> **Line[{p1, …, pn}]**   A broken line through points **p1, …, pn**
> **Line[{{p11, …, p1n}, {p21, …, p2n}, … }]** (⌘6)   Several broken lines

With **Line**, we can easily mimic **ListLinePlot**:

```
t = Table[{x, Sin[x]}, {x, 0, 2 π, π / 5}];
```

```
{ListLinePlot[t],
 Graphics[Line[t], Axes → True, AspectRatio → 1 / GoldenRatio]}
```

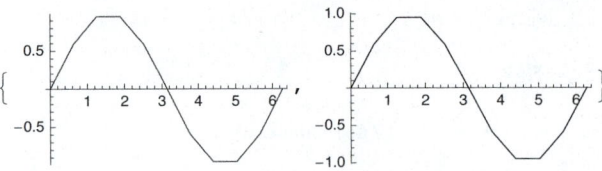

With **Point** and **Line**, we can easily show both the points and the connecting lines between them:

```
{ListLinePlot[t, Mesh → All],
 Graphics[{Point[t], Line[t]}, Axes → True, AspectRatio → 1 / GoldenRatio]}
```

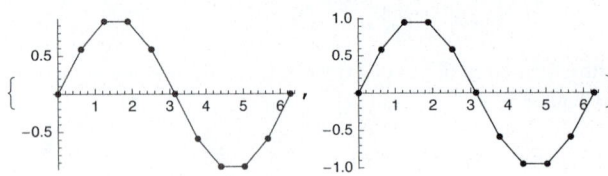

■ **Thickness**

`Thickness[d]`	The thickness of a line is a fraction **d** of the width of the graph
`AbsoluteThickness[d]`	The thickness of a line is a multiple **d** of 1/72 inch
`Tiny, Small, Medium, Large` (⌘6)	Special values of **d** (they correspond approximately to absolute thickness 0.2, 0.5, 1.1, and 1.9)
`Thin, Thick` (⌘6)	Special thickness specifications (they correspond approximately to absolute thickness 0.2 and 1.9)

The default thickness is `AbsoluteThickness[0.5]`. Thickness defined by `Thickness` depends on the size of the plot: The larger the plot, the thicker the curves. Thickness defined with `AbsoluteThickness` does not depend on the size of the plot. In the following plot, the absolute thickness varies from 0 to 3 in steps of 0.5:

```
Graphics[{Table[{AbsoluteThickness[d], Line[{{0, d}, {1, d}}]}, {d, 0, 3, 0.5}],
    Line[{{0, 0}, {0, 3}}], Line[{{1, 0}, {1, 3}}]}, AspectRatio → 0.5]
```

As can be seen, the ends of a line float somewhat outside of the intended interval and do so more as the line becomes thicker.

■ **Dashing**

`Dashing[{d1, d2, … }]`	The length of a segment is a fraction **di** of the width of the graph
`AbsoluteDashing[{d1, d2, … }]`	The length of a segment is a multiple **di** of 1/72 inch
`Tiny, Small, Medium, Large` (⌘6)	Special arguments of `Dashing` and `AbsoluteDashing` (they correspond approximately to absolute dashing {1, 3}, {2.5, 4.5}, {5, 7}, and {10, 13})
`Dotted, Dashed, DotDashed` (⌘6)	Special dashing specifications (they correspond approximately to absolute dashing {1, 4}, {4, 4}, and {1, 4, 4, 4})

In the dashing style, a line consists of small segments, and they alternate between black and white. The lengths of the segments are defined with the dashing directive. Usually, only a few segments are defined, and they are used cyclically. For example,

`Dashing[{}]`	No dashing is used; lines are solid (this is the default)
`Dashing[{d, d}]` or `Dashing[{d}]` or `Dashing[d]`	Black and white segments, each of length **d**, alternate
`Dashing[{d1, d2}]`	Black and white segments of lengths **d1** and **d2**, respectively, alternate

Dashing defined with `Dashing` depends on the size of the figure: The larger the figure, the longer the segments. Dashing defined with `AbsoluteDashing` does not depend on the size of the figure. Next, we show absolute dashings between 2 and 4 in steps of 0.5:

```
Graphics[Table[{AbsoluteDashing[d], Line[{{0, d}, {1, d}}]}, {d, 2, 4, 0.5}],
  AspectRatio → 0.5]
```

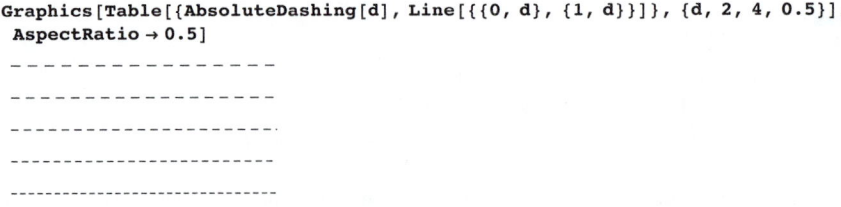

Note that the dashing is not exactly what we would expect: The white segments are smaller than they should be. The thicker the dashed curve, the smaller the white segments. This is a consequence of the fact that the ends of lines float outside of the desired interval, as we saw previously when we considered thickness. By the way, this problem with dashing may actually be a useful feature: The dashing looks better when the white parts are shorter than the black ones.

If the length of a segment is 0, such a segment is drawn as a dot whose diameter is the thickness of the line:

```
Graphics[{AbsoluteDashing[{4, 2, 0, 2}], Line[{{0, 1.5}, {1, 1.5}}]}],
  AspectRatio → 0.15]
```

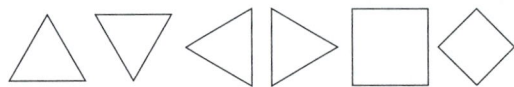

■ Regular Polygons

With **Line**, we can easily create regular polygons. Here is a function for doing this:

```
regularPolygon[n_, x0_, y0_, r_, θ_, opts___] :=
  Graphics[Line[{r Cos[#] + x0, r Sin[#] + y0} & /@ Range[θ, θ + 2 π, 2 π / n]], opts]
```

The arguments of this function are the order **n** of the polygon, the center point (**x0**, **y0**), the radius **r** (from the center point to a vertex), an angle θ (one of the vertices is at this angle), and zero or more options. To show, for example, a triangle, type **regularPolygon[3, 0, 0, 1, π/2]**. Here are some examples:

```
g1 = regularPolygon[3, 0, 0, 1, π / 2]; g2 = regularPolygon[3, 0, 0, 1, -π / 2];
g3 = regularPolygon[3, 0, 0, 1, -π]; g4 = regularPolygon[3, 0, 0, 1, 0];
g5 = regularPolygon[4, 0, 0, 1, π / 4]; g6 = regularPolygon[4, 0, 0, 1, 0];
  GraphicsRow[{g1, g2, g3, g4, g5, g6}, ImageSize → 240]
```

With the Polytopes` package, we can get information about regular polygons, from **Digon** to **Dodecagon**.

6.2.3 Circle and Ellipse

> **Circle[p]**   Circle with center **p** and radius 1 (**Circle[ ]** means **Circle[{0,0}]**)
>
> **Circle[p, r]**   Circle with center **p** and radius **r**
>
> **Circle[p, Offset[{r, r}]]**   Circle with center **p** and radius **r/72** inch
>
> **Circle[p, r, {θ1, θ2}]**   Circular arc
>
> **Circle[p, {rx, ry}]**   Ellipse with center **p** and semi-axes of lengths **rx** and **ry**, oriented parallel to the coordinate axes
>
> **Circle[p, {rx, ry}, {θ1, θ2}]**   Elliptical arc

Consider the following circles:

```
c1 = Graphics[Circle[]];
c2 = Graphics[Circle[], AspectRatio → 0.5];
c3 = Graphics[Circle[{0, 0}, Offset[{18, 18}]], AspectRatio → 0.5];
GraphicsRow[{c1, c2, c3}, ImageSize → 200]
```

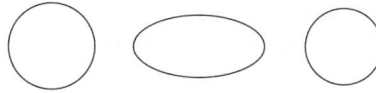

In the first plot, the aspect ratio has the default value **Automatic**; the circle looks like a circle. In the second plot, the aspect ratio is 0.5; the circle looks like an ellipse. In the third plot, the aspect ratio is again 0.5, but now the radius is defined with **Offset**; the circle again looks like a circle.

Thus, to get true circles, either use **AspectRatio → Automatic** or define the radius with **Offset**. The difference between these methods is that if we change the size of a plot containing a circle, the size of the circle also changes in the first method but remains fixed in the second method.

Circles can be used in plotting data:

```
data = Table[{i, RandomReal[]}, {i, 0, 40}];

Graphics[{Line[data], Circle[#, Offset[{1.5, 1.5}]] & /@ data},
  Axes → True, AspectRatio → 0.25, ImageSize → 300]
```

Later, we will use **Disk** in the same way. The use of **Disk** has the advantage that it enables us to hide the lines inside the circles.

Circular arcs can be used to denote angles:

```
p =.; Graphics[{PointSize[Medium], Point[{3, 1}],
   Line[{{0, 0}, {3, 1}}], Circle[{0, 0}, 0.8, {0, ArcTan[1 / 3]}], Dashed,
   Line[{{0, 1}, {3, 1}, {3, 0}}], Text[p, {3.15, 1}], Text[α, {0.67, 0.1}]},
 Axes → True,  Ticks → None, ImageSize → 150]
```

6.2.4 Transformations

Rotate[prim, θ] (⌘6)   Rotate the primitive counterclockwise by θ radians about the center point
Rotate[prim, θ, p]   Rotate about the given point
GeometricTransformation[prim, transf] (⌘6)   Apply the given transformation function
GeometricTransformation[prim, mat]   Multiply by the given matrix

In addition to **Rotate**, we also have **Translate** and **Scale**. Graphics primitives can be transformed in many ways. We have such transformation functions as **Affine-**, **LinearFractional-**, **Reflection-**, **Rescaling-**, **Rotation-**, **Scaling-**, **Shearing-**, and **TranslationTransform**. Examples:

```
{Graphics[Circle[{0, 0}, {4, 1}], Axes → True],
 Graphics[Rotate[Circle[{0, 0}, {4, 1}], π / 6], Axes → True],
 Graphics[GeometricTransformation[Circle[{0, 0}, {4, 1}],
   RotationTransform[π / 6, {0, 0}]], Axes → True]}
```

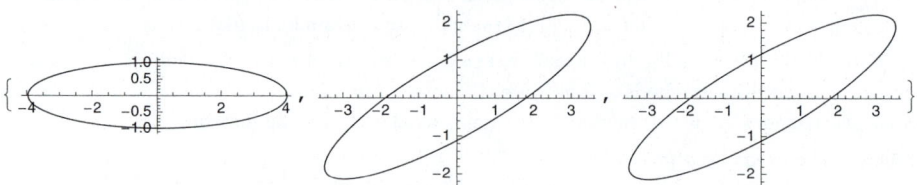

The next example shows how we can manipulate a circle by two vectors (**Manipulate** is considered in Chapter 10):

```
Manipulate[
 Graphics[{PointSize[Medium], Point[{0, 0}], Circle[], Blue, Line[{{0, 0}, v1}],
   Line[{{0, 0}, v2}], Red, GeometricTransformation[Circle[], {v1, v2}ᵀ]},
  PlotRange → 5, ImageSize → 130], {{v1, {3, 4}}, Locator}, {{v2, {-3, -1}}, Locator}]
```

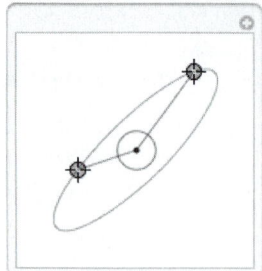

In the RegressionCommon` package, we also have the primitive **Ellipsoid**.

6.2.5 Arrow and Spline

■ **Arrow**

Arrow[{p1, p2}] (❀6) Arrow from point **p1** to point **p2**

Arrow[{p1, p2}, d] The ends of the arrow are set back from **p1** and **p2** by a distance **d**

Arrow[{p1, p2}, {d1, d2}] Sets back by **d1** from **p1** and by **d2** from **p2**

A directive:

Arrowheads[s] (❀6) Length of the arrowhead is a fraction **s** of the width of the whole plot

Arrowheads[{{s, pos}}] Arrowhead is of size **s** at position **pos**

Arrowheads[{{s, pos, g}}] Arrowhead is graphic **g**

Arrowheads[{{s, pos, {g, δ}}}] Shorten the shaft by δ at the arrowhead end

Arrowheads[{-s, s}] Double-headed arrow with heads of size **s**

Arrowheads[{{-s1, pos1}, {s2, pos2}}] Double-headed arrow with custom sizes and positions

---

Note that by choosing **Graphics ▷ Graphics Inspector**, we get a window for specifying various styles. To give an arrow a suitable form, select the arrow on your plot and choose a suitable arrowhead from the 2D graphics inspector.

The default length of the arrowhead (the default value of **s**) is 0.04. The head size **s** can also be **Tiny**, **Small**, **Medium**, or **Large**. The position of the vertex of the arrowhead runs from 0 to 1.

```
Graphics[{Arrow[{{0, 0}, {0, 1}}],
  {Arrowheads[0.1], Arrow[{{1, 0}, {1, 1}}]},
  {Arrowheads[{{0.1, 0.8}}], Arrow[{{2, 0}, {2, 1}}]},
  {Arrowheads[{-0.1, 0.1}], Arrow[{{3, 0}, {3, 1}}]}},
 AspectRatio → 0.3, ImageSize → 200, PlotRange → {{-0.5, 6.5}, Automatic}]
```

The first arrow is in the default form, the second has a modified head size, and the third has a modified head size and position. The fourth arrow is double-headed.

Now we create some custom arrowheads:

```
GraphicsRow[{h1 = Graphics[Line[{{-1, 1 / 3}, {0, 0}, {-1, -1 / 3}}]],
  h2 = Graphics[Polygon[{{-1, 1 / 3}, {0, 0}, {-1, -1 / 3}, {-0.4, 0}}]],
  h3 = Graphics[Line[{{-1, 1 / 3}, {0, 0}, {-1, -1 / 3}, {-0.4, 0}, {-1, 1 / 3}}]]}]
```

A custom arrowhead is placed so that the origin {0, 0} of its coordinates lies at the given position **pos**.

```
Graphics[{{Arrowheads[{{0.1, 1, h1}}], Arrow[{{0, 0}, {0, 1}}]},
  {Arrowheads[{{0.1, 1, h2}}], Arrow[{{1, 0}, {1, 1}}]},
  {Arrowheads[{{0.1, 1, {h3, 1 / 3}}}], Arrow[{{2, 0}, {2, 1}}]}},
  AspectRatio → 0.3, ImageSize → 200, PlotRange → {{-0.5, 6.5}, Automatic}]
```

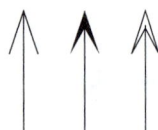

If we connect circles with arrows, it may be helpful to set back the ends of the arrow by a distance that is the radius of the circle:

```
{Graphics[{Circle[{0, 0}, 1], Circle[{4, 0}, 1], Arrow[{{0, 0}, {4, 0}}]}],
  Graphics[{Circle[{0, 0}, 1], Circle[{4, 0}, 1], Arrow[{{0, 0}, {4, 0}}, 1]}]}
```

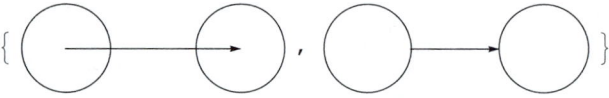

An arrow may sometimes be useful in pointing out some properties of a plot:

```
Plot[Sin[x], {x, 0, 2 π}, Ticks → {{π, 2 π}, {-1, 1}}, PlotRange → {Full, All},
  ImageSize → 140, Epilog → {Text["point of inflection", {3.7, 0.45}, {-1, -1}],
    Arrow[{{3.6, 0.4}, {π + 0.05, 0.05}}]}]
```

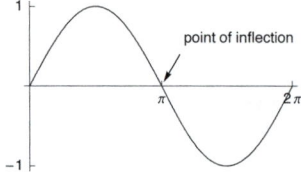

With the following program, we can easily define labeled arrows.

```
labeledArrow[{p1_, p2_}, label_, d_: 0, s_: 0.04] :=
 With[{h = Graphics[Text[Style[label, Small]], {0, 0}, {0, -1.5}]]},
  {Arrowheads[{{-s, 0}, {s, .5, h}, {s, 1}}], Arrow[{p1, p2}, d]}]
```

Note that here we defined that in the middle of the arrow we have a custom "arrowhead," the label. The arguments **d** and **s** have the default values 0 and 0.04; if these values are suitable, the arguments **d** and **s** need not be written.

```
Graphics[{Circle[{0, 0}, 1], Circle[{8, 3}, 1],
  labeledArrow[{{0, 0}, {8, 3}}, "label", 1, 0.07]}]
```

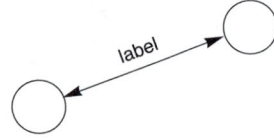

■ **Spline**

In the **Splines`** *package:*

Spline[{p1, p2, … }, type] Spline of **type** through (or controlled by) **p1, p2,** …

An option:

SplineDots The style used for the given points; examples of values: **None** (points are not plotted),
Automatic (points are red and of size 0.03)

For the **Splines`** package, look at Splines/guide/SplinesPackage` in the Documentation Center.

Splines are considered in detail in Section 24.3, p. 803. Here, we only note that a spline is a smooth curve through all or some of the points and that three types of splines can be used: **Cubic** (goes through all points), **Bezier** (goes through the end points), and **CompositeBezier** (goes through every other point). The default is that the given points are not shown. The splines are drawn by an adaptive method, which can be controlled with the additional options **SplinePoints**, **MaxBend**, and **SplineDivision**. Splines may be useful, for example, for drawing some arcs:

```
<< Splines`

Graphics[{Circle[{0, 0}, 1], Circle[{3, 0}, 1],
  Circle[{6, 0}, 1], Text[1, {0, 0}], Text[2, {3, 0}], Text[3, {6, 0}],
  Spline[{{0, 1}, {0.7, 1.7}, {2.3, 1.7}, {3, 1}}, Cubic],
  Spline[{{3, 1}, {3.7, 1.7}, {5.3, 1.7}, {6, 1}}, Cubic],
  Spline[{{0, -1}, {0.7, -1.7}, {2.3, -1.7}, {3, -1}}, Cubic],
  Spline[{{3, -1}, {3.7, -1.7}, {5.3, -1.7}, {6, -1}}, Cubic]},
 BaseStyle → {FontSize → 11}]
```

6.2.6 Text and Coordinates

■ **Text**

Text[expr, {x, y}] The center of the text **expr** is at the point **{x, y}**
Text[expr, {x, y}, {u, v}] The point **{u, v}**, expressed in text coordinates, of the text is at the
point **{x, y}**

The expression to be printed can be a mathematical expression such as **Sin[x]** or a string such as **"Here is text"**. The expression in printed, by default, in **TraditionalForm**. The default is that the text is centered at the given point. However, a text element has its own coordinate system ranging from −1 to 1 in both the x and y directions. For example, the point **{-1, -1}** in text coordinates represents the bottom left-hand corner of the text element and **{1, 1}** the top right-hand corner. With the help of the text coordinates, we can place the text element in other ways. For example,

Text[expr, {3, 4}, {-1, 0}]

Here, the vertical middle of the left-hand end of **expr** is at the point **{3, 4}**; that is, the text starts from **{3, 4}**. The following table explains the most frequently used text coordinates:

Text[t,{x,y},{-1,1}] The top left corner of t is at {x,y}	Text[t,{x,y},{0,1}] t is centered below {x,y}	Text[t,{x,y},{1,1}] The top right corner of t is at {x,y}
Text[t,{x,y},{-1,0}] t starts from {x,y}	Text[t,{x,y},{0,0}] t is centered at {x,y}	Text[t,{x,y},{1,0}] t ends at {x,y}
Text[t,{x,y},{-1,-1}] The bottom left corner of t is at {x,y}	Text[t,{x,y},{0,-1}] t is centered above {x,y}	Text[t,{x,y},{1,-1}] The bottom right corner of t is at {x,y}

An example:

```
Plot[Sin[x], {x, 0, 2 π}, PlotRange → {-1.2, 1.2},
  Ticks → {{π, 2 π}, {-1, 1}}, Epilog → {Point[{π / 2, 1}], Point[{3 π / 2, -1}],
    Text["maximum", {π / 2, 1}, {-1, -1}], Text["minimum", {3 π / 2, -1}, {-1, 1}]}]
```

Note that if a part of a text goes outside the ordinary plot region, that part is not shown in the plot. Without the explicit value of the **PlotRange** option, this would be the case in our example (even **PlotRange → All** does not help). Enlarging the figure with the mouse may also help to see all **Text** primitives.

Text[expr, {x, y}, {u, v}, {r, s}]   Text is rotated to have slope **s / r**

The following plot shows some slope definitions **{r, s}**:

```
Graphics[{Text["bottom to top: {0,1}", {-1, 0}, {0, 0}, {0, 1}],
  Text["left to right: {1,0}", {0, .6}, {0, 0}, {1, 0}],
  Text["top to bottom: {0,-1}", {1, 0}, {0, 0}, {0, -1}],
  Text["right to left: {-1,0}", {0, -0.7}, {0, 0}, {-1, 0}],
  Text["ascending: {2,1}", {-.4, .1}, {0, 0}, {2, 1}],
  Text["descending: {2,-1}", {.3, -.2}, {0, 0}, {2, -1}]}]
```

■ Style of Text

We can adjust the style of text items with **Style**. It accepts both font directives and font options.

Text[Style[expr, dirsAndOpts], {x, y}]   Text has the given directives and options

---

Font directives:

n, **Tiny, Small, Medium, Large, Smaller, Larger**

Bold, *Italic*, Underlined, Plain

Red, LightBlue, GrayLevel[0.3], etc.

Font options:

FontFamily; examples of values: **"Times"**, **"Helvetica"**, **"Courier"**

FontColor; examples of values: **Red, Gray**

FontTracking; examples of values: **"Narrow"**, **"Condensed"**, **"SemiCondensed"**, **"Extended"**, **"Wide"**

Background; examples of values: **LightGray, LightBlue**

The default size of text in graphics is 10. The size can be any number such as 9; or **Tiny** (means size 6), **Small** (9), **Medium** (12), or **Large** (24); or **Smaller** (8) or **Larger** (10). This kind of absolute font size does not change if the size of the plot is changed with the mouse. If the size is defined with, for example, **Scaled[0.05]**, then the size of the font is 0.05 times the width of the plot.

In addition to the options we have shown previously for **Style**, we can use many other options, such as **LineSpacing**, **TextAlignment**, and **Magnification**.

```
Plot[Sin[x], {x, 0, 2 π}, PlotRange → {-1.2, 1.4}, Epilog → {Point[{π / 2, 1}],
    Text[Style["maximum", Small, Blue, Bold, Italic, Underlined,
        FontFamily → "Arial", Background → Yellow], {π / 2, 1}, {-1, -1}]}]
```

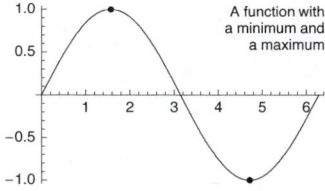

■ Coordinates

{x, y}   A point in the original coordinates of the plot

Scaled[{sx, sy}]   A point in the scaled coordinates

Thus far, we have defined the positions of graphics primitives with the normal coordinates used in the plot. Another way is to use scaled coordinates that run from 0 to 1 in both directions. An example:

```
Plot[Sin[x], {x, 0, 2 π}, Epilog → {Point[{π / 2, 1}], Point[{3 π / 2, -1}],
    Text[Style["A function with\na minimum and\na maximum", TextAlignment → Right],
        Scaled[{1, 1}], {1, 1}]}, ImageSize → 150]
```

Here, we told *Mathematica* that the top right corner of the text is at the point **Scaled[{1, 1}]**, which is at the top right corner of the plot (note that in the text string, **\n** defines a new line). Using scaled coordinates for primitives may be useful, for example, when the original coordinates vary from figure to figure but we want a primitive at the same position in each figure.

> `Scaled[{sdx, sdy}, {x, y}]`   Scaled offset `{sdx, sdy}` from `{x, y}`
>
> `Offset[{adx, ady}, {x, y}]`   Absolute offset `{adx, ady}` from `{x, y}`
>
> `Offset[{adx, ady}, Scaled[{sx, sy}]]`   Absolute offset from `Scaled[{sx, sy}]`

The position of a primitive can also be expressed as an offset from a given point. In this way, we can define a position relative to another position. The offset can be set either in scaled coordinates or in absolute units. The absolute unit is one printer's point (1/72 inch). We already used **Offset** in a special way when we considered circles: For this primitive, we can give the radius in absolute units with **Offset[{adx, ady}]**.

6.2.7 Rectangle, Polygon, and Disk

■ **Filled Primitives**

Mathematica has three filled primitives: **Rectangle**, **Polygon**, and **Disk**.

> `Rectangle[p]`   Filled unit square with two opposite corners at **p** and **p** + 1 (`Rectangle[]` means
>   `Rectangle[{0,0}]`)
>
> `Rectangle[p1, p2]`   Filled rectangle with two opposite corners at **p1** and **p2**
>
> `Polygon[{p1, …, pn}]`   Filled polygon with vertices **p1**, …, **pn**
>
> `Disk[p]`   Filled disk with center **p** and radius 1 (`Disk[]` means `Disk[{0,0}]`)
>
> `Disk[p, r]`   Filled disk with center **p** and radius **r**
>
> `Disk[p, Offset[{r, r}]]`   Filled disk with center **p** and radius **r**/72 inch
>
> *Two directives:*
>
> `EdgeForm[`styles`]` (✑6)   The styles (color, opacity, thickness, dashing) used for the edges
>
> `FaceForm[`styles`]` (✑6)   The styles (color, opacity) used for the faces (or insides)

EdgeForm[] means no edges (this is also the default). **FaceForm[]** means no face (meaning an empty inside).

Disk is used in the same way as **Circle**. Note that although we have a primitive corresponding to an unfilled disk, namely **Circle**, we do not have a primitive corresponding to an unfilled rectangle or unfilled polygon. However, as we will see, the **EdgeForm** and **FaceForm** directives enable us to create unfilled primitives. Unfilled rectangles and polygons can also be easily generated with **Line**:

> `Line[{p1, p2, p3, p4, p1}]`   Rectangle with vertices **p1**, …, **p4**
>
> `Line[{p1, …, pn, p1}]`   Polygon with vertices **p1**, …, **pn**

Here are simple examples:

```
Graphics[{Rectangle[{0, 0}, {2.5, 1}], Rotate[Rectangle[{3, 0}, {5.5, 1}], π / 6],
  Polygon[{{6, 0}, {7, 0.5}, {8, 0}, {7, 1.5}}], Disk[{9, 0.5}, 0.7],
  Disk[{10, 0}, 1.5, {0, Pi / 4}]}, ImageSize → 200]
```

Next, we use various directives. The style of the edge is defined with **EdgeForm**. The style of the inside can be defined directly, but **FaceForm** can also be used. **FaceForm[]** is needed if we want an empty inside:

```
Graphics[{{Red, Disk[{0, 0}, 0.8]},
  {Yellow, EdgeForm[Black], Disk[{2, 0}, 0.8]},
  {LightGreen, EdgeForm[{Blue, Thick, Dashed}]], Disk[{4, 0}, 0.8]},
  {White, EdgeForm[Black], Disk[{6, 0}, 0.8], Disk[{6.5, 0}, 0.8]},
  {FaceForm[], EdgeForm[Black], Disk[{8.5, 0}, 0.8], Disk[{9, 0}, 0.8]}},
 ImageSize → 300]
```

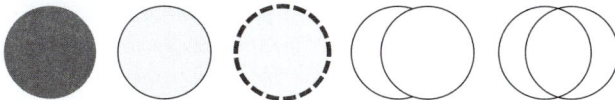

In the fourth figure, there are two white disks partly on top of each other. In the last figure, there are two empty disks or, which is the same, two circles. The last figure shows how to get a simple unfilled primitive: Use **FaceForm[]** and **EdgeForm[Black]**.

Next, we use a yellow filled rectangle with a black edge and a white text on a red background:

```
Graphics[{Yellow, EdgeForm[{Black, Thick}], Rectangle[{0, 0}, {2, 0.7}],
  White, Text[Style["STOP", 40, Bold, Background → Red], {1, 0.31}]}]
```

■ Histograms and Pie Charts

Toss a dice 20 times and plot the frequencies as a histogram:

```
SeedRandom[1]; data = RandomInteger[{1, 6}, 20]
```

{5, 3, 5, 1, 2, 1, 1, 3, 1, 1, 4, 6, 3, 1, 4, 5, 5, 2, 4, 4}

```
fr = Tally[data] // Sort
```

{{1, 6}, {2, 2}, {3, 3}, {4, 4}, {5, 4}, {6, 1}}

```
Graphics[{Red, Rectangle[{#[[1]] - .4, 0}, {#[[1]] + .4, #[[2]]}]} & /@ fr,
 AspectRatio → 1 / GoldenRatio, Frame → True]
```

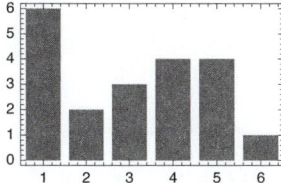

Plot the frequencies as a pie chart:

```
fr2 = fr[[All, 2]]
```

{6, 2, 3, 4, 4, 1}

```
Module[{t = 0, n = Length[fr2], α = 2 π fr2 / Total[fr2]},
  Graphics[Table[{Hue[i / n], EdgeForm[Black], Disk[{0, 0}, 1, {t, t = t + α[[i]]}]},
    {i, n}], ImageSize → 110]]
```

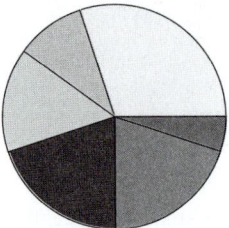

■ Plotting Data

A useful application of **Disk** is to plot data. Whereas **ListPlot** uses black points or disks, with **Disk** we can get disks with white inside:

```
data = Table[{i, RandomReal[]}, {i, 0, 40}];

Graphics[{Line[data], White,
  EdgeForm[Black], Disk[#, Offset[{1, 1}]] & /@ data},
 Axes → True, AspectRatio → 0.25, ImageSize → 300]
```

Here, we first plotted a line connecting the points. Then we plotted white disks with black edges. These disks hide the lines inside the disks. In Section 6.2.3, p. 159, we used circles and the result was not very good because the lines can be seen inside the circles.

6.2.8 Colors and Raster

■ Color Schemes

Here are the four basic color schemes:

> **GrayLevel[g]**   Gray level **g** between 0 (black) and 1 (white)
> **Hue[h]**   Color with hue **h** between 0 and 1 (with maximum saturation and brightness)
> **Hue[h, s, b]**   Color with hue **h**, saturation **s**, and brightness **b**, each between 0 and 1
> **RGBColor[r, g, b]**   Color with specified red **r**, green **g**, and blue **b** components, each between 0 and 1
> **CMYKColor[c, m, y, k]**   Color with specified cyan **c**, magenta **m**, yellow **y**, and black **k** components, each between 0 and 1 (used in four-color printing)

Later, we will study a palette that gives us many additional color schemes.

Colors can be modified in the following ways:

Opacity[o] (⌘6) Opacity **o** (0 means complete transparency, 1 complete opacity)

Blend[{col1, col2}, x] (⌘6) Blend a fraction 1 - **x** of **col1** and **x** of **col2**

Lighter[col] or **Lighter[Red, f]** (⌘6) A lighter color (lightened by a fraction **f**)

Darker[col] or **Darker[Red, f]** (⌘6) A darker color (darkened by a fraction **f**)

Opacity can be defined with **Opacity**, but opacity specification can also be added as a last argument to each of the color systems **GrayLevel**, **Hue**, **RGBColor**, and **CMYKColor**. The default lightening or darkening is 1/3 (**Lighter[col, 0]** and **Darker[col, 0]** give **col**, **Lighter[col, 1]** gives white, and **Darker[col, 1]** gives black).

For some basic colors, we also have ready-to-use names.

Grays: **Black, Gray, White**

Hues: **Red, Yellow, Green, Cyan, Blue, Magenta**

Special colors: **Pink, Orange, Purple, Brown**

Each color, except **Black** and **White**, can be preceded by **Light** (**LightRed**, etc.)

All primitives except **Raster** can be modified with a color. Here are some disks:

```
{Graphics[{RGBColor[0, 1, 0], Disk[{0, 0}],
   {Opacity[0.5], Red, Disk[{1, 0}]}, LightRed, Disk[{3, 0}]}],
 Graphics[Table[{Blend[{Red, Green}, x], Disk[{4 x, 0}]}, {x, 0, 1, 1 / 4}]]}
```

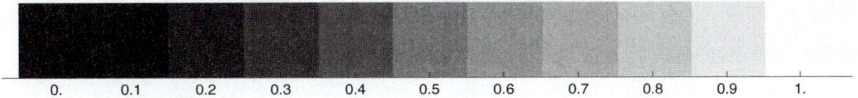

Later, we will mention a palette that helps in choosing a suitable color.

■ Raster

With **Raster** we can form raster images.

Raster[{{g11, … }, … }] An array of gray cells

Raster[{{h11, … }, … }, ColorFunction → Hue] An array of hue cells

Raster[{{{r11, g11, b11}, … }, … }] (⌘6) An array of RGB cells

---

With a second argument of **Raster** we can inform the rectangle that the raster image will occupy. Next, we demonstrate the color systems by using **Raster**.

■ Gray Level

Gray level 0 corresponds to black and 1 to white (contrary perhaps to what you might expect). Here is a sequence of grays from 0 to 1 in steps of 0.1 (the last rectangle, which is white, is not visible):

```
Graphics[Raster[{Table[g, {g, 0, 1, 1 / 10}]}, {{-0.05, 0}, {1.05, 0.1}}],
  ImageSize → 400, Axes → {True, False}, Ticks → {Range[0, 1, 0.1], None}]
```

| 0. | 0.1 | 0.2 | 0.3 | 0.4 | 0.5 | 0.6 | 0.7 | 0.8 | 0.9 | 1. |

■ **Hue**

When the argument in the hue specification ranges from 0 to 1/6, 2/6, 3/6, 4/6, 5/6, and 1, the color changes from red to yellow, green, cyan, blue, magenta, and back to red. In the following, we have several hues:

```
Graphics[Raster[{Table[g, {g, 0, 1, 1 / 12}]},
  {{-1 / 24, 0}, {25 / 24, 0.1}}, ColorFunction → Hue], ImageSize → 400,
  Axes → {True, False}, Ticks → {Range[0, 1, 1 / 12], None}]
```

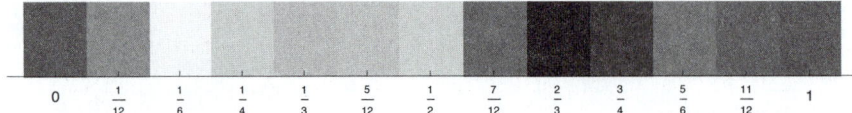

Next, we show a color wheel showing how the hue changes (the wheel is implemented by plotting 300 narrow sectors of a disk):

```
d = π / 150;
Graphics[{Hue[# / (2 π - d)], Disk[{0, 0}, 1, {#, # + d}]} & /@ Range[0, 2 π - d, d],
  ImageSize → 120]
```

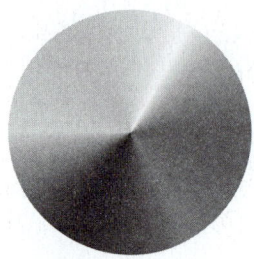

■ **RGB Color**

In the RGB system, we specify the intensity of red, green, and blue. For example, **RGBColor[1, 0, 0]** is red, **RGBColor[0, 1, 0]** is green, and **RGBColor[0, 0, 1]** is blue. This color system corresponds to the one used in color monitors. In the following three tables, we have various RGB colors when the blue component has the value 0, 0.5, and 1.

```
Graphics[Raster[Table[{r, g, #}, {g, 0, 1, .5}, {r, 0, 1, .5}],
  {{-1 / 4, -1 / 4}, {5 / 4, 5 / 4}}], ImageSize → 120, Axes → True,
  AxesLabel → {r, g}, AxesOrigin → {-1 / 4, -1 / 4}, PlotLabel → Row[{"b = ", #}],
  Ticks → {{0, .5, 1}, {0, .5, 1}}] & /@ {0, .5, 1}
```

■ Summary

Here is a summary of how the basic colors can be obtained with the four color schemes:

		GrayLevel	Hue	RGBColor	CMYKColor
Black		0	0, 0, 0	0, 0, 0	0, 0, 0, 1
Gray		.5	0, 0, .5	.5, .5, .5	0, 0, 0, .5
White		1	0, 0, 1	1, 1, 1	0, 0, 0, 0
Red			0	1, 0, 0	0, 1, 1, 0
Yellow			1/6	1, 1, 0	0, 0, 1, 0
Green			2/6	0, 1, 0	1, 0, 1, 0
Cyan			3/6	0, 1, 1	1, 0, 0, 0
Blue			4/6	0, 0, 1	1, 1, 0, 0
Magenta			5/6	1, 0, 1	0, 1, 0, 0

■ Interactive Choice of Colors

For help choosing a color, there is a **Color** item in the **Insert** menu. This item gives you various ways to select colors interactively. On a Macintosh, some of the ways are as follows:

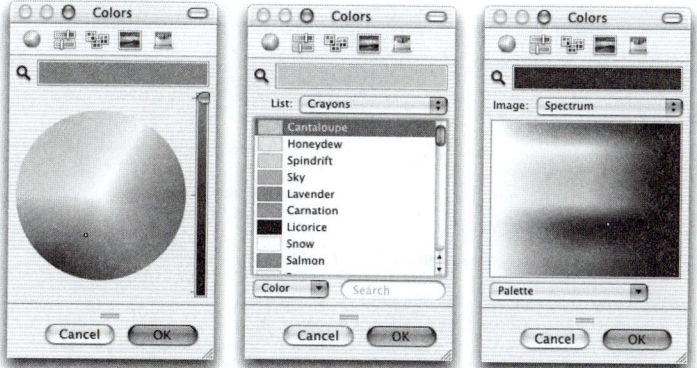

Choose a color from a window by clicking with the mouse and then clicking *OK*. The chosen color definition appears at the location of the cursor as an RGB color definition. For example, type:

```
Plot[Sin[x], {x, 0, 2 π}, PlotStyle →]
```

Place the cursor after →, then select a color by the color selector.

With `ColorSlider`, we get a panel in which we can click on a color. The selected color is shown on the square on the left and shown as an RGB color on the right:

```
{ColorSlider[Dynamic[cl]], Dynamic[cl]}
```

{  , RGBColor[0., 0., 0.]}

■ **Interactive Choice of Color Schemes**

The basic color schemes are **GrayLevel**, **Hue**, **RGBColor**, and **CMYKColor**. Many additional color schemes can be used with the **ColorSchemes** palette from the **Palette** menu.

The schemes are classified to gradients and to physical, named, and indexed color schemes. One use of the gradients and physical color schemes is as values of the **ColorFunction** option. In the first plot below, we have selected, from the palette, the gradient with the name *Rainbow*. The gradient then appears in our notebook, at the cursor position, in the form **ColorData["Rainbow"]**. Similarly, in the second plot, we use the *TemperatureMap* gradient. In the third plot, we use the physical color scheme *VisibleSpectrum*.

```
{Plot3D[Sin[x y], {x, 0, π}, {y, 0, π},
  ColorFunction → (ColorData["Rainbow"][#3] &), Mesh → False],
 ContourPlot[Sin[x y], {x, 0, π}, {y, 0, π},
  ColorFunction → ColorData["TemperatureMap"]],
 DensityPlot[x, {x, 380, 750}, {y, 0, 50},
  ColorFunction → ColorData["VisibleSpectrum"],
  ColorFunctionScaling → False, AspectRatio → Automatic,
  PlotRangePadding → 5, FrameTicks → {{None, None}, {Automatic, None}}]}
```

Indexed and named color schemes may be useful as values of **PlotStyle** in commands such as **Plot** or **ListPlot** or as color directives in **Graphics**:

```
data = Table[(2 - n / 6) x + 0.4 RandomReal[], {n, 1, 6}, {x, 0, 5, 0.1}];

{Plot[Evaluate[Table[Sin[n x], {n, 1, 3}]],
  {x, 0, 2 π}, PlotStyle → ColorData[16, "ColorList"]],
 ListLinePlot[data, PlotStyle → ColorData[22, "ColorList"], InterpolationOrder → 2],
 Graphics[{ColorData["HTML"]["Maroon"], Disk[],
   ColorData["HTML"]["OliveDrab"], Disk[{1, 0}]}]}
```

In the first plot, the palette only gives **ColorData[16]**; we have to add **"ColorList"** after the index of the color scheme to get a list of colors. In the second plot, we have done similarly. The color definitions in the third plot are directly from the palette.

The color schemes of the palette can also be studied and used directly with the **ColorData** command; see Section 9.3.3, p. 304.

6.2.9 GraphicsComplex

■ **Building Plots from Primitives**

Often in building a plot from primitives, we have a set of points and then we draw various primitives, many of them depending on the same points. Of course, we can use the points as such in the primitives, as in the following:

```
p = Table[{x, Sin[x]}, {x, 0., 2 π, π / 10}];

Graphics[{Point[p], Line[p]}, Axes → True]
```

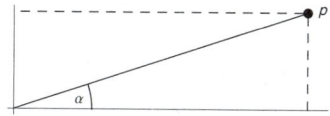

However, we have a special command to help build a plot from primitives.

> **GraphicsComplex[{p1, p2, … }, data]** (⌗6) In the nested list **data** of directives and primitives,
> refer to point **pi** by **i**

Here, we first give the list of points. Then we can refer to the points simply by their ordinal numbers: Point **pi** can be referred to by **i**. This means that the set of all points can be referred to by **Range[n]**, where **n** is the number of points. Let us now redraw the previous plot:

```
With[{rp = Range[Length[p]]},
  Graphics[GraphicsComplex[p, {Point[rp], Line[rp]}], Axes → True]]
```

GraphicsComplex is treated like a single primitive in **Graphics**. The advantage of **GraphicsComplex** is not very clear from this simple example, but in more complex plots the advantage may become more prominent.

Next, using **GraphicsComplex**, we redraw a plot we considered in Section 6.2.3, p. 159:

```
Graphics[GraphicsComplex[p = {{0, 0}, {3, 0}, {3, 1}, {0, 1}},
  {PointSize[Medium], Point[3], Line[{1, 3}], Circle[1, 0.8, {0, ArcTan[1 / 3]}],
   Dashed, Line[{2, 3, 4}], Text["p", {3.15, 1}], Text[α, {0.67, 0.1}]}],
 Axes → True, Ticks → None, ImageSize → 150]
```

With **Normal** we get the usual graphics primitives:

```
GraphicsComplex[p = {{0, 0}, {3, 0}, {3, 1}, {0, 1}},
  {Point[3], Line[{1, 3}]}] // Normal
{{Point[{3, 1}]}, Line[{{0, 0}, {3, 1}}]}
```

Most surface and region plots produce **GraphicsComplex**:

```
Plot3D[Sin[x y], {x, 0, π}, {y, 0, π}] // InputForm // Shallow
Graphics3D[GraphicsComplex[<< 3 >>], {<< 7 >>}]
```

■ **Network Plot**

Let us write a program that draws circles with labels and connects some of the circles with arrows. Let the data be as follows:

```
nodes = {{0, 2}, {2, 4}, {2, 3}, {2, 2}, {2, 1}, {2, 0}, {4, 3}, {4, 1}};
labels = CharacterRange["a", "h"];
arcs = {{1, 2}, {1, 3}, {1, 4}, {1, 5}, {1, 6}, {2, 7},
   {2, 8}, {3, 7}, {3, 8}, {4, 7}, {4, 8}, {5, 7}, {5, 8}, {6, 7}, {6, 8}};
```

We want to draw circles of radius **r** at **nodes**, and **labels** are written inside the circles. The list **arcs** gives the pairs of node numbers between which arrows are drawn; node numbers are 1, 2, and so on in the order they are given in **nodes**. First, define that **rn** contains the ordinal numbers of the nodes. Then generate the circles, labels, and arrows:

```
rn = Range[Length[nodes]]
```
```
{1, 2, 3, 4, 5, 6, 7, 8}
```

```
Circle[#, r] & /@ rn // Short
```
```
{Circle[1, r], <<6>>, Circle[8, r]}
```

```
MapThread[Text, {labels, rn}] // Short
```
```
{Text[a, 1], Text[b, 2], <<5>>, Text[h, 8]}
```

```
Arrow[#, r] & /@ arcs // Short
```
```
{Arrow[{1, 2}, r], <<13>>, Arrow[{6, 8}, r]}
```

Thus, a program could be as follows:

```
networkPlot[nodes_, labels_, arcs_, r_, opts___] :=
 With[{rn = Range[Length[nodes]]}, Graphics[GraphicsComplex[nodes,
    {Circle[#, r] & /@ rn, MapThread[Text, {labels, rn}],
     Arrowheads[{{0.04, 0.8}}], Arrow[#, r] & /@ arcs}], opts]]
```

```
networkPlot[nodes, labels, arcs, 0.3]
```

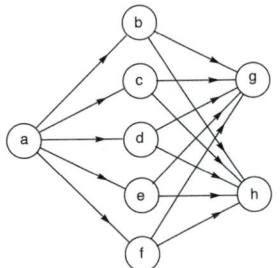

We also have the built-in command **GraphPlot** that we consider in Section 8.5, p. 267:

```
arcs2 = {1 → 2, 1 → 3, 1 → 4, 1 → 5, 1 → 6, 2 → 7,
   2 → 8, 3 → 7, 3 → 8, 4 → 7, 4 → 8, 5 → 7, 5 → 8, 6 → 7, 6 → 8};
```

```
GraphPlot[arcs2, VertexCoordinateRules → nodes,
  VertexLabeling → True, DirectedEdges → True]
```

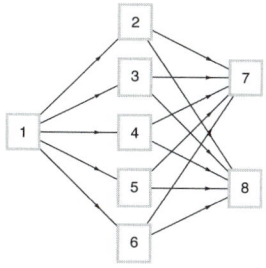

6.2.10 Inset

Previously, we have considered primitives such as **Point**, **Line**, and **Text**. With **Inset**, we can create, from an arbitrary graphics, an object that can be used like a primitive.

Inset[gr] (⌘6)	Inset graphics **gr** at the center of the enclosing graphics
Inset[gr, pos]	Put the center of **gr** at **pos** of the enclosing graphics
Inset[gr, pos, opos]	Put point **opos** of **gr** at **pos** of the enclosing graphics
Inset[gr, pos, opos, size]	Define the size of **gr** in units of the enclosing graphics
Inset[gr, pos, opos, size, dir]	Define the direction of **gr**

Although here we use **Inset** for a graphics **gr**, we can inset any expression. In the positions, we can also use *x*-positions **Automatic**, **Left**, **Center**, **Right**, and **Axis** and *y*-positions **Automatic**, **Bottom**, **Center**, **Top**, **Axes**, or **Baseline**. With the **Alignment** option (**Left**, **Center**, or **Right**) we can define how to align the contents of the inset.

In the following figures, we use the plot of sin(*x*) as a primitive:

```
p = Plot[Sin[x], {x, 0, 2 π}];

p1 = Graphics[{Circle[{1, 0}, 1], Inset[p]}, ImageSize → 90];
p2 = Graphics[{Circle[{1, 0}, 1], Inset[p, {1, 0.5}]}, ImageSize → {90, 90}];
p3 = Graphics[{Circle[{1, 0}, 1], Inset[p, {1, 0}, {0, -1}, 1.2]}, ImageSize → 90];
p4 = Graphics[
    {Circle[{1, 0}, 1], Inset[p, {1, 0}, {0, -1}, 1.2, {2, 1}]}, ImageSize → 90];
Row[{p1, p2, p3, p4}, "   "]
```

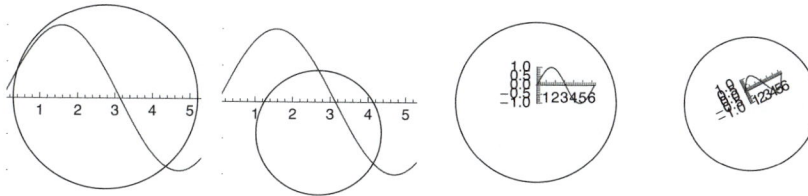

- In the first plot, the center of the plot of sin(x) is at the center of the enclosing graphics.
- In the second plot, the center of the plot of sin(x) is at the point (1, 0.5) of the enclosing graphics.
- In the third plot, the point (0, −1) of the plot of sin(x) is at the point (1, 0) of the enclosing graphics and the size of the plot of sin(x) is 1.2 units in the enclosing graphics.
- In the fourth plot, the slope of the *x* axis of the plot of sin(x) is 1/2.

Next, we use **Inset** in an **Epilog** option:

```
{Plot[Sin[x], {x, 0, 2 π},
  Epilog → Inset[Plot[Cos[x], {x, 0, 2 π}], {2 π, 1}, {Right, Top}, 3.5]],
  Plot[Sin[x], {x, 0, 2 π}, Epilog → Inset[Framed["The graph of the sine function"],
    {2 π, 1}, {Right, Top}, 3.5, Alignment → Right]],
  Plot[Sin[x], {x, 0, 2 π}, Epilog → Text[Framed[Style[
    "The graph of\nthe sine function", TextAlignment → Right]], {2 π, 1}, {1, 1}]]]}
```

- In the first plot, we inset the plot of cos(x) with size 3.5 so that its right top point is at the point (2 π, 1) of the enclosing graphics.
- In the second plot, we inset a text with right alignment.
- In the third plot, we use, instead of **Inset**, the **Text** primitive. Now we have to define the line breaks with **\n** and use **Style** to define the alignment.

GraphicsGroup[{gr1, gr2, … }] (❀6) Represents a collection of graphics objects grouped together

Grouping of graphics objects may be useful in interactive selection of objects. Grouping can also be done by using the menu command **Graphics ▷ Operations ▷ Group**; see Section 5.1.3, p. 131.

6.2.11 3D Primitives and Directives

■ The Structure of 3D Graphics

Thus far, we have considered 2D graphics primitives and directives. Similarly, we have 3D primitives and directives. As for 2D graphics, the importance of the 3D primitives and directives derives from three facts:

- *Mathematica* uses the primitives and directives in the construction of all plots.
- We can construct a plot directly from the primitives and directives.
- The directives can be used to modify plots with options.

The construction of 3D graphics from primitives is done with **Graphics3D**:

Graphics3D[{directives and primitives}, options] Create and show a graphics object from directives and primitives, using some options

Graphics3D has mostly the same options as **Plot3D**, but the default values of **Axes** and **BoxRatios** are **False** and **Automatic**, whereas the default values for **Plot3D** are **True** and **{1, 1, 0.4}**.

Here are two examples. The second example shows that we can use **GraphicsComplex** (see Section 6.2.9, p. 173) in the same way as for 2D graphics:

```
{Graphics3D[{Cuboid[{0, 0, 0}, {1, 1, 1}], Thick, Line[{{0, 0, 1}, {1, 1, 1}}],
    AbsolutePointSize[5], Point[{{0, 0, 1}, {1, 1, 1}}]}, ImageSize → 100],
 Graphics3D[GraphicsComplex[{{0, 0, 0}, {0, 0, 1}, {1, 1, 1}}, {Cuboid[1, 3],
    Thick, Line[{2, 3}], AbsolutePointSize[5], Point[{2, 3}]}], ImageSize → 100]}
```

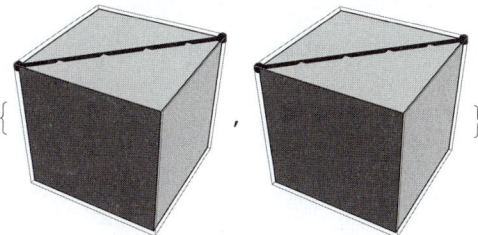

■ 3D Graphics Primitives

The following are all the 3D graphics primitives. With **p**, **p1**, **p2**, and **pn**, we denote a point that consists of the x, y, and z coordinates; an example is **{1, 4, 3}**.

Point[p]   Point at **p**

Point[{p1, …, pn}]   Points at **p1**, …, **pn**

Line[{p1, …, pn}]   Line through points **p1**, …, **pn**

Text[expr, p]   Text **expr** centered at **p**

Text[expr, p, {u, v}]   Text **expr** placed so that the point **{u, v}**, expressed in text coordinates, of **expr** is at the point **p**

Cuboid[p]   Cube with opposite corners **p** and **p** + 1 (**Cuboid[]** means **Cuboid[{0, 0, 0}]**)

Cuboid[p1, p2]   Rectangular parallelepiped with opposite corners **p1** and **p2**

Polygon[{p1, …, pn}]   Polygon with vertices **p1**, …, **pn**

Sphere[p] (❀6)   Sphere of radius 1 centered at **p** (**Sphere[]** means **Sphere[{0, 0, 0}]**)

Sphere[p, r]   Sphere of radius **r** centered at **p**

Cylinder[{p1, p2}] (❀6)   Filled cylinder of radius 1 around the line from **p1** to **p2** (**Cylinder[]** means **Cylinder[{{0, 0, -1}, {0, 0, 1}}]**)

Cylinder[{p1, p2}, r]   Filled cylinder of radius **r** around the line from **p1** to **p2**

For **Text**, note that the point **p** at which the text is placed is a 3D point, but the text coordinates **{u, v}** are 2D. The faces of a **Cuboid** are parallel to the axes. To get other cuboids, use **Rotate**. To get ellipsoids, apply **Scale** to a **Sphere**.

With **Graphics3D**, we can use **GraphicsComplex**, **GraphicsGroup**, and **Inset** as with **Graphics** (see Sections 6.2.9, p. 173, and 6.2.10, p. 175).

Note that the 3D primitives cannot be used in the **Prolog** and **Epilog** options. In these options, we can only use 2D primitives.

■ **3D Graphics Directives**

For 3D primitives, we have all of the directives of point size, thickness, dashing, and color as we do for 2D primitives:

```
PointSize[d], AbsolutePointSize[d]
Thickness[d], AbsoluteThickness[d]
Dashing[{d1, d2, … }], AbsoluteDashing[{d1, d2 ,… }]
GrayLevel[g], Hue[h], RGBColor[r, g, b], CMYKColor[c, m, y, k]
```

In addition, the style of 3D polygons can be adjusted with the following directives:

```
EdgeForm[styles]   Styles of the edges (EdgeForm[] means no edges)
FaceForm[styles]   Styles of both front and back faces (FaceForm[] means no faces)
FaceForm[front styles, back styles]   Styles of front and back faces

Opacity[o] (⌘6)   The opacity
Specularity[col, n] (⌘6)   The specularity (a color and a specular exponent)
Glow[col] (⌘6)   The glow
```

In **FaceForm**, the front face of a polygon is defined to be the one for which the corners, as you specify them, are in counterclockwise order. Here are some examples of how the style of polygons can be adjusted:

```
SetOptions[Graphics3D, ImageSize → 90];

{Graphics3D[Cuboid[]],
 Graphics3D[{Opacity[0.4], Cuboid[]}],
 Graphics3D[{EdgeForm[{Gray, AbsoluteThickness[4]}], Cuboid[]}],
 Graphics3D[{Opacity[0.5], FaceForm[Green, Red], Cuboid[]}]}
```

■ **Polyhedrons**

With **PolyhedronData**, we can get information about many polyhedrons. This command is considered in Section 9.3.1, p. 300. Here are images of some well-known polyhedrons:

```
Show[PolyhedronData[#, "Image"], Boxed → False, ImageSize → 70] & /@
  {"Tetrahedron", "Cube", "Octahedron", "Dodecahedron", "Icosahedron"}
```

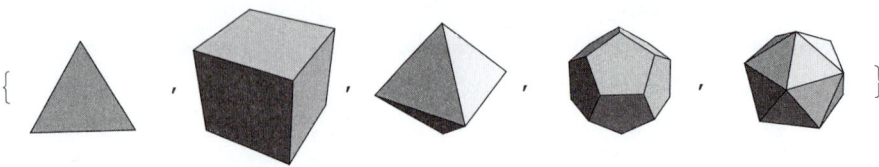

Graphics Options

Introduction

> *Isaac Newton, it seems, was one of the original absentminded mathematicians. He once cut a hole in the bottom of the door of an outhouse to allow his favorite cat easy access. When the cat had kittens, he added a small hole next to the big one.*

We shall now explore in detail all the options of the main 2D plotting commands **Plot**, **ParametricPlot**, and **Graphics** and of the main 3D commands **Plot3D**, **ParametricPlot3D**, **Graphics3D**, **ContourPlot**, and **DensityPlot**.

In Sections 7.1.2 and 7.5.1, we present summaries of the options. The options are classified into *global options*, which modify the plot as a whole, *local options*, which modify separate components of the figure, and *options for the curve or surface*. The varying importance of the options is also shown.

Many options relating to styles such as **PlotStyle**, **BaseStyle**, and **FrameStyle** use *graphics directives*, and the options **Prolog** and **Epilog** also use *graphics primitives*. You may want to study them in Chapter 6.

7.1 Introduction to Options

7.1.1 Using Options

■ **Four Ways to Adjust Options**

In Section 5.1.1, p. 118, we explained how options can be used to modify a plot. Here, we summarize four methods of using options. We use **Plot** as an example of a plotting command.

(a) Setting options in **Plot**:

    Plot[expr, {x, a, b}, opt1 → val1, opt2 → val2, …]

(b) Setting options in **Show**:

    p = Plot[expr, {x, a, b}]
    Show[p, opt1 → val1, opt2 → val2, …]

(c) Giving a name to the options:

    opts = Sequence[opt1 → val1, opt2 → val2, …]
    Plot[expr, {x, a, b}, Evaluate[opts]]

(d) Setting options with **SetOptions**:

    SetOptions[Plot, opt1 → val1, opt2 → val2, …]
    Plot[expr, {x, a, b}]

(a) The first method may be the most convenient in a notebook environment (as in Windows and on a Macintosh). You can first plot the function without any options. If the result is not satisfactory, add an option or several options to the original plotting command and then execute the command anew. Continue adding and modifying the options and executing the plotting command until you are satisfied with the result.

(b) The second method has a certain advantage. Indeed, **Show** does not execute the plotting command anew; only the appearance of the figure is changed. Thus, if the plotting command is time-consuming, it is better to use **Show** to avoid executing the plotting command anew every time.

(c) The third method may be useful if you use the same options (**opts**) for several plots: Write the options once and use the name of the set of options in the subsequent plots.

(d) The fourth method is useful if, during a session, you continuously use certain values for some options. Before using this method, it may be useful to look at the default values with **Options[Plot]** for the case in which you want to return to the default values. The default values can then be set with another **SetOptions** command.

■ **Notes about Show**

If you combine two or more figures with **Show**, you should be aware of the fact that if the figures have different values for the same option, then **Show** takes the value given in the *first* figure; the values given in later figures for this option are disregarded (see Section 5.1.2, p. 125).

Also, with **Show** we can adjust most of the options *but not all of them*. Indeed, with **Show** we can only adjust the options of **Graphics**. For example, **Plot** has the following options that **Graphics** does not have:

ClippingStyle, ColorFunction, ColorFunctionScaling, EvaluationMonitor, Exclusions, ExclusionsStyle, Filling, FillingStyle, MaxRecursion, Mesh, MeshFunctions, MeshShading, MeshStyle, PerformanceGoal, PlotPoints, PlotStyle, RegionFunction, WorkingPrecision.

These options control the sampling algorithm and the style of the curve. Thus, the options mentioned cannot be adjusted with **Show**. Note especially that **PlotStyle** cannot be used with **Show**.

■ **Information about Options**

> **Options[comm]** Give the options and their default values of a command **comm**
> **Options[comm, opt]** Give the default value of an option **opt** of a command **comm**
> **AbsoluteOptions[p]** Give the detailed values of the options used in a plot **p**, even if a value is **Automatic** or **All**
> **AbsoluteOptions[p, opt]** Give the detailed value of an option **opt** used in a plot **p**

Note that for a plot produced by **Plot**, the values of the special options mentioned previously cannot be asked.

Here are all the 55 options and their default values for **Plot**:

```
Style[Options[Plot], 7]
```

$$\Big\{\text{AlignmentPoint} \to \text{Center}, \text{AspectRatio} \to \frac{1}{\text{GoldenRatio}}, \text{Axes} \to \text{True}, \text{AxesLabel} \to \text{None},$$

AxesOrigin → Automatic, AxesStyle → {}, Background → None, BaselinePosition → Automatic,
BaseStyle → {}, ClippingStyle → None, ColorFunction → Automatic, ColorFunctionScaling → True,
ColorOutput → Automatic, ContentSelectable → Automatic, DisplayFunction :→ $DisplayFunction, Epilog → {},
Evaluated → Automatic, EvaluationMonitor → None, Exclusions → Automatic, ExclusionsStyle → None,
Filling → None, FillingStyle → Automatic, FormatType :→ TraditionalForm, Frame → False,
FrameLabel → None, FrameStyle → {}, FrameTicks → Automatic, FrameTicksStyle → {}, GridLines → None,
GridLinesStyle → {}, ImageMargins → 0., ImagePadding → All, ImageSize → Automatic, LabelStyle → {},
MaxRecursion → Automatic, Mesh → None, MeshFunctions → {#1 &}, MeshShading → None, MeshStyle → Automatic,
Method → Automatic, PerformanceGoal :→ $PerformanceGoal, PlotLabel → None, PlotPoints → Automatic,
PlotRange → {Full, Automatic}, PlotRangeClipping → True, PlotRangePadding → Automatic, PlotRegion → Automatic,
PlotStyle → Automatic, PreserveImageOptions → Automatic, Prolog → {}, RegionFunction → (True &),
RotateLabel → True, Ticks → Automatic, TicksStyle → {}, WorkingPrecision → MachinePrecision}

The default value of a certain option or a list of options can also be displayed:

```
Options[Plot, PlotRange]
```

{PlotRange → {Full, Automatic}}

Remember also that we can ask for information about an option by typing, for example, **?PlotRange** or using the Documentation Center (see Sections 1.4.2, p. 17, and 1.4.3, p. 20).

Some common values of options are as follows. Most default values of the options of **Plot** are **Automatic**. This value means that there is a special value chosen by **Plot** according to certain rules. Some options, such as **AxesLabel**, have the value **None**, meaning that the plot does not have the corresponding components. Some options, such as **Frame**, have the value **True** or **False**, which tells whether or not something is present. **PlotRange** can have the value **All**, meaning that the whole plot is shown.

We can also ask for the options of a given plot:

```
p = Plot[Sin[x], {x, 0, 2 π}]
```

```
AbsoluteOptions[p, AxesStyle]
```

```
{AxesStyle → {{GrayLevel[0.], AbsoluteThickness[0.25]},
    {GrayLevel[0.], AbsoluteThickness[0.25]}}}
```

■ Comparing Options

We can easily compare the options of various plotting commands. For example, **LogPlot** has exactly the same options and default values as **LogLinearPlot**:

```
Options[LogPlot] == Options[LogLinearPlot]
```

```
True
```

The options of **Plot** and **Graphics** have some differences:

```
c = Options[Plot]; g = Options[Graphics];
```

```
Style[Intersection[c, g], 7]
```

```
{AlignmentPoint → Center, AxesLabel → None, AxesOrigin → Automatic, AxesStyle → {}, Background → None,
 BaselinePosition → Automatic, BaseStyle → {}, ColorOutput → Automatic, ContentSelectable → Automatic,
 Epilog → {}, Frame → False, FrameLabel → None, FrameStyle → {}, FrameTicks → Automatic,
 FrameTicksStyle → {}, GridLines → None, GridLinesStyle → {}, ImageMargins → 0., ImagePadding → All,
 ImageSize → Automatic, LabelStyle → {}, Method → Automatic, PlotLabel → None, PlotRangePadding → Automatic,
 PlotRegion → Automatic, PreserveImageOptions → Automatic, Prolog → {}, RotateLabel → True,
 Ticks → Automatic, TicksStyle → {}, DisplayFunction ⧴ $DisplayFunction, FormatType ⧴ TraditionalForm}
```

```
Style[Complement[c, g], 7]
```

```
                      1
{AspectRatio → ──────────── , Axes → True, ClippingStyle → None, ColorFunction → Automatic,
                 GoldenRatio
 ColorFunctionScaling → True, Evaluated → Automatic, EvaluationMonitor → None, Exclusions → Automatic,
 ExclusionsStyle → None, Filling → None, FillingStyle → Automatic, MaxRecursion → Automatic,
 Mesh → None, MeshFunctions → {#1 &}, MeshShading → None, MeshStyle → Automatic, PlotPoints → Automatic,
 PlotRange → {Full, Automatic}, PlotRangeClipping → True, PlotStyle → Automatic,
 RegionFunction → (True &), WorkingPrecision → MachinePrecision, PerformanceGoal ⧴ $PerformanceGoal}
```

```
Style[Complement[g, c], 7]
```

```
{AspectRatio → Automatic, Axes → False, PlotRange → All, PlotRangeClipping → False}
```

Complement[c, g] gives the elements of **c** that are not in **g**. Thus, the default values of **AspectRatio**, **Axes**, **PlotRange**, and **PlotRangeClipping** are different for the two plotting commands, and **Graphics** does not have the special options of **Plot** we mentioned previously.

The options of **Plot** and **ParametricPlot** also have some differences:

```
c = Options[Plot]; p = Options[ParametricPlot];
```

```
Complement[c, p]
```

$$\left\{\text{AspectRatio} \to \frac{1}{\text{GoldenRatio}}, \text{ClippingStyle} \to \text{None},\right.$$
$$\text{Filling} \to \text{None}, \text{FillingStyle} \to \text{Automatic}, \text{Frame} \to \text{False},$$
$$\left.\text{Mesh} \to \text{None}, \text{MeshFunctions} \to \{\#1 \&\}, \text{PlotRange} \to \{\text{Full}, \text{Automatic}\}\right\}$$

```
Complement[p, c]
```

$$\{\text{AspectRatio} \to \text{Automatic}, \text{BoundaryStyle} \to \text{Automatic}, \text{Frame} \to \text{Automatic},$$
$$\text{Mesh} \to \text{Automatic}, \text{MeshFunctions} \to \text{Automatic}, \text{PlotRange} \to \text{Automatic}\}$$

Thus, the default values of **AspectRatio**, **Frame**, **Mesh**, **MeshFunctions**, and **PlotRange** are different, and **ParametricPlot** does not have the options **ClippingStyle**, **Filling**, and **FillingStyle**.

■ Example

With the help of options we can get interesting results, such as the following. This plot is overdone, but our aim is simply to show what can be done with options.

```
Plot[Sin[x], {x, 0, 2 π}, PlotRange → {{-0.7, 2 π + 0.7}, {-1.6, 1.6}},
  PlotRegion → {{-0.05, 1}, {0.04, 0.94}}, ImageSize → {280, 190},
  Background → Black, BaseStyle → {9, White, FontFamily → "Helvetica"},
  PlotStyle → Directive[White, Thick],
  PlotLabel → Style["An interesting wave", Bold, 12], Frame → True,
  FrameLabel → {Style[x, White], Style[Sin[x], White]},
  RotateLabel → False, FrameStyle → Directive[Gray, Thick, Bold],
  FrameTicks → {{0, {π / 2, "π/2"}, π, {3 π / 2, "3π/2"}, 2 π}, {-1, 0, 1}, None, None},
  FrameTicksStyle → Directive[White, Thin], Epilog → {
    Text[Style["Maximum point", Bold, 7], {π / 2, 1}, {-1, -1.3}],
    Text[Style["Minimum point", Bold, 7], {3 π / 2, -1}, {1, 2}],
    Text[Style["Point of inflection", Bold, 7], {3.5, 0.4}, {-1, -1}],
    Arrowheads[0.03], Arrow[{{3.55, 0.4}, {π + 0.1, 0.1}}],
    PointSize[Medium], Gray, Point[{{π / 2, 1}, {π, 0}, {3 π / 2, -1}}]}]
```

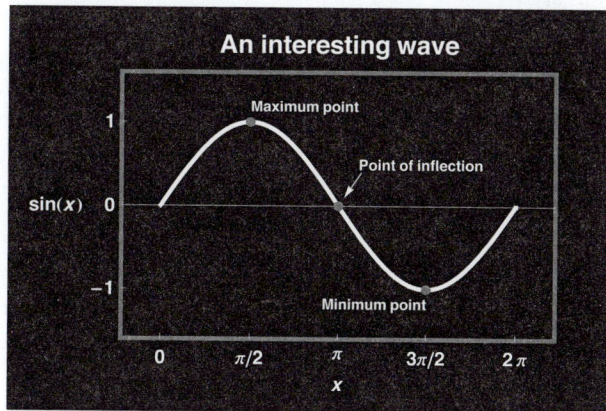

7.1.2 Summary

■ **Introduction**

Here, we list all the options of **Plot** (see Section 5.1.1, p. 116), **ParametricPlot** (see Section 5.2.1, p. 132), and **Graphics** (see Section 6.1.1, p. 153), with short descriptions and some common values. The default value of an option is mentioned first, and after that we mention either all other possible values or some examples of other values (the examples are simple; more advanced forms may exist).

The options are divided into three groups:

- *Global options:* These are options to modify global aspects of the plot—that is, to adjust how the plot looks in general. This means, for example, adjusting the form, size, plot range, various margins, or fonts.
- *Local options:* These are options to modify local components of a plot such as plot label, axes, ticks, frame, grid lines, or add-ons.
- *Options for the curve:* These are options to modify the curve produced by **Plot** or **ParametricPlot**. This means, for example, adjusting the style, filling, mesh, or exclusions or controlling the plotting algorithm.

All of the options in the first two groups are common to all three plotting commands mentioned. However, options in the third group are applicable only with **Plot** or **ParametricPlot**. Options and their default values applicable only to certain commands are expressed in the lists in this section by a superscript after the option name or value:

^{c}: applicable to **Plot** (shorthand for *c*urve)
^{p}: applicable to **ParametricPlot** (shorthand for *p*arametric)
^{g}: applicable to **Graphics** (shorthand for *g*raphics)

For example, **PlotStyle** ^{cp} means that this option applies to **Plot** and **ParametricPlot** but not to **Graphics**, and **False** ^{g} means that the default value of the option in question (**Axes**) is **False** for **Graphics**.

Three options assumed to be the most important are marked with two asterisks (**). Nine options assumed to be less important are marked with one asterisk (*). A single parenthesis indicates options that most of us will seldom need. The remaining options, without any special markings, may sometimes be useful. Of course, the given classification of the options according to importance reflects my personal impression. You may well have a different classification.

■ **Global Options**

The following options can be used to adjust some global aspects of a plot.

Options for form and size:

** **AspectRatio** Ratio of height to width of the plotting rectangle; examples of values:
  1/GoldenRatio ^c, **Automatic** ^{g p}, **Full**, **0.4**

ImageSize The absolute size (in printer's points) of the plot; examples of values: **Automatic**, **Full**,
  All, **width**, **{width, height}**, **{maxsize}**, **{{maxwidth}, {maxheight}}**

Options for plot range:

** **PlotRange** Ranges for *x* and *y* in the plot; examples of values:

Automatic ^p	**{ymin, Automatic}**	**{Full, Automatic}** ^c
All ^g	**{ymin, All}**	**{{xmin, xmax}, Automatic}**
Full	**{All, ymax}**	**{Full, {ymin, Automatic}}**
5	**{ymin, ymax}**	**{{xmin, xmax}, {ymin, ymax}}**

(**PlotRangeClipping** (⌘6) Whether graphics objects should be clipped at the edge of the region
  defined by **PlotRange** or should be allowed to extend to the selection rectangle; possible values:
  True ^{c p}, **False** ^g)

Options for margins:

PlotRegion Specifies margins around the plot inside the selection rectangle; examples of values:
  Automatic (means **{{0, 1}, {0, 1}}**), **{{xmin, xmax}, {ymin, ymax}}**

(**ImageMargins** Specifies margins (in printer's points) outside the selection rectangle; examples of
  values: **0**, **Automatic**, **15**, **{{left, right}, {bottom, top}}**)

Background Color of the background; examples of values: **None**, **LightGray**

(**ImagePadding** (⌘6) Extra space (in printer's points) for objects such as thick lines and tick and axes
  labels; examples of values: **All**, **None**, **15**, **{{left, right}, {bottom, top}}**)

(**PlotRangePadding** (⌘6) How much farther axes etc. should extend beyond the range of coordinates specified by **PlotRange**; examples of values: **Automatic**, **None**, **{0.2, 0.5}**)

Options for fonts and formatting:

* **BaseStyle** (⌘6) Style of all texts; examples of values: **{}**, **{9, "Bold", FontFamily → "Arial"}**

LabelStyle (⌘6) Style of all labels; examples of values: **{}**, **{9, "Bold", FontFamily → "Arial"}**

(**FormatType** Format type of text used in a plot; examples of values: **TraditionalForm**,
  StandardForm, **InputForm**, **OutputForm**)

Miscellaneous options:

(**BaselinePosition** (⌘6) Where the baseline of a plot should be if the plot is combined with other
  plots or text; examples of values: **Automatic**, **Axis**, **Bottom**, **Top**, **Center**, **Baseline**)

(**AlignmentPoint** (⌘6) How objects should by default be aligned when they appear in **Inset**;
  default value: **Center**)

(**ContentSelectable** (⌘6) Whether and how content of a plot should be selectable; possible values:
  Automatic (double-click allows content selection), **True** (single clicks immediately select content
  objects), **False** (content objects cannot be selected))

(**PreserveImageOptions** (⌘6) Whether the size and margins of a plot should remain the same if the
  plotting command is executed anew; examples of values: **Automatic** (the properties should remain
  the same if not explicitly otherwise stated), **True** (the properties should remain the same), **False**
  (the previous properties are ignored))

(**DisplayFunction** Function to apply to a graphic; default value: **$DisplayFunction**)

■ **Local Options**

The following options can be used to adjust some local components of a plot.

An option for plot label:

PlotLabel   Label of the plot; examples of values: **None**, **Sin[x/2]**, **"Function** $g(x)$**"**

Options for axes and ticks:

* **Axes**   Whether to draw the axes; examples of values: **True** ^{cp}, **False** ^{g}, **{True, False}**

* **AxesOrigin**   Point where the axes cross; examples of values: **Automatic**, **{0, 0}**

AxesLabel   Labels for the axes; examples of values: **None**, **y**, **{x, None}**, **{x, y}**

(**AxesStyle**   Style of the axes, axes labels, tick marks, and tick labels; examples of values: **{}**, **Thick**,
      Blue, **Arrowheads[0.07]**, **Directive[Thick, Blue, 12, Italic]**

** **Ticks**   Tick on the axes; simple examples of values: **Automatic**, **None**, **{{π, 2π}, Automatic}**,
      {Automatic, {-1, 0, 1}}, **{{π, 2π}, {-1, 0, 1}}**

TicksStyle (⌘6)   Style of tick marks and tick labels; examples of values: **{}**, **Blue**,
      Directive[Thick, Blue, 12]

Options for frame and frame ticks:

* **Frame**   Whether to draw a frame; examples of values: **False**, **True**, **{True, True, False, False}**

* **FrameLabel**   Labels for the frame; examples of values: **None**, **{x, y}**, **{"bottom", "left", "top",**
      "right"}

* **RotateLabel**   Whether to rotate the labels for the vertical edges; possible values: **True**, **False**

FrameStyle   Style of the frame, frame labels, frame tick marks, and frame tick labels; examples of
      values: **Automatic**, **Blue**, **Directive[Thick, Red, 12, Italic]**

* **FrameTicks**   Frame tick marks on the frame; simple examples of values: **Automatic**, **None**, **All** (on
      all edges), **{{0, π, 2π}, {-1, 1}, None, None}**

FrameTicksStyle (⌘6)   Style of frame tick marks and frame tick labels; examples of values: **{}**, **Red**,
      Directive[Thick, Red, 12, Bold]

Options for grid lines:

GridLines   How the grid lines are drawn; simple examples of values: **None**, **Automatic**, **{None,**
      Automatic}, **{None, {-1,0,1}}**, **{{0,1,2}, {-1,0,1}}**

(**GridLinesStyle** (⌘6)   Style of the grid lines; examples of values: **{}**, **Dashed**,
      Directive[LightGray, Dashed])

Options for primitives:

(**Prolog**   Graphics primitives to be plotted before the main plot; examples of values: **{}**, **{Red,**
      PointSize[Medium], Point[{3, 2}]})

* **Epilog**   Graphics primitives to be plotted after the main plot; examples of values: **{}**, **{Red,**
      PointSize[Medium], Point[{3, 2}]}

■ **Options for the Curve**

With **Plot** and **ParametricPlot**, the following options can be used to adjust the curve to be plotted.

Options for plot style:

* **PlotStyle** ^{cp} Style(s) of the curve(s); examples of values: **Automatic, Thickness[Medium]**,
  Directive[Thick, Red, Dashed]

(**ClippingStyle** ^{c} (⌘6) How to indicate clipped parts that fall outside of the plot range; examples
  of values: **None, Automatic, Red, {Blue, Red}**)

Options for color function:

(**ColorFunction** ^{cp} (⌘6) Function that determines the color of the curve; examples of values:
  Automatic, (Hue[#2] &), (RGBColor[#2, 0, 1 - #2] &), "Rainbow")

(**ColorFunctionScaling** ^{cp} (⌘6) Whether arguments to a color function should be scaled to lie
  between 0 and 1; examples of values: **True, False**)

Options for filling:

Filling ^{c} (⌘6) Type of filling to use; examples of values: **None, Axis, Bottom, Top, 0.3, True**

(**FillingStyle** ^{c} (⌘6) Style of filling; examples of values: **Automatic, Red, {Blue, Red}** (different
  style for negative and positive values), **Directive[Opacity[0.3], Blue]**)

Options for mesh:

Mesh ^{cp} How many mesh points should be drawn; examples of values: **None** ^{c}, **Automatic** ^{p}, **10**,
  Full, All, {{0}}

MeshFunctions ^{cp} (⌘6) How to determine the placement of the mesh points; examples of values:
  {#1 &} ^{c}, **Automatic** ^{p}, **{#2 &}**

MeshStyle ^{cp} The style of mesh points; examples of values: **Automatic, PointSize[Small]**,
  Directive[Red, PointSize[Medium]]

(**MeshShading** ^{cp} (⌘6) How to shade regions between mesh points; examples of values: **None, {Red,
  Blue}**)

Options for exclusions:

Exclusions ^{cp} (⌘6) The *x* points that are excluded in plotting; examples of values: **Automatic,
  None, {π/2, 3 π/2}, {Cos[x] == 0}**

ExclusionsStyle ^{cp} (⌘6) What to draw at excluded points; examples of values: **None,
  Directive[Blue, Dashed], {None, Directive[Red, PointSize[Medium]]}**

(**RegionFunction** ^{cp} (⌘6) Specifies the region to include in the plot drawn; examples of values:
  (True &), (Abs[#2] > 0.7&))

Options for plotting algorithm:

PlotPoints ^{cp} Number of initial sampling points; examples of values: **Automatic, 100**

MaxRecursion ^{cp} (⌘6) The maximum number of recursive subdivisions allowed; examples of
  values: **Automatic, 8**

WorkingPrecision ^{cp} (⌘6) The precision used in computations; examples of values:
  MachinePrecision, 20

(**Evaluated** ^{cp} (⌘6) Whether the expression to be plotted is evaluated before the expression is
  sampled; possible values: **Automatic, True, False**)

(**PerformanceGoal** ^{cp} (⌘6) What aspect of performance to try to optimize; examples of values:
  $PerformanceGoal, "Quality", "Speed")

(**EvaluationMonitor** ^{cp} (⌘6) Expression to evaluate at every function evaluation; examples of
  values: **None, Sow[{x, Sin[x]}]**)

In various style options we can use graphics primitives such as `Point`, `Line`, and `Text` and graphics directives such as `AbsolutePointSize`, `Thick`, `Dashed`, `Gray`, or `Red`. These are explained in Chapter 6.

All of the options are explained in detail in the remaining sections of this chapter, after we have considered the combination of various styles. Note that in the **PlotLegends`** package we have options for legends.

■ Combining Styles

The styles of various components of a plot can be defined with several options: `PlotStyle`, `ClippingStyle`, `FillingStyle`, `MeshStyle`, `ExclusionsStyle`, `LabelStyle`, `AxesStyle`, `TicksStyle`, `FrameStyle`, `FrameTicksStyle`, and `GridLinesStyle`. As an example, consider `PlotStyle`:

```
GraphicsRow[{Plot[{Sin[x], Cos[x]}, {x, 0, 2 π}, PlotStyle → Thick],
   Plot[{Sin[x], Cos[x]}, {x, 0, 2 π}, PlotStyle → {Thick, Blue}],
   Plot[{Sin[x], Cos[x]}, {x, 0, 2 π}, PlotStyle → {{Thick, Blue}}],
   Plot[{Sin[x], Cos[x]}, {x, 0, 2 π},
    PlotStyle → {{Thick, Blue}, {Thick, Red}}]}], ImageSize → 420]
```

In these four plots, examine the use of braces in the value of `PlotStyle`. In the first plot, we have only one directive, `Thick`, and it is applied for both curves. In the second plot, we have two directives inside single braces, and now the first directive, `Thick`, defines the style of the first curve and the second directive, `Blue`, the style of the second curve. If we want both curves to be blue and thick, we have to use double braces, as is done in the third plot. If we want unique styles for both curves, we use a nested list for the directives, as can be seen in the fourth plot.

It may be difficult to remember when to use single braces and when to use double braces. The use of `Directive` may then be helpful.

> `Directive[dir1, dir2, … ]` (⌘6)   Represents a single graphics directive composed of the given directives

Thus, with `Directive` we can collect together several directives and use the collection like a single directive. The third and fourth plots can now be written as follows:

```
{Plot[{Sin[x], Cos[x]}, {x, 0, 2 π}, PlotStyle → Directive[Blue, Thick]],
  Plot[{Sin[x], Cos[x]}, {x, 0, 2 π},
   PlotStyle → {Directive[Blue, Thick], Directive[Red, Thick]}]}
```

As can be seen, with `Directive` we need single braces only if we define different styles for each of several curves, and double braces are not needed at all. As another example, consider defining the style of a frame:

```
{Plot[Sin[x], {x, 0, 2 π}, Frame → True,
  FrameStyle → {{Thick, Blue}, {Thick, Blue}, {Thick, Blue}, {Thick, Blue}}],
  Plot[Sin[x], {x, 0, 2 π}, Frame → True, FrameStyle → Directive[Thick, Blue]]}
```

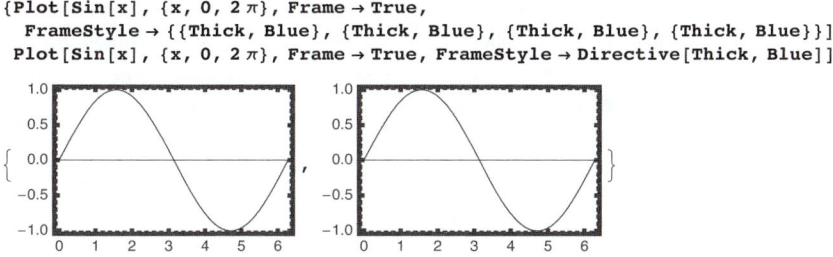

In the first plot, we had to separately define the style of each edge of the frame, whereas in the second plot the use of **Directive** enables us to define the style only once.

We will use **Directive** in this book to collect several directives together, but if you are comfortable with the braces, they can often also be used.

7.2 Options for Form, Ranges, and Fonts

7.2.1 Form and Size

> ** **AspectRatio** Ratio of height to width of the plotting rectangle; examples of values:
> **1/GoldenRatio** ^{c}, **Automatic** ^{gp}, **Full**, **0.4**

The default value **1/GoldenRatio** = 0.618 of **AspectRatio** for **Plot** gives an aesthetically pleasing form. The default value **Automatic** of **AspectRatio** for **Graphics** and **ParametricPlot** sets one unit on the x axis to have the same length as one unit on the y axis. The aspect ratio can be any positive real number. The value **Full** means that the graphic should be stretched so as to fill out its enclosing region in **Grid** or related construct. Next, we plot a circle with each of the three commands. The curve produced by **Plot** looks like an ellipse because of the aspect ratio 0.618:

```
GraphicsRow[{Plot[{Sqrt[1 - x^2], -Sqrt[1 - x^2]}, {x, -1, 1}],
  ParametricPlot[{Cos[t], Sin[t]}, {t, 0, 2 π}],
  Graphics[Circle[], Axes → True]}, ImageSize → 330]
```

> **ImageSize** The absolute size (in printer's points) of the plot; examples of values: **Automatic**, **Full**,
> **All**, **width**, **{width, height}**, **{maxsize}**, **{{maxwidth}, {maxheight}}**

The size of the plot is easy to change with the mouse, but we can also use the **ImageSize** option. It determines the absolute size of the plot in units of printer's points (1/72 inch). One number as the value of the option defines the width, and a list of two numbers determines both the width and the height. Note that if both the width and the height are specified, the plot fills this area only if the aspect ratio is exactly **height/width**. The default size can be changed with the Option Inspector.

The value **Full** for **ImageSize** means that on the screen, the size of the graphic is automatically adjusted to fit the window and, when printed, the graphic has the full width of the content area of the page. The aspect ratio of a plot is kept fixed when resizing, unless **AspectRatio** is **Full**. The value of **ImageSize** can also be **Tiny**, **Small**, **Medium**, or **Large**.

7.2.2 Plot Range

** **PlotRange** Ranges for x and y in the plot; examples of values:		
Automatic [p]	**{ymin, Automatic}**	**{Full, Automatic}** [c]
All [g]	**{ymin, All}**	**{{xmin, xmax}, All}**
Full	**{All, ymax}**	**{Full, {ymin, Automatic}}**
5	**{ymin, ymax}**	**{{xmin, xmax}, {ymin, ymax}}**

When showing a plot, *Mathematica* normally displays all values of the function in the given interval. However, if the function takes on very small or very large values on a small interval, *Mathematica* may decide to cut such values away from the plot so that the remaining parts of the function can be seen more clearly; this may happen if the value **Automatic** is used for **PlotRange**.

To see the whole function, use the value **All**. The value **Full** also causes the whole function to be plotted but, in addition, the whole plotting range on x axis is included in the plot. A constant value such as **5** means, for **Plot**, the plot range **{Full, {-5, 5}}** and, for **Graphics** and **ParametricPlot**, the plot range **{{-5, 5}, {-5, 5}}**.

```
GraphicsRow[{Plot[Exp[x], {x, -1, 10}],
  Plot[Exp[x], {x, -1, 10}, PlotRange → All], Plot[Sqrt[x], {x, -1, 1}],
  Plot[Sqrt[x], {x, -1, 1}, PlotRange → All]}, ImageSize → 420]
```

In the first plot, we use the default value **{Full, Automatic}** of **PlotRange** for **Plot**. The y range definition **Automatic** has, in this example, caused the function to be shown only up to approximately $x = 9$. The x range definition **Full** means that the whole x range (0, 10) is shown. In the second example, the value **All** is used to show the whole function up to $x = 10$.

In the third plot, we again use the default value **{Full, Automatic}**. Thus, the whole x plotting range is included, although the function is not defined on $[-1, 0)$. In the fourth plot, we use the value **All**, and then all the points where the function is defined (and only these points) are shown.

The **ClippingStyle** option can be used to define the style for how the clipped parts are displayed; see Section 7.4.1, p. 203.

(**PlotRangeClipping** (⌘6)   Whether graphics objects should be clipped at the edge of the region defined by **PlotRange** or should be allowed to extend to the selection rectangle; possible values: **True** [cp], **False** [g])

7.2.3 Margins and Background

> **PlotRegion** Specifies margins around the plot inside the selection rectangle; examples of values:
>   **Automatic** (means **{{0, 1}, {0, 1}}**), **{{xmin, xmax}, {ymin, ymax}}**
> (**ImageMargins** Specifies margins (in printer's points) outside the selection rectangle; examples of
>   values: **0**, **Automatic**, **15**, **{{left, right}, {bottom, top}}**)
> **Background** Color of the background; examples of values: **None**, **LightGray**

Each plot has a *display area* that can be seen by clicking the plot: A rectangle around the plot appears. The curve normally fills this area, but with **PlotRegion** we can define other ways for the curve to be placed in the display area. The plot region is given in scaled coordinates ranging from 0 to 1 in each direction. The default setting **Automatic** is the same as **{{0, 1}, {0, 1}}**, which is the whole display area. By specifying other values (values less than 0 and greater than 1 are allowed), we can adjust the margins around the curve in the display area. For example, plots with gray or colored backgrounds often look better with a somewhat reduced plot region (i.e., with larger margins). The second plot shown here has a wider margin around the plot than the first plot. In the third plot, the margin is outside the selection rectangle.

```
GraphicsRow[
  {Plot[Sin[x], {x, 0, 2 π}, Background → LightGray, PlotRegion → Automatic],
   Plot[Sin[x], {x, 0, 2 π}, Background → LightGray,
    PlotRegion → {{0.1, 0.9}, {0.1, 0.9}}], Plot[Sin[x], {x, 0, 2 π},
    Background → LightGray, ImageMargins → 10]}, ImageSize → 420]
```

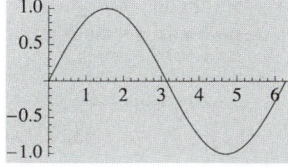 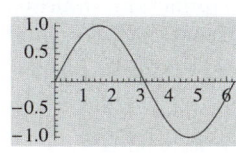 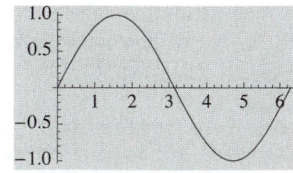

> (**ImagePadding** (⌘6) Extra space (in printer's points) for objects such as thick lines and tick and axes
>   labels; examples of values: **All**, **None**, **15**, **{{left, right}, {bottom, top}}**)
> (**PlotRangePadding** (⌘6) How much further axes etc. should extend beyond the range of coordi-
>   nates specified by **PlotRange**; examples of values: **Automatic**, **None**, **{0.2, 0.5}**)

The first plot below shows that usually the axes are a little extended (2% in each direction) from the given *x* range and the range of *y* values. In the second plot, we use no padding, and in the third plot we use unusually wide padding. No padding is used if an explicit plot range is used.

```
Plot[Sin[x], {x, 0, 2 π}, PlotRangePadding → #] & /@ {Automatic, None, {0.5, 0.2}}
```

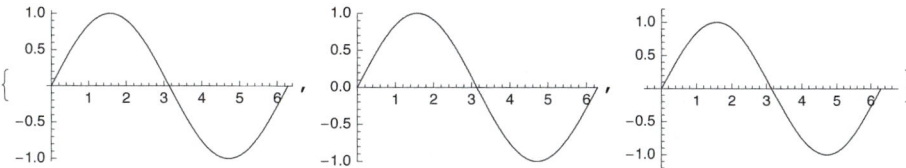

7.2.4 Fonts

All text in a plot is, by default, written with the Times font of size 10. We can change the font properties at various levels: for all plots during a session, for all texts in a single plot, for all labels in a single plot, and for a text item inside a plot. Here is a summary:

Setting font properties at various levels:

For all plots in a session: `SetOptions[Plot, BaseStyle` → {font directives and options}`]`
For all texts in a plot: `BaseStyle` → {font directives and options}
For all labels in a plot: `LabelStyle` → {font directives and options}
For a text item: `Style[expr, {font directives and options}]`

Note that the more specific style options `AxesStyle`, `TicksStyle`, `FrameStyle`, and `FrameTicksStyle` can be used to adjust fonts in axes labels, tick labels, frame labels, and frame tick labels. See Section 7.3.5, p. 200, for a discussion of the hierarchy of styles on plots. Next, we consider `BaseStyle`, `LabelStyle`, `SetOptions`, and `Style` in detail.

■ **Setting Font Properties for All Texts or Labels in a Plot**

---

* **BaseStyle** → {font directives and options} (✿6)

The font styles given in `BaseStyle` apply for all texts in a plot: for plot label, axes labels, tick labels, frame labels, frame tick labels, and `Text` primitives. In defining the style of font, we can use font directives such as **7** (font size), `Bold`, and `Italic` and font options such as `FontFamily` and `FontColor`. Examples:

```
BaseStyle → 7
BaseStyle → {7, Bold, Italic}
BaseStyle → {7, Bold, FontFamily → "Helvetica"}
```

In the next example, we use bold Helvetica in size 7:

```
Plot[Sin[x], {x, 0, 2 π}, AxesLabel → {x, Sin[x]},
  BaseStyle → {7, Bold, FontFamily → "Helvetica"}]
```

The next boxes give font directives and font options.

Font directives:

n, `Tiny, Small, Medium, Large, Smaller, Larger`

`Bold, Italic, Underlined, Plain`

`Red, LightBlue, GrayLevel[0.3]`, etc.

> *Font options:*
>
> **FontFamily**; examples of values: **"Times"**, "Helvetica", "Courier"
>
> **FontColor**; examples of values: **Red, Gray**
>
> **FontTracking**; examples of values: **"Narrow"**, **"Condensed"**, **"SemiCondensed"**, **"Extended"**, **"Wide"**
>
> **Background**; examples of values: **LightGray, LightBlue**

Instead of the directives, we can use the options **FontSize**, **FontWeight**, **FontSlant**, and **FontColor**. If we want to define a color for all text with **BaseStyle**, a definition such as **BaseStyle → Red** causes the axes and labels to also be colored. Thus, font color should be defined by using **FontColor**, as in **BaseStyle → {FontColor → Red}**.

If you export *Mathematica* figures in, for example, EPS form into another application such as TeX and encounter problems with special characters appearing in the plots, one solution may be to render the document in the front end (and not in the printer). To do this, choose **Format ▷ Option Inspector...**, set *Show option values for* to be *notebook*, then go to **Notebook Options ▷ Printing Options ▷ PrintingOptions** and change the value of **GraphicsPrintingFormat** from **Automatic** to **RenderInFrontEnd**. Now when you print a document, it is rendered in the front end.

> **LabelStyle → {**font directives and options**}** (⌘6)

LabelStyle applies for all labels in a plot (i.e., for plot labels, axes labels, tick labels, frame labels, and frame tick labels) but not for **Text** primitives.

■ Setting Font Properties for a Session

Usually, the same style of text is used for all plots in a document or in a session. The style definitions need not be done for each plot separately because we can set, with **SetOptions**, the value of **BaseStyle** so that the given value is used during the rest of the current session.

> **SetOptions[Plot, BaseStyle → {**font directives and options**}]**

For example, to use the Helvetica font of size 8 during a session, simply execute, at the beginning of the session, the following command:

    SetOptions[Plot, BaseStyle → {8, FontFamily → "Helvetica"}]

If you want to use the same text style for several plotting commands, write, for example,

    SetOptions[{Plot, ListPlot, ListLinePlot},
      BaseStyle → {8, FontFamily → "Helvetica"}]

The font properties can also be defined in the **init.m** file; the definitions then hold automatically for all sessions, unless you change the definitions (see Section 4.1.1, p. 96).

■ Setting Font Properties for a Text Item

Sometimes we want to use varying font properties for several text items in a plot. We can then define the font used most often with the **BaseStyle** and use **Style** in places where we want to use special font properties.

> **Style[expr, {**font directives and options**}]** (⌘6)

The font directives and options need not even be given as a list; they can simply be written separated by commas. For colors, we need not use the **FontColor** option; just write the color specification. Here are some examples:

```
f = Exp[x] ArcTan[x] / Sqrt[x];
```

```
{Plot[f, {x, 0, 5}, PlotLabel → Framed[Style[f, 8, Blue]], ImageSize → 110],
 Plot[f, {x, 0, 5}, AxesLabel → (Style[#, 8, Red] & /@ {x, f})],
 Plot[f, {x, 0, 5}, Epilog → Text[Style[f, 8, Background → LightGray], {2.7, 73}]]}
```

■ Setting Font Properties in a Style Sheet

The font style in graphics can also be defined in a style sheet. In Section 3.2.2, p. 63, we showed how to change styles. Using this technique, we can change the style of Graphics cells.

- Choose **Edit Stylesheet…** from the **Format** menu. A notebook appears with the title *Private Style Definitions for ….*
- In the *Enter a style name* input field, write Graphics; the corresponding style definition cell appears in the document. Edit this cell with the **Format** menu by defining, for example, a font.
- Thus far, the new Graphics style holds for the current notebook. You can also save the style sheet so that you can use it for any notebook (see Section 3.2.2).

■ Formatting

(**FormatType** Format type of text used in a plot; examples of values: **TraditionalForm**,
StandardForm, **InputForm**, **OutputForm**)

These types of formatting are considered in Section 3.3.1, p. 70. The default is to use traditional formatting—that is, the type of formatting used in traditional mathematical typesetting: The font is Times, all variables are in italic, and all formulas are typeset according to traditional mathematical notation. These properties can be seen in the first plot:

```
Plot[f, {x, 0, 5}, AxesLabel → {x, y}, PlotLabel → f,
    FormatType → #, ImageSize → 110] & /@ {TraditionalForm, StandardForm}
```

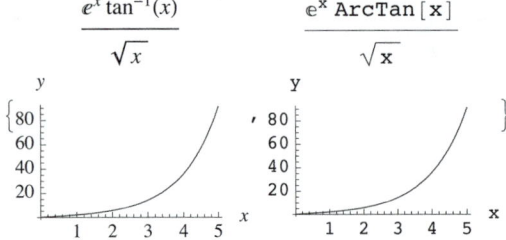

7.2.5 Miscellaneous Options

(**BaselinePosition** (⌘6)   Where the baseline of a plot should be if the plot is combined with other
  plots or text; examples of values: **Automatic, Axis, Bottom, Top, Center, Baseline**)
(**AlignmentPoint** (⌘6)   How objects should by default be aligned when they appear in **Inset**;
  default value: **Center**)
(**ContentSelectable** (⌘6)   Whether and how content of a plot should be selectable; possible values:
  Automatic (double-click allows content selection), **True** (single clicks immediately select content
  objects), **False** (content objects cannot be selected))
(**PreserveImageOptions** (⌘6)   Whether the size and margins of a plot should remain the same if the
  plotting command is executed anew; examples of values: **Automatic** (the properties should remain
  the same if not explicitly otherwise stated), **True** (the properties should remain the same), **False**
  (the previous properties are ignored))
(**DisplayFunction**   Function to apply to a graphic; default value: **$DisplayFunction**

With **BaselinePosition** we can ask to align several plots according to the x axis:

```
Row[{Plot[Sin[x], {x, 0, 2 π}, BaselinePosition → Axis, ImageSize → 120],
    Plot[Exp[x], {x, 0, 3}, BaselinePosition → Axis, ImageSize → 120],
    Plot[Log[x], {x, 0, 3}, BaselinePosition → Axis, ImageSize → 120]}, "    "]
```

7.3 Options for Axes, Frames, and Primitives

7.3.1 Plot Label

PlotLabel   Label of the plot; examples of values: **None, Sin[x/2], "Function g(x)"**

With **Style** we can modify the style of the plot label:

```
{Plot[Sin[x / 2], {x, 0, 4 π}, PlotLabel → Sin[x / 2]], Plot[Sin[x / 2],
    {x, 0, 4 π}, PlotLabel → Style[Sin[x / 2], 9, Red, FontFamily → "Helvetica"]]}
```

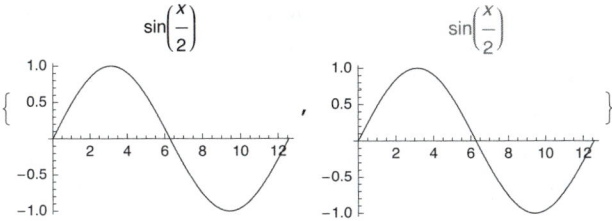

Sometimes we want to vary a parameter in the label:

```
Table[Plot[Sin[n x], {x, 0, 2 π},
    PlotLabel → Row[{"The graph of ", Sin[n x]}]], {n, 3}]
```

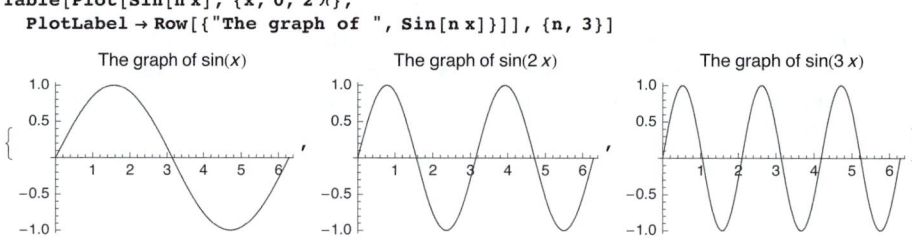

Next, we give a plot a more complete caption:

```
{Plot[Sin[x / 2], {x, 0, 4 π}, ImageSize → 160, PlotLabel →
    Style[Row[{Style["Figure 1.1  ", Bold], "The graph of ", Sin[x / 2]}], 9]],
  Labeled[Plot[Sin[x / 2], {x, 0, 4 π}, ImageSize → 160],
    Style[Row[{Style["Figure 1.1  ", Bold], "The graph of ",
        Sin[x / 2] // TraditionalForm}], 10, FontFamily → "Times"]]}
```

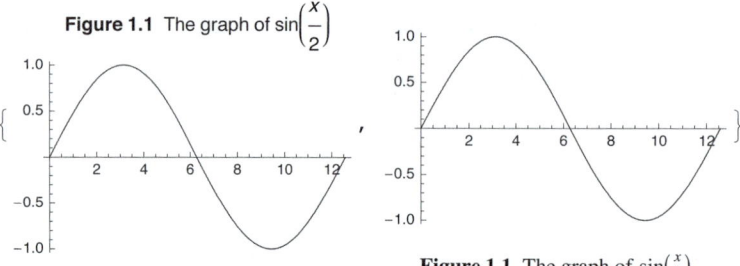

Figure 1.1 The graph of $\sin\left(\frac{x}{2}\right)$

The **PlotLabel** option puts the caption on top of the plot. However, the traditional position of a figure caption is on the bottom of the plot; this can be done with the **Labeled** command, as can be seen from the second plot. For more about **Labeled**, see Section 3.3.1, p. 72.

7.3.2 Axes and Ticks

■ **Axes**

---

* **Axes** Whether to draw the axes; examples of values: **True** [cp], **False** [8], **{True, False}**

* **AxesOrigin** Point where the axes cross; examples of values: **Automatic**, **{0, 0}**

AxesLabel Labels for the axes; examples of values: **None**, **y**, **{x, None}**, **{x, y}**

(**AxesStyle** Style of the axes, axes labels, tick marks, and tick labels; examples of values: **{}**, **Thick**, **Blue**, **Arrowheads[0.07]**, **Directive[Thick, Blue, 12, Italic]**

The point where the axes cross is **AxesOrigin**. Its default value is determined by an algorithm; this method usually chooses the point **{0, 0}** if it is within or close to the region defined by **PlotRange**.

Axes labels are placed at the ends of the axes (frame labels, instead, are in the middle of the frame edges). If the axes labels are long, consider using a frame instead of axes (see Section 7.3.3, p. 198).

The default thickness of axes and the major tick marks is **AbsoluteThickness[0.25]**, and that for the minor tick marks is **AbsoluteThickness[0.125]**. With **Arrowheads**, we can draw arrows at the ends of the axes.

```
{Plot[2 + Sin[x], {x, 1, 6}, AxesLabel → {x, y}],
 Plot[2 + Sin[x], {x, 1, 6}, AxesOrigin → {0, 0}],
 Plot[2 + Sin[x], {x, 1, 6}, AxesStyle → Arrowheads[0.07]]}
```

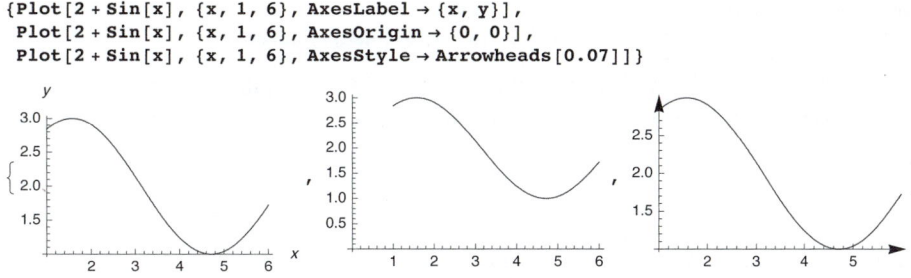

■ **Defining Positions of Ticks**

** Ticks Tick on the axes; simple examples of values: **Automatic**, **None**, **{{π, 2π}, Automatic}**,
{Automatic, {-1, 0, 1}}, {{π, 2π}, {-1, 0, 1}}

TicksStyle (✿6) Style of tick marks and tick labels; examples of values: **{}**, **Blue**,
Directive[Thick, Blue, 12]

The general form of the value of **Ticks** is **{xticks, yticks}**. We can define ticks on both axes or let *Mathematica* choose the ticks on one of the axes. The automatic algorithm often uses appropriate ticks but frequently also too many ticks, especially if the plots are scaled to be small (as in this book). In particular, the minor tick marks (without labels) between the major tick marks (with labels) are often unnecessary. A few carefully selected tick marks often suffice. Indeed, **Ticks** is perhaps the option I use most often. For example,

```
{Plot[Sin[x], {x, 0, 2 π}],
 Plot[Sin[x], {x, 0, 2 π}, Ticks → {{0, π, 2 π}, {-1, 0, 1}}]}
```

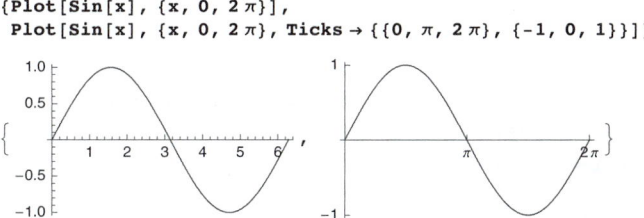

In the second plot, we asked for only six ticks. Note, however, that we got only four tick labels: The labels on the *x* and *y* axis at 0 were not drawn. Indeed, tick labels on the axes origin are not displayed.

■ **Defining Labels, Lengths, and Styles of Ticks**

Ticks has more advanced forms, where we can define—in addition to the positions of the tick marks— the tick labels and the length and style of the tick marks. In the next box, we give forms of a *single* tick and examples in which each form is used to define two ticks on the *x* axis (on the *y* axis, the default ticks are used).

---

position
Example: **{{π, 2π}, Automatic}**

{position, label}
Example: **{{{π, a}, {2π, b}}, Automatic}**

{position, label, {poslength, neglength}}
Example: **{{{π, a, {0.05, 0}}, {2π, b, {0.05, 0}}}, Automatic}**

```
{position, label, {poslength, neglength}, style}
```
Example: {{{π , a, {0.05, 0}, Red}, {2π, b, {0.05, 0}, Red}}, Automatic}

Often, when plotting mathematical figures, we do not want numerical tick labels but, rather, a few symbolic labels. Then we can define, with **Ticks**, the positions of the labels and the labels. The next figure is an example:

```
aa = {0.2, 0}; bb = {0.8, 0}; cc = {0, Exp[0.2]};
dd = {0, Exp[0.8]}; ee = {0.2, Exp[0.2]}; ff = {0.8, Exp[0.8]};
Plot[Exp[x], {x, 0, 1}, AxesOrigin → {0, 0},
  Ticks → {{{0.2 , a}, {0.8, b}}, {{Exp[0.2], c}, {Exp[0.8], d}}},
  Epilog → {Line[{cc, ee, aa}], Line[{dd, ff, bb}], Point[{ee, ff}]}]
```

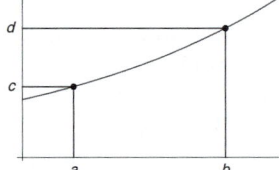

Tick labels are, by default, written in traditional format. This may sometimes make the labels somewhat large relative to the figure. Defining the labels as strings may be a solution:

```
{Plot[Sin[x], {x, 0, 2 π}, Ticks → {{π / 2, π, 3 π / 2, 2 π}, {-1, 1}}],
 Plot[Sin[x], {x, 0, 2 π}, Ticks → {{{π / 2, "π/2"}, π, {3 π / 2, "3π/2"}, 2 π}, {-1, 1}}]}
```

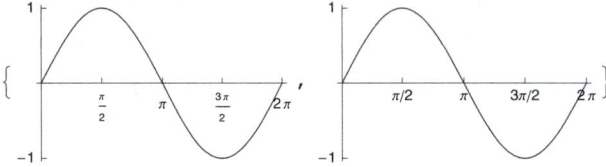

7.3.3 Frame and Frame Ticks

* **Frame** Whether to draw a frame; examples of values: **False, True, {True, True, False, False}**
* **FrameLabel** Labels for the frame; examples of values: **None, {x, y}, {"bottom", "left", "top", "right"}**
* **RotateLabel** Whether to rotate the labels for the vertical edges; possible values: **True, False**
  FrameStyle Style of the frame, frame labels, frame tick marks, and frame tick labels; examples of values: **Automatic, Blue, Directive[Thick, Red, 12, Italic]**
* **FrameTicks** Frame tick marks on the frame; simple examples of values: **Automatic, None, All** (on all edges), **{{0, π, 2π}, {-1, 1}, None, None}**
  FrameTicksStyle (✤6) Style of frame tick marks and frame tick labels; examples of values: **{}, Red, Directive[Thick, Red, 12, Bold]**

For most frame options (not for **RotateLabel**), we can define the properties separately on each of the four edges of the frame. In the previous box, the properties were given in the form {bottom, left, top, right}. The properties can also be given in the form {{left, right}, {bottom, top}}.

Here are some examples:

```
{Plot[Sin[x], {x, -π/2, 5π/2}, PlotRange → {-1.3, 1.3},
  Frame → True, FrameLabel → {x, y}, RotateLabel → False],
 Plot[Sin[x], {x, -π/2, 5π/2}, AxesLabel → {x, y}, PlotRange → {-1.3, 1.3},
  Frame → True, FrameTicks → {{0, π, 2π}, {-1, 0, 1}, None, None}],
 Plot[Sin[x], {x, -π/2, 5π/2}, AxesLabel → {x, y},
  Ticks → {{π, 2π}, {-1, 0, 1}}, ImageSize → 110] // Framed}
```

As the first plot shows, frame labels are placed midway on the edges (axes labels are placed at the ends of the axes). The default value of **RotateLabel** is **True**, which means that the frame labels on the vertical parts of the frame are rotated so that they read from bottom to top. A short label such as y looks better and is easier to read when not rotated. If we want a frame, then the figure often looks better if there is somewhat more space around the curve. This can be done with **PlotRange**.

In the second plot, we have defined our own ticks. The default is that there are tick marks on all edges but tick labels only on the bottom and left edges. If we define our own x and y ticks, they are used on all four edges, but with **None** we can remove the ticks from the edges on which we do not want them. Frame ticks are defined in the same way as axes ticks (see Section 7.3.2, p. 196).

In the third plot, we used the **Framed** command instead of the **Frame** option; for **Framed**, see Section 3.3.1, p. 72.

7.3.4 Grid Lines

> **GridLines** How the grid lines are drawn; simple examples of values: **None**, **Automatic**, **{None,**
> **Automatic}**, **{None, {-1,0,1}}**, **{{0,1,2}, {-1,0,1}}**
> (**GridLinesStyle** (✥6) Style of the grid lines; examples of values: **{}**, **Dashed**,
> **Directive[LightGray, Dashed]**)

Ticks often suffice to give information about the values of the coordinates. However, if we want to read approximate coordinates from a figure, then grid lines may help us. The value **Automatic** of **GridLines** draws lines on the major ticks and colors them gray. For example,

```
Plot[Sin[x], {x, 0, 2π}, Ticks → {{π, 2π}, {-1, 1}}, GridLines → #] & /@ {
  Automatic,
  {None, Automatic},
  {{π/2, π, 3π/2, 2π}, {-1, -0.5, 0.5, 1}}}
```

7.3.5 Hierarchy of Styles

For all of the components of a plot, we have options to give the components the styles we like most. We have very general options such as **BaseStyle** that affect most components of a plot. We have options such as **AxesStyle** that affect many components. We have special options such as **FrameTicksStyle** that affect only a few components. Finally, with either the **Style** command or special inputs we can control the style of single components. Let us look at the various components of a plot and how they can be styled with options. First, the curve in a plot can be styled with **PlotStyle** (**ColorFunction** can also be used); the **BaseStyle** option has no effect on the curve.

■ **Plot Label**

BaseStyle affects the style of a plot label. However, if **LabelStyle** is used, it overrides possible **BaseStyle** and so has a higher priority. Finally, if we use the **Style** command, it overrides possible other options. The following examples illustrate this hierarchy of the options:

```
{Plot[Sin[x], {x, 0, 2 π}, PlotLabel → Sin[x], BaseStyle → Blue],
 Plot[Sin[x], {x, 0, 2 π},
  PlotLabel → Sin[x], BaseStyle → Blue, LabelStyle → Green],
 Plot[Sin[x], {x, 0, 2 π}, BaseStyle → Blue, LabelStyle → Green,
  PlotLabel → Style[Sin[x], Red]]}
```

■ **Axes**

Next, we consider options and commands that can be used to adjust the styles of components related to axes. We prepare a table:

```
Grid[{Join[{Style["Component", Bold]}, Style[#, "Input", FontFamily → "Courier"] & /@
    {BaseStyle, LabelStyle, AxesStyle, TicksStyle, Style}],
   {"Axes", "×", "", "×", "", ""}, {"Axes labels", "×", "×", "×", "", "×"},
   {"Axes tick marks", "×", "", "×", "×", ""},
   {"Axes tick labels", "×", "×", "×", "×", "×"}},
  Dividers → {{False, True}, {{True}}}, Alignment → {{Left, Center}}] // Text
```

Component	BaseStyle	LabelStyle	AxesStyle	TicksStyle	Style
Axes	×		×		
Axes labels	×	×	×		×
Axes tick marks	×		×	×	
Axes tick labels	×	×	×	×	×

With × we have denoted the components that each option or command can control. **BaseStyle**, **LabelStyle**, **AxesStyle**, and **TicksStyle** are options, and **Style** is a command.

The options and commands are mentioned in the table in the order of increasing priority. **BaseStyle** affects all the components mentioned in the table. However, if **LabelStyle** is used, it has higher priority for the labels. If **AxesStyle** is used, it has still higher priority over **BaseStyle** and **LabelStyle**. **TicksStyle** has the highest priority for axes tick marks and tick labels. However, if we use the **Style** command for axes or tick labels, it overrides possible options.

7.3.6 Primitives

■ Adding Primitives with Prolog and Epilog

Occasionally, we may want to add some components to a plot. We may want to have clarifying text, an important point, an arrow, or a line. Such small additions may considerably improve the quality of a plot: They guide the eyes of the reader to important aspects of the plot. The options **Prolog** and **Epilog** are the tools for making such additions.

(**Prolog** Graphics primitives to be plotted before the main plot; examples of values: **{}**, **{Red,**
    PointSize[Medium], Point[{3, 2}]})
** **Epilog** Graphics primitives to be plotted after the main plot; examples of values: **{}**, **{Red,**
    PointSize[Medium], Point[{3, 2}]}

The values of **Prolog** and **Epilog** are lists of graphics directives and primitives. Directives and primitives were explained in Chapter 6. For easy reference, we list here the four primitives that are most useful when using **Prolog** and **Epilog**.

Point[p] Point at **p**
Line[{p1, …, pn}] Line through points **p1**, …, **pn**
Text[expr, p, q] The point **q**, expressed in text coordinates, of **expr** is at the point **p**
Arrow[{p1, p2}] Arrow from **p1** to **p2**

■ Example

We have already used some primitives with the **Epilog** option in a few examples; see Sections 6.2.6, p. 163 (**Point**, **Text**), 6.2.10, p. 175 (**Inset**), and 7.1.1, p. 183 (**Point**, **Text**, **Arrow**). Many more examples are in the forthcoming chapters. In the next example, we use the primitives **Point**, **Line**, **Text**, and **Arrow**:

```
Plot[{Log[x] + 1, Sqrt[x]}, {x, 0, 1.5}, PlotRange → {{0, 1.9}, {0, 1.55}},
  ImageSize → 220, Ticks → {{0.5, 1, 1.5}, {0.5, 1, 1.5}}, Epilog →
   {Text[Log[x] + 1, {1.55, 1.42}, {-1, 0}], Text[Sqrt[x], {1.55, 1.22}, {-1, 0}],
    Text[Log[x] + 1 == Sqrt[x], {1.2, 0.85}, {-1, 1}], Blue, Arrowheads[0.03], Arrow[
      {{1.19, 0.82}, {1.05, 0.95}}], Green, Dashed, Line[{{0, 1}, {1, 1}, {1, 0}}],
    Red, PointSize[Medium], Point[{1, 1}]}]
```

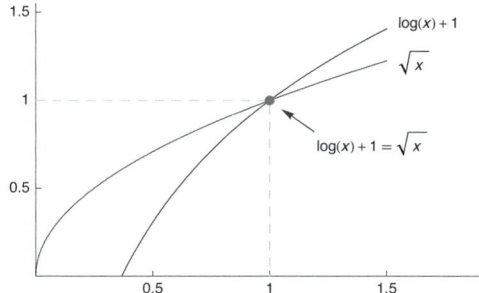

Here, we defined a large enough `PlotRange` to include all the primitives in the plot. Indeed, the default value of `PlotRange` does not take into account possible added primitives.

■ Disappearing Primitives

We may also have the problem of disappearing primitives. Remember what we said in Section 5.1.2, p. 125, about options with `Show`: If plots to be combined with `Show` have different values for some options, the values of the first plot are applied. Thus, if we combine two or more plots with `Show`—each having an `Epilog` option—then `Show` takes on the value of `Epilog` given in the first figure. This means that all of the graphics primitives in the other figures are left out. You may think the problem is solved by writing an `Epilog` in `Show` to add the lacking primitives. In reality, the primitives of the first figure are now lacking.

Thus, if you intend to combine several figures with `Show`, prepare to use `Epilog` also in `Show`: Write in the `Epilog` option *all* the primitives of the component figures. If you do not need the intermediate figures as such, do not add primitives into them but only into the final combined figure. However, if you plot the primitives with `Graphics`, then the problem does not occur; this method is considered next.

■ Adding Primitives with Graphics

`Prolog` and `Epilog` are not the only ways to add graphics primitives; we can also use `Graphics` (see Section 6.1.1, p. 152). In fact, we can separately plot the main figure with `Plot` (or with another command) and the primitives with `Graphics` and combine the two plots with `Show`. For example,

```
{p1 = Plot[{Log[x] + 1, Sqrt[x]}, {x, 0, 1.5},
   PlotRange → {{0, 1.9}, {0, 1.55}}, Ticks → {{0.5, 1, 1.5}, {0.5, 1, 1.5}}],
 p2 = Graphics[{Text[Log[x] + 1, {1.55, 1.42}, {-1, 0}], Text[Sqrt[x],
     {1.55, 1.22}, {-1, 0}], Text[Log[x] + 1 == Sqrt[x], {1.2, 0.85}, {-1, 1}],
   Blue, Arrowheads[0.03], Arrow[{{1.19, 0.82}, {1.05, 0.95}}],
   Green, Dashed, Line[{{0, 1}, {1, 1}, {1, 0}}],
   Red, PointSize[Medium], Point[{1, 1}]}]}
```

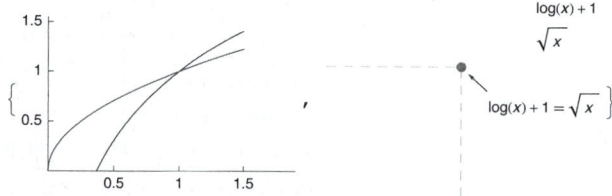

`Show[p1, p2, ImageSize → 220]`

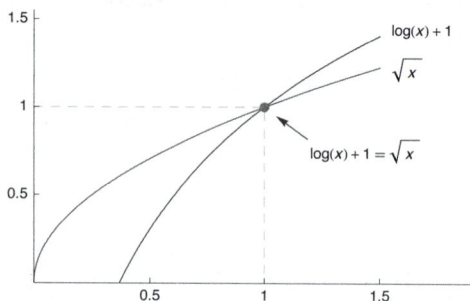

7.4 Options for the Curve

7.4.1 Plot Style

> * **PlotStyle** ^{cp} Style(s) of the curve(s); examples of values: **Automatic**, **Thickness[Medium]**,
> **Directive[Thick, Red, Dashed]**

When plotting 2D curves, **PlotStyle** defines the thickness, dashing, and color of the curves. The default value **Automatic** means a thin, nondashed curve whose color is **Hue[0.67, 0.6, 0.6]**. The style can be defined with *graphics directives*. Thickness can be defined, among others, with **Thickness** and **AbsoluteThickness**, dashing with **Dashing** and **AbsoluteDashing**, and color with, for example, **Red**. For thickness and dashing, see Section 6.2.2, p. 156, and for colors, see Section 6.2.8, p. 168.

PlotStyle is handy for distinguishing different curves in the same plot. Note that a good way to identify the curves is with the use of **Tooltip** (see Section 5.1.2, p. 122). We can also add a legend telling which style belongs to which curve (see Section 7.4.7, p. 208).

With **Directive** (see Section 6.1.2, p. 155), the **PlotStyle** option can be written in the following forms:

> *Style definitions for one curve:*
>
> **PlotStyle → Directive[s1, s2, …]**
>
> *Style definitions for several curves:*
>
> **PlotStyle → Directive[s1, s2, …]** Same style for each curve
> **PlotStyle → {Directive[s11, s12, …], Directive[s21, s22, …], … }** Different styles for the curves

Here, each **si** or **sij** is a graphics directive such as **Red**, **Thick**, or **Dashed**. The default style is denoted by the empty list **{}**. If a style is defined only with a single directive, then **Directive** is not needed. Some examples:

```
GraphicsRow[{Plot[Sin[x], {x, 0, 2 π}, PlotStyle → Red],
  Plot[Sin[x], {x, 0, 2 π}, PlotStyle → Directive[Red, Thick]],
  Plot[{Sin[x], Cos[x]}, {x, 0, 2 π}, PlotStyle → Directive[Red, Thick]],
  Plot[{Sin[x], Cos[x]}, {x, 0, 2 π},
    PlotStyle → {Directive[Red, Thick], Directive[Blue, Thick]}]}, ImageSize → 420]
```

> (**ClippingStyle** ^{c} (✻6) How to indicate clipped parts that fall outside of the plot range; examples of values: **None**, **Automatic**, **Red**, **{Blue, Red}**)

Parts of a plot may be clipped as discussed previously. The **ClippingStyle** option can be used to define the way that clipped parts of the plot are shown:

```
GraphicsRow[Plot[Sin[x], {x, 0, 2 π}, PlotRange → 0.5, ClippingStyle → #] & /@
   {None, Automatic, Red, {Blue, Red}}, ImageSize → 420]
```

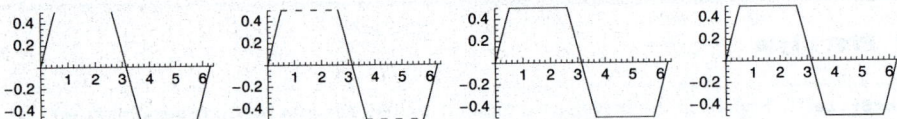

The first plot shows the default way that clipped parts are indicated: Nothing is plotted. With the value **Automatic**, we get a dashed line. We can give the line some directives such as a color or two colors: The first color is used at the bottom and the second at the top.

7.4.2 Color Function

> (**ColorFunction** ^{cp} (⌘6) Function that determines the color of the curve; examples of values:
>   **Automatic, (Hue[#2] &), (RGBColor[#2, 0, 1 - #2] &), "Rainbow")**
> (**ColorFunctionScaling** ^{cp} (⌘6) Whether arguments to a color function should be scaled to lie
>   between 0 and 1; examples of values: **True, False**)

With **ColorFunction**, we can get colors that vary according to the values of x and y. The color function is a pure function whose arguments are **#1** (corresponds to x) and **#2** (y); in **ParametricPlot**, an argument can also be **#3** (the parameter). Many gradients (e.g., **"Rainbow"**) useful as a color function can be found from the *ColorSchemes* palette. Usually, it is useful for the arguments of a color function to be scaled to (0, 1) because the arguments of all color schemes can be in this interval.

```
Plot[Sin[x], {x, 0, 2 π}, PlotStyle → Thick, ColorFunction → #] & /@
   {(Hue[#2] &), (RGBColor[#2, 0, 1 - #2] &), "Rainbow"}
```

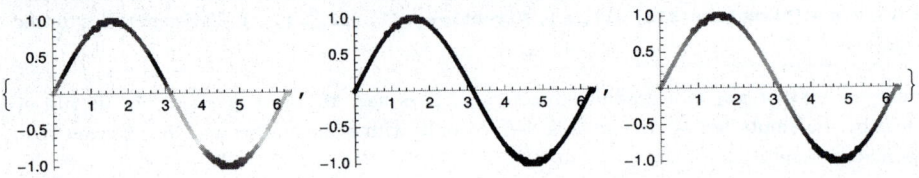

7.4.3 Filling

> **Filling** ^{c} (⌘6) Type of filling to use; examples of values: **None, Axis, Bottom, Top, 0.3, True**
> (**FillingStyle** ^{c} (⌘6) Style of filling; examples of values: **Automatic, Red, {Blue, Red}** (different
>   style for negative and positive values), **Directive[Opacity[0.3], Blue]**)

Filling was considered in Section 5.2.4, p. 135.

```
{Plot[BesselJ[1, x], {x, 0, 20}, Filling → Axis],
 Plot[{Sqrt[x], Log[x + 1]}, {x, 0, 100}, Filling → True]}
```

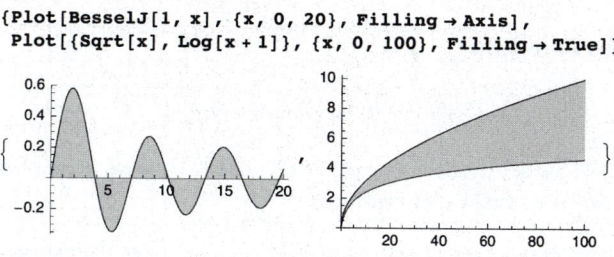

7.4.4 Mesh Points

Mesh ^{cp} How many mesh points should be drawn; examples of values: **None** ^{c}, **Automatic** ^{p}, **10**,
  Full, **All**, **{{0}}**

MeshFunctions ^{cp} (✻6) How to determine the placement of the mesh points; examples of values:
  {#1 &} ^{c}, **Automatic** ^{p}, **{#2 &}**,

MeshStyle ^{cp} The style of mesh points; examples of values: **Automatic**, **PointSize[Small]**,
  Directive[Red, PointSize[Medium]]

(**MeshShading** ^{cp} (✻6) How to shade regions between mesh points; examples of values: **None**, **{Red,
  Blue}**)

By a mesh for a curve, we mean points on the curve. With **Mesh**, we can ask a given number of points
that correspond to equally spaced x points. The value **Full** means that the initial sample points are
drawn. With the value **All**, we get the final sample points (for sampling, see Section 7.4.6, p. 207):

```
Plot[Sin[x], {x, 0, 2 π}, Mesh → #,
   MeshStyle → Directive[Red, PointSize[Small]]] & /@ {15, Full, All}
```

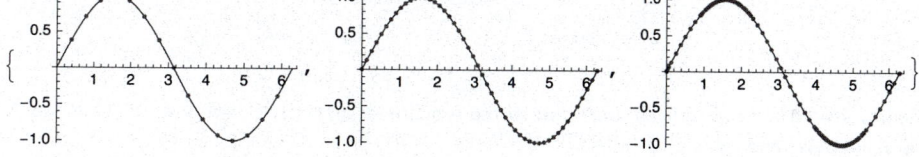

With **Mesh**, we can show points at which a function or a difference of two functions has a given
value, for example, 0. Another way is to use **Epilog**:

```
f = x^5 - 7 x^3 + 9 x; g = 2 x;
{Plot[f, {x, -2.5, 2.5}, Mesh → {{0}},
  MeshFunctions → {#2 &}, MeshStyle → PointSize[Medium]],
 Plot[{f, g}, {x, -2.5, 2.5}, Mesh → {{0}}, MeshFunctions → {(f - g) /. x → #1 &}],
 Plot[{f, g}, {x, -2.5, 2.5},
  Epilog → {Point[{#, f /. x → #}] & /@ (x /. NSolve[f == g, x])}]}
```

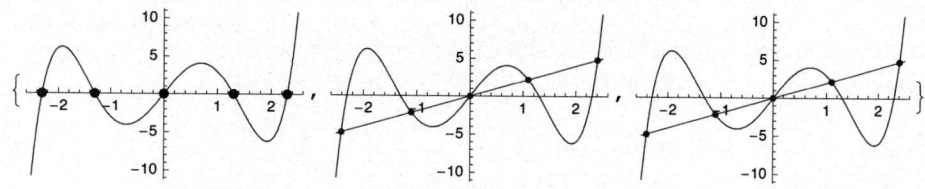

7.4.5 Exclusions and Region Function

> **Exclusions** [cp] (⌘6) The *x* points that are excluded in plotting; examples of values: `Automatic`,
> `None`, `{π/2, 3 π/2}`, `{Cos[x] == 0}`
> **ExclusionsStyle** [cp] (⌘6) What to draw at excluded points; examples of values: `None`, `Directive[`
> `Blue, Dashed]`, `{None, Directive[Red, PointSize[Medium]]}`

Possible steps in functions are automatically plotted discontinuously, but by giving the `Exclusions` style the value `None`, we can also get a vertical line:

```
Plot[Piecewise[{{-1, x < 1}, {1, x ≥ 1}}], {x, 0, 2}, Exclusions → #] & /@
  {Automatic, None}
```

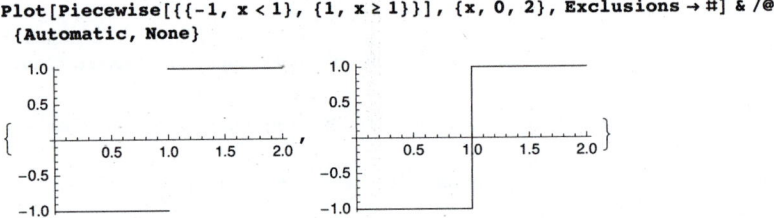

We can give the vertical line a suitable style and we can also get points at the ends of the step:

```
Plot[Piecewise[{{-1, x < 1}, {1, x ≥ 1}}], {x, 0, 2},
  PlotRangePadding → {Automatic, 0.1}, ExclusionsStyle → #] & /@ {
  Directive[Blue, Dashed],
  {None, Directive[Red, PointSize[Medium]]},
  {Directive[Blue, Dashed], Directive[Red, PointSize[Medium]]} }
```

Sometimes we want to indicate the value of the function at a point of discontinuity. One way to do this is to use disks and circles:

```
Plot[Piecewise[{{-1, x < 1}, {1, x ≥ 1}}], {x, 0, 2},
  PlotRangePadding → {Automatic, 0.1}, Epilog → {Disk[{1, 1}, Offset[{2.2, 2.2}]],
    White, EdgeForm[Black], Disk[{1, -1}, Offset[{2, 2}]]}]
```

For infinite jumps, a vertical line is automatically drawn. To get rid of the line, define, with **Exclusions**, points to avoid either explicitly as a list or implicitly by an equation:

```
{Plot[Tan[x], {x, 0, 2 π}], Plot[Tan[x], {x, 0, 2 π}, Exclusions → {π / 2, 3 π / 2}],
  Plot[Tan[x], {x, 0, 2 π}, Exclusions → Cos[x] == 0,
    ExclusionsStyle → Directive[Blue, Dashed]]}
```

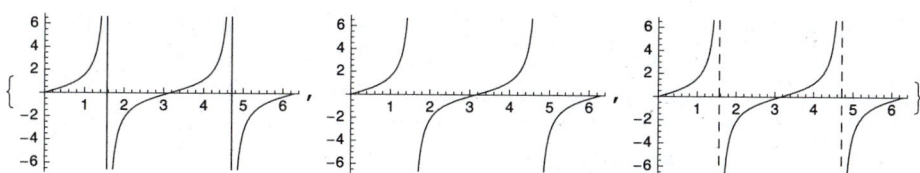

(**RegionFunction** ^{cp} (⌘6) Specifies the region to include in the plot drawn; examples of values:
(**True &**), (**Abs[#2] > 0.7&**))

The basic way to control the x values is to give a suitable interval in the plotting command. The basic way to control the y values is to use **PlotRange**. More complex rules that determine the values of x and y for which the curve is plotted can be given with **RegionFunction**. A point is included in the plot if the region function at that point gives True. The arguments of the pure function are **#1** (corresponds to x) and **#2** (y); in **ParametricPlot**, an argument can also be **#3** (the parameter).

```
Plot[Sin[x], {x, 0, 2 π}, RegionFunction → (Abs[#2] < 0.6 || Abs[#2] > 0.8 &)]
```

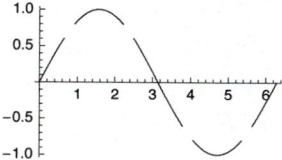

7.4.6 Plotting Algorithm

Mostly, the default values of the following options give very good results. In special cases, the use of some of these options may improve the quality of the plot.

PlotPoints ^{cp} Number of initial sampling points; examples of values: **Automatic, 100**

MaxRecursion ^{cp} (⌘6) The maximum number of recursive subdivisions allowed; examples of values: **Automatic, 8**

WorkingPrecision ^{cp} (⌘6) The precision used in computations; examples of values: **MachinePrecision, 20**

(**Evaluated** ^{cp} (⌘6) Whether the expression to be plotted is evaluated before the expression is sampled; possible values: **Automatic, True, False**)

(**PerformanceGoal** ^{cp} (⌘6) What aspect of performance to try to optimize; examples of values: **$PerformanceGoal, "Quality", "Speed"**)

(**EvaluationMonitor** ^{cp} (⌘6) Expression to evaluate at every function evaluation; examples of values: **None, Sow[{x, Sin[x]}]**)

Plot works by first sampling the function at some equally spaced points specified by **PlotPoints**. If the function changes rapidly between two points, the interval is divided into two smaller intervals. This subdivision is continued until the curve resulting by joining the sample points with lines is smooth enough. However, subdivisions are done at most **MaxRecursion** times. In this way, **Plot** is adaptive, and, accordingly, the resulting plot is usually sufficiently smooth and accurate.

As an example, we plot a function and collect all the points where the function is evaluated:

```
a = Reap[Plot[Sin[x], {x, 0, 2 π}, EvaluationMonitor :> Sow[{x, Sin[x]}]]];
```

(Note the delayed rule :> that is written as :>; for **Sow** and **Reap**, see Section 18.2.3, p. 564.) The next plot shows all the points (more than 400) where the function was evaluated to get the curve:

```
ListPlot[a[[2, 1]], PlotStyle → PointSize[Small],
  Filling → Axis, AspectRatio → 0.2, ImageSize → 420]
```

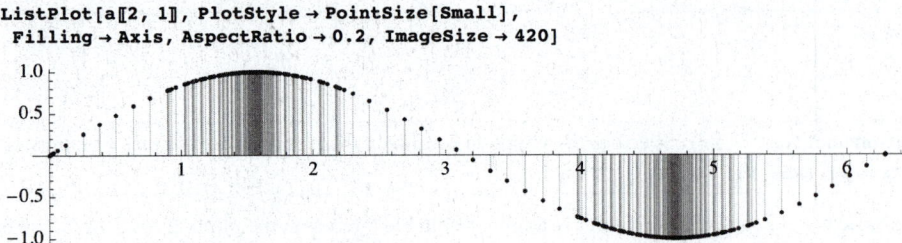

We see that where the function changes rapidly, more points are calculated than where the function behaves almost linearly.

Next, we plot a function that behaves wildly near the origin:

```
{Plot[Sin[1 / x], {x, 0, 0.1}, PerformanceGoal → "Speed", AspectRatio → 0.4],
  Plot[Sin[1 / x], {x, 0, 0.1}, AspectRatio → 0.4],
  Plot[Sin[1 / x], {x, 0, 0.1}, MaxRecursion → 8, AspectRatio → 0.4]}
```

The numbers of sampled points are as follows:

```
Length[%[[#]][[1, 1, 3, 2, 1]]] & /@ {1, 2, 3}
{182, 1669, 4105}
```

In the first figure, we rejected the goal of quality and, instead, asked to get a plot rapidly; only 182 sample points were calculated. In the second plot, **Plot** gradually increased the number of sampled points up to 1669 (**MaxRecursion** was 6), but by comparing the two figures, we see that even this is not sufficient. In the third figure, we allowed 8 recursive subdivisions and ended with 4105 points, and now the result is good.

The **WorkingPrecision** option can be useful if the computation of the values of the function involves round-off errors. See an example in Section 12.2.3, p. 407.

When **Plot** samples the expression to be plotted, the expression is held in an unevaluated form. However, sometimes it is necessary for the expression to be evaluated before sampling. This can be achieved with the option **Evaluated → True**; this is equivalent to enclosing the expression with the **Evaluate** command.

7.4.7 Legends

■ Legends with Primitives

If we have several curves in the same plot, it is useful to identify the curves in some way. On the screen, a good way is to use **Tooltip**; see Section 5.1.2, p. 122.

Another good way is to place a suitable **Text** primitive near each curve; see an example in Section 7.3.6, p 201. This method has several advantages. It is simple to implement, we need not look at a separate legend, and we can use the same style for each curve (as was done in the example in Section 7.3.6).

Still another way to identify the curves is the use of a legend. This can easily be done with graphics primitives in an **Epilog** option, but we also have a package for legends. Here is an example of the use of **Epilog**:

```
Plot[{Sin[x], Cos[x]}, {x, 0, 2 π}, AspectRatio → 0.4, PlotRange → {{-0.1, 9.3}, All},
  Ticks → {{π, 2 π}, {-1, 1}}, PlotStyle → {Black, {Black, AbsoluteDashing[{1.3}]}},
  AxesStyle → {White, Black}, TicksStyle → Black,
  Epilog → {Line[{{-0.1, 0}, {2 π + 0.1, 0}}], Line[{{7.0, 0.8}, {7.8, 0.8}}],
    AbsoluteDashing[{1.3}], Line[{{7.0, 0.4}, {7.8, 0.4}}], Text[Sin[x],
    {8.1, 0.8}, {-1, 0}], Text[Cos[x], {8.1, 0.4}, {-1, 0}]}, ImageSize → 160]
```

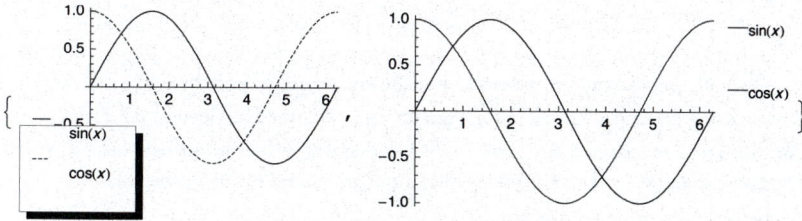

If the legend is placed outside the *x* range of the curves as above, then **PlotRange** must be used to extend the *x* range so that the legend is within it. In the previous example, the extended *x* range causes the *x* axis to continue from 2π to 9.6. To stop the *x* axis at 2π, we have defined white as the style of the *x* axis and drawn the *x* axis ourselves with the **Line** primitive.

■ Legends with a Package

The **PlotLegends`** package (see PlotLegends/guide/PlotLegendsPackage`) adds some new options for **Plot** enabling the use of legends. Note that this holds only for **Plot** and not for any other plotting command. For other commands, the package defines the **ShowLegend** command.

Before listing the legend options, we present two examples. In the first example, we have used the default style of the legend.

```
<< PlotLegends`
```

```
{Plot[{Sin[x], Cos[x]}, {x, 0, 2 π}, PlotStyle → {{}, AbsoluteDashing[{1.3}]},
  PlotLegend → {Sin[x], Cos[x]}, ImageSize → 150],
  Plot[{Sin[x], Cos[x]}, {x, 0, 2 π}, PlotStyle → {Sin[x], Cos[x]},
  LegendPosition → {1, -0.15}, LegendSize → {0.4, 0.8}, LegendShadow → None,
  LegendBorder → None, LegendTextOffset → {-1.1, 0}, ImageSize → 200]}
```

Options for the legend and its position, size, and orientation:

** `PlotLegend` The texts in a legend (mandatory legend option with `Plot`); an example of value: `{Sin[x], Cos[x]}`

** `LegendPosition` Position of the lower left corner of the legend box (in a coordinate system in which the center of the figure is {0, 0} and the longest side of the plot runs from −1 to 1); default value: `{-1.2, -0.82}`

** `LegendSize` Size of the legend box (in the same units as `LegendPosition`); examples of values: `Automatic, 0.6, {width, height}`

`LegendOrientation` Direction of the legend; possible values: `Vertical, Horizontal`

Options for the style of the legend box and its shadow:

`LegendBorder` Style of the border line; default value: `Automatic`

`LegendBorderSpace` Space between the border and its content; default value: `Automatic`

`LegendBackground` Color of the background; default value: `Automatic`

** `LegendShadow` Offset of the shadow from the legend box (in the same units as `LegendPosition`); examples of values: `Automatic, None` (no border and no shadow), `{0, 0}` (no shadow), `{xoffset, yoffset}`

`ShadowBackground` Color of the shadow; default value: `GrayLevel[0]`

Options for the key boxes, texts, and legend label:

`LegendSpacing` Space around each key box; default value: `Automatic`

`LegendTextSpace` The width of the space of each text; default value: `Automatic`

`LegendTextOffset` Offset of a text from a key box; default value: `Automatic`

`LegendTextDirection` Direction of text; default value: `Automatic`

`LegendLabel` Label of the legend box; default value: `None`

`LegendLabelSpace` Space above and below the label; default value: `Automatic`

`ShowLegend[graphic, {{{key1, text1}, {key2, text2}, … }, opts}]` Show `graphic` with a legend containing the key graphics or key colors `key1, key2, …` and texts `text1, text2, …`; modify the legend with the legend options `opts`

7.5 Options for Surface Plots

7.5.1 Summary

■ Introduction

In this section, we list all the options of `Plot3D`, `ParametricPlot3D`, and `Graphics3D`, with short descriptions and with some common values (`Graphics3D` was explained in Section 6.2.11, p. 176). The default value of an option is mentioned first, and then we mention either all other possible values or some examples of other values (the examples are simple; more advanced forms may exist).

The options are divided into three groups:

• *Global options:* These are options to modify global aspects of the plot—that is, to adjust how the plot looks in general. This means, for example, adjusting the form, size, plot range, various margins, fonts, or viewpoint.

- *Local options:* These are options to modify local components of a plot such as plot label, axes, ticks, bounding box, face grids, or add-ons.

- *Options for the surface:* These are options to modify the curve produced by **Plot3D** or **ParametricPlot3D**. This means, for example, adjusting the style, filling, mesh, or exclusions or controlling the plotting algorithm.

All of the options in the first two groups are common to all three plotting commands mentioned. However, options in the third group are applicable only to **Plot3D** or **ParametricPlot3D**. Options and their default values applicable only to certain commands are expressed in the following lists by a superscript after the option name or value:

^{s}: applicable to **Plot3D** (shorthand for *surface*)
^{p}: applicable to **ParametricPlot3D** (shorthand for *parametric*)
^{g}: applicable to **Graphics3D** (shorthand for *graphics*)

For example, **PlotStyle** ^{sp} means that this option applies to **Plot3D** and **ParametricPlot3D** but not to **Graphics3D**, and **False** ^{g} means that the default value of the option in question (**Axes**) is **False** for **Graphics3D**.

Ten options assumed to be somewhat important are marked with an asterisk (*). A single parenthesis indicates several options that most of us will seldom need. The remaining options, without any special markings, may sometimes be useful. Of course, the given classification of the options according to importance reflects my personal impression. You may well have a different classification.

■ Global Options

With **Plot3D**, **ParametricPlot3D**, and **Graphics3D**, the following options can be used to adjust some global aspects of a plot.

Options for form and size:

* BoxRatios Ratios of side lengths of the bounding box; default values: **{1, 1, 0.4}** ^{s}, **Automatic** ^{pg}

(**AspectRatio** Ratio of height to width of the plotting rectangle; default value: **Automatic**)

ImageSize The absolute size (in printer's points) of the plot; examples of values: **Automatic**, **Full**, **All**, width, **{width, height}**, **{maxsize}**, **{{maxwidth}, {maxheight}}**

An option for plot range:

* PlotRange Ranges for *x*, *y*, and *z* in the plot; examples of values:

Automatic ^{p},	**{zmin,Automatic}**,	**{Full, Full, Automatic}** ^{s},
All ^{g},	**{zmin,All}**,	**{Full, Full, All}**,
Full,	**{All,zmax}**,	**{{xmin, xmax}, {ymin, ymax}, All}**,
5,	**{zmin,zmax}**,	**{{xmin, xmax}, {ymin, ymax}, {zmin, zmax}}**

Options for margins and background:

PlotRegion Specifies margins around the plot inside the selection rectangle; examples of values: **Automatic** (means **{{0, 1}, {0, 1}}**), **{{xmin, xmax}, {ymin, ymax}}**

SphericalRegion Whether to leave room in the plotting rectangle for a sphere enclosing the bounding box; possible values: **False**, **True**

(**ImageMargins** Specifies margins (in printer's points) outside the selection rectangle; examples of values: **0**, **Automatic**, **15**, **{{left, right}, {bottom, top}}**)

(**ImagePadding** (※6) Extra space (in printer's points) for objects such as thick lines and tick and axes labels; examples of values: **All**, **None**, **15**, **{{left, right}, {bottom, top}}**)

(**PlotRangePadding** (⌘6)   How much farther axes etc. should extend beyond the range of coordinates specified by **PlotRange**; examples of values: **Automatic**, **None**, **{0.2, 0.5}**)

Background   Color of the background; examples of values: **None**, **LightGray**

Options for fonts and formatting:

* **BaseStyle** (⌘6)   Style of texts; examples of values: **{}**, **{9, Bold, FontFamily → "Helvetica"}**

LabelStyle (⌘6)   Style of all labels; examples of values: **{}**, **{9, Bold, FontFamily → "Helvetica"}**

(**FormatType**   Format type of text used in a plot; examples of values: **TraditionalForm**, **StandardForm**, **InputForm**, **OutputForm**)

Options for viewpoint and lighting:

* **ViewPoint**   Point (in a scaled coordinate system) from which the surface is viewed; default value: **{1.3, -2.4, 2}**

RotationAction (⌘6)   How to render the plot if rotated with the mouse; examples of values: **"Fit"** (after rotation, the plot is rescaled to fit into the image region), **"Clip"** (after rotation, the plot is not rescaled and so may be clipped)

(**ViewAngle** (⌘6)   Angle of the field of view; default value: **Automatic**)

(**ViewCenter**   Point (in scaled coordinates) to display at the center; default value: **{1/2, 1/2, 1/2}**)

(**ViewMatrix** (⌘6)   Explicit transformation matrix; default value: **Automatic**)

(**ViewRange** (⌘6)   Range of viewing distances to include; default value: **All**)

(**ViewVector** (⌘6)   Position and direction of a simulated camera; default value: **Automatic**)

(**ViewVertical**   Direction to make vertical; default value: **{0, 0, 1}**)

Lighting   What simulated lighting to use; simple examples of values: **Automatic**, **"Neutral"**, **None**

Miscellaneous options:

(**BaselinePosition** (⌘6)   Where the baseline of a plot should be if the plot is combined with other plots or text; examples of values: **Automatic**, **Axis**, **Bottom**, **Top**, **Center**, **Baseline**)

(**AlignmentPoint** (⌘6)   How objects should by default be aligned when they appear in **Inset**; default value: **Center**)

(**ContentSelectable** (⌘6)   Whether and how content of a plot should be selectable; possible values: **Automatic** (double-click allows content selection), **True** (single clicks immediately select content objects), **False** (content objects cannot be selected))

(**PreserveImageOptions** (⌘6)   Whether the size and margins of a plot should remain the same if the plotting command is executed anew; examples of values: **Automatic** (the properties should remain the same if not explicitly otherwise stated), **True** (the properties should remain the same), **False** (the previous properties are ignored))

(**DisplayFunction**   Function to apply to a graphic before returning it; default value: **$DisplayFunction** (is normally **Identity**))

The main difference between the options for the three commands is in the default value of **BoxRatios**: It is **{1, 1, 0.4}** for **Plot3D** and **Automatic** for **ParametricPlot3D** and **Graphics3D**. Note that in the previous box, we do not mention the **ControllerLinking**, **ControllerMethod**, and **ControllerPath** options.

■ **Local Options**

With **Plot3D**, **ParametricPlot3D**, and **Graphics3D**, the following options can be used to adjust some local components of a plot:

An option for plot label:
PlotLabel   Label of the plot; examples of values: **None**, **Sin[x y]**, **"A surface plot"**

Options for axes and ticks:
* **Axes**   Whether to draw the axes; examples of values: **True** [sp], **False** [g], **{True, True, False}**
AxesEdge   Where to draw the axes; examples of values: **Automatic**, **{{1,1}, {-1,1}, {1,1}}**, **{ Automatic, {-1,1}, {1,1}}**, **{{1,1}, {-1,1}, None}**
* **AxesLabel**   Labels for the axes; examples of values: **None**, **z**, **{x, y}**, **{x, y, z}**
AxesStyle   Style of the axes; examples of values: **{}**, **Directive[Thick, Gray]**
* **Ticks**   Ticks on the axes; simple examples of values: **Automatic**, **None**, **{{0,1,2}, {0,1}, Automatic}**, **{{0,1,2}, {0,1}, {-1,0,1}}**
TicksStyle (⌗6)   Style of tick marks and tick labels; examples of values: **{}**, **Blue**, **Directive[Thick, Blue, 12]**

Options for bounding box:
* **Boxed**   Whether to draw a bounding box around the surface; possible values: **True**, **False**
* **BoxRatios**   Ratios of side lengths of the bounding box; default values: **{1, 1, 0.4}** [s], **Automatic** [pg]
(**BoxStyle**   Style of the bounding box; examples of values: **{}**, **Directive[Thick, Gray]**)

Options for face grids:
FaceGrids   Grid lines drawn on the faces of the bounding box; examples of values: **None**, **All**, **{{-1,0,0}, {0,1,0}, {0,0,-1}}**, **{{{-1,0,0}, {ygrid, zgrid}}, {{0,1,0}, {xgrid, zgrid}}, {{0,0,-1}, {xgrid, ygrid}}}**
(**FaceGridsStyle**   Style of the grid lines; examples of values: **{}**, **Directive[Thick, Gray]**)

Options for primitives:
(**Prolog**   2D graphics primitives to be plotted before the main plot; examples of values: **{}**, **{PointSize[Medium], Red, Point[{0.4, 0.7}]}**)
Epilog   2D graphics primitives to be plotted after the main plot; examples of values: **{}**, **{PointSize[Medium], Red, Point[{0.4, 0.7}]}**

The main difference between the options for the three commands is in the default value of **Axes**: It is **True** for **Plot3D** and **ParametricPlot3D** and **False** for **Graphics3D**.

■ **Options for the Surface**

With **Plot3D** and **ParametricPlot3D**, the following options can be used to adjust the surface to be plotted:

Options for plot style:
PlotStyle [sp] (⌗6)   Style(s) of the surface(s); examples of values: **Automatic**, **Yellow**, **Directive[Yellow, Specularity[Red, 5]]**, **Opacity[0.3]**, **None**
(**BoundaryStyle** [sp] (⌗6)   How to draw boundary lines for surfaces; examples of values: **Automatic** [s], **None** [p], **Thick**)
(**ClippingStyle** [s] (⌗6)   How to draw clipped parts of the surface; examples of values: **Automatic**, **None**, **Red**, **{Green, Red}**)
(**NormalsFunction** [sp] (⌗6)   How to determine effective surface normals; default value: **Automatic**)

Options for color function:
ColorFunction [sp]   What colors are used to color the surface, if simulated illumination is not used;

examples of values: `Automatic, GrayLevel, "Rainbow", (Hue[1 - #3] &), (RGBColor[#3, 1 - #3, 0] &)`

(`ColorFunctionScaling` *sp* Whether values provided for a color function are scaled to lie between 0 and 1 (`True`) or left as such (`False`))

Options for filling:

(`Filling` *s* (⌘6) Type of filling to use; examples of values: `None, Axis, Bottom, Top, 0.3, True`)

(`FillingStyle` *s* (⌘6) Style of filling; examples of values: `Opacity[0.5], Red`)

Options for mesh lines:

* `Mesh` *sp* How many mesh lines should be drawn; examples of values: `Automatic, 10, {5, 10}, Full, All, {{0, 1, 2}, {1, 2}}, None`

`MeshFunctions` *sp* (⌘6) How to determine the placement of the mesh lines; examples of values: `{#1 &, #2 &}` *s*, `Automatic` *p*, `{#3 &}`,

`MeshStyle` *sp* Style of mesh lines; examples of values: `Automatic, Directive[Red, Thick]`

(`MeshShading` *sp* (⌘6) How to shade regions between mesh lines; examples of values: `None, {{Red,Green}, {Green, Red}}`)

Options for exclusions and region function:

`Exclusions` *sp* (⌘6) The curves on the (x, y) surface that are excluded in plotting; examples of values: `Automatic, None, {x + y == 0}`

(`ExclusionsStyle` *sp* (⌘6) What to draw at excluded curves; examples of values: `None, Opacity[0.5], {Opacity[0.5], Directive[Red, Thick]}`)

`RegionFunction` *sp* (⌘6) Specifies the region to include in the plot drawn; examples of values: `(True &), (#1^2 + #2^2 < 1 &)`

Options for plotting algorithm:

`PlotPoints` *sp* Number of initial sampling points; examples of values: `Automatic, 30, {30, 50}`

`MaxRecursion` *sp* (⌘6) The maximum number of recursive subdivisions allowed; examples of values: `Automatic, 3`

`WorkingPrecision` *sp* (⌘6) The precision used in computations; examples of values: `MachinePrecision, 20`

(`PerformanceGoal` *sp* (⌘6) What aspect of performance to try to optimize; examples of values: `$PerformanceGoal, "Quality", "Speed"`)

(`EvaluationMonitor` *sp* (⌘6) Expression to evaluate at every function evaluation; examples of values: `None, Sow[{x, y, Sin[x y]}]`)

■ **Example**

First, we define the grid lines for *x*, *y*, and *z*:

`xg = {π / 4, π / 2, 3 π / 4}; yg = {π / 4, π / 2, 3 π / 4}; zg = {0};`

In the plot, we color the surface with the `ColorFunction` option according to the height (to see the colors, look at the plot in the Help Browser if you have installed the CD-ROM of this book):

```
Plot3D[Cos[x y] Cos[x], {x, 0, π}, {y, 0, π}, BoxRatios → Automatic,
  Background → Black, PlotRegion → {{0.05, 0.95}, {0.03, 0.95}},
  ColorFunction → (Hue[1 - #3] &), PlotPoints → 50,
  BaseStyle → {FontFamily → "Helvetica"}, MeshStyle → Gray, MeshFunctions → (#3 &),
  AxesLabel → {x, y}, AxesStyle → Directive[White, 9, Bold], BoxStyle → White,
  Ticks → {{0, {π / 2, "π/2"}, π}, {0, {π / 2, "π/2"}, π}, {-1, 0, 1}},
  FaceGrids → {{{-1, 0, 0}, {yg, zg}}, {{0, 1, 0}, {xg, zg}}, {{0, 0, -1}, {xg, yg}}},
  PlotLabel → Style[Cos[x y] Cos[x], "Bold", 11, White], ImageSize → 250] // Framed
```

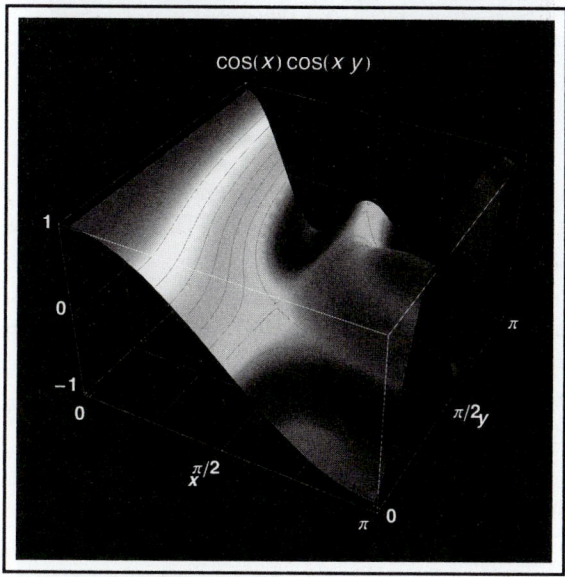

7.5.2 Global Options

■ Form and Size

> * **BoxRatios**   Ratios of side lengths of the bounding box; default values: **{1, 1, 0.4}** [s],
>   **Automatic** [p g]
>
> (**AspectRatio**   Ratio of height to width of the plotting rectangle; default value: **Automatic**)
>
> **ImageSize**   The absolute size (in printer's points) of the plot; examples of values: **Automatic, Full,**
>   **All, width, {width, height}, {maxsize}, {{maxwidth}, {maxheight}}**

(Remember that options and values denoted by s, p, and g apply for **Plot3D**, **ParametricPlot3D**, and **Graphics3D**, respectively.)

BoxRatios determines the form of the 3D bounding box. The default value **{1, 1, 0.4}** for **Plot3D** means that the x and y axes have the same length and that the z axis is 0.4 times this length. The default value **Automatic** for **ParametricPlot3D** and **Graphics3D** tells us that one unit in all three axes has the same length. Thus, the setting **BoxRatios → Automatic** for 3D graphics corresponds to the setting **AspectRatio → Automatic** for 2D graphics.

```
Plot3D[Sin[x y], {x, 0, π}, {y, 0, π}, BoxRatios → #] & /@
  {{1, 1, 0.4}, Automatic, {1, 1, 1}}
```

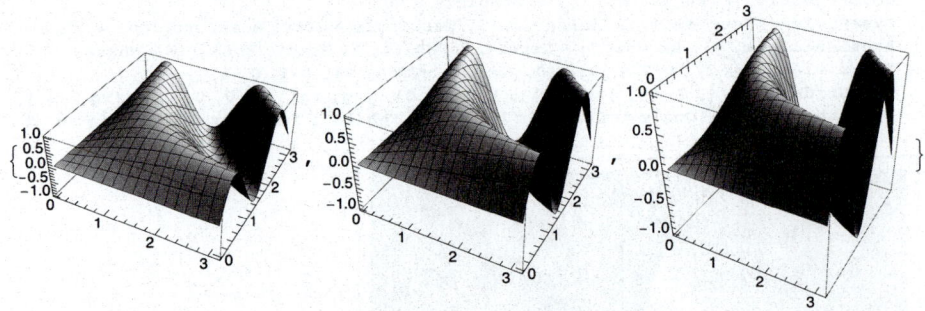

For surfaces produced by **Plot3D**, it may be worth setting the **BoxRatios** such that one unit on the x and y axes has the same length to get the correct impression of the form of the (x, y) region and setting the third component such that we get a good impression of the form of the surface.

AspectRatio defines the 2D form of the plot. The default value **Automatic** leaves unchanged the form of the graphics determined by the **BoxRatios** option. **AspectRatio** is seldom used for 3D graphics. Instead, **BoxRatios** is the option for defining the 3D form of the surface.

For **ImageSize**, see Section 7.2.1, p. 189.

■ **Plot Range**

* PlotRange Ranges for x, y, and z in the plot; examples of values:		
Automatic [p],	{zmin,Automatic},	{Full, Full, Automatic} [s],
All [g],	{zmin,All},	{Full, Full, All},
Full,	{All,zmax},	{{xmin, xmax}, {ymin, ymax}, All},
5,	{zmin,zmax},	{{xmin, xmax}, {ymin, ymax}, {zmin, zmax}}

The values in the first two columns specify the range for z only. The values in the third column also specify the ranges for x and y. The value **Automatic** determines the z range with an algorithm that may clip some high or low parts of the surface. Giving the value **All** ensures that the surface is wholly plotted without clipping. The value **Full** also causes the whole function to be plotted but, in addition, the whole plotting range on x and y axes is included in the plot. A constant value such as **5** means, for **Plot3D**, the plot range {Full, Full, {-5, 5}} and, for **Graphics3D** and **ParametricPlot3D**, the plot range {{-5, 5}, {-5, 5}, {-5, 5}}. In the first plot shown next, high values near the point (3, 3) are clipped:

```
{Plot3D[Exp[x y], {x, 0, 3}, {y, 0, 3}],
  Plot3D[Exp[x y], {x, 0, 3}, {y, 0, 3}, PlotRange → All]}
```

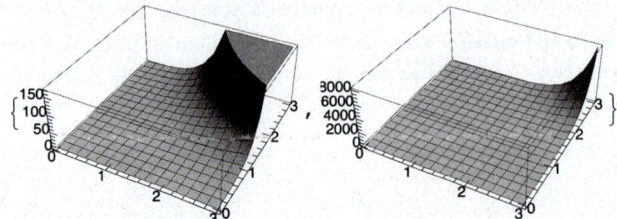

(ClippingStyle ^{s} (⌘6)   How to draw clipped parts of the surface; examples of values: Automatic,
   None, Red, {Green, Red})

■ **Margins and Background**

PlotRegion   Specifies margins around the plot inside the selection rectangle; examples of values:
   Automatic (means {{0, 1}, {0, 1}}), {{xmin, xmax}, {ymin, ymax}}
SphericalRegion   Whether to leave room in the plotting rectangle for a sphere enclosing the
   bounding box; possible values: False, True
(ImageMargins   Specifies margins (in printer's points) outside the selection rectangle; examples of
   values: 0, Automatic, 15, {{left, right}, {bottom, top}})
(ImagePadding (⌘6)   Extra space (in printer's points) for objects such as thick lines and tick and axes
   labels; examples of values: All, None, 15, {{left, right}, {bottom, top}})
(PlotRangePadding (⌘6)   How much farther axes etc. should extend beyond the range of coordinates specified by PlotRange; examples of values: Automatic, None, {0.2, 0.5})
Background   Color of the background; examples of values: None, LightGray

PlotRegion determines the 2D margins around the plot in the usual way, as it does for 2D graphics
(see Section 7.2.3, p. 191). The default value Automatic means {{0, 1}, {0, 1}}: The plot fills the entire
display area. A value such as {{0.05, 0.95}, {0.05, 0.95}} leaves wider margins; this may be useful
when using a colored background (see the example of Section 7.5.1, p. 214).

SphericalRegion determines whether to leave room in the plotting rectangle for a sphere enclosing
the bounding box. The value True is useful when rotating a graphic with animation because the graphic
has the same size from all viewpoints (this is important in animation).

■ **Fonts and Formatting**

* BaseStyle (⌘6)   Style of texts; examples of values: {}, {9, Bold, FontFamily → "Helvetica"}
LabelStyle (⌘6)   Style of all labels; examples of values: {}, {9, Bold, FontFamily → "Helvetica"}
(FormatType   Format type of text used in a plot; examples of values: TraditionalForm,
   StandardForm, InputForm, OutputForm)

For fonts and formatting, see Section 7.2.4, p. 192.

■ **Viewpoint and Lighting**

* ViewPoint   Point (in a scaled coordinate system) from which the surface is viewed; default value:
   {1.3, -2.4, 2}
RotationAction (⌘6)   How to render the plot if rotated with the mouse; examples of values: "Fit"
   (after rotation, the plot is rescaled to fit into the image region), "Clip" (after rotation, the plot is
   not rescaled and so may be clipped)
(ViewAngle (⌘6)   Angle of the field of view; default value: Automatic)
(ViewCenter   Point (in scaled coordinates) to display at the center; default value: {1/2, 1/2, 1/2})
(ViewMatrix (⌘6)   Explicit transformation matrix; default value: Automatic)
(ViewRange (⌘6)   Range of viewing distances to include; default value: All)
(ViewVector (⌘6)   Position and direction of a simulated camera; default value: Automatic)
(ViewVertical   Direction to make vertical; default value: {0, 0, 1})

We plot the same surface from three different viewpoints:

```
Plot3D[Sin[x y], {x, 0, π}, {y, 0, π}, Ticks → None, ViewPoint → #] & /@
   {{1.3, -2.4, 2}, {1.3, -2.4, 0}, {1.3, -2.4, -1}}
```

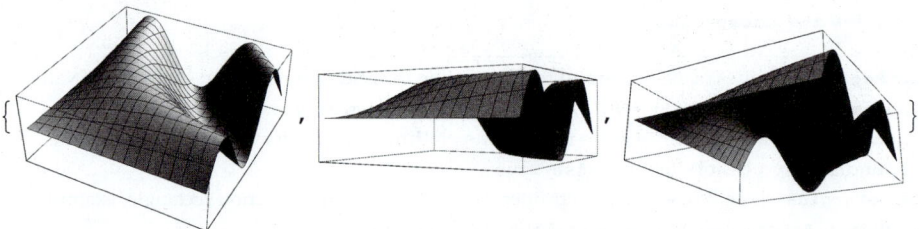

The viewpoint coordinates are not coordinates of the surface; the viewpoint has its own coordinate system. The origin of this system is at the center of the bounding box, and the longest side of the box runs from −0.5 to 0.5. The other sides then run so that the lengths of the sides satisfy the proportions expressed in **BoxRatios**. Thus, if we have **BoxRatios → {1, 1, 0.4}**, then the viewpoint coordinates of the box run from −0.5 to 0.5 for x and y and from −0.2 to 0.2 for z. The value of the **ViewPoint** option can also be one of **Above, Below, Front, Back, Left,** or **Right**. One of the coordinates can also be infinite, such as in **{0, 0, ∞}**.

Recall from Section 5.3.1, p. 139, that we can rotate a 3D plot by dragging with the mouse so that we can easily see the surface from various viewpoints. A 3D plot can also be zoomed with the mouse by holding down the ⌨CTRL (Windows) or ⌘ (Macintosh) key and dragging upward (to zoom in) or downward (to zoom out).

After rotating with the mouse, the plot is rescaled to fit into the image region. This causes a "jump" when the mouse is released. Give the **RotationAction** option the value **"Clip"** if you do not want the plot rescaled; now the plot may partly fall out of the image region and thus be clipped, or extra space may appear around the plot.

The current viewpoint of a rotated plot can be seen by pasting the figure as the argument to **Options**:

ViewPoint /. Options [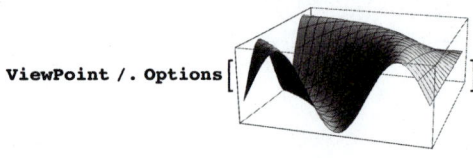]

{-0.972355, 3.08212, 1.00253}

Lighting What simulated lighting to use; simple examples of values: **Automatic, "Neutral", None**

The automatic lighting uses ambient light together with four light sources. Neutral lighting uses white light sources. If we use no lighting, the result is a black surface. For more about lighting, see the Documentation Center.

```
Plot3D[Sin[x y], {x, 0, π}, {y, 0, π}, Ticks → None, Lighting → #] & /@
  {Automatic, "Neutral", None}
```

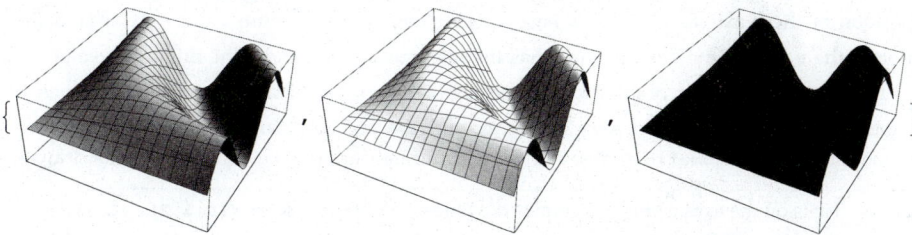

■ Miscellaneous Options

(**BaselinePosition** (🏶6) Where the baseline of a plot should be if the plot is combined with other
plots or text; examples of values: **Automatic, Axis, Bottom, Top, Center, Baseline**)

(**AlignmentPoint** (🏶6) How objects should by default be aligned when they appear in **Inset**;
default value: **Center**)

(**ContentSelectable** (🏶6) Whether and how content of a plot should be selectable; possible values:
Automatic (double-click allows content selection), **True** (single clicks immediately select content
objects), **False** (content objects cannot be selected))

(**PreserveImageOptions** (🏶6) Whether the size and margins of a plot should remain the same if the
plotting command is executed anew; examples of values: **Automatic** (the properties should remain
the same if not explicitly otherwise stated), **True** (the properties should remain the same), **False**
(the previous properties are ignored))

(**DisplayFunction** Function to apply to a graphic before returning it; default value:
$DisplayFunction (is normally **Identity**))

7.5.3 Local Options

■ Plot Label

PlotLabel Label of the plot; examples of values: **None, Sin[x y], "A surface plot"**

PlotLabel was considered in Section 7.3.1, p. 195; there is also an example in Section 7.5.1, p. 214.

■ Axes and Ticks

* **Axes** Whether to draw the axes; examples of values: **True** [sp], **False** [g], {True, True, False}
AxesEdge Where to draw the axes; examples of values: **Automatic**, {{1,1}, {-1,1}, {1,1}}, {
  Automatic, {-1,1}, {1,1}}, {{1,1}, {-1,1}, **None**}
* **AxesLabel** Labels for the axes; examples of values: **None, z, {x, y}, {x, y, z}**
AxesStyle Style of the axes; examples of values: {}, **Directive[Thick, Gray]**

The default value of **Axes** is **True** for **Plot3D** and **ParametricPlot3D** and **False** for **Graphics3D**. If
the plot has the bounding box, some edges of the box are selected as the axes. Note, however, that axes
and the box can be drawn independently: You can have axes without the box, the box without the axes,
or neither the axes nor the box; see forthcoming examples.

The edges of the box that are used as axes are, by default, determined automatically. However, each axis can be drawn on any of four edges with **AxesEdge**. In the next example, the first **{-1, -1}** defines the position of the x axis as the edge where y and z have the minimum values. **{1, -1}** defines the position of the y axis as the edge where x has the maximum value and z the minimum value. The last **{-1, -1}** defines the position of the z axis as the edge where x and y have the minimum values. (By the way, these definitions are unnecessary in this example because they are the default definitions.) Any definition can also be **Automatic** (that axis is chosen automatically) or **None** (that axis is not drawn).

---

* **Ticks** Ticks on the axes; simple examples of values: **Automatic**, **None**, **{{0,1,2}, {0,1},**
  Automatic}, {{0,1,2}, {0,1}, {-1,0,1}}
TicksStyle (⌘6) Style of tick marks and tick labels; examples of values: **{}**, **Blue**,
  Directive[Thick, Blue, 12]

```
Plot3D[Sin[x y], {x, 0, π}, {y, 0, π},
 AxesLabel → {x, y, z}, AxesEdge → {{-1, -1}, {1, -1}, {-1, -1}},
 AxesStyle → AbsoluteThickness[1], Ticks → {{0, π}, {0, π}, {-1, 1}}]
```

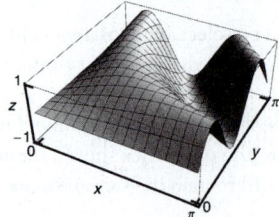

AxesLabel, **AxesStyle**, **Ticks**, and **TicksStyle** are used as they are with 2D plots (see Section 7.3.2, p. 196). We have marked **AxesLabel** with an asterisk because in 3D plots, the labels are useful: It is not obvious which one of the two horizontal axes is the x axis and which is the y axis.

■ **Bounding Box**

---

* **Boxed** Whether to draw a bounding box around the surface; possible values: **True**, **False**
* **BoxRatios** Ratios of side lengths of the bounding box; default values: **{1, 1, 0.4}** [s],
  Automatic [p8]
(**BoxStyle** Style of the bounding box; examples of values: **{}**, **Directive[Thick, Gray]**)

BoxRatios was considered in Section 7.5.2, p. 215. For example,

```
p1 = Plot3D[Sin[x y], {x, 0, π}, {y, 0, π}, Ticks → {{0, π}, {0, π}, {-1, 1}}];
{Show[p1, Boxed → False],
Show[p1, Axes → False, BoxStyle → Directive[Thick, Gray]],
Show[p1, Axes → False, Boxed → False]}
```

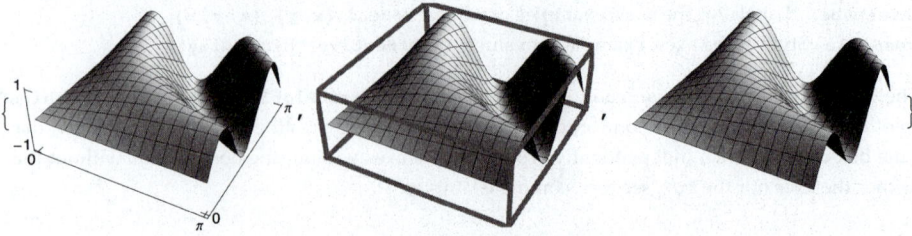

■ **Face Grids**

> **FaceGrids** Grid lines drawn on the faces of the bounding box; examples of values: **None**, **All**,
>   **{{-1,0,0}, {0,1,0}, {0,0,-1}}, {{{-1,0,0}, {ygrid, zgrid}}, {{0,1,0}, {xgrid, zgrid}},**
>   **{{0,0,-1}, {xgrid, ygrid}}}**
> (**FaceGridsStyle** Style of the grid lines; examples of values: **{}**, **Directive[Thick, Gray]**)

The value **All** draws grids on all six faces. We can define the faces where we want the grids. For example, **{-1,0,0}** is the face where x has the smallest value, **{0,1,0}** is the face where y has the largest value, and **{0,0,-1}** is the face where z has the smallest value. The example in Section 7.5.1, p. 214, contains grid lines.

■ **Primitives**

> (**Prolog** 2D graphics primitives to be plotted before the main plot; examples of values: **{}**,
>   **{PointSize[Medium], Red, Point[{0.4, 0.7}]}**)
> **Epilog** 2D graphics primitives to be plotted after the main plot; examples of values: **{}**,
>   **{PointSize[Medium], Red, Point[{0.4, 0.7}]}**

For **Prolog** and **Epilog**, we refer to Section 7.3.6, p. 201. These options can be used to add 2D graphics primitives to the plot. Note that the primitives cannot be 3D. In addition, the coordinates used in the primitives have to be given with scaled coordinates, which run from 0 to 1 in both horizontal and vertical directions.

If you want to add 3D primitives (see Section 6.2.11, p. 176), plot the primitives with **Graphics3D** and then use **Show** to combine the main plot and the primitives. For example,

```
p1 = ParametricPlot3D[{Cos[s] (3 + Cos[t]), Sin[t], Sin[s] (3 + Cos[t])},
    {s, 0, 2 π}, {t, 0, 2 π}, Boxed → False, Axes → False];
p2 = Graphics3D[{Thick, Line[{{-5, 0, 0}, {5, 0, 0}}], Line[{{0, -5, 0}, {0, 6, 0}}],
    Line[{{0, 0, -5.5}, {0, 0, 5}}], PointSize[Large], Red, Point[{0, 0, 0}]}];
Show[p1, p2, PlotRange → All, PlotRegion → {{-0.2, 1.2}, {-0.2, 1.2}}]
```

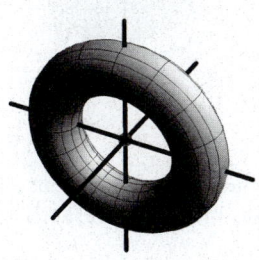

7.5.4 Options for the Surface

■ **Plot Style**

> **PlotStyle** ^{sp} (✻6) Style(s) of the surface(s); examples of values: **Automatic**, **Yellow**,
>   **Directive[Yellow, Specularity[Red, 5]]**, **Opacity[0.3]**, **None**
> (**BoundaryStyle** ^{sp} (✻6) How to draw boundary lines for surfaces; examples of values:
>   **Automatic** ^{s}, **None** ^{p}, **Thick**)
> (**ClippingStyle** ^{s} (✻6) How to draw clipped parts of the surface; examples of values: **Automatic**,
>   **None**, **Red**, **{Green, Red}**)
> (**NormalsFunction** ^{sp} (✻6) How to determine effective surface normals; default value: **Automatic**)

With **PlotStyle**, we can change the colors of a surface:

Plot3D[Sin[x y], {x, 0, π}, {y, 0, π}, Ticks → None, ImageSize → 90, PlotStyle → #] & /@
{Automatic, Yellow, Opacity[0.3], None}

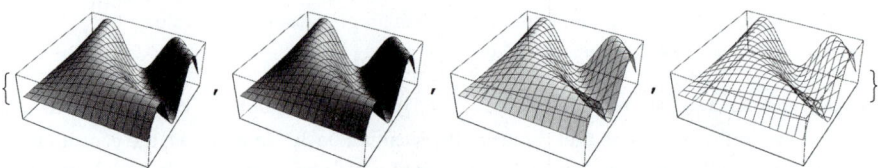

The default value **Automatic** for **ClippingStyle** produces plateaus at the top and bottom of the surface. The plateaus have no mesh lines; see the first plot shown next. The value **None** leaves the clipped parts empty, and this reveals the clipped parts very clearly; see the second plot. The clipped parts can also be shown with colors; see the third and fourth plots.

Plot3D[Sin[x y], {x, 0, π}, {y, 0, π}, Ticks → None, PlotRange → {-0.7, 0.85},
ImageSize → 90, ClippingStyle → #] & /@ {Automatic, None, Red, {Green, Red}}

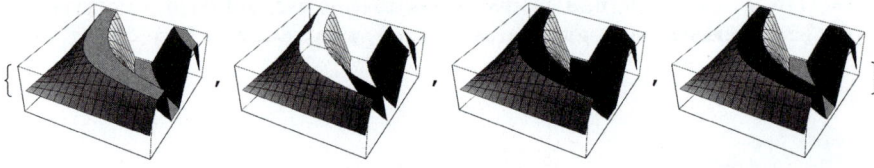

■ **Color Function**

> **ColorFunction** ^{sp} What colors are used to color the surface, if simulated illumination is not used;
>   examples of values: **Automatic**, **GrayLevel**, **"Rainbow"**, **(Hue[1 - #3] &)**, **(RGBColor[#3, 1 - #3,**
>   **0] &)**
> (**ColorFunctionScaling** ^{sp} Whether values provided for a color function are scaled to lie between
>   0 and 1 (**True**) or left as such (**False**))

Instead of simulated illumination, we can use a color function to color the surface. The function is expressed as a pure function (for pure functions, see Section 2.2.2, p. 38). In the function, we can use **#1**, **#2**, and **#3** to denote the *x*, *y*, and *z* coordinates, respectively. In a parametric surface, we can also use **#4** and **#5** to denote the parameters *u* and *v*, respectively. A useful coloring is obtained by coloring according to the value of *z*:

```
Plot3D[Sin[x y], {x, 0, π}, {y, 0, π}, Ticks → None, ColorFunction → #] & /@
 {Automatic, GrayLevel, "Rainbow"}
```

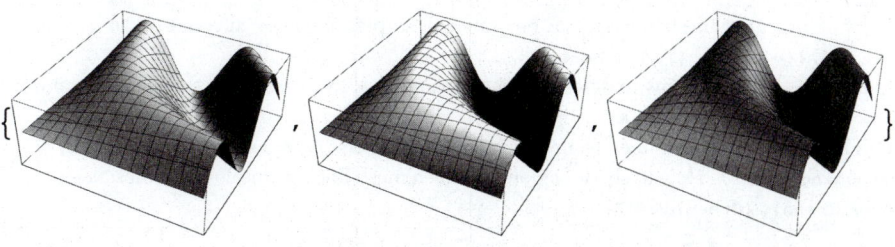

```
{Plot3D[Sin[x y], {x, 0, π}, {y, 0, π}, Ticks → None, ColorFunction → (Hue[1 - #3] &)],
 Plot3D[Sin[x y], {x, 0, π}, {y, 0, π},
  Ticks → None, ColorFunction → (RGBColor[#3, 1 - #3, 0] &)],
 ParametricPlot3D[{s Cos[t] Sin[s], s Cos[s] Cos[t], -s Sin[t]}, {s, 0, 2 π},
  {t, 0, π}, Mesh → False, ColorFunction → (Hue[#4 #5] &), Boxed → False, Axes → False]}
```

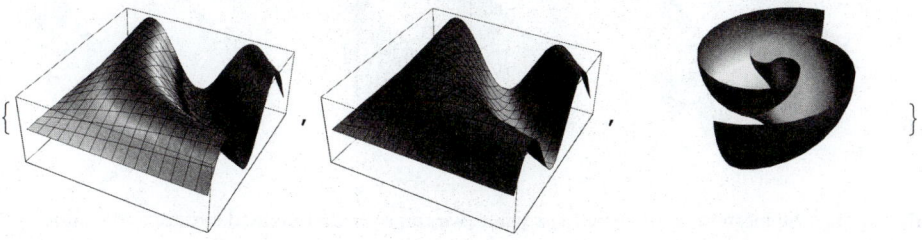

Here, we used the special color function **"Rainbow"**. To see a list of other color functions, execute **ColorData["Gradients"]**. For **ColorData**, see Section 9.3.3, p. 304.

■ Filling

(**Filling** ^{s} (❖6) Type of filling to use; examples of values: **None, Axis, Bottom, Top, 0.3, True**)
(**FillingStyle** ^{s} (❖6) Style of filling; examples of values: **Opacity[0.5], Red**)

Filling may be useful in connection with **RegionFunction** to make an unusual plotting region clearer:

```
{Plot3D[Sin[x y], {x, 0, π}, {y, 0, π}, Ticks → None, Filling → Bottom],
 Plot3D[Sin[x y], {x, 0, π}, {y, 0, π}, Ticks → None, Filling → Bottom,
  RegionFunction → ((#1 - π / 2) ^ 2 + (#2 - π / 2) ^ 2 ≤ (π / 2) ^ 2 &)]}
```

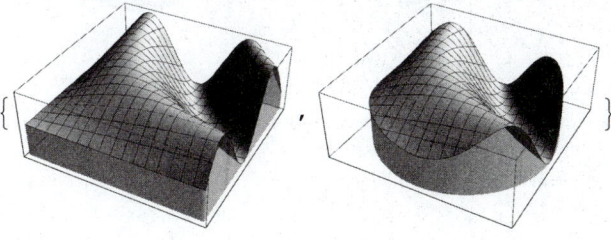

■ **Mesh Lines**

> * **Mesh** ^{sp} How many mesh lines should be drawn; examples of values: **Automatic**, 10, {5, 10},
> **Full, All, {{0, 1, 2}, {1, 2}}, None**
> **MeshFunctions** ^{sp} (⊞6) How to determine the placement of the mesh lines; examples of values: {#1
> **&, #2 &}** ^{s}, **Automatic** ^{p}, {#3 &}
> **MeshStyle** ^{sp} Style of mesh lines; examples of values: **Automatic, Directive[Red, Thick]**
> (**MeshShading** ^{sp} (⊞6) How to shade regions between mesh lines; examples of values: **None**,
> **{{Red,Green}, {Green, Red}})**

If the value of **Mesh** is **Full**, we get the initial sampling mesh. Value **All** gives the final sampling
mesh.

```
Plot3D[Sin[x y], {x, 0, π}, {y, 0, π}, Ticks → None, Mesh → #] & /@ {Full, All, None}
```

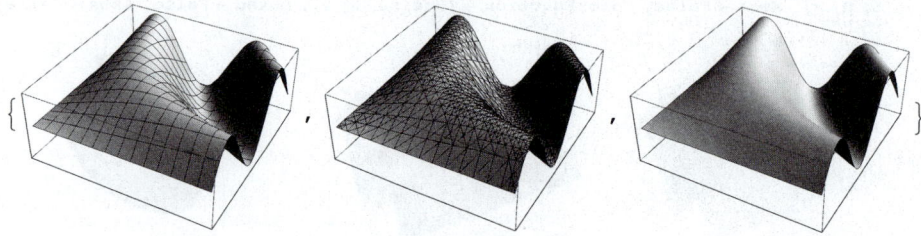

By defining the mesh function to be the z value, we can draw a combined surface and contour plot, as
is shown in the first plot below.

```
{Plot3D[Sin[x y], {x, 0, π}, {y, 0, π}, Ticks → None,
  MeshFunctions → {#3 &}, Mesh → {Range[-0.9, 0.9, 0.2]}],
 Plot3D[Sin[x y], {x, 0, π}, {y, 0, π}, Ticks → None,
  MeshShading → {{Red, Green}, {Green, Red}}],
 ParametricPlot3D[{Sin[s] Cos[t], Sin[s] Sin[t], 1 + Cos[s]}, {s, 0, π}, {t, 0, 2 π},
  Axes → False, Boxed → False, MeshStyle → Directive[Thickness[Small], White]]}
```

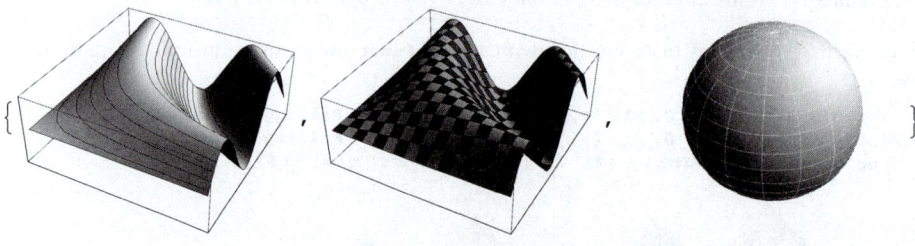

Next, we see three norm functions. Each plot contains a mesh curve where the norm gets the value 1:

```
Plot3D[Norm[{x, y}, #], {x, -1.5, 1.5}, {y, -1.5, 1.5},
   PlotRange → {0, 3}, MeshFunctions → {#3 &}, Mesh → {{1}},
   BoxRatios → Automatic, ViewPoint → {1.1, -2.3, 2.4}] & /@ {1, 2, ∞}
```

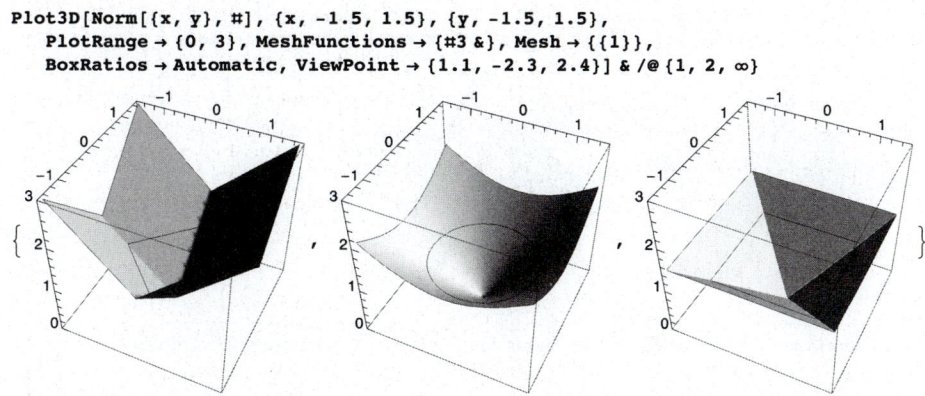

■ Exclusions and Region Function

Exclusions ^{sp} (✿6) The curves on the (x, y) surface that are excluded in plotting; examples of values: **Automatic, None, {x + y == 0}**

(**ExclusionsStyle** ^{sp} (✿6) What to draw at excluded curves; examples of values: **None, Opacity[0.5], {Opacity[0.5], Directive[Red, Thick]}**)

The default is that nothing is drawn at excluded curves, as can be seen from the first plot shown next. In the second plot, we ask not to exclude anything. In the third plot, we use a custom style to indicate the exclusions. In the fourth plot, we use a special style for the boundary of the surface at the excluded curve.

```
GraphicsRow[{Plot3D[Im[Log[x + I y]], {x, -1, 1}, {y, -1, 1}, Ticks → None],
   Plot3D[Im[Log[x + I y]], {x, -1, 1}, {y, -1, 1}, Ticks → None, Exclusions → None],
   Plot3D[Im[Log[x + I y]], {x, -1, 1},
     {y, -1, 1}, Ticks → None, ExclusionsStyle → Opacity[0.5]],
   Plot3D[Im[Log[x + I y]], {x, -1, 1}, {y, -1, 1}, Ticks → None,
     ExclusionsStyle → {None, Directive[Red, Thick]}]}, ImageSize → 420]
```

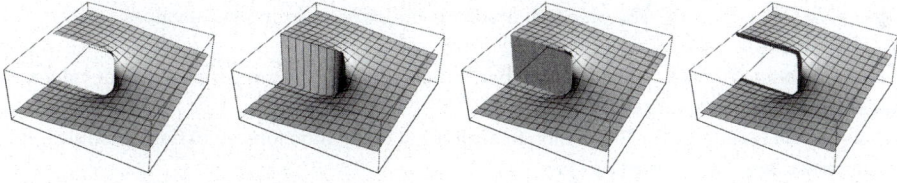

Exclusions can be defined with equations:

```
{Plot3D[1 / (x + y) ^ 3, {x, -1, 1}, {y, -1, 1}, Ticks → None],
 Plot3D[1 / (x + y) ^ 3, {x, -1, 1}, {y, -1, 1}, Ticks → None, Exclusions → {x + y == 0}]}
```

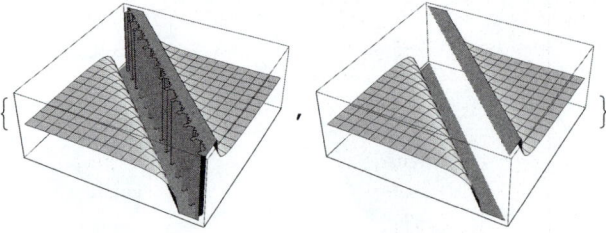

RegionFunction [sp] (⌘6) Specifies the region to include in the plot drawn; examples of values:

  (True &), (#1^2 + #2^2 < 1 &)

With **RegionFunction**, we can show a surface at a region defined by a logical combination of inequalities:

```
Plot3D[Sin[x y], {x, 0, π}, {y, 0, π}, Ticks → None,
  RegionFunction → ((π / 3) ^ 2 < (#1 – π / 2) ^ 2 + (#2 – π / 2) ^ 2 ≤ (π / 2) ^ 2 &)]
```

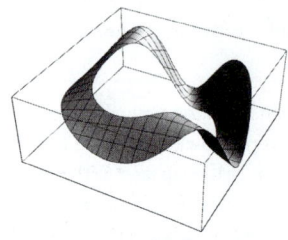

■ Plotting Algorithm

PlotPoints [sp]   Number of initial sampling points; examples of values: **Automatic**, **30**, **{30, 50}**

MaxRecursion [sp] (⌘6)   The maximum number of recursive subdivisions allowed; examples of
  values: **Automatic**, **3**

WorkingPrecision [sp] (⌘6)   The precision used in computations; examples of values:
  MachinePrecision, **20**

(**PerformanceGoal** [sp] (⌘6)   What aspect of performance to try to optimize; examples of values:
  $PerformanceGoal, **"Quality"**, **"Speed"**)

(**EvaluationMonitor** [sp] (⌘6)   Expression to evaluate at every function evaluation; examples of
  values: **None**, **Sow[{x, y, Sin[x y]}]**)

7.6 Options for Contour and Density Plots

7.6.1 Special Options

Contour and density plots describe 3D functions, but the plots are like 2D plots. Indeed, most options and their default values of these commands are the same as those of **Plot**. Thus, we can mainly refer to Sections 7.1 through 7.4 for the options. Here, it may suffice to mention the differences between the default values of a few options and some new options of **ContourPlot**.

■ **Aspect Ratio**

The default value of **AspectRatio** is **1/GoldenRatio** for **Plot** but **1** for **ContourPlot** and **DensityPlot**. Thus, the form of a contour or density plot is, by default, a square. If you want one unit in the x and y axes to have the same length, give the value **Automatic** for this option.

■ **Plot Range**

Note that the value of **PlotRange** in a contour or density plot is given as for **Plot3D**. Remember that high or low z values may be clipped. In a contour or density plot, the clipped parts appear as white or dark blue regions and may easily remain unobserved. Use **PlotRange → All** to see all z values:

```
{ContourPlot[Exp[x y], {x, 0, 1.5}, {y, 0, 1.5}],
 ContourPlot[Exp[x y], {x, 0, 1.5}, {y, 0, 1.5}, PlotRange → All]}
```

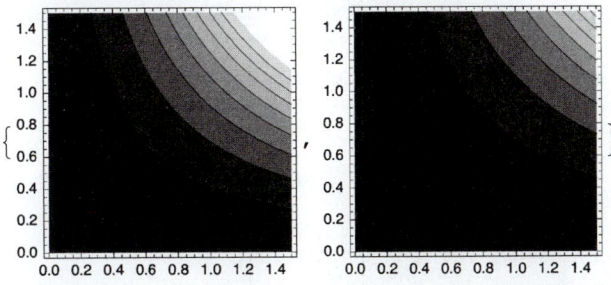

■ **Axes, Frame, and Ticks**

A contour or density plot has, by default, a frame and no axes. Thus, ticks are set with **FrameTicks** and not with **Ticks**, and x and y labels are set with **FrameLabel** and not with **AxesLabel**. If the y label is short, the rotating of the label should be avoided by setting **RotateLabel → False**. Axes can be useful either in addition to the frame or in place of the frame:

```
{p1 = ContourPlot[-x^4 - 3 x^2 y - 5 y^2 - x - y, {x, -1.6, 0.6},
   {y, -1.1, 0.6}, AspectRatio → Automatic, PlotRange → All, Contours → 24],
 Show[p1, Axes → True],
 Show[p1, Axes → True, Frame → False]}
```

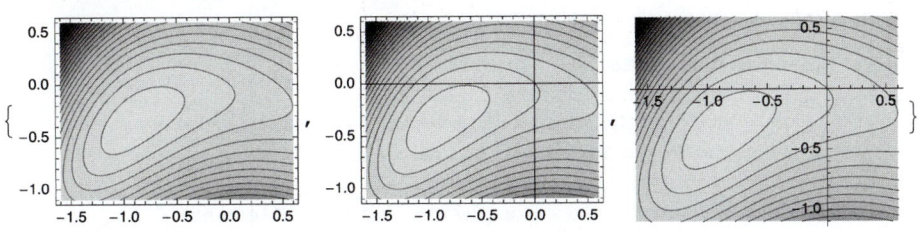

■ **Color Function**

ColorFunction How to color the regions between the contour lines (**ContourPlot**) or how to color various values of the function (**DensityPlot**); examples of values: **Automatic**, **GrayLevel**, **"Rainbow"**, **(Hue[1 - #] &)**, **(RGBColor[#, 0, 1 - #] &)**

(**ColorFunctionScaling** Whether values provided for a color function are scaled to lie between 0 and 1 (**True**) or left as such (**False**))

`ContourPlot` and `DensityPlot` do not have the `PlotStyle` option. Instead, we can use the `ColorFunction` option:

`ContourPlot[Cos[x y] Cos[x], {x, 0, π}, {y, 0, π}, ColorFunction → #] & /@`
`{Automatic, GrayLevel, (RGBColor[#, 0, 1 - #] &)}`

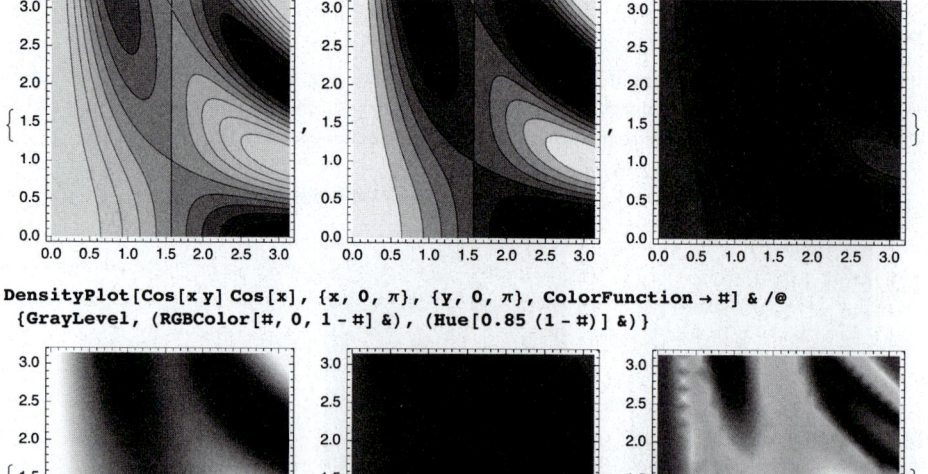

`DensityPlot[Cos[x y] Cos[x], {x, 0, π}, {y, 0, π}, ColorFunction → #] & /@`
`{GrayLevel, (RGBColor[#, 0, 1 - #] &), (Hue[0.85 (1 - #)] &)}`

From the last density plot we can see that the colors are not smooth enough. A setting of, for example, `PlotPoints → 50` would give a good result. As an aside, from the first density plot I get the strong impression of a surface seen above and lighted from northeast, but this interpretation is wrong: The gray parts are not shadows but instead describe deep parts of the surface.

`ContourPlot` and `DensityPlot` have the `BoundaryStyle` option (with default value `None`) to define the style of the boundary of the plotted region.

■ Special Options of ContourPlot

* **Contours** Number of z values for which contour lines are to be drawn or a list of z values; examples of values: `Automatic, 10, {-2, -1, 0, 1, 2}`
* **ContourLines** Whether to draw the contour lines; possible values: `True, False`
  ContourStyle Style of the contour lines; examples of values: `Automatic, Black`
  ContourLabels (⌘6) How to label contour lines; examples of values: `None, Automatic`
* **ContourShading** How to shade regions between the contour lines; examples of values: `Automatic, True, False`

In the first plot that follows, we ask for five equally paced contours, whereas in the second plot we define the contours explicitly. In the third plot, we ask not to draw the contour lines.

```
{ContourPlot[Cos[x y] Cos[x], {x, 0, π}, {y, 0, π}, Contours → 5],
 ContourPlot[Cos[x y] Cos[x], {x, 0, π}, {y, 0, π}, Contours → Range[-1, 1, 0.25]],
 ContourPlot[Cos[x y] Cos[x], {x, 0, π}, {y, 0, π}, ContourLines → False]}
```

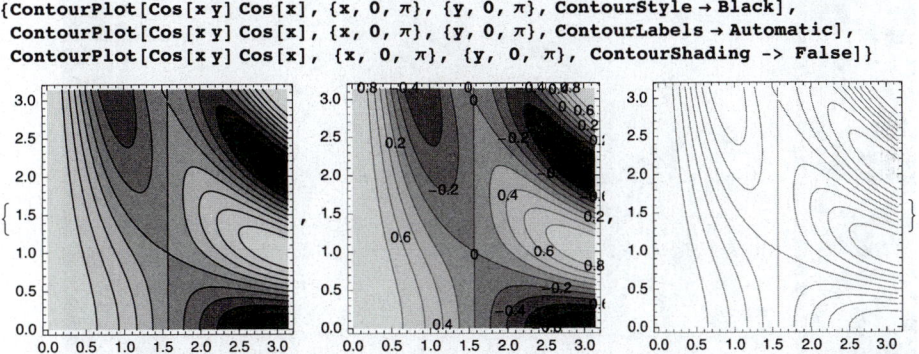

In the first plot that follows, we define black to be the style of the contours. In the second plot, we ask to label the contours. In the third plot, we ask not to use shading between the contour lines.

```
{ContourPlot[Cos[x y] Cos[x], {x, 0, π}, {y, 0, π}, ContourStyle → Black],
 ContourPlot[Cos[x y] Cos[x], {x, 0, π}, {y, 0, π}, ContourLabels → Automatic],
 ContourPlot[Cos[x y] Cos[x], {x, 0, π}, {y, 0, π}, ContourShading -> False]}
```

Remember that by moving the mouse pointer over a contour plot (without pressing the mouse button), we can see the constants corresponding to the contours.

Using evenly spaced contours, it may sometimes be difficult to get a good description of a function, even with a large number of contours. The value of the **Contours** option can also be a function of the minimum and maximum values of the function, and with such a function we can get better contours:

```
{ContourPlot[(x - 1)^2 + 100 (y - x^2)^2, {x, -1, 1},
 {y, -1, 1}, ContourShading → False, Contours → 30, PlotRange → All],
 ContourPlot[(x - 1)^2 + 100 (y - x^2)^2, {x, -1, 1},
 {y, -1, 1}, ContourShading → False, PlotRange → All,
 Contours → Function[{min, max}, Exp[Range[0, Log[max], Log[max] / 10]]]]}
```

■ **Example**

For the following plot, we use several options to enhance it.

```
f = -x^4 - 3 x^2 y - 5 y^2 - x - y;
p = {x, y} /. FindMaximum[f, {x, 0}, {y, 0}][[2]]
{-0.886324, -0.335671}

ContourPlot[f, {x, -1.3, 0.1}, {y, -0.8, 0.1}, Contours → 19,
  AspectRatio → Automatic, PlotRegion → {{0.01, 0.94}, {0.04, 0.96}},
  Background → Black, BaseStyle → {10, White}, ColorFunction → "Rainbow",
  PlotLabel → Style[-x^4 - 3 x^2 y - 5 y^2 - x - y, Bold, 14],
  FrameTicks → {{p[[1]], 0}, {p[[2]], 0}, None, None},
  FrameStyle → Thickness[Medium], ImageSize → 400,
  Epilog → {PointSize[Medium], Point[p], Text["Maximum point", p, {-1.1, -1.4}]}]
```

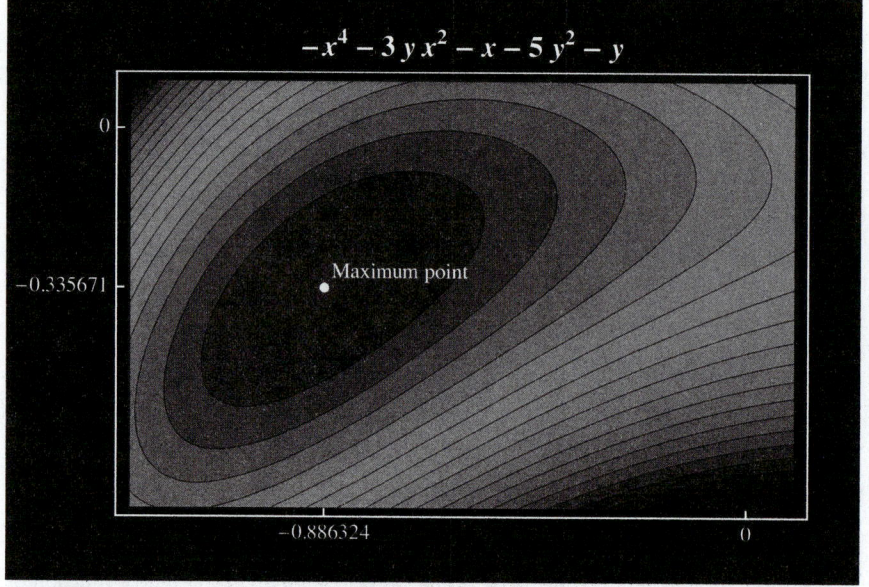

Graphics for Data

Introduction

> *"Data! data! data!" he cried impatiently. "I can't make bricks without clay."—Sherlock Holmes*

In this chapter, we present ways to illustrate data with 2D and 3D graphics. A basic situation encountered when plotting data is one or more time series: data sets with one dependent variable and time as the independent variable. Plotting such data is considered in Section 8.1. The basic commands are **ListPlot** and **ListLinePlot**, but we can also easily use **Graphics**.

When we want to study the relationships between two or more dependent variables, the scatter plot methods of Section 8.2 are useful. The basic approach, with **ListPlot** or **PairwiseScatterPlot**, is to plot one variable against another variable, thereby yielding a scatter plot or, more generally, a scatter plot matrix. This method of pairing observations is also used in quantile–quantile plots.

Other plotting commands for data include **BarChart**, **Histograms**, **dotPlot**, **BoxWhiskerPlot**, and **PieChart**. These are considered in Sections 8.3 and 8.4. New in *Mathematica* 6 are **GraphPlot**, **LayeredGraphPlot**, and **TreePlot**; they are addressed in Section 8.5.

For 3D graphics, **ListPlot3D**, **ListContourPlot**, and **ListDensityPlot** make up one collection of commands to be tried. An effective plot may be a 3D bar chart done with **BarChart3D**. Also, a scatter plot made with **ListPointPlot3D** may sometimes be effective.

In Chapter 9, we consider the data sets that come with *Mathematica*. There, we use some of the plotting methods presented here. The data sets also contain some original graphics such as chemical diagrams, photos of astronomical objects, shapes of countries, flags, polyhedrons, graphs, lattices, knots, and various test images.

Chapters 10 and 11 contain some manipulations of data. For example, we show how to interactively and graphically study interpolation and approximation.

8.1 Basic Plots

8.1.1 Built-in Plots

■ Basic Plots

ListPlot[data]   Plot points

ListLinePlot[data] (⌘6)   Plot joining lines

ListLinePlot[data, Mesh → All]   Plot joining lines and points

ListPlot[data, Filling → Axis]   Plot points and vertical lines (stems)

ListLinePlot[data, Filling → Axis]   Plot joining lines; fill area below the curve

ListLinePlot[data, Mesh → All, Filling → Axis]   Plot joining lines and points; fill area below the curve

Data can be given in either of the following forms:

$\{y_1, y_2, ...\}$ Plot the points $\{1, y_1\}, \{2, y_2\}, ...$

$\{\{x_1, y_1\}, \{x_2, y_2\}, ...\}$ Plot the given points

Instead of **ListLinePlot[data]**, we can apply **ListPlot[data, Joined → True]**. As an example, we plot 30 random numbers in all six ways:

```
SeedRandom[1]; data = Table[{x, 0.2 x + 2 RandomReal[]}, {x, 0, 30}];
```

```
{ListPlot[data],
 ListLinePlot[data],
 ListLinePlot[data, Mesh → All]}
```

```
{ListPlot[data, Filling → Axis],
 ListLinePlot[data, Filling → Axis],
 ListLinePlot[data, Mesh → All, Filling → Axis]}
```

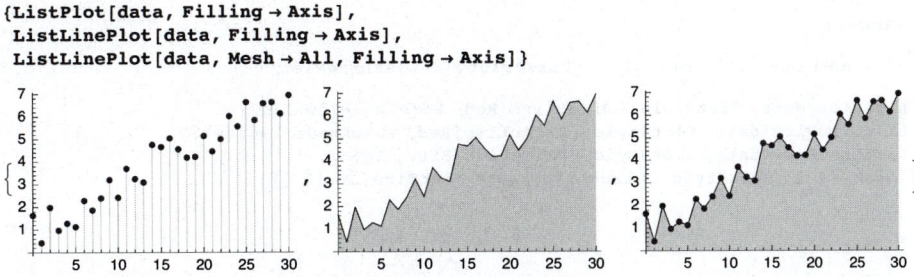

The first of these six plots shows clearly the points and the second plot shows the path, but the third shows clearly both the points and the path. The plots in the second row make the data somewhat more concrete with the vertical lines or fills. A plot with vertical lines is able to clearly reveal possible lacking observations because the lacking lines will be so remarkable. A plot with a fill makes the shape of the curve very clear.

■ Defining Styles

Defining styles in **ListPlot**:
PlotStyle Style of points
FillingStyle (⌘6) Style of vertical lines

Defining styles in **ListLinePlot**:
PlotStyle Style of joining lines
MeshStyle Style of points
FillingStyle (⌘6) Style of fills

A single style such as **Red** can be written as such, but several styles need either braces or **Directive** (see Section 7.1.2, p. 188).

The default size of points is **AbsolutePointSize[3]**, and the default thickness of lines is **AbsoluteThickness[0.5]**. The default color is **Hue[0.67, 0.6, 0.6]** (a modified blue); in vertical lines, this color is modified with **Opacity[0.2]** (thus, the blue color is very transparent and light).

Remember that we considered the styles of points and lines in Sections 6.2.1, p. 155, and 6.2.2, p. 156. Styles are defined with so-called style directives. The following is a summary:

- For colors, we have directives such as **Hue**, **RGBColor**, **GrayLevel**, and **Opacity**; for some colors, we also have ready-to-use names such as **Red** or **LightBlue**.
- Point size can be defined with **PointSize** or **AbsolutePointSize**.
- Thickness of lines can be defined with **Thickness** or **AbsoluteThickness**; we also have special thickness definitions **Thin** and **Thick**.
- Dashing of lines can be defined with **Dashing** or **AbsoluteDashing**; we also have special dashing definitions **Dotted**, **Dashed**, and **DotDashed**.
- The argument of the directives defining point size, thickness, or dashing can also be one of the special symbols **Tiny**, **Small**, **Medium**, or **Large**.

■ **Examples**

Here are some examples of using styles in `ListPlot` and `ListLinePlot`:

```
{ListPlot[data, PlotStyle → Directive[Red, PointSize[Small]]],
 ListLinePlot[data, PlotStyle → Directive[Red, Thickness[Medium]]],
 ListLinePlot[data, PlotStyle → Directive[Blue, Thin],
  Mesh → All, MeshStyle → Directive[Red, PointSize[Small]]]}
```

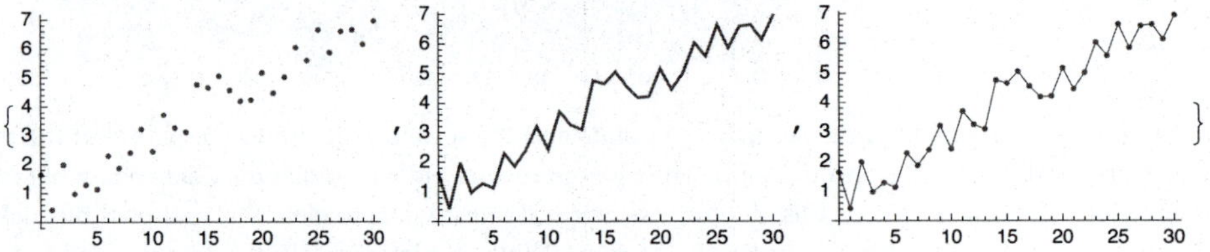

```
{ListPlot[data, PlotStyle → Directive[Red, PointSize[Small]],
  Filling → Axis, FillingStyle → Directive[Blue, Thickness[Small]]],
 ListLinePlot[data, PlotStyle → Directive[Red, Thickness[Medium]],
  Filling → Axis, FillingStyle → Lighter[Blue, 0.6]],
 ListLinePlot[data, PlotStyle → Directive[Blue, Thin],
  Mesh → All, MeshStyle → Directive[Red, PointSize[Small]],
  Filling → Axis, FillingStyle → Directive[Opacity[0.3], Blue]]}
```

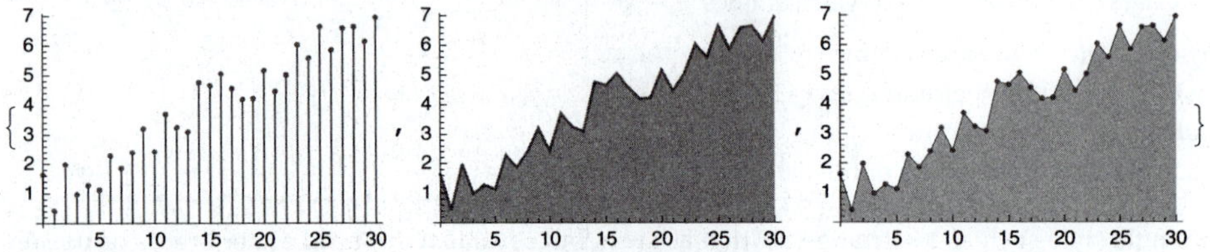

If the data contain both negative and positive values, we can define a different filling style for the negative and positive values:

```
ListLinePlot[Table[{x, -4 + 0.2 x + 2 RandomReal[]}, {x, 0, 30}],
 Filling → Axis, FillingStyle → {LightBlue, LightRed}]
```

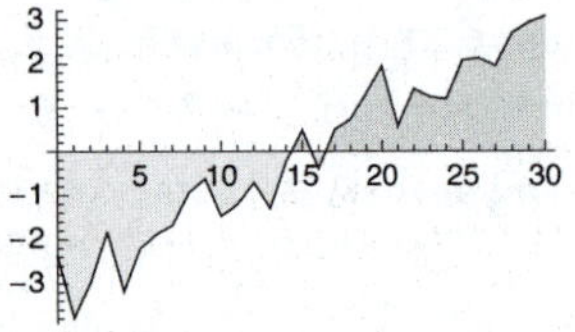

■ **Options**

The options and their default values for `ListPlot` are mostly the same as for `Plot`. However, `ListPlot` has five options that `Plot` does not have. On the other hand, `ListPlot` does not have some options (e.g., `PlotPoints` and `MaxRecursion`) of `Plot` that control the sampling of the function to be plotted.

The options of `ListPlot` and `ListLinePlot` are otherwise the same, but the default value of `Joined` is `False` for `ListPlot` and `True` for `ListLinePlot`.

Thus, we can mainly refer to Chapter 7 for the options of `ListPlot` and `ListLinePlot`. However, here are the five options that `Plot` does not have.

DataRange   The range of x values to assume; examples of values: **Automatic, {0, 1}**

InterpolationOrder   The degree of the polynomials joining the points; examples of values: **None, 0, 1, 2, 3**

Joined (❀6)   Whether to join the points; possible values: **False** (the default for **ListPlot**), **True** (the default for **ListLinePlot**)

MaxPlotPoints   The maximum number of points plotted; examples of values: ∞, **100**

PlotMarkers (❀6)   Markers to use for the points; examples of values: **None, Automatic**

In our previous examples, we had points with both x and y coordinates. If the data do not contain the x coordinates, the x coordinates 1, 2, … are used. If we want to determine the x coordinates, we can either add the coordinates to the data or use the **DataRange** option. Here are data without the x coordinates:

```
SeedRandom[1]; data2 = Table[0.2 x + 2 RandomReal[], {x, 0, 30}];
```

If we plot these data as such, the x coordinates will be 1, …, 31, not 0, …, 30. However, we can add the x coordinates:

```
data3 = {Range[0, 30], data2}ᵀ;
```

(Here, ᵀ means transpose; it can be written as ⎡ESC⎤tr⎡ESC⎤.) We can also use the **DataRange** option:

```
{ListPlot[data3], ListPlot[data2, DataRange → {0, 30}]}
```

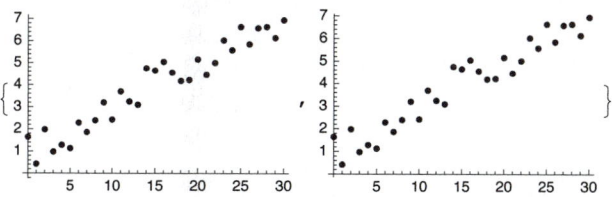

With the **InterpolationOrder** option we can get piecewise constant, linear (the default), etc. curves between the points. With the value 3, we get third-order spline interpolation:

```
t = RandomInteger[{1, 6}, {20}]
```

{4, 6, 3, 2, 1, 3, 1, 6, 5, 1, 1, 6, 5, 1, 1, 4, 1, 2, 3, 4}

```
ListLinePlot[t, InterpolationOrder → #, Mesh → Full] & /@ {0, 1, 3}
```

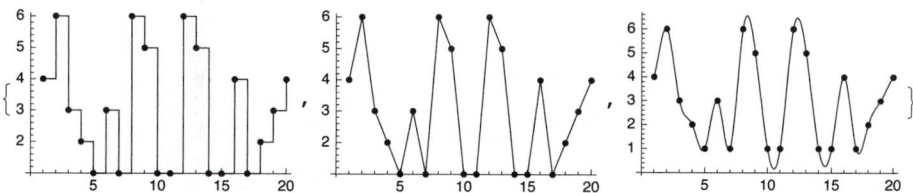

For large data sets, the **MaxPlotPoints** option may be used to reduce the number of points plotted. The **PlotMarkers** option is useful in plotting several data sets, and so it is considered in Section 8.1.3, p. 244.

■ **Tooltips**

Tooltip[data] (⌘6) Show the coordinates of **data** as a tooltip when the mouse pointer is moved
over the points of **data**
Tooltip[{xi, yi}, label] Show **label** as a tooltip for point **{xi, yi}**

In the following plot, we can see the coordinates of a point if we move the mouse pointer over that
point:

ListLinePlot[Tooltip[data], Mesh → All]

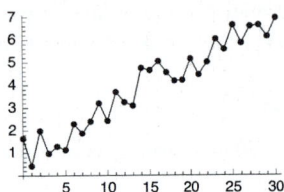

In the next plot, we can see labels of the form $f[k]$ when we move the mouse pointer over the points:

dist = BinomialDistribution[20, 1 / 2];
t = Table[Tooltip[{k, PDF[dist, k]}, f[k]], {k, 0, 20}];
ListPlot[t, Filling → Axis]

■ **Example**

Generate 100 random points on the unit square:

SeedRandom[2]; xy = RandomReal[1, {100, 2}];

The task is to visit all of the points once in such a way that the length of the tour is the minimum. With
the following command, we can try to solve this problem:

tour = FindShortestTour[xy]

{7.86367, {1, 16, 39, 32, 59, 68, 6, 96, 75, 29, 80, 19, 85, 38, 78, 21, 65, 92, 97, 57,
52, 53, 84, 17, 7, 70, 83, 89, 88, 26, 11, 47, 5, 95, 86, 79, 82, 30, 93, 48,
9, 77, 94, 55, 71, 10, 12, 41, 27, 31, 2, 22, 58, 14, 99, 91, 42, 64, 36, 33,
72, 50, 28, 69, 45, 40, 87, 60, 15, 25, 63, 34, 67, 13, 61, 20, 24, 4, 43, 46,
51, 23, 54, 100, 62, 98, 90, 8, 74, 56, 18, 66, 73, 44, 81, 76, 3, 35, 37, 49}}

The length of the tour find is 7.86. The tour goes through the 1st, 16th, ..., 49th point. We show both the
points and the (at least near optimal) tour:

```
{ListPlot[xy, AspectRatio → 1],
 ListLinePlot[xy[[tour // Last]], Mesh → All, AspectRatio → 1]}
```

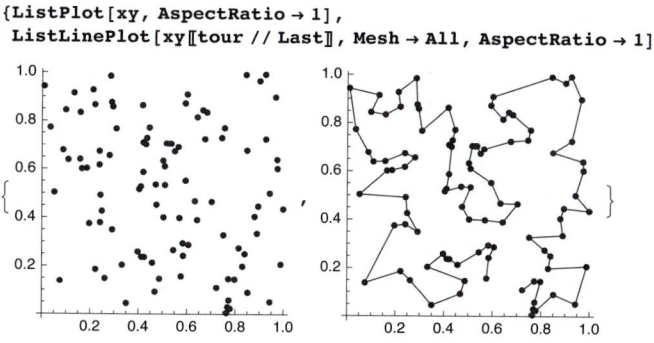

■ **Other Plots**

> **ListLogPlot[data]** (⌘6)   A plot of **Log[yi]** as a function of **xi**
> **ListLogLinearPlot[data]** (⌘6)   A plot of **yi** as a function of **Log[xi]**
> **ListLogLogPlot[data]** (⌘6)   A plot of **Log[yi]** as a function of **Log[xi]**

The data are given in the same form as they are for **ListPlot**. The option **Joined** can be used. In a logarithmic scale, exponential data are close to a line:

```
SeedRandom[4];
data = Table[{x, Exp[x] + RandomReal[{-0.5, 0.5}]}, {x, 0.1, 2, 0.1}];
ListLogPlot[data]
```

> *In the* **ErrorBarPlots`** *package:*
>
> **ErrorListPlot[data]**   Show data by points with error bars
>
> Examples of data forms:
>
> **{{y₁, yerr₁}, … }**
>
> **{{x₁, y₁, yerr₁}, … }**
>
> **{{{x₁, y₁}, ErrorBar[{negerr₁, poserr₁}]}, … }**

In the first two data forms, the error is shown by a vertical bar centered at **{xᵢ, yᵢ}** and having a total length of two times the given error. In the third data form, there are different errors in the negative and positive directions. For example,

```
data = Table[
   {x, 1.5 + Sin[x] + RandomReal[{-0.2, 0.2}], RandomReal[{0, 0.3}]}, {x, 0, 7, 0.2}];
<< ErrorBarPlots`
```

```
ErrorListPlot[data, AspectRatio → 0.4,
  PlotRange → All, AxesOrigin → {0, 0}, ImageSize → 230]
```

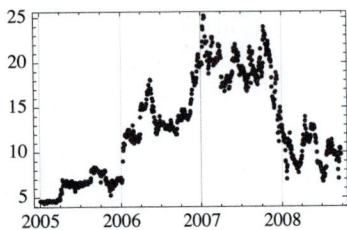

`DateListPlot[{{date₁, y₁}, … }]` (⌘6) Plot points with dates as *x* coordinates

To try this command, let us examine financial data; see Section 9.2.2, p. 299. The stock price of ADY from January 3, 2005, on has been as follows:

```
FinancialData["ADY", "Jan. 3, 2005"] // Short
```

`{{{2005, 1, 3}, 4.61}, {{2005, 1, 4}, 4.59}, ≪937≫, {{2008, 9, 25}, 9.75}}`

Here is a plot of the price until now:

```
DateListPlot[FinancialData["ADY", "Jan. 1, 2005"],
  ImageSize → 160, PlotStyle → PointSize[Small]]
```

Look at the Documentation Center for more information about **DateListPlot**.

`ListPolarPlot[radii]` (⌘6) Plot points equally spaced in angle at `radii`

8.1.2 Self-Made Plots

■ **Using Graphics**

We have **ListPlot** and **ListLinePlot** to plot data. However, it is very easy to plot data directly with the graphics primitives **Point** and **Line** (see Sections 6.2.1, p. 155, and 6.2.2, p. 156). With **Graphics** we collect the primitives together and show the resulting plot (see Section 6.1.1, p. 153). Here are ways to prepare four kinds of plots we encountered previously plus a plot where the points are shown as circles.

`Graphics[Point[data]]` Points
`Graphics[Line[data]]` Joining lines
`Graphics[{Line[data], Point[data]}]` Joining lines and points
`Graphics[{ Line[{{#⟦1⟧, 0}, #}] & /@ data, Point[data]}]` Vertical lines and points
`Graphics[{Line[data], White, EdgeForm[Black],
   Disk[#, Offset[{1.5, 1.5}]] & /@ data}]` Joining lines and circles
---

These commands require that the data points contain both the x and the y coordinate. We can add some directives into the commands to get suitable styles for the plots. Also, we have to add the options **Axes → True** and **AspectRatio → 1/GoldenRatio** to get similar plots as we get with **ListPlot** and **ListLinePlot** (remember that **Graphics** has, by default, **Axes → False** and **AspectRatio → Automatic**). In the following, we first save the default values of the options of **Graphics** and then set some options with **SetOptions**:

```
gropts = Options[Graphics];
SetOptions[Graphics, Axes → True,
   AspectRatio → 1 / GoldenRatio, AxesOrigin → {0, 0}];
```

Now we try using **Graphics**:

```
SeedRandom[1]; data = Table[{x, 0.2 x + 2 RandomReal[]}, {x, 0, 30}];
```

```
{Graphics[{Red, PointSize[Small], Point[data]}],
 Graphics[{Blue, Thickness[Medium], Line[data]}]}
```

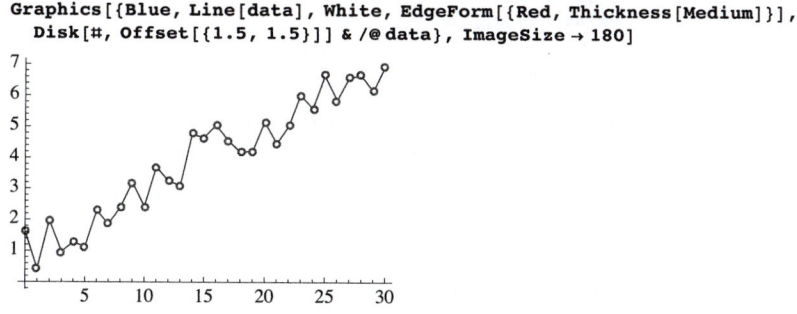

```
{Graphics[
  {Blue, Thickness[Small], Line[data], Red, PointSize[Small], Point[data]}],
 Graphics[{Blue, Thickness[Small], Line[{{#[[1]], 0}, #}] & /@ data,
   Red, PointSize[Small], Point[data]}]}
```

Similarly, we can plot the data with circles [circles are popular in scientific publications; for example, Cleveland (1993) mainly uses circles]. If we used **Circle**, then the joining lines could be seen inside the circles, which is not beautiful. Therefore, we use **Disk** because it has a filled inside (see Section 6.2.7, p. 166). In the following example, we define the disks to be white with red edges:

```
Graphics[{Blue, Line[data], White, EdgeForm[{Red, Thickness[Medium]}],
   Disk[#, Offset[{1.5, 1.5}]] & /@ data}, ImageSize → 180]
```

Here we expressed the radius of the disks with **Offset** to get true circles and not ellipses.

■ An Example

In this example, we consider the numbers of hare pelts sold to the Hudson Bay Trading Company in Canada from 1844 to 1934. These observations are from Burghes and Borrie (1981) (reproduced with the permission of the authors). First, we read the numbers of hare pelts from a file. I have the data in a text file called **hare** in a folder called **MNData**:

```
haredata = Import["/Users/heikki/Documents/MNData/hare", "Table"]
```

```
{{1844, 30}, {1845, 25}, {1847, 25}, {1848, 15}, {1849, 30}, {1850, 55}, {1851, 80},
 {1852, 80}, {1853, 90}, {1854, 70}, {1855, 80}, {1856, 95}, {1857, 75},
 {1858, 30}, {1859, 15}, {1860, 20}, {1861, 40}, {1862, 5}, {1863, 155},
 {1864, 140}, {1865, 105}, {1866, 45}, {1867, 20}, {1868, 5}, {1869, 5}, {1870, 10},
 {1871, 10}, {1872, 60}, {1873, 50}, {1874, 50}, {1875, 105}, {1876, 85},
 {1877, 60}, {1878, 15}, {1879, 10}, {1880, 15}, {1881, 10}, {1882, 10},
 {1883, 40}, {1884, 50}, {1885, 135}, {1886, 135}, {1887, 90}, {1888, 30},
 {1889, 20}, {1890, 50}, {1891, 55}, {1892, 60}, {1893, 55}, {1894, 80},
 {1895, 95}, {1896, 50}, {1897, 15}, {1898, 5}, {1899, 5}, {1900, 15}, {1901, 5},
 {1902, 10}, {1903, 50}, {1904, 70}, {1906, 20}, {1909, 25}, {1910, 50},
 {1911, 55}, {1912, 75}, {1913, 70}, {1914, 55}, {1915, 30}, {1916, 20},
 {1917, 15}, {1918, 15}, {1919, 20}, {1920, 35}, {1921, 60}, {1922, 80},
 {1923, 85}, {1924, 60}, {1925, 30}, {1926, 20}, {1927, 10}, {1928, 5},
 {1929, 5}, {1930, 10}, {1931, 30}, {1932, 80}, {1933, 100}, {1934, 80}}
```

The file **hare** can be found on the CD-ROM that comes with this book. For **Import**, see Section 4.2.1, p. 100. The numbers are in thousands and are quoted to the nearest 5000. Observations regarding the number of hare pelts are lacking for the years 1846, 1905, 1907, and 1908. We take the last 31 data points:

```
hdata = Take[haredata, -31]
```

```
{{1901, 5}, {1902, 10}, {1903, 50}, {1904, 70}, {1906, 20}, {1909, 25}, {1910, 50},
 {1911, 55}, {1912, 75}, {1913, 70}, {1914, 55}, {1915, 30}, {1916, 20},
 {1917, 15}, {1918, 15}, {1919, 20}, {1920, 35}, {1921, 60}, {1922, 80},
 {1923, 85}, {1924, 60}, {1925, 30}, {1926, 20}, {1927, 10}, {1928, 5},
 {1929, 5}, {1930, 10}, {1931, 30}, {1932, 80}, {1933, 100}, {1934, 80}}
```

Here are the data plotted in eight ways:

```
SetOptions[Graphics, Axes → True,
  AspectRatio → 1/GoldenRatio, AxesOrigin → {1900, 0}, Ticks → None];
circles = {White, EdgeForm[Black], Disk[#, Offset[{1.3, 1.3}]]} & /@ hdata};
verticalLines = Line[{{#[[1]], 0}, #}] & /@ hdata;
p1 = Graphics[{}, Axes → False];
p2 = Graphics[circles];
p3 = Graphics[Point[hdata]];
p4 = Graphics[Line[hdata]];
p5 = Graphics[{Line[hdata], circles}];
p6 = Graphics[{Line[hdata], Point[hdata]}];
p7 = Graphics[verticalLines];
p8 = Graphics[{verticalLines, circles}];
p9 = Graphics[{verticalLines, Point[hdata]}];
```

Next, we show the plots as a grid. Note how only the plots with vertical lines are able to clearly reveal the missing observations:

```
GraphicsGrid[{{p1, p2, p3}, {p4, p5, p6}, {p7, p8, p9}}, ImageSize → 420, Axes → False]
```

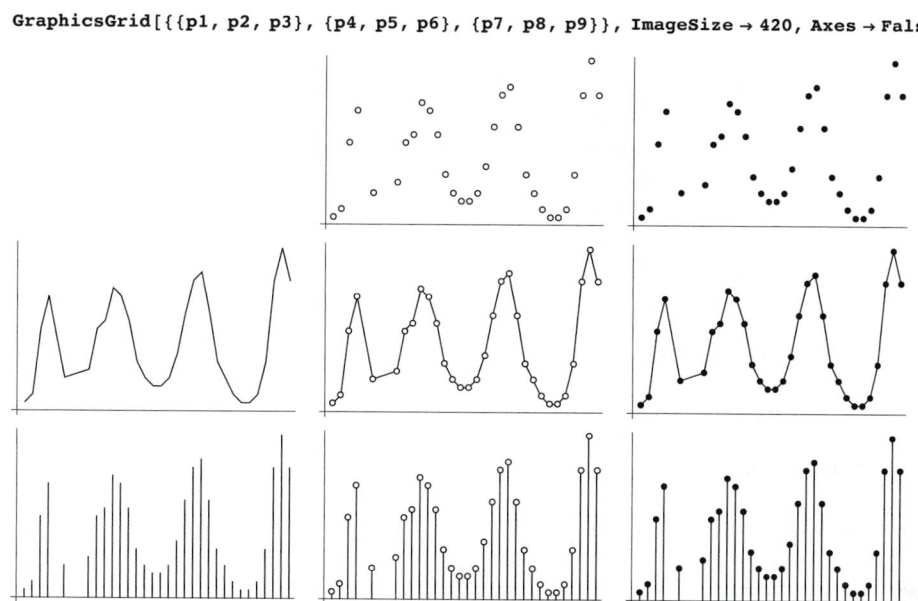

■ **A Quality Plot**

To get a plot of good quality for all of the hare pelt data, we define a smaller aspect ratio and our own ticks:

```
p1 = Graphics[{Line[haredata], Point[haredata]},
   AspectRatio → 0.2, Axes → True, AxesOrigin → {1843, 0},
   Ticks → {Range[1850, 1930, 10], Range[20, 140, 20]}, ImageSize → 420]
```

Using a fill shows very clearly the shape of the data:

```
ListLinePlot[haredata, Mesh → All,
   Filling → Axis, AspectRatio → 0.2, AxesOrigin → {1843, 0},
   Ticks → {Range[1850, 1930, 10], Range[20, 140, 20]}, ImageSize → 420]
```

Generally, a plot can be considered to consist of some *line segments* that go up or down. The slope of a segment indicates the *orientation* of the segment. For example, if the slope is 1, we say that the line segment has an orientation of 45°. The orientations of the segments have an effect on how well the information contained in a plot can be perceived. Typically, the judgments of a curve are optimized when the absolute values of the orientations of the line segments that make up the curve are approximately 45° (see Cleveland 1993, p. 89). The orientations can be adjusted by the aspect ratio of the plot. Choosing the aspect ratio to center the absolute orientations on 45° is *banking to 45°*.

Banking to 45° in the previous plot would require an aspect ratio of approximately 0.05, but then the plot becomes very low. As a compromise, we have chosen the value 0.2.

We go back to the default values of the options of **Graphics**:

```
SetOptions[Graphics, gropts];
```

8.1.3 Plots of Several Data Sets

■ **Basic Ways to Plot Data**

`ListPlot[{data1, data2, … }]` (⌘6)	Points
`ListLinePlot[{data1, data2, … }]` (⌘6)	Joining lines
`ListLinePlot[{data1, data2, … }, Mesh → All]` (⌘6)	Joining lines and points
`ListPlot[{data1, data2, … }, Filling → True]` (⌘6)	Points and vertical lines
`ListLinePlot[{data1, data2, … }, Filling → True]` (⌘6)	Points and fills between the curves
`ListLinePlot[{data1, data2, … }, Filling → Axis]` (⌘6)	Points and fills between the curves and *x* axis
`Joining → {True, False, … }` (⌘6)	Points for some data sets are joined, for others not

ListPlot and **ListLinePlot** are also suitable for plotting several data sets. By default, the data sets are identified with different colors; otherwise, each data set is plotted in the same way. The first, second, third, and fourth data sets are plotted with colors resembling blue, purple, brown, and green, respectively. After that, the same colors are used slightly modified. (Note that in plots having joining lines and points, the points are blue for all data sets; only the color of the joining lines changes.) Here are some examples:

```
SeedRandom[3]; data1 = Table[{x, 2 + 0.2 x - 3 RandomReal[]}, {x, 0, 30}];
data2 = Table[{x, 2.5 + 0.4 x + 3 RandomReal[]}, {x, 0, 30}];
{ListPlot[{data1, data2}],
 ListLinePlot[{data1, data2}],
 ListLinePlot[{data1, data2}, Mesh → All]}
```

```
{ListPlot[{data1, data2}, Filling → True],
 ListLinePlot[{data1, data2}, Filling → True],
 ListLinePlot[{data1, data2}, Filling → Axis, Mesh → All]}
```

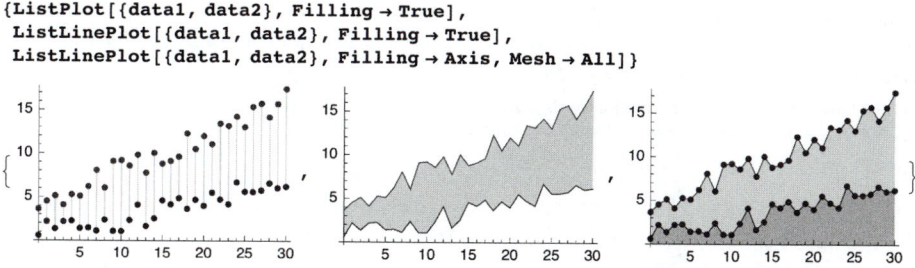

■ Defining Styles

Defining styles in **ListPlot**:

PlotStyle Style of points

PlotMarkers (⌘6) Markers to use for the data sets (enabling other markers besides points)

FillingStyle (⌘6) Style of vertical lines

Defining styles in **ListLinePlot**:

PlotStyle Style of joining lines

MeshStyle Style of points (same style for all data sets)

PlotMarkers (⌘6) Markers to use for the data sets (enabling other markers besides points and
  allowing different styles for the data sets)

FillingStyle (⌘6) Style of fills

First, we consider defining styles of points and lines with **PlotStyle** and **MeshStyle**. Later, we
consider the use of **PlotMarkers** and **FillingStyle**.

■ Styles of Points and Lines

In our first example, we show how to define styles of points in **ListPlot**:

```
GraphicsRow[ListPlot[{data1, data2}, PlotStyle → #] & /@ {
   Black,
   {Gray, Black},
   Directive[Gray, PointSize[Small]],
   {Directive[Gray, PointSize[Small]], Directive[Black, PointSize[Small]]}
   }, ImageSize → 400]
```

Similarly, we can define the style of joining lines with **PlotStyle**:

```
GraphicsRow[ListLinePlot[{data1, data2}, PlotStyle → #] & /@ {
   Thick, {Black, Thick}, Directive[Black, Thick],
   {Directive[Gray, Thick], Directive[Black, Thick]}}, ImageSize → 400]
```

To give styles for points in plots containing both points and joining lines, we can use **MeshStyle**; unfortunately, however, with this option we cannot define different styles of points for each data set. Indeed, the directives given apply for all data sets:

```
ListLinePlot[{data1, data2}, Mesh → All, MeshStyle → #] & /@
 {Black, Directive[Black, PointSize[Small]]}
```

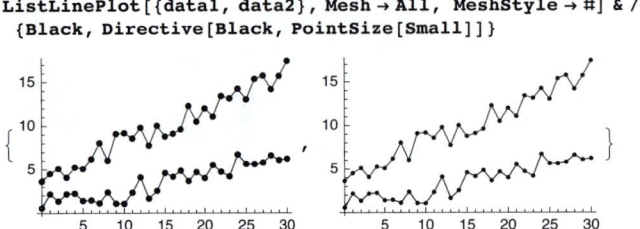

To get different styles of points for each data set, we need the **PlotMarkers** option. This option is explained next.

■ Plot Markers

Normally, if we use **ListPlot** or if we use **ListLinePlot** with **Mesh → All**, each data set is plotted with the same symbol, point; only the color of the points differs between the data sets. This is a good way to show the data sets on the screen and when printed with a color printer. When printed with a grayscale printer, the colors may not be easily distinguished. Then we can use different symbols for the data sets with the **PlotMarkers** option.

PlotMarkers (✻6)   Markers to use for the points; examples of values: **None**, **Automatic**,
 {Automatic, Tiny}, **{Automatic, 10}**, **{"●", "○"}**, **{{"●", 10}, {"○", 10}}**, **{{gr1, 0.08},**
 {gr2, 0.08}}

The default value of **PlotMarkers** is **None**, meaning that all data sets are plotted with points. If we use the value **Automatic**, then the first five data sets are plotted with the following filled symbols: blue disk, purple square, brown diamond, green upward triangle, and blue downward triangle. If more data sets are plotted, the same symbols are used unfilled. The value of the option can also be of the form **{Automatic, size}**, where the size of the symbol refers to font size; for the size, we can also use the special symbols **Tiny**, **Small**, **Medium**, and **Large**.

The value of the option can also be a list of symbols such as **{"●", "○"}**. Such symbols can be picked from the *SpecialCharacters* palette or typed, for example, as \[FilledCircle] or \[EmptyCircle]. The size of the symbols can be told as in **{{"●", 10}, {"○", 10}}**.

In addition, a symbol can be a graphic we have created ourselves. Now the size of the symbol is a fraction of the width of the plot.

First, we use the automatic symbols:

```
SeedRandom[3]; data1 = Table[{x, 2 + 0.2 x - 3 RandomReal[]}, {x, 0, 10}];
data2 = Table[{x, 2.5 + 0.4 x + 3 RandomReal[]}, {x, 0, 10}];
```

```
ListPlot[{data1, data2}, PlotMarkers → #] & /@
  {Automatic, {Automatic, Tiny}, {Automatic, 10}}
```

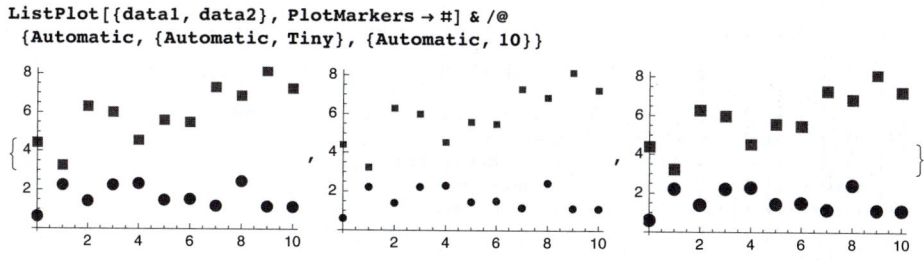

The size of the symbol is defined with the **PlotMarkers** option, but the color can be defined with the **PlotStyle** option:

```
ListPlot[{data1, data2}, PlotMarkers → Automatic, PlotStyle → #] & /@
  {Black, {Red, Blue}}
```

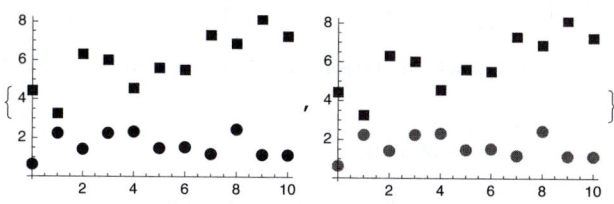

Next, we use special symbols:

```
ListPlot[{data1, data2}, PlotMarkers → #] & /@
  {{"O", "●"}, {{"◆", 8}, {"●", 8}}, {"1", "2"}}
```

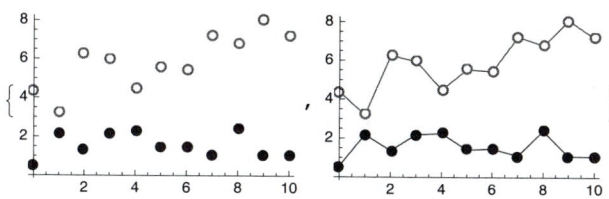

Lastly, we define two symbols with **Graphics**:

```
gr1 = Graphics[{Blue, Disk[]}];
gr2 = Graphics[{EdgeForm[{Red, Thickness[Medium]}], White, Disk[]}];
{ListPlot[{data1, data2}, PlotMarkers → {{gr1, 0.07}, {gr2, 0.07}}],
 ListLinePlot[{data1, data2}, PlotMarkers → {{gr1, 0.07}, {gr2, 0.07}}]}
```

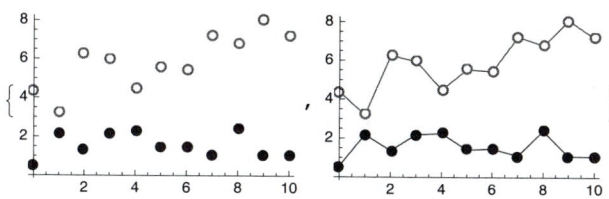

■ **Styles of Fills**

To give styles for the vertical lines or for the fill between two data sets, use `FillingStyle`:

```
{ListPlot[{data1, data2}, Filling → True, FillingStyle → Black],
 ListPlot[{data1, data2}, Filling → True,
   FillingStyle → Directive[Black, Thickness[Tiny]]],
 ListLinePlot[{data1, data2}, Filling → True,
   FillingStyle → Directive[Opacity[0.3], Brown]]}
```

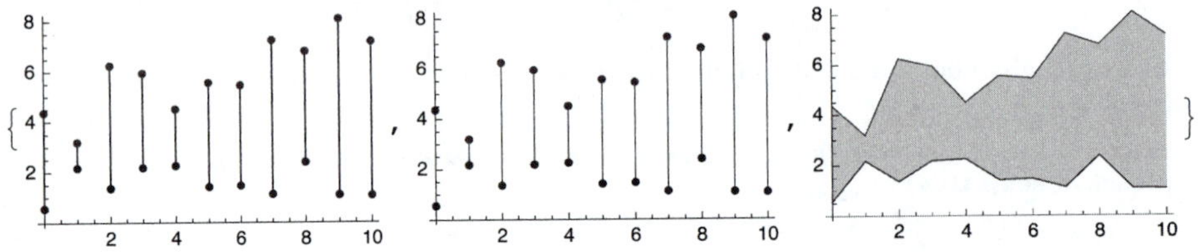

If we want to define styles for more than one filling, the styles have to be defined within the `Filling` option:

```
Filling → {{1 → {Axis, style1}, {2 → {{1}, style2}, … }
```

```
{ListPlot[{data1, data2}, Filling → {1 → {Axis, Green}, 2 → {{1}, Blue}}],
 ListLinePlot[{data1, data2}, Filling → {1 → {Axis, Directive[Opacity[0.3], Green]},
   2 → {{1}, Directive[Opacity[0.3], Blue]}}]}
```

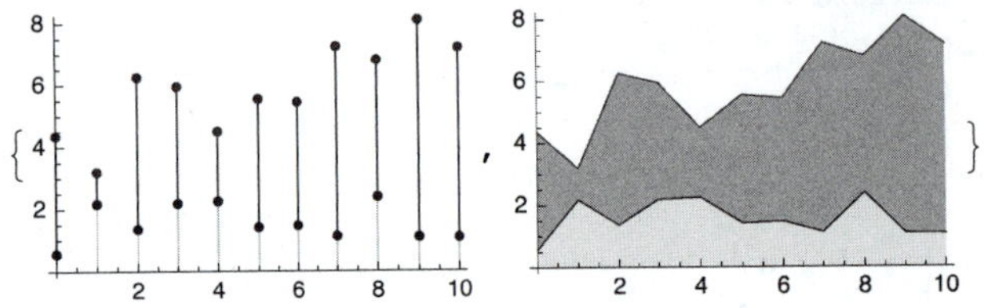

Next, we plot the cumulative sums of three data sets:

```
data1 = RandomReal[{1, 2}, {20}];
data2 = RandomReal[{1, 2}, {20}];
data3 = RandomReal[{1, 2}, {20}];
ListLinePlot[Accumulate[{data1, data2, data3}],
  Mesh → All, Filling → #, AxesOrigin → {0, 0}] & /@
 {Axis, {1 → {Axis, LightBlue}, 2 → {{1}, LightGreen}, 3 → {{2}, LightRed}}}
```

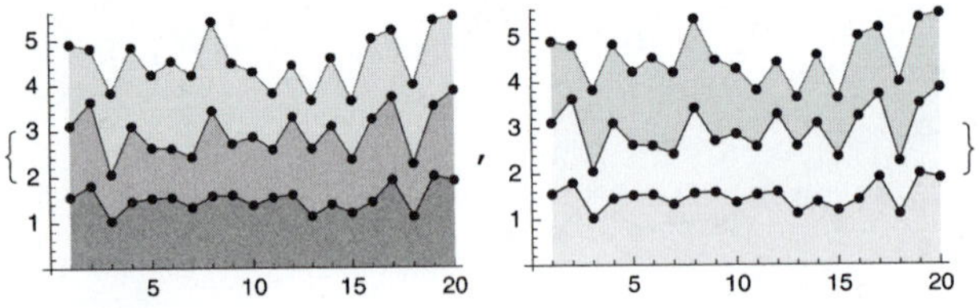

■ **Example**

We consider again the numbers of hare pelts sold to the Hudson Bay Trading Company in Canada from 1844 to 1934; this example was examined in Section 8.1.2, p. 240. We first use the default symbols and line styles but somewhat reduce the size of the points:

```
haredata = Import["/Users/heikki/Documents/MNData/hare", "Table"];
lynxdata = Import["/Users/heikki/Documents/MNData/lynx", "Table"]
{{1844, 6}, {1845, 14}, {1846, 22}, {1847, 36}, {1848, 29}, {1849, 7},
 {1850, 2}, {1851, 1}, {1852, 1}, {1853, 1}, {1854, 5}, {1855, 13}, {1856, 16},
 {1857, 25}, {1858, 14}, {1859, 8}, {1860, 3}, {1861, 2}, {1862, 1}, {1863, 3},
 {1864, 10}, {1865, 27}, {1866, 58}, {1867, 30}, {1868, 26}, {1869, 9},
 {1870, 4}, {1871, 2}, {1872, 2}, {1873, 6}, {1874, 10}, {1875, 26}, {1876, 29},
 {1877, 21}, {1878, 11}, {1879, 10}, {1880, 5}, {1881, 3}, {1882, 5},
 {1883, 16}, {1884, 42}, {1885, 64}, {1886, 63}, {1887, 32}, {1888, 15},
 {1889, 7}, {1890, 3}, {1891, 4}, {1897, 15}, {1898, 7}, {1899, 2}, {1900, 3},
 {1901, 5}, {1902, 14}, {1903, 27}, {1904, 47}, {1905, 54}, {1906, 29},
 {1907, 7}, {1908, 2}, {1909, 2}, {1910, 4}, {1911, 10}, {1912, 14},
 {1913, 19}, {1915, 8}, {1916, 9}, {1917, 2}, {1918, 1}, {1919, 1}, {1920, 2},
 {1921, 4}, {1922, 4}, {1923, 8}, {1924, 7}, {1925, 9}, {1926, 7}, {1927, 4},
 {1928, 3}, {1929, 2}, {1930, 3}, {1931, 3}, {1932, 5}, {1933, 7}, {1934, 7}}

opts = Sequence[{AspectRatio → 0.2, AxesOrigin → {1843, 0},
    Ticks → {Range[1850, 1930, 10], Range[20, 140, 20]}, ImageSize → 420}];

ListLinePlot[{haredata, lynxdata},
  Mesh → All, MeshStyle → AbsolutePointSize[2.3], opts]
```

Both hare and lynx seem to have a cycle of approximately 10 years. More than 90% of the diet of the lynx is hare. When there are few hares available, lynx starve rather than eat other species. The two data sets are not very clearly distinguished in the previous plot because the color and the symbol of the points are the same for both data sets (the color of the lines is different, however). Next, we use our own symbols:

```
gr1 = Graphics[{Black, Disk[]}];
gr2 = Graphics[{EdgeForm[Black], White, Disk[]}];

ListLinePlot[{haredata, lynxdata}, Mesh → All, opts,
  PlotStyle → Black, PlotMarkers → {{gr1, 0.038}, {gr2, 0.036}}]
```

Now the two data sets can be seen more clearly (note that we have made the size of the black disks slightly larger than the size of the circles so that both seem to be approximately the same size). A filled plot also gives a good illustration:

```
ListLinePlot[{haredata, lynxdata}, Mesh → All, opts, PlotStyle → Black,
 Filling → Axis, PlotMarkers → {{gr1, 0.038}, {gr2, 0.036}}]
```

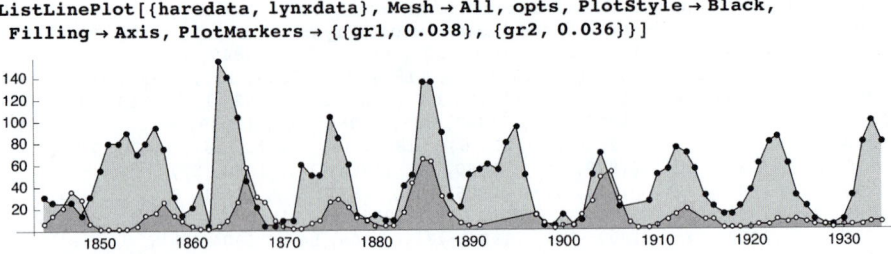

■ Self-Made Plotting

Plotting several data sets with **Graphics**, using the primitives **Line**, **Point**, and **Disk** with suitable directives, is very straightforward:

```
Graphics[{Line[haredata], Line[lynxdata],
  AbsolutePointSize[2.5], Point[haredata], White,
  EdgeForm[Black], Disk[#, Offset[{1, 1}]] & /@ lynxdata, Black,
  Text["hare", {1936, 80}, {-1, 0}], Text["lynx", {1936, 9}, {-1, 0}]},
 Axes → True, opts, PlotRange → {{1837, 1940}, All}]
```

■ A Phase Plot

Until now, we have plotted two data sets—$(x_1, x_2, \ldots)$ and $(y_1, y_2, \ldots)$—as time series. Another kind of plot is obtained by plotting the pairs (x_i, y_i). In this way, we get a plot that is analogous to a phase plot of the solution of a pair of differential equations (see Section 26.3.2, p. 855). For the years 1909 to 1934, the numbers of hare and lynx pelts are as follows (the observation for the year 1914 is lacking):

```
harelynx = {{25, 2}, {50, 4}, {55, 10}, {75, 14}, {70, 19}, {30, 8}, {20, 9}, {15, 2},
  {15, 1}, {20, 1}, {35, 2}, {60, 4}, {80, 4}, {85, 8}, {60, 7}, {30, 9}, {20, 7},
  {10, 4}, {5, 3}, {5, 2}, {10, 3}, {30, 3}, {80, 5}, {100, 7}, {80, 7}};
```

We construct a phase plot:

```
ListLinePlot[harelynx, Mesh → All,
 Epilog → {AbsolutePointSize[5], Hue[0], Point[First[harelynx]]},
 AxesLabel → {"hare", "lynx"}, ImageSize → 200]
```

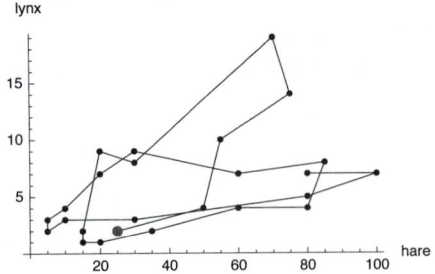

The starting point (25, 2) is red and larger than the other points. We can see a counterclockwise cycle; this pattern is typical for a predator–prey system.

8.2 Scatter Plots

8.2.1 Scatter Plots

Thus far, we have presented plotting methods for showing each dependent variable separately, typically as time series. Now we present some plotting methods that show two or even more dependent variables in the same plot. Such plots are useful when studying relationships among a number of dependent variables. We study scatter plots and quantile–quantile plots.

In a *scatter plot*, we plot one variable against another variable. Such a plot may yield valuable information about the connections between the variables. A pairwise scatter plot, which is also called a scatter plot matrix or a correlation plot, is a collection of plots in which each plot shows one variable against another variable. A scatter plot matrix is among the best ways to illustrate multidimensional data.

■ **Example 1**

As an example, we consider the data file **environmental**, which contains 111 observations of ozone, radiation, temperature, and wind in New York City from May to September of 1973. The data are from a collection of data sets in http://lib.stat.cmu.edu/S/visualizing.data. All of the data are visualized in Cleveland (1993), and the data sets are also on the CD-ROM that comes with this book (the data sets are reproduced with the permission of the publisher, Hobart Press). We also consider the environmental data set in Section 30.5.3, p. 1038, when presenting local regression. On my computer, the environmental data are in a text file **environmental** in a folder **visdata**. First, we read the data (**Rest** drops the first row containing the headings of the columns; for reading data, see Section 4.2.1, p. 100):

```
Style[data = Rest[
   Import["/Users/heikki/Documents/MNData/visdata/environmental", "Table"]], 6]
{{1, 41, 190, 67, 7.4}, {2, 36, 118, 72, 8.}, {3, 12, 149, 74, 12.6}, {4, 18, 313, 62, 11.5}, {5, 23, 299, 65, 8.6},
{6, 19, 99, 59, 13.8}, {7, 8, 19, 61, 20.1}, {8, 16, 256, 69, 9.7}, {9, 11, 290, 66, 9.2}, {10, 14, 274, 68, 10.9},
{11, 18, 65, 58, 13.2}, {12, 14, 334, 64, 11.5}, {13, 34, 307, 66, 12.}, {14, 6, 78, 57, 18.4}, {15, 30, 322, 68, 11.5},
{16, 11, 44, 62, 9.7}, {17, 1, 8, 59, 9.7}, {18, 11, 320, 73, 16.6}, {19, 4, 25, 61, 9.7}, {20, 32, 92, 61, 12.}, {21, 23, 13, 67, 12.},
{22, 45, 252, 81, 14.9}, {23, 115, 223, 79, 5.7}, {24, 37, 279, 76, 7.4}, {25, 29, 127, 82, 9.7}, {26, 71, 291, 90, 13.8},
{27, 39, 323, 87, 11.5}, {28, 23, 148, 82, 8.}, {29, 21, 191, 77, 14.9}, {30, 37, 284, 72, 20.7}, {31, 20, 37, 65, 9.2},
{32, 12, 120, 73, 11.5}, {33, 13, 137, 76, 10.3}, {34, 135, 269, 84, 4.}, {35, 49, 248, 85, 9.2}, {36, 32, 236, 81, 9.2},
{37, 64, 175, 83, 4.6}, {38, 40, 314, 83, 10.9}, {39, 77, 276, 88, 5.1}, {40, 97, 267, 92, 6.3}, {41, 97, 272, 92, 5.7},
{42, 85, 175, 89, 7.4}, {43, 10, 264, 73, 14.3}, {44, 27, 175, 81, 14.9}, {45, 7, 48, 80, 14.3}, {46, 48, 260, 81, 6.9},
{47, 35, 274, 82, 10.3}, {48, 61, 285, 84, 6.3}, {49, 79, 187, 87, 5.1}, {50, 63, 220, 85, 11.5}, {51, 16, 7, 74, 6.9},
{52, 80, 294, 86, 8.6}, {53, 108, 223, 85, 8.}, {54, 20, 81, 82, 8.6}, {55, 52, 82, 86, 12.}, {56, 82, 213, 88, 7.4},
{57, 50, 275, 86, 7.4}, {58, 64, 253, 83, 7.4}, {59, 59, 254, 81, 9.2}, {60, 39, 83, 81, 6.9}, {61, 9, 24, 81, 13.8},
{62, 16, 77, 82, 7.4}, {63, 122, 255, 89, 4.}, {64, 89, 229, 90, 10.3}, {65, 110, 207, 90, 8.}, {66, 44, 192, 86, 11.5},
{67, 28, 273, 82, 11.5}, {68, 65, 157, 80, 9.7}, {69, 22, 71, 77, 10.3}, {70, 59, 51, 79, 6.3}, {71, 23, 115, 76, 7.4},
{72, 31, 244, 78, 10.9}, {73, 44, 190, 78, 10.3}, {74, 21, 259, 77, 15.5}, {75, 9, 36, 72, 14.3}, {76, 45, 212, 79, 9.7},
{77, 168, 238, 81, 3.4}, {78, 73, 215, 86, 8.}, {79, 76, 203, 97, 9.7}, {80, 118, 225, 94, 2.3}, {81, 84, 237, 96, 6.3},
{82, 85, 188, 94, 6.3}, {83, 96, 167, 91, 6.9}, {84, 78, 197, 92, 5.1}, {85, 73, 183, 93, 2.8}, {86, 91, 189, 93, 4.6},
{87, 47, 95, 87, 7.4}, {88, 32, 92, 84, 15.5}, {89, 20, 252, 80, 10.9}, {90, 23, 220, 78, 10.3}, {91, 21, 230, 75, 10.9},
{92, 24, 259, 73, 9.7}, {93, 44, 236, 81, 14.9}, {94, 21, 259, 76, 15.5}, {95, 28, 238, 77, 6.3}, {96, 9, 24, 71, 10.9},
{97, 13, 112, 71, 11.5}, {98, 46, 237, 78, 6.9}, {99, 18, 224, 67, 13.8}, {100, 13, 27, 76, 10.3}, {101, 24, 238, 68, 10.3},
{102, 16, 201, 82, 8.}, {103, 13, 238, 64, 12.6}, {104, 23, 14, 71, 9.2}, {105, 36, 139, 81, 10.3}, {106, 7, 49, 69, 10.3},
{107, 14, 20, 63, 16.6}, {108, 30, 193, 70, 6.9}, {109, 14, 191, 75, 14.3}, {110, 18, 131, 76, 8.}, {111, 20, 223, 68, 11.5}}
```

Extract the columns of the data:

```
{no, ozone, radiation, temperature, wind} = data⊤;
```

We are interested in how ozone depends on the other variables. Thus, we plot ozone against radiation, temperature, and wind:

```
GraphicsRow[{ListPlot[{radiation, ozone}ᵀ, PlotStyle → PointSize[Small],
    Frame → True, FrameLabel → {"radiation", "ozone"}],
  ListPlot[{temperature, ozone}ᵀ, PlotStyle → PointSize[Small],
    Frame → True, FrameLabel → {"temperature", None}],
  ListPlot[{wind, ozone}ᵀ, PlotStyle → PointSize[Small], Frame → True,
    FrameLabel → {"wind", None}, Axes → False]}, ImageSize → 420, Spacings → -40]
```

We can see that ozone values are high when radiation or temperature is high (a positive correlation) or wind is low (a negative correlation); however, a high value of radiation does not necessarily mean a high ozone value.

■ Example 2

With `CountryData` (see Section 9.2.1, p. 293), we can study many properties of countries. Here, we investigate life expectancy against birth rate fraction and literacy fraction:

```
co = CountryData["Countries"];
birth = Tooltip[{CountryData[#, "BirthRateFraction"],
      CountryData[#, "LifeExpectancy"]}, #] & /@ co;
literacy = Tooltip[{CountryData[#, "LiteracyFraction"],
      CountryData[#, "LifeExpectancy"]}, #] & /@ co;
```

The corresponding scatter plots are as follows:

```
GraphicsRow[{ListPlot[birth, Frame → True, AxesOrigin → {0, 30},
    FrameLabel → {"Birth rate fraction", "Life expectancy"}],
  ListPlot[literacy, Frame → True, AxesOrigin → {0, 30},
    FrameLabel → {"Literacy fraction", None}]}, ImageSize → 420, Spacings → -10]
```

The life expectancy is higher the lower the birth rate or the higher the literacy fraction. By using **Tooltip**, the plots have the property that the names of the countries can be seen by moving the mouse (without pressing a button) over the points.

■ **Example 3**

In Example 2, we used **Tootip** to show the countries corresponding to the points. Such labels cannot be seen in a printed document. Here, we show how we can easily produce a plot with explicit labels.

As an example, we plot the body and brain weights of some animals. The data are from the same **visdata** collection as we considered in Example 1; the collection also comes on the CD-ROM of this book. On my computer, the animal data are in a text file **modAnimal** in a folder **visdata**. (Note that the original file **animal** contained spaces in the names of the animals, but we have now deleted the spaces.) We read the file and take only rows 38 through 48:

```
data = Import["/Users/heikki/Documents/MNData/visdata/modAnimal", "Table"];
```

```
{no, name, body, brain} = Take[data, {38, 48}]ᵀ
```
{{37, 38, 39, 40, 41, 42, 43, 44, 45, 46, 47},
 {RoeDeer, Goat, Kangaroo, GrayWolf, Sheep, GiantArmadillo,
  GraySeal, Jaguar, BrazilianTapir, Donkey, Pig}, {14 830, 27 660,
  35 000, 36 330, 55 500, 60 000, 85 000, 100 000, 160 000, 187 100, 192 000},
 {98.2, 115., 56., 119.5, 175., 81., 325., 157., 169., 419., 180.}}

We use **Graphics** to get a plot with labels for the points. As can be seen, overlapping labels may be a problem with labeled plots:

```
Graphics[{Point[{body, brain}ᵀ],
    MapThread[Text[#1, #2, {-1.3, 0}] &, {name, {body, brain}ᵀ}]}, Frame → True,
  FrameLabel → {"Body weight", "Brain weight"}, AspectRatio → 1 / GoldenRatio,
  PlotRange → {{0, 225 000}, {0, 460}}, ImageSize → 250]
```

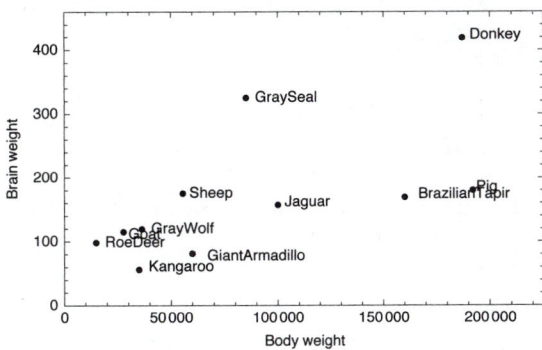

■ **Scatter Plot Matrix**

In the **StatisticalPlots`** *package:*

PairwiseScatterPlot[data]   Plot multidimensional data as a pairwise scatter plot

Options:

DataLabels   Labels for the variables; examples of values: **None**, {"X", "Y", "Z"}

DataTicks   Ticks for the variables; examples of values: **None**, **Automatic**

DataSpacing   Space between the subgraphs; examples of values: **0**, **0.05**

DataRanges   Ranges for the data; default value: **All**

PlotDirection   Direction in which scatter plots are generated; examples of values: {**Right**, **Down**}, {**Right**, **Up**}

PlotStyle   Style of the points; examples of values: **Automatic**, **PointSize[Small]**

As an example, we consider the data of Example 1:

```
<< StatisticalPlots`
```

```
PairwiseScatterPlot[
  {radiation, temperature, wind, ozone}ᵀ, PlotStyle → PointSize[Small],
  DataLabels → {"radiation", "temperature", "wind", "ozone"},
  PlotDirection → {Right, Up}, ImageSize → 330]
```

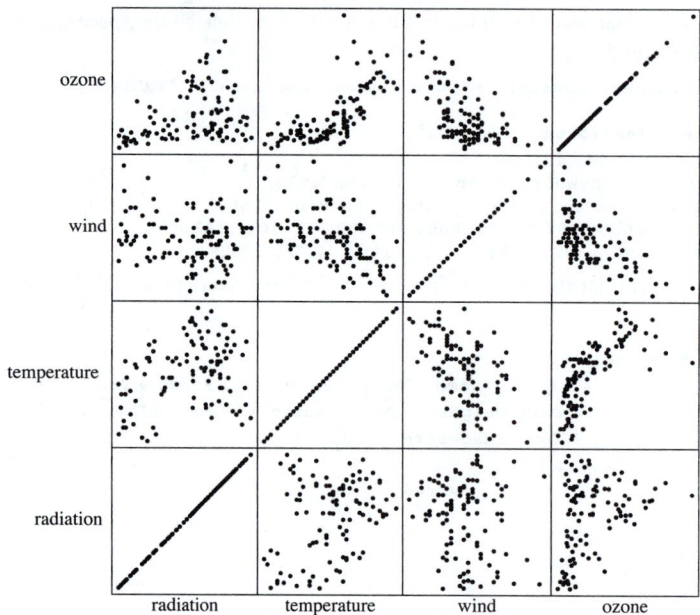

Of these 16 plots, we plotted, in Example 1, the first three in the top row. For more about the statistical plots package, look at StatisticalPlots/guide/StatisticalPlotsPackage.

8.2.2 Quantile–Quantile Plots

In the **StatisticalPlots`** *package:*

QuantilePlot[data1, data2]   Create a quantile–quantile plot

Options:

PlotMarkers   Markers for the points; examples of values: **Automatic**, **{Automatic, 3}**, **None**
ReferenceLineStyle   Style of the reference line; examples of values: **Automatic**, **None**
Joined   Whether the points are joined with lines; possible values: **False**, **True**
PlotStyle   Style of the joining line; default value: **Automatic**

A *quantile–quantile plot* or a *q-q plot* (see Cleveland 1993, p. 21) is a powerful method for comparing the distributions of two or more sets of univariate data. The plot is a special scatter plot: It shows the quantiles of one data set against the quantiles of another data set. If the resulting points are close to a line with a slope of 1, this supports the hypothesis that the distributions of the two data sets are the same.

QuantilePlot first determines the interpolated quantiles of the shorter of the two data sets at the equivalent positions in the longer data set. It then plots the two sets of quantiles against each other.

As an example, generate data sets from a Student t-distribution with parameter 10 and from the standard normal distribution:

```
SeedRandom[2]; ran1 = RandomReal[StudentTDistribution[10], {2000}];
ran2 = RandomReal[NormalDistribution[0, 1], {2000}];
```

How close are the two distributions? Prepare a q-q plot:

```
QuantilePlot[ran1, ran2, PlotMarkers → {Automatic, 2}, ImageSize → 200]
```

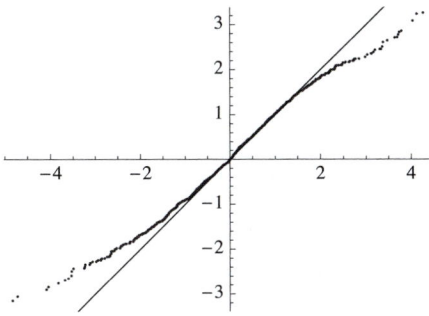

We see that the points are not sufficiently close to the reference line for supporting the hypothesis that the distributions are the same. Indeed, although we know that the t-distribution approaches the normal distribution as the parameter approaches infinity, the value 10 simply is not large enough. The tails of the t-distribution are fatter than the tails of the normal distribution.

8.3 Bar Charts

8.3.1 Bar Charts

■ Bar Charts

The **BarCharts`** package contains several commands for bar charts. **BarChart**, **StackedBarChart**, and **PercentileBarChart** are suitable for charts in which the bars are simply drawn side by side (without having specified positions) and in which there are labels (not coordinates) under the bars. **GeneralizedBarChart** is designed for charts in which the bars have specified positions and widths. **Histogram** (see Section 8.3.2, p. 258) is a special command to calculate frequencies and plot them as bar charts. We also have **BarChart3D**, **GeneralizedBarChart3D**, and **Histogram3D** (see Section 8.6.1, p. 275).

In the **BarCharts`** *package:*

BarChart[{y_1, y_2, … }] Plot bars of heights y_1, y_2, … and label the bars by **1**, **2**, …

```
data = {4, 7, 6, 3, 5, 4, 8, 7};

<< BarCharts`
```

`BarChart[data]`

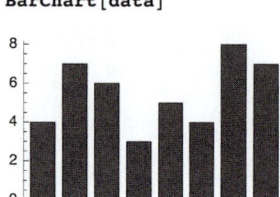

■ **Options**

`BarChart` accepts the following options and the options used with `Graphics`. For `BarChart`, the default value of `AspectRatio` is `1/GoldenRatio` and that of `Axes` is `True`.

Options of `BarChart`:

`BarLabels` Labels under the bars; examples of values: `Automatic`, `{"A", "B", "C"}`

`BarValues` Whether to write the values on top of the bars; possible values: `False`, `True`

`BarOrientation` Orientation of the bars; possible values: `Vertical`, `Horizontal`

`BarGroupSpacing` The space between each group of bars as a fraction of the width of one bar; examples of values: `Automatic` (means `0.2`), `0`

`BarSpacing` The space between the bars within a group as a fraction of the width of one bar; examples of values: `Automatic` (means `0`), `0.1`, `-0.4`

`BarStyle` Style inside the edges of the bars; examples of values for one data set: `Automatic` (means `Hue[0.67, 0.45, 0.65]`), `Gray`; an example for two data sets: `{Gray, GrayLevel[0.9]}`

`BarEdgeStyle` Style of the edges of the bars; examples of values for one data set: `Opacity[0.5]`, `Black`, `Directive[Gray, Thickness[Medium]]`; an example for two data sets: `{Thickness[Medium], Directive[Gray, Thickness[Medium]]}`

`BarEdges` Whether edges are drawn for the bars; possible values: `True`, `False`

`BarGroupSpacing` defines the space between each group of bars. Indeed, `BarChart` can generate bars for multiple data sets, and then the bars are collected into groups; each group contains as many bars as there are data sets. Thus, if we only have one data set, then `BarGroupSpacing` simply defines the space between the bars. The default value `Automatic` means the value `0.2`. If you want no space between the bars, give the value `0`.

`BarSpacing` defines the space between bars within a group of bars. The default value `Automatic` means the value `0`: There is no space between the bars within a group. If you want a small space between the bars, give a small positive value for the option, and if you want the bars to overlap, give a small negative value.

```
{BarChart[data, BarLabels → CharacterRange["a", "h"]],
 BarChart[data, BarValues → True],
 BarChart[data, BarOrientation → Horizontal]}
```

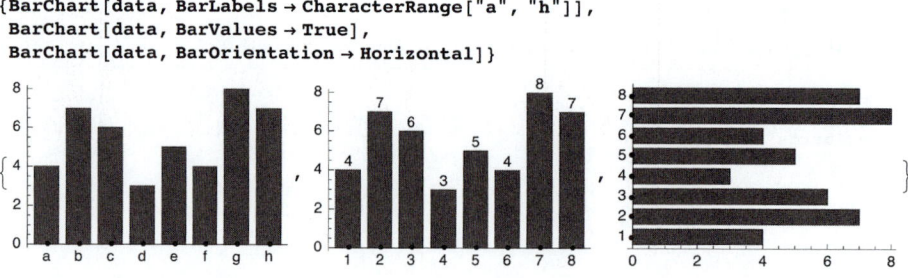

```
{BarChart[data, BarGroupSpacing → 0],
 BarChart[data, BarStyle → Lighter[Blue, 0.4]],
 BarChart[data, Ticks → {{Range[8], CharacterRange["a", "h"]}ᵀ, Range[8]}]}
```

The last example is about defining ticks (the symbol ᵀ means a transpose; type it as [ESC]tr[ESC]). Suppose we want to make a bar chart like the first plot shown previously, except that we want to define new *y* ticks. A problem is that giving a value to **Ticks** causes **BarLabels** to no longer be taken into account. Thus, if we use **Ticks**, we also have to redefine the labels with the aid of **Ticks**; this is done in the last example.

■ Example 1

As an example, we plot the body and brain weights of some animals. This data set was also considered in Example 3 of Section 8.2.1, p. 251. We read the file and take only rows 38 through 48:

```
data0 = Import["/Users/heikki/Documents/MNData/visdata/modAnimal", "Table"];
```

```
{no, name, body, brain} = Take[data0, {38, 48}]ᵀ
{{37, 38, 39, 40, 41, 42, 43, 44, 45, 46, 47},
 {RoeDeer, Goat, Kangaroo, GrayWolf, Sheep, GiantArmadillo,
  GraySeal, Jaguar, BrazilianTapir, Donkey, Pig}, {14 830, 27 660,
  35 000, 36 330, 55 500, 60 000, 85 000, 100 000, 160 000, 187 100, 192 000},
 {98.2, 115., 56., 119.5, 175., 81., 325., 157., 169., 419., 180.}}
```

Next, we form pairs of brain weight and animal name and sort the pairs in ascending order according to brain weight:

```
data = Sort[{brain, name}ᵀ]
{{56., Kangaroo}, {81., GiantArmadillo}, {98.2, RoeDeer},
 {115., Goat}, {119.5, GrayWolf}, {157., Jaguar}, {169., BrazilianTapir},
 {175., Sheep}, {180., Pig}, {325., GraySeal}, {419., Donkey}}
```

The bar chart is given as follows:

```
p1 = BarChart[data, AspectRatio → 0.25, ImageSize → 420]
```

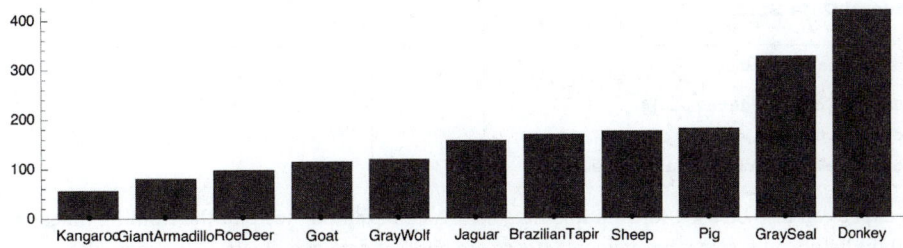

Here, we have used a large image size to avoid the overlapping of the labels. In the next example, we present other solutions to this problem.

■ **Example 2**

A solution to overlapping labels is to use horizontal bars:

BarChart[data, BarOrientation → Horizontal, ImageSize → 250]

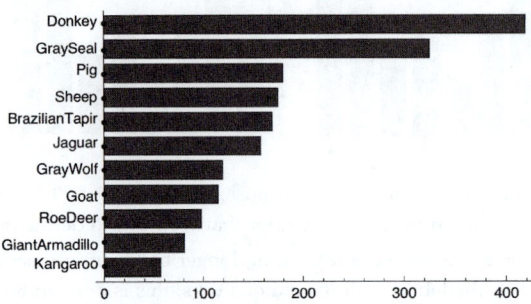

Another solution is to give a suitable slope to the labels. Remember from Section 6.2.6, p. 163, that **Text** primitives can have any slope. We define the slope as **{4, 3}**; that is, the slope is 3/4. We also adjust the positioning of the labels: The old position in text coordinates is **{0, 1}** so that the labels are centered below the bars, but the new position is **{0.8, 0.8}** so that the labels are to the left of the bars:

Show[FullGraphics[p1] /. Text[a_String, b_, c_] → Text[a, b, {0.8, 0.8}, {4, 3}],
Axes → False, AspectRatio → 1 / GoldenRatio, ImageSize → 250]

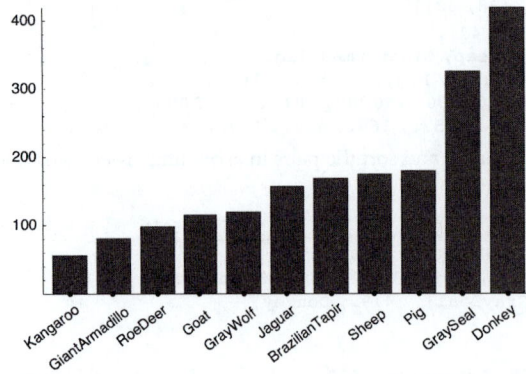

■ **Several Data Sets**

> *In the **BarCharts`** package:*
>
> **BarChart[{data1, data2, … }]**
> **StackedBarChart[{data1, data2, … }]**
> **PercentileBarChart[{data1, data2, … }]**

BarChart can also be used for multiple data sets. As an example, we plot the running times of four algorithms before and after improvements (these are not real data; we illustrate the same data sets with a dot plot in Section 8.4.1, p. 260). The default is that the bars are side by side, but we produce a plot in which the bars somewhat overlap:

```
p = BarChart[{{8, 9, 11, 10}, {5, 7, 6, 8}},
   BarLabels → {"Alg. 1", "Alg. 2", "Alg. 3", "Alg. 4"}, BarGroupSpacing → 0.6,
   BarSpacing → -0.4, BarStyle → {GrayLevel[0.5], GrayLevel[0.8]},
   PlotLabel → Style["Running times before (dark) and\nafter (light) improvements",
   8], ImageSize → 200]
```

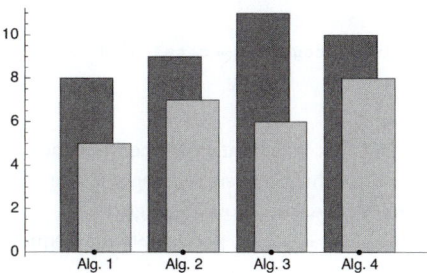

Bars with Positions

> *In the* **BarCharts`** *package:*
>
> **GeneralizedBarChart[{{x_1, y_1, w_1}, {x_2, y_2, w_2}, ... }]** Plot bars of heights y_1, y_2, ... and widths w_1, w_2, ... at the positions x_1, x_2, ...

Recall that **BarChart** produces bars side by side, with possible labels. With **GeneralizedBarChart** we can define the positions of the bars.

GeneralizedBarChart has the same options as **BarChart** except for the options **BarLabels**, **BarGroupSpacing**, and **BarSpacing**. The options are in fact unnecessary because the labels can be given with the **Ticks** option, and the spacing can be adjusted with the widths of the bars.

In this example, we toss a die 100 times and calculate and plot the frequencies. Note that for each frequency, we add the width 1 of the bar as the third component:

```
SeedRandom[1];
data = RandomInteger[{1, 6}, 100];
freq = {#, Count[data, #], 1} & /@ Range[6]
{{1, 17, 1}, {2, 14, 1}, {3, 22, 1}, {4, 16, 1}, {5, 14, 1}, {6, 17, 1}}

GeneralizedBarChart[freq, AxesOrigin → {0.5, 0}, ImageSize → 120]
```

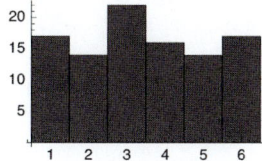

In this example, **Histogram** would be the correct command, but we used **GeneralizedBarChart** just to illustrate this command. **Histogram** is considered in the next section.

8.3.2 Histograms

With **Histogram** we can plot frequencies as a bar chart; the data can be either raw data or frequencies. In the former case, **Histogram** first calculates the frequencies.

In the **Histograms`** *package:*

Histogram[{x$_1$, x$_2$, … }] Plot the frequencies of the given raw data

Histogram[{f$_1$, …, f$_n$}, FrequencyData → True, HistogramCategories → cats] Plot the given frequencies

Options:

HistogramCategories How the data is categorized—that is, for which intervals the frequencies are calculated; possible values: **Automatic** (use an internal algorithm), a positive integer **n** (use exactly **n** categories of equal width, if **ApproximateIntervals → False**, and about **n** categories, if **ApproximateIntervals → True**), or a list of cutoff values {c$_0$, c$_1$, …, c$_n$} (calculate the frequencies in the intervals [c$_0$, c$_1$), …, [c$_{n-1}$, c$_n$))

ApproximateIntervals Whether interval boundaries should be approximated by simple numbers; possible values: **Automatic** (usually means **True**), **True**, **False**

HistogramScale Whether to scale the heights of the bars; examples of values: **Automatic** (means **False** for categories with equal widths and **True** for categories with unequal widths), **False** (no scaling: plot frequencies as such), **True** (scale by dividing the heights by the widths of the bars to get a frequency density), **1** (scale to get the sum of the areas of the bars equal to 1 so that the histogram approximates the probability density function of the data; other constants can also be used)

HistogramRange Range of data to be included in the histogram; examples of values: **Automatic** (means that all data are included), **{0, 10}**

BarOrientation, **BarStyle**, **BarEdgeStyle**, **BarEdges** (see Section 8.3.1, p. 254)

Histogram also has the options of **Graphics**. For **Histogram**, the default value of **AspectRatio** is **1/GoldenRatio** and that of **Axes** is **True**.

We plot the frequencies of the same data that were used in Section 8.3.1, p. 257. In the first plot that follows, we used raw data. In the second plot, we used the frequencies we had calculated ourselves.

```
<< Histograms`
```

```
SeedRandom[1];
data = RandomInteger[{1, 6}, 100];
freq = Count[data, #] & /@ Range[6]
{17, 14, 22, 16, 14, 17}
```

```
{Histogram[data, HistogramCategories → Range[0.5, 6.5, 1]],
 Histogram[freq, HistogramCategories → Range[0.5, 6.5, 1], FrequencyData → True]}
```

With `ParetoPlot` from the **StatisticalPlots`** package, we can plot bars for the frequencies together with a line plot for the cumulative frequencies.

In Section 30.2, p. 1011, we consider the calculation of frequencies and the plotting of histograms in more detail.

8.3.3 Stem-and-Leaf Plots

In the **StatisticalPlots`** *package:*

`StemLeafPlot[vector]`   Create a stem-and-leaf plot for one data set

`StemLeafPlot[vector1, vector2]`   Create a stem-and-leaf plot for two data sets

Options:

`Leaves` How leaves are represented; examples of values: `"Digits"`, `"Tallies"`, `{"Tallies"`,
  `"TallySymbol" → ●, "LeafWrapping" → 20}`, `None`

`IncludeStemCounts` Whether to include column(s) for counts; possible values: `False`, `True`

`IncludeEmptyStems` Whether stems within the data range without leaves should be included;
  possible values: `False`, `True`

`ColumnLabels` Labels for the columns; examples of values: `Automatic`, `{"Values", "Tallies"}`

`StemExponent` If the value is x, the stem unit is 10^x; examples of values: `Automatic` (the exponent is
  chosen based on the magnitudes of the data), `2`

`IncludeStemUnits` Whether a reminder of the stem units should be included; possible values: `True`,
  `False`

The `StemExponent` and `Leaves` options have a number of suboptions (some of them are shown above); see StatisticalPlots/tutorial/StatisticalPlots in the Documentation Center. Options of `GridBox` (e.g., `RowLines → True`) can also be used.

A stem-and-leaf plot is like a histogram: It shows how many observations fall into some categories. An example:

```
data = {3.26, 1.4, 4.33, 3.6, 1.27, 3.5};
```

```
<< StatisticalPlots`
<< Histograms`
{StemLeafPlot[data, IncludeEmptyStems → True], Histogram[data, ImageSize → 130]}
```

The first "plot" shows that two values have integer part 1; their first (rounded) decimals are 3 and 4. We have no values that have integer part 2. Three values have integer part 3; their first decimals are 3, 5, and 6. Lastly, one value has integer part 4; its first decimal is 3.

One advantage of a stem-and-leaf plot over a histogram is that we can also read the approximate values of the individual observations. In addition, we can easily compare two data sets. In the following, we compare a binomial distribution with the corresponding approximate Poisson distribution:

```
SeedRandom[2];
data1 = RandomInteger[BinomialDistribution[50, 0.05], 50];
data2 = RandomInteger[PoissonDistribution[50 × 0.05], 50];
```

```
StemLeafPlot[data1, data2, Leaves → {"Tallies", "TallySymbol" → ●},
  ColumnLabels → {"Counts", "Tallies", "Values", "Tallies", "Counts"},
  IncludeStemCounts → True, IncludeStemUnits → False]
```

Counts	Tallies	Values	Tallies	Counts
4	●●●●	0	●●	2
8	●●●●●●●●	1	●●●●●●●●●●●	11
14	●●●●●●●●●●●●●●	2	●●●●●●●●●●●●●●●●	16
12	●●●●●●●●●●●●	3	●●●●●●●●●●	10
7	●●●●●●●	4	●●●●●●●●●	9
3	●●●	5		0
1	●	6	●	1
1	●	7	●	1

8.4 Other Plots

8.4.1 Dot Plots

A dot plot can be drawn with the following program:

```
dotPlot[values_, labels_, styles_, {xmin_, xmax_}, xticks_, grid_, opts___] :=
 Module[{n = Length[labels], vlines, hlines, points},
  vlines = If[grid == {}, {},
    {Gray, Line[Table[{{i, 0.3}, {i, n + 0.7}}, {i, grid[[1]], grid[[2]], grid[[3]]}]]}];
  hlines = {Thin, AbsoluteDashing[{1, 1.5}],
    Line[{{xmin, #}, {xmax, #}}] & /@ Range[n]]};
  points = Table[If[styles == {},
      {PointSize[Medium], Black, Point[#]} &,
      {styles[[i, 1]], styles[[i, 2]], Point[#]} &] /@
    ({values[[i]], Range[n]}ᵀ), {i, Length[values]}];
  Graphics[{vlines, hlines, points}, PlotRange → {{xmin, xmax}, {0.3, n + 0.7}},
   Frame → True, FrameTicks → {xticks, {Range[n], labels}ᵀ, None, None}, opts]]
```

Here, **values** is a list of one or more data sets, with each data set being a list of numbers; **labels** is a list containing the labels for the y axis. The variable **styles** is a list of styles for the points: The list has as many components as there are data sets, with each component being a list of two elements giving the point size and color of the points; **styles** can also be an empty list **{}**, and then default styles are used. The variables **xmin** and **xmin** define the x range; **xticks** defines the ticks on the x axis; and **grid** defines a list of three numbers giving the position of the first vertical grid line, position of the last vertical line, and the increment of the grid lines (an empty list **{}** can also be given). In addition, we can give options of **Graphics**.

■ **Example 1**

To illustrate dot plots, we use the same animal brain weight data we considered, with horizontal bar charts, in Section 8.3.1, p. 256:

```
data = Import["/Users/heikki/Documents/MNData/visdata/modAnimal", "Table"];
```

```
{no, name, body, brain} = Take[data, {38, 48}]ᵀ;
```

Sort according to the brain weights:

```
{values, labels} = Sort[{brain, name}ᵀ]ᵀ
{{56., 81., 98.2, 115., 119.5, 157., 169., 175., 180., 325., 419.},
 {Kangaroo, GiantArmadillo, RoeDeer, Goat, GrayWolf,
  Jaguar, BrazilianTapir, Sheep, Pig, GraySeal, Donkey}}
```

Prepare a dot plot:

```
dotPlot[{values}, labels, {}, {0, 460}, Automatic, {100, 400, 100},
 PlotLabel → Style["Brain weights for some animals", 10, Bold],
 AspectRatio → 1 / GoldenRatio, ImageSize → 300]
```

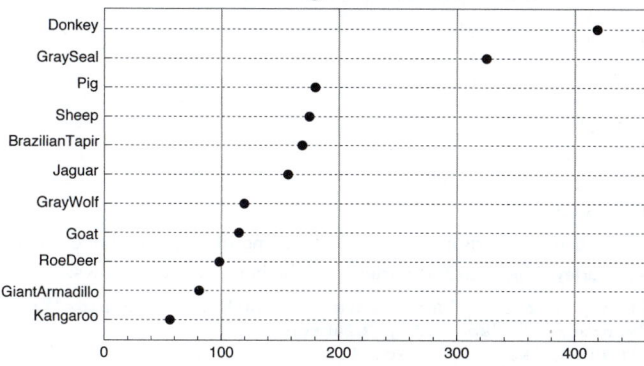

Brain weights for some animals

■ Example 2

With a dot plot we can also compare two or more data sets (another way is to use a bar chart; see Section 8.3.1, p. 256). As an example, we plot the running times of four algorithms before and after improvements (these are not real data):

```
labels = {"Algorithm 1", "Algorithm 2", "Algorithm 3", "Algorithm 4"};
times1 = {8, 9, 11, 10}; times2 = {5, 7, 6, 8};
style1 = {PointSize[Large], Black};
style2 = {PointSize[Large], Gray};

dotPlot[{times1, times2}, labels, {style1, style2},
 {0, 12}, Range[11], {}, AspectRatio → 1 / GoldenRatio,
 PlotLabel → Style["Running times before (black) and\nafter (gray) improvements",
   9, Bold], ImageSize → 200]
```

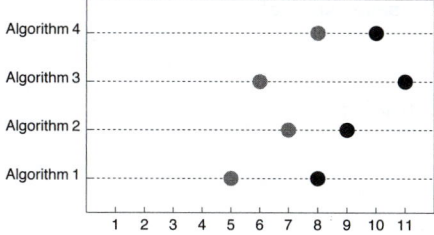

Running times before (black) and after (gray) improvements

■ **Multiway Dot Plots**

An effective way to illustrate 3D data is by using a *multiway dot plot* (Cleveland, 1993). As an example, we consider the data in the file **modBarley**. This file is from Cleveland (1993) and can be found on the CD-ROM accompanying this book. The file contains barley yields at six sites for 10 varieties in 1931. On my computer, the file is in a folder **visdata** in the folder **MNata**:

```
(data = Import["/Users/heikki/Documents/MNData/visdata/modBarley", "Table"]) //
  TableForm
47.3  40.5  35.   35.1  25.7  29.7
48.9  39.9  34.4  27.   29.   33.
46.8  44.1  44.2  24.7  33.1  19.7
55.2  38.1  35.1  43.1  29.7  29.1
50.2  41.3  38.8  39.9  26.3  23.
48.6  41.6  43.2  32.8  32.   34.7
63.8  46.9  46.6  36.6  33.9  29.8
65.8  48.6  47.   36.6  28.1  24.9
58.1  45.7  43.5  43.3  33.6  32.2
58.8  49.9  47.2  39.3  31.6  34.5
```

This file is a slightly modified version of the original file **barley**. In **modBarley**, we have somewhat rearranged the rows and columns of **barley**. The sites and varieties of the barley are as follows:

```
sites = {"Waseca", "Crookston", "Morris", "Univ. Farm", "Duluth", "Gr. Rapids"};
varieties = {"Svansota", "Manchuria", "No. 475", "Glabron",
    "Velvet", "Peatland", "Trebi", "No. 462", "No. 457", "Wisconsin"};
```

Calculate the total mean of all 60 yields:

```
tmean = Mean[Flatten[data]]
```

```
39.1183
```

We produce two multiway dot plots with the program **dotPlot** presented previously. First, we make a multiway dot plot showing the yields of the 10 varieties at each of the six sites:

```
p1 = GraphicsColumn[
    dotPlot[{#[[1]]}, varieties, {{AbsolutePointSize[2.5], Black}}, {0, 72},
      Automatic, {tmean, tmean, tmean}, BaseStyle → {5, FontFamily -> "Helvetica"},
      AspectRatio → 0.6, ImagePadding → 15, PlotLabel → Style[#[[2]], 7]] & /@
    ({dataᵀ, sites}ᵀ), Spacings → 0];
```

Then we make a multiway dot plot showing the yields at the six sites of each of the 10 varieties:

```
p2 = GraphicsColumn[
    dotPlot[{#[[1]]}, Reverse[sites], {{AbsolutePointSize[2.5], Black}}, {0, 72},
      Automatic, {tmean, tmean, tmean}, BaseStyle → {5, FontFamily -> "Helvetica"},
      AspectRatio → 0.4, ImagePadding → 15, PlotLabel → Style[#[[2]], FontSize → 7]] & /@
    Reverse[{Map[Reverse, data], varieties}ᵀ], Spacings → 0];
```

Now we show both multiway dot plots side by side:

```
GraphicsRow[{p1, p2}, AspectRatio → 3.5,
  ImageSize → 400, Spacings → -20, PlotRangePadding → 0]
```

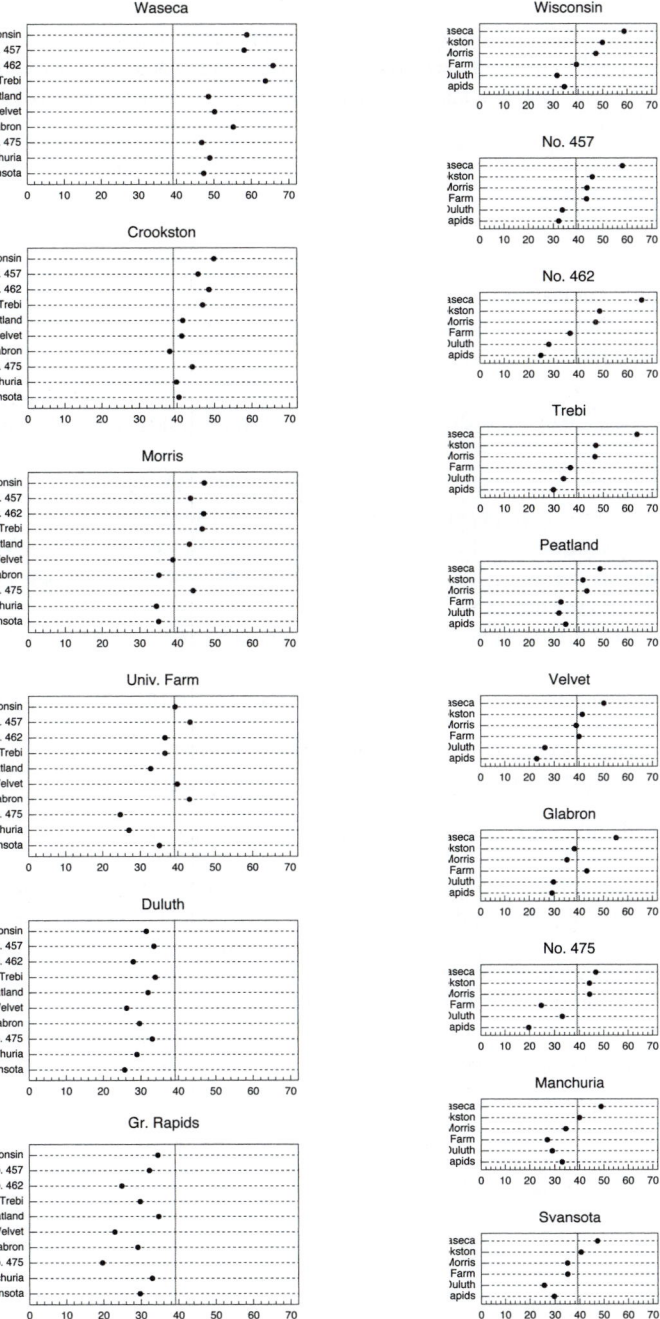

The multiway dot plot of the first column can be used to infer how the yields of the varieties vary within each site and how the yields vary among sites in general. The multiway dot plot of the second column can be used to infer how the yield of each variety varies among the sites. The gray line in both plots is the total mean. If we want to perform detailed comparisons with the data, a multiway dot plot is among the best ways to show the data. The barley data are also considered in the next section and in Section 8.6.1, p. 280.

8.4.2 Box-and-Whisker Plots

In the **StatisticalPlots`** *package:*

BoxWhiskerPlot[vector]   Plot **vector**

BoxWhiskerPlot[matrix]   Plot the columns of **matrix**

BoxWhiskerPlot[vector1, vector2, …]   Plot each vector

Options:

BoxQuantile   If set to a ($0 < a < 0.5$), a box shows data from ($0.5 - a$)-quantile to ($0.5 + a$)-quantile; examples of values: **0.25, 0.4**

BoxLabels   Labels for the boxes; examples of values: **Automatic, {"X", "Y", "Z"}**

BoxOrientation   Orientation of the graph; possible values: **Vertical, Horizontal**

BoxOutliers   Whether to indicate outliers; possible values: **None** (whiskers are drawn to cover the entire data set), **All** (outliers are shown separately), **Automatic** (near and far outliers can be drawn differently by using **BoxOutlierMarkers**)

BoxOutlierMarkers   Markers for the outliers; examples of values: **Automatic, {●, ○}, {{Automatic, 10}}**

BoxFillingStyle   Styles of the boxes; examples of values: **Automatic, Hue[0], {Hue[0], Hue[1/3], Hue[2/3]}**

BoxLineStyle   Style of all lines; examples of values: **Automatic, Hue[2/3]**

BoxMedianStyle   Additional styles for the median line; examples of values: **Automatic, Thickness[Medium]**

BoxExtraSpacing   Extra space between the boxes; examples of values: **0, 0.1**

The following plots illustrate the first two forms of data mentioned previously. In the first case, we have a vector of 100 observations. In the second case, we have a matrix with 20 rows and three columns.

```
SeedRandom[3];
data1 = RandomReal[GammaDistribution[5, 1], 100];
data2 = RandomReal[GammaDistribution[8, 1], {20, 3}];
<< StatisticalPlots`

{BoxWhiskerPlot[data1], BoxWhiskerPlot[data2]}
```

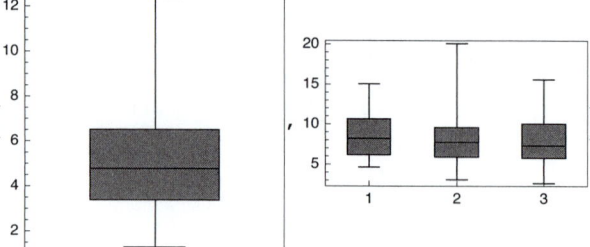

A box-and-whisker plot is simply a way to show the quartiles and the minimum and maximum of the data. In the first example, these statistics are as follows:

```
{Min[data1], Quantile[data1, 0.25],
 Median[data1], Quantile[data1, 0.75], Max[data1]}
{1.27455, 3.39826, 4.78942, 6.52369, 12.4698}
```

The horizontal line inside the box is the median or the 0.5 quantile. Both below and above the median, we have 50% of the data. The bottom and top of the box are at the 0.25 and 0.75 quantiles, so inside the box we have 50% of the data. Both below and above the box, we have 0.25% of the data. The bottom and top horizontal lines of the "whiskers" are at the minimum and maximum of the data. This kind of plot gives a quick overview of the extent of a data set.

For more about the statistical plots package, look at StatisticalPlots/guide/StatisticalPlotsPackage.

■ Example 1

As an example, consider the same barley data we investigated, by using multiway dot plots, in Section 8.4.1, p. 262:

```
data = Import["/Users/heikki/Documents/MNData/visdata/modBarley", "Table"];

BoxWhiskerPlot[data, AspectRatio → 0.4,
  BoxOrientation → Horizontal, BoxLabels → {"Waseca", "Crookston",
    "Morris", "Univ. Farm", "Duluth", "Gr. Rapids"}, ImageSize → 400]
```

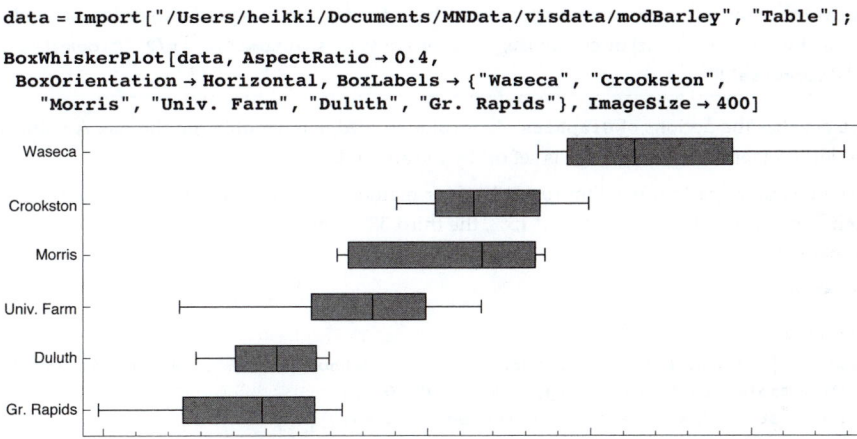

■ Example 2

A *near outlier* is a value beyond 1.5 times the interquantile range from the edge of the box. A *far outlier* is a value beyond 3 times the interquantile range. If the value of **BoxOutliers** is **All**, then all outliers are plotted in the same way; if the value is **Automatic**, then near and far outliers are plotted differently if so determined by the **BoxOutlierMarkers** option. The markers are defined in the same way as for **PlotMarkers** (see Section 8.1.3, p. 244).

```
BoxWhiskerPlot[{6, 3, 8, 5, 2, 7, 10, 4, 3, 5, 16, 24},
  BoxOutliers → Automatic, BoxOutlierMarkers → {●, ○},
  BoxOrientation → Horizontal, AspectRatio → 0.2, ImageSize → 200]
```

8.4.3 Pie Charts

A pie chart illustrates how a total amount is made up from certain components. The purpose is to give the reader an impression of the relative magnitudes of the components.

---

*In the **PieCharts`** package:*

`PieChart[{y₁, y₂, … }]` Plot a pie chart from the positive numbers $y_1, y_2, \ldots$

Options:

`PieLabels` Labels in the wedges; examples of values: `Automatic`, `{"A", "B", "C"}`

`PieStyle` Style(s) inside the borders of the wedges; examples of values: `Automatic`,
   `Table[GrayLevel[p], {p, 0.7, 1, 0.1}]`

`PieEdgeStyle` Style of the border of the wedges; examples of values: `Automatic`,
   `Thickness[Medium]`, `Directive[Thickness[Medium], Blue]`

`PieExploded` Whether some wedges are exploded; examples of values: `None`, `All`, `{4}`, `{4, 0.2}`,
   `{4, 5}`, `{{4, 0.2}, {5, 0.2}}`

`PieOrientation` Starting angle of the first wedge (the default is 0) and whether to order the wedges
   counterclockwise (the default) or clockwise; examples of values: `Automatic`, `-π/2`, `"Clockwise"`,
   `{π/2, "Clockwise"}`

`PieChart` also has the options of `Graphics`. An exploded wedge is set off from the pie. A value such as `{4, 0.2}` defines that the fourth wedge is set off by the amount 0.2.

As an example, an algorithm was improved by four methods. Of the total savings in running time, the first method contributed 12%, the second 15%, the third 38%, and the fourth 35%. The corresponding pie chart is shown here:

```
<< PieCharts`

PieChart[{12, 15, 38, 35},
  PieLabels → {"Method 1\n12%", "Method 2\n15%", "Method 3\n38%", "Method 4\n35%"},
  PieStyle → Table[GrayLevel[p], {p, 0.65, 0.95, 0.1}],
  PlotLabel → Style["Savings by four methods", 9, Bold]]
```

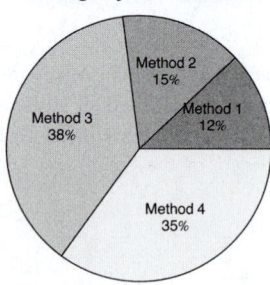

8.4.4 Vector Fields

---

*In the **VectorFieldPlots`** package:*

`ListVectorFieldPlot[data]` Plot the given array of vectors

```
data = Table[{i, Sqrt[j]}, {i, 0, 1, 0.2}, {j, 0, 1, 0.2}];

<< VectorFieldPlots`

ListVectorFieldPlot[data, ImageSize → 100]
```

8.5 Graph Plots

8.5.1 Graph Plots

■ **Graph Plots**

> **GraphPlot[{v1 → v2, v3 → v4, … }]** (❀6)   Connect vertices **v1** and **v2**, …
> **GraphPlot[{{v1 → v2, lb1}, {v3 → v4, lb2}, … }]**   Edges have the given labels
> **GraphPlot[m]**   Plot the graph represented by the adjacency matrix **m**
>
> *Some options:*
>
> **DirectedEdges**   Whether to show edges as directed arrows; possible values: **False** (edges are shown as lines), **True**, **{True, "ArrowheadsSize"→ s}**
>
> **VertexLabeling**   Whether to show vertex names as labels; examples of values: **Automatic** (show labels as tooltips if the graph is small), **Tooltip** (show labels as tooltips), **True** (show labels explicitly), **False** (do not show labels at all), **All** (show labels both explicitly and as tooltips)

Graphs are useful in illustrating connections and flows between points. The points are called *vertices* (or nodes), and the lines or arrows between the vertices are called *edges*. In using **GraphPlot**, we only indicate, by using some names for the vertices, which vertices should be connected; the coordinates of the vertices are automatically chosen by **GraphPlot** using some algorithms that try to obtain a clear graph. However, we can also define the coordinates with an option, if the automatic coordinates do not satisfy us. Here are simple examples of graphs:

```
{GraphPlot[{1 → 2, 1 → 3, 2 → 3, 2 → 4, 3 → 4}],
 GraphPlot[{{1 → 2, c₁₂}, {1 → 3, c₁₃}, {2 → 3, c₂₃}, {2 → 4, c₂₄}, {3 → 4, c₃₄}}],
 GraphPlot[{{0, 1, 1, 0}, {0, 0, 1, 1}, {0, 0, 0, 1}, {0, 0, 0, 0}}]}
```

By default, vertices are shown as points and edges as lines. In the previous plots, the names 1, 2, 3, and 4 of the vertices can be seen as tooltips: The name of a vertex appears when the mouse cursor is moved (without pressing the button) above the vertex. In the second plot, we have labels of edges. In the third example, we represented the graph with an adjacency matrix (the matrix can also be a sparse array; see Section 21.2.1, p. 689). Some more examples:

```
{GraphPlot[{1 → 2, 1 → 3, 2 → 3, 2 → 4, 3 → 4}, DirectedEdges → True],
 GraphPlot[{1 → 2, 1 → 3, 2 → 3, 2 → 4, 3 → 4},
   DirectedEdges → {True, "ArrowheadsSize" → 0.06}, VertexLabeling → True],
 GraphPlot[{a → b, a → c, b → c, b → d, c → d}, VertexLabeling → True]}
```

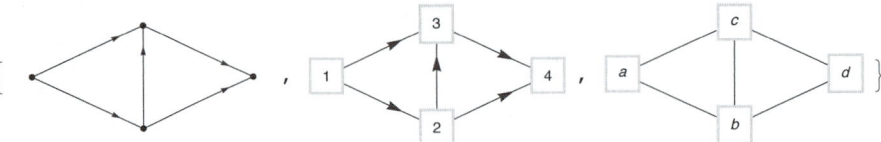

In the first plot, we have arrows. In the second plot, we have labels for the vertices and the size of the arrowheads is made larger (the size and other properties of the arrowheads can also be adjusted with the **PlotStyle** option; see General Options). The third plot shows that the names of the vertices can be any expression.

Note that we also have the **GraphData** command with which we can plot and study many graphs (see Section 9.3.1, p. 301). See also *Combinatorica*/guide/CombinatoricaPackage in the Documentation Center for information about the **Combinatorica`** package.

Next, we study all the special options of **GraphPlot**. The options are grouped into options for edges, options for vertices, and general options. In addition to these options, we have the options of **Graphics**.

In addition to **GraphPlot**, we have **LayeredGraphPlot** and **TreePlot**. These are considered later.

■ Options for Edges

DirectedEdges Whether to show edges as directed arrows; possible values: **False** (edges are shown as lines), **True**, **{True, "ArrowheadsSize"→ s}**

MultiedgeStyle How to draw multiple edges between vertices; examples of values: **Automatic** (multiple edges are shown; however, if the graph is defined by an adjacency matrix, multiple edges are not shown), **0.2** (the distance between the outermost edges is a fraction of 0.2 of the distance of the corresponding vertices), **True** (show multiple edges), **False** (do not show multiple edges)

SelfLoopStyle How to draw edges linking a vertex to itself; examples of values: **Automatic** (self-loops are shown; however, if the graph is defined by an adjacency matrix, self-loops are not shown), **0.5** (the diameter of a self-loop circle is a fraction of 0.5 of the average edge length), **True** (show self-loops), **False** (do not show self-loops)

EdgeLabeling Whether to include the given labels of the edges; examples of values: **True** (show edge labels explicitly), **Automatic** (show edge labels as tooltips), **All** (show edge labels both explicitly and as tooltips), **False** (do not show edge labels)

EdgeRenderingFunction Function to give explicit graphics for edges; examples of values: **Automatic** (edges are dark red lines), **None** (no edges), **({Red, Arrow[#1]} &)**

The **DirectedEdges** option was explained previously. Here are examples of **MultiedgeStyle**:

```
GraphPlot[{1 → 2, 2 → 3, 3 → 2}, MultiedgeStyle → #] & /@ {Automatic, 0.5, False}
```

With the **SelfLoopStyle** we can adjust the size of a circle:

```
GraphPlot[{1 → 2, 2 → 3, 3 → 3}, SelfLoopStyle → #] & /@ {Automatic, 0.3, 1}
```

Edge labels are normally drawn explicitly, as is shown in the first plot that follows. If the **EdgeLabeling** option has the value **Automatic**, the edge labels are only shown as tooltips. The value **All** causes the labels to be shown both explicitly and as tooltips.

```
GraphPlot[{{1 → 2, c₁₂}, {2 → 3, c₂₃}, {3 → 3, c₃₃}}, EdgeLabeling → #] & /@
  {True, Automatic, All}
```

Edges are normally lines (as can be seen from the first plot that follows) or arrows, with the arrow-head somewhat back from the top of the arrow. If the **EdgeRenderingFunction** option has the value **None**, the edges are not drawn, as can be seen from the second plot. We can give the option a value as a pure function, where **#1** refers to the list of coordinates of the vertices at the ends of an edge. (The function can also have arguments **#2** and **#3** that refer to the names of the vertices and to the label of the edge, respectively.) In the third plot, we ask to draw red arrows between the vertices.

```
GraphPlot[{1 → 2, 2 → 3, 3 → 3}, EdgeRenderingFunction → #] & /@
  {Automatic, None, ({Red, Arrow[#1]} &)}
```

For more about arrows, see Section 6.2.5, p. 161. Next, we adjust the arrows in various ways. In the first plot, the ends of the arrows are set back from the end points by a small amount. In the second plot, we use the **Arrowheads** directive to insert arrowheads at both ends of the arrows. In the third plot, we define custom size and position for the arrowheads.

```
GraphPlot[{1 → 2, 2 → 3, 3 → 3}, EdgeRenderingFunction → #] & /@
  {(Arrow[#1, 0.1] &),
   ({Arrowheads[{-0.05, 0.05}], Arrow[#1, 0.1]} &),
   ({Arrowheads[{{0.06, 0.9}}], Arrow[#1, 0.1]} &)}
```

■ **Options for Vertices**

VertexLabeling Whether to show vertex names as labels; examples of values: **Automatic** (show labels as tooltips if the graph is small), **Tooltip** (show labels as tooltips), **True** (show labels explicitly), **False** (do not show labels at all), **All** (show labels both explicitly and as tooltips)

VertexCoordinateRules Explicit vertex coordinates as a complete list of $\{x, y\}$ pairs or as a complete or incomplete list of rules for the $\{x, y\}$ pairs; examples of values: **Automatic**, {coordinates}, {1 → {0, 0}, 4 → {1, 0.5}}

VertexRenderingFunction Function to give explicit graphics for vertices; examples of values: **Automatic** (vertices are blue points with tooltips showing the names), **None** (no vertices), ({Yellow, EdgeForm[Black], Disk[#1, 0.2], Black, Text[#2, #1]} &) (yellow disks with black edges and black labels)

The first option was explained previously. If the graph given by `GraphPlot` does not satisfy us, even after trying several different methods (see the `Method` option discussed later), we can ourselves define the coordinates of the vertices. In the first plot that follows, we define the coordinates as a list of $\{x, y\}$ pairs. In the second plot, we define the coordinates as rules. In the third plot, we only define the coordinates of two vertices.

```
GraphPlot[{1 → 2, 1 → 3, 2 → 3, 2 → 4, 3 → 4},
    VertexLabeling → True, VertexCoordinateRules → #] & /@
{{{0, 0}, {1, 0.5}, {1, -0.5}, {2, 0}},
    {1 → {0, 0}, 2 → {1, 0.5}, 3 → {1, -0.5}, 4 → {2, 0}},
    {1 → {0, 0}, 4 → {1, 0.5}}}}
```

Note that coordinates of vertices can only be defined by rules if the `Method` option has the default value `"SpringElectricalEmbedding"`.

The value of the `VertexRenderingFunction` option can be a pure function where `#1` refers to the coordinates and `#2` to the label of a vertex. Here are examples of this option:

```
GraphPlot[{1 → 2, 2 → 3, 3 → 3}, VertexRenderingFunction → #] & /@
    {Automatic,
    None,
    ({Yellow, EdgeForm[Black], Disk[#1, 0.2], Black, Text[#2, #1]} &)}
```

■ General Options

> **Method**   Method used to lay out the graph; possible values: **Automatic** (means
>   `"SpringElectricalEmbedding"`; however, `"RadialDrawing"` is used for trees),
>   `"CircularEmbedding"`, `"HighDimensionalEmbedding"`, `"LinearEmbedding"`,
>   `"RandomEmbedding"`, `"SpiralEmbedding"`, `"SpringEmbedding"`, `"SpringElectricalEmbedding"`,
>   `"LayeredDrawing"`, `"LayeredDigraphDrawing"`, `"RadialDrawing"`, {`"meth"`, `"Rotation"` → θ}
>   (rotate the graph θ radians clockwise)
>
> **PackingMethod**   Method used to lay out a graph with disconnected components; possible values:
>   `Automatic`, `"ClosestPacking"`, `"ClosestPackingCenter"`, `"Layered"`, `"LayeredLeft"`,
>   `"LayeredTop"`, `"NestedGrid"`
>
> **PlotStyle**   Overall graphics directives for vertices and edges; examples of values: **Automatic**, {Red,
>   PointSize[Medium], Thickness[Medium], Dashed, Arrowheads[0.05]}

Each value of `Method` has some options. The `"Rotation"` option is one. Consider the following three plots:

```
edges = {1 → 2, 1 → 3, 1 → 4, 1 → 5, 1 → 6, 2 → 7,
    2 → 8, 3 → 7, 3 → 8, 4 → 7, 4 → 8, 5 → 7, 5 → 8, 6 → 7, 6 → 8};
vertices = {{0, 2}, {2, 4}, {2, 3}, {2, 2}, {2, 1}, {2, 0}, {4, 3}, {4, 1}};
```

```
{GraphPlot[edges, VertexLabeling → True, AspectRatio → 1],
 GraphPlot[edges, VertexLabeling → True,
   Method → {"LayeredDigraphDrawing", "Rotation" → -π/2}, AspectRatio → 1],
 GraphPlot[edges, VertexLabeling → True,
   VertexCoordinateRules → vertices, AspectRatio → 1]}
```

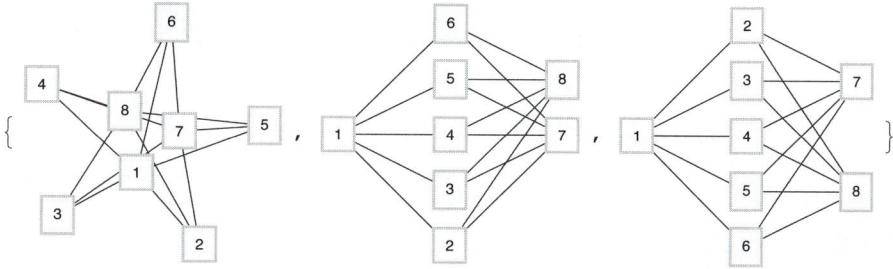

With the default **"SpringElectricalEmbedding"** method, the plot is not very clear. With the **"LayeredDigraphDrawing"** method, with a rotation of $-\pi/2$, we get a good result. By defining the coordinates of the vertices, we get just the result we like most. (We also considered this example in Section 6.2.9, p. 174, by using our own plotting program.)

The methods used by **GraphPlot** are explained in the Documentation Center at tutorial/GraphDrawingIntroduction and tutorial/GraphDrawing. The default method, **"SpringElectrical-Embedding"**, is described as follows: "Invoke the spring embedding method, in which a vertex is subject to either attractive or repulsive force from another vertex, as though they are connected by a spring; the spring has an ideal length equal to the graph distance between the vertices; the total spring energy is minimized."

Note that the plot given by **GraphPlot** can also be adjusted manually with the mouse. Suppose that, in the second plot shown previously, we want to move the vertex 7 somewhat downwards. Just click on the graph and on the vertex 7 a sufficient number of times until the vertex 7 is selected. Then drag the vertex downwards. For more about manipulating plots with the mouse, see Section 5.1.3, p. 126.

If the graph contains disconnected parts, the **PackingMethod** option can be used to adjust the way the graph is displayed:

```
GraphPlot[{1 → 2, 1 → 3, 2 → 3, 2 → 4, 3 → 4, 5 → 6}, PackingMethod → #] & /@
   {Automatic, "ClosestPackingCenter", "NestedGrid"}
```

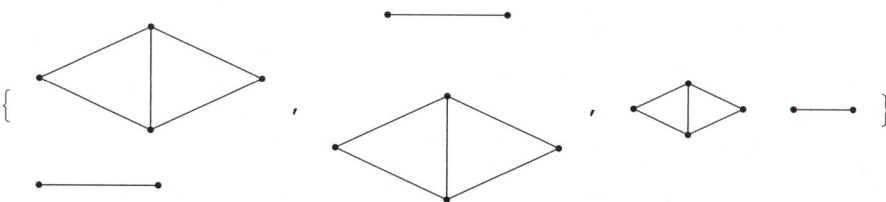

With **PlotStyle**, we can adjust the size of the points representing the vertices, the thickness and dashing of the edges, and the size and position of the arrowheads. In addition, we can define an overall color that is used both for the edges and for either the points representing the vertices or the text in the vertex labels. However, edge and vertex rendering functions have higher priority than plot style.

```
{GraphPlot[{1 → 2, 2 → 3, 3 → 3}, VertexLabeling → True, PlotStyle → Red],
 GraphPlot[{1 → 2, 2 → 3, 3 → 3}, DirectedEdges → True,
  PlotStyle → {Red, Arrowheads[{{0.05, 0.8}}]},
  VertexRenderingFunction → ({Blue, PointSize[Medium], Point[#1]} &)],
 GraphPlot[{1 → 2, 2 → 3, 3 → 3},
  PlotStyle → {Red, PointSize[Medium], Thickness[Medium], Dashed}]}
```

{ [1]——[2]——[3]⟲ , •——▸•——▸◯⟲ , •----•----•⟲ }

■ Applications

Here is a rate diagram for a birth–death process:

```
GraphPlot[{{0 → 1, λ₀}, {1 → 2, λ₁}, {2 → 3, λ₂}, {3 → 2, μ₃}, {2 → 1, μ₂}, {1 → 0, μ₁}},
  DirectedEdges → True, MultiedgeStyle → 0.5, VertexLabeling → True,
  ImageSize → 350, Epilog → Text[Style[". . .", 12], {3.25, 0.03}]]
```

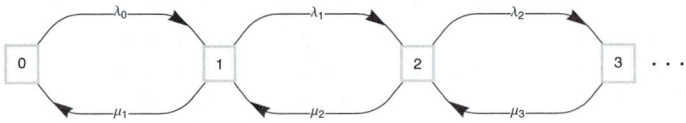

Next, we draw a graph describing the process of modeling (see Giordano, Weir, and Fox, 1997, p. 33):

```
b1 = "Real-world data"; b2 = "Model";
b3 = "Mathematical\nconclusions"; b4 = "Predictions/\nexplanations";
GraphPlot[{{b1 → b2, "Formulation"}, {b2 → b3, "Analysis"},
   {b3 → b4, "Interpretation"}, {b4 → b1, "Test"}}, DirectedEdges → True,
  VertexLabeling → True, VertexCoordinateRules → {{0, 1}, {1, 1}, {1, 0}, {0, 0}},
  ImageSize → 200, AspectRatio → 0.5]
```

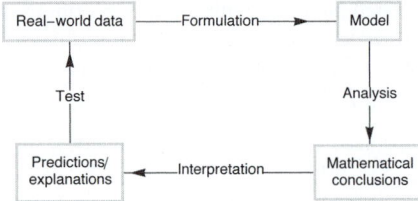

Now we draw relationships between some probability distributions:

$$\{ber = Ber[p], hyp = Hyp[n, K, N], bin = Bin[n, p],$$

$$po = Po[\lambda], normal = "N(\mu, \sigma)", gamma = Gamma[\alpha, \lambda], sum = \sum_{i=1}^{n} x_i\};$$

```
GraphPlot[{{hyp → bin, Column[{p ≡ K / N, N → ∞}, Center]},
  {bin → bin, sum}, {bin → ber, n ≡ 1}, {ber → bin, sum},
  {bin → po, Column[{λ ≡ n p, n → ∞, p → 0}, Center]},
  {bin → normal, Column[{μ ≡ n p, σ² ≡ n p q, n → ∞}, Center]},
  {po → po, sum}, {po → normal, Column[{μ ≡ λ, σ² ≡ λ, λ → ∞}, Center]},
  {normal → normal, sum}, {gamma → gamma, sum},
  {gamma → normal, Column[{μ ≡ α / λ, σ² ≡ α / λ², α → ∞}, Center]}}, ImageSize → 400,
 DirectedEdges → {True, "ArrowheadsSize" → 0.02}, MultiedgeStyle → 0.3,
 SelfLoopStyle → 0.4, VertexLabeling → True, VertexCoordinateRules →
  {hyp → {0, 1}, bin → {1, 1}, ber → {2, 1}, po → {0, 0}, normal → {1, 0}, gamma → {2, 0}}]
```

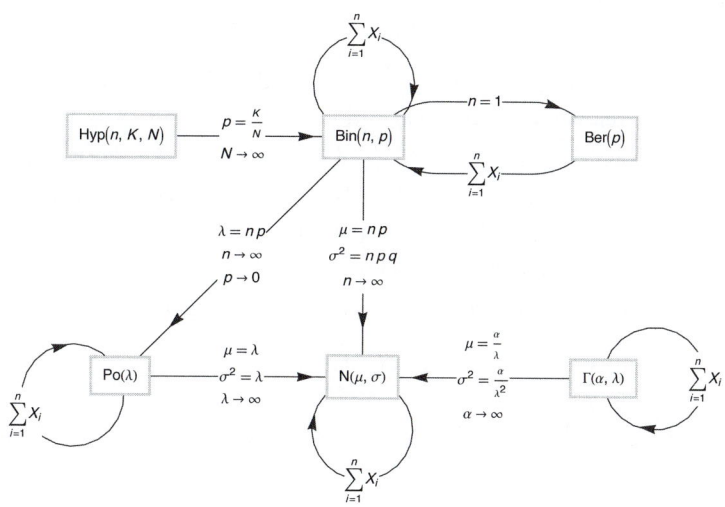

For example, the sum of n independent variables having the Ber(p) distribution has the binomial distribution. The sum of n independent variables having the Po(λ) distribution also has the Poisson distribution [more exactly, Po$(n\lambda)$ distribution]. The binomial distribution Bin(n, p) can be approximated, for large n, by the normal distribution $N\left(np, \sqrt{npq}\right)$.

Next, we draw a social network of my wife:

```
relationships = {"Marjatta" → "Heikki", "Marjatta" → "Hanna",
    "Hanna" → "Kerttu", "Hanna" → "BO", "Marjatta" → "Kerttu",
    "Kerttu" → "Raimo", "Marjatta" → "Eine", "Eine" → "Jussi", "Eine" → "Tuula",
    "Marjatta" → "Marjukka", "Marjukka" → "Antti", "Marjatta" → "Ilse",
    "Ilse" → "Hanna", "Marjatta" → "Seija", "Seija" → "Pirkko", "Seija" → "Eino"};
GraphPlot[relationships, VertexLabeling → True, ImageSize → 230]
```

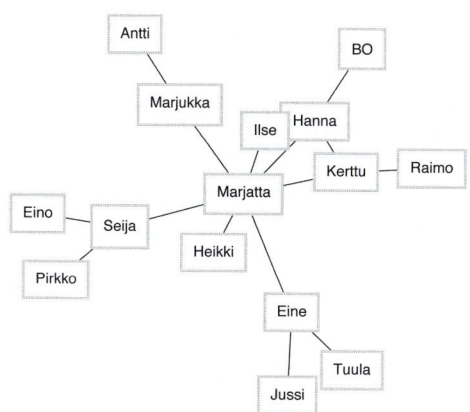

■ **Layered Graph Plots**

`LayeredGraphPlot[{v1 → v2, v3 → v4, … }]` (✿6)	Connect vertices **v1** and **v2**, …
`LayeredGraphPlot[{v1 → v2, v3 → v4, … }, pos]`	Place the dominant vertices at position **pos**;
possible values of **pos: Top, Bottom, Left, Right**	

This command is used in the same way as `GraphPlot`; edge labels can be given and the graph can be defined by an adjacency matrix. The options are almost the same as the ones of `GraphPlot`; the default value of `DirectedEdges` is now `True`, and the `Method` option does not exist.

A layered graph plot shows the vertices at several layers or levels:

`LayeredGraphPlot[relationships, VertexLabeling → True, ImageSize → 260]`

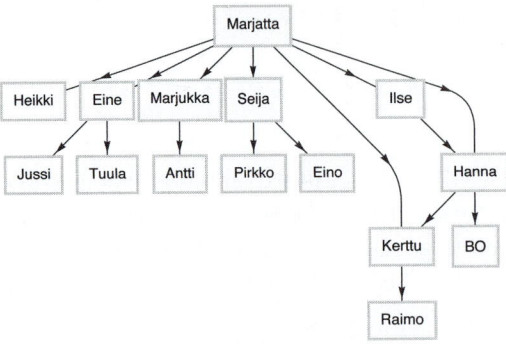

If vertex coordinates are not given, some edges may be curved. If vertex coordinates are given, the edges are straight lines.

8.5.2 Tree Plots

`TreePlot[{v1 → v2, v3 → v4, … }]` (✿6)	Connect vertices **v1** and **v2**, …
`TreePlot[{v1 → v2, v3 → v4, … }, pos]`	Place the dominant vertices at position **pos**; possible values
of **pos: Top, Bottom, Left, Right, Center**	
`TreePlot[{v1 → v2, v3 → v4, … }, pos, vk]`	Use **vk** as the root node

This command, too, is used in the same way as `GraphPlot`; edge labels can be given and the graph can be defined by an adjacency matrix. For `TreePlot`, we have the new option `LayerSizeFunction` with the default value `(1 &)` (the function defines the height of the layers), whereas the `Method` option does not exist. The next plot describes the dependencies of the chapters of a book:

`TreePlot[{1 → 3, 1 → 2, 2 → 4, 2 → 6, 4 → 5, 6 → 8}, Top, 1,`
`    VertexLabeling → True, LayerSizeFunction → (0.3 &), ImageSize → 100]`

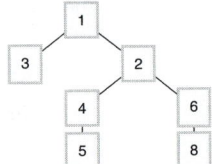

In the following plot, we show methods used in modeling (see Giordano *et al.*, 1997, p. 34):

```
TreePlot[{{1 → 2, "Phenomenon\nof interest"},
   {2 → 3, "Replication\nof behavior"}, {2 → 4, "Mathematical\nrepresentation"},
   {3 → 5, "Simulation"}, {3 → 6, "Experimentation"},
   {4 → 7, "Model selection"}, {4 → 8, "Model construction"}}, Left, 1,
  VertexRenderingFunction → None, LayerSizeFunction → (8 &), ImageSize → 350]
```

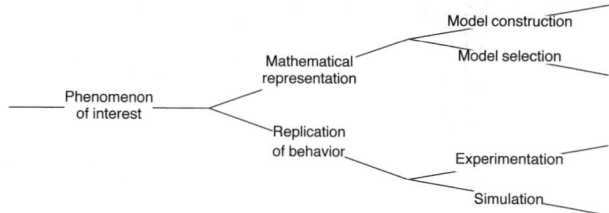

8.6 Plots for 3D Data

8.6.1 Plots for 3D Data

■ **Regular Data**

Often, 3D data are in a form of a matrix:

$$\begin{pmatrix} z_{11} & \cdots & z_{1n} \\ & \vdots & \\ z_{m1} & \cdots & z_{mn} \end{pmatrix}$$

We call such data regular: Each row has the same number of values. Also, only the z values are given; the x and y values are not given at all. Indeed, it is assumed that the z values are given for equally spaced x and y coordinates. In such a situation, we simply need to be able to tell the ranges of the x and y coordinates, and this can be done with the **DataRange** option, which is explained soon. In the next box, we have commands to plot regular 3D data. Irregular data are considered later.

ListPlot3D[data] 3D surface plot
ListPointPlot3D[data, Filling → Bottom] (❀6) 3D points, possibly with stems
BarChart3D[data] 3D bar chart (in the **BarCharts`** package)

ArrayPlot[data] Grayscale squares
MatrixPlot[data] (❀6) Color squares
ListDensityPlot[data] A density plot
ListContourPlot[data] A contour plot

Data are given in the matrix form:
{{z₁₁, …, z₁ₙ}, …, {zₘ₁, …, zₘₙ}} (each row corresponds to a fixed value of y)

We also have **GeneralizedBarChart3D** (in the **BarCharts`** package), **ListVectorFieldPlot3D** (in the **VectorFieldPlots`** package), and **ReliefPlot**.

■ **Examples**

To illustrate these commands, we define a very simple data set:

```
data1 = {{0, 0, 1, 1, 0}, {1, 0, 1, 1, 1}, {2, 1, 2, 2, 1}, {2, 1, 2, 2, 2}};
```

First we show six plots that have a true 3D nature. Here are three surface plots:

```
{ListPlot3D[data1, BoxRatios → Automatic, AxesLabel → {"x", "y", ""}],
 ListPlot3D[data1, Mesh → Full, BoxRatios → Automatic],
 ListPlot3D[data1, Mesh → Full, PlotStyle → None, BoxRatios → Automatic]}
```

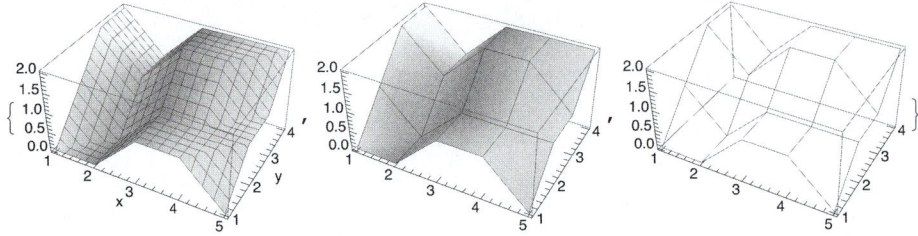

In the first plot, the given points are connected by surface pieces and a mesh is drawn on the resulting surface. The first row {0, 0, 1, 1, 0} of the data is in front (parallel to the x axis), the second row {1, 0, 1, 1, 1} is next, and so on. In the second plot, we ask for a mesh corresponding to the data points: The points are at the corner points of the surface pieces. In the third plot, we only have the mesh lines.

```
<< BarCharts`
```

```
{ListPointPlot3D[data1, BoxRatios → Automatic],
 ListPointPlot3D[data1, Filling → Bottom, BoxRatios → Automatic],
 BarChart3D[data1ᵀ, BoxRatios → Automatic,
  Ticks → {{1, 2, 3, 4, 5}, {1, 2, 3, 4}, Automatic}]}
```

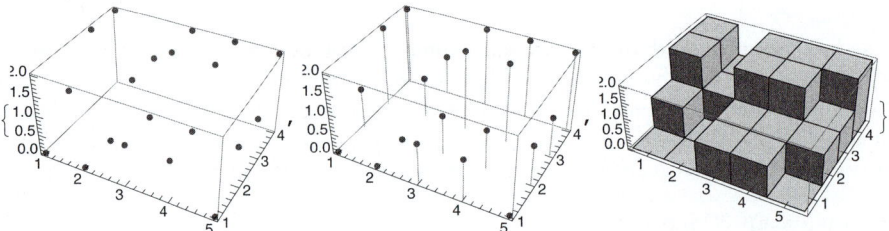

In the first plot, we have 3D points (a scatter plot). Without vertical lines, it is difficult to infer the positions of the points in the space. In the second plot, we have the vertical lines to the (x, y) plane, and now the plot is much clearer. The third plot shows the points as a bar chart; this plot gives a very clear view for the data. In a bar chart, we have to transpose the data to get a plot comparable with the other plots.

Note that all six 3D plots shown previously can be rotated with the mouse. Rotation dramatically improves the illusion of space.

Next, we draw other kinds of plots:

```
GraphicsRow[{ArrayPlot[data1, Mesh → True, Frame → True,
    FrameTicks → {True, True, None, None}, DataReversed → True],
  MatrixPlot[data1, Mesh → True, Frame → True,
    FrameTicks → {True, True, None, None}, DataReversed → True],
  ListDensityPlot[data1, AspectRatio → Automatic,
    FrameTicks → {{1, 2, 3, 4, 5}, {1, 2, 3, 4}, None, None}],
  ListContourPlot[data1, AspectRatio → Automatic,
    FrameTicks → {{1, 2, 3, 4, 5}, {1, 2, 3, 4}, None, None}]}, ImageSize → 420]
```

In the first plot, the values are plotted as grayscale squares, and in the second plot they are plotted as color squares. The density plot shows the data values somewhat smoothed: high values as light and low values as dark. In the contour plot, there are 10 contours that correspond to 10 equally spaced values between the minimum and maximum values. The constants that correspond to the contours can be seen by moving the mouse over the contour plot. With the option **ContourLabels → Automatic** we get some explicit labels in the plot.

■ Options

The options of **ListPlot3D**, **ListDensityPlot**, and **ListContourPlot** are almost the same as the ones of **Plot3D**, **DensityPlot**, and **ContourPlot**. Thus, we can refer to Chapter 7 for the options. Some options, such as **MaxRecursion** and **PlotPoints**, are lacking, but we have the following new options:

Common special options of **ListPlot3D**, **ListDensityPlot**, *and* **ListContourPlot**:

DataRange   The range of x and y values to assume for data; examples of values: **Automatic**, **{{xmin, xmax}, {ymin, ymax}}**

InterpolationOrder   The polynomial degree (in each variable) of surfaces used in joining data points; default value: **None**

MaxPlotPoints   The maximum number of points to include; default value: **Automatic**

As stated previously, the data points contain only the z values. To plot ticks on the axes, *Mathematica* assumes that the x and y values are evenly spaced and are, in fact, the integers 1, 2, 3, …; you can see this from the previous plots. If the true x and y values are not these integers, the option **DataRange** should be used to input the true ranges within which the points lie. For example, suppose that x values are 0, 1, 2, 3, and 4 and y values 0, 2, 4, and 6; then the x range is {0, 4} and the y range is {0, 6}. We can get the correct x and y coordinates as follows:

ListContourPlot[data1, DataRange → {{0, 4}, {0, 6}}, ImageSize → 100]

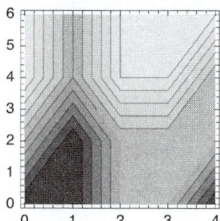

The options of **ListPointPlot3D** are mostly the same as the ones of **Graphics3D**. **Axes** and **BoxRatios** now have the default values **True** and **{1, 1, 0.4}**, respectively. The following new options are available:

Special options of **ListPointPlot3D**:

DataRange The range of *x* and *y* values to assume for data; examples of values: **Automatic**, **{{xmin, xmax}, {ymin, ymax}}**

PlotStyle Style of the points; default value: **Automatic**

ColorFunction How to determine the colors of points; default value: **Automatic**

ColorFunctionScaling Whether to scale arguments to **ColorFunction**; possible values: **True**, **False**

The options of **BarChart3D** also are mostly the same as the ones of **Graphics3D**. **Axes** and **BoxRatios** now have the default values **Automatic** and **{1, 1, 1}**, respectively. The following new options are available:

Special options of **BarChart3D**:

BarEdges Whether to draw the edges of the bars; possible values: **True**, **False**

BarEdgeStyle Style of the edges; default value: **GrayLevel[0]**

BarSpacing Space between the bars in the *x* and *y* directions; default value: **0**

BarStyle Style of the faces of the bars; default value: **GrayLevel[1]**

ArrayPlot and **MatrixPlot** and their options are considered in Section 21.2.1, p. 690.

■ Coloring

We used the **ColorFunction** option in Sections 7.5.1, p. 214, 7.5.4, p. 222, and 7.6.1, pp. 227 and 230. Now we use this option to color data plots according to the values of the data points. In **BarChart3D**, the colors have to be attached to each data value.

```
data2 = Table[RandomInteger[{8 i + 5 j, 10 i + 7 j}], {i, 10}, {j, 15, 1, -1}];

data3 = Partition[{#, Hue[1 - # / Max[data2]]} & /@ Flatten[data2], 15];

{ListPlot3D[data2, Ticks → None,
   BoxRatios → {15, 10, 10}, ColorFunction → (Hue[1 - #3] &)],
  ListPointPlot3D[data2, Ticks → None, BoxRatios → {15, 10, 10},
   ColorFunction → (Hue[1 - #3] &)],
  BarChart3D[data3ᵀ, BoxRatios → {15, 10, 10}, Ticks → None]}
```

```
{ArrayPlot[data2, DataReversed → True, ColorFunction → (Hue[1 - #] &)],
 ListContourPlot[data2, AspectRatio → Automatic,
  FrameTicks → None, ColorFunction → (Hue[1 - #] &)],
 ListDensityPlot[data2, AspectRatio → Automatic,
  ColorFunction → (Hue[1 - #] &), FrameTicks → None]}
```

{ 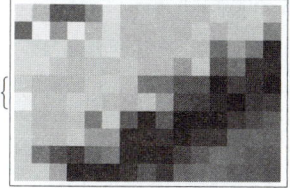 , 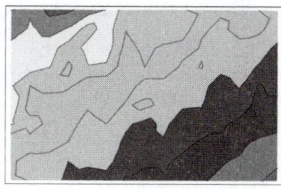 , 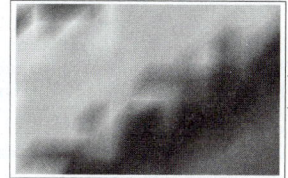 }

■ Example: Galaxy

We have in a text file **galaxy** various data about NGC 7531, a spiral galaxy in the Northern Hemisphere. The data are, again, from Cleveland (1993) and can be found on the CD-ROM accompanying this book.

```
data6 = Rest[Import["/Users/heikki/Documents/MNData/visdata/galaxy", "Table"]];
```

(**Rest** drops the first row, which contains the headings of the columns.) The file contains 323 rows. The first row is as follows:

```
data6[[1]]
```

```
{3, 8.46279, -38.1732, 102.5, 39.1, 1769}
```

The first item is the observation number (ranging from 3 to 417 but having missing observations), the second and third items are the coordinates of a point of the galaxy, and the sixth item is the velocity of the galaxy at the given point. We extract the columns from the data:

```
{no, eastwest, southnorth, slitangle, radialposition, velocity} = data6ᵀ;
```

The velocity varies between the following numbers (given in kilometers per second):

```
{Min[velocity], Max[velocity]}
```

```
{1409, 1775}
```

Then we plot the velocities. Note that the plot can be rotated with the mouse.

```
ListPointPlot3D[{eastwest, southnorth, velocity}ᵀ, BoxRatios → {6, 10, 10},
 PlotStyle → PointSize[Small], ViewPoint → {-2.9, 1, 1.2},
 ImageSize → 200, AxesLabel → {"sn", "ew", "   velocity"}]
```

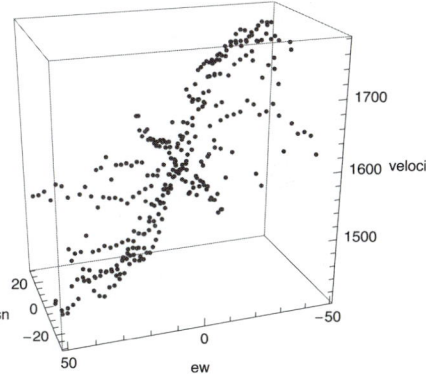

■ **Example: Barley**

In Sections 8.4.1, p. 262, and 8.4.2, p. 265, we considered barley yields data. Now we illustrate the same data with a bar chart:

```
data4 = Import["/Users/heikki/Documents/MNData/visdata/modBarley", "Table"];

sites = {"Waseca", "Crookston", "Morris", "Univ. Farm", "Duluth", "Gr. Rapids"};
varieties = {"Svansota", "Manchuria", "No. 475", "Glabron",
    "Velvet", "Peatland", "Trebi", "No. 462", "No. 457", "Wisconsin"};

xticks = {Range[6], sites}ᵀ;
yticks = {Range[10], varieties}ᵀ;
xgrid = Range[1.5, 5.5, 1]; ygrid = Range[1.5, 9.5, 1]; zgrid = Range[10, 60, 10];
grids = {{{-1, 0, 0}, {ygrid, zgrid}},
    {{0, 1, 0}, {xgrid, zgrid}}, {{0, 0, -1}, {xgrid, ygrid}}};

BarChart3D[data4ᵀ, BoxRatios → {6, 10, 7},
  ViewPoint → {1.8, -2.4, 1.6}, AxesEdge → {Automatic, {1, -1}, Automatic},
  Ticks → {xticks, yticks, Range[10, 60, 10]}, FaceGrids → grids, ImageSize → 280]
```

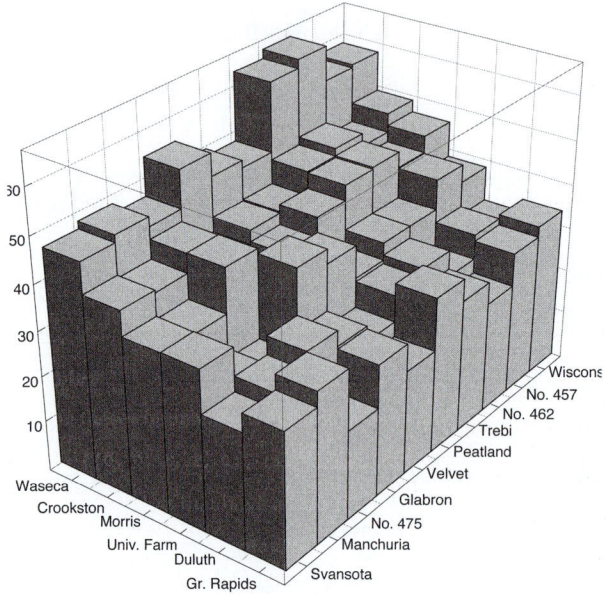

■ **Histograms**

> *In the* **Histograms`** *package:*
>
> **Histogram3D[data]**   Plot the frequencies of the given raw data

With the option **FrequencyData → True**, we can plot given frequencies.

```
<< MultivariateStatistics`
<< Histograms`

SeedRandom[1]; data =
  Table[RandomReal[MultinormalDistribution[{0, 0}, {{1, 0.6}, {0.6, 1}}]], {2000}];
```

```
Histogram3D[data, ImageSize → 260]
```

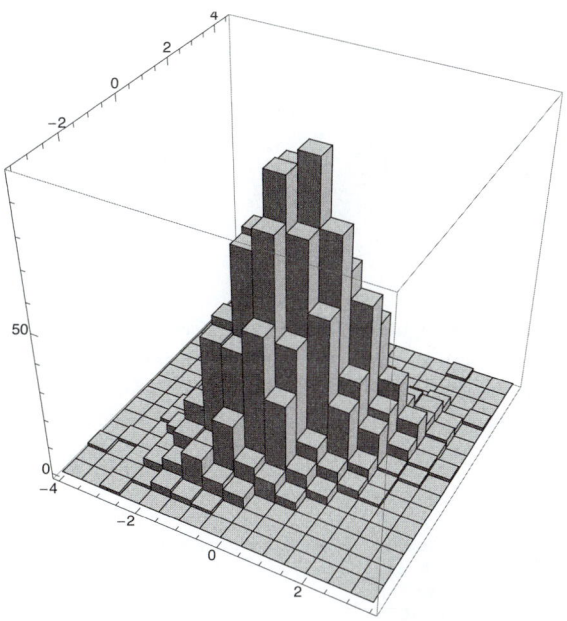

See Section 30.2, p. 1011, for more information about frequencies and histograms.

■ Irregular Data

If the data are irregular, we have to tell, in addition to the z values, the x and y values of the data points; that is, the data is of the form $\{\{x_1, y_1, z_1\}, ..., \{x_n, y_n, z_n\}\}$. For irregular data, the following commands presented previously still work: **ListPlot3D**, **ListPointPlot3D**, **ListDensityPlot**, and **ListContourPlot**. However, **BarChart3D**, **ArrayPlot**, and **MatrixPlot** are no longer available. In addition, we have the following new commands:

ListSurfacePlot3D[data] (✻6) Find a surface that approximates the given points

TriangularSurfacePlot[data] Plot a triangular surface plot according to the Delaunay triangulation (in the **ComputationalGeometry`** package)

To illustrate plots of irregular data, we first generate such data:

```
SeedRandom[2];
data5 =
  Table[{x = RandomReal[{0, π}], y = RandomReal[{0, π}], Sin[x] + Sin[y]}, {30}];
```

Here are some surface plots:

```
<< ComputationalGeometry`
```

```
GraphicsRow[{ListPlot3D[data5, BoxRatios → Automatic, AxesLabel → {"x", "y", ""}],
  ListPlot3D[data5, Mesh → All, BoxRatios → Automatic],
  TriangularSurfacePlot[data5],
  ListSurfacePlot3D[data5, Ticks → None]}, ImageSize → 420]
```

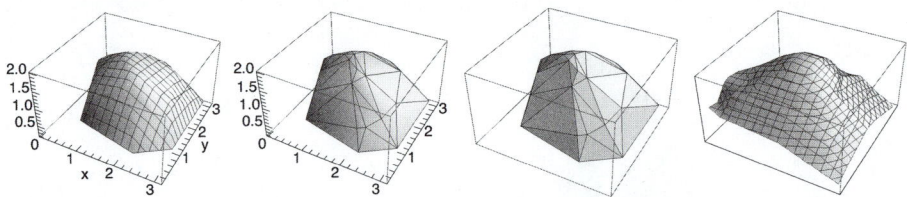

If we use **Mesh → All** in **ListPlot3D**, we see the triangularization of the surface, although the surface is colored smoothly. With **TriangularSurfacePlot** we get a surface consisting of triangles. **ListSurfacePlot3D** finds a surface that approximates the points. For more information about the computational geometry package, see ComputationalGeometry/guide/ComputationalGeometryPackage.

Next, we show two scatter plots and a density and contour plot. The second plot shows how easy it is to plot points with **Graphics3D**.

```
GraphicsRow[
  {ListPointPlot3D[data5, Filling → Bottom, BoxRatios → Automatic, Ticks → None],
   Graphics3D[{Red, AbsolutePointSize[3.5], Point[data5],
     Black, Line[{{#[[1]], #[[2]], 0}, #}] & /@ data5}],
   ListDensityPlot[data5, AspectRatio → Automatic,
    FrameTicks → {{1, 2, 3, 4, 5}, {1, 2, 3, 4}, None, None}],
   ListContourPlot[data5, AspectRatio → Automatic,
    FrameTicks → {{1, 2, 3, 4, 5}, {1, 2, 3, 4}, None, None}]}, ImageSize → 420]
```

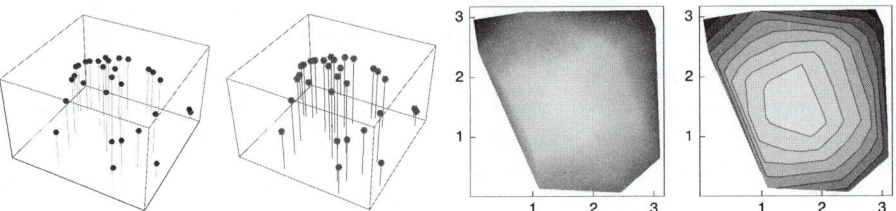

With **ListContourPlot3D** we can plot surfaces of constant value; see **ContourPlot3D** in Section 5.4.2, p. 149.

Data

Introduction

> *Laplace was once asked who was the greatest mathematician in Germany.*
> *He replied "Pfaff." "But what about Gauss?" asked the questioner.*
> *"Gauss," said Laplace, "is the greatest mathematician in the world."*

New in *Mathematica 6* are built-in data over various fields. Here is a list of the data sets with short descriptions:

Chemical data:
- **ElementData**: bulk, atomic, chemical, etc. properties of chemical elements
- **ChemicalData**: structural, physical, and other properties of chemical compounds
- **IsotopeData**: static and decay properties of all nuclear isotopes

Physical data:
- **ParticleData**: properties of stable, unstable, and resonance particles
- **AstronomicalData**: properties of stars, planets, and other objects

Geographical data:
- **CountryData**: many categories of data about countries, groups, etc.
- **CityData**: properties of cities throughout the world

Financial data:
- **FinancialData**: current and historical stock, fund, index, currency, etc. data

Mathematical data:
- **PolyhedronData**: geometry and properties of polyhedra
- **GraphData**: properties of named and enumerated graphs
- **LatticeData**: properties of named lattices
- **KnotData**: properties of enumerated knots

Word data:
- **WordData**: properties of words and network of relations between them
- **DictionaryLookup**: use string patterns to look up words in the dictionary

Color data:
- **ColorData**: various color schemes such as gradients

Example data:
- **ExampleData**: many types of standard test and example data

Here, we can only give short introductions to these data sets. If you are interested in a particular data set, please look at the corresponding document from **Help ▷ Documentation Center**.

Note that we also have the following physical packages:

- **PhysicalConstants`**, **Units`** (see Section 12.1.3, p. 402)
- **StandardAtmosphere`**, **ResonanceAbsorptionLines`**, **BlackBodyRadiation`**, **Geodesy`**

Regarding mathematics, note the following packages:

- **PolyhedronOperations`**, **Polytopes`**
- **GraphUtilities`**, **Combinatorica`**, **ComputationalGeometry`**

See guide/DatabaseConnectivity and *DatabaseLink*/tutorial/Overview in the Documentation Center for tools to work with databases.

9.1 Chemical and Physical Data

9.1.1 Element, Chemical, and Isotope Data

■ ElementData

`ElementData[]` (✻6)	Give a list of all standard chemical elements sorted by atomic number
`ElementData[n]`	Give the full name of the nth element
`ElementData[patt]`	Give a list of all elements matching the string pattern
`ElementData["Properties"]`	Give a list of all properties
`ElementData["elem", "prop"]`	Give the value of the property for the element
`ElementData["elem", "prop", "Units"]`	Give the units of the property
`ElementData["elem", "prop", "ann"]`	Give the specified annotation
`ElementData["Classes"]`	Give a list of all classes
`ElementData["class"]`	Give a list of elements in the given class
`ElementData["elem", "class"]`	Give True if **elem** belongs to **class**

An element can be specified by the full name such as **"Iron"**, by the abbreviation such as **"Fe"**, or by the atomic number such as **26**.

Typical annotations include **"Units"**, **"UnitsName"**, **"UnitsStandardName"**, **"UnitsNotation"**, **"Description"**, **"LongDescription"**, **"Interval"** (uncertainty), and **"Note"**.

Here are all the elements, properties, and classes:

Style[ElementData[], 6]

{Hydrogen, Helium, Lithium, Beryllium, Boron, Carbon, Nitrogen, Oxygen, Fluorine, Neon, Sodium, Magnesium, Aluminum, Silicon, Phosphorus, Sulfur, Chlorine, Argon, Potassium, Calcium, Scandium, Titanium, Vanadium, Chromium, Manganese, Iron, Cobalt, Nickel, Copper, Zinc, Gallium, Germanium, Arsenic, Selenium, Bromine, Krypton, Rubidium, Strontium, Yttrium, Zirconium, Niobium, Molybdenum, Technetium, Ruthenium, Rhodium, Palladium, Silver, Cadmium, Indium, Tin, Antimony, Tellurium, Iodine, Xenon, Cesium, Barium, Lanthanum, Cerium, Praseodymium, Neodymium, Promethium, Samarium, Europium, Gadolinium, Terbium, Dysprosium, Holmium, Erbium, Thulium, Ytterbium, Lutetium, Hafnium, Tantalum, Tungsten, Rhenium, Osmium, Iridium, Platinum, Gold, Mercury, Thallium, Lead, Bismuth, Polonium, Astatine, Radon, Francium, Radium, Actinium, Thorium, Protactinium, Uranium, Neptunium, Plutonium, Americium, Curium, Berkelium, Californium, Einsteinium, Fermium, Mendelevium, Nobelium, Lawrencium, Rutherfordium, Dubnium, Seaborgium, Bohrium, Hassium, Meitnerium, Darmstadtium, Roentgenium, Ununbium, Ununtrium, Ununquadium, Ununpentium, Ununhexium, Ununseptium, Ununoctium}

Style[ElementData["Properties"], 6]

{Abbreviation, AbsoluteBoilingPoint, AbsoluteMeltingPoint, AdiabaticIndex, AllotropeNames, AllotropicMultiplicities, AlternateNames, AlternateStandardNames, AtomicNumber, AtomicRadius, AtomicWeight, Block, BoilingPoint, BrinellHardness, BulkModulus, CASNumber, Color, CommonCompoundNames, CovalentRadius, CriticalPressure, CriticalTemperature, CrustAbundance, CrystalStructure, CuriePoint, DecayMode, Density, DiscoveryCountries, DiscoveryYear, ElectricalConductivity, ElectricalType, ElectronAffinity, ElectronConfiguration, ElectronConfigurationString, Electronegativity, ElectronShellConfiguration, FusionHeat, GasAtomicMultiplicities, Group, HalfLife, HumanAbundance, IconColor, IonizationEnergies, IsotopeAbundances, KnownIsotopes, LatticeAngles, LatticeConstants, Lifetime, LiquidDensity, MagneticType, MassMagneticSusceptibility, MeltingPoint, Memberships, MeteoriteAbundance, MohsHardness, MolarMagneticSusceptibility, MolarVolume, Name, NeelPoint, NeutronCrossSection, NeutronMassAbsorption, OceanAbundance, Period, Phase, PoissonRatio, QuantumNumbers, Radioactive, RefractiveIndex, Resistivity, ShearModulus, SolarAbundance, SoundSpeed, SpaceGroupName, SpaceGroupNumber, SpecificHeat, StableIsotopes, StandardName, SuperconductingPoint, ThermalConductivity, ThermalExpansion, UniverseAbundance, Valence, VanDerWaalsRadius, VaporizationHeat, VickersHardness, VolumeMagneticSusceptibility, YoungModulus}

Style[ElementData["Classes"], 6]

{Actinide, AlkaliMetal, AlkalineEarthMetal, Antiferromagnetic, Conductor, Diamagnetic, Ferromagnetic, Gas, Halogen, Insulator, Lanthanide, Liquid, Metal, Metalloid, Natural, NobleGas, Nonmetal, Paramagnetic, PoorMetal, Radioactive, Semiconductor, Solid, Stable, Synthetic, TransitionMetal}

Length /@ {ElementData[], ElementData["Properties"], ElementData["Classes"]}

{118, 86, 22}

In the Documentation Center, under **ElementData**, the properties are classified as follows: basic properties, material, thermodynamic, electromagnetic and optical, abundance, periodic table, basic chemical, crystallographic, atomic, nuclear, names-related, and historical and commercial properties.

A property that is not applicable to an element has the value Missing["NotApplicable"]. A property that is not available for an element has the value Missing["NotAvailable"]. A property that is unknown for an element has the value Missing["Unknown"].

■ **Example 1**

An element can be specified by standard name, abbreviation, or atomic number:

ElementData["Fe", "StandardName"]

Iron

ElementData["Iron", "Abbreviation"]

Fe

ElementData["Iron", "AtomicNumber"]

26

Any of these specifications can be used when asking for properties:

ElementData["Iron", "Density"]

7874.

ElementData["Fe", "Density"]

7874.

```
ElementData[26, "Density"]
```

7874.

An element is found in Finland:

```
ElementData["Yttrium", "DiscoveryCountries"]
```

{Finland}

```
ElementData["Yttrium", "DiscoveryYear"]
```

1794

Find all elements beginning with H:

```
ElementData["H*"]
```

{Hydrogen, Helium, Holmium, Hafnium, Hassium}

Ask for some annotations:

```
ElementData["Fe", "Density", "Units"]
```

KilogramsPerCubicMeter

```
ElementData["Fe", "Density", "UnitsName"]
```

kilograms per cubic meter

```
ElementData["Fe", "Density", "UnitsNotation"]
```

kg/m^3

```
ElementData["Fe", "Density", "LongDescription"]
```

density at standard temperature and pressure

Consider the class of liquids:

```
ElementData["Liquid"]
```

{Bromine, Mercury}

```
ElementData["Bromine", "Liquid"]
```

True

■ Example 2

Here are all boiling points:

```
Style[t1 = Table[ElementData[n, "BoilingPoint"], {n, 118}], 7]
```

{-252.87, -268.93, 1342., 2470., 4000., 4027., -195.79, -182.9, -188.12, -246.08, 883., 1090.,
2519., 2.9×10^3, 280.5, 444.72, -34.04, -185.8, 759., 1484., 2830., 3287., 3407., 2671., 2061.,
2861., 2927., 2913., 2927., 907., 2204., 2820., 614., 685., 59., -153.22, 688., 1382., 3345.,
4409., 4744., 4639., 4265., 4150., 3695., 2963., 2162., 767., 2072., 2602., 1587., 988., 184.3,
-108., 671., 1870., 3464., 3360., 3290., 3.1×10^3, 3.0×10^3, 1803., 1527., 3250., 3230., 2567.,
2700., 2868., 1950., 3402., 4603., 5458., 5555., 5596., 5012., 4428., 3825., 2856., 356.73,
1473., 1749., 1564., 962., Missing[NotAvailable], -61.7, Missing[NotAvailable], 1737., 3200.,
4820., 4000., 3927., 4.0×10^3, 3230., 2011., 3110., Missing[NotAvailable], Missing[NotAvailable],
Missing[NotAvailable], Missing[NotAvailable], Missing[NotAvailable], Missing[NotAvailable],
Missing[NotAvailable], Missing[NotAvailable], Missing[NotAvailable], Missing[NotAvailable],
Missing[NotAvailable], Missing[NotAvailable], Missing[NotAvailable], Missing[Unknown],
Missing[NotAvailable], Missing[NotAvailable], Missing[Unknown], Missing[NotAvailable],
Missing[NotAvailable], Missing[Unknown], Missing[NotAvailable], Missing[NotAvailable]}

Many items are missing or unknown; nevertheless, we can plot the data:

```
ListLinePlot[t1, Mesh → All, ImageSize → 200]
```

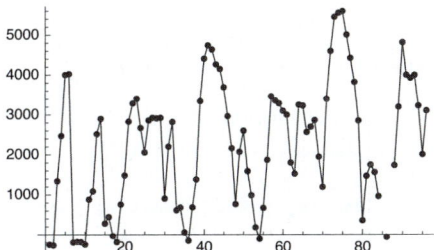

List all the melting points:

```
t2 = Table[ElementData[n, "MeltingPoint"], {n, 118}];
```

Plot pairs of boiling points and melting points:

```
ListPlot[{t1, t2}ᵀ, AxesLabel → {"Boiling", "Melting"}, ImageSize → 200]
```

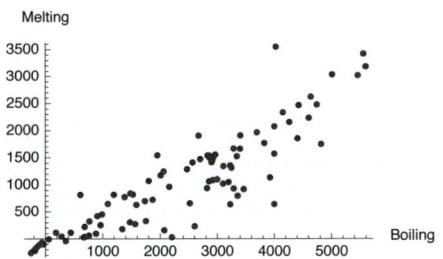

We can assign a tooltip for each point that gives the abbreviation of the element:

```
t1t2 =
  Table[Tooltip[{ElementData[n, "BoilingPoint"], ElementData[n, "MeltingPoint"]},
    ElementData[n, "Abbreviation"]], {n, 118}];
```

By moving the mouse over the points in the following plot, we can see the abbreviations:

```
ListPlot[t1t2, ImageSize → 300]
```

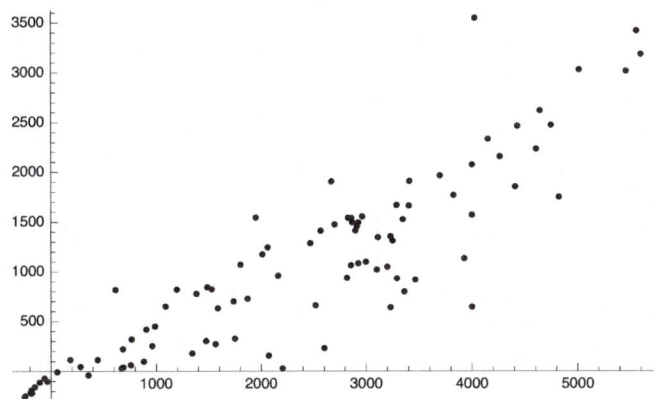

■ **Example 3**

Prepare a table of boiling and melting points of all alkaline earth metals:

```
t = {#, ElementData[#, "BoilingPoint"], ElementData[#, "MeltingPoint"]} & /@
  ElementData["AlkalineEarthMetal"]
```
$\{\{Beryllium, 2470., 1287.\}, \{Magnesium, 1090., 650.\}, \{Calcium, 1484., 842.\},$
$\{Strontium, 1382., 777.\}, \{Barium, 1870., 727.\}, \{Radium, 1737., 7.0 \times 10^2\}\}$

```
Text@
  TableForm[t, TableHeadings → {None, {"Element", "Boiling point", "Melting point"}}]
```

Element	Boiling point	Melting point
Beryllium	2470.	1287.
Magnesium	1090.	650.
Calcium	1484.	842.
Strontium	1382.	777.
Barium	1870.	727.
Radium	1737.	7.0×10^2

■ **Example 4**

In the following example, some data are missing:

```
t = {#, ElementData[#, "BoilingPoint"], ElementData[#, "MeltingPoint"]} & /@
  ElementData["NobleGas"]
```
$\{\{Helium, -268.93, Missing[NotApplicable]\}, \{Neon, -246.08, -248.59\},$
$\{Argon, -185.8, -189.3\}, \{Krypton, -153.22, -157.36\}, \{Xenon, -108., -111.8\},$
$\{Radon, -61.7, -71.\}, \{Ununoctium, Missing[NotAvailable], Missing[NotAvailable]\}\}$

Before tabulating the data, we can delete all missing items:

```
t1 = DeleteCases[t, {_, _, _Missing}]
```
$\{\{Neon, -246.08, -248.59\}, \{Argon, -185.8, -189.3\},$
$\{Krypton, -153.22, -157.36\}, \{Xenon, -108., -111.8\}, \{Radon, -61.7, -71.\}\}$

```
Text@Grid[Prepend[t1, {"Element", "Boiling point", "Melting point"}],
  Dividers → {False, {False, True}},
  Alignment → {{Left, ".", "."}, Baseline, {{1, 2} → Right, {1, 3} → Right}},
  ItemStyle → {Automatic, {Bold}}]
```

Element	**Boiling point**	**Melting point**
Neon	−246.08	−248.59
Argon	−185.8	−189.3
Krypton	−153.22	−157.36
Xenon	−108.	−111.8
Radon	−61.7	−71.

■ **ChemicalData**

ChemicalData is used in the same way as ElementData:

ChemicalData[] (✸6)   Give a list of all available chemicals in order of increasing molecular weight
ChemicalData["chem"]   Give a structure diagram for the chemical
ChemicalData[patt]   Give a list of all chemicals matching the string pattern

ChemicalData["Properties"]   Give a list of all properties
ChemicalData["chem", "prop"]   Give the value of the property for the chemical
ChemicalData["chem", "prop", "Units"]   Give the units of the property

---

```
ChemicalData["chem", "prop", "ann"]   Give the specified annotation

ChemicalData["Classes"]   Give a list of all classes
ChemicalData["class"]   Give a list of chemicals in the given class
ChemicalData["chem", "class"]   Give True if chem belongs to class

ChemicalData[{"elem", "Compound"}]   Give a list of available chemicals that contain the given
   element
```

Typical annotations include **"Units"**, **"UnitsName"**, **"UnitsStandardName"**, **"UnitsNotation"**, **"Description"**, **"LongDescription"**, **"Interval"** (uncertainty), and **"Note"**.

Here are sizes of some lists:

```
Length /@ {ChemicalData[], ChemicalData["Properties"], ChemicalData["Classes"]}
```
{18 179, 68, 20}

Some properties of the aspirin are as follows:

```
ChemicalData["Aspirin"]
```

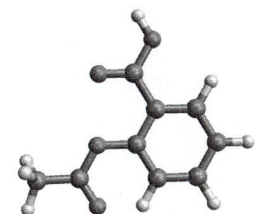

```
ChemicalData["Aspirin", "MoleculePlot"]
```

```
ChemicalData["Aspirin", "FormulaString"]
```
C9H8O4

```
ChemicalData["Aspirin", "AlternateStandardNames"]
```
{2-(Acetyloxy)BenzoicAcid, AcetylsalicylicAcid,
 2-AcetoxybenzoicAcid, Acenterine, Acetosal}

Ask for chemicals that contain gold:

```
ChemicalData[{"Gold", "Compound"}] // Short
```
{Gold, ≪13≫, Bis(Propane-1, … o-C)Aurate(1-)]}

Here are chemicals whose names begin with J:

```
ChemicalData["J*"]
```

```
{Jasmone, Jacobine, Javanicin, Jervine,
 Julolidine, Juglone, JunipericAcid, JanusGreenB, Josamycin}
```

Consider the melting points:

```
t = DeleteCases[ChemicalData[#, "MeltingPoint"] & /@ ChemicalData[], _Missing];
```

```
{Min[t], Max[t]}
```

```
{-259.14, 3550.}
```

Plot a part of the frequency distribution of the melting points:

```
ListPlot[{Range[-260, 500, 10], BinCounts[t, {-265, 505, 10}]}ᵀ,
  Filling → Axis, ImageSize → 320]
```

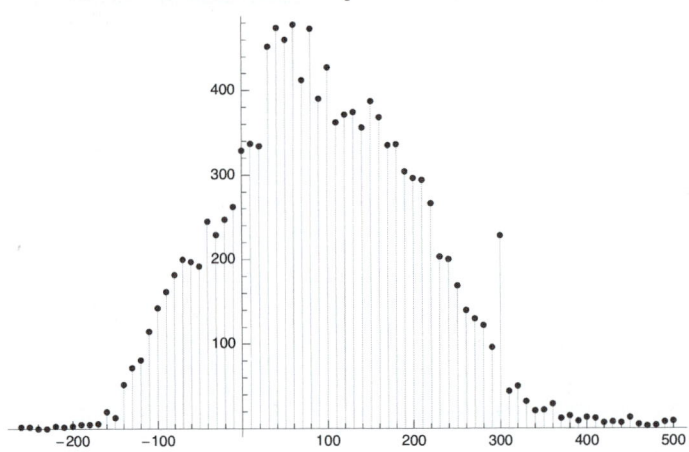

■ **IsotopeData**

`IsotopeData[]` (⌘6)	Give a list of all known isotopes sorted by atomic number and mass number
`IsotopeData["elem"]`	Give the known isotopes of the element
`IsotopeData[patt]`	Give a list of all isotopes matching the string pattern
`IsotopeData["Properties"]`	Give a list of all properties
`IsotopeData["isot", "prop"]`	Give the value of the property for the isotope
`IsotopeData["isot", "prop", "Units"]`	Give the units of the property
`IsotopeData["isot", "prop", "ann"]`	Give the specified annotation
`IsotopeData["Classes"]`	Give a list of all classes
`IsotopeData["class"]`	Give a list of isotopes in the given class
`IsotopeData["isot", "class"]`	Give True if `isot` belongs to `class`

   Typical annotations include `"Units"`, `"UnitsName"`, `"UnitsStandardName"`, `"UnitsNotation"`, `"Description"`, `"LongDescription"`, `"Interval"` (uncertainty), and `"Note"`.

   Here are sizes of some lists:

```
Length /@ {IsotopeData[], IsotopeData["Properties"], IsotopeData["Classes"]}
```

```
{3182, 33, 47}
```

Isotopes of uranium are as follows:

```
IsotopeData["Uranium"] // Short
{Uranium217, ≪24≫, Uranium242}
```

Isotopes can be referred to in several ways:

```
IsotopeData["Uranium217", "Lifetime"]
0.038
```

```
IsotopeData["U217", "Lifetime"]
0.038
```

```
IsotopeData[{92, 217}, "Lifetime"]
0.038
```

9.1.2 Particle and Astronomical Data

■ ParticleData

ParticleData[] (❀6)	Give a list of all known particles sorted by mass
ParticleData[patt]	Give a list of all particles matching the string pattern
ParticleData["Properties"]	Give a list of all properties
ParticleData["part", "prop"]	Give the value of the property for the particle
ParticleData[{"part", q}, "prop"]	Give the value of the property for the particle with charge **q**
ParticleData["part", "prop", "Units"]	Give the units of the property
ParticleData["part", "prop", "ann"]	Give the specified annotation
ParticleData["Classes"]	Give a list of all classes
ParticleData["class"]	Give a list of isotopes in the given class
ParticleData["part", "class"]	Give True if **part** belongs to **class**

Typical annotations include **"Units"**, **"UnitsName"**, **"UnitsStandardName"**, **"UnitsNotation"**, **"Description"**, **"LongDescription"**, **"Interval"** (uncertainty), and **"Note"**.

Here are sizes of some lists:

```
Length /@ {ParticleData[], ParticleData["Properties"], ParticleData["Classes"]}
{1002, 35, 24}
```

The mass of electron is as follows:

```
ParticleData["Electron", "Mass"]
0.51099892
```

```
ParticleData["Electron", "Mass", "Units"]
MegaelectronVoltsPerSpeedOfLightSquared
```

■ AstronomicalData

AstronomicalData has information about planets, stars, galaxies, etc.

AstronomicalData[] (❀6)	Give a list of all available astronomical objects
AstronomicalData["tag"]	Give the standardized name or list of names of **tag**
AstronomicalData[patt]	Give a list of all objects matching the string pattern

AstronomicalData["Properties"]   Give a list of all properties

AstronomicalData["obj", "prop"]   Give the value of the property for the object

AstronomicalData["obj", "prop", "Units"]   Give the units of the property

AstronomicalData["obj", "prop", "ann"]   Give the specified annotation

AstronomicalData["obj", "Image"]   Give a picture of the object

AstronomicalData["obj", "ObjectType"]   Give the basic type of the object

AstronomicalData["Classes"]   Give a list of all classes

AstronomicalData["class"]   Give a list of objects in the given class

AstronomicalData["obj", "Classes"]   Give the classes in which **obj** occurs

AstronomicalData["obj", "class"]   Give True if **obj** belongs to **class**

Typical annotations include **"Units"**, **"UnitsName"**, **"UnitsStandardName"**, **"UnitsNotation"**, **"Description"**, **"LongDescription"**, and **"Note"**.

Possible object types are as follows:

Union[AstronomicalData[#, "ObjectType"] & /@ AstronomicalData[]]

{BrightHIIRegion, DwarfPlanet, Galaxy, GlobularCluster, Nebula, OpenCluster, Planet, PlanetaryMoon, PlanetaryNebula, Star, Missing[NotAvailable]}

Here are sizes of some lists:

Length /@
{AstronomicalData[], AstronomicalData["Properties"], AstronomicalData["Classes"]}
{100 910, 71, 48}

Here are planets:

AstronomicalData["Planet"]

{Mercury, Venus, Earth, Mars, Jupiter, Saturn, Uranus, Neptune, Pluto}

Pluto is considered to be a dwarf planet:

AstronomicalData["Pluto", "ObjectType"]

DwarfPlanet

The classes and picture of sun are as follows:

AstronomicalData["Sun", "Classes"]

{Star, MainSequenceStar, ClassGStar, StarNearest100, StarBrightest100, StarNearest10, StarBrightest10}

AstronomicalData["Sun", "Image"]

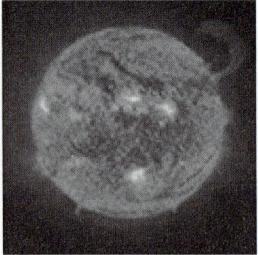

We have 162 planetary moons:

AstronomicalData["PlanetaryMoon"] // Length

162

Consider the density and radius of them (many of these data are missing):

```
t = Tooltip[{AstronomicalData[#, "Density"], AstronomicalData[#, "Radius"]},
    AstronomicalData[#, "Name"]] & /@ AstronomicalData["PlanetaryMoon"];
```

By moving the mouse over the points in the following plot, we can see the names of the moons:

```
ListPlot[t, PlotRange → All, ImageSize → 280]
```

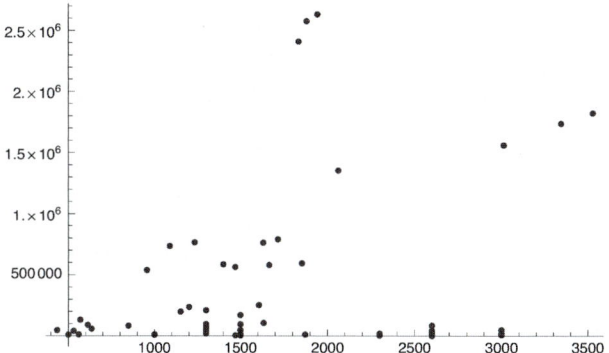

9.2 Geographical and Financial Data

9.2.1 Country and City Data

■ **CountryData**

CountryData[] (✦6) Give a list of all ordinary countries and dependencies
CountryData["tag"] Give the standardized name or list of names of the tag

CountryData["Properties"] Give a list of all properties
CountryData["tag", "prop"] Give the value of the property for the tag
CountryData["tag", {"prop", y}] Give the value of the property for year **y**
CountryData["tag", {"prop", All}] All available years
CountryData["tag", {"prop", {y1, y2}}] Years from **y1** to **y2**
CountryData["tag", {{"prop", "curr"}, y}] Give values in currency "**curr**"
CountryData["tag", "prop", "Units"] Give the units of the property
CountryData["tag", "prop", "ann"] Give the specified annotation
CountryData["tag", "Shape"] Give the shape of the tag

CountryData["class"] Give a list of tags in the given class; possible classes: "**Countries**"
    (ordinary countries and dependencies), "**Groups**" (groups of countries), "**Continents**", "**Oceans**"
CountryData["tag", "Classes"] Give the classes and groups in which **tag** appears
CountryData["tag", "class"] Give True if **tag** belongs to the given class or group

Special values of the currency "**curr**" are as follows:

"**USDollars**": nominal value in U.S. dollars (the default)

"**Local**": nominal value in local currency

"**Adjusted**": adjusted to give a real value in current U.S. dollars

"**AdjustedLocal**": adjusted to give a real value in current local currency

Typical annotations include "**Units**", "**UnitsName**", "**UnitsStandardName**", "**UnitsNotation**", "**Description**", "**Date**", "**DateValue**", and "**Note**".

Here are numbers of countries and other tags:

```
Length /@ {CountryData[], CountryData["Groups"],
   CountryData["Continents"], CountryData["Oceans"], CountryData["Properties"]}
{237, 305, 7, 5, 220}
```

■ Example 1

As an example, here are some properties of Finland:

```
CountryData["Finland", #] & /@ {"Shape", "Flag", "NativeName", "IndependenceYear"}
```

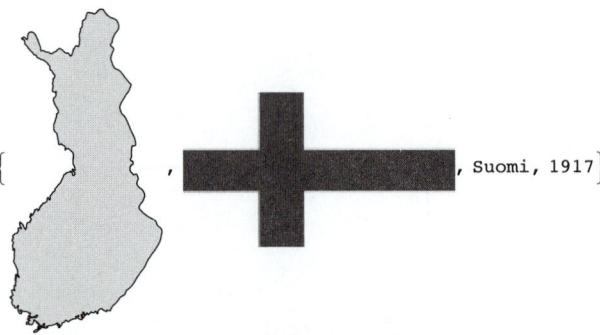
{ , , Suomi, 1917}

Here is a link to a web satellite image for Finland:

```
CountryData["Finland", "CenterLocationLink"]
```

http://maps.google.com/maps?q=+64+26&z=6&t=h

We plot the GDP (real value in current local currency) of Finland:

```
DateListPlot[
  CountryData["Finland", {{"GDP", "LocalAdjusted"}, All}], Filling → Bottom]
```

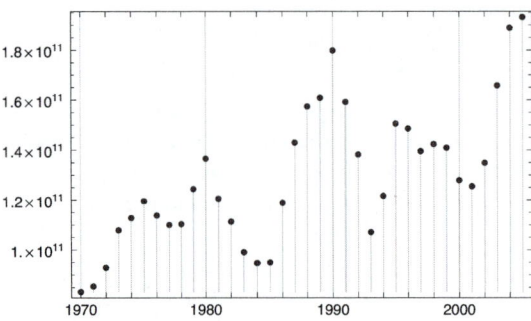

■ **Example 2**

Properties of the G8 countries:

> **CountryData["G8"]**

> {Canada, France, Germany, Italy, Japan, Russia, UnitedKingdom, UnitedStates}

> **CountryData["G8", "Population"]**

> $\{3.22682 \times 10^7, 6.23121 \times 10^7, 8.26892 \times 10^7, 5.80927 \times 10^7,$
> $1.28085 \times 10^8, 1.43202 \times 10^8, 5.96678 \times 10^7, 2.98213 \times 10^8\}$

Find the 10 largest populations:

> **Reverse@Take[**
> ** SortBy[{CountryData[#], CountryData[#, "Population"]} & /@ CountryData[], Last],**
> ** -10] // TableForm**

China	1.29299×10^9
India	1.10337×10^9
UnitedStates	2.98213×10^8
Indonesia	2.22781×10^8
Brazil	1.86405×10^8
Pakistan	1.57935×10^8
Russia	1.43202×10^8
Bangladesh	1.41822×10^8
Nigeria	1.3153×10^8
Japan	1.28085×10^8

Show infant mortatility fraction against literacy fraction:

> **ListPlot[Tooltip[{CountryData[#, "LiteracyFraction"],**
> ** CountryData[#, "InfantMortalityFraction"]},**
> ** CountryData[#]] & /@ CountryData[], ImageSize → 420]**

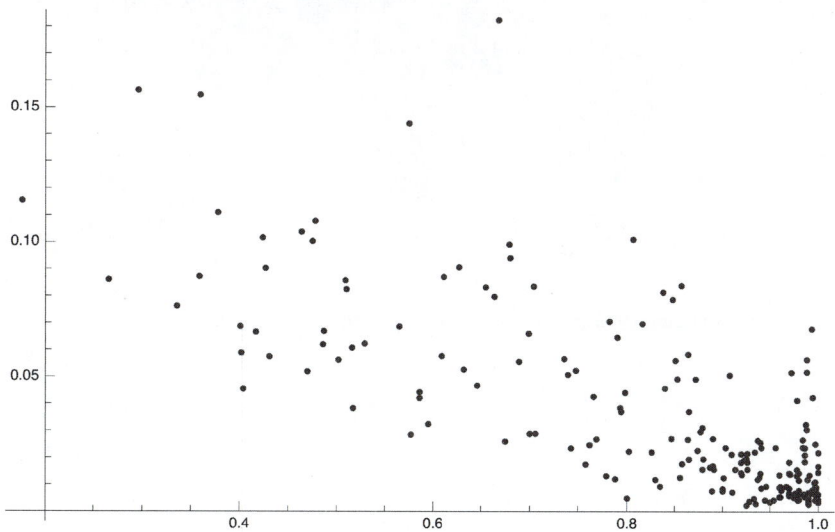

■ **Example 3**

Plot the world:

```
Graphics[{Darker[Green, 0.6],
  CountryData[#, {"SchematicPolygon", "Mollweide"}] & /@ CountryData[]},
 Background → LightBlue, ImageSize → 420]
```

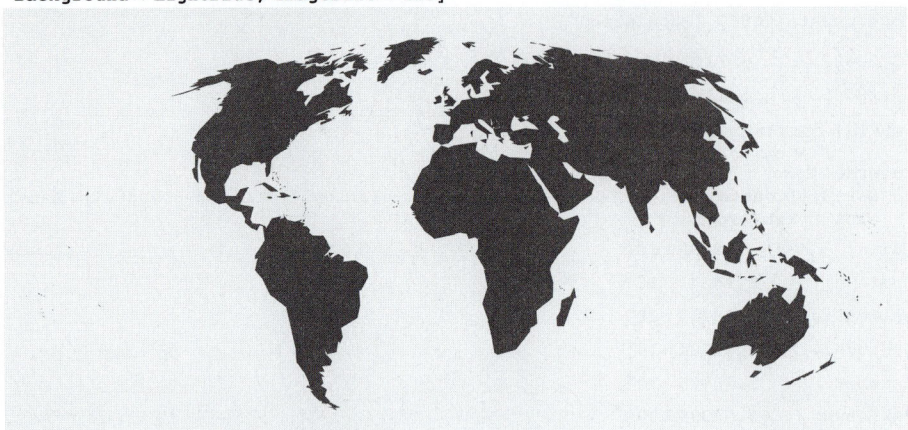

Show the name of each country by a tooltip:

```
Graphics[{Darker[Orange, 0.1], EdgeForm[Black],
  Tooltip[CountryData[#, {"SchematicPolygon", "Mollweide"}], #] & /@ CountryData[]},
 Background → LightBlue, ImageSize → 420]
```

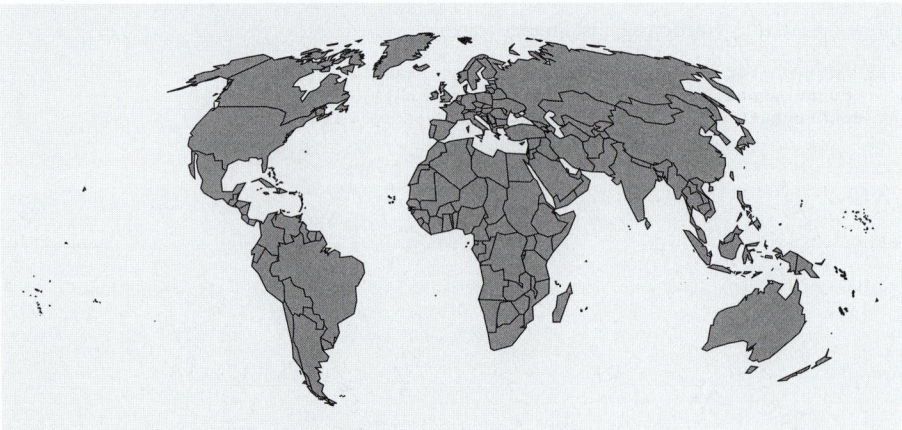

Note that in addition to **CountryData**, we also have the **WorldPlot`** package.

■ **CityData**

CityData contains information about more than 160,000 cities:

CityData[] (⌘6)   Give a list of all cities in the world sorted by population

CityData["city"]   Give a list of the full specifications of cities with the given name

CityData[{"city", "country"}]   Give a list of the full specifications of cities with the given name in the given country

CityData["Properties"]   Give a list of all properties

CityData["city", "prop"]   Give the value of the property for the given city

CityData[{"city", "region", "country"}, "prop"]   Give the value of the property for the given city in the given region of the given country

CityData["city", "prop", "Units"]   Give the units of the property

CityData["city", "prop", "ann"]   Give the specified annotation

CityData[{All, "country"}]   Give a list of all available cities in the given country

CityData[{Large, "country"}]   Give a list of all large cities in the given country

CityData[{All, "region", "country"}]   Give a list of all available cities in the given region of the given country

CityData[{Large, "region", "country"}]   Give a list of all large cities in the given region of the given country

---

Typical annotations include **"Units"**, **"UnitsName"**, **"UnitsStandardName"**, **"UnitsNotation"**, **"Description"**, **"Date"**, and **"Note"**.

Here are sizes of some lists:

Length /@ {CityData[], CityData["Properties"]}

{163 428, 14}

We have several cities with the name Paris:

Style[CityData["Paris"], 7]

```
{{Paris, IleDeFrance, France}, {Paris, Texas, UnitedStates},
 {Paris, Ontario, Canada}, {Paris, Tennessee, UnitedStates}, {Paris, Kentucky, UnitedStates},
 {Paris, Illinois, UnitedStates}, {Paris, Maine, UnitedStates}, {Paris, NewYork, UnitedStates},
 {Paris, Arkansas, UnitedStates}, {Paris, Wisconsin, UnitedStates}, {Paris, Missouri, UnitedStates},
 {ParisGrant, Wisconsin, UnitedStates}, {Paris, Idaho, UnitedStates}}
```

CityData["Paris", "Population"]

2 138 551

CityData[{"Paris", "Ontario", "Canada"}, "Population"]

10 437

The 10 cities with the largest populations:

Take[CityData[], 10]

```
{{Beijing, Beijing, China}, {Shanghai, Shanghai, China},
 {Bombay, Maharashtra, India}, {Karachi, Sind, Pakistan},
 {BuenosAires, BuenosAires, Argentina}, {Delhi, Delhi, India},
 {Manila, Manila, Philippines}, {Moscow, Moscow, Russia},
 {Seoul, SoulTvkpyolsi, SouthKorea}, {SaoPaulo, SaoPaulo, Brazil}}
```

```
TableForm[{#〚1〛, CityData[#, "Population"]} & /@ %]
```

```
Beijing       14 930 000
Shanghai      14 608 512
Bombay        12 691 836
Karachi       11 624 219
BuenosAires   11 574 205
Delhi         10 927 986
Manila        10 444 527
Moscow        10 381 222
Seoul         10 349 312
SaoPaulo      10 021 295
```

Here is a link to a map of Bombay:

```
CityData["Bombay", "LocationLink"]
```

```
http://maps.google.com/maps?q=+18.96+72.82&z=12&t=h
```

The next plot shows the largest cities of India:

```
Graphics[{Darker[Green, 0.4], CountryData["India", "Polygon"],
  PointSize[Medium], Lighter[Orange], DeleteCases[
   Tooltip[Point[Reverse[CityData[#, "Coordinates"]]], CityData[#, "Name"]] & /@
    CityData[{Large, "India"}],
   Tooltip[Point[{_Missing, _Missing}], _]]}, ImageSize → 380]
```

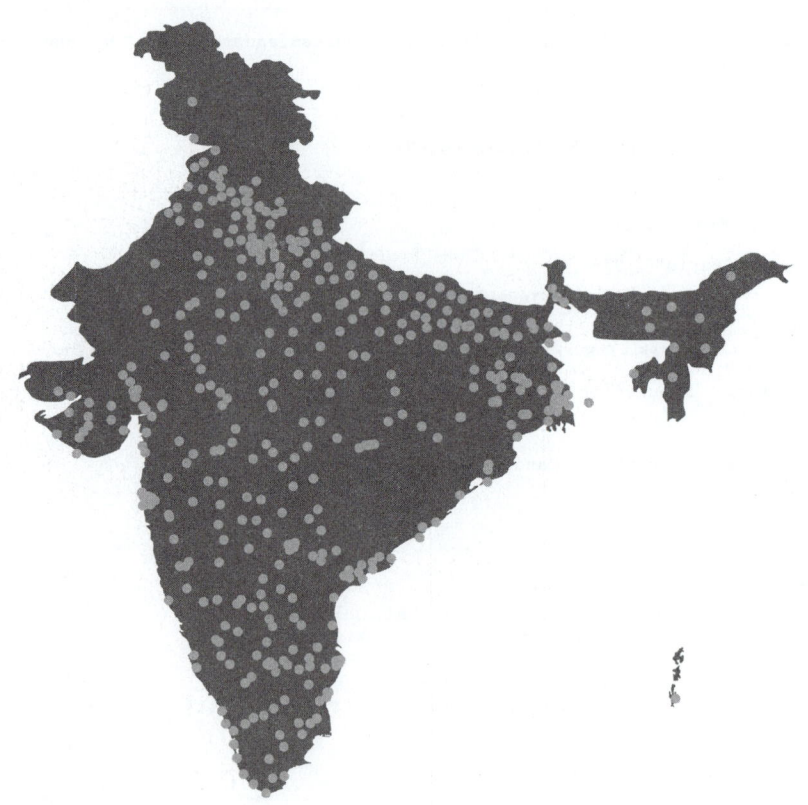

9.2.2 Financial Data

`FinancialData[]` (✻6)   Give a list of all financial instruments sorted by ticker symbols

`FinancialData[patt]`   Give a list of entities matching the string pattern

`FinancialData["name"]`   Give the last known price or value of the financial entity

`FinancialData["name", {start}]`   Give a list of dates and daily closing values from **start**

`FinancialData["name", {start, end}]`   Give a list of dates and daily closing values from **start** to **end**

`FinancialData["name", {start, end, period}]`   Give a list of dates and prices for the specified periods (`"Day"`, `"Week"`, `"Month"`, `"Year"`) lying between **start** and **end**

`FinancialData["Properties"]`   Give a list of all properties

`FinancialData["name", "prop"]`   Give the value of the property for the entity

`FinancialData["name", "prop", {start, end, … }]`   Give a list of dates and values of the property for a sequence of dates or periods

`FinancialData["name", "prop", …, "Value"]`   Give the value

`FinancialData["name", "prop", …, "DateValue"]`   Give a list of date and value

`FinancialData["name", "prop", …, "Units"]`   Give the units of the property

`FinancialData["name", "prop", …, "ann"]`   Give the specified annotation

`FinancialData["Classes"]`   Give a list of all available classes

`FinancialData["class"]`   Give a list of entities in the given class

Typical annotations include `"Units"`, `"UnitsName"`, `"UnitsStandardName"`, `"UnitsNotation"`, `"Description"`, `"LongDescription"`, and `"Currency"`.

Here are sizes of some lists:

`Length /@ {FinancialData[], FinancialData["Properties"], FinancialData["Classes"]}`

`{186 127, 71, 8}`

Here are the various classes of financial entities:

`FinancialData["Classes"]`

```
{Currencies, Exchanges, ExchangeTradedFunds,
 Futures, Indices, MutualFunds, Sectors, Stocks}
```

Next, we show financial entities beginning with NASDAQ:AA:

`FinancialData["NASDAQ:AA*"]`

```
{NASDAQ:AACC, NASDAQ:AAME, NASDAQ:AANB, NASDAQ:AAON,
 NASDAQ:AAPL, NASDAQ:AATI, NASDAQ:AATK, NASDAQ:AAUK, NASDAQ:AAWW}
```

Here, NASDAQ : AAPL is the Apple company:

`FinancialData["NASDAQ:AAPL", "Name"]`

`Apple Inc`

Ask the current value:

`FinancialData["NASDAQ:AAPL"]`

`169.26`

Plot the value from 2004 on:

```
DateListPlot[FinancialData["NASDAQ:AAPL", {2004}]]
```

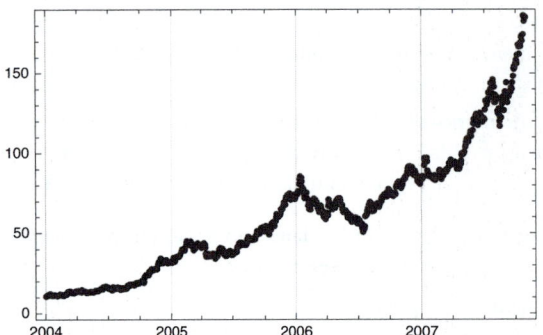

Find the current exchange rate between euros and U.S. dollars:

```
FinancialData["EUR/USD"]
```
```
1.4205
```

9.3 Mathematical and Other Data

9.3.1 Mathematical Data

■ **PolyhedronData**

PolyhedronData[] (❖6) Give a list of all available polyhedra
PolyhedronData[n] Give a list of all polyhedra with **n** faces
PolyhedronData[patt] Give a list of all polyhedra matching the string pattern
PolyhedronData["poly"] Give an image of the given polyhedron
PolyhedronData["Properties"] Give a list of all properties available
PolyhedronData["poly", "prop"] Give the value of the property for the polyhedron
PolyhedronData["prop", "ann"] Give an annotation of the property
PolyhedronData["Classes"] Give a list of all classes available
PolyhedronData["class"] Give a list of the polyhedra in the given class
PolyhedronData["poly","Classes"] Give the classes in which **poly** occurs
PolyhedronData["poly","class"] Give True if **poly** belongs to **class**

Typical annotations include **"Description"**, **"LongDescription"**, and **"Note"**.

Here are sizes of some lists:

```
Length /@
 {PolyhedronData[], PolyhedronData["Properties"], PolyhedronData["Classes"]}
 {147, 85, 27}
```

All Platonic polyhedra are as follows:

```
PolyhedronData["Platonic"]
```
```
{Cube, Dodecahedron, Icosahedron, Octahedron, Tetrahedron}
```

Ask for some properties of a dodecahedron:

```
{Show[PolyhedronData["Dodecahedron"], ImageSize → 70],
 PolyhedronData["Dodecahedron", "NetImage"],
 PolyhedronData["Dodecahedron", "Volume"]}
```

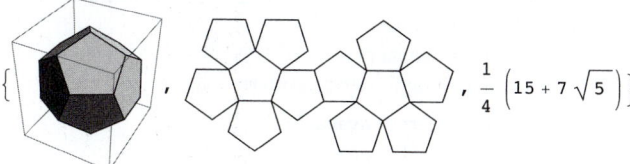

```
PolyhedronData["Volume", "LongDescription"]
```
enclosed volume assuming unit smallest edge length

■ **GraphData**

`GraphData[]` (⚛6)	Give a list of all standard named graphs
`GraphData[All]`	Give a list of all available graphs
`GraphData[n]`	Give a list of all named graphs with **n** vertices
`GraphData[patt]`	Give a list of all graphs matching the string pattern
`GraphData["graph"]`	Give an image of the graph
`GraphData["Properties"]`	Give a list of all properties available
`GraphData["graph", "prop"]`	Give the value of the property for the graph
`GraphData[{n, i}, …]`	Give data for **i**th simple graph with **n** vertices
`GraphData[{"type", id}, …]`	Give data for the graph of **type** with identifier **id**
`GraphData["prop", "ann"]`	Give an annotation of the property
`GraphData["Classes"]`	Give a list of all classes available
`GraphData["class"]`	Give a list of the graphs in the given class
`GraphData["class", n]`	Give a list of graphs with **n** vertices in the given class
`GraphData["graph", "Classes"]`	Give the classes in which **graph** occurs
`GraphData["graph", "class"]`	Give True if **graph** belongs to **class**

Typical annotations include **"Description"**, **"LongDescription"**, and **"Note"**. Here are sizes of some lists:

```
Length /@
  {GraphData[], GraphData[All], GraphData["Properties"], GraphData["Classes"]}
{735, 1939, 121, 71}
```

Ask for some properties of the cuboctahedral graph:

```
{Show[GraphData["CuboctahedralGraph"], ImageSize → 80],
 GraphData["CuboctahedralGraph", "VertexCount"],
 GraphData["CuboctahedralGraph", "EdgeCount"]}
```

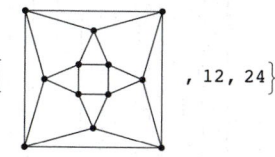 , 12, 24}

For graphs, look also at *Combinatorica*/tutorial/*Combinatorica* in the Documentation Center. Also remember that we have `GraphPlot` (see Section 8.5, p. 267).

■ **LatticeData**

`LatticeData[]` (⌘6) Give a list of classical named lattices
`LatticeData[n]` Give a list of named lattices of dimension **n**
`LatticeData[patt]` Give a list of all lattices matching the string pattern

`LatticeData["Properties"]` Give a list of all properties available
`LatticeData["lattice", "prop"]` Give the value of the property for the lattice
`LatticeData[{"type", id}, … ]` Give data for the lattice of **type** with identifier **id**

`LatticeData["Classes"]` Give a list of all classes available
`LatticeData["class"]` Give a list of the graphs in the given class
`LatticeData["lattice", "Classes"]` Give the classes in which **lattice** occurs
`LatticeData["lattice", "class"]` Give True if **lattice** belongs to **class**

Here are sizes of some lists:

`Length /@ {LatticeData[], LatticeData["Properties"], LatticeData["Classes"]}`
`{21, 37, 8}`

As an example, here is the body-centered cubic lattice:

`Show[LatticeData["BodyCenteredCubic", "Image"], ImageSize → 140]`

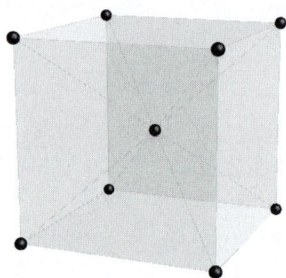

■ **KnotData**

`KnotData[]` (⌘6) Give a list of classical named knots
`KnotData[All]` Give a list of knots that have Alexander-Briggs notations
`KnotData["knot"]` Give an image of the knot

`KnotData["Properties"]` Give a list of all properties available
`KnotData["knot", "prop"]` Give the value of the property for the knot
`KnotData[{"type", id}, … ]` Give data for the knot of **type** with identifier **id**

`KnotData["Classes"]` Give a list of all classes available
`KnotData["class"]` Give a list of the knots in the given class
`KnotData["knot", "Classes"]` Give the classes in which **knot** occurs
`KnotData["knot", "class"]` Give True if **knot** belongs to **class**

Here are sizes of some lists:

```
Length /@ {KnotData[], KnotData[All], KnotData["Properties"], KnotData["Classes"]}
{6, 250, 63, 14}
```

Classical knots are as follows:

```
KnotData[]
{Unknot, Trefoil, FigureEight, SolomonSeal, Stevedore, PerkoPair}
```

Here are their images (we do not have an image for **PerkoPair**):

```
Show[KnotData[#], ImageSize → 70] & /@ Most@KnotData[]
```

9.3.2 Word Data

■ **WordData**

WordData[] (❀6)   Give a list of all words and phrases
WordData["word"]   Give a list of full word specifications representing possible uses and senses of the given word
WordData[patt, "Lookup"]   Give a list of all words matching the string pattern
WordData["Properties"]   Give a list of all properties
WordData[wordspec, "prop"]   Give the value of the property for the given word specification
WordData[wordspec, "prop", "form"]   Give the value in the given form; possible forms: **"List"**, **"Rules"**, **"ShortRules"**
WordData[All, "PartOfSpeech"]   Give a list of parts of speech
WordData[All, "part"]   Give a list of words of a given part of speech
WordData[All, "Stopwords"]   Give a list of words typically ignored in text comparisons

WordData contains approximately 150,000 words and phrases, and we have 37 properties:

```
Length /@ {WordData[], WordData["Properties"]}
{149 191, 37}
```

Parts of speech are as follows:

```
WordData[All, "PartsOfSpeech"]
{Noun, Verb, Adjective, Adverb, Preposition,
 Conjunction, Pronoun, Determiner, Interjection}
```

Most of the words are nouns:

```
WordData[All, "Noun"] // Length
119 034
```

Let us look at what **WordData** knows about words or phrases beginning with matrix:

```
WordData["matrix*", "Lookup"]
```
{matrix, matrix addition, matrix algebra, matrix inversion,
 matrix multiplication, matrix operation, matrix printer, matrix transposition}

Here is what **WordData** knows about matrix:

```
WordData["matrix"]
```
{{matrix, Noun, Mold}, {matrix, Noun, AnimalTissue}, {matrix, Noun, BodySubstance},
 {matrix, Noun, Array}, {matrix, Noun, Enclosure}, {matrix, Noun, Stone}}

The definition of the fourth meaning is as follows:

```
WordData["matrix", "Definitions"]〚4〛
```
{matrix, Noun, Array} →
   (mathematics) a rectangular array of quantities or expressions set out by rows
   and columns; treated as a single element and manipulated according to rules

■ **DictionaryLookup**

DictionaryLookup[] (⌘6)	Give a list of all words in an English dictionary
DictionaryLookup[patt]	Find all words that match the string pattern **patt**
DictionaryLookup[patt, n]	Give only the first **n** words found
DictionaryLookup[patt, IgnoreCase → True]	Do not take the case of words into account

We consider this command in Section 16.2.1, p. 505. However, the following is an example:

```
DictionaryLookup["math" ~~ ___]
```
{math, mathematical, mathematically, mathematician, mathematicians, mathematics}

9.3.3 Color Data

■ **ColorData**

ColorData[] (⌘6)	Give a list of named collections of color schemes
ColorData["collection"]	Give a list of color schemes in the given collection
ColorData["scheme"]	Give a function that generates colors in the given color scheme
ColorData["scheme"][par]	Give the RGB color that corresponds to the parameter value
ColorData["scheme", par]	Give the RGB color that corresponds to the parameter value
ColorData["Properties"]	Give a list of all properties
ColorData["scheme", "prop"]	Give the value of the property for the color scheme

We have four collections of color schemes:

```
ColorData[]
```
{Gradients, Indexed, Named, Physical}

Properties are as follows:

```
ColorData["Properties"]
```
{ColorFunction, ColorList, ColorRules, Image, Name, Panel, ParameterCount, Range}

Note that the color schemes can be used more easily with the **ColorSchemes** palette (see Section 6.2.8, p. 172).

■ **Gradients**

We have 51 gradients:

```
ColorData["Gradients"] // Short
```
{DarkRainbow, Rainbow, ≪48≫, DarkBands}

Here is the color function of the rainbow color scheme:

```
ColorData["Rainbow"]
```

ColorDataFunction$\left[\{0, 1\}, \right.$ 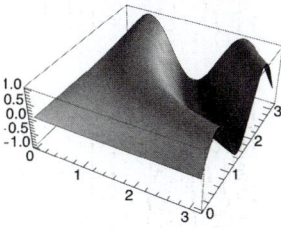 $\left.\right]$

A value of this function is

```
ColorData["Rainbow"][0.5]
```
RGBColor[0.513417, 0.72992, 0.440682]

In the next plot, we color the surface by using the rainbow color scheme to give the surface different colors according to the height of the surface:

```
Plot3D[Sin[x y], {x, 0, π}, {y, 0, π},
  ColorFunction → (ColorData["Rainbow"][#3] &), Mesh → False]
```

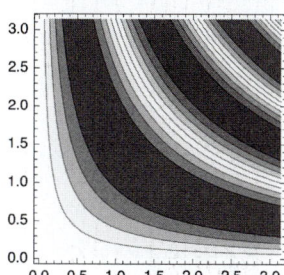

In density plots, we do not use arguments for the color function:

```
ContourPlot[Sin[x y], {x, 0, π}, {y, 0, π},
  ColorFunction → ColorData["TemperatureMap"]]
```

■ **Physical Color Schemes**

We have three physical color schemes:

```
ColorData["Physical"]
```
{BlackBodySpectrum, HypsometricTints, VisibleSpectrum}

Here is the color function of the visible spectrum color scheme:

```
ColorData["VisibleSpectrum"]
```

ColorDataFunction[{380, 750}, 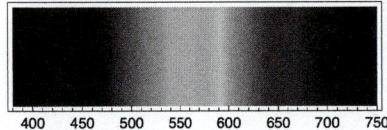]

One of these colors is as follows:

```
ColorData["VisibleSpectrum"][400]
```

RGBColor[0.263347, 0, 0.632745]

In the following plot, we use this color scheme:

```
DensityPlot[x, {x, 380, 750}, {y, 0, 100},
 ColorFunction → ColorData["VisibleSpectrum"],
 ColorFunctionScaling → False, AspectRatio → Automatic, ImageSize → 180,
 PlotRangePadding → 5, FrameTicks → {{None, None}, {Automatic, None}}]
```

■ **Indexed Color Schemes**

We have 43 indexed color schemes:

```
ColorData["Indexed"] // Short
```

{1, 2, 3, 4, 5, 6, 7, 8, 9, ≪25≫, 35, 36, 37, 38, 39, 40, 41, 42, 43}

Here is the color function of the 16th indexed color scheme:

```
ColorData[16]
```

ColorDataFunction[{1, 9, 1}, 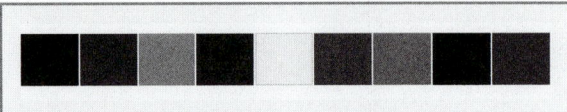]

By clicking on a color in the following panel, we get the corresponding RGB color:

```
ColorData[16, "Panel"]
```

The nine colors of this color scheme are as follows:

```
Short[ColorData[16, "ColorList"], 4]
```

{RGBColor[0.454902, 0.0509804, 0.0235294],
  ≪7≫, RGBColor[0.941176, 0., 0.00784314]}

The ninth color is

```
ColorData[16][9]
```

RGBColor[0.941176, 0., 0.00784314]

The following plot uses the first three colors of the color scheme:

```
Plot[Evaluate[Table[Sin[n x], {n, 3}]],
 {x, 0, 2 π}, PlotStyle → ColorData[16, "ColorList"]]
```

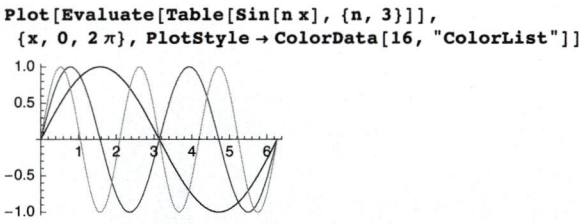

■ Named Color Schemes

We have five named color schemes:

```
ColorData["Named"]
```

{Atoms, GeologicAges, HTML, Legacy, WebSafe}

Here is the color function of the HTML color scheme:

```
ColorData["HTML"]
```

ColorDataFunction[{AliceBlue, ≪146≫},

By moving the mouse over the following panel, we can see the name of the color as a tooltip. By clicking on a color, we get the corresponding RGB color:

```
ColorData["HTML", "Panel"]
```

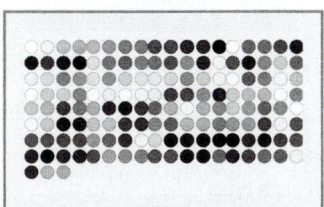

This color scheme contains 147 colors with names:

```
ColorData["HTML", "Range"] // Short
```

{{AliceBlue, AntiqueWhite, ≪144≫, YellowGreen}}

In the following plot, we use two of these colors:

```
Graphics[{ColorData["HTML"]["Maroon"],
 Disk[], ColorData["HTML"]["OliveDrab"], Disk[{1, 0}]}]
```

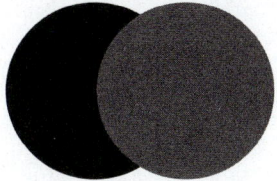

The image of the atoms color scheme gives a special plot:

```
Show[ColorData["Atoms", "Image"], ImageSize → 380]
```

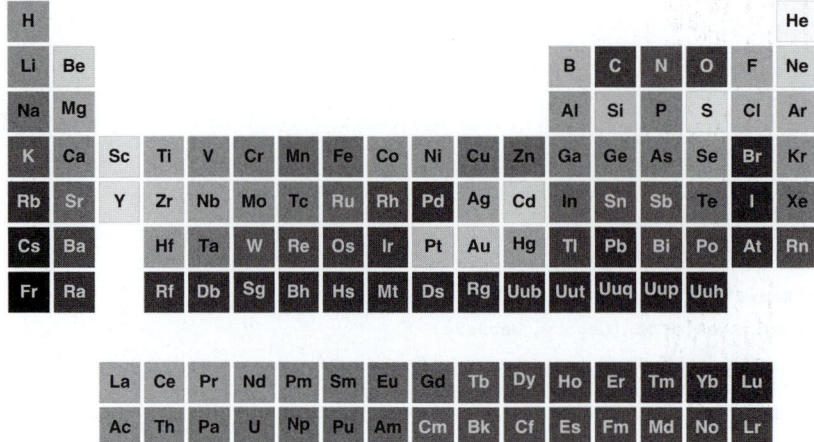

9.3.4 Example Data

■ **ExampleData**

ExampleData[] (✻6)   Give a list of all types of examples

ExampleData["type"]   Give a list of examples of the given type

ExampleData[{"type", "name"}]   Show the named example of the given type

ExampleData[{"type", "name"}, "Properties"]   Give a list of properties available for the example

ExampleData[{"type", "name"}, "prop"]   Give the value of the given property

---

Here are all the types of examples and the numbers of examples in each type:

```
Grid[{#, Length@ExampleData[#]} & /@ ExampleData[], Alignment → {{Left, Right}}]
```

```
AerialImage          38
Geometry3D           27
LinearProgramming   138
Matrix             2338
Sound                63
TestAnimation         4
TestImage            44
Text                 50
Texture              64
```

Short descriptions of the example types are as follows:

Graphics and sound examples:

"TestImage": test images for image processing

"AerialImage": sample aerial photography images

"Texture": sample textures

"Geometry3D": 3D geometry data for models and shapes

"TestAnimation": test animations for image processing

"Sound": sample audio clips

Mathematical examples:

"Matrix": sparse and dense matrices

"LinearProgramming": linear programming problems

Text examples:

"Text": sample text pieces

■ **TestImage, AerialImage, and Texture**

Here are examples of test images, aerial images, and textures:

```
{Show[ExampleData[{"TestImage", "Sailboat"}], ImageSize → 130],
 Show[ExampleData[{"AerialImage", "Earth"}], ImageSize → 130],
 Show[ExampleData[{"Texture", "Bark"}], ImageSize → 130]}
```

{ ,  , }

For these examples, we can ask for the following properties:

```
ExampleData[{"TestImage", "Sailboat"}, "Properties"]
```

{BitDepth, ColorSpace, Data, DataType, Graphics,
 GrayLevels, Image, ImageSize, Name, RGBColorArray}

■ **Geometry3D, TestAnimation, and Sound**

Next, we show an example of a 3D geometry. The image can be rotated with the mouse.

```
Show[ExampleData[{"Geometry3D", "Galleon"}], ImageSize → 200, ImagePadding → 0]
```

```
ExampleData[{"Geometry3D", "Galleon"}, "Properties"]
```

{Graphics3D, GraphicsComplex, Name, PolygonCount,
 PolygonData, PolygonObjects, VertexData, VertexNormals}

The four test animation examples animate a series of photographs. These examples have two properties:

```
ExampleData[{"TestAnimation", "WalterCronkite"}, "Properties"]
{Animation, Frames}
```

Now we can hear a bassoon scale:

```
ExampleData[{"Sound", "BassoonScale"}]
```

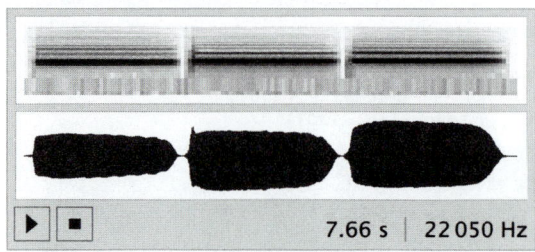

```
ExampleData[{"Sound", "BassoonScale"}, "Properties"]
{Channels, Data, Duration, SampledSoundList, SampleRate, Sound}
```

■ Matrix

One of the more than 2000 example matrices is FIDAP007:

```
m = ExampleData[{"Matrix", "FIDAP007"}]
SparseArray[<46 570>, {1633, 1633}]
```

```
MatrixPlot[m, ImageSize → 220]
```

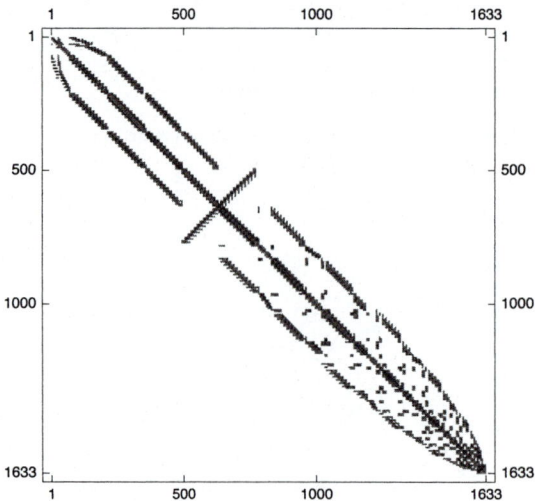

Matrices have the following properties:

```
ExampleData[{"Matrix", "FIDAP007"}, "Properties"]
```

{AverageEntriesPerColumn, AverageEntriesPerRow, Bandwidth, Collection,
 Dimensions, Entries, Format, ID, LowerBandwidth, Matrix, MatrixStructure,
 Name, PatternSymmetry, PositiveDefiniteQ, Source, StrongComponents,
 StructuralFullRankQ, StructuralRank, Symmetry, Type, UpperBandwidth, URL}

Our matrix has, for example, the following properties:

```
ExampleData[{"Matrix", "FIDAP007"}, #] & /@
 {"Type", "Bandwidth", "AverageEntriesPerRow", "Source"}
```
{Real, 277, 28.5181,
 ftp://math.nist.gov/pub/MatrixMarket2/SPARSKIT/fidap/fidap007.mtx.gz}

■ LinearProgramming

One of the smallest linear programming examples is afiro:

```
ExampleData[{"LinearProgramming", "afiro"}, "Dimensions"]
```
{27, 32}

This example has 27 constraints and 32 variables. The form of the example is as follows:

```
Shallow[ExampleData[{"LinearProgramming", "afiro"}], 3]
```
{{ ≪32≫ }, SparseArray[≪4≫], { ≪27≫ }, { ≪32≫ }}

Thus, the example has four components. Let us give the components the names **c**, **m**, **b**, and **n**:

```
{c, m, b, n} = ExampleData[{"LinearProgramming", "afiro"}];
```

Here are the four components:

```
c
```
{0, -0.4, 0, 0, 0, 0, 0, 0, 0, 0, 0, 0, -0.32, 0,
 0, 0, -0.6, 0, 0, 0, 0, 0, 0, 0, 0, 0, 0, -0.48, 0, 0, 10.}

```
m
```
SparseArray[<83>, {27, 32}]

```
Short[m // Normal, 4]
```
{{-1., 1., 1., 0, 0, 0, 0, 0, 0, 0, 0, 0, 0, 0, 0, 0, 0, 0, 0, 0, 0,
  0, 0, 0, 0, 0, 0, 0, 0, 0, 0, 0}, ≪25≫, {0, 0, 0, 0, 0, 0, 0, 0, 0,
  0, 0, 0, 0, 0, 1., 0, 0, 0, 0, 0, 0, 0, 0, 0, 0, 0, 0, 0, 0, 1., 0}}

```
b
```
{{0., 0}, {0., 0}, {80., -1}, {0., -1}, {0., 0}, {0., 0}, {80., -1},
 {0., -1}, {0., -1}, {0., -1}, {0., 0}, {0., 0}, {500., -1}, {0., -1},
 {0., 0}, {44., 0}, {500., -1}, {0., -1}, {0., -1}, {0., -1}, {0., -1},
 {0., -1}, {0., -1}, {0., -1}, {0., -1}, {310., -1}, {300., -1}}

```
n
```
{{0, ∞}, {0, ∞}, {0, ∞}, {0, ∞}, {0, ∞}, {0, ∞}, {0, ∞}, {0, ∞}, {0, ∞}, {0, ∞}, {0, ∞},
 {0, ∞}, {0, ∞}, {0, ∞}, {0, ∞}, {0, ∞}, {0, ∞}, {0, ∞}, {0, ∞}, {0, ∞}, {0, ∞}, {0, ∞},
 {0, ∞}, {0, ∞}, {0, ∞}, {0, ∞}, {0, ∞}, {0, ∞}, {0, ∞}, {0, ∞}, {0, ∞}, {0, ∞}}

The components correspond with the linear programming problem where the coefficient vector of the objective function is **c**, the constraint matrix is **m**, the right-hand side of the constraints is **b**, and the usual nonnegativity constraints apply. So, we can solve the problem as follows (for linear programming, see Section 23.2.1, p. 753):

```
sol = LinearProgramming[c, m, b]
```
{80., 25.5, 54.5, 84.8, 18.2143, 0., 0., 0., 0.,
 0., 0., 0., 18.2143, 0., 19.3071, 500., 475.92, 24.08, 0.,
 215., 0., 0., 0., 0., 0., 0., 0., 0., 339.943, 383.943, 0., 0.}

```
c.sol
```

```
-464.753
```

We could also write simply as follows:

```
ExampleData[{"LinearProgramming", "afiro"}];
```

```
LinearProgramming @@ %
```

```
{80., 25.5, 54.5, 84.8, 18.2143, 0., 0., 0., 0.,
 0., 0., 0., 18.2143, 0., 19.3071, 500., 475.92, 24.08, 0.,
 215., 0., 0., 0., 0., 0., 0., 0., 0., 339.943, 383.943, 0., 0.}
```

Linear programming examples have the following properties:

```
ExampleData[{"LinearProgramming", "afiro"}, "Properties"]
```

```
{Collection, ConstraintMatrix, Dimensions,
 Equations, LinearProgrammingData, Name, Source}
```

Next we ask for the explicit equations:

```
{{obj, cons}, vars} = ExampleData[{"LinearProgramming", "afiro"}, "Equations"];
```

For example, here are the objective function and the variables:

```
obj
```

$-0.4\,X02_{MPS} - 0.32\,X14_{MPS} - 0.6\,X23_{MPS} - 0.48\,X36_{MPS} + 10.\,X39_{MPS}$

```
vars
```

$\{X01_{MPS}, X02_{MPS}, X03_{MPS}, X04_{MPS}, X06_{MPS}, X07_{MPS}, X08_{MPS}, X09_{MPS}, X10_{MPS}, X11_{MPS}, X12_{MPS},$
$X13_{MPS}, X14_{MPS}, X15_{MPS}, X16_{MPS}, X22_{MPS}, X23_{MPS}, X24_{MPS}, X25_{MPS}, X26_{MPS}, X28_{MPS}, X29_{MPS},$
$X30_{MPS}, X31_{MPS}, X32_{MPS}, X33_{MPS}, X34_{MPS}, X35_{MPS}, X36_{MPS}, X37_{MPS}, X38_{MPS}, X39_{MPS}\}$

The constraints form a large expression that we do not show here. Now we can use **Minimize**:

```
Minimize[{obj, cons}, vars]
```

$\{-464.753, \{X01_{MPS} \rightarrow 80., X02_{MPS} \rightarrow 25.5, X03_{MPS} \rightarrow 54.5, X04_{MPS} \rightarrow 84.8, X06_{MPS} \rightarrow 18.2143,$
$X07_{MPS} \rightarrow 0., X08_{MPS} \rightarrow 0., X09_{MPS} \rightarrow 0., X10_{MPS} \rightarrow 0., X11_{MPS} \rightarrow 0., X12_{MPS} \rightarrow 0.,$
$X13_{MPS} \rightarrow 0., X14_{MPS} \rightarrow 18.2143, X15_{MPS} \rightarrow 0., X16_{MPS} \rightarrow 19.3071, X22_{MPS} \rightarrow 500.,$
$X23_{MPS} \rightarrow 475.92, X24_{MPS} \rightarrow 24.08, X25_{MPS} \rightarrow 0., X26_{MPS} \rightarrow 215., X28_{MPS} \rightarrow 0.,$
$X29_{MPS} \rightarrow 0., X30_{MPS} \rightarrow 0., X31_{MPS} \rightarrow 0., X32_{MPS} \rightarrow 0., X33_{MPS} \rightarrow 0., X34_{MPS} \rightarrow 0.,$
$X35_{MPS} \rightarrow 0., X36_{MPS} \rightarrow 339.943, X37_{MPS} \rightarrow 383.943, X38_{MPS} \rightarrow 0., X39_{MPS} \rightarrow 0.\}\}$

We obtained the same solution as with **LinearProgramming**.

■ Text

One of the text examples is an excerpt from Hamlet. We only show the text of the first four rows:

```
StringTake[ExampleData[{"Text", "ToBeOrNotToBe"}], 173]
```

```
To be, or not to be,--that is the question:-- Whether
   'tis nobler in the mind to suffer The slings and arrows of
   outrageous fortune Or to take arms against a sea of troubles
```

This is a single string, as can be seen with **InputForm**:

```
% // InputForm
```

```
"To be, or not to be,--that is the question:-- \
Whether 'tis nobler in the mind to suffer The \
slings and arrows of outrageous fortune Or to take \
arms against a sea of troubles"
```

For text examples, we can ask for the following properties:

```
ExampleData[{"Text", "ToBeOrNotToBe"}, "Properties"]
```

```
{Author, FormattedText, FullTitle, Language,
 Lines, NotebookExpression, String, Title, Words}
```

The property **"String"** gives the same result as we presented previously. With **"FormattedText"** we get correct lines:

```
StringTake[ExampleData[{"Text", "ToBeOrNotToBe"}, "FormattedText"], 173]
```

```
To be, or not to be,--that is the question:--
Whether 'tis nobler in the mind to suffer
The slings and arrows of outrageous fortune
Or to take arms against a sea of troubles
```

With **InputForm** we can see the newline characters **\n**:

```
% // InputForm
```

```
"To be, or not to be,--that is the \
question:--\nWhether 'tis nobler in the mind to \
suffer\nThe slings and arrows of outrageous \
fortune\nOr to take arms against a sea of \
troubles"
```

With **"Lines"** we get the text as a list of lines (each line is a string):

```
Take[ExampleData[{"Text", "ToBeOrNotToBe"}, "Lines"], 4]
```

```
{To be, or not to be,--that is the question:--,
 Whether 'tis nobler in the mind to suffer,
 The slings and arrows of outrageous fortune,
 Or to take arms against a sea of troubles,}
```

With **"Words"** we get the text as a list of words (each word is a string):

```
Take[ExampleData[{"Text", "ToBeOrNotToBe"}, "Words"], 33]
```

```
{To, be,, or, not, to, be,--that, is, the, question:--, Whether,
 'tis, nobler, in, the, mind, to, suffer, The, slings, and, arrows, of,
 outrageous, fortune, Or, to, take, arms, against, a, sea, of, troubles,}
```

With **"NotebookExpression"** we can put the text into a separate notebook (the notebook is not shown here):

```
NotebookPut[ExampleData[{"Text", "ToBeOrNotToBe"}, "NotebookExpression"]]
```

```
NotebookObject[ [□] Untitled-4 ]
```

For more about strings, see Sections 13.3.6, p. 433, and 16.2, p. 505.

■ **Additional Examples**

Mathematica also has more than 100 additional example files:

```
SetDirectory[$InstallationDirectory <>
  "/Documentation/English/System/ExampleData"]; Short[FileNames[], 1]
{100d.pdb, 1PPT.pdb, <<133>>, wr.rss}
```

These examples are mainly intended for internal use of *Mathematica*. The files can be imported with **Import**:

```
Short[Import["ExampleData/financialtimeseries.csv"], 2]
{{Jan 03 2006, 11.82}, <<249>>, {Dec 29 2006, 13.91}}
```

Manipulations

Introduction

> *The story is told about the mathematician Littlewood that he was lecturing to a class one day and remarked about some step in a mathematical argument that it was obvious. Then he stepped back from the board and said, "Mmm, I wonder if it is obvious?" He spent about half an hour doing various calculations and finally declared with a smile, "Yes, it is obvious."*

Mathematica 6 brings to us *dynamic interactivity*—that is, ways to create interfaces where the user can explore the output by some controls such as sliders or popup menus. Such interfaces range from interactive data viewers to interactive applications.

The main command, considered in this chapter, is **Manipulate**. It creates an interface where an expression containing some parameters can easily be studied or animated by adjusting the parameters with some controls. We also have the lower-level command **Dynamic** that can be used for more special or more advanced cases or for more detailed formatting of the output. In addition, we have **Animate** for animations and commands such as **MenuView**, **TabView**, and **OpenerView** to view data in various ways. The viewers are capable of creating hierarchical views, suitable for representing large data sets. We address **Dynamic**, **Animator**, and the views in Chapter 11. The commands presented in this and the next chapter are new in *Mathematica* 6.

Wolfram Research has a special Demonstrations Project at http://demonstrations.wolfram.com to advance the use of **Manipulate**. Users of *Mathematica* 6 can submit applications of this command for publication on the website. Currently, the site contains nearly 4000 demonstrations in subject areas such as mathematics; computation; physical sciences; life sciences; business and social systems; systems, models, and methods; engineering and technology; our world; creative arts; kids and fun; and *Mathematica* functionality. The demonstrations and their source codes can be freely loaded.

Although interesting and useful as such, the demonstrations are also valuable in studying the use of **Manipulate**. The interested reader is encouraged to study the applications in the Demonstrations Project because here we only have room for some basic examples.

Regarding dynamic features, also remember palettes, hyperlinks, and slide shows that we considered in Section 3.1.3, p. 56.

To avoid some printing problems, all dynamic outputs shown in this book are GIF images of the original outputs. Thus, each dynamic output was exported by **Export["dyn.gif", %]** and then imported by **Import[%]**.

10.1 Basic Manipulation

10.1.1 Introduction

■ Manipulation

Here is a simple **Manipulate** command and its output panel:

```
Manipulate[Plot[Cos[a x], {x, 0, 3 π}, ImageSize → 200], {a, 1, 10}]
```

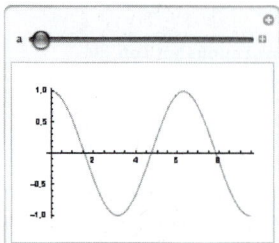

We can change the value of the parameter **a** from 1 to 10 by moving the slider with the mouse; the plot of cos(*a x*) is redrawn in real time.

In general, **Manipulate** has two arguments: the *expression to be manipulated* (the plotting command in the previous example) and the *specifications of the parameters* (**{a, 1, 10}**). The expression to be manipulated depends on the parameters. The output panel contains a *control* (e.g., a slider) for each parameter and a *content area*. The expression can be manipulated with the controls, and the content area shows the current state of the expression.

The panel also has two special buttons: the *animation button* ⊞ and the *utility button* ⊕. Clicking the animation button ⊞ opens *animation controls* and a *value field* showing the current value of the parameter. Clicking the utility button ⊕ opens the *utility menu*. We discuss these buttons later.

Mathematica contains several types of controls. **Manipulate** often automatically chooses a suitable control type based on the form of the parameter specification. The automatically chosen control can be one of the following: a manipulator, a 2D slider, a setter bar, a popup menu, a checkbox, an input field, or a color slider. We can override the automatic control by using the **ControlType** option. With this option we can also ask for still other control types, such as a slider, a vertical slider, an animator, a trigger, a locator, a radio button bar, a toggler, a checkbox bar, a toggler bar, or a color setter. We consider this option in the next subsection.

Manipulate is an extremely powerful way to explore how various objects change as we interactively change one or more parameters.

■ **Control Types and Placements**

`Manipulate` chooses the type of control from the form of the parameter specification. With the following option we can override the automatic setting. With this option we have access to all of the various control types.

> `ControlType` The type of control to produce in the output; possible values: `Automatic, Animator,`
> `Checkbox, CheckboxBar, ColorSetter, ColorSlider, InputField, Locator, Manipulator` (slider;
> animator under the ⊞ icon), `PopupMenu, RadioButton, RadioButtonBar, Setter, SetterBar`
> (tabs), `Slider, Slider2D, Toggler, TogglerBar, Trigger` (like animator), `VerticalSlider, None`

The option can be used like other options, but a simpler way is to add the control type as an additional list element of the control specification. For example, the automatic control in the following is a slider:

```
Manipulate[Prime[n], {n, 1, 200, 1}]
```

We can ask, for example, for an input field with the option:

```
Manipulate[Prime[n], {n, 1, 200, 1}, ControlType → InputField]
```

A simpler way is to add the control type to the parameter specification:

```
Manipulate[Prime[n], {n, 1, 200, 1, InputField}]
```

The default is that the controls are placed at the top of the panel. With the following option we can adjust the placement of the controls.

> `ControlPlacement` Placement of controls in the panel; examples of values: `Automatic, Left, Right,`
> `Bottom, Top`

```
Manipulate[Prime[n], {n, 1, 200, 1}, ControlPlacement → Bottom]
```

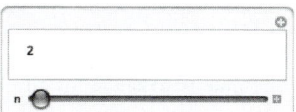

Next, we discuss the various controls, which have been grouped as sliders, locators, and other controls.

10.1.2 Sliders

■ Manipulator, Slider, and VerticalSlider

The following box gives the two most common forms of **Manipulate**.

To create an interface enabling the interactive choice of the value of the parameter **u** with a slider and
  showing the corresponding value of **expr**:

Manipulate[expr, {u, u$_{min}$, u$_{max}$}] (❀6) u can have any value between u$_{min}$ and u$_{max}$

Manipulate[expr, {u, u$_{min}$, u$_{max}$, du}] u can have any value between u$_{min}$ and u$_{max}$ in steps of **du**

For these kinds of parameter specifications, **Manipulate** automatically chooses the control to be a
manipulator. A manipulator means a slider together with animation controls under the ⊞ button
(animation is considered later). In the first case, the slider is continuous, whereas in the latter case it is
discrete.

As an example, we create a panel where we can show the probability density function (PDF) of the
normal distribution with mean 0 and standard deviation σ for any value of σ in the interval [0.2, 5].
Here, we have a continuous slider.

```
Manipulate[Plot[PDF[NormalDistribution[0, σ], x],
  {x, -5, 5}, PlotRange → {0, 2.05}, ImageSize → 200], {σ, 0.2, 5}]
```

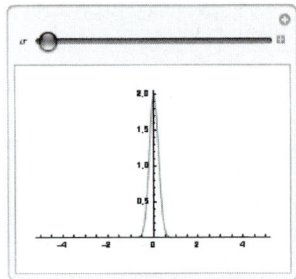

Next, we show the PDFs of $\chi^2(n)$ distributions with $n = 1, \ldots, 15$. Now we have a discrete slider.

```
Manipulate[Plot[PDF[ChiSquareDistribution[n], x],
  {x, 0, 20}, PlotRange → {0, 0.51}, ImageSize → 200], {n, 1, 15, 1}]
```

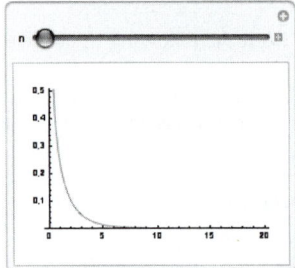

Note that often, as in the previous examples, it is useful to define the same **PlotRange** for all plots.
Then the plots can be more easily compared.

Here are the probability functions of the Poisson distribution with parameter in the interval [0.7, 10]:

```
Manipulate[ListPlot[Table[{x, PDF[PoissonDistribution[λ], x]}, {x, 0, 20}],
    Filling → Axis, PlotRange → {-0.02, 0.51}, ImageSize → 200], {λ, 0.7, 10}]
```

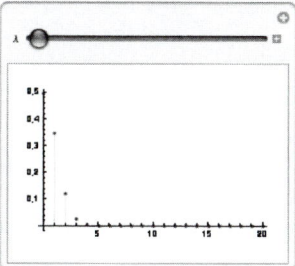

Note that **Manipulate** does its work in real time: It does not precompute all the possible outputs. Accordingly, for the manipulation to be practical, the computation caused by moving a slider should not take more than, for example, 1 second. Fortunately, nowadays quite a lot can be computed within 1 second.

```
Manipulate[expr, {u, umin, umax, Manipulator}]

Manipulate[expr, {u, umin, umax, Slider}]

Manipulate[expr, {u, umin, umax, VerticalSlider}]
```

A manipulator can also be asked for by specifying the type of the control to be **Manipulator**. If we define the control to be **Slider**, we only get a slider without the animation controls. With **VerticalSlider**, we get a slider running from bottom to top (a good location for this control is on the left; use the **ControlPlacement** option). Here, we ask for a slider:

```
Manipulate[Prime[n], {n, 1, 200, 1, Slider}]
```

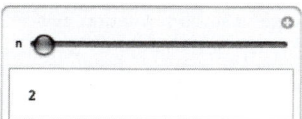

■ **More Examples of Manipulate**

Besides plots, we can manipulate any other expression. Below we can see any of the first 200 primes. We have used the **Alignment** option to place the primes at the right of the panel.

```
Manipulate[Prime[n], {n, 1, 200, 1}, Alignment → Right]
```

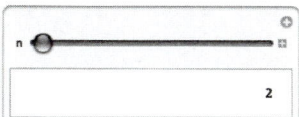

Next, we tabulate the probabilities of getting k sixes when tossing a die n times. We can adjust the value of n.

```
Manipulate[TableForm[Table[{k, PDF[BinomialDistribution[n, 1/6.], k]}, {k, 0, n}],
    TableSpacing → {1, 3},
    TableHeadings → {None, {"k", "P(X=k)"}}] // Framed, {n, 1, 10, 1}]
```

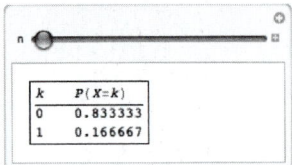

3D plots can be rotated with the mouse. Another way to rotate is to use **Rotate** with **Manipulate**. To get a smooth rotation, we define a large enough **PlotRange**:

```
Manipulate[
    Graphics3D[Rotate[Cuboid[{-1, -1, -1}, {1, 1, 1}], θ, {0, 0, 1}], Boxed → False,
    PlotRange → {{-1.5, 1.5}, {-1.5, 1.5}, All}, ImageSize → 200], {θ, 0, 2 π}]
```

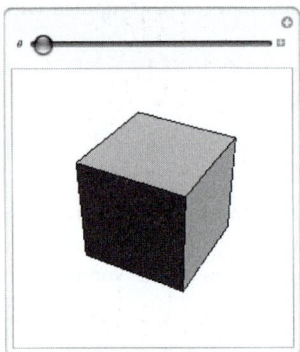

Now we rotate a parametric surface by changing the viewpoint. To get a smooth rotation, we defined a large enough **PlotRange** and set **SphericalRegion** to be True:

```
Manipulate[ParametricPlot3D[{s Cos[t] Sin[s], s Cos[s] Cos[t], -s Sin[t]},
    {s, 0, 2 π}, {t, 0, π}, ViewPoint → {Cos[θ], Sin[θ], 1},
    Boxed → False, Axes → False, PlotRange → {{-8, 8}, {-8, 8}, All},
    SphericalRegion → True, PlotPoints → 20, ImageSize → 200], {θ, 0, π}]
```

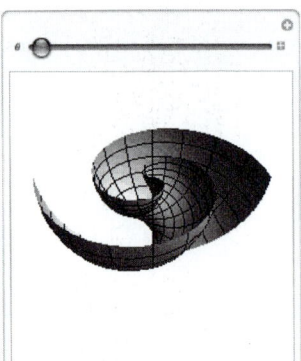

Next, we show how a small sphere moves around a larger sphere:

```
Manipulate[Graphics3D[{Sphere[], Sphere[1.8 {Cos[t], Sin[t], 0}, 0.2]},
  Boxed → False, ViewPoint → {1.3, -2.4, 0.8},
  PlotRange → {{-2, 2}, {-2, 2}, {-1, 1}}, ImageSize → 200], {t, 0, 2 π}]
```

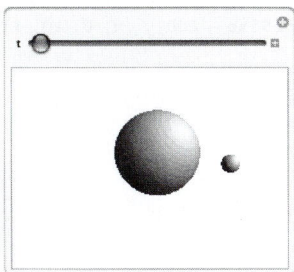

■ Animation

We have several ways to run an animation:

```
Animate[expr, {u, u_min, u_max}]

Manipulate[expr, {u, u_min, u_max}]   Click ⊞ and use the animation controls

Manipulate[expr, {u, u_min, u_max}]   Click ⊕ and choose Autorun

Manipulate[expr, {u, u_min, u_max, Animator}]

Manipulate[expr, {u, u_min, u_max, Trigger} ]
```

Animate is considered in Section 11.1.2, p. 365. Here is an example:

```
Animate[Plot[ChebyshevT[n, x], {x, -1, 1}, PlotRange → 1.05, ImageSize → 130],
  {n, 0, 10, 1}, AnimationRunning → False]
```

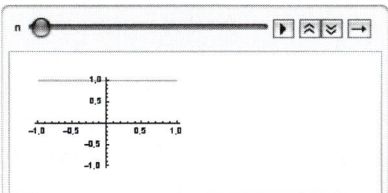

We can run the animation by clicking the play button. We can also drag the slider with the mouse to show a particular plot. **Manipulate** produces very much the same view:

```
Manipulate[
  Plot[ChebyshevT[n, x], {x, -1, 1}, PlotRange → 1.05, ImageSize → 200], {n, 0, 10, 1}]
```

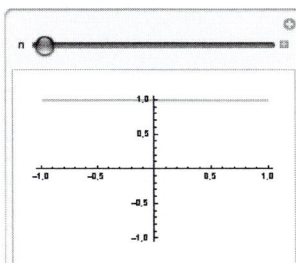

Here, we can drag the slider. However, by clicking the animation button ⊞ next to the slider, we also get controls to play an animation and a field showing the current value of the parameter:

```
Manipulate[
 Plot[ChebyshevT[n, x], {x, -1, 1}, PlotRange → 1.05, ImageSize → 200], {n, 0, 10, 1}]
```

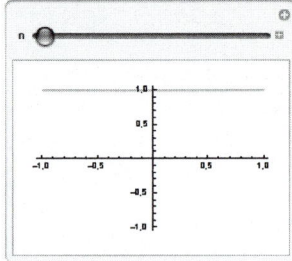

Clicking the utility ⊕ button, we get a menu, with one of its items being **Autorun**. If we choose this item, an animation is run automatically (for each parameter in turn, if we have several parameters). The animation can be stopped by clicking *close*.

```
Manipulate[
 Plot[ChebyshevT[n, x], {x, -1, 1}, PlotRange → 1.05, ImageSize → 200], {n, 0, 10, 1}]
```

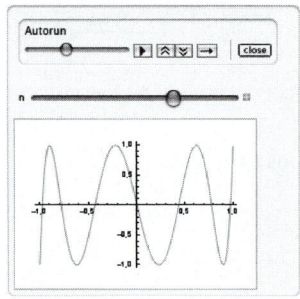

A fourth way to get an animation is to ask **Manipulate** to produce an animator:

```
Manipulate[Plot[ChebyshevT[n, x], {x, -1, 1}, PlotRange → 1.05, ImageSize → 200],
 {n, 0, 1×0, 1, Animator}]
```

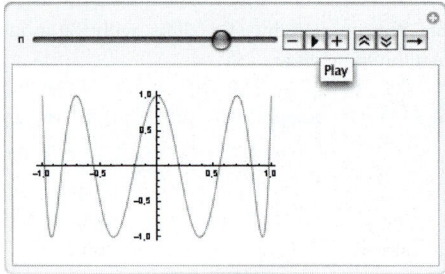

A fifth way is to ask for a trigger. It is like an animator but does not have a slider:

```
Manipulate[Plot[ChebyshevT[n, x], {x, -1, 1}, PlotRange → 1.05, ImageSize → 160],
  {n, 0, 10, 1, Trigger}]
```

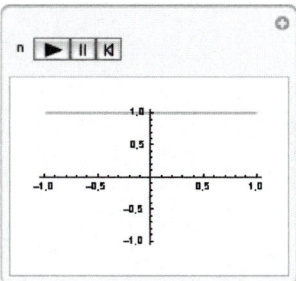

In summary, **Animate** is intended to automatically show a series of plots from the beginning to the end (although we can also drag the slider by ourselves). **Manipulate** produces an interface where we can adjust the parameter in question by ourselves (although we can also run an animation). If we only have one parameter, there is not much difference between **Animate** and **Manipulate**.

■ Initial Values, Labels, and Current Values

Manipulate[expr, {{u, u$_{init}$}, u$_{min}$, u$_{max}$}] The initial value of **u** is u$_{init}$
Manipulate[expr, {{u, u$_{init}$, u$_{lbl}$}, u$_{min}$, u$_{max}$}] The initial value of **u** is u$_{init}$ and the label of the slider is u$_{lbl}$
Manipulate[expr, {u, u$_{min}$, u$_{max}$, Appearance → "Labeled"}] Show the current value of **u**

Previously, we created a panel showing the plot of some Chebyshev polynomials. The starting value 1 of the parameter n produced a somewhat noninteresting plot of a constant. It would be good if the panel, immediately after executing the **Manipulate** command, would show something more interesting. This can be done by giving an initial value for the parameter:

```
Manipulate[Plot[ChebyshevT[n, x], {x, -1, 1},
  PlotRange → 1.05, ImageSize → 200], {{n, 5}, 0, 10, 1}]
```

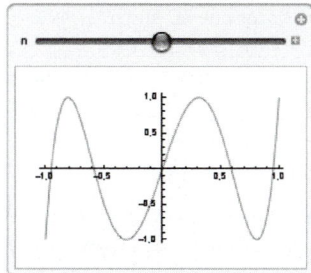

The label on the left-hand side of the slider is, by default, the corresponding parameter. We can give another label. As an example, next we write the parameter of the exponential distribution as $1/\lambda$. Then the expected value of the corresponding random variable is λ. Hence, we can label the slider as $E(X)$:

```
Manipulate[Plot[PDF[ExponentialDistribution[1 / λ], x], {x, 0, 3},
   PlotRange → {0, 5.1}, ImageSize → 200], {{λ, 0.25, "E(X)"}, 0.2, 10}]
```

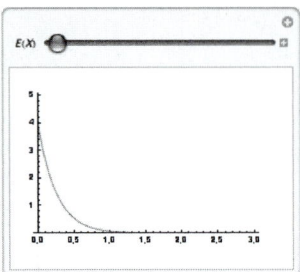

By clicking the animation button ⊞, we can see the current value of the parameter; this is shown in the previous panel. Another way to see this value is to use the **Appearance** option. The value of the parameter can be seen next to the ⊞ icon:

```
Manipulate[Plot[PDF[ExponentialDistribution[1 / λ], x],
   {x, 0, 3}, PlotRange → {0, 5.1}, ImageSize → 200],
   {{λ, 0.25, "E(X)"}, 0.2, 10, Appearance → "Labeled"}]
```

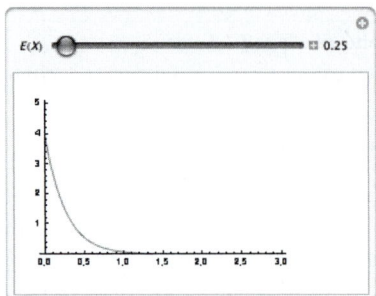

■ **Two Manipulators and Slider2D**

If we have two parameters, we can ask either for two usual 1D sliders or for one 2D slider:

`Manipulate[expr, {x, xmin, xmax}, {y, ymin, ymax}]`	Control the two parameters with two sliders
`Manipulate[expr, {u, {xmin, ymin}, {xmax, ymax}}]`	Control the two parameters with a 2D slider

A 2D slider can also be asked for by specifying the type of the control to be **Slider2D**.

With a manipulation, we can investigate the PDF of a beta distribution with respect to its two parameters. Here, we use the **LabelStyle** option to make the labels α and β bold and somewhat larger.

```
Manipulate[Plot[PDF[BetaDistribution[α, β], x], {x, 0, 1}, PlotRange → {0, 4.1},
   ImageSize → 200], {{α, 2.5}, 0.01, 3, Appearance → "Labeled"},
 {{β, 1.5}, 0.01, 3, Appearance → "Labeled"}, LabelStyle → {Bold, 12}]
```

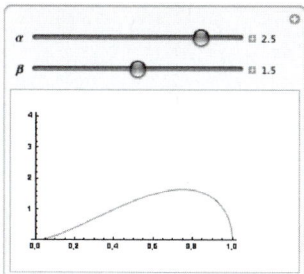

We can also ask for a 2D slider that allows simultaneous changes of the parameters. In the next example, the 2D slider can be moved on the region $(0.01, 3) \times (0.01, 3)$. With the **ControlPlacement** option, we asked to put the 2D slider at the left of the panel.

```
Manipulate[Plot[PDF[BetaDistribution[α[[1]], α[[2]]], x],
   {x, 0, 1}, PlotRange → {0, 4.1}, ImageSize → 200],
  {{α, {2.5, 1.5}}, {0.01, 0.01}, {3, 3}}, ControlPlacement → Left]
```

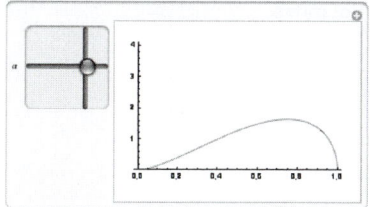

In the 2D slider, the values of α go from left bottom to right bottom, and the values of β go from bottom left to top left. For example, by moving the slider from bottom left to top right we see that for these values of the parameters the PDF is symmetric, one of the PDF's being the one of the uniform distribution.

■ Several Manipulators

We can have an arbitrary amount of parameters. In the following example, we ask for three sliders. The plot describes the evolution of the solution of a logistic differential equation. We can adjust the initial value and the two parameters of the model.

```
Manipulate[Plot[M / (1 + (M / y0 - 1) Exp[-r M t]), {t, 0, 20},
  PlotRange → {0, All}, ImagePadding → {{15, 2}, {15, 5}}, ImageSize → 200],
 {{y0, 0.2, "y(0)"}, 0.001, 5, Appearance → "Labeled"},
 {{r, 0.15, "r"}, 0, 2, Appearance → "Labeled"},
 {{M, 4, "M"}, 0, 10, Appearance → "Labeled"}]
```

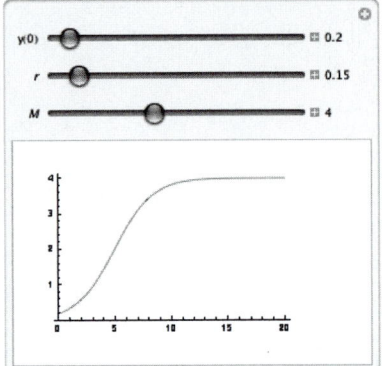

Note that in this example, we used the **ImagePadding** option to make suitable space around the plot so that even if the y tick labels change, the axes of the plot do not move.

10.1.3 Locators

■ **Locator**

In the following example, we create a point that we can move with a 2D slider:

```
Manipulate[Graphics[{PointSize[Large], Point[p]},
  Frame → True, FrameTicks → None, PlotRange → 1.2, ImageSize → 60],
 {{p, {0, 0}}, {-1, -1}, {1, 1}},
 ControlPlacement → Left]
```

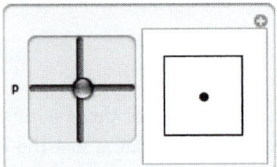

However, we have a special control, called a *locator,* that is specifically designed for points that can be moved.

Manipulate[expr, {pt, Locator}]   Create an interface enabling the interactive movement of a
point **pt** (displayed as a locator) and showing the corresponding value of **expr**

Manipulate[expr, {{pt, {x1, y1}}, Locator}]   The initial position of the locator is (x_1, y_1)

We do not get a locator automatically; we have to define the control type as a **Locator**.

Here is a simple example:

```
Manipulate[Graphics[{}, PlotLabel → NumberForm[pt // Chop, {3, 2}], PlotRange → 1.1,
   Frame → True, FrameTicks → None, ImageSize → 100], {{pt, {0, 0}}, Locator}]
```

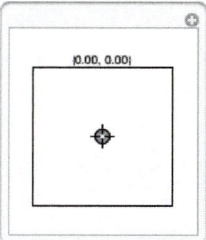

We created an empty graphics, containing only a frame and a plot label. However, the plot also contains a locator that can be moved within the frame. The position of the locator is shown in the plot label. For a locator, we do not need to define ranges because a locator takes suitable ranges from the ranges of the plot.

Now we create three locators:

```
Manipulate[Graphics[{Orange, Polygon[{p1, p2, p3}]},
   PlotRange → 1.1, ImageSize → 120], {{p1, {-0.9, -0.9}}, Locator},
   {{p2, {0.9, -0.9}}, Locator}, {{p3, {0, 0.9}}, Locator}]
```

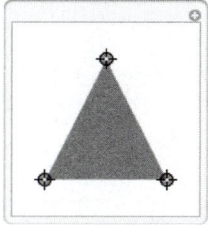

By dragging on any one of the locators, the plot changes accordingly. You can also click anywhere in the plot and the nearest locator moves to that point.

■ More about Locator

It is almost always useful to define initial values for the locators. If we have several locators, we can treat them as a whole, as is shown in the second item here:

`{{pt, {x1, y1}}, Locator}`	Represent **pt** as a locator whose initial position is (x_1, y_1)
`{{pt, {{x1, y1}, … , {xn, yn}}}, Locator}`	Represent **pt** as a set of locators whose initial positions are $(x_1, y_1), …, (x_n, y_n)$

We can now write the preceding example as follows:

```
Manipulate[Graphics[{Orange, Polygon[p]}, PlotRange → 1.1, ImageSize → 120],
  {{p, {{-0.9, -0.9}, {0.9, -0.9}, {0, 0.9}}}, Locator}]
```

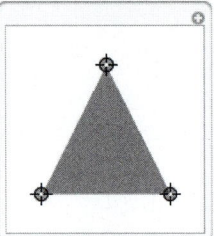

With the **Appearance** option we can define other symbols for the locators.

{{pt, {x1, y1}}, Locator, Appearance → obj}   Use **obj** as a locator object; examples of values of
  obj: **Automatic** (the default "crosshairs" appearance), **None** (display nothing visible), **"*"**, **"●"**,
  Graphics[{PointSize[Large], Point[{0, 0}]}]

Now, we show 50 random locators as colored points (the symbol ● can be typed as \[FilledSmallCircle]). Each point can be moved separately.

```
Manipulate[Graphics[{}, PlotRange → 1.1, ImageSize → 150],
  {{pt, RandomReal[{-1, 1}, {50, 2}]}, Locator,
   Appearance → Table[Style["●", Hue[h / 50]], {h, 50}]}]
```

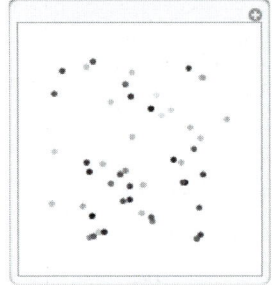

■ **Adding and Removing Locators**

The number of locators need not be fixed. With an option, we can allow more locators to be created or old locators to be removed.

{{pt, {x1, y1}}, Locator, LocatorAutoCreate → True}   Allow autocreation and autodeletion of
  locators

Now, if you hold down the ⒜ⓁⓉ key (Windows) or ⌘ key (Macintosh) and then click on the plot, a new locator is created. If you hold down the ⒜ⓁⓉ or ⌘ key and then click on an existing locator, that locator is removed.

Consider again the familiar triangle and try adding and removing locators:

```
Manipulate[Graphics[{Orange, Polygon[p]}, PlotRange → 1, ImageSize → 120],
  {{p, {{-0.9, -0.9}, {0.9, -0.9}, {0, 0.9}}}, Locator, LocatorAutoCreate → True}]
```

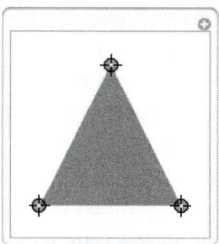

■ Example: Interactive Interpolation and Curve Fitting

The document tutorial/IntroductionToManipulate in the Documentation Center contains interesting examples of interactive interpolating and curve fitting. First, we consider interpolation:

```
Manipulate[Plot[InterpolatingPolynomial[points, x],
  {x, -2, 2}, PlotRange → 10.5, ImageSize → 230],
  {{points, RandomReal[{-2, 2}, {2, 2}]}, Locator, LocatorAutoCreate → True}]
```

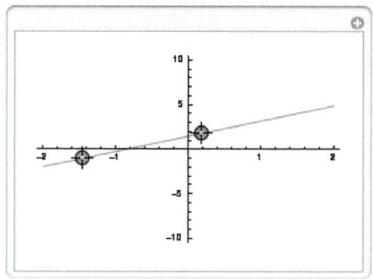

Here, the points to be interpolated are represented as locators. The figure shows the interpolating polynomial. Points can be added and removed by ⎇- or ⌘-clicking. Now, move the locators and add new locators to see how the interpolating polynomial changes. For more about interpolating polynomials, see Section 24.1.1, p. 792.

Similarly, we can interactively investigate polynomial curve fitting:

```
Manipulate[DynamicModule[{x, fi}, fi = Fit[points, x^Range[0, order], x];
  Plot[fi, {x, -2, 2}, PlotRange → 5.1, ImageSize → 230]],
  {{order, 1}, 1, 10, 1, Appearance → "Labeled"},
  {{points, RandomReal[{-2, 2}, {3, 2}]}, Locator, LocatorAutoCreate → True}]
```

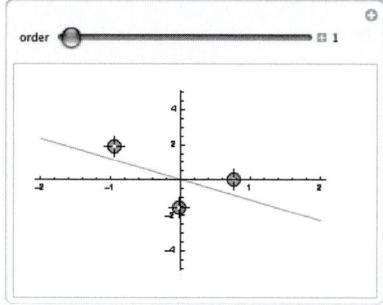

Again, the points to be fitted are represented as locators. The figure shows the fitted polynomial. Points can be added and removed by ⎇- or ⌘-clicking. Now, move the locators, add new locators, and change the order of the polynomial to see how the fit changes. Note that in the previous command we used **DynamicModule** to make the **x** variable local; now the command works even if **x** happens to have a value. **DynamicModule** is considered in Section 11.2.1, p. 371. It suffices here to say that it is used like **Module** and is designed for dynamic calculations. For more on curve fitting, see Section 25.1.1, p. 812.

■ **Example: Interactive Differential Equation Plotting**

Next, we study a differential equation model of competing species. In the initial position of the panel, we show four trajectories of the solution of the equations, corresponding to four starting points. The four points are represented by locators so that you can move them with the mouse. You can also create new trajectories and delete old ones. This model is also studied in Section 26.3.2, p. 855.

```
Manipulate[DynamicModule[{x, y, t, sol},
  sol = {x[t], y[t]} /. NDSolve[{x'[t] == x[t] (2 - 2 / 3 x[t] - 2 y[t]),
      y'[t] == y[t] (2 - 4 / 3 x[t] - y[t]), x[0] == #[[1]], y[0] == #[[2]]},
    {x[t], y[t]}, {t, 0, 10}] & /@ start;
  ParametricPlot[sol, {t, 0, 10}, PlotRange → {0, 2.5}, ImageSize → 230]],
 {{start, {{0.2, 0.1}, {0.3, 0.3}, {2.5, 1.5}, {2.5, 2}}},
  Locator, LocatorAutoCreate → True}]
```

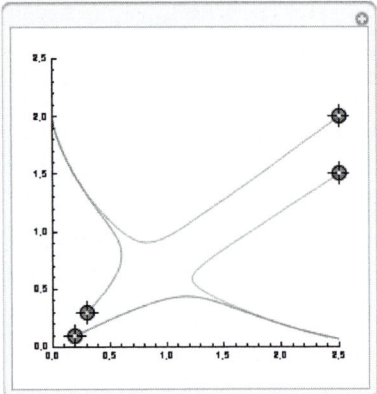

The examples of interactive interpolation, fitting, and differential equation plotting begin to give an impression of the power of **Manipulate**. In a few lines of easy code we can create impressive interactive applications.

■ **Geometric Constraints on Points**

In the next examples, we create points that are constrained on a curve. The points can be moved with locators. In the first example, we can freely move a locator, but it in turn moves an arrow whose head is on a circle:

```
Manipulate[Graphics[{Circle[], Arrowheads[0.1], Arrow[{{0, 0}, Normalize[pt]}]},
  PlotRange → 1.2, ImageSize → 100], {{pt, {1, 0}}, Locator}]
```

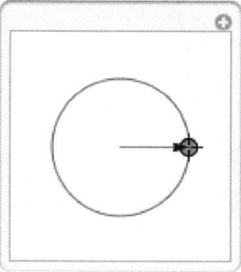

Here, the **pt** is a locator. The position of the arrowhead is obtained from the locator position by normalizing the latter (i.e., by dividing with the norm). Thus, the arrowhead always has norm 1, and this means that the point is on a circle.

Now, we create a locator that moves a point on a curve. With the **Appearance** option, we have asked not to show the locator:

```
Manipulate[Plot[Sin[x], {x, 0, 10},
  Epilog → {PointSize[Large], Point[{First[pt], Sin[First[pt]]}]}], ImageSize → 200],
  {{pt, {2, Sin[2]}}, Locator, Appearance → None}]
```

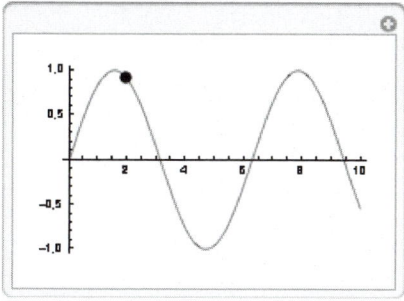

10.1.4 Other Controls

■ SetterBar and RadioButtonBar

> **Manipulate[expr, {u, {u$_1$, u$_2$, … }}]**   Create an interface enabling the interactive choice of the value of the parameter **u** from the values u$_1$, u$_2$, … with a setter bar or popup menu and showing the corresponding value of **expr**

For this type of parameter specification, we automatically get either a setter bar or a popup menu. Indeed, if we have at most five values for **u**, we get a setter bar (a row of tabs); for at least six values we get a popup menu.

A setter bar or popup menu can also be asked for by specifying the control type to be **SetterBar** (or **Setter**) or **PopupMenu**. The type **RadioButtonBar** (or **RadioButton**) produces a set of radio buttons.

```
Manipulate[expr, {u, {u₁, u₂, … }, SetterBar}]
Manipulate[expr, {u, {u₁, u₂, … }, PopupMenu}]
Manipulate[expr, {u, {u₁, u₂, … }, RadioButtonBar}]
```

Setter bars and popup menus are useful in situations in which we are interested in an irregular set of values of the parameter that cannot be represented with a regular iteration specification. We can then define the list of values in which we are interested. In the following example, we can ask for some quantiles of the standard normal distribution:

```
Manipulate[Quantile[NormalDistribution[0, 1], q], {q, {0.9, 0.95, 0.99, 0.999}}]
```

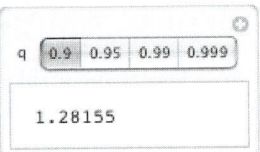

We got a setter bar or a set of tabs. A **TabView**, considered in Section 11.1.1, p. 360, produces a very similar view:

```
TabView[# → Quantile[NormalDistribution[0, 1], #] & /@ {0.9, 0.95, 0.99, 0.999}]
```

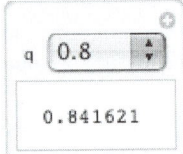

We can also use radio buttons:

```
Manipulate[Quantile[NormalDistribution[0, 1], q],
  {q, {0.9, 0.95, 0.99, 0.999}, RadioButtonBar}]
```

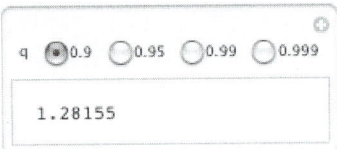

■ **PopupMenu**

We continue the preceding examples. In the next example, we have more than five items so that now we get a popup menu:

```
Manipulate[Quantile[NormalDistribution[0, 1], q],
  {q, {0.8, 0.85, 0.9, 0.95, 0.99, 0.999}}]
```

A **MenuView**, considered in Section 11.1.1, p. 358, produces a very similar view:

```
MenuView[
  # → Quantile[NormalDistribution[0, 1], #] & /@ {0.8, 0.85, 0.9, 0.95, 0.99, 0.999}]
```

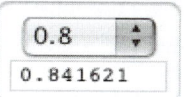

Now we ask for the plots of the basic trigonometric functions:

```
Manipulate[Plot[f[x], {x, -π, π}, PlotRange → {-5.1, 5.1}],
  {{f, Sin, ""}, {Sin, Cos, Tan, Cot, Sec, Csc}}]
```

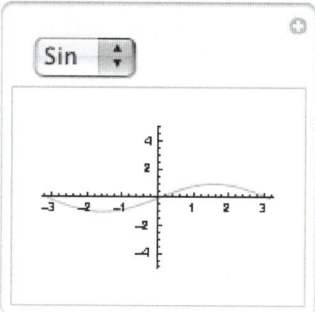

■ **Checkbox and Toggler**

A special case of the control specification $\{u, \{u_1, u_2, \dots \}\}$ presented for setter bars is the one in which the possible values are **True** or **False**:

Manipulate[expr, {u, {True, False}}]   u takes on the values **True** and **False**

For this kind of parameter specification, **Manipulate** automatically chooses the control to be a checkbox. A checkbox can also be asked for by specifying the control type to be **Checkbox**.

In the next example, we get the plot of the probability density function of various Poisson distributions. In addition, we can ask, with a checkbox, to also plot the cumulative distribution function.

```
Manipulate[dist = PoissonDistribution[λ]; ListPlot[
  If[cdf, Flatten[Table[{{x, PDF[dist, x]}, {x, CDF[dist, x]}}, {x, 0, 15}], 1],
    Table[{x, PDF[dist, x]}, {x, 0, 15}]], PlotRange → {0, If[cdf, 1, 0.51]},
  ImageSize → 200], {{λ, 3}, 0.7, 10}, {{cdf, True, "CDF"}, {True, False}}]
```

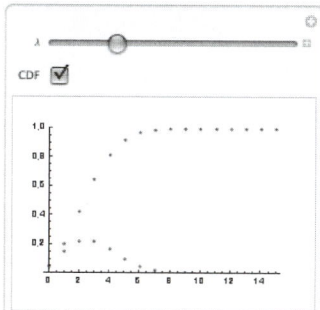

```
Manipulate[expr, {u, {u1, u2, … }, Checkbox}]
Manipulate[expr, {u, {u1, u2, … }, Toggler}]
```

If we have more than two choices, clicking the checkbox goes through the alternatives:

```
Manipulate[Quantile[NormalDistribution[0, 1], q],
 {q, {.9, .95, .99, .999}, Checkbox}]
```

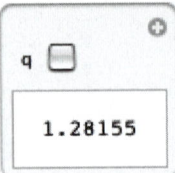

In the next output, we have a toggler. When clicking the number next to *q*, the number goes through the given alternatives:

```
Manipulate[Quantile[NormalDistribution[0, 1], q], {q, {.9, .95, .99, .999}, Toggler}]
```

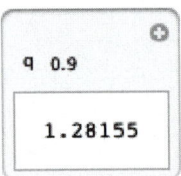

■ **CheckboxBar and TogglerBar**

```
Manipulate[expr, {u, {u1, u2, … }, CheckboxBar}]
Manipulate[expr, {u, {u1, u2, … }, TogglerBar}]
```

With checkbox bars and toggler bars we can choose a *list* of values from a given list (the `Initialization` option is explained in Section 10.2.3, p. 349):

```
Manipulate[Quantile[NormalDistribution[0, 1], q],
 {q, {.9, .95, .99, .999}, CheckboxBar}, Initialization :> (q = {})]
```

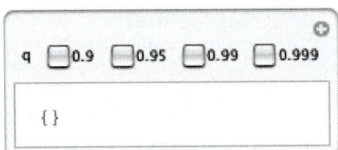

```
Manipulate[Quantile[NormalDistribution[0, 1], q],
 {q, {.9, .95, .99, .999}, TogglerBar}, Initialization :> (q = {})]
```

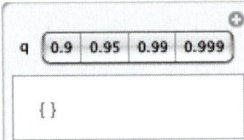

■ **InputField**

> To create an interface enabling the interactive choice of the value of the parameter **u** with an input
> field and showing the corresponding value of **expr**:
>
> **Manipulate[expr, {u}]**   The input field is initially empty
>
> **Manipulate[expr, {u, u$_0$}]**   The input field initially contains **u$_0$**

For these kinds of parameter specifications, **Manipulate** automatically chooses the control to be an input field. An input field can also be asked for by specifying the control type to be **InputField**. Here is an example:

Manipulate[Plot[f, {x, a, b}, ImageSize → 130], {f, Sin[x]}, {a, 0}, {b, 2 π}]

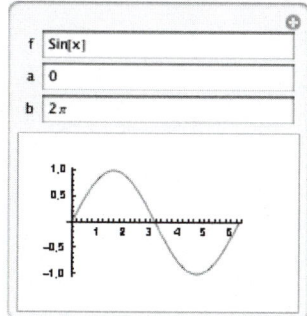

We got three input fields—one each for the function and the starting and ending points. After executing the **Manipulate** command, the input fields contain the values of parameters we gave in the command. These values should be chosen to be typical: With the aid of the example, the user of the panel should be able to produce similar outputs. Indeed, we can now write new functions to be plotted and new starting and ending points; press ⌷TAB⌷ to go from one field to the next. Once the new inputs are ready, press the ⌷RET⌷ key to get the new plot.

As another example, we create an interface where we can ask for binomial probabilities:

```
Manipulate[
  Column[{Grid[{{"P(X = k):", PDF[BinomialDistribution[n, p], k]}, {"P(X ≤ k):",
      CDF[BinomialDistribution[n, p], k]}}, Spacings → 2],
    ListPlot[Table[{x, PDF[BinomialDistribution[n, p], x]}, {x, 0, n}],
      Filling → Axis, ImageSize → 150]}, Center, 2],
  {n, 10}, {p, 0.5}, {k, 3}] // TraditionalForm
```

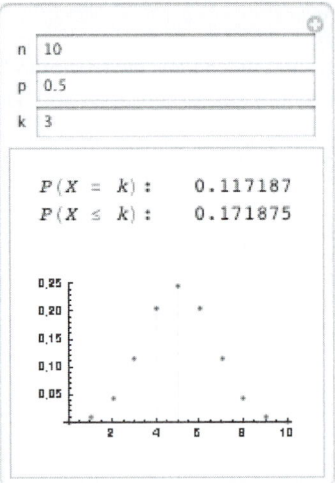

Once you have set the parameters *n*, *p*, and *k*, press the ⎉ key. In Section 11.2.3, p. 384, we create a similar panel by using **Dynamic**.

■ ColorSlider and ColorSetter

> **Manipulate[expr, {u, col}]**   Create an interface enabling, with a color slider, the interactive choice of a color as the value of the parameter **u** and showing the corresponding value of **expr**; the initial value of the color is **col**

For this kind of parameter specification, **Manipulate** automatically chooses the control to be a color slider. A color slider can also be asked for by specifying the control type to be **ColorSlider**. The color can be selected by clicking on the slider. We can also ask for a **ColorSetter**. When we click on a color setter, a separate window appears where we can choose the color. Here we have both a color slider and a color setter:

```
{Manipulate[Graphics[{col, Disk[]}, ImageSize → 30], {col, Red}],
 Manipulate[Graphics[{col, Disk[]}, ImageSize → 30], {col, Red, ColorSetter}]}
```

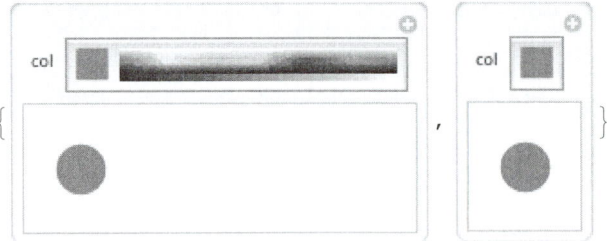

Thus far, we have manipulated plots produced by commands such as `Plot` or `ListPlot`. Interesting manipulations can also be done with `Graphics`. Here is a simple example:

```
Manipulate[Graphics[{col1, PointSize[Large], Point[p], col2, Circle[p, r]},
  Frame → True, FrameTicks → None, PlotRange → 2.05, ImageSize → 130],
 {{p, {0, 0}, "Center point"}, {-1, -1}, {1, 1}},
 {{r, 0.8, "Radius"}, 0, 1},
 {{col1, Blue, "Point color"}, Blue},
 {{col2, Red, "Circle color"}, Red}, Alignment → Center, ControlPlacement → Left]
```

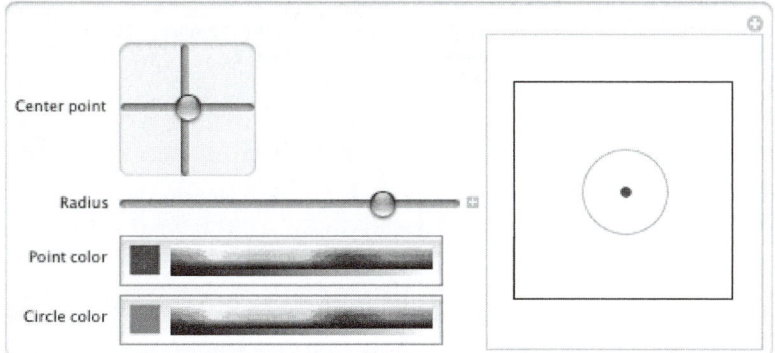

Note that in manipulating graphics, it is almost always useful to give a fixed `PlotRange`. Previously, we gave the value 2.05 for this option; it means that the plot range is (−2.05, 2.05) in both the *x* and the *y* direction. Without a fixed plot range, the plot range varies according to the plotted figure and we cannot get a good idea of the relative sizes and positions of the manipulated plots.

As another example of using colors, next we create an interface that enables us to study how RGB colors change when we adjust the amount of red, green, and blue:

```
Manipulate[Graphics[{RGBColor[red, green, blue], Disk[]}, ImageSize → 60],
 {{red, 0.75, Graphics[{Red, Rectangle[]}, ImageSize → 20]},
  0, 1, Appearance → "Labeled"},
 {{green, 0.5, Graphics[{Green, Rectangle[]}, ImageSize → 20]},
  0, 1, Appearance → "Labeled"},
 {{blue, 0.25, Graphics[{Blue, Rectangle[]}, ImageSize → 20]},
  0, 1, Appearance → "Labeled"}, ControlPlacement → Left]
```

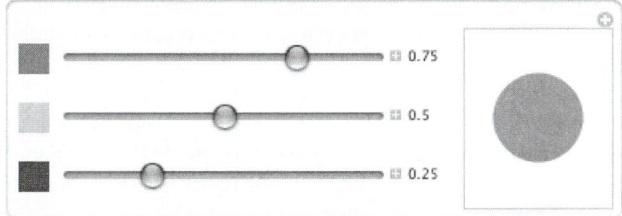

10.2 Advanced Manipulation

10.2.1 More about Controls

■ **Slowing Down the Speed of a Slider**

Consider again the plots of the PDF of the beta distribution:

```
Manipulate[Plot[PDF[BetaDistribution[α, β], x], {x, 0, 1}, PlotRange → {0, 4.1},
  ImageSize → 200], {{α, 2.5}, 0.01, 3, Appearance → "Labeled"},
  {{β, 1.0}, 0.01, 3, Appearance → "Labeled"}]
```

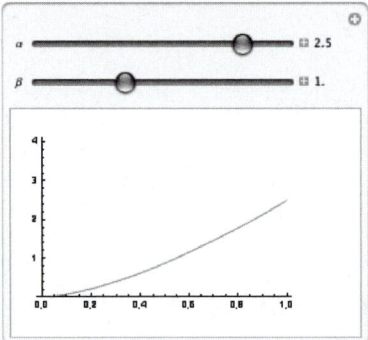

If you let the slider of α be in the default position and slightly move the slider of β, the plot changes so rapidly that we lose the smooth deformation of the curve; we get a series of separate plots that differ clearly from each other.

Fortunately, we can fine-tune the speed of the slider. Hold down the [ALT] key on a Windows machine or the [OPTION] key on a Macintosh machine, and then move the slider. Now the action of the slider is slowed down by a factor of 20 relative to the movements of the mouse. With this method, we can see in our example more clearly what happens near the value $\beta = 1$. While pressing the special keys, we can also move the mouse pointer outside the slider; then the value of the parameter keeps changing.

The speed of the slider can still be slowed down. In addition to the [ALT] or [OPTION] key, if we also hold down the [CTRL] key, the speed is slowed down by an additional factor of 20. By holding down [ALT], [CTRL], and [SHIFT] keys, the speed is still slowed down by a factor of 20. These techniques apply, besides for sliders, for 2D sliders and locators.

■ **Enhancing the Controls**

We can fine-tune the controls by defining their placement and style and by adding dividers and text among the controls. Often the best place for the controls is at the top of the panel. However, we can also ask to put the controls at the left, bottom, or right by using the **ControlPlacement** option.

ControlPlacement Placement of controls; examples of values: **Automatic, Left, Right, Bottom, Top**

A single value such as **Left** defines that all controls are at the left. The placement of the controls can also be defined differently for the variables. Indeed, we can either define the value of **ControlPlacement** to be a list of placements or define the placement for each parameter separately. We do the latter in the next example:

```
Manipulate[ListPlot[Table[{x, PDF[HypergeometricDistribution[n, W, T], x]},
   {x, Max[0, W + n - T], Min[n, W]}], PlotRange → {-0.04, 1.04},
  PlotStyle → PointSize[Medium], ImageSize → 200], {{T, 20}, 1, 20, 1},
 {{W, 12}, 0, T, 1}, {{n, 14}, 1, T, 1, ControlPlacement → Bottom}]
```

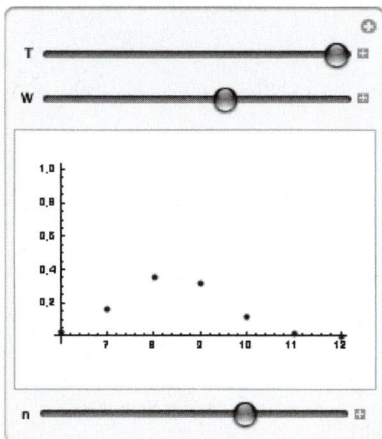

The controls can be grouped by using **Delimiter** among the parameter specifications. A delimiter is shown by a line between the controls:

```
Manipulate[ListPlot[Table[{x, PDF[HypergeometricDistribution[n, W, T], x]},
   {x, Max[0, W + n - T], Min[n, W]}], PlotRange → {-0.04, 1.04},
  PlotStyle → PointSize[Medium], ImageSize → 200], {{T, 20}, 1, 20, 1},
 {{W, 12}, 0, T, 1}, Delimiter, {{n, 14}, 1, T, 1}]
```

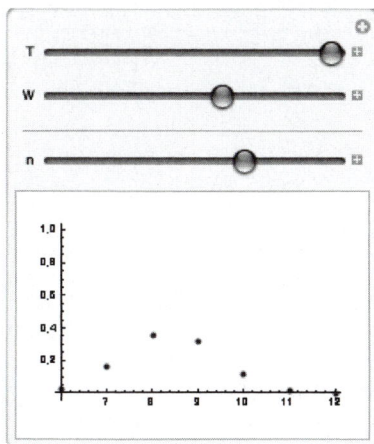

We can also add text among the controls. The texts can be pure strings or styled with **Style**:

```
Manipulate[Plot[PDF[ChiSquareDistribution[n], x], {x, 0, 20}, PlotRange → {0, 0.51},
    ImageSize → 200], Style["Degrees of Freedom", Bold, 10], {{n, 4}, 1, 15, 1}]
```

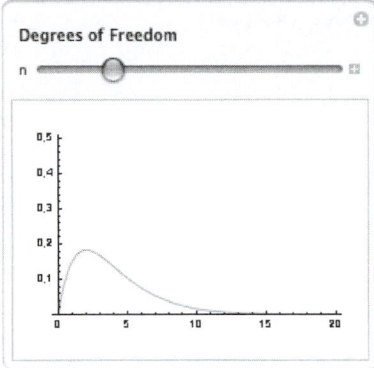

Using **Item**, we can define different alignments:

```
Manipulate[Plot[PDF[ChiSquareDistribution[n], x],
    {x, 0, 20}, PlotRange → {0, 0.51}, ImageSize → 200],
    Item[Style["Degrees of Freedom", Bold], Alignment → Center], {{n, 4}, 1, 15, 1}]
```

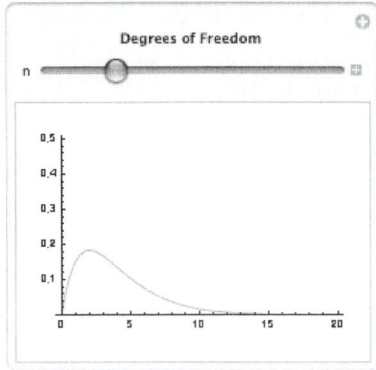

■ Interdependent Controls

In the next example, we again plot the PDF of a hypergeometric distribution. The distribution arises when we draw, without replacement, n balls from an urn containing a total of T balls, of which W are white. A random variable with this distribution gives the number of white balls in the sample.

```
Manipulate[ListPlot[Table[{x, PDF[HypergeometricDistribution[n, W, T], x]},
   {x, Max[0, W + n - T], Min[n, W]}], PlotRange → {-0.04, 1.04},
  PlotStyle → PointSize[Large], ImageSize → 200],
 {{T, 20}, 1, 20, 1, Appearance → "Labeled"},
 {{W, 12}, 0, T, 1, Appearance → "Labeled"},
 {{n, 14}, 1, T, 1, Appearance → "Labeled"}]
```

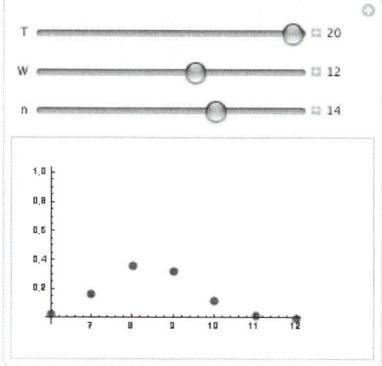

Here, first let T, the total number of balls, be at its maximum. Then try adjusting the value of n, the sample size. Then, leave n somewhere in the middle and try adjusting the value of W, the number of white balls. Then, leave W somewhere in the middle and try adjusting the value of T.

Try again giving T increasingly smaller values. You observe that the sliders of W and n also change, in the opposite direction! Click on the ⊞ buttons next to the right of the sliders of W and n. Now you can see the actual values of these parameters. As you now move the slider of T, you see that the *value* of W or n does not change. Note that the upper limits of W and n are T. Thus, the smaller is T, the smaller are the upper limits of W and n so that fixed values of W or n are increasingly nearer the upper limit. This causes the movement of the sliders of W and n.

When you move the slider of T to the left, at some point one of the sliders of W and n reaches its right end; let it be the slider of W. If you continue moving the slider of T to the left, a red background appears for the slider of W. Also, the figure disappears and we only get the code of the plot with a pink background and we get an error message above the plot. This means that something is wrong with this parameter. Indeed, in our example, the cause is that we cannot have more white balls in the urn than we have total balls. The list of values to be plotted became empty.

If you still continue moving the slider of T to the left, at some point another slider also gets a red background.

We can add the conditions that if T becomes smaller than W, then W should be replaced by T, and if T becomes smaller than n, then n should be replaced by T:

```
Manipulate[If[T < W, W = T]; If[n > T, n = T];
  ListPlot[Table[{x, PDF[HypergeometricDistribution[n, W, T], x]},
    {x, Max[0, W + n - T], Min[n, W]}], PlotRange → {-0.04, 1.04},
  PlotStyle → PointSize[Large], ImageSize → 200],
  {{T, 20}, 1, 20, 1, Appearance → "Labeled"},
  {{W, 12}, 0, T, 1, Appearance → "Labeled"},
  {{n, 14}, 1, T, 1, Appearance → "Labeled"}]
```

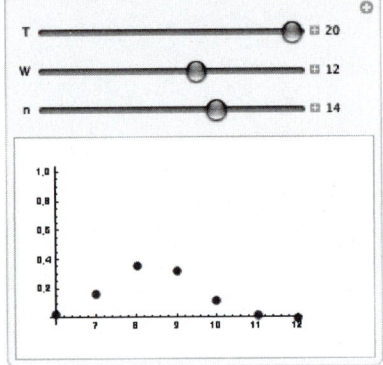

Now the pink background only flashes and the values of *W* and *n* are automatically made smaller if we make *T* small enough.

■ Using Gamepads and Joysticks

Usually, the controls of a panel created by **Manipulate** are used with the mouse. However, there are other input devices such as gamepads and joysticks. Devices that enable us to use the controls are called *controllers*. Indeed, gamepads and joysticks can also be used to interact with the outputs of **Manipulate**. These devices have the advantage that they have several buttons and joysticks so that we can use several controls at the same time. To use a gamepad or joystick, simply plug it in and select (highlight) with the mouse the cell bracket containing the **Manipulate** output you want to control. *Mathematica* automatically detects a gamepad or joystick, and **Manipulate** automatically links as many parameters as possible with the available joysticks and buttons. We do not go into further detail; see the document tutorial/IntroductionToManipulate in the Documentation Center.

10.2.2 Handling Slow Manipulations

■ Manipulating 3D Graphics

We can also manipulate 3D graphics:

```
Manipulate[ParametricPlot3D[2 {Sin[u] Cos[v], Sin[u] Sin[v], Cos[u]},
   {u, π, a}, {v, -π, π}, ImageSize → 150, PlotRange → 2], {{a, 1}, 0, π - 0.01}]
```

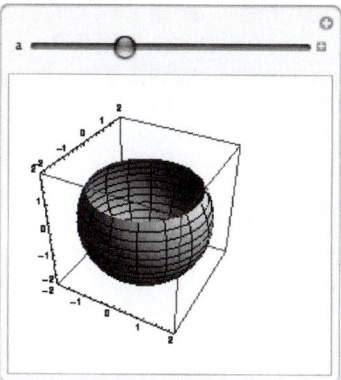

When you try the slider of this panel, you will see that when you drag the slider, the quality of the plot is not as high as usual. As soon as you release the mouse button, a high-quality plot is rendered. This behavior is designed to speed up the manipulation. Indeed, a 3D plot requires a lot of computation; if all the plots during manipulation were of high quality, the manipulation would often be too slow.

Manipulated 3D plots can also be rotated with the mouse.

■ Showing Degraded Output during Manipulation

When we previously considered the manipulation of 3D plots, we noted that the quality of the plot is automatically downgraded when moving a slider to make the manipulation speedier. Similarly, we can do this in any manipulation: With the **ControlActive** command, we can ask for simpler and faster computations while a slider is moved and produce a quality output when the mouse is released.

ControlActive[act, norm] (⌘6) Evaluates to **act** if a control is actively being used, and to **norm** otherwise

---

In plotting, a useful way to speed up the manipulation is to use fewer plot points when moving the mouse and use more points to produce the plot when the mouse is released:

```
Manipulate[DensityPlot[Sin[Sqrt[n] x y], {x, 0, π}, {y, 0, π},
   ImageSize → 150, PlotPoints → ControlActive[10, 50]], {n, 1, 5}]
```

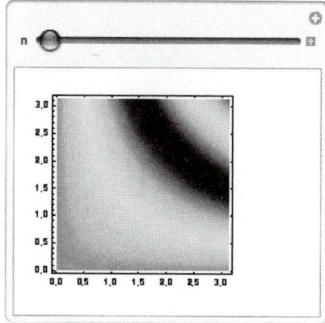

Here, we used 10 as the value of `PlotPoints` when the slider is moved and the value 50 to produce the final plot. In fact, this kind of functionality is the default in density plots, although the final plot is produced with a smaller amount of points. The default functionality corresponds with the following use of the `PerformanceGoal` option:

```
Manipulate[DensityPlot[Sin[Sqrt[n] x y], {x, 0, π}, {y, 0, π}, ImageSize → 150,
    PerformanceGoal → ControlActive["Speed", "Quality"]], {n, 1, 5}]
```

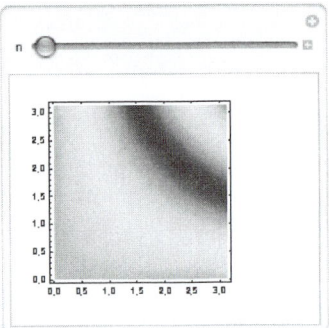

The `ControlActive` command can also be used to control the value of the `MaxRecursion` option. This option defines the maximum number of recursive subdivisions allowed in improving the quality of a plot. An application of `ControlActive` to `MaxRecursion` can be seen in a manipulation in Section 26.4.3, p. 876.

■ Showing Only the Final Output of Manipulation

Another solution to slow manipulations is to arrange so that during movement of the slider the output is not updated. Only when we release the mouse is the output updated.

> `ContinuousAction` Whether to update the panel continuously when controls are changed; possible
>   values: `Automatic`, `True`, `False` (updating only when mouse is released)

Usually, the output is continuously updated. If we ask not to update continuously, then we can rapidly move the slider. Here is an example:

```
Manipulate[ParametricPlot3D[{2 Sin[u] Cos[v], Sin[u] Sin[v], Cos[u]},
    {u, π, a}, {v, -π, π}, ImageSize → 100, PlotRange -> 2],
    {a, 0, π - 0.01}, ContinuousAction → False]
```

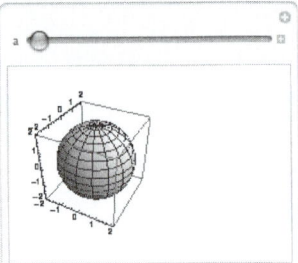

Usually, if an update takes more than 5 seconds, the calculation is aborted. With the previous option, there is no 5-second limit.

■ Example: Two-Dimensional Normal Distribution

The following panel shows the contour plot of the probability density function of the 2D normal distribution with means μ_1 and μ_2, standard deviations σ_1 and σ_2, and correlation ρ. The panel also shows the corresponding marginal distributions. To speed up the computations, we have asked for only five contours. If the manipulation seems to be too slow, you can add the option **ContinuousAction →**
False.

```
<< MultivariateStatistics`
Manipulate[
 Grid[{{Plot[PDF[NormalDistribution[μ2, σ2], y], {y, -6, 6}, PlotRange → All,
     ImageSize → 150], ContourPlot[PDF[
       MultinormalDistribution[{μ1, μ2}, {{σ1^2, ρ σ1 σ2}, {ρ σ1 σ2, σ2^2}}]], {x, y}],
      {x, -6, 6}, {y, -6, 6}, Contours → 5, PlotRange → All, ImageSize → 150]},
    {Null, Plot[PDF[NormalDistribution[μ1, σ1], x], {x, -6, 6},
      PlotRange → All, ImageSize → 150]}}],
 {{μ1, 0}, -2, 2}, {{μ2, 0}, -2, 2}, {{σ1, 1.5}, 0.5, 2},
 {{σ2, 1.5}, 0.5, 2},
 {{ρ, 0.5}, -0.9, 0.9, Appearance → "Labeled"},
 SaveDefinitions → True]
```

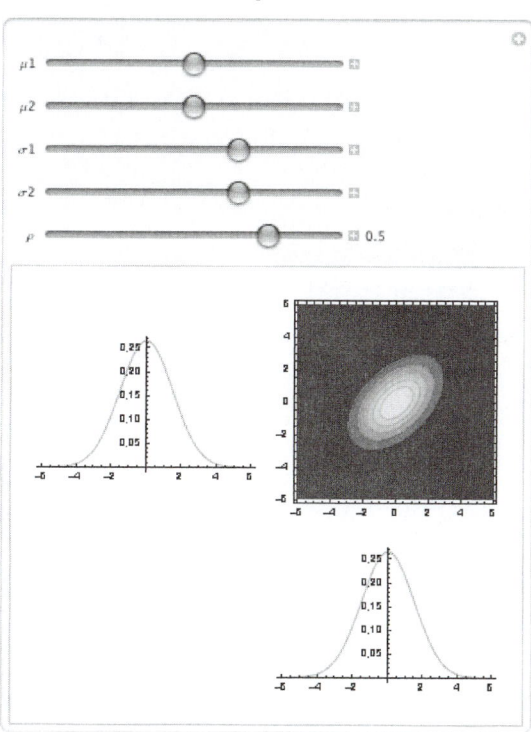

We can also put the marginal densities in the control area. This requires the plotting commands for the marginals to be enclosed by **Dynamic** so that the plots are dynamically updated; **Dynamic** is considered in Section 11.2, p. 369.

```
<< MultivariateStatistics`
Manipulate[ContourPlot[
  PDF[MultinormalDistribution[{μ1, μ2}, {{σ1^2, ρ σ1 σ2}, {ρ σ1 σ2, σ2^2}}]], {x, y}],
  {x, -6, 6}, {y, -6, 6}, Contours → 5, PlotRange → All, ImageSize → 210],
  {{μ1, 0}, -2, 2}, {{σ1, 1.5}, 0.5, 2},
  Dynamic[Plot[PDF[NormalDistribution[μ1, σ1], x],
    {x, -6, 6}, PlotRange → All, ImageSize → 100]],
  {{μ2, 0}, -2, 2}, {{σ2, 1.5}, 0.5, 2}, Dynamic[Plot[
    PDF[NormalDistribution[μ2, σ2], y], {y, -6, 6}, PlotRange → All, ImageSize → 100]],
  {{ρ, 0.5}, -0.9, 0.9}, SaveDefinitions → True]
```

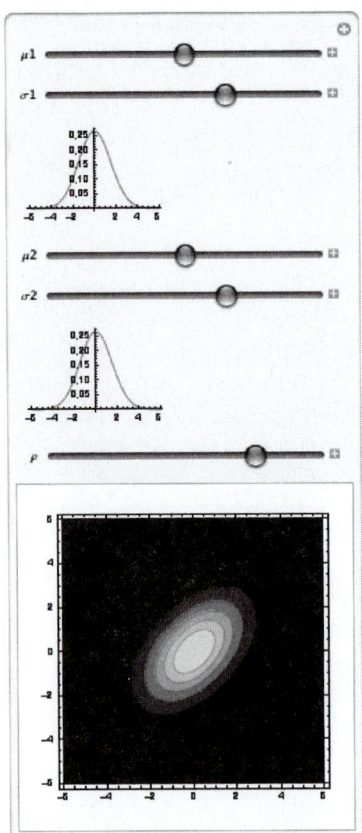

■ Updating the Output Only Partly

Each time we manipulate an expression, the entire first argument of **Manipulate** is evaluated anew to reflect the new values of the parameters. This is suitable in most cases. However, there are situations in which we would like only some parts of the first expression to be evaluated anew. One situation is when we specifically want only some aspects of the output to be updated, to be able to see the effect of only some parameters. Another is when it would be advantageous not to evaluate the entire first argument to reduce computing time.

In the following example, we generate a sample of 10,000 random numbers from the standard normal distribution and show a histogram. We use the **SaveDefinitions** option (see Section 10.2.3, p. 348).

```
<< Histograms`
Manipulate[DynamicModule[{r},
  r = RandomReal[NormalDistribution[0, 1], 10 000];
  Histogram[r, HistogramCategories → n, ApproximateIntervals → False,
    ImageSize → 200, HistogramRange → {-5, 5}, Ticks → {Range[-4, 4], Automatic}]],
 {{n, 20}, 1, 50, 1}, SaveDefinitions → True]
```

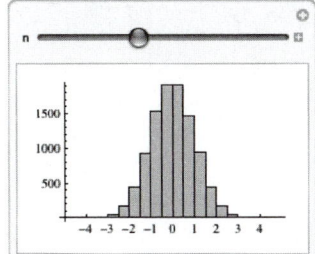

The adjustable parameter **n** gives the number of histogram categories. The panel works but has a drawback: Each time we adjust **n**, the first argument of **Manipulate** is evaluated anew, causing a new sample to be drawn from the normal distribution. We would instead like to see the same sample displayed with various values of **n**.

To solve the problem, calculate the sample outside of **Manipulate**:

```
With[{s = RandomReal[NormalDistribution[0, 1], 10^5]},
  Manipulate[Histogram[s, HistogramCategories → n,
    ApproximateIntervals → False, ImageSize → 200, HistogramRange → {-5, 5},
    Ticks → {Range[-4, 4], Automatic}], {{n, 20}, 1, 50, 1}]]
```

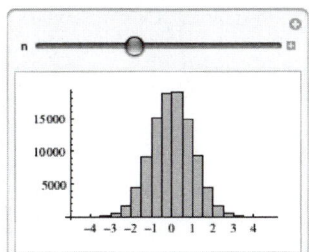

When we now adjust **n**, the sample will remain the same. To investigate a different sample, execute the command again.

Another solution is to enclose the command that calculates the histogram with **Dynamic**:

```
Manipulate[DynamicModule[{t},
    t = RandomReal[NormalDistribution[0, 1], 10^5];
    Dynamic[Histogram[t, HistogramCategories → n,
        ApproximateIntervals → False, ImageSize → 200, HistogramRange → {-5, 5},
        Ticks → {Range[-4, 4], Automatic}]]], {{n, 20}, 1, 50, 1}]
```

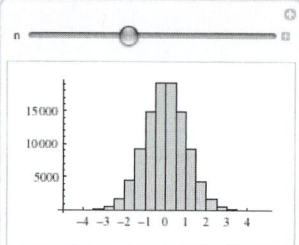

Again, when we now adjust **n**, the sample will remain the same. To investigate a different sample, execute the command anew. The use of **Dynamic** has the effect that when we change the value of **n**, only the value of the expression inside **Dynamic** will be updated. The essential point here is that **n** only occurs inside the expression enclosed by **Dynamic**. However, note that generally it is recommended that **Dynamic** not be used inside **Manipulate**.

10.2.3 Bookmarks and More

■ **Saving Definitions**

Suppose we have defined the probability function of a binomial distribution and then tabulated it for various values of *n*. Lastly, we have saved and closed this notebook and opened it again. This is what we then get:

```
f[k_, n_] := PDF[BinomialDistribution[n, 1 / 6.], k]
```

```
Manipulate[TableForm[Table[{k, f[k, n]}, {k, 0, n}],
    TableSpacing → {1, 3}, TableHeadings → {None, {"k", "P(X=k)"}}], {n, 1, 10, 1}]
```

```
n  ●────────────────────────────────

   k      P(X=k)
   0      f[0, 1]
   1      f[1, 1]
```

As we see, the table in the panel does not know about the *f* function. We have to execute the definition of *f* to see the values of the probabilities.

To avoid this problem, we have the **SaveDefinitions** option.

SaveDefinitions   Whether to save with the output all definitions associated with the first argument of **Manipulate**; possible values: **False**, **True**

With this option, we can save all the definitions used within the manipulation so that when we open the notebook, the panel works without executing the definitions. Here is an example:

```
g[k_, n_] := PDF[BinomialDistribution[n, 1 / 6.], k]
```

```
Manipulate[TableForm[Table[{k, g[k, n]}, {k, 0, n}], TableSpacing → {1, 3},
  TableHeadings → {None, {"k", "P(X=k)"}}], {n, 1, 10, 1}, SaveDefinitions → True]
```

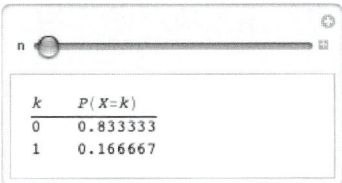

Now we see that the table in the panel knows the *g* function. We do not need to execute the definition of *g* to see the values of the probabilities.

■ Initialization

Consider using data in manipulations. To get an example, we tabulate some binomial probabilities and save them into files:

```
t = Table[Table[{k, g[k, n]}, {k, 0, n}], {n, 1, 10}];
Do[Export["bin" <> ToString[n] <> ".dat", t[[n]]], {n, 1, 10}]
```

Suppose we now would like to create a panel showing the binomial probabilities. We can first import the files:

```
s = Table[Import["bin" <> ToString[n] <> ".dat"], {n, 1, 10}];
```

To also show the probabilities in future sessions, we can use **SaveDefinitions** as above:

```
Manipulate[TableForm[s[[n]], TableSpacing → {1, 3},
  TableHeadings → {None, {"k", "P(X=k)"}}], {n, 1, 10, 1}, SaveDefinitions → True]
```

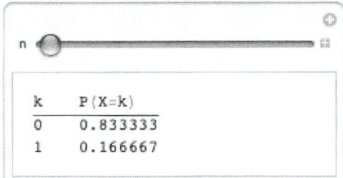

The probabilities are now saved in the current notebook. However, if the data sets are large, saving the definitions could create a large notebook.

An alternative way is the use of the **Initialization** option to import the data sets.

Initialization An expression to be evaluated before the main body of **Manipulate** is executed or when the output of **Manipulate** is first displayed in a particular session; default value: **None**

The value of this option can consist of several commands. In the next example, we use the option to import the probabilities:

```
Manipulate[TableForm[r[[n]], TableSpacing → {1, 3},
  TableHeadings → {None, {"k", "P(X=k)"}}], {n, 1, 10, 1},
  Initialization :→ (r = Table[Import["bin" <> ToString[n] <> ".dat"], {n, 1, 10}])]
```

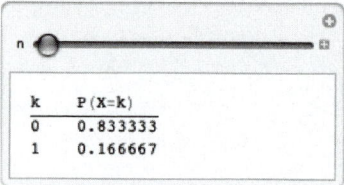

Now the probabilities are not saved within the notebook. Instead, they are imported when the previous table is first displayed in the current session.

■ Snapshots

If we click on the utility button , we get the following utility menu:

If we have found, by manipulating the parameters, a particularly interesting value of a parameter or a particularly interesting combination of values of parameters, we can produce a snapshot of the situation by letting the controllers be at the interesting values and then choosing **Paste Snapshot** from the menu. For example, a special case of the beta distribution is the arcsine distribution. This distribution is obtained with the parameters $\alpha = 1/2$ and $\beta = 1/2$:

```
Manipulate[Plot[PDF[BetaDistribution[α, β], x], {x, 0, 1},
  PlotRange → {0, 4.1}, ImageSize → 200], {{α, 0.5}, 0.01, 3}, {{β, 0.5}, 0.01, 3}]
```

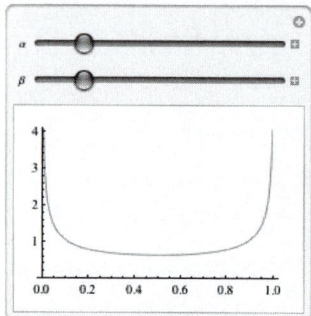

By choosing **Paste Snapshot**, we get, in a separate cell below the panel, the following input:

```
DynamicModule[{α = 0.5`, β = 0.5`}, Plot[PDF[BetaDistribution[α, β], x],
  {x, 0, 1}, PlotRange → {0, 4.1`}, ImageSize → 200]]
```

The snapshot is in the form of a dynamic module (discussed in Section 11.2.1, p. 371). By executing this command, we get a plot of the special case:

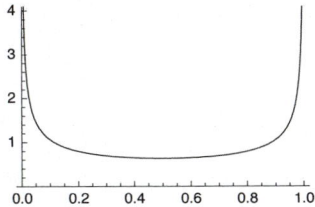

■ **Bookmarks**

With the utility menu ✛ we can also create a list of interesting values. They are stored into the memory as so-called *bookmarks*. When you have found an interesting combination of values of parameters, choose **Add To Bookmarks...** from the menu. At the top of the panel, there is a field in which you can write a name for the bookmark; then click *add* (click *close* if you do not want to create a bookmark).

In the following example, we have created four bookmarks:

```
Manipulate[Plot[PDF[BetaDistribution[α, β], x], {x, 0, 1},
   PlotRange → {0, 4.1}, ImageSize → 200], {{α, 0.5}, 0.01, 3}, {{β, 0.5}, 0.01, 3}]
```

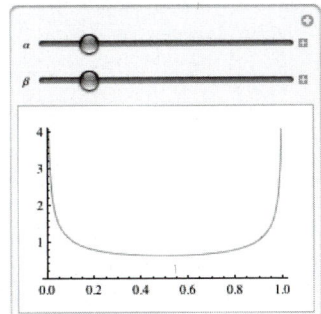

The bookmarks can be seen from the same menu:

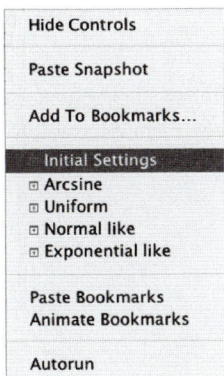

Here, we initially only have one bookmark, **Initial Settings**; choosing this bookmark gives us the plot with the initial values used for the parameters. Similarly, we can choose other bookmarks to show interesting situations corresponding to special values of the parameters.

■ **Pasting and Animating Bookmarks**

With **Paste Bookmarks** from the utility menu 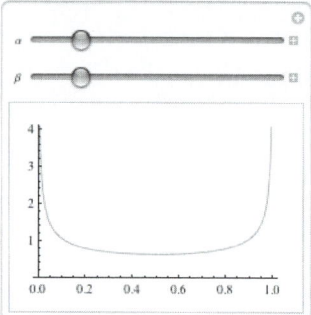 we get a list of rules for the bookmarks. In our example, we get the following list:

```
{"Arcsine" :→ {α = 0.5`, β = 0.5`}, "Uniform" :→ {α = 1.`, β = 1.`},
  "Normal like" :→ {α = 3.`, β = 3.`}, "Exponential like" :→ {α = 1.`, β = 3.`}}
```

This list is of an appropriate form to be used as the value of the **Bookmarks** option. In this way, we can get the bookmarks to a new panel. Now, they are also easy to edit:

```
Manipulate[Plot[PDF[BetaDistribution[α, β], x], {x, 0, 1},
  PlotRange → {0, 4.1}, ImageSize → 200], {{α, 0.5}, 0.01, 3}, {{β, 0.5}, 0.01, 3},
  Bookmarks → {"Arcsine" :→ {α = 0.5`, β = 0.5`}, "Uniform" :→ {α = 1.`, β = 1.`},
    "Normal like" :→ {α = 3.`, β = 3.`}, "Exponential like" :→ {α = 1.`, β = 3.`}}]
```

The utility menu also has the **Animate Bookmarks** command. Now the animation goes through the bookmarks. However, the animation does not simply show each bookmark as such but, rather, shows a smooth animation where (by default) quadratic interpolation is used to get curves between the bookmarks (the order of the interpolation can be set with the **InterpolationOrder** option). Try the animation with the previous panel.

10.2.4 Options of Manipulate

Manipulate has many options. However, the command works so well that we seldom need options. To help find suitable options, we have grouped them into categories and marked noticeable options with an asterisk *. Some of the options were considered previously.

■ **Controls**

> * `ControlType`   The type of control to produce in the output; possible values: `Automatic`, `Animator`,
>   `Checkbox`, `CheckboxBar`, `ColorSetter`, `ColorSlider`, `InputField`, `Locator`, `Manipulator` (slider;
>   animator under the ⊞ icon), `PopupMenu`, `RadioButton`, `RadioButtonBar`, `Setter`, `SetterBar`
>   (tabs), `Slider`, `Slider2D`, `Toggler`, `TogglerBar`, `Trigger` (like animator), `VerticalSlider`, `None`
> * `ControlPlacement`   Placement of controls in the panel; examples of values: `Automatic`, `Left`,
>   `Right`, `Bottom`, `Top`
> `AutoAction`   Whether to allow changing the controls by moving the mouse over them (without
>   pressing the mouse button); possible values: `False`, `True`

In the next example, we can change a control simply by moving the mouse over the control area, without pressing the mouse button:

```
Manipulate[Cos[x], {x, 0, 1}, AutoAction → True]
```

■ **Initialization**

> * `Initialization`   An expression to be evaluated before the main body of `Manipulate` is executed
>   or when the output of `Manipulate` is first displayed in a particular session; default value: `None`
> `SynchronousInitialization`   Whether to perform initialization synchronously; possible values:
>   `True` (notebook editing impossible during initialization), `False` (notebook editing possible during
>   initialization)
> `Deinitialization`   An expression to be evaluated if the output from `Manipulate` is deleted; default
>   value: `None`
> * `SaveDefinitions`   Whether to save with the output all definitions associated with the first
>   argument of `Manipulate`; possible values: `False`, `True`

■ **Updating**

> * `ContinuousAction`   Whether to update the panel continuously when controls are changed;
>   possible values: `Automatic`, `True`, `False` (updating only when mouse is released)
> `SynchronousUpdating`   Whether to update synchronously; possible values: `Automatic` (update
>   synchronously when controls are used, asynchronously otherwise), `True` (notebook editing
>   impossible during updating), `False` (notebook editing possible during updating)
> `TrackedSymbols`   Symbols whose changes trigger updates in the output; examples of values: `Full`
>   (only symbols that appear explicitly in the first argument of `Manipulate` are tracked), `All` (output
>   is updated whenever any symbol encountered in its evaluation is changed), `{x, y}`
> `PreserveImageOptions`   Whether to preserve image size and other options when updating graphics;
>   possible values: `True`, `False`
> `ShrinkingDelay`   How long to delay before shrinking if the displayed object becomes smaller;
>   default value: `0`

■ **Bookmarks**

> `Bookmarks` Bookmarks of the panel; examples of values: `{}`, `{"Arcsine" :→ {α = 0.5, β = 0.5,`
>   `"Uniform" :→ {α = 1., β = 1.}}`
> `InterpolationOrder` The order of interpolation used to get expressions between bookmarks;
>   examples of values: `Automatic, 3`

■ **Styling**

> `BaseStyle` Base style specifications for the panel; default value: `{}`
> `DefaultBaseStyle` Default base style of the panel; default value: `"Manipulate"`
> `LabelStyle` Style specifications for the labels in the control area and for the labels of the panel;
>   examples of values: `{}`, `{14, Red, Bold, Italic}`
> `DefaultLabelStyle` Default label style of the panel; default value: `"ManipulateLabel"`
> `Alignment` How to align the output in the content area of the panel; examples of values: `Automatic,`
>   `Left, Right, Bottom, Top, Center`
> `BaselinePosition` Alignment of the panel relative to surrounding text; examples of values:
>   `Automatic, Bottom, Center, Top`
> `Paneled` Whether to put the displayed output in a panel; possible values: `True, False`

The default is that the output of `Manipulate` is put into a panel. We can ask not to use a panel:

`Manipulate[Cos[x], {x, 0, 1}, Paneled → False]`

x ⊙━━━━━━━━━━

1

■ **Frames and Margins**

> `Frame` Whether to draw a frame around the controls; possible values: `False, True`
> `FrameMargins` Margins inside the content area; examples of values: `Automatic, Large, 10`
> `FrameLabel` Labels for each side of the panel; examples of values: `None, {"bottom", "left",`
>   `"top", "right"}`
> `RotateLabel` Whether to rotate *y* labels on the panel; possible values: `False, True`
> `ImageMargins` Margins outside the panel; examples of values: `0, Large, 10`

We show two outputs that enable us to see the effect of these options. Here is a basic output:

`Manipulate[Cos[x], {x, 0, 1}]`

Now we show an enhanced panel:

```
Manipulate[Cos[x], {x, 0, 1}, Frame → True,
  FrameMargins → 20, FrameLabel → {"bottom", "left", "top", "right"},
  RotateLabel → True, ImageMargins → 20] // Framed
```

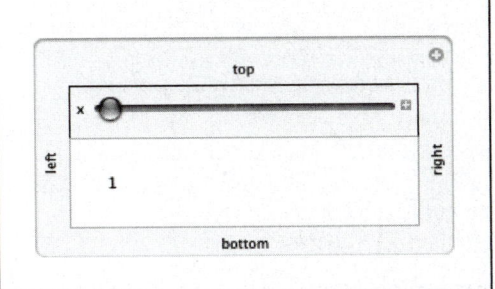

Thus, with **Frame** we can get a frame around the controls (the frame in a way continues the frame around the content area). **FrameMargins** changes the white space around the object in the content area. With **FrameLabel** we can add labels at each side of the panel. With **RotateLabel** we can ask to rotate the labels on the vertical sides of the panel. **ImageMargins** adjusts the margins outside the panel (we have used **Framed** to make these margins more concrete).

■ **Controllers**

ControllerLinking When to activate links to external controllers; examples of values: **Automatic**, **Full**, **All**, **True**, **False**

ControllerMethod How external controllers should operate; examples of values: **Automatic**, **"Absolute"**

ControllerPath What external controllers to try to use; examples of values: **Automatic**, **"Gamepad"**, **"Joystick"**, **"Multi-Axis Controller"**, **"Detachable"**, **"BuiltIn"**

■ **Miscellaneous**

AppearanceElements Buttons to include at the top right corner of the panel; examples of values: **Automatic** (means the utility menu), **All**; possible elements: **"HideControlsButton"**, **"SnapshotButton"**, **"ResetButton"**, **"UpdateButton"**

AutorunSequencing How autorun (from the menu) of the panel should use the controls; examples of values: **Automatic** (run one parameter at a time), **All** (run all parameters simultaneously), **{3, 1, 2}** (run the parameters in the given order), **{{3, 7}, {1, 4}, {2, 9}}** (run the parameters in the given order, each a given amount of seconds; the default is 5 seconds)

Deployed Whether to restrict interactivity to the controls (then the output cannot be manipulated in other ways); possible values: **False**, **True**

Evaluator The kernel to use for evaluations; default value: **Automatic**

LocalizeVariables Whether to localize the parameters; possible values: **True**, **False** (parameters are treated as global)

By default, we only have the utility button at the top right corner of the panel. We can ask for more elements:

```
Manipulate[Cos[x], {x, 0, 1}, AppearanceElements → All]
```

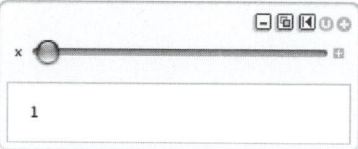

Now we have the following buttons: *Hide Controls* (the controls disappear; the controls again appear by clicking on the panel), *Paste Snapshot* (takes a snapshot of the current state), *Reset* (sets the panel into its initial state), *Update* (updates the content area), and the default utility menu button.

11

Dynamics

Introduction

> *Bertrand Russell was once standing on the platform at Oxford railway station having just missed his train to London. Suddenly, an express train to London made a unscheduled stop there, so Russell hopped on board. "I'm sorry, sir," said a porter. "You'll have to get off, because this train doesn't stop here." "That's all right," said Russell, "because in that case I'm not on it."*

In the preceding chapter, we considered **Manipulate**, the main command for creating interactive interfaces. Now we explore other commands to create dynamic interfaces.

Viewers such as **MenuView**, **TabView**, and **OpenerView** provide useful ways to represent data. As we have seen, animations can be done with **Manipulate**, but they can also be done using the special command **Animate**.

Most dynamic interfaces can be created with **Manipulate**, but **Dynamic** may be valuable in more special or more advanced cases or for more detailed formatting of the output.

The commands presented in this chapter are all new in *Mathematica* 6.

11.1 Views and Animations

11.1.1 Views

■ **Introduction**

Tables are good representations for various data. Relatively small tables can be easily shown both on the screen and printed. However, on the screen we have many other possibilities to show data when using *Mathematica* 6.

Indeed, if the amount of information is large, it may be advantageous to allow the user to interactively ask for just the information he or she needs. This can be done with *dynamic views* such as menu view, tab view, or opener view.

Before explaining the dynamic views, we show, as a comparison, how to use **Grid** (see Section 15.2, p. 470) and **Panel** (see Section 3.3.1, p. 72) to display tables. As an example, we show the probability density function of some continuous distributions:

```
dist =
  {ExponentialDistribution[λ], NormalDistribution[μ, σ], GammaDistribution[α, λ]};
t = {#, PDF[#, x]} & /@ dist
```

$$\left\{\left\{\text{ExponentialDistribution}[\lambda],\ e^{-x\lambda}\,\lambda\right\},\right.$$

$$\left\{\text{NormalDistribution}[\mu,\ \sigma],\ \frac{e^{-\frac{(x-\mu)^2}{2\sigma^2}}}{\sqrt{2\pi}\ \sigma}\right\},\ \left\{\text{GammaDistribution}[\alpha,\ \lambda],\ \frac{e^{-\frac{x}{\lambda}}x^{-1+\alpha}\,\lambda^{-\alpha}}{\text{Gamma}[\alpha]}\right\}\right\}$$

```
Labeled[t // Grid // Framed, "Some discrete distributions", Top] // TraditionalForm
```

Some discrete distributions

ExponentialDistribution[λ]	$e^{-x\lambda}\,\lambda$
NormalDistribution[μ, σ]	$\dfrac{e^{-\frac{(x-\mu)^2}{2\sigma^2}}}{\sqrt{2\pi}\ \sigma}$
GammaDistribution[α, λ]	$\dfrac{e^{-\frac{x}{\lambda}}x^{\alpha-1}\,\lambda^{-\alpha}}{\Gamma(\alpha)}$

```
Panel[t // Grid, "Some discrete distributions"] // TraditionalForm
```

Some discrete distributions

ExponentialDistribution[λ]	$e^{-x\lambda}\,\lambda$
NormalDistribution[μ, σ]	$\dfrac{e^{-\frac{(x-\mu)^2}{2\sigma^2}}}{\sqrt{2\pi}\ \sigma}$
GammaDistribution[α, λ]	$\dfrac{e^{-\frac{x}{\lambda}}x^{\alpha-1}\,\lambda^{-\alpha}}{\Gamma(\alpha)}$

■ **MenuView**

MenuView[{lbl1 → expr1, lbl2 → expr2, … }] (✿6) Create a popup menu where selecting the
  menu item with label **lbli** displays **expri**
MenuView[{lbl1 → expr1, lbl2 → expr2, … }, i] Show initially the **i**th item
MenuView[{expr1, expr2, … }] The labels are consecutive integers

---

A menu view is able to show large amounts of data in a very condensed way. A label is chosen from a popup menu and the panel then shows the corresponding expression:

```
dist =
  {ExponentialDistribution[λ], NormalDistribution[μ, σ], GammaDistribution[α, λ]};
```

```
MenuView[# → PDF[#, x] & /@ dist, 2]
```

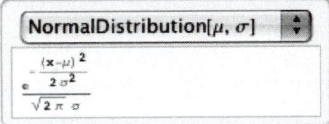

With **Manipulate** we can get a similar view:

```
Manipulate[PDF[d, x], {d, dist, PopupMenu}]
```

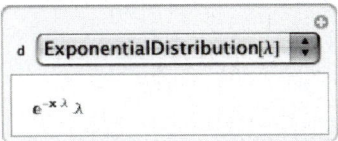

The menus can be hierarchical:

```
MenuView[# → MenuView[{PDF → PDF[#, x], CDF → CDF[#, x],
        Mean → Mean[#], Variance → Variance[#]}, 2] & /@ dist, 2]
```

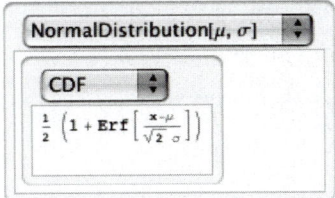

A menu view is useful for showing a set of plots:

```
MenuView[
  Table[n → Plot[ChebyshevT[n, x], {x, -1, 1}, PlotRange → 1.05, ImageSize → 200],
    {n, 0, 10}], 11]
```

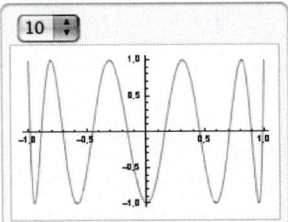

Next, we show the complete name of the function plotted:

```
MenuView[Table[Row[{"ChebyshevT[", n, ", x]"}] → Plot[ChebyshevT[n, x],
    {x, -1, 1}, PlotRange → 1.05, ImageSize → 200], {n, 0, 10}], 11]
```

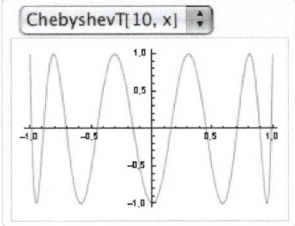

We consider **MenuView** again in Section 11.2.5, p. 388.

■ **TabView**

> **TabView[{lbl1 → expr1, lbl2 → expr2, … }]** (⌘6) Create a bar of tabs where selecting the tab with
>   label **lbli** displays **expri**
> **TabView[{lbl1 → expr1, lbl2 → expr2, … }, i]** Show initially the **i**th item
> **TabView[{expr1, expr2, … }]** The labels are consecutive integers

In a tab view, all the labels are shown on the top of the panel as a row of tabs. Clicking a tab shows
the corresponding expression:

```
Labeled[
  TabView[# → Quantile[NormalDistribution[0, 1], #] & /@ {0.9, 0.95, 0.99, 0.999}, 3],
  "Quantiles", Top]
```

Quantiles

| 0.9 | 0.95 | 0.99 | 0.999 |

2.32635

With **Manipulate** we can get a similar view:

```
Manipulate[Quantile[NormalDistribution[0, 1], q],
  {q, {0.9, 0.95, 0.99, 0.999}, SetterBar}]
```

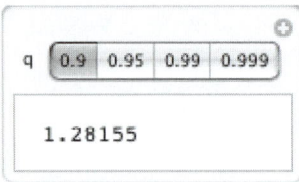

Tab views can be hierarchical:

```
TabView[# → TabView[{PDF → PDF[#, x],
      CDF → CDF[#, x], Mean → Mean[#], Variance → Variance[#]}, 1] & /@
  {NormalDistribution[μ, σ], GammaDistribution[α, λ]}, 2]
```

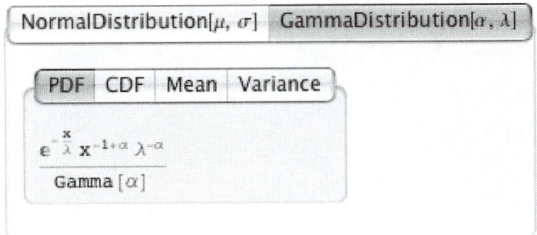

A tab view is also useful in presenting a set of plots:

```
TabView[
  Table[n → Plot[ChebyshevT[n, x], {x, -1, 1}, PlotRange → 1.05, ImageSize → 200],
   {n, 0, 10}], 7]
```

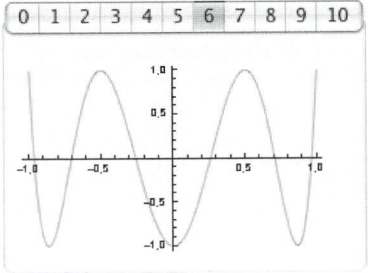

■ **OpenerView**

`OpenerView[{lbl, expr}]` (✿6) Create an opener where label `lbl` is displayed if the opener is closed and both `lbl` and `expr` are displayed if the opener is open	
`OpenerView[{lbl, expr}, open]` Show the item open if `open` is `True`, closed if `open` is `False`	

An opener view shows a triangle before each label. Clicking a triangle opens the corresponding expression. We have to separately create an opener for each label–expression pair. In the following example, we use **Map** to get a list of openers and **Column** to show the list as a column:

```
dist =
  {ExponentialDistribution[λ], NormalDistribution[μ, σ], GammaDistribution[α, λ]};
OpenerView[{#, PDF[#, x]}] & /@ dist // Column // TraditionalForm
```

▶ ExponentialDistribution[λ]

▼ NormalDistribution[μ, σ]

$$\frac{e^{-\frac{(x-\mu)^2}{2\sigma^2}}}{\sqrt{2\pi}\ \sigma}$$

▶ GammaDistribution[α, λ]

We can create hierarchical displays:

```
OpenerView[{#,
       {OpenerView[{PDF, PDF[#, x]}],
          OpenerView[{CDF, CDF[#, x]}],
          OpenerView[{Mean, Mean[#]}],
          OpenerView[{Variance, Variance[#]}]} // Column}] & /@ dist // Column //
  TraditionalForm
```

▶ ExponentialDistribution[λ]

▼ NormalDistribution[μ, σ]

    ▶ PDF

    ▼ CDF

$$\frac{1}{2}\left(\mathrm{erf}\left(\frac{x-\mu}{\sqrt{2}\ \sigma}\right)+1\right)$$

    ▶ Mean

    ▶ Variance

▶ GammaDistribution[α, λ]

An advantage of an opener view over menu view or tab view is that more than one item can be open so that comparing two or more items is easier.

■ PopupView, SlideView, and FlipView

> To create a popup/slide/flip view that shows the given expressions:
> ```
> PopupView[{expr1, expr2, … }] (⌘6)
> SlideView[{expr1, expr2, … }] (⌘6)
> FlipView[{expr1, expr2, … }] (⌘6)
> ```

With a second argument, we can tell which item we would initially like to see shown.

Popup, slide, and flip views are restricted in that they merely show a chosen expression. They do not have the ability to show a label attached to an expression (the expression shown can, however, be a list containing as much information as we want; we do this in the following examples). In a popup view, an expression is chosen from a popup menu:

```
t = {#, PDF[#, x]} & /@ dist
```

$$\left\{\left\{\texttt{ExponentialDistribution}[\lambda],\ e^{-x\lambda}\lambda\right\},\right.$$

$$\left\{\texttt{NormalDistribution}[\mu,\ \sigma],\ \frac{e^{-\frac{(x-\mu)^2}{2\sigma^2}}}{\sqrt{2\pi}\ \sigma}\right\},\ \left\{\texttt{GammaDistribution}[\alpha,\ \lambda],\ \frac{e^{-\frac{x}{\lambda}}x^{-1+\alpha}\lambda^{-\alpha}}{\texttt{Gamma}[\alpha]}\right\}\right\}$$

```
PopupView[t, 2]
```

$$\left\{\texttt{NormalDistribution}[\mu,\ \sigma],\ \frac{e^{-\frac{(x-\mu)^2}{2\sigma^2}}}{\sqrt{2\pi}\ \sigma}\right\}$$

We consider `PopupView` again in Section 11.2.5, p. 389.

With a slide view, we can show the expressions like in a slide show by using the four buttons:

```
SlideView[t, 2]
```

$$\left\{ \text{NormalDistribution}[\mu, \sigma], \frac{e^{-\frac{(x-\mu)^2}{2\sigma^2}}}{\sqrt{2\pi}\ \sigma} \right\}$$

In a flip view, the output moves to the next expression each time we click on the output:

```
FlipView[t, 2]
```

$$\left\{ \text{NormalDistribution}[\mu, \sigma], \frac{e^{-\frac{(x-\mu)^2}{2\sigma^2}}}{\sqrt{2\ \pi}\ \sigma} \right\}$$

■ PopupWindow, Tooltip, Mouseover, StatusArea, and Annotation

```
PopupWindow[lbl, expr] (⌘6)
Tooltip[lbl, expr] (⌘6)
Mouseover[lbl, expr] (⌘6)
StatusArea[lbl, expr] (⌘6)
Annotation[lbl, expr, "Mouse"] (⌘6)
```

The methods presented here produce a list a labels. In a special way, we can see the corresponding expression. By clicking a label in the following output, the corresponding expression appears in a separate window:

```
PopupWindow[#, PDF[#, x]] & /@ dist // Column
```

```
ExponentialDistribution[λ]
NormalDistribution[μ, σ]
GammaDistribution[α, λ]
```

Next, when the mouse pointer is over a label, a tooltip appears showing the corresponding expression:

```
Tooltip[#, PDF[#, x]] & /@ dist // Column
```

```
ExponentialDistribution[λ]
NormalDistribution[μ, σ]
GammaDistribution[α, λ]
```

By moving the mouse pointer over a label (without clicking), the corresponding expression replaces the expression:

```
Mouseover[#, PDF[#, x]] & /@ dist // Column
```

```
ExponentialDistribution[λ]
NormalDistribution[μ, σ]
GammaDistribution[α, λ]
```

Now the expression appears at the lower left corner of the window when the mouse pointer is over a label:

```
StatusArea[#, PDF[#, x]] & /@ dist // Column
```

```
ExponentialDistribution[λ]
NormalDistribution[μ, σ]
GammaDistribution[α, λ]
```

Lastly, in the following way we get an expression in place of the Null output of the **Dynamic** command when we move the mouse over the **Annotation** output that contains the labels:

```
Column[
 {Column[Annotation[#, PDF[#, x], "Mouse"] & /@ dist], Dynamic[MouseAnnotation[]]}]
```

ExponentialDistribution[λ]
NormalDistribution[μ, σ]
GammaDistribution[α, λ]
Null

The document tutorial/FormattedOutput in the Documentation Center gives the following application of **Tooltip**. By moving the mouse over a country on the map, we can see the name of the country and its flag.

```
Graphics[{LightBlue, EdgeForm[Gray],
  Dynamic[Tooltip[CountryData[#, {"SchematicPolygon", "Mollweide"}],
    Panel[CountryData[#, "Flag"], #]]] & /@ CountryData[]}, ImageSize → 420]
```

Here, we used **Dynamic** to reduce the size of the output (see more in Section 11.2.5, p. 388).

▪ Options

The dynamic views have several options, but we rarely need them; the default views are very good. For example, here are the options of **MenuView**:

```
Options[MenuView]
```

{Alignment → {Left, Top}, Background → None, BaselinePosition → Automatic,
 BaseStyle → {}, ControlPlacement → {Top, Left}, DefaultBaseStyle → MenuView,
 DefaultLabelStyle → MenuViewLabel, Deployed → False, Enabled → Automatic,
 FrameMargins → Automatic, ImageMargins → Automatic, ImageSize → All, LabelStyle → {}}

With **ControlPlacement** we can ask to put the controls, for example, at the bottom or left:

```
MenuView[(# → PDF[#, x] &) /@
   {ExponentialDistribution[λ], NormalDistribution[μ, σ], GammaDistribution[α, λ]},
   ControlPlacement → {Bottom, Center}]
```

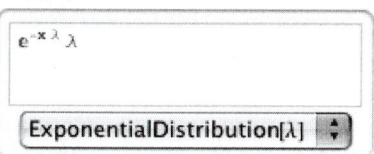

```
TabView[# → Quantile[NormalDistribution[0, 1], #] & /@ {0.9, 0.95, 0.99, 0.999},
   ControlPlacement → Left]
```

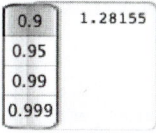

The default value of **ImageSize** is **All** for most views. The default value is **Automatic** for **OpenerView** and **FlipView**. The value **All** means that the output is large enough to contain the largest of the items. In this case, the size of the output remains the same for all items. On the other hand, the value **Automatic** means that the size of the output is set separately for each item so that the size is just large enough to show the current item. In this case, the size of the output changes according to the size of the current item. For example, in the following menu view, the size changes according to the item shown:

```
MenuView[(# → PDF[#, x] &) /@ {ExponentialDistribution[λ],
   NormalDistribution[μ, σ], GammaDistribution[α, λ]}, ImageSize → Automatic]
```

11.1.2 Animations

■ Basics of Animation

Here is a simple **Animate** command and its output panel:

```
Animate[Plot[Sin[x - a], {x, 0, 4 π}, ImageSize → 200], {a, 0, 2 π}]
```

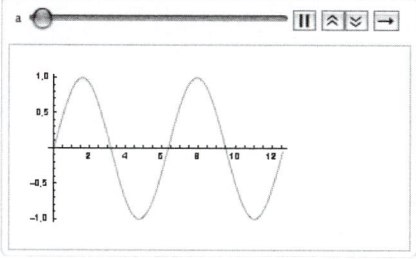

The animation starts immediately after we have executed the command: The value of the parameter **a** is changed and the plot is redrawn in real time. Thus, a sequence of plots is shown rapidly one after the other, giving the impression of a continuously changing curve. We can control the animation with the following icons at the top right of the panel: *Play/Pause*, *Faster*, *Slower*, and *Forward/Backward/Forward and Backward*. We can also manually change the value of the parameter **a** by moving the slider with the mouse; the plot of $\sin(x - a)$ is again redrawn in real time.

In general, **Animate** has two arguments: the *expression to be animated* (the plotting command in the previous example) and the *specifications of the parameters* (**{a, 0, 2π}**). In addition, we can have options. The expression to be animated depends on one or more parameters. The output panel contains a *slider* and *animation controls* for each parameter and a *content area* for the expression to be animated.

Recall from the previous chapter that animations can also be run with **Manipulate**.

■ **Forms of Animation**

To create an interface enabling the animation of **expr** by changing the value of a parameter **u**:

Animate[expr, {u, u$_{min}$, u$_{max}$}] (❀6) u can have any value between u$_{min}$ and u$_{max}$

Animate[expr, {u, u$_{max}$, u$_{max}$, du}] u can have any value between u$_{min}$ and u$_{max}$ in steps of **du**

Animate[expr, {{u, u$_{init}$}, u$_{min}$, u$_{max}$}] The initial value of u is u$_{init}$

Animate[expr, {u, {u$_1$, u$_2$, ... }}] u takes on the values u$_1$, u$_2$, ...

AnimationRunning Whether the animation is running after execution of **Animate** or after opening a notebook containing the animation; possible values: **True, False**

We will use the **AnimationRunning** option to ask that the animation not be run automatically; this saves computer load.

The typical use of **Animate** is to animate a series of plots by changing a continuous parameter (note, however, that **expr** in the previous box need not be a plotting command; it can be any expression). This was the case in the previous example, in which we animated the graph of the function $\sin(x - a)$.

Sometimes we want to show only the animation for a given discrete set of values of the parameter. Next, we show some Chebyshev polynomials:

```
Animate[Plot[ChebyshevT[n, x], {x, -1, 1}, PlotRange → 1.05, ImageSize → 200],
   {n, 0, 10, 1}, AnimationRunning → False]
```

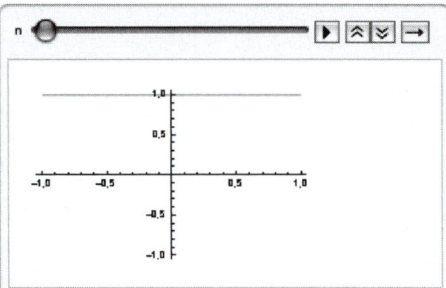

Note that often, as in the previous example, it is useful to define the same **PlotRange** for all of the plots. Then the plots can be more easily compared. Sometimes **ImagePadding** is a useful option for **Plot** to ensure that the same space is reserved for tick labels in all plots so that, again, the positions of the axes do not change during the animation.

With the third form of **Animate** given previously, we can define an initial value of the parameter. The corresponding plot is shown immediately after the animation command is executed. This could be useful if the minimum value of the parameter gives an uninteresting plot.

The maximum value of the parameter can be infinite. In this case, the animation can run forever.

The fourth form of **Animate** provides the possibility to give a list of (irregular) values for the parameter.

We can also animate with respect to several parameters. However, such an animation is probably not very useful. Use **Manipulate** instead; see Chapter 12.

3D plots can be animated in the same way as 2D plots:

```
Animate[Plot3D[Sin[x y - a], {x, 0, π}, {y, 0, π}, ImageSize → 160],
  {a, 0, 2 π}, AnimationRunning → False]
```

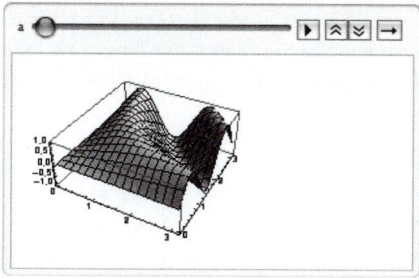

Here is an example in which we animate a plot produced by **Graphics**:

```
Animate[Graphics[
  {Translate[Rotate[{Thick, Green, Line[{{-1, 0}, {1, 0}}], Line[{{0, -1}, {0, 1}}],
      Red, Circle[], Blue, PointSize[Large], Point[{0, 0}]}, -t], {t, 0}],
   Line[{{-1, -1.03}, {3 π + 1, -1.03}}]}], PlotRange → {{-1.1, 3 π + 1}, {-1.3, 1.3}},
  ImageSize → 250], {t, 0, 3 π}, AnimationRunning → False]
```

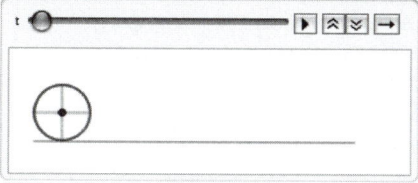

■ **Animating a Given List of Plots**

Thus far, we have considered animations where the sequence of plots is produced during the animation. Now we consider animating an existing list of plots.

ListAnimate[plots] (⌘6)	Animate the given list of plots
ListAnimate[plots, rate]	Animate the given list of plots, displaying **rate** plots per second

Previously, we animated Chebyshev polynomials with **Animate**. The same animation can also be obtained by precomputing the plots and asking for an animation:

```
t = Table[Plot[ChebyshevT[n, x],
    {x, -1, 1}, PlotRange → 1.05, ImageSize → 200], {n, 0, 10, 1}];
```

`ListAnimate[t, AnimationRunning → False]`

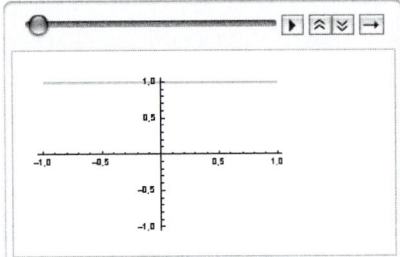

■ **Options of Animate**

`Animate` has many options. Most of them are the same as the options of `Manipulate`. Indeed, `Animate` can be seen as a special case of `Manipulate`. The latter command is considered in Chapter 12. Here are special options of `Animate`; we also give one option (`AppearanceElements`) whose default value is different than the default value of the same option for `Manipulate`.

> `DefaultDuration` The default duration (in seconds) of one run of the animation; default value: `5.`
>
> `AnimationRate` The rate at which the animation should run; examples of values: `Automatic, 0.1`
>
> `RefreshRate` The default number of times per second to refresh; default value: `Automatic`
>
> `AnimationRunning` Whether the animation is running after execution of `Animate` or after opening a notebook containing the animation; possible values: `True, False`
>
> `AnimationDirection` The direction of the animation; possible value: `Forward, Backward, ForwardBackward`
>
> `AnimationRepetitions` How many times to run before stopping; default value: ∞
>
> `DisplayAllSteps` Whether to force all discrete steps to be displayed; possible values: `False, True`
>
> `Exclusions` Specific values to be excluded; default value: `{}`
>
> `AnimatorElements` Animation control elements to include; examples of values: `Automatic, All`; default elements: "`ProgressSlider`", "`PlayPauseButton`", "`FasterSlowerButtons`", "`DirectionButton`"; additional elements: "`StepLeftButton`", "`StepRightButton`", "`ResetButton`", "`PlayButton`", "`ResetPlayButton`"
>
> `AppearanceElements` Utility buttons to include at the top right corner of the panel; examples of values: `None, Automatic` (means the utility menu ⊙), `All`; possible elements: "`HideControlsButton`", "`SnapshotButton`", "`ResetButton`", "`UpdateButton`"

In addition, `Animate` has the following options:

> `Alignment, AppearanceElements, BaselinePosition, BaseStyle, ControlAlignment, ControllerLinking, ControllerMethod, ControllerPath, ControlPlacement, Deinitialization, Deployed, FrameMargins, ImageMargins, Initialization, LabelStyle, PreserveImageOptions, SaveDefinitions, ShrinkingDelay, SynchronousInitialization, SynchronousUpdating, TrackedSymbols`

For these options, see the same options of `Manipulate` in Section 10.2.4, p. 352.

In animating a continuous variable such as in `Animate[expr, {u, u`$_{\min}$`, u`$_{\max}$`}]`, a discrete set of plots are shown by producing plots for which the value of the parameter **u** differs by an amount of **du**. The value of **du** is determined from the value of **RefreshRate**. For example, if the value of this option is 3, then 3 plots are produced in a second, and because the animation runs for 5 seconds (by default), then we get 15 plots in a run so that **du** should be approximately $(u_{\max} - u_{\min})/15$ (the actual value may differ somewhat).

An explicit setting for **AnimationRate** takes precedence over the setting for **DefaultDuration**. For **AppearanceElements**, see the options of **Manipulate** in Section 10.2.4, p. 352.

ListOptions has almost the same options as **Animate**; the default value of **DisplayAllSteps** is now **True**.

11.2 Advanced Dynamics

11.2.1 Dynamic Expressions

■ **Introduction**

Recall that for graphics we have such high-level commands as **Plot** but also the lower-level command **Graphics**. Mostly high-level plotting commands suffice for us, but sometimes we need **Graphics** to create advanced or special plots. Actually, commands such as **Plot** utilize the lower-level command **Graphics**, as can be seen from the following output:

```
Plot[Sin[x], {x, 0, π}] // InputForm // Short
Graphics[{{{}, {}, {<<2>>}}}, {<<6>>}]
```

For dynamic interfaces, the situation is similar. We have the high-level command **Manipulate**, but we also have the lower-level command **Dynamic**. **Manipulate** usually suffices for us, but sometimes we need **Dynamic** to create advanced or special dynamic interfaces. Actually, **Manipulate** utilizes the lower-level command **Dynamic**. This cannot be directly demonstrated, but consider the following example:

```
Manipulate[Sin[x], {x, 0, π}]
```

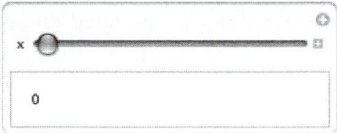

A similar output can be obtained with **Dynamic** by using the **Manipulator** control:

```
Panel@DynamicModule[{x = 0},
  Column[{Manipulator[Dynamic[x], {0, π}], Dynamic[Sin[x]]}]]
```

Manipulate works as follows: It applies **Dynamic** to allow dynamic changes of values, adds controls such as sliders to create an interactive interface, and adds formatting constructs to arrange the components of the panel.

We now begin to study **Dynamic** in more detail.

■ **Dynamic Expressions**

Calculate the value of the factorial function $n!$:

    n = 3;

    n!

    6

Then change the value of n:

    n = 4;

The value of $n!$ did not change. This is the normal way *Mathematica* works: New values of variables do not change the values of old expressions. With **Dynamic** we get a different working.

> **Dynamic[expr]** (⌘6) An object whose value is updated whenever any of the parameters in **expr** get a new value

Define a dynamic expression:

    Dynamic[n!]

    n!

We do not have any unusual features here: The value of this expression reflects the current value of n. However, change the value of n:

    n = 5;

Here, the value of **Dynamic[n!]** changed to 5!. In this way, a dynamic expression always changes its value if any of its variables gets a new value. Note that here we did not execute the command **n = 5** in order not to destroy the initial value of **Dynamic[n!]**. You should perform the computations presented here on your computer to actually see what happens.

■ **Adjusting Parameters**

The heart of dynamics is the ability to interactively change the values of some parameters to see what effect this has on an expression. To change parameters, we have many types of controls, as we saw when we studied **Manipulate**. The basic control is a slider. We consider sliders and other controls in forthcoming sections, but here we present some simple examples.

For convenience, we again present the dynamic expression we considered previously:

    Dynamic[n!]

    n!

Now define a slider:

    Slider[Dynamic[n], {1, 100, 1}]

To dynamically change the value of n, we enclosed it with **Dynamic**. Try adjusting the value of n with the slider. Observe how the value of **Dynamic[n!]** changes accordingly. In fact, any dynamic expression containing **n** would change accordingly.

Note that only dynamic outputs currently visible on the screen are updated. Dynamic outputs outside the current state of the screen are not updated; they are updated when we scroll to them.

Often, it is useful to gather together a slider and an expression to be manipulated:

```
{Slider[Dynamic[n1], {1, 100, 1}], Dynamic[n1]}
```

Again, to show the current value of n_1, we enclosed it with **Dynamic**. A more polished version of this output can be obtained with **Row**:

```
Row[{"n ", Slider[Dynamic[n2], {1, 100, 1}], " ", Dynamic[n2]}]
```

n ●━━━━━━━━━━━━━━━ 1

Create an interface to find the nth prime:

```
Column[{Row[{"n ", Slider[Dynamic[n3], {1, 100, 1}], " ", Dynamic[n3]}],
  Row[{"Prime[n]: ", Dynamic[Prime[n3]]}]}]
```

n ●━━━━━━━━━━━━━━━ 1
Prime[n]: 2

Finally, add a panel:

```
Panel@Column[{Row[{"n ", Slider[Dynamic[n4], {1, 100, 1}], " ", Dynamic[n4]}],
    Row[{"Prime[n]: ", Dynamic[Prime[n4]]}]}]
```

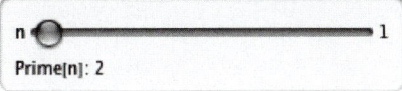

Thus, we have obtained a similar output as is given by **Manipulate**:

```
Manipulate[Prime[n], {n, 1, 100, 1}, FrameLabel → "Prime[n]"]
```

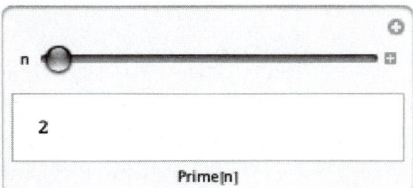

In summary, with **Dynamic** we can obtain similar interfaces as with **Manipulate**, but usually the latter command is easier to apply.

■ Dynamic Modules

Previously, we observed that adjusting n with a slider changes all dynamic expressions containing n. This may not be what we want. Indeed, often we want to restrict the effect of changing parameters into a single dynamic expression. This can be achieved with a dynamic module.

DynamicModule[{x = x0, y = y0, … }, expr] (❀6) An object that maintains the same *local* instance
of the variables **x**, **y**, … in the course of all evaluations of **Dynamic** objects in **expr**; initial values **x0**,
y0, … are used for the variables (initial values need not be given)

Consider now the following example:

```
DynamicModule[{n = 5}, {Slider[Dynamic[n], {1, 100, 1}], Dynamic[n]}]
```

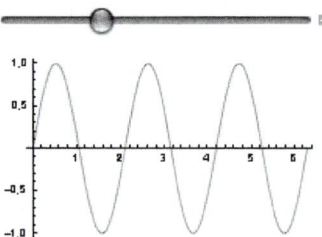

When moving the slider, other dynamic objects having the parameter n do not change because this object has a local version of n.

You may observe that **DynamicModule** and **Module** are formally similar. However, behind the scenes they work differently. **Module** does its work in the kernel, and **DynamicModule** does its work in the front end. The latter command maintains, in the output cell, the current values of the parameters so that the values are saved when the notebook is saved. For example, suppose we have a dynamic module and we have set the slider to a specific position. Then we save the notebook, quit *Mathematica*, start *Mathematica* again, and open the same notebook. The slider still has the same position. If we had used **Module**, the slider would instead be in the default position.

■ **Dynamic Graphics**

Like **DynamicModule**, **Dynamic** is an exceptional command in that it does its work entirely in the front end, not in the kernel. This is important to keep in mind when we design dynamic interfaces.

For example, in the following we have enclosed the plotting command with **Dynamic** to get it updated every time the value of n is changed:

```
DynamicModule[{n = 3}, Column[{Manipulator[Dynamic[n], {0, 10, 1}],
    Dynamic[Plot[Sin[n x], {x, 0, 2 π}, PlotRange → 1.05, ImageSize → 200]]}]]
```

Note that here **Dynamic** is used *inside* **Manipulator**, around **n**. Why did we use **Dynamic** *outside* of **Plot**? Why did we not write **Plot[Sin[Dynamic[n] x]]**…? The latter does not work. This is because the kernel needs a specific value of **n** to plot the function, but **Dynamic[n]** does not evaluate to anything in the kernel—it remains **Dynamic[n]**. Thus, the plotting does not succeed.

We can also use **Dynamic** to adjust the value of some options. Here, we adjust the font size:

```
DynamicModule[{s = 5},
  Column[{Manipulator[Dynamic[s], {5, 16, 1}], Plot[Sin[3 x], {x, 0, 2 π},
    LabelStyle → {FontSize → Dynamic[s]}, ImageSize → 200]}]]
```

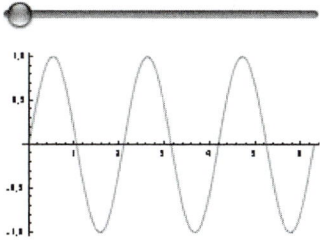

Indeed, an option such as **FontSize** takes its effect in the front end so that we can write **FontSize →
Dynamic[s]** to be able to change **s** (without recomputing the curve).

However, there are options whose values cannot be adjusted in this way. For example, the kernel
needs   the   value   of   **PlotStyle**   so   that   we   cannot   write   **PlotStyle →
AbsoluteThickness[Dynamic[th]]**. In this case, we have to enclose the entire plotting command with
Dynamic:

```
DynamicModule[{th = 1},
  Column[{Manipulator[Dynamic[th], {0, 6}], Dynamic[Plot[Sin[3 x],
    {x, 0, 2 π}, PlotStyle → AbsoluteThickness[th], ImageSize → 200]]}]]
```

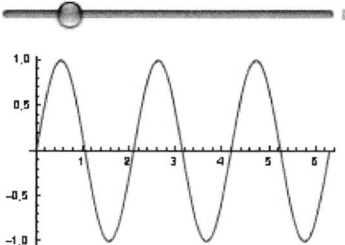

Now, each time we adjust the thickness of the curve, the kernel produces a new plot.

■ Options of Dynamic

Initialization An expression to be evaluated before the main body of **Dynamic** is executed or
    when the output of **Dynamic** is first displayed in a particular session; default value: **None**
Deinitialization An expression to be evaluated if the output from **Dynamic** is deleted; default
    value: **None**
TrackedSymbols Symbols whose changes trigger updates in the output; examples of values: **All**
    (output is updated whenever any symbol encountered in its evaluation is changed), **Full** (only
    symbols that appear explicitly in the first argument of **Dynamic** are tracked), **{x, y}**
UpdateInterval Time interval (in seconds) at which to do updates; examples of values: ∞, 1
ShrinkingDelay How long to delay before shrinking if the displayed object gets smaller; default
    value: **0**.
Editable Whether to allow the textual display of **Dynamic** to be edited; possible values: **False, True**
Evaluator The kernel to use for evaluations; default value: **Automatic**

Options of **Dynamic** are seldom needed; more often, we need the options of the various controls such as sliders, locators, or input fields.

Sometimes the update or refreshing of a dynamic object does not work unless we specifically ask to refresh. For example, the following does not continuously generate random numbers; it only generates one random number:

Dynamic[RandomReal[]]

0.753821

To continuously get new random numbers, define how often the update should occur:

Dynamic[RandomReal[], UpdateInterval → 1]

0.256163

We can also use the **Refresh** command and its options to define when an update should be made.

Dynamic[Refresh[expr, opts]] (⌘6) Update according to the options

Options of **Refresh**:
TrackedSymbols Symbols whose changes trigger an update; examples of values: **Automatic**, **{x, y}**
UpdateInterval Time interval (in seconds) at which to do updates; examples of values: ∞, **1**

■ Common Options of Controls

We will soon study various controls such as **Manipulator**, **Slider**, **Locator**, and **InputField**. Before we do so, we present the options that are shared with most of the controls so that these options need not be repeated for all the controls. In addition to these common options, each control has a few special options; they are mentioned for each control separately.

We have divided the common options into two categories:

Options relating to style:
ImageSize The overall size of the control; examples of values: **Automatic**, **All** (default for
  PopupMenu, **Setter**, and **Toggler**), **Tiny**, **Small**, **Medium**, **Large**, **20**, **{100, 20}** (the option is not
  for **Checkbox** or **Opener**)
ImageMargins Margins around the control; examples of values: **0**, **Tiny**, **Small**, **Medium**, **Large**, **10**
Background The background color of the control; examples of values: **Automatic**, **Red** (the option is
  not for **Manipulator**)
BaselinePosition Alignment of the control relative to surrounding text; examples of values:
  Automatic, **Bottom**, **Center**, **Top** (the option is not for **Locator**)
BaseStyle Base style specifications for the control; examples of values: **{}**, **{Red, Bold}** (the option
  is not for **Manipulator**)
DefaultBaseStyle Default base style of the control; examples of values: **{}**, **"InputField"** (the
  option is not for **Manipulator**)

Options relating to controls:
Enabled Whether adjusting the control is enabled; possible values: **Automatic**, **True**, **False**
AutoAction Whether to allow adjusting the control by moving the mouse over it (without pressing
  the mouse button); possible values: **False**, **True** (the option is not for **InputField**)
ContinuousAction Whether to update continuously when the control is adjusted; possible values:
  True, **False** (updating only when mouse is released; default for **InputField**, **Checkbox**, **Opener**)

Previously, we indicated some controls that do not have some of the options. In addition, the options **ImageSize**, **ImageMargins**, **AutoAction**, and **ContinuousAction** are not for **SetterBar**, **RadioButtonBar**, **CheckboxBar**, or **TogglerBar**.

For example, here are sliders and locators of varying sizes:

Slider[3, {0, 10, 1}, ImageSize → #] & /@ {Tiny, Small, Medium}

Graphics[Locator[{0, 0}, ImageSize → #], ImageSize → 25] & /@ {Tiny, Small, Medium}

$$\{\ \blacklozenge\ ,\ \Diamondblack\ ,\ \bigoplus\ \}$$

Next, we discuss the various controls and how they are used with **Dynamic**. We consider the controls approximately in the same order they were presented for **Manipulate** in the previous chapter.

11.2.2 Sliders and Locators

■ Manipulator

With the **Manipulator** control we can get similar results as with **Manipulate**.

To create a manipulator enabling the interactive choice of the value of the parameter **u**:

Manipulator[Dynamic[u]] (✻6) **u** can have any value between 0 and 1

Manipulator[Dynamic[u], {u$_{min}$, u$_{max}$}] **u** can have any value between u$_{min}$ and u$_{max}$

Manipulator[Dynamic[u], {u$_{min}$, u$_{max}$, du}] **u** can have any value between u$_{min}$ and u$_{max}$ in steps
  of **du**

Special options:

Appearance The appearance of the manipulator; examples of values: **Automatic**, **"Closed"**, **"Open"**,
  "Labeled", **{"Open", "Labeled"}**

AppearanceElements Animation controls to include; examples of values: **Automatic**, **All**; default
  elements: "**ProgressSlider**", **"InputField"**, **"StepLeftButton"**, **"PlayPauseButton"**,
  "StepRightButton", **"FasterSlowerButtons"**, **"DirectionButton"**; additional elements:
  "InlineInputField", **"ResetButton"**, **"PlayButton"**, **"ResetPlayButton"**

AnimationRate The rate at which the animation should run; examples of values: **Automatic**, **0.1**

AnimationDirection The direction of the animation; possible values: **Forward**, **Backward**,
  ForwardBackward

Exclusions Specific values to be excluded; default value: **{}**

With the **Appearance** option we can ask to show the animation control opened and to display the current value of the parameter next to the slider. The current value of the parameter can also be obtained with the appearance element **"InlineInputField"**. Note that the animation controls can be adjusted with **AppearanceElements**; recall that for **Animate**, the animation controls can be adjusted with **AnimatorElements**. The size of the slider can be adjusted with **ImageSize**.

```
DynamicModule[{a = 3},
 Column[{Manipulator[Dynamic[a], {1, 10}, Appearance → {"Open", "Labeled"}],
   Dynamic[Plot[Cos[a x], {x, 0, 2 π}, ImageSize → 200]]}]]
```

■ Animator and Trigger

To create an animator enabling the interactive choice of the value of the parameter **u**:

Animator[Dynamic[u]] (⌘6) u can have any value between 0 and 1

Animator[Dynamic[u], {u$_{min}$, u$_{max}$}] u can have any value between u$_{min}$ and u$_{max}$

Animator[Dynamic[u], {u$_{min}$, u$_{max}$, du}]   u can have any value between u$_{min}$ and u$_{max}$ in steps of **du**

Special options:

Appearance The appearance of the manipulator; examples of values: **Automatic**, **Tiny**, **Small**,
  Medium, **Large**

AppearanceElements Animation controls to include; examples of values: **Automatic**, **All**; default
  elements: **"ProgressSlider"**, **"StepLeftButton"**, **"PlayPauseButton"**, **"StepRightButton"**,
  "FasterSlowerButtons", **"DirectionButton"**; additional elements: **"ResetButton"**,
  "PlayButton", **"ResetPlayButton"**

Exclusions Specific values to be excluded; default value: **{}**

In addition to the options mentioned in the box, **Animator** has the following options familiar from
Animate: **AnimationDirection**, **AnimationRate**, **AnimationRepetitions**, **AnimationRunning**,
DefaultDuration, **DisplayAllSteps**, and **RefreshRate**. Note again that the animation controls can be
adjusted with **AppearanceElements**; recall that for **Animate**, the animation controls can be adjusted
with **AnimatorElements**.

With **Animator** we get a similar result as with **Animate**:

```
DynamicModule[{a = 3},
 Column[{Animator[Dynamic[a], {1, 10}, AnimationRunning → False],
   Dynamic[Plot[Cos[a x], {x, 0, 2 π}, ImageSize → 200]]}]]
```

With **Trigger** we get a simplified animation:

```
DynamicModule[{a = 3}, Column[{Trigger[Dynamic[a], {1, 10}],
    Dynamic[Plot[Cos[a x], {x, 0, 2 π}, ImageSize → 200]]}]]
```

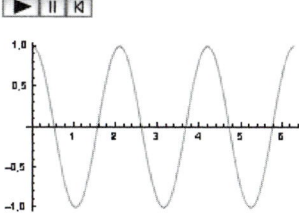

■ Slider and VerticalSlider

> To create a slider enabling the interactive choice of the value of the parameter **u**:
>
> **Slider[Dynamic[u]]** (⌘6)   **u** can have any value between 0 and 1
>
> **Slider[Dynamic[u], {u_{min}, u_{max}}]**   **u** can have any value between u_{min} and u_{max}
>
> **Slider[Dynamic[u], {u_{min}, u_{max}, du}]**   **u** can have any value between u_{min} and u_{max} in steps of **du**
>
> *Special options*:
>
> **Appearance**   The appearance of the slider; examples of values: **Automatic**, **"UpArrow"**, **"DownArrow"**
>
> **Exclusions**   Specific values to be excluded; default value: **{}**

Here is an example showing the thumb of a slider as a down arrow:

```
DynamicModule[{n = 3},
  Slider[Dynamic[n], {0, 10, 1}, Appearance → "DownArrow", Background → LightPurple]]
```

Sometimes we would like to create a set of sliders with **Table**. The following does not work:

```
DynamicModule[{data = {3, 3, 3, 3}},
  Column[{Table[Slider[Dynamic[data[[i]]], {0, 10}], {i, 4}], Dynamic[data]}]]
```

{3, 3, 3, 3}

This does not work because **Dynamic** has the attribute **HoldFirst** so that **Dynamic** does not evaluate its first argument. Thus, **data[[i]]** remains as such for each value of **i**. A trick to get the command to work is the use of **With**:

```
DynamicModule[{data = {3, 3, 3, 3}}, Column[
  {Table[With[{i = i}, Slider[Dynamic[data[[i]]], {0, 10}]], {i, 4}], Dynamic[data]}]]
```

{3, 3, 3, 3}

Here is an example of a vertical slider:

```
<< MultivariateStatistics`
DynamicModule[{σ1 = 1, σ2 = 1, x, y},
 Grid[{{VerticalSlider[Dynamic[σ2], {0.5, 2}, ImageSize → Small],
    Dynamic[ContourPlot[PDF[MultinormalDistribution[{0, 0},
       {{σ1^2, 0.5 σ1 σ2}, {0.5 σ1 σ2, σ2^2}}], {x, y}], {x, -6, 6},
      {y, -6, 6}, Contours → 5, PlotRange → All, ImageSize → 150]]},
   {Null, Slider[Dynamic[σ1], {0.5, 2}, ImageSize → Small]}}]]
```

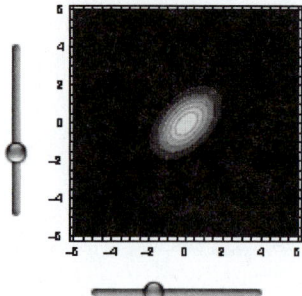

■ Slider2D

To create a 2D slider enabling the interactive choice of the value of the point **pt**:

Slider2D[Dynamic[pt]] (⌘6) x and y coordinates can have any value between 0 and 1

Slider2D[Dynamic[pt], {min, max}] The coordinates are between **min** and **max**

Slider2D[Dynamic[pt], {min, max, d}] The coordinates jump in steps of **d**

Define different ranges in x and y directions:

Slider2D[Dynamic[pt], {{x$_{min}$, y$_{min}$}, {x$_{max}$, y$_{max}$}}]

Slider2D[Dynamic[pt], {{x$_{min}$, y$_{min}$}, {x$_{max}$, y$_{max}$}, {dx, dy}}]

A special option:

Exclusions Specific values to be excluded; default value: **{}**

In the following, we use a 2D slider to move a **Point**:

```
DynamicModule[{pt = {0, 0}}, Row[{Slider2D[Dynamic[pt], {-1, 1}],
  " ", Graphics[{PointSize[Large], Point[Dynamic[pt]]},
   PlotRange → 1.1, Frame → True, FrameTicks → None, ImageSize → 80]}]]
```

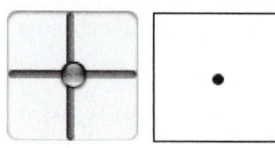

With **Manipulate** we get a similar result:

```
Manipulate[Graphics[{PointSize[Large], Point[pt]},
   PlotRange → 1.1, Frame → True, FrameTicks → None, ImageSize → 80],
  {{pt, {0, 0}}, {-1, -1}, {1, 1}}, ControlPlacement → Left]
```

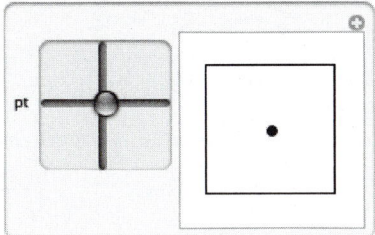

■ Locator

> Locator[Dynamic[pt]] (⌘6)   Create a locator enabling the interactive choice of the point pt
>
> Locator[Dynamic[pt], obj]   Use obj as the locator object; examples of values of obj: Automatic
> (the default "crosshairs" appearance), None (display nothing visible), "*",
>   Graphics[{PointSize[Large], Point[{0, 0}]}]
>
> *A special option*:
>
> LocatorRegion   Where the locator is allowed to go; examples of values: Automatic (within plot
> range), Full (all of the graphics)

Recall that a locator can be moved with the mouse. An example:

```
DynamicModule[{pt = {0, 0}},
  Graphics[{Locator[Dynamic[pt]]}, PlotLabel → Dynamic[NumberForm[pt, {3, 2}]],
   PlotRange → 1, Frame → True, FrameTicks → None, ImageSize → 100]]
```

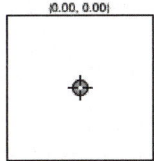

With **Manipulate** we get a similar result (we already presented this example in Section 10.1.3, p. 326):

```
Manipulate[Graphics[{}, PlotLabel → NumberForm[pt, {3, 2}], PlotRange → 1.1,
   Frame → True, FrameTicks → None, ImageSize → 100], {{pt, {0, 0}}, Locator}]
```

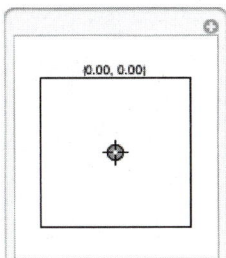

11.2.3 Other Controls

■ **SetterBar and RadioButtonBar**

> `SetterBar[Dynamic[x], list]` (⌘6) Create a setter bar (a set of tabs) enabling the interactive choice of the value of the parameter **x** from `list`
>
> `RadioButtonBar[Dynamic[x], list]` (⌘6) Create a radio button bar enabling the interactive choice of the value of the parameter **x** from `list`
>
> *A special option*:
>
> `Appearance` The appearance of the bar; examples of values: `Automatic`, `"Horizontal"`, `"Vertical"`, `"Row"`

An example:

```
DynamicModule[{q = 0.95},
 Column[{ SetterBar[Dynamic[q], {0.9, 0.95, 0.99, 0.999}],
   Dynamic[Quantile[NormalDistribution[0, 1], q]]}]]
```

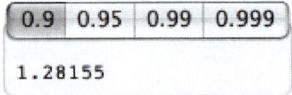

```
1.64485
```

The same kind of interface can be more easily obtained with `TabView`:

```
TabView[# → Quantile[NormalDistribution[0, 1], #] & /@ {0.9, 0.95, 0.99, 0.999}]
```

0.9	0.95	0.99	0.999

```
1.28155
```

An example of a radio button bar:

```
DynamicModule[{q = 0.95},
 Row[{RadioButtonBar[Dynamic[q], {0.9, 0.95, 0.99, 0.999}, Appearance → "Vertical"],
   " Quantile: ", Dynamic[Quantile[NormalDistribution[0, 1], q]]}]]
```

⊙0.9

⦿0.95
       Quantile: 1.64485
⊙0.99

⊙0.999

■ **PopupMenu**

`PopupMenu[Dynamic[x], {val1, val2, … }]` (⌘6) Create a popup menu enabling the interactive choice of the value of the parameter **x** from the given list

`PopupMenu[Dynamic[x], {val1 → lbl1, val2 → lbl2, … }]` Show the values as the given labels in the popup menu

Special options:

`FieldSize` The size of the menu; examples of values: `{{1, 50}, {1, 10}}`, `width`, `{width, height}`, `{{wmin, wmax}, {{hmin, hmax}}`

`FrameMargins` Margins inside the menu; examples of values: `Automatic`, `2`

Here is an example of the first form of **PopupMenu**:

```
DynamicModule[{q},
 Column[{Row[{PopupMenu[Dynamic[q], {0.9, 0.95, 0.99, 0.999}], "-quantile:"}],
   Dynamic[Quantile[NormalDistribution[0, 1], q]]}]]
```

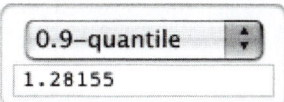

-quantile:

1.28155

With the second form we can get arbitrary labels in the menu:

```
DynamicModule[{r},
 Column[{PopupMenu[Dynamic[r],
   Quantile[NormalDistribution[0, 1], #] → Row[{#, "-quantile"}] & /@
    {0.9, 0.95, 0.99, 0.999}], Dynamic[r]}]]
```

> 0.9–quantile

1.28155

With **MenuView** we get a similar result:

```
MenuView[Row[{#, "-quantile"}] → Quantile[NormalDistribution[0, 1], #] & /@
 {0.9, 0.95, 0.99, 0.999}]
```

> 0.9–quantile

1.28155

In the next example, we use **PopupMenu** to show plots:

```
DynamicModule[{y},
  Column[{PopupMenu[Dynamic[y], Table[Plot[ChebyshevT[n, x], {x, -1, 1}] →
    Row[{"ChebyshevT[", n, ", x]"}], {n, 0, 10}]], Dynamic[y]}]]
```

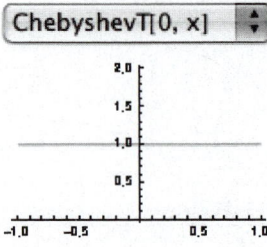

Now the popup menu appears as the plot label:

```
DynamicModule[{n = 5}, Dynamic[Plot[ChebyshevT[n, x], {x, -1, 1}, ImageSize → 200,
  PlotLabel → Row[{"n = ", PopupMenu[Dynamic[n], Range[0, 10]]}]]]]
```

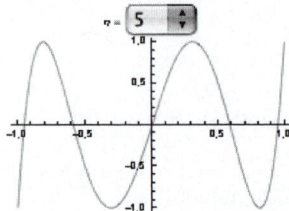

■ Checkbox and Toggler

Checkbox[Dynamic[x]] (⌘6) Create a checkbox enabling the interactive choice of either **False**

(☐ , the default) or **True** (☑) as the value of the parameter **x**

Checkbox[Dynamic[x], list] Create a checkbox enabling the interactive choice of an element from

list as the value of the parameter **x**; the values cycle through the elements of **list** (the first

element is shown as ☐ , the second as ☑ , and next ones as ⊟)

A special option:

Appearance The appearance of the input field; examples of values: **Automatic, Tiny, Small, Medium, Large**

In the following, the variable is either **False** or **True**:

```
DynamicModule[{x}, Row[{Checkbox[Dynamic[x]], Dynamic[x]}]]
```

⊟False

Now the values go through the list {1, 2, 3, 4, 5, 6}:

```
DynamicModule[{q = 0.95},
  Column[{ Row[{Checkbox[Dynamic[q], {0.9, 0.95, 0.99, 0.999}], " ", Dynamic[q]}],
    Dynamic[Quantile[NormalDistribution[0, 1], q]]}]]
```

☑ 0.95

1.64485

Toggler[Dynamic[x]] (⌘6)   Create a toggler enabling the interactive choice of **True** or **False** as the
value of the parameter **x**

Toggler[Dynamic[x], list]   Create a toggler enabling the interactive choice of an element from
list as the value of the parameter **x**

```
DynamicModule[{q = 0.95},
  Row[{Toggler[Dynamic[q], {0.9, 0.95, 0.99, 0.999}],
    " ", Dynamic[Quantile[NormalDistribution[0, 1], q]]}]]
```

0.95   1.64485

Here, by clicking the first number, the second number goes through the various quantiles of the
standard normal distribution.

■ **CheckboxBar and TogglerBar**

CheckboxBar[Dynamic[x], list] (⌘6)   Create a checkbox bar enabling the interactive choice of a
list of values from **list** as the value of the parameter **x**

TogglerBar[Dynamic[x], list] (⌘6)   Create a toggler bar enabling the interactive choice of a list of
values from **list** as the value of the parameter **x**

A special option:

Appearance   The appearance of the setter bar; examples of values: **Automatic**, **"Horizontal"**,
"Vertical", **"Row"**

With checkbox bars and toggler bars we can choose a list of values from a given list:

```
DynamicModule[{q = {}},
  Column[{CheckboxBar[Dynamic[q], {0.9, 0.95, 0.99, 0.999}], Dynamic[q]}]]
```

☐0.9 ☑0.95 ☑0.99 ☐0.999
{0.95, 0.99}

```
DynamicModule[{q = {}},
  Column[{TogglerBar[Dynamic[q], {0.9, 0.95, 0.99, 0.999}], Dynamic[q]}]]
```

| 0.9 | 0.95 | 0.99 | 0.999 |

{0.9, 0.95, 0.999}

■ **InputField**

`InputField[Dynamic[u]]` (✻6)   Create an input field enabling the interactive input of the value of
the parameter **u**
`InputField[Dynamic[u], Number]`   The input can only contain number characters
`InputField[Dynamic[u], String]`   The input is converted into a string
Special options:
`FieldSize`   The size of the input field; examples of values: `{{20, 20}, {1, ∞}}`, `width`, `{width,
height}`, `{{wmin, wmax}, {{hmin, hmax}}`
`Appearance`   The appearance of the input field; examples of values: `Automatic`, `"Framed"`,
`"Frameless"`
`FrameMargins`   Margins inside the frame; examples of values: `Automatic, 3`

Tab can be used to move from one input field to the next. Once all fields are filled in, press the ⌅ key
to get the corresponding result. Note that if the second argument is **Number**, we can only input number
characters 0, …, 9 and a decimal point; we cannot input, for example, fractions such as $1/6$.

We return to an example discussed in Section 10.1.4, p. 335, in which we used **Manipulate** to create a
panel to calculate binomial probabilities. Now we use **Dynamic** and **InputField**:

```
DynamicModule[{n = 10, p = 0.5, k = 3, x, left, right},
  left = Style[Grid[{{"n", InputField[Dynamic[n]]},
      {"p", InputField[Dynamic[p]]},
      {"k", InputField[Dynamic[k]]}, {"", ""}, {"P(X = k)",
       InputField[Dynamic[PDF[BinomialDistribution[n, p], k]], Enabled → False]},
      {"P(X ≤ k)", InputField[Dynamic[CDF[BinomialDistribution[n, p], k]],
        Enabled → False]}}], DefaultOptions → {InputField → {FieldSize → 8}}];
  right = InputField[Dynamic[ListPlot[Table[{x, PDF[BinomialDistribution[n, p], x]},
      {x, 0, n}], Filling → Axis, PlotRange → All,
     ImagePadding → {{25, 15}, {10, 10}}, ImageSize → 180]], Enabled → False];
  Deploy@Panel@TraditionalForm@Row[{left, right}, Spacer[14]]]
```

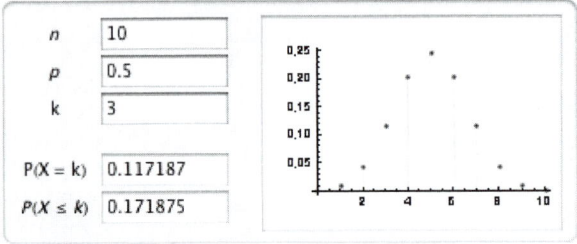

Here, we used **InputField** for all the numbers and also for the plot. Using **Enabled → False**, we
specified that the probabilities or the plot cannot be edited. We have also used **Deploy** to disable the
edition of the panel. In **Style**, we defined with **DefaultOptions** the default value of the **FieldSize**
option of **InputField**.

■ **ColorSlider and ColorSetter**

To create an interface enabling the interactive choice of a color as the value of the parameter **x**:

ColorSlider[Dynamic[x]] (⌘6)

ColorSetter[Dynamic[x]] (⌘6)

A special option of **ColorSlider**:

AppearanceElements Elements to include in the slider; examples of values: **Automatic**, **"Swatch"**, **"Spectrum"**, **"SwatchSpectrum"**

In the following way, we can get the RGB color corresponding to a given color:

DynamicModule[{x}, Column[{ColorSlider[Dynamic[x]], Dynamic[x]}]]

RGBColor[0., 0., 0.]

A color setter is a special case of a color slider. Indeed, a color setter only contains the "swatch" part of a standard color slider:

DynamicModule[{x}, Row[{ColorSetter[Dynamic[x]], Dynamic[x]}]]

RGBColor[0., 0., 0.]

11.2.4 Special Controls

■ **Opener**

Row[{Opener[Dynamic[x]], PaneSelector[{False → expr1, True → expr2}, Dynamic[x]]}] (⌘6)

Create an opener enabling the interactive choice of either **False** (opener is displayed ▶ and **expr1** is evaluated) or **True** (opener is displayed ▼ and **expr2** is evaluated) as the value of the parameter **x**

A special option:

Appearance The appearance of the opener; examples of values: **Automatic**, **Tiny**, **Small**, **Medium**, **Large**

An opener can be used to hide and expose additional controls. In the next example, with the opener we get animation controls:

```
DynamicModule[{n = 2, x}, Column[
  {Row[{Opener[Dynamic[x]], PaneSelector[{False → Slider[Dynamic[n], {1, 6, 1}],
    True → Animator[Dynamic[n], {1, 6, 1}, AnimationRunning → False,
      AppearanceElements → {"StepLeftButton", "PlayPauseButton",
        "StepRightButton", "FasterSlowerButtons", "DirectionButton"}]},
      Dynamic[x]]}], Dynamic[Plot[Sin[n x], {x, 0, 2 π}]]}]]
```

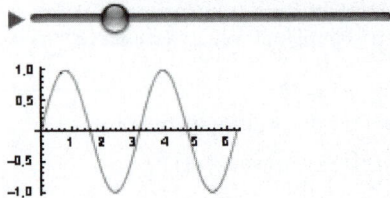

■ Button and PasteButton

> **Button[label, action]** (⌘6) Create a button that does **action** when the button is pressed; put
>   **label** on the button
> **PasteButton[label, expr]** (⌘6) Create a button that pastes **expr** when the button is pressed; put
>   **label** on the button
>
> *Special options:*
> **Appearance** The appearance of the button; examples of values: **Automatic** (default for **Button**),
>   **"DialogBox"**, **"Palette"** (default for **PasteButton**), **"Frameless"**, **"AbuttingLeftRight"**,
>   **"AbuttingRight"**, **"Pressed"**, **"None"**
> **FrameMargins** Margins inside the frame; examples of values: **Automatic**, 3
> **Method** The evaluation method to use; examples of values: **"Preemptive"** (default for **Button**),
>   **"Queued"** (default for **PasteButton**)

With many views and controls we can easily choose items from a given list. If we would like to do a more general action, then **Button** (or **ActionMenu**) can be useful.

In the following example, we take a sample from the standard normal distribution. With four buttons, we can double the sample size, keep it unchanged, or halve it; resetting the sample size to the default 100 is also possible. The two lines in the plot give the 95% and 5% quantiles; thus, approximately 10% of the points lie outside the region given by the two lines.

```
DynamicModule[{n = 100, q = Quantile[NormalDistribution[0, 1], 0.95]},
  Row[{Column[{Button["Double", n = 2 n], Button["Keep", ++n; --n],
     Button["Halve", n = n / 2], Button["Reset", n = 100], Dynamic[n]}, Center, 2],
    Dynamic[ListPlot[Sort[RandomReal[NormalDistribution[0, 1], n]],
     PlotStyle → {Black, AbsolutePointSize[1]},
     PlotRange → {-5.1, 5.1}, AxesOrigin → {-n / 20, 0}, ImageSize → 300,
     Epilog → {Line[{{0, q}, {n, q}}], Line[{{0, -q}, {n, -q}}]}]]}]]
```

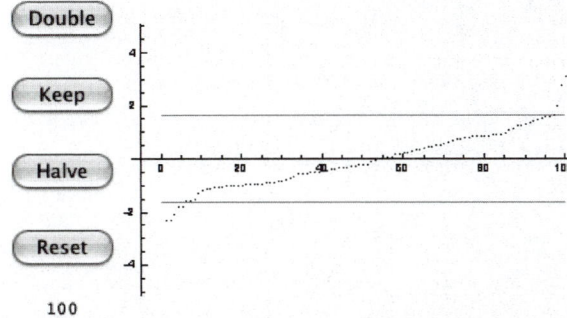

Note that to keep the value of *n* unchanged, we have added 1 to *n* and then subtracted 1 from the result with ++*n* and --*n*. This is a trick: We do something with *n* to trigger the update of the **Dynamic** part of the code.

With **FileNameSetter** we get a *Browse* button. Clicking the button opens a standard window where we can choose a file; the complete name of that file is set as the value of the given variable:

```
DynamicModule[{f = ""}, Column[{FileNameSetter[Dynamic[f]], Dynamic[f]}]]
```

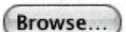

/Users/heikki/Mathematica Navigator 3/Dynamic.nb

■ ActionMenu

ActionMenu[name, {lbl1 :> act1, lbl2 :> act2, … }] (⌘6) Create a menu with **name**, containing items with the given labels; each item does the given action

Special options:

Appearance The appearance of the menu; examples of values: **Automatic**, **"PopupMenu"**, **"Button"**, **"None"**

FieldSize The size of the field for menu items; default value: **{{1, 50}, {1, 10}}**

FrameMargins Margins inside the frame; examples of values: **Automatic**, **3**

We continue the normal distribution example we presented previously. Now we apply an action menu:

```
DynamicModule[{n = 100, q = Quantile[NormalDistribution[0, 1], 0.95]}, Row[
  {Column[{ActionMenu["Number of Points", {"Double" :> (n = 2 n), "Keep" :> (++n; --n),
      "Halve" :> (n = n / 2), "Reset" :> (n = 100)}], Dynamic[n]}, Center, 2],
   Dynamic[ListPlot[Sort[RandomReal[NormalDistribution[0, 1], n]],
     PlotStyle -> {Black, AbsolutePointSize[1]},
     PlotRange -> {-5.1, 5.1}, AxesOrigin -> {-n / 20, 0}, ImageSize -> 250,
     Epilog -> {Line[{{0, q}, {n, q}}], Line[{{0, -q}, {n, -q}}]}]}]]]
```

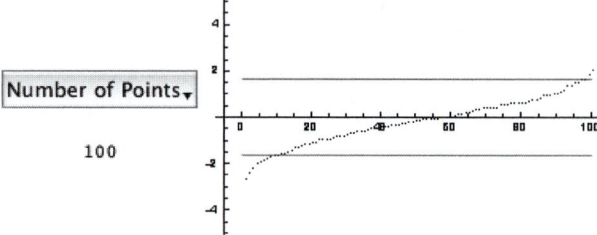

11.2.5 More about Dynamics

■ MenuView Revisited

Now that we know about **Dynamic**, we can consider again some of the views we presented in Section 11.1.1, p. 357. Indeed, as we will see, using **Dynamic** with the views may have advantages and give new possibilities to show data.

> **MenuView[{lbl1 → expr1, lbl2 → expr2, … }]** Create a popup menu where selecting the menu item with label **lbl1** displays **expr1**; compute all the values in advance
>
> **MenuView[{lbl1 → Dynamic[expr1], lbl2 → Dynamic[expr2], … }]** Compute the values of the expressions on the fly

The first item is familiar from Section 11.1.1, p. 358. The second item is useful to show large data sets. As an example, consider the following, in which we can choose a country and the menu view then shows the size of the population from 1970 to 2005:

```
MenuView[
  # → Dynamic[DateListPlot[CountryData[#, {{"Population"}, {1970, 2005}}]]] & /@
    CountryData[], ImageSize → 250]
```

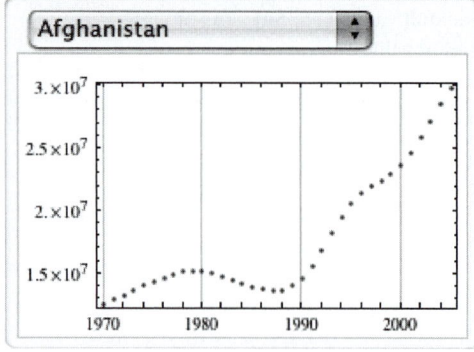

Why have we enclosed the plotting command with **Dynamic**? It is not necessary; indeed, the following command works (although it generates a series of warnings):

```
MenuView[# → DateListPlot[CountryData[#, {{"Population"}, {1970, 2005}}]] & /@
    CountryData[], ImageSize → 250]
```

The point here is that this command computes in advance all the plots corresponding to the 237 countries of the world. This takes a lot of time and memory. Instead, when we use **Dynamic**, each plot is generated on demand, not in advance. Accordingly, the view appears almost instantly and takes very little memory.

MenuView[{v1, lbl1 → expr1}, {v2, lbl2 → expr2}, … }, Dynamic[x]] If **lbli** is selected,
 display **expri** and give **x** the value **vi**; if **x** has the value **vi**, select **lbli** and display **expri**

This form of **MenuView** is useful if we need the value chosen from the popup menu outside **MenuView**. As an example, consider again the population data. Because the list of countries is very long, it may take some time to scroll to a specific country. Thus, it may be useful to be able to simply write the name of the country. Thus, we add an input field:

```
DynamicModule[{country = "Finland"},
    Column[{Row[{InputField[Dynamic[country], String, FieldSize → 16], Dynamic[
        Show[CountryData[country, "Shape"], ImageSize → {50, 50}]]}], MenuView[{#, # →
            Dynamic[DateListPlot[CountryData[#, {{"Population"}, {1970, 2005}}]]]} & /@
        CountryData[], Dynamic[country], ImageSize → 250]}]]
```

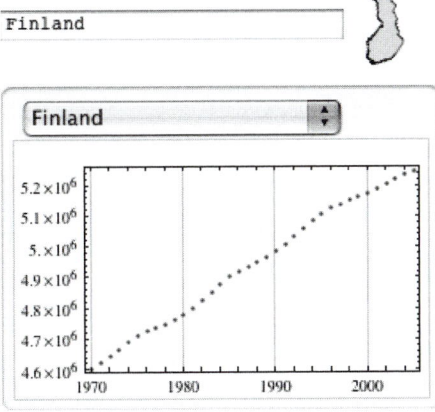

■ PopupView Revisited

PopupView[{expr1, expr2, … }] Create a popup menu whose items are the given expressions
PopupView[{expr1, expr2, … }, Dynamic[n]] If the *i*th expression is chosen, give **n** the value *i*; if **n**
 has value *i*, select the *i*th expression

First, we choose quantiles of the standard normal distribution:

```
DynamicModule[{a = {0.9, 0.95, 0.99, 0.999}, n},
 Column[{PopupView[Row[{#, "-quantile"}] & /@ a, Dynamic[n]],
  Dynamic[Quantile[NormalDistribution[0, 1], a[[n]]]]}]]
```

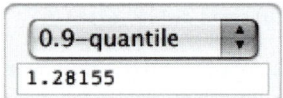

1.28155

A menu view is easier to build:

```
MenuView[Row[{#, "-quantile"}] → Quantile[NormalDistribution[0, 1], #] & /@
 {0.9, 0.95, 0.99, 0.999}]
```

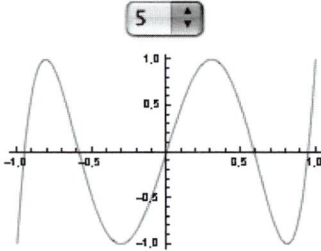

Next, we show Chebyshev polynomials:

```
DynamicModule[{n = 6},
 Column[{PopupView[Range[0, 10], Dynamic[n]], Dynamic[Plot[ChebyshevT[n - 1, x],
  {x, -1, 1}, PlotRange → 1.05, ImageSize → 200]]}, Center]]
```

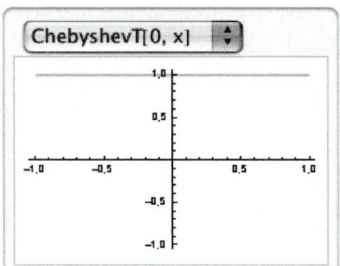

Again, menu view is easier:

```
MenuView[Table[Row[{"ChebyshevT[", n, ", x]"}] →
 Plot[ChebyshevT[n, x], {x, -1, 1}, PlotRange → 1.05, ImageSize → 200], {n, 0, 10}]]
```

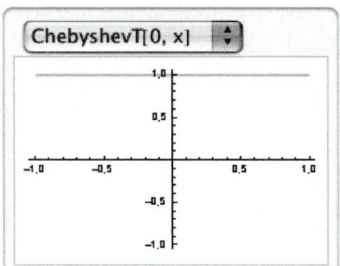

Here are some country data:

```
Panel@DynamicModule[{c = CountryData[], n}, Column[{PopupView[c, Dynamic[n]],
    Dynamic[Framed@DateListPlot[CountryData[c〚n〛, {{"Population"}, {1970, 2005}}],
      ImageSize → 250, Background → White]]}, Right]]
```

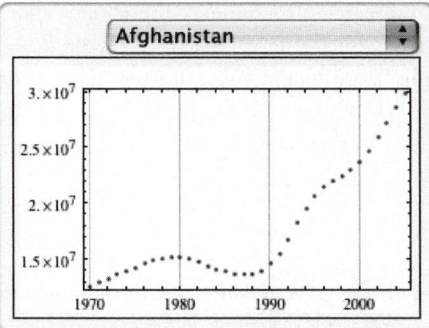

■ What to Do When Moving a Slider

In simple cases, when we move the slider, the thumb of the slider continuously follows the mouse and the value of the parameter is directly based on the position of the mouse:

```
{Slider[Dynamic[m], {0, 10}], Dynamic[m]}
```

This is the automatic working of a slider. We can add a second argument **Automatic** into **Dynamic** to get the same functioning:

```
{Slider[Dynamic[n, Automatic], {0, 10}], Dynamic[n]}
```

However, the second argument can also be an arbitrary pure function. In interactive mouse operations, this function defines what is done while we move the slider with the mouse. The argument # of the pure function is the current position of the mouse.

> **Dynamic[u, (…)&]** The pure function (…)& defines what is done during mouse movement; the argument # of the function is the position of the mouse

Typically, the pure function is used to define the current value of the parameter **u**. The default functioning of a slider can also be obtained as follows:

```
{Slider[Dynamic[p, (p = #) &], {0, 10}], Dynamic[p]}
```

Here, during moving of the slider, the value of the parameter **p** is defined to be the position # of the mouse.

Next, the value of **q** is constrained to be the integer part of the position # of the mouse:

```
{Slider[Dynamic[q, (q = IntegerPart[#]) &], {0, 10}], Dynamic[q]}
```

We can see that although we can drag the thumb with the mouse continuously, the thumb moves in steps of unit and the value of **q** shown at the end of the slider also shows only integer values.

In the next example, the slider shows directly the position of the mouse, but from the position we extract both the integer and the fractional part and show them at the end of the slider:

```
{Slider[Dynamic[u, (u = #; v = IntegerPart[#]; w = FractionalPart[#]) &], {0, 10}],
  Dynamic[{v, w}]}
```

If the second argument is **Temporary**, the value of the parameter is updated after the mouse button is released:

```
{Slider[Dynamic[q, Temporary], {0, 10}], Dynamic[q]}
```

We can also define what to do when the mouse button is first pressed and lastly released:

Dynamic[u, (during)&]   The given pure function defines what is done during mouse movement

Dynamic[u, {(during)&, (end)&}]   The given pure functions define what is done during mouse movement and when the mouse button is released

Dynamic[u, {(start)&, (during)&, (end)&}]   The given pure functions define what is done when the mouse button is first pressed, during mouse movement, and when the mouse button is released

---

For example, in the following, the background of the frame is yellow when the slider is not moved but green when the slider is moved:

```
{Slider[Dynamic[q, {(q = IntegerPart[#]; col = Green) &, (col = Yellow) &}], {0, 10}],
  Framed[Dynamic[q], Background → Dynamic[col]]}
```

{ ●━━━━━━━━━━━━━ , ⌷ 0. ⌷ }

■ **Geometric Constraints on Points**

In Section 10.1.3, p. 330, we demonstrated how we can constrain the movement of a point with **Manipulate**. Now we show how the second argument of **Dynamic** can be used to define geometric constraints to a 2D slider or locator.

In the following example, the position **pt** of the slider is obtained from the mouse position by normalizing it (i.e., by dividing with the norm). Thus, the point corresponding to the slider always has norm 1, and this means that the point is on a circle.

```
DynamicModule[{pt = {1, 0}},
  {Slider2D[Dynamic[pt, (pt = Normalize[#]) &], {-1, 1}], Dynamic[pt]}]
```

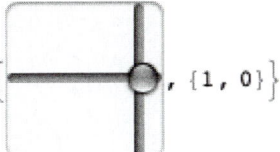

In the same way, we can restrict a locator to move on a circle:

```
DynamicModule[{pt = {1, 0}},
  Graphics[{Circle[], Locator[Dynamic[pt, (pt = Normalize[#]) &]]},
    PlotRange → 1.2, ImageSize → 100]]
```

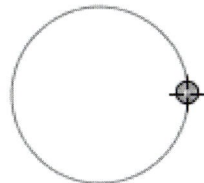

■ Locator Panes and Click Panes

LocatorPane[Dynamic[pt], back] (⌘6) Create a pane with a locator enabling the interactive choice of the point **pt**; the background of the pane is **back**

LocatorPane[Dynamic[{pt1, pt2, … }], back] Create locators at the given points

A special option:

LocatorAutoCreate Whether to allow additions and deletions of locators by ALT- or ⌘-clicking; possible values: **False**, **True**

In addition to **Locator**, locators can also be created with **LocatorPane**. In the following, we create a locator both with **Locator** and with **LocatorPane**:

```
{DynamicModule[{pt = {0, 0}},
   Graphics[Locator[Dynamic[pt]], PlotRange → 1, ImageSize → 100]],
  LocatorPane[Dynamic[ps], Graphics[{}, PlotRange → 1, ImageSize → 100]]}
```

ClickPane[image, (…) &] (⌘6) Display **image** and apply the pure function (…) **&** to the x and y coordinates of each click on the pane

---

In the following, we create the same kind of plot both with **ClickPane** and with **Manipulate**:

```
{DynamicModule[{pt = {0, 0}},
  Framed@ClickPane[Graphics[{Red, PointSize[Large], Dynamic[Point[pt]]},
    PlotRange → 1, ImageSize → 90], (pt = #) &]], Manipulate[
  Graphics[{Red, PointSize[Large], Point[pt]}, PlotRange → 1, ImageSize → 60],
  {{pt, {0, 0}}, Locator, Appearance → None}]}
```

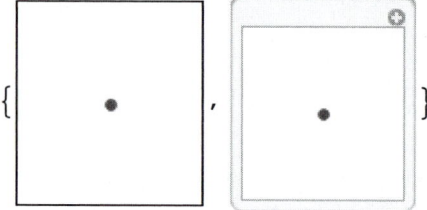

In the first plot, each time we click on the plot, the red point moves to the corresponding point. The same happens in the second plot but, in addition, we can drag the point with the mouse.

■ Event Handler and Mouse Position

> **EventHandler[expr, "MouseDown" :> action]** (⌘6) If the mouse button is pressed down, do
>   **action** and then evaluate **expr**
> **MousePosition["Graphics"]** (⌘6) Give the mouse position in the coordinates of the current
>   graphic

As an example of handling mouse-related events, we create an interface in which the position of the point moves according to where we click with the mouse:

```
DynamicModule[{pt = {0, 0}}, EventHandler[Framed@
  Graphics[{PointSize[Large], Point[Dynamic[pt]]}, PlotRange → 1, ImageSize → 70],
  "MouseDown" :> (pt = MousePosition["Graphics"])]]
```

Numbers

Introduction

> *As Ramanujan, the great Indian mathematical genius, lay dying, he was visited by his friend and mentor G. H. Hardy. To make conversation, Hardy mentioned the number of the taxi he had arrived in, 1729, and remarked that it seemed a very uninteresting number. Ramanujan is reputed to have raised himself from his deathbed and said feebly, "On the contrary, my dear Hardy. It is the smallest number expressible as the sum of two cubes in two different ways."*

Numbers are clear for us, but the representation and interaction of various kinds of numbers in the computer involve aspects worth careful study. Also, *Mathematica* has a richer assortment of numbers than is usually found in computer applications. We can use arbitrarily large integers, we can calculate with exact rational numbers, and we can ask to calculate with real numbers containing as many digits as we want.

We also study the precision and accuracy of real numbers. These relate to the relative and absolute error in the result. In Section 12.3, we study how to control the precision and accuracy of numerical routines in *Mathematica*.

Note that although Section 12.1.1 contains commands relating to number theory, we do not consider number theory with *Mathematica* at any length. The interested reader should consult the following pages in the Documentation Center:

- guide/NumberTheory • guide/AlgebraicNumberTheory • guide/CryptographicNumberTheory
- tutorial/IntegerAndNumberTheoreticalFunctions • PrimalityProving/guide/PrimalityProvingPackage

See also Ruskeepää (2008a, b).

12.1 Introduction to Numbers

12.1.1 Integers

■ Four Types of Numbers

Mathematica has four types of numbers: *integers* such as 38254, *rationals* such as 41/7, *reals* such as 58.723, and *complexes* such as 9.45 + 3 *i*. The type of the number can be asked with **Head**:

```
Head /@ {38 254, 41/7, 58.723, 9.45 + 3 I}
{Integer, Rational, Real, Complex}
```

Although *Mathematica* has the four basic types of numbers, it recognizes more types with special tests. For example, *Mathematica* knows that $\sqrt{2}$ is an algebraic number:

```
Sqrt[2] ∈ Algebraics      True
```

These tests are often used in simplifying expressions, and so we consider them in Section 13.2.1, p. 419, in which we study simplification of expressions.

■ Primes

```
Prime[n]   nth prime (the first prime is 2)
Prime[{m, n, … }]   List of mth, nth, … primes
PrimePi[x]   The number of primes π(x) less than or equal to x
NextPrime[x]   The next prime larger than x
NextPrime[x, -1]   The largest prime smaller than x
RandomPrime[{imin, imax}]   A random prime in the range imin to imax

PrimeQ[n]   Test whether n is a prime
PrimeQ[n, GaussianIntegers → True]   Test whether n is a Gaussian prime
PrimePowerQ[n]   Test whether n is a power of a prime

CoprimeQ[m, n]   Test whether m and n are relatively prime
EulerPhi[n]   The number of integers ≤ n which are relatively prime to n
```

Find the seven first primes:

```
Prime[Range[7]]      {2, 3, 5, 7, 11, 13, 17}
```

Check integers up to 17 for primality and show each prime framed:

```
If[PrimeQ[#], Framed[#], #] & /@ Range[17]
```

$$\left\{1, \boxed{2}, \boxed{3}, 4, \boxed{5}, 6, \boxed{7}, 8, 9, 10, \boxed{11}, 12, \boxed{13}, 14, 15, 16, \boxed{17}\right\}$$

■ Factors and Divisors of Integers

```
FactorInteger[n]   List of the prime factors of n together with their exponents
FactorInteger[n, GaussianIntegers → True]   Factor over Gaussian integers
```

```
fa = FactorInteger[3 361 743]      {{3, 4}, {7, 3}, {11, 2}}
```

With **Apply** we can check that the factors give the original number:

```
Apply[Times, Apply[Power, fa, {1}]]     3 361 743
```

This can also be written as

```
Times @@ Power @@@ fa     3 361 743
```

We can also show the factorization as follows:

```
CenterDot @@ Superscript @@@ fa     3⁴ · 7³ · 11²
```

$3^4 \cdot 7^3 \cdot 11^2$

Next, we factor over Gaussian integers (complex numbers with integer real and imaginary parts):

```
FactorInteger[13, GaussianIntegers → True]
{{-i, 1}, {2 + 3 i, 1}, {3 + 2 i, 1}}

Times @@ Power @@@ %     13
```

For an application of **FactorInteger**, see Section 1.1.1, p. 2.

Divisors[n]   List of the integers that divide **n**

Divisors[n, GaussianIntegers → True]   Include Gaussian integers

DivisorSigma[k, n]   The sum of the **k**th powers of the divisors of **n**

Divisible[m, n] (✱6)   Test whether **m** is divisible by **n**

IntegerExponent[n, b]   The highest power of **b** that divides **n**

GCD[m, n, …]   The greatest common divisor of the integers

LCM[m, n, …]   The least common multiple of the integers

A perfect number is a number whose sum of divisors is two times the number. We find all perfect numbers less than or equal to 10^4:

```
Select[Range[10 000], DivisorSigma[1, #] == 2 # &]
{6, 28, 496, 8128}
```

■ **More about Integers**

Quotient[m, n]   The integer quotient of **m** and **n**

Mod[m, n]   **m** modulo **n** or the remainder on division of **m** by **n**

PowerMod[a, b, n]   **a^b** modulo **n**

QuotientRemainder[m, n]   The quotient and remainder from division of **m** by **n**

IntegerLength[n]   Number of digits in **n**

IntegerDigits[n]   List of digits of **n**

FromDigits[list]   Construct an integer from the list of its digits

DigitCount[n]   List of the numbers of the digits 1, 2, …, 9, 0 in **n**

EvenQ[n]   Test whether **n** is even

OddQ[n]   Test whether **n** is odd

IntegerPartitions[n] (✱6)   Give all ways to partition **n** into sum of integers

IntegerPartitions[n, k]   Give partitions into at most **k** integers

IntegerPartitions[n, {k}]   Give partitions into exactly **k** integers

PartitionsP[n]   Give the number of partitions

PowersRepresentations[n, k, p] (⌘6) Give distinct sets of **k** integers whose sum of **p**th powers is **n**: $n = n_1{}^p + \ldots + n_k{}^p$

Throw three dice. In how many ways can we get 7 as the sum?

```
IntegerPartitions[7, {3}]
```

$\{\{5, 1, 1\}, \{4, 2, 1\}, \{3, 3, 1\}, \{3, 2, 2\}\}$

We now return to the story about Ramanujan at the beginning of this chapter. Are there numbers n_1 and n_2 such that $1729 = n_1{}^3 + n_2{}^3$?

```
pr = PowersRepresentations[1729, 2, 3]      {{1, 12}, {9, 10}}
```

Thus, both $1729 = 1^3 + 12^3$ and $1729 = 9^3 + 10^3$. Verify this as follows:

```
Total /@ (pr^3)     {1729, 1729}
```

Next, we find all integers less than or equal to 21,000 that can be represented as the sum of two cubes in at least two ways:

```
(pr = Select[PowersRepresentations[#, 2, 3] & /@ Range[21 000], Length[#] ≥ 2 &]) //
  Timing
{70.1857,
 {{{1, 12}, {9, 10}}, {{2, 16}, {9, 15}}, {{2, 24}, {18, 20}}, {{10, 27}, {19, 24}}}}
Map[Total, pr^3, {2}]
{{1729, 1729}, {4104, 4104}, {13 832, 13 832}, {20 683, 20 683}}
```

We see that 1729 is the smallest of such numbers, as Ramanujan said. The next ones are 4104, 13,832, and 20,683.

RamanujanTau[n] (⌘6) Ramanujan tau function
RamanujanTauL[s] (⌘6) Ramanujan tau Dirichlet L-function
RamanujanTauZ[t] (⌘6) Ramanujan tau Z-function
RamanujanTauTheta[t] (⌘6) Ramanujan tau theta function

12.1.2 Real and Complex Numbers

■ Asking a Decimal Value

If an expression does not contain a decimal point, the result is an exact expression:

```
3 + (3 / 5) ^ 2 - Sin[π / 3] + Log[2]
```
$$\frac{84}{25} - \frac{\sqrt{3}}{2} + \text{Log}[2]$$

If an expression contains a decimal point, the whole result is a real number:

```
3. + (3 / 5) ^ 2 - Sin[π / 3] + Log[2]     3.18712
```

To get a decimal result, we can also use **N**.

N[expr] or **expr//N** Calculate decimal value of **expr**
N[expr, n] Calculate decimal value of **expr** to **n**-digit precision

---

First, we calculate a numerical value in the usual way:

```
Sin[2] // N     0.909297
```

The result was calculated by using the normal floating-point numbers with 16 decimal digits of precision. Then we ask for a numerical value to 30-digit precision:

N[Sin[2], 30]   0.909297426825681695396019865912

Note that **expr** in **N[expr, n]** must contain exact numbers or numbers of sufficiently high precision for the **n** to have an effect. Asking for **N[2.3 Pi, 20]**, for example, is useless because the expression **2.3 Pi** contains a low-precision number; the result is the same as with **N[2.3 Pi]**. Indeed, normal 16-digit real numbers are used in the evaluation of an expression as soon as such a number is encountered in the expression. Ask for **N[23/10 Pi, 20]** or **N[2.3`20 Pi, 20]** instead (with **2.3`20** we give 2.3 to 20 digits of precision, as explained in Section 12.2.2, p. 406).

If you are trying to find an *answer* satisfying a given requirement of precision, **N** can use extra precision in the *calculations*. However, the extra precision cannot exceed **$MaxExtraPrecision**, which has a default value of 50. This means that if you ask for a value to 20-digit precision, *Mathematica* can use at most 70-digit precision during the computation. If a calculation does not succeed within this limit, we can increase the value of this constant (by typing, for example, **$MaxExtraPrecision = 100**) and then retry the calculation.

■ **Adjusting the Number of Digits Shown**

Mathematica normally shows six digits of decimal numbers; this is suitable in most cases. Sometimes we want to show more or fewer digits, and then we can use **NumberForm**.

NumberForm[expr, n]   Show **expr** using **n** digits
NumberForm[expr, {n, f}]   Show **expr** using **n** digits, of which **f** digits are to the right of the decimal point

Exp[3.2]   24.5325

NumberForm[Exp[3.2], 4]   24.53

NumberForm[Exp[3.2], {4, 1}]   24.5

The way numbers are displayed can be controlled by opening the preferences from **Mathematica ▷ Preferences** and going to **Appearance ▷ Numbers**. In addition to **NumberForm**, we have other formatting commands: **ScientificForm**, **EngineeringForm**, and **AccountingForm**.

■ **Manipulating Real Numbers**

Round[x]   Integer closest to **x**
Floor[x] or ⌊x⌋   Greatest integer less than or equal to **x**
Ceiling[x] or ⌈x⌉   Smallest integer greater than or equal to **x**
IntegerPart[x]   Integer part of **x**
FractionalPart[x]   Fractional part of **x**
Chop[expr]   Replace all real numbers in **expr** with magnitude less than 10^{-10} with 0
Chop[expr, dx]   Replace all real numbers in **expr** with magnitude less than **dx** with 0

Note that the result of **Round[x.5]** is the *even* integer nearest to **x**. Thus, **Round[2.5]** gives 2, but **Round[3.5]** gives 4. Symbols ⌊ and ⌋ can be written as ESC**lf**ESC and ESC**rf**ESC, and ⌈ and ⌉ can be written as ESC**lc**ESC and ESC**rc**ESC, respectively.

> **Rationalize[x]**   Rational number close to **x** with small denominator
>
> **Rationalize[x, dx]**   Rational number within **dx** of **x** with the smallest denominator
>
> **ContinuedFraction[x, n]**   Continued fraction representation $\{a, b, c, ...\}$ of **x** with **n** terms: $x = a + 1/(b + 1/(c + ...$
>
> **Convergents[list]** (✿6)   Give the convergents $\{a, a + 1/b, a + 1/(b + 1/c), ...\}$ from the continued fraction representation
>
> **FromContinuedFraction[list]**   Reconstruct a number from its continued fraction representation
>
> **RealDigits[x]**   List of the digits in **x** together with the number of digits to the left of the decimal point
>
> **FromDigits[list]**   Construct a real number from the list of its digits
>
> **MantissaExponent[x]**   Give the mantissa and the exponent of **x**
>
> **Rescale[x, {A, B}, {a, b}]**   Rescale **x** so that for **x** = **A** the rescaled expression is **a**, and for **x** = **B** the rescaled expression is **b**

```
r = Rescale[x, {A, B}, {a, b}] // Simplify
```

$$\frac{A\,b - a\,B + a\,x - b\,x}{A - B}$$

```
{r /. x → A, r /. x → B} // Simplify
```

$\{a, b\}$

■ Manipulating Complex Numbers

> **I** $\sqrt{-1}$
>
> **Re[z]**   Real part
>
> **Im[z]**   Imaginary part
>
> **Conjugate[z]** or **z***   Complex conjugate
>
> **Abs[z]**   Absolute value
>
> **Arg[z]**   Argument ϕ such that $z = |z|\,e^{i\phi}$
>
> **ComplexExpand[z]**   Expand **z** to real and imaginary parts

The imaginary unit can be written as **I** or **i**. The latter form can be written as ⎡ESC⎤**ii**⎡ESC⎤. The symbol * can be written as ⎡ESC⎤**co**⎡ESC⎤.

```
z = 3 - 2 I; {Re[z], Im[z], Conjugate[z], Abs[z], Arg[z]}
```

$$\left\{3, -2, 3 + 2\,i, \sqrt{13}, -\text{ArcTan}\left[\frac{2}{3}\right]\right\}$$

```
ComplexExpand[(-1) ^ (1 / 3)]
```

$$\frac{1}{2} + \frac{i\,\sqrt{3}}{2}$$

```
ComplexExpand[Log[2 I]]
```

$$\frac{i\,\pi}{2} + \text{Log}[2]$$

For more information on **ComplexExpand**, see Section 13.3.4, p. 430.

12.1.3 Constants and Units

■ **Mathematical Constants**

Mathematica has the following mathematical constants:

> `const = Select[Names["*"], MemberQ[Attributes[#], Constant] &]`

```
{Catalan, Degree, E, EulerGamma, Glaisher, GoldenRatio, Khinchin,
  MachinePrecision, Pi}
```

They have the following numerical values, respectively:

> `const // ToExpression // N`
>
> `{0.915966, 0.0174533, 2.71828, 0.577216, 1.28243, 1.61803, 2.68545, 15.9546, 3.14159}`

Of these constants, **Degree**, **E**, and **Pi** can also be entered as ESCdegESC, ESCeeESC, and ESCpESC, and the results are °, e, and π. In traditional form, **Catalan**, **Glaisher**, **GoldenRatio**, and **Khinchin** are written as C, A, ϕ, and K, respectively:

> `const // ToExpression // TraditionalForm`
>
> $\{C, °, e, \gamma, A, \phi, K, \text{MachinePrecision}, \pi\}$

Note that **Catalan** C is $\sum_{k=0}^{\infty} (-1)^k (2k+1)^{-2}$, **Degree** ° is $\pi/180$ (the degrees-to-radians conversion factor), **EulerGamma** γ is $\lim_{m\to\infty}\left(\sum_{k=1}^{m} \frac{1}{k} - \log(m)\right)$, **Glaisher** A satisfies the equation $\log(A) = \frac{1}{12} - \zeta'(-1)$, **GoldenRatio** ϕ is $\frac{1}{2}\left(1 + \sqrt{5}\right)$, and **Khinchin** K is $\Pi_{s=1}^{\infty}\left(1 + \frac{1}{s(s+2)}\right)^{\log_2 s}$. **MachinePrecision** is considered in Section 12.2.2, p. 405.

■ **Infinite and Indeterminate Quantities**

> `Infinity` ∞
>
> `ComplexInfinity` An infinite quantity with an undetermined direction
>
> `DirectedInfinity[z]` An infinite quantity in the direction of the complex number **z**
>
> `Indeterminate` An indeterminate numerical result

Of these, **Infinity** can be written as ESCinfESC and the result is ∞. Internally, *Mathematica* transforms **Infinity** to **DirectedInfinity[1]**, **-Infinity** to **DirectedInfinity[-1]**, and **ComplexInfinity** to **DirectedInfinity[]**. To demonstrate these special symbols, write the following:

> `{Sqrt[-1], I^2, Exp[-π I]}`   `{i, -1, -1}`
>
> $\left\{\text{Limit}\left[\frac{1}{x}, x \to 0, \text{Direction} \to -1\right], \text{Limit}\left[\frac{1}{x}, x \to 0, \text{Direction} \to 1\right], \frac{1}{0}\right\}$
>
> Power::infy : Infinite expression $\frac{1}{0}$ encountered. ≫
>
> `{∞, -∞, ComplexInfinity}`

Next, we get several indeterminate quantities:

```
{0 / 0, 0^0, ∞ - ∞}
```

Power::infy : Infinite expression $\dfrac{1}{0}$ encountered. ≫

∞::indet : Indeterminate expression 0 ComplexInfinity encountered. ≫

Power::indet : Indeterminate expression 0^0 encountered. ≫

∞::indet : Indeterminate expression $-\infty + \infty$ encountered. ≫

```
{Indeterminate, Indeterminate, Indeterminate}
```

Note that the limit of x^x exists at 0:

```
Limit[x^x, x → 0]      1
```

The result of some calculations may also be an interval (for interval arithmetic, see Section 12.2.3, p. 408):

```
Limit[Sin[1 / x], x → 0]     Interval[{-1, 1}]
```

■ Physical Constants

The **PhysicalConstants`** package contains the values of 51 physical constants. To get a list of the constants, type the following:

```
<< PhysicalConstants`
```

```
Style[Names["PhysicalConstants`*"], 7]
```

```
{AccelerationDueToGravity, AgeOfUniverse, AvogadroConstant, BohrRadius, BoltzmannConstant,
 ClassicalElectronRadius, CosmicBackgroundTemperature, DeuteronMagneticMoment, DeuteronMass, EarthMass,
 EarthRadius, ElectronCharge, ElectronComptonWavelength, ElectronGFactor, ElectronMagneticMoment,
 ElectronMass, FaradayConstant, FineStructureConstant, GalacticUnit, GravitationalConstant,
 HubbleConstant, IcePoint, MagneticFluxQuantum, MolarGasConstant, MolarVolume, MuonGFactor,
 MuonMagneticMoment, MuonMass, NeutronComptonWavelength, NeutronMagneticMoment, NeutronMass,
 PlanckConstant, PlanckConstantReduced, PlanckMass, ProtonComptonWavelength, ProtonMagneticMoment,
 ProtonMass, QuantizedHallConductance, RydbergConstant, SackurTetrodeConstant, SolarConstant,
 SolarLuminosity, SolarRadius, SolarSchwarzschildRadius, SpeedOfLight, SpeedOfSound,
 StefanConstant, ThomsonCrossSection, VacuumPermeability, VacuumPermittivity, WeakMixingAngle}
```

Here are some examples:

```
? AccelerationDueToGravity
```

AccelerationDueToGravity is the acceleration
    of a body freely falling in a vacuum on Earth at sea level. ≫

```
{IcePoint, SpeedOfLight, SpeedOfSound, AccelerationDueToGravity}
```

$$\left\{273.15 \, \text{Kelvin}, \frac{299\,792\,458 \, \text{Meter}}{\text{Second}}, \frac{340.292 \, \text{Meter}}{\text{Second}}, \frac{9.80665 \, \text{Meter}}{\text{Second}^2}\right\}$$

See PhysicalConstants/guide/PhysicalConstantsPackage for more information about physical constants.

■ Physical Units

In the **Units`** *package:*

Convert[expr, newunits]   Convert **expr** to **newunits**

ConvertTemperature[temp, oldunits, newunits]   Convert temperature

SI[expr]   Convert to SI units

MKS[expr]   Convert to MKS units (meter/kilogram/second)

CGS[expr]   Convert to CGS units (centimeter/gram/second)

This package has information on approximately 250 units related to temperature, electricity, length, information, time, mass, weight, force, inverse length, volume, viscosity, luminous energy and intensity, radiation, angles, power, area, amounts of substances, acceleration due to gravity, magnetism, pressure, energy, frequency, speed, and fineness for yam or thread. Here are some examples of conversion:

```
<< Units`
```

$$\mathtt{Convert[60\,Mile\,/\,Hour,\,Kilo\,Meter\,/\,Hour]} \quad \frac{301\,752\,\mathtt{Kilo\,Meter}}{3125\,\mathtt{Hour}}$$

```
ConvertTemperature[100, Celsius, Fahrenheit]    212
```

See Units/guide/UnitsPackage for more information on physical constants.

12.2 Real Numbers

12.2.1 Precision and Accuracy

■ Precision and Accuracy

The following definitions are used for real numbers:

> *Precision:* The total number of significant decimal digits
> *Accuracy:* The number of significant decimal digits to the right of the decimal point

The precision and accuracy of a number can be asked for using **Precision** and **Accuracy**. Before presenting examples, we define a function that gives the precision and accuracy of a number:

```
pa[x_] := {Precision[x], Accuracy[x]}
```

An example:

```
pa[11 111.222223333344444]     {19.0458, 15.}
```

The precision of the number is approximately 20 and the accuracy is 15. Some other examples:

```
pa[0.00000111111222223333344444]     {19.0458, 25.}
```

```
pa[111 112 222 233 333.44444 × 10^10]     {19.0458, -5.}
```

Thus, the accuracy can be a negative number; in this example, it tells us that there are five insignificant digits (zeros in this case) between the least significant digit and the decimal point. The precision and accuracy of exact numbers (e.g., integers, rational numbers, and special constants) are infinity:

```
{pa[7], pa[3 / 4], pa[Pi], pa[Sin[2]]}
```
```
{{∞, ∞}, {∞, ∞}, {∞, ∞}, {∞, ∞}}
```

■ Relative and Absolute Errors

Precision and accuracy have interpretations in terms of relative and absolute errors.

> Relative error $\simeq 10^{-\text{precision}}$
> Absolute error $\simeq 10^{-\text{accuracy}}$

Thus, if the precision of a result is p, the relative error of the result is of the order 10^{-p}. Similarly, if the accuracy of a result is a, the absolute error of the result is of the order 10^{-a}.

Note that precision is approximately $-\log_{10}$(relative error) and accuracy is approximately $-\log_{10}$(absolute error). For example, the absolute error in 11111.22222 33333 44444 can be considered to be approximately 0.00000 00000 00000 5, and the precision and accuracy thus have the following approximate values:

```
-Log[10, 0.0000000000000005 / 11 111.222223333344444]    19.3468
```

```
-Log[10, 0.0000000000000005]    15.301
```

These are close to the values 19.0458 and 15. given by **Precision** and **Accuracy**. We can also see the precision of our number from the **InputForm**:

```
11 111.222223333344444 // InputForm
11111.222223333344444`19.04576183352721
```

12.2.2 Two Types of Real Numbers

■ Two Types of Arithmetic

Mathematica has two types of floating-point arithmetic.

> *Fixed-precision arithmetic:* Implemented in the hardware
> *Variable-precision arithmetic:* Implemented in *Mathematica*

There are several ways to guide *Mathematica* to use the arithmetic we want, as we soon discuss. As an example, we calculate a decimal value by fixed-precision arithmetic and ask the precision of the result:

```
N[Sin[2]]    0.909297
```

```
Precision[%] // N    15.9546
```

Then we use variable-precision arithmetic:

```
N[Sin[2], 30]    0.909297426825681695396019865912
```

```
Precision[%]    30.
```

Usually, fixed-precision arithmetic is used. The name of this system comes from the representation of real numbers in the computer hardware. Real numbers have a mantissa and an exponent, with the mantissa always containing a fixed number of bits; this usually means 16 decimal digits. There is no way to determine how precise such a number is, and *Mathematica* has adopted the convention that if you ask for the precision of such a number, the maximum precision (usually approximately 16, as in the previous example) is given as the answer, independent of the true precision. Thus the name fixed-precision arithmetic: All numbers have the same precision, independent of what can be justified for them (i.e., whether or not all the digits in the result can be determined to be correct on the basis of the numbers in the input). Therefore, the results you get with this arithmetic can contain insignificant digits because *Mathematica* cannot determine which digits are significant and which are insignificant.

Variable-precision arithmetic is what arithmetic should be: The precision of the result is what can be justified from the input and calculations. Only significant digits are then included in the result. The precision of such numbers varies—thus the name variable-precision arithmetic. This arithmetic is implemented in the software of *Mathematica*. Variable-precision arithmetic has two remarkable properties. First, all digits returned by *Mathematica* are correct if this arithmetic is used. Second, we can ask for the result to whatever precision we want.

■ Two Types of Real Numbers

There are two types of real numbers, which correspond with the two types of arithmetic:

> *Machine-precision numbers:* Numbers produced by fixed-precision arithmetic
> *Arbitrary-precision numbers:* Numbers produced by variable-precision arithmetic

Machine-precision numbers correspond to double-precision floating-point numbers in the underlying computer system. Arbitrary-precision numbers are handled with the software of *Mathematica*. Usually, machine-precision numbers are used. Arbitrary-precision numbers can be formed in some special ways, which we consider later.

The precision of machine-precision numbers is indicated by the special symbol `MachinePrecision`.

> `MachinePrecision`   The precision specification used to indicate machine-precision numbers
> `$MachinePrecision`   Numerical value of `MachinePrecision`

When we calculate a numerical value with `N[expr]`, we actually ask the result as a machine-precision number. We could equally well write `N[expr, MachinePrecision]`:

    {N[Pi], N[Pi, MachinePrecision]}

    {3.14159, 3.14159}

The precision of these numbers is, indeed, `MachinePrecision`:

    Precision /@ %

    {MachinePrecision, MachinePrecision}

The numerical value of `MachinePrecision` is `$MachinePrecision`, which is approximately 16:

    {MachinePrecision // N, $MachinePrecision}

    {15.9546, 15.9546}

The fixed precision used in machine-precision numbers may vary between computer systems, but usually it is approximately 16. *Mathematica* knows the precision: It is the value of the constant `$MachinePrecision`.

■ Machine-Precision Numbers

Here are examples of machine-precision numbers:

    {2.2, N[Pi], 1.2345678901234567, N[Sin[2.2], 20]}

    {2.2, 3.14159, 1.23457, 0.808496}

The internal representations in *Mathematica* are as follows:

    InputForm[%]

    {2.2, 3.141592653589793, 1.2345678901234567, 0.8084964038195901}

For example, `N[Pi]` is internally calculated with all of the standard 16 digits, but normally only 6 digits are shown. Note especially that in `N[Sin[2.2], 20]`, the number 2.2 is a machine-precision number; this causes all calculations in this expression to be done with fixed-precision arithmetic. Thus, asking for the value of `Sin[2.2]` to 20-digit precision does not have the desired effect; the result is a 16-digit machine-precision number.

■ **Arbitrary-Precision Numbers**

Ways to form an arbitrary-precision number:

**2.2`9**  Use ` to write a number with any precision (here with precision 9)

N[22/10, 20] Use **N** to form a number with any precision (here with precision 20)

2.20000000000000000 Write at least 18 significant digits (in most computers)

SetPrecision[2.2, 9] Use **SetPrecision** to write a number with any precision

Note that *machine-precision numbers are used in a calculation as soon as a machine-precision number is encountered.* Thus, to use arbitrary-precision numbers in a computation, *all* numbers in the input have to be exact quantities or arbitrary-precision numbers.

Here are examples of arbitrary-precision numbers and their internal representations:

```
{2.2`9, N[Pi, 17], 1.234567890123456789, N[Sin[22 / 10], 20]}
{2.20000000, 3.1415926535897932, 1.234567890123456789, 0.80849640381959018430}
```

```
InputForm[%]
{2.2`9.000000000000002,
 3.14159265358979323846264338358`17.,
 1.234567890123456789`18.091514977212704,
 0.808496403819590184304040369104161190646`20.}
```

We see that the internal representation of arbitrary-precision numbers contains, after the mark `, the precision of the number. The input forms of the second and fourth numbers contain even more digits than were requested. The accuracy of a number can be set with ``.

■ **Printing**

For machine-precision numbers, usually 6 digits are printed (with **InputForm** all 16 digits are shown).
For arbitrary-precision numbers, *significant* digits are printed.

Machine-precision numbers are usually printed with 6 digits (however, trailing zeros, even significant ones, are not printed). If you want to see all of the internal 16 digits, apply **InputForm** to the result. If you want to adjust the number of digits shown, use **NumberForm** (see Section 12.1.2, p. 399).

An excellent aspect of arbitrary-precision numbers is that *Mathematica* shows for them only the digits for which it can be sure are significant. Thus, we can trust that all digits in such a result are correct.

■ **Advantages and Disadvantages**

The primary advantage of fixed-precision arithmetic is that it is fast; this is because the calculations are done by the hardware in the floating-point unit. Conversely, the primary disadvantage of variable-precision arithmetic is that it is slow; this is because the arithmetic is implemented in the software of *Mathematica*.

Advantages of variable-precision arithmetic include the following:

• we can use arbitrary precision in the calculations;
• no round-off errors are introduced by the arithmetic itself; and
• results contain only correct digits.

Conversely, disadvantages of fixed-precision arithmetic include the following:

- we cannot do high-precision calculations;
- round-off errors are introduced by the arithmetic (see the next section); and
- results may contain insignificant digits.

An additional disadvantage of variable-precision arithmetic has to be mentioned. It is good that the result contains only correct digits, but it is not so good that the rules used to determine the precision of the result may yield an overly pessimistic precision. This means that the result often has a better precision than the one given by *Mathematica*. The cause for this pessimism is the assumption that all errors are independent. For example, let *a* be a given arbitrary-precision number. Then *a* − *a* should be exactly 0, but it is not:

```
a = N[Pi, 20];
```

```
a - a        0. × 10^-20
```

The errors in the two *a*'s are considered as independent instead of equal, and therefore *a* − *a* cannot be assigned the value 0.

12.2.3 Round-off Errors and Interval Arithmetic

■ Round-off Errors

It is well-known how round-off errors affect the results of fixed-precision arithmetic. Consider the following sum:

```
1.234567890123456 + 10.^-16 // InputForm
1.234567890123456
```

We see that the value of the sum is the same as the first summand; the second summand had no effect. The fixed-precision 16-digit system could not represent the result adequately. Such round-off errors are the primary sources of errors in this system. Round-off errors are also called representation errors. Instead, if we use variable-precision arithmetic, then no round-off error is introduced:

```
1.234567890123456`17 + 10.`17^-16
1.2345678901234561
```

As another example, define the following function:

```
f[x_] := (Log[1 - x] + x Exp[x / 2]) / x^3
```

Here is its limit at the origin (we use **Quiet** to drop any messages):

```
Limit[f[x], x → 0.] // Quiet        -0.208333
```

Calculating values of the function near the origin results in huge errors:

```
Table[f[10.^-n], {n, 4, 8}]
{-0.208345, -0.162823, -28.9641, 52 635.4, -5.02476 × 10^7}
```

To obtain correct values, use arbitrary-precision numbers:

```
Table[f[10.`23^-n], {n, 4, 8}]
{-0.20835625197412, -0.20833562502, -0.208333563, -0.2083334, -0.20833}
```

Notice how we get increasingly fewer digits the nearer we get to the origin. This is because significant digits are lost increasingly more. If we try to plot the function near the origin, the result is far from good:

```
Plot[f[x], {x, 0, 0.001}, PlotPoints → 200, Ticks → {{0.0005}, Automatic}]
```

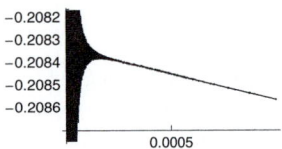

The plot shows clearly how wildly the machine-precision values of the function vary near the origin. If we use arbitrary-precision numbers, we get the correct plot:

```
Plot[f[x], {x, 0, 0.001}, WorkingPrecision → 25, Ticks → {{0.0005}, Automatic}]
```

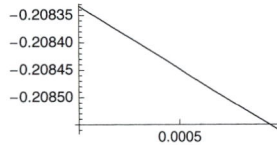

In Section 18.1.1, p. 543, there is a similar example.

Sometimes *Mathematica* may give numbers so near to 0 that we prefer to replace them with an exact 0; the nonzero digits are possibly only the result of round-off and other errors. We can use **Chop**, which is explained in Section 12.1.2, p. 399. **Chop[expr]** replaces all real numbers in **expr** with a magnitude of less than 10^{-10} with 0:

```
Exp[N[Pi / 2 I]]      6.12323 × 10⁻¹⁷ + 1. i
Chop[%]      1. i
```

■ Interval Arithmetic

We can do interval arithmetic with *Mathematica*. **Interval[{min, max}]** represents the range of values between **min** and **max**. As an example, we consider linear algebra. First, we form a square matrix that has intervals as elements, which reflects the uncertainty we have about these numbers:

```
a = {{2, -1}, {3, -5}};

aa = Map[Interval[{# - 0.01, # + 0.01}] &, a, {2}]
{{Interval[{1.99, 2.01}], Interval[{-1.01, -0.99}]},
 {Interval[{2.99, 3.01}], Interval[{-5.01, -4.99}]}}
```

Then we calculate the determinant and solve a system of linear equations:

```
Det[aa]      Interval[{-7.11, -6.89}]

Solve[aa.{x, y} == {4, 1}, {x, y}]
{{x → Interval[{2.67932, 2.75036}], y → Interval[{1.39944, 1.45864}]}}
```

The results belong to the intervals shown.

■ More about Computer Arithmetic

Not all machine-precision numbers can be distinguished because of the limited precision available. **$MachineEpsilon** is the smallest machine-precision number such that 1.0 + **$MachineEpsilon** is not equal to 1.0. **$MaxMachineNumber** is the largest positive machine-precision number and **$MinMachineNumber** the smallest positive machine-precision number:

{$MachineEpsilon, $MaxMachineNumber, $MinMachineNumber}

$\{2.22045 \times 10^{-16}, 1.79769 \times 10^{308}, 2.22507 \times 10^{-308}\}$

Note that if the result of a calculation is a number outside the range specified by **$MinMachineNumber** and **$MaxMachineNumber**, the result is automatically converted to arbitrary-precision form.

The **ComputerArithmetic`** package can be used to investigate floating-point systems with various rounding rules, bases, and precisions; see ComputerArithmetic/guide/ComputerArithmeticPackage.

12.3 Options of Numerical Routines

12.3.1 Options for Precision

■ Three Options

Although the main scope of *Mathematica* is symbolic calculation, sooner or later we encounter a problem for which *Mathematica* cannot find a solution in a symbolic and exact form. Then we can resort to numerical routines such as **NIntegrate**, **FindRoot**, and **NDSolve** and obtain an approximate numerical solution.

The numerical routines have several options with which we can control and modify the routines. Most of the routines include one or more of the following three options. We consider them here for two reasons. First, they are closely related to round-off errors, precision, and accuracy, which we considered in Section 12.2. Second, the three options do not then need to be separately considered in detail for each routine.

WorkingPrecision → w Calculations are done using numbers with **w**-digit precision
PrecisionGoal → p Result should have **p**-digit precision
AccuracyGoal → a Result should have **a**-digit accuracy

WorkingPrecision affects the precision of the *calculations*, whereas **PrecisionGoal** and **AccuracyGoal** affect the precision of the *result*. With **WorkingPrecision**, we can control the effect of *round-off error*. With **PrecisionGoal** and **AccuracyGoal**, we can control the effect of *truncation error*.

■ Precision and Accuracy Goals

Truncation errors are caused by the iterative method used—that is, by approximating the original problem with another, simpler one. Typically, the iterative methods rely on calculating the value of a function at a finite set of points, whereas an infinite set would be required for the exact solution. For example, in calculating the value of an infinite sum by numerical methods, only a finite number of terms are summed, and the rest are estimated by various methods.

Precision and accuracy were considered in Section 12.2.1, p. 403. There, we noted that if the precision of a result is p, the relative error of the result is of the order 10^{-p}, and if the accuracy of a result is a, the absolute error of the result is of the order 10^{-a}. So we get the following interpretations of the two options:

PrecisionGoal → p The *relative error* of the result should be at most of the order 10^{-p}.
AccuracyGoal → a The *absolute error* of the result should be at most of the order 10^{-a}.

The goals are used in the stopping criteria for the iterative methods: *Iterations are stopped as soon as either the accuracy goal or the precision goal is satisfied*. Note that the precision and accuracy goals are only goals; this is because the true relative and absolute errors are unknown and have to be estimated with the iterative methods. The true relative or absolute error may be much larger or much smaller than the given goal.

Note that **p** and **a** can also be infinite. If **p** is infinity, the precision goal will never be satisfied, and so only absolute error is used as the criterion. Similarly, if **a** is infinity, the accuracy goal will never be satisfied, and so only relative error is used as the criterion.

■ **Example**

We calculate an integral:

```
f = Sin[4 x] Exp[-x];
```

```
Integrate[f, {x, 0, 50}]
```

$$\frac{1}{17} \left(4 - \frac{4 \, \text{Cos}[200] + \text{Sin}[200]}{e^{50}} \right)$$

```
N[%, 20]
```
```
0.23529411764705882353
```

Numerical integration gives the following result:

```
NIntegrate[f, {x, 0, 50}] // InputForm
```
```
0.23529411799402022
```

The result has 9 correct decimals, so it is very good. However, we try to get a still better result by using 30 digits during the calculations (so arbitrary-precision numbers are used) and asking for a result with an absolute error of at most 10^{-20}:

```
NIntegrate[f, {x, 0, 50}, WorkingPrecision → 30, AccuracyGoal → 20]
```
```
0.235294117647058823529399530360
```

The result has, indeed, at least 20 correct digits. If we are satisfied with a lower precision and accuracy, we can write the following:

```
NIntegrate[f, {x, 0, 10}, PrecisionGoal → 2, AccuracyGoal → 2] // InputForm
```
```
0.23529925221643583
```

■ **Default Values of Options**

The three options mentioned previously usually have the following default values:

```
WorkingPrecision → MachinePrecision   (= 16 in most computers)
PrecisionGoal → Automatic   (= 8 in most computers)
AccuracyGoal → Automatic   (= 8 in most computers)
```

The default value **MachinePrecision** of **WorkingPrecision** means that normal machine-precision numbers are used in the calculations. As we discussed in Section 12.2.2, p. 405, the numerical value of **MachinePrecision** is **$MachinePrecision**, which usually means a number close to 16.

The default values of the two goals mean that iterations are normally stopped when *either* the relative *or* the absolute error is less than 10^{-8}. More detailed information about the default stopping criteria is as follows:

Default stopping criteria:

FindMinimum, FindMaximum The relative or absolute error of *the optimum point and of the value of the function at the optimum point* is less than 10^{-8}.

NMinimize, NMaximize The relative or absolute error of *the optimum point and of the value of a penalty function at the optimum point* is less than 10^{-8}.

FindFit The relative or absolute error of *the optimum point and of a norm function* (e.g., squared residuals) is less than 10^{-8}.

FindRoot The relative or absolute error of *the root* is less than 10^{-8} and the absolute value of *the function at the root* is less than 10^{-8}.

NDSolve The relative or absolute error of *the solution of the differential equation at each chosen point* is less than 10^{-8}.

NIntegrate The relative error of *the integral* is less than 10^{-6}.

NSum The relative error of *the sum* is less than 10^{-6}.

NProduct The relative error of *the product* is less than 10^{-6}.

■ Adjusting the Options

With the default value **MachinePrecision** (which is approximately 16 in most computers) of **WorkingPrecision**, the usual fixed-precision numbers are used and the calculations are fast, but the precision may not suffice in a critical or ill-conditioned case (see Section 12.2.2, p. 406, for advantages and disadvantages of fixed-precision numbers).

By specifying a value for **WorkingPrecision** other than **MachinePrecision**, the calculations are done with variable-precision arithmetic. Using a high value such as 20 or more generally gives more accurate results; the disadvantage is that the calculations take more time. Note that **WorkingPrecision** affects only the calculations; the result probably has a lower precision. To control the precision and accuracy of the result, use **PrecisionGoal** and **AccuracyGoal**.

The default value **Automatic** of **PrecisionGoal** and **AccuracyGoal** usually means 8 if **WorkingPrecision** has its default value. If **WorkingPrecision** has another value, then the default value of the precision and accuracy goals is generally **WorkingPrecision**/2. For example, if **WorkingPrecision** is 30, the default value of the precision and accuracy goals is usually 15. (For **NIntegrate, NSum,** and **NProduct**, the default value of **PrecisionGoal** is 6 or **WorkingPrecision** - 10.)

If you increase the precision or accuracy goal from the default value 8, you often also have to increase the value of **WorkingPrecision**; give it a value that is at least a few digits larger than the goal.

12.3.2 Other Common Options

■ StepMonitor and EvaluationMonitor

Here are two options for various iterative numerical methods:

StepMonitor An option that gives a command to be executed after each step; examples of values:
   None, Sow[x], ++n, AppendTo[iters, x]
EvaluationMonitor An option that gives a command to be executed after each evaluation of functions derived from the input; examples of values: None, Sow[x], ++n, AppendTo[points, x]

StepMonitor and **EvaluationMonitor** are useful when investigating how a numerical method proceeds (e.g., how many iterations are needed and what are all the points that a numerical method generates). The following commands have both of these options: **FindFit**, **FindMaximum**, **FindMinimum**, **FindRoot**, **NMaximize**, **NMinimize**, and **NDSolve**. In addition, **NIntegrate**, **NProduct**, **NSum**, and many plotting commands have the **EvaluationMonitor** option.

The values of these two options are set with a delayed setting by using **:>** instead of **->** (to avoid the immediate evaluation of the command given). Note that *Mathematica* automatically replaces **:>** with the special symbol ⧴. The following is a typical example:

```
f = Exp[-x] - x^2;

n = 0; FindRoot[f, {x, -1}, StepMonitor ⧴ ++n]
{x → 0.703467}

n        6
```

We needed six iterations to find the root. Here is another example:

```
iters = {}; FindRoot[f, {x, -1}, StepMonitor ⧴ AppendTo[iters, x]]
{x → 0.703467}

iters
{1.39221, 0.835088, 0.709834, 0.703483, 0.703467, 0.703467}
```

These are the six points generated by the iterative method used by **FindRoot**. We could also use **Sow** and **Reap** (see Section 18.2.3, p. 564):

```
Reap[FindRoot[f, {x, -1}, StepMonitor ⧴ Sow[x]]]
{{x → 0.703467}, {{1.39221, 0.835088, 0.709834, 0.703483, 0.703467, 0.703467}}}
```

Reap[expr] returns a list of two components: the value of **expr** (here, the result of **FindRoot**) together with a list of values of the expression to which **Sow** has been applied during the calculation of **expr**.

■ **Compiled**

> **Compiled** An option for various numerical methods that indicates whether to compile the expression with which they are working; possible values: **Automatic**, **True**, **False**

The default value **Automatic** of **Compiled** usually means **True**; this means that the command *compiles* the expression to be manipulated. Compilation transforms the expression to a kind of pseudocode that contains simple instructions for evaluating the expression. This speeds up the computations.

Compiled expressions use only normal machine-precision numbers. If you want the command to use arbitrary-precision numbers when calculating the value of an expression, you have to turn the compiling off by giving the option **Compiled → False**.

Compiling is considered in more detail in Section 17.2.3, p. 528.

13

Expressions

Introduction

> *Algebra is generous: She often gives more than is asked for.* —J. d'Alembert

Section 13.1 contains some very basic techniques for expressions: assigning values for variables, clearing values of variables, inserting a value of a variable into an expression, and picking parts of an expression. Read this section carefully.

Mathematica often writes the result in a good form, but occasionally we will want to perform some manipulations on an expression; see Sections 13.2 and 13.3. *Mathematica* has many commands to this end, although a little experimenting is sometimes needed to get the desired result.

You will probably find in *Mathematica* all of the usual and special functions you need; Section 13.4 gives some lists of these functions. *Mathematica* also uses the functions effectively by itself; for example, many integrals can be written only in terms of certain special functions. Note that http://functions.wolfram.com/ is an excellent website for information about functions.

Lists are very important expressions in *Mathematica*; they are considered in Chapter 14. Chapter 15 is devoted to converting lists to tables. *Patterns*, considered in Chapter 16, represent classes of expressions.

13.1 Basic Techniques

13.1.1 Assigning and Clearing Values

For referring to earlier results, you have at least three methods at your disposal: use `%`, use `Out[n]`, or give a name to the result to which you want to refer.

Using `%` (see Section 1.2.2, p. 7) is convenient provided that you refer to a very recent result and you do not need to execute the command containing `%` several times. If you have to execute several times until you obtain the desired result, then you need to use `%%`, `%%%`, and so on in the succeeding executions so that you refer to the correct result.

A better possibility to refer to an earlier result may be the use of `Out[n]` (see Section 1.2.2) because `n` remains the same even if you execute several times. (By the way, to start the numbering of results from 1 again, execute `$Line = 0`.)

Giving a name to the result and using it later may be the best method. This is considered next.

■ Assigning Values to Symbols

> `x = a`   Assign the value **a** to **x**
> `x = y = a`   Assign the same value to several symbols
> `{x, y, …} = {a, b, …}`   Assign **a** to **x**, **b** to **y**, …

Assigning values to symbols with `=` (or `Set`) has a drawback. Suppose you want to consider the following expression:

> `p = x + Sin[y];`

You want to calculate the value of **p** when **x** is 3.7 and **y** is 1.2. One possibility is to assign the values to **x** and **y** and then ask the value of **p**:

> `x = 3.7; y = 1.2; p     4.63204`

This is a straightforward method but has a drawback: **x** and **y** have from here on the values 3.7 and 1.2, and **p** has from here on the value 4.63204, unless you assign a new value for **x** or **y** or clear the values of **x** and **y**. *The values of* **x** *and* **y** *are applied in all expressions in which* **x** *or* **y** *appear*. This can cause trouble later when you have perhaps forgotten that you assigned a value for **x** and **y**. Thus, remember to clear or remove a symbol when you no longer need it so that the value of the symbol does not cause trouble in later calculations. If the symbol is **x**, type `x =.`, `Clear[x]`, or `Remove[x]`; these commands are explained soon.

A recommended rule is to assign values for variables sparingly. The preferred method for calculating values of expressions for particular values of variables is to use transformation rules (see Section 13.1.2, p. 416).

■ **Asking Information about Symbols**

> **x** Give the value of symbol **x**
> **Definition[x]** Give the definitions associated with **x**
> **?x** Give the context of and definitions associated with symbol **x**
> **?x*** Print all symbols beginning with **x**
>
> **?Global`*** Print all user-defined symbols
> **Names["Global`*"]** Give a list of all user-defined symbols
> **ToExpression[%]** Give a list of the values of the symbols given by **Names**

These commands are useful if you have forgotten what symbols you have used and what values they have been given. We ask the current value and the original definition of **p**:

    p   4.63204

    Definition[p]
    p = x + Sin[y]

    ?p
_____

    Global`p

    p = x + Sin[y]

Here are all the symbols we have used in this section:

    ?Global`*

▼ Global`

    p   x   y

Clicking the name of a symbol prints its definition. Next, we ask for our symbols and their values in a list form:

    Names["Global`*"]   {p, x, y}

    ToExpression[%]   {4.63204, 3.7, 1.2}

We consider contexts such as **Global`** in Section 17.3.1, p. 531. It may suffice here to mention that the **Global`** context is an environment that contains all the user-defined symbols.

■ **Removing Symbols**

> **x =.** Clear the value of **x**
> **Clear[x]** Clear the value of **x**
> **Remove[x]** Remove **x** completely
> **Remove[x, y , …]** Remove **x**, **y**, ...
> **Remove["x*"]** Remove all symbols that start with **x**
> **Remove["Global`*"]** Remove all user-defined symbols

Clear and **Remove** are used in the same way. **Clear** as well as **=.** (or **Unset**) only clears the *value* of a symbol. **Remove** not only clears the value of a symbol but also removes the entire symbol.

Occasionally, we may observe that *Mathematica* is not working as expected. With high probability, we have forgotten that some symbols have specified values, and this causes the trouble. One way to "clear the table" is to use the command **Remove["Global`*"]**; it removes all user-defined symbols. Another, even easier, way is to quit the kernel with the menu command **Evaluation ▷ Quit Kernel** (the kernel can be started anew by executing a command or with the menu command **Evaluation ▷ Start Kernel**).

13.1.2 Inserting Values

■ Using Transformation Rules

If we want to calculate the value of an expression **expr** for a specific value **a** of a variable **x**, we can use two methods. One method is to assign the value **a** to **x** and then ask the value of **expr**:

```
x = a
expr
```

Another method is to apply a *transformation rule*:

```
expr /. x → a
```

The latter method is better. The first method has the drawback that **x** has the value **a** during the rest of the session, and this may cause trouble.

Getting used to the transformation technique may take some time. However, it is worth spending the time on this topic because it is very important in *Mathematica*. First, you can use the technique to insert values into expressions. Second, some important commands, such as **Solve**, **NSolve**, **FindRoot**, **FindMinimum**, **Minimize**, **NMinimize**, **FindFit**, **DSolve**, and **NDSolve**, give the result in the form of a transformation rule, so you have to know how to handle such results. Some examples were already worked in Sections 2.3.3, p. 43, and 2.3.6, p. 48. Transformation rules are also considered in Sections 16.1 p. 494, 18.4.1, p. 584, and 18.5.3, p. 610.

Note that the arrow (→) can be written as -> because *Mathematica* automatically replaces these two marks with a genuine arrow. The arrow can also be written as ⎡ESC⎤->⎡ESC⎤.

■ Simple Rules

expr /. x → a Replace **x** with **a** in **expr**
f = expr /. x → a Replace **x** with **a** and assign the result to **f**

In place of **expr /. x → a**, we could use **ReplaceAll[expr, x → a]**. Here is an expression:

```
Clear[x, y]; p = x + Sin[y];
```

Here is a value of the expression:

```
p /. x → 3.7     3.7+Sin[y]
```

Note that after the transformation **p /. x → 3.7**, the variable **p** still has its original value **x + Sin[y]** and **x** has no value, as can be seen by asking for the value of **{p, x}**:

```
{p, x}    {x + Sin[y], x}
```

In general, after a transformation such as **expr /. x → a**, **x** has no value and the value of **expr** has not changed. *Mathematica* only calculates and shows the value of **expr** after the transformation is done; no assignment is made for **expr**. Write **f = expr /. x → a** if you want to store the result as a value of a variable.

■ **Several Rules**

---

expr /. {x → a, y → b, … }   Replace **x** with **a**, **y** with **b**, …

expr /. Thread[{x, y, … } → {a, b, … }]   Replace **x** with **a**, **y** with **b**, …

expr /. Thread[vars → vals]   Replace variables in the list **vars** with the corresponding values in the list **vals**

expr /. x → a /. y → b   Replace **x** with **a**; in the result, replace **y** with **b**

Try two rules:

   p /. {x → 3.7, y → 1.2}   4.63204

Suppose we have some variables and some values:

   vars = {x, y}; vals = {3.7, 1.2};

If we want to replace **vars** with **vals** in **expr**, we cannot write **expr /. vars → vals**. We have to apply **Thread** to obtain the list of rules:

   Thread[vars → vals]   {x → 3.7, y → 1.2}

Thus, we can write the following:

   p /. Thread[vars → vals]   4.63204

Note that when calculating a transformation such as **expr /. {x → a, y → b, … }**, *Mathematica* looks at each part of the expression, tries all the rules on it, and then goes on to the next part. The first rule that applies to a particular part is used; no other rules are tried on that part or on any of its subparts. For example,

   {x, y} /. {x → y, y → x}   {y, x}

Here, *Mathematica* tries the rules that may apply to **x**, finds that **x → y** is suitable, and applies it (**y → x** is no longer applied). Then it goes to **y**, finds that **y → x** is suitable, and applies it (**x → y** is no longer applied). On the other hand, consider the next example:

   {x, y} /. x → y /. y → x   {x, x}

Mathematica first replaces **x** with **y** and then, in the resulting list **{y, y}**, replaces **y** with **x**.

■ **Forming a List of Values**

---

expr /. {{x → a}, {x → b}, … }   Replace **x** with **a**, **b**, …; the result is a list

expr /. {{x → a, y → b, … }, {x → c, y → d, … }, … }   Apply several sets of rules; the result is a list

---

We calculate the value of **p** with several values of **x**:

   p /. {{x → 3.7}, {x → 3.8}, {x → 3.9}}
   {3.7 + Sin[y], 3.8 + Sin[y], 3.9 + Sin[y]}

The result is a list. We can apply several sets of rules:

   p /. {{x → 3.7, y → 1.2}, {x → 3.8, y → 1.3}, {x → 3.9, y → 1.4}}
   {4.63204, 4.76356, 4.88545}

Solve gives several sets of rules as the answer. As an example, we solve the following equation:

   eqn = 6 − 5 x + x^2 == 0;

```
sol = Solve[eqn, x]     {{x → 2}, {x → 3}}
```

The result is a list of lists. Each sublist gives one solution of the equation. The form of the solution is useful for continuing the calculations. Because **sol** now has the value {{x → 2}, {x → 3}}, to get a list of values (instead of a list of transformation rules), we can write the following:

```
x /. sol      {2, 3}
```

We can check that the solution is correct by inserting it into the equation:

```
eqn /. sol     {True, True}
```

Mathematica could conclude that both solutions are correct—that is, that both solutions simplify the equation into the form 0 == 0, which is true. Similarly, we can calculate the value of any expressions with the solutions:

$$1 / (1 + x) /. \text{sol} \qquad \left\{ \frac{1}{3}, \frac{1}{4} \right\}$$

We could also ask directly for the values of **x**:

```
sol2 = x /. Solve[eqn, x]      {2, 3}
```

If we now want to check the solution, we must directly write the necessary transformation rules:

```
eqn /. {{x → sol2[[1]]}, {x → sol2[[2]]}}     {True, True}
```

Compare this with the simple command **eqn /. sol**. We see that the transformation rules given by **Solve** are handy when we continue the calculations.

■ **Using Rules Repeatedly**

> **expr //. {x → a, y → b, … }** Apply the rules repeatedly until the result no longer changes

Consider the following example:

```
p /. {x → y, y → a}      y + Sin[a]
```

At each part of **p**, the rules are applied only once. Write the following:

```
p /. x → y /. y → a     a + Sin[a]
```

First, **x** is replaced with **y**, and then **y** is replaced with **a**. *Mathematica* also has the command **ReplaceRepeated**, which is formed by **//.**. It applies the rules until the result no longer changes:

```
p //. {x → y, y → a}      a + Sin[a]
```

13.1.3 Picking Parts

> **expr[[i]]** or **expr⟦i⟧** The *i*th part of **expr**
> **expr[[i, j]]** or **expr⟦i, j⟧** The *j*th part of the *i*th part of **expr**

Picking parts of expressions can be done with double brackets. We can also use the special symbols ⟦ and ⟧; they can be written as ESC[[ESC and ESC]]ESC. We will use the symbols ⟦ and ⟧ because they are simpler and shorter than the double brackets. Consider the following expression:

```
p = (y^2 + 2 y - 8) / (y + 4)^2 + 2 / (x + 3)
```

$$\frac{2}{3 + x} + \frac{-8 + 2 y + y^2}{(4 + y)^2}$$

We pick several parts:

`{p[[1]], p[[2]], p[[1, 1]], p[[1, 2]], p[[1, 2, 1]], p[[1, 2, 2]]}`

$$\left\{ \frac{2}{3+x}, \frac{-8+2\,y+y^2}{(4+y)^2}, 2, \frac{1}{3+x}, 3+x, -1 \right\}$$

The result is not always the one you expect. For example, you might expect that $p[[1, 2]]$ is $3+x$ instead of $1/(3+x)$. Given that $p[[1, 2]]$ is $1/(3+x)$, you might expect that $p[[1, 2, 1]]$ is 1 instead of $3+x$. The explanation is that the parts are picked from the representation *Mathematica* uses internally. You can look at this representation with **FullForm**:

`FullForm[p[[1]]]`

`Times[2, Power[Plus[3, x], -1]]`

From this we see that $p[[1, 2]]$ is in fact **Power[Plus[3, x], -1]**—that is, $1/(3+x)$—and that $p[[1, 2, 1]]$ is **Plus[3, x]**—that is, $3+x$.

■ Example

Let us calculate the area enclosed by two curves. Here are the curves:

`f1 = x^2 / 2 + x / 2 + 1;`
`f2 = x^2 - 1 / 2;`

We want to calculate the filled area between the points of intersection:

`Plot[{f1, f2}, {x, -1.5, 2.5}, Filling → True]`

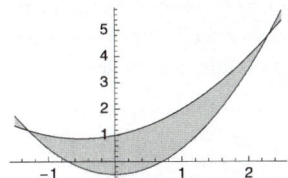

The points of intersection are as follows:

`sol = x /. Solve[f1 == f2]` $\left\{ \frac{1}{2} \left(1 - \sqrt{13} \right), \frac{1}{2} \left(1 + \sqrt{13} \right) \right\}$

Then we integrate the difference of the curves between these points:

`Integrate[f1 - f2, {x, sol[[1]], sol[[2]]}]` $\dfrac{13\,\sqrt{13}}{12}$

13.2 Manipulating Expressions

13.2.1 Simplifying

Simplify[expr]	Simplify by trying algebraic and trigonometric transformations
FullSimplify[expr]	Try a much wider range of transformations

These commands can also be used in the form **expr // Simplify** as all commands with one argument. Often, **Simplify** does a good job, but if we are not satisfied, we can use **FullSimplify**. It tries many more transformations, and it knows much about special functions; however, calculations can be long. As an example, **Simplify** does not do anything to the following expressions:

(n + 3) ! / n ! // Simplify $\dfrac{(3 + n)\,!}{n\,!}$

Sqrt[5 + 2 Sqrt[6]] // Simplify $\sqrt{5 + 2\sqrt{6}}$

FullSimplify succeeds in the simplification:

(n + 3) ! / n ! // FullSimplify $(1 + n)\,(2 + n)\,(3 + n)$

Sqrt[5 + 2 Sqrt[6]] // FullSimplify $\sqrt{2} + \sqrt{3}$

Next, we consider the use of assumptions with simplification. Then, we examine the options of the two commands.

■ Using Assumptions

Simplify[expr, ass]   Use assumptions **ass** during the simplification
FullSimplify[expr, ass]   Use assumptions **ass** during the simplification

Simplify and **FullSimplify** (and also **FunctionExpand**; see Section 13.2.2, p. 424) accept assumptions. The assumptions can be given by declaring domains of variables and by specifying equations and inequalities (note that a variable declared to satisfy an inequality is automatically assumed to be real). Various assumptions can be combined with logical operators such as **&&** (and), **||** (or), and **!** (not) (see Section 13.3.5, p. 431). Domains can be declared as follows:

x ∈ dom   Declare that **x** is an element of domain **dom**
{x, y, … } ∈ dom or **(x|y| …) ∈ dom**   Declare that **x, y**, ... are elements of domain **dom**
Possible domains:
Complexes, Reals, Algebraics, Rationals, Integers, Primes, Booleans

Write ∈ by pressing ⎋el⎋. In place of **x ∈ dom** and **{x, y, … } ∈ dom**, we can write **Element[x, dom]** and **Element[{x, y, … }, dom]**.

Domains are often used in simplification, but we can also ask whether a given expression belongs to a certain domain:

{Sqrt[5 + 2 Sqrt[7 ^ (3 / 8)]] ∈ Algebraics,
 E ∈ Algebraics, Pi ∈ Algebraics, E + Pi ∈ Algebraics}
 {True, False, False, $e + \pi \in$ Algebraics}

(Note that it is not yet known whether $e + \pi$ is an algebraic number.)

In simplifying expressions, note the following: Every variable is assumed to be a generic *complex* quantity. This causes *Mathematica* sometimes not to do the simplifications we expect. However, by specifying suitable domains or other assumptions, we can get the results we want. Next, we consider several examples.

■ Square Roots

Mathematica does not automatically simplify $\sqrt{x^2}$ to $|x|$:

    `Sqrt[x^2] // Simplify` $\sqrt{x^2}$

The reason for this is that this simplification does not hold for complex numbers:

    `{Sqrt[(2 + 3 I)^2], Abs[2 + 3 I]}` $\left\{2 + 3\,i,\ \sqrt{13}\right\}$

However, if we assume that x is real, then the simplification is valid:

    `Simplify[Sqrt[x^2], x ∈ Reals]`   `Abs[x]`

If we assume that $x < 0$, then we get $-x$:

    `Simplify[Sqrt[x^2], x < 0]`   `-x`

In the following, we also need various assumptions:

    `Simplify[Sqrt[x] Sqrt[y], x > 0 && y > 0]` $\sqrt{x\,y}$

In the following, we need the assumption that x is positive:

    `Simplify[Sqrt[1 / x], x > 0]` $\dfrac{1}{\sqrt{x}}$

■ Powers of Powers

The expression $\left(a^b\right)^c$ cannot always be simplified as $a^{b\,c}$:

    `Simplify[(a^b)^c]` $\left(a^b\right)^c$

If c is an integer or if a and b are positive, then we get the simplified form:

    `Simplify[(a^b)^c, c ∈ Integers]` $a^{b\,c}$

    `Simplify[(a^b)^c, a > 0 && b > 0]` $a^{b\,c}$

■ Logarithmic Expressions

Simplification of logarithmic expressions often also requires suitable assumptions:

    `Simplify[Log[a] + Log[b], a > 0 && b > 0]`   `Log[a b]`

    `Simplify[Log[a^2], a > 0]`   `2 Log[a]`

    `Simplify[Log[1 / a], a > 0]`   `-Log[a]`

■ Trigonometric Expressions

Trigonometric expressions often become simpler if some parameters are integers:

    `Integrate[Sin[m x] Cos[n x], {x, 0, Pi}]`

$$\frac{m - m\,\mathrm{Cos}[m\,\pi]\,\mathrm{Cos}[n\,\pi] - n\,\mathrm{Sin}[m\,\pi]\,\mathrm{Sin}[n\,\pi]}{m^2 - n^2}$$

    `Simplify[%, {m, n} ∈ Integers]` $-\dfrac{\left(-1 + (-1)^{m+n}\right)\,m}{m^2 - n^2}$

Note, however, that this result holds for generic values of m and n. For the special case of $m = n$, we get another result:

`Integrate[Sin[m x] Cos[m x], {x, 0, Pi}]` $\dfrac{\text{Sin}[m\,\pi]^2}{2\,m}$

`Simplify[%, m ∈ Integers]`   0

■ Inequalities

If the expression to be simplified is an inequality, then **Simplify**

* gives `True` if the inequality holds for all values of variables in the specified domain;
* gives `False` if the inequality does not hold at any point in the specified domain; and
* does not claim anything if the inequality holds for some points but does not hold for some others, or if the validity of the inequality cannot be decided.

Here are some examples:

`Simplify[x ≤ x + a, a ≥ 0]`   True

`Simplify[x ≤ x + a, a < 0]`   False

`Simplify[x ≤ x + a, a ≤ 0]` $a ≥ 0$

Here is a less trivial example:

`Simplify[Exp[x] > Log[x] + 2, x > 0]`   True

■ Other Expressions

Many kinds of expressions can be simplified with assumptions. Here, we demonstrate that *Mathematica* knows about Fermat's Last Theorem:

`FullSimplify[x^n + y^n == z^n, {x, y, z, n} ∈ Integers && n > 2 && x y z ≠ 0]`   False

■ Specifying Default Assumptions

Assumptions can be used with some commands, such as **Simplify**, **FullSimplify**, **FunctionExpand**, **Limit**, and **Integrate**. Actually, we can distinguish two types of assumptions: default and specific. When we previously wrote **Simplify[Sqrt[x^2], x < 0]**, we used a specific assumption. The default assumptions are represented by the variable **$Assumptions**. Its initial value is true:

`$Assumptions`   True

This has no effect, and so this means, in practice, that there are no default assumptions.

Default assumptions can be added by **Assuming**. As an example, we note that in place of **Simplify[Sqrt[x^2], x < 0]**, we can write the following:

`Assuming[x < 0, Simplify[Sqrt[x^2]]]` $-x$

Here, we added the assumption $x < 0$ into the current value of **$Assumptions**, giving the default assumption True && $(x < 0)$—that is, $x < 0$.

Assuming can be used to define local environments with given default assumptions in a way that is similar to the way **Module** and **Block** (see Section 17.1.4, p. 520) are used to define local environments with given variables or their values. Inside **Assuming**, we can write several commands: **Assuming[assum, expr1; expr2; …]**. Thus, we do not need to write the assumptions in each command. Each command, however, can have its specific assumptions.

■ **Options**

Options of `Simplify` *and* `FullSimplify`:

`Assumptions` Default assumptions; default value: `$Assumptions`

`TimeConstraint` Time in seconds to try a particular transformation; default values: `300` (`Simplify`), ∞ (`FullSimplify`)

`ComplexityFunction` Function used in assessing the complexity of the transformed expression; examples of values: `Automatic`, `(LeafCount[#]&)`

`Trig` Whether to also use trigonometric transformations; possible values: `True, False`

`TransformationFunctions` Functions tried in transforming the expression; examples of values: `Automatic, {Automatic, f}`

`ExcludedForms` Forms of subexpressions not touched in simplification; examples of values: `{}, Gamma[_]`

In addition to using `Assuming`, default assumptions can be given with the `Assumptions` option.

`Simplify` and `FullSimplify` do a sequence of transformations when searching for the simplest form of the expression. If `Simplify` does not complete a particular transformation in 300 seconds, it gives up, prints a warning, and goes to the next transformation. In such a case, you can consider trying the simplification anew with a larger value of `TimeConstraint`. `FullSimplify`, by default, does all transformations until completion.

Normally, `Simplify` and `FullSimplify` assess the complexity of the transformed expressions primarily according to their `LeafCount` (the total number of indivisible subexpressions), with corrections to treat integers with more digits as more complex. For example, $4\log(6)$ is considered the simplest form, but if only `LeafCount` is used, then the simplest form is the following:

```
Simplify[4 Log[6], ComplexityFunction → (LeafCount[#] &)]
Log[1296]
```

Normally, both algebraic and trigonometric transformations are tried. Set `Trig` to `False` if you want trigonometric transformations not to be tried.

`Simplify` and, in particular, `FullSimplify` have large collections of transformations to try. However, sometimes they are not able to simplify an expression. We can then help them by adding one or more transformation functions. For example, without the function `comp`, `FullSimplify` does not succeed in performing the following simplification:

```
comp[x_ ^ (2 n_)] := Expand[x ^ 2] ^ n

FullSimplify[(1 + Sqrt[2]) ^ (2 n) - (3 + 2 Sqrt[2]) ^ n,
  n ∈ Integers, TransformationFunctions → {Automatic, comp}]
0
```

With an option we can exclude some forms of subexpressions in the simplification. For example, if all subexpressions are taken into account in the simplification, we get the following:

```
FullSimplify[(n + 3) ! Gamma[n + 1] / (n ! Gamma[n + 2])]
(2 + n) (3 + n)
```

However, if gamma functions are not touched, we get the following:

```
FullSimplify[(n + 3) ! Gamma[n + 1] / (n ! Gamma[n + 2]), ExcludedForms → Gamma[_]]
```

$$\frac{(1 + n) \ (2 + n) \ (3 + n) \ \text{Gamma}[1 + n]}{\text{Gamma}[2 + n]}$$

■ **Refine**

> **Refine[expr, ass]**   Give the form of **expr** that would be obtained if symbols in it were replaced
>   with explicit numerical expressions satisfying assumptions **ass**

Refine is useful for simplifying or developing expressions under given assumptions. For example, in the following cases, **Simplify** does nothing, but **Refine** gives us a modified expression:

```
{Refine[Abs[x], x < 0], Refine[Log[x], x < 0],
  Refine[Sin[x + n π], n ∈ Integers], Refine[Sqrt[x y], x > 0]}
```
$$\left\{-x, \ i\,\pi + \text{Log}[-x], \ (-1)^n \,\text{Sin}[x], \ \sqrt{x}\ \sqrt{y}\right\}$$

13.2.2 Expanding

> **Expand[expr]**   Expand products and positive integer powers in the top level
>
> **Expand[expr, patt]**   Do not expand any parts that are free of pattern **patt**
>
> **ExpandAll[expr]**   Expand products and positive integer powers in all levels
>
> **ExpandAll[expr, patt]**   Do not expand any parts that are free of pattern **patt**
>
> **FunctionExpand[expr]**   Expand special functions
>
> **FunctionExpand[expr, ass]**   Use assumptions **ass** during the expansion
>
> **PowerExpand[expr]**   Expand powers of products and powers of powers
>
> **PowerExpand[expr, Assumptions → ass]**   Use assumptions **ass**. The default value **Automatic** does
>   not check the validity of the expansion. The value **True** gives a universally valid expansion.
>
> **PowerExpand[expr, {x, y, …}]**   Expand only with respect to the given variables

■ **Expand and ExpandAll**

In the following example, we do not expand terms not containing **x**:

```
Expand[(1 + a) ^2 (1 + x) ^2, x]
```
$$(1 + a)^2 + 2\,(1 + a)^2\, x + (1 + a)^2\, x^2$$

Expand does not expand denominators, but **ExpandAll** does:

```
Expand[(1 + a) ^2 / (1 + x) ^2]
```
$$\frac{1}{(1 + x)^2} + \frac{2\,a}{(1 + x)^2} + \frac{a^2}{(1 + x)^2}$$

```
ExpandAll[(1 + a) ^2 / (1 + x) ^2]
```
$$\frac{1}{1 + 2\,x + x^2} + \frac{2\,a}{1 + 2\,x + x^2} + \frac{a^2}{1 + 2\,x + x^2}$$

■ **FunctionExpand**

FunctionExpand does not automatically write $\log(x\,y) = \log(x)\log(y)$:

```
Log[x y] // FunctionExpand      Log[x y]
```

This happens because $\log(x\,y) = \log(x)\log(y)$ is not always correct. If x and y are positive, then the expansion can be made:

```
FunctionExpand[Log[x y], x > 0 && y > 0]      Log[x] + Log[y]
```

FunctionExpand is sometimes able to convert expressions to forms in which the arguments and functions are simpler:

$$\texttt{Cos[ArcSin[x / 3]] // FunctionExpand} \qquad \frac{1}{3} \sqrt{3 - x} \; \sqrt{3 + x}$$

$$\texttt{ChebyshevT[n, x] // FunctionExpand} \qquad \texttt{Cos[n ArcCos[x]]}$$

$$\texttt{Log[ProductLog[3, z]] // FunctionExpand}$$
$$6 \, i \, \pi + \texttt{Log[z]} - \texttt{ProductLog[3, z]}$$

■ PowerExpand

PowerExpand does, by default, brutal expansions without taking into account the appropriate assumptions. For example,

$$\texttt{\{PowerExpand[Sqrt[x\^2]], PowerExpand[(a\^b)\^c], PowerExpand[Log[x y]]\}}$$

$$\left\{ \texttt{x, a}^{\texttt{bc}}, \texttt{Log[x] + Log[y]} \right\}$$

As previously noted, these results are not always correct. In general, $\left(a^b\right)^c = a^{bc}$ and $(a\,b)^c = a^c b^c$ hold only if c is an integer or a and b are positive real numbers. However, using assumptions we get correct results:

$$\texttt{PowerExpand[Sqrt[x\^2], Assumptions → True]}$$

$$e^{i \, \pi \, \text{Floor}\left[\frac{1}{2} - \frac{\text{Arg}[x]}{\pi}\right]} \, x$$

$$\texttt{\{Simplify[\%, x < 0], Simplify[\%, x > 0]\}} \qquad \{-x, \, x\}$$

$$\texttt{PowerExpand[Sqrt[x\^2], Assumptions → x < 0]} \qquad -x$$

13.2.3 More about Expressions

■ Short Forms for Results

Sometimes the result of a computation is so long and complicated that we do not want to see it completely. On the other hand, we might sometimes be interested in the structure of the complicated expression. The following commands are useful in these cases:

Length[expr]	Give the number of topmost parts of **expr**
Short[expr]	Print **expr** in a shortened form that is less than approximately one line
Short[expr, n]	Print **expr** in a shortened form that is approximately **n** lines long
Shallow[expr]	Show a shallow form of **expr**
Shallow[expr, depth]	Show all parts below **depth** in skeleton form
Shallow[expr, {depth, length}]	Show all parts below **depth** or longer than **length** in skeleton form

Shallow[expr] means **Shallow[expr, {4, 10}]**. Here is a long expression:

$$\texttt{q = Integrate[x\^2 Sqrt[1 + x\^2] / (1 - x + x\^2), x] // Simplify;}$$

We ask for some short and shallow forms:

$$\texttt{Short[q]}$$

$$\frac{1}{12} \left(12 \, \sqrt{1 + x^2} \; + \; \ll 11 \gg \right)$$

```
Short[q, 2]
```

$$\frac{1}{12}\left(12\sqrt{1+x^2}+6x\sqrt{1+x^2}+\ll 10\gg\right)$$

```
Shallow[q]
```

$$\frac{1}{12}\,(\text{Times}[\ll 2\gg]+\text{Times}[\ll 3\gg]+\text{Times}[\ll 2\gg]+\text{Times}[\ll 3\gg]+$$
$$\text{Times}[\ll 3\gg]+\text{Times}[\ll 3\gg]+\text{Times}[\ll 3\gg]+\text{Times}[\ll 3\gg]+\text{Times}[\ll 3\gg])$$

```
Shallow[q, {∞, 3}]
```

$$\frac{1}{12}\left(12\sqrt{1+x^2}+6x\sqrt{1+x^2}+6\,\text{ArcSinh}[x]+\ll 6\gg\right)$$

If an output is very long, *Mathematica* may show only a short form of it:

```
Range[10^5]
```

> ## A very large output was generated. Here is a sample of it:
>
> {1, 2, 3, 4, 5, 6, 7, 8, 9, 10, 11, 12, 13, 14, 15, 16, 17, 18, 19, ≪99 962≫,
> 99 982, 99 983, 99 984, 99 985, 99 986, 99 987, 99 988, 99 989, 99 990, 99 991,
> 99 992, 99 993, 99 994, 99 995, 99 996, 99 997, 99 998, 99 999, 100 000}
>
> | Show Less | Show More | Show Full Output | Set Size Limit... |

■ Everything Is an Expression

In previous sections, we considered expressions in the ordinary mathematical sense. In *Mathematica*, however, an expression is a much wider concept. In fact, in *Mathematica*, *everything is an expression*. This fact is most clearly seen from the internal representation of the various objects of *Mathematica*. A typical expression is of the form **head[arguments]**. The internal form can be seen with **FullForm**, and the head can be seen with **Head**. Here are simple examples:

```
FullForm /@ {x + y, x y, x^y, {x, y}}
{Plus[x, y], Times[x, y], Power[x, y], List[x, y]}
```

```
Head /@ {x + y, x y, x^y, {x, y}}
{Plus, Times, Power, List}
```

Here are some other examples of expressions:

```
FullForm /@ {x == y, x < y, x → y}
{Equal[x, y], Less[x, y], Rule[x, y]}
```

```
FullForm /@ {x - y, x / y, Sqrt[x]}
{Plus[x, Times[-1, y]], Times[x, Power[y, -1]], Power[x, Rational[1, 2]]}
```

It is sometimes useful to know the internal form, particularly if we want to pick a part of an expression, as we saw in Section 13.1.3, p. 418. The parts of an expression are decided from the internal representation and not from the normal form.

Mathematica has exactly six basic expressions, which are called *atoms*. All expressions are made up of these basic elements. Here are examples of all of them:

```
Head /@ {2, 2 / 5, 3.7, 6 + 2 I, x, "message"}
{Integer, Rational, Real, Complex, Symbol, String}
```

■ Levels of Expressions

Here is a list:

    t = {{0, 1}, {1, p}, {2, p^2}, {3, p^3}}

$$\big\{\{0, 1\}, \{1, p\}, \{2, p^2\}, \{3, p^3\}\big\}$$

To illustrate the *levels* of an expression, we first define that the zeroth level of an expression is the expression itself. So, the zeroth level of **t** is **t**. At the first level, we have the four lists **{0, 1}**, ..., **{3, p^3}**. At the second level, we have the elements of the lists: **0, 1, 1, p**, ..., **p^3**. At the third level, we have the components of the powers: **p, 2, p, 3**. Expressed in another way, the level of a part is the number of indices needed to show the position. For example,

    {t[[4]], t[[4, 2]], t[[4, 2, 1]]}

$$\big\{\{3, p^3\}, p^3, p\big\}$$

The levels of the expressions **{3, p^3}**, **p^3**, and **p** (the **p** in **p^3**) in **t** are thus 1, 2, and 3, respectively. A list of the parts of an expression at a given level can be seen with **Level**:

    Level[t, {3}]   {p, 2, p, 3}

In some commands, such as **Apply** and **Map** (Section 14.2, p. 459) and **Position** and **Cases** (Section 16.1.1, p. 493), we can use a *level specification*. It defines the level or levels of the expression toward which the operation of the command is directed. Levels are specified as follows:

0	The expression itself
n	Levels 1 through n
∞	All levels 1, 2, ...
{n}	Level n only
{n, m}	Levels n through m
{-1}	The lowest level

For example, the position of 2 in **t** at level 3 is as follows:

    Position[t, 2, {3}]   {{3, 2, 2}}

13.3 Manipulating Special Expressions

13.3.1 Rational Expressions

■ Factoring

Factor[expr]	Factor numerator and denominator
Factor[expr, Extension → {a, b, …}]	Allow coefficients that are rational combinations of the algebraic numbers **a**, **b**, …
Factor[expr, Extension → Automatic]	Allow coefficients that are rational combinations of the algebraic numbers in **expr**
Factor[expr, GaussianIntegers → True]	Factor over Gaussian integers (complex numbers with integer real and imaginary parts)
FactorList[expr]	Give the factors in a list form

```
a = (x^2 + 2 x - 8) / (x + 4) ^2 + 2 / (x + 3)
```

$$\frac{2}{3+x} + \frac{-8+2x+x^2}{(4+x)^2}$$

```
Factor[a]
```
$$\frac{(1+x)\ (2+x)}{(3+x)\ (4+x)}$$

Next, we try the **Extension** option:

```
Factor[1 + x^4]      1 + x^4
```

```
Factor[1 + x^4, Extension → I]      (-i + x^2) (i + x^2)
```

```
Factor[1 + x^4, Extension → Sqrt[2]]
```

$$-\left(-1 + \sqrt{2}\ x - x^2\right)\left(1 + \sqrt{2}\ x + x^2\right)$$

```
Factor[1 + x^4, Extension → {I, Sqrt[2]}]
```

$$\frac{1}{4}\left(\sqrt{2} - (1+i)\ x\right)\left(\sqrt{2} - (1-i)\ x\right)\left(\sqrt{2} + (1-i)\ x\right)\left(\sqrt{2} + (1+i)\ x\right)$$

```
Factor[1 + x^4, GaussianIntegers → True]
```

$$\left(-i + x^2\right)\left(i + x^2\right)$$

With **FactorList** we get a list of the factors:

```
FactorList[a]
```
```
{{1, 1}, {1 + x, 1}, {2 + x, 1}, {3 + x, -1}, {4 + x, -1}}
```

From this list, we get the factored expression as follows, by using **Apply**:

```
Times @@ Power @@@ %
```
$$\frac{(1+x)\ (2+x)}{(3+x)\ (4+x)}$$

■ **Other Manipulations**

Together[expr] Put terms over a common denominator and cancel factors	
Cancel[expr] Cancel out common factors in the numerator and denominator	
Apart[expr] Give the partial fraction expansion	
Apart[expr, x] Treat only **x** as a variable (treat other variables as constants)	
HornerForm[expr] (✳6) Put the numerator and denominator in Horner form	
ExpandNumerator[expr] Expand the numerator	
ExpandDenominator[expr] Expand the denominator	
Numerator[expr] Give the numerator	
Denominator[expr] Give the denominator	
Variables[expr] Give a list of all variables	

Together and **Cancel** also have the **Extension** option mentioned previously. Remember also the commands **Simplify** and **FullSimplify** (see Section 13.2.1, p. 419). We try some commands:

```
{Simplify[a], Factor[a], Together[a], Cancel[a], Apart[a]}
```

$$\left\{\frac{2+3x+x^2}{12+7x+x^2}, \frac{(1+x)\ (2+x)}{(3+x)\ (4+x)}, \frac{2+3x+x^2}{(3+x)\ (4+x)}, \frac{2}{3+x} + \frac{-2+x}{4+x}, 1 + \frac{2}{3+x} - \frac{6}{4+x}\right\}$$

13.3.2 Polynomial Expressions

`Factor[poly]` Factor **poly**; see other forms of **Factor** in Section 13.3.1, p. 427

`FactorTerms[poly]` Pull out any overall numerical factor
`FactorTerms[poly, {x, y, …}]` Pull out any overall factor that does not depend on **x, y**, …

`Expand[poly]` Expand out products and powers
`Expand[poly, x]` Avoid expanding parts not containing **x**

`Collect[poly, x]` Collect together terms involving the same powers of **x**
`Collect[poly, x, h]` Apply function **h** to the coefficients of the powers of **x**

`Decompose[poly, x]` Decompose **poly** into a composition of simpler polynomials
`HornerForm[poly, x]` (🎖6) Put **poly** in Horner form

`Coefficient[poly, expr]` Give the coefficient of **expr**
`Coefficient[poly, expr, n]` Give the coefficient of **expr^n**
`CoefficientList[poly, x]` Give a list of coefficients of powers of **x**

`Exponent[poly, x]` Give the maximum power of **x**
`Exponent[poly, x, h]` Apply **h** to the set of exponents of **x** (the default of **h** is **Max**)
`Variables[poly]` Give a list of all variables

`PolynomialQuotient[p, q, x]` Give the result of dividing **p** by **q**, with any remainder dropped
`PolynomialRemainder[p, q, x]` Give the remainder from dividing **p** by **q**

Remember also the commands **Simplify** and **FullSimplify**. Consider the following polynomial:

$\quad$ `r = c^2 + 8 c x + 16 x^2 + 2 c x^2 + 8 x^3 + x^4;`

We collect terms with respect to **c** and factor the coefficients:

$\quad$ `Collect[r, c, Factor]` $\quad$ $c^2 + 2 c x (4 + x) + x^2 (4 + x)^2$

The Horner form is efficient and stable in numerical computations:

$\quad$ `HornerForm[r /. c → 8]` $\quad$ $64 + x (64 + x (32 + x (8 + x)))$

Here are the coefficients of various powers of **x**:

$\quad$ `CoefficientList[r, x]` $\quad$ $\{c^2, 8 c, 16 + 2 c, 8, 1\}$

Here are the maximum and minimum exponents of **x** and all the exponents with which **x** appears:

$\quad$ `{Exponent[r, x], Exponent[r, x, Min], Exponent[r, x, List]}`
$\quad$ `{4, 0, {0, 1, 2, 3, 4}}`

13.3.3 Trigonometric and Hyperbolic Expressions

`TrigExpand[expr]` Expand sums and multiple angles in arguments into sums of powers of trigonometric functions
`TrigFactor[expr]` Factor **expr** in a product form, converting powers into multiple angles
`TrigReduce[expr]` Write **expr** into a sum with no products or powers, using combined arguments
`TrigToExp[expr]` Write trigonometric functions in terms of complex exponentials
`ExpToTrig[expr]` Write complex exponentials in terms of trigonometric functions

---

These commands work for both trigonometric and hyperbolic functions. Remember that `Simplify` and `FullSimplify` also simplify, by default, trigonometric expressions (write the option `Trig → False` if you do not want do trigonometric simplifications):

```
{Simplify[1 - Sin[x]^2], Simplify[(1 - Cos[x]) / Sin[x]]}
```

$$\left\{ \text{Cos}[x]^2, \ \text{Tan}\left[\frac{x}{2}\right] \right\}$$

Remember also that `Simplify` and `FullSimplify` accept assumptions:

```
Integrate[x^2 Sin[n x], {x, 0, Pi}]
```

$$\frac{-2 + \left(2 - n^2 \pi^2\right) \text{Cos}[n \pi] + 2 n \pi \text{Sin}[n \pi]}{n^3}$$

```
FullSimplify[%, n ∈ Integers]
```

$$\frac{-2 + (-1)^n \left(2 - n^2 \pi^2\right)}{n^3}$$

Here are some examples of the other commands:

```
{TrigReduce[2 Sin[x]^2], TrigReduce[2 Sin[x] Cos[y]]}
```
$$\{1 - \text{Cos}[2 x], \ \text{Sin}[x - y] + \text{Sin}[x + y]\}$$

```
{TrigFactor[Sin[3 x]], TrigExpand[Sin[3 x]]}
```
$$\left\{ (1 + 2 \text{Cos}[2 x]) \text{Sin}[x], \ 3 \text{Cos}[x]^2 \text{Sin}[x] - \text{Sin}[x]^3 \right\}$$

```
{TrigToExp[2 Sin[x]], ExpToTrig[Exp[I x] + Exp[-I x]]}
```
$$\left\{ i\, e^{-i x} - i\, e^{i x}, \ 2 \text{Cos}[x] \right\}$$

With `TrigExpand`, we can calculate Chebyshev polynomials from a trigonometric expression:

```
Table[TrigExpand[Cos[n ArcCos[x]]], {n, 0, 4}]
```
$$\left\{ 1, \ x, \ -1 + 2 x^2, \ -3 x + 4 x^3, \ 1 - 8 x^2 + 8 x^4 \right\}$$

The same commands we considered previously for trigonometric functions also work for hyperbolic functions. The following are examples:

```
{Simplify[1 + Sinh[x]^2], TrigReduce[2 Sinh[x]^2], TrigFactor[Sinh[3 x]]}
```
$$\left\{ \text{Cosh}[x]^2, \ -1 + \text{Cosh}[2 x], \ (1 + 2 \text{Cosh}[2 x]) \text{Sinh}[x] \right\}$$

```
ExpToTrig[Exp[x] - Exp[-x]]
```
```
2 Sinh[x]
```

13.3.4 Complex Expressions

`ComplexExpand[expr]`	Expand to real and imaginary parts assuming all variables are real
`ComplexExpand[expr, {x, y, … }]`	Expand assuming **x**, **y**, … are complex
`ComplexExpand[expr, TargetFunctions → list]`	Try to expand in terms of functions in **list**

Possible target functions are `Re`, `Im`, `Abs`, `Arg`, `Conjugate`, and `Sign` (for these functions, see Section 12.1.2, p. 400). The default is typically to give results in terms of `Re` and `Im`. The following are examples:

```
(-8)^(1/3)       2 (-1)^{1/3}
```

```
ComplexExpand[%]        1 + i √3
```
```
ComplexExpand[Sin[x + I y]]        Cosh[y] Sin[x] + i Cos[x] Sinh[y]
```

We assume that **z** is complex:

ComplexExpand[z, z]   i Im[z] +Re[z]

The result is written, by default, in terms of **Re** and **Im**. Next, we ask for the result in a polar form—that is, in terms of **Abs** and **Arg**:

ComplexExpand[z, z, TargetFunctions → {Abs, Arg}]

Abs[z] Cos[Arg[z]] + i Abs[z] Sin[Arg[z]]

Note that in Section 5.2.6, p. 137, we considered plotting complex-valued functions. In Section 13.4.1, p. 435, we consider roots such as $(-8)^{1/3}$.

13.3.5 Logical Expressions

■ Logical Tests

For testing whether an expression has a given property, *Mathematica* has several built-in tests. The following is a collection of such tests. Tests are used, for example, with **Select** (see Section 14.1.7, p. 45? with patterns (see Section 16.1.2, p. 496), and with **If** and **Which** (see Section 18.2.2, p. 556).

== (Equal), != (Unequal), === (SameQ), =!= (UnsameQ), PossibleZeroQ (🖐6)

< (Less), ≤ (LessEqual), > (Greater), ≥ (GreaterEqual)

Negative, NonPositive, NonNegative, Positive

NumericQ, NumberQ, IntegerQ, EvenQ, OddQ, PrimeQ

ListQ, VectorQ, MatrixQ, TensorQ, PositiveDefiniteMatrixQ (🖐6)

PolynomialQ, StringQ, OptionQ

FreeQ, MemberQ, MatchQ

To get all names that end with **Q**, execute **?*Q**.

A logical statement gives a result of **True** or **False** or, if *Mathematica* cannot decide the validity of the property, the test as such. Here are some examples:

{2 < 3, 4 < 3, x < 3}

{True, False, x < 3}

{Positive[-3], IntegerQ[3], EvenQ[3], PrimeQ[3]}

{False, True, False, True}

{ListQ[3], ListQ[{}], ListQ[{3, 5, 2}]}

{False, True, True}

IntegerQ[2] && EvenQ[2] && PrimeQ[2]   True

NumericQ tests whether the expression has a numerical value, whereas **NumberQ** tests whether the expression is a number. Recall from Section 12.1.1, p. 396, that *Mathematica* has four kinds of number: integers, rationals, reals, and complexes. *Mathematica* does not consider exact expressions such as **Pi**, **Sqrt[2]**, or **Sin[5]** as numbers, but they do have numeric values:

{NumberQ[Pi], NumericQ[Pi]}   {False, True}

VectorQ and **MatrixQ** accept a second argument defining a test to be satisfied by the elements (the test is written as a pure function; see Section 2.2.2, p. 38):

VectorQ[{2, a, Sqrt[5]}]   True

```
VectorQ[{2, -3, Sqrt[5]}, NumericQ]     True

VectorQ[{2, 3, Sqrt[5]}, NumericQ[#] && Positive[#] &]     True
```

The following second-order polynomial does not contain **y**:

```
FreeQ[a + b x + c x^2, y]     True
```

The attributes of **Pi** are as follows:

```
Attributes[Pi]     {Constant, Protected, ReadProtected}
```

This means that **Constant** is a member of **Attributes[Pi]**:

```
MemberQ[Attributes[Pi], Constant]     True
```

■ Testing Equality

Now we compare the tests **expr1 == expr2** (**Equal**) and **expr1 === expr2** (**SameQ**). In general, the latter test is more demanding. Both tests give **True** if the expressions are identical and **False** if they are not identical. If *Mathematica* cannot decide whether the expressions are identical, == gives the original test **expr1 == expr2** as such, but === gives **False**. The expression **expr1 == expr2** returned by == can be considered as an equation from which we can perhaps solve a variable. For example,

```
{2 == 2, 2 === 2, 2 - x == 0, 2 - x === 0}
{True, True, 2 - x == 0, False}
```

The tests differ somewhat in the way they treat numerical expressions. The test == gives **True** if the numerical values of the expressions differ in at most their 8 binary digits, which correspond roughly to their last two decimal digits of the 16 standard digits. On the other hand, the test === gives **True** if the difference of the expressions is less than the uncertainty of either of them, which means in practice that the expressions must be equal to the last digit and that exact numbers are not considered equal to their decimal values. Here are some examples:

```
{2 == 2., 2 === 2.}
{True, False}

{2. + 10^-13 == 2., 2. + 10^-14 == 2.}
{False, True}

{2. + 10^-15 === 2., 2. + 10^-16 === 2.}
{False, True}
```

In general, use **Equal** to form equations and to test the equality of numbers and strings. Use **SameQ** to test arbitrary expressions for equality in structure.

Note that **Equal**, **Unequal**, **SameQ**, and **UnsameQ** accept more than two expressions as arguments. For example,

```
UnsameQ[1, 2, 3, 4, 5, 2]     False
```

■ Logical Expressions

The logical tests can be combined with the following logical operations to form more complex logical expressions:

---

p && q True if both **p** and **q** are true (AND)

p || q True if one or both of **p** and **q** are true (OR)

!p True if **p** is false (NOT)

Xor[p, q] True if one and only one of **p** and **q** is true (exclusive OR)

Nand[p, q] Means **Not[And[p, q]]** (true if **p** or **q** is false; false if they are both true)

Nor[p, q] Means **Not[Or[p, q]]** (true if both **p** and **q** are false; false if either is true)

LogicalExpand[expr] Expand a logical statement

We can write **&&**, **||**, and **!** in the form ∧, ∨, ¬ by writing ⎡ESC⎤and⎡ESC⎤, ⎡ESC⎤or⎡ESC⎤, and ⎡ESC⎤not⎡ESC⎤, respectively.

LogicalExpand[Nand[p, q]]   ! p || ! q

LogicalExpand[(p || q) && (r || s)]

(p && r) || (p && s) || (q && r) || (q && s)

Quantifiers **ForAll** and **Exists** and **Resolve** are considered in Section 22.2.5, p. 728.

13.3.6 Strings

■ Introduction

A string is an expression written inside quotation marks, such as **"Here is a message."**. The quotation marks do not appear in the output:

"Here is a message."

Here is a message.

However, the quotation marks can be seen with **InputForm**:

% // InputForm

"Here is a message."

In a string, a new line can be defined by **\n** and a tab with **\t**:

"\tHere is a message\n\tin two rows."
    Here is a message
    in two rows.

A string consists of characters:

Characters["A message"]

{A, , m, e, s, s, a, g, e}

Characters also have quotation marks:

% // InputForm

{"A", " ", "m", "e", "s", "s", "a", "g", "e"}

Mathematica has powerful string and character manipulation commands. Here, we consider the basic manipulation of strings. In Section 16.2, p. 505, we consider string patterns. We denote strings by symbols such as **s**, **s1**, and **s2**.

■ **Characters**

Characters[s] Convert a string into a list of characters
CharacterRange["c1", "c2"] Generate a list of all characters from **c1** to **c2**
ToCharacterCode[s] Give a list of the integer codes of the characters in **s**
FromCharacterCode[{n1, n2, … }] Convert the character codes into a string

---

CharacterRange["a", "f"]   {a, b, c, d, e, f}

% // InputForm   {"a", "b", "c", "d", "e", "f"}

ToCharacterCode["Message"]   {77, 101, 115, 115, 97, 103, 101}

FromCharacterCode[%]   Message

■ **Basics of Strings**

StringLength[s] Give the number of characters (\n and \t count as one character)
StringLength[{s1, s2, …}] Give the number of characters of each of the strings

ToString[expr] Convert **expr** into a string
ToExpression[s] Convert **s** into an expression

ToLowerCase[s] Change all uppercase letters to lowercase
ToUpperCase[s] Change all lowercase letters to uppercase

StringQ[expr] Give **True** is **expr** is a string, **False** otherwise
LetterQ[s] Give **True** if all the characters in **s** are letters, **False** otherwise
DigitQ[s] Give **True** if all the characters in **s** are digits (0–9), **False** otherwise

---

ToString[2 + 3]   5

% // InputForm   "5"

ToExpression["2+3"]   5

■ **String Manipulation**

StringJoin[s1, s2, …] or **StringJoin[{s1, s2, …}]** or **s1 <> s2 <> …** Join the strings together
Sort[{s1, s2, … }] Sort the strings into standard order
StringReverse[s] Reverse the order of the characters of **s**

StringTake[s, n] Give the string containing the first **n** characters of **s**
StringDrop[s, n] Give the string **s** with its first **n** characters dropped
StringInsert[s, s1, n] Give a string with **s1** inserted starting at position **n** in **s**
StringReplacePart[s, s1, {m, n}] Replace characters at positions **m**, …, **n** in **s** by **s1**

For **StringTake** and **StringDrop**, we can use the same kinds of part specifications as with **Take** and **Drop** (see Section 14.1.2, p. 449). Thus, in place of **n** we can have **-n** (take/drop the last **n** characters), **{n}** (take/drop the **n**th character), **{-n}** (take/drop the **n**th character from the end), **{m, n}** (take/drop characters **m** through **n**), or **{m, n, d}** (take/drop characters **m** through **n** in steps of **d**). Here is an example of **StringJoin**. Note that the same kind of output can also be obtained with **Row**:

"abc" <> " " <> "def"   abc def

Row[{"abc", " ", "def"}]   abc def

13.4 Mathematical Functions

13.4.1 Basic Functions

Some basic functions are **Sqrt**, **Exp**, and **Log**, together with the trigonometric and hyperbolic functions:

```
trig = {Sin, Cos, Tan, Cot, Sec, Csc};
invtrig = {ArcSin, ArcCos, ArcTan, ArcCot, ArcSec, ArcCsc};
hyp = {Sinh, Cosh, Tanh, Coth, Sech, Csch};
invhyp = {ArcSinh, ArcCosh, ArcTanh, ArcCoth, ArcSech, ArcCsch};
```

We also have **Sinc[x]** or $\sin(x)/x$; $\mathrm{sinc}(0) = 1$. Note that **Sqrt[x]** can also be written as **x^(1/2)** or $\sqrt{x}$ and **Exp[x]** as **E^x** or **e^x**. Also note that **Log[x]** is the natural logarithm and **Log[b, x]** the logarithm to base **b**. The argument of the trigonometric functions is in radians.

For all mathematical functions of *Mathematica,* note the following:

- For integers, rational numbers, and special symbols, the functions give an exact result: the expression as such if it cannot be simplified or a simplified expression.
- A decimal value is calculated if the argument contains a decimal number.
- Values can be calculated to any numerical precision by giving a high-precision argument or with **N[expr, n]** (see Sections 12.1.2, p. 398, and 12.2.2, p. 404).
- Arguments can be complex numbers.
- Arguments can be lists.

For example,

```
{Exp[0.3`18], Exp[3.5 + I], Exp[{1., 2., 3.}]}
```
```
{1.349858807576003104, 17.8924 + 27.8657 i, {2.71828, 7.38906, 20.0855}}
```

Note that to calculate exp(0.3) to high precision, the argument has to be written in a high-precision form (see Section 12.2.2, p. 406). We could also have written **N[Exp[3/10], 19]**.

Some notes about roots and inverse trigonometric functions follow.

■ Roots

Sqrt[x] is not a true inverse function of the function **x^2**. The true inverse is two-valued: The square root of 4 is a number x such that $x^2 = 4$, and there are two solutions for this equation: $x = 2$ and $x = -2$. However, **Sqrt[4]** gives only the positive value 2. If we want both the positive and the negative value for a square root, one possibility is to use **Solve**:

```
Solve[x^2 == 4]      {{x → -2}, {x → 2}}
```

The situation is similar for other roots. For example, the third root of 8 is a number x such that $x^3 = 8$. There are three solutions to this equation, but *Mathematica* gives only one—the *principal root* (the root with the last positive argument):

```
{8^(1/3), (-8)^(1/3)}     {2, 2 (-1)^{1/3}}
```

```
% // ComplexExpand     {2, 1 + i √3}
```

With **Solve**, we get all roots:

```
Solve[x^3 == 8]     {{x → 2}, {x → -1 - i √3}, {x → -1 + i √3}}
```

```
Solve[x^3 == -8]      {{x → -2}, {x → 1 - i √3}, {x → 1 + i √3}}
```

If we only want a real root, we can use **Reduce**:

```
Reduce[x^3 == -8, x, Reals]      x == -2
```

■ Inverse Trigonometric and Hyperbolic Functions

Inverse trigonometric functions and **ArcCosh** and **ArcSech** are also multiple-valued. *Mathematica* gives the principal value for them. For easy reference, here are plots of all of the trigonometric and hyperbolic functions and their inverse functions:

```
trigPlot[functs_, interval_, ranges_, ticks_] :=
  Plot[#[x], interval, PlotLabel → #, Ticks → ticks,
    PlotRange → ranges, AspectRatio → Automatic] & /@ functs;
```

```
r = Range[-3, 3]; t = {-3.2, 3.2}; p1 = {-π / 2, "-π/2"}; p2 = {π / 2, "π/2"};
```

```
g1 = trigPlot[trig, {x, -π, π}, t, {{-π, p1, p2, π}, r}];
g2 = trigPlot[invtrig, {x, -3.2, 3.2}, {t, {-1.8, 3.4}}, {r, {p1, p2, π}}];
GraphicsGrid[{g1, g2}, Spacings → 0, ImageSize → 420]
```

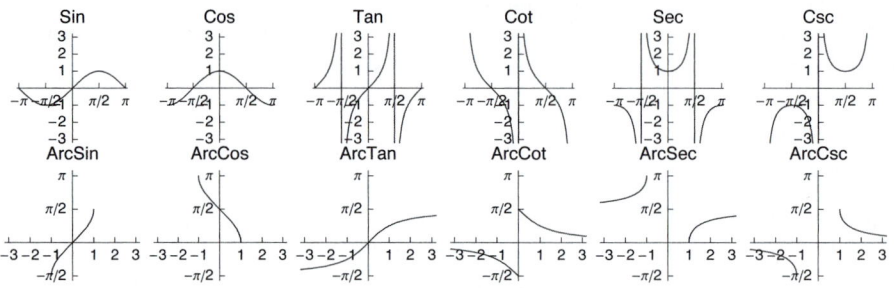

```
g3 = trigPlot[hyp, {x, -3.2, 3.2}, t, {r, r}];
g4 = trigPlot[invhyp, {x, -3, 3}, {t, t}, {r, r}];
GraphicsGrid[{g3, g4}, Spacings → 0, ImageSize → 420]
```

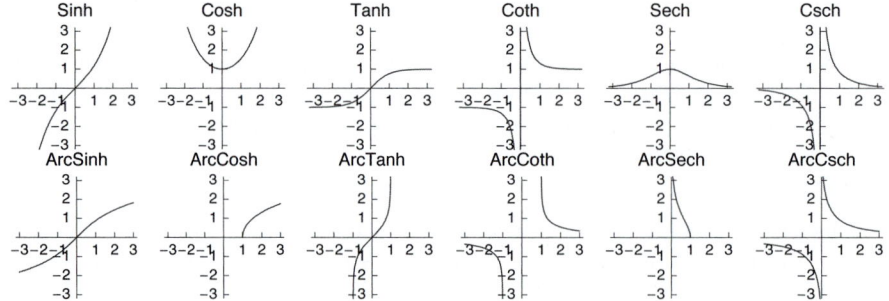

13.4.2 More Functions

■ Combinatorial Functions

n! or **Factorial[n]**   Factorial $n(n-1)(n-2)\dots 1$

n!! or **Factorial2[n]**   Double factorial $n(n-2)(n-4)\dots$

Subfactorial[n] (✏6)   Subfactorial $!\,n = n!\sum_{k=0}^{n}\frac{(-1)^k}{k!}$ (the number of derangements of n objects—

that is, the number of ways to permute n objects so that none is in its natural place)

Pochhammer[n, m]   Pochhammer symbol $(n)_m = n(n+1)(n+2)\dots(n+m-1)$

Binomial[n, m]   Binomial coefficient $\binom{n}{m} = \frac{n!}{m!\,(n-m)!}$

CatalanNumber[n] (✏6)   Catalan number $C_n = \frac{1}{n+1}\binom{2n}{n}$

Multinomial[n1, …, nk]   Multinomial coefficient $\left(N; n_1, \dots, n_k\right) = \frac{N!}{n_1!\dots n_k!}$, $N = \sum_{i=1}^{k} n_i$

BellB[n] (✏6)   Bell number B_n

BernoulliB[n]   Bernoulli number B_n

EulerE[n]   Euler number E_n

StirlingS1[n, m]   Stirling number of the first kind $S_n^{(m)}$

StirlingS2[n, m]   Stirling number of the second kind $\mathcal{S}_n^{(m)}$

Fibonacci[n]   Fibonacci number F_n: $F_n = F_{n-1} + F_{n-2}$, $F_1 = F_2 = 1$

LucasL[n] (✏6)   Lucas number L_n: $L_n = L_{n-1} + L_{n-2}$, $L_1 = 1$, $L_2 = 3$

HarmonicNumber[n]   Harmonic number $H_n = \sum_{i=1}^{n}\frac{1}{i}$

HarmonicNumber[n, k]   Harmonic number $H_n^{(k)} = \sum_{i=1}^{n}\frac{1}{i^k}$

Note that with the exception of subfactorial, Bell, Bernoulli, Euler, and Stirling numbers, the numbers in the previous boxes can also be calculated for noninteger arguments.

■ Nonsmooth Functions

Max[x, y, …], **Max[list]**   The maximum of x, y, … or of the elements in **list**

Min[x, y, …], **Min[list]**   The minimum of x, y, … or of the elements in **list**

Clip[x]   Give x clipped to be within $[-1, 1]$: -1 if $x < -1$, 1 if $x > 1$, and x if $|x| \le 1$

Abs[x] $|x|$: $-x$ if $x < 0$, and x if $x \ge 0$

`Plot[Abs[Max[Sin[x], Cos[x]]], {x, 0, 2 Pi}]`

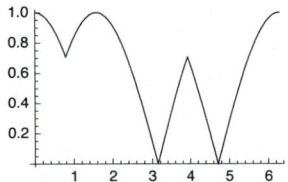

`Boole[expr]`	1 if `expr` is `True`, 0 if `expr` is `False`
`Piecewise[{{val1, cond1}, {val2, cond2}, …}]`	Give the first `vali` for which `condi` is `True`

We also consider `Boole` in Section 20.1.4, p. 642, in the context of integration. `Piecewise` is considered in Section 17.1.2, p. 516, in the context of piecewise functions. Here are some examples:

`Integrate[Boole[x^2 + y^2 ≤ 1] x y^2, {x, 0, 1}, {y, -1, 1}]`

$$\frac{2}{15}$$

`g[x_] = Piecewise[{{Cos[x], -Pi ≤ x ≤ 0}, {-Cos[x], 0 < x ≤ Pi}}, 1/2]`

$$\begin{cases} \text{Cos}[x] & -\pi \le x \le 0 \\ -\text{Cos}[x] & 0 < x \le \pi \\ \frac{1}{2} & \text{True} \end{cases}$$

`Plot[g[x], {x, -5, 5}]`

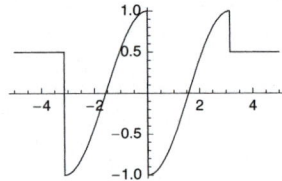

`Sign[x]`	sgn(x): −1, 0, or 1 if $x < 0$, $x = 0$, or $x > 0$, respectively
`Unitize[x]` (✎6)	0 if $x = 0$, and 1 for any other numerical x
`UnitStep[x]`	$\theta(x)$: 0 if $x < 0$, and 1 if $x \ge 0$
`UnitStep[x, y, … ]`	0 if any of $x, y, …$ is negative, and 1 if $x, y, …$ are all nonnegative
`DiscreteDelta[m, n, … ]`	$\delta(m, n, …)$: 1 if $m, n, …$ are all 0, and 0 otherwise
`KroneckerDelta[m, n, … ]`	$\delta_{m, n, …}$: 1 if $m, n, …$ are all equal, and 0 otherwise

`Plot[UnitStep[Sin[2 x]], {x, 0, 10}, AxesOrigin → {-0.3, -0.1}]`

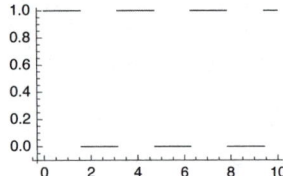

■ **Generalized Functions**

HeavisideTheta[x] (※6) $\theta(x)$: 0 if $x < 0$, and 1 if $x > 0$

HeavisideTheta[x, y, …]   0 if any of $x, y, \ldots$ is negative, and 1 if $x, y, \ldots$ are all positive

DiracDelta[x] $\delta(x)$: 0 if $x \neq 0$

DiracDelta[x, y, …]   0 if any of $x, y, \ldots$ is not 0

Heaviside theta and Dirac delta are generalized functions. They can be used, for example, with integrals, integral transformations, and differential equations:

```
D[HeavisideTheta[x] , x]      DiracDelta[x]

Integrate[DiracDelta[x] , x]      HeavisideTheta[x]

Integrate[DiracDelta[x], {x, -∞, ∞}]      1

Integrate[DiracDelta[x - 3] f[x], {x, -∞, ∞}]      f[3]
```

■ **Orthogonal Polynomials**

The following are some orthogonal polynomials; for more, see tutorial/OrthogonalPolynomials. If $P_n(x)$ is an orthogonal polynomial on (a, b) with respect to weight function $w(x)$, it satisfies the orthogonality condition $\int_a^b P_n(x)\, P_m(x)\, w(x)\, dx = 0$, $m \neq n$.

LegendreP[n,x]   Orthogonal in $(-1, 1)$ with respect to 1

ChebyshevT[n,x]   Orthogonal in $(-1, 1)$ with respect to $\left(1 - x^2\right)^{-1/2}$

ChebyshevU[n,x]   Orthogonal in $(-1, 1)$ with respect to $\left(1 - x^2\right)^{1/2}$

HermiteH[n,x]   Orthogonal in $(-\infty, \infty)$ with respect to e^{-x^2}

LaguerreL[n,x]   Orthogonal in $(0, \infty)$ with respect to e^{-x}

```
t1 = Table[ChebyshevT[n, x], {n, 0, 4}]
```
$\left\{1,\ x,\ -1 + 2\,x^2,\ -3\,x + 4\,x^3,\ 1 - 8\,x^2 + 8\,x^4\right\}$
```
Plot[t1, {x, -1, 1}]
```

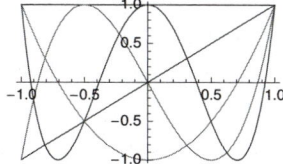

13.4.3 Special Functions

The following are some lists of special functions in *Mathematica*; for details and a complete list, see guide/SpecialFunctions. We consider probability distribution functions in Section 29.1.

Gamma[z]   Gamma function $\Gamma(z)$

Gamma[a, z]   Incomplete gamma function $\Gamma(a, z)$

Gamma[a, z0, z1]   Generalized incomplete gamma function $\Gamma(a, z_0) - \Gamma(a, z_1)$

GammaRegularized[a, z]   Regularized incomplete gamma function $Q(a, z)$

`InverseGammaRegularized[a, s]`   Inverse gamma function

`PolyGamma[z]`   Digamma function $\psi(z)$

`PolyGamma[n, z]`   The nth derivative of $\psi(z)$

$$\Gamma(z) = \int_0^\infty t^{z-1}\, e^{-t}\, dt, \ \ \Gamma(a, z) = \int_z^\infty t^{a-1}\, e^{-t}\, dt, \ \ Q(a, z) = \frac{\Gamma(a, z)}{\Gamma(a)}, \ \ \psi(z) = \frac{d}{dz}\log\Gamma(z) = \frac{\Gamma'(z)}{\Gamma(z)}.$$

`Beta[a, b]`   Beta function $B(a, b)$

`Beta[z, a, b]`   Incomplete beta function $B_z(a, b)$

`BetaRegularized[z, a, b]`   Regularized incomplete beta function $I_z(a, b)$

`InverseBetaRegularized[s, a, b]`   Inverse beta function

`Hypergeometric1F1[a, b, z]`   Confluent hypergeometric function $_1F_1(a; b; z)$

`Hypergeometric2F1[c, a, b, z]`   Hypergeometric function $_2F_1(c, a; b; z)$

$$B(a, b) = \frac{\Gamma(a)\,\Gamma(b)}{\Gamma(a+b)} = \int_0^1 t^{a-1}\,(1-t)^{b-1}\, dt, \ \ B_z(a, b) = \int_0^z t^{a-1}\,(1-t)^{b-1}\, dt, \ \ I_z(a, b) = \frac{B_z(a, b)}{B(a, b)},$$

$$_1F_1(a; b; z) = \frac{\int_0^1 t^{a-1}\,(1-t)^{b-a-1}\, e^{zt}\, dt}{B(a, b-a)}, \ \ _2F_1(c, a; b; z) = \frac{\int_0^1 t^{a-1}\,(1-t)^{b-a-1}\,(1-tz)^{-c}\, dt}{B(a, b-a)}.$$

`Erf[z]; Erfc[z]`   Error function $\mathrm{erf}(z)$; complementary error function $\mathrm{erfc}(z)$

`InverseErf[s]; InverseErfc[s]`   Inverse error and complementary error function

`ExpIntegralE[n, z]; ExpIntegralEi[z]`   Exponential integrals $E_n(z)$ and $\mathrm{Ei}(z)$

`LogIntegral[z]; PolyLog[n, z]`   Logarithmic integral $\mathrm{li}(z)$; polylogarithm function $\mathrm{Li}_n(z)$

`Zeta[s]`   Riemann zeta function $\zeta(s)$

`SinIntegral[z]; CosIntegral[z]`   Sine and cosine integrals $\mathrm{Si}(z)$ and $\mathrm{Ci}(z)$

`SinhIntegral[z]; CoshIntegral[z]`   Hyperbolic sine and cosine integrals $\mathrm{Shi}(z)$ and $\mathrm{Chi}(z)$

`ProductLog[z]`   Product log function $W(z)$ (solution for w of $z = w\, e^w$)

$$\mathrm{erf}\, z = \frac{2}{\sqrt{\pi}} \int_0^z e^{-t^2}\, dt, \ \ \mathrm{erfc}(z) = 1 - \mathrm{erf}(z), \ \ E_n(z) = \int_1^\infty \frac{e^{-zt}}{t^n}\, dt, \ \ \mathrm{Ei}(z) = -\int_{-z}^\infty \frac{e^{-t}}{t}\, dt,$$

$$\mathrm{li}(z) = \int_0^z \frac{dt}{\log(t)}\, dt, \ \ \mathrm{Li}_2(z) = \int_z^0 \frac{\log(1-t)}{t}\, dt, \ \ \mathrm{Li}_n(z) = \sum_{k=1}^\infty \frac{z^k}{k^n}, \ \ \zeta(s) = \sum_{k=1}^\infty \frac{1}{k^s} \ (s > 1),$$

$$\mathrm{Si}(z) = \int_0^z \frac{\sin(t)}{t}\, dt, \ \ \mathrm{Ci}(z) = -\int_z^\infty \frac{\cos(t)}{t}\, dt,$$

$$\mathrm{Shi}(z) = \int_0^z \frac{\sinh(t)}{t}\, dt, \ \ \mathrm{Chi}(z) = \gamma + \log(z) + \int_0^z \frac{\cosh(t) - 1}{t}\, dt.$$

LegendreP[n, z]; LegendreQ[n, z] Legendre function $P_n(z)$; Legendre function of the second kind $Q_n(z)$

LegendreP[n, m, z]; LegendreQ[n, m, z] Associated Legendre function $P_n^m(z)$; associated Legendre function of the second kind $Q_n^m(z)$

BesselJ[n, z]; BesselY[n, z] Bessel functions $J_n(z)$ and $Y_n(z)$

BesselI[n, z]; BesselK[n, z] Modified Bessel functions $I_n(z)$ and $K_n(z)$

AiryAi[z]; AiryBi[z] Airy functions $\mathrm{Ai}(z)$ and $\mathrm{Bi}(z)$

$P_n(z)$, $Q_n(z)$: independent solutions of $\left(1 - z^2\right) y'' - 2\,z\,y' + n(n+1)\,y = 0$

$P_n^m(z)$, $Q_n^m(z)$: independent solutions of $\left(1 - z^2\right) y'' - 2\,z\,y' + \left[n(n+1) - m^2 \big/ \left(1 - z^2\right)\right] y = 0$

$J_n(z)$, $Y_n(z)$: independent solutions of $z^2\,y'' + z\,y' + \left(z^2 - n^2\right) y = 0$

$I_n(z)$, $K_n(z)$: independent solutions of $z^2\,y'' + z\,y' - \left(z^2 + n^2\right) y = 0$

$\mathrm{Ai}(z)$, $\mathrm{Bi}(z)$: independent solutions of $y'' - z\,y = 0$

```
Plot[BesselJ[5, x], {x, 0, 50}]
```

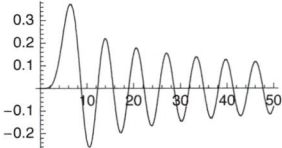

Remember that **FullSimplify** and **FunctionExpand** (see Sections 13.2.1, p. 419, and 13.2.2, p. 424) know much about special functions. For more information about special functions, see Trott (2006b).

Lists

Introduction

> *Wiener once went to a doctor and told him that his memory was terrible and that he couldn't remember anything from one minute to the next. "How long has this been going on?" asked the doctor. "How long has what been going on?" said Wiener.*

A list is *Mathematica*'s way of storing information so that the pieces of information are well arranged and can, at any time, be easily "remembered" or retrieved. Lists are the bread and butter of *Mathematica*— you simply cannot live without them.

    Lists are used as the basic method of collecting numbers, symbols, and other objects. In addition, vectors and matrices are, in fact, lists. *Mathematica* contains a rich collection of tools to work with lists, as can be seen from this chapter. In the next chapter, we show how to form tables from lists. By using patterns we get powerful tools for working with lists and other expressions (see Chapter 16). Vectors and matrices are considered in Chapter 21. In Section 29.1, we discuss the commands **RandomChoice** and **RandomSample** for taking samples; in Section 30.2, we discuss **Tally** and **BinCounts** for calculating frequencies; and in Section 30.6.1, we discuss **ListCorrelate** and **ListConvolve** for smoothing.

14.1　Basic List Manipulation

14.1.1　Forming Lists

■　Lists of Various Dimensions

A list is an ordered collection of zero or more elements. Here are some examples:

```
m1 = {32, 214, 5};
m2 = {{32.7, 8.39, -412.64}, {4.5, -56.2163, -7.606}};
m3 = {{{1, a}, {2, b}, {3, c}}, {{4, d}, {5, e}, {6, f}}};
```

The ordering means that, for example, lists `{3, 2, 5}` and `{2, 3, 5}` are not the same lists. An empty list is `{}`. A 1D list such as `m1` can also be considered as a vector and a 2D list such as `m2` as a matrix (each row is a list and so `m2` is a 2×3 matrix). Higher-dimensional lists such as `m3` are tensors.

With the menu command **Insert ▷ Table/Matrix ▷ New** we can ask for an empty 2D table that contains placeholders for the entries:

```
       □  □  □
m4 =
       □  □  □
```

We can then fill in the placeholders with the ⊞ key to get a 2D list:

```
       3  2  7
m4 =
       6  3  5
{{3, 2, 7}, {6, 3, 5}}
```

2D tables can also be entered with the keyboard. To start, type ⊞,⏎ (i.e., keep the Control key pressed and type the comma); two placeholders of a 1×2 table appear. To get more columns, type ⊞,⏎ several times. To get more rows, type ⊞⏎ several times. Then fill in the table.

Length[list]　Number of elements at the top level of **list**

Dimensions[list]　Dimensions of **list**

ArrayDepth[list]　The depth to which **list** is a full array

ArrayQ[list]　Test whether **list** is a full array (i.e., all parts at a particular level are lists of the same length)

```
Length /@ {m1, m2, m3}       {3, 2, 2}

Dimensions /@ {m1, m2, m3}       {{3}, {2, 3}, {2, 3, 2}}

ArrayDepth /@ {m1, m2, m3}       {1, 2, 3}

ArrayQ[m3]     True
```

Manipulating lists is considered in detail in forthcoming sections, but here we present some important commands that you may need soon.

---

list[[i]], list[[i, j]]　ith and (i, j)th part of **list**

Transpose[list] or **list^T**　Transpose the first two levels of **list**

Recall that `list[[i]]` can also be written as `list⟦i⟧`, where ⟦ and ⟧ can be written as ⎋[[⎋ and ⎋]]⎋. In `list`^{T}, the ^{T} can be written as ⎋tr⎋.

Picking of parts was considered in Section 13.1.3, p. 418.

```
m2⟦1⟧     {32.7, 8.39, -412.64}
```

    m3⟦1, 2, 2⟧   **b**

Transposing a 2D list means converting columns to rows:

    m2ᵀ   {{32.7, 4.5}, {8.39, −56.2163}, {−412.64, −7.606}}

Next, we consider various commands to generate lists. Note that sparse arrays are considered in Section 21.2.1, p. 689.

■ Forming Lists with Table

Table is one of the most useful commands in *Mathematica*. If the elements of a list can be obtained from a formula, then **Table** is the correct tool to use to form the list.

Table[expr, iterspec] Form a list from **expr** according to the iteration specification **iterspec**

The following forms can be used for **iterspec**:

{b} Make a list of **b** copies of **expr**

{i, b} Make a list of the values of **expr** when **i** runs from **1** to **b**

{i, a, b} **i** runs from **a** to **b**

{i, a, b, d} **i** runs from **a** to **b** in steps of **d**

Iteration specifications extend to more indices. For example, the specification **{i, a, b}, {j, c, d}** means that **i** runs from **a** to **b** and, for each **i**, **j** runs from **c** to **d** (**c** and **d** can contain **i**). Here are some examples:

    Table[1, {3}]   {1, 1, 1}

    Table[Cos[n Pi / 2], {n, 0, 6, 2}]   {1, −1, 1, −1}

    Table[Integrate[1 / x^i, x], {i, 0, 4}]

$$\left\{x, \text{Log}[x], -\frac{1}{x}, -\frac{1}{2\,x^2}, -\frac{1}{3\,x^3}\right\}$$

    Table[1 / (i + j − 1), {i, 3}, {j, 3}]

$$\left\{\left\{1, \frac{1}{2}, \frac{1}{3}\right\}, \left\{\frac{1}{2}, \frac{1}{3}, \frac{1}{4}\right\}, \left\{\frac{1}{3}, \frac{1}{4}, \frac{1}{5}\right\}\right\}$$

The index can also have decimal values, and the expression to be tabulated may be a list:

    Table[Log[x], {x, 1, 2, 0.2}]

    {0, 0.182322, 0.336472, 0.470004, 0.587787, 0.693147}

    Table[{x, Log[x]}, {x, 1, 2, 0.2}]

    {{1, 0}, {1.2, 0.182322}, {1.4, 0.336472},
    {1.6, 0.470004}, {1.8, 0.587787}, {2., 0.693147}}

Table[expr, {x, {x1, x2, …}}] (✸6) Form a list by giving, in **expr**, **x** the values **x1, x2, …**

Map[expr &, {x1, x2, …}] Form a list by giving, in **expr**, **#** the values **x1, x2, …**

expr & /@ {x1, x2, …} Form a list by giving, in **expr**, **#** the values **x1, x2, …**

Sometimes we want to tabulate an expression for such irregular values of a variable that cannot be formed by an iteration specification; instead, we have the values of the variable as a list. **Table** can also be used in such cases, although **Map** is an alternative. **Map** can also be used with **/@**:

    Table[x^2, {x, {1, 3, 4, 9}}]   {1, 9, 16, 81}

    Map[#^2 &, {1, 3, 4, 9}]   {1, 9, 16, 81}

```
#^2 & /@ {1, 3, 4, 9}    {1, 9, 16, 81}
```

Here we form a 2D table:

```
Table[x^2 + y^2, {x, {1, 3, 4, 9}}, {y, {2, 5, 7}}]
{{5, 26, 50}, {13, 34, 58}, {20, 41, 65}, {85, 106, 130}}
```

■ Forming Lists with Range

With **Range**, it is easy to form lists of consecutive numbers.

Range[n]	$\{1, 2, 3, ..., n\}$
Range[m, n]	$\{m, m + 1, m + 2, ..., n\}$
Range[m, n, d]	$\{m, m + d, m + 2d, ..., n\}$

```
{Range[4], Range[0, 4], Range[0, 4, 2]}
{{1, 2, 3, 4}, {0, 1, 2, 3, 4}, {0, 2, 4}}
```

Range even works with real numbers:

```
Range[2.6, 5.4, 0.5]
{2.6, 3.1, 3.6, 4.1, 4.6, 5.1}
```

Note that **Range** stopped at 5.1 because the next number, 5.6, would be larger than the upper bound given—5.4.

■ Forming Lists with Array

With **Table** we can also generate lists of indexed variables:

```
vv = Table[v[i], {i, 0, 5}]
{v[0], v[1], v[2], v[3], v[4], v[5]}
```

```
ww = Table[w[i, j], {i, 2}, {j, 2}]
{{w[1, 1], w[1, 2]}, {w[2, 1], w[2, 2]}}
```

This can also be done with **Array**.

Array[f, n]	$\{f[1], ..., f[n]\}$
Array[f, {m, n}]	$\{\{f[1, 1], ..., f[1, n]\}, ..., \{f[m, 1], ..., f[m, n]\}\}$
ConstantArray[c, n] (✤6)	$\{c, ..., c\}$
ConstantArray[c, {m, n}]	$\{\{c, ...,c\}, ..., \{c, ..., c\}\}$

A possible third argument gives the index origin (the default is 1). For example,

```
rr = Array[r, 5]
{r[1], r[2], r[3], r[4], r[5]}
```

```
ss = Array[s, 6, 0]
{s[0], s[1], s[2], s[3], s[4], s[5]}
```

```
tt = Array[t, {2, 3}]
{{t[1, 1], t[1, 2], t[1, 3]}, {t[2, 1], t[2, 2], t[2, 3]}}
```

For the elements, we can define a function:

```
r[i_] := 2 + i 0.2
```

Now all elements of **rr** have a value:

    rr   {2.2, 2.4, 2.6, 2.8, 3.}

The function can also be given directly in the **Array** command as a pure function:

    Array[2 + # 0.2 &, 5]

    {2.2, 2.4, 2.6, 2.8, 3.}

    Array[#1^#2 &, {3, 4}]

    {{1, 1, 1, 1}, {2, 4, 8, 16}, {3, 9, 27, 81}}

Here are constant arrays:

    Table[4, {6}]   {4, 4, 4, 4, 4, 4}

    Array[4 &, 6]   {4, 4, 4, 4, 4, 4}

    ConstantArray[4, 6]   {4, 4, 4, 4, 4, 4}

■ Some Notes about Indexed Variables

Forming a series of values is important in many mathematical calculations. Often, it suffices to generate a list of values, for example, **xx**, and then to refer to its components with **xx[[i]]**:

    xx = Table[2 + i 0.2, {i, 0, 5}]

    {2, 2.2, 2.4, 2.6, 2.8, 3.}

    xx[[1]]   2

We can also use indexed variables such as **x[i]**:

    Do[x[i] = 2 + i 0.2, {i, 0, 5}]

    x[0]   2

If we need both the indexed variables and a symbol for the whole set of variables, we can do as follows:

    xx = Table[x[i] = 2 + i 0.2, {i, 0, 5}]

    {2, 2.2, 2.4, 2.6, 2.8, 3.}

If we want to see the values of all indexed variables with a certain name, for example, **x**, then we type **?x**. To clear a single value, we type **x[i] =.**, and to clear all values, we type **Clear[x]**.

For indexed variables, we can also use subscripts. They can be made with **Subscript** or by entering the subscripts in a 2D form. 2D subscripts can be written with the **BasicMathInput** palette (the second button in the next-to-last row) and with CTRL[_] or CTRL[-] (see Section 3.3.3, p. 76; to get out of the subscript position, press CTRL[_]). Here, we use both **Subscript** and 2D input:

    vv = Table[Subscript[v, i], {i, 0, 5}]

    {v_0, v_1, v_2, v_3, v_4, v_5}

    ww = Table[$w_{i,j}$, {i, 2}, {j, 3}]

    {{$w_{1,1}$, $w_{1,2}$, $w_{1,3}$}, {$w_{2,1}$, $w_{2,2}$, $w_{2,3}$}}

    Array[$f_{\#\#}$ &, {3, 3}]

    {{$f_{1,1}$, $f_{1,2}$, $f_{1,3}$}, {$f_{2,1}$, $f_{2,2}$, $f_{2,3}$}, {$f_{3,1}$, $f_{3,2}$, $f_{3,3}$}}

14.1.2 Selecting Parts

■ **Taking Parts of Lists**

> list⟦i⟧   ith part of **list**
> list⟦-i⟧   ith part counted from the end
> list⟦i, j⟧   (i, j)th part
> list⟦i, j, k⟧   (i, j, k)th part
> list⟦{i, j, … }⟧   Parts i, j, ...

Instead of **list⟦i⟧**, we can write **list[[i]]**. Recall that ⟦ and ⟧ can be written as ᴱˢᶜ[[ᴱˢᶜ and ᴱˢᶜ]]ᴱˢᶜ, respectively. We will use the symbols ⟦ and ⟧ because they are simpler and shorter than the double brackets. Instead of **list⟦i⟧**, we can write **Part[list, i]**. Examples:

    m = {{11, 12, 13}, {21, 22, 23}, {31, 32, 33}};

    m⟦2⟧   {21, 22, 23}

    m⟦2, 3⟧   23

    m⟦{2, 1, 3}⟧   {{21, 22, 23}, {11, 12, 13}, {31, 32, 33}}

> list⟦i ;; j⟧ (⌘6) Take parts i through j
> list⟦i ;; ⟧ Take parts from i to the end
> list⟦-i ;; ⟧ Take the last i parts
> list⟦ ;; j⟧ Take the first j parts
> list⟦ ;; -j⟧ Take parts from the beginning to the jth part from the end
> list⟦i ;; j ;; d⟧ Take parts i through j in steps of d

Internally, *Mathematica* uses **Span** for **;;**:

    2 ;; 6 // FullForm   Span[2, 6]

Examples:

    m⟦2 ;;⟧   {{21, 22, 23}, {31, 32, 33}}

    m⟦ ;; -2⟧   {{11, 12, 13}, {21, 22, 23}}

■ **Resetting Parts**

Parts specified by ⟦ ⟧ can be used to set new values for parts:

> list⟦i⟧ = a Set the ith part to **a**
> list⟦i ;; j⟧ = a Set parts i through j to **a**
> list⟦i ;; j⟧ = {a, …, b} Set part i to **a**, ..., part j to **b**
> list⟦{i, j, … }⟧ = {a, b, …} Set parts i, j, ... to **a, b, ...**

Note that these commands modify the original list:

    m⟦3, 3⟧ = 333   333

    m   {{11, 12, 13}, {21, 22, 23}, {31, 32, 333}}

■ **Taking and Dropping Parts**

> `First[list]`, `Rest[list]` Take/drop the first part
> `Last[list]`, `Most[list]` Take/drop the last part
>
> `Take[list, n]`, `Drop[list, n]` Take/drop the first **n** parts
> `Take[list, -n]`, `Drop[list, -n]` Take/drop the last **n** parts
>
> `Take[list, {n}]`, `Drop[list, {n}]` Take/drop the nth part
> `Take[list, {-n}]`, `Drop[list, {-n}]` Take/drop the nth part from the end
>
> `Take[list, {m, n}]`, `Drop[list, {m, n}]` Take/drop parts **m** through **n**
> `Take[list, {m, n, d}]`, `Drop[list, {m, n, d}]` Take/drop parts **m** through **n** in steps of **d**

Note that the commands in the previous box do not modify the original list. For example, if we want to replace **list** with a list in which the last part is dropped, we have to write **list = Most[list]**.

> `TakeWhile[list, crit]` (❀6) Take elements from **list** as long as **crit** gives **True**
> `LengthWhile[list, crit]` (❀6) Give the length of the list given by **TakeWhile**

The criterion is written as a pure function:

```
t = {4, 7, 2, 4, 8, 6, 10, 3, 9};

TakeWhile[t, # < 10 &]      {4, 7, 2, 4, 8, 6}

LengthWhile[t, # < 10 &]      6
```

14.1.3 Inserting and Deleting

> `Prepend[list, elem]` or `Join[{elem}, list]` Insert **elem** at the beginning of **list**
> `Append[list, elem]` or `Join[list, {elem}]` Insert **elem** at the end of **list**
>
> `PrependTo[list, elem]` Insert **elem** at the beginning of **list** and reset **list** to the result
> `AppendTo[list, elem]` Insert **elem** at the end of **list** and reset **list** to the result
>
> `Insert[list, elem, i]` Insert **elem** at position **i** in **list**
> `ReplacePart[list, i → elem]` Replace the **i**th element of **list** with **elem**
>
> `Delete[list, i]` Delete the **i**th element of **list**
> `Delete[list, {{i}, {j}, … }]` Delete elements **i**, **j**, …

The argument **i** in **Insert**, **ReplacePart**, and **Delete** can also be negative, meaning that the position is counted from the end of the list. The last argument can also be a more detailed definition of the position, such as **{i, j}**.

Note that with the exception of **PrependTo** and **AppendTo**, the commands in the previous box do not modify the original value of **list**:

```
Append[m, {41, 42, 43}]

{{11, 12, 13}, {21, 22, 23}, {31, 32, 333}, {41, 42, 43}}

m    {{11, 12, 13}, {21, 22, 23}, {31, 32, 333}}
```

However, **PrependTo[list, a]** and **AppendTo[list, a]** do modify the original list. The same effect can also be obtained by **list = Prepend[list, a]** and **list = Append[list, a]**:

```
AppendTo[m, {41, 42, 43}]

{{11, 12, 13}, {21, 22, 23}, {31, 32, 333}, {41, 42, 43}}

m    {{11, 12, 13}, {21, 22, 23}, {31, 32, 333}, {41, 42, 43}}
```

Riffle[{a, b, c, … }, x] (✿6) {a, x, b, x, c, x, …}

Riffle[{a, b, c, … }, x, d] Every **d**th element is **x**

Riffle[{a, b, c, … }, x, {p0, p1, d}] **x** appears at positions **p0**, **p0 + d**, **p0 + 2d**, … (a negative **p1** counts from the end)

Riffle[{a, b, c, … }, {x, y, z, … }] {a, x, b, y, c, z, …}

Note that

Riffle[list, x] is equivalent to **Riffle[list, x, {2, -2, 2}]**;

Riffle[list, x, d] is equivalent to **Riffle[list, x, {d, -2, d}]**.

With **Riffle** we can intersperse additional elements:

```
Riffle[{1, 2, 3, 4}, x]     {1, x, 2, x, 3, x, 4}

Riffle[{1, 2, 3, 4, 5}, x, 3]     {1, 2, x, 3, 4, x, 5}

Riffle[{1, 2, 3, 4}, x, {2, -2, 2}]     {1, x, 2, x, 3, x, 4}

Riffle[{1, 2, 3, 4}, x, {1, -1, 2}]     {x, 1, x, 2, x, 3, x, 4, x}
```

We can also merge two lists. If the latter list is shorter than the first, the elements of the latter list are used cyclically:

```
Riffle[{1, 2, 3, 4}, {x, y}]     {1, x, 2, y, 3, x, 4}

Riffle[{1, 2, 3, 4}, {u, v, w, x}]     {1, u, 2, v, 3, w, 4, x}
```

PadRight[v, n] Pad vector **v** with zeros on the right to make a vector of length **n**

PadRight[v, n, a] Pad vector **v** with replicates of **a**

PadRight[m, {n_1, n_2}] Pad matrix **m** with zeros to an $n_1 \times n_2$ matrix

PadRight[m, {n_1, n_2}, a] Pad matrix **m** with replicates of matrix **a**

PadLeft works similarly.

14.1.4 Ungrouping and Grouping

Flatten[list] Flatten out all levels in **list** by deleting all inner braces

Flatten[list, n] Flatten out the top **n** levels in **list**

Partition[list, n] Partition **list** into nonoverlapping sublists of **n** elements

Partition[list, n, d] Generate sublists with offset **d**

Split[list] Split **list** into pieces consisting of runs of identical elements

Split[list, test] Consider adjacent elements as identical if **test** gives **True**

■ **Flatten**

We first make a table:

```
a = Table[{x, y, x y, x + y}, {x, 1, 2}, {y, 2, 4}]
{{{1, 2, 2, 3}, {1, 3, 3, 4}, {1, 4, 4, 5}}, {{2, 2, 4, 4}, {2, 3, 6, 5}, {2, 4, 8, 6}}}
```

Flatten removes all inner curly braces { } and so ungroups the list:

```
at = Flatten[a]
{1, 2, 2, 3, 1, 3, 3, 4, 1, 4, 4, 5, 2, 2, 4, 4, 2, 3, 6, 5, 2, 4, 8, 6}
```

Next, we flatten only the first level:

```
Flatten[a, 1]
{{1, 2, 2, 3}, {1, 3, 3, 4}, {1, 4, 4, 5}, {2, 2, 4, 4}, {2, 3, 6, 5}, {2, 4, 8, 6}}
```

We will use this kind of flattening several times in later chapters.

■ **Partition**

Now we partition the flattened list into sublists of four elements:

```
b = Range[12]     {1, 2, 3, 4, 5, 6, 7, 8, 9, 10, 11, 12}

Partition[b, 3]     {{1, 2, 3}, {4, 5, 6}, {7, 8, 9}, {10, 11, 12}}
```

If we partition into sublists of five elements, the remaining two elements (11 and 12) are dropped:

```
Partition[b, 5]     {{1, 2, 3, 4, 5}, {6, 7, 8, 9, 10}}
```

We then partition with offset 1:

```
Partition[{1, 2, 3, 4, 5, 6}, 3, 1]     {{1, 2, 3}, {2, 3, 4}, {3, 4, 5}, {4, 5, 6}}
```

As we saw, a possible incomplete sublist at the end will, by default, be dropped. However, we have more advanced forms of **Partition** that allow us to define what to do with an incomplete sublist. In particular, we can define how to pad an incomplete sublist.

How to pad a possible incomplete sublist:

Partition[list, n]   Do not pad: drop an incomplete sublist
Partition[list, n, n, 1]   Pad with elements from the beginning of **list**
Partition[list, n, n, 1, a]   Pad with repetitions of element **a**
Partition[list, n, n, 1, padlist]   Pad cyclically with **padlist**
Partition[list, n, n, 1, {}]   Do not pad: allow an incomplete sublist at the end

---

```
Partition[b, 5, 5, 1]
{{1, 2, 3, 4, 5}, {6, 7, 8, 9, 10}, {11, 12, 1, 2, 3}}

Partition[b, 5, 5, 1, 0]
{{1, 2, 3, 4, 5}, {6, 7, 8, 9, 10}, {11, 12, 0, 0, 0}}

Partition[b, 5, 5, 1, {x, y}]
{{1, 2, 3, 4, 5}, {6, 7, 8, 9, 10}, {11, 12, x, y, x}}

Partition[b, 5, 5, 1, {}]
{{1, 2, 3, 4, 5}, {6, 7, 8, 9, 10}, {11, 12}}
```

■ **Split**

Split finds runs of identical elements:

```
at     {1, 2, 2, 3, 1, 3, 3, 4, 1, 4, 4, 5, 2, 2, 4, 4, 2, 3, 6, 5, 2, 4, 8, 6}
```

```
Sort[at]    {1, 1, 1, 2, 2, 2, 2, 2, 2, 3, 3, 3, 3, 4, 4, 4, 4, 4, 4, 5, 5, 6, 6, 8}

Split[%]
{{1, 1, 1}, {2, 2, 2, 2, 2, 2}, {3, 3, 3, 3}, {4, 4, 4, 4, 4, 4}, {5, 5}, {6, 6}, {8}}
```

If we also count the number of elements in each sublist, we get the so-called run-length encoding:

```
{First[#], Length[#]} & /@ %
{{1, 3}, {2, 6}, {3, 4}, {4, 6}, {5, 2}, {6, 2}, {8, 1}}
```

To get the original list, do as follows:

```
ConstantArray[#⟦1⟧, #⟦2⟧] & /@ % // Flatten
{1, 1, 1, 2, 2, 2, 2, 2, 2, 3, 3, 3, 3, 4, 4, 4, 4, 4, 4, 5, 5, 6, 6, 8}
```

The default test used by **Split** is **SameQ** or, to write it more completely, **SameQ[#1, #2]&**. Here, **#1** and **#2** represent two consecutive elements. Thus, two consecutive elements are put into the same run, if they are considered the same. Consider some other test functions. First, define a pair of elements to belong to the same run if the absolute value of their difference is at most 2:

```
Split[at, Abs[#1 - #2] ≤ 2 &]
{{1, 2, 2, 3, 1, 3, 3, 4}, {1}, {4, 4, 5}, {2, 2, 4, 4, 2, 3}, {6, 5}, {2, 4}, {8, 6}}
```

Then, consider elements to belong to the same run if they are different:

```
Split[at, UnsameQ]
{{1, 2}, {2, 3, 1, 3}, {3, 4, 1, 4}, {4, 5, 2}, {2, 4}, {4, 2, 3, 6, 5, 2, 4, 8, 6}}
```

Now we consider a pair of elements to belong to different runs if the previous element is at most 4 and the next element is greater than 4:

```
Split[at, ! (#1 ≤ 4 && #2 > 4) &]
{{1, 2, 2, 3, 1, 3, 3, 4, 1, 4, 4}, {5, 2, 2, 4, 4, 2, 3}, {6, 5, 2, 4}, {8, 6}}
```

Split after each element that equals 4:

```
Split[at, #1 != 4 &]
{{1, 2, 2, 3, 1, 3, 3, 4}, {1, 4}, {4}, {5, 2, 2, 4}, {4}, {2, 3, 6, 5, 2, 4}, {8, 6}}

a =.; at =.
```

14.1.5 Reordering

Transpose[list] or **listᵀ** Transpose the first two levels of **list** (ᵀ can be written as ⎣ESC⎦**tr**⎣ESC⎦)

Sort[list] Sort the elements of **list** into a standard order
Sort[list, p] Sort the elements of **list** using the ordering function **p**
SortBy[list, f] (🟦6) Sort the elements of **list** in the order defined by applying **f** to each of them
Union[list] Sort the elements and remove any duplicates

Reverse[list] Reverse the order of elements
Reverse[list, n] Reverse the order of elements at level **n**

RotateLeft[list] Rotate the elements one position to the left; similarly for **RotateRight**
RotateLeft[list, n] Rotate the elements **n** positions to the left
RotateLeft[list, {n1, n2, … }] Rotate the elements to the left **n1** positions at the first level, **n2** positions at the second level, …

■ Sort and Union

Consider the following list:

```
r = {3, 2, C, b, A, 1, a, 2, B, A, c}
{3, 2, C, b, A, 1, a, 2, B, A, c}
```

We sort the elements into standard order:

```
r // Sort     {1, 2, 2, 3, a, A, A, b, B, c, C}
```

Then we sort the elements but also remove any duplicates:

```
r // Union     {1, 2, 3, a, A, b, B, c, C}
```

■ Advanced Sorting

By default, the ordering function that **Sort** uses is **OrderedQ[{#1, #2}]&**. *Mathematica* uses this function to decide whether two elements are in order. For example,

```
{OrderedQ[{2, 3}], OrderedQ[{3, 2}]}
{True, False}
```

We can define another ordering function. In the following examples, we order into descending order:

```
s = {3, 2, 4, 5, 1};

s // Sort // Reverse     {5, 4, 3, 2, 1}

Sort[s, OrderedQ[{#2, #1}] &]     {5, 4, 3, 2, 1}

Sort[s, Greater]     {5, 4, 3, 2, 1}

Sort[s, Greater[#1, #2] &]     {5, 4, 3, 2, 1}

Sort[s, #1 > #2 &]     {5, 4, 3, 2, 1}
```

Next, we order according to absolute value. We can use both **Sort** and **SortBy**, the latter being easier:

```
Sort[{-4, 6, -2, 3}, Abs[#1] ≤ Abs[#2] &]     {-2, 3, -4, 6}

SortBy[{-4, 6, -2, 3}, Abs]     {-2, 3, -4, 6}
```

Now we order according to real part:

```
SortBy[{3 - 2 I, 4 I, -5 + 3 I}, Re]     {-5 + 3 i, 4 i, 3 - 2 i}
```

Consider a nested list:

```
t = {{30, 2}, {10, 3}, {20, 1}};

Sort[t]     {{10, 3}, {20, 1}, {30, 2}}
```

By default, the sublists are ordered according to the first element. Then we order the sublists such that the second elements of the sublists appear in ascending order. Again, **SortBy** is easier:

```
Sort[t, #1[[2]] ≤ #2[[2]] &]     {{20, 1}, {30, 2}, {10, 3}}

SortBy[t, Last]     {{20, 1}, {30, 2}, {10, 3}}
```

Note that special symbols and structures are not ordered according to their numerical values:

```
u = {-∞, Sin[2], e, π, √10 , ∞};

u // Sort     {√10 , e, π, -∞, ∞, Sin[2]}
```

To get the list ordered according to size, do as follows:

Sort[u, Less] $\{-\infty, \text{Sin}[2], e, \pi, \sqrt{10}, \infty\}$

14.1.6 Combinatorial Operations

Permutations[list] Give all possible permutations having all the elements of **list** (repeated elements are treated as identical)
Permutations[list, {k}] Give all permutations having exactly **k** elements (**k**-permutations)
Permutations[list, k] Give all permutations having at most **k** elements

Subsets[list] Give all possible subsets (the power set) (repeated elements are treated as distinct)
Subsets[list, {k}] Give all subsets having exactly **k** elements (**k**-combinations)
Subsets[list, k] Give all subsets having at most **k** elements

Tuples[list, k] Give all **k**-tuples (repeated elements are treated as distinct)
Tuples[{list1, list2, … }] Give all tuples whose *i*th element is from **list***i*

See also *Combinatorica*/guide/CombinatoricaPackage in the Documentation Center for information about the **Combinatorica`** package.

▪ Permutations

Here are all permutations of the elements of a list:

r = {1, 2, 3};

Permutations[r]
{{1, 2, 3}, {1, 3, 2}, {2, 1, 3}, {2, 3, 1}, {3, 1, 2}, {3, 2, 1}}

For a list with n elements, we have $n!$ permutations. Generate all k-permutations—that is, permutations having length k, $k = 0, 1, 2, 3$:

Permutations[r, {0}]   {{}}

Permutations[r, {1}]   {{1}, {2}, {3}}

Permutations[r, {2}]
{{1, 2}, {1, 3}, {2, 1}, {2, 3}, {3, 1}, {3, 2}}

Permutations[r, {3}]
{{1, 2, 3}, {1, 3, 2}, {2, 1, 3}, {2, 3, 1}, {3, 1, 2}, {3, 2, 1}}

For a list with n elements, the number of k-permutations is $n!/(n-k)!$ or $(n-k+1)_k$ or **Pochhammer[n – k + 1, k]**. Generate all permutations having lengths 0, 1, 2, and 3:

Permutations[r, 3]
{{}, {1}, {2}, {3}, {1, 2}, {1, 3}, {2, 1}, {2, 3}, {3, 1},
  {3, 2}, {1, 2, 3}, {1, 3, 2}, {2, 1, 3}, {2, 3, 1}, {3, 1, 2}, {3, 2, 1}}

Consider a list with repeated elements:

s = {1, 1, 2};

Now we get only three permutations because repeated elements are treated as identical:

Permutations[s]   {{1, 1, 2}, {1, 2, 1}, {2, 1, 1}}

■ **Subsets**

For subsets, the order of the elements does not matter. Thus, for **r**, we get the following different subsets:

```
r = {1, 2, 3};

Subsets[r]
{{}, {1}, {2}, {3}, {1, 2}, {1, 3}, {2, 3}, {1, 2, 3}}
```

For a set with n elements, we have 2^n different subsets; all these together form the so-called power set of the given set. Next, generate all k-combinations—that is, all subsets having k elements, $k = 0, 1, 2, 3$:

```
Subsets[r, {0}]     {{}}

Subsets[r, {1}]     {{1}, {2}, {3}}

Subsets[r, {2}]     {{1, 2}, {1, 3}, {2, 3}}

Subsets[r, {3}]     {{1, 2, 3}}
```

For a set with n elements, the number of k-combinations is $\binom{n}{k}$. Repeated elements are treated as distinct:

```
Subsets[{1, 1, 2}]
{{}, {1}, {1}, {2}, {1, 1}, {1, 2}, {1, 2}, {1, 1, 2}}
```

As an example, generate six points and plot the points and their connecting lines:

```
Table[{Cos[2 Pi i / 6], Sin[2 Pi i / 6]}, {i, 6}];

Subsets[%, {2}];

Graphics[{Point[%%], Line[%]}]
```

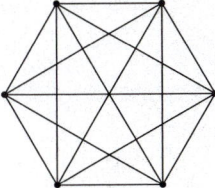

■ **Tuples**

Generate, for **r**, tuples of sizes 0, 1, 2, and 3:

```
r = {1, 2, 3};

Tuples[r, 0]     {{}}

Tuples[r, 1]     {{1}, {2}, {3}}

Tuples[r, 2]
{{1, 1}, {1, 2}, {1, 3}, {2, 1}, {2, 2}, {2, 3}, {3, 1}, {3, 2}, {3, 3}}

Tuples[r, 3]
{{1, 1, 1}, {1, 1, 2}, {1, 1, 3}, {1, 2, 1}, {1, 2, 2}, {1, 2, 3},
 {1, 3, 1}, {1, 3, 2}, {1, 3, 3}, {2, 1, 1}, {2, 1, 2}, {2, 1, 3}, {2, 2, 1},
 {2, 2, 2}, {2, 2, 3}, {2, 3, 1}, {2, 3, 2}, {2, 3, 3}, {3, 1, 1}, {3, 1, 2},
 {3, 1, 3}, {3, 2, 1}, {3, 2, 2}, {3, 2, 3}, {3, 3, 1}, {3, 3, 2}, {3, 3, 3}}
```

For a list with n elements, we have n^k tuples of size k. Here are all possible results with two dice:

```
Tuples[Range[6], {2}]
```

```
{{1, 1}, {1, 2}, {1, 3}, {1, 4}, {1, 5}, {1, 6}, {2, 1}, {2, 2}, {2, 3},
 {2, 4}, {2, 5}, {2, 6}, {3, 1}, {3, 2}, {3, 3}, {3, 4}, {3, 5}, {3, 6},
 {4, 1}, {4, 2}, {4, 3}, {4, 4}, {4, 5}, {4, 6}, {5, 1}, {5, 2}, {5, 3},
 {5, 4}, {5, 5}, {5, 6}, {6, 1}, {6, 2}, {6, 3}, {6, 4}, {6, 5}, {6, 6}}
```

Next, we generate, from two letters, all words of length three:

```
StringJoin /@ Tuples[{"a", "b"}, {3}]
```

```
{aaa, aab, aba, abb, baa, bab, bba, bbb}
```

Repeated elements are treated as distinct:

```
Tuples[{1, 1, 2}, 2]
```

```
{{1, 1}, {1, 1}, {1, 2}, {1, 1}, {1, 1}, {1, 2}, {2, 1}, {2, 1}, {2, 2}}
```

As an example, generate the corner points of a unit cube, from all pairs of the points, and draw the points and the connecting lines between points:

```
Tuples[{0, 1}, 3]
```

```
{{0, 0, 0}, {0, 0, 1}, {0, 1, 0}, {0, 1, 1}, {1, 0, 0}, {1, 0, 1}, {1, 1, 0}, {1, 1, 1}}
```

```
Subsets[%, {2}];
```

```
Graphics3D[{AbsolutePointSize[4], Point[%%], Line[%]}, Boxed → False]
```

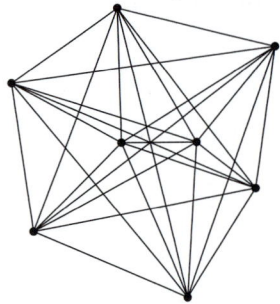

Form all possible tuples from two or three lists:

```
Tuples[{{1, 2, 3}, {a, b}}]
```

```
{{1, a}, {1, b}, {2, a}, {2, b}, {3, a}, {3, b}}
```

```
Tuples[{{1, 2}, {a, b}, {A, B}}]
```

```
{{1, a, A}, {1, a, B}, {1, b, A}, {1, b, B}, {2, a, A}, {2, a, B}, {2, b, A}, {2, b, B}}
```

We could also use **Distribute**:

```
Distribute[{{1, 2, 3}, {a, b}}, List]
```

```
{{1, a}, {1, b}, {2, a}, {2, b}, {3, a}, {3, b}}
```

With **Outer** we get a more structured list:

```
Outer[List, {1, 2, 3}, {a, b}]
```

```
{{{1, a}, {1, b}}, {{2, a}, {2, b}}, {{3, a}, {3, b}}}
```

14.1.7 Searching Elements

■ Searching with a Test

> `Select[list, test]` Select the elements of `list` for which `test` gives `True`
>
> `Select[list, test, n]` Select the first `n` elements

With `Select`, we can search for elements that satisfy a logical test. Logical tests were mentioned in Section 13.3.5, p. 431. Tests useful with `Select` include the following:

`==, !=, <, ≤, >, ≥, Negative, Nonnegative, Positive,`

`NumericQ, IntegerQ, EvenQ, OddQ, FreeQ, MemberQ`.

In general, a test used in `Select` is written as a pure function, such as `Select[c, EvenQ[#]&]` (for pure functions, see Sections 2.2.2, p. 38, and 17.1.4, p. 520). The argument of a pure function is written as `#`, and at the end we write `&`. However, simple built-in tests with one argument can be written without `#` and `&` so that we can write simply `Select[c, EvenQ]`. More complicated built-in tests that use two arguments have to be written as pure functions; for example, `Select[c, # < 2 &]`.

The built-in tests can be combined with logical operations such as `&&` (and), `||` (or), and `!` (not) (see Section 13.3.5, p. 431). Such combined tests also have to be written as pure functions; for example, `Select[c, EvenQ[#] && Positive[#] &]`.

To demonstrate, we throw a die 20 times and select the results that satisfy various tests:

```
a = RandomInteger[{1, 6}, {20}]
{2, 4, 3, 5, 4, 1, 4, 5, 6, 2, 4, 6, 4, 1, 3, 2, 4, 2, 1, 3}

Select[a, EvenQ]    {2, 4, 4, 4, 6, 2, 4, 6, 4, 2, 4, 2}

Select[a, EvenQ[#] && # ≥ 4 &]    {4, 4, 4, 6, 4, 6, 4, 4}

Select[a, 2 ≤ # ≤ 4 &]    {2, 4, 3, 4, 4, 2, 4, 4, 3, 2, 4, 2, 3}

Select[a, # == 1 || # == 6 &]    {1, 6, 6, 1, 1}
```

In nested lists, a part specification may be needed:

```
b = Table[{i, RandomReal[]}, {i, 5}]
{{1, 0.179938}, {2, 0.0702301}, {3, 0.900999}, {4, 0.274275}, {5, 0.799332}}

Select[b, #[[2]] ≤ 0.5 &]
{{1, 0.179938}, {2, 0.0702301}, {4, 0.274275}}
```

■ Searching with a Pattern

> `Count[list, pattern]` Give the number of elements in `list` that match `pattern`
>
> `Cases[list, pattern]` Give the elements of `list` that match `pattern`
>
> `DeleteCases[list, pattern]` Remove the elements of `list` that match `pattern`

These commands are considered in Section 16.1.1, p. 493, so here we consider them only briefly. A true pattern is formed with the underscore (`_`), but simple expressions such as `6`, `x`, or `{1, 2}` can also be considered as patterns: They are degenerate patterns. We count the number of sixes and results that are at least 4:

```
Count[a, 6]      2

Cases[a, x_ /; x ≥ 4]      {4, 5, 4, 4, 5, 6, 4, 6, 4, 4}
```

14.1.8 Searching Positions

■ Searching Positions with a Pattern

> **Position[list, pattern]** Give the positions at which objects matching **pattern** occur in **list**
> **Extract[list, positions]** Extract the parts at the given positions

Position is considered in more detail in Section 16.1.1, p. 493. Here, we give only one example. In t[
sequence of tosses presented in Section 14.1.7, sixes occurred at the following times:

```
Position[a, 6]      {{9}, {12}}

Extract[a, %]      {6, 6}

Clear[a, b]
```

■ Searching the Positions of the Smallest through Largest Elements

> **Ordering[list]** Give the positions of all elements of **Sort[list]** in **list**
> **Ordering[list, n]** Give the positions of the first **n** elements of **Sort[list]** in **list**
> **Ordering[list, -n]** Give the positions of the last **n** elements in **Sort[list]** in **list**
> **Ordering[list, 1]** Give the position of the smallest element in **list**
> **Ordering[list, -1]** Give the position of the largest element in **list**

Consider the following list and its sorted version:

```
c = {13, 16, 14, 15, 11, 12};

sc = Sort[c]      {11, 12, 13, 14, 15, 16}
```

Ordering gives the following list:

```
Ordering[c]      {5, 6, 1, 3, 4, 2}
```

This means that the first element of **sc** (the smallest element of **c**) is the fifth element of **c**, the second-
smallest element of **c** is the sixth element of **c**, and so on, with the largest element of **c** being the second
element of **c**. The sorted list can be obtained as follows:

```
c[[%]]      {11, 12, 13, 14, 15, 16}
```

The position of the smallest element and the smallest element itself are as follows:

```
{minpos = Ordering[c, 1][[1]], c[[minpos]]}      {5, 11}
```

Another way to get the position of the smallest element is to write the following:

```
Position[c, Min[c]][[1, 1]]      5
```

The position of the largest element and the largest element itself are as follows:

```
{maxpos = Ordering[c, -1][[1]], c[[maxpos]]}      {2, 16}
```

Another way to get the position of the largest element is to write the following:

```
Position[c, Max[c]][[1, 1]]      2
```

```
c =.
```

14.1.9 Operations on Several Lists

Join[list1, list2, …] Concatenate the lists together

Union[list1, list2, …] or **list1 ∪ list2 ∪ …** Give a sorted list of all distinct elements that
   appear in any of the lists

Intersection[list1, list2, …] or **list1 ∩ list2 ∩ …** Give a sorted list of all distinct elements
   common to all lists

Complement[list0, list1, list2, …] Give a sorted list of all distinct elements of **list0** that are
   not in any of the other lists

Here, ∪ can be written as ESC**un**ESC and ∩ as ESC**inter**ESC. With **Union**, **Complement**, and **Intersection**
we can use the option **SameTest** to define a test to be used for deciding whether two elements should be
considered the same.

```
e = {r, 3, 1, p, 2, r, 2, q};
f = {A, C, p, A, B, 3};
Join[e, f]      {r, 3, 1, p, 2, r, 2, q, A, C, p, A, B, 3}

Union[e, f]      {1, 2, 3, A, B, C, p, q, r}

Complement[e, f]      {1, 2, q, r}

Union[{3, -2, -4, -3, 2}, SameTest → (Abs[#1] == Abs[#2] &)]

{-4, -3, -2}

Clear[e, f]
```

14.2 Advanced List Manipulation

14.2.1 Mapping the Elements

■ **Basic Mapping**

Suppose we have a list such as

```
t = {a, b, c};
```

and we want to map each element of the list with a function. One way to do this is to use **Table**:

```
Table[f[t〚i〛], {i, 3}]      {f[a], f[b], f[c]}
```

However, we can also use **Table** in a special way:

Table[expr, {x, {x1, x2, … }}] (✾6) Form a list by giving, in **expr**, **x** the values **x1, x2**, …

```
Table[f[x], {x, t}]      {f[a], f[b], f[c]}
```

In addition to **Table**, we also have **Map** and its shortened form **/@**:

Map[expr &, {x1, x2, … }] Form a list by giving, in **expr**, **#** the values **x1, x2**, …

expr & /@ {x1, x2, … } Form a list by giving, in **expr**, **#** the values **x1, x2**, …

Map or **/@** is one of the most useful commands of *Mathematica*. With map, we can apply a given function to each element (at the first level) of a list. The function to be mapped is expressed as a pure function (for pure functions, see Sections 2.2.2, p. 38, and 17.1.4, p. 520). The argument of the function is **#** and at the end of the function we have **&**. We can map the elements of **t** in either of the following ways:

    Map[f[#] &, t]   {f[a], f[b], f[c]}

    f[#] & /@ t   {f[a], f[b], f[c]}

As another example, take the logarithm of each element:

    Map[Log[#] &, t]   {Log[a], Log[b], Log[c]}

    Log[#] & /@ t   {Log[a], Log[b], Log[c]}

However, when the function to be mapped is a built-in function, such as **Log**, the argument of the pure function need not be written:

    Log /@ t   {Log[a], Log[b], Log[c]}

Note that often we can calculate with lists very simply without **Map** because *Mathematica* does all operations automatically element by element:

    Log[t]   {Log[a], Log[b], Log[c]}

    t^2 + 1 $\{1 + a^2, 1 + b^2, 1 + c^2\}$

    Map is very useful for more complicated list manipulation, and we will use it frequently in this book. Next, we consider some examples.

■ Matrix Manipulation

Consider the matrix

    m = {{1, 2, 3}, {a, b, c}, {A, B, C}};

Reverse reverses the order of the rows:

    Reverse[m]   {{A, B, C}, {a, b, c}, {1, 2, 3}}

However, to reverse each of the rows, write

    Reverse /@ m   {{3, 2, 1}, {c, b, a}, {C, B, A}}

To sort the elements of each row of the previous matrix, write

    Sort /@ %   {{1, 2, 3}, {a, b, c}, {A, B, C}}

The first row of **m** can be taken with **m[[1]]**, but to take the first column, pick the first element of each row with

    #[[1]] & /@ m   {1, a, A}

or with **m[[All, 1]]**.

■ Gradient

To calculate the gradient of a function, we have **D**:

    D[x y z, {{x, y, z}}]   {y z, x z, x y}

However, we can also write the following:

    D[x y z, #] & /@ {x, y, z}   {y z, x z, x y}

■ Frequencies

To calculate frequencies, we have **Tally**:

```
u = RandomInteger[{1, 6}, {20}]
```

{4, 1, 4, 5, 6, 3, 3, 6, 6, 5, 6, 6, 2, 2, 5, 5, 1, 5, 6, 1}

```
Tally[u] // Sort
```

{{1, 3}, {2, 2}, {3, 2}, {4, 2}, {5, 5}, {6, 6}}

However, we can also write the following:

```
{#, Count[u, #]} & /@ Range[6]
```

{{1, 3}, {2, 2}, {3, 2}, {4, 2}, {5, 5}, {6, 6}}

■ Run-length Encoding

Consider the following list:

```
u = RandomInteger[1, {20}]
```

{1, 0, 0, 1, 1, 1, 0, 0, 0, 0, 1, 0, 1, 1, 1, 0, 0, 0, 0, 0}

Split it into runs of identical elements:

```
Split[u]
```

{{1}, {0, 0}, {1, 1, 1}, {0, 0, 0, 0}, {1}, {0}, {1, 1, 1}, {0, 0, 0, 0, 0}}

Then form the run-length encoding of **u** by forming a list of pairs of numbers where the first number is the element of the run in consideration and the second number the length of the run:

```
{First[#], Length[#]} & /@ %
```

{{1, 1}, {0, 2}, {1, 3}, {0, 4}, {1, 1}, {0, 1}, {1, 3}, {0, 5}}

The decoding can be done as follows:

```
ConstantArray[#[[1]], #[[2]]] & /@ % // Flatten
```

{1, 0, 0, 1, 1, 1, 0, 0, 0, 0, 1, 0, 1, 1, 1, 0, 0, 0, 0, 0}

■ Special Mappings

Map[f, list, levspec]	Apply **f** to each element at the specified levels of **list**
MapAt[f, list, parts]	Apply **f** to the specified parts of **list**
MapAll[f, list]	Apply **f** to all parts of **list**

These special mapping commands are not at all important. Level specifications were considered in Section 13.2.3, p. 427. The default level in **Map** is {1}, which means that each element at the first level is mapped.

```
m = {{1, 2, 3}, {a, b, c}, {A, B, C}};
```

```
Map[ToString, m, {2}] // InputForm
```

{{"1", "2", "3"}, {"a", "b", "c"}, {"A", "B", "C"}}

```
MapAt[#^2 &, m, 3]
```

$\left\{\{1, 2, 3\}, \{a, b, c\}, \left\{A^2, B^2, C^2\right\}\right\}$

> **MapIndexed[f, list]** Apply **f** to the elements of **list**, putting the part specification of each
>   element as a second argument to **f**
> **Scan[f, list]** Apply **f** to each element in **list**, but do not form a list from the results

MapIndexed[f[#1, #2] &, {a, b, c}]

{f[a, {1}], f[b, {2}], f[c, {3}]}

MapIndexed[1 / (1 + #1) ^ #2[[1]] &, {1, 2, 3}]

$$\{\frac{1}{2}, \frac{1}{9}, \frac{1}{64}\}$$

Scan[Print["The divisors of ", #, " are ", Divisors[#]] &, {15, 16, 17}]

The divisors of 15 are {1, 3, 5, 15}

The divisors of 16 are {1, 2, 4, 8, 16}

The divisors of 17 are {1, 17}

14.2.2 Changing the Head

■ **How to Change the Head**

Consider the internal forms of a list, sum, and product:

{a, b, c} // FullForm   List[a, b, c]

a + b + c // FullForm   Plus[a, b, c]

a b c // FullForm   Times[a, b, c]

We can see that the internal forms are very similar: Only the head of the expression is different (for heads, see Section 13.2.3, p. 426). Therefore, if we had a command to change the head of an expression, we would easily do various changes to expressions. **Apply** is such a command.

> **Apply[head, list]** or **head @@ list** Replace the head **List** of **list** with **head**

The only thing **Apply** does is change the head:

t = {a, b, c};

Apply[head, t]   head[a, b, c]

head @@ t   head[a, b, c]

Then change the head to **Plus**, **Times**, and **And**:

{Plus @@ t, Times @@ t, And @@ t}

{a + b + c, a b c, a && b && c}

In the following table, we have collected useful applications of **Apply**:

> **Apply[Plus, list]** Calculate the sum of the elements of **list**; also **Total[list]**
> **Apply[Times, list]** Calculate the product of the elements of **list**
> **Apply[And, list]** Calculate the logical AND of the elements of **list**
> **Apply[List, sum]** Form a list from the terms of **sum**

■ Level Specifications

> `Apply[head, list, {1}]` or `head @@@ list` Replace the head at level 1
> `Apply[head, list, levspec]` Replace the head at the specified levels

Level specifications were considered in Section 13.2.3, p. 427. For example, level specification {0} means the whole expression, {1} means the parts of the expression at level 1, and {0, 1} means the parts at levels 0 and 1. The default level of `Apply` is {0}, which means that `Apply` replaces the head of the whole expression. As an example, we calculate column and row sums and the sum of all elements of a matrix:

```
m = {{1, 2, 3}, {a, b, c}, {A, B, C}};

Apply[Plus, m]    {1 + a + A, 2 + b + B, 3 + c + C}

Apply[Plus, m, {1}]    {6, a + b + c, A + B + C}

Apply[Plus, m, {0, 1}]    6 + a + A + b + B + c + C
```

As another example, factor an integer:

```
fac = FactorInteger[1 234 800]

{{2, 4}, {3, 2}, {5, 2}, {7, 3}}
```

This means that $1\,234\,800 = 2^4\,3^2\,5^2\,7^3$. Suppose we are given this factorization and we have to construct the original number. Let us see how *Mathematica* represents a power:

```
a^b // FullForm    Power[a, b]
```

Thus, the head is `Power`. So, we have to change the head `List` of the pairs of numbers (which are at level 1) to the head `Power` and then multiply the powers:

```
Power @@@ fac    {16, 9, 25, 343}

Times @@ %    1 234 800
```

By the way, a formula such as $2^4\,3^2\,5^2\,7^3$ can be obtained as follows:

```
CenterDot @@ (Superscript @@@ fac)    2^4 · 3^2 · 5^2 · 7^3
```

14.2.3 Sequences

■ Sequences

Suppose we want to calculate, with `Plus`, the sum of **a**, **b**, and **c**. Compare the following two commands:

```
Plus[{a, b, c}]    {a, b, c}

Plus[a, b, c]    a + b + c
```

We see that the arguments of `Plus` cannot be within a list: The arguments have to be loose. In *Mathematica*, it is said that the arguments **a**, **b**, and **c** in `Plus[a, b, c]` form a *sequence*. Thus, a sequence is like a list but the braces `{ }` are lacking. We can form a sequence with `Sequence`:

```
Sequence[a, b, c]    Sequence[a, b, c]
```

However, the result is not **a**, **b**, **c** but `Sequence[a, b, c]`. We get the result **a**, **b**, **c** only inside another expression:

```
{v, x, Sequence[a, b, c], y, z}    {v, x, a, b, c, y, z}
```

Suppose, in general, that we have a multivariate function `f` where the arguments have to be loose—that is, they form a sequence. Thus, the function has to be called like `f[x, y, … ]`, not like `f[{x, y, … }]`. `Apply` is a command that enables us to calculate the value of the multivariate function even if the arguments are supplied in the form of a list. Indeed, a more general syntax of `Apply` is the following:

> `Apply[f[##]&, {x, y, … }]` Calculate `f[x, y, … ]`

Here, `##` represents all of the arguments of `f`. A simple example:

> `Plus[##] & @@ {a, b, c}`   a + b + c

■ Example: Multiple Iteration Specifications

An application of the more general syntax of `Apply` is with commands such as `Table`, `Do`, `Sum`, or `Integrate` that sometimes have several iteration specifications. Suppose we want to calculate

$$\sum_{i_1=1}^{3}\sum_{i_2=1}^{3}\sum_{i_3=1}^{3}(i_1 + i_2 + i_3)$$

In this example, it is simple to write

> `Sum[i₁ + i₂ + i₃, {i₁, 3}, {i₂, 3}, {i₃, 3}]`   162

Note that the three iteration specifications have to form a sequence—that is, they cannot be within a list:

> `Sum[i₁ + i₂ + i₃, {{i₁, 3}, {i₂, 3}, {i₃, 3}}]`
> `{(1 - i₂ + i₃) (i₁ + i₂ + i₃), i₁ + i₂ + i₃}`

Suppose now that we have the iteration specifications as a list:

> `iter = Table[{i_j, 3}, {j, 3}]`
> `{{i₁, 3}, {i₂, 3}, {i₃, 3}}`

Note again that we cannot write

> `Sum[i₁ + i₂ + i₃, iter]`
> Sum::itform : Argument iter at position 2 does not have the correct form for an iterator. ≫
> `Sum[i₁ + i₂ + i₃, iter]`

Instead, we can use `Apply` and write

> `Sum[i₁ + i₂ + i₃, ##] & @@ iter`   162

because now the iteration specifications appear as a sequence in `Sum`. So we get the following command:

> `Sum[i₁ + i₂ + i₃, {i₁, 3}, {i₂, 3}, {i₃, 3}]`   162

We could also write

> `Sum[i₁ + i₂ + i₃, Evaluate[Sequence @@ iter]]`   162

Here, we changed the *list* of iteration specifications into a *sequence* with `Apply` by changing the head `List` of `iter` to the head `Sequence`. In addition, `Sum` requires the use of `Evaluate` to get explicit iteration specifications.

■ Example: Stirling Numbers

Stirling numbers of the second kind $S_n^{(k)}$ can be calculated with `StirlingS2[n, k]`:

StirlingS2[7, 3]   301

Let us try to calculate these numbers from (see Trott, 2004b, p. 717)

$$S_n^{(k)} = \frac{n!}{k!} \sum_{r_1=1}^{n} \sum_{r_2=1}^{n-r_1} \cdots \sum_{r_k=1}^{n-r_1-\ldots-r_{k-1}} \frac{\delta_{n, \sum_{j=1}^{k} r_j}}{\prod_{j=1}^{k} r_j!}$$

Here, δ is the Kronecker δ: It is 1 if all the arguments are equal and 0 otherwise. Note that we have k indices of summation. How can we write such a general number of summation indices? The solution is to form a *list* of summation specifications with **Table** and then feed the specifications into **Sum** in a *sequence* form with **Apply**:

$$\text{stirlingS2}[n\_, k\_] :=$$
$$\frac{n!}{k!} \text{ Sum}\left[\frac{\text{KroneckerDelta}\left[n, \sum_{j=1}^{k} r_j\right]}{\prod_{j=1}^{k} r_j!}, \text{ \#\#}\right] \& @@ \text{Table}\left[\left\{r_i, n - \sum_{j=1}^{i-1} r_j\right\}, \{i, k\}\right]$$

stirlingS2[7, 3]   301

14.2.4 Mapping Two Lists

■ Threading

Thread[f[{a, b, c}, {A, B, C}]]   {f[a, A], f[b, B], f[c, C]}

MapThread[f, {{a, b, c}, {A, B, C}}]   {f[a, A], f[b, B], f[c, C]}

The results of **Thread** and **MapThread** are the same, but the way the two lists are inputted differs. Some examples follow. **Thread** also has applications in manipulating equations (see Section 22.2.3, p. 725).

■ Constructing Explicit Lists of Rules

Note that the following does not work:

a + b + c /. {a, b, c} → {A, B, C}   a + b + c

We have to use an explicit list of rules:

a + b + c /. {a → A, b → B, c → C}   A + B + C

With **Thread** we can construct such a list of substitutions; we could also use **MapThread**, but **Thread** is simpler:

Thread[{a, b, c} → {A, B, C}]   {a → A, b → B, c → C}

MapThread[Rule, {{a, b, c}, {A, B, C}}]   {a → A, b → B, c → C}

Thus, write as follows:

a + b + c /. Thread[{a, b, c} → {A, B, C}]   A + B + C

■ Constructing Explicit Lists of Equations

Note that **Solve** does not need an explicit list of equations:

```
Solve[{x + y, y + z, x + z} == {1, 0, 1}]
```

$\{\{x \to 1, y \to 0, z \to 0\}\}$

However, an explicit list of equations can be obtained with **Thread** or **MapThread**:

```
Thread[{x + y, y + z, x + z} == {1, 0, 1}]
```

$\{x + y = 1, y + z = 0, x + z = 1\}$

```
MapThread[Equal, {{x + y, y + z, x + z}, {1, 0, 1}}]
```

$\{x + y = 1, y + z = 0, x + z = 1\}$

■ **Other Examples**

Form pairs with a constant second element:

```
Thread[{{a, b, c}, 1}]        {{a, 1}, {b, 1}, {c, 1}}
```

Calculate the pairwise maximums of two lists:

```
MapThread[Max, {{2, 3, 1, 4}, {3, 1, 2, 3}}]        {3, 3, 2, 4}
```

■ **Inner and Outer Products**

```
Inner[f, {a, b, c}, {A, B, C}]  f[a, A] + f[b, B] + f[c, C]

Inner[f, {a, b, c}, {A, B, C}, g]  g[f[a, A], f[b, B], f[c, C]]

Outer[f, {a, b, c}, {A, B, C}]  {{f[a, A], f[a, B], f[a, C]},
                                 {f[b, A], f[b, B], f[b, C]},
                                 {f[c, A], f[c, B], f[c, C]}}
```

With **Inner**, we can form the inner product of two vectors. For a simple function such as **Times**, the name of the function suffices, although we could also write two arguments for the pure function:

```
Inner[Times, {a, b, c}, {A, B, C}]      a A + b B + c C

Inner[Times[#1, #2] &, {a, b, c}, {A, B, C}]      a A + b B + c C
```

An easier way, however, is to use the dot:

```
{a, b, c}.{A, B, C}      a A + b B + c C
```

Inner can have a function as the fourth argument. The default of this function is **Plus**:

```
Inner[f, {a, b, c}, {A, B, C}, Plus]
```

f[a, A] + f[b, B] + f[c, C]

With **Outer**, we can calculate the outer product of two vectors:

```
Outer[Times, {a, b, c}, {A, B, C}]
```

$\{\{a A, a B, a C\}, \{A b, b B, b C\}, \{A c, B c, c C\}\}$

We then form all pairs of the elements of two lists:

```
Outer[List, {a, b, c}, {A, B, C}]
```

$\{\{\{a, A\}, \{a, B\}, \{a, C\}\}, \{\{b, A\}, \{b, B\}, \{b, C\}\}, \{\{c, A\}, \{c, B\}, \{c, C\}\}\}$

With **Tuples** of **Distribute** we get a nonnested list of pairs:

```
Tuples[{{a, b, c}, {A, B, C}}]
```

$\{\{a, A\}, \{a, B\}, \{a, C\}, \{b, A\}, \{b, B\}, \{b, C\}, \{c, A\}, \{c, B\}, \{c, C\}\}$

Tables

Introduction

> *Why are numbers beautiful? It is like asking why is Beethoven's Ninth Symphony beautiful.*
> *If you don't see why, someone can't tell you.*
> *I know numbers are beautiful. If they aren't beautiful, nothing is.*—Paul Erdös

Tabular representation of data has proved to be an efficient way to transmit information. *Mathematica* has good tools for preparing tables. **TableForm** is familiar from earlier versions of *Mathematica*, whereas **Grid**, **Column**, and **Row** are new in version 6. **Grid** is versatile and powerful for even complex tables, giving the possibility of detailed formatting.

15.1 Basic Tabulating

15.1.1 TableForm

The main commands used to tabulate lists are **TableForm** and **Grid**. Of these, **TableForm** is an elementary tabulating command that can be used to adjust the spacings between the items, to give headings to rows and columns, and to also tabulate three- and higher-dimensional lists. **Grid** is an advanced tabulating command with which we can adjust the table in numerous ways, for example, with dividers and frames. First, we consider **TableForm**; **Grid** is addressed in Section 15.2, p. 470.

TableForm[list]   Form a table from **list**

Options:

TableAlignments   Alignment of elements in horizontal and vertical directions; examples of values: **Automatic** (means **{Left, Baseline}**), **Right**, **Decimal**, **"."**, **{Center, Baseline}**, **{Right, Bottom}**

TableSpacing   Space between rows and columns; examples of values: **Automatic** (means, for a 2D list, **{1, 1}**), **{1, 2}**

TableHeadings   Labels for rows and columns; examples of values: **{None, None, None, None, … }**, **Automatic** (means consecutive integers), **{None, {"Col1", "Col2", "Col3"}}**

TableDepth   Up to what level the tabular form is used; examples of values: ∞, **2**

TableDirections   How to arrange eacn dimension (as a row or column); examples of values:

{Column, Row, Column, Row, … }, {Row, Column}

We try **TableForm** for a 1D, 2D, and 3D list:

```
m1 = {32, 214, 5};
m2 = {{32.7, 8.39, -412.64}, {4.5, -56.2163, -7.606}};
m3 = {{{1, a}, {2, b}, {3, c}}, {{4, d}, {5, e}, {6, f}}};
{TableForm[m1], TableForm[m2], TableForm[m3]}
```

$$\left\{ \begin{matrix} 32 \\ 214 \\ 5 \end{matrix} \right. , \quad \begin{matrix} 32.7 & 8.39 & -412.64 \\ 4.5 & -56.2163 & -7.606 \end{matrix} , \quad \left. \begin{matrix} 1 & 2 & 3 \\ a & b & c \\ 4 & 5 & 6 \\ d & e & f \end{matrix} \right\}$$

For decimal numbers, alignment according to the decimal point is useful:

TableForm[m2, TableAlignments → Decimal]

```
32.7     8.39     -412.64
 4.5   -56.2163    -7.606
```

For integers, right alignment is suitable:

TableForm[Table[{n!, (2 n) !, (3 n) !}, {n, 4}], TableAlignments → Right]

```
 1      2        6
 2     24      720
 6    720   362 880
24  40 320  479 001 600
```

For a higher-dimensional table, it may be advantageous to adjust the spacing and directions:

```
{TableForm[m3, TableSpacing → {2, 3, 0}],
 TableForm[m3, TableSpacing → {3, 3, 1}, TableDirections → {Column, Row, Row}]}
```

$$\left\{ \begin{matrix} 1 & 2 & 3 \\ a & b & c \\ 4 & 5 & 6 \\ d & e & f \end{matrix} , \quad \begin{matrix} 1 & a & 2 & b & 3 & c \\ 4 & d & 5 & e & 6 & f \end{matrix} \right\}$$

Here are some examples of **TableHeadings**:

TableForm[m2, TableHeadings → Automatic]

	1	2	3
1	32.7	8.39	-412.64
2	4.5	-56.2163	-7.606

TableForm[m2, TableHeadings → {None, {"Col. 1 ", "Col. 2 ", "Col. 3"}}]

Col. 1	Col. 2	Col. 3
32.7	8.39	-412.64
4.5	-56.2163	-7.606

TableForm[m2, TableHeadings → {Automatic, {"Col. 1 ", "Col. 2 ", "Col. 3"}}]

	Col. 1	Col. 2	Col. 3
1	32.7	8.39	-412.64
2	4.5	-56.2163	-7.606

```
TableForm[m2,
   TableHeadings → {{"Row 1", "Row 2"}, {"Col. 1 ", "Col. 2 ", "Col. 3"}}] // Text
```

	Col. 1	Col. 2	Col. 3
Row 1	32.7	8.39	−412.64
Row 2	4.5	−56.2163	−7.606

PaddedForm[TableForm[list], {n, f}]   Align all numbers right; reserve space for **n** digits for all numbers, with **f** of them for decimals

---

To align columns with the decimal point, use **PaddedForm** to define a fixed space for the decimals:

```
{TableForm[m2], PaddedForm[TableForm[m2], {6, 3}]}
```

$$\left\{ \begin{array}{lll} 32.7 & 8.39 & -412.64 \\ 4.5 & -56.2163 & -7.606 \end{array} \right. , \left. \begin{array}{lll} 32.700 & 8.390 & -412.640 \\ 4.500 & -56.216 & -7.606 \end{array} \right\}$$

If necessary, the decimal digits are shortened and zeros are added to fill the fixed space of the decimals; the decimal point and the possible sign are not counted in the total space. **PaddedForm** has several options that we do not explore here. We only note that if you do not want the filling zeros after short decimal parts, use the option **NumberPadding → {" ", " "}**.

15.1.2 Column and Row

Column[v] (✿6)   Form a column from a list
Column[v, alignment]   Align columns in the specified way
Column[v, alignment, spacings]   Leave the specified spaces between the rows

---

Column has the same options as **Grid**, with the following exceptions. For **Grid**, the default value of **Alignment** is **{Center, Baseline}**, but for **Column** the default value is **{Left, Baseline}**. Thus, the elements in the column are, by default, aligned left. In addition, **Column** also has the special option **ColumnAlignments**, with default value **Left**. However, usually we do not need either **Alignment** or **ColumnAlignments** because we can simply add to the command a second argument that indicates the alignment. The alignment can be **Left**, **Center**, **Right**, and **"c"** (where **c** is a character).

```
v = Table[Binomial[n, i], {n, 0, 3}, {i, 0, n}]
```

```
{{1}, {1, 1}, {1, 2, 1}, {1, 3, 3, 1}}
```

```
{Grid[v], Column[v], Column[v, Center, 0.6]}
```

$$\left\{ \begin{array}{llll} 1 \\ 1 & 1 \\ 1 & 2 & 1 \\ 1 & 3 & 3 & 1 \end{array} \right. , \begin{array}{l} \{1\} \\ \{1, 1\} \\ \{1, 2, 1\} \\ \{1, 3, 3, 1\} \end{array} , \left. \begin{array}{l} \{1\} \\ \{1, 1\} \\ \{1, 2, 1\} \\ \{1, 3, 3, 1\} \end{array} \right\}$$

```
{Column[Table[n!, {n, 10, 12}], Right, Frame → True],
   Column[Table[Exp[x], {x, 2., 6, 2}], ".",
      Dividers → All, Background → Lighter[Green, 0.7]]} // Text
```

$$\left\{ \begin{array}{|r|} \hline 3\,628\,800 \\ \hline 39\,916\,800 \\ \hline 479\,001\,600 \\ \hline \end{array} , \begin{array}{|r|} \hline 7.38906 \\ \hline 54.5982 \\ \hline 403.429 \\ \hline \end{array} \right\}$$

Row[v] (✿6)   Form a row from a list (word wrap if necessary)
Row[v, separator]   Insert the given separator between elements

---

Row does not have any options. Without a separator the elements are written side by side with no space in between. The separator is typically a string such as " " or ", "; it can also be given in printer's points with **Spacer**:

```
v = Table[Exp[x], {x, 2., 6, 2}]
```

{7.38906, 54.5982, 403.429}

```
Row[v] // Text
```

7.3890654.5982403.429

```
Row[v, ", "] // Text
```

7.38906, 54.5982, 403.429

```
Row[v, "     "] // Text
```

7.38906   54.5982   403.429

```
Row[v, Spacer[8]] // Text
```

7.38906   54.5982   403.429

15.2 Advanced Tabulating

15.2.1 Introduction to Grid and Its Options

■ Creating Tables

> **Grid[m]** (⌘6)   Form a table from a 2D list
> **Grid[m] // Text**   Format the items as text

With **Grid**, we can show 2D lists in a tabular form:

```
m = {{6, 21.2, 3.05, 64.2}, {34, 9.582, 143.17, 8.702}, {985, 0.6914, 70.4, 126.6}};
```

```
m2 = Table[Binomial[n, i], {n, 0, 3}, {i, 0, n}]
```

{{1}, {1, 1}, {1, 2, 1}, {1, 3, 3, 1}}

```
{Grid[m], Grid[m] // Text, Grid[m2]}
```

$$\left\{ \begin{matrix} 6 & 21.2 & 3.05 & 64.2 \\ 34 & 9.582 & 143.17 & 8.702 \\ 985 & 0.6914 & 70.4 & 126.6 \end{matrix} \right., \begin{matrix} 6 & 21.2 & 3.05 & 64.2 \\ 34 & 9.582 & 143.17 & 8.702, \\ 985 & 0.6914 & 70.4 & 126.6 \end{matrix} \left. \begin{matrix} 1 \\ 1 & 1 \\ 1 & 2 & 1 \\ 1 & 3 & 3 & 1 \end{matrix} \right\} $$

The first two examples show that the default is that the items in each column are centered. The last example shows that the rows of the list need not be of the same length.

Next, we begin the study of the many options of **Grid**. We list the options and show some general rules for using them. In Section 15.2.2, p. 475, we study each option in detail.

■ Options of Grid

Here are the options of **Grid**. We have tried to classify the options with respect to importance: Options with ** are assumed to be the most important, options with * are assumed to be not as important, and options without an asterisk are assumed to be seldom used.

Options of `Grid`:

** `Alignment` Horizontal and vertical alignment of items; examples of values: `{Center, Baseline}` (columns are centered, rows are at baseline), `Left` (columns are aligned left), `Right` (columns are aligned right), `"."` (columns are aligned at the decimal point). In horizontal (or column) alignment, we can use `Left`, `Center`, `Right`, and `"c"` (where `c` is a character). In vertical (or row) alignment, we can use `Bottom`, `Center`, `Baseline`, and `Top`.

** `Dividers` Where to draw lines; examples of values: `None`, `All` (all items become boxed), `Center` (all interior dividers), `{None, All}` (no column lines, all row lines, lines also before the first row and after the last row), `{2 → True}` (a line after the first column), `{None, 2 → True}` (a line below the first row), `{2 → True, 2 → True}` (a line after the first column and below the first row)

** `Spacings` Space between columns and rows (in units of the current font size); examples of values: `Automatic` (usually means `{0.7, 0.4}`: the space between columns is 0.7 and that between rows is 0.4), `1.2` (the space between columns is 1.2 and that between rows is the default 0.4)

** `ItemStyle` Styles of columns and rows; examples of values: `None`, `Blue` (all items are blue), `{1 → Red}` (the first column is red), `{Automatic, 1 → Bold}` (the first row is bold), `{1 → Directive[Red, Bold, 14]}` (the first column is red, bold, and size 14), `{1 → Bold, 1 → Bold}` (the first column and row are bold)

* `Background` Colors of the background; examples of values: `None`, `GrayLevel[0.9]` (all items are gray), `{Automatic, {{ White, LightGray}}}` (columns are as default, rows alternate between white and gray)

** `Frame` Where to draw frames; examples of values: `None`, `True` (frame around the whole grid), `All` (all items become boxed), `{All}` (frame around each column), `{None, All}` (frame around each row)

`FrameStyle` Style of frames and dividers; examples of values: `Automatic`, `Red`, `Directive[Gray, Thickness[2]]`

`ItemSize` Width and height of each item; examples of values: `Automatic` (separately size items to fit within the total formatting width; long items may take several lines), `All` (make all items the same width and height), `Full` (allow each item its full width and height; long items are not divided into several lines), `w` (give all items width `w`, measured in ems), `{w, h}` (give all items width `w` and height `h`, with `h` measured in line heights)

`BaseStyle` Base style specifications for the grid; examples of values: `{}`, `Blue` (all items are blue), `Directive[Red, Bold, Italic]` (all items are red, bold, and italic)

`BaselinePosition` How the grid should be positioned inside text; examples of values: `Automatic` (means `Axis`), `Axis` (axis of the middle row in the grid), `Baseline` (baseline of the middle row in the grid), `Bottom` (bottom of the whole grid), `Center` (halfway from top to bottom), `Top` (top of the whole grid)

Note that the examples of values of the options in the previous table are very simple. More advanced forms for the options exist, as is explained next.

■ General Forms of the Values of Options

With the exception of `BaselinePosition`, `BaseStyle`, `FrameStyle`, and `Spacings`, the values of the options can be of the forms given in the next box.

Values of options can be given as follows:

`val` Apply `val` for all items

`{valc}` Apply `valc` for columns

`{valc, valr}` Apply `valc` for columns and `valr` for rows

The values `valc` and `valr` can be of the forms given in the following box:

Values `valc` and `valr` can be a single value or a list of values of the following general form:

$\{a_1, ..., a_k, \{b_1, ..., b_l\}, c_1, ..., c_m\}$ Apply $a_1, ..., a_k$ at the beginning, $c_1, ..., c_m$ at the end, and cyclically the sequence $b_1, ..., b_l$ in between

• Some typical special cases are as follows:

$\{a_1, ..., a_k, \{ \}, c_1, ..., c_m\}$ Apply the default value in between

$\{a_1, ..., a_k, \{b_1, ..., b_l\}\}$ Apply cyclically $b_1, ..., b_l$ through the end

$\{a_1, ..., a_k\}$ Apply $a_1, ..., a_k$ at the beginning and then the default value

$\{a_1\}$ Apply a_1 at the beginning and then the default value

$\{\{b_1, ..., b_l\}\}$ Apply cyclically $b_1, ..., b_l$

$\{\{b_1, b_2\}\}$ Apply cyclically b_1 and b_2—that is, alternate between b_1 and b_2

$\{\{b_1\}\}$ or b_1 Apply repeatedly b_1

The forms of the options are very flexible but, as such, it requires some time to get used to them. In the following examples, we illustrate the various forms of the options. After that, we study, in some detail, most of the options.

■ Example 1

To practice the forms of the options, we consider the following matrix:

```
w = {{1, 2, 3, 4, 5, 6, 7}, {8, 9, 10, 11, 12, 13, 14}};
```

Here are three different tables for the matrix:

```
Row[{Grid[w, Background → {{Red, Blue, {Gray}, Green, Yellow}}],
    Grid[w, Background → {{Red, Blue, {}, Green, Yellow}}],
    Grid[w, Background → {{Red, Blue, {White, Gray}}}]}, ", "]
```

In these examples, the value of the `Background` option is of the form `{valc}` mentioned previously and so the value is applied for columns. In the first example, `valc` is of the form $\{a_1, a_2, \{b_1\}, c_1, c_2\}$ mentioned previously and so the first column is red, the second column is blue, the next-to-last column is green, the last column is yellow, and the columns in between are gray. In the second example, we use the default background (white) for the columns in between. In the third example, beginning from the third column, white and gray are used cyclically.

Here are further examples:

```
Row[{Grid[w, Background → {{Red, Blue}}],
  Grid[w, Background → {{Red}}],
  Grid[w, Background → {{{White, Gray}}}]}], ", "]
```

In these examples, the value of the option is again of the form **{valc}**. In the first example, **valc** is of the form $\{a_1, a_2\}$ and so the first two columns are red and blue, respectively. In the second example, **valc** is of the form $\{a_1\}$ and so the first column is red. In the third example, **valc** is of the form $\{\{b_1, b_2\}\}$ mentioned previously and so white and gray are used cyclically.

■ **Example 2**

Now we draw various frames:

```
v = {{1, 2, 3, 4}, {5, 6, 7, 8}, {9, 1, 11, 12}};
```

```
Row[{Grid[v, Frame → All],
  Grid[v, Frame → {All}],
  Grid[v, Frame → {{All}}],
  Grid[v, Frame → {{{All}}}]}], ", "]
```

In the first example, the value **All** is of the general form **val** mentioned previously; thus, the value **All** is used for all items so that all items have a frame.

In the second example, the value **{All}** is of the form **{valc}** mentioned previously; thus, the value **All** is used for all columns.

In the third example, the value **{{All}}** is again of the form **{valc}**; however, now **valc** is **{All}**, and this is of the general form $\{a_1\}$ mentioned previously. Thus, the value **All** is used for the first column; the default (i.e., no frame) is used for other columns.

In the fourth example, the value **{{{All}}}** is again of the form **{valc}**; however, now **valc** is **{{All}}**, and this is of the general form $\{\{b_1\}\}$ mentioned previously. Thus, the value **All** is used cyclically for columns, and this means that the value **All** is used repeatedly for all columns.

Next, we draw frames for rows:

```
Row[{Grid[v, Frame → {None, All}],
  Grid[v, Frame → {None, {All}}],
  Grid[v, Frame → {None, {{All}}}]}], ", "]
```

In this example, the value of the **Frame** option is of the form **{valc, valr}** so that **valc** or **None** is used for columns and **valr** for rows.

■ **Exceptional Values**

If we have some exceptional formatting for some columns, rows, or items, we can add definitions for these cases as rules.

Exceptional values can be given as rules (a single rule or a list of rules):

{rulesj} Apply **rulesj** for specific columns

{rulesj, rulesi} Apply **rulesj** for specific columns and **rulesi** for specific rows

{valc, rulesj} Apply **valc** for columns but **rulesj** to specific columns

{{valc, rulesj}, {valr, rulesi}} Apply **valc** for columns and **valr** for rows but **rulesj** to specific columns and **rulesi** to specific rows

{valc, valr, rulesij} Apply **valc** and **valr** for columns and rows but **rulesij** for specific items (not applicable for **Dividers** and **Spacings**)

Here, **rulesj** may be, for example, **4 → True**, meaning that the value **True** is used for the fourth column. Similarly, **rulesi** may be, for example, **3 → Green**, meaning that the value **Green** is used for the third row. Also, **rulesij** may be, for example, **{2, 5} → Red**, meaning that the value **Red** is used for the item at position (2, 5). In addition, **rulesij** may define a range of items. An example is **{{2, 4}, {1, 3}} → Red**, meaning that the value **Red** is used for items with row index in {2, 4} and column index in {1, 3}.

■ Example 3

As an example, consider again the **Frame** option:

```
Row[{Grid[v, Frame → {1 → True}],
  Grid[v, Frame → {{1 → True, 3 → True}}], Grid[v, Frame → {1 → True, 3 → True}]}, ", "]
```

In the first example, the value **{1 → True}** is of the form **{rulesj}** so that a frame is drawn for the first column. In the second example, the value **{{1 → True, 3 → True}}** is again of the form **{rulesj}** so that a frame is drawn for the first and third columns. In the third example, the value **{1 → True, 3 → True}** is of the form **{rulesj, rulesi}** so that a frame is drawn for the first column and the third row.

Here are further examples:

```
Row[{Grid[v, Frame → {{1 → True, 3 → True}, 3 → True}],
  Grid[v, Frame → {None, 3 → True}],
  Grid[v, Frame → {None, None, {3, 3} → True}],
  Grid[v, Frame → {None, None, {{2, 3}, {2, 4}} → True}]}, ", "]
```

In the first example, the value **{{1 → True, 3 → True}, 3 → True}** is again of the form **{rulesj, rulesi}** so that a frame is drawn for the first and third columns and the third row.

In the second example, the value **{None, 3 → True}** is of the form **{{valc, rulesj}, {valr, rulesi}}** or, actually, the special case **{valc, rulesi}** of this form, so that frames are not drawn for columns but a frame is drawn for the third row.

In the third example, the value **{None, None, {3, 3} → True}** is of the form **{valc, valr, rulesij}** so that a frame is not drawn for the columns or rows but a frame is drawn for the (3, 3)th item.

In the fourth example, we draw a frame around items with row index in {2, 3} and column index in {2, 4}.

Next, we study the various options in more detail with the aid of an example. For other examples, see Sections 23.2.1, p. 755, 29.2.1, p. 968, 29.3.2, p. 980, and 29.4.2, p. 992.

■ **Example 4**

The grid may also contain text and graphics:

```
t1 = "Here is the cumulative distribution
        function (CDF) of the standard normal distribution with mean

      0 and standard deviation 1.\nThe CDF is  ────── ∫  e⁻ᵗ²/²dt.";
                                                 √2π

t2 = "Here is the probability density function (PDF) of the
        standard normal distribution with mean 0 and

      standard deviation 1.\nThe PDF is  ────── e⁻ˣ²/² . ";
                                          √2π

g1 = Plot[CDF[NormalDistribution[0, 1], x], {x, -3, 3}, ImageSize → 120];
g2 = Plot[PDF[NormalDistribution[0, 1], x], {x, -3, 3}, ImageSize → 120];
Grid[{{t1, g1}, {t2, g2}}, Alignment → {Left, Center}, ItemSize → {{20, 20}}] // Text
```

Here is the *cumulative distribution function* (CDF) of the standard normal distribution with mean 0 and standard deviation 1.

The CDF is $\frac{1}{\sqrt{2\pi}}\int_{-\infty}^{x} e^{-t^2/2}dt$.

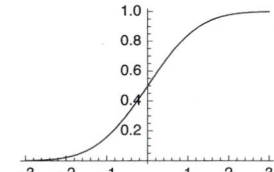

Here is the *probability density function* (PDF) of the standard normal distribution with mean 0 and standard deviation 1.

The PDF is $\frac{1}{\sqrt{2\pi}}e^{-x^2/2}$.

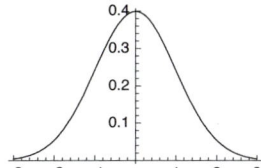

15.2.2 Options of Grid

■ **Example: Row and Column Sums**

In Section 17.2.1, we listed the options of **Grid** and discussed general rules of how they are used. Now we explore in detail each of the options. However, first we generate an example. Consider the matrix

```
m = {{6, 21.2, 3.05, 64.2}, {34, 9.582, 143.17, 8.702}, {985, 0.6914, 70.4, 126.6}};
```

Let us prepare a table in which we have the column sums, row sums, and the total sum. First, calculate the sums (see Section 21.2.3, p. 697):

```
colSums = Total[m]        {1025, 31.4734, 216.62, 199.502}

rowSums = Total /@ m       {94.45, 195.454, 1182.69}

totSum = Total[rowSums]    1472.6
```

Then form the rows of the table:

```
firstRow = {{"Rows", "Col 1", "Col 2", "Col 3", "Col 4", "Sums"}}
```
{{Rows, Col 1, Col 2, Col 3, Col 4, Sums}}

```
mainRows = Join[{{"Row 1", "Row 2", "Row 3"}}ᵀ, m, {rowSums}ᵀ, 2]
```
{{Row 1, 6, 21.2, 3.05, 64.2, 94.45}, {Row 2, 34, 9.582, 143.17, 8.702, 195.454},
 {Row 3, 985, 0.6914, 70.4, 126.6, 1182.69}}

```
lastRow = {Join[{"Sums"}, colSums, {totSum}]}
```
{{Sums, 1025, 31.4734, 216.62, 199.502, 1472.6}}

(Here, ^{T} means a transpose; it can be written as ⎣ESC⎦tr⎣ESC⎦.) Join the rows:

```
t = Join[firstRow, mainRows, lastRow]
```
{{Rows, Col 1, Col 2, Col 3, Col 4, Sums},
 {Row 1, 6, 21.2, 3.05, 64.2, 94.45}, {Row 2, 34, 9.582, 143.17, 8.702, 195.454},
 {Row 3, 985, 0.6914, 70.4, 126.6, 1182.69},
 {Sums, 1025, 31.4734, 216.62, 199.502, 1472.6}}

A first draft for a table is as follows:

```
Grid[t]
```

Rows	Col 1	Col 2	Col 3	Col 4	Sums
Row 1	6	21.2	3.05	64.2	94.45
Row 2	34	9.582	143.17	8.702	195.454
Row 3	985	0.6914	70.4	126.6	1182.69
Sums	1025	31.4734	216.62	199.502	1472.6

Next, we try several options to enhance the table.

■ **Alignment**

> ** **Alignment** Horizontal and vertical alignment of items; examples of values: {Center, Baseline}
> (columns are centered, rows are at baseline), **Left** (columns are aligned left), **Right** (columns are
> aligned right), "**.**" (columns are aligned at the decimal point). In horizontal (or column) alignment,
> we can use **Left**, **Center**, **Right**, and "**c**" (where **c** is a character). In vertical (or row) alignment,
> we can use **Bottom**, **Center**, **Baseline**, and **Top**.

As an example, we consider the table **t** we formed previously. First, we align the columns at the right:

```
Grid[t, Alignment → Right] // Text
```

Rows	Col 1	Col 2	Col 3	Col 4	Sums
Row 1	6	21.2	3.05	64.2	94.45
Row 2	34	9.582	143.17	8.702	195.454
Row 3	985	0.6914	70.4	126.6	1182.69
Sums	1025	31.4734	216.62	199.502	1472.6

Right alignment is good for integers. Then we align the columns at the decimal point:

```
Grid[t, Alignment → "."] // Text
```

Rows	Col 1	Col 2	Col 3	Col 4	Sums
Row 1	6	21.2	3.05	64.2	94.45
Row 2	34	9.582	143.17	8.702	195.454
Row 3	985	0.6914	70.4	126.6	1182.69
Sums	1025	31.4734	216.62	199.502	1472.6

This alignment is also well suited for integers. Note that the column headers are aligned such that they end at the position of the decimal point, and this causes the headers to be too far to the left. For the row headers, the right alignment is not good.

Now we align the first column left, the second column right, and the rest of the columns at the decimal point:

```
Grid[t, Alignment → {{Left, Right, {"."}}}] // Text
```

Rows	Col 1	Col 2	Col 3	Col 4	Sums
Row 1	6	21.2	3.05	64.2	94.45
Row 2	34	9.582	143.17	8.702	195.454
Row 3	985	0.6914	70.4	126.6	1182.69
Sums	1025	31.4734	216.62	199.502	1472.6

Note that here the value of the option is of the form **{valc}** mentioned previously in a box. This form of the option means that **valc** is applied for columns. Here, **valc** is **{Left, Right, {"."}}** and this is of the general form $\{a_1, a_2, \{b_1\}\}$ mentioned in a box in Section 17.2.1. This form of the option means that a_1 is applied for the first column, a_2 for the second column, and b_1 for the rest of the columns.

The previous table is quite good. However, the column headers should be aligned better. We define these exceptional alignments as rules:

```
Grid[t, Alignment → {{Left, Right, {"."}}, Baseline, {{1, 1}, {2, 6}} → Center}] //
   Text
```

Rows	Col 1	Col 2	Col 3	Col 4	Sums
Row 1	6	21.2	3.05	64.2	94.45
Row 2	34	9.582	143.17	8.702	195.454
Row 3	985	0.6914	70.4	126.6	1182.69
Sums	1025	31.4734	216.62	199.502	1472.6

Here, the value of the option is of the form **{valc, valr, rulesij}** mentioned previously in a box. This form of the option means that **valc** is applied for columns, **valr** for rows, and **rulesij** for special items. Actually, **valr** or **Baseline** is the default for rows but we have to define something for the rows so that we can define rules for single items. The rule **{{1, 1}, {2, 6}} → Center** defines that the column alignment should be **Center** for items with row indices in the range {1, 1} and column indices in the range {2, 6}.

■ Dividers

> ** **Dividers** Where to draw lines; examples of values: **None**, **All** (all items become boxed), **Center** (all interior dividers), **{None, All}** (no column lines, all row lines, lines also before the first row and after the last row), **{2 → True}** (a line after the first column), **{None, 2 → True}** (a line below the first row), **{2 → True, 2 → True}** (a line after the first column and below the first row)

Note that if we give **Dividers** a list of values, the first value corresponds with the divider *before* the first column or *above* the first row. The last value that can be given concerns the divider *after* the last column or *below* the last row.

First, we draw all dividers:

`Grid[t, Dividers → All] // Text`

Rows	Col 1	Col 2	Col 3	Col 4	Sums
Row 1	6	21.2	3.05	64.2	94.45
Row 2	34	9.582	143.17	8.702	195.454
Row 3	985	0.6914	70.4	126.6	1182.69
Sums	1025	31.4734	216.62	199.502	1472.6

Then we draw all interior dividers:

`Grid[t, Dividers → Center] // Text`

Rows	Col 1	Col 2	Col 3	Col 4	Sums
Row 1	6	21.2	3.05	64.2	94.45
Row 2	34	9.582	143.17	8.702	195.454
Row 3	985	0.6914	70.4	126.6	1182.69
Sums	1025	31.4734	216.62	199.502	1472.6

Next, we ask for no column dividers and all row dividers:

`Grid[t, Dividers → {None, All}] // Text`

Rows	Col 1	Col 2	Col 3	Col 4	Sums
Row 1	6	21.2	3.05	64.2	94.45
Row 2	34	9.582	143.17	8.702	195.454
Row 3	985	0.6914	70.4	126.6	1182.69
Sums	1025	31.4734	216.62	199.502	1472.6

Next, we add, for both the columns and the rows, the second divider:

`Grid[t, Dividers → {2 → True, 2 → True}] // Text`

Rows	Col 1	Col 2	Col 3	Col 4	Sums
Row 1	6	21.2	3.05	64.2	94.45
Row 2	34	9.582	143.17	8.702	195.454
Row 3	985	0.6914	70.4	126.6	1182.69
Sums	1025	31.4734	216.62	199.502	1472.6

Now we add, for columns, the second divider and the next-to-last divider:

`Grid[t, Dividers → {{2 → True, -2 → True}}] // Text`

Rows	Col 1	Col 2	Col 3	Col 4	Sums
Row 1	6	21.2	3.05	64.2	94.45
Row 2	34	9.582	143.17	8.702	195.454
Row 3	985	0.6914	70.4	126.6	1182.69
Sums	1025	31.4734	216.62	199.502	1472.6

Next, we add, for rows, the second divider and the next-to-last divider:

`Grid[t, Dividers → {None, {2 → True, -2 → True}}] // Text`

Rows	Col 1	Col 2	Col 3	Col 4	Sums
Row 1	6	21.2	3.05	64.2	94.45
Row 2	34	9.582	143.17	8.702	195.454
Row 3	985	0.6914	70.4	126.6	1182.69
Sums	1025	31.4734	216.62	199.502	1472.6

Now we combine the column and row dividers:

`Grid[t, Dividers → {{2 → True, -2 → True}, {2 → True, -2 → True}}] // Text`

Rows	Col 1	Col 2	Col 3	Col 4	Sums
Row 1	6	21.2	3.05	64.2	94.45
Row 2	34	9.582	143.17	8.702	195.454
Row 3	985	0.6914	70.4	126.6	1182.69
Sums	1025	31.4734	216.62	199.502	1472.6

Instead of the value **True**, we can also define colors:

`Grid[t, Dividers → {{2 → Blue, -2 → Blue}, {2 → Red, -2 → Red}}] // Text`

Rows	Col 1	Col 2	Col 3	Col 4	Sums
Row 1	6	21.2	3.05	64.2	94.45
Row 2	34	9.582	143.17	8.702	195.454
Row 3	985	0.6914	70.4	126.6	1182.69
Sums	1025	31.4734	216.62	199.502	1472.6

The style of the dividers can also be defined by the **FrameStyle** option; this option is considered later.

For tables with many rows, it may be useful to define a row divider after, for example, five rows:

```
Style[Grid[RandomReal[1, {15, 5}],
  Dividers → {None, {False, {False, False, False, False, True}}}], 7]
```

0.351699	0.0975843	0.920731	0.328851	0.278958
0.300958	0.75641	0.456994	0.0566562	0.822485
0.721353	0.894927	0.3933	0.263716	0.858048
0.371972	0.656398	0.835972	0.473315	0.657873
0.402682	0.126343	0.923583	0.605294	0.26255
0.78003	0.125083	0.49996	0.812553	0.864977
0.440816	0.29955	0.845526	0.201106	0.0514095
0.0339601	0.153938	0.322785	0.68958	0.228968
0.739786	0.0508808	0.453728	0.908037	0.573318
0.761619	0.0938451	0.199507	0.199536	0.602789
0.259405	0.742096	0.143953	0.699057	0.277642
0.0825613	0.354035	0.223007	0.0525714	0.057686
0.562795	0.470452	0.137385	0.31363	0.548075
0.778625	0.868359	0.735533	0.414053	0.288367
0.345906	0.826166	0.425129	0.877104	0.756518

Here, we defined that there should not be a divider above the first row, and after that we have, repeatedly, four rows with no divider and one row with a divider.

■ **Spacings**

> ** **Spacings** Space between columns and rows (in units of the current font size); examples of values: **Automatic** (usually means **{0.7, 0.4}**: the space between columns is 0.7 and that between rows is 0.4), **1.2** (the space between columns is 1.2 and the space between rows is the default 0.4)

Note that if we give **Spacings** a list of values, the first value corresponds with the space *before* the first column or *above* the first row, and the last value that can be given concerns the space *after* the last column or *below* the last row. However, the first or last value does not have any effect unless there is a frame line or divider line before the first column or row or after the last column or row.

We make the space between columns 2 (the space between rows is the default 0.4):

```
Grid[t, Spacings → 2] // Text
```

Rows	Col 1	Col 2	Col 3	Col 4	Sums
Row 1	6	21.2	3.05	64.2	94.45
Row 2	34	9.582	143.17	8.702	195.454
Row 3	985	0.6914	70.4	126.6	1182.69
Sums	1025	31.4734	216.62	199.502	1472.6

Now make the space between columns 2 and the space between rows 0.1:

```
Grid[t, Spacings → {2, 0.1}] // Text
```

Rows	Col 1	Col 2	Col 3	Col 4	Sums
Row 1	6	21.2	3.05	64.2	94.45
Row 2	34	9.582	143.17	8.702	195.454
Row 3	985	0.6914	70.4	126.6	1182.69
Sums	1025	31.4734	216.62	199.502	1472.6

Next, we fine-tune the spacings:

```
Grid[t, Spacings → {1.5, {0, 1, {0.5}, 1, 0}}] // Text
```

Rows	Col 1	Col 2	Col 3	Col 4	Sums
Row 1	6	21.2	3.05	64.2	94.45
Row 2	34	9.582	143.17	8.702	195.454
Row 3	985	0.6914	70.4	126.6	1182.69
Sums	1025	31.4734	216.62	199.502	1472.6

Here, the space between columns is 1.5. The space below the first row and above the last row is 1; for the rows in between, we use the space of 0.5.

Note that *Mathematica* assumes that the spacings begin above the first row and extend below the last row. Thus, to get the space 1 after the first row and above the last row, we have to define a space also above the first row and below the last row. We used the value 0 for these spaces, but in this example these values do not have any effect.

If we have a frame or dividers above the first row and below the last row, then the first and last values of **Spacings** do have an effect because these spacings define how much space should be around the first and last divider. Here, we use the value 1 for these spacings:

```
Grid[t, Spacings → {1.5, {1, 1, {0.5}, 1, 1}},
  Dividers → {None, {1 → True, 2 → True, -2 → True, -1 → True}}] // Text
```

Rows	Col 1	Col 2	Col 3	Col 4	Sums
Row 1	6	21.2	3.05	64.2	94.45
Row 2	34	9.582	143.17	8.702	195.454
Row 3	985	0.6914	70.4	126.6	1182.69
Sums	1025	31.4734	216.62	199.502	1472.6

■ **ItemStyle**

> ** **ItemStyle** Styles of columns and rows; examples of values: **None**, **Blue** (all items are blue), **{1 →** **Red}** (the first column is red), **{Automatic, 1 → Bold}** (the first row is bold), **{1 → Directive[Red,** **Bold, 14]}** (the first column is red, bold, and size 14), **{1 → Bold, 1 → Bold}** (the first column and row are bold)

First, we define bold style for the first row:

Grid[t, ItemStyle → {Automatic, 1 → Bold}] // Text

Rows	Col 1	Col 2	Col 3	Col 4	Sums
Row 1	6	21.2	3.05	64.2	94.45
Row 2	34	9.582	143.17	8.702	195.454
Row 3	985	0.6914	70.4	126.6	1182.69
Sums	1025	31.4734	216.62	199.502	1472.6

Then we use the bold style for both the first column and the first row:

Grid[t, ItemStyle → {1 → Bold, 1 → Bold}] // Text

Rows	Col 1	Col 2	Col 3	Col 4	Sums
Row 1	6	21.2	3.05	64.2	94.45
Row 2	34	9.582	143.17	8.702	195.454
Row 3	985	0.6914	70.4	126.6	1182.69
Sums	1025	31.4734	216.62	199.502	1472.6

Now the first and last column and row are bold:

Grid[t, ItemStyle → {{1 → Bold, -1 → Bold}, {1 → Bold, -1 → Bold}}] // Text

Rows	Col 1	Col 2	Col 3	Col 4	Sums
Row 1	6	21.2	3.05	64.2	**94.45**
Row 2	34	9.582	143.17	8.702	**195.454**
Row 3	985	0.6914	70.4	126.6	**1182.69**
Sums	**1025**	**31.4734**	**216.62**	**199.502**	**1472.6**

The following is another way:

Grid[t, ItemStyle → {{Bold, {}, Bold}, {Bold, {}, Bold}}] // Text

Rows	Col 1	Col 2	Col 3	Col 4	Sums
Row 1	6	21.2	3.05	64.2	**94.45**
Row 2	34	9.582	143.17	8.702	**195.454**
Row 3	985	0.6914	70.4	126.6	**1182.69**
Sums	**1025**	**31.4734**	**216.62**	**199.502**	**1472.6**

Next, we use both the bold style and a color:

Grid[t, ItemStyle → {1 → Directive[Bold, Blue], 1 → Directive[Bold, Red]}] // Text

Rows	Col 1	Col 2	Col 3	Col 4	Sums
Row 1	6	21.2	3.05	64.2	94.45
Row 2	34	9.582	143.17	8.702	195.454
Row 3	985	0.6914	70.4	126.6	1182.69
Sums	1025	31.4734	216.62	199.502	1472.6

Styles can also be defined for a range of items or for a singe item:

```
Grid[t, ItemStyle →
   {Automatic, Automatic, {{2, 4}, {2, 5}} → Directive[Bold, Blue]}] // Text
```

Rows	Col 1	Col 2	Col 3	Col 4	Sums
Row 1	6	**21.2**	**3.05**	**64.2**	94.45
Row 2	34	**9.582**	**143.17**	**8.702**	195.454
Row 3	985	**0.6914**	**70.4**	**126.6**	1182.69
Sums	1025	31.4734	216.62	199.502	1472.6

```
Grid[t, ItemStyle → {Automatic, Automatic, {5, 6} → Directive[Bold, Red]}] // Text
```

Rows	Col 1	Col 2	Col 3	Col 4	Sums
Row 1	6	21.2	3.05	64.2	94.45
Row 2	34	9.582	143.17	8.702	195.454
Row 3	985	0.6914	70.4	126.6	1182.69
Sums	1025	31.4734	216.62	199.502	**1472.6**

■ Background

> ** **Background** Colors of the background; examples of values: **None**, **GrayLevel[0.9]** (all items are gray), **{Automatic, {{ White, LightGray}}}** (columns are default, rows alternate between white and gray)

Define a gray background for the table:

```
Grid[t, Background → LightGray] // Text
```

Rows	Col 1	Col 2	Col 3	Col 4	Sums
Row 1	6	21.2	3.05	64.2	94.45
Row 2	34	9.582	143.17	8.702	195.454
Row 3	985	0.6914	70.4	126.6	1182.69
Sums	1025	31.4734	216.62	199.502	1472.6

Next, we define a gray background for the first row:

```
Grid[t, Background → {Automatic, 1 → GrayLevel[0.9]}] // Text
```

Rows	Col 1	Col 2	Col 3	Col 4	Sums
Row 1	6	21.2	3.05	64.2	94.45
Row 2	34	9.582	143.17	8.702	195.454
Row 3	985	0.6914	70.4	126.6	1182.69
Sums	1025	31.4734	216.62	199.502	1472.6

Now we define an alternating background for the rows:

```
Grid[t, Background → {Automatic, {{White, LightGray}}}] // Text
```

Rows	Col 1	Col 2	Col 3	Col 4	Sums
Row 1	6	21.2	3.05	64.2	94.45
Row 2	34	9.582	143.17	8.702	195.454
Row 3	985	0.6914	70.4	126.6	1182.69
Sums	1025	31.4734	216.62	199.502	1472.6

Define then the first and last rows to be light red:

`Grid[t, Background → {Automatic, {LightRed, {White, LightGray}, LightRed}}] // Text`

Rows	Col 1	Col 2	Col 3	Col 4	Sums
Row 1	6	21.2	3.05	64.2	94.45
Row 2	34	9.582	143.17	8.702	195.454
Row 3	985	0.6914	70.4	126.6	1182.69
Sums	1025	31.4734	216.62	199.502	1472.6

Next, we define the background for a range of items or for a singe item:

`Grid[t, Background → {Automatic, Automatic, {{2, 4}, {2, 5}} → LightGray}] // Text`

Rows	Col 1	Col 2	Col 3	Col 4	Sums
Row 1	6	21.2	3.05	64.2	94.45
Row 2	34	9.582	143.17	8.702	195.454
Row 3	985	0.6914	70.4	126.6	1182.69
Sums	1025	31.4734	216.62	199.502	1472.6

`Grid[t, Background → {Automatic, Automatic, {{{5, 5}, {2, 5}} → Darker[Yellow, 0.1], {{2, 4}, {6, 6}} → Darker[Yellow, 0.1], {5, 6} → Red}}] // Text`

Rows	Col 1	Col 2	Col 3	Col 4	Sums
Row 1	6	21.2	3.05	64.2	94.45
Row 2	34	9.582	143.17	8.702	195.454
Row 3	985	0.6914	70.4	126.6	1182.69
Sums	1025	31.4734	216.62	199.502	1472.6

`Grid[t, Background → {Automatic, Automatic, {5, 6} → Pink}] // Text`

Rows	Col 1	Col 2	Col 3	Col 4	Sums
Row 1	6	21.2	3.05	64.2	94.45
Row 2	34	9.582	143.17	8.702	195.454
Row 3	985	0.6914	70.4	126.6	1182.69
Sums	1025	31.4734	216.62	199.502	1472.6

■ **Frame**

> ** **Frame** Where to draw frames; examples of values: **None**, **True** (frame around the whole grid), **All** (all items become boxed), **{All}** (frame around each column), **{None, All}** (frame around each row)

Add a frame for the whole table:

`Grid[t, Frame → True] // Text`

Rows	Col 1	Col 2	Col 3	Col 4	Sums
Row 1	6	21.2	3.05	64.2	94.45
Row 2	34	9.582	143.17	8.702	195.454
Row 3	985	0.6914	70.4	126.6	1182.69
Sums	1025	31.4734	216.62	199.502	1472.6

Add a frame for all items (this can also be obtained with **Dividers → All**):

`Grid[t, Frame → All] // Text`

Rows	Col 1	Col 2	Col 3	Col 4	Sums
Row 1	6	21.2	3.05	64.2	94.45
Row 2	34	9.582	143.17	8.702	195.454
Row 3	985	0.6914	70.4	126.6	1182.69
Sums	1025	31.4734	216.62	199.502	1472.6

Add a frame for each row:

`Grid[t, Frame → {None, All}] // Text`

Rows	Col 1	Col 2	Col 3	Col 4	Sums
Row 1	6	21.2	3.05	64.2	94.45
Row 2	34	9.582	143.17	8.702	195.454
Row 3	985	0.6914	70.4	126.6	1182.69
Sums	1025	31.4734	216.62	199.502	1472.6

Next, we use both dividers and a frame:

`Grid[t, Dividers → {{2 → True, -2 → True}, {2 → True, -2 → True}}, Frame → True] // Text`

Rows	Col 1	Col 2	Col 3	Col 4	Sums
Row 1	6	21.2	3.05	64.2	94.45
Row 2	34	9.582	143.17	8.702	195.454
Row 3	985	0.6914	70.4	126.6	1182.69
Sums	1025	31.4734	216.62	199.502	1472.6

Now we define a frame for items with row index in {2, 4} and column index in {2, 5}:

`Grid[t, Frame → {None, None, {{2, 4}, {2, 5}} → True}] // Text`

Rows	Col 1	Col 2	Col 3	Col 4	Sums
Row 1	6	21.2	3.05	64.2	94.45
Row 2	34	9.582	143.17	8.702	195.454
Row 3	985	0.6914	70.4	126.6	1182.69
Sums	1025	31.4734	216.62	199.502	1472.6

Lastly, we give a frame for a single item:

`Grid[t, Frame → {None, None, {5, 6} → True}] // Text`

Rows	Col 1	Col 2	Col 3	Col 4	Sums
Row 1	6	21.2	3.05	64.2	94.45
Row 2	34	9.582	143.17	8.702	195.454
Row 3	985	0.6914	70.4	126.6	1182.69
Sums	1025	31.4734	216.62	199.502	1472.6

■ **FrameStyle**

FrameStyle Style of frames and dividers; examples of values: **Automatic**, **Red**, **Directive[Gray, Thickness[2]]**

Here, we define the style of a frame:

`Grid[t, Frame → True, FrameStyle → Directive[Red, Thickness[2], Dashing[3]]] // Text`

Rows	Col 1	Col 2	Col 3	Col 4	Sums
Row 1	6	21.2	3.05	64.2	94.45
Row 2	34	9.582	143.17	8.702	195.454
Row 3	985	0.6914	70.4	126.6	1182.69
Sums	1025	31.4734	216.62	199.502	1472.6

We could also define the style in the **Frame** option:

`Grid[t, Frame → Directive[Red, Thickness[2], Dashing[3]]] // Text`

Rows	Col 1	Col 2	Col 3	Col 4	Sums
Row 1	6	21.2	3.05	64.2	94.45
Row 2	34	9.582	143.17	8.702	195.454
Row 3	985	0.6914	70.4	126.6	1182.69
Sums	1025	31.4734	216.62	199.502	1472.6

Next, we define a style for dividers:

`Grid[t, Dividers → {{2 → True, -2 → True}, {2 → True, -2 → True}},`
`  FrameStyle → Directive[Blue, Thickness[1]]] // Text`

Rows	Col 1	Col 2	Col 3	Col 4	Sums
Row 1	6	21.2	3.05	64.2	94.45
Row 2	34	9.582	143.17	8.702	195.454
Row 3	985	0.6914	70.4	126.6	1182.69
Sums	1025	31.4734	216.62	199.502	1472.6

We could also define the style in the **Dividers** option using **Directive**.

■ **ItemSize**

> **ItemSize** Width and height of each item; examples of values: **Automatic** (separately size items to fit within the total formatting width; long items may take several lines), **All** (make all items the same width and height), **Full** (allow each item its full width and height; long items are not divided into several lines), **w** (give all items width **w**, measured in ems), **{w, h}** (give all items width **w** and height **h**, with **h** measured in line heights; the default value of **h** is 1)

The default is to adjust the width of each column separately:

`Grid[t, ItemSize → Automatic, Frame → {All}]`

Rows	Col 1	Col 2	Col 3	Col 4	Sums
Row 1	6	21.2	3.05	64.2	94.45
Row 2	34	9.582	143.17	8.702	195.454
Row 3	985	0.6914	70.4	126.6	1182.69
Sums	1025	31.4734	216.62	199.502	1472.6

Next, we ask for the same width for all columns:

```
Grid[t, ItemSize → All, Frame → {All}]
```

Rows	Col 1	Col 2	Col 3	Col 4	Sums
Row 1	6	21.2	3.05	64.2	94.45
Row 2	34	9.582	143.17	8.702	195.454
Row 3	985	0.6914	70.4	126.6	1182.69
Sums	1025	31.4734	216.62	199.502	1472.6

■ **Example: Combining the Options**

Now we use many of the options we considered previously:

```
Labeled[
  Grid[t, Alignment → {{Left, Right, {"."}}, Baseline, {{1, 1}, {2, 6}} → Center},
    Frame → True, Dividers → {{2 → True, -2 → True}, {2 → True, -2 → True}},
    Spacings → {1.5, {1.5, 1, {0.5}, 1, 1}}, ItemStyle → {1 → Bold, 1 → Bold},
    Background → {Automatic, Automatic, {{2, 4}, {2, 5}} → GrayLevel[0.9]}],
  Style["Row and column sums", Bold], Top] // Text
```

Row and column sums

Rows	**Col 1**	**Col 2**	**Col 3**	**Col 4**	**Sums**
Row 1	6	21.2	3.05	64.2	94.45
Row 2	34	9.582	143.17	8.702	195.454
Row 3	985	0.6914	70.4	126.6	1182.69
Sums	1025	31.4734	216.62	199.502	1472.6

```
% // Panel
```

Row and column sums

Rows	**Col 1**	**Col 2**	**Col 3**	**Col 4**	**Sums**
Row 1	6	21.2	3.05	64.2	94.45
Row 2	34	9.582	143.17	8.702	195.454
Row 3	985	0.6914	70.4	126.6	1182.69
Sums	1025	31.4734	216.62	199.502	1472.6

■ **Example: Chemical Elements**

Consider the following chemical elements:

```
e = Table[ElementData[i], {i, 1, 51, 10}]
```

```
{Hydrogen, Sodium, Scandium, Gallium, Niobium, Antimony}
```

For these elements, we gather the element abbreviations, atomic numbers, atomic weights, boiling points, melting points, and stable isotopes:

```
ea = Table[ElementData[i, "Abbreviation"], {i, 1, 51, 10}]
```

```
{H, Na, Sc, Ga, Nb, Sb}
```

```
a = Table[ElementData[i, "AtomicNumber"], {i, 1, 51, 10}]
```

```
{1, 11, 21, 31, 41, 51}
```

```
w = Table[ElementData[i, "AtomicWeight"], {i, 1, 51, 10}]
{1.00794, 22.989770, 44.955910, 69.723, 92.90638, 121.760}

b = Table[ElementData[i, "BoilingPoint"], {i, 1, 51, 10}]
{-252.87, 883., 2830., 2204., 4744., 1587.}

m = Table[ElementData[i, "MeltingPoint"], {i, 1, 51, 10}]
{-259.14, 97.72, 1541., 29.76, 2477., 630.63}

s = Table[ElementData[i, "StableIsotopes"], {i, 1, 51, 10}]
{{1, 2}, {23}, {45}, {69, 71}, {93}, {121, 123}}
```

We form the rows of the table:

```
data0 = {e, ea, a, w, b, m, s}ᵀ
{{Hydrogen, H, 1, 1.00794, -252.87, -259.14, {1, 2}},
 {Sodium, Na, 11, 22.989770, 883., 97.72, {23}},
 {Scandium, Sc, 21, 44.955910, 2830., 1541., {45}},
 {Gallium, Ga, 31, 69.723, 2204., 29.76, {69, 71}},
 {Niobium, Nb, 41, 92.90638, 4744., 2477., {93}},
 {Antimony, Sb, 51, 121.760, 1587., 630.63, {121, 123}}}
```

We form the headings of the table to be formed, and we add these headings as the first row of the table:

```
h = {"Element", "Abbr.", "No.", "Weight    ", "Boiling ", "Melting", "Isotopes"};

data = Prepend[data0, h]

{{Element, Abbr., No., Weight    , Boiling , Melting, Isotopes},
 {Hydrogen, H, 1, 1.00794, -252.87, -259.14, {1, 2}},
 {Sodium, Na, 11, 22.989770, 883., 97.72, {23}},
 {Scandium, Sc, 21, 44.955910, 2830., 1541., {45}},
 {Gallium, Ga, 31, 69.723, 2204., 29.76, {69, 71}},
 {Niobium, Nb, 41, 92.90638, 4744., 2477., {93}},
 {Antimony, Sb, 51, 121.760, 1587., 630.63, {121, 123}}}
```

Here is a table for the data:

```
Grid[data, Alignment → {{Left, Left, Right, {"."}, Left}, Baseline,
    Thread[Table[{1, j}, {j, 7}] → {Left, Left, Right, Right, Right, Right, Left}]},
  Dividers → {None, {1 → True, 2 → True, -1 → True}},
  Spacings → {{0, 1.8, 1.1, -0.4, 1, 0.4, 2.6}, {1.2, 0.8, {0.5}, 1}},
  ItemStyle → {Automatic, 1 → Bold, {{2, 7}, {1, 1}} → Italic},
  Background → {None, {{White, LightGray}}}] // Text
```

Element	**Abbr.**	**No.**	**Weight**	**Boiling**	**Melting**	**Isotopes**
Hydrogen	H	1	1.00794	−252.87	−259.14	{1, 2}
Sodium	Na	11	22.989770	883.	97.72	{23}
Scandium	Sc	21	44.955910	2830.	1541.	{45}
Gallium	Ga	31	69.723	2204.	29.76	{69, 71}
Niobium	Nb	41	92.90638	4744.	2477.	{93}
Antimony	Sb	51	121.760	1587.	630.63	{121, 123}

Here, we used **Thread** to form a list of rules:

```
Thread[Table[{1, j}, {j, 7}] → {Left, Left, Left, Right, Right, Right, Left}]
{{1, 1} → Left, {1, 2} → Left, {1, 3} → Left,
 {1, 4} → Right, {1, 5} → Right, {1, 6} → Right, {1, 7} → Left}
```

Note also that when we defined the list **h** of headings, we added two spaces after "Weight" and one space after "Boiling" to further fine-tune the alignments of the headings.

■ Item

Previously, we found that we can format, with the options of **Grid**, even single items: We can, for example, draw a frame around or give a color or background to a single item. Sometimes it is useful to format single items with **Item**.

Item[expr, opts]   Apply options **opts** to the given item

Options:

Alignment   Alignment of the item; examples of values: **{}, Left, Right**

Background   Background color of the item; examples of values: **Automatic, Yellow**

Frame   Whether to draw a frame around the item; examples of values: **{}, True**

FrameStyle   Style of the frame; examples of values: **{}, Blue**

ItemSize   Size of the item; examples of values: **{}, 1.5**

Toss a die 30 times:

```
SeedRandom[2]; t2 = RandomInteger[{1, 6}, {30}]
```
```
{6, 2, 3, 3, 6, 3, 2, 6, 6, 1, 1, 5, 4, 5, 1, 2, 2, 6, 2, 6, 5, 5, 1, 1, 5, 5, 2, 3, 4, 4}
```

We give each 6 a red color, a yellow background, and a frame:

```
t3 = If[# == 6, Item[Style[#, Red], Background → Yellow, Frame → True], #] & /@ t2
```
```
{6, 2, 3, 3, 6, 3, 2, 6, 6, 1, 1, 5, 4, 5, 1, 2, 2, 6, 2, 6, 5, 5, 1, 1, 5, 5, 2, 3, 4, 4}
```

Then we ask a grid:

```
Grid[Partition[t3, 10]]
```

```
6  2  3  3  6  3  2  6  6  1
1  5  4  5  1  2  2  6  2  6
5  5  1  1  5  5  2  3  4  4
```

Notice the high quality of the frames: Frames that are side by side have common edges.

■ Spanning Elements and Nested Grids

With **SpanFromLeft** or ⋯ (ESC**sfl**ESC), **SpanFromAbove** or ⋮ (ESC**sfa**ESC), and **SpanFromBoth** or ⋱ (ESC**sfb**ESC), we get results such as the following:

```
Grid[{
  {4, 3, ⋯, 5, 6},
  {6, 5, 2, 9, ⋯},
  {⋮, 6, 8, ⋮, ⋱},
  {⋮, 4, 1, 7, 5}}, Frame → All]
```

```
4     3    5  6
6  5  2    9
   6  8
   4  1  7  5
```

Spanning elements can also be specified with the menu command **Insert ▷ Table/Matrix ▷ Make Spanning**. As an example, write first

```
Grid[{{4, 3, x, 5, 6},
   {6, 5, 2, 9, x}, {x, 6, 8, x, x}, {x, 4, 1, 7, 5}}, Frame → All]
```

4	3	x	5	6
6	5	2	9	x
x	6	8	x	x
x	4	1	7	5

Then select, with the mouse, the elements 3 and x from the first row and choose the menu command. Then select the elements 6, x, and x from the first column and choose the menu command. Lastly, select the elements 9, x, x, and x and again choose the menu command. The result is

4	3 x	5	6
6	5	2	9 x
x	6	8	x x
x	4	1	7	5

Then delete the x's to get

4	3	5	6
6	5	2	9
	6	8	
	4	1	7	5

One use of spanning elements is to write a common explanation for the columns:

```
r0 = {"", "Columns", ⋯, ⋯, ⋯, ""};
```

```
Grid[Prepend[t, r0],
   Alignment → {{Left, Right, {"."}}, Baseline, {{1, 2}, {2, 6}} → Center},
   Dividers → {{2 → True, -2 → True}, {2 → True, 3 → True, -2 → True}},
   Spacings → {1.5, {0, 0.7, {0.5}, 0.7, 0}},
   ItemStyle → {1 → Bold, {1 → Bold, 2 → Bold}},
   Background → {Automatic, Automatic, {{3, 5}, {2, 5}} → GrayLevel[0.9]}] // Text
```

	Columns				
Rows	Col 1	Col 2	Col 3	Col 4	**Sums**
Row 1	6	21.2	3.05	64.2	94.45
Row 2	34	9.582	143.17	8.702	195.454
Row 3	985	0.6914	70.4	126.6	1182.69
Sums	1025	31.4734	216.62	199.502	1472.6

Grids can also be nested:

```
Grid[{{Grid[{{""}}], Grid[{{"j"}}]},
   {Grid[{{"i"}}], Grid[{{a, b, c, d}, {e, f, g, h}, {i, j, k, l}}]}}, Frame → All]
```

	j			
	a	b	c	d
i	e	f	g	h
	i	j	k	l

Patterns

Introduction

> *Mathematics is the art of giving the same name to different things.* —Henri Poincaré
> *Poetry is the art of giving different names to the same thing.* —Anonymous poet

A pattern represents *a class of expressions*. Expressions matching a pattern all have the *structure* given by the pattern. Using patterns we get powerful methods to manipulate lists and other expressions. We can also use patterns with transformation rules, and patterns are used with the arguments of functions to restrict the arguments to suitable expressions. Here, patterns are considered in two parts. Section 16.1 is devoted to general patterns and Section 16.2 to special patterns for strings.

Patterns constitute one of the two main components of rule-based programming. The other component is the use of rules. We consider rule-based and other programming styles in Chapter 18.

For more information about patterns, see tutorial/PatternsOverview.

16.1 Patterns

16.1.1 Introduction to Patterns

■ Applications of Patterns

A pattern represents *a class of expressions*—that is, all expressions having a given *structure*. Patterns are used

    • in function definitions to specify the class of arguments;
    • in transformation rules to specify the class of subexpressions to be transformed; and
    • in several list manipulation commands.

Let us first consider functions.

```
f[pattern] := expr
```

The pattern in **f** defines the class of arguments that the function is designed to accept. Often, the pattern is simply, for example, **x_**, and then any expression is accepted as an argument. However, we can also form more restrictive patterns such as the one in the following example:

```
f[n_Integer ? Positive] := Prime[n]
```

This function only accepts positive integers as arguments; these numbers *match* the given pattern. For other types of arguments, nothing is done:

```
{f[5], f[7.4], f[-8]}
```
```
{11, f[7.4], f[-8]}
```

In transformation rules, patterns are used as follows:

```
expr /. pattern → value
```

The pattern in the transformation rule defines the class of subexpressions of **expr** that are replaced with **value**. Often, the pattern is a degenerate pattern such as **x**, and then **x** is replaced with **value** in all places **x** appears in **expr**. However, we can also form more restrictive patterns such as **x_ /; x > 0.5**, and now the numbers that are greater than 0.5 are replaced with a given value:

```
{0.2, 0.6, 0.8} /. (x_ /; x > 0.5) → a
```
```
{0.2, a, a}
```

Patterns are also used in some commands, such as **Position**, **Cases**, **DeleteCases**, **Count**, and **Switch** (for **Switch**, see Section 18.2.2, p. 556) and in some tests such as **FreeQ**, **MemberQ**, and **MatchQ**.

■ **Special Patterns**

Two patterns are very special: a degenerate pattern and a pattern representing anything. A degenerate pattern is a given expression such as 6. This represents a class of expressions consisting of only one element: the element given. Later, we consider commands such as **Count**. In the following example, we count all the elements of a given list that match the degenerate pattern 6—that is, we count all 6's:

```
Count[{3, 2, 6, 4, 6, 1}, 6]      2
```

Now we replace all 6's with a string:

```
{3, 2, 6, 4, 6, 1} /. 6 → "win"      {3, 2, win, 4, win, 1}
```

Next, we consider a pattern representing anything.

```
_   A pattern representing any expression
x_  A pattern representing any expression; the expression is referred to with the name x
```

The most general pattern is **_**. It matches anything, and its internal name is **Blank**:

```
FullForm[_]      Blank[]
```

In Section 18.2.2, p. 556, we present the following example of **Switch**:

```
volume[name_] := Switch[name, cylinder, Pi r^2 h,
  sphere, 4 Pi r^3 / 3, ellipsoid, 4 Pi a b c / 3, _, unknown]
```

The blank **_** matches anything, and so the result of **volume** is **unknown** if **name** is not cylinder, sphere, or ellipsoid.

The pattern **x_** also represents any expression, but now we give a name **x** for the expression so that we can refer to it later. The most important use of this pattern is in defining the arguments of a function (see Section 17.1.1, p. 512). An example is **f[x_] := x Sin[x]**. In the next example, we form powers:

{{a, b}, {c, d}, {c, d}} /. {x_, y_} → x^y

$\{a^b, c^d, c^d\}$

■ Searching Expressions Using Patterns

> **Position[list, pattern]**   Give the positions of the parts of **list** that match **pattern**
> **Position[list, pattern, {1}]**   Give the positions of the elements of **list** that match **pattern**
> **Count[list, pattern]**   Give the number of the elements of **list** that match **pattern**
> **Cases[list, pattern]**   Give the elements of **list** that match **pattern**
> **Cases[list, pattern → value]**   Apply the given transformation to the found elements
> **DeleteCases[list, pattern]**   Delete the elements of **list** that match **pattern**
> **Pick[list, sel, pattern]**   Pick out those elements of **list** for which the corresponding element of
>   **sel** matches **pattern**

It is useful to compare **Cases** with **Select**. Note that **Select** does not use patterns: It uses tests.

> **Select[list, test]**   Select the elements of **list** that satisfy **test**

First, we generate a list of 20 tosses of a die:

SeedRandom[7]; t = RandomInteger[{1, 6}, 20]
{6, 1, 3, 3, 6, 4, 2, 1, 4, 5, 3, 4, 3, 2, 4, 2, 2, 6, 4, 3}

Then we search the positions of all 6's, the number of 6's, and the cases of 6's. We also delete all 6's. In these examples, we use the degenerate pattern 6:

Position[t, 6]   {{1}, {5}, {18}}

Count[t, 6]   3

Cases[t, 6]   {6, 6, 6}

DeleteCases[t, 6]
{1, 3, 3, 4, 2, 1, 4, 5, 3, 4, 3, 2, 4, 2, 2, 4, 3}

To give an example of **Pick**, generate a second list:

SeedRandom[1]; u = RandomInteger[{0, 1}, 20]
{1, 1, 0, 1, 0, 0, 0, 1, 0, 1, 0, 0, 0, 0, 0, 0, 0, 1, 1, 1}

Now we pick all the elements of **t** for which the corresponding element in **u** is 1:

Pick[t, u, 1]   {6, 1, 3, 1, 5, 6, 4, 3}

To use **Select**, we write a test to select all 6's:

Select[t, # == 6 &]   {6, 6, 6}

■ Levels

Position, **Count**, **Cases**, and **DeleteCases** accept a third argument that specifies the level from which matching objects are searched. For example, positions can be searched from given levels:

> **Position[expr, pattern, levspec]** Give positions at levels specified by **levspec** at which objects matching **pattern** occur in **expr**

Similarly, a level specification in, for example, **Cases** directs the search of elements into the specified levels. Level specifications are considered in Section 13.2.3, p. 427. Recall that

- a level specification **n** means all levels 1, 2, …, *n*;
- a level specification **{n}** means only the level *n*; and
- a level specification **{n, m}** means levels **n** through **m**.

If a level specification is not given in **Position**, the specification is assumed to be **{0, ∞}**, which means all levels 0, 1, 2, and so on. In **Count**, **Cases**, and **DeleteCases**, the default level is **{1}**. As an example, consider the following list:

    t = {{a, Sin[b]}, {c, Sin[d]}};

Find out all positions where the sin function appears:

    Position[t, Sin[_]]   {{1, 2}, {2, 2}}

This means that sin appears as the second element of both the first and the second element. Note that sin does not appear as an element of **t**—that is, at the first level of **t**:

    Position[t, Sin[_], {1}]   {}

By default, **Cases** searches from the first level so that no cases are found:

    Cases[t, Sin[_]]   {}

If we search from the second level, some cases are found:

    Cases[t, Sin[_], {2}]   {Sin[b], Sin[d]}

■ Limiting the Number of Parts Searched For

Position, **Cases**, and **DeleteCases** also accept a fourth argument that limits the number of parts to search for. For example, we can ask for the first **n** positions:

> **Position[expr, pattern, levspec, n]** Give positions of first **n** parts at levels specified by **levspec** at which objects matching **pattern** occur in **expr**

    Position[t, Sin[_], ∞, 1]   {{1, 2}}

■ Transforming Expressions Using Patterns

> **expr /. pattern → value** Apply the transformation once to each part of **expr**
> **expr //. pattern → value** Apply the transformation until the result no longer changes

(The internal names of **/.** and **//.** are **ReplaceAll** and **ReplaceRepeated**, respectively.) Applying transformation rules is a powerful way to manipulate lists. Here are some examples:

Take square roots from the second elements:

    {{1, 10}, {2, 11}, {3, 12}} /. {x_, y_} → {x, Sqrt[y]}

$$\left\{\left\{1, \sqrt{10}\right\}, \left\{2, \sqrt{11}\right\}, \left\{3, 2\sqrt{3}\right\}\right\}$$

Define an expression containing square roots:

```
t1 = Sqrt[x] + 1 / Sqrt[x];
```

If we want to transform the expression by replacing $\sqrt{x}$ with y, we do not get what we want:

t1 /. Sqrt[x] → y $\dfrac{1}{\sqrt{x}} + y$

Indeed, *Mathematica* uses internally the **FullForm** in transformation rules. Here is this form of our expression:

```
FullForm[t1]
Plus[Power[x, Rational[-1, 2]], Power[x, Rational[1, 2]]]
```

We see that whereas the internal representation of $\sqrt{x}$ is $x^{1/2}$, that of $1\big/\sqrt{x}$ is not $1\big/x^{1/2}$ but rather $x^{-1/2}$, which does not match the pattern $x^{1/2}$. For this reason, we have to transform $\sqrt{x}$ and $1\big/\sqrt{x}$ separately:

t1 /. {Sqrt[x] → y, 1 / Sqrt[x] → 1 / y} $\dfrac{1}{y} + y$

Mathematica does not automatically expand logarithms such as $\log(x^2)$ to $2\log(x)$:

```
t2 = Log[x^2] + Log[1 / Sqrt[y]];
```

t2 // Expand $\mathrm{Log}\big[x^2\big] + \mathrm{Log}\Big[\dfrac{1}{\sqrt{y}}\Big]$

This is because such an expansion is not always correct:

{Log[(-1)^2], 2 Log[-1]} $\{0, 2\,i\,\pi\}$

PowerExpand does the transformation, but we have to consider the conditions under which the transformation is correct:

t2 // PowerExpand $2\,\mathrm{Log}[x] - \dfrac{\mathrm{Log}[y]}{2}$

The same effect can be obtained with a transformation rule:

t2 /. Log[a_^b_] → b Log[a] $2\,\mathrm{Log}[x] - \dfrac{\mathrm{Log}[y]}{2}$

ReplaceList[expr, pattern → value] Find all ways **expr** can match **pattern**; apply the transformation for each way

We do a transformation in all possible ways and form a list from the results:

ReplaceList[a + b + c, x_ + y_ → x y]
$\{a\,(b+c),\ b\,(a+c),\ (a+b)\,c,\ (a+b)\,c,\ b\,(a+c),\ a\,(b+c)\}$

With **Union**, we get the results that are distinct:

% // Union $\{(a+b)\,c,\ b\,(a+c),\ a\,(b+c)\}$

An application of **ReplaceList** can be found in Section 29.4.2, p. 994.

■ Testing Expressions Using Patterns

> **MatchQ[expr, pattern]**   Give **True** if **expr** matches **pattern** and **False** otherwise
>
> **FreeQ[expr, pattern]**   Give **True** if no subexpression of **expr** matches **pattern** and **False** otherwise
>
> **MemberQ[list, pattern]**   Give **True** if an element of **list** matches **pattern** and **False** otherwise

MatchQ can be used to check whether a given expression matches a given pattern. In the following examples, we demonstrate that any expression matches the most general pattern _:

```
{MatchQ[5, _], MatchQ[anything, _], MatchQ[a + b x, _]}
```
{True, True, True}

With **FreeQ** we can test whether an expression is free from a pattern:

```
{FreeQ[a, x], FreeQ[a + b x, x]}
```
{True, False}

MemberQ enables us to test whether an expression is an element of a list:

```
{MemberQ[{4, 2, 1, 5}, 2], MemberQ[{4, 2, 1, 5}, 6]}
```
{True, False}

16.1.2 Patterns with Restrictions

In Section 16.1.1, we introduced the most general pattern _. Next, we consider ways to restrict this pattern. A restriction can be formed with heads, tests, and conditions. Note that whereas a pattern represents a *structural* constraint, a restriction represents a *mathematical* constraint. A restricted pattern first accepts expressions that match the pattern but then does further mathematical tests.

■ Restricting Patterns with Heads

> **x_head**   Any expression having the head **head**

One type of restriction is to require that the expression must have a certain head. When we considered heads in Section 13.2.3, p. 426, we noted that every expression has a head, and we encountered such heads as **Integer**, **Rational**, **Real**, **Complex**, **Symbol**, **String**, **List**, **Plus**, **Times**, and **Power**. The head can be asked with **Head**:

```
Head /@ {4, 3 / 5, 2.1, 4 + 5 I, a, "xyz"}
```
{Integer, Rational, Real, Complex, Symbol, String}

```
Head /@ {{a, b, c}, a + b, a b, a^b}
```
{List, Plus, Times, Power}

The head can also be seen with **FullForm**:

```
FullForm /@ {{a, b, c}, a + b, a b, a^b}
```
{List[a, b, c], Plus[a, b], Times[a, b], Power[a, b]}

As an example, we form a function that calculates the square of an integer:

```
f1[n_Integer] := n^2
```

```
{f1[3], f1[0], f1[-3], f1[2.7], f1[2 + 5 I], f1[a]}
{9, 0, 9, f1[2.7], f1[2 + 5 i], f1[a]}
```

As can be seen, the function **f1** is applied only if the argument is an integer. The following function picks the first element of a list:

```
f2[x_List] := First[x]

{f2[{1, 2, 3}], f2[{{1, 2}, {3, 4}}], f2[8]}
{1, {1, 2}, f2[8]}
```

The expression 8 is not a list, and so **f2** is not applied. Next, we find out all lists:

```
Cases[{a, {b, c, d}, e, {f, g}}, x_List]
{{b, c, d}, {f, g}}
```

We could also use **Select**:

```
Select[{a, {b, c, d}, e, {f, g}}, ListQ]
{{b, c, d}, {f, g}}
```

Find the positions of all integers:

```
Position[{7, 2 + 4 I, 5 / 3, 6, 1.78}, x_Integer]
{{1}, {4}}
```

Count the number of complex numbers:

```
t = {2 + 6 I, 7, 3 - I, 2, 5 + 4 I};

Count[t, z_Complex]    3
```

Replace all integers with the corresponding prime:

```
t /. n_Integer → Prime[n]
{2 + 6 i, 17, 3 - i, 3, 5 + 4 i}
```

Replace all complex numbers with a list of the real and imaginary parts:

```
t /. z_Complex → {Re[z], Im[z]}
{{2, 6}, 7, {3, -1}, 2, {5, 4}}
```

Pick all complex numbers and replace them with a list of the real and imaginary parts:

```
Cases[t, z_Complex → {Re[z], Im[z]}]
{{2, 6}, {3, -1}, {5, 4}}
```

■ Restricting Patterns with Tests

> **x_ ? test** Any expression giving **True** from **test**
> **x_head ? test** Any expression having the head **head** and giving **True** from **test**

(The internal name of **?** is **PatternTest**.) Tests were considered in Section 13.3.5, p. 431. We have tests such as ==, !=, ===, =!=, <, ≤, >, ≥, **NonNegative**, **Positive**, **NumericQ**, **EvenQ**, **OddQ**, **VectorQ**, **MatrixQ**, and **OptionQ**. We can form more complex tests with operations such as **&&** (AND), **||** (OR), and **!** (NOT).

As an example, we program the factorial function. We require that the argument **n** must be a positive integer:

```
fac[0] = 1;
fac[n_Integer?Positive] := n fac[n - 1]
{fac[5], fac[2.7], fac[-2]}
{120, fac[2.7], fac[-2]}
```

Indeed, 2.7 or −2 does not match the given pattern:

```
MatchQ[2.7, n_Integer?Positive]
False
```

Transpose a matrix:

```
f3[x_?MatrixQ] := Transpose[x]

{f3[{{1, 2}, {3, 4}}], f3[{1, 2, 3}]}
{{{1, 3}, {2, 4}}, f3[{1, 2, 3}]}
```

Replace all even numbers with a string:

```
{4, 1, 3, 5, 3, 6, 2} /. _?EvenQ → "even"
{even, 1, 3, 5, 3, even, even}
```

Solve an equation and find all positive solutions:

```
Solve[49 - 22 x^2 + x^4 == 0]
```

$$\left\{\left\{x \to -3 - \sqrt{2}\right\}, \left\{x \to 3 - \sqrt{2}\right\}, \left\{x \to -3 + \sqrt{2}\right\}, \left\{x \to 3 + \sqrt{2}\right\}\right\}$$

```
Cases[%, {x → _?Positive}]
```

$$\left\{\left\{x \to 3 - \sqrt{2}\right\}, \left\{x \to 3 + \sqrt{2}\right\}\right\}$$

We could also use **Select**:

```
Select[%%, #[[1, 2]] > 0 &]
```

$$\left\{\left\{x \to 3 - \sqrt{2}\right\}, \left\{x \to 3 + \sqrt{2}\right\}\right\}$$

The built-in tests can be used without an argument. If an argument is written, the test is written as a pure function in which the argument is **#**: `f3[x_?(MatrixQ[#] &)] := ...`; note that parentheses are needed to enclose the pure function. User-defined tests are normally written as pure functions. As an example, we require that the argument is in the interval (0, 1):

```
f4[x_?(0 < # < 1 &)] := ArcSin[x]

{f4[0.3], f4[1.3]}
{0.304693, f4[1.3]}
```

Find all elements whose absolute value is less than 5:

```
Cases[{2 + 6 I, 7, 3 - I, 2, 5 + 4 I}, _?(Abs[#] < 5 &)]
{3 - i, 2}
```

We could also use **Select**:

```
Select[{2 + 6 I, 7, 3 - I, 2, 5 + 4 I}, Abs[#] < 5 &]
{3 - i, 2}
```

Require that argument is a matrix whose elements are numeric:

```
f4[m_?(MatrixQ[#, NumericQ] &)] := Det[m]
```

```
f4[{{3, 5}, {2, 1}}]
```

-7

We write a program for Newton's method and restrict the arguments with suitable heads and tests:

```
newton9[f_, x_Symbol, x0_?NumericQ, max_Integer?Positive] :=
 With[{df = D[f, x]}, FixedPointList[(x - f / df) /. x → # &, N[x0], max]]
```

```
newton9[3 x^3 - E^x, x, 2, 20]
```

```
{2., 1.41942, 1.1019, 0.975117, 0.953089, 0.952446, 0.952446, 0.952446, 0.952446}
```

The program does not accept, for example, a nonnumeric initial value:

```
newton9[3 x^3 - E^x, x, x0, 20]
```

$\text{newton9}\left[-e^x + 3\,x^3, x, x0, 20\right]$

■ Restricting Patterns with Conditions

> **x_ /; cond** Any expression giving **True** from **cond**
>
> **x_head /; cond** Any expression having the head **head** and giving **True** from **cond**
>
> **x_ ? test /; cond** Any expression giving **True** from **test** and from **cond**
>
> **x_head ? test /; cond** Any expression having the head **head** and giving **True** from **test** and from **cond**

(The internal name of **/;** is **Condition**.) As an example, select all numbers that are on a given interval:

```
Cases[{4, 1, 3, 5, 3, 6, 2}, x_ /; 2 ≤ x ≤ 4]
```

```
{4, 3, 3, 2}
```

We could also use a test or **Select**:

```
Cases[{4, 1, 3, 5, 3, 6, 2}, _? (2 ≤ # ≤ 4 &)]
```

```
{4, 3, 3, 2}
```

```
Select[{4, 1, 3, 5, 3, 6, 2}, 2 ≤ # ≤ 4 &]
```

```
{4, 3, 3, 2}
```

Replace all big even numbers with a string:

```
SeedRandom[2]; t = RandomInteger[{1, 20}, 20]
```

```
{4, 19, 12, 11, 18, 9, 4, 9, 1, 20, 10, 15, 5, 4, 5, 13, 13, 9, 9, 19}
```

```
t /. x_ /; EvenQ[x] && x > 10 → "bigEven"
```

```
{4, 19, bigEven, 11, bigEven, 9, 4, 9, 1, bigEven, 10, 15, 5, 4, 5, 13, 13, 9, 9, 19}
```

The following function accepts only real numbers as the argument since the imaginary part of the argument is required to be 0:

```
f5[x_ /; Im[x] == 0] := x^2
```

```
{f5[-3], f5[Sqrt[2 - Sqrt[2]]], f5[Sqrt[1 - Sqrt[2]]], f5[2 + 5 I], f5[a]}
```

$\left\{9,\ 2 - \sqrt{2}\ ,\ \text{f5}\left[i\,\sqrt{-1 + \sqrt{2}}\ \right],\ \text{f5}[2 + 5\,i],\ \text{f5}[a]\right\}$

The next function requires that the argument be positive:

```
f6[x_ /; Im[x] == 0 && x > 0] := x^2
```

```
{f6[3], f6[-7], f6[2 + I], f6[a]}
```

```
{9, f6[-7], f6[2 + i], f6[a]}
```

Now we require that **x** is a numerical matrix:

```
f7[x_ /; MatrixQ[x, NumericQ]] := Det[x]
```

```
{f7[{{3, 2}, {5, 1}}], f7[{{a, 2}, {5, 1}}]}
```

```
{-7, f7[{{a, 2}, {5, 1}}]}
```

Select lists with length 3:

```
Cases[{{1, 2}, {1, 2, 3}, {1, 2, 3, 4}}, x_List /; Length[x] == 3]
```

```
{{1, 2, 3}}
```

```
Select[{{1, 2}, {1, 2, 3}, {1, 2, 3, 4}}, ListQ[#] && Length[#] == 3 &]
```

```
{{1, 2, 3}}
```

All the examples of tests presented previously can also be written by using conditions. Indeed, the use of conditions may often be easier than the use of tests:

```
fac[0] = 1;
fac[n_Integer /; n > 0] := n fac[n - 1]
```

```
{fac[5], fac[2.7], fac[-2]}
```

```
{120, fac[2.7], fac[-2]}
```

```
{4, 1, 3, 5, 3, 6, 2} /. x_ /; EvenQ[x] → "even"
```

```
{even, 1, 3, 5, 3, even, even}
```

```
Solve[49 - 22 x^2 + x^4 == 0]
```

$$\left\{\left\{x \to -3 - \sqrt{2}\right\}, \left\{x \to 3 - \sqrt{2}\right\}, \left\{x \to -3 + \sqrt{2}\right\}, \left\{x \to 3 + \sqrt{2}\right\}\right\}$$

```
Cases[%, {x → x_ /; x > 0}]
```

$$\left\{\left\{x \to 3 - \sqrt{2}\right\}, \left\{x \to 3 + \sqrt{2}\right\}\right\}$$

```
f4[x_ /; 0 < x < 1] := ArcSin[x]
```

```
{f4[0.3], f4[1.3]}
```

```
{0.304693, f4[1.3]}
```

```
Cases[{2 + 6 I, 7, 3 - I, 2, 5 + 4 I}, x_ /; Abs[x] < 5]
```

```
{3 - i, 2}
```

16.1.3 More about Patterns

■ Variable Number of Arguments

x_	Any single expression
x__	Any sequence of *one* or more expressions
x___	Any sequence of *zero* or more expressions

Sometimes it is useful to form a function in which the number of arguments is unspecified. Such functions can be formed with a double underscore (__) (**BlankSequence**) or a triple underscore (___) (**BlankNullSequence**).

The functions **f1**, **f2**, and **f3** accept one, one or more, and zero or more arguments, respectively:

```
f1[x_] := Apply[Plus, {x}]
f2[x__] := Apply[Plus, {x}]
f3[x___] := Apply[Plus, {x}]
```

We try the functions for several numbers of arguments:

```
{f1[], f1[a], f1[a, b], f1[a, b, c]}     {f1[], a, f1[a, b], f1[a, b, c]}

{f2[], f2[a], f2[a, b], f2[a, b, c]}     {f2[], a, a + b, a + b + c}

{f3[], f3[a], f3[a, b], f3[a, b, c]}     {0, a, a + b, a + b + c}
```

We see that **f1** accepts only one argument, **f2** does not accept zero arguments, and **f3** accepts zero or more arguments.

Next, we search lists of exactly one element, at least one element, and zero or more elements:

```
t = {{3, 7}, {5}, {}, {2, 6, 1}, 7, {a, b}};

Cases[t, {x_}]      {{5}}

Cases[t, {x__}]      {{3, 7}, {5}, {2, 6, 1}, {a, b}}

Cases[t, {x___}]      {{3, 7}, {5}, {}, {2, 6, 1}, {a, b}}
```

The triple underscore is used especially for options. In Section 18.3.4, p. 579, we present the following example:

```
newton8[f_, x_, x0_, max_, opts___] := With[{df = D[f, x]},
   FixedPointList[(x - f / df) /. x → # &, N[x0], max, opts]]
```

This function accepts zero or more options so that we can write, for example, the following commands:

```
newton8[f, x, 2, 20]

newton8[f, x, 2, 20, SameTest → (Abs[#1 - #2] < 10^-3 &)]
```

■ Default Values

> **x_.** Any expression; if it does not occur explicitly, use a built-in default value
> **x_:v** Any expression; if it does not occur explicitly, use the default value **v**
> **x_head:v** Any expression with the head **head**; if it does not occur explicitly, use the default value **v**

(These constructs are internally formed with **Optional**.) Default values are useful, for example, in such expressions as **a^n_**. Consider the following three examples. In the first example, we do not have a default value for the exponent, and **a** or **a^1** is not transformed:

```
{a, a^2, a^3} /. a^n_ → b^(2 n)     {a, b^4, b^6}
```

In the following example, we use the built-in default value 1 for the exponent, and now **a**, too, is transformed:

```
{a, a^2, a^3} /. a^n_. → b^(2 n)     {b^2, b^4, b^6}
```

We can also define our own default value:

```
{a, a^2, a^3} /. a^(n_: 1) → b^(2 n)     {b^2, b^4, b^6}
```

In general, the default value of **y** in **x_ + y_.** is 0. In **x_ y_.** the default value of **y** is 1, and in **x_^y_.** it is also 1. As an example, the function **coeff** extracts the coefficients from a linear expression:

```
coeff[a_. + b_. x_, x_] := {a, b}
coeff[a_, x_] := {a, 0}
```

We try the function:

```
{coeff[u, z], coeff[u + z, z], coeff[v z, z], coeff[u + v z, z]}
{{u, 0}, {u, 1}, {0, v}, {u, v}}
```

■ Optional Arguments

A default value such as **v** in **x_:v** may be useful in function definitions. Here is an example:

```
newtonSolve[f_, x_, x0_, d_: 1, n_: 20, opts___?OptionQ] :=
 With[{df = D[f, x]}, FixedPointList[(x - d f / df) /. x → # &, N[x0], n, opts]]
```

If we give only the first three arguments, as in **newtonSolve[f, x, 3]**, then the default value 1 is used for **d** and 20 for **n**. On the other hand, we can define **d** as in **newtonSolve[f, x, 3, 2]**, and then 2 is used as the damping factor. We can also define both **d** and **n** as in **newtonSolve[f, x, 3, 2, 25]**. In this way, we can define and use optional arguments.

An optional argument can sometimes be handy at the end of the parameter list. Another approach, which is useful in more complicated situations, is to define the function several times for several forms of arguments. Still another approach is to use options (see Section 17.3.4, p. 538). Options are good for situations in which the function has many adjustable features but the default settings of the features work in most cases.

■ Alternative Patterns

```
pattern1 | pattern2 | …   Represents any expression matching any of the patterns
```

Toss a die 10 times:

```
t = {2, 6, 4, 3, 3, 6, 3, 1, 6, 5};
```

Find all 1's and 6's:

```
Cases[t, 1 | 6]     {6, 1, 6}
```

We could also use **Select**:

```
Select[t, # == 1 || # == 6 &]     {6, 1, 6}
```

Replace all 1's and 6's with a string:

```
t /. 1 | 6 → "onesix"
{2, onesix, 4, 3, 3, onesix, 3, onesix, onesix, 5}
```

Of course, we could also write

```
t /. {1 → "onesix", 6 → "onesix"}
{2, onesix, 4, 3, 3, onesix, 3, onesix, onesix, 5}
```

Define a function with alternative arguments:

```
die[1 | 6] := win
die[2 | 3 | 4 | 5] := lose
die /@ Range[6]
{win, lose, lose, lose, lose, win}
```

■ A More Complete Definition of a Pattern with a Name

Thus far, we have considered patterns of the form **x_** (added with tests and conditions). To form more complicated patterns, the following more general definition of a pattern is useful:

> **x : pattern** Represents any expression matching **pattern**; the expression is referred to with the name **x**

(The internal name of this construct is **Pattern**.) The pattern **x_** is a special case of this general definition. Indeed, **x_** is equivalent to **x:_**, and **x_head** is equivalent to **x:_head**.

In the next example, the argument of the function has to be a positive integer or a positive rational number:

```
f4[x : (_Integer | _Rational) /; x > 0] := x^2
```

```
{f4[3], f4[2/3], f4[1.6], f4[-5]}
```

$$\left\{9, \frac{4}{9}, f4[1.6], f4[-5]\right\}$$

■ Repeated Patterns

> **pattern ..** Represents any sequence of one or more expressions, each matching **pattern**
> **pattern ...** Represents any sequence of zero or more expressions, each matching **pattern**

Generate random strings:

```
SeedRandom[1]; t = RandomChoice[{"a", "b"}, 20]
```

```
{b, b, a, b, a, a, a, b, a, b, a, a, a, a, a, a, a, b, b, b}
```

Split the list:

```
t2 = Split[%]
```

```
{{b, b}, {a}, {b}, {a, a, a}, {b}, {a}, {b}, {a, a, a, a, a, a, a}, {b, b, b}}
```

Find all lists of one or more a's:

```
Cases[t2, {"a" ..}]
```

```
{{a}, {a, a, a}, {a}, {a, a, a, a, a, a, a}}
```

Replace all lists of one or more a's with A:

```
t2 /. {"a" ..} → "A"
```

```
{{b, b}, A, {b}, A, {b}, A, {b}, A, {b, b, b}}
```

The function **f5** accepts as an argument a nonempty list of two-element lists and then transposes the list:

```
f5[x : {{_, _} ..}] := Transpose[x]
```

```
f5[{{1, a}, {2, b}, {3, c}}]
```

```
{{1, 2, 3}, {a, b, c}}
```

We could also write

```
f5a[x_ ?MatrixQ /; Dimensions[x][[2]] == 2] := Transpose[x]
```

```
f5a[{{1, a}, {2, b}, {3, c}}]
```

```
{{1, 2, 3}, {a, b, c}}
```

With `Repeated` we can control in more detail the number of times a pattern is allowed to repeat.

`Repeated[pattern]`   A pattern repeated at least once (the same as `pattern..`)
`Repeated[pattern, {n}]`   A pattern repeated exactly `n` times
`Repeated[pattern, max]`   A pattern repeated at most `max` times
`Repeated[pattern, {min, max}]`   A pattern repeated between `min` and `max` times

Consider the list `t2` generated above. Instead of the pattern `"a"..` we can also write `Repeated["a"]`:

```
Cases[t2, {Repeated["a"]}]
```
```
{{a}, {a, a, a}, {a}, {a, a, a, a, a, a, a}}
```

Find all lists with at most three a's:

```
Cases[t2, {Repeated["a", 3]}]
```
```
{{a}, {a, a, a}, {a}}
```

We could also write

```
Cases[t2, {"a"} | {"a", "a"} | {"a", "a", "a"}]
```
```
{{a}, {a, a, a}, {a}}
```

■ Exceptions

`Except[c]`   A pattern matching any expression except `c`
`Except[c, pattern]`   A pattern matching `pattern` but not `c`

Toss a die 20 times:

```
SeedRandom[2]; t = RandomInteger[{1, 6}, 20]
```
```
{6, 2, 3, 3, 6, 3, 2, 6, 6, 1, 1, 5, 4, 5, 1, 2, 2, 6, 2, 6}
```

Pick all but 6's or all results that are not 1's or 6's:

```
Cases[t, Except[6]]
```
```
{2, 3, 3, 3, 2, 1, 1, 5, 4, 5, 1, 2, 2, 2}
```

```
Cases[t, Except[1 | 6]]
```
```
{2, 3, 3, 3, 2, 5, 4, 5, 2, 2, 2}
```

We could also write

```
Cases[t, x_ /; x ≠ 6]
```
```
{2, 3, 3, 3, 2, 1, 1, 5, 4, 5, 1, 2, 2, 2}
```

Pick all even results that are not 6:

```
Cases[t, Except[6, _?EvenQ]]      {2, 2, 4, 2, 2, 2}
```

We could also write

```
Cases[t, x_?EvenQ /; x ≠ 6]      {2, 2, 4, 2, 2, 2}
```

■ Longest and Shortest Sequences

`Longest[pattern]` (✻6)   The longest sequence matching `pattern`
`Shortest[pattern]` (✻6)   The shortest sequence matching `pattern`

Find the longest sequence of the digit 9 appearing in the first 1000 digits of π:

```
d = RealDigits[N[π, 1000]] // First;
```

```
d /. {___, x : Longest[9 ..], ___} → {x}
```

{9, 9, 9, 9, 9, 9}

Another way:

```
d /. {___, Longest[x__ ? (# == 9 &)], ___} → {x}
```

{9, 9, 9, 9, 9, 9}

This sequence of six 9's begins from the 763th digit:

```
Position[Partition[d, 6, 1], {9, 9, 9, 9, 9, 9}]
```

{{763}}

■ Pattern Sequences

PatternSequence[pattern1, pattern2, …] (✿6) A sequence of objects, with the first object matching **pattern1**, the second matching **pattern2**, …

What is the longest sequence of the digits 1 and 2 appearing in the first 1000 digits of π:

```
d /. {___, x : Longest[PatternSequence[1, 2] ..], ___} :→ {x}
```

{1, 2, 1, 2}

16.2 String Patterns

16.2.1 String Patterns

■ String Patterns

Strings were considered in Section 13.3.6, p. 433. String patterns are very useful in manipulating strings. **StringPosition** (to be considered soon) is one of the commands that support string patterns. The simplest form of this command searches the start and end positions of an explicitly given string:

```
StringPosition["mathematics", "mat"]
```

{{1, 3}, {6, 8}}

However, we can also ask the positions of a more complex pattern. Here we ask, in two ways, the positions of "math" or "mati":

```
StringPosition["mathematics", {"math", "mati"}]
```

{{1, 4}, {6, 9}}

```
StringPosition["mathematics", "math" | "mati"]
```

{{1, 4}, {6, 9}}

Mathematica has a built-in English dictionary. With the following command we can search words from the dictionary. The search is based on string patterns.

`DictionaryLookup[]` (⌘6)   Give a list of all words in an English dictionary

`DictionaryLookup[patt]`   Find all words that match the string pattern **patt**

`DictionaryLookup[patt, n]`   Give only the first **n** words found

`DictionaryLookup[patt, IgnoreCase → True]`   Do not take the case of words into account

The dictionary contains almost 100,000 words:

```
DictionaryLookup[] // Length        92 518
```

In Section 16.1.3, p. 500, we studied pattern objects such as _, __, and ___. For strings, the pattern object _ represents any single character, __ represents a sequence of one or more characters, and ___ represents a sequence of zero or more characters. Next, we search, from the dictionary of the English language, all palindromes—that is, words with one or more characters that are the same if read from the end to the beginning:

```
Style[DictionaryLookup[x__ /; x === StringReverse[x]], 8]
```

{a, aha, aka, bib, bob, boob, bub, CFC, civic, dad, deed, deified, did, dud, DVD,
 eke, ere, eve, ewe, eye, gag, gig, huh, I, kayak, kook, level, ma'am, madam, mam,
 MGM, minim, mom, mum, nan, non, noon, nun, oho, pap, peep, pep, pip, poop, pop,
 pup, radar, redder, refer, repaper, reviver, rotor, sagas, sees, seres, sexes,
 shahs, sis, solos, SOS, stats, stets, tat, tenet, TNT, toot, tot, tut, wow, WWW}

Find all words containing one or more times the letters h, e, i, and k:

```
DictionaryLookup[("h" | "e" | "i" | "k") .., IgnoreCase → True]
```

{eek, eh, eke, he, hi, hie, hike, I, Ike}

■ **String Expressions**

In addition to standard pattern objects such as _ or __, we can form more complex patterns by combining, with ~~, strings and standard pattern objects. Such patterns constructs are called *string expressions*. For example, next we ask

- the positions of "mat" followed by any single character;
- the positions of "mat" followed by any single character followed by "cs"; and
- the positions of "mat" followed by any single character not equal to "h":

```
StringPosition["mathematics", "mat" ~~ _]
```

{{1, 4}, {6, 9}}

```
StringPosition["mathematics", "mat" ~~ _ ~~ "cs"]
```

{{6, 11}}

```
StringPosition["mathematics", "mat" ~~ (c_ /; c ≠ "h")]
```

{{6, 9}}

Next, we search, from the dictionary of the English language, all words starting with "x"

```
DictionaryLookup["x" ~~ ___]
```

{xenon, xenophobe, xenophobes, xenophobia, xenophobic,
 xerographic, xerography, xerox, xeroxed, xeroxes, xeroxing, xi, xis,
 xylem, xylene, xylophone, xylophones, xylophonist, xylophonists}

A simpler way is the following:

```
DictionaryLookup["x*"]
```

{xenon, xenophobe, xenophobes, xenophobia, xenophobic,
 xerographic, xerography, xerox, xeroxed, xeroxes, xeroxing, xi, xis,
 xylem, xylene, xylophone, xylophones, xylophonist, xylophonists}

Now we calculate the number of words beginning with a, ..., z (case ignored):

```
freq =
 Length[DictionaryLookup[# ~~ ___, IgnoreCase → True]] & /@ CharacterRange["a", "z"]
```
{5305, 5510, 8726, 5647, 3607, 3757, 3096, 3466, 3569, 1017, 928, 2913, 5190,
 2021, 2320, 7130, 450, 5572, 10474, 4664, 2647, 1405, 2492, 46, 331, 224}

```
ListPlot[freq, Filling → Axis, ImageSize → 220,
 Ticks → {Transpose[{Range[26], CharacterRange["a", "z"]}], Automatic}]
```

Here is the general syntax of a string expression:

StringExpression[s1, s2, …] or **s1 ~~ s2 ~~ …** A string expression—that is, a sequence of strings and pattern objects

Next, we consider commands that support string expressions. Note that the following option can be used for all the commands we will consider:

An option for commands using string patterns:

IgnoreCase Whether lower- and uppercase letters should be treated as equivalent; possible values:
False, True

The default is that lower- and uppercase letters are not treated as equivalent.

■ **Searching Positions and Cases**

Here, we again denote by **s** a given string. Also, we denote by **patt** a string pattern—that is, a string expression.

StringPosition[s, patt] Give positions of substrings of **s** matching **patt**
StringCases[s, patt] Give a list of substrings in **s** that match **patt**
StringCases[s, patt → val] Replace each case of **patt** with **val**
StringCount[s, patt] Count how many substrings of **s** match **patt**
StringMatchQ[s, patt] Test whether **s** matches the string pattern **patt**
StringFreeQ[s, patt] Test whether no substring of **s** matches the string pattern **patt**

```
StringCases["mathematics", "mat"]
```
{mat, mat}

```
StringCases["mathematics", "mat" ~~ _]
```
{math, mati}

```
StringCases["mathematics", "mat" ~~ _ ~~ "cs"]
```
{matics}

```
StringCases["mathematics", "mat" ~~ c_ ~~ "cs" → "MAT" ~~ c ~~ "CS"]
```
{MATiCS}

Here are the frequencies of all letters (case ignored) in the built-in dictionary:

```
dl = DictionaryLookup[__, IgnoreCase → True];
```

```
{#, StringCount[StringJoin[dl], #]} & /@ CharacterRange["a", "z"]
```
{{a, 59 823}, {b, 14 496}, {c, 30 209}, {d, 28 950}, {e, 88 302}, {f, 10 232}, {g, 22 567},
 {h, 17 621}, {i, 66 782}, {j, 1322}, {k, 7188}, {l, 40 655}, {m, 20 161},
 {n, 55 210}, {o, 47 069}, {p, 21 188}, {q, 1403}, {r, 55 984}, {s, 66 601},
 {t, 51 121}, {u, 25 717}, {v, 7694}, {w, 6768}, {x, 2078}, {y, 12 372}, {z, 3322}}

In place of, for example, **StringCases**, we can use **Pick** with **StringMatchQ**. Here, we search for all words starting with "aa" (for **StartOfString**, see Special Positions):

```
StringCases[dl, StartOfString ~~ "aa" ~~ ___] // Flatten
```
{aah, aardvark, aardvarks}

```
Pick[dl, StringMatchQ[dl, "aa" ~~ ___]]
```
{aah, aardvark, aardvarks}

An option for **StringPosition**, **StringCases**, *and* **StringCount**:

Overlaps   How to treat overlapping substrings; possible values: **False** (allow no overlaps; the default for **StringCases** and **StringCount**), **All** (allow all overlaps), **True** (allow overlaps starting at different positions; the default for **StringPosition**)

StringCases and **StringCount** do not, by default, allow overlaps. Here, we search for a sequence of one or more characters:

```
StringCases["abcd", __]
```
{abcd}

Next, we allow all overlaps:

```
StringCases["abcd", __, Overlaps → All]
```
{abcd, abc, ab, a, bcd, bc, b, cd, c, d}

Now we only allow overlaps at different positions:

```
StringCases["abcd", __, Overlaps → True]
```
{abcd, bcd, cd, d}

StringPosition, by default, allows overlaps at different positions:

```
StringPosition["abcd", __]
```
{{1, 4}, {2, 4}, {3, 4}, {4, 4}}

■ **Replacing**

StringReplace[s, patt → val] Replace every occurrence of string **patt** in **s** with **val**

StringReplace[{s, t, … }, patt → val] Replace in each of the strings **s, t** …

StringReplaceList[s, patt → val] Do the replacement in all possible ways

---

```
StringReplace["aaag aabg aacg", "aa" → "AA"]
```
AAag AAbg AAcg

```
StringReplace["aaag aabg aacg", "aa" ~~ _ → "AA"]
```
AAg AAg AAg

```
StringReplace[{"aaag", "aabg", "aacg"}, "aa" ~~ _ → "AA"]
```
{AAg, AAg, AAg}

16.2.2 More about String Patterns

■ **Splitting**

StringSplit[s] Split **s** into sublists at every whitespace (space, newline, tab)

StringSplit[s, patt] Split **s** into sublists at every delimiter matching **patt**

StringSplit[s, {patt1, patt2, …}] Split at points matching any of the patterns

StringSplit[s, patt → val] Insert **val** at the position of each delimiter

---

Note that the delimiters (whitespace, etc.) are dropped from the splitted string.

```
StringSplit["Here is a hat"]
```
{Here, is, a, hat}

```
StringSplit["aababaaaabba", "b"]
```
{aa, a, aaaa, , a}

```
StringSplit["aababaaaabba", "b" → "b"]
```
{aa, b, a, b, aaaa, b, , b, a}

Generate a random sequence of letters a and b. Then split the sequence such that every occurrence of "ayya", where "y" can be any character, is picked and colored red:

```
SeedRandom[1]; s = StringJoin[RandomChoice[{"a", "b", "c"}, 50]]
```
babbaaabaaaacabcaabbaabacabbccabababaacaccabcbcaaa

```
StringSplit[s, x : ("a" ~~ y_ ~~ y_ ~~ "a") ⧴ Style[x, Red]]
```
{b, abba, aab, aaaa, cabca, abba, abacabbccabababaac, acca, bcbcaaa}

■ **Special Patterns**

NumberString A number in a string form

DigitCharacter A digit character (0-9)

LetterCharacter A letter character

WordCharacter A letter or a digit character

Whitespace A sequence of space, newline, or tab characters

WhitespaceCharacter A single space, newline, or tab character

Except[c] Represents any character except ones matching **c**

Replace every sequence of one or more digits with "D":

```
StringReplace["ab123cde45f", DigitCharacter .. → "D"]
abDcdeDf
```

Replace every sequence of one or more characters that are not digits with "D":

```
StringReplace["ab123cde45f", Except[DigitCharacter] .. → "D"]
D123D45D
```

Bertrand Russell has defined mathematics as

```
t = "The subject in which we never know what we
        are talking about nor whether what we are saying is true." ;
```

(See http://mathworld.wolfram.com/Mathematics.html.) Search all words, sort the list, split the resulting list, and calculate the length of each run. Lastly, calculate how many words we have that occur once, twice, thrice, …:

```
words = ToLowerCase[StringCases[t, WordCharacter ..]]
{the, subject, in, which, we, never, know, what, we, are,
 talking, about, nor, whether, what, we, are, saying, is, true}
```

```
words // Sort // Split
{{about}, {are, are}, {in}, {is}, {know}, {never}, {nor}, {saying}, {subject},
 {talking}, {the}, {true}, {we, we, we}, {what, what}, {whether}, {which}}
```

```
Length /@ %    {1, 2, 1, 1, 1, 1, 1, 1, 1, 1, 1, 1, 3, 2, 1, 1}
```

```
% // Tally    {{1, 13}, {2, 2}, {3, 1}}
```

■ Special Positions

StartOfString, EndOfString	The start/end of string
StartOfLine, EndOfLine	The start/end of a line
WordBoundary	The boundary between word characters and others
Except[StartOfString]	Anywhere except at the start of the string, etc.

Change the first character of each word to uppercase:

```
StringReplace["Here is a hat", WordBoundary ~~ x_ → ToUpperCase[x]]
Here Is A Hat
```

■ Regular Expressions

RegularExpression[s]	The generalized regular expression specified by **s**

String patterns can be defined, besides with string expressions and patterns objects, with a pattern notation called *regular expression*. We give only one example here. We can write

```
StringReplace["aaebf cadbb", "a" | "b" | "c" → "C"]
CCeCf CCdCC
```

but we can also use a regular expression:

```
StringReplace["aaebf cadbb", RegularExpression["[abc]"] → "C"]
CCeCf CCdCC
```

Functions

Introduction

> *Teacher: "If Tom gave you three apples and Bill gave you two apples, how many apples would you have then?" Mary: "Seven apples, teacher." Teacher: "Wrong, Mary, 3 + 2 = 5." Mary: "I know that, teacher, but I have two apples already."*

A function defined by **f[x_] := expr** is usually not needed to do calculations with expressions. Instead, give a name for the expression, such as **f = expr**, and use this name. On the other hand, the syntax of a function is needed to write more complex functions, such as recursive functions or programs. Recursive and periodic functions are addressed in Chapter 18 in the context of recursive programming.

Pure functions are important in *Mathematica*, however odd they may seem at first sight. With *tracing*, we can see how a calculation proceeds inside *Mathematica*; this may be useful to correct programs. *Mathematica* 6 has a special *debugger* to detect errors in programs. By *compiling* a function, we can get a more efficient function and save computing time. We can also assign some *attributes* to a function and so give it some desired properties.

Functions are used in programming, and a collection of related functions can be gathered together into a *package*. To understand the structure of a package, we have to consider *contexts*.

Sections 17.1.1 and 17.1.4 are the most important ones in this chapter.

17.1 User-Defined Functions

17.1.1 Defining a Function

■ A Name or a Function?

Suppose we want to consider the function **x Cos[x]**. We want to calculate its value at some points and to differentiate, integrate, and plot it. To avoid having to write the expression again and again or use %, %%, %%%, and so on, we can either give a name to the expression or define a function:

1. Naming the expression	2. Defining a Function
`f = x Cos[x]`	`f[x_] := x Cos[x]`
`f /. x → 2`	`f[2]`
`Table[f, {x, 0, 1, .1}]`	`Table[f[x], {x, 0, 1, .1}]`
`D[f, x]`	`f'[x]`
`Integrate[f, x]`	`Integrate[f[x], x]`
`Plot[f, {x, 0, Pi}]`	`Plot[f[x], {x, 0, Pi}]`

The first method is often the most appropriate. With it, we give a name to the expression, and later we need only type this name. The second method is often not needed to perform calculations with an expression. However, function definitions are used to define more complicated functions and also programs.

■ Functions

> `f[x_] := expr` A function of one variable
> `f[x_, y_] := expr` A function of two variables
>
> `?f` Show the context and definition of `f`
> `Definition[f]` Show the definition of `f`
>
> `Clear[f]` Clear all definitions for `f`
> `Remove[f]` Remove `f` completely

Functions are defined with `:=`, and each argument is followed by the underscore (_) (named "blank"). For example,

```
f1[x_] := x Sin[x]
```

$$\text{f1[Pi / 3]} \qquad \frac{\pi}{2\sqrt{3}}$$

```
Definition[f1]
f1[x_] := x Sin[x]
```

```
? f1
```

```
Global`f1
```

```
f1[x_] := x Sin[x]
```

(The full name of `f1` is `Global`f1`, where `Global`` is the context of all user-defined symbols.) If you forget the underscore, the function does not work: *Mathematica* knows the value of the function only for the argument you have used in the definition and not for any other argument:

> `f2[x] := x Cos[x]`
>
> `f2[x]`   x Cos[x]
>
> `f2[Pi]`   f2[π]

Here is a function with two arguments:

> `charpoly[m_, x_] := Det[m - x IdentityMatrix[Length[m]]]`
>
> `charpoly[{{2, 5}, {6, 1}}, x]` $-28 - 3 x + x^2$

One or more arguments can be used to index a function:

> `f3[n_, x_] := (n - 1) x^n`
>
> `Table[f3[n, y], {n, 4}]` $\{0, y^2, 2 y^3, 3 y^4\}$

■ Details

At this point, there are two things that require explanation: the underscore and the colon.

The underscore in `x_` makes the argument a *pattern*. The pattern `x_` is matched with *anything*: The value of the function is calculated using whatever you give as an argument. Indeed, this is what we want of a function. Patterns were considered in more detail in Chapter 16. As we previously stated, if we define a function such as `f2` without the underscore, the definition is nearly useless because the value of the function is known only for the argument given in the definition and not for any other argument. Thus, *the crucial point to remember when defining a function is to write the underscore.*

Recall that assigning values can be done as follows: `f = x Sin[x]`. This means that the value of `x Sin[x]` is *immediately* calculated and then assigned to `f` (if `x` happens to have a value when you define `f`, then `f` gets the corresponding special value of `x Sin[x]`). If you define a function `f[x_] := x Sin[x]`, the value of the right-hand side is not evaluated *until* you ask for the value of the function with a specific argument. Indeed, the effect of the colon in `:=` is only to *delay* the evaluation of the right-hand side. This may seem somewhat unimportant now, but later you will encounter more complex functions in which the right-hand side simply cannot be evaluated or its value is extremely complex and useless unless the argument has a numerical value. Hence, it makes sense to routinely use `:=` in function definitions (although `=` could work for simpler functions).

■ Example 1

We try to use `=` in the definition of a function:

> `g1[x_] = NIntegrate[Sin[t^2], {t, 0, x}]`
>
> NIntegrate::nlim : t = x is not a valid limit of integration. ≫
>
> NIntegrate::nlim : t = x is not a valid limit of integration. ≫
>
> NIntegrate[Sin[t²], {t, 0, x}]

Mathematica immediately tries to evaluate the right-hand side of `g1`, but it cannot be evaluated: In `NIntegrate`, the limits of integration must be numerical (and not `x`, for example), and so we get the error messages. However, we can calculate the value of `g1` at a point:

> `g1[2]`   0.804776

Then we use `:=` in the definition:

```
g2[x_] := NIntegrate[Sin[t^2], {t, 0, x}]
```

Now the right-hand side is not evaluated until a specific value of **x** is given:

```
g2[2]      0.804776
```

■ Example 2

In this example, we demonstrate why it may be dangerous not to use the colon. Suppose we have defined a value for a variable **x** and then we define, without the colon, a function using this variable:

```
x = 3;
```

```
g3[x_] = x + 10      13
```

The value of the variable **x** was immediately used. From the definition of **g3**, we see that our function does not work because its value is 13 for all arguments:

```
Definition[g3]      g3[x_] = 13
```

If we instead use the colon, we get a working function:

```
g4[x_] := x + 10
```

```
Definition[g4]      g4[x_] := x + 10
```

```
x =.
```

■ Example 3

Sometimes we have to use = and not :=. For example, we define the following:

```
g5[x_] = D[x Sin[x], x]      x Cos[x] + Sin[x]
```

The important point is that the derivative was calculated immediately because we did not write the colon. Now **g5** contains the derivative, and we can calculate its value:

```
g5[2.]      0.0770038
```

Let us then try := in the definition:

```
g6[x_] := D[x Sin[x], x]
```

```
g6[2.]
```
General::ivar : 2.` is not a valid variable. ≫
$\partial_{2.} 1.81859$

We got an error message. The right-hand side of **g6** was not evaluated when **g6** was defined, and when we then asked for the value at 2, we ended up with the impossible expression **D[1.81859, 2.]**; clearly, we cannot differentiate with respect to 2. Therefore, in this example, we should not use the colon in the definition of the function.

■ Example 4

Sometimes we may want to define a function from the result of a computation. For example, we may have calculated the following:

```
D[x Sin[x] + Sinh[x], x]      x Cos[x] + Cosh[x] + Sin[x]
```

Now we want to define this result as the value of a function **g7** at **x**. This is another case in which we have to use = and not :=:

```
g7[x_] = %      x Cos[x] + Cosh[x] + Sin[x]
```

g7[1]   Cos[1] +Cosh[1] +Sin[1]

■ **A Tip**

When you define a function, you often need to do some experimenting to get the correct and best form. During the experimenting, it is possible that the definition of the function will not remain in the form you really want, and you may get odd or erroneous results. To keep the definition in the desired form, ask frequently, during the experimenting, for the definition of the function with **?f** or **Definition[f]**. If the definition is not what you want, remove it with **Remove[f]** and redefine. You can also simply develop the habit of removing the old definition with **Remove[f]** every time you modify the definition.

Before the next section, we remove all of our current definitions:

Remove["Global`*"]

■ **Syntax Coloring**

SyntaxInformation[f] (❀6) Define syntax information for **f** used in syntax coloring

Syntax coloring is used with the built-in functions to show that the number of arguments is wrong:

{Sin[], Sin[x], Sin[x, y]}

To get similar colorings for user-defined functions, use **SyntaxInformation**. In the next example, we define that **g8** is a function of two arguments:

SyntaxInformation[g8] = {"ArgumentsPattern" → {_, _}}

{ArgumentsPattern → {_, _}}

If we write **g8** with less than two or more than two arguments, we get colored information about the wrong number of arguments:

{g8[], g8[x], g8[x, y], g8[x, y, z]}

■ **Some Special Functions**

Identity[x] Identity function
InverseFunction[f][x] Inverse function
Composition[f, g][x] or **f[g[x]]** Composite function

Here are some examples of inverse functions:

{InverseFunction[f][x], InverseFunction[Sin][x]} $\left\{ f^{(-1)}[x], ArcSin[x] \right\}$

If **Solve** uses inverse functions, it gives a warning:

Solve[h[x] ⩵ 2, x]

InverseFunction::ifun :
  Inverse functions are being used. Values may be lost for multivalued inverses. ≫
  $\left\{ \left\{ x \rightarrow h^{(-1)}[2] \right\} \right\}$

Values of composite functions can be calculated naturally by writing **f[g[x]]**:

f[x_] := Sin[x]
g[x_] := x^2

{f[g[a]], g[f[a]]} $\left\{ Sin\left[a^2\right], Sin[a]^2 \right\}$
Remove["Global`*"]

17.1.2 Piecewise-Defined Functions

■ **Piecewise-Defined Functions**

> **Piecewise[{{val$_1$, cond$_1$}, {val$_2$, cond$_2$}, … }]**   A piecewise function with values **val$_i$** in regions
> defined by conditions **cond$_i$**
> **Piecewise[{{val$_1$, cond$_1$}, {val$_2$, cond$_2$}, … }, val]**   Use value **val** if none of the **cond$_i$** apply;
> the default of **val** is 0

Mathematica has advanced functionality for piecewise-defined functions. Such functions can be used in many kinds of computations, such as differentiating, integrating, minimizing, and solving differential equations. Here is an example:

```
f = Piecewise[{{1 + x, -1 < x ≤ 0}, {1 - x, 0 < x ≤ 1}}]
```

$$\begin{cases} 1 + x & -1 < x \le 0 \\ 1 - x & 0 < x \le 1 \end{cases}$$

Mathematica shows piecewise-defined functions in a form that is familiar from traditional mathematical notation.

Note that it is not necessary to define the value of the function when $x \le -1$ or $x > 1$ because *Mathematica* assumes that the value of a piecewise-defined function at points that are not explicitly included in the definition of the function is zero; this default value suits us in this example.

Piecewise functions can also be written as follows. First, type ⎣ESC⎦**pw**⎣ESC⎦ to get **{**. Then type ⎣CTRL⎦**,** to get

$$\begin{cases} \square & \square \\ \square & \square \end{cases}$$. Then fill in the boxes. If you want more rows, type ⎣CTRL⎦⎣RET⎦. In this way, we get the following

definition:

$$f = \begin{cases} 1 + x & -1 < x \le 0 \\ 1 - x & 0 < x \le 1 \end{cases}$$

■ **Example 1**

The function defined previously is the triangular probability density function (PDF):

```
Plot[f, {x, -1.5, 1.5}, Ticks → {{-1, 1}, {1}}, AspectRatio → Automatic]
```

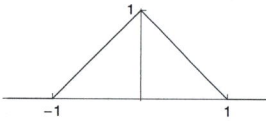

Check that the integral of the PDF is 1:

```
Integrate[f, {x, -∞, ∞}]      1
```

To calculate the cumulative distribution function (CDF) $F(x) = \int_{-\infty}^{x} f(t)\,dt$, we could try the following:

```
F = Integrate[f /. x → t, {t, -∞, x}]
```

$$\int_{-\infty}^{x} \begin{cases} 1 + t & -1 < t \le 0 \\ 1 - t & 0 < t \le 1 \end{cases} dt$$

This did not succeed. We have to declare that x is real:

```
F = Integrate[f /. x → t, {t, -∞, x}, Assumptions → x ∈ Reals]
```

$$\begin{cases} 1 & x > 1 \\ \frac{1}{2}\left(1 + 2x - x^2\right) & 0 < x \le 1 \\ \frac{1}{2}\left(1 + 2x + x^2\right) & -1 < x \le 0 \end{cases}$$

Note that *Mathematica* did not explicitly show the value of the CDF when $x \le -1$; in this interval, the value of the CDF is the default value 0. Plot the CDF:

```
Plot[F, {x, -1.5, 1.5}, Ticks → {{-1, 1}, {1}}, AspectRatio → Automatic]
```

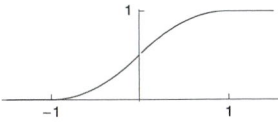

Check that $F'(x) = f(x)$:

```
D[F, x] // Simplify
```

$$\begin{cases} 1 & x == 0 \\ 1 - x & 0 < x < 1 \\ 1 + x & -1 < x < 0 \end{cases}$$

Again, *Mathematica* did not explicitly show the regions $x \le -1$ and $x \ge 1$ where the value of the derivative is the default value 0.

Note that *Mathematica* has the triangular distribution:

```
PDF[TriangularDistribution[{-1, 1}], x]
```

$$\begin{cases} 1 + x & -1 \le x \le 0 \\ 1 - x & 0 < x \le 1 \end{cases}$$

■ Example 2

Define another PDF:

$$f2 = \begin{cases} \frac{x}{ab} & 0 < x \le b \\ \frac{1}{a} & b < x \le a \\ \frac{a+b-x}{ab} & a < x \le a + b \end{cases} ;$$

```
Plot[f2 /. {b → 2, a → 4}, {x, 0, 6}]
```

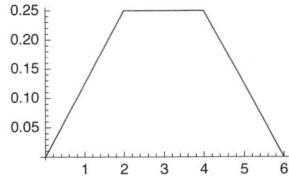

This is the PDF of the random variable $X + Y$ when X is uniform on $(0, a)$, Y is uniform on $(0, b)$, X and Y are independent, and $a > b$. Check that the integral is 1:

```
Integrate[f2, {x, -∞, ∞}, Assumptions → {a > 0, b > 0, a > b}]     1
```

To calculate the CDF, the following, which worked in Example 1, now gives a very complex, useless result (we do not show the output here):

```
F2 = Integrate[f2 /. x → t, {t, -∞, x}, Assumptions → {a > 0, b > 0, a > b, x ∈ Reals}]
```

Instead, do as follows. Here we calculate, with **Map** or **/@**, the integral separately for each interval:

```
F2 = Piecewise[
  {Integrate[f2 /. x → t, {t, -∞, x}, Assumptions → {a > 0, b > 0, a > b, #}], #} & /@
  {x ≤ 0, 0 < x ≤ b, b < x ≤ a, a < x ≤ a + b, x > a + b}]
```

$$
\begin{cases}
0 & x \le 0 \\
\frac{x^2}{2\,a\,b} & 0 < x \le b \\
\frac{-b+2\,x}{2\,a} & b < x \le a \\
\frac{-a^2-b^2+2\,a\,x+2\,b\,x-x^2}{2\,a\,b} & a < x \le a + b \\
1 & x > a + b
\end{cases}
$$

■ **Expanding Piecewise Functions**

> **PiecewiseExpand[expr]** Expand nested piecewise functions in **expr** to give a single piecewise function
>
> **PiecewiseExpand[expr, ass]** Use assumptions **ass**

We may have a function consisting of nested piecewise functions, such as the following:

$$
\begin{cases}
\begin{cases}
1 & 0 < x < 1 \\
2 & 1 < x < 2
\end{cases} & 0 < x < 2 \\
3 & 2 < x < 3
\end{cases} ;
$$

With **PiecewiseExpand** we get a single piecewise function:

```
% // PiecewiseExpand
```

$$
\begin{cases}
1 & 0 < x < 1 \\
2 & 1 < x < 2 \\
3 & 2 < x < 3
\end{cases}
$$

Some functions are implicitly piecewise. They can also be expanded into one explicit piecewise function:

```
PiecewiseExpand[Abs[x], x ∈ Reals]
```

$$
\begin{cases}
-x & x < 0 \\
x & \text{True}
\end{cases}
$$

```
PiecewiseExpand[Max[x, y]]
```

$$
\begin{cases}
x & x - y \ge 0 \\
y & \text{True}
\end{cases}
$$

```
PiecewiseExpand[If[x < 0, Cos[x], Sin[x]]]
```

$$
\begin{cases}
Cos[x] & x < 0 \\
Sin[x] & \text{True}
\end{cases}
$$

Here, the condition True is used to express the "otherwise" situation.

17.1.3 Implicit Functions

Implicit functions are defined via equations: The solution of an equation defines the value of the function. For example, the equation $x^3 - y^4 + x\,y^3 - y + 1 = 0$ implicitly defines a function y of x. Calculating derivatives of an implicit function is considered in Section 19.1.4, p. 622. Implicit functions can be plotted with **ContourPlot**, as shown in Section 5.2.3, p. 134. Here are two additional examples.

■ **Example 1**

Consider the following polynomial expression:

```
expr = x^3 - y^4 + x y^3 - y + 1;
```

In using `ContourPlot` to plot an implicit function defined by `expr == 0`, we ask only for the contour in which `expr` takes on the value `0`:

```
ContourPlot[expr == 0, {x, -2, 2}, {y, -2, 2},
  Frame → False, Axes → True, AspectRatio → Automatic]
```

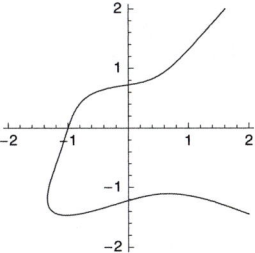

We can find values of `y` for a fixed `x` with `NSolve`:

```
Cases[y /. NSolve[expr == 0 /. x → 0, y], _Real]
{-1.22074, 0.724492}
```

■ **Example 2**

The implicit function can be quite complex, as in $\int_0^2 \sin(\sin(x\, y\, t))\, dt = 1$, and we can still succeed in plotting it (we drop some messages by `Off`):

```
Off[NIntegrate::inumr]
ContourPlot[NIntegrate[Sin[Sin[x y t]], {t, 0, 2}] == 1, {x, 0, 4}, {y, 0, 3},
  Frame → False, Axes → True, AspectRatio → Automatic, Ticks → {{1, 2, 3, 4}, {1, 2, 3}}]
```

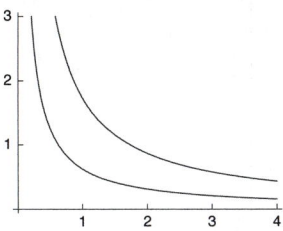

To calculate values of `y` for a fixed `x`, we can define a function:

```
f[x_, y0_, y1_] :=
  y /. FindRoot[NIntegrate[Sin[Sin[x y t]], {t, 0, 2}] == 1, {y, y0, y1}]
```

Here, `y0` and `y1` are the starting values for the iterative method used by `FindRoot`. Giving different starting values may result in different values if the function has multiple values. We ask for the value of `y` when `x` is 1. First, we start from 0.5 and 1, and then from 1 and 2 (we get a warning message but the result is correct):

```
{f[1, 0.5, 1], f[1, 1, 2]} // Quiet
{0.622871, 1.73409}

Remove["Global`*"]
```

17.1.4 Pure Functions and Scoping Constructs

■ Introduction

We already encountered *pure functions* in Sections 2.2.2, p. 38 (when considering **Map**), 14.1.5, p. 452 (when considering **Sort**), 14.1.7, p. 457 (the selecting criterion of **Select** was a pure function), 14.2.1, p. 459 (when considering **Map** once again), 14.2.3, p. 463 (when considering **Sequence**), and 16.1.2, p. 497 (when considering tests in patterns).

Mathematica has a number of *scoping constructs*, in which certain names are made local. The most important of these constructs is **Module**, but we also have **With** and **Block**. We have already used **Module** several times in this book.

■ Pure Functions

We illustrate pure functions using **Select**. First, we generate a table of 100 random numbers:

```
SeedRandom[2]; ran = RandomReal[1, {100}];
```

Suppose we are interested in picking the numbers that are either less than 0.02 or greater than 0.98. **Select[list, test]** is the appropriate command. It selects from **list** all the elements for which **test** gives **True**. One possibility is to define a function to this end:

```
f[x_] := x < 0.02 || x > 0.98
```

The value of this function is **True** if $x < 0.02$ or $x > 0.98$ and **False** otherwise. For example,

```
{f[0.01], f[0.03]}
{True, False}
```

Then we can use this function in **Select**:

```
Select[ran, f]
{0.00984124, 0.0154057, 0.986939, 0.986734, 0.0113432}
```

This works fine. However, we used the function **f** only once, and it seems too ceremonious to define a function for such a temporary use. Pure functions are handy for temporary use. They need not be defined beforehand but instead only exactly where they are used. In our example, we can use a pure function as follows:

```
Select[ran, Function[x, x < 0.02 || x > 0.98]]
{0.00984124, 0.0154057, 0.986939, 0.986734, 0.0113432}
```

There is even a simpler construction in which **#** means the argument and the ampersand **&** is the mark of a pure function:

```
Select[ran, # < 0.02 || # > 0.98 &]
{0.00984124, 0.0154057, 0.986939, 0.986734, 0.0113432}
```

In general, there are two ways of writing a pure function to perform a sequence **body** of commands for the arguments:

Function[x, body] A pure function with the argument **x**

Function[{x, y, … }, body] A pure function with the arguments **x**, **y**, …

body & A pure function with the argument **#** or with the arguments **#1**, **#2**, … (**##** means the sequence of all arguments supplied to a pure function)

Additional examples of pure functions are presented in Section 20.3 (**Nest**, **FixedPoint**, **Fold**) and throughout the following chapters.

■ Module: Local Variables

Often the definition of a function consists of a sequence of operations. One possibility is to separate the steps with semicolons. Consider the following example, in which we calculate the sum and product of integers $1, 2, ..., n$:

```
f1[n_] := (s = 0; p = 1; Do[s = s + i; p = i p, {i, n}]; {s, p})
```

```
f1[7]     {28, 5040}
```

One drawback is that the temporary variables **s** and **p** retain their values outside of the function:

```
{s, p}     {28, 5040}
```

```
Remove[s, p]
```

This can cause confusion later. It would be useful if the variables **s** and **d** were local to the function: Outside the function they had no values. This can be achieved with a module.

```
f[x_] := Module[{local variables}, body]
```

Note that

- local variables are separated by commas;
- initial values can be given for the local variables (the initial values cannot depend on each other);
- commands in the body are separated by semicolons;
- the result of the last command of the module is printed automatically; and
- to print intermediate values (even graphics), use **Print**.

The preceding example is now as follows:

```
f2[n_] := Module[{s = 0, p = 1},
   Do[s = s + i; p = i p, {i, n}];
   {s, p}]
```

As we see, initial values of the local variables can be given at the same time as they are made local. Often, each command in the module is written in its own row to make the module more readable. Before we use **f2**, we give some values for **s** and **p**:

```
s = -1; p = -2;
```

```
f2[7]     {28, 5040}
```

Now **s** and **p** have their old values:

```
{s, p}     {-1, -2}
```

This means that **s** and **p** in the module are not the same as **s** and **p** outside the module (indeed, inside the module *Mathematica* uses names of the form **s$nnn** and **p$nnn**, where **nnn** is increased incrementally by 1 every time we use the module).

The result of the last command of the module is automatically printed; in the example, the last command was **{s, p}**. If we want to print some intermediate results, we can use **Print** (see Section 18.2.3, p. 562).

Programming with *Mathematica* is essentially done by writing functions, often with `Module`. The programs can use all the commands of *Mathematica*. In Chapter 18, we consider some styles of programming and some special commands that are useful in these styles. You will encounter modules frequently in this book.

■ With: Local Constants

```
f[x_] := With[{x = x0, y = y0, … }, body]
```

The `With` construction is formally similar to a `Module` but is actually more restricted: `With` is used to define local *constants;* these constants cannot be changed later within the `With` construction. Note that a `Module` contains local *variables,* and their values can be changed in the module. `With` can be used inside a `Module` to define local constants. For example,

```
f3[n_] := With[{c = Range[6]},
   {RandomChoice[c, n], RandomSample[c, n]}]
f3[6]      {{1, 6, 6, 5, 6, 3}, {1, 4, 3, 6, 5, 2}}
```

■ Block: Local Values of Variables

```
Block[{x = x0, y = y0, … }, body]
```

In a `Block`, the variables within curly braces have only *local values.* In Section 18.5, when we consider recursive programming, we will show how to use `Block` to give temporary values for `$RecursionLimit` and `$IterationLimit`.

Blocks are also important because iteration commands such as `Table`, `Sum`, `Product`, and `Do` localize the values of the iterators with `Block`. For example,

```
i = 10; Sum[2^i, {i, 3}]; i      10
```

We see that `i` still has its original value of 10.

However, there is still a possibility of confusion with iterator commands. Consider the following function:

```
i =.

f5[x_, n_] := Sum[x^i, {i, n}]

f5[x, 3]      x + x^2 + x^3
```

If we ask for `f5[i, 3]`, we do not get the desired result:

```
f5[i, 3]      32
```

We get the number $1^1 + 2^2 + 3^3$ instead of $i^1 + i^2 + i^3$. We can see this with `Trace` (see Section 17.2.1, p. 523):

```
Trace[f5[i, 3], TraceDepth → 1]
```

$$\left\{ f5[i, 3], \sum_{i=1}^{3} i^i, 32 \right\}$$

For this reason, it is safe to use a module to localize the iterator `i`:

```
f6[x_, n_] := Module[{i}, Sum[x^i, {i, n}]]

f6[i, 3]      i + i^2 + i^3
```

17.2 More about Functions

17.2.1 Tracing

■ Tracing Expressions

If we are interested in seeing in detail how *Mathematica* calculates an expression, we can use tracing. When we write our own functions or programs, tracing may be useful for detecting coding errors and for making a function effective. Here, we present a short introduction to tracing; for more information, see tutorial/TracingEvaluation.

Trace[expr]	Generate a list of all expressions used in the evaluation of **expr**
Trace[expr, TraceDepth → n]	Ignore steps that lead to lists nested more than **n** levels deep

For example,

```
Trace[5 (3 + 1) - 4]
```
```
{{{3 + 1, 4}, 5 × 4, 20}, 20 - 4, 16}
```

We see that *Mathematica* starts the calculation from 3 + 1, gets 4, forms the product 5×4, gets 20, forms the difference 20 - 4, and finally gets 16. As another example, consider the following function:

```
f[x_] := Module[{c = 2 x}, Cos[c]]
```
```
f[2. Pi]
1.
```

Let us see how *Mathematica* got the result 1.:

```
f[2. Pi] // Trace // Column
```
```
{2. π, 6.28319}
f[6.28319]
Module[{c = 2 × 6.28319}, Cos[c]]
{2 × 6.28319, 12.5664}
{c$127 = 12.5664, 12.5664}
{{c$127, 12.5664}, Cos[12.5664], 1.}
1.
```

Here we see how *Mathematica* used a local variable **c$nnn**.

■ Tracing Assignments

Trace[expr, var = _]	Trace assignments to **var**
Trace[expr, _ = _]	Trace all assignments

Often, it is useful to see all assignments to variables to check that they are done correctly. Remember that the underscore (_) means anything in *Mathematica* and so, for example, **var = _** means all assignments to **var** and **_ = _** means all assignments to all variables.

Here is a procedural program for calculating a factorial:

```
g[n_] := Module[{fac, i}, For[fac = 1; i = 1, i ≤ n, ++i, fac = fac i]; fac]
```

We trace all assignments for **fac**:

```
Trace[g[3], fac = _]
{{{{{fac$224 = 1}}, {fac$224 = 1}, {fac$224 = 2}, {fac$224 = 6}}}}
```

■ Tracing Functions

Trace[expr, f]	Trace all calls to function **f**
Trace[expr, f[_]]	Give a list of intermediate expressions of the form **f[_]**
Trace[expr, f[_], TraceForward → True]	Also show the values of the function
Trace[expr, f[x_] → x]	Give a list of all values of the argument of **f**
On[f]	Switch tracing on for the symbol **f**
Off[f]	Switch tracing off for the symbol **f**

Here is a function for calculating Fibonacci numbers (we do not use dynamic programming; see Section 18.5.1, p. 597):

```
fib[1] = 1;
fib[2] = 1;
fib[n_] := fib[n - 1] + fib[n - 2]
```

It is convenient to turn tracing on and off, particularly for recursive functions:

```
On[fib]

fib[3]
fib::trace : fib[3] --> fib[3 - 1] + fib[3 - 2].
fib::trace : fib[3 - 1] --> fib[2].
fib::trace : fib[2] --> 1.
fib::trace : fib[3 - 2] --> fib[1].
fib::trace : fib[1] --> 1.
2

Off[fib]
```

Then we trace all calls to **fib**:

```
Trace[fib[3], fib]
{fib[3], fib[3 - 1] + fib[3 - 2], {fib[2], 1}, {fib[1], 1}}

Remove["Global`*"]
```

17.2.2 Debugging

■ Introduction

In debugging programs, tracing can be useful. However, we also have a special *Debugger*. The idea of the debugger is that we select, from the program, a set of *breakpoints*—that is, points where the program stops so that we can see the current status of the computation. In this way, we can see, step by step, how the values of variables used in the program change, and this can help us in debugging the program.

■ An Example

As an example, we write a procedural program that, from a given set **a**, chooses a given number **n** of elements without replacement:

```
SWOR[a_List, n_Integer] /; 0 < n ≤ Length[a] :=
 Module[{aa = a, b = Length[a], pos, sample = {}},
  Do[pos = Random[Integer, {1, b}];
    sample = {sample, aa[[pos]]};
    aa = Delete[aa, pos];
    b = b - 1, {n}];
   Sort[Flatten[sample]]]
```

Here, we wrote a **Do** loop to pick the elements in turn. The variable **aa** contains the list of currently available elements and **b** its length. At each step, we choose a random position from the remaining elements, add the corresponding element to the sample (which initially is an empty list), delete this element from the list of available elements, and subtract one from the length of this list. After doing this loop, we output the sample as a flattened and sorted list.

For example, from the numbers 0, 1, …, 9, we choose, without replacement, five numbers:

```
SeedRandom[1];
SWOR[Range[0, 9], 5]        {2, 4, 6, 7, 8}
```

■ **The Debugger Palette**

Next, we use the *Debugger* to see how the program proceeds. We select **Evaluation** ▷ **Debugger**. The following palette appears:

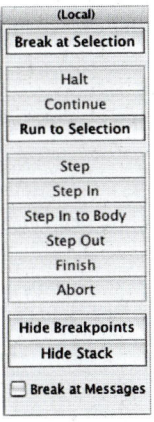

This palette can be used to debug the program. Our program is already debugged, but we can use the *Debugger* to see how the program works. First, we define the breakpoints. To set a breakpoint, select an expression from the program and, on the palette, click *Break at Selection*. In our example, we select all four commands of the **Do** loop as breakpoints. The breakpoints appear, in the program code, inside red frames:

```
SWOR[a_List, n_Integer] /;
  0 < n ≤ Length[a] :=
 Module[{aa = a, b = Length[a], pos,
    sample = {}},
  Do[pos = Random[Integer, {1, b}];
    sample = {sample, aa[[pos]]};
    aa = Delete[aa, pos];
    b = b - 1, {n}];
   Sort[Flatten[sample]]]
```

From the palette, we can click the *Show Breakpoints* to get a list of current breakpoints:

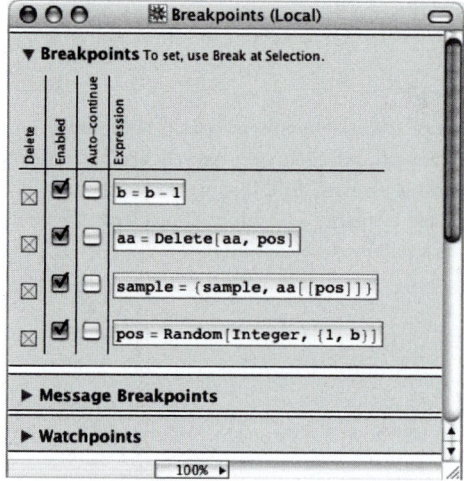

From this list, we can delete, enable, and disable breakpoints.

■ Debugging a Program

Now, execute the program code again and ask a sample from a set:

```
SeedRandom[1];
SWOR[Range[0, 9], 5]
```

We do not get a result since the program stopped on the first breakpoint. The program code now appears as follows:

```
SWOR[a_List, n_Integer] /; 0 < n ≤ Length[a] :=
Module[{aa = a, b = Length[a], pos, sample = {}},
  Do[pos = Random[Integer, {1, b}];
    sample = {sample, aa[[pos]]};
    aa = Delete[aa, pos];
    b = b - 1, {n}];
  Sort[Flatten[sample]]]
```

We can see, on a green background, the expression at which the program broke. Click on the *Show Stack* button on the palette. The following window appears:

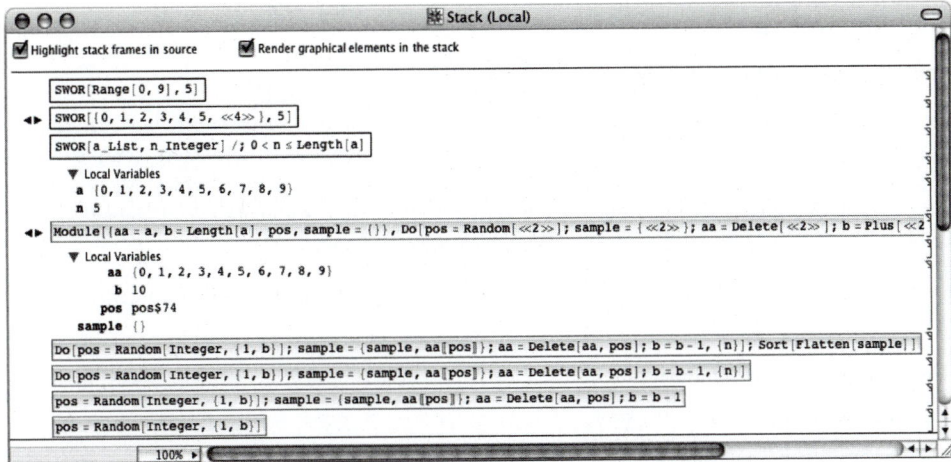

Here we can see the current status of the program. First, we see the command we executed and its local variables **a** and **n**. Then we see program code and—the most important part—the current values of local variables. Thus far, the program has not executed commands in the **Do** loop so that the variable **pos** does not yet have a value. At the bottom of the window we see various parts of the program. They are hyperlinks to the program, and in this way we can easily correct a part of the program.

To advance the program to the next breakpoints, repeatedly click on the *Continue* button on the palette. Observe, from the stack window, how the value of a variable changes each time you click *Continue*: First **pos** gets the value 8, then **sample** the value {{}, 7}, **aa** the value {0, 1, 2, 3, 4, 5, 6, 8, 9} (the element 7 has been removed), and **b** the value 9. Now a step of the **Do** loop is ready. By repeatedly clicking the *Continue* button, we could see how the program proceeds until completion. In this way, we can easily see how the program works and then easily correct possible problem points.

■ More about the Debugger

In the **Debugger** palette, we have the following buttons:

Break at Selection: make the selected expression to a breakpoint
Halt: interrupt the debugger and display the evaluation stack
Continue: resume debugging from one breakpoint to the next breakpoint
Run to Selection: continue until reaching selected expression
Step: stop at the beginning of the next expression
Step In: stop at the next possible stopping point
Step In to Body: step over arguments into body of function
Step Out: stop after exiting current stack frame
Finish: run through the entire evaluation, ignoring any breakpoints
Abort: exit evaluation without finishing
Show/Hide Breakpoints: toggles the breakpoints window on and off
Show/Hide Stack: toggles the stack window on and off

Note that with **Evaluation ▷ Debugger Controls**, we can do most of these tasks.

At the bottom of the *Debugger* palette, we can ask to break at all messages. In the *Breakpoints* window, we also have the possibility to ask the program to stop at specific messages.

17.2.3 Compiling

■ **Defining a Compiled Function**

With compiling, we can speed up functions and programs. Normally, *Mathematica* must handle many different kinds of arguments for functions (e.g., numbers, symbols, lists, and more complex expressions), and this makes the evaluation process slower than in a situation in which *Mathematica* can assume that all arguments are numbers. In compiled functions, the arguments are, by default, assumed to be real numbers, and this simplifies the evaluation process and speeds it up.

> **f = Compile[{x, y, … }, expr]**   Create a compiled function **f** to evaluate **expr** for numerical values of the real variables **x, y**, …

Here is a simple compiled function:

```
f1 = Compile[{x}, x + 7.]
CompiledFunction[{x}, x + 7., -CompiledCode-]
```

The result of compiling is an object called `CompiledFunction`. The compiled code is not shown; `-CompiledCode-` represents it. We try the compiled function:

```
f1[10]    17.
```

With **InputForm**, we can see the compiled code:

```
f1 // InputForm
CompiledFunction[{_Real},
  {{3, 0, 0}, {3, 0, 2}}, {0, 0, 3, 0, 0},
  {{1, 5}, {8, 7., 1}, {18, 0, 1, 2},
    {2}}, Function[{x}, x + 7.], Evaluate]
```

The code is not readable for us, but for *Mathematica* it gives clear instructions about how to effectively calculate the value of the function. For example, the instructions specify the types of arguments; the version of the compiler; the numbers of needed logical, integer, real, complex, and tensor registers; and what to load into these registers when the calculation proceeds. Thus, *Mathematica* compiles the function to form a kind of pseudocode that contains simple instructions for evaluating the compiled function. At the end of the code, we see the function as a pure function. This pure function is used if for some reason the compiled code cannot be used.

As another example, suppose that we have to calculate the value of the seventh Chebyshev polynomial many times. To speed up the computations, we form a compiled function:

```
ch = Compile[{x}, ChebyshevT[7, x] // Evaluate]
```

$$\text{CompiledFunction}\left[\{x\},\ -7\,x + 56\,x^3 - 112\,x^5 + 64\,x^7,\ \text{-CompiledCode-}\right]$$

To get the explicit expression, we needed **Evaluate**. By comparing execution times, we see that the compilation really speeds up the computation:

```
Do[ch[3.], {10^5}] // Timing    {0.471302, Null}
```

```
Do[ChebyshevT[7, 3.], {10^5}] // Timing    {1.56084, Null}
```

Compile handles numerical functions, list manipulation functions, matrix operations, procedural and functional programming constructs, and so on. Compiling is primarily designed for heavy numerical computations with machine-precision numbers.

Note, however, that with compiled code we generally cannot obtain commands that are faster than the built-in commands of *Mathematica*. Also, a compiled code does not handle numerical precision in the same way as ordinary *Mathematica* code and may use somewhat restricted algorithms. Thus, use the built-in commands as much as possible, but remember compilation as a way to speed up your own heavy numerical programs.

Note that many built-in *Mathematica* commands, such as **NIntegrate** or **NDSolve**, automatically compile the function to be processed. If we do not want to compile the function, we can set **Compiled →** **False**.

■ Integer, Complex, and Logical Arguments

The default is that all arguments of compiled functions are real numbers. We can also specify integer, complex, and logical arguments.

> **f = Compile[{{x1, t1}, {x2, t2}, … }, expr]** Arguments **x1**, **x2**, … are of types **t1**, **t2**, …; allowed
> types: **_Integer, _Real, _Complex, True|False**

The basic compiled function **Compile[{x, y, … }, expr]** considered previously is the same as the following:

> **Compile[{{x, _Real}, {y, _Real}, … }, expr]**

Here is a compiled function with one real and one integer argument (the default real type need not be declared):

> **f2 = Compile[{x, {n, _Integer}}, x^n]**
>
> CompiledFunction$\left[\{x, n\}, x^n, \text{-CompiledCode-}\right]$
>
> **Table[f2[1.5, n], {n, 4}]**
>
> {1.5, 2.25, 3.375, 5.0625}

What if the value of **n** was 1751?

> **f2[1.5, 1751]**
>
> CompiledFunction::cfn :
>   Numerical error encountered at instruction 2; proceeding with uncompiled evaluation. ≫
> $2.166679161861306 \times 10^{308}$

We got an error message because the number exceeded the limit allowed for machine-precision numbers. The maximum machine-precision number is shown here:

> **$MaxMachineNumber** 1.79769×10^{308}

The value of the function was thus calculated with the usual uncompiled methods.

■ Tensor Arguments

> **f = Compile[{{x1, t1, r1}, {x2, t2, r2}, … }, expr]** Arguments **x1**, **x2**, … are tensors, having
> elements of type **t1**, **t2**, … and ranks **r1**, **r2**, …

If an argument of a compiled function is a list, matrix, or, in general, a tensor, we define the type of its elements and also its rank. The rank of a tensor is the number of indices needed to specify each element. So the rank of a vector is 1 and that of a matrix is 2.

We write a compiled function to calculate the product of a matrix and a vector:

```
prod = Compile[{{a, _Real, 2}, {b, _Real, 1}}, a.b]
CompiledFunction[{a, b}, a.b, -CompiledCode-]

prod[{{6, 8, 3}, {5, 1, 7}, {4, 2, 9}}, {2, 3, 1}]
{39., 20., 23.}
```

■ **Types of Subexpressions**

> **f = Compile[vars, expr, {{s1, t1}, {s2, t2}, … }]** Subexpressions that match **s1, s2**, … are of
> type **t1, t2**, …

Mathematica knows the type of result from a calculation if the calculation only contains standard arithmetic operations. Thus, the result of $2 + 3$ is an integer, and the result of $2.0 + 3$ is a real number. If we use other functions, *Mathematica* may not know the type of an intermediate result and has to assume that the result is a real number. If we know that the result is not a real number, we can give this information at the end of the definition of the compiled function.

In the following example, the compiled code is not even used unless we declare that the result of **Eigenvalues[_]** is a complex tensor of rank 1 (i.e., a complex vector):

```
f3 = Compile[{{a, _Real, 2}}, Eigenvalues[a.a], {{Eigenvalues[_], _Complex, 1}}];

f3[{{1., 2}, {-4, 1}}]
{-7. + 5.65685 i, -7. - 5.65685 i}
```

17.2.4 Attributes

Sometimes we may want a function to have a special property, such as being able to accept a list as an argument and calculate the value of the function for each element of that list. We have observed that the built-in functions have this property. For example,

```
Sin[{-2, -1, 0, 1}]     {-Sin[2], -Sin[1], 0, Sin[1]}
```

For this reason, the function **Sin** is said to have the **Listable** property or *attribute*. This can also be seen by asking for the attributes of **Sin**:

```
Attributes[Sin]
{Listable, NumericFunction, Protected}
```

If you form your own function, it may not necessarily be listable. The function **f** in the following example is not listable:

```
f[x_] := If[x < 0, Cos[x], Sin[x]]

f[{-2, -1, 0, 1}]
If[{-2, -1, 0, 1} < 0, Cos[{-2, -1, 0, 1}], Sin[{-2, -1, 0, 1}]]
```

One possibility for overcoming this problem is to use **Map**:

```
Map[f, {-2, -1, 0, 1}]     {Cos[2], Cos[1], 0, Sin[1]}
```

We can also assign the **Listable** attribute to **f**:

```
SetAttributes[f, Listable]
```

Now we can give a list as an argument:

```
f[{-2, -1, 0, 1}]     {Cos[2], Cos[1], 0, Sin[1]}
```

> **Attributes[f]** Give the attributes of **f**
>
> **SetAttributes[f, attr]** Add **attr** to the attributes of **f**
>
> **ClearAttributes[f, attr]** Remove **attr** from the attributes of **f**

The different attributes of the built-in commands are as follows (in addition, we have the **Stub** attribute):

```
Attributes /@ Names["*"] // Flatten // Union
```
```
{Constant, Flat, HoldAll, HoldAllComplete, HoldFirst, HoldRest,
 Listable, Locked, NHoldAll, NHoldFirst, NHoldRest, NumericFunction,
 OneIdentity, Orderless, Protected, ReadProtected, SequenceHold, Temporary}
```

We do not explain all of the attributes; for more information, read tutorial/Attributes. We explain only the **HoldAll** attribute. Many plotting commands have this attribute:

```
Select[Names["*Plot*"], MemberQ[Attributes[#], HoldAll] &]
```
```
{ContourPlot, ContourPlot3D, DensityPlot, LogLinearPlot,
 LogLogPlot, LogPlot, ParametricPlot, ParametricPlot3D, Plot, Plot3D,
 PolarPlot, RegionPlot, RegionPlot3D, RevolutionPlot3D, SphericalPlot3D}
```

This means that the expression to be plotted is not evaluated first but only at specific numerical sampling points as the plotting proceeds. On the other hand, we can force the expression to be evaluated first by enclosing the expression with **Evaluate**.

The **HoldAll** attribute is also a property of numerical commands such as **FindMinimum**, **FindRoot**, and **NIntegrate**, and so the expression is not evaluated first but only at the sampling points. In addition, iteration commands such as **Product**, **NProduct**, **Sum**, **NSum**, **Table**, and **Do** have the **HoldAll** attribute. This again means that the expression to be processed is evaluated only at the specific values of the iteration variable.

17.3 Contexts and Packages

17.3.1 What Is a Context?

■ **Four Basic Contexts**

Simply stated, contexts are like directories. They enable us to arrange material hierarchically and to keep content from blending. Normally, we do not need to concern ourselves with contexts; *Mathematica* takes care of them. However, in some cases contexts help us understand the way *Mathematica* works. Also, in writing packages, we need contexts.

We start a new session and ask what contexts we have right now:

```
$ContextPath
```
```
{PacletManager`, WebServices`, System`, Global`}
```

The **System`** context contains *Mathematica*'s built-in definitions and the **Global`** context contains our own definitions; in addition, we have the **PacletManager`** and **WebServices`** contexts. Here are the numbers of symbols these contexts contain:

```
Length[#] & /@ {Names["System`*"],
  Names["PacletManager`*"], Names["WebServices`*"], Names["Global`*"]}
{2964, 37, 9, 0}
```

The **Global`** context currently contains no symbols because thus far we have defined no symbols.

- **Complete Names**

We define some symbols:

```
a = 5; b = 8;
```

Now the **Global`** context contains these symbols:

```
Names["Global`*"]     {a, b}
```

We can ask for information about **a**:

```
? a
```

```
Global`a

a = 5
```

We see the context where **a** is defined and we also see its value. In fact, we see the complete name of the symbol **a**, which is **Global`a**. In this way, all symbols have complete names that indicate the contexts in which they are defined. We can use the complete names if we want:

```
{Global`a, System`Exp[1]}     {5, e}
```

However, this is normally not necessary.

- **Adding Contexts**

Then we load a package:

```
<< PhysicalConstants`
```

The list of contexts is now as follows:

```
$ContextPath
{PhysicalConstants`, Units`, PacletManager`, WebServices`, System`, Global`}
```

(Typically, new contexts are added to the beginning of the list of contexts.) Basically, packages have their own contexts: We have the new **PhysicalConstants`** context and, in addition, the **Units`** context. Indeed, a package may need some other packages, and so they are also loaded. The package of physical constants contains 51 symbols:

```
Names["PhysicalConstants`*"] // Short
{AccelerationDueToGravity, «49», We … gle}
```

We can ask for the context of a particular name:

```
{Context[Integrate], Context[a], Context[IcePoint]}
{System`, Global`, PhysicalConstants`}
```

- **Commands Relating to Contexts**

$ContextPath   Give a list of all currently existing contexts
$Context   Give the current context
Context[name]   Give the context of **name**

Contexts[]   Give a list of all contexts in *Mathematica*
Contexts["string"]   Give a list of contexts that match the given string

---

`Names[]`   Give a list of all symbols in all contexts of *Mathematica*

`Names["*"]`   Give a list of all symbols in all currently existing contexts

`Names["cont`*"]`   Give a list of all symbols in context **cont**

`?cont`*`   Give a table of all symbols in context **cont**

`Remove["Global`*"]`   Remove all user-defined symbols

■ Searching for a Name in the Contexts

Mathematica always has a current context. This is the context in which we are currently working and in which all new names are stored. Normally, it is **Global`**:

   `$Context`   `Global``

Within a package, the current context changes, as we will see in Section 17.3.3, p. 535.

When we have entered a short name (i.e., a name without the context), *Mathematica* searches the name from the contexts in the list **$ContextPath** from left to right, but first from the current context **$Context**. If the entered name is found, its value or definition is used. If the name is not found, the name is created in the current context. (If you enter a complete name—that is, a name with the context—then *Mathematica* searches for the name only in the given context.)

For example, write the following:

   `sin[π]`   `sin[π]`

Mathematica could not find the name **sin** anywhere and so creates it. From now on, the name **sin** stays in the **Global`** context:

   `Names["Global`*"]`   `{a, b, sin}`

You did not intend to create a new name **sin**, but *Mathematica* had no other choice. If you want to remove **sin**, you can write **Remove[sin]**. If you want to remove all user-defined symbols, write the following:

   `Remove["Global`*"]`

■ Notebook-Specific Contexts

If we work simultaneously with several notebooks, we may want to use the same symbols in several notebooks. For example, suppose we would like to define **a = 1** in the first notebook and **a = 2** in the second notebook. Normally, each symbol has the value most recently defined. Thus, if we first execute **a = 1** and then **a = 2**, the variable **a** also has the value **2** if we do a calculation in the first notebook.

With **Evaluation ▷ Notebook's Default Context ▷ Unique to This Notebook** we can ask to use a unique default context for a notebook. If we do this for all open notebooks, we can use the same symbols in the notebooks, and the values of the symbols are notebook-specific. Thus, **a** has the value **1** only in the first notebook and the value **2** in the second notebook.

17.3.2 Forgetting to Load: Once Again

■ Problem

What to do when you forget to load a package was addressed in Section 4.1.2, p. 96, but let us now see what actually happens. We start a new session, and we want to ask the value of **IcePoint** but have forgotten to load the **PhysicalConstants`** package:

`IcePoint`   IcePoint

Mathematica searched the name **IcePoint** from all contexts, could not find such a name, and so created the name in the context **Global`**:

`Names["Global`*"]`   {IcePoint}

Next we load the package in the hope that we can use the command:

`<< PhysicalConstants``

IcePoint::shdw :
   Symbol IcePoint appears in multiple contexts {PhysicalConstants`, Global`}; definitions in
      context PhysicalConstants` may shadow or be shadowed by other definitions. ≫

We get a warning message and **IcePoint** becomes red.

■ **Shadowing**

The previous message warns about multiple **IcePoint**: We have an **IcePoint** in the **PhysicalConstants`** context and another in the **Global`** context:

`Names["*`IcePoint"]`    {Global`IcePoint, IcePoint}

One of the multiple names *shadows* the other, which means that *Mathematica* has to use one of the names and disregard the other. Indeed, the name in the **PhysicalConstants`** context seems to shadow the name in the **Global`** context (in earlier versions of *Mathematica*, the situation was the reverse) because **IcePoint** now, after loading the package, works:

`IcePoint`   273.15 Kelvin

   However, the red color stays with **IcePoint**. To resolve the situation, we remove our own, unintentionally created symbol:

`Remove[Global`IcePoint]`

Now the red color changes to the normal black color and shadowing has disappeared.

■ **Summary**

Let us now discuss how to proceed if we forget to load a package. We again start a new session and try to ask the value of **IcePoint**:

`IcePoint`   IcePoint

We observe that we have forgotten to load the package. Now we simply remove our own **IcePoint**:

`Remove[IcePoint]`

(Now we do not need to write **Remove[Global`IcePoint]** because we have only one **IcePoint**.) Then we load the package and use **IcePoint**:

`<< PhysicalConstants``

`IcePoint`   273.15 Kelvin

   Here is a summary:

If you forget to load a package before using one of its commands,
• remove the name you have tried to use;
• load the package; and
• use the command of the package again.

However, note that even without removing the name, the package works. The two minor drawbacks of not removing the name are that the loading of the package generates a warning message and the name remains red.

Another solution is to quit the kernel from **Evaluation ▷ Quit Kernel** and then restart the kernel from **Evaluation ▷ Start Kernel** (or simply by executing a command), but then you may need to do some calculations again. Loading packages was considered in Section 4.1.

17.3.3 Writing a Package

■ **Forming a Simple Package**

You may have developed some useful functions or programs that you may need later. A simple way to reuse them is to open the notebook that contains them and execute the cells that contain the definitions of the programs. Another way is to save the useful programs as a package; to use the programs later, you only need to load in the package. To create a package, do as follows:

- Select **File ▷ New ▷ Package**. In the package file, write useful functions that you intend to use later. These cells automatically are so-called *initialization cells*. You can also copy, from an existing notebook, useful functions and paste them into the package file. If you copy the *content* of a cell (by selecting the content with the mouse), then the cell created by pasting will automatically be an initialization cell. On the other hand, if you copy a whole cell (by selecting the cell bracket), then set the pasted cell, in the package file, as an initialization cell by choosing the cell bracket and then choosing **Cell ▷ Cell Properties ▷ Initialization Cell**.

- Save the document as a *Mathematica* package by choosing **File ▷ Save As...**. In the opening dialog, give the file a suitable name that ends with **.m** and choose a location you prefer. Install the package in the correct location by choosing **File ▷ Install...**. A dialog opens. Set *Type of Item to Install* to be *Package*; in *Source*, select the package you saved; in *Install Name*, give a suitable name for the package; and click *Finish*.

Suppose the name of the notebook saved is **programs.m**. To load the package and use the functions defined in it, write **<<programs`**. In this way, you can create notebooks that you can use like packages.

What is an initialization cell? This is a cell that, when the file is read in, is automatically executed. In this way, the definitions of the programs in the package are automatically executed when the package is loaded.

The procedure previously described creates a file that can be used like a package. However, a true package has a special structure that we now begin to study.

■ **An Example of a Package**

A package has a special structure in which contexts play a key role. In a package, we use the following commands:

```
BeginPackage["packageName"]  Begin a package
Begin["`Private`"]  Begin the code of a package after usage messages
End[]  End the code of a package
EndPackage[]  End the package
```

To give an elementary example, we have written the following code in a new notebook:

```
BeginPackage["Own`newton`"]

newton::usage = "newton[f,x,x0,n] calculates a
   zero of f starting from x0 and using at most n iterations"

Begin["`Private`"]
(* one step *)
newtonStep[f_, df_, x_, xi_] := (x - f / df) /. x -> N[xi]
(* iterate the step *)
newton[f_, x_, x0_, n_] :=
 With[{df = D[f, x]}, FixedPointList[newtonStep[f, df, x, #] &, N[x0], n]]
End[]

EndPackage[]
```

This is a single cell that we have set as an initialization cell (**Cell** ▷ **Cell Properties** ▷ **Initialization Cell**), and we have saved the notebook, by **File** ▷ **Save As**, with the name **newton.m** as a *Mathematica* package (select this from "Format" in the save dialog) into a new folder called **Own** in the *Mathematica* **Applications** folder. To try the package, we first load it:

```
<< Own`newton`
```

If we have already forgotten how **newton** is used, we can ask for information:

```
? newton
```

newton[f,x,x0,n] calculates a zero of f starting from x0 and using at most n iterations

We calculate a zero of an expression, starting from point 2:

```
newton[3 x^3 - E^x, x, 2, 20]
```

{2., 1.41942, 1.1019, 0.975117, 0.953089, 0.952446, 0.952446, 0.952446, 0.952446}

■ **The Structure of a Package**

The definition of the package begins with **BeginPackage**, in which the package name is the argument. If the package needs other packages, these packages can also be mentioned as arguments of **BeginPackage** (an example is given later).

Also coming into play are the usage messages. These messages give information about the usage of the various functions. This information is displayed when the user writes **?name**, where **name** is a name defined in the package. *The user of the package can only use the names for which a usage message exists.* In this way, you can restrict the set of names available for use. In the previous example, we did not define a usage message for the function **newtonStep**, and so this function cannot be used (it is used only within the package).

The program begins with the command **Begin["`Private`"]** and ends with the command **End[]**. Note that *Mathematica* commands can be annotated by inserting comments into the code. A comment starts with **(*** and ends with ***)**; a comment can be placed anywhere. Comments are especially useful in packages and in other longer codes to help with the reading of the code. The whole package ends with the command **EndPackage[]**.

The commands **BeginPackage**, **Begin**, **End**, and **EndPackage** affect the contexts. This topic is discussed next.

■ Contexts in Packages

Packages are normally loaded or read in by << in one step, and we do not see what actually happens during the loading. Let us now investigate what happens when we load a package. We specifically discuss how the context changes during the reading. We start a new session and first load a package:

```
<< ComputerArithmetic`
```

Then we consider what happens when we load the following package:

```
BeginPackage["Own`newton`", "NumericalCalculus`"]
newton::usage = "newton[f,x,x0,n] calculates a
    zero of f starting from x0 and using at most n iterations"
Begin["`Private`"]
newtonStep[f_, df_, x_, xi_] := (x - f / df) /. x → N[xi]
newton[f_, x_, x0_, n_] :=
  With[{df = D[f, x]}, FixedPointList[newtonStep[f, df, x, #] &, N[x0], n]]
End[]
EndPackage[]
```

This is almost the same example as before. Now, however, we only assumed that we also need the package **NumericalCalculus`** (although we actually do not need it), so this package is mentioned in **BeginPackage**. We proceed step by step and observe how **$Context** (the current context) and **$ContextPath** (a list of contexts from which information is searched) change from the initial state (before loading the package) to the state after loading. The current context and context path change after the commands **BeginPackage**, **Begin**, **End**, and **EndPackage**:

```
{$Context, $ContextPath}
{Global`, {ComputerArithmetic`, PacletManager`, WebServices`, System`, Global`}}

BeginPackage["Own`newton`", "NumericalCalculus`"];
{$Context, $ContextPath}
{Own`newton`, {Own`newton`, NumericalCalculus`, System`}}

Begin["`Private`"];
{$Context, $ContextPath}
{Own`newton`Private`, {Own`newton`, NumericalCalculus`, System`}}

End[];
{$Context, $ContextPath}
{Own`newton`, {Own`newton`, NumericalCalculus`, System`}}

EndPackage[];
{$Context, $ContextPath}
{Global`, {Own`newton`, NumericalCalculus`,
   ComputerArithmetic`, PacletManager`, WebServices`, System`, Global`}}
```

With regard to the current context, initially it is the usual **Global`**, where all user-defined names are stored. When the reading of the package begins, the context changes to **Own`newton`**. Thus, all names defined in the package are stored in this context. In particular, the names with a usage message are in this context. When the ordinary program begins, the current context gets a subcontext called **Private`**. Names defined in this context without usage messages are not available to the user. After the program ends, the current context changes back to **Own`newton`**; when the whole package is read in, we are again in the **Global`** context.

From the context paths, we see that several things happens when **BeginPackage** is read. First, the **ComputerArithmetic`** package that was loaded before **Own`newton`** disappears and thus is not available for the package. Consequently, a package can only use the packages declared in **BeginPackage**. The **PacletManager`** context also disappears from the context path.

Second, the **Global`** context disappears from the context path; this means that the names in **Global`** are not available in the package. This is safe: We cannot accidentally use or change, within the package, the values of names in **Global`**. In fact, we can have the same names in **Global`** as we have in the package. After loading the package, the names in **Global`** still have their old values and not the values defined in the package.

Third, the **Own`newton`** and **NumericalCalculus`** contexts are added to the context path so that new definitions made in the package are placed in the former context and the latter context makes available the commands in the corresponding package.

After **EndPackage**, we have access to the contexts that were available before reading the package and to the contexts **Own`newton`** and **NumericalCalculus`**.

Maeder (1997) is an excellent source of information about programming and package development.

17.3.4 Handling Options and Messages

■ **Handling Options**

Now we want to add some options to the **newton** program presented in Section 19.3.3. Let the new program be **newton2**. For example, we want to be able to call this program as follows:

```
newton2[x^3, x, 2, 20, dampingConstant → 3]
```

To write such a program, we can use the following commands:

Options[progr] = {opt1 → val1, opt2 → val2, …}   Define that the options of program or function **progr** are **opt1, opt2**, … with default values **val1, val2**, …

OptionsPattern[] (❀6)   Represents a collection of options given as rules; the values of the options can be accessed using **OptionValue**

OptionValue[opt] (❀6)   Give the value of option **opt** in options represented by **OptionsPattern**

In the following package, we have used these commands:

```
BeginPackage["Own`newton2`"]
newton2::usage = "newton2[f,x,x0,n,opts] calculates
   a zero of f starting from x0 and using at most n iterations."
dampingConstant::usage = "dampingConstant -> d is an option for newton2 that
      gives the damping factor. Default value: 1. For example, if the zero is of
      multiplicity 2, define dampingConstant -> 2 to accelerate the convergence."
stoppingCriterion::usage = "stoppingCriterion -> (pure function) is an option
      for newton2 that gives the stopping criterion for the iteration. Default
      value: (#1 === #2&). Examples of values: (Abs[#1 - #2] < 10^-8 &), (Abs[f /.
      x->#2] < 10^-6 &). Here #1 is the next to last point and #2 the last point."

Begin["`Private`"]
Options[newton2] = {dampingConstant → 1, stoppingCriterion → (#1 === #2 &)}
newtonStep2[f_, df_, x_, d_, xi_] := (x - d f / df) /. x → N[xi]
newton2[f_, x_, x0_?NumericQ, n_Integer, OptionsPattern[]] :=
  Module[{df = D[f, x]},
    FixedPointList[newtonStep2[f, df, x, OptionValue[dampingConstant], #] &,
    N[x0], n, SameTest → OptionValue[stoppingCriterion]]]
End[]
EndPackage[]
```

Here, we used the three commands declared previously:

- We defined, with **Options**, that the function **newton2** has the two options **dampingConstant** and **stoppingCriterion**, with the default values **1** and **(#1 === #2 &)**, respectively.
- The last argument of **newton2** is **OptionsPattern[]**. This allows us to call the function with some options.
- Inside **newton2**, we use the values of the options with **OptionValue[dampingConstant]** and **OptionValue[stoppingCriterion]**.

■ Example

We defined the single cell containing the program **newton2** to be an initialization cell (**Cell ▷ Cell Properties ▷ Initialization Cell**). Then we saved the notebook, by **File ▷ Save As**, with the name **newton2.m** as a *Mathematica* Package (we selected this from "Format" in the save dialog) into a folder **Own** in the *Mathematica* **Applications** folder. To try the package, we start a new session and first load the package:

```
<< Own`newton2`
```

We can ask for information about the package:

```
Names["Own`newton2`*"]
```

```
{dampingConstant, newton2, stoppingCriterion}
```

```
? dampingConstant
```

---

dampingConstant –> d is an option for newton2 that
    gives the damping factor. Default value: 1. For example, if the zero is of
    multiplicity 2, define dampingConstant –> 2 to accelerate the convergence.

We can use **newton2** with no options, with one option, or with two options:

```
newton2[x^2, x, 2, 10]
```

```
{2., 1., 0.5, 0.25, 0.125, 0.0625,
 0.03125, 0.015625, 0.0078125, 0.00390625, 0.00195313}
```

```
newton2[x^2, x, 2, 10, dampingConstant → 2]
```

```
{2., 0, 0}
```

```
newton2[x^2, x, 2, 10, dampingConstant → 3,
 stoppingCriterion → (Abs[#1 - #2] < 10^-6 &)]
```

```
{2., -1., 0.5, -0.25, 0.125, -0.0625,
 0.03125, -0.015625, 0.0078125, -0.00390625, 0.00195313}
```

If we call the program with arguments that do not satisfy the given restrictions (e.g., **n** has to be an integer), the program does nothing:

```
newton2[3 x^3 - E^x, x, 2, 20.]
```

$$\text{newton2}\left[-e^x + 3\,x^3,\ x,\ 2,\ 20.\right]$$

■ Filtering Options

In a program, you may use several commands with options. The question then arises as to how to pick the suitable options for the various commands. This can be done with **FilterRules**.

FilterRules[opts, Options[command]] (⌘6) From options **opts**, pick options of **command**

For example, suppose your program uses **FindRoot** and **Plot**. Let **opts** contain all of the options given by the user of your program; suppose that

```
opts = {DampingFactor → 2, AspectRatio → Automatic};
```

From this list, you can pick the options belonging to **FindRoot** and the options belonging to **Plot** by entering the following:

```
frOpts = FilterRules[opts, Options[FindRoot]]
```
{DampingFactor → 2}

```
plOpts = FilterRules[opts, Options[Plot]]
```
{AspectRatio → Automatic}

In your program you could then write, for example,

```
FindRoot[x^2, {x, 1}, Evaluate[frOpts]]
```
{x → 0.}

```
Plot[Sin[x], {x, 0, 2 Pi}, Evaluate[plOpts]]
```

■ Handling Messages

A well-designed package prints messages if problems are encountered during the execution of the functions in the package. Messages can be printed with **Print** (see Section 18.2.3, p. 562), but the use **Message** is a better method.

---

program::messageName = " … `` `1` `` … `` `2` `` … "   For **program**, define a message with the name **messageName** as a string that possibly contains slots for the values of some variables

Message[program::messageName, val1, val2, …]   Print the given message of **program**, inserting the given values into the slots of the message

Quiet[expr] (❄6)   Do not print any messages possibly generated during the evaluation of **expr**

Check[expr, failexpr]   Evaluate and return **expr**, unless messages were generated, in which case evaluate and return **failexpr**

Suppose an algorithm does not converge for the given values of **maxit** and **eps**. For this situation, the package contains, after the usage messages, the following message template:

```
myPackage::nonconv =
  "The algorithm did not converge with maxit = `1` and eps = `2`"
```

The program also has, in the correct place, this command:

```
Message[myPackage::nonconv, maxit, eps]
```

The result is a message such as the following (if **maxit** and **eps** are 50 and 0.0001):

myPackage::nonconv: The algorithm did not converge with maxit = 50
and eps = 0.0001`

18

Programs

Introduction

> There is a square room of side twenty feet with a pure mathematician in one corner and an applied mathematician in the opposite corner. In a third corner is a delicious apple. The mathematicians are allowed to approach the apple in bounces along the sides of the square, the first bounce a maximum of ten feet and every subsequent bounce a maximum of half the previous bounce. The pure mathematician, well versed in limits, quickly calculates that no matter how many bounces he takes he can never reach the apple, so he doesn't even begin to bounce. The applied mathematician sets off at once because he realizes that after five or six bounces he will be close enough to take the apple.

Although *Mathematica* has a great many ready-to-use commands for almost all kinds of mathematical problems, occasionally in this book we present some short programs for doing calculations we find interesting, pedagogically worthwhile, or sometimes even practically useful. The same reasons may give you motivation to study programming with *Mathematica*. Although *Mathematica* is a kind of interactive calculator that does calculations step by step, by combining the steps into one or a few logical blocks, a program may make the calculation vastly more effective and even simpler.

Mathematica supports many styles of programming, such as *procedural*, *functional*, *rule-based*, and *recursive*. In addition, the use of graphics primitives and directives leads us to *graphics* programming (see Chapter 6). Even *object-oriented* programming can be approached with *Mathematica*. Functional and rule-based programming make up the heart of programming in *Mathematica*. Let us briefly introduce the main programming styles.

Programs built with traditional programming languages such as Fortran and C are called procedures, and from here comes the term procedural programming. *Mathematica* also has similar commands, such as **For**, **While**, and **If**, as are used in procedural programming, and so we can program in the procedural style with *Mathematica*.

With functional programming, we apply functions to arguments. The functions can be built-in functions or functions we have defined, and they can be applied in a nested way. The functions are applied to the arguments by special powerful commands, such as **Map**, **Apply**, **Nest**, **FixedPoint**, and **Fold**. Functional programming is effective especially in list manipulations and iterative calculations.

Rule-based programming uses *rules* and *patterns*. A function definition **f[x_] := expr** is an example of a *global rule:* Whenever **f** is encountered with a specified argument, for example, **a**, this rule replaces **f[a]** with the value of **expr**, where **x** is replaced with **a**. In **expr /. x → a** we apply a *local rule:* Whenever **x** is encountered in **expr**, replace it with **a**. The argument **x_** of a function is an example of a pattern. The pattern **x_** is very general and is, in fact, matched by anything; the name of the pattern is **x**. We can form more restrictive patterns in many ways. Generally, in rule-based programming we give several rules for the same function. These rules cover several different situations or several patterns of argument.

Recursive programming naturally arises by programming recursive mathematical formulas: The program calls itself with another argument. Many list manipulation tasks can also be programmed recursively: A transformation is made for the list until the list no longer changes.

However, before discussing procedural, functional, rule-based, and recursive programming, we first present examples of simple programs where we do not need any special programming commands but use familiar commands such as **Table**, **N**, **/.**, and the many list manipulation commands. Remember also that with **Manipulate** and **Dynamic** we can get interactive applications (see Chapters 12 and 13).

For more information about programming, see Gray (1997), Maeder (1997), Trott (2004b), and Wellin, Gaylord, and Kamin (2005).

18.1 Simple Programming

18.1.1 Numerical Methods

Before discussing various special programming styles, let us present some examples of simple programs in which we do not use any special programming commands but, rather, familiar commands such as **/.** (see Section 13.1.2, p. 416), **Table** (see Section 14.1.1, p. 445), and various list manipulation commands. First, we consider examples from numerical analysis.

■ Approximating a Derivative

Because the derivative $f'(a)$ is defined to be $\lim_{h\to 0}\left[f(a+h)-f(a)\right]/h$, an approximation of the derivative is $\left[f(a+h)-f(a)\right]/h$ for a small h. We write a program for this approximation:

```
der[f_, x_, a_, h_] := (N[f /. x → a + h] - N[f /. x → a]) / h
```

Note that we have used **N** to calculate the decimal value of the approximation because in numerical calculations we are not interested in "exact approximations." Furthermore, the results obtained when calculating with exact quantities grow easily to huge expressions, in iterative calculations especially, and the computation time may become very long. Thus, use decimal numbers from the start in a numerical program.

We try the program and also calculate the true derivative:

```
f = x Sin[x];
```

```
{der[f, x, 1, 10^-4], df = D[f, x] /. x → 1.}
{1.38179, 1.38177}
```

The approximation is quite good.

Next, we use several values of h and investigate the error:

```
t1 = Table[{h = 10.^-n, d = der[f, x, 1, h], Abs[d - df]}, {n, 1, 15}];
```

```
TableForm[t1,
  TableHeadings → {None, {"h", "approx.", "error"}}, TableSpacing → {1, 3}]
```

h	approx.	error
0.1	1.38857	0.00679782
0.01	1.38292	0.00114453
0.001	1.38189	0.000119056
0.0001	1.38179	0.0000119516
0.00001	1.38177	1.19562×10^{-6}
$1. \times 10^{-6}$	1.38177	1.1946×10^{-7}
$1. \times 10^{-7}$	1.38177	1.28788×10^{-8}
$1. \times 10^{-8}$	1.38177	5.995×10^{-9}
$1. \times 10^{-9}$	1.38177	7.17206×10^{-8}
$1. \times 10^{-10}$	1.38177	2.93765×10^{-7}
$1. \times 10^{-11}$	1.38177	8.16458×10^{-7}
$1. \times 10^{-12}$	1.38189	0.000121308
$1. \times 10^{-13}$	1.38001	0.00176607
$1. \times 10^{-14}$	1.38778	0.00600549
$1. \times 10^{-15}$	1.55431	0.172539

The error is smallest when $h = 10^{-8}$; after that, the error grows due to increasing rounding errors.

We plot the **der** as a function of h for small h:

```
Plot[der[f, x, 1, h], {h, 10^-15, 2×10^-7}, PlotPoints → 2000, ImageSize → 200,
  PlotRange → {1.381773255`10, 1.381773335`10}, AxesOrigin → {0, 1.381773255`10},
  Ticks → {{0, 2.×10^-7}, {1.38177327`9, 1.38177330`9, 1.38177333`9}}]
```

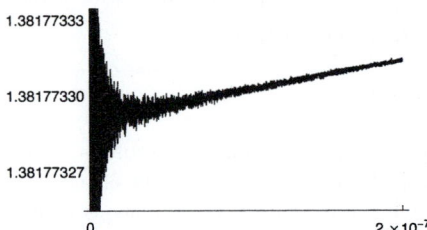

The plot shows the increasing difficulties seen when calculating the approximation as h becomes smaller: The values of **der** vary wildly. The usual fixed-precision decimal numbers simply cannot do better.

If we use arbitrary-precision numbers, we have no problems (a similar example was presented in Section 12.2.3, p. 407):

```
der2[f_, x_, a_, h_] := (N[f /. x → a + h, 20] - N[f /. x → a, 20]) / h
```

```
Plot[der2[f, x, 1, SetPrecision[h, 20]], {h, 10^-15, 2×10^-7}, ImageSize → 200,
  PlotRange → {1.381773255`10, 1.381773335`10}, AxesOrigin → {0, 1.381773255`10},
  Ticks → {{0, 2.×10^-7}, {1.38177327`9, 1.38177330`9, 1.38177333`9}}]
```

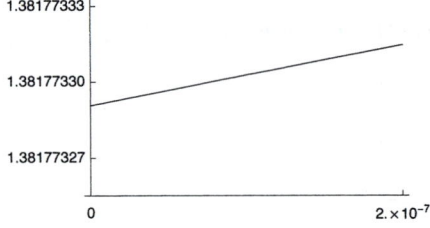

■ Approximating an Integral

The trapezoidal rule to approximate an integral is as follows:

$$\int_a^b f(x)\, dx \simeq \frac{h}{2}\left(f(a) + 2\sum_{i=1}^{n-1} f(a + i\,h) + f(b) \right).$$

Here, $h = (b - a)/n$. We write a program for this rule:

$$\text{trapez}[f\_, x\_, a\_, b\_, n\_] := \text{With}\left[\left\{h = \frac{b-a}{n}\right\},\right.$$
$$\left.\frac{h}{2}\left(N[f /. x \to a] + 2\sum_{i=1}^{n-1} N[f /. x \to a + i\,h] + N[f /. x \to b]\right)\right]$$

Using the **With** scoping construct, we made h a local constant (see Section 17.1.4, p. 522). For example,

```
f = x Sin[x];
```

```
{trapez[f, x, 0, 2, 50], int = Integrate[f, {x, 0., 2}]}
{1.7416, 1.74159}
```

The following table shows how the error depends on the number of steps:

```
t2 = Table[{10^n, InputForm[i = trapez[f, x, 0, 2, 10^n]], Abs[i - int]}, {n, 1, 4}];

TableForm[t2,
  TableHeadings → {None, {"n", "approx.", "error"}}, TableSpacing → {1, 3}]
```

n	approx.	error
10	1.741851999412873	0.000260899
100	1.7415936671330097	2.56721×10^{-6}
1000	1.7415911255879262	2.5668×10^{-8}
10 000	1.7415911001766418	2.56675×10^{-10}

```
h =.
```

■ **Approximating a Zero**

Newton's method for solving an equation $f(x) = 0$ uses the following recursion formula by starting from a given point x_0:

$$x_{i+1} = x_i - \frac{f(x_i)}{f'(x_i)}, \ i = 0, 1, \dots$$

A program for it is as follows:

```
newton[f_, x_, x0_, n_] := Module[{newx = x - f/D[f, x], xi = x0},
    Table[xi = N[newx /. x → xi], {n}]]
```

The scoping construct **Module** makes **newx** and **xi** local variables (see Section 17.1.4, p. 521). The starting values of **newx** and **xi** are the right-hand side of the recursion formula and the starting point **x0**. **Table** then does the iteration **n** times and stores the results in a list. At each iteration, the new value of **xi** is computed by inserting the old value of **xi** into **newx**. In this way, **Table** can be used in iterative methods, although we have more powerful functional iteration commands such as **FixedPoint**. For example, define the following function:

```
g = 3 x^3 - E^x;

Plot[g, {x, -1, 2}]
```

One zero seems to be near $x = 1$. We start from $x = 2$ and do seven iterations:

```
zero = newton[g, x, 2, 7]
{1.41942, 1.1019, 0.975117, 0.953089, 0.952446, 0.952446, 0.952446}
```

(Note that the starting point 2 is lacking from the list.) The function is zero very accurately at the last point:

```
g /. x → Last[zero]       4.44089 × 10⁻¹⁶
```

We investigate how the error in the zero and in the value of the function evolve as we do increasingly more iterations. As can be seen from the following table, Newton's method converges quickly:

```
t3 = {Range[7], zero, Abs[zero - Last[zero]], g /. x → zero}ᵀ;

TableForm[t3,
 TableHeadings → {None, {"n", "zero", "error", "g[zero]"}}, TableSpacing → {1, 3}]
```

n	zero	error	g[zero]
1	1.41942	0.466974	4.44462
2	1.1019	0.149457	1.00387
3	0.975117	0.0226711	0.1301
4	0.953089	0.000643299	0.00358769
5	0.952446	5.39696×10^{-7}	3.00737×10^{-6}
6	0.952446	3.80362×10^{-13}	2.11964×10^{-12}
7	0.952446	0.	4.44089×10^{-16}

■ Approximating the Solution of a Differential Equation

Consider solving, with Euler's method, a differential equation $y'(x) = f[x, y(x)]$ with the initial value $y(x_0) = y_0$. The recursive formulas are as follows:

$$x_{n+1} = x_n + h,$$
$$y_{n+1} = y_n + h\, f(x_n, y_n).$$

Here, h is a given step size. We again use **Table** to do the iterations:

```
euler[f_, x_, y_, x0_, y0_, x1_, n_] :=
 Module[{xi = x0, yi = y0, h = N[(x1 - x0) / n]},
  Prepend[Table[{xi, yi} = N[{x + h, y + h f} /. {x → xi, y → yi}], {n}], {x0, y0}]]
```

The solution is calculated from **x0** to **x1** using **n** steps. With **Prepend**, we add the starting point to the list. As an example, we solve the equation $y' = x - y^2$, $y(0) = 1$ in the interval [0, 1] using 10 steps:

```
eu = euler[x - y^2, x, y, 0, 1, 1, 10]

{{0, 1}, {0.1, 0.9}, {0.2, 0.829}, {0.3, 0.780276}, {0.4, 0.749393}, {0.5, 0.733234},
 {0.6, 0.729471}, {0.7, 0.736258}, {0.8, 0.75205}, {0.9, 0.775492}, {1., 0.805354}}

ListLinePlot[eu, Mesh → All]
```

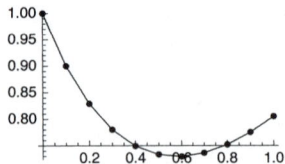

The following is a very accurate value of the solution at $x = 1$:

```
soln = y[1] /. NDSolve[{y'[x] == x - y[x]^2, y[0] == 1}, y, {x, 0, 1},
    WorkingPrecision → 25, PrecisionGoal → 20, AccuracyGoal → 20][[1]]
0.8333833914643544468327539
```

We compare this value with various Euler approximations:

```
t4 = Table[{10^n,
    eu = Last[euler[x - y^2, x, y, 0, 1, 1, 10^n]][[2]], Abs[eu - soln]}, {n, 1, 4}];
TableForm[t4,
 TableHeadings → {None, {"n", "approx.", "error"}}, TableSpacing → {1, 3}]
```

n	approx.	error
10	0.805354	0.0280298
100	0.830746	0.0026377
1000	0.833121	0.000262254
10 000	0.833357	0.0000262104

The number of steps n has to be very large or the step size h has to be very small to get good precision with Euler's method. In fact, this method has almost only pedagogical value.

18.1.2 List Manipulation

Mathematica has very powerful list manipulation commands (see Chapter 14). They enable us to do complicated calculations without any special programming commands. Programming where manipulation of lists is central is sometimes called *list-based programming*.

■ Interleaving Lists

Suppose we want to interleave two lists of the same length. For example, from the lists {1, 2, 3} and {a, b, c}, we would like to get the list {1, a, 2, b, 3, c}. This is easy. First, transpose the matrix having the two lists as rows:

```
{{1, 2, 3}, {a, b, c}}ᵀ    {{1, a}, {2, b}, {3, c}}
```

(Here, the transpose ᵀ can be written as ⎡ESC⎤tr⎡ESC⎤.) Then flatten this list:

```
% // Flatten    {1, a, 2, b, 3, c}
```

Thus, a program is as follows:

```
interleave[list1_, list2_] := Flatten[{list1, list2}ᵀ]
```

An example:

```
interleave[{1, 2, 3}, {a, b, c}]    {1, a, 2, b, 3, c}
```

Note that we also have a ready-to-use command:

```
Riffle[{1, 2, 3}, {a, b, c}]    {1, a, 2, b, 3, c}
```

■ Hamming Distance

Consider two lists of the same length containing only zeros and ones. The Hamming distance between them is the number of element positions having nonmatching elements (see Wellin *et al.*, 2005, p. 107). For example, the Hamming distance between the lists {1, 0, 0, 1} and {0, 1, 0, 1} is 2 because at positions 1 and 2 we have nonmatching elements. First, we form the pairs of elements:

```
{{1, 0, 0, 1}, {0, 1, 0, 1}}ᵀ    {{1, 0}, {0, 1}, {0, 0}, {1, 1}}
```

Then we count the number of pairs that are either {0, 1} or {1, 0}:

```
Count[%, {0, 1} | {1, 0}]    2
```

(With | we can express alternative expressions; see Section 16.1.3, p. 502.) Thus, we can write the following program:

```
hamming1[list1_, list2_] := Count[{list1, list2}ᵀ, {0, 1} | {1, 0}]
```

```
hamming1[{1, 0, 0, 1}, {0, 1, 0, 1}]    2
```

The following is a different approach. First, add the two lists:

```
{1, 0, 0, 1} + {0, 1, 0, 1}    {1, 1, 0, 2}
```

Then divide each element of the sum by 2 and consider the remainder:

```
Mod[%, 2]    {1, 1, 0, 0}
```

We see that the remainder is 1 if and only if the corresponding elements are nonmatching, and the remainder is 0 if the corresponding elements are matching. It then suffices to calculate the sum of the remainders:

```
hamming2[list1_, list2_] := Total[Mod[list1 + list2, 2]]
```

```
hamming2[{1, 0, 0, 1}, {0, 1, 0, 1}]    2
```

To investigate the speed of the two programs, generate two random lists of 10^6 elements and calculate the Hamming distance with both programs:

```
SeedRandom[1];
list1 = RandomInteger[{0, 1}, 10^6];
list2 = RandomInteger[{0, 1}, 10^6];
```

```
hamming1[list1, list2] // Timing    {1.97515, 500 821}
```

```
hamming2[list1, list2] // Timing    {0.106975, 500 821}
```

The speed of the first program is approximately 17 times slower than the speed of the second program.

■ **Finding a Subsequence**

Let us write a program that searches, from a given list of digits, a given sequence of digits. For example, consider the digits of a decimal representation of the number π:

```
N[π, 30]    3.14159265358979323846264338328
```

We would like to check whether the sequence 626 appears somewhere within the decimals. First, extract the digits:

```
RealDigits[%]
{{3, 1, 4, 1, 5, 9, 2, 6, 5, 3, 5, 8,
  9, 7, 9, 3, 2, 3, 8, 4, 6, 2, 6, 4, 3, 3, 8, 3, 2, 8}, 1}
```

Here we have a list of the digits together with the number of digits to the left of the decimal point. Pick the list:

```
seq = %[[1]]
{3, 1, 4, 1, 5, 9, 2, 6, 5, 3, 5, 8, 9, 7, 9, 3, 2, 3, 8, 4, 6, 2, 6, 4, 3, 3, 8, 3, 2, 8}
```

Define the subsequence we are interested in:

```
subseq = {6, 2, 6};
```

To search the subsequence, partition the list of digits into sublists having the same length as the subsequence. Each sublist moves one element to the right (the offset is 1):

```
partitionedSeq = Partition[seq, 3, 1]
```

```
{{3, 1, 4}, {1, 4, 1}, {4, 1, 5}, {1, 5, 9}, {5, 9, 2}, {9, 2, 6}, {2, 6, 5},
 {6, 5, 3}, {5, 3, 5}, {3, 5, 8}, {5, 8, 9}, {8, 9, 7}, {9, 7, 9}, {7, 9, 3},
 {9, 3, 2}, {3, 2, 3}, {2, 3, 8}, {3, 8, 4}, {8, 4, 6}, {4, 6, 2}, {6, 2, 6},
 {2, 6, 4}, {6, 4, 3}, {4, 3, 3}, {3, 3, 8}, {3, 8, 3}, {8, 3, 2}, {3, 2, 8}}
```

Now we are ready to check whether the given subsequence is one of these subsequences:

```
Position[partitionedSeq, subseq]      {{21}}
```

We found one hit. It begins from the 21st digit of π:

```
Take[seq, {21, 23}]      {6, 2, 6}
```

Let us then collapse this calculation into a program:

```
subsequence[seq_, subseq_] :=
  Position[Partition[seq, Length[subseq], 1], subseq]
```

Check that the program works:

```
subsequence[seq, subseq]      {{21}}
```

As a more advanced example, let us check whether the sequence of digits 314159 appears somewhere else within the first 200,000 digits of π besides at the beginning of the digits of π:

```
seq = First[RealDigits[N[π, 200 000]]];
```

```
subsequence[seq, {3, 1, 4, 1, 5, 9}]      {{1}, {176 452}}
```

We found a second hit! It begins from the 176,452nd digit of π:

```
Take[seq, {176 452, 176 457}]      {3, 1, 4, 1, 5, 9}
```

■ Perfect Numbers

A positive integer is a perfect number if it is equal to the sum of its proper divisors. A proper divisor is a divisor that is not the number itself. For example, because $6 = 1 \times 2 \times 3$ and $6 = 1 + 2 + 3$, 6 is a perfect number. We would like to develop a program to search perfect numbers.

Mathematica has the **Divisors** command to calculate the divisors:

```
Divisors[6]      {1, 2, 3, 6}
```

The command also gives the number itself. Thus, we can say that a number is perfect if the sum of all divisors is equal to two times the number. The following function tests whether a number is perfect:

```
perfectQ[n_] := Total[Divisors[n]] == 2 n
```

For example, 6 is perfect but 7 is not:

```
{perfectQ[6], perfectQ[7]}      {True, False}
```

Then we write a program to find perfect numbers from integers being at most a given number:

```
findPerfect[n_] := Select[Range[n], perfectQ[#] &]
```

For example, here are the perfect numbers among the first 10,000 integers:

```
findPerfect[10 000]      {6, 28, 496, 8128}
```

Other perfect numbers are very large. Here is the next:

```
perfectQ[33 550 336]      True
```

Note that with **DivisorSigma** we can also easily find perfect numbers (see Section 12.1.1, p. 396):

```
Select[Range[10 000], DivisorSigma[1, #] == 2 # &]
```

```
{6, 28, 496, 8128}
```

18.1.3 String Manipulation

Mathematica has advanced string manipulation commands; see Sections 13.3.6, p. 433, and 16.2, p. 505. As examples, we consider palindromes and simple cryptography.

■ **Palindromes**

A palindrome is a word or sentence that is the same whether the characters are read from the beginning to the end or from the end to the beginning. For example, "reviver" is a palindrome. We would like to write a program to test whether a word or a sentence is a palindrome.

In a palindrome, we do not distinguish between lowercase and uppercase letters. So, we change all letters to lowercase:

```
ToLowerCase["A man, a plan, a canal—Panama!"]
```

```
a man, a plan, a canal—panama!
```

Also, we do not take care of spaces and other special characters—that is, we are only interested in letter characters. With **StringCases** we can pick up all letters:

```
StringCases[%, LetterCharacter]
```

```
{a, m, a, n, a, p, l, a, n, a, c, a, n, a, l, p, a, n, a, m, a}
```

We have a palindrome if these characters are the same as the reversed characters:

```
% == Reverse[%]      True
```

Thus, we arrive at the following test:

```
palindromeQ[string_] :=
 With[{c = StringCases[ToLowerCase[string], LetterCharacter]},
  c == Reverse[c]]
```

An example:

```
palindromeQ["A man, a plan, a canal—Panama!"]      True
```

Here is a Finnish palindrome I have discovered:

```
palindromeQ["Leseidolin iski niksin ilo: diesel!"]      True
```

(The sentence in English is approximately as follows: "An idol of bran was hit by the joy of a gimmick: diesel!")

■ **Cryptography**

A simple cryptographic method is to replace each character by another character. Here, we consider only letters. The usual alphabet of lowercase letters is as follows:

```
alphabet = CharacterRange["a", "z"]
```

{a, b, c, d, e, f, g, h, i, j, k, l, m, n, o, p, q, r, s, t, u, v, w, x, y, z}

We create a new alphabet by forming a random permutation of the usual alphabet:

```
<< Combinatorica`
```

```
newAlphabet = RandomPermutation[alphabet]
```

{j, h, x, o, f, s, c, p, q, v, i, g, b, m, u, n, k, d, t, r, z, y, a, l, w, e}

A message is encrypted by replacing the original characters with the corresponding characters of the new alphabet—that is, by applying the following rules:

```
encryptRules = Thread[alphabet → newAlphabet]
```

{a → j, b → h, c → x, d → o, e → f, f → s, g → c, h → p, i → q, j → v, k → i, l → g, m → b, n → m, o → u, p → n, q → k, r → d, s → t, t → r, u → z, v → y, w → a, x → l, y → w, z → e}

An encrypted message is decrypted with the reverse transformation:

```
decryptRules = Thread[newAlphabet → alphabet]
```

{j → a, h → b, x → c, o → d, f → e, s → f, c → g, p → h, q → i, v → j, i → k, g → l, b → m, m → n, u → o, n → p, k → q, d → r, t → s, r → t, z → u, y → v, a → w, l → x, w → y, e → z}

As an example, we encrypt the word "mathematics":

```
Characters["mathematics"]    {m, a, t, h, e, m, a, t, i, c, s}
```

```
% /. encryptRules    {b, j, r, p, f, b, j, r, q, x, t}
```

```
% // StringJoin    bjrpfbjrqxt
```

A program for encryption could be as follows:

```
encrypt[string_] := StringJoin[Characters[string] /. encryptRules]
```

```
m = encrypt["mathematics"]    bjrpfbjrqxt
```

Similarly, a program for decryption could be as follows:

```
decrypt[string_] := StringJoin[Characters[string] /. decryptRules]
```

```
decrypt[m]    mathematics
```

18.1.4 Mathematical Formulas

Mathematical formulas are often easy to transform into *Mathematica* code because *Mathematica* has almost all the traditional mathematical notations and we can write a formula into a 2D form with palettes or by direct typing. Two examples are presented here. In Section 14.2.3, p. 463, we considered mathematical formulas where we needed **Apply** to form multiple iteration specifications. In Section 18.5.1, p. 596, we consider recursive mathematical formulas.

■ Day of Week

Consider a date such as 12.25.2010. What day of week may this be? Let $m = 12$, $d = 25$, and $y = 2010$ be the month, day, and year of the date, respectively. The day of week can be computed as follows (see Trott, 2006a, p. 1065). Let

$$a = \left\lfloor \frac{23\,m}{9} \right\rfloor + d + y + 4 + \left\lfloor \frac{z}{4} \right\rfloor - \left\lfloor \frac{z}{100} \right\rfloor + \left\lfloor \frac{z}{400} \right\rfloor - \delta,$$

where

$$z = \begin{cases} y - 1 & \text{if } m < 3 \\ y & \text{otherwise} \end{cases}, \qquad \delta = \begin{cases} 2 & \text{if } m \geq 3 \\ 0 & \text{otherwise} \end{cases}.$$

Let $b = (a \bmod 7) + 1$. The day of week is the bth element of the list {Sunday, Monday, Tuesday, Wednesday, Thursday, Friday, Saturday}. These formulas are easy to transform into a *Mathematica* program:

```
dayOfWeek[m_, d_, y_] :=
  Module[{δ = If[m ≥ 3, 2, 0], z = If[m < 3, y - 1, y], a,
    days = {Sunday, Monday, Tuesday, Wednesday, Thursday, Friday, Saturday}},
    a = ⌊ 23 m/9 ⌋ + d + y + 4 + ⌊ z/4 ⌋ - ⌊ z/100 ⌋ + ⌊ z/400 ⌋ - δ;
    days[[Mod[a, 7] + 1]]]
```

The floor function $\lfloor\ \rfloor$ can be written with ⎋lf⎋ and ⎋rf⎋. For example,

```
dayOfWeek[12, 25, 2010]     Saturday
```

■ **Number of Primes**

The number of primes at most x can be calculated from the following formula (see Trott, 2006a, p. 1068):

$$\pi(x) = -\sum_{k=1}^{\lfloor \log_2(x) \rfloor} \left(\mu(k) \sum_{n=2}^{\lfloor x^{1/k} \rfloor} \mu(n)\,\Omega(n) \left\lfloor \frac{x^{1/k}}{n} \right\rfloor \right).$$

Here, $\Omega(n)$ is the number of prime factors of n, and $\mu(n)$ is the Möbius μ function: $\mu(n) = 1$ if n is the product of an even number of distinct primes, $\mu(n) = -1$ if n is the product of an odd number of distinct primes, and $\mu(n) = 0$ if n has a multiple prime factor.

This formula is easy to program. First, we write the $\Omega(n)$ function. As an example, here are the factors of 1120:

```
FactorInteger[1120]     {{2, 5}, {5, 1}, {7, 1}}
```

Now $\Omega(1120)$ is the number of factors: $5 + 1 + 1 = 7$. These numbers can be extracted as follows:

```
%[[All, 2]]     {5, 1, 1}
```

Thus, we can write the following:

```
Ω[n_] := Total[FactorInteger[n][[All, 2]]]
```

```
Ω[1120]     7
```

Mathematica has the Möbius μ function but we define a shorter name for this function:

```
μ[n_] := MoebiusMu[n]
```

The formula for $\pi(x)$ is now as follows:

$$\text{primePi}[x\_] := - \sum_{k=1}^{\lfloor \text{Log}[2,x] \rfloor} \left(\mu[k] \sum_{n=2}^{\lfloor x^{1/k} \rfloor} \mu[n]\, \Omega[n] \left\lfloor \frac{x^{1/k}}{n} \right\rfloor \right)$$

Here, $\lfloor$ and $\rfloor$ can be written as ESClfESC and ESCrfESC. As an example, we calculate the number of primes at most 100,000:

> primePi[100 000]   9592

This is the same result we get with the built-in **PrimePi**:

> PrimePi[100 000]   9592

18.2 Procedural Programming

18.2.1 Doing

■ Procedural Programming

The procedural style is familiar from ordinary programming languages such as Fortran and C. You probably know such structures as **For**, **While**, and **If**; they all exist in *Mathematica*, too. However, as you study other styles of programming in *Mathematica*, you will find that they are often more effective than the procedural style. A large procedural program may be rewritten in a few lines of code of functional or rule-based programming. Thus, before you begin to code your problem, study in detail whether you can use functional or rule-based programming. Of course, these styles first require serious study, but the time you spend on them is very interesting and saves you time as you progress.

■ Commands for Doing

> Do[body, {i, min, max}]   Do **body** while **i** goes from **min** to **max**
> While[test, body]   Check **test**, then repeat **body** until **test** fails to give **True**
> For[start, test, incr, body]   Do **start**; do **test**, **body**, and **incr** until **test** fails to give **True**

The iteration specification in **Do** can have all of the same forms as the one in **Table** (see Section 14.1.1, p. 445). Thus, the specification can also be of the forms {n} (**body** is done **n** times); {i, max} (**i** goes from 1 to **max**); and {i, min, max, step}. In addition, multiple specifications such as {i, imax}, {j, jmax} can be used.

If **body**, **test**, **start**, or **incr** consists of a sequence of commands, they are separated by semicolons. An index **i** can be incremented with ++i; it is equivalent to i = i + 1.

Note that **Do**, **While**, and **For** do not print anything as an answer; they only do what they are asked to do. Thus, after the calculation, remember to ask for the value of the variable in which you are interested. A prototype **Do** calculation could be as follows:

```
x = init                (*give an initial value for x*)
Do[body, {i, max}]      (*do something for x*)
x                       (*ask the final value of x*)
```

■ Simple Examples

As an example, we calculate the sum of the first 10 integers with all three commands:

```
s = 0; Do[s = s + i, {i, 10}]; s      55

s = 0; i = 1; While[i ≤ 10, s = s + i; i = i + 1]; s      55

For[s = 0; i = 1, i ≤ 10, i = i + 1, s = s + i]; s      55
```

However, the following are better ways to calculate the sum:

```
Sum[i, {i, 10}]      55

Total[Range[10]]      55
```

Do and **Table** are very similar commands; both do what we ask, but whereas **Do** does not print anything, **Table** gathers the results into a list and prints it. For example,

```
s = 0; Table[s = s + i, {i, 10}]
{1, 3, 6, 10, 15, 21, 28, 36, 45, 55}
```

■ Example: Newton's Method

In Section 18.1.1, p. 545, we used **Table** to implement Newton's method. Now we simply replace **Table** with **Do** and add the command **xi** at the end of the code:

```
newton2[f_, x_, x0_, n_] := Module[{newx = x - f / D[f, x], xi = x0},
  Do[xi = N[newx /. x → xi], {n}];
  xi]
```

```
newton2[3 x^3 - E^x, x, 2, 4]      0.953089
```

Next, we use **While** and **For**:

```
newton3[f_, x_, x0_, eps_] := Module[{newx = x - f / D[f, x], xi = x0},
  While[Abs[f /. x → xi] > eps, xi = N[newx /. x → xi]];
  xi]
```

```
newton3[3 x^3 - E^x, x, 2, 10^-2]      0.953089
```

```
newton4[f_, x_, x0_, eps_] := Module[{newx = x - f / D[f, x], xi},
  For[xi = x0,
    Abs[f /. x → xi] > eps,
    xi = N[newx /. x → xi]];
  xi]
```

```
newton4[3 x^3 - E^x, x, 2, 10^-2]      0.953089
```

■ Example: Sampling without Replacement

We would like to take a sample of n elements from a given set. The sampling is done without replacement (so n must be at most the number of elements of the set). As an example, let the set be the integers 1, 2, ..., 10. We have a ready-to-use command:

```
RandomSample[Range[10], 4]      {6, 4, 7, 1}
```

However, now we would like to write a procedural program for taking a sample without replacement.

Let us gather the sample into the variable **sample**, and let its initial value be the empty list **{}**. After each new sample element, for example, **c**, we update the sample with **sample = {sample, c}**. Suppose we have already sampled the elements 7, 4, and 8. The list of elements still available, its length, and the current sample are thus as follows:

```
a = {1, 2, 3, 5, 6, 9, 10};
b = 7;
sample = {{{{}, 7}, 4}, 8};
```

To take the next random element, choose a random integer from 1, ..., 7, add the corresponding element to the sample, delete the element from the list of available elements, and decrease the length of the list by one:

```
pos = RandomInteger[{1, b}]        6

sample = {sample, a[[pos]]}        {{{{{}, 7}, 4}, 8}, 9}

a = Delete[a, pos]        {1, 2, 3, 5, 6, 10}

b = b - 1        6
```

In this way, we arrive at the following program:

```
SWOR[list_, n_] := Module[{a = list, b = Length[list], pos, sample = {}},
  Do[pos = RandomInteger[{1, b}];
   sample = {sample, a[[pos]]};
   a = Delete[a, pos];
   b = b - 1, {n}];
  Sort[Flatten[sample]]]
```

At the end of the program, we remove all unnecessary braces by flattening the sample. We also sort the sample. For example,

```
SWOR[Range[10], 4]        {3, 4, 7, 9}
```

■ Example: Sieve of Eratosthenes

A very old method to find all primes at most n is the sieve of Eratosthenes (ca. 276-194 BC; see Wellin *et al.*, 2005, p. 142). First, write down the integers 1, 2, ..., n. Let the initial value of p be 2. Repeat the following until $p > \sqrt{n}$: Cross out all integer multiples of p on the interval $[2p, n]$ and increase the value of p by 1.

Suppose we would like to find all primes at most 20:

```
p = 2; list = Range[20]
{1, 2, 3, 4, 5, 6, 7, 8, 9, 10, 11, 12, 13, 14, 15, 16, 17, 18, 19, 20}
```

Write down all integer multiples of 2:

```
Range[2 p, 20, p]        {4, 6, 8, 10, 12, 14, 16, 18, 20}
```

Instead of crossing out, we mark integers to be deleted by 1:

```
list[[%]] = 1;
```

The list of integers is now

```
list        {1, 2, 3, 1, 5, 1, 7, 1, 9, 1, 11, 1, 13, 1, 15, 1, 17, 1, 19, 1}
```

Increase the value of *p*:

```
p = 3;
```

Write down all integer multiples of 3:

```
Range[2 p, 20, p]      {6, 9, 12, 15, 18}
```

Set these integers to 1:

```
list[[%]] = 1;
```

The list of integers is now

```
list     {1, 2, 3, 1, 5, 1, 7, 1, 1, 1, 11, 1, 13, 1, 1, 1, 17, 1, 19, 1}
```

We see that two iterations suffice in our example. Indeed, deleting all 1's, we get the list of primes:

```
DeleteCases[list, 1]      {2, 3, 5, 7, 11, 13, 17, 19}
```

In the following program, we use **While** to do the iteration.

```
eratosthenes[n_] := Module[{list = Range[n], p = 2},
  While[p ≤ Sqrt[n],
   list[[Range[2 p, n, p]]] = 1;
   p++];
  DeleteCases[list, 1]]
```

An example:

```
eratosthenes[100]
{2, 3, 5, 7, 11, 13, 17, 19, 23, 29, 31,
 37, 41, 43, 47, 53, 59, 61, 67, 71, 73, 79, 83, 89, 97}
```

To check the result, use **Prime**:

```
Table[Prime[i], {i, 25}]
{2, 3, 5, 7, 11, 13, 17, 19, 23, 29, 31,
 37, 41, 43, 47, 53, 59, 61, 67, 71, 73, 79, 83, 89, 97}
```

18.2.2 Branching

■ Commands for Branching

If[test, then, else, otherwise] If **test** gives **True**, do **then**; else, if **test** gives **False**, do **else**; otherwise (if **test** is neither **True** nor **False**), do **otherwise**.

Which[test1, then1, test2, then2, ... , True, otherwise] Evaluate each of the **testi** in turn and do the **theni** corresponding to the first test giving **True**. If all **testi** give **False**, do **otherwise**. If, in searching the first test giving **True**, a test is encountered giving neither **True** nor **False**, the searching is stopped and the **Which** command is returned with the remaining arguments unevaluated.

Switch[expr, form1, value1, form2, value2, ... , _, otherwise] Compare **expr** with each of the **formi** in turn and return the **valuei** corresponding to the first match. If no matches are found, return **otherwise**.

The box contains the most complete forms of the three commands; shorter forms exist. Regarding **If**, it also accepts the following form:

```
If[test, then, else]
```

Then the `If` command is returned as such if **test** gives neither **True** nor **False**. Here is a still shorter form:

```
If[test, then]
```

This does nothing (meaning that the value of the command is **Null**) if **test** gives **False** and returns the `If` command if **test** gives neither **True** nor **False**.

The **Which** command has the following shorter form:

```
Which[test1, then1, test2, then2, … ]
```

Now nothing is done (meaning that the value of the command is **Null**) if all tests give **False**.

The **Switch** command has the following shorter form:

```
Switch[expr, form1, value1, form2, value2, … ]
```

Now the **Switch** command is returned if no matches are found. In the complete form, remember that _ means anything (see Section 16.1.1, p. 491) so that it matches all expressions.

The complete forms are safe to use because you have specified what to do in all possible cases (with the exception that in **Which** we cannot define what to do if a test is encountered that gives neither **True** nor **False**).

In the tests, we can apply operators such as ==, !=, <, ≤, >, and ≥; use tests such as **IntegerQ** and **OddQ**; and form more complex logical expressions with && (AND), || (OR), and ! (NOT) (see Section 13.3.5, p. 431).

■ **Simple Examples**

Here are some examples of `If`:

```
{If[2 < 3, Yes], If[4 < 3, Yes], If[x < 3, Yes]}
{Yes, Null, If[x < 3, Yes]}
```

In the second example, the test $4 < 3$ gives **False**, so nothing is done. In the third example, we cannot say whether $x < 3$ or not, so the test gives $x < 3$, which is neither **True** nor **False** and, consequently, the `If` command is returned as such.

The following function gives the integral of x^n for each n:

```
f[n_] := If[n == -1, Log[x], x ^ (n + 1) / (n + 1)]
```

Examples:

```
{f[-2], f[-1], f[0], f[1]}
```

$$\left\{ -\frac{1}{x}, \ \text{Log}[x], \ x, \ \frac{x^2}{2} \right\}$$

Next, we use **Which**:

```
g[n_] := Which[
    n == -1, Log[x],
    True, x ^ (n + 1) / (n + 1)]
```

Examples:

```
{g[-2], g[-1], g[0], g[1]}
```

$$\left\{-\frac{1}{x}, \text{Log}[x], x, \frac{x^2}{2}\right\}$$

We could also write **n != -1** in place of **True**. Here is an example of **Switch**:

```
volume[name_] := Switch[name,
  cylinder, Pi r^2 h,
  sphere, 4 Pi r^3 / 3,
  ellipsoid, 4 Pi a b c / 3,
  _, unknown]
```

Examples:

```
{volume[cylinder], volume[sphere], volume[cone]}
```

$$\left\{h \pi r^2, \frac{4 \pi r^3}{3}, \text{unknown}\right\}$$

■ Piecewise Functions

Remember that in Section 17.1.2, p. 516, we considered piecewise-defined functions:

```
Piecewise[{{val₁, cond₁}, {val₂, cond₂}, …}]   A piecewise function with values valᵢ in regions
  defined by conditions condᵢ
Piecewise[{{val₁, cond₁}, {val₂, cond₂}, …}, val]   Use value val if none of the condᵢ apply;
  the default of val is 0
```

This is the preferred method to form functions defined piecewise. Previously, we defined, with **If** and **Which**, a function to give the integral of x^n for each n. Such a function could also be defined as follows:

```
h[n_] = Piecewise[{{x ^ (n + 1) / (n + 1), n ≠ -1}, {Log[x], n == 1}}]
```

$$\begin{cases} \frac{x^{1+n}}{1+n} & n \neq -1 \\ \text{Log}[x] & \text{True} \end{cases}$$

With **PiecewiseExpand** we can develop expressions containing **If** and **Which** into a piecewise expression:

```
PiecewiseExpand[f[n]]
```

$$\begin{cases} \frac{x^{1+n}}{1+n} & n \neq -1 \\ \text{Log}[x] & \text{True} \end{cases}$$

■ Example: Collatz Sequences

A Collatz sequence starts from a given positive integer n and proceeds iteratively. At each step, if the current value n is even, the next value is $n/2$; if the current value n is odd, the next value is $3n + 1$. The hypothesis has been presented that all Collatz sequences sooner or later get the value 1. Here is a program to generate a Collatz sequence:

```
collatzSequence[n_] := Module[{a = n, seq = {n}},
  While[a ≠ 1, a = If[EvenQ[a], a / 2, 3 a + 1]; seq = {seq, a}];
  Flatten[seq]]
```

```
Table[collatzSequence[n], {n, 1, 10}] // Column
{1}
{2, 1}
{3, 10, 5, 16, 8, 4, 2, 1}
{4, 2, 1}
{5, 16, 8, 4, 2, 1}
{6, 3, 10, 5, 16, 8, 4, 2, 1}
{7, 22, 11, 34, 17, 52, 26, 13, 40, 20, 10, 5, 16, 8, 4, 2, 1}
{8, 4, 2, 1}
{9, 28, 14, 7, 22, 11, 34, 17, 52, 26, 13, 40, 20, 10, 5, 16, 8, 4, 2, 1}
{10, 5, 16, 8, 4, 2, 1}
```

■ Example: Switching Fleas

Two dogs, A and B, share n fleas (see Trott, 2006a, p. 1064). Initially, m fleas are on dog A and $n - m$ fleas on dog B. Consider an iteration, where in each step a random flea switches its dog. We would like to write a program that gives the number of fleas on dog A during t iterations.

Let the initial value of the list containing the number of fleas on dog A be $\{m\}$. Let the current number of fleas on dog A be a (initially $a = m$). In each step, choose a random integer from the set 1, 2, ..., n. If this integer is larger than a, this corresponds to the situation in which a flea on dog B switches to dog A, but if the integer is at most a, this corresponds to the situation in which a flea on dog A switches to dog B.

```
fleas[n_, m_, t_] := Module[{a = m, iters = {m}},
  Do[If[RandomInteger[{1, n}] > a, a = a + 1, a = a - 1];
   iters = {iters, a}, {t}];
  Flatten[iters]]
```

Assume that we have 100 fleas. Initially, they are all on dog B. We do 500 steps:

```
ListLinePlot[fleas[100, 0, 500], PlotRange → All, Epilog → Line[{{0, 50}, {500, 50}}]]
```

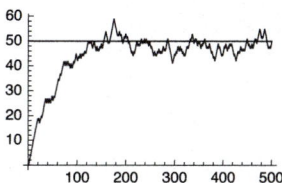

The number of fleas on dog A seems to settle down near 50 in approximately 100–200 steps. Next, we do 51,000 steps, drop the first 1000 steps, and plot the frequencies of the rest of the steps:

```
ListPlot[Tally[Drop[fleas[100, 0, 51 000], 1000]]]
```

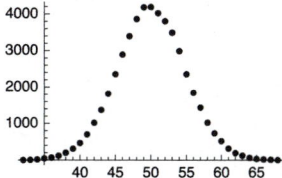

In the long range, the number of fleas seems to be, with high probability, from 40 to 60.

■ Example: Coding Lotto Results

In the lotto game in Finland, each player guesses seven numbers from the numbers 1, 2, ..., 39. Let us study the frequencies of results having a variable number of consecutive numbers. First, we write a program that codes lotto results in the following way. If the result has, for example, five consecutive numbers, the result of the program is 5. If the result contains five consecutive numbers and also two consecutive numbers, the result is 52. The set of possible results is as follows: 7, 6, 52, 5, 43, 42, 4, 322, 33, 32, 3, 222, 22, 22, 2, and 1 (the last result means that the result does not have consecutive numbers). The following program codes a lotto result. The program assumes that the lotto result is sorted in ascending order.

```
lottoCodes[{a_, b_, c_, d_, e_, f_, g_}] := Which[
  (g == a + 6), 7,
  (f == a + 5 || g == b + 5), 6,
  (e == a + 4 && g == f + 1) || (b == a + 1 && g == c + 4), 52,
  (e == a + 4 || f == b + 4 || g == c + 4), 5,
  (d == a + 3 && g == e + 2) || (c == a + 2 && g == d + 3), 43,
  (d == a + 3 && f == e + 1) || (d == a + 3 && g == f + 1) || (e == b + 3 && g == f + 1) ||
   (b == a + 1 && f == c + 3) || (b == a + 1 && g == d + 3) || (c == b + 1 && g == d + 3), 42,
  (d == a + 3 || e == b + 3 || f == c + 3 || g == d + 3), 4,
  (c == a + 2 && e == d + 1 && g == f + 1) ||
   (b == a + 1 && e == c + 2 && g == f + 1) || (b == a + 1 && d == c + 1 && g == e + 2), 322,
  (c == a + 2 && f == d + 2) || (c == a + 2 && g == e + 2) || (d == b + 2 && g == e + 2), 33,
  (c == a + 2 && e == d + 1) || (c == a + 2 && f == e + 1) || (c == a + 2 && g == f + 1) ||
   (d == b + 2 && f == e + 1) || (d == b + 2 && g == f + 1) ||
   (b == a + 1 && e == c + 2) || (e == c + 2 && g == f + 1) ||
   (b == a + 1 && f == d + 2) || (c == b + 1 && f == d + 2) || (b == a + 1 && g == e + 2) ||
   (c == b + 1 && g == e + 2) || (d == c + 1 && g == e + 2), 32,
  (c == a + 2 || d == b + 2 || e == c + 2 || f == d + 2 || g == e + 2), 3,
  (b == a + 1 && d == c + 1 && f == e + 1) || (b == a + 1 && d == c + 1 && g == f + 1) ||
   (b == a + 1 && e == d + 1 && g == f + 1) || (c == b + 1 && e == d + 1 && g == f + 1), 222,
  (b == a + 1 && d == c + 1) || (b == a + 1 && e == d + 1) ||
   (b == a + 1 && f == e + 1) || (b == a + 1 && g == f + 1) ||
   (c == b + 1 && e == d + 1) || (c == b + 1 && f == e + 1) || (c == b + 1 && g == f + 1) ||
   (d == c + 1 && f == e + 1) || (d == c + 1 && g == f + 1) || (e == d + 1 && g == f + 1), 22,
  (b == a + 1 || c == b + 1 || d == c + 1 || e == d + 1 || f == e + 1 || g == f + 1), 2,
  True, 1]
```

The program is based on the assumption that the lotto result is sorted in ascending order. For example, in order for $\{a, b, c, d, e, f, g\}$ to have seven consecutive numbers, it suffices that $g = a + 6$. To get five and two consecutive numbers, it suffices that either $e = a + 4$ and $g = f + 1$ or $b = a + 1$ and $g = c + 4$. Note that here we need not explicitly exclude the cases of seven or six consecutive numbers because if one of these is the case, the program codes the result correctly as these cases are covered earlier in the program.

To test the program, we give here a set of 25 artificial lotto results:

```
lottoResults = {{1, 2, 3, 4, 5, 6, 7}, {1, 2, 3, 4, 5, 6, 11},
  {1, 11, 12, 13, 14, 15, 16}, {1, 2, 11, 12, 13, 14, 15}, {1, 2, 3, 4, 5, 11, 12},
  {1, 2, 3, 4, 5, 11, 21}, {1, 11, 12, 13, 14, 15, 21}, {1, 11, 21, 22, 23, 24, 25},
  {1, 2, 3, 11, 12, 13, 14}, {1, 2, 3, 4, 11, 12, 13}, {1, 2, 11, 12, 13, 14, 21},
  {1, 2, 11, 21, 22, 23, 24}, {1, 2, 3, 4, 11, 12, 21}, {1, 2, 3, 4, 11, 21, 31},
  {1, 11, 12, 13, 14, 21, 31}, {1, 11, 21, 22, 23, 24, 31},
  {1, 11, 21, 31, 32, 33, 34}, {1, 2, 11, 12, 21, 22, 23}, {1, 11, 12, 13, 21, 22, 23},
  {1, 11, 12, 21, 31, 32, 33}, {1, 11, 12, 13, 21, 31, 39}, {1, 2, 11, 12, 21, 22, 31},
  {1, 5, 6, 11, 12, 21, 31}, {1, 5, 6, 11, 21, 31, 39}, {1, 3, 5, 7, 9, 11, 13}};
```

Here, the first result has seven consecutive numbers, the next two have six consecutive numbers, and so on. Now we code the results:

```
lottoCodes /@ lottoResults
```

{7, 6, 6, 52, 52, 5, 5, 5, 43, 43, 42, 42, 42, 4, 4, 4, 4, 322, 33, 32, 3, 222, 22, 2, 1}

■ Example: A Lotto Simulation

We continue the preceding example and calculate approximative probabilities of lotto results. First, we generate a set of lotto results. Use **RandomSample** to sample without replacement seven numbers from the integers 1, ..., 39 and sort the results:

```
RandomSample[Range[39], 7] // Sort    {6, 16, 17, 20, 24, 31, 39}
```

Generate 1 million sorted lotto results:

```
SeedRandom[1];
lottoResults = Table[RandomSample[Range[39], 7] // Sort, {10^6}];
```

First code the results:

```
(c = lottoCodes /@ lottoResults;) // Timing
```

{288.415, Null}

The coding may take some minutes. Then calculate the frequencies:

```
Tally[c]
```

{{3, 77127}, {2, 431162}, {22, 154770}, {1, 278198},
 {32, 31921}, {4, 10591}, {42, 2124}, {222, 10634}, {322, 1067},
 {33, 1119}, {5, 1086}, {52, 61}, {43, 61}, {6, 78}, {7, 1}}

Sort the frequencies in descending order:

```
fr = Reverse@SortBy[%, Last]
```

{{2, 431162}, {1, 278198}, {22, 154770}, {3, 77127},
 {32, 31921}, {222, 10634}, {4, 10591}, {42, 2124}, {33, 1119},
 {5, 1086}, {322, 1067}, {6, 78}, {52, 61}, {43, 61}, {7, 1}}

Check that the sum of the frequencies is 1 million:

```
Total[fr[[All, 2]]]    1 000 000
```

Then form a table from the results:

```
Grid[Prepend[fr, {"Case", "Frequency"}], Spacings → {0.7, 0.2},
 Dividers → {False, {False, True}}, Alignment → {{Left, Right}}]
```

Case	Frequency
2	431 162
1	278 198
22	154 770
3	77 127
32	31 921
222	10 634
4	10 591
42	2124
33	1119
5	1086
322	1067
6	78
52	61
43	61
7	1

Approximate probabilities are obtained by dividing the frequencies by 1 million. Thus, the most probable event is that the lotto result has one times two consecutive numbers (and no other consecutive numbers); an example of this kind of result is {4, 11, 12, 19, 25, 31, 35}. The probability of such an event is approximately 0.43. Thus, almost every other lotto result has one times two consecutive numbers. The second largest probability, approximately 0.28, is with the case of no consecutive numbers. Two times two consecutive numbers occur with a probability of approximately 0.15; an example of this kind of result is {4, 11, 12, 19, 25, 26, 35}. It is highly unlikely to get seven consecutive numbers as in {1, 2, 3, 4, 5, 6, 7}. The simulation gives for this kind of event only an approximate probability of $1. \times 10^{-6}$ (this

happens to be quite the same as the exact probability, $33 \big/ \binom{39}{7}$).

In Section 18.4.2, p. 592, we use rule-based programming to calculate the frequencies of lotto results.

18.2.3 Communicating

■ **Printing Values**

`Print[expr1, expr2, … ]`	Print the values of the expressions

Often, it suffices that we only get the final result when all the computations have been completed. In the following example, we calculate the number of different factors of $2^n - 1$ for some values of n. We get the result once all the numbers have been factored:

```
Table[Length[FactorInteger[2^n - 1]], {n, 135, 145}]
```

{10, 10, 2, 8, 2, 15, 6, 6, 6, 17, 5}

In long calculations, it may be useful to see in real time how the calculations proceed. This can be done by printing intermediate results as soon as they become complete or by printing the current value of an iterator:

```
Table[fi = Length[FactorInteger[2^n - 1]]; Print[fi]; fi, {n, 135, 140}]
```

10

10

2

8

2

15

{10, 10, 2, 8, 2, 15}

```
Table[Print[n]; Length[FactorInteger[2^n - 1]], {n, 135, 140}]
```

135

136

137

138

139

140

{10, 10, 2, 8, 2, 15}

As we noted in Section 17.1.4, p. 521, a module automatically prints the result of its last statement. If we want to print some intermediate results, we can use **Print**, although **Message** is the preferred command (see Section 17.3.4, p. 540).

Print is useful for debugging a larger program. You can print the values of some variables that play a central role in a susceptible block of the program. This may help you to infer what the program actually does, and then you can more easily correct the code. However, remember that we also have a special debugger (see Section 17.2.2, p. 524).

Another way to print intermediate results is with the use of **Monitor** or **PrintTemporary**, as is explained next.

■ **Monitoring Values**

> **Monitor[expr, mon]** (⌘6) Show, in a temporary cell, the current value of **mon** during the evaluation of **expr**
>
> **Monitor[expr, ProgressIndicator[var, {min, max}]]** Show, in a temporary cell, the progress of the computation of **expr** as **var** takes values from **min** to **max**
>
> **PrintTemporary[expr]** (⌘6) Print, in a series of temporary cells, the value of **expr** during the execution of the current command

These commands print the value of a variable of interest into temporary cells. The cells disappear once the computations are complete.

When we execute the following command, we get a temporary cell showing the current value of **n**. The cell disappears when the computation becomes ready so that we cannot show the temporary cell here:

```
Monitor[Table[Length[FactorInteger[2^n - 1]], {n, 135, 145}], n]
```
```
{10, 10, 2, 8, 2, 15, 6, 6, 6, 17, 5}
```

During the execution of the following command, we get a temporary progress indicator showing the growth of the value of **n** during the computation. The cell again disappears when the computation becomes ready so that we cannot show the temporary cell here:

```
Monitor[Table[Length[FactorInteger[2^n - 1]], {n, 135, 145}],
 ProgressIndicator[n, {135, 145}]]
{10, 10, 2, 8, 2, 15, 6, 6, 6, 17, 5}
```

In the following way, we can show a permanent progress indicator:

```
ProgressIndicator[Dynamic[n], {135, 145}]
```

```
Table[(n = k; Length[FactorInteger[2^n - 1]]), {k, 135, 145}]
{10, 10, 2, 8, 2, 15, 6, 6, 6, 17, 5}
```

The following command creates a series of temporary cells showing the current value of **n**:

```
Table[Length[PrintTemporary[n]; FactorInteger[2^n - 1]], {n, 135, 145}]
{10, 10, 2, 8, 2, 15, 6, 6, 6, 17, 5}
```

■ **Sowing and Reaping Values**

> `Sow[val]` Sow **val** for the nearest enclosing **Reap**
> `Reap[expr]` Evaluate **expr**; return a list of values sown by **Sow**

During a calculation, we often would like to collect some values into a list. As an example, in the following calculation we collect all values of n for which the number of factors of $2^n - 1$ is one:

```
Table[fi = Length[FactorInteger[2^n - 1]]; If[fi == 1, Print[n]]; fi, {n, 2, 10}]
```

2

3

5

7

```
{1, 1, 2, 1, 2, 1, 3, 2, 3}
```

We can also gather the interesting values into a list:

```
a = {};
Table[fi = Length[FactorInteger[2^n - 1]]; If[fi == 1, a = {a, n}]; fi, {n, 2, 10}]
{1, 1, 2, 1, 2, 1, 3, 2, 3}

Flatten[a]     {2, 3, 5, 7}
```

However, a simpler way is to use **Sow** and **Reap**:

```
Reap[Table[fi = Length[FactorInteger[2^n - 1]]; If[fi == 1, Sow[n]]; fi, {n, 2, 10}]]
{{1, 1, 2, 1, 2, 1, 3, 2, 3}, {{2, 3, 5, 7}}}
```

The result first shows the value of the expression inside **Reap** and then a list of values sown by **Sow**.

In Section 12.3.2, p. 411, we presented the following example:

```
iters = {};
FindRoot[Exp[-x] - x^2, {x, -1}, StepMonitor :> AppendTo[iters, x]]
{x → 0.703467}

iters
{1.39221, 0.835088, 0.709834, 0.703483, 0.703467, 0.703467}
```

The **StepMonitor** option defines a command executed after each step so that we get in the list **iters** the value of **x** after each step. Again, it is simpler to use **Sow** and **Reap**:

```
Reap[FindRoot[Exp[-x] - x^2, {x, -1}, StepMonitor :> Sow[x]]]
{{x → 0.703467}, {{1.39221, 0.835088, 0.709834, 0.703483, 0.703467, 0.703467}}}
```

■ **Inputting Values**

> `x = Input["prompt"]` Print the prompt; then read an expression as a value for **x**

With **Input** we can build an interactive program that asks for some values. As an example, the function **div** asks for an integer and then prints its divisors. This is continued until the integer is negative:

```
div := Module[{n, cont = True},
  While[cont == True, n = Input["Give an integer"];
   If[n ≥ 0, Print["The divisors of ", n, " are ", Divisors[n]], cont = False]]]
```

```
div
```

The divisors of 1111 are {1, 11, 101, 1111}

The divisors of 11 111 are {1, 41, 271, 11 111}

The divisors of 111 111 are
  {1, 3, 7, 11, 13, 21, 33, 37, 39, 77, 91, 111, 143, 231, 259, 273, 407, 429, 481, 777,
   1001, 1221, 1443, 2849, 3003, 3367, 5291, 8547, 10 101, 15 873, 37 037, 111 111}

When **div** is entered, a new window appears in which we can enter the requested input. The result is then printed in the notebook.

18.2.4 Controlling

■ **Commands for Controlling**

Continue[]	Go to the next step in the current loop of **Do**, **While**, or **For**
Break[]	Exit the nearest enclosing loop of **Do**, **While**, or **For**
Return[expr]	Return **expr**, exiting all procedures and loops in a function
Goto[name]	Go to **Label[name]**

These commands are used to perform an exceptional operation. As an example, we program the tossing of an *n*-face die until the result is 1 or *n*. The following program is very clumsy:

```
die[n_] := Module[{r, i = 1},
  Label[start];
  r = RandomInteger[{1, n}];
  If[r == 1 || r == n,
   Print["We got ", r, " after ", i, " tosses"]; Goto[finish],
   ++i; Goto[start]];
  Label[finish];]
```

An example:

```
die[6]

We got 6 after 4 tosses
```

We write a better program (here we assume that we get 1 or *n* in at most 100 tosses):

```
die2[n_] := Module[{r, i = 1},
  Do[r = RandomInteger[{1, n}];
   If[r == 1 || r == n, Return[{r, i}], ++i], {100}]]
```

An example:

```
die2[6]     {6, 3}
```

Here is a still better program:

```
die3[n_] := Module[{r = 0, i = 0},
  While[r != 1 && r != n, r = RandomInteger[{1, n}]; ++i];
  {r, i}]
```

An example:

```
die3[6]     {1, 4}
```

■ **Example: Newton's Method**

When we programmed Newton's method in Section 18.2.1, p. 554, with **Do**, we did a fixed, sufficiently large number of iterations to obtain the zero. Normally, the convergence is controlled with the program: Iterations are stopped once the present approximation to the solution is accurate enough. The following module checks the convergence:

```
newton5[f_, x_, x0_, eps_, max_] :=
 Module[{xi = x0, fi, df = D[f, x], dfi, iters = {x0}},
  Do[{fi, dfi} = N[{f, df} /. x → xi];
   If[Abs[fi] ≥ eps, xi = xi - fi / dfi; iters = {iters, xi},
    Return[Flatten[iters]]], {max}]]
```

The starting value is **x0**. If the value of the function is at least **eps**, iterations are continued; otherwise, iterations are stopped and the program returns the flattened list of iterations. However, iterations are done at most **max** times (since we have written **Do[… , {max}]**). For example,

```
f = 3 x^3 - E^x;
```

```
newton5[f, x, 2, 10^-14, 20]
```

```
{2, 1.41942, 1.1019, 0.975117, 0.953089, 0.952446, 0.952446, 0.952446}
```

```
f /. x → Last[%]    4.44089 × 10^-16
```

We see that the value of the function is, in fact, less than 10^{-14} at the last point. We can also find complex zeros if we start at a complex point:

```
newton5[f, x, -0.5 + 0.5 I, 10^-14, 20]
```

```
{-0.5 + 0.5 i, -0.400328 + 0.465629 i, -0.384087 + 0.473305 i,
 -0.38428 + 0.473739 i, -0.38428 + 0.473739 i, -0.38428 + 0.473739 i}
```

```
Remove["Global`*"]
```

In Section 22.3.4, p. 739, we write a similar program for the secant method.

Next, we present some questions regarding the previous program.

■ **Some Questions**

1. Why have we written **iters = {iters, xi}** and not **AppendTo[iters, xi]**? The latter command would directly give a list of the simple form $\{x_0, x_1, x_2, …\}$ so that at the end of the program we could simply write **iters** (flattening would not be needed).

The answer is that **AppendTo[iters, xi]** is slower than **iters = {iters, xi}**, even when we take into account the time required for the flattening:

Use **iters = {iters, xi}** *and lastly* **Flatten[iters]** *instead of* **AppendTo[iters, xi]**.

The time saved may not be noticeable in the small iteration of **newton5**, but for longer iterations it may be considerable. If each iteration gives two numbers **{xi, yi}**, we can in the same way write **iters = {iters, {xi, yi}}** and lastly **Flatten[iters]**, but if the result must consist of the pairs **{xi, yi}**, the last step is **Partition[Flatten[iters], 2]**.

2. Instead of **iters = {iters, xi}**, could we write **iters = {iters, %}**, as **xi** is calculated in the preceding command?

The answer is no:

We cannot use **%** *in programs.*

The symbol **%** is intended to be used only in interactive calculations.

3. Why have we introduced the variable **xi** in the program and given it the starting value **x0**? Why have we not directly used the variable **x0** containing the starting value?

An important point to note is the following:

The arguments of a program cannot be changed inside the program.

We might assume that we do not need the variable **xi** and instead write directly **x0 = x0 - fi/dfi**. This does not work, however, because when we use the program, **x0** has a specific numerical value such as 2, and we cannot make an assignment such as **2 = …** . We would obtain an error message such as the following:

Set::setraw : Cannot assign to raw object 2. ≫

Thus, **x0** can be used only as the starting value, and the iterations have to be stored in another variable; we have used **xi**.

4. In defining the starting value of **iters**, we have written **iters = {x0}**. Could we write **iters = {xi}**? Remember that we previously defined that **xi = x0**.

Note the following:

In giving starting values for local variables in a module, previously defined local variables cannot be used.

If we write **iters = {xi}**, we get the following result:

{xi, 1.41942, 1.1019, 0.975117, 0.953089, 0.952446, 0.952446}

We see that the starting value **xi** of **iters** was unknown to *Mathematica*.

■ **Example: Removing Repetitive Elements**

We write a procedural program that, from a given list, removes all elements that are the same as the preceding element. For example, the list {0, 1, 1, 2, 2, 2, 1, 1} becomes {0, 1, 2, 1}.

```
removeRepetitions[list_] :=
 Module[{result = {First[list]}, current = First[list], next},
  Do[next = list[[i]];
   If[next == current, Continue[],
    result = {result, next}; current = next],
   {i, 2, Length[list]}];
  Flatten[result]]
```

Here, we proceed element by element. If the next element is the same as the current element, we continue to the next step. If the next element is different than the current element, it is added to the list to be outputted.

```
removeRepetitions[{0, 1, 1, 2, 2, 2, 1, 1}]
```

{0, 1, 2, 1}

Note that we can write a much better and simpler program using high-level list manipulation commands:

```
removeRepetitions2[list_] := First /@ Split[list]
```

To understand this program, consider an example:

```
Split[{0, 1, 1, 2, 2, 2, 1, 1}]

{{0}, {1, 1}, {2, 2, 2}, {1, 1}}

First /@ %    {0, 1, 2, 1}
```

To compare the computing times of the two programs, generate a list of 100,000 random 0's and 1's and then remove repetitions with both programs:

```
list = RandomInteger[{0, 1}, 100 000];

removeRepetitions[list]; // Timing    {1.54165, Null}

removeRepetitions2[list]; // Timing    {0.143988, Null}
```

The procedural program is approximately 10 times slower than the list-based program.

The two programs describe the properties of procedural and list-based/functional programs in general:

- Procedural programs apply an element-by-element approach, whereas list-based/functional programs use high-level commands to avoid explicit treatment of each element.
- Procedural programs typically have longer code and slower speed than list-based/functional programs.

18.3 Functional Programming

18.3.1 Introduction

Functional programming may be the most important programming style in *Mathematica*. It is suitable in many problems in which we *manipulate lists* or *iterate functions*. In both types of tasks, we *apply functions to arguments*. The functions are often written as pure functions (see Section 17.1.4, p. 520).

In list manipulation, we typically modify the elements by applying a function to them so that, for example, we transform the list $\{a, b, c\}$ into $\{f(a), f(b), f(c)\}$ with **Map** or into $f(a, b, c)$ with **Apply**. In iterations, we start from a given value and then iterate it with a function so that, for example, from the starting value x_0 we get new values with the iteration formula $x_{i+1} = f(x_i)$; this can be done with **Nest**. Next, we consider list manipulation and function iteration in more detail.

▪ List Manipulation

For list manipulation, the most important functional style programming commands are **Map** and **Apply**, but **Map** also has the variations **MapAt**, **MapAll**, **MapIndexed**, and **MapThread**; we also have **Thread**, **Inner**, and **Outer**. All these commands were considered in Section 14.2, p. 459. We could have considered these commands here in the context of programming, but we believed it was more suitable to consider them as one group of list manipulation commands because they—or at least **Map** and **Apply**—are very useful in everyday calculations with *Mathematica*, not only in ordinary programs.

The key in functional list manipulation is that we do not *explicitly* treat each element of a list separately; we only indicate what we want do with the elements. Of course, ultimately *Mathematica* has to do the operations for each element, but the essential point is that the list manipulation routines inside the kernel of *Mathematica* are much faster than any code we are able to write with *Mathematica*.

For example, to calculate the sum of the elements of a list, we do not use **Sum**:

```
t = {a, b, c};
Sum[t[[i]], {i, 1, 3}]     a + b + c
```

A still worse way is the use of **Do**:

```
s = 0; Do[s = s + t[[i]], {i, 1, 3}]; s     a + b + c
```

Instead, the command is simply the following:

```
Total[t]     a + b + c
```

Alternatively, we can use the following:

```
Apply[Plus, t]     a + b + c
```

To square the elements, we do not use **Table**:

```
Table[t[[i]]^2, {i, 1, 3}]     {a², b², c²}
```

Instead, we enter the following simple command:

```
t^2     {a², b², c²}
```

To calculate the row sums of a matrix, we do not use **Table** and **Sum**:

```
m = {{1, 2, 3}, {a, b, c}, {A, B, C}};
Table[Sum[m[[i, j]], {j, 1, 3}], {i, 1, 3}]
{6, a + b + c, A + B + C}
```

Instead, we write the following:

```
Map[Total, m]     {6, a + b + c, A + B + C}
```

Alternatively, we can use the following:

```
Apply[Plus, m, 1]     {6, a + b + c, A + B + C}
```

The functional list manipulation commands have two advantages. First, once you have learned them, they are short to write and thus shorten the code needed to do a calculation. Second, they are fast. As an example, we calculate the row sums of a 1000×1000 matrix by three methods:

```
n = 1000; r = RandomReal[{0, 1}, {n, n}];

Table[s = 0; Do[s = s + r[[i, j]], {j, 1, n}]; s, {i, n}]; // Timing
{6.65994, Null}

Table[Sum[r[[i, j]], {j, n}], {i, n}]; // Timing
{0.746879, Null}

Map[Total, r]; // Timing
{0.030034, Null}
```

■ Function Iteration

Many mathematical methods are iterative or recursive, having the typical form $x_{i+1} = f(x_i)$. Examples are several iterative numerical methods for nonlinear equations, nonlinear optimization, and differential, partial differential, and difference equations. *Mathematica* has special functional type commands such as **Nest**, **FixedPoint**, and **Fold** for iterative calculations.

Nest is the basic iterating command, which does the iteration $x_{i+1} = f(x_i)$ a fixed number of times:

```
Nest[f, x0, 3]     f[f[f[x0]]]
```

FixedPoint does the iteration until the result does not change (the stopping criterion can be given). **Fold** also iterates a function, but now the function has two variables and, at each iteration, it takes one element from a given list as the second argument. We also have the commands **NestList**, **FixedPointList**, and **FoldList**, which also print all intermediate steps.

For example, Newton's method can be written as follows:

```
newton[f_, x_, x0_, max_, opts___] := With[{df = D[f, x]},
  FixedPointList[(x - f / df) /. x → # &, N[x0], max, opts]]
```

Here is an application of it:

```
newton[3 x^3 - E^x, x, 2, 20, SameTest → (Abs[#1 - #2] < 10^-6 &)]
{2., 1.41942, 1.1019, 0.975117, 0.953089, 0.952446, 0.952446}
```

The first advantage of function iteration commands is that they shorten the code compared with procedural programming. Compare the **newton** program discussed here with the **newton5** program discussed in Section 18.2.4, p. 566. Indeed, in **newton** we do not need to be concerned with setting suitable initial values for variables, implementing the stopping of the iteration according to the result of the stopping criterion, or adding iteration counters. We simply input the three or four items needed in the iteration: the function to be iterated, the starting point, the maximum number of steps, and (if the default stopping criterion is not suitable) a stopping criterion (as a pure function).

The second advantage is that the function iteration commands are fast. As an example, we compare two programs:

```
x = 1.; Do[x = Cos[x] + RandomReal[], {500 000}] // Timing
{3.71337, Null}
```

```
Nest[Cos[#] + RandomReal[] &, 1., 500 000]; // Timing
{0.417417, Null}
```

The time needed by the procedural program is approximately 10 times the time needed by the functional program.

In Sections 18.3.2–18.3.6, we study functional list manipulation and function iteration.

```
x = .
```

18.3.2 List Manipulation

■ Useful Commands

Recall from Section 14.2, p. 459, the two most useful functional list manipulation commands—**Map** and **Apply**.

Map[f[#]&, list] or **f[#]& /@ list** Apply **f** to each element at the first level of **list**

An example:

```
Map[#^2 &, {a, b, c}]     {a², b², c²}
```

Apply[head, list] or **head @@ list** Replace the head **List** of **list** with **head**
Apply[head, list, {1}] or **head @@@ list** Replace the head at level 1

Recall that the head of a list is **List** and that with **Apply** we can change the head:

```
{a, b, c} // FullForm      List[a, b, c]

Apply[Plus, {a, b, c}]      a + b + c

% // FullForm      Plus[a, b, c]
```

Next, we present some examples of programs that use these commands.

■ Harmonic Numbers

The sum of the first n terms of the harmonic series $1 + \frac{1}{2} + \frac{1}{3} + \dots$ is the nth harmonic number. We have a ready-to-use command for it:

$$\texttt{HarmonicNumber[10]} \qquad \frac{7381}{2520}$$

To build a program for the harmonic number, we could use **Sum**:

$$\texttt{Sum[1 / i, \{i, 10\}]} \qquad \frac{7381}{2520}$$

However, in functional programming we try to not calculate with the elements of lists. A better program can be obtained by first observing that the first, for example, 10 terms of the series can be calculated as follows:

$$\texttt{1 / Range[10]} \qquad \left\{ 1, \frac{1}{2}, \frac{1}{3}, \frac{1}{4}, \frac{1}{5}, \frac{1}{6}, \frac{1}{7}, \frac{1}{8}, \frac{1}{9}, \frac{1}{10} \right\}$$

This is due to the fact that *Mathematica* automatically performs calculations with lists term by term. Now we only have to sum these terms with **Total**:

```
harmonicNumber[n_] := Total[1 / Range[n]]
```

$$\texttt{harmonicNumber[10]} \qquad \frac{7381}{2520}$$

■ Harmonic and Geometric Means

The harmonic mean of numbers $x_1, \dots, x_n$ is $n \Big/ \left(\frac{1}{x_1} + \dots + \frac{1}{x_n} \right)$. Again, we could use **Sum**:

```
x = {a, b, c, d};

Length[x] / Sum[1 / x[[i]], {i, Length[x]}]
```

$$\frac{4}{\frac{1}{a} + \frac{1}{b} + \frac{1}{c} + \frac{1}{d}}$$

However, we do otherwise:

```
harmonicMean[x_] := Length[x] / Total[1 / x]
```

```
harmonicMean[x]
```

$$\frac{4}{\frac{1}{a} + \frac{1}{b} + \frac{1}{c} + \frac{1}{d}}$$

The geometric mean of numbers $x_1, \dots, x_n$ is $(x_1 \cdots x_n)^{1/n}$. We could use **Product**:

```
Product[x〚i〛, {i, Length[x]}] ^ (1 / Length[x])
```

$(a\,b\,c\,d)^{1/4}$

However, we again do otherwise:

```
geometricMean[x_] := Apply[Times, x] ^ (1 / Length[x])
```

```
geometricMean[x]        (a b c d)^(1/4)
```

Note that we have the built-in **HarmonicMean** and **GeometricMean**.

■ Constructing a Number from the Factors

In Section 11.1.1, we considered the factorization of integers. Here is an example:

```
fa = FactorInteger[3 361 743]        {{3, 4}, {7, 3}, {11, 2}}
```

This means that 3,361,734 can be written as

```
3 ^ 4 × 7 ^ 3 × 11 ^ 2        3 361 743
```

Suppose that the result of the factorization is as follows:

```
factors = {{f1, p1}, {f2, p2}, {f3, p3}};
```

The head of the lists **{fi, pi}** is **List**:

```
% // FullForm
List[List[f1, p1], List[f2, p2], List[f3, p3]]
```

The head of a power is **Power**:

```
fi ^ p1 // FullForm        Power[fi, p1]
```

Thus, we change, with **Apply**, the head **List** at the level 1 of **fa** to the head **Power**:

```
Apply[Power, factors, {1}]        {f1^p1, f2^p2, f3^p3}
```

Then we form the product of the elements of this list:

```
Apply[Times, %]        f1^p1 f2^p2 f3^p3
```

Now we can write the following program:

```
numberFromFactors[factors_] := Apply[Times, Apply[Power, factors, {1}]]
```

```
numberFromFactors[fa]        3 361 743
```

The program can also be written in the shorter form

```
numberFromFactors2[factors_] := Times @@ Power @@@ factors
```

■ Frequencies

We use **Tally** for frequencies:

```
SeedRandom[2]; u = RandomInteger[{1, 6}, {20}]
{6, 2, 3, 3, 6, 3, 2, 6, 6, 1, 1, 5, 4, 5, 1, 2, 2, 6, 2, 6}
```

```
Tally[u] // Sort
{{1, 3}, {2, 5}, {3, 3}, {4, 1}, {5, 2}, {6, 6}}
```

To write a program for frequencies, note that if we know that the elements of the list are from the set $\{m, m+1 \ldots, n\}$, we can write the following (remember that **Map** can also be written as **/@**):

```
frequencies1[u_, m_, n_] := {#, Count[u, #]} & /@ Range[m, n]
```

```
frequencies1[u, 1, 6]
{{1, 3}, {2, 5}, {3, 3}, {4, 1}, {5, 2}, {6, 6}}
```

This program has the advantage that it also shows possible zero frequencies. If we do not know from what set the elements derive, we can write

```
frequencies2[u_] := {#, Count[u, #]} & /@ Union[u]
```

```
frequencies2[u]
{{1, 3}, {2, 5}, {3, 3}, {4, 1}, {5, 2}, {6, 6}}
```

■ Frequencies of Characters

Mathematica contains a dictionary of nearly 100,000 English words (see Section 16.2.1, p. 505):

```
words = DictionaryLookup[];
```

```
words // Length    92 518
```

We would like to calculate the frequencies of letters in this dictionary. As an example, take the first five words:

```
Take[words, 5]    {a, Aachen, aah, Aaliyah, aardvark}
```

In counting the frequencies, we do not distinguish between lower- and uppercase letters, and so we change all words to lowercase:

```
ToLowerCase[%]    {a, aachen, aah, aaliyah, aardvark}
```

Then we extract the characters of the words:

```
Characters[%]
{{a}, {a, a, c, h, e, n}, {a, a, h}, {a, a, l, i, y, a, h}, {a, a, r, d, v, a, r, k}}
```

Once flattened, we get a list from which we can calculate the frequencies. Thus, we write the following:

```
chars = Flatten[Characters[ToLowerCase[words]]];
```

```
chars // Length    776 570
```

The frequencies of the characters of the usual English alphabet are as follows:

```
fr = {Count[chars, #], #} & /@ CharacterRange["a", "z"]
```

```
{{60 670, a}, {15 303, b}, {31 145, c}, {29 423, d}, {88 677, e}, {10 556, f}, {23 047, g},
 {18 143, h}, {67 014, i}, {1630, j}, {7550, k}, {41 195, l}, {21 179, m},
 {55 550, n}, {47 311, o}, {21 777, p}, {1448, q}, {56 424, r}, {67 506, s},
 {51 649, t}, {25 806, u}, {7900, v}, {7062, w}, {2108, x}, {12 457, y}, {3410, z}}
```

Plot the frequencies in decreasing order as a bar chart:

```
<< BarCharts`
```

```
BarChart[Reverse[Sort[fr]], ImageSize → 300]
```

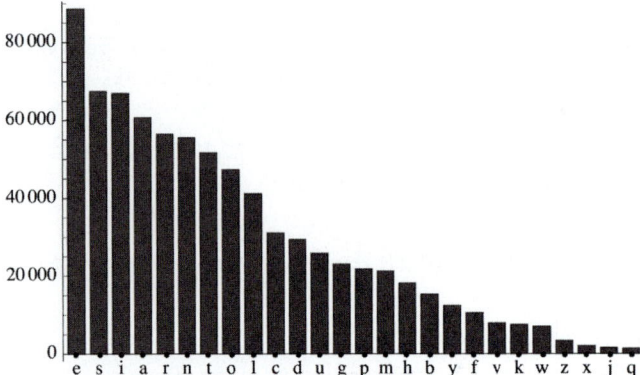

■ **Run-Length Encoding and Decoding**

Consider the following list:

```
SeedRandom[1]; u = RandomInteger[1, {20}]
```

{1, 1, 0, 1, 0, 0, 0, 1, 0, 1, 0, 0, 0, 0, 0, 0, 0, 1, 1, 1}

Split it into runs of identical elements:

```
Split[u]
```

{{1, 1}, {0}, {1}, {0, 0, 0}, {1}, {0}, {1}, {0, 0, 0, 0, 0, 0, 0}, {1, 1, 1}}

Then form the run-length encoding of **u** by forming a list of pairs of numbers where the first number is the element of the run in consideration and the second number the length of the run:

```
en = {First[#], Length[#]} & /@ %
```

{{1, 2}, {0, 1}, {1, 1}, {0, 3}, {1, 1}, {0, 1}, {1, 1}, {0, 7}, {1, 3}}

One way to do the decoding is to use **Map** as follows:

```
ConstantArray[#[[1]], #[[2]]] & /@ en
```

{{1, 1}, {0}, {1}, {0, 0, 0}, {1}, {0}, {1}, {0, 0, 0, 0, 0, 0, 0}, {1, 1, 1}}

This list can then be flattened. A simpler way is to apply **ConstantArray** at the first level:

```
Apply[ConstantArray[##] &, en, {1}]
```

{{1, 1}, {0}, {1}, {0, 0, 0}, {1}, {0}, {1}, {0, 0, 0, 0, 0, 0, 0}, {1, 1, 1}}

An even simpler way is to write

```
ConstantArray @@@ en
```

{{1, 1}, {0}, {1}, {0, 0, 0}, {1}, {0}, {1}, {0, 0, 0, 0, 0, 0, 0}, {1, 1, 1}}

```
runLengthEncoding[u_] := {First[#], Length[#]} & /@ Split[u]
runLengthDecoding[u_] := Flatten[ConstantArray @@@ u]
```

```
runLengthEncoding[u]
```

{{1, 2}, {0, 1}, {1, 1}, {0, 3}, {1, 1}, {0, 1}, {1, 1}, {0, 7}, {1, 3}}

```
runLengthDecoding[%]
```

{1, 1, 0, 1, 0, 0, 0, 1, 0, 1, 0, 0, 0, 0, 0, 0, 0, 1, 1, 1}

18.3.3 Iterating a Mapping

NestList[f, x0, n]   Do n times the iteration $x_{i+1} = f(x_i)$, starting from x_0; give all iterations

Nest[f, x0, n]   Give only the last iteration

NestList gives the starting point x_0 and all the iterations (thus, a list of $n + 1$ elements). **Nest** only gives the final result. The function to be iterated is most naturally written as a pure function. With an unspecified function, we see a general result:

NestList[f[#] &, x0, 3]   {x0, f[x0], f[f[x0]], f[f[f[x0]]]}

Nest[f[#] &, x0, 3]   f[f[f[x0]]]

Note that **ComposeList** does not iterate the same function but possibly a different function each time:

ComposeList[{f, g, h}, x0]   {x0, f[x0], g[f[x0]], h[g[f[x0]]]}

- **Simple Examples**

We do the iteration $x_{i+1} = \cos(x_i)$ four times by starting from $x_0 = 1$:

NestList[Cos[#] &, 1, 4]

{1, Cos[1], Cos[Cos[1]], Cos[Cos[Cos[1]]], Cos[Cos[Cos[Cos[1]]]]}

NestList[Cos[#] &, 1., 4]

{1., 0.540302, 0.857553, 0.65429, 0.79348}

Next, we calculate continued fractions $x_{i+1} = \frac{1}{1+x_i}$:

NestList[1 / (1 + #) &, x, 4]

$$\left\{ x, \; \frac{1}{1+x}, \; \frac{1}{1+\frac{1}{1+x}}, \; \frac{1}{1+\frac{1}{1+\frac{1}{1+x}}}, \; \frac{1}{1+\frac{1}{1+\frac{1}{1+\frac{1}{1+x}}}} \right\}$$

If the starting value is numerical, with **HoldForm** we can get the unsimplified results:

NestList[1 / (1 + #) &, 3, 4]

$$\left\{ 3, \; \frac{1}{4}, \; \frac{4}{5}, \; \frac{5}{9}, \; \frac{9}{14} \right\}$$

NestList[HoldForm[1 / (1 + #)] &, 3, 4]

$$\left\{ 3, \; \frac{1}{1+3}, \; \frac{1}{1+\frac{1}{1+3}}, \; \frac{1}{1+\frac{1}{1+\frac{1}{1+3}}}, \; \frac{1}{1+\frac{1}{1+\frac{1}{1+\frac{1}{1+3}}}} \right\}$$

Then we calculate successive derivatives of 2^{x^2} by applying the iteration formula $f_{i+1}(x) = \partial_x f_i(x)$, $f_0(x) = 2^{x^2}$:

NestList[D[#, x] &, 2 ^ (x^2), 3] // Simplify

$$\left\{ 2^{x^2}, \; 2^{1+x^2} \, x \, \text{Log}[2], \; 2^{1+x^2} \, \text{Log}[2] \left(1 + x^2 \, \text{Log}[4] \right), \; 2^{2+x^2} \, x \, \text{Log}[2]^2 \left(3 + x^2 \, \text{Log}[4] \right) \right\}$$

We use **Differences** to calculate differences:

```
Table[Differences[{a, b, c, d}, n], {n, 4}]
```

$\{\{-a + b, -b + c, -c + d\}, \{a - 2 b + c, b - 2 c + d\}, \{-a + 3 b - 3 c + d\}, \{\}\}$

The same can also be done with **NestList**:

```
NestList[Rest[#] - Most[#] &, {a, b, c, d}, 4]
```

$\{\{a, b, c, d\}, \{-a + b, -b + c, -c + d\}, \{a - 2 b + c, b - 2 c + d\}, \{-a + 3 b - 3 c + d\}, \{\}\}$

■ Newton's Method

```
newton6[f_, x_, x0_, n_] := With[{df = D[f, x]},
  NestList[(x - f / df) /. x → # &, N[x0], n]]
```

```
newton6[f[x], x, x0, 2]
```

$$\left\{x0, x0 - \frac{f[x0]}{f'[x0]}, x0 - \frac{f[x0]}{f'[x0]} - \frac{f\left[x0 - \frac{f[x0]}{f'[x0]}\right]}{f'\left[x0 - \frac{f[x0]}{f'[x0]}\right]}\right\}$$

```
newton6[3 x^3 - E^x, x, 2, 7]
```

$\{2., 1.41942, 1.1019, 0.975117, 0.953089, 0.952446, 0.952446, 0.952446\}$

We can also define a separate function for a single step of Newton's method.

```
newtonStep7[f_, df_, x_, xi_] := (x - f / df) /. x → N[xi]
newton7[f_, x_, x0_, n_] := With[{df = D[f, x]},
  NestList[newtonStep7[f, df, x, #] &, N[x0], n]]
```

Here, **newtonStep7** gives the operation performed in a single step, and **newton7** nests this operation **n** times starting from **x0**. The function **newtonStep7** has several arguments, and so we have to express, in **newton7**, the argument with respect to which the iterations are to be done: This argument is **#** in the pure function. This method of defining separately a single step and the nesting of this step is often used in later chapters. Here are some examples:

```
newton7[f[x], x, x0, 2]
```

$$\left\{x0, x0 - \frac{f[x0]}{f'[x0]}, x0 - \frac{1. f[x0]}{f'[x0]} - \frac{f\left[x0 - \frac{1. f[x0]}{f'[x0]}\right]}{f'\left[x0 - \frac{1. f[x0]}{f'[x0]}\right]}\right\}$$

```
newton7[3 x^3 - E^x, x, 2, 7]
```

$\{2., 1.41942, 1.1019, 0.975117, 0.953089, 0.952446, 0.952446, 0.952446\}$

■ Euler's Method

```
eulerStep2[f_, x_, y_, {xi_, yi_}, h_] := {x + h, y + h f} /. {x → xi, y → yi}
euler2[f_, x_, y_, x0_, y0_, h_, n_] :=
  NestList[eulerStep2[f, x, y, #, h] &, N[{x0, y0}], n]
```

Note that because the pure function in **NestList** has to have only one argument (the **#**), we have to group the variables **xi** and **yi** into a list **{xi, yi}** in **eulerStep2**. We check **euler2** with the problem $y' = f(x, y), y(x_0) = y_0$:

```
euler2[f[x, y], x, y, x0, y0, h, 2] // Column
```

$\{x0, y0\}$
$\{h + x0, y0 + h f[x0, y0]\}$
$\{2 h + x0, y0 + h f[x0, y0] + h f[h + x0, y0 + h f[x0, y0]]\}$

Here are some numerical values:

```
euler2[x - y^2, x, y, 0, 1, 0.1, 10]
```
```
{{0., 1.}, {0.1, 0.9}, {0.2, 0.829}, {0.3, 0.780276},
  {0.4, 0.749393}, {0.5, 0.733234}, {0.6, 0.729471},
  {0.7, 0.736258}, {0.8, 0.75205}, {0.9, 0.775492}, {1., 0.805354}}
```

■ Sampling without Replacement

In Section 18.2.1, p. 554, we presented a procedural program for sampling without replacement. Now we apply the functional programming style. First, we write a function to delete a random element from a given set:

```
del[list_] := Delete[list, RandomInteger[{1, Length[list]}]]
```

If we then iterate this function, we get a set, from which random elements are removed:

```
Nest[del, Range[10], 4]     {1, 2, 3, 6, 9, 10}
```

However, we are interested in the removed elements so that lastly we take a complement:

```
Complement[Range[10], %]     {4, 5, 7, 8}
```

```
SWOR2[list_, n_] := Complement[list, Nest[del, list, n]]
```

```
SWOR2[Range[10], 4]     {1, 4, 7, 9}
```

■ Josephus Problem

We have n persons in a queue. We do the following operation repeatedly: The first person of the current queue moves to the end and the next person leaves the queue. After a number of iterations, only one person is left; let us call him or her the winner. Who is the winner? This is the so-called Josephus problem (see Wellin *et al.*, 2005, p. 109). Suppose, for example, that we have the queue {a, b, c, d}, with a being the first person. After the first step we have the queue {c, d, a}, after the second step the queue {a, c}, and after the third step the queue {a}. In this example, the person a is the winner. We would like to write a program that, for a given queue, reveals the winner.

Suppose, as above, that the initial queue is {a, b, c, d}. To move the first person to the end, use **RotateLeft**:

```
RotateLeft[{a, b, c, d}]     {b, c, d, a}
```

Then drop the first person of this queue:

```
Rest[%]     {c, d, a}
```

We do this operation repeatedly with **Nest**:

```
josephus[list_] := Nest[Rest[RotateLeft[#]] &, list, Length[list] - 1][[1]]
```

```
josephus[{a, b, c, d}]     a
```

Next, we denote people by 1, 2, … and calculate the winning person for various lengths of the queue:

```
t = Table[{n, josephus[Range[n]]}, {n, 1, 15}]
```
```
{{1, 1}, {2, 1}, {3, 3}, {4, 1}, {5, 3}, {6, 5}, {7, 7}, {8, 1},
  {9, 3}, {10, 5}, {11, 7}, {12, 9}, {13, 11}, {14, 13}, {15, 15}}
```

18.3.4 Iterating until Convergence

> **FixedPointList[f, x0]**   Do the iteration $x_{i+1} = f(x_i)$, starting from x_0, until convergence; give all
> iterations
> **FixedPoint[f, x0]**   Give only the last iteration
> **FixedPointList[f, x0, max]**   Iterate until convergence but at most **max** times
> **FixedPoint[f, x0, max]**   Give only the last iteration
>
> *An option:*
> **SameTest**   The test used as the stopping criterion; examples of values: **Automatic** (means **(#1 === #2**
> **&)**), **(Abs[#1 – #2] < 10^-10 &)**, **(Abs[f /. x → #2] < 10^-5 &)**

FixedPoint is similar to **Nest**, but it applies the given function iteratively until the result no longer changes. In particular, **FixedPoint[f, x0, max]** is similar to **Nest[f, x0, n]**, although **FixedPoint** can stop before **max** iterations are done. If both **max** and a stopping criterion are given, then iterations are stopped as soon as the criterion gives **True**, but in all cases at most **max** iterations are done. The third argument **max** prevents infinite calculations.

With an unspecified function, we see a general result:

    FixedPointList[f[#] &, x0, 3]   {x0, f[x0], f[f[x0]], f[f[f[x0]]]}

    FixedPoint[f[#] &, x0, 3]   f[f[f[x0]]]

The stopping criterion is a pure function of the last two iterations; the next-to-last iteration is denoted by **#1** and the last iteration by **#2**. The default stopping criterion is **(SameQ[#1, #2] &)**—that is, **(#1 === #2 &)**. This means that the last two iterations are the same to 16-digit precision (in most computers) (see Section 13.3.5, p. 432). This is a tight condition, but we can formulate milder criteria.

■ The Fixed-Point Method

If we have a nonlinear equation of the form $x = f(x)$, one possibility to solve the equation is to apply the *fixed-point method* by doing the iterations $x_{i+1} = f(x_i)$. **FixedPoint** does exactly the iterations of this method. The method converges at least when $|f'(x)| < 1$ near the solution. We try to solve the equation $x = \cos(x)$:

    t = FixedPointList[Cos[#] &, 0.05];

The last point and the difference of the left- and right-hand sides of the equation at this point are as follows:

    {Last[t], x - Cos[x] /. x → Last[t]}
    {0.739085, 0.}

The iterations proceed as follows:

    ListLinePlot[{Range[0, Length[t] - 1], t}ᵀ,
      PlotRange → {0, 1.05}, AspectRatio → 0.15, ImageSize → 350]

In approximately 15 iterations, we are already near the solution. However, to get it to the last decimal, we still need approximately 75 iterations. Next, we use a custom stopping criterion:

```
FixedPointList[Cos[#] &, 0.05, SameTest → (Abs[#1 - #2] < 10^-2 &)]
```
```
{0.05, 0.99875, 0.541354, 0.857012, 0.654699, 0.793231,
 0.701546, 0.763845, 0.722182, 0.750365, 0.73144, 0.744213, 0.735621}
```

With the following program, we can illustrate the search of a fixed point (see also Sections 28.1.3, p. 932, and 28.2.1, p. 939):

```
cobwebPlot[f_, x_, x0_, n_, a_, b_, opts___] := Module[{xi = x0, t},
  t = Table[{{xi, xi = f /. x → xi}, {xi, xi}}, {n}];
  Plot[{x, f}, {x, a, b}, Epilog → Line[Prepend[Flatten[t, 1], {x0, 0}]], opts]];
```

```
cobwebPlot[Cos[x], x, 0.05, 12, 0, 1.03, AspectRatio → Automatic]
```

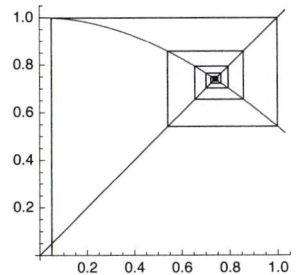

■ Newton's Method

```
newton8[f_, x_, x0_, max_, opts___] := With[{df = D[f, x]},
  FixedPointList[(x - f / df) /. x → # &, N[x0], max, opts]]
```

This formula is the same as that for **newton6** in Section 18.3.3, p. 576, except that we now use **FixedPointList** instead of **NestList** and we have also added the possibility of writing a stopping criterion as an option. The three underscores (___) after **opts** mean zero or more options (see Section 16.1.3, p. 500). Here is an application:

```
f = 3 x^3 - E^x;
```

```
newton8[f, x, 2, 20]
```
```
{2., 1.41942, 1.1019, 0.975117, 0.953089, 0.952446, 0.952446, 0.952446, 0.952446}
```

The last two iterations are as follows:

```
Take[%, -2] // InputForm
```
```
{0.9524456794271664, 0.9524456794271663}
```

They are considered to be the same. Next, we use a loose stopping criterion:

```
newton8[f, x, 2, 20, SameTest → (Abs[#1 - #2] < 10^-3 &)]
```
```
{2., 1.41942, 1.1019, 0.975117, 0.953089, 0.952446}
```

Now we require that the value of the function at the root should be sufficiently small at the last point calculated:

```
newton8[f, x, 2, 20, SameTest → (Abs[f /. x → #2] < 10^-6 &)]
```
```
{2., 1.41942, 1.1019, 0.975117, 0.953089, 0.952446, 0.952446}
```

A modification of **newton8** is also considered in Section 22.3.4, p. 737. There, we also generalize the function to simultaneous nonlinear equations.

```
f =.
```

■ Collatz Sequences

In Section 18.2.2, p. 558, we wrote a procedural program to calculate Collatz sequences. Now we use **FixedPointList**:

```
collatz2[n_] := If[EvenQ[n], n / 2, 3 n + 1]
collatzSequence2[n_] := FixedPointList[collatz2, n, SameTest → (#2 == 1 &)]
```

We stop the iteration once we get the value 1. For example,

```
ListLinePlot[collatzSequence2[27], PlotRange → All, ImageSize -> 210]
```

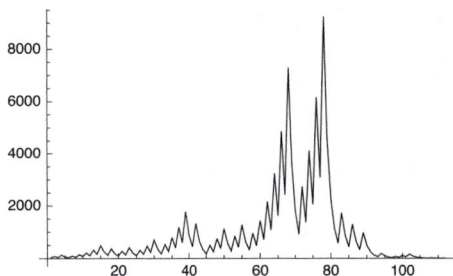

■ Stopping Computation

Throw[val]	Stop computation and return **val** as the value of the nearest enclosing **Catch**
Catch[expr]	Return the argument of the first **Throw** generated in the evaluation of **expr**

Once a **Throw** is encountered, the current calculation is stopped and **Catch** returns the argument of **Throw**; the computation continues after **Catch**. In this way, we are able to stop a whole sequence of nested functions in the case of an error, for example. Recall that **Return** only stops the current **Do**, **While**, or **For** loop.

For example, if we do the recursion $x_{i+1} = e^{x_i}$, $x_0 = 1$, and want the calculation to stop if the result goes over 100, we can use **If**:

```
x = 1; res = {1};
Do[AppendTo[res, x = Exp[x]]; If[x > 100, Return[res]], {10}]
```
$\{1, e, e^e, e^{e^e}\}$

We can also use **Throw** and **Catch**:

```
x = 1; res = {1};
Catch[Do[AppendTo[res, x = Exp[x]]; If[x > 100, Throw[res]], {10}]]
```
$\{1, e, e^e, e^{e^e}\}$

However, in procedural programming the use of **Return** is often easier than the use of **Throw** and **Catch**.

In functional programming **Throw** and **Catch** may have some use. For example, if we use **NestList**, we can stop the iteration at the point we want:

```
Catch[NestList[If[# > 100, Throw[#], Exp[#]] &, 1, 50]]       e^{e^e}
```

However, now we only get the last iteration, not all the iterations. If we use **FixedPointList**, we can use the **SameTest** option, and in this way we get all the iterations:

> **FixedPointList[Exp[#] &, 1, 50, SameTest → (#2 > 100 &)]** $\{1, e, e^e, e^{e^e}\}$

18.3.5 More General Testing of Convergence

NestWhile[f, x0, test]   While **test** gives **True**, do the iteration $x_{i+1} = f(x_i)$, starting from x_0; give the first x_i, $i = 0, 1, \ldots$, for which **test** does not give **True**

NestWhile[f, x0, test, m]   Use the most recent m results as arguments to **test**

NestWhile[f, x0, test, All]   Use all results so far as arguments to **test**

NestWhile[f, x0, UnsameQ, All]   Iterate until a previous result appears the second time

NestWhile[f, x0, test, m, max]   Do at most **max** iterations

NestWhile[f, x0, test, m, max, n]   Apply **f** an additional **n** times after **test** fails or **max** iterations have already been done

NestWhile[f, x0, test, m, max, -n]   Give the result found when **f** had been applied **n** fewer times

NestWhile[f, x0, test, m, ∞, -1]   Give the last result for which **test** gives **True**

We also have **NestWhileList**, which gives all iterations. **NestWhile** is similar to **FixedPoint**, but the tests used in these commands are opposites of each other: Iterations stop in **FixedPoint** once the test (the default of **SameTest** is **SameQ**) gives **True**, whereas in **NestWhile**, they stop when the test no longer gives **True**. Indeed, the following commands are equivalent:

> **FixedPoint[f, x0, max]**

> **NestWhile[f, x0, UnsameQ, 2, max]**

■ **Simple Examples**

We do the same calculation in two ways:

> **FixedPointList[Cos[#] &, 0.05, 10, SameTest → (Abs[#1 - #2] < 0.1 &)]**
> $\{0.05, 0.99875, 0.541354, 0.857012, 0.654699, 0.793231, 0.701546\}$

> **NestWhileList[Cos[#] &, 0.05, Abs[#1 - #2] ≥ 0.1 &, 2, 10]**
> $\{0.05, 0.99875, 0.541354, 0.857012, 0.654699, 0.793231, 0.701546\}$

Newton's method can be written as follows:

```
newton9[f_, x_, x0_, max_, test_] := With[{df = D[f, x]},
  NestWhileList[(x - f / df) /. x → # &, N[x0], test, 2, max]]
```

> **newton9[3 x^3 - E^x, x, 2, 20, UnsameQ]**
> $\{2., 1.41942, 1.1019, 0.975117, 0.953089, 0.952446, 0.952446, 0.952446\}$

NestWhile is useful in finding numbers of given properties. We find the first integer n that is prime and also $n - 2$ and $n - 6$ are primes:

> **NestWhile[# + 1 &, 1, ! (PrimeQ[#1] && PrimeQ[#5] && PrimeQ[#7]) &, 7]**
> 13

Thus, the number is 13; indeed, 13, 11, and 7 are all primes. Note that above **#7** represents the last iteration, **#5** the third-to-last iteration, and **#1** the seventh-to-last iteration.

■ **Tossing a Die**

If we want to use more than the last two iterations in the test, then **NestWhile** is very useful. As an example, we toss a die until we get a result the second time:

```
SeedRandom[3]; NestWhileList[RandomInteger[{1, 6}] &, 0, UnsameQ, All] // Rest
{4, 6, 1, 2, 3, 1}
```

Here, we use all iterations so far as arguments to the test; remember that **UnsameQ** accepts several arguments. In place of **UnsameQ**, we could also have written the more complete pure function **UnsameQ[##]&**, where **##** represents the whole sequence of results thus far computed.

■ **Power Sum of Digits**

In the next example, we calculate the sum of the third powers of the digits of a number. To the result we apply the same operation; this is continued until we get a result for the second time.

```
powerSumOfDigits[n_] := NestWhileList[Total[IntegerDigits[#] ^ 3] &, n, UnsameQ, All]
```

```
Table[powerSumOfDigits[i], {i, 1, 10}] // Column
{1, 1}
{2, 8, 512, 134, 92, 737, 713, 371, 371}
{3, 27, 351, 153, 153}
{4, 64, 280, 520, 133, 55, 250, 133}
{5, 125, 134, 92, 737, 713, 371, 371}
{6, 216, 225, 141, 66, 432, 99, 1458, 702, 351, 153, 153}
{7, 343, 118, 514, 190, 730, 370, 370}
{8, 512, 134, 92, 737, 713, 371, 371}
{9, 729, 1080, 513, 153, 153}
{10, 1, 1}
```

We see that either a single number begins to repeat or we end up with a cycle of several numbers. We could find that, for example, numbers 1, 153, 370, 371, and 407 repeat themselves and sequences of numbers {55, 250, 133}, {136, 244}, {160, 217, 352}, {919, 1459} form cycles.

18.3.6 Iterating with a Resource

FoldList[f, x0, {y1, y2, …, yn}] Do the iteration $x_{i+1} = f(x_i, y_{i+1})$, $i = 0, …, n - 1$, starting from
x_0, and give x_n; give all iterations
Fold[f, x0, {y1, y2, …, yn}] Give only the last iteration

Fold is similar to **Nest**, but whereas **Nest** uses a function of one variable, **Fold** uses a function of two variables. The second argument of **Fold** is the starting value. The third argument is a list from which one element at a time is fed in as the second argument of the function. The list can be considered as a resource from which a new element is drawn at each iteration. With an unspecified function, we see a general result:

```
FoldList[f[#1, #2] &, x_0, {y_1, y_2, y_3}]
{x_0, f[x_0, y_1], f[f[x_0, y_1], y_2], f[f[f[x_0, y_1], y_2], y_3]}
```

```
Fold[f[#1, #2] &, x_0, {y_1, y_2, y_3}]
f[f[f[x_0, y_1], y_2], y_3]
```

■ **Cumulative Sums, Products, and Maximums**

Consider the following list:

```
t = {a, b, c};
```

To calculate the cumulative sums of the elements of this list, we have **Accumulate**:

```
Accumulate[t]     {a, a + b, a + b + c}
```

As an exercise, we also use **FoldList**:

```
FoldList[Plus[#1, #2] &, 0, t]     {0, a, a + b, a + b + c}
```

With **Rest** we can drop the starting value 0:

```
% // Rest     {a, a + b, a + b + c}
```

The arguments in the pure function **Plus[#1, #2] &** are not necessary, and so we can simply write

```
FoldList[Plus, 0, t] // Rest     {a, a + b, a + b + c}
```

In this example, **FoldList** works as follows:

```
{x_0 = 0, x_1 = Plus[x_0, a], x_2 = Plus[x_1, b], x_3 = Plus[x_2, c]}
{0, a, a + b, a + b + c}
```

Similarly, we can calculate cumulative products and maximums:

```
FoldList[Plus, 0, {a, b, c, … }] //Rest  Compute the cumulative sums
FoldList[Times, 1, {a, b, c, … }] //Rest  Compute the cumulative products
FoldList[Max, 0, {a, b, c, … }] //Rest  Compute the cumulative maximums
```

Here are the first few factorials:

```
FoldList[Times, 1, Range[8]] // Rest
{1, 2, 6, 24, 120, 720, 5040, 40 320}
```

Next, we compute cumulative maximums—that is, record values:

```
FoldList[Max, 0,
  {16, 12, 8, 14, 5, 10, 3, 2, 17, 7, 15, 18, 6, 1, 20, 13, 9, 11, 4, 19}] // Rest
{16, 16, 16, 16, 16, 16, 16, 16, 17, 17, 17, 18, 18, 18, 20, 20, 20, 20, 20, 20}
```

The different record values are

```
% // Union     {16, 17, 18, 20}
```

■ **Other Examples**

Construct a number from given digits:

```
FoldList[10 #1 + #2 &, 0, {2, 9, 3, 6, 1, 5}]
{0, 2, 29, 293, 2936, 29 361, 293 615}
```

Write a third-degree polynomial $a x^3 + b x^2 + c x + d$ in Horner form:

```
FoldList[#2 + #1 x &, 0, {a, b, c, d}]
{0, a, b + a x, c + x (b + a x), d + x (c + x (b + a x))}
```

Write a continued fraction:

```
FoldList[1 / (#2 + #1) &, x, {a, b, c}]
```

$$\left\{x, \ \frac{1}{a + x}, \ \frac{1}{b + \frac{1}{a+x}}, \ \frac{1}{c + \frac{1}{b+\frac{1}{a+x}}}\right\}$$

18.4 Rule-Based Programming

18.4.1 Rules

■ **Rule-Based Programming**

Thus far, we have encountered operations such as the following:

```
Grid[{{Text["Operation"], Text["Meaning"], Text["Example"]},
  {Style["=", Bold], Text["assign a value"], Style["x = 3", Bold]},
  {Style[":=", Bold], Text["define a function"],
   Style["f[x_] := x Sin[x]", Bold]}, {Style["→", Bold],
   Text["make a transformation"], Style["a x + b /. x → 3", Bold]}},
  Alignment → Left, Dividers → {False, 2 → True}, Spacings → {1.5, 0.5}]
```

Operation	Meaning	Example
=	assign a value	`x = 3`
:=	define a function	`f[x_] := x Sin[x]`
→	make a transformation	`a x + b /. x → 3`

All of these operations can be seen as *rules*. For example, **x = 3** is the rule that whenever **x** appears (in any expression during the rest of the current session), it will be replaced with 3. The rule **f[x_] := x Sin[x]** tells *Mathematica* that whenever **f[anything]** occurs during the rest of the current session, it should be replaced with **anything Sin[anything]**. The rule **a + b x /. x → 3** asks that wherever **x** appears in the expression **a + b x** (and only in this particular occurrence of the expression), it should be replaced with 3. *Rule-based programming* uses such rules.

In addition, an important concept in rule-based programming is a *pattern*. For example, in **f[x_]**, the argument is given in the form of a pattern: **x_** matches anything, and so this is a very general pattern. We can form a more restrictive pattern such as the one in **f[x_^y_] := …**, in which the argument of **f** can be anything in the form **x^y**.

Thus, central in rule-based programming are minor symbols such as → and _ that relate to rules and patterns. In addition to these, rule-based programming does not introduce any new programming commands.

Here, we consider rules; the treatment is short because we have already studied most rules. Patterns, on the other hand, were considered in Chapter 16. Thus, we already have in our hands the tools of rule-based programming. Accordingly, in Section 18.4.2 we can concentrate on examples of rule-based programming.

■ **Global and Local Rules**

	Global rules	Local rules
Evaluate the right-hand side	= (**Set**)	→ (**Rule**)
Delay the evaluation of the right-hand side	:= (**SetDelayed**)	:→ (**RuleDelayed**)

These are the four most common rules. Some rules evaluate the right-hand side, whereas others do not; the latter delay the evaluation until the rule is actually applied. Global rules are applied whenever the left-hand side is encountered. Local rules are applied only to the given expression.

We have already considered =, →, and := in Sections 13.1.1, p. 414, 13.1.2, p. 416, and 17.1.1, p. 512. These are the most important rules. The delayed local rule :> is quite seldom used. Some examples of rules follow.

■ **Examples**

With the rule **f = Integrate[1 + x, x]**, the right-hand side is evaluated so that **f** gets the value $x + \frac{1}{2} x^2$. The rule is global: The value $x + \frac{1}{2} x^2$ is used for **f** whenever **f** is encountered in the current session.

With the rule **g[y_, x_] := Integrate[y, x]**, the right-hand side is not evaluated at the time **g** is defined; it is evaluated only when we ask for the value of **g** for specific **y** and **x**. This rule, too, is a global rule and is used whenever **g** is encountered.

We then consider local rules. Here is an example:

```
a + b x /. x → Integrate[Sin[x], {x, 0, Pi}]
a + 2 b
```

The integral of the right-hand side of the rule is evaluated and gives 2, and this value is then substituted for **x** in **a + b x**. The rule **x → …** is local and is applied only to the present occurrence of the expression **a + b x**.

To see the effect of a delayed rule **:>** (which *Mathematica* transforms to the form :→), we consider the following more complex examples:

```
1 + f[a + 2 b x] /. f[y_] → f[Integrate[y, x]]
1 + f[x (a + 2 b x)]

1 + f[a + 2 b x] /. f[y_] :→ f[Integrate[y, x]]
1 + f[a x + b x²]
```

In these examples, our aim is to replace the arguments of all functions **f[…]** with their integrals. The first example seems not to work. The reason is that the right-hand side **f[Integrate[y, x]]** of the rule is evaluated and gives **f[x y]**, and when **f[anything]** is replaced with **f[x anything]** in **1 + f[a + 2 b x]**, we get the result shown.

On the other hand, the second example works because the right-hand side of the rule is not evaluated until the rule is first applied. Thus, we first get **1 + f[Integrate[a + 2 b x, x]]** and then the result shown.

The following is another example:

```
{a, a} /. a → RandomReal[]      {0.192013, 0.192013}

{a, a} /. a :→ RandomReal[]     {0.153438, 0.351603}
```

In the first case, the right-hand side **RandomReal[]** of the rule is first computed and then this value is substituted for **a** in **{a, a}**. Thus, we get the same random number twice. In the second case, **RandomReal[]** is first substituted for **a** in **{a, a}** and then the resulting expression is evaluated. Thus, we get two different random numbers.

■ Giving Several Rules

When defining a function, we often have to consider several special cases. As an example, here is a function to define an absolute value:

```
abs[x_] := Which[
  ! NumericQ[x], x,
  x ∈ Complexes, Sqrt[Re[x]^2 + Im[x]^2],
  x < 0, -x,
  x = 0, 0,
  x > 0, x]
```

```
abs /@ {4, -3, 0, 2 + 3 I, a}
```

$$\left\{4,\ 3,\ 0,\ \sqrt{13}\ ,\ a\right\}$$

Another way to define the same function is to write a separate definition or rule for each case by using conditions with /; (see Section 16.1.2, p. 499):

```
abs2[x_ /; ! NumericQ[x]] := x
abs2[x_ /; x ∈ Complexes] := Sqrt[Re[x]^2 + Im[x]^2]
abs2[x_ /; x < 0] := -x
abs2[0] := 0
abs2[x_ /; x > 0] := x
```

```
abs2 /@ {4, -3, 2 + 3 I, a}
```

$$\left\{4,\ 3,\ \sqrt{13}\ ,\ a\right\}$$

When we call **abs2**, *Mathematica* will check through the rules to find a rule whose pattern allows the use of the given argument.

A condition is often written next to a variable. A condition can also be written in other places, for example, at the end of the rule. The function **int** gives the integral of $\sin(m\,x)\cos(n\,x)$ from 0 to π:

```
int[m_Integer, n_Integer] := 0 /; EvenQ[m + n]
int[m_Integer, n_Integer] := 2 m / (m^2 - n^2) /; OddQ[m + n]
```

In this case, the conditions depend on both argument **m** and argument **n**, and in such a case a condition cannot be written next to a variable; a condition has to be written after the left-hand side or at the end of the rule. Here are some examples:

```
{int[0, 1], int[1, 0], int[1, 1], int[1, 2]}
```

$$\left\{0,\ 2,\ 0,\ -\frac{2}{3}\right\}$$

■ Rule Base

For a given symbol, *Mathematica* maintains a *rule base* that contains all the rules defined for the symbol. With a question mark we can ask *Mathematica* to show the rule base:

```
? abs2
```

```
Global`abs2

abs2[0] := 0

abs2[x_ /; ! NumericQ[x]] := x

abs2[x_ /; x ∈ Complexes] := √(Re[x]² + Im[x]²)

abs2[x_ /; x > 0] := x

abs2[x_ /; x < 0] := -x
```

When we compare the original definition of **abs2** with the rule base of **abs2**, we can find some differences in the order of the rules. Indeed, *Mathematica* will order the rules according to a certain principle. The principle is that more specific rules should appear earlier in the rule base than more general rules; that is, *Mathematica* tries to order the rules *from more specific to more general*. Regarding **abs2**, we can see that the special rule for **abs2[0]** is at the beginning of the rule base. Whenever the appropriate ordering is not clear, *Mathematica* stores rules in the order we give them.

When we write rules for a symbol, often we try to give an exhaustive set of rules so that in all cases we can find a suitable rule to apply. In this case, the last rule of the rule base can be written without any conditions because if the earlier rules have not been suitable, the last rule has to be suitable. Thus, in the rules defining **abs2**, we can leave out the condition $x < 0$:

```
abs3[x_ /; ! NumericQ[x]] := x
abs3[x_ /; x ∈ Complexes] := Sqrt[Re[x] ^ 2 + Im[x] ^ 2]
abs3[x_] := -x
abs3[0] := 0
abs3[x_ /; x > 0] := x
```

```
abs3 /@ {4, -3, 2 + 3 I, a}
```

$$\left\{4, 3, \sqrt{13}, a\right\}$$

In rare cases, it may happen that the rules are not in appropriate order in the rule base. We can then change the order. For example, if we would like the rules of **abs2** for the cases $x < 0$ and $x > 0$ to be in the opposite order in the rule base, we could write as follows:

```
DownValues[abs2] = DownValues[abs2][[{1, 2, 3, 5, 4}]]
```

$\big\{$HoldPattern[abs2[0]] :→ 0, HoldPattern[abs2[x_ /; ! NumericQ[x]]] :→ x,

HoldPattern[abs2[x_ /; x ∈ Complexes]] :→ $\sqrt{Re[x]^2 + Im[x]^2}$,

HoldPattern[abs2[x_ /; x < 0]] :→ -x, HoldPattern[abs2[x_ /; x > 0]] :→ x$\big\}$

Now, the order of the two rules has changed:

```
? abs2
```

Global`abs2

abs2[0] := 0

abs2[x_ /; ! NumericQ[x]] := x

$$abs2[x_ /; x \in Complexes] := \sqrt{Re[x]^2 + Im[x]^2}$$

abs2[x_ /; x < 0] := -x

abs2[x_ /; x > 0] := x

■ **Downvalues and Upvalues**

	Downvalues	Upvalues
Evaluate the right-hand side	= (**Set**)	^= (**UpSet**)
Delay the evaluation of the right-hand side	:= (**SetDelayed**)	^:= (**UpSetDelayed**)

Setting a value such as **x = 3** is also called defining a *downvalue* for **x**. Similarly, a function definition such as **f[x_] := x Sin[x]** defines a downvalue for **f**. We also have *upvalues*.

In the following example, we consider objects **sin**, **cos**, and **tan** and associate two properties or upvalues with each object, namely its derivative and integral:

```
der[sin] ^= cos; int[sin] ^= -cos;
der[cos] ^= -sin; int[cos] ^= sin;
der[tan] ^= sec^2; int[tan] ^= -ln[cos];
```

For example, we can ask information about the **tan** object:

```
? tan
```

Global`tan

$$der[tan] \; \hat{} = sec^2$$

$$int[tan] \; \hat{} = -ln[cos]$$

Upvalues can also be defined with **/:** (**TagSet**):

```
sin /: der[sin] = cos;
sin /: int[sin] = -cos;
cos /: der[cos] = -sin;
cos /: int[cos] = sin;
tan /: der[tan] = sec^2;
tan /: int[tan] = -ln[cos];
```

This is a brief example of object-oriented programming. For more information about object-oriented programming, see Gray (1997) or Maeder (1994).

■ **Dispatching**

dispatchrules = Dispatch[rules]   Generate an optimized dispatch table representation of a list of rules

expr /. dispatchrules   Apply the dispatch rules

Dispatching allows **/.** to "dispatch" to potentially applicable rules immediately rather than testing all of the rules in turn. If the list of rules is long, this may save a significant amount of time. As an example, we form a list of rules and dispatch them:

```
rules = Table[f[i] → i^2, {i, 3000}];

dispatchrules = Dispatch[rules];
```

Then we do the same calculation with both sets of rules:

```
Do[f[i] /. rules, {i, 3000}] // Timing
{1.76769, Null}

Do[f[i] /. dispatchrules, {i, 3000}] // Timing
{0.015387, Null}
```

18.4.2 Examples of Rule-Based Programming

■ Collatz Sequences

In Section 18.2.2, p. 558, we presented a procedural program to calculate Collatz sequences. In Section 18.3.4, p. 580, we wrote the following functional program:

```
collatz2[n_] := If[EvenQ[n], n / 2, 3 n + 1]
collatzSequence2[n_] := FixedPointList[collatz2, n, SameTest → (#2 == 1 &)]
```

Now we slightly change this program in that we write separate rules for even and odd *n*:

```
collatz3[n_ ? EvenQ] := n / 2
collatz3[n_ ? OddQ] := 3 n + 1
collatzSequence3[n_Integer ? Positive] :=
 FixedPointList[collatz3, n, SameTest → (#2 == 1 &)]
```

```
collatzSequence3[19]
{19, 58, 29, 88, 44, 22, 11, 34, 17, 52, 26, 13, 40, 20, 10, 5, 16, 8, 4, 2, 1}
```

■ Interchanging Two Elements

We have a list for which we would like to change the order of the first two elements. One approach is the following:

```
interchange[x_List /; Length[x] > 1] := Join[{x[[2]], x[[1]]}, Drop[x, 2]]
interchange[x_List /; Length[x] ≤ 1] := x
```

Thus, if the list is empty or only has one element, then return the list as such; otherwise, change the order of the first two elements. Try the program:

```
{interchange[{}], interchange[{1}], interchange[{1, 2, 3, 4}]}
{{}, {1}, {2, 1, 3, 4}}
```

Check the rule base:

```
? interchange
```

Global`interchange

```
interchange[x_List /; Length[x] > 1] := Join[{x〚2〛, x〚1〛}, Drop[x, 2]]
```

```
interchange[x_List /; Length[x] ≤ 1] := x
```

As noted in Section 18.4.1, p. 586, if the rules defined for a function cover all cases exhaustively, in the last rule we do not need any conditions, because it is used if other rules are not applicable. Thus, we could also write the following:

```
interchange2[x_List /; Length[x] > 1] := Join[{x〚2〛, x〚1〛}, Drop[x, 2]]
interchange2[x_List] := x
```

```
{interchange2[{}], interchange2[{1}], interchange2[{1, 2, 3, 4}]}
```
```
{{}, {1}, {2, 1, 3, 4}}
```

Another, simpler approach is the following:

```
interchange3[{a_, b_, c___}] := {b, a, c}
interchange3[x_List] := x
```

The first rule again covers the cases in which the list has two or more elements. The second rule is applied if the list is empty or only has one element. The rule base is as follows:

```
? interchange2
```

Global`interchange2

```
interchange2[x_List /; Length[x] > 1] := Join[{x〚2〛, x〚1〛}, Drop[x, 2]]
```

```
interchange2[x_List] := x
```

Again, in the second rule we do not need any conditions. Try the program:

```
{interchange3[{}], interchange3[{1}], interchange3[{1, 2, 3, 4}]}
```
```
{{}, {1}, {2, 1, 3, 4}}
```

■ **Classifying Poker Hands 1**

We would like to write a function that classifies hands of poker. A hand contains five cards. Each card has one of the values 2, 3, …, 10, J, Q, K, and A. Each card belongs to one of the four suits—spades ♠, clubs ♣, hearts ♡, and diamonds ◊. We use the following terminology:

- A hand containing one value four times is *four of a kind*.
- A hand containing one value three times and another value two times is *full house*.
- A hand containing one value three times is *three of a kind*.
- A hand containing one value two times and another value two times is *two pair*.
- A hand containing one value two times is *one pair*.
- A hand containing five consecutive values is *straight*.
- Other hands are called *nothing*.

We assume that the hands are in ascending order according to the value. One way to classify poker hands is to write down all the cases that can appear:

```
poker[{a_, a_, a_, a_, b_}] := "Four of a kind"
poker[{a_, b_, b_, b_, b_}] := "Four of a kind"
poker[{a_, a_, a_, b_, b_}] := "Full house"
poker[{a_, a_, b_, b_, b_}] := "Full house"
poker[{a_, a_, a_, b_, c_}] := "Three of a kind"
poker[{a_, b_, b_, b_, c_}] := "Three of a kind"
poker[{a_, b_, c_, c_, c_}] := "Three of a kind"
poker[{a_, a_, b_, b_, c_}] := "Two pair"
poker[{a_, a_, b_, c_, c_}] := "Two pair"
poker[{a_, b_, b_, c_, c_}] := "Two pair"
poker[{a_, a_, b_, c_, d_}] := "Pair"
poker[{a_, b_, b_, c_, d_}] := "Pair"
poker[{a_, b_, c_, c_, d_}] := "Pair"
poker[{a_, b_, c_, d_, d_}] := "Pair"
poker[hand_ /; MemberQ[straights, hand]] := "Straight"
poker[hand_] := "Nothing"
straights = Partition[Join[Range[2, 10], {J, Q, K, A}], 5, 1];
```

For example, in `poker[{a_, a_, a_, b_, c_}]` we have three times the same value in succession and two other values. Note that we do not need to constrain **a**, **b**, and **c** to be different because if some of these values are the same, a previously presented case of `poker` applies. Indeed, we could see with `?poker` that *Mathematica* applies the rules in the same order that we have presented them here. Thus, the order in which we write the previous rules is significant.

To check whether we have a straight, we have, in the previous box, formed a list of all straights:

```
straights = Partition[Join[Range[2, 10], {J, Q, K, A}], 5, 1]
```

```
{{2, 3, 4, 5, 6}, {3, 4, 5, 6, 7}, {4, 5, 6, 7, 8}, {5, 6, 7, 8, 9}, {6, 7, 8, 9, 10},
 {7, 8, 9, 10, J}, {8, 9, 10, J, Q}, {9, 10, J, Q, K}, {10, J, Q, K, A}}
```

Here is test material containing 16 artificial poker hands:

```
hands = {{2, 2, 2, 2, 3}, {2, 3, 3, 3, 3},
    {2, 2, 3, 3, 3}, {2, 2, 2, 3, 3},
    {2, 2, 2, 3, 4}, {2, 3, 3, 3, 4}, {2, 3, 4, 4, 4},
    {2, 2, 3, 3, 4}, {2, 2, 3, 4, 4}, {2, 3, 3, 4, 4},
    {2, 2, 3, 4, 5}, {2, 3, 3, 4, 5}, {2, 3, 4, 4, 5}, {2, 3, 4, 5, 5},
    {2, 3, 4, 5, 6},
    {2, 4, 5, 6, 7}};
```

These hands contain the following numbers of various hands: four of a kind, 2; full house, 2; three of a kind, 3; two pair, 3; pair, 4; straight, 1; and nothing, 1. We then ask our program to classify the hands:

```
poker /@ hands
```

```
{Four of a kind, Four of a kind, Full house, Full house,
 Three of a kind, Three of a kind, Three of a kind, Two pair,
 Two pair, Two pair, Pair, Pair, Pair, Pair, Straight, Nothing}
```

■ Classifying Poker Hands 2

Now we use another approach to classify poker hands. In four of a kind, require four cards of the same value and allow zero or more cards at the beginning and end:

```
MatchQ[#, {___, a_, a_, a_, a_, ___}] & /@ {{2, 2, 2, 2, 3}, {2, 3, 3, 3, 3}}
```

```
{True, True}
```

In full house, we have two possibilities:

```
MatchQ[#, {a_, a_, b_, b_, b_} | {a_, a_, a_, b_, b_}] & /@
  {{2, 2, 3, 3, 3}, {2, 2, 2, 3, 3}}
{True, True}
```

Here, | is a pattern meaning alternatives. In three of a kind, require three cards of the same value:

```
MatchQ[#, {a___, b_, b_, b_, c___}] & /@
  {{2, 2, 2, 3, 4}, {2, 3, 3, 3, 4}, {2, 3, 4, 4, 4}}
{True, True, True}
```

In a two pair, require that we have two times two cards of the same value:

```
MatchQ[#, {a___, b_, b_, c___, d_, d_, e___}] & /@
  {{2, 2, 3, 3, 4}, {2, 2, 3, 4, 4}, {2, 3, 3, 4, 4}}
{True, True, True}
```

In one pair, require that we have two cards of the same value:

```
MatchQ[#, {a___, b_, b_, c___}] & /@
  {{2, 2, 3, 4, 5}, {2, 3, 3, 4, 5}, {2, 3, 4, 4, 5}, {2, 3, 4, 5, 5}}
{True, True, True, True}
```

To treat the case of a straight, we use the same list of all straights formed previously:

```
straights = Partition[Join[Range[2, 10], {J, Q, K, A}], 5, 1];
```

Then we can ask whether a given hand is a straight:

```
MemberQ[straights, {2, 3, 4, 5, 6}]

True
```

Now we form a program that classifies a given poker hand:

```
poker2[h_ ? VectorQ /; Length[h] == 5 && OrderedQ[h]] :=
 With[{straights = Partition[Join[Range[2, 10], {J, Q, K, A}], 5, 1]},
  Switch[h,
   {___, a_, a_, a_, a_, ___}, "Four of a kind",
   {a_, a_, b_, b_, b_} | {a_, a_, a_, b_, b_}, "Full house",
   {a___, b_, b_, b_, c___}, "Three of a kind",
   {a___, b_, b_, c___, d_, d_, e___}, "Two pair",
   {a___, b_, b_, c___}, "Pair",
   hand_ /; MemberQ[straights, hand], "Straight",
   _, "Nothing"]]
```

We use the same test material presented previously:

```
poker2 /@ hands
{Four of a kind, Four of a kind, Full house, Full house,
  Three of a kind, Three of a kind, Three of a kind, Two pair,
  Two pair, Two pair, Pair, Pair, Pair, Pair, Straight, Nothing}
```

Again note that, for example, in {a___, b_, b_, b_, c___}, representing three of a kind, we do not need to constrain **a**, **b**, and **c** to be different because if some of these values are the same, a previously presented case applies. Thus, the order in which we write the cases in **Switch** is significant.

■ Consecutive Numbers in Lotto

In an example in Section 18.2.2, p. 560, we considered the lotto game with procedural programming. Let us now try rule-based programming. Recall that in the lotto game in Finland, each player guesses 7 numbers from the numbers 1, 2, ..., 39. We would like to study the frequencies of results having a variable number of consecutive numbers. To test our procedures, we give here a set of 25 artificial lotto results:

```
lottoResults = {{1, 2, 3, 4, 5, 6, 7}, {1, 2, 3, 4, 5, 6, 11},
    {1, 11, 12, 13, 14, 15, 16}, {1, 2, 3, 4, 5, 11, 21}, {1, 11, 12, 13, 14, 15, 21},
    {1, 11, 21, 22, 23, 24, 25}, {1, 2, 11, 12, 13, 14, 15}, {1, 2, 3, 4, 5, 11, 12},
    {1, 2, 3, 4, 11, 21, 31}, {1, 11, 12, 13, 14, 21, 31}, {1, 11, 21, 22, 23, 24, 31},
    {1, 11, 21, 31, 32, 33, 34}, {1, 2, 11, 12, 13, 14, 21}, {1, 2, 11, 21, 22, 23, 24},
    {1, 2, 3, 4, 11, 12, 21}, {1, 2, 3, 11, 12, 13, 14}, {1, 2, 3, 4, 11, 12, 13},
    {1, 11, 12, 13, 21, 31, 39}, {1, 11, 12, 21, 31, 32, 33},
    {1, 2, 11, 12, 21, 22, 23}, {1, 11, 12, 13, 21, 22, 23}, {1, 5, 6, 11, 21, 31, 39},
    {1, 5, 6, 11, 12, 21, 31}, {1, 2, 11, 12, 21, 22, 31}, {1, 3, 5, 7, 9, 11, 13}};
```

Here, the first result has seven consecutive numbers, the next two have six consecutive numbers, and so on. As for the previous test material, we assume that all lotto results have been ordered into ascending order.

We proceed in two phases. In the first phase, we only code the lotto results in a suitable way. In the second phase, we then calculate the frequencies.

First, we mark all sequences having seven consecutive numbers with a string:

```
Take[12 = lottoResults /. {a_, b_, c_, d_, e_, f_, g_} /; g == a + 6 → {"seven"}, 2]

{{seven}, {1, 2, 3, 4, 5, 6, 11}}
```

Later, we will count all results containing, for example, "seven". Next, we mark all sequences having six consecutive numbers:

```
Take[13 = 12 /. {a___, b_, c_, d_, e_, f_, g_, h___} /; g == b + 5 → {a, "six", h}, 4]

{{seven}, {six, 11}, {1, six}, {1, 2, 3, 4, 5, 11, 21}}
```

We can continue in this way:

```
Take[14 = 13 /. {a___, b_, c_, d_, e_, f_, g___} /; f == b + 4 → {a, "five", g}, 9]

{{seven}, {six, 11}, {1, six}, {five, 11, 21}, {1, five, 21},
 {1, 11, five}, {1, 2, five}, {five, 11, 12}, {1, 2, 3, 4, 11, 21, 31}}
```

```
Take[15 = 14 /. {a___, b_, c_, d_, e_, f___} /; e == b + 3 → {a, "four", f}, 18]

{{seven}, {six, 11}, {1, six}, {five, 11, 21}, {1, five, 21}, {1, 11, five},
 {1, 2, five}, {five, 11, 12}, {four, 11, 21, 31}, {1, four, 21, 31},
 {1, 11, four, 31}, {1, 11, 21, four}, {1, 2, four, 21}, {1, 2, 11, four},
 {four, 11, 12, 21}, {1, 2, 3, four}, {four, 11, 12, 13}, {1, 11, 12, 13, 21, 31, 39}}
```

```
16 = 15 /. {a___, b_, c_, d_, e___} /; d == b + 2 → {a, "three", e}

{{seven}, {six, 11}, {1, six}, {five, 11, 21}, {1, five, 21}, {1, 11, five},
 {1, 2, five}, {five, 11, 12}, {four, 11, 21, 31}, {1, four, 21, 31}, {1, 11, four, 31},
 {1, 11, 21, four}, {1, 2, four, 21}, {1, 2, 11, four}, {four, 11, 12, 21},
 {three, four}, {four, three}, {1, three, 21, 31, 39}, {1, 11, 12, 21, three},
 {1, 2, 11, 12, three}, {1, three, 21, 22, 23}, {1, 5, 6, 11, 21, 31, 39},
 {1, 5, 6, 11, 12, 21, 31}, {1, 2, 11, 12, 21, 22, 31}, {1, 3, 5, 7, 9, 11, 13}}
```

Now we observe that in the fifth result from the end, we have another sequence with three consecutive numbers. Thus, we have to make the same transformation again, or we can use the repeated transformation //.:

```
16 = 15 //. {a___, b_, c_, d_, e___} /; d == b + 2 → {a, "three", e}

{{seven}, {six, 11}, {1, six}, {five, 11, 21}, {1, five, 21}, {1, 11, five},
 {1, 2, five}, {five, 11, 12}, {four, 11, 21, 31}, {1, four, 21, 31}, {1, 11, four, 31},
 {1, 11, 21, four}, {1, 2, four, 21}, {1, 2, 11, four}, {four, 11, 12, 21},
 {three, four}, {four, three}, {1, three, 21, 31, 39}, {1, 11, 12, 21, three},
 {1, 2, 11, 12, three}, {1, three, three}, {1, 5, 6, 11, 21, 31, 39},
 {1, 5, 6, 11, 12, 21, 31}, {1, 2, 11, 12, 21, 22, 31}, {1, 3, 5, 7, 9, 11, 13}}
```

Similarly, we also apply the repeated transformation in the last case:

```
16 //. {a___, b_, c_, d___} /; c == b + 1 → {a, "two", d}
```

{{seven}, {six, 11}, {1, six}, {five, 11, 21}, {1, five, 21}, {1, 11, five},
{two, five}, {five, two}, {four, 11, 21, 31}, {1, four, 21, 31},
{1, 11, four, 31}, {1, 11, 21, four}, {two, four, 21}, {two, 11, four},
{four, two, 21}, {three, four}, {four, three}, {1, three, 21, 31, 39},
{1, two, 21, three}, {two, two, three}, {1, three, three}, {1, two, 11, 21, 31, 39},
{1, two, two, 21, 31}, {two, two, two, 31}, {1, 3, 5, 7, 9, 11, 13}}

Now we have coded all the test results. We put the commands into a program that codes one lotto result:

```
lottoCodes[l_?VectorQ] := Block[{a, b, c, d, e, f, g, h}, l /.
      {a_, b_, c_, d_, e_, f_, g_} /; g == a + 6 → {"seven"} /.
    {a___, b_, c_, d_, e_, f_, g_, h___} /; g == b + 5 → {a, "six", h} /.
   {a___, b_, c_, d_, e_, f_, g___} /; f == b + 4 → {a, "five", g} /.
  {a___, b_, c_, d_, e_, f___} /; e == b + 3 → {a, "four", f} //.
 {a___, b_, c_, d_, e___} /; d == b + 2 → {a, "three", e} //.
{a___, b_, c_, d___} /; c == b + 1 → {a, "two", d}]
```

With **Map** (or /@) we can code several lotto results:

```
c = lottoCodes /@ lottoResults
```

{{seven}, {six, 11}, {1, six}, {five, 11, 21}, {1, five, 21}, {1, 11, five},
{two, five}, {five, two}, {four, 11, 21, 31}, {1, four, 21, 31},
{1, 11, four, 31}, {1, 11, 21, four}, {two, four, 21}, {two, 11, four},
{four, two, 21}, {three, four}, {four, three}, {1, three, 21, 31, 39},
{1, two, 21, three}, {two, two, three}, {1, three, three}, {1, two, 11, 21, 31, 39},
{1, two, two, 21, 31}, {two, two, two, 31}, {1, 3, 5, 7, 9, 11, 13}}

■ Frequencies of Coded Lotto Results

We continue the preceding lotto example in which we coded the lotto results in a special way. Now we count the coded results with, for example, the code "seven" (meaning seven consecutive numbers) or the code "six". Consider the following program.

```
lottoFrequencies[c_List] := With[{alt = Alternatives, per = Permutations},
  {{7, Count[c, {"seven"}]},
   {6, Count[c, {"six", _} | {_, "six"}]},
   {52, Count[c, alt @@ per[{"five", "two"}]]},
   {5, Count[c, alt @@ per[{"five", _, _}]]},
   {43, Count[c, alt @@ per[{"four", "three"}]]},
   {42, Count[c, alt @@ per[{"four", "two", _}]]},
   {4, Count[c, alt @@ per[{"four", _, _, _}]]},
   {322, Count[c, alt @@ per[{"three", "two", "two"}]]},
   {33, Count[c, alt @@ per[{"three", "three", _}]]},
   {32, Count[c, alt @@ per[{"three", "two", _, _}]]},
   {3, Count[c, alt @@ per[{"three", _, _, _, _}]]},
   {222, Count[c, alt @@ per[{"two", "two", "two", _}]]},
   {22, Count[c, alt @@ per[{"two", "two", _, _, _}]]},
   {2, Count[c, alt @@ per[{"two", _, _, _, _, _}]]},
   {1, Count[c, {_, _, _, _, _, _, _}]}}]
```

Here are the frequencies of our test example:

```
lottoFrequencies[c]
```

{{7, 1}, {6, 2}, {52, 2}, {5, 3}, {43, 2}, {42, 3}, {4, 4},
 {322, 1}, {33, 1}, {32, 1}, {3, 1}, {222, 1}, {22, 1}, {2, 1}, {1, 1}}

The program takes a list of coded lotto results as an input and gives a matrix as the output. The second element of each row of the output is the frequency of a result type, and the first element is a code for the result type. For example, the code 52 means a result with five and two consecutive numbers and the code 5 a result with only five consecutive numbers.

Note that we could compute the number of results having, for example, the code "five" as follows:

```
Count[c, {"five", _, _} | {_, "five", _} | {_, _, "five"}]
3
```

This kind of an alternative pattern can also be formed by first forming all permutations,

```
Permutations[{"five", _, _}]
{{five, _, _}, {_, five, _}, {_, _, five}}
```

and then changing, with **Apply** or **@@**, the head **List** into the head **Alternatives**:

```
Alternatives @@ %
{five, _, _} | {_, five, _} | {_, _, five}
```

Thus, we can simply write

```
Count[c, Alternatives @@ Permutations[{"five", _, _}]]
3
```

■ **A Lotto Simulation**

To calculate approximative probabilities of lotto results, first generate a set of 1 million lotto results:

```
SeedRandom[1]; lottoResults = Table[RandomSample[Range[39], 7] // Sort, {10^6}];
```

Code the results:

```
c = lottoCodes /@ lottoResults; // Timing
{651.003, Null}
```

The coding takes more than twice the amount of time than the procedural program of Section 18.2.2 required. Then calculate the frequencies:

```
(fr = lottoFrequencies[c]) // Timing
{34.1913, {{7, 1}, {6, 78}, {52, 61}, {5, 1086}, {43, 61},
  {42, 2124}, {4, 10 591}, {322, 1067}, {33, 1119}, {32, 31 921},
  {3, 77 127}, {222, 10 634}, {22, 154 770}, {2, 431 162}, {1, 278 198}}}
```

Check that the sum of the frequencies is 1 million:

```
Total[fr[[All, 2]]]    1 000 000
```

Sort the frequencies into descending order:

```
fr2 = Reverse@SortBy[fr, Last]
{{2, 431 162}, {1, 278 198}, {22, 154 770}, {3, 77 127},
  {32, 31 921}, {222, 10 634}, {4, 10 591}, {42, 2124}, {33, 1119},
  {5, 1086}, {322, 1067}, {6, 78}, {52, 61}, {43, 61}, {7, 1}}
```

Then form a table from the results:

```
Grid[Prepend[fr2, {"Case", "Frequency"}], Spacings → {0.7, 0.2},
  Dividers → {False, {False, True}}, Alignment → {{Left, Right}}]
```

Case	Frequency
2	431 162
1	278 198
22	154 770
3	77 127
32	31 921
222	10 634
4	10 591
42	2124
33	1119
5	1086
322	1067
6	78
52	61
43	61
7	1

18.5 Recursive Programming

18.5.1 Indexed Recursive Formulas

■ An Introduction to Recursive Programming

Now we proceed to a programming technique called *recursive programming*. In this programming style, we can distinguish two methods: the use of recursive functions and the use of recursive transformations.

> *Recursive function:* The function repeatedly calls itself until the result no longer changes:
>
>   `f[expr1_] := ... f[expr2] ...`
>
> *Recursive transformation rule:* A rule is repeatedly applied until the result no longer changes:
>
>   `expr //. pattern :> value`

Recursive functions apply global rules, whereas recursive transformation rules apply local rules.

We can distinguish two types of recursive functions: indexed and nonindexed recursive functions.

Many mathematical formulas are expressed with indexed recursive functions. Often, an object with an index n is calculated by using objects with indices $n-1$, $n-2$, ... An example is the formulas defining a Fibonacci sequence. Some numerical methods are also expressed as indexed recursive formulas. Such recursive formulas are very easy to program in *Mathematica*: Just write the formulas almost as such.

The calculation of some mathematical objects can be written as nonindexed recursive formulas. In such a formula, the value of a function with an argument is an expression containing the same function with a simpler argument. Such recursive calls are continued until the new expression no longer contains a call of the function. Examples are formulas for the calculation of derivatives, integrals, and determinants.

Recursive transformation rules can in particular be applied for list manipulation, although the computation time often becomes large.

We will first consider examples of indexed recursive formulas. Additional examples can be found in Section 23.5.2, p. 780, where we consider an optimization method called dynamic programming.

■ **Indexed Recursive Formulas**

Many mathematical formulas are of the recursive form $x_0 = a$, $x_n = f(x_{n-1})$, $n = 1, 2, \ldots$. Also, many objects whose basic definitions are not recursive can also be calculated recursively. For example, if $f(n)$ is the factorial of n, we can calculate $f(n)$ from the recursive formulas $f(0) = 1$, $f(n) = n\,f(n-1)$, $n = 1, 2,$ $\ldots$. Although we can calculate factorials with **Factorial[n]** or **n!**, we can also write our own program:

```
f[0] = 1;
f[n_] := n f[n - 1]
```

   f[4]   24

Let us study, with **Trace**, how the computation proceeds:

   f[4] // Trace
```
{f[4], 4 f[4 - 1],
  {{4 - 1, 3}, f[3], 3 f[3 - 1], {{3 - 1, 2}, f[2], 2 f[2 - 1], {{2 - 1, 1}, f[1],
    1 f[1 - 1], {{1 - 1, 0}, f[0], 1}, 1×1, 1}, 2 × 1, 2}, 3 × 2, 6}, 4 × 6, 24}
```

Thus, *Mathematica* observes that

$f(4) = 4\,f(3)$, where

$f(3) = 3\,f(2)$, where

$f(2) = 2\,f(1)$, where

$f(1) = 1\,f(0)$, where

$f(0) = 1$.

Now, $f(1) = 1 \times 1 = 1$, $f(2) = 2 \times 1 = 2$, $f(3) = 3 \times 2 = 6$, and $f(4) = 4 \times 6 = 24$. Thus, $f(4)$ is first reduced back to $f(0)$ whose value we know. Then we go forward and calculate values of $f(i)$, $i = 1, 2, \ldots$ until we have the value of $f(4)$.

Let us check what *Mathematica* knows about the **f** function:

   ? f

Global`f

f[0] = 1

f[n_] := n f[n - 1]

The rule base contains our definitions as such.

If we now would like to calculate, for example, 5!, we have to again calculate 1!, 2!, 3!, and 4! that we actually calculated when we computed $f(4)$. So we observe that it would be useful if we could store the values of the recursive function into the memory of the computer so that they need not be recalculated. *Dynamic programming* is a technique that enables us to store the values of indexed recursive functions.

■ **Dynamic Programming**

We write a new version of our factorial function:

```
g[0] = 1;
g[n_] := (g[n] = n g[n - 1])
```

(Here, the parentheses `( )` are not necessary.) The difference with the **f** function is that we have added **g[n] =** into the definition of **g**. Now, when we call this function with an argument, the command **g[n] = n g[n - 1]** causes the value of **g[n]** to be stored into memory. With **Trace** we can again check to see how the computation proceeds:

```
g[4] // Trace
{g[4], g[4] = 4 g[4 - 1],
  {{{4 - 1, 3}, g[3], g[3] = 3 g[3 - 1], {{{3 - 1, 2}, g[2], g[2] = 2 g[2 - 1],
    {{{2 - 1, 1}, g[1], g[1] = 1 g[1 - 1], {{{1 - 1, 0}, g[0], 1}, 1×1, 1},
    2×1, 2}, g[2] = 2, 2}, 3×2, 6}, g[3] = 6, 6}, 4×6, 24}, g[4] = 24, 24}
```

Mathematica again observes that $g(4) = 4 g(3)$, where $g(3) = 3 g(2)$, where $g(2) = 2 g(1)$, where $g(1) = 1 g(0)$, where $g(0) = 1$. Now $g(1) = 1 \times 1 = 1$, and this value is stored. Furthermore, $g(2) = 2 \times 1 = 2$, and $g(2)$ is again stored. In the same way, the values $g(3) = 3 \times 2 = 6$ and $g(4) = 4 \times 6 = 24$ are stored. Let us now check what *Mathematica* knows about the **g** function:

```
?g
```

```
Global`g

g[0] = 1

g[1] = 1

g[2] = 2

g[3] = 6

g[4] = 24

g[n_] := g[n] = n g[n - 1]
```

We see that the rule base contains, in addition to our original definition, all the values we actually computed while computing $g(4)$.

Let us see how *Mathematica* calculates $g(5)$:

```
g[5] // Trace
{g[5], g[5] = 5 g[5 - 1], {{{5 - 1, 4}, g[4], 24}, 5×24, 120}, g[5] = 120, 120}
```

From the rule base, *Mathematica* observed that $g(4) = 24$ so that it sufficed to compute $5 g(4)$. There was no need to calculate the values $g(1)$, $g(2)$, $g(3)$, and $g(4)$ anew. In this way, the **g** function saves computing time when comparing with the **f** function. When we calculated $g(5)$, this value was again stored into the rule base of **g**.

The technique we used with **g** is known as *dynamic programming*. This terminology derives from the fact that the rule base is updated during the computation—that is, dynamically. In the next box, we present a general formulation of dynamic programming.

Dynamic programming: The function repeatedly calls itself and all values of the function are stored into the rule base: **f[n_] := f[n] = ... f[n - 1] ...**

Saving the values of the function with **f[n_] := f[n] = ...** is also known as *caching*.

```
Remove[f, g]
```

■ Fibonacci Numbers

Fibonacci numbers are defined with the formulas $F_1 = 1$, $F_2 = 1$, $F_n = F_{n-1} + F_{n-2}$, $n = 3, 4, \ldots$. Although we have the ready-to-use command **Fibonacci**, we now present our own function. The function uses dynamic programming:

```
fib[1] = 1;
fib[2] = 1;
fib[n_] := fib[n] = fib[n - 1] + fib[n - 2]
```

```
fib[30]    832 040
```

■ Legendre Polynomials

Legendre orthogonal polynomials are defined with the formulas $P_0(x) = 1$, $P_1(x) = x$, $P_n(x) = \frac{1}{n}[(2n - 1)\, x\, P_{n-1}(x) - (n - 1)\, P_{n-2}(x)]$, $n = 3, 4, \ldots$ Although we have **LegendreP[n, x]**, we write our own program:

```
leg[0, x_] := 1
leg[1, x_] := x
leg[n_, x_] := leg[n, x] = Simplify[((2 n - 1) x leg[n - 1, x] - (n - 1) leg[n - 2, x]) / n]
```

```
leg[6, x]    1/16 (-5 + 105 x^2 - 315 x^4 + 231 x^6)
```

■ A Prime-Generating Recurrence

Rowland (2008) has proved that if $a(n) = a(n - 1) + \gcd(n, a(n - 1))$ with the initial condition $a(1) = 7$, then for all n, the difference $a(n) - a(n - 1)$ is either 1 or a prime; see also http://demonstrations.wolfram.com/PrimeGeneratingRecurrence/ . First, define the recurrence formula by using dynamic programming:

```
a[1] = 7;
a[n_] := a[n] = a[n - 1] + GCD[n, a[n - 1]]
```

Calculate 2 million values:

```
aa = Table[a[n], {n, 2×10^6}]; // Timing
{60.4718, Null}
```

The first 50 values are the following:

```
Take[aa, 50]
{7, 8, 9, 10, 15, 18, 19, 20, 21, 22, 33, 36, 37, 38, 39, 40, 41,
 42, 43, 44, 45, 46, 69, 72, 73, 74, 75, 76, 77, 78, 79, 80, 81, 82,
 83, 84, 85, 86, 87, 88, 89, 90, 91, 92, 93, 94, 141, 144, 145, 150}
```

Then, calculate the differences:

```
d = Differences[aa];
```

The first 50 of these are the following:

```
Take[d, 50]
{1, 1, 1, 5, 3, 1, 1, 1, 1, 11, 3, 1, 1, 1, 1, 1, 1, 1, 1, 1, 1, 23, 3, 1,
 1, 1, 1, 1, 1, 1, 1, 1, 1, 1, 1, 1, 1, 1, 1, 1, 1, 1, 47, 3, 1, 5, 3}
```

Delete all the uninteresting 1's:

```
dd = DeleteCases[d, 1]
```

{5, 3, 11, 3, 23, 3, 47, 3, 5, 3, 101, 3, 7, 11, 3, 13, 233, 3, 467, 3, 5, 3, 941, 3,
  7, 1889, 3, 3779, 3, 7559, 3, 13, 15 131, 3, 53, 3, 7, 30 323, 3, 60 647, 3, 5,
  3, 101, 3, 121 403, 3, 242 807, 3, 5, 3, 19, 7, 5, 3, 47, 3, 37, 5, 3, 17, 3, 199,
  53, 3, 29, 3, 486 041, 3, 7, 421, 23, 3, 972 533, 3, 577, 7, 1 945 649, 3, 163, 7}

Prepare a logarithmic plot for these numbers:

```
ListLogPlot[dd, Filling → Axis, FillingStyle → Black,
  AxesOrigin → {0, 0}, PlotRange → All, ImageSize → 250]
```

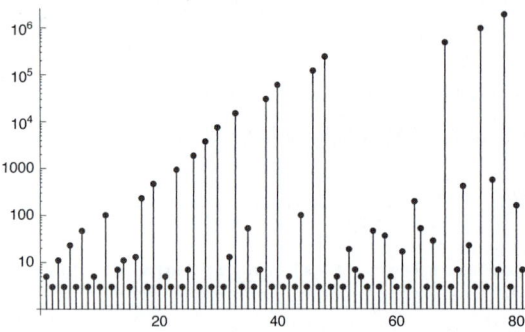

The different numbers in **dd** are the following:

```
Union[dd]
```

{3, 5, 7, 11, 13, 17, 19, 23, 29, 37, 47, 53, 101, 163, 199, 233, 421, 467, 577, 941,
  1889, 3779, 7559, 15 131, 30 323, 60 647, 121 403, 242 807, 486 041, 972 533, 1 945 649}

As Rowland has proved, they are all primes:

```
Union[PrimeQ[%]]     {True}
```

■ Pascal's Triangle

Pascal's triangle is easy to form by using the binomial coefficient:

```
Column[Table[Binomial[n, k], {n, 0, 4}, {k, 0, n}], Center]
```

```
         {1}
        {1, 1}
       {1, 2, 1}
      {1, 3, 3, 1}
     {1, 4, 6, 4, 1}
```

The elements of one row of the triangle can also be obtained by pairwise adding the elements of the previous row ($1 + 3 = 4, 3 + 3 = 6, 3 + 1 = 4$) and adding 1's at the ends of the new row. We would like to write this kind of recursive function to calculate rows of Pascal's triangle.

We number the rows as 0, 1, 2, …; thus, row 0 is {1}. To calculate, for example, the fourth row, first partition the third row to sublists of two elements, then calculate the pairwise sums, and lastly add 1's at the ends:

```
Partition[{1, 3, 3, 1}, 2, 1]     {{1, 3}, {3, 3}, {3, 1}}
```

```
Total /@ %     {4, 6, 4}
```

```
Join[{1}, %, {1}]     {1, 4, 6, 4, 1}
```

```
pascal[0] := {1}
pascal[n_] := pascal[n] = Join[{1}, Total /@ Partition[pascal[n - 1], 2, 1], {1}]
```

```
Column[Table[pascal[n], {n, 0, 6}], Center]
```

```
        {1}
       {1, 1}
      {1, 2, 1}
     {1, 3, 3, 1}
    {1, 4, 6, 4, 1}
   {1, 5, 10, 10, 5, 1}
  {1, 6, 15, 20, 15, 6, 1}
```

■ Newton's Method

We would like to find a zero of the following function:

```
h[x_] := 3 x^3 - E^x;
```

Newton's method uses the recursive formula $x_{i+1} = x_i - \dfrac{f(x_i)}{f'(x_i)}$, $i = 0, 1, \ldots$, by starting from a given value x_0. We write this method as a recursive function:

```
Remove[x];
x[0] = 2;
x[n_] := x[n] = N[x[n - 1] - h[x[n - 1]] / h'[x[n - 1]]]
```

At the beginning, we have written **Remove[x]** because if we change the initial value, the previously calculated values have first to be removed; otherwise, the old stored values are used. For example,

```
Table[x[n], {n, 0, 6}]
```

```
{2, 1.41942, 1.1019, 0.975117, 0.953089, 0.952446, 0.952446}
```

■ Euler's Method

Euler's method to solve a differential equation $y'(x) = f[x, y(x)]$ with the initial value $y(x_0) = y_0$ uses the following recursive formulas:

$$x_{n+1} = x_n + h,$$
$$y_{n+1} = y_n + h\, f(x_n, y_n).$$

Here, h is the step size. In Section 18.1.1, p. 546, we used **Table** to program Euler's method. Now we use a recursive function. We would like to solve the problem $y' = x - y^2$, $y(0) = 1$ on the interval $[0, 1]$ by using 10 steps:

```
v[x_, y_] := x - y^2
```

```
Remove[x, y];
x[0] = 0; y[0] = 1; h = 0.1;
x[n_] := x[n] = N[x[n - 1] + h]
y[n_] := y[n] = N[y[n - 1] + h v[x[n - 1], y[n - 1]]]
```

```
Table[{x[n], y[n]}, {n, 0, 10}]
```

```
{{0, 1}, {0.1, 0.9}, {0.2, 0.829}, {0.3, 0.780276}, {0.4, 0.749393}, {0.5, 0.733234},
 {0.6, 0.729471}, {0.7, 0.736258}, {0.8, 0.75205}, {0.9, 0.775492}, {1., 0.805354}}
```

■ Periodic Functions

We can define a periodic function very simply. If, for example, the function has a period of 2, simply define, with **If** or **Which**, the function in a basic interval, for example, [0, 2], and add the condition that the function is $f(x - 2)$ for $x > 2$:

```
f[x_] := If[0 ≤ x ≤ 2, x^2, f[x - 2]]
```

The value of the function at a given point is calculated by using the recursion $f(x) = f(x - 2)$ until the argument $x - 2$ reduces onto the interval $0 \le x \le 2$. For example, $f(5) = f(3) = f(1) = 1^2 = 1$. Here is a plot:

```
Plot[f[x], {x, 0, 10}]
```

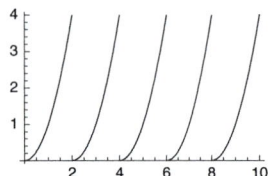

Note that calculus with this function is very limited. If you want to do calculus with a periodic function, then it may be useful to generate the function as an explicit piecewise function for a specific set of intervals:

```
f2[x_] = Piecewise[Table[{(x - a)^2, a < x < a + 2}, {a, 0, 8, 2}]]
```

$$
\begin{cases}
x^2 & 0 < x < 2 \\
(-2 + x)^2 & 2 < x < 4 \\
(-4 + x)^2 & 4 < x < 6 \\
(-6 + x)^2 & 6 < x < 8 \\
(-8 + x)^2 & 8 < x < 10
\end{cases}
$$

```
Plot[f2[x], {x, 0, 10}]
```

■ Continuous Convolutions

Let X_1, X_2, ..., be independent, identically distributed, continuous random variables, with a common probability density function (PDF) $f(x)$. Consider the sum $S_n = X_1 + ... + X_n$ of the variables, and let $f_{S_n}(x)$ be its PDF. It can be shown that $f_{S_2}(x) = \int_{-\infty}^{\infty} f(v) f(x - v)\, dv$. This integral is also called a *convolution* of the PDF $f(x)$. Furthermore, it can be shown that $f_{S_3}(x) = \int_{-\infty}^{\infty} f_{S_2}(v) f(x - v)\, dv$; this is called the twofold convolution. Generally, $f_{S_n}(x) = \int_{-\infty}^{\infty} f_{S_{n-1}}(v) f(x - v)\, dv$. This formula is easily written using dynamic programming:

```
convolution[1] = f[x];
convolution[n_] :=
  convolution[n] = Integrate[(convolution[n - 1] /. x → v) f[x - v], {v, -∞, ∞}]
```

As an example, let the variables X_i have the uniform density on $(-1, 1)$:

```
f[x_] = PDF[UniformDistribution[{0, 1}], x]
```

$$\left\{ 1 \quad 0 \le x \le 1 \right.$$

Calculate the PDFs of S_2, S_3, and S_4:

```
Table[convolution[n], {n, 2, 4}] // FullSimplify
```

$$\left\{ \left[\begin{array}{ll} 2-x & 1<x<2 \\ x & 0<x\le 1 \end{array} \right., \left[\begin{array}{ll} \frac{1}{2} & x==2 \\ \frac{x^2}{2} & 0<x\le 1 \\ -\frac{3}{2}-(-3+x)\,x & 1<x<2 \\ \frac{1}{2}\,(-3+x)^2 & 2<x<3 \end{array} \right., \left[\begin{array}{ll} \frac{1}{6} & x==3 \\ \frac{2}{3} & x==2 \\ \frac{x^3}{6} & 0<x\le 1 \\ \frac{2}{3}-\frac{1}{2}\,(-2+x)^2\,x & 1<x<2 \\ -\frac{1}{6}\,(-4+x)^3 & 3<x<4 \\ -\frac{22}{3}+\frac{1}{2}\,x\,(20+(-8+x)\,x) & 2<x<3 \end{array} \right. \right\}$$

According to the central limit theorem, the sum of independent, identically distributed random variables converges in distribution to the normal disribution. Let us compare the previous densities to the densities of the corresponding normal disributions. The mean and variance of the uniform distribution are

```
{m, s2} = #[UniformDistribution[{0, 1}]] & /@ {Mean, Variance}
```

$$\left\{ \frac{1}{2}, \frac{1}{12} \right\}$$

Thus, the mean and variance of the sum of n such independent random variables are 0 and $n/3$, respectively. Therefore, we use the following normal densities:

```
fn[n_] := PDF[NormalDistribution[n m, Sqrt[n s2]], x]
```

Next, we plot the true densities of S_2, S_3, and S_4 and compare them with the corresponding normal densities:

```
Table[Plot[{convolution[n], fn[n]}, {x, 0, n}, Exclusions → None], {n, 2, 4}]
```

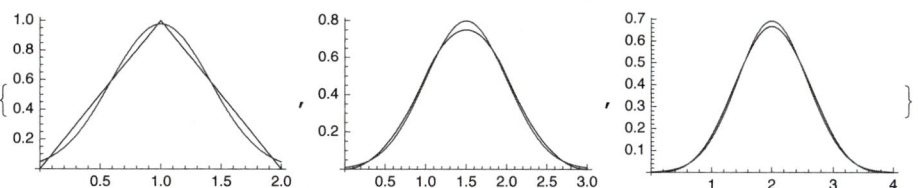

We can see that, in this example, the convergence to the normal distribution is fast: The density of S_4 is already very near to the corresponding normal density.

■ **Discrete Convolutions**

Let X_1, X_2, ..., be independent, identically distributed, discrete random variables, with a common probability mass function (PMF) $p(x)$. Consider the sum $S_n = X_1 + ... + X_n$ of the variables, and let $p_{S_n}(k)$ be its PMF. It can be shown that $p_{S_2}(k) = \sum_i p(i)\, p(k - i)$. This sum is also called a *convolution* of the PMF $p(k)$. Furthermore, it can be shown that $p_{S_3}(k) = \sum_i p_{S_2}(i)\, p(k - i)$; this is called the twofold convolution. Generally, $p_{S_n}(x) = \sum_i p_{S_{n-1}}(i)\, p(k - i)$. This formula is easily written using dynamic programming:

```
convolution2[1, k_] := p[k];
convolution2[n_, k_] :=
 convolution2[n, k] = Sum[(convolution2[n - 1, i]) p[k - i], {i, 0, 100}]
```

As an example, let the variables X_i have the discrete uniform density on {1, 2, 3, 4, 5, 6} (i.e., we toss a die):

```
p[1 | 2 | 3 | 4 | 5 | 6] = 1 / 6;
p[k_] = 0;
```

Calculate the PMFs of S_2, S_3, and S_4:

```
c[2] = Table[{k, convolution2[2, k]}, {k, 2, 12}]
```

$$\left\{\left\{2, \frac{1}{36}\right\}, \left\{3, \frac{1}{18}\right\}, \left\{4, \frac{1}{12}\right\}, \left\{5, \frac{1}{9}\right\}, \left\{6, \frac{5}{36}\right\},\right.$$

$$\left.\left\{7, \frac{1}{6}\right\}, \left\{8, \frac{5}{36}\right\}, \left\{9, \frac{1}{9}\right\}, \left\{10, \frac{1}{12}\right\}, \left\{11, \frac{1}{18}\right\}, \left\{12, \frac{1}{36}\right\}\right\}$$

```
c[3] = Table[{k, convolution2[3, k]}, {k, 3, 18}]
```

$$\left\{\left\{3, \frac{1}{216}\right\}, \left\{4, \frac{1}{72}\right\}, \left\{5, \frac{1}{36}\right\}, \left\{6, \frac{5}{108}\right\}, \left\{7, \frac{5}{72}\right\}, \left\{8, \frac{7}{72}\right\}, \left\{9, \frac{25}{216}\right\}, \left\{10, \frac{1}{8}\right\},\right.$$

$$\left.\left\{11, \frac{1}{8}\right\}, \left\{12, \frac{25}{216}\right\}, \left\{13, \frac{7}{72}\right\}, \left\{14, \frac{5}{72}\right\}, \left\{15, \frac{5}{108}\right\}, \left\{16, \frac{1}{36}\right\}, \left\{17, \frac{1}{72}\right\}, \left\{18, \frac{1}{216}\right\}\right\}$$

```
c[4] = Table[{k, convolution2[4, k]}, {k, 4, 24}]
```

$$\left\{\left\{4, \frac{1}{1296}\right\}, \left\{5, \frac{1}{324}\right\}, \left\{6, \frac{5}{648}\right\}, \left\{7, \frac{5}{324}\right\}, \left\{8, \frac{35}{1296}\right\}, \left\{9, \frac{7}{162}\right\}, \left\{10, \frac{5}{81}\right\},\right.$$

$$\left\{11, \frac{13}{162}\right\}, \left\{12, \frac{125}{1296}\right\}, \left\{13, \frac{35}{324}\right\}, \left\{14, \frac{73}{648}\right\}, \left\{15, \frac{35}{324}\right\}, \left\{16, \frac{125}{1296}\right\}, \left\{17, \frac{13}{162}\right\},$$

$$\left.\left\{18, \frac{5}{81}\right\}, \left\{19, \frac{7}{162}\right\}, \left\{20, \frac{35}{1296}\right\}, \left\{21, \frac{5}{324}\right\}, \left\{22, \frac{5}{648}\right\}, \left\{23, \frac{1}{324}\right\}, \left\{24, \frac{1}{1296}\right\}\right\}$$

According to the central limit theorem, the sum of independent, identically distributed random variables converges in distribution to the normal disribution. Let us compare the previous PMFs to the densities of the corresponding normal disributions. The mean and variance of the uniform distribution are

```
{m, s2} = #[DiscreteUniformDistribution[{1, 6}]] & /@ {Mean, Variance}
```

$$\left\{\frac{7}{2}, \frac{35}{12}\right\}$$

Thus, the mean and variance of the sum of n such independent random variables are $7n/2$ and $35n/12$, respectively. Therefore, we use the following normal densities:

```
fn[n_] := PDF[NormalDistribution[n m, Sqrt[n s2]], x]
```

Next, we plot the true PMFs of S_2, S_3, and S_4 and compare them with the corresponding normal densities:

```
Table[Show[ListPlot[c[n], AxesOrigin → {n, 0}], Plot[fn[n], {x, n, 6 n}]], {n, 2, 4}]
```

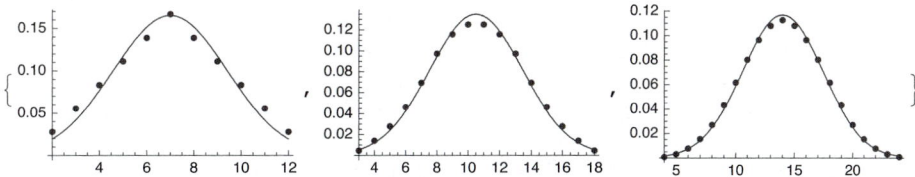

We can see that, in this example, the convergence to the normal distribution is fast: The PMF of S_4 is already very near the corresponding normal density.

■ Recursion Limit

Consider the following recursive function to calculate the sum of the first n positive integers:

```
s[0] = 0;
s[n_] := s[n - 1] + n
```

Here is the $s(254)$:

```
s[254]        32 385
```

However, we do not get a complete result for $s(255)$:

```
s[255]
```

$RecursionLimit::reclim : Recursion depth of 256 exceeded. ≫

$RecursionLimit::reclim : Recursion depth of 256 exceeded. ≫

32 640 + Hold[Hold[s[1 - 1]]]

We got the message that the default value 256 of the **$RecursionLimit** constant has been exceeded. Indeed, *Mathematica* uses this constant to prevent infinite recursion. Mostly, infinite recursion is the result of improper programming. However, in running correct recursive programs we can also reach the recursion limit.

To check the value of the limit, write

```
$RecursionLimit      256
```

If a program calls itself more than this, the computation is interrupted and a message is shown. However, we can continue the computation with **ReleaseHold**:

```
%% // ReleaseHold     32 640 +Hold[s[1 - 1]]
```

```
% // ReleaseHold     32 640
```

> **$RecursionLimit**   The largest allowed number of levels of recursion during one computation

■ Adjusting the Recursion Limit

We can give the **$RecursionLimit** constant the value we want (also the value ∞):

```
$RecursionLimit = 500      500
```

Now we calculate the value of our function up to $n = 498$:

```
s[498]      124 251
```

We go back to the default value:

```
$RecursionLimit = 256;
```

A better way to adjust the value of the constant is the use of a **Block** construct:

Block[{$RecursionLimit = n}, ...]   Temporarily change the value of the recursion limit

The advantage of this construct is that after the computation the constant again has the default value. Let us try:

```
Block[{$RecursionLimit = 500}, s[497]]      123 753
```

Note that if we use dynamic programming, we get speedier code but we may also sometimes avoid the recursion limit by computing in suitable pieces. To calculate the value of, for example, $s(500)$, calculate first, say, $s(250)$ and then $s(500)$:

```
Remove[s]
s[0] = 0;
s[n_] := s[n] = s[n - 1] + n
s[250]      31 375

s[500]      125 250
```

We did not get a message about recursion limit. Indeed, when we calculated $s(500)$, we could start the calculation from the value $s(250)$ since this value is in the memory.

18.5.2 Nonindexed Recursive Formulas

■ **Logarithms**

To expand some expressions containing the logarithm function, we need some special commands:

```
{FunctionExpand[Log[3 a]], FullSimplify[Log[a^3], a > 0]}
```
$\{Log[3] + Log[a], 3 Log[a]\}$

PowerExpand also makes these kinds of expansions (without checking their validity):

```
{PowerExpand[Log[3 a]], PowerExpand[Log[a^3]]}
```
$\{Log[3] + Log[a], 3 Log[a]\}$

Now we write our own logarithm function that automatically expands products and powers:

loga[x_ y_] := loga[x] + loga[y] **loga[x_^y_] := y loga[x]**

In the first definition, the argument has to be in a product form, whereas the second definition applies for arguments of a power form. The definitions are recursive in that they call themselves. When these definitions are done, then, every time **loga** appears in an expression, the given rules are applied until the result no longer changes. Here is a simple example:

```
loga[a b c^2] // Trace
```
$\left\{loga\left[a\,b\,c^2\right],\; loga[a] + loga\left[b\,c^2\right],\right.$
$\quad \left\{loga\left[b\,c^2\right],\; loga[b] + loga\left[c^2\right],\; \left\{loga\left[c^2\right],\; 2\,loga[c]\right\},\; loga[b] + 2\,loga[c]\right\},$
$\quad \left. loga[a] + (loga[b] + 2\,loga[c]),\; loga[a] + loga[b] + 2\,loga[c]\right\}$

A more advanced example:

```
loga[a^2 b^3 Sqrt[c] / d^4]
```

$$2 \, \text{loga}[a] + 3 \, \text{loga}[b] + \frac{\text{loga}[c]}{2} - 4 \, \text{loga}[d]$$

■ Derivatives

Mathematica knows all the rules of differentiating:

```
D[#, x] & /@ {f[x] + g[x], f[x] g[x], f[x]^g[x]}
```

$$\left\{ f'[x] + g'[x], \, g[x] \, f'[x] + f[x] \, g'[x], \, f[x]^{g[x]} \left(\frac{g[x] \, f'[x]}{f[x]} + \text{Log}[f[x]] \, g'[x] \right) \right\}$$

Next, we write a recursive function that calculates derivatives of rational expressions:

```
der[a_, x_] := 0 /; FreeQ[a, x]
der[x_, x_] := 1
der[a_ f_, x_] := a der[f, x] /; FreeQ[a, x]
der[f_ + g_, x_] := der[f, x] + der[g, x]
der[f_ g_, x_] := der[f, x] g + f der[g, x]
der[f_^a_, x_] := a f^(a - 1) der[f, x] /; FreeQ[a, x]
der[f_^g_, x_] := f^g (der[f, x] g / f + der[g, x] Log[f])
```

For the derivative of a quotient, we do not need a special rule because *Mathematica* writes quotients as products of powers:

```
FullForm[f[x] / g[x]]
```

```
Times[f[x], Power[g[x], -1]]
```

An example:

```
der[(1 + a / (b + x)) ^ (2 + x^2), x]
```

$$\left(1 + \frac{a}{b+x}\right)^{2+x^2} \left(-\frac{a\left(2 + x^2\right)}{(b+x)^2 \left(1 + \frac{a}{b+x}\right)} + 2 \, x \, \text{Log}\left[1 + \frac{a}{b+x}\right] \right)$$

The **D** command gives the same result:

```
D[(1 + a / (b + x)) ^ (2 + x^2), x]
```

$$\left(1 + \frac{a}{b+x}\right)^{2+x^2} \left(-\frac{a\left(2 + x^2\right)}{(b+x)^2 \left(1 + \frac{a}{b+x}\right)} + 2 \, x \, \text{Log}\left[1 + \frac{a}{b+x}\right] \right)$$

■ Integrals

From the Documentation Center, we can find the following recursive function that calculates indefinite integrals; see tutorial/AnExampleDefiningYourOwnIntegrationFunction. The function covers simple rational and exponential expressions.

```
int[a_, x_] := a x /; FreeQ[a, x]
int[a_ f_, x_] := a int[f, x] /; FreeQ[a, x]
int[f_ + g_, x_] := int[f, x] + int[g, x]
int[x_^a_., x_] := x^(a + 1) / (a + 1) /; FreeQ[a, x] && a != -1
int[1 / (a_. x_ + b_.), x_] := Log[a x + b] / a /; FreeQ[{a, b}, x]
int[Exp[a_. x_ + b_.], x_] := Exp[a x + b] / a /; FreeQ[{a, b}, x]
```

```
int[a x^2 + Exp[-bx] + c / (d - e x), x]
```

$$e^{-bx} x + \frac{a x^3}{3} - \frac{c \, Log[d - e x]}{e}$$

```
Integrate[a x^2 + Exp[-bx] + c / (d - e x), x]
```

$$e^{-bx} x + \frac{a x^3}{3} - \frac{c \, Log[d - e x]}{e}$$

■ Determinants

Consider an $(n \times n)$ matrix whose (i, j)th element is m_{ij}. The determinant of the matrix can be computed recursively from the formula $\sum_{j=1}^{n} (-1)^{j+1} m_{1j} c_j$, where c_j is the determinant of the matrix that is obtained by deleting the first row and jth column. The determinant of a matrix consisting of only one element is the element itself.

Consider the matrix

```
m = {{a, b, c}, {d, e, f}, {g, h, i}};
```

To delete the first row, write

```
Rest[m]      {{d, e, f}, {g, h, i}}
```

To delete, for example, the first column, write

```
Drop[%, {}, {1}]      {{e, f}, {h, i}}
```

To test whether an $(n_1 \times n_2)$ matrix is a square matrix, change, with **Apply** or **@@**, the head **List** of the dimensions to the head **Equal**:

```
Equal @@ {n1, n2}      n1 == n2
```

Thus, we can write the following program:

```
deter[{{m_}}] := m
deter[m_ ?MatrixQ] /; Equal @@ Dimensions[m] := deter[m]
 = Sum[(-1) ^ (1 + j) m[[1, j]] deter[Drop[Rest[m], {}, {j}]], {j, Length[m]}]
```

To speed up the program, we have used dynamic programming. An example:

```
m = RandomReal[{0, 1}, {15, 15}];
```

```
{deter[m] // Timing, Det[m] // Timing}
  {{12.9303, 0.136651}, {0.000115, 0.136651}}
```

Our program is very slow for larger matrices.

18.5.3 Recursive List Manipulation

■ Removing Zeros: Built-in Commands

From a given vector, we would like to remove all zeros. We will apply built-in commands and procedural, functional, and rule-based programming to solve this problem. First, we take a sample with
replacement from the set {−1, 0, 1}:

```
t = RandomChoice[{-1, 0, 1}, 10^5];
```

In solving a problem, we should first investigate if we have ready-to-use commands because they are
fast. In our example, we have **Cases**:

```
Cases[t, -1 | 1]; // Timing      {0.086028, Null}

Cases[t, x_ /; x ≠ 0]; // Timing      {0.203768, Null}

Cases[t, x_ /; x == -1 || x == 1]; // Timing      {0.461625, Null}
```

We also have **Select**:

```
Select[t, # ≠ 0 &]; // Timing      {0.255259, Null}

Select[t, # == -1 || # == 1 &]; // Timing      {0.524612, Null}
```

■ Removing Zeros: Functional Programming

Functional programming also often gives fast code. In our example, we can use **Fold**:

```
removeZeros1[x_ ?VectorQ] := Fold[If[#2 == 0, #1, {#1, #2}] &, {}, x] // Flatten
```

Here, the starting point is the empty list. From the **x** list we take one element at a time. With **#1** we
denote the list to be outputted and with **#2** the next element of **x**. If the next element is zero, we do not
change **#1**. If the next element is not zero, we add the element at the end of the output list with **{#1,
#2}**; thus, the output list will be a nested list so that at the end we flatten the list. We try our program:

```
removeZeros1[t]; // Timing      {0.554869, Null}
```

The program is fast.

■ Removing Zeros: Procedural Programming

A procedural program is often easy to write and the code is frequently quite fast. In the following
program, we test whether the next element is nonzero. If so, we add it at the end of the output list.

```
removeZeros2[x_ ?VectorQ] := Module[{nonzero = {}, next},
  Do[If[(next = x[[i]]) ≠ 0, nonzero = {nonzero, next}], {i, Length[x]}];
  Flatten[nonzero]]
```

```
removeZeros2[t]; // Timing      {0.873215, Null}
```

The program is fast. Note that we have written **nonzero = {nonzero, next}**. Indeed, we should not
use **AppendTo[nonzero, next]**. To demonstrate why, write a second version of the program, now using
AppendTo:

```
removeZeros3[x_ ?VectorQ] := Module[{nonzero = {}, next},
  Do[If[(next = x[[i]]) ≠ 0, AppendTo[nonzero, next]], {i, Length[x]}]; nonzero]
```

Now we do not need **Flatten** at the end of the program. An example:

```
removeZeros3[t]; // Timing     {67.1121, Null}
```

Thus, the program using **AppendTo** is very slow.

■ **Removing Zeros: Recursive Function**

To remove zero elements with a recursive function, write the following definition:

```
removeZeros4[{a___, 0, b___}] := removeZeros4[{a, b}]
```

Here, if a zero is found somewhere in the given list, it is removed and the same function is called again with the resulting vector as the argument. An example:

```
removeZeros4[{-1, 0, -1, 1, 1, 0, 1, 0}]
removeZeros4[{-1, -1, 1, 1, 1}]
```

We see that the program works, but to get the result, we need a definition that simply gives its argument as such:

```
removeZeros4[x_] := x
```

Now the function works:

```
removeZeros4[{-1, 0, -1, 1, 1, 0, 1, 0}]
{-1, -1, 1, 1, 1}
```

The program also works with an empty list:

```
removeZeros4[{}]     {}
```

Thus, we can write the following function:

```
removeZeros4[{a___, 0, b___}] := removeZeros4[{a, b}]
removeZeros4[x_] := x
```

This recursive program is so slow that we have to try it to a smaller vector:

```
t0 = RandomChoice[{-1, 0, 1}, 7 × 10^3];
```

```
removeZeros4[t0]; // Timing     {1.5609, Null}
```

The slowness of recursive functions in list manipulation is typical. Thus, resort to other programming styles to get fast code.

Why is the program so slow? To find a zero from the given list, elements are tested from the beginning until a zero is found; it is removed. Then, elements are again tested from the beginning until a zero is found. Thus, each time a zero is found, the search for a zero starts from the beginning of the list instead of starting after the last found zero. Accordingly, much time is spent in testing the elements.

■ **Removing Zeros: Recursive Transformation Rule**

The idea of the preceding program can also be applied with a recursive transformation rule:

```
removeZeros5[x_] := x //. {a___, 0, c___} :> {a, c}
```

Recall that //. applies a rule until the result no longer changes. We try this program for the smaller test example:

```
removeZeros5[t0]; // Timing     {1.55814, Null}
```

The computing time is the same as with the preceding program. Note that in the previous program we used :→ (write is as :>), not → (or ->). The arrow :→ means that transformation is delayed: The evaluation of the right-hand side {a, c} is delayed until the rule is applied. In this way, the program works even if a or c happens to have some values.

■ Record Values: Functional and Procedural Programs

In Section 18.3.6, p. 582, we calculated cumulative maximums or record values with **FoldList**. Now we write the procedure as a program:

```
records1[x_?VectorQ] := Union[Rest[FoldList[Max, 0, x]]]
```

To try the program, generate a random permutation of the integers 1, 2, ..., 20:

```
SeedRandom[1]; t = RandomSample[Range[20], 20]
```
```
{6, 1, 8, 19, 2, 17, 13, 16, 11, 9, 4, 15, 20, 18, 12, 14, 7, 5, 10, 3}
```
```
records1[t]    {6, 8, 19, 20}
```

Next, we write a procedural program.

```
records2[x_?VectorQ] := Module[{result = {First[x]}, rec = First[x], next},
  Do[If[(next = x[[i]]) > rec, result = {result, next}; rec = next],
    {i, 2, Length[x]}];
  Flatten[result]]
records2[{}] := {}
```

```
records2[t]    {6, 8, 19, 20}
```

■ Record Values: Recursive Programs

A wholly different approach is the use of a recursive transformation rule (see Trott, 2004b, p. 634).

```
records3[x_?VectorQ] := x //. {a___, b_, c_, d___} /; c ≤ b :→ {a, b, d}
```

In the program, we repeatedly remove one element from the list; this element is denoted by c. If a c is found that is at most the preceding element b, then c is deleted because it cannot be a record value. Before b and after c there can be zero or more elements.

```
records3[t]    {6, 8, 19, 20}
```

In the following program, we use a recursive function:

```
records4[{a___, b_, c_, d___}] /; c ≤ b := records4[{a, b, d}]
records4[x_] := x
```

```
records4[t]    {6, 8, 19, 20}
```

To test the speed of the programs, generate a random permutation of the first 4000 integers:

```
SeedRandom[1]; t = RandomSample[Range[4000], 4000];
```

```
records1[t] // Timing
```
```
{0.013578, {3349, 3810, 3858, 3924, 3967, 3998, 4000}}
```

```
records2[t] // Timing
```
```
{0.029646, {3349, 3810, 3858, 3924, 3967, 3998, 4000}}
```

```
records3[t] // Timing
```
{1.31727, {3349, 3810, 3858, 3924, 3967, 3998, 4000}}

```
records4[t] // Timing
```
{1.37323, {3349, 3810, 3858, 3924, 3967, 3998, 4000}}

The functional and procedural programs are fast, whereas the recursive programs are quite slow.

■ Record Values: Faster Recursive Programs

By careful design, we can get faster recursive programs for finding record values; see Wellin *et al.* (2005, pp. 184, 191). Here is the first faster program:

```
records5[{a_, b___}] := Join[{a}, Select[records5[{b}], # > a &]]
records5[{}] := {}
```

```
Block[{$RecursionLimit = 4004}, records5[t]] // Timing
```
{0.757501, {3349, 3810, 3858, 3924, 3967, 3998, 4000}}

In the previous program, we first set the first element, **a**, as the first record value. Then we find the record values from among the rest of the elements, **b**, and select the record values that are greater than the first element, **a**.

A still faster recursive function is as follows:

```
records6[x_List] := records6[First[x], Rest[x]]
records6[a_, {b_, c___}] /; a ≥ b := records6[a, {c}]
records6[a_, {b_, c___}] /; a < b := Join[{a}, records6[b, {c}]]
records6[a_, {}] := {a}
records6[{}] := {}
```

```
records6[t] // Timing
```
{0.455719, {3349, 3810, 3858, 3924, 3967, 3998, 4000}}

The idea of the program is to use two arguments for the function: The first is the current record value and the second a list of all remaining elements. The second rule given previously states that if the first element, **b**, of the remaining elements is at most the current record value, **a**, then **b** is not a record value so that we can call the function with arguments where **b** is dropped. The third rule applies in the other case: **b** is larger than **a** so that **b** is a new record value; in this case, we save **a** and call the function with **b** as the new current record value. The first rule is only used at the beginning: The first element must be the first record value. The last two rules cover some special cases.

■ Iteration Limit

Consider again **records4**:

```
records4[{a___, b_, c_, d___}] /; c ≤ b := records4[{a, b, d}]
records4[x_] := x
```

We try to find the record values from a list of length 5000:

```
SeedRandom[1]; t = RandomSample[Range[5000], 5000];
```

```
records4[t];
```

$IterationLimit::itlim : Iteration limit of 4096 exceeded. ≫

The procedure stopped because an internal limit of iterations, 4096, had been exceeded.

> **$IterationLimit** The maximum allowed amount of iterations during one computation: at most this number of times *Mathematica* tries to simplify an expression

We can ask for the current value of this constant:

```
$IterationLimit    4096
```

Similarly as for **$RecursionLimit** (see Section 18.5.1, p. 605), a good way to temporarily change the value of **$IterationLimit** is with the use of **Block**:

```
Block[{$IterationLimit = 4992}, records4[t]]

{449, 1073, 4053, 4345, 4403, 4902, 4965, 4994, 4999, 5000}
```

■ Sorting

The following recursive program sorts the elements of a list into the standard order:

```
sort[x_?VectorQ] := x //. {a___, b_, c_, d___} /; ! OrderedQ[{b, c}] :> {a, c, b, d}
```

The program repeatedly finds pairs of elements {*b*, *c*} that are not in standard order and then puts them into order. The program is very slow.

```
sort[{3, b, 5, a, c, 2}]    {2, 3, 5, a, b, c}
```

■ Run-Length Encoding and Decoding

In Section 18.3.2, p. 574, we presented the following programs for run-length encoding and decoding:

```
runLengthEncoding[x_?VectorQ] := {First[#], Length[#]} & /@ Split[x]
runLengthDecoding[x_?MatrixQ] := Flatten[ConstantArray @@@ x]
```

```
u = {1, 0, 1, 0, 1, 0, 0, 1, 1, 1, 1};

runLengthEncoding[u]

{{1, 1}, {0, 1}, {1, 1}, {0, 1}, {1, 1}, {0, 2}, {1, 4}}

runLengthDecoding[%]

{1, 0, 1, 0, 1, 0, 0, 1, 1, 1, 1}
```

Here are recursive programs (Trott, 2004b, p. 630):

```
runLengthEncoding2[x_?VectorQ] :=
 x /. {a_Integer :> {a, 1}} //. {a___, {b_, i_}, {b_, j_}, c___} :> {a, {b, i + j}, c}
runLengthDecoding2[x_?MatrixQ] :=
 x //. {y___, {a_, k_}, z___} :> Join[{y}, ConstantArray[a, k], {z}]
```

```
runLengthEncoding2[u]

{{1, 1}, {0, 1}, {1, 1}, {0, 1}, {1, 1}, {0, 2}, {1, 4}}

runLengthDecoding2[%]

{1, 0, 1, 0, 1, 0, 0, 1, 1, 1, 1}
```

In encoding, we replace all elements *a* with pairs {*a*, 1}. Then we repeatedly find pairs {*b*, *i*} and {*b*, *j*} having the same first element. Such two pairs are replaced with {*b*, *i* + *j*}. In decoding, we repeatedly find pairs {*a*, *k*}. They are replaced with a list having *k* times the element *a*. The programs are very slow.

Differential Calculus

Introduction

> *Analysis takes back with one hand what it gives with the other. I recoil with fear and loathing from that deplorable evil, continuous functions with no derivatives.—Hermite*

Traditional differential calculus includes derivatives, Taylor series, and limits; the corresponding *Mathematica* commands are **D**, **Series**, and **Limit**. Later, derivatives play a central role when we, for example, solve nonlinear equations by numerical methods (Chapter 22) and solve optimization problems (Chapter 23). Of course, derivatives are essential in differential and partial differential equations (Chapters 26 and 27).

Because this chapter begins the mathematical part of the book, we note that http://mathworld.wolfram.com is a place to find information about mathematical topics such as calculus, algebra, applied mathematics, discrete mathematics, and probability and statistics.

19.1 Derivatives

19.1.1 Partial Derivatives

In the following box are some examples of calculating partial derivatives:

`D[f, x]`	$\partial f/\partial x$	`D[f, x, y]`	$\partial^2 f/(\partial x\,\partial y)$
`D[f, x, x]`	$\partial^2 f/\partial x^2$	`D[f, x, y, y, y]`	$\partial^4 f/(\partial x\,\partial y^3)$
`D[f, x, x, x]`	$\partial^3 f/\partial x^3$	`D[f, x, y, z]`	$\partial^3 f/(\partial x\,\partial y\,\partial z)$

You can write in **D** the whole expression to be differentiated, as in `D[a + b x, x]`, or you can first give a name to the expression, such as `h = a + b x`, and then use that name, such as in the expression `D[h, x]`. Functions such as `f[x]` can be differentiated simply by `f'[x]` (this is considered in more detail later).

Another way to calculate partial derivatives is to use the buttons $\partial_\square\,\blacksquare$ and $\partial_{\square,\square}\,\blacksquare$ of the **BasicMathInput** palette. Alternatively, you can do it yourself: Write ∂ as ESC pd ESC, go to a subscript by pressing CTRL -, write the variable, go out of the subscript position with CTRL ␣ (␣ means the space key), and write the expression (see Section 3.3.3, p. 76):

> ∂_x (a + b x)   b

The command **D** also has another form in which the order of differentiation is expressed explicitly. For example, instead of `D[f, x]`, `D[f, x, x]`, and `D[f, x, y, y, y]`, we can write as follows:

```
D[f, {x, 1}],   D[f, {x, 2}],   D[f, {x, 1}, {y, 3}]
```

■ Example 1: Tangent

Consider the following function:

```
f = (x + 2) (x^2 + 1) x (x - 1) (x - 2);
```

We calculate and plot a tangent for it at $x = 1.6$. First, we calculate the corresponding values of the function and its derivative:

```
x1 = 1.6;

f1 = f /. x → x1     -4.92134

df1 = D[f, x] /. x → x1     -4.76544
```

Then we form the tangent and plot it together with the function:

```
tangent = f1 + df1 (x - x1)     -4.92134 - 4.76544 (-1.6 + x)

Plot[{f, tangent}, {x, 0, 2.1}, PlotStyle → {{}, Blue},
  Epilog → {PointSize[Medium], Red, Point[{x1, f1}]}]
```

■ Example 2: Critical Points

Let us now examine the critical points of the function we considered in Example 1:

```
p = Plot[f, {x, -2.05, 2.1}]
```

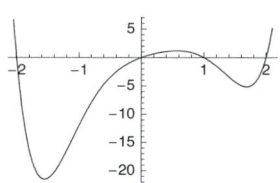

Critical points are characterized by the property that the first derivative is zero at these points. Critical points contain points of local maximum and minimum values and saddle points. Our function seems to have three critical points. To find them, we calculate the derivative and its zeros:

```
df = D[f, x] // Simplify
```

$4 - 8 x + 9 x^2 - 12 x^3 - 5 x^4 + 6 x^5$

```
c = NSolve[df == 0, x]
```

$\{\{x \to -1.55135\}, \{x \to 0.065021 - 0.666791\,i\},$
$\{x \to 0.065021 + 0.666791\,i\}, \{x \to 0.567487\}, \{x \to 1.68715\}\}$

We are interested only in real critical points, and so we select such points:

```
crit = c[[{1, 4, 5}]]
```

$\{\{x \to -1.55135\}, \{x \to 0.567487\}, \{x \to 1.68715\}\}$

Then we form the corresponding points on the curve:

```
points = {x, f} /. crit
```

$\{\{-1.55135, -21.4839\}, \{0.567487, 1.19346\}, \{1.68715, -5.14393\}\}$

Lastly, we show the points on the curve:

```
Show[p, Epilog → {PointSize[Medium], Red, Point[points]}]
```

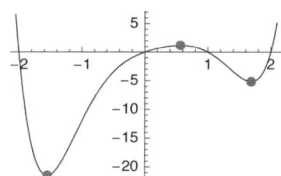

```
Remove["Global`*"]
```

■ **Derivatives of Functions of One Variable**

If you have defined a function `f[x]` of one variable, derivatives can be calculated with primes (this resembles the usual mathematical notation).

`f'[x], f''[x], f'''[x], ...` The first, second, third, ... derivative of a function `f` at `x`

Consider, for example, the following function:

```
f[x_] := a + b x + c x^2
```

The first, second, and third derivatives are as follows:

```
{f'[x], f''[x], f'''[x]}     {b + 2 c x, 2 c, 0}
```

Note that we can at the same time also specify the point at which the derivative is calculated:

```
f'[3]     b + 6 c
```

However, if the function has several arguments, such as **g** in this example,

```
g[x_, y_] := Sin[x] Cos[y]
```

then derivatives are again calculated with **D** (or **Derivative**):

```
D[g[x, y], y]     -Sin[x] Sin[y]
```

For some special functions, we do not get an explicit expression for the derivative but we can ask numerical values:

```
{Zeta'[x], Zeta'[2.]}     {Zeta'[x], -0.937548}
```

■ Derivatives of Functions of Several Variables

Consider the function **g** defined previously. If we want to calculate the value of its derivative at a point, we first have to calculate its derivative and then ask its value at the point:

```
D[g[x, y], x, x, y] /. {x → 4, y → 5}     Sin[4] Sin[5]
```

However, *Mathematica* still has one way to represent derivatives, which is **Derivative**, and with this command we can specify the orders and the point at the same time:

```
Derivative[2, 1][g][4, 5]     Sin[4] Sin[5]
```

`Derivative[m, n, … ][f][a, b, … ]` Differentiate function `f[x, y , … ]` **m** times with respect to **x**, **n** times with respect to **y**, … and evaluate the derivative at the point **x** = **a**, **y** = **b**, …

When differentiating unspecified functions, *Mathematica* shows primes for functions of one variable:

```
D[r[s[x]], x]     r'[s[x]] s'[x]
```

Superscripts are used for functions of several variables:

```
D[r[x, y], x, x, y]     r^{(2,1)}[x, y]
```

However, in internal representations, *Mathematica* uses **Derivative**:

```
D[r[s[x]], x] // InputForm
Derivative[1][r][s[x]]*Derivative[1][s][x]

D[r[x, y], x, x, y] // InputForm
Derivative[2, 1][r][x, y]
```

■ NonConstants

If a variable depends on another variable, it is simple to explicitly denote the dependency. In the next example, both **r** and **a** depend on **x**:

```
D[a[x] r[x], x]
r[x] a'[x] + a[x] r'[x]
```

Another way to express the dependency is to use the **NonConstants** option:

```
D[a r[x], x, NonConstants → a]
D[a, x, NonConstants → {a}] r[x] + a r'[x]
```

`D[f, x, NonConstants → {a, b, … }]` Declare that **a**, **b**, … depend on **x**

19.1.2 Vector Analysis

■ Gradient, Hessian, Jacobian, and More

For a scalar-valued function of several variables, we may want to calculate

the *gradient* vector (vector of the first derivatives);
the *Hessian* matrix (matrix of the pure and mixed second derivatives); and
the *Laplacian* scalar (sum of the unmixed second derivatives).

For a vector-valued function of several variables, we may want to calculate

the *Jacobian* matrix (the *i*th row is the gradient of the *i*th function); and
the *divergence* scalar (sum in which the *i*th term is the derivative of the *i*th function with respect to the *i*th variable).

These can be calculated as given in the following box. Here, **f** is a scalar-valued and **fs** a vector-valued function, and **vars** is a list of variables.

D[f, {vars}] The gradient of **f** with respect to **vars**
D[f, {vars, 2}] The Hessian of **f** with respect to **vars**
Inner[D, D[f, {vars}], vars] The Laplacian of **f** with respect to **vars**

D[fs, {vars}] The Jacobian of **fs** with respect to **vars**
Inner[D, fs, vars] The divergence of **fs** with respect to **vars**

Note that in calculating a gradient, Hessian, or Jacobian, the variables are within double braces such as in **{{x, y, z}}** or **{{x, y, z}, 2}**. In calculating the Laplacian and divergence, we used **Inner** (see Section 14.2.4, p. 465). Here is a general example of this command:

Inner[f, {a, b, c}, {A, B, C}]

f[a, A] + f[b, B] + f[c, C]

As an example, here is a gradient:

vars = {x, y, z}; p = f[x, y, z];

D[p, {vars}]

$\left\{ f^{(1,0,0)}[x, y, z], f^{(0,1,0)}[x, y, z], f^{(0,0,1)}[x, y, z] \right\}$

Next, we calculate a Hessian:

D[p, {vars, 2}] // MatrixForm

$$\begin{pmatrix} f^{(2,0,0)}[x, y, z] & f^{(1,1,0)}[x, y, z] & f^{(1,0,1)}[x, y, z] \\ f^{(1,1,0)}[x, y, z] & f^{(0,2,0)}[x, y, z] & f^{(0,1,1)}[x, y, z] \\ f^{(1,0,1)}[x, y, z] & f^{(0,1,1)}[x, y, z] & f^{(0,0,2)}[x, y, z] \end{pmatrix}$$

Here is a Laplacian:

Inner[D, D[p, {vars}], vars]

$f^{(0,0,2)}[x, y, z] + f^{(0,2,0)}[x, y, z] + f^{(2,0,0)}[x, y, z]$

Calculate the Jacobian of a vector-valued function:

fs = {f[x, y, z], g[x, y, z], h[x, y, z]};

```
D[fs, {vars}] // MatrixForm
```

$$\begin{pmatrix} f^{(1,0,0)}[x, y, z] & f^{(0,1,0)}[x, y, z] & f^{(0,0,1)}[x, y, z] \\ g^{(1,0,0)}[x, y, z] & g^{(0,1,0)}[x, y, z] & g^{(0,0,1)}[x, y, z] \\ h^{(1,0,0)}[x, y, z] & h^{(0,1,0)}[x, y, z] & h^{(0,0,1)}[x, y, z] \end{pmatrix}$$

Lastly, calculate a divergence:

```
Inner[D, fs, vars]
```

$h^{(0,0,1)}[x, y, z] + g^{(0,1,0)}[x, y, z] + f^{(1,0,0)}[x, y, z]$

Here are more specific examples of a gradient, Hessian, and Jacobian:

```
f = x^2 + x y^2 + x y z^2; g = Exp[x y z]; h = Sin[x y z];
```

```
D[f, {{x, y, z}}]
```

$\{2x + y^2 + yz^2, 2xy + xz^2, 2xyz\}$

```
D[f, {{x, y, z}, 2}]
```

$\{\{2, 2y + z^2, 2yz\}, \{2y + z^2, 2x, 2xz\}, \{2yz, 2xz, 2xy\}\}$

```
D[{f, g, h}, {{x, y, z}}]
```

$\{\{2x + y^2 + yz^2, 2xy + xz^2, 2xyz\},$
$\{e^{xyz}yz, e^{xyz}xz, e^{xyz}xy\}, \{yz\cos[xyz], xz\cos[xyz], xy\cos[xyz]\}\}$

■ A Package for Vector Analysis

In the **VectorAnalysis`** package, there are many more commands for vector analysis. With this package, we can do calculations in various 3D coordinate systems. We will not give a thorough presentation of this package but, rather, a quick overview. For details, see VectorAnalysis/tutorial/VectorAnalysis. First, load the package:

```
<< VectorAnalysis`
```

Then we can look at the names of this package with the following command:

```
? VectorAnalysis`*
```

The resulting table of 47 names is not presented here. We can use 14 coordinate systems (e.g., **Cartesian**, **Cylindrical**, and **Spherical**) and calculate **Grad**, **Laplacian**, **Biharmonic**, **Div**, **Curl**, **DotProduct**, **CrossProduct**, and **ScalarTripleProduct**, among others.

Let us use some of the commands. First, we ask for the current coordinate system and the default coordinates:

```
{CoordinateSystem, Coordinates[]}      {Cartesian, {Xx, Yy, Zz}}
```

Now we can calculate, for example, the gradient of a function:

```
Grad[Xx + Sin[Yy Zz]]      {1, Zz Cos[Yy Zz], Yy Cos[Yy Zz]}
```

We can set the coordinates:

```
SetCoordinates[Cartesian[x, y, z]]      Cartesian[x, y, z]
```

Now we can use **x**, **y**, and **z**:

```
Div[{x y, x y z, Sin[x y z]}]      y + x z + x y Cos[x y z]
```

Let us then move to spherical coordinates and ask for some information about this system:

```
SetCoordinates[Spherical[r, θ, φ]]      Spherical[r, θ, φ]
```

    CoordinateRanges[] $\{0 \le r < \infty,\ 0 \le \theta \le \pi,\ -\pi < \phi \le \pi\}$

    CoordinatesToCartesian[{r, θ, ϕ}]

    $\{r\,\text{Cos}[\phi]\,\text{Sin}[\theta],\ r\,\text{Sin}[\theta]\,\text{Sin}[\phi],\ r\,\text{Cos}[\theta]\}$

    jdet = JacobianDeterminant[] $r^2\,\text{Sin}[\theta]$

As an application, we calculate the area and volume of a sphere of radius R:

    Integrate[jdet, {θ, 0, Pi}, {ϕ, -Pi, Pi}] /. r $\to$ R $4\,\pi\,R^2$

    Integrate[jdet, {θ, 0, Pi}, {ϕ, -Pi, Pi}, {r, 0, R}] $\dfrac{4\,\pi\,R^3}{3}$

19.1.3 Numerical Derivatives

■ Numerical Derivatives of Functions

Sometimes it may be too difficult or impossible to calculate a derivative symbolically. You can then use **ND** from a package (or the program we presented in Section 18.1.1, p. 543).

In the **NumericalCalculus`** *package:*

ND[f, x, a] First derivative of **f** with respect to **x** at **a**

ND[f, {x, n}, a] nth derivative of **f** with respect to **x** at **a**

Options:

WorkingPrecision Precision used in internal computations; examples of values:
  MachinePrecision, **20**

Method Extrapolation method; possible values: **EulerSum**, **NIntegrate**

Scale Initial step size (**EulerSum**) or the radius of the circle of integration (**NIntegrate**); default value: **1**

Terms Number of divided differences calculated (**EulerSum**); default value: **7**

ND has two methods. If **EulerSum** is used, then **ND** forms a sequence of divided differences with successively smaller step sizes and then extrapolates to the limit by calculating a numerical limit of the divided differences. The initial step size is **Scale**, and a total of **Terms** difference quotients is calculated by successively halving the previous step size. If **NIntegrate** is used, then **ND** applies Cauchy's integral formula, and **Scale** is the radius of the circle of integration.

    As an example, we calculate the first, second, and third derivatives of an expression numerically with **ND** using both methods and compare the results with the numerical values of the exact derivatives given by **D**. The approximations are very good, particularly with the **NIntegrate** method:

```
<< NumericalCalculus`

f = Exp[-x^2];
a1 = Table[ND[f, {x, i}, 1, Method → EulerSum], {i, 3}];
a2 = Table[ND[f, {x, i}, 1, Method → NIntegrate], {i, 3}];
b = Table[D[f, {x, i}], {i, 3}] /. x → 1.;
{a1 - b, a2 - b} // Chop
```

$\left\{\left\{1.76167 \times 10^{-9},\ 7.28006 \times 10^{-7},\ -0.0000482498\right\},\ \{0, 0, 0\}\right\}$

```
Remove["Global`*"]
```

■ **Numerical Derivatives of Data**

If we have values of a function only at some given points, we can still calculate approximations to derivatives at most of the points. As an example, here are two different kinds of approximations to the first derivative:

    data = {a, b, c, d, e};

    Differences[data, 1] / h

$$\left\{ \frac{-a+b}{h}, \frac{-b+c}{h}, \frac{-c+d}{h}, \frac{-d+e}{h} \right\}$$

    (data[[3 ;;]] - data[[;; -3]]) / (2 h)

$$\left\{ \frac{-a+c}{2h}, \frac{-b+d}{2h}, \frac{-c+e}{2h} \right\}$$

Here, **h** is the step size between the points. In the first case, we did not get the approximation at the last point and in the second case at the first and last points. Remember that, for example, **data[[3 ;;]]** takes elements from the third to the end and **data[[;; -3]]** from the first to the third last (see Section 14.1.2, p. 448).

To calculate an approximation to the second derivative, do as follows:

    Differences[data, 2] / h^2

$$\left\{ \frac{a-2b+c}{h^2}, \frac{b-2c+d}{h^2}, \frac{c-2d+e}{h^2} \right\}$$

Here, we did not get the approximation at the first and last points. Note that all these calculations rely on the fact that *Mathematica* automatically does all calculations with lists component by component.

To calculate finite difference approximations of derivatives, we present an advanced command, **NDSolve`FiniteDifferenceDerivative**, in Section 27.3.2, p. 914.

19.1.4 Total Derivatives

Dt calculates total derivatives in which all variables in the expression are assumed to depend on all of the variables with respect to which the total derivative is calculated. The form of **Dt** is the same as the form of **D**. Here are some examples:

    Dt[f, x], Dt[f, x, x], Dt[f, x, y]

We can also use forms such as **Dt[f, {x, 2}]**. Here are two examples of total derivatives:

    Dt[x y, x] $y + x\, Dt[y, x]$

    Dt[x y, x, x] $2\, Dt[y, x] + x\, Dt[y, \{x, 2\}]$

■ **Example: Differentiating an Implicit Function**

We define a function of two variables:

    f = x^2 + 3 y^2 - x y - 1;

Then we define an equation in which this function is equal to zero:

    eqn = f == 0 $-1 + x^2 - x y + 3 y^2 == 0$

The equation **eqn** implicitly defines a function $y(x)$. We considered the plotting of implicit functions in Section 17.1.3, p. 518:

```
p1 = ContourPlot[f == 0, {x, -1.1, 1.1}, {y, -0.7, 0.7},
    Frame → False, Axes → True, AspectRatio → Automatic]
```

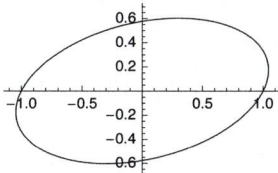

Now we calculate an equation for the derivative of the implicitly defined function:

```
deqn = Dt[eqn, x]       2 x -y -x Dt[y, x] +6 y Dt[y, x] == 0
```

Solving **Dt[y, x]** from here, we get the derivative $y'(x)$, and we give it the name **dy**:

$$\texttt{Solve[deqn, Dt[y, x]]} \qquad \left\{\left\{\texttt{Dt[y, x]} \rightarrow \frac{2\,x - y}{x - 6\,y}\right\}\right\}$$

$$\texttt{dy = Dt[y, x] /. \%[\![1]\!]} \qquad \frac{2\,x - y}{x - 6\,y}$$

Another way to calculate the derivative is to use a result of analysis:

$$\texttt{-D[f, x] / D[f, y]} \qquad \frac{-2\,x + y}{-x + 6\,y}$$

Let us consider the derivative for $x = 0.5$. First, we solve the corresponding values of y:

```
x1 = 0.5;
```

```
y1 = y /. Solve[eqn /. x → x1, y]      {-0.423564, 0.59023}
```

Then we calculate the corresponding values of the derivative:

```
dy1 = dy /. {x → x1, y → y1}      {0.468065, -0.134731}
```

Next, we form the tangents at these points:

```
tan1 = y1 + dy1 (x - x1)
{-0.423564 + 0.468065 (-0.5 + x), 0.59023 - 0.134731 (-0.5 + x)}
```

(Note how easily we obtained both of the tangents with one command: *Mathematica* automatically does vector operations component by component.) Lastly, we show the function, the tangents, and the points:

```
p2 = Plot[tan1, {x, -1.1, 1.1}];
```

```
Show[p1, p2, Epilog → {Point[{x1, y1[[1]]}], Point[{x1, y1[[2]]}]}]
```

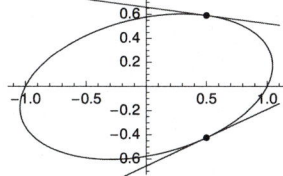

■ **Constants**

> `Dt[f, x, Constants → {a, b, … }]` Declare that **a**, **b**, … do not depend on **x**

In this way, we can tell that some symbols are treated as constants when we calculate a total derivative. For example,

> `Dt[a x^m, x, Constants → {a, m}]` $a m x^{-1+m}$

If we want to define some symbols permanently as constants, we can give them the attribute **Constant** (see Section 17.2.4, p. 530):

> `SetAttributes[{a, m}, Constant]`

Now **a** and **m** are treated as constants:

> `Dt[a x^m, x]` $a m x^{-1+m}$

The attribute can be removed by writing `Remove[a, m]`.

We can define a certain derivative of a given symbol as having a certain value:

> `n /: Dt[n, x] = 0`   0

This defines **n** as having the property that it does not depend on **x**. Now we get the following:

> `Dt[b x^n, x]` $b n x^{-1+n} + x^n \, Dt[b, x]$

The property `Dt[n, x] = 0` can be removed by writing `Remove[n]`.

■ **Total Differentials**

> `Dt[f]` Total differential of **f**

> `Dt[x^2 y^2]` $2 x y^2 \, Dt[x] + 2 x^2 y \, Dt[y]$
> `Remove["Global`*"]`

19.2 Taylor Series

19.2.1 Taylor Series and Polynomials

■ **Taylor Series**

> `Series[f, {x, a, n}]` Taylor series of **f** with respect to **x** about the point **a** with terms up to the **n**th power of **x** – **a**

> `Series[Exp[c x], {x, 1, 2}]`
>
> $$\mathrm{e}^c + c \, \mathrm{e}^c \, (x - 1) + \frac{1}{2} c^2 \, \mathrm{e}^c \, (x - 1)^2 + O[x - 1]^3$$

The remainder is in the form of a capital O. In normal mathematical notation, the remainder here would be written as $O\big((x - 1)^3\big)$, indicating terms where the power of $x - 1$ is at least 3. Here is another example:

```
t = Series[1 / Sqrt[1 + x], {x, 0, 4}]
```

$$1 - \frac{x}{2} + \frac{3\,x^2}{8} - \frac{5\,x^3}{16} + \frac{35\,x^4}{128} + O[x]^5$$

We can calculate with a Taylor series expansion:

```
t^2
```
$$1 - x + x^2 - x^3 + x^4 + O[x]^5$$

Here, all terms of an order higher than four are gathered in the remainder. We can also calculate derivatives and integrals (note the change in the remainder):

```
D[t, x]
```
$$-\frac{1}{2} + \frac{3\,x}{4} - \frac{15\,x^2}{16} + \frac{35\,x^3}{32} + O[x]^4$$

```
Integrate[t, x]
```
$$x - \frac{x^2}{4} + \frac{x^3}{8} - \frac{5\,x^4}{64} + \frac{7\,x^5}{128} + O[x]^6$$

Additional functions are automatically expanded if they occur together with a series:

```
t + Sin[c x]
```
$$1 + \left(-\frac{1}{2} + c\right) x + \frac{3\,x^2}{8} + \left(-\frac{5}{16} - \frac{c^3}{6}\right) x^3 + \frac{35\,x^4}{128} + O[x]^5$$

A series expansion is calculated automatically if we add a remainder:

```
Sin[c x] + O[x]^7
```
$$c\,x - \frac{c^3\,x^3}{6} + \frac{c^5\,x^5}{120} + O[x]^7$$

Internally, power series are **SeriesData** objects:

```
t // InputForm

SeriesData[x, 0, {1, -1/2, 3/8, -5/16, 35/128}, 0, 5, 1]
```

■ Taylor Polynomial

Normal[series]	Delete the remainder from the Taylor series

If we remove the remainder, we get the Taylor polynomial:

```
Normal[t]
```
$$1 - \frac{x}{2} + \frac{3\,x^2}{8} - \frac{5\,x^3}{16} + \frac{35\,x^4}{128}$$

The result is an ordinary expression, and now all calculations are done as with usual expressions. The following animation shows Taylor polynomials of BesselJ(5, x) at $x = 10$ up to degree 15:

```
Animate[Plot[Evaluate[{BesselJ[5, x], Normal[Series[BesselJ[5, x], {x, 10, n}]]}],
  {x, 0, 20}, PlotRange → {-0.41, 0.41}, ImageSize → 200],
  {n, 0, 15, 1}, AnimationRunning → False]
```

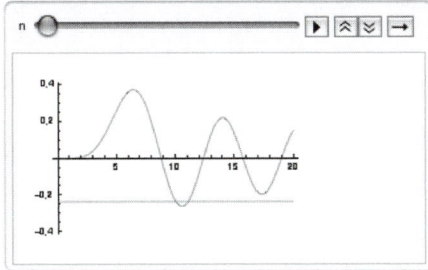

■ **Special Power Series**

Series can also calculate power series that contain negative and fractional powers:

Series[1 / Sin[x]^3, {x, 0, 6}]

$$\frac{1}{x^3} + \frac{1}{2\,x} + \frac{17\,x}{120} + \frac{457\,x^3}{15\,120} + \frac{3287\,x^5}{604\,800} + O[x]^7$$

Series[Sin[Sqrt[x]], {x, 0, 4}]

$$\sqrt{x} - \frac{x^{3/2}}{6} + \frac{x^{5/2}}{120} - \frac{x^{7/2}}{5040} + O[x]^{9/2}$$

Sometimes an essential singularity is encountered, and the series cannot be calculated:

Series[Sin[1 / x], {x, 0, 3}] $\mathrm{Sin}\!\left[\dfrac{1}{x}\right]$

However, this series can be calculated about infinity:

Series[Sin[1 / x], {x, ∞, 8}]

$$\frac{1}{x} - \frac{1}{6}\left(\frac{1}{x}\right)^3 + \frac{1}{120}\left(\frac{1}{x}\right)^5 - \frac{\left(\frac{1}{x}\right)^7}{5040} + O\!\left[\frac{1}{x}\right]^9$$

Unspecified functions are treated correctly:

Series[h[x], {x, 0, 4}]

$$h[0] + h'[0]\,x + \frac{1}{2}\,h''[0]\,x^2 + \frac{1}{6}\,h^{(3)}[0]\,x^3 + \frac{1}{24}\,h^{(4)}[0]\,x^4 + O[x]^5$$

By default, it is assumed that unspecified functions are analytic. With an option, we can tell that analyticity is not assumed:

Series[h[x] Exp[x], {x, 0, 2}]

$$h[0] + (h[0] + h'[0])\,x + \frac{1}{2}\,(h[0] + 2\,h'[0] + h''[0])\,x^2 + O[x]^3$$

Series[h[x] Exp[x], {x, 0, 2}, Analytic → False]

$$h[x]\left(1 + x + \frac{x^2}{2} + O[x]^3\right)$$

■ **More about Power Series**

If we want to develop a function of several variables into a series, we give the information about each variable in turn:

Series[f, {x, a, m}, {y, b, n}]

Series[Sin[x] Cos[y], {x, 0, 5}, {y, 0, 5}]

$$\left(1 - \frac{y^2}{2} + \frac{y^4}{24} + O[y]^6\right)x + \left(-\frac{1}{6} + \frac{y^2}{12} - \frac{y^4}{144} + O[y]^6\right)x^3 + \left(\frac{1}{120} - \frac{y^2}{240} + \frac{y^4}{2880} + O[y]^6\right)x^5 + O[x]^6$$

InverseSeries[series, x]   Find the inverse series of **series**
ComposeSeries[series1, series2]   Replace the variable in **series1** with **series2**
Residue[f, {x, a}]   Residue of **f** when **x** equals **a**

If $s(y)$ is a series expansion of $f(y)$, then `InverseSeries[s, x]` gives the series expansion of the inverse function of $f(y)$—that is, for y such that $f(y) = x$. Calculate the inverse series of a general function:

`InverseSeries[Series[f[y], {y, a, 2}], x]`

$$a + \frac{x - f[a]}{f'[a]} - \frac{f''[a] \, (x - f[a])^2}{2 \, f'[a]^3} + O[x - f[a]]^3$$

This is a series expansion of y such that $f(y) = x$. Thus, if we give x the value 0, we get a series expansion of y such that $f(y) = 0$—that is, a series expansion for a zero of $f(y)$:

`Normal[%] /. x → 0`

$$a - \frac{f[a]}{f'[a]} - \frac{f[a]^2 \, f''[a]}{2 \, f'[a]^3}$$

Here, the first two terms correspond with Newton's method. Taking more terms, we get higher-order methods.

With `NSeries` from the **NumericalCalculus`** package, we can calculate a numerical approximation to a power series expansion, and with `NResidue` from the same package we can calculate a numerical approximation to a residue.

19.2.2 Coefficients

> `SeriesCoefficient[series, n]`　Give the coefficient of the **n**th-order term of **series**
> `SeriesCoefficient[f, {x, a, n}]`　Give the coefficient of the **n**th-order term of the series expansion of **f** (here, **n** can be symbolic)
> `CoefficientList[poly, var]`　Give a list of coefficients of powers of **var** in **poly**, starting with power 0

■ Specific Coefficients

Form a series expansion:

`s = Series[Log[1 + x], {x, 1, 4}]`

$$\mathrm{Log}[2] + \frac{x - 1}{2} - \frac{1}{8} \, (x - 1)^2 + \frac{1}{24} \, (x - 1)^3 - \frac{1}{64} \, (x - 1)^4 + O[x - 1]^5$$

Find the coefficients:

`Table[SeriesCoefficient[s, n], {n, 0, 4}]`

$$\left\{ \mathrm{Log}[2], \, \frac{1}{2}, \, -\frac{1}{8}, \, \frac{1}{24}, \, -\frac{1}{64} \right\}$$

The list of coefficients of powers of x is found from the expanded expression:

`CoefficientList[s, x]`

$$\left\{ -\frac{131}{192} + \mathrm{Log}[2], \, \frac{15}{16}, \, -\frac{11}{32}, \, \frac{5}{48}, \, -\frac{1}{64} \right\}$$

`Expand@Normal@s`

$$-\frac{131}{192} + \frac{15 \, x}{16} - \frac{11 \, x^2}{32} + \frac{5 \, x^3}{48} - \frac{x^4}{64} + \mathrm{Log}[2]$$

■ **General Coefficients**

Now we ask for the general nth-order coefficient:

```
SeriesCoefficient[Log[1 + x], {x, 1, n}]
```

$$\text{If}\left[n == 0, \text{Log}[2], \frac{\text{If}\left[-1 + n \geq 0, \left(-\frac{1}{2}\right)^{-1+n}, 0\right]}{2\,n}\right]$$

```
Table[%, {n, 0, 4}]
```

$$\left\{\text{Log}[2], \frac{1}{2}, -\frac{1}{8}, \frac{1}{24}, -\frac{1}{64}\right\}$$

As another example, consider the following series:

```
Series[1 / (1 + x^4), {x, 0, 16}]
```

$$1 - x^4 + x^8 - x^{12} + x^{16} + O[x]^{17}$$

In this example, it is quite easy to write a general formula for the coefficients, but we try the command:

```
SeriesCoefficient[1 / (1 + x^4), {x, 0, n}]
```

$$\frac{1}{4}\,(-1)^{n/4}\left(1 + (-1)^n + i^n + i^{3\,n}\right)$$

```
Table[%, {n, 0, 16}]
```

$$\{1, 0, 0, 0, -1, 0, 0, 0, 1, 0, 0, 0, -1, 0, 0, 0, 1\}$$

■ **Generating Functions**

The probability-generating function of a discrete random variable X is defined to be $G(z) = E(z^X)$. Thus, $G(z)$ can be calculated from $G(z) = \sum_n P(X = n)\,z^n$. Inversely, if we know a closed-form expression for a probability generating function $G(z)$, then $P(X = n)$ can be calculated as the coefficient of z^n in the power series expansion of $G(z)$. As an example, if $G(z) = e^{\lambda(z-1)}$, the coefficient of z^n is

```
SeriesCoefficient[Exp[λ (z - 1)], {z, 0, n}]
```

$$\frac{e^{-\lambda}\,\lambda^n}{n!}$$

This is the probability function of a Poisson random variable.

Toss a die n times. Let X_i be the ith result and S_n the sum of the results. The probability-generating function of X_i is $G_{X_i}(z) = \sum_{k=1}^{6}\frac{1}{6}z^k$, and that of S_n is $G_{S_n}(z) = \prod_{i=1}^{n} G_{X_i}(z)$. Thus, $G_{S_n}(z) = \frac{1}{6^n}\left(\sum_{k=1}^{6} z^k\right)^n$. We can now get the probabilities of S_n from the power series expansion of $G_{S_n}(z)$. For example,

```
GS[n_] := Sum[z^k, {k, 1, 6}]^n / 6^n
```

```
CoefficientList[GS[3], z]
```

$$\left\{0, 0, 0, \frac{1}{216}, \frac{1}{72}, \frac{1}{36}, \frac{5}{108}, \frac{5}{72}, \frac{7}{72}, \frac{25}{216}, \frac{1}{8}, \frac{1}{8}, \frac{25}{216}, \frac{7}{72}, \frac{5}{72}, \frac{5}{108}, \frac{1}{36}, \frac{1}{72}, \frac{1}{216}\right\}$$

```
ListPlot[%, Filling → Axis]
```

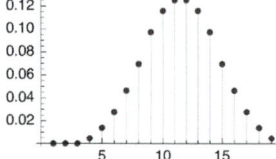

It is known that if $P_n(x)$ is the nth Legendre polynomial, then $\sum_{n=0}^{\infty} P_n(x) t^n = 1 \Big/ \sqrt{1 - 2 t x + t^2}$. Thus,

```
SeriesCoefficient[1 / Sqrt[1 - 2 t x + t^2], {t, 0, n}]
```

LegendreP[n, x]

With special values of n we get explicit polynomials:

```
SeriesCoefficient[1 / Sqrt[1 - 2 t x + t^2], {t, 0, 5}]
```

$$\frac{1}{8} \left(15 x - 70 x^3 + 63 x^5 \right)$$

LegendreP[5, x] $\frac{1}{8} \left(15 x - 70 x^3 + 63 x^5 \right)$

19.2.3 Equations

Sometimes we want to determine conditions under which two power series are equivalent. For example, consider the following series:

```
s = Series[y[x], {x, 0, 4}] /. y[0] → 1
```

$$1 + y'[0] x + \frac{1}{2} y''[0] x^2 + \frac{1}{6} y^{(3)}[0] x^3 + \frac{1}{24} y^{(4)}[0] x^4 + O[x]^5$$

We want conditions for the derivatives of y at 0 under which the following equation is true (up to the remainder):

```
eqn = D[s, x] + s == Exp[x]
```

$$(1 + y'[0]) + (y'[0] + y''[0]) x +$$
$$\left(\frac{y''[0]}{2} + \frac{1}{2} y^{(3)}[0] \right) x^2 + \left(\frac{1}{6} y^{(3)}[0] + \frac{1}{6} y^{(4)}[0] \right) x^3 + O[x]^4 == e^x$$

The expansion of e^x is as follows:

```
Series[Exp[x], {x, 0, 3}]
```
$1 + x + \frac{x^2}{2} + \frac{x^3}{6} + O[x]^4$

We see that for the equation **eqn** to be true, we must have $1 + y'(0) = 1$, and so on. With **LogicalExpand**, we can easily form such conditions.

LogicalExpand[series1 == series2] Find conditions that make the equation true

We try this command with the preceding example:

```
LogicalExpand[eqn]
```

$$y'[0] == 0 \, \&\& \, -1 + y'[0] + y''[0] == 0 \, \&\&$$
$$-\frac{1}{2} + \frac{y''[0]}{2} + \frac{1}{2} y^{(3)}[0] == 0 \, \&\& \, -\frac{1}{6} + \frac{1}{6} y^{(3)}[0] + \frac{1}{6} y^{(4)}[0] == 0$$

Now we can solve the equations:

```
sol = Solve[%]
```

$$\left\{ \left\{ y'[0] \to 0, \, y''[0] \to 1, \, y^{(3)}[0] \to 0, \, y^{(4)}[0] \to 1 \right\} \right\}$$

We can also directly apply **Solve** to the equation **eqn** (without first using **LogicalExpand**):

```
Solve[eqn]
```

$$\{\{y'[0] \to 0, y''[0] \to 1, y^{(3)}[0] \to 0, y^{(4)}[0] \to 1\}\}$$

If we now insert these values into **s**, we get, in fact, a (crude) series solution for the differential equation $y'(x) + y(x) = e^x$ with the initial value $y(0) = 1$:

```
s /. sol[[1]] // Normal
```
$$1 + \frac{x^2}{2} + \frac{x^4}{24}$$

The exact solution of the equation is

```
DSolve[{y'[x] + y[x] == Exp[x], y[0] == 1}, y[x], x]
```
$$\left\{\left\{y[x] \to \frac{1}{2} e^{-x} \left(1 + e^{2x}\right)\right\}\right\}$$

with a series expansion

```
Series[y[x] /. %[[1]], {x, 0, 4}]
```
$$1 + \frac{x^2}{2} + \frac{x^4}{24} + O[x]^5$$

For more information about series solutions of differential equations, see Section 26.2.2, p. 843.

19.3 Limits

19.3.1 Symbolic Limits

> **Limit[f, x → a]**   Limit of **f** as **x** approaches **a**
>
> *Options:*
> **Direction**   Direction from which **a** is approached; possible values: **Automatic** (usually means **-1**),
>   **-1** (from above), **1** (from below)
> **Assumptions**   Assumptions for parameters; examples of values: **$Assumptions, a > 0**
> **Analytic**   Whether unknown functions are treated as analytic; possible values: **False, True**

Write the arrow as **->** (*Mathematica* then replaces it with a true arrow →). The default value **Automatic** for **Direction** means **Direction → -1** (i.e., from above or from larger values) except for limits at infinity, where it means **Direction → 1**. For example,

```
Limit[(Cos[x] - 1) / x^2, x → 0]
```
$$-\frac{1}{2}$$

```
Limit[(1 + c / x)^x, x → ∞]        e^c
```

```
Limit[x^y, x → ∞, Assumptions → y < 0]        0
```

The following limit is not unique, and we get a whole interval:

```
Limit[Sin[1 / x], x → 0]    Interval[{-1, 1}]
```

■ **Derivatives**

By definition, derivatives are obtained from limits:

```
Limit[(Sin[x + h] - Sin[x]) / h, h → 0]    Cos[x]
```

If we have an expression containing unknown functions, the default is that **Limit** does not assume that they are analytic and, consequently, the limit is not calculated:

```
Limit[(f[x + h] - f[x]) / h, h → 0]
```

$$\text{Limit}\left[\frac{-f[x] + f[h + x]}{h}, h \to 0\right]$$

Assuming that **f** is analytic, we get a result:

```
Limit[(f[x + h] - f[x]) / h, h → 0, Analytic → True]      f'[x]
```

```
Limit[(f[x + 2 h] - 2 f[x + h] + f[x]) / h^2, h → 0, Analytic → True]      f''[x]
```

■ Integrals

By definition, integrals are obtained from the limit of a Riemann sum.

```
riemannSum[f_, x_, a_, b_, n_] := With[{h = (b - a) / n},
  h Sum[f /. x → a + h k, {k, 0, n - 1}]]
```

Calculate a Riemann sum and compare its limit with the corresponding integral:

```
riemannSum[1 + x + x^2, x, 0, 1, n]
```
$$\frac{1 - 6 n + 11 n^2}{6 n^2}$$

```
Limit[%, n → ∞]
```
$$\frac{11}{6}$$

```
Integrate[1 + x + x^2, {x, 0, 1}]
```
$$\frac{11}{6}$$

■ Direction

Consider the following discontinuous function:

```
g = 1 / (2 ^ (1 / x) + 1);
```

```
Plot[g, {x, -3, 3}]
```

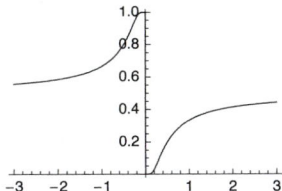

We see that the function has two different limits—1 and 0—at $x = 0$. *Mathematica* gives, by default, the limit from above:

```
Limit[g, x → 0]      0
```

If we suspect that the limit may be different depending on the direction, we can also calculate the limit from below:

```
Limit[g, x → 0, Direction → 1]      1
```

19.3.2 Numerical Limits

In the **NumericalCalculus`** *package:*

NLimit[f, x → a]　Numerical limit of **f** as **x** approaches **a**

Options:

Direction　Direction from which **a** is approached; possible values: **Automatic** (usually means **-1**),
　-1 (from above), **1** (from below)

WorkingPrecision　Precision used in internal computations; examples of values:
　MachinePrecision, **20**

Scale　Initial step size; default value: **1**

Terms　Number of values calculated; default value: **7**

Method　Extrapolation method; possible values: **EulerSum**, **SequenceLimit**

WynnDegree　Degree to use in **SequenceLimit**; default value: **1**

NLimit works by calculating a sequence of values for the function with successively smaller step sizes and then extrapolating to the limit. The initial step size is **Scale** (default is 1), and a total of **Terms** (default is 7) values is calculated by successively dividing the previous step size by 10. The sequence of values is then extrapolated by applying either a generalized Euler transformation (**Method → EulerSum**; this is the default) or Wynn's ϵ-algorithm (**Method → SequenceLimit**).

It turns out that the default method does not calculate the limit of the first example in Section 19.3.1, p. 630, but the other method works:

```
<< NumericalCalculus`

NLimit[(Cos[x] - 1) / x^2, x → 0, Method → SequenceLimit]

-0.5
```

Integral Calculus

Introduction

> *Riemann consulted a doctor about his diet. He was told to reduce the amount of food he ate at each meal but to increase the number of meals. He proceeded to do so and ultimately ate infinitesimal amount infinitely often, and he found that his weight did not change. Shortly after this, he gave a precise definition of a definite integral.*

Mathematica's symbolic integration methods have such power and knowledge that in the Documentation Center it is said that **Integrate** can evaluate essentially all indefinite integrals and most definite integrals listed in standard books of tables.

In the case in which we do not get an exact answer for a definite integral, we can resort to numerical methods (numerical quadrature) and get an approximate answer with `NIntegrate`.

Mathematica has excellent adaptive methods for numerical integration. The default methods often work very well, but we also have extensive possibilities to choose a method we think should suit our problem. We can choose between global and local adaptive *integration strategy* and between many particular *integration rules*. For a detailed account of numerical integration, see tutorial/NIntegrateOverview.

In this chapter, we also consider sums, products, and transforms. Many transforms of functions are based on integrals or sums. The best-known transform is the Laplace transform. Solving differential and partial differential equations with this transform is considered in Sections 26.2.1, p. 841, and 27.1.2, p. 891. In Section 28.1.3, p. 932, we demonstrate how to use the Z-transform to solve difference equations. In Section 30.6.2, p. 1045, we use the discrete Fourier transform to smooth data.

20.1 Integration

20.1.1 Indefinite Integration

> `Integrate[f, x]`   Indefinite integral of `f` with respect to `x`

Note that the constant of integration is not shown:

`Integrate[x / (a + b x), x]`

$$\frac{x}{b} - \frac{a \, \text{Log}[a + b \, x]}{b^2}$$

If the derivative of the integral is the same as the integrand, this supports the correctness of the integral:

`D[%, x] // Simplify` $\dfrac{x}{a + b \, x}$

Remember that integrals can also be entered with the aid of the **BasicMathInput** palette (see Section 1.4.1, p. 15). Still another way to enter integrals is to write 2D input directly with the keyboard (see Section 3.3.3, p. 76): The integral sign $\int$ can be written as ⎋int⎋, and the d appearing before the variable of the integration is entered as ⎋dd⎋:

$\int (a + x\,\text{Exp}[x])\, dx$ $e^x\,(-1 + x) + a\,x$

`Derivative` can also be used to calculate integrals: `Derivative[-n][f][x]` gives the nth antiderivative (or indefinite integral) of `f` at `x` (recall from Section 19.1.1, p. 618, that `Derivative[n][f][x]` gives the nth derivative of `f` at `x`). For example,

`f[x_] := 1`

`Table[Derivative[-n][f][x], {n, 0, 3}]`

$$\left\{1,\, x,\, \frac{x^2}{2},\, \frac{x^3}{6}\right\}$$

■ Special Values of Parameters

An important point is that *Mathematica* assumes that all parameters in the integrand have *generic* values. For example, the first integral given previously is correct only if b is not 0. If $b = 0$, the integral is, of course, $x^2/(2a)$. *Mathematica* does not tell you for what values the result holds; you have to check special cases directly. Here is another example:

Integrate[Log[x]^n / x, x] $\quad \dfrac{\text{Log}[x]^{1+n}}{1+n}$

This holds for a generic n—that is, for an n not equal to -1. If $n = -1$, the result is different:

Integrate[Log[x]^(-1) / x, x] $\quad$ Log[Log[x]]

However, for *definite* integrals, *Mathematica* can give conditions under which the integral converges (see Section 20.1.3, p. 639).

■ Special Functions

Many functions do not have integrals in terms of elementary functions. However, the integral may be representable in terms of some special functions:

Integrate[Exp[-x^2], x] $\quad \dfrac{1}{2}\sqrt{\pi}\ \text{Erf}[x]$

The result of the following integral contains an elliptic integral:

Integrate[1 / Sqrt[1 + x^3], x]

$$\dfrac{1}{3^{1/4}\sqrt{1+x^3}}\ 2\,(-1)^{1/6}\sqrt{-(-1)^{1/6}\left((-1)^{2/3}+x\right)}$$

$$\sqrt{1+(-1)^{1/3}\,x+(-1)^{2/3}\,x^2}\ \text{EllipticF}\!\left[\text{ArcSin}\!\left[\dfrac{\sqrt{-(-1)^{5/6}\,(1+x)}}{3^{1/4}}\right],\,(-1)^{1/3}\right]$$

■ Piecewise Functions

Piecewise functions can also be integrated:

g[x_] = Piecewise[{{Sin[x], x^2 < 1}, {Cos[x], x^2 ≥ 1}}]

$$\begin{cases}\text{Sin}[x] & x^2 < 1 \\ \text{Cos}[x] & x^2 \geq 1\end{cases}$$

Integrate[g[x], x]

$$\begin{cases}\text{Sin}[x] & x \leq -1 \\ \text{Cos}[1] - \text{Cos}[x] - \text{Sin}[1] & -1 < x \leq 1 \\ -2\,\text{Sin}[1] + \text{Sin}[x] & \text{True}\end{cases}$$

Integrals containing absolute value are not done unless the variable is real:

Integrate[Abs[x] x, x]

$$\int x\,\text{Abs}[x]\,\mathrm{d}x$$

Integrate[Abs[x] x, x, Assumptions → x ∈ Reals]

$$\begin{cases}-\dfrac{x^3}{3} & x \leq 0 \\ \dfrac{x^3}{3} & \text{True}\end{cases}$$

■ Root Sums

In integrating rational functions, we often get an explicit result:

```
int = Integrate[1 / (2 + 3 x - x^3), x]
```

$$\frac{1}{9} \left(-\frac{3}{1+x} - \text{Log}[-2 + x] + \text{Log}[1 + x] \right)$$

Often, the result also contains a **RootSum** object:

```
int = Integrate[1 / (2 + 2 x - x^3), x]
```

$$-\text{RootSum}\left[-2 - 2 \#1 + \#1^3 \, \&, \ \frac{\text{Log}[x - \#1]}{-2 + 3 \#1^2} \, \& \right]$$

Thus, the integral is $-\sum_{i=1}^{3} \log(x - r_i)/(-2 + 3 r_i^2)$, where r_1, r_2, and r_3 are the roots of $-2 - 2x + x^3$. We can see the sum with **Normal**:

```
int // Normal
```

$$-\frac{\text{Log}\left[x - \text{Root}\left[-2 - 2\#1 + \#1^3 \, \&, \ 1\right]\right]}{-2 + 3 \, \text{Root}\left[-2 - 2\#1 + \#1^3 \, \&, \ 1\right]^2} -$$

$$\frac{\text{Log}\left[x - \text{Root}\left[-2 - 2\#1 + \#1^3 \, \&, \ 2\right]\right]}{-2 + 3 \, \text{Root}\left[-2 - 2\#1 + \#1^3 \, \&, \ 2\right]^2} - \frac{\text{Log}\left[x - \text{Root}\left[-2 - 2\#1 + \#1^3 \, \&, \ 3\right]\right]}{-2 + 3 \, \text{Root}\left[-2 - 2\#1 + \#1^3 \, \&, \ 3\right]^2}$$

We can also ask for the explicit expression to be shown:

```
int // ToRadicals
```

$$-\frac{\text{Log}\left[-\frac{1}{3}\left(27 - 3\sqrt{57}\right)^{1/3} - \frac{\left(9+\sqrt{57}\right)^{1/3}}{3^{2/3}} + x\right]}{-2 + 3 \left(\frac{1}{3}\left(27 - 3\sqrt{57}\right)^{1/3} + \frac{\left(9+\sqrt{57}\right)^{1/3}}{3^{2/3}}\right)^2} -$$

$$\frac{\text{Log}\left[\frac{1}{6}\left(1 + i\sqrt{3}\right)\left(27 - 3\sqrt{57}\right)^{1/3} + \frac{\left(1-i\sqrt{3}\right)\left(9+\sqrt{57}\right)^{1/3}}{2 \cdot 3^{2/3}} + x\right]}{-2 + 3 \left(-\frac{1}{6}\left(1 + i\sqrt{3}\right)\left(27 - 3\sqrt{57}\right)^{1/3} - \frac{\left(1-i\sqrt{3}\right)\left(9+\sqrt{57}\right)^{1/3}}{2 \cdot 3^{2/3}}\right)^2} -$$

$$\frac{\text{Log}\left[\frac{1}{6}\left(1 - i\sqrt{3}\right)\left(27 - 3\sqrt{57}\right)^{1/3} + \frac{\left(1+i\sqrt{3}\right)\left(9+\sqrt{57}\right)^{1/3}}{2 \cdot 3^{2/3}} + x\right]}{-2 + 3 \left(-\frac{1}{6}\left(1 - i\sqrt{3}\right)\left(27 - 3\sqrt{57}\right)^{1/3} - \frac{\left(1+i\sqrt{3}\right)\left(9+\sqrt{57}\right)^{1/3}}{2 \cdot 3^{2/3}}\right)^2}$$

Thus, when asking the value of the integral, *Mathematica* decided not to show the explicit expression because it is quite involved. Anyway, we can do usual calculations with root sums:

```
Plot[int, {x, 2, 10}]
```

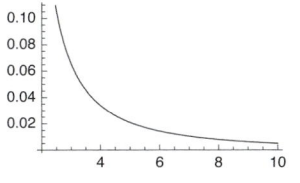

```
D[int, x] // Simplify
```
$$\frac{1}{2 + 2\,x - x^3}$$

20.1.2 More about Indefinite Integration

■ Difficult Integrals

If *Mathematica* does not do the integration, there are two possibilities: The integral really cannot be calculated in terms of any of the built-in elementary and special functions of *Mathematica*, or the integral can be calculated but *Mathematica* did not succeed at doing so. If you think the integral could be calculated, try helping *Mathematica*. For example, try special values for the parameters.

If *Mathematica* cannot calculate the integral, it simply writes the given command as such:

```
Integrate[Log[x^x]^n, x]
```
$$\int \text{Log}\left[x^x\right]^n\, dx$$

Mathematica can calculate this integral for any given positive integer value of n. For example,

```
Integrate[Log[x^x]^2, x] // Simplify
```
$$\frac{1}{54}\,x\left(4\,x^2 + 18\,x^2\,\text{Log}[x]^2 + 3\,x\,\text{Log}[x]\left(5\,x - 18\,\text{Log}\left[x^x\right]\right) - 27\,x\,\text{Log}\left[x^x\right] + 54\,\text{Log}\left[x^x\right]^2\right)$$

You can also try to write the integrand in another form (e.g., with **Apart** you get partial fraction expansions), or try integration by parts or change of variable. If you have a definite integral, you can also try numerical quadrature (see Section 20.2).

If the integrand contains unknown functions, *Mathematica* may not be able to calculate the integral:

```
Integrate[p[x] p''[x], x]
```
$$\int p[x]\,p''[x]\, dx$$

However, sometimes it succeeds:

```
Integrate[2 p[x] p'[x], x]
```
$$p[x]^2$$

It is always good to check the result given by *Mathematica*. For an indefinite integral, we can differentiate the result. For a definite integral, we can also use numerical quadrature (for specific values of the parameters) and compare the results.

■ Integration by Parts

Mathematica can do the following integral:

```
Integrate[Log[x] x^n, x]
```
$$\frac{x^{1+n}\,(-1 + (1 + n)\,\text{Log}[x])}{(1 + n)^2}$$

However, let us try integration by parts; our example is from Wrede and Spiegel (2002, p. 106). The rule can be written as follows:

```
Integrate[u dv, x] = u v - Integrate[v du, x]
```

Here, **du** and **dv** are derivatives of **u** and **v**. Define **u** and **dv**:

```
u = Log[x]; dv = x^n;
```

Then calculate the derivative of **u** and the integral of **dv**:

```
{du = D[u, x], v = Integrate[dv, x]}
```
$$\left\{ \frac{1}{x}, \frac{x^{1+n}}{1+n} \right\}$$

The original integral is then as follows:

```
u v - Integrate[v du, x]
```
$$-\frac{x^{1+n}}{(1+n)^2} + \frac{x^{1+n} \, Log[x]}{1+n}$$

■ Change of Variable

Let us integrate, by change of variable, the following function:

```
f = a^Sqrt[b + c x];
```

This example is a generalization of an example in Wrede and Spiegel (2002, p. 107). Denote $y = \sqrt{b + c x}$ and express this as an equation:

```
eqn = y == Sqrt[b + c x]
```
$$y == \sqrt{b + c \, x}$$

Solve this for **x**:

```
xx = x /. Solve[eqn, x][[1]]
```
$$\frac{-b + y^2}{c}$$

Insert this into the integrand and into the differential:

```
g = PowerExpand[(f /. x → xx) D[xx, y]]
```
$$\frac{2 \, a^y \, y}{c}$$

(Here, we used **PowerExpand** to simplify $\sqrt{y^2}$ to y, which is true if $y \geq 0$.) Let us now try integration:

```
iy = Integrate[g, y]
```
$$\frac{2 \, a^y \, (-1 + y \, Log[a])}{c \, Log[a]^2}$$

Lastly, we go back to the variable x:

```
ix = iy /. ToRules[eqn]
```
$$\frac{2 \, a^{\sqrt{b+c x}} \left(-1 + \sqrt{b + c \, x} \, Log[a] \right)}{c \, Log[a]^2}$$

(Here, **ToRules[eqn]** writes the equation **eqn** as the rule **{y → Sqrt[b + c x]}**.) By the way, *Mathematica* does this integral, too:

```
Integrate[f, x]
```
$$\frac{2 \, a^{\sqrt{b+c x}} \left(-1 + \sqrt{b + c \, x} \, Log[a] \right)}{c \, Log[a]^2}$$

20.1.3 Definite Integration

Integrate[f, {x, a, b}]   Definite integral of **f** when **x** varies from **a** to **b**

```
Integrate[Exp[-x^2], {x, -∞, ∞}]     √π
```

(Recall that ∞ can be written as ⎡ESC⎤inf⎡ESC⎤.) Definite integrals can also be calculated by the **BasicMathIn-put** palette, or you can write 2D input directly with the keyboard (see Section 3.3.3, p. 76). For example, consider the following integral:

$$\int_0^\pi (\text{Sin}[x] + \text{Log}[x])\, dx$$

To write this expression, type ⎡ESC⎤int⎡ESC⎤⎡CTRL⎤-⎤0⎡CTRL⎤%⎤⎡ESC⎤p⎡ESC⎤⎡CTRL⎤␣⎤(Sin[x]+Log[x])⎡ESC⎤dd⎡ESC⎤x.

If a definite integral is not calculated, we can ask for a numerical value (see Section 20.2.1, p. 644):

```
Integrate[Sin[Sin[x]], {x, 0, π / 3}]
```
$$\int_0^{\frac{\pi}{3}} \text{Sin}[\text{Sin}[x]]\, dx$$

```
% // N    0.466185
```

■ **Simplifying the Result**

Let us try to derive formula 18.26 in Spiegel (1999, p. 107). This formula states that the integral of $\sin(m\,x)\sin(n\,x)$ (m and n are integers) over $(0, \pi)$ is zero unless $m = n$, in which case the integral is $\pi/2$. First, we compute the integral with the general m and n:

```
int = Integrate[Sin[m x] Sin[n x], {x, 0, π}]
```
$$\frac{n\,\text{Cos}[n\,\pi]\,\text{Sin}[m\,\pi] - m\,\text{Cos}[m\,\pi]\,\text{Sin}[n\,\pi]}{m^2 - n^2}$$

Then we assume that m and n are integers:

```
Simplify[int, {m, n} ∈ Integers]    0
```

However, the result obtained is a generic result that is valid only for most values of m and n. Specifically, it does not hold for the case $m = n$. We calculate and simplify this integral separately:

```
int2 = Integrate[Sin[m x]^2, {x, 0, π}]
```
$$\frac{\pi}{2} - \frac{\text{Sin}[2\,m\,\pi]}{4\,m}$$

```
Simplify[int2, m ∈ Integers]
```
$$\frac{\pi}{2}$$

■ **Conditions of Convergence**

For definite integrals, *Mathematica* can give conditions under which the integral converges. The following integral converges only if the real part of **a** is positive:

```
Integrate[Exp[-x / a], {x, 0, ∞}]
```
$$\text{If}\left[\text{Re}[a] > 0,\, a,\, \text{Integrate}\left[e^{-\frac{x}{a}},\, \{x, 0, \infty\},\, \text{Assumptions} \to \text{Re}[a] \le 0\right]\right]$$

Here is another example (the definition of the beta function):

```
Integrate[t^(a - 1) (1 - t)^(b - 1), {t, 0, 1}]
```
$$\text{If}\left[\text{Re}[a] > 0\,\&\&\,\text{Re}[b] > 0,\, \frac{\pi\,\text{Csc}[b\,\pi]\,\text{Gamma}[a]}{\text{Gamma}[1 - b]\,\text{Gamma}[a + b]},\, \text{Integrate}\left[\right.\right.$$
$$\left.\left. (1 - t)^{-1+b}\,t^{-1+a},\, \{t, 0, 1\},\, \text{Assumptions} \to \text{Re}[a] \le 0\,||\,(\text{Re}[a] > 0\,\&\&\,\text{Re}[b] \le 0)\right]\right]$$

The integral converges only if the real parts of **a** and **b** are positive. Actually, we did not get the usual form $\Gamma(a)\,\Gamma(b)/\Gamma(a + b)$ for the beta function, but we can show that the result given is correct:

$$\frac{\pi\,\text{Csc}[b\,\pi]\,\text{Gamma}[a]}{\text{Gamma}[1-b]\,\text{Gamma}[a+b]} - \frac{\text{Gamma}[a]\,\text{Gamma}[b]}{\text{Gamma}[a+b]} \; \text{// FullSimplify}$$
0

Here is a third example:

`Integrate[1 / x^2, {x, a, 1}, Assumptions → a ∈ Reals]`

$$\text{If}\left[a > 1 \;||\; 0 < a < 1,\; -1 + \frac{1}{a},\; \text{Integrate}\left[\frac{1}{x^2},\; \{x, a, 1\},\; \text{Assumptions} \to a \le 0 \;||\; a == 1\right]\right]$$

Thus, a should be positive (the result $-1 + \frac{1}{a}$ actually also holds for $a = 1$).

■ Options

Options of `Integrate` *in definite integration:*

`Assumptions`   Assumptions about parameters; examples of values: `$Assumptions, n > 0, n ∈`
  `Integers, n > 0 && n ∈ Integers`

`GenerateConditions`   Whether to generate answers containing conditions for the parameters;
    possible values: `Automatic` (usually means `True`), `True`, `False`

`PrincipalValue`   Whether to find the Cauchy principal value; possible values: `False`, `True`

The assumptions can be equations, inequalities, and domain specifications (see Section 13.2.1, p. 420, for various domains) and their logical combinations. An assumption could be, for example, `x ∈ Reals` (write ∈ as ⓔsⓒelemⓔsⓒ). The default value `Automatic` for `GenerateConditions` essentially means `True`.

We try some assumptions for the integrals we calculated previously in **Conditions of Convergence**:

`Integrate[Exp[-x / a], {x, 0, ∞}, Assumptions → a > 0]`   a

`Integrate[t^(a - 1) (1 - t)^(b - 1), {t, 0, 1}, Assumptions → a > 0 && b > 0]`

$$\frac{\pi\,\text{Csc}[b\,\pi]\,\text{Gamma}[a]}{\text{Gamma}[1-b]\,\text{Gamma}[a+b]}$$

We can ask not to print conditions of convergence:

`Integrate[Exp[-x / a], {x, 0, ∞}, GenerateConditions → False]`   a

Some integrals that do not have a finite value in the usual (Riemannian) sense over intervals containing a point of singularity may have a finite Cauchy principal value. This value for the integral is obtained by deleting a small interval centered at the singular point and then taking the limit of the integral as the length of the interval goes to zero.

■ Advanced Integrals

The integrand may contain unknown functions:

`Integrate[f[x, t], {x, a[t], b[t]}]` $\displaystyle\int_{a[t]}^{b[t]} f[x, t]\, dx$

We can calculate the derivative of this with respect to t:

`D[%, t]` $\displaystyle\int_{a[t]}^{b[t]} f^{(0,1)}[x, t]\, dx - f[a[t], t]\, a'[t] + f[b[t], t]\, b'[t]$

Here, $f^{(0,1)}[x, t]$ means derivative with respect to t.

The integrand may contain such functions as `Abs`, `Sign`, `UnitStep`, `Min`, and `Max`:

```
Integrate[Abs[x] + Sign[x] + Min[1/2, Sin[x]], {x, -1, 1}]
```

$$\frac{1}{12}\left(18 - 6\sqrt{3} - \pi + 12\,\text{Cos}[1]\right)$$

Piecewise defined functions are also integrated:

```
Integrate[Piecewise[{{x^2, x^2 > x^3 - x}, {x^3 - 1, True}}], {x, 0, 3}]
```

$$\frac{1}{24}\left(421 + 11\sqrt{5}\right)$$

```
Integrate[Which[x^2 > x^3 - x, x^2, True, x^3 - 1], {x, 0, 3}]
```

$$\frac{1}{24}\left(421 + 11\sqrt{5}\right)$$

```
Integrate[If[x^2 > x^3 - x, x^2, x^3 - 1], {x, 0, 3}]
```

$$\frac{1}{24}\left(421 + 11\sqrt{5}\right)$$

```
Integrate[Max[x, Cos[x]], {x, 0, π}]
```
$$\frac{\pi^2}{2}$$

■ Change of Variable

Let us calculate the integral of the following function over $(0, \pi)$:

```
f = x Sin[x] / (1 + Cos[x]^2);
```

According to Wrede and Spiegel (2002, p. 108), the integral has the value $\pi^2/4$. For this integral, *Mathematica* gives a long expression of approximately 1 page containing **ArcCos**, **ArcTan**, **Log**, and **PolyLog** (we do not show the result here):

```
int = Integrate[f, {x, 0, π}]; // Timing
{221.194, Null}
```

The simplified result is correct:

```
int // FullSimplify
```
$$\frac{\pi^2}{4}$$

Let us try the same technique as Spiegel used to directly obtain the simple result $\pi^2/4$. Do a change of variable $y = \pi - x$:

```
g = (f /. x → π - y) D[π - y, y]
```
$$-\frac{(\pi - y)\,\text{Sin}[y]}{1 + \text{Cos}[y]^2}$$

Expand this expression:

```
g = Expand[g]
```
$$-\frac{\pi\,\text{Sin}[y]}{1 + \text{Cos}[y]^2} + \frac{y\,\text{Sin}[y]}{1 + \text{Cos}[y]^2}$$

The last term is in the same form as the original function. If the original integral is **int**, then the integral of the last term is **-int** because the integration with respect to y goes from π to 0. Integrate the first term:

```
i1 = Integrate[First[g], {y, π, 0}]
```
$$\frac{\pi^2}{2}$$

We now have the equation **int == i1 - int**. Solving the equation for **int**, we get $\pi^2/4$.

20.1.4 Integration over Regions

■ **Basic Multiple Integrals**

Consider the following multiple integral:

$$\int_a^b \left(\int_x^{x+1} 12\,x\,y\,dy \right) dx$$

$$-3\,a^2 - 4\,a^3 + 3\,b^2 + 4\,b^3$$

This can also be calculated with nested **Integrate** commands:

```
Integrate[Integrate[12 x y, {y, x, x + 1}], {x, a, b}]
```

$$-3\,a^2 - 4\,a^3 + 3\,b^2 + 4\,b^3$$

However, **Integrate** also has special forms for multiple integrals:

`Integrate[g, {x, a, b}, {y, c, d}]`	Calculate $\int_a^b \left(\int_c^d g\,dy \right) dx$
`Integrate[g, {x, a, b}, {y, c, d}, {z, e, f}]`	Calculate $\int_a^b \left[\int_c^d \left(\int_e^f g\,dz \right) dy \right] dx$

In the first command, the integral is first calculated with respect to **y** and then with respect to **x**. The end points **c** and **d** may be functions of **x**. For example,

```
Integrate[12 x y, {x, a, b}, {y, x, x + 1}]
```

$$-3\,a^2 - 4\,a^3 + b^2\,(3 + 4\,b)$$

■ **Advanced Multiple Integrals**

Recall that **Boole[ineqs]** is the characteristic function of the set defined by inequalities **ineqs**; that is, **Boole[True]** is 1 and **Boole[False]** is 0. With **Boole** we can integrate over various regions.

`Integrate[f Boole[cond], {x, a, b}, {y, c, d}]`	Integrate **f** over the region where **cond** is True

In the first example, we integrate the function 1 over a circle, that is, we calculate the area of the circle:

```
Integrate[Boole[x^2 + y^2 ≤ r^2], {x, -∞, ∞}, {y, -∞, ∞}, Assumptions → r > 0]
```

$$\pi\,r^2$$

Calculate then the volume of a sphere:

```
Integrate[Boole[x^2 + y^2 + z^2 ≤ r^2],
 {x, -∞, ∞}, {y, -∞, ∞}, {z, -∞, ∞}, Assumptions → r > 0]
```

$$\frac{4\,\pi\,r^3}{3}$$

Define two inequalities that determine the area of integration:

```
ineqs = x^2 + y^2 < 1 && (x - 1 / 2)^2 + (y - 1 / 2)^2 > 1 / 30
```

$$x^2 + y^2 < 1 \,\&\&\, \left(-\frac{1}{2} + x\right)^2 + \left(-\frac{1}{2} + y\right)^2 > \frac{1}{30}$$

```
RegionPlot[ineqs , {x, 0, 1}, {y, 0, 1}]
```

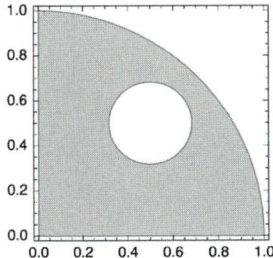

Now, integrate $x^3 + y^3$ over this region. First, define and plot the function:

```
f = (x^3 + y^3) Boole[ineqs];
```

```
Plot3D[f, {x, 0, 1}, {y, 0, 1}, Ticks → None]
```

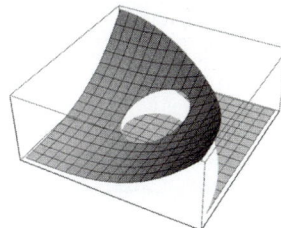

Then calculate the integral:

```
Integrate[f, {x, 0, 1}, {y, 0, 1}] // FullSimplify // Timing
```

$$\left\{51.2201, \frac{4}{15} - \frac{11\,\pi}{1200}\right\}$$

In the following example, we get different results for different values of the parameter a:

```
f = (x^3 + y^3) Boole[x^2 + y^2 < a];
```

```
Integrate[f, {x, 0, 1}, {y, 0, 1}]
```

$$\begin{cases} \dfrac{1}{2} & a \geq 2 \\[2mm] \dfrac{4\,a^{5/2}}{15} & 0 < a \leq 1 \\[2mm] \dfrac{1}{30}\left(3 + 12\sqrt{-1+a} - 10\,a - 4\sqrt{-1+a}\;a + 15\,a^2 - 8\sqrt{-1+a}\;a^2\right) & 1 < a < 2 \end{cases}$$

■ A Probability Example

Suppose that the joint probability density function (PDF) of random variables X and Y is a constant c in the triangular region defined by the inequalities $0 \leq x \leq 1$, $0 \leq y \leq 1$, and $x + y \leq 1$. What is the value of c? Integrate the density function over the region:

```
Integrate[c Boole[x + y ≤ 1], {x, 0, 1}, {y, 0, 1}]
```
$$\frac{c}{2}$$

Because this has to be equal to 1, we get that $c = 2$. What is the cumulative distribution function (CDF) of the random variable $Z = Y - X$? The CDF is $F_Z(z) = P(Z \leq z) = P(Y \leq X + z)$. This probability can be computed by integrating the PDF over the region $y \leq x + z$:

```
cdf = Integrate[2 Boole[x + y ≤ 1 && y ≤ x + z], {x, 0, 1}, {y, 0, 1}]
```

$$\begin{cases} 1 & z > 1 \\ \frac{1}{2}\left(1 + 2z - z^2\right) & 0 < z \le 1 \\ \frac{1}{2}\left(1 + 2z + z^2\right) & -1 < z \le 0 \end{cases}$$

The corresponding PDF is

```
pdf = D[cdf, z] // Simplify
```

$$\begin{cases} 1 & z == 0 \\ 1 - z & 0 < z < 1 \\ 1 + z & -1 < z < 0 \end{cases}$$

```
{Plot[cdf, {z, -1, 1}], Plot[pdf, {z, -1, 1}]}
```

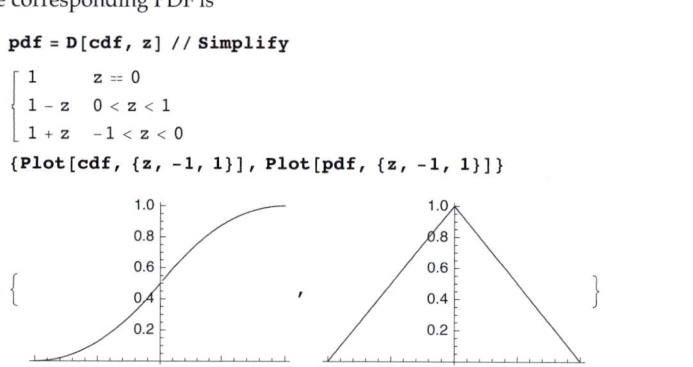

20.2 Numerical Quadrature

20.2.1 Introduction

■ **Numerical Integration**

If **Integrate** does not calculate your definite integral, apply numerical integration. You can first try exact integration, and if a result is not obtained, ask for a numerical value.

Integrate[f, {x, a, b}]	Try exact integration
% // N	If exact integration did not succeed, use numerical methods

We try the following integral:

```
i = Integrate[Exp[x^x], {x, 0, 1}]
```
$$\int_0^1 e^{x^x}\, dx$$

Exact integration did not succeed, and so we ask for a numerical value:

```
N[i]      2.19754
```

We can also directly resort to numerical methods.

NIntegrate[f, {x, a, b}]	Calculate the integral using numerical methods

```
NIntegrate[Exp[x^x], {x, 0, 1}]      2.19754
```

If we use the method **% // N**, eventually **NIntegrate** will be applied.

The integration interval can be infinite, and the end points can be singular:

```
NIntegrate[Exp[-x^2] Log[x], {x, 0, ∞}]      -0.870058
```

Multiple integrals are calculated in the familiar way.

`NIntegrate[f, {x, a, b}, {y, c, d}]`   Integrate first with respect to **y** and then with respect to **x**

`NIntegrate[Sin[x y], {x, 0, π}, {y, 0, x}]`   1.45034

Mathematica automatically takes into account possible singularities at the end points of the integration interval. If we have singularities within the interval, then we can integrate in several pieces, as follows:

`NIntegrate[f, {x, a, s1, s2, …, sn, b}]`   Integrate in several pieces

Here, `NIntegrate` integrates separately on each of the intervals (a, s_1), (s_1, s_2), ..., (s_n, b). This forces *Mathematica* to take into account the possible singularity of the points s_1, s_2, ..., s_n in addition to the possible singularity of the points a and b. Another application of the piecewise integration is to divide very long intervals into smaller pieces to get a more precise result. The technique can also be used to specify a piecewise linear integration contour in the complex plane.

■ Options

Options of `NIntegrate`:

`WorkingPrecision`   Precision used in internal computations; examples of values:
   `MachinePrecision`, 20

`PrecisionGoal`   If the value of the option is **p**, the relative error of the integral should be of the order
   10^{-p}; examples of values: `Automatic` (usually means **6**; is **2** for Monte Carlo methods), **10**

`AccuracyGoal`   If the value of the option is **a**, the absolute error of the integral should be of the order
   10^{-a}; examples of values: ∞, **10**

`Method`   Method or strategy to use; default value: `Automatic`

`MinRecursion`   Minimum number of recursive subdivisions; examples of values: **0**, **3**

`MaxRecursion`   Maximum number of recursive subdivisions; examples of values: `Automatic`, 9, 12

`MaxPoints`   Maximum total number of sample points; default value: `Automatic`

`Exclusions`   Parts of the integration region to exclude; examples of values: `None`, `{2, 3}`, `x^2 + y^2`
   `== 1`

`EvaluationMonitor`   Command to be executed after each evaluation of the function to be integrated;
   examples of values: `None`, `Sow[x]`, `Sow[{x, y}]`, `++n`, `AppendTo[points, x]`

`Compiled`   Whether the integrand should be compiled or not; possible values: `Automatic`, `True`,
   `False`

Next, we study the use of some of the options. However, the `Method` option will be considered in Sections 20.2.2 to 20.2.9. Note that `NIntegrate` automatically chooses a suitable method. For example, `NIntegrate` automatically handles singularities at end points and oscillatory integrands. Thus, we rarely need to use the `Method` option. Likewise, the other options are seldom needed.

■ WorkingPrecision

The default value of `WorkingPrecision` is `MachinePrecision` so that `NIntegrate` uses, by default, the usual fixed-precision arithmetic with 16-digit precision. Give another value for this option if you want the calculations to be done with variable-precision arithmetic (see Section 12.3.1, p. 409). In the following examples, we get a very good result with fixed-precision arithmetic, but with variable-precision arithmetic we can improve the result:

```
NIntegrate[Exp[-x], {x, 0, ∞}] // InputForm
1.000000000053296

NIntegrate[Exp[-x], {x, 0, ∞}, WorkingPrecision → 25] // InputForm
0.9999999999999999999999997192550370090045`25.
```

■ PrecisionGoal

The method used by **NIntegrate** stops improving the result as soon as either **PrecisionGoal** or **AccuracyGoal** is met.

The default value **Automatic** of **PrecisionGoal** means 6 so that **NIntegrate** tries to give you an answer having a *relative error* of the order 10^{-6} (this is only a goal; the actual relative error can be smaller or larger). If you have defined your own value for **WorkingPrecision**, then **PrecisionGoal** has the default value **WorkingPrecision - 10**. In the following integral, we have increased the value of the precision goal to get a better result:

```
NIntegrate[Exp[-x], {x, 0, ∞}, PrecisionGoal → 10] // InputForm
1.0000000000000013
```

As another example, calculate an integral both exactly and numerically:

```
i1 = Integrate[Sin[1 / x], {x, π / 90, π / 4}];
i2 = NIntegrate[Sin[1 / x], {x, π / 90, π / 4}];
```

Here are the absolute and relative errors of the numerical integral:

```
{Abs[i1 - i2], Abs[i1 - i2] / i1}
```

$\left\{1.08247 \times 10^{-15}, 3.46795 \times 10^{-15}\right\}$

In this example, the actual relative error is much better than its goal of 10^{-6}.

Note that if we increase **PrecisionGoal**, we often also have to increase **WorkingPrecision**; otherwise, convergence may not be reached.

■ AccuracyGoal

The default value of **AccuracyGoal** is ∞. Such a goal will never be satisfied, so the method actually stops as soon as **PrecisionGoal** is met. In practice, this means that the *absolute error* is, by default, not used as a goal. If you define a finite value of, for example, 6 for **AccuracyGoal**, then the method stops if it gets a result having an absolute error at most of the order 10^{-6}. Next, we use a very low accuracy goal:

```
NIntegrate[Exp[-x], {x, 0, ∞}, AccuracyGoal → 1] // InputForm
0.999917271688709
```

If the true value of the integral is zero, **NIntegrate** tells us that it cannot reach convergence:

```
NIntegrate[Sin[x] - 2 / π, {x, 0, π}]
```

NIntegrate::ncvb :
    NIntegrate failed to converge to prescribed accuracy after 9 recursive bisections in
        x near {x} = {≪83≫}. NIntegrate obtained 5.898059818321144`*^-17
        and 3.55359807013137`*^-13 for the integral and error estimates. ≫
$0. \times 10^{-13}$

If we want to get rid of the warning, we can set the value of **AccuracyGoal** to be smaller than ∞:

```
NIntegrate[Sin[x] - 2 / π, {x, 0, π}, AccuracyGoal → 10]
```

$0. \times 10^{-11}$

The three options `WorkingPrecision`, `PrecisionGoal`, and `AccuracyGoal` are explained in detail in Section 12.3.1, p. 409. If you skipped this section, now is a good time to read it.

■ MinRecursion, MaxRecursion, and MaxPoints

To achieve the precision or accuracy goal, the integration region is recursively bisected into smaller regions. The `MinRecursion` option defines the minimum number of recursive bisections of the interval before the integration starts; the default value is 0. This option can be used, for example, to ensure that a narrow spike on the integrand is not missed.

The `MaxRecursion` option is the largest number of recursive bisections allowed; the default value `Automatic` usually means 9. This option stops the recursive bisections at singular points so that special methods can be applied to handle the singularity.

With the `MaxPoints` option we can restrict the total number of sample points.

As an example, the following integral did not converge after nine recursive bisections:

```
NIntegrate[Sin[x^2], {x, 0, 23 Pi}]
```

NIntegrate::ncvb :
   NIntegrate failed to converge to prescribed accuracy after 9 recursive bisections in
      x near {x} = {0.}. NIntegrate obtained 0.620550468258415` and
      0.0059618314998086345` for the integral and error estimates. ≫
0.62055

We can solve the problem by increasing either the `MinRecursion` or the `MaxRecursion` option:

```
NIntegrate[Sin[x^2], {x, 0, 23 Pi}, MinRecursion → 1]       0.620059

NIntegrate[Sin[x^2], {x, 0, 23 Pi}, MaxRecursion → 10]      0.620059
```

20.2.2 Strategies and Rules

■ Strategies

The working of `NIntegrate` is governed by *strategies* and *rules*. The main strategies are `"GlobalAdaptive"` and `"LocalAdaptive"`. These strategies indicate the general way in which the adaptive method proceeds (as is explained later). With these strategies, we can use several quadrature rules, such as the Gauss–Kronrod rule or the Newton–Cotes rule. In addition to the two main strategies, there are special strategies. The strategies are defined with the `Method` option of `NIntegrate`:

`Method` Method or strategy to use:
- default value: `Automatic`
- general adaptive strategies: `"GlobalAdaptive"`, `"LocalAdaptive"`
- for singular integrands: `"DoubleExponential"`, `"DuffyCoordinates"`
- for Cauchy principal value: `"PrincipalValue"`
- for periodic integrands over one period: `"Trapezoidal"`
- for periodizing multidimensional integrands: `"MultiPeriod"`
- for oscillating integrands on infinite or semi-infinite intervals: `"DoubleExponentialOscillatory"`, `"ExtrapolatingOscillatory"`
- preprocessor strategies: `"SymbolicPiecewiseSubdivision"`, `"EvenOddSubdivision"`, `"OscillatorySelection"`, `"UnitCubeRescaling"`
- Monte Carlo strategies: `"MonteCarlo"`, `"AdaptiveMonteCarlo"`, `"QuasiMonteCarlo"`, `"AdaptiveQuasiMonteCarlo"`

The special strategies mentioned in the box are of three kinds: Some strategies are for special types of integrals, some define preprocessing techniques, and some are for Monte Carlo integration.

■ Rules

As stated previously, with the two general adaptive strategies we can ask to use a specific integration rule, as follows:

`Method → {"strategy", Method → "rule"}` Use the `"GlobalAdaptive"` or `"LocalAdaptive"` strategy with the given rule; possible rules:

- for open Gaussian quadrature: `"GaussBerntsenEspelidRule"`, `"GaussKronrodRule"`
- for closed Gaussian quadrature: `"LobattoKronrodRule"`, `"LobattoPeanoRule"`
- for interpolatory type quadrature: `"NewtonCotesRule"`
- for periodic integrands over one period: `"TrapezoidalRule"`
- for oscillatory integrands on finite intervals: `"ClenshawCurtisRule"`
- for multipanel (or compounded or composite) rules: `"MultiPanelRule"`
- for multidimensional integrals: `"CartesianRule"`, `"MultiDimensionalRule"`

As an example, we can use the `"GaussKronrodRule"` within either the `"GlobalAdaptive"` or the `"LocalAdaptive"` strategy. All the names of the rules end with `Rule`. (Note that we have both the `"Trapezoidal"` strategy and the `"TrapezoidalRule"` rule.)

The default value `Automatic` of the `Method` option actually means that the strategy is `"GlobalAdaptive"` and the rule is `"GaussKronrodRule"` for one-dimensional integrals and `"MultiDimensionalRule"` for multidimensional integrals.

■ Method-Specific Options

In addition to the general options of `NIntegrate`, each of the integration strategies and rules has its own special options. For example, here are the special options of the global adaptive strategy and of the Gauss–Kronrod rule:

`Options[NIntegrate`GlobalAdaptive]`

{Method → Automatic, MinRecursion → 0, MaxRecursion → 9, MaxPoints → ∞, SingularityDepth → Automatic, MaxErrorIncreases → Automatic, SingularityHandler → Automatic, SymbolicProcessing → Automatic}

`Options[NIntegrate`GaussKronrodRule]`

{Points → Automatic, SymbolicProcessing → Automatic}

Some of the special options are mentioned later, but generally we do not consider the special options.

In Sections 20.2.3 through 20.2.9, we study the strategies and rules in more detail. Note again that `NIntegrate` automatically chooses a suitable method. Thus, we rarely need to use the `Method` option to specify an integration strategy or rule.

20.2.3 General Adaptive Strategies

■ Global and Local Adaptive Strategies

Numerical integration works by sampling the function to be integrated at some points and then computing a weighted sum of sampled values. *Mathematica*, by default, uses *adaptive* sampling methods; that is, more points are sampled in subregions where the current estimated error is relatively large. In nonadaptive methods, more points are sampled in the *whole* integration interval if the error estimate is too large.

In adaptive integration, two main strategies can be used: global and local adaptive strategies. The default strategy is `"GlobalAdaptive"`. In this strategy, the subinterval with the largest current error estimate is bisected and more points are sampled from the two smaller intervals. This kind of bisection is continued recursively. After each bisection, new global estimates of the integral and the error are calculated over the whole integration interval. The method stops if the current global absolute or relative error satisfies the accuracy or precision goal.

In `"LocalAdaptive"` strategy, the error is considered on each subinterval (instead of the whole integration interval). If the error on a subinterval is too large compared to an initial estimate of the integral, that subinterval is divided into more subintervals. This is continued until the error estimate is acceptable on each of the current subintervals. In general, global adaptive strategies work better than the local ones.

Whichever strategy we choose, we can choose from several quadrature rules that were mentioned previously. The default rule is `"GaussKronrodRule"` for one-dimensional integrals and `"MultiDimensionalRule"` for multidimensional integrals.

Next, we study the global and local adaptive strategies in more detail. The various rules are addressed in Section 20.2.4.

As stated previously, `NIntegrate` automatically chooses a suitable method. Thus, mostly we do not need to use the `Method` option.

■ Global Adaptive Strategy

To show the points where `NIntegrate` evaluates the function, use the `EvaluationMonitor` option:

```
f = Sin[1 / (x + π / 10)];
```

```
{int, {xpoints}} =
    Reap[NIntegrate[f, {x, 0, π}, EvaluationMonitor :> Sow[x], PrecisionGoal → 9]];
xpoints // Length        201
```

The 201 points where the function was evaluated are as follows:

```
Graphics[Line[{{#, 0}, {#, f /. x → #}}] & /@ xpoints,
  Axes → True, AspectRatio → 0.2, ImageSize → 420]
```

This plot demonstrates the *adaptivity* property of the integration method: More points are sampled where the function changes rapidly.

In the following way, we see the order in which the points were sampled:

```
ListPlot[{xpoints, Range[201]}ᵀ, ImageSize → 420,
  Ticks → {{π / 16, π / 8, π / 4, 3 π / 8, π / 2, 3 π / 4, π}, Automatic},
  Epilog → {Line[{{π, 0}, {π, 201}}], Line[{{π / 2, 6}, {π / 2, 201}}],
    Line[{{π / 4, 17}, {π / 4, 201}}], Line[{{π / 8, 39}, {π / 8, 201}}],
    Line[{{π / 16, 61}, {π / 16, 201}}], Line[{{π / 32, 83}, {π / 32, 201}}],
    Line[{{3 π / 8, 50}, {3 π / 8, 201}}], Line[{{3 π / 4, 28}, {3 π / 4, 201}}]}]
```

First, the function is sampled at 11 points. Because the global error estimate is found to be too large, the interval is bisected and the function is sampled at 11 points on both subintervals. The error estimate on the first subinterval is the largest and is found to be too large. Thus, this interval is bisected and the function is sampled on both subintervals. This recursive bisection is continued several times until it is observed that the largest error estimate is on $\left(\frac{\pi}{4}, \frac{\pi}{2}\right)$; this interval is bisected. The error on $\left(\frac{\pi}{2}, \pi\right)$ is then the largest so that this interval is bisected. Now, the global error estimate satisfies the given precision or accuracy goal.

In the following example, we give the **MinRecursion** option the value 2, and now two recursive subdivisions are done before the integration starts, resulting in four initial subintervals:

```
{int, {xpoints}} = Reap[NIntegrate[f, {x, 0, π},
    EvaluationMonitor :→ Sow[x], PrecisionGoal → 9, MinRecursion → 2]];
```

```
ListPlot[{xpoints, Range[Length[xpoints]]}ᵀ,
  ImageSize → 420, Ticks → {{π / 4, π / 2, 3 π / 4, π}, Automatic},
  Epilog → {Line[{{π, 0}, {π, 201}}], Line[{{3 π / 4, 0}, {3 π / 4, 201}}],
    Line[{{π / 2, 0}, {π / 2, 201}}], Line[{{π / 4, 0}, {π / 4, 201}}]}]
```

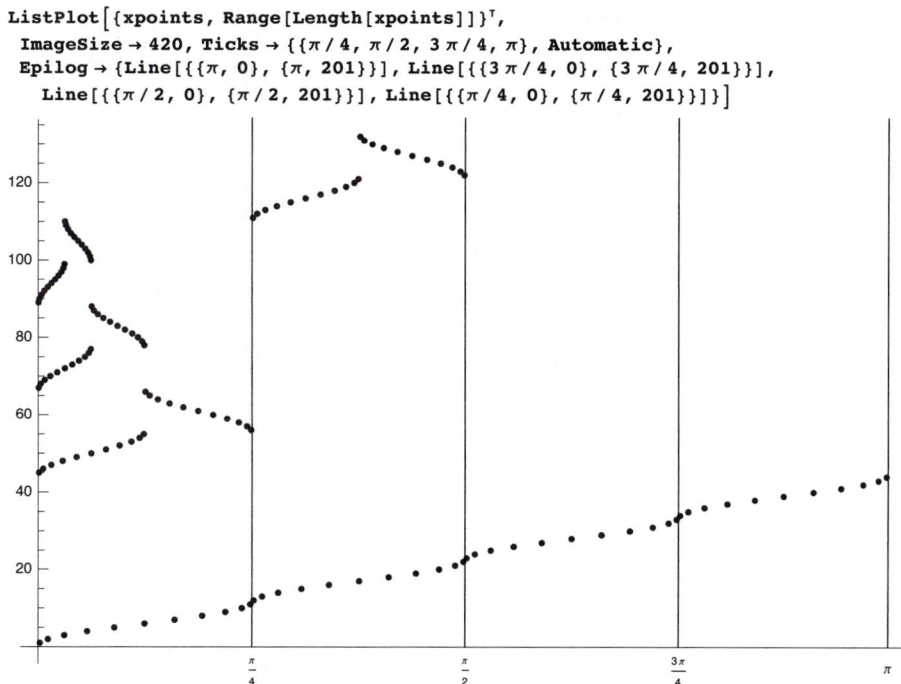

■ **Local Adaptive Strategy**

Now we use the local adaptive strategy. To show the points where **NIntegrate** evaluates the function, do again as follows:

```
f = Sin[1 / (x + π / 10)];

{int, {xpoints}} = Reap[
    NIntegrate[f, {x, 0, π}, Method → "LocalAdaptive", EvaluationMonitor :> Sow[x]]];
xpoints // Length      177
```

The 177 points where the function was evaluated are as follows:

```
Graphics[Line[{{#, 0}, {#, f /. x → #}}] & /@ xpoints,
  Axes → True, AspectRatio → 0.2, ImageSize → 420]
```

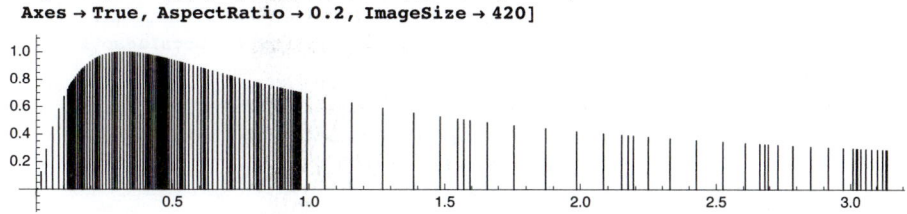

This plot again demonstrates the adaptivity property of the integration method: More points are sampled where the function changes rapidly.

In the following way, we see the order in which the points are sampled:

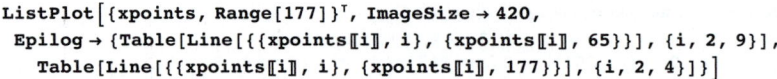

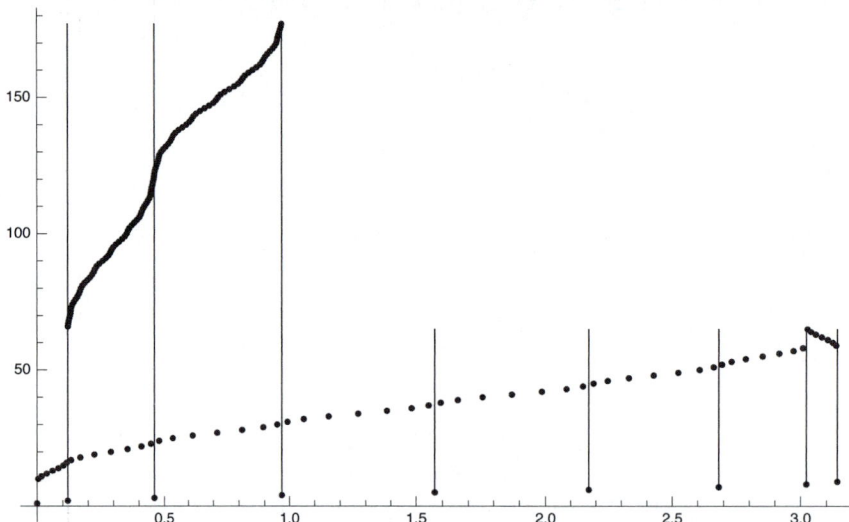

The function is first sampled at nine points; this gives an initial estimate of the integral. The error estimates to be calculated are based on this initial estimate of the integral. The error estimate on the first subinterval is found to be too large so that the function is sampled at seven new points between the first two original points. It is found that the same has to be done for each of the eight subintervals. After that, because the error estimate on the second and third subintervals is still too large, more sample points are taken from each subinterval of these two subintervals. Now the local error estimate on each subinterval is small enough compared to the initial estimate of the integral.

20.2.4 Quadrature Rules

■ Using the Rules

In Section 20.2.2, we mentioned that with the **"GlobalAdaptive"** and **"LocalAdaptive"** strategies we can ask to use a specific integration rule, as follows:

Method → {"strategy", Method → "rule"} Use the "GlobalAdaptive" or "LocalAdaptive"
   strategy with the given rule; possible rules:
- for open Gaussian quadrature: "**GaussBerntsenEspelidRule**", "**GaussKronrodRule**"
- for closed Gaussian quadrature: "**LobattoKronrodRule**", "**LobattoPeanoRule**"
- for interpolatory type quadrature: "**NewtonCotesRule**"
- for periodic integrands over one period: "**TrapezoidalRule**"
- for oscillatory integrands on finite intervals: "**ClenshawCurtisRule**"
- for multipanel (or compounded or composite) rules: "**MultiPanelRule**"
- for multidimensional integrals: "**CartesianRule**", "**MultiDimensionalRule**"

As previously discussed, the default value **Automatic** of the **Method** option actually means that the strategy is **"GlobalAdaptive"** and the rule is **"GaussKronrodRule"** for one-dimensional integrals and **"MultiDimensionalRule"** for multidimensional integrals.

To use, for example, the Newton–Cotes rule with the global and local adaptive strategies, write as follows:

```
NIntegrate[Sin[x], {x, 0, π},
 Method → {"GlobalAdaptive", Method → "NewtonCotesRule"}]
2.

NIntegrate[Sin[x], {x, 0, π}, Method → {"LocalAdaptive", Method → "NewtonCotesRule"}]
2.
```

Because the default is to use the global adaptive strategy, we can also only write the name of the rule if we would like to use the global strategy:

```
NIntegrate[Sin[x], {x, 0, π}, Method → "NewtonCotesRule"]
2.
```

We would again like to note that **NIntegrate** automatically chooses a suitable method. Thus, we rarely need to use the **Method** option.

■ One-Dimensional Integrals

Most of the integration rules have the **"Points"** option. Its default value is usually 5; however, for the Newton–Cotes rule the default value is 3, and for the Gauss–Berntsen–Espelid rule and Gauss–Kronrod rule the default value is **Automatic**.

The Gauss–Kronrod rule uses Gaussian quadrature with error estimation based on evaluation at Kronrod points. Here is an example of the use of the Gauss–Kronrod rule:

```
NIntegrate[Sin[x], {x, 0, π}, Method → {"GaussKronrodRule", "Points" → 7}]
2.
```

The Newton–Cotes rule is, by default, of the closed type. Next, we ask to apply an open rule:

```
NIntegrate[Sin[x], {x, 0, π},
 Method → {"NewtonCotesRule", "Points" → 7, "Type" → Open}]
2.
```

The trapezoidal rule, by default, applies Romberg quadrature. We can also ask to apply the pure trapezoidal rule:

```
NIntegrate[Sin[x], {x, 0, π},
 Method → {"TrapezoidalRule", "Points" → 7, "RombergQuadrature" → False}]
2.
```

The Clenshaw–Curtis rule is useful in integrating oscillatory functions (see Section 20.2.6, p. 658).

■ Multidimensional Integrals

The **"MultiDimensionalRule"** is a fully symmetric cubature for multidimensional integrals. This is the default rule for multidimensional integrals.

A d-dimensional **"CartesianRule"** has sampling points that are a Cartesian product of the sampling points of d one-dimensional rules. The weight associated with a Cartesian rule sampling point is the product of the one-dimensional rule weights that correspond to its coordinates. The Cartesian rule, by default, uses the Gauss–Kronrod rule.

As an example, we calculate the same 2D integral with both methods:

```
{{int1, {xypoints1}}, {int2, {xypoints2}}} =
  Reap[NIntegrate[Exp[x y], {x, 0, 2}, {y, 0, 2}, EvaluationMonitor :→ Sow[{x, y}],
    Method → #]] & /@ {"MultiDimensionalRule", "CartesianRule"};
```

```
{{int1, xypoints1 // Length},
 {int2, xypoints2 // Length}}
{{17.6674, 1071}, {17.6674, 121}}
```

Here are plots of the sampled points:

```
ListPlot[#, AspectRatio → 1, PlotStyle → {Black, PointSize[Small]},
  ImageSize → 180] & /@ {xypoints1, xypoints2}
```

We can also define the rule to be used with the Cartesian rule:

```
Method → {"CartesianRule", Method → {"GaussKronrodRule", "Points" → 7}}
```

■ **Rule Data**

For all but the Monte Carlo rules we can ask to show the data of the rule; that is,

- the abscissas: points at which the integrand is to be evaluated;
- the function weights: weights of the values of the function for the integral; and
- the error weights: weights of the values of the function for the error estimate.

For example, here are data for a Gauss–Kronrod rule:

```
intdata = NIntegrate`GaussKronrodRuleData[2, MachinePrecision]
{{0.03709, 0.211325, 0.5, 0.788675, 0.96291},
 {0.0989899, 0.245455, 0.311111, 0.245455, 0.0989899},
 {0.0989899, -0.254545, 0.311111, -0.254545, 0.0989899}}
```

The data give the points on the interval (0, 1). To integrate over other intervals, the integral must first be transformed onto (0, 1). Thus, here is a program to calculate an integral with the Gauss–Kronrod method:

```
gaussKronrod[f_, {x_, a_, b_}, n_] := Module[{y, g},
  g = (b - a) f /. x → a + (b - a) y; Total@MapThread[{(g /. y → #1) #2, (g /. y → #1) #3} &,
    NIntegrate`GaussKronrodRuleData[n, MachinePrecision]]]
```

Here, with **MapThread** we get a matrix where the first column contains the weighted function values for the integral and the second column contains the weighted function values for the error estimate:

```
MapThread[{(g[x] /. x → #1) #2, (g[x] /. x → #1) #3} &, intdata]
{{0.0989899 g[0.03709], 0.0989899 g[0.03709]},
 {0.245455 g[0.211325], -0.254545 g[0.211325]},
 {0.311111 g[0.5], 0.311111 g[0.5]}, {0.245455 g[0.788675], -0.254545 g[0.788675]},
 {0.0989899 g[0.96291], 0.0989899 g[0.96291]}}
```

With `Total` we get the column sums we need:

```
Total@%
```
```
{0.0989899 g[0.03709] + 0.245455 g[0.211325] + 0.311111 g[0.5] + 0.245455 g[0.788675] +
    0.0989899 g[0.96291], 0.0989899 g[0.03709] - 0.254545 g[0.211325] +
    0.311111 g[0.5] - 0.254545 g[0.788675] + 0.0989899 g[0.96291]}
```

As an example, compute an integral first exactly and then with a Gauss–Kronrod rule:

```
Integrate[Exp[x], {x, -1, 2}] // N // InputForm
```
```
7.0211766577592085
```

```
gaussKronrod[Exp[x], {x, -1, 2}, 4] // InputForm
```
```
{7.021176657759308, 0.000019289921552423372}
```

The rule we used here is exact for all polynomials of at most degree 13:

```
f = Sum[aᵢ xⁱ, {i, 0, 13}];
Integrate[f, {x, p, q}] - gaussKronrod[f, {x, p, q}, 4]〚1〛 // Simplify // Chop
0
```

20.2.5 Singular Integrands

■ Telling Singularities

As mentioned previously, *Mathematica* automatically takes into account possible singularities at the end points of the integration interval. If we have singularities within the interval, we can integrate in several pieces or tell the points to exclude:

`NIntegrate[f, {x, a, s1, s2, …, sn, b}]`	Integrate in several pieces
`Exclusions` Parts of the integration region to exclude; examples of values: `None`, `{2, 3}`, `x^2 + y^2` `== 1`	

In the first case, `NIntegrate` integrates separately on each of the intervals (a, s_1), (s_1, s_2), ..., (s_n, b). This forces *Mathematica* to take into account the possible singularity of the points $s_1, s_2, ..., s_n$ in addition to the possible singularity of the points a and b. In the second case, we tell the points, curves, or surfaces to exclude in the integration.

■ Example 1

Here we do not tell the singularities at π and 2π:

```
NIntegrate[1 / Sqrt[Sin[x]], {x, 0, 7}]
```

NIntegrate::ncvb :

  NIntegrate failed to converge to prescribed accuracy after 9 recursive bisections in
      x near {x} = {3.13097}. NIntegrate obtained 6.92914 - 5.16748 i and
      0.21240735884114376` for the integral and error estimates. ≫
6.92914 - 5.16748 i

We had problems with the convergence. If we proceed piecewise, *Mathematica* properly handles the singularities at π and 2π:

```
NIntegrate[1 / Sqrt[Sin[x]], {x, 0, π, 2 π, 7}]
```
```
6.95223 - 5.24412 i
```

The singular points can also be told with the `Exclusions` option:

```
NIntegrate[1 / Sqrt[Sin[x]], {x, 0, 7}, Exclusions → {π, 2 π}]
6.95223 - 5.24412 i
```

■ **Example 2**

In multidimensional integrals, we may have, in addition to points of singularity, curves or surfaces of singularity. Here is an example:

```
NIntegrate[1 / Sqrt[x^2 + y^2 - 1], {x, -2, 2}, {y, -2, 2}]
```

NIntegrate::slwcon :

   Numerical integration converging too slowly; suspect one of the following: singularity, value of
      the integration is 0, highly oscillatory integrand, or WorkingPrecision too small. ≫

NIntegrate::ncvb :

   NIntegrate failed to converge to prescribed accuracy after 18 recursive bisections in
      x near {x, y} = {0.600259, 0.799133}. NIntegrate obtained $12.5906 - 6.26927 i$
      and 0.10860320865547311` for the integral and error estimates. ≫

```
12.5906 - 6.26927 i
```

By telling the curve of singularity with the **Exclusions** option, we can resolve the problems of convergence:

```
NIntegrate[1 / Sqrt[x^2 + y^2 - 1], {x, -2, 2}, {y, -2, 2}, Exclusions → x^2 + y^2 == 1]
12.6049 - 6.28319 i
```

Another solution is the use of **Boole** to make the singular curve as a boundary; singularities at boundaries are detected by **NIntegrate**:

```
NIntegrate[1 / Sqrt[x^2 + y^2 - 1] Boole[x^2 + y^2 - 1 > 0], {x, -2, 2}, {y, -2, 2}] +
  NIntegrate[1 / Sqrt[x^2 + y^2 - 1] Boole[x^2 + y^2 - 1 < 0], {x, -2, 2}, {y, -2, 2}]
12.6049 - 6.28319 i
```

■ **Special Methods**

Mathematica automatically chooses a suitable method to handle singularities. We can also ask to use a specified method, as follows:

Method → {"GlobalAdaptive", "SingularityHandler" → sh}   Use a special singularity handler **sh**; possible values of **sh**: {IMT, "TuningParameters" → {a, p}}, "DoubleExponential", "DuffyCoordinates"
Method → "DoubleExponential"   Use the double-exponential singularity handler
Method → "DuffyCoordinates"   Use the Duffy's coordinates singularity handler

To achieve the prescribed precision and accuracy goals, subintervals of the integration interval are bisected and more points are sampled. At a singularity, it may be difficult to get sufficient accuracy so that a large amount of recursive bisections are needed. After a certain amount of bisections, it becomes clear that we have a singular point and so continuing the bisections may not be the best way to proceed. Instead, we should resort to a *singularity handler*.

To handle singularities, the **"GlobalAdaptive"** and **"LocalAdaptive"** strategies use the IMT transformation of variables (published by Iri, Moriguti, and Takasawa in 1970). The **"DoubleExponential"** strategy uses the trapezoidal rule with a special variable transformation. The **"DuffyCoordinates"** strategy simplifies or eliminates certain types of singularities in multidimensional integrals. Recall also the **"PrincipalValue"** strategy for computing the Cauchy principal value.

■ Demonstrating Singularity Handling

In the following example, *Mathematica* uses a singularity handler (the IMT transformation):

```
NIntegrate[1 / Sqrt[x], {x, 0, 1}]
2.
```

Compare this with the following output. Here, we use the **"SingularityDepth"** option to tell how many recursive bisections to try before applying a singularity handler. We ask for an unlimited number of bisections so that a singularity handler will actually not be used:

```
NIntegrate[1 / Sqrt[x], {x, 0, 1},
 Method → {"GlobalAdaptive", "SingularityDepth" → ∞}]
```
NIntegrate::ncvb :

  NIntegrate failed to converge to prescribed accuracy after 9 recursive bisections in
    x near {x} = {0.00193758}. NIntegrate obtained 1.997237951127453`
    and 0.004238324551439297` for the integral and error estimates. ≫
```
1.99724
```

Nine bisections did not suffice to handle the singularity. Another way to demonstrate the effect of a singularity handler is to ask not to use any handler:

```
NIntegrate[1 / Sqrt[x], {x, 0, 1},
 Method → {"GlobalAdaptive", "SingularityHandler" → None}]
```
NIntegrate::ncvb :

  NIntegrate failed to converge to prescribed accuracy after 9 recursive bisections in
    x near {x} = {0.00193758}. NIntegrate obtained 1.997237951127453`
    and 0.004238324551439297` for the integral and error estimates. ≫
```
1.99724
```

■ IMT Transformation

Consider again the integral of $1 \big/ \sqrt{x}$ over (0, 1). Do the IMT transformation of variable $x = \exp\!\left(1 - \frac{1}{t}\right)$. Then $dx = \exp\!\left(1 - \frac{1}{t}\right)\frac{1}{t^2}\,dt$, $t = 1\big/(1 - \log x)$, and the integration interval is again (0, 1). Thus, we arrive at the integral of $\left(\exp\!\left(1 - \frac{1}{t}\right)\right)^{1/2}\frac{1}{t^2}$. Compare the plots of the original and the transformed integrand:

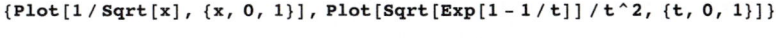

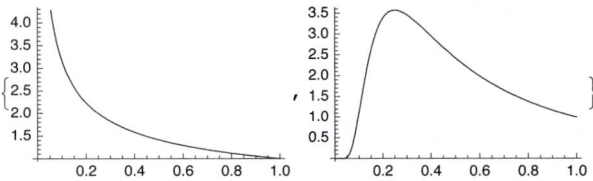

The original function has singularity at $x = 0$, whereas the transformed function does not have any singularities.

A more general IMT transformation is $x = a\exp\!\left(1 - \frac{1}{t^p}\right)$, where a and p are so-called tuning parameters. The default values are $a = 10$ and $p = 1$. For example,

```
NIntegrate[1 / Sqrt[x], {x, 0, 1}, Method →
  {"GlobalAdaptive", "SingularityHandler" → {IMT, "TuningParameters" → {10, 2}}}]
2.
```

■ Double-Exponential Method

In the double-exponential method (Mori and Takahasi, 1974), first a transformation of variable is applied and then the trapezoidal rule is used. The name of the transformation derives from the fact that the derivative of the transformation decreases double-exponentially at the ends of the integration interval. Here, we apply the double-exponential singularity handler:

```
NIntegrate[1 / Sqrt[x], {x, 0, 1},
  Method → {"GlobalAdaptive", "SingularityHandler" → "DoubleExponential"}]
2.
```

We can also use the double-exponential strategy:

```
NIntegrate[1 / Sqrt[x], {x, 0, 1}, Method → "DoubleExponential"]
2.
```

■ Duffy's Coordinates

For multidimensional integrals with singularities, a method called Duffy's coordinates can be tried. This is a technique that transforms an integrand over a square, cube, or hypercube with a singular point in one of the corners into an integrand with a singularity over a line, which might be easier to integrate.

In the following integral, we have a singular corner point and *Mathematica* automatically handles it correctly:

```
NIntegrate[1 / Sqrt[x^2 + y^2], {x, 0, 1}, {y, 0, 1}]
1.76275
```

We can also explicitly ask to use Duffy's coordinates as the singularity handler:

```
NIntegrate[1 / Sqrt[x^2 + y^2], {x, 0, 1}, {y, 0, 1},
  Method → {"GlobalAdaptive", "SingularityHandler" → "DuffyCoordinates"}]
1.76275
```

Still another way is to use the strategy of Duffy's coordinates:

```
NIntegrate[1 / Sqrt[x^2 + y^2], {x, 0, 1}, {y, 0, 1}, Method → "DuffyCoordinates"]
1.76275
```

20.2.6 Oscillatory Integrands

■ Methods

Mathematica automatically detects an oscillatory integrand and chooses a suitable method based on the integration interval and the form of the integrand. We can also ask to use a specified method, as follows:

`Method → "ClenshawCurtisRule"`   For oscillatory integrands of the form $k(x) f(x)$, where $k(x)$ is of the type $\sin(\omega x^p + c)$, $\cos(\omega x^p + c)$, or $\exp(i \omega x^p + c)$; integration interval is finite

`Method → "DoubleExponentialOscillatory"`   For slowly decaying oscillatory integrands of the form $k(x) f(x)$, where $k(x)$ is of the type $\sin(\omega x^p + c)$, $\cos(\omega x^p + c)$, or $\exp(i \omega x^p)$; integration interval is infinite or semi-infinite

`Method → "ExtrapolatingOscillatory"`   For oscillatory integrands of the form $k(x) f(x)$, where $k(x)$ is of the type $\sin(\omega x^p + c)$, $\cos(\omega x^p + c)$, $J_v(\omega x^p + c)$, $Y_v(\omega x^p + c)$, $H_v^{(1)}(\omega x^p + c)$, $H_v^{(2)}(\omega x^p + c)$, $j_v(\omega x^p + c)$, or $y_v(\omega x^p + c)$; integration interval is infinite or semi-infinite

---

Here, ω, c, and v are real constants, and p is a positive integer. Furthermore, the special functions are as follows:

```
{BesselJ[v, x], BesselY[v, x], HankelH1[v, x], HankelH2[v, x],
  SphericalBesselJ[v, x], SphericalBesselY[v, x]} // TraditionalForm
```

$$\{J_v(x),\ Y_v(x),\ H_v^{(1)}(x),\ H_v^{(2)}(x),\ j_v(x),\ y_v(x)\}$$

The `"ClenshawCurtisRule"` uses Chebyshev expansions of the integrand and the global adaptive integration strategy. `"DoubleExponentialOscillatory"` uses a modification of the double-exponential algorithm and `"ExtrapolatingOscillatory"` a sequence summation acceleration. In the last method, a sequence of integrals is calculated between the zeros of the integrand; this sequence is then summed (`NSum`) with Wynn's extrapolation method.

■ **Example 1**

Consider the following highly oscillatory function:

```
f = x Sin[x] / (x^2 + 1);

Plot[f, {x, 0, 100}]
```

Its integral from 0 to 100 is as follows:

```
(i1 = Integrate[f, {x, 0, 100}] // N // Chop) // InputForm
0.5692936610326388
```

Numerical integration gives a very good result:

```
NIntegrate[f, {x, 0, 100}] // InputForm
0.5692936604752128

i1 - %
5.57426 × 10^-10
```

Here, the Clenshaw–Curtis rule was applied, as can be seen if we explicitly ask to use this rule:

```
NIntegrate[f, {x, 0, 100}, Method → "ClenshawCurtisRule"] // InputForm
0.5692936604752128
```

■ **Example 2**

Integrate the same function on a semi-infinite interval:

```
(i2 = Integrate[f, {x, 0, ∞}] // N) // InputForm
0.5778636748954609
```

Again, numerical integration gives a very good result:

```
NIntegrate[f, {x, 0, ∞}] // InputForm
0.5778636820019379

i2 - %
-7.10648 × 10^-9
```

Here, the double-exponential oscillatory strategy was applied:

```
NIntegrate[f, {x, 0, ∞}, Method → "DoubleExponentialOscillatory"] // InputForm
0.5778636820019379
```

The extrapolating oscillatory strategy also gives a good result:

```
NIntegrate[f, {x, 0, ∞}, Method → "ExtrapolatingOscillatory"] // InputForm
0.5778636315399948

i2 - %

4.33555 × 10⁻⁸
```

20.2.7 Symbolic Preprocessing

■ Symbolic Preprocessing Methods

By a symbolic preprocessing, *Mathematica* is able to detect special types of integrands such as piecewise, even, odd, and oscillatory functions. Symbolic preprocessing is done automatically to get more precise results more rapidly, but we can also ask to apply a special preprocessing method:

> `Method → "SymbolicPiecewiseSubdivision"` Divide an integral with a piecewise integrand into integrals with disjoint integration regions
> `Method → "EvenOddSubdivision"` Reduce the integration region if the region is symmetric around the origin and the integrand is even or odd
> `Method → "OscillatorySelection"` Select specialized algorithms for efficient evaluation of one-dimensional oscillating integrals
> `Method → "UnitCubeRescaling"` Transform the integration region into a unit cube or hypercube

■ Example 1

Consider the following function:

```
Plot3D[Boole[x^2 + y^2 ≤ 1] (x^2 + y^2), {x, -1, 1}, {y, -1, 1}]
```

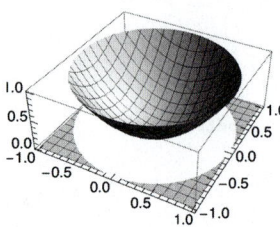

Check how many sample points are needed and where they are located:

```
{int, {xypoints}} = Reap[NIntegrate[Boole[x^2 + y^2 ≤ 1] (x^2 + y^2),
    {x, -1, 1}, {y, -1, 1}, EvaluationMonitor :> Sow[{x, y}]]];
{int, xypoints // Length}
{1.5708, 476}
```

```
ListPlot[xypoints, PlotStyle → {Black, PointSize[Tiny]}, AspectRatio → 1]
```

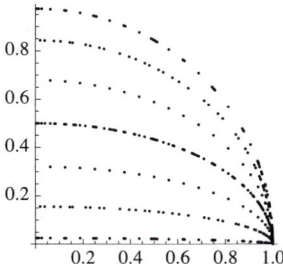

Thus, only 476 sample point were needed. The integrand was only sampled in the first quadrant and in the region where the integrand is nonzero.

■ Example 2

With the `"SymbolicProcessing"` option we can tell how many seconds to do symbolic preprocessing. Giving the value 0, no symbolic preprocessing is done. In this way, we can see the effect of the preprocessing:

```
{int, {xypoints}} = Reap[NIntegrate[Boole[x^2 + y^2 ≤ 1] (x^2 + y^2),
    {x, -1, 1}, {y, -1, 1}, EvaluationMonitor :→ Sow[{x, y}],
    Method → {Automatic, "SymbolicProcessing" → 0}]];
```

NIntegrate::slwcon :

    Numerical integration converging too slowly; suspect one of the following: singularity, value of
        the integration is 0, highly oscillatory integrand, or WorkingPrecision too small. ≫

NIntegrate::eincr :

    The global error of the strategy GlobalAdaptive has increased more than 2000 times. The
        global error is expected to decrease monotonically after a number of integrand
        evaluations. Suspect one of the following: the working precision is insufficient
        for the specified precision goal; the integrand is highly oscillatory or it is not
        a (piecewise) smooth function; or the true value of the integral is 0. Increasing
        the value of the GlobalAdaptive option MaxErrorIncreases might lead to a
        convergent numerical integration. NIntegrate obtained 1.5691392282066634`
        and 0.0008974905096775965` for the integral and error estimates. ≫

Thus, without symbolic preprocessing, we have difficulties with convergence, the result is not very accurate, and a huge amount of approximately 200,000 sampling points were needed:

```
{int, xypoints // Length}
```

```
{1.56914, 206 805}
```

```
ListPlot[xypoints, PlotStyle → {Black, PointSize[Tiny]}, AspectRatio → 1]
```

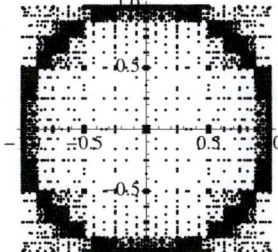

We see that points were sampled on the whole integration region instead of only in the first quadrant, and an enormous effort was needed to track the circle-form boundary of the surface.

20.2.8 Monte Carlo Methods

■ Monte Carlo and Quasi-Monte Carlo Methods

`Method → "MonteCarlo"`　Usual Monte Carlo method
`Method → "AdaptiveMonteCarlo"`　Adaptive Monte Carlo method
`Method → "QuasiMonteCarlo"`　Quasi-Monte Carlo method
`Method → "AdaptiveQuasiMonteCarlo"`　Adaptive quasi-Monte Carlo method

In the Monte Carlo methods, the function is sampled at a large number of points. Usually, Monte Carlo methods use uniformly distributed random points in the integral's region. The number of points is incremented until the estimated standard deviation is small enough to satisfy the specified precision or accuracy goal. However, a quasi-Monte Carlo algorithm uses equidistributed, deterministically generated sequences of points.

The adaptive versions of the Monte Carlo methods apply recursive stratified sampling: They recursively bisect the subregion with the largest variance estimate into two halves, and they compute integral and variance estimates for each half (with the nonadaptive Monte Carlo methods).

The maximum number of points is determined by `MaxPoints`; its default value `Automatic` means 50,000 for nonadaptive Monte Carlo methods. The default value of `PrecisionGoal` is 2 for Monte Carlo methods, which means that the result will not be very precise. Monte Carlo methods are suited for multidimensional integrals.

■ Examples

By applying the Monte Carlo method several times, each time we get a slightly different result because the method is based on random numbers:

```
Table[NIntegrate[Sin[x y], {x, 0, π}, {y, 0, π}, Method → "MonteCarlo"], {5}]
{2.90471, 2.88406, 2.88166, 2.90878, 2.86634}
```

One of the special options of the Monte Carlo method is `"RandomSeed"`. By using it, we get the same result each time:

```
Table[NIntegrate[Sin[x y], {x, 0, π},
  {y, 0, π}, Method → {"MonteCarlo", "RandomSeed" → 1}], {5}]
{2.91815, 2.91815, 2.91815, 2.91815, 2.91815}
```

A quasi-Monte Carlo method uses deterministic points so that each time we get the same result:

```
Table[NIntegrate[Sin[x y], {x, 0, π}, {y, 0, π},
  Method → "QuasiMonteCarlo", MaxPoints → 60 000], {5}]
{2.90061, 2.90061, 2.90061, 2.90061, 2.90061}
```

Next, we use the adaptive Monte Carlo method:

```
Table[NIntegrate[Sin[x y], {x, 0, π}, {y, 0, π}, Method → "AdaptiveMonteCarlo"], {5}]
{2.9067, 2.88096, 2.88858, 2.88829, 2.95755}
```

Lastly, we apply the adaptive quasi-Monte Carlo method:

```
Table[NIntegrate[Sin[x y], {x, 0, π},
  {y, 0, π}, Method → "AdaptiveQuasiMonteCarlo"], {5}]
{2.89929, 2.90143, 2.8997, 2.90235, 2.89995}
```

■ A Comparison

Next we compare the four Monte Carlo methods:

```
i = Integrate[Sin[x y], {x, 0, π}, {y, 0, π}]
```

$$\text{EulerGamma} - \text{CosIntegral}\left[\pi^2\right] + 2\,\text{Log}[\pi]$$

```
methods = {"MonteCarlo", "QuasiMonteCarlo",
    "AdaptiveMonteCarlo", "AdaptiveQuasiMonteCarlo"};
```

```
ints = Reap[NIntegrate[Sin[x y], {x, 0, π}, {y, 0, π}, MaxPoints → 70 000,
        Method → #, EvaluationMonitor :→ Sow[{x, y}]]] & /@ methods;
```

The values of the integral, the error, and the number of sample points are as follows:

```
TableForm[{#[[1]], i - #[[1]], Length@#[[2, 1]]} & /@ ints,
  TableHeadings → {methods, { "Integral", "Error", "Points"}}]
```

	Integral	Error	Points
MonteCarlo	2.89243	0.00825584	43 600
QuasiMonteCarlo	2.90061	0.0000693623	52 399
AdaptiveMonteCarlo	2.89471	0.00597262	9700
AdaptiveQuasiMonteCarlo	2.89889	0.00179695	65 900

The adaptive Monte Carlo method uses many fewer points than the usual Monte Carlo method.

■ Stratified Monte Carlo

It can be shown that the performance of Monte Carlo methods can be improved by partitioning the integration region into smaller parts or strata and applying the Monte Carlo method separately for each part. In this way, we get what is called stratified Monte Carlo integration. If the number of strata is s, the standard deviation of the stratified Monte Carlo estimate is s times less the standard deviation of the crude Monte Carlo estimate.

Stratified Monte Carlo integration can be applied either by specifying intermediate points for the integration variables or by using the **"Partitioning"** option:

```
NIntegrate[f, {x, a0, a1, …, am}, {y, b0, b1, …, bn}, Method → "MonteCarlo"]
NIntegrate[f, {x, a, b}, {y, c, d}, Method → {"MonteCarlo", "Partitioning" → {m, n}}]
```

With the first method, we can use strata of unequal sizes. With the second method, the first dimension is stratified into **m** parts, the second dimension into **n** parts, etc.; if the value of the option is a single number, then each dimension is stratified into the same number of strata.

Note that, as mentioned previously, the adaptive Monte Carlo methods automatically use stratified sampling. For these methods, the **"Partitioning"** option gives the *initial* partitioning. The default value of this option is **Automatic** for the adaptive methods, whereas it is 1 for the nonadaptive methods.

■ Examples

Now we compare all four Monte Carlo methods and their stratified versions. A low value for **MaxPoints** is used to make the plots of the sampled points clearer.

```
Plot3D[Sin[x y], {x, 0, π}, {y, 0, π}]
```

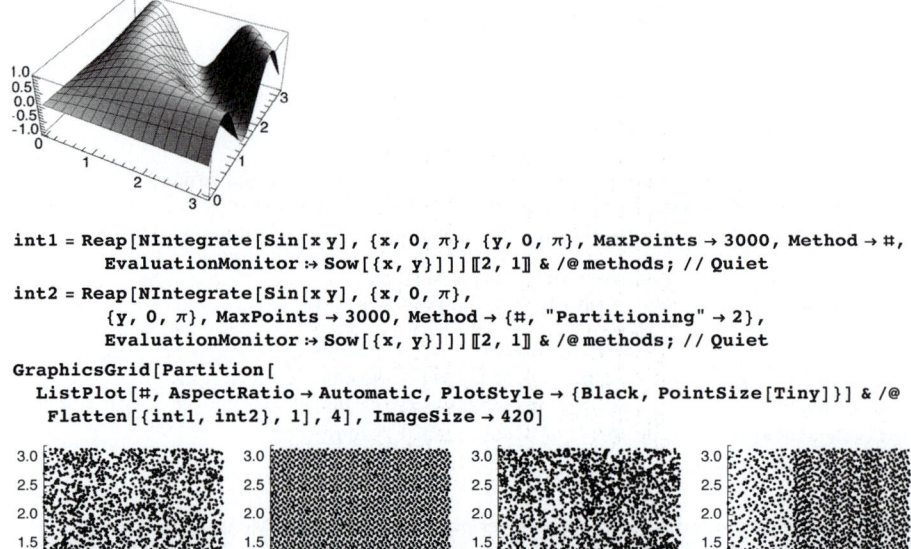

```
int1 = Reap[NIntegrate[Sin[x y], {x, 0, π}, {y, 0, π}, MaxPoints → 3000, Method → #,
        EvaluationMonitor :→ Sow[{x, y}]]][[2, 1]] & /@ methods; // Quiet

int2 = Reap[NIntegrate[Sin[x y], {x, 0, π},
        {y, 0, π}, MaxPoints → 3000, Method → {#, "Partitioning" → 2},
        EvaluationMonitor :→ Sow[{x, y}]]][[2, 1]] & /@ methods; // Quiet

GraphicsGrid[Partition[
    ListPlot[#, AspectRatio → Automatic, PlotStyle → {Black, PointSize[Tiny]}] & /@
    Flatten[{int1, int2}, 1], 4], ImageSize → 420]
```

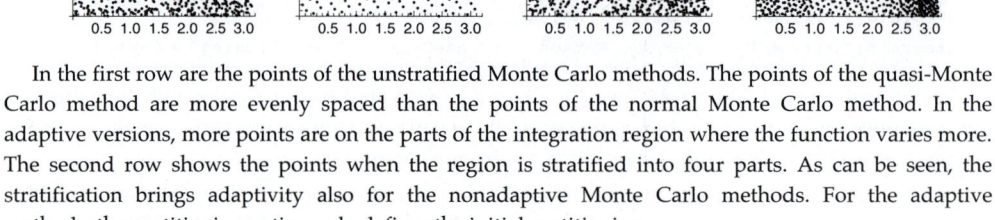

In the first row are the points of the unstratified Monte Carlo methods. The points of the quasi-Monte Carlo method are more evenly spaced than the points of the normal Monte Carlo method. In the adaptive versions, more points are on the parts of the integration region where the function varies more. The second row shows the points when the region is stratified into four parts. As can be seen, the stratification brings adaptivity also for the nonadaptive Monte Carlo methods. For the adaptive methods, the partitioning option only defines the initial partitioning.

20.2.9 More about Quadrature

■ Newton–Cotes Rule

With a package, we can derive various Newton–Cotes and Gaussian quadrature formulas. First, let us ask for information about a two-point Newton–Cotes rule:

```
<< NumericalDifferentialEquationAnalysis`
```

```
{NewtonCotesWeights[2, a, a + h], NewtonCotesError[2, f, a, a + h]}
```

$$\left\{\left\{\left\{a, \frac{h}{2}\right\}, \left\{a + h, \frac{h}{2}\right\}\right\}, \frac{h^3 f''}{12}\right\}$$

Thus, $\int_a^{a+h} f(x)\,dx \simeq \frac{h}{2} f(a) + \frac{h}{2} f(a + h)$. The error term indicates that the rule gives exact results for all linear polynomials. To make the method more useful, we create a multipanel (or compounded or composite) rule:

```
nc = Table[NewtonCotesWeights[2, a + i h, a + (i + 1) h], {i, 0, 3}]
```

$$\left\{\left\{\left\{a, \frac{h}{2}\right\}, \left\{a + h, \frac{h}{2}\right\}\right\}, \left\{\left\{a + h, \frac{h}{2}\right\}, \left\{a + 2 h, \frac{h}{2}\right\}\right\},\right.$$

$$\left.\left\{\left\{a + 2 h, \frac{h}{2}\right\}, \left\{a + 3 h, \frac{h}{2}\right\}\right\}, \left\{\left\{a + 3 h, \frac{h}{2}\right\}, \left\{a + 4 h, \frac{h}{2}\right\}\right\}\right\}$$

```
FactorTerms[Total[#[[2]] f[#[[1]]] & /@ Flatten[nc, 1]], h] /. a + 4 h → b
```

$$\frac{1}{2} h (f[a] + f[b] + 2 f[a + h] + 2 f[a + 2 h] + 2 f[a + 3 h])$$

Thus, we arrive at the trapezoidal rule we programmed in Section 18.1.1, p. 544:

$$\int_a^b f(x)\,dx \simeq \frac{h}{2}\left(f(a) + 2\sum_{i=1}^{n-1} f(a + i h) + f(b)\right).$$

An illustration of the trapezoidal rule is as follows:

```
g = Sin[x] + 1.2;
points = Table[{x, g}, {x, 0.5, 5.5}];
Show[Plot[g, {x, 0, 6}], ListPlot[points, Filling → Axis, FillingStyle → Black],
  ListLinePlot[points, Filling → Axis], Ticks → {{{0.5, a}, {5.5, b}}, None}]
```

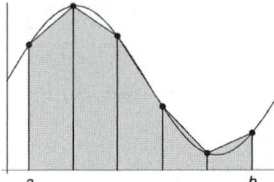

■ Gaussian Rule

Now we ask for information about a three-point Gaussian quadrature in the interval $(-1, 1)$:

```
<< NumericalDifferentialEquationAnalysis`
```

```
g = GaussianQuadratureWeights[3, -1, 1]
```
```
{{-0.774597, 0.555556}, {0., 0.888889}, {0.774597, 0.555556}}
```

```
GaussianQuadratureError[3, f, -1, 1]
```

$$-0.0000634921\ f^{(6)}$$

Thus, the integral of f over $(-1, 1)$ is approximated by the following:

```
Total[#[[2]] f[#[[1]]] & /@ g]
```
```
0.555556 f[-0.774597] + 0.888889 f[0.] + 0.555556 f[0.774597]
```

The formula gives correct results for all fifth-order polynomials. The following program calculates an integral using n-point Gaussian quadrature:

```
gaussianQuadrature[f_, {x_, a_, b_}, n_] :=
 Total[#[[2]] f /. x → #[[1]] & /@ GaussianQuadratureWeights[n, a, b]]
```

We try the same example as presented previously but with **gaussianQuadrature**:

```
Integrate[Sin[x], {x, 0, 2}] - gaussianQuadrature[Sin[x], {x, 0, 2}, 3]
-0.0000518162
```

■ Quadrature for Data

If we do not know the whole function to be integrated but only its values at a collection of points, we have some possibilities. If the data contain observational errors, then a good method is first to fit a function to the data using **Fit** or **FindFit** (see Section 25.1) and then integrate this function (using **Integrate** or **NIntegrate**). Otherwise, we can form a piecewise-interpolating polynomial using **Interpolation** or **ListInterpolation** (see Section 24.2) and then use **NIntegrate**. Here is an example:

```
data = Table[{x, y, BesselI[2, x y]}, {x, 0, 2, 0.2}, {y, 0, 2, 0.2}];

int = Interpolation[Flatten[data, 1]]
InterpolatingFunction[{{0., 2.}, {0., 2.}}, <>]

NIntegrate[Cos[int[x, y]], {x, 0, 2}, {y, 0, 2}]
3.37132
```

20.3 Sums and Products

20.3.1 Exact Sums

Sum[expr, {i, a, b}]   Sum of the values of **expr** when **i** varies from **a** to **b**

The iteration specification can also be of the form **{b}** if we sum **b** copies of **expr**, **{i, b}** if the starting value of **i** is 1, and **{i, a, b, d}** if **i** goes from **a** to **b** in steps of **d**.

Remember that sums can also be entered with the aid of the **BasicInput** palette (see Section 1.4.1, p. 15). The keyboard can also be used (see Section 3.3.3, p. 76). Consider the following sum:

$$\sum_{i=1}^{10} 2^i$$

To write this expression, type ESC sum ESC CTRL + i=1 CTRL % 10 CTRL _ 2 CTRL ^ i CTRL _.

Here are some examples of infinite sums:

```
Sum[1 / i^2, {i, ∞}]            π²/6
```

```
Sum[(1 + i) / (2 + i)^3, {i, 0, ∞}]      1/6 (π² - 6 Zeta[3])
```

```
Sum[(-1)^(n - 1) x^n / n, {n, ∞}]      Log[1 + x]
```

In the last example, we succeeded in obtaining the function that corresponds to a given power series.

Next, we calculate symbolic sums—that is, sums in which the upper bound is a symbol:

`Sum[i^2 + 2^i, {i, n}]` $2\left(-1 + 2^n\right) + \dfrac{1}{6} n\,(1 + n)\,(1 + 2\,n)$

`Sum[i^2 Binomial[m, i], {i, m}]` $2^{-2+m}\, m\,(1 + m)$

`Sum[1 / i, {i, n}]`   `HarmonicNumber[n]`

`Sum[c^i, {i, 0, ∞}]` $\dfrac{1}{1 - c}$

For the last result, note that the result holds if $|c| < 1$.

What is the expectation of X^2 when X has a binomial distribution with parameters n and p?

`pdf = PDF[BinomialDistribution[n, p], k]`

$(1 - p)^{-k+n}\, p^k\, \text{Binomial}[n, k]$

`Simplify[Sum[k^2 pdf, {k, 0, n}], n ∈ Integers]`

$n\, p\,(1 + (-1 + n)\, p)$

What is the probability-generating function of a negative binomial distribution with parameters n and p?

`pdf = PDF[NegativeBinomialDistribution[n, p], k]`

$(1 - p)^k\, p^n\, \text{Binomial}[-1 + k + n,\ -1 + n]$

`Sum[z^k pdf, {k, 0, ∞}]`

$p^n\,(1 - z + p\, z)^{-n}$

Here is an example in which we get an incorrect result:

`Sum[i Binomial[2 n - i, n] / 2^(2 n - i), {i, 0, n}]`

$\left((-1)^n\, 2^{-2 n}\, \pi\, \text{Csc}[n\,\pi]\, \text{HypergeometricPFQRegularized}[\{2,\ 1 - n\},\ \{1 - 2\,n\},\ 2]\right) /$
$(\text{Gamma}[n]\, \text{Gamma}[1 + n])$

The true value of the sum is $\dbinom{2\,n}{n} \dfrac{2\,n+1}{2^{2\,n}} - 1.$

`Sum[expr, {i, a, b}, {j, c, d}]`

Multiple sums are calculated in the same way as multiple integrals. Here, **c** and **d** may depend on **i**; that is, the ranges of the indices are given in the familiar mathematical notation, with the outer index given first.

`Sum[x^i y^j, {i, 1, 3}, {j, 1, i}]`

$x\, y + x^2\, y + x^3\, y + x^2\, y^2 + x^3\, y^2 + x^3\, y^3$

We can also calculate a sum where the summing index gets irregular values:

`Sum[f[n], {n, {1, 4, 5, 8, 10}}]`

$f[1] + f[4] + f[5] + f[8] + f[10]$

20.3.2 Numerical Sums

■ Calculating Sums Numerically

`NSum[expr, {i, a, b}]`   Calculate the sum with numerical methods

The idea of **NSum** is to sum a certain number of the first terms and then estimate accurately the sum of the terms neglected. Numerical summation is useful for such infinite sums that cannot be calculated with **Sum**. Also, sums with exceptionally many terms can effectively be calculated using numerical methods. One application of numerical summation is in numerical integration of oscillatory functions (see Section 20.2.5, p. 659).

We calculate several partial sums of the harmonic series with both **Sum** and **NSum**:

(s1 = Table[Sum[N[1 / i], {i, 10^n}], {n, 1, 6}]) // Timing

{7.02223, {2.92897, 5.18738, 7.48547, 9.78761, 12.0901, 14.3927}}

(s2 = Table[NSum[1 / i, {i, 10^n}], {n, 1, 6}]) // Timing

{0.110215, {2.92897, 5.18738, 7.48547, 9.78761, 12.0901, 14.3927}}

The difference in computing times is considerable, but the differences in the results are very small:

s1 – s2

$\{0., 2.35502 \times 10^{-10}, 2.35556 \times 10^{-10}, 2.34898 \times 10^{-10}, 2.35461 \times 10^{-10}, 2.34811 \times 10^{-10}\}$

Here is a sum for which **Sum** does not give a result but **NSum** works:

NSum[Log[i^2] / (2^i i!), {i, ∞}]   0.227205

We can also apply **% // N** to the result of **Sum** if **Sum** does not succeed. The sum is then actually calculated by **NSum**.

■ **Options**

Options of **NSum**:

WorkingPrecision Precision used in internal computations; examples of values:
  MachinePrecision, 20

PrecisionGoal If the value of the option is **p**, the relative error of the sum should be of the order
  10^{-p}; examples of values: **Automatic** (usually means **6**), **10**; this option is applicable if **Method** is
  "EulerMaclaurin"

AccuracyGoal If the value of the option is **a**, the relative error of the integral should be of the order
  10^{-a}; examples of values: ∞, **10**; this option is applicable if **Method** is **"EulerMaclaurin"**

Method Method to use; possible values: **Automatic, "AlternatingSigns", "EulerMaclaurin"**,
  "WynnEpsilon"

NSumTerms Number of terms summed explicitly; examples of values: **15, 20**

VerifyConvergence Whether to test explicitly for convergence of infinite sums; possible values:
  True, False

EvaluationMonitor Command to be executed after each evaluation of the expression to be
  summed; examples of values: **None, Sow[i], ++i, AppendTo[points, i]**

Compiled Whether the summand should be compiled or not; possible values: **Automatic, True,**
  False

■ **Methods**

NSum first calculates **NSumTerms** (default is 15) terms of the sum and then estimates the rest by a suitable method. There are three methods.

The **"EulerMaclaurin"** method estimates the value of the sum of the neglected terms by integration. Using this method, we can set our own **PrecisionGoal** and **AccuracyGoal**; the current estimate of the sum is accepted if either of these goals is satisfied.

The `"WynnEpsilon"` method calculates a number of extra terms and then tries to fit them to a polynomial multiplied by a decaying exponential. The `"AlternatingSigns"` method also samples a number of additional terms and approximates their sum by the ratio of two polynomials (Padé approximation). If the method is not specified, **NSum** tries to decide between the Euler–Maclaurin and Wynn's epsilon methods.

Here we calculate the same sum with two methods:

```
{sum1, {points1}} = Reap[NSum[Log[i^2] / (2^i i!),
    {i, ∞}, Method → "WynnEpsilon", EvaluationMonitor :→ Sow[i]]];
{sum2, {points2}} = Reap[NSum[Log[i^2] / (2^i i!),
    {i, ∞}, Method → "EulerMaclaurin", EvaluationMonitor :→ Sow[i]]];
{sum1 - sum2, points1 // Length, points2 // Length}
{0., 30, 70}
```

The values of the sum are the same. Next, we show the values of *i* used by these methods:

```
{ListPlot[{points1, Range[Length[points1]]}ᵀ, ImageSize → 200],
 ListPlot[{points2, Range[Length[points2]]}ᵀ, ImageSize → 200]}
```

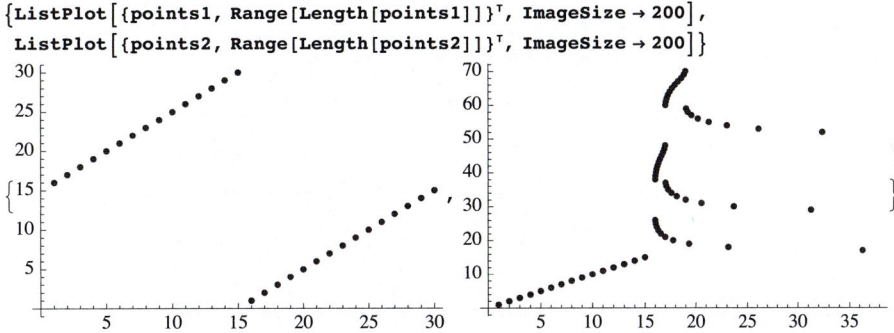

The first plot contains the 30 values of *i* for the term to be summed using Wynn's epsilon method. We see that first terms for $i = 16, \ldots, 30$ were calculated; with these terms the sum from $i = 16$ to $i = \infty$ was estimated. Then the sum of the first 15 terms was calculated.

The second plot is for the 70 points used by the Euler–Maclaurin method. The first 15 terms are again summed explicitly. Then the rest is estimated by numerical integration. Note that the plot only shows the values of *i* at most approximately 36; the largest value of *i* for which the term is evaluated is

```
Max[points2]
517.682
```

■ Euler–Maclaurin Formula

We would like to calculate the sum of $1/n^2$ for *n* from 1 to ∞. Calculate explicitly the sum of the first 15 terms and use the Euler–Maclaurin formula to approximate the rest. The four terms of the formula are as follows (this example is from the Documentation Center, under **NIntegrate**):

```
f[n_] := 1 / n^2; n0 = 1; n1 = 15; n2 = ∞; m = 5;
```

$$\left\{ \sum_{n=n0}^{n1} f[n], \text{Integrate}[f[n], \{n, n1, n2\}], \right.$$

$$\left. -\frac{f[n1] + f[n2]}{2}, \sum_{i=1}^{m} \frac{\text{BernoulliB}[2\,i]}{(2\,i)!} \left(f^{(2\,i-1)}[n2] - f^{(2\,i-1)}[n1] \right) \right\} // N$$

```
{1.58044, 0.0666667, -0.00222222, 0.000049339}
```

Here, **Integrate** can also be replaced with **NIntegrate**. The sum of the four terms is

```
Total@%
1.64493
```

This is a very accurate approximation, as can be seen when comparing with the exact result:

```
Sum[1 / n^2, {n, 1, ∞}] - %
```
-4.44089×10^{-16}

20.3.3 Products

```
Product[expr, {i, a, b}]
NProduct[expr, {i, a, b}]
```

Products can also be entered with the aid of the **BasicInput** palette or with the keyboard ($\prod$ is entered as ⎋prod⎋). Here are some examples:

```
Product[(a + i), {i, 4}]      (1 + a) (2 + a) (3 + a) (4 + a)
```

```
Product[1 - 1 / (2 i^2), {i, ∞}]
```
$$\frac{\sqrt{2} \; \mathrm{Sin}\left[\frac{\pi}{\sqrt{2}}\right]}{\pi}$$

NProduct has the same options and default values as **NSum** except that **NSumTerms** is replaced with **NProductFactors**.

20.4 Transforms

20.4.1 Laplace Transforms

```
LaplaceTransform[F, t, s]   Laplace transform of F (a function of t); the transform will be a
   function of s
InverseLaplaceTransform[f, s, t]   Inverse Laplace transform of f (a function of s); the original
   function will be a function of t
```

The Laplace transform of a function $F(t)$ is $f(s) = \int_0^\infty F(t)\, e^{-st}\, dt$. For example,

```
LaplaceTransform[Sin[3 t], t, s]
```
$$\frac{3}{9 + s^2}$$

```
InverseLaplaceTransform[%, s, t]      Sin[3 t]
```

Here is another example:

```
LaplaceTransform[Exp[a t] Cosh[b t], t, s]
```
$$\frac{-a + s}{-b^2 + (a - s)^2}$$

```
InverseLaplaceTransform[%, s, t]
```
$$\frac{1}{2}\, e^{-(-a-b)\,t} + \frac{1}{2}\, e^{-(-a+b)\,t}$$

```
FullSimplify[%]      e^{a t} Cosh[b t]
```

Another example:

```
LaplaceTransform[UnitStep[t - a], t, s]
```

$$\frac{\text{UnitStep}[-a] + e^{-as}\,\text{UnitStep}[a]}{s}$$

```
InverseLaplaceTransform[%, s, t]
```

UnitStep[-a] + HeavisideTheta[-a + t] UnitStep[a]

And another example:

```
LaplaceTransform[DiracDelta[t], t, s]      1
```

```
InverseLaplaceTransform[%, s, t]      DiracDelta[t]
```

We can also calculate transforms of some expressions that contain unspecified functions. Here is an example:

```
LaplaceTransform[F'[t], t, s]
```

-F[0] + s LaplaceTransform[F[t], t, s]

```
InverseLaplaceTransform[%, s, t]      F'[t]
```

Another example:

```
LaplaceTransform[Integrate[F[u], {u, 0, t}], t, s]
```

$$\frac{\text{LaplaceTransform}[F[t], t, s]}{s}$$

The inversion of this does not succeed. A convolution:

```
LaplaceTransform[Integrate[F[u] G[t - u], {u, 0, t}], t, s]
```

LaplaceTransform[F[t], t, s] LaplaceTransform[G[t], t, s]

```
InverseLaplaceTransform[%, s, t]
```

$$\int_0^t F[K\$154]\, G[-K\$154 + t]\, dK\$154$$

The original function may be an infinite sum (see Spiegel, 1999, p. 187). In this case, *Mathematica* cannot calculate the inverse transform:

```
InverseLaplaceTransform[Sinh[s x] / (s Sinh[s a]), s, t]
```

$$2\ \text{InverseLaplaceTransform}\left[\frac{-\frac{1}{2}e^{-sx} + \frac{e^{sx}}{2}}{(-e^{-as} + e^{as})\,s},\ s,\ t\right]$$

With **LaplaceTransform**, we can use the same **Assumptions**, **GenerateConditions**, and **PrincipalValue** options as we did with **Integrate** (see Section 20.1.3, p. 640) and also the **Analytic** option we encountered with **Limit** (see Sections 19.3.1, p. 630).

Multidimensional Laplace transforms and their inverse transforms can also be calculated.

For application of the Laplace transform to the solution of ordinary differential equations, integral equations, and partial differential equations, see Sections 26.2.1, p. 841, 26.2.4, p. 847, and 27.1.2, p. 891.

■ **Numerical Inversion**

For the numerical inversion of Laplace transforms, see Cheng, Sidauruk, and Abousleiman (1994). Here is one such method—the Stehfest method:

$$c[n\_, i\_] := (-1)^{i+\frac{n}{2}} \sum_{k=\left\lceil \frac{i+1}{2} \right\rceil}^{\text{Min}\left[i, \frac{n}{2}\right]} \frac{N\left[k^{\frac{n}{2}} (2k)!\right]}{N\left[\left(\frac{n}{2}-k\right)! k! (k-1)! (i-k)! (2k-i)!\right]}$$

$$\text{stehfest}[f\_, s\_, t\_, n\_?\text{EvenQ}] := \frac{\text{Log}[2]}{t} \sum_{i=1}^{n} c[n, i] \left(f /. s \to \frac{i \, \text{Log}[2]}{t}\right)$$

Compare the exact value $\sin(1)$ and the numerical inverse of $1/(1+s^2)$ at $t = 1$:

```
Sin[1.] - stehfest[1 / (1 + s^2), s, 1, 24]
```
-0.00122814

The numerical inverse is quite good. We can plot the inverse transform:

```
Plot[{Sin[t], Evaluate[stehfest[1 / (1 + s^2), s, t, 24]]}, {t, 0, 2 π}]
```

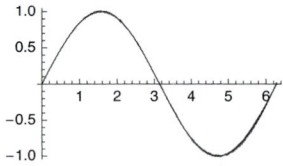

20.4.2 Z-Transforms

ZTransform[f, n, z]   Z-transform of **f** (a function of **n**); the transform will be a function of **z**
InverseZTransform[g, z, n]   Inverse Z-transform of **g** (a function of **z**); the original function will be a function of **n**

The Z-transform is defined by $g(z) = \sum_{m=0}^{\infty} f(n) z^{-n}$. For example,

```
ZTransform[a^n n^2, n, z]
```
$$-\frac{a z (a + z)}{(a - z)^3}$$

```
InverseZTransform[%, z, n]
```
$a^n n^2$

The Z-transform can be used to solve difference equations (see Section 28.1.3, p. 932).

20.4.3 Fourier Transforms and Series

■ **Fourier Transforms**

FourierTransform[f, t, w]   Fourier transform of **f** (a function of **t**); the transform will be a function of **w**
InverseFourierTransform[F, w, t]   Inverse Fourier transform of **F** (a function of **w**); the original function will be a function of **t**

An option:

FourierParameters   Parameters of the Fourier transform; examples of values: **{0, 1}, {1, -1}, {-1, 1}, {0, -2π}**

The Fourier transform of $f(t)$ in *Mathematica* is $F(w) = \frac{1}{\sqrt{2\pi}} \int_{-\infty}^{\infty} f(t) e^{iwt} dt$ if the option

FourierParameters has the default value **{0, 1}**. Various other Fourier transforms are applied in different disciplines. They can be used by giving **FourierParameters** a suitable value. If the value is

{a, b}, the transform is $\left(\frac{|b|}{(2\pi)^{1-a}}\right)^{1/2} \int_{-\infty}^{\infty} f(t) e^{ibwt} dt$. The **Assumptions** and **GenerateConditions** options

can also be used.

Here is an example:

```
FourierTransform[Exp[-t^2] Sinh[t], t, w]
```

$$\frac{(-1 + \text{Cos}[w] + i\, \text{Sin}[w]) \left(\text{Cosh}\left[\frac{1}{4}(i+w)^2\right] - \text{Sinh}\left[\frac{1}{4}(i+w)^2\right]\right)}{2\sqrt{2}}$$

```
InverseFourierTransform[%, w, t] // FullSimplify
```

e^{-t^2} Sinh[t]

```
FourierSinTransform[f, t, w],   InverseFourierSinTransform[f, t, w]
FourierCosTransform[f, t, w],   InverseFourierCosTransform[f, t, w]
```

The default Fourier sine transform is $\sqrt{\frac{2}{\pi}} \int_0^\infty f(t) \sin(wt) dt$. Other Fourier sine transforms can be

obtained from the most general form, $2\left(\frac{|b|}{(2\pi)^{1-a}}\right)^{1/2} \int_0^\infty f(t) \sin(bwt) dt$, by giving the parameters a and b

suitable values with **FourierParameters**. Similarly, we get Fourier cosine transforms.

The **FourierSeries`** package also has the commands **DTFourierTransform** and **InverseDTFourierTransform** and numerical versions of all of the eight commands mentioned previously (e.g., **NFourierTransform**).

■ **Fourier Series**

In the **FourierSeries`** *package:*

FourierTrigSeries[f, t, k]   Fourier trigonometric series expansion to order **k** of **f**; by default, **f** is treated as a periodic function of **t** with one period on $(-1/2, 1/2)$

FourierCosCoefficient[f, t, k]   Coefficient of a cos term

FourierSinCoefficient[f, t, k]   Coefficient of a sin term

An option:

FourierParameters   Parameters of the series expansion; default value: **{0, 1}**

The package also defines the commands **FourierSeries** (Fourier exponential series) and **FourierCoefficient**.

If the option **FourierParameters** has the default value **{0, 1}**, the Fourier trigonometric series is $c_0 + \sum_{n=1}^{k} [c_n \cos(2\pi n t) + d_n \sin(2\pi n t)]$ for a periodic function with one period on $\left(-\frac{1}{2}, \frac{1}{2}\right)$. If **FourierParameters** is given the value **{0, b}**, then the series is (if b is positive)

$$\sqrt{b}\,\left\{c_0 + \sum_{n=1}^{k} [c_n \cos(2\pi b n t) + d_n \sin(2\pi b n t)]\right\}$$

for a periodic function with one period on $\left(-\frac{1}{2b}, \frac{1}{2b}\right)$. The coefficients c_0, c_n, and d_n are as follows:

$$c_0 = \sqrt{b} \int_{-\frac{1}{2b}}^{\frac{1}{2b}} f(t)\,dt, \quad c_n = 2\sqrt{b}\int_{-\frac{1}{2b}}^{\frac{1}{2b}} f(t)\cos(2\pi b n t)\,dt, \quad d_n = 2\sqrt{b}\int_{-\frac{1}{2b}}^{\frac{1}{2b}} f(t)\sin(2\pi b n t)\,dt.$$

As an example, we calculate the third-order Fourier trigonometric series of $(t-1)^2$ when this function is treated as periodic with one period on $(-2, 2)$. Because we want $\frac{1}{2b}$ to be 2, we choose b to be $1/4$:

```
<< FourierSeries`
```

```
ser = FourierTrigSeries[(t - 1)^2, t, 3, FourierParameters → {0, 1/4}]
```

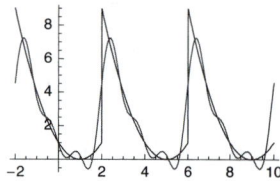

To compare this function with the original periodic function, we define the periodic function (see Section 18.5.1, p. 602):

```
f[t_] := If[-2 ≤ t < 2, (t - 1)^2, f[t - 4]]
```

```
Plot[{f[t], ser}, {t, -2, 10}]
```

We can also calculate the general coefficients of the Fourier series:

```
c0 = FourierCosCoefficient[(t - 1)^2, t, 0, FourierParameters → {0, 1/4}]
```

$$\frac{14}{3}$$

```
cn = FourierCosCoefficient[(t - 1)^2, t, n, FourierParameters → {0, 1/4}]
```

$$\frac{32\,(-1)^n}{n^2\,\pi^2}$$

```
dn = FourierSinCoefficient[(t - 1)^2, t, n, FourierParameters → {0, 1/4}]
```

$$\frac{16\,(-1)^n}{n\,\pi}$$

So we see that in the interval $(-2, 2)$, we have

$$(t-1)^2 = \frac{1}{2}\left[\frac{14}{3} + \sum_{n=1}^{\infty}\left(\frac{32}{n^2\pi^2}(-1)^n\cos\left(\frac{n\pi t}{2}\right) + \frac{16}{n\pi}(-1)^n\sin\left(\frac{n\pi t}{2}\right)\right)\right].$$

For more information about the **FourierSeries`** package, see FourierSeries/tutorial/FourierSeries.

20.4.4 Discrete Fourier Transforms

■ Discrete Fourier Transforms

> **Fourier[data]** Discrete Fourier transform
> **InverseFourier[data]** Discrete inverse Fourier transform
>
> *An option:*
> **FourierParameters** Parameters of the transform; examples of values: **{0, 1}**, **{-1, 1}**, **{1, -1}**

Data are often analyzed by calculating the discrete Fourier transform or the spectrum. For data $\{u_1, \ldots, u_n\}$, the transform is $n^{-1/2} \sum_{r=1}^{n} u_r \, e^{2\pi i (r-1)(s-1)/n}$ if the option **FourierParameters** has the default value **{0, 1}**. If **FourierParameters** is given the value **{a, b}**, then the transform is $n^{-(1-a)/2} \sum_{r=1}^{n} u_r \, e^{2\pi i b (r-1)(s-1)/n}$. To ensure a unique inverse discrete Fourier transform, $|b|$ must be relatively prime to n (greatest common divisor of $|b|$ and n is 1). **Fourier** can also find the transform for higher-dimensional data.

As a simple example, we calculate the Fourier transform of a list having three elements; **Chop** can be used to replace near-zero real or imaginary parts with an exact zero:

  Fourier[{1, 0, 2}]   {1.73205 + 0. i, 0. - 1. i, 0. + 1. i}

  Chop[%]   {1.73205, -1. i, 1. i}

Then we calculate the inverse transform:

  InverseFourier[%] $\{1., 2.56395 \times 10^{-16}, 2.\}$
  Chop[%]   {1., 0, 2.}

In Section 30.6.2, p. 1045, we present an example that shows how the discrete Fourier transform can be used to smooth or filter data.

■ Fourier Discrete Sine and Cosine Transforms

> **FourierDST[data, m]** (✻6) Fourier discrete sine transform of type **m**
> **FourierDCT[data, m]** (✻6) Fourier discrete cosine transform of type **m**

The parameter **m** can have the values 1, 2, 3, and 4. The value 2 is the default. The inverse of type 1, 2, 3, or 4 transform is type 1, 3, 2, and 4 transform, respectively.

  FourierDCT[{1, 0, 2}]   {1.73205, -0.5, 0.866025}

  FourierDCT[%, 3] // Chop   {1., 0, 2.}

 **21**

Matrices

Introduction

> *In some colleges of music, part of the doctoral requirement is to compose an original full-length symphony. Because modern music sounds so weird, a good ploy is to take a well-known classical symphony, write it backwards, and submit it as an original work. One student took the daring step of taking his professor's doctoral symphony and reversing it. He failed to receive his degree, the examiners remarking that he had reproduced Sibelius' Fourth Symphony with not a single note changed.*

In *Mathematica*, vectors and matrices are represented as lists. Accordingly, in this chapter we encounter many of the commands that are familiar from Chapter 14. Note that systems of linear equations are considered in Section 22.1, and linear programming is addressed in Section 23.2.1. For more information about linear algebra with *Mathematica*, see tutorial/LinearAlgebraOverview. See also Szabo (2000, 2001), and Ruskeepää (2007).

21.1 Vectors

21.1.1 Basics of Vectors

■ **Displaying Vectors**

{a, b, c, … }   A vector with elements **a**, **b**, **c**, …

A vector is a one-dimensional list:

    v = {4, 3, 7}   {4, 3, 7}

This is the usual form of a vector in *Mathematica*. Vectors can also be displayed in other forms:

> **MatrixForm[v]** Show vector **v** as a column, with parentheses at left and right
> **MatrixForm[{v}]** Show vector **v** as a row, with parentheses at left and right
> **Column[v]** (⌘6) Show vector **v** as a column without parentheses
> **Row[v, s]** (⌘6) Show vector **v** as a row without parentheses; separate elements with **s**

    {MatrixForm[v], MatrixForm[{v}], Column[v], Row[v, "   "]}

$$\left\{\begin{pmatrix}4\\3\\7\end{pmatrix},\ (4\ \ 3\ \ 7),\ \begin{matrix}4\\3\\7\end{matrix},\ 4\ \ 3\ \ 7\right\}$$

For more information about **Column** and **Row**, see Section 15.1.2, p. 469. For more information about **MatrixForm**, see Section 21.2.1, p. 686. Note that these commands are only used to *display* vectors: Vectors displayed with these commands cannot be used in any *calculations*.

■ **Generating Vectors**

Vectors that arise from a systematic scheme can be input with special commands such as **Table**, **Range**, and **Array**, familiar from Section 14.1.1, p. 444. Here, we recall these commands, together with some new commands.

> **Range[n]** Create the vector $\{1, 2, …, n\}$; **Range[n$_0$, n]** creates the vector $\{n_0, n_0 + 1, …, n\}$ and **Range[n$_0$, n, d]** the vector of numbers from **n$_0$** to **n** in steps of **d**
> **CharacterRange["c$_1$", "c$_2$"]** Create a vector of characters from **c$_1$** to **c$_2$**
> **Table[expr, {i, n}]** Create a vector from the values of **expr** when **i** goes from 1 to **n**; write **{i, n$_0$, n}** if **i** goes from **n$_0$** to **n** and **{i, n$_0$, n, d}** if **i** goes from **n$_0$** to **n** in steps of **d**; write **{n}** to create **n** copies of **expr**
> **Array[f, n]** Create a vector from the values of **f[i]** when **i** goes from 1 to **n**; write **Array[f, n, n$_0$]** if **i** goes from **n$_0$** to **n + n$_0$ - 1**
> **ConstantArray[c, n]** (⌘6) Create a vector of **n** copies of element **c**
> **SparseArray[rules, dims, default]** Create a sparse array (see Section 21.2.1, p. 689)
> **UnitVector[n, k]** (⌘6) The **n**-dimensional unit vector in the **k**th direction

    Range[10]   {1, 2, 3, 4, 5, 6, 7, 8, 9, 10}

    CharacterRange["a", "g"]   {a, b, c, d, e, f, g}

    Table[Prime[n], {n, 6}]   {2, 3, 5, 7, 11, 13}

    Array[f, {4}]   {f[1], f[2], f[3], f[4]}

    ConstantArray[1, 5]   {1, 1, 1, 1, 1}

    Table[UnitVector[3, i], {i, 3}]   {{1, 0, 0}, {0, 1, 0}, {0, 0, 1}}

21.1.2 Manipulating Vectors

■ **Taking Parts of Vectors**

Because vectors are lists, we can use all the list manipulation commands presented in Chapter 14. Here, we recall these commands. First, we show ways to take elements by using **Part** or **[[ ]]**. Note that the two characters **[[** can also be replaced with the single character **〚**; write it as ESC **[[** ESC. Similarly, **]]** can be replaced with **〛**; write it as ESC **]]** ESC. Still another way to take parts is to use the **□〚□〛** button in the **BasicMathInput** palette.

---

v〚i〛 Take the **i**th element

v〚-i〛 Take the **i**th element from the end

v〚i ;; j〛 (⌘6) Take elements **i** through **j**

v〚i ;;〛 Take elements from **i** to the end

v〚-i ;;〛 Take the last **i** elements

v〚 ;; j〛 Take the first **j** elements

v〚 ;; -j〛 Take elements from the beginning to the **j**th element from the end

v〚i ;; j ;; d〛 Take elements **i** through **j** in steps of **d**

v〚{i, j, … }〛 Take elements **i, j, …**

---

    v = Range[6]   {1, 2, 3, 4, 5, 6}

    {v〚4〛, v〚3 ;; 6〛, v〚{2, 3, 6}〛}

    {4, {3, 4, 5, 6}, {2, 3, 6}}

■ **Resetting Elements**

---

v〚i〛 = a Set the **i**th element to scalar **a**

v〚i ;; j〛 = a Set elements **i** through **j** to scalar **a**

v〚i ;; j〛 = {a, …, b} Set element **i** to **a**, …, element **j** to **b**

v〚{i, j, … }〛 = {a, b, … } Set elements **i, j, …** to **a, b, …**

Parts specified by **〚 〛** can be used to set new values for elements:

    v〚4〛 = 44   44

    v   {1, 2, 3, 44, 5, 6}

■ **Taking and Dropping Elements**

First[v], **Rest[v]** Take/drop the first element

Last[v], **Most[v]** Take/drop the last element

Take[v, i], **Drop[v, i]** Take/drop the first **i** elements

Take[v, -i], **Drop[v, -i]** Take/drop the last **i** elements

Take[v, {i, j}], **Drop[v, {i, j}]** Take/drop elements **i, …, j**

Take[v, {i, j, d}], **Drop[v, {i, j, d}]** Take/drop elements in steps of **d**

TakeWhile[list, crit] (⌘6) Take elements from **list** as long as **crit** gives **True**

LengthWhile[list, crit] (⌘6) Give the length of the list given by **TakeWhile**

Note that these commands do not modify the original value of **v**:

```
Drop[v, -3]     {1, 2, 3}

v     {1, 2, 3, 44, 5, 6}
```

■ Inserting and Deleting Elements

Among the following commands, `PrependTo` and `AppendTo` are special, as explained later.

`Prepend[v, a]` or `Join[{a}, v]` Insert element **a** at the beginning of **v**

`Append[v, a]` or `Join[v, {a}]` Insert element **a** at the end of **v**

`PrependTo[v, a]` Insert element **a** at the beginning of **v** and reset **v** to the result

`AppendTo[v, a]` Insert element **a** at the end of **v** and reset **v** to the result

`Insert[v, a, i]` Insert element **a** between elements **i** - 1 and **i**

`ReplacePart[m, a, i]` Replace i̇th element with **a**

`Delete[v, i]`, `Delete[v, -i]` Delete the i̇th element, counting from the beginning/end

`Delete[v, {{i}, {j}, … }]` Delete elements **i, j**, …

`Riffle[{a, b, c, … }, {x, y, z, … }]` {a, x, b, y, c, z, ...}

`Riffle[{a, b, c, … }, x, {p0, p1, d}]` (❀6) **x** appears at positions **p0, p0 + d, p0 + 2d**, … (a negative **p1** counts from the end)

`PadRight[v, n]` Pad vector **v** with zeros on the right to make a vector of length **n**

`PadRight[v, n, a]` Pad vector **v** cyclically with **a** (**a** may be a scalar or list)

Note that with the exception of `PrependTo` and `AppendTo`, these commands do not modify the original value of **v**:

```
Prepend[v, 0]     {0, 1, 2, 3, 44, 5, 6}

v     {1, 2, 3, 44, 5, 6}
```

However, `PrependTo[v, a]` and `AppendTo[v, a]` do modify the original vector. The same effect can also be obtained by **v = Prepend[v, a]** and **v = Append[v, a]**:

```
PrependTo[v, 0]     {0, 1, 2, 3, 44, 5, 6}

v     {0, 1, 2, 3, 44, 5, 6}

v = Prepend[v, -1]     {-1, 0, 1, 2, 3, 44, 5, 6}

v     {-1, 0, 1, 2, 3, 44, 5, 6}
```

With `PadRight` or `PadLeft` we can pad a vector to form a longer vector:

```
PadRight[{1, 2, 3}, 7]     {1, 2, 3, 0, 0, 0, 0}
```

■ Selecting Elements

`Select[v, test]` Select the elements that give True for **test**

`Cases[v, pattern]` Select the elements that match **pattern**

`DeleteCases[v, pattern]` Delete the elements that match **pattern**

`Count[v, pattern]` Give the number of the elements that match **pattern**

`Position[v, pattern]` Give the positions at which elements match **pattern**

`Extract[v, pos]` Extract the elements that are at the positions given by `Position`

```
SeedRandom[1]; v = RandomInteger[6, 10]
{6, 4, 2, 4, 0, 1, 6, 0, 0, 2}

Select[v, # ≤ 3 &]    {2, 0, 1, 0, 0, 2}

Select[v, # == 1 || # == 6 &]    {6, 1, 6}

Cases[v, x_ /; x ≤ 3]    {2, 0, 1, 0, 0, 2}

Cases[v, 1 | 6]    {6, 1, 6}

Position[v, x_ /; x ≤ 3]    {{3}, {5}, {6}, {8}, {9}, {10}}

Extract[v, %]    {2, 0, 1, 0, 0, 2}
```

Select is considered in Section 14.1.7, p. 457, and most other commands are discussed in Section 16.1.1, p. 493.

■ Reordering Elements

Sort[v] Sort the elements into a standard order
SortBy[list, f] (✿6) Sort the elements of **list** in the order defined by applying **f** to each of them
Union[v] Sort the elements into a standard order and remove any duplicates
Join[u, v, …] Join the given vectors into one vector
Partition[v, n] Partition the vector into subvectors of **n** elements
Split[v] Split the vector into subvectors consisting of runs of identical elements

21.1.3 Vector Calculus

■ Properties of Vectors

Length[v] The number of elements of vector **v**
VectorQ[v] Test whether **v** is a vector
VectorQ[v, test] Test whether **v** is a vector with elements that satisfy **test**

```
{VectorQ[{4, 3, 7}], VectorQ[{{4, 3}, {2, 8}}], VectorQ[{}]}
{True, False, True}

{VectorQ[{4, 3, 7}, NumericQ], VectorQ[{4, 2, 8}, EvenQ]}
{True, True}
```

We consider tests in more detail in Section 13.3.5, p. 431.

■ Arithmetic with Vectors

Arithmetic with vectors is easy because *Mathematica* automatically does all operations element by element. Here are some calculations with a single vector:

a + v Add scalar **a** to each element of vector **v**
a v Multiply each element of vector **v** with scalar **a**
1/v Calculate the reciprocal of each element of vector **v**
v^a Calculate the **a**th power (**a** is a scalar) of each element of vector **v**
a^v Calculate the powers of scalar **a** that are given in vector **v**

```
v = {x, y, z};
```

```
{1 + v, 5 v, 1 / v, v^2, 2^v}
```

$$\left\{ \{1 + x, 1 + y, 1 + z\}, \{5\,x, 5\,y, 5\,z\}, \left\{ \frac{1}{x}, \frac{1}{y}, \frac{1}{z} \right\}, \{x^2, y^2, z^2\}, \{2^x, 2^y, 2^z\} \right\}$$

Built-in functions of vectors are also calculated elementwise:

Log[v]   {Log[x], Log[y], Log[z]}

Here are some calculations with two vectors of the same size:

---

u + v Add two vectors **u** and **v**
u v Multiply the corresponding elements of **u** and **v** (use **u.v** for an inner product)
u/v Divide the corresponding elements of vectors **u** and **v**
u^v Calculate powers of corresponding elements of vectors **u** and **v**

---

u = {4, 3, 7}; v = {x, y, z};

{u + v, u v, u / v, u^v}

$$\left\{ \{4 + x, 3 + y, 7 + z\}, \{4\,x, 3\,y, 7\,z\}, \left\{ \frac{4}{x}, \frac{3}{y}, \frac{7}{z} \right\}, \{4^x, 3^y, 7^z\} \right\}$$

■ Sum, Product, Minimum, and Maximum of Elements

Total[v] The sum of the elements of vector **v**
Total[v, Method → CompensatedSummation] Use compensated summation
Apply[Times, v] The product of the elements of vector **v**
Min[v], Max[v] The smallest/largest element

---

v = {a, b, c, d};

Total[v]   a + b + c + d

Apply[Times, v]   a b c d

In summing long lists, round-off error may accumulate. Using an option resolves the problem:

t = ConstantArray[2.9, 10^6];

Total[t] − 2.9 × 10^6   −0.0000535152

Total[t, Method → CompensatedSummation] − 2.9 × 10^6   0.

■ Cumulative Sums and Differences of Elements

Accumulate[v] (✿6) The list of successive cumulative sums of elements
Differences[v] (✿6) The list of successive differences of elements
Differences[v, n] (✿6) The list of **n**th differences

---

v = {a, b, c, d};

Accumulate[v]   {a, a + b, a + b + c, a + b + c + d}

Differences[v]   {−a + b, −b + c, −c + d}

Let us calculate several differences in two ways:

Table[Differences[v, n], {n, 0, 4}]

{{a, b, c, d}, {−a + b, −b + c, −c + d}, {a − 2 b + c, b − 2 c + d}, {−a + 3 b − 3 c + d}, {}}

```
NestList[Differences, v, 4]
```

{{a, b, c, d}, {-a + b, -b + c, -c + d}, {a - 2 b + c, b - 2 c + d}, {-a + 3 b - 3 c + d}, {}}

Cumulative sums are needed in simulation of random walks:

```
ListLinePlot[Accumulate[RandomChoice[{-1, 1}, 50]], Mesh → All]
```

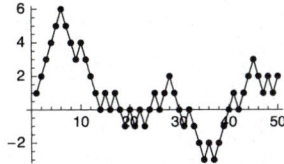

■ **Products**

u v	Element-by-element product (the result is a vector)
u.v or **Dot[u, v]**	Scalar (or inner) product (the result is a scalar)
u×v or **Cross[u, v]**	Cross product (the result is a vector)
KroneckerProduct[u, v] (✻6)	Kronecker (or outer) product (the result is a matrix)

The cross (×) can be written as ESC cross ESC. Examples:

```
u = {a, b, c}; v = {P, Q, R};
```

```
u v      {a P, b Q, c R}
```

```
u.v      a P + b Q + c R
```

```
u × v    {-c Q + b R, c P - a R, -b P + a Q}
```

Calculate the Kronecker product in two ways:

```
KroneckerProduct[u, v]      {{a P, a Q, a R}, {b P, b Q, b R}, {c P, c Q, c R}}
```

```
Outer[Times, u, v]      {{a P, a Q, a R}, {b P, b Q, b R}, {c P, c Q, c R}}
```

Note especially that the normal product (written as a space) cannot be used to calculate the scalar product; we have to use the dot.

Note also that a vector such as {a, b, c} looks like a *row* vector. The truth is, however, that *Mathematica* does not distinguish between row and column vectors; all vectors are written in the same way. *Mathematica* can work in this way because it is almost always clear to *Mathematica* what should be done in a computation that contains vectors and matrices. To illustrate this further, we introduce a matrix together with two vectors:

```
T = {{3, 1}, {4, 6}};
r = {x, y}; s = {3, 8};
```

To multiply **T** and **r**, write the following:

```
T.r      {3 x + y, 4 x + 6 y}
```

To multiply **r** and **T**, write

```
r.T      {3 x + 4 y, x + 6 y}
```

However, if we want the product of **T**, **s** (considered as a column vector), **s** (considered as a row vector), and **T**, we have to be careful and use **KroneckerProduct**:

```
T.KroneckerProduct[s, s].T      {{697, 867}, {2460, 3060}}
```

■ **Norms**

`Norm[v]`	The 2-norm of vector **v**
`Norm[v, p]`	The p-norm of vector **v**, $1 \le p \le \infty$

The p-norm of a vector is $\left(\sum |v_i|^p\right)^{1/p}$ for $1 \le p < \infty$, and the ∞-norm is the maximum of the absolute values of the elements.

```
v = {a, b, c};
```

```
Norm[v, #] & /@ {1, 2, p, ∞}
```

$$\left\{ \mathrm{Abs}[a] + \mathrm{Abs}[b] + \mathrm{Abs}[c], \; \sqrt{\mathrm{Abs}[a]^2 + \mathrm{Abs}[b]^2 + \mathrm{Abs}[c]^2}, \right.$$
$$\left. \left(\mathrm{Abs}[a]^p + \mathrm{Abs}[b]^p + \mathrm{Abs}[c]^p\right)^{\frac{1}{p}}, \; \mathrm{Max}[\mathrm{Abs}[a], \mathrm{Abs}[b], \mathrm{Abs}[c]]\right\}$$

The 1-, 2-, and ∞-norms are easy to program:

```
vectorNorm[v_, p_] := Switch[p,
   1, Total[Abs[v]],
   2, Sqrt[Abs[v].Abs[v]],
   ∞, Max[Abs[v]]]
```

```
vectorNorm[v, 2]
```
$$\sqrt{\mathrm{Abs}[a]^2 + \mathrm{Abs}[b]^2 + \mathrm{Abs}[c]^2}$$

■ **Distances**

`EuclideanDistance[u, v]` (✻6)	The same as `Norm[u - v]`
`HammingDistance[u, v]` (✻6)	The number of elements whose values disagree in **u** and **v**

```
EuclideanDistance[{2, 3}, {x, y}]
```
$$\sqrt{\mathrm{Abs}[2 - x]^2 + \mathrm{Abs}[3 - y]^2}$$

```
HammingDistance[{1, 2, 3, 4, 5}, {2, 2, 3, 5, 4}]        3
```

Mathematica has many other distances: `BrayCurtisDistance`, `CanberraDistance`, `ChebyshevDistance` (the same as `Norm[u - v, ∞]`), `CorrelationDistance`, `CosineDistance`, `EditDistance`, `ManhattanDistance` (the same as `Norm[u - v, 1]`), and `SquaredEuclideanDistance`. These distances can be used as a value for the `DistanceFunction` option of such commands as `FindShortestTour`, `FindClusters`, and `Nearest`; see Sections 23.5.1, p. 777, and 30.1.2, p. 1009.

As an example, let us draw 100 random lines in unit square:

```
SeedRandom[1];
tt = RandomReal[{0, 1}, {100, 2, 2}];
Graphics[{Line /@ tt}, Axes → True]
```

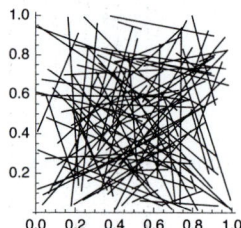

We calculate the lengths of the lines and the mean and variance of the lengths:

```
lengths = EuclideanDistance[#[[1]], #[[2]]] & /@ tt;
```

```
{Mean[lengths], Variance[lengths]}
```

{0.532353, 0.0602447}

■ Orthogonalization

Orthogonalize[{v1, v2, … }] (❀6) Generate an orthonormal set from the vectors
Projection[u, v] (❀6) Orthogonal projection of **u** onto **v**
Normalize[v] (❀6) Normalize the vector **v**

Here is an example of Gram–Schmidt orthonormalization:

```
u = {5, 1}; v = {3, 4};
```

```
{o1, o2} = Orthogonalize[{u, v}]
```
$$\left\{\left\{\frac{5}{\sqrt{26}}, \frac{1}{\sqrt{26}}\right\}, \left\{-\frac{1}{\sqrt{26}}, \frac{5}{\sqrt{26}}\right\}\right\}$$

The inner product of these vectors is 0 and their norm is 1:

```
{o1.o2, Norm[o1], Norm[o2]}
```
{0, 1, 1}

Next, we calculate and show the orthogonal projection of **v** onto **u**:

```
pr = Projection[v, u]
```
$$\left\{\frac{95}{26}, \frac{19}{26}\right\}$$

```
Graphics[{Arrow[{{0, 0}, u}], Arrow[{{0, 0}, v}], Arrow[{{0, 0}, pr}],
  Line[{v, pr}], Text["u", u, {-2, 0}], Text["v", v, {-1, -1}]}, Axes → True]
```

Orthogonalize and **Projection**, by default, use the usual inner product (**#1.#2 &**), but another inner product can be defined by writing a pure function as an additional argument to these commands.

■ Angles and Rotation

VectorAngle[u, v] (❀6) The angle between the two vectors
RotationMatrix[θ] (❀6) The matrix to rotate in 2D by θ radians counterclockwise about the origin

---

```
VectorAngle[o1, o2]
```
$$\frac{\pi}{2}$$

```
RotationMatrix[θ]
```
{{Cos[θ], -Sin[θ]}, {Sin[θ], Cos[θ]}}

```
RotationMatrix[π / 2].{1, 0}
```
{0, 1}

For more information about these commands, see the *Mathematica* documentation.

■ **Transforms**

Vectors can be transformed with **Affine-**, **LinearFractional-**, **Reflection-**, **Rescaling-**, **Rotation-**, **Scaling-**, **Shearing-**, and **TranslationTransform**.

```
ParametricPlot[RotationTransform[Pi / 4][{2 Cos[t], Sin[t]}] // Evaluate,
  {t, 0, 2 π}, Ticks → {{-1, 1}, {-1, 1}}, ImageSize → 90]
```

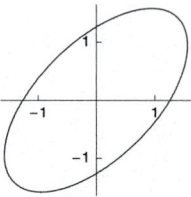

21.2 Matrices

21.2.1 Basics of Matrices

■ **Writing Matrices**

{{a, b, … }, {c, d, … }, … }   A matrix with rows {a, b, …}, {c, d, …}, …

A matrix is represented as a list; each row of the matrix is a separate sublist. Here is a matrix with two rows and three columns:

```
m = {{4, 2, 7}, {3, 5, 1}}     {{4, 2, 7}, {3, 5, 1}}
```

The menu command **Insert ▷ Table/Matrix ▷ New** generates an empty matrix such as the one in the following:

$$m = \begin{pmatrix} \square & \square & \square \\ \square & \square & \square \end{pmatrix}$$

We can then fill in the placeholders:

$$m = \begin{pmatrix} 4 & 2 & 7 \\ 3 & 5 & 1 \end{pmatrix} \qquad \{\{4, 2, 7\}, \{3, 5, 1\}\}$$

Matrices can also be entered with the keyboard. To start, type CTRL{,] to get two placeholders of a (1 × 2) matrix. To get more columns, type CTRL{,] several times. To get more rows, type CTRL{↵] several times. Then fill in the matrix.

■ **Displaying Matrices**

MatrixForm[m]   Write matrix **m** in a 2D form, with parentheses at left and right
TraditionalForm[m]   Write matrix **m** in a 2D form, with parentheses at left and right
Grid[m] (✤6)   Write matrix **m** in a 2D form, without parentheses

---

```
{MatrixForm[m], TraditionalForm[m], Grid[m]}
```

$$\left\{ \begin{pmatrix} 4 & 2 & 7 \\ 3 & 5 & 1 \end{pmatrix}, \begin{pmatrix} 4 & 2 & 7 \\ 3 & 5 & 1 \end{pmatrix}, \begin{matrix} 4 & 2 & 7 \\ 3 & 5 & 1 \end{matrix} \right\}$$

Note that **MatrixForm**, **TraditionalForm**, and **Grid** are only used to *display* matrices: Matrices displayed with these commands cannot be used in any *calculations*. For example, write the following:

q = {{3, 1}, {2, 5}} // MatrixForm $\begin{pmatrix} 3 & 1 \\ 2 & 5 \end{pmatrix}$

We defined **q** to be not the matrix itself but its matrix form. *Mathematica* will not do any calculations with such a form:

{2 + q, 3 q, Transpose[q], Inverse[q]}

$\left\{2 + \begin{pmatrix} 3 & 1 \\ 2 & 5 \end{pmatrix},\ 3\begin{pmatrix} 3 & 1 \\ 2 & 5 \end{pmatrix},\ \text{Transpose}\left[\begin{pmatrix} 3 & 1 \\ 2 & 5 \end{pmatrix}\right],\ \text{Inverse}\left[\begin{pmatrix} 3 & 1 \\ 2 & 5 \end{pmatrix}\right]\right\}$

Therefore, when defining **q**, we have to be careful so that the value of **q** will be the matrix itself and not its matrix form. The simplest way is to use two separate commands: First define the matrix and then ask for its matrix form:

q = {{3, 1}, {2, 5}}   {{3, 1}, {2, 5}}

q // MatrixForm $\begin{pmatrix} 3 & 1 \\ 2 & 5 \end{pmatrix}$

If we want, in one command, to define and display a matrix, we can do so in either of the following ways:

MatrixForm[q = {{3, 1}, {2, 5}}] $\begin{pmatrix} 3 & 1 \\ 2 & 5 \end{pmatrix}$

(q = {{3, 1}, {2, 5}}) // MatrixForm $\begin{pmatrix} 3 & 1 \\ 2 & 5 \end{pmatrix}$

A good way to show all matrices in the matrix form—without problems in calculations—is to define **$PrePrint = If[MatrixQ[#], MatrixForm[#], #]&**. Now all matrices are shown in the matrix form. By defining **$PrePrint =.**, we can cancel the definition.

For **MatrixForm**, we have the same five options as for **TableForm**; the options of the latter command were considered in Section 15.1.1, p. 467. The default value **Automatic** of the **TableAlignments** option now means **{Center, Baseline}** (the default value is **{Left, Baseline}** for **TableForm**). The default value **Automatic** of the **TableSpacing** option (defining the space between rows and columns) actually means **{1, 1}**.

TraditionalForm is considered in Section 3.3.1, p. 70, and **Grid** in Section 15.2, p. 470.

■ **Generating Matrices**

Matrices that arise from a systematic scheme can be input with special commands such as **Table**, **Range**, and **Array**, familiar from Section 14.1.1, p. 444. Here, we recall these commands, together with some new commands.

Table[expr, {i, m}, {j, n}] Create an (**m**×**n**) matrix; other forms of the iteration specifications are **{i, m₀, m}**, **{i, m₀, m, d}**, and **{m}**

Array[f, {m, n}] Create an (**m**×**n**) matrix with elements **f[i, j]**; add a third argument **{m₀, n₀}** if index origins are **m₀** and **n₀**

ConstantArray[c, {m, n}] (✦6) Create an (**m**×**n**) matrix, all elements being **c**

SparseArray[rules, dims, default] Create a sparse matrix

IdentityMatrix[n]   An (n×n) identity matrix

DiagonalMatrix[v]   A diagonal matrix with diagonal elements from vector **v**

HilbertMatrix[n]   The (n×n) Hilbert matrix

Partition[v, n]   From vector **v**, form a matrix with **n** columns and as many rows as become complete

---

Here is a Hilbert matrix, calculated in three ways:

```
Table[1 / (i + j - 1), {i, 3}, {j, 3}]
```

$$\left\{\left\{1, \frac{1}{2}, \frac{1}{3}\right\}, \left\{\frac{1}{2}, \frac{1}{3}, \frac{1}{4}\right\}, \left\{\frac{1}{3}, \frac{1}{4}, \frac{1}{5}\right\}\right\}$$

```
Array[1 / (#1 + #2 - 1) &, {3, 3}]
```

$$\left\{\left\{1, \frac{1}{2}, \frac{1}{3}\right\}, \left\{\frac{1}{2}, \frac{1}{3}, \frac{1}{4}\right\}, \left\{\frac{1}{3}, \frac{1}{4}, \frac{1}{5}\right\}\right\}$$

```
HilbertMatrix[3]
```

$$\left\{\left\{1, \frac{1}{2}, \frac{1}{3}\right\}, \left\{\frac{1}{2}, \frac{1}{3}, \frac{1}{4}\right\}, \left\{\frac{1}{3}, \frac{1}{4}, \frac{1}{5}\right\}\right\}$$

Here is matrix with general elements:

```
Array[f, {2, 2}]
```

```
{{f[1, 1], f[1, 2]}, {f[2, 1], f[2, 2]}}
```

Here is a zero matrix in three ways:

```
ConstantArray[0, {3, 3}]      {{0, 0, 0}, {0, 0, 0}, {0, 0, 0}}
```

```
Array[0 &, {3, 3}]       {{0, 0, 0}, {0, 0, 0}, {0, 0, 0}}
```

```
Table[0, {3}, {3}]       {{0, 0, 0}, {0, 0, 0}, {0, 0, 0}}
```

Other examples:

```
IdentityMatrix[3]      {{1, 0, 0}, {0, 1, 0}, {0, 0, 1}}
```

```
DiagonalMatrix[{a, b, c}]      {{a, 0, 0}, {0, b, 0}, {0, 0, c}}
```

```
Partition[Range[10], 3]      {{1, 2, 3}, {4, 5, 6}, {7, 8, 9}}
```

Next, we use **If**, **Which**, and **Switch** (see Section 18.2.2, p. 556):

```
Table[If[i ≤ j, i + j, 0], {i, 3}, {j, 3}] // MatrixForm
```

$$\begin{pmatrix} 2 & 3 & 4 \\ 0 & 4 & 5 \\ 0 & 0 & 6 \end{pmatrix}$$

```
Table[Which[i > j, 0, i == j, 1, i == j - 1, 2, i ≤ j - 2, 3], {i, 3}, {j, 3}] // MatrixForm
```

$$\begin{pmatrix} 1 & 2 & 3 \\ 0 & 1 & 2 \\ 0 & 0 & 1 \end{pmatrix}$$

```
Table[Switch[i - j, -1, a, 0, b, 1, c, _, 0], {i, 3}, {j, 3}] // MatrixForm
```

$$\begin{pmatrix} b & a & 0 \\ c & b & a \\ 0 & c & b \end{pmatrix}$$

Here is a Vandermonde matrix:

```
Outer[Power, {x, y, z}, Range[0, 2]] // MatrixForm
```

$$\begin{pmatrix} 1 & x & x^2 \\ 1 & y & y^2 \\ 1 & z & z^2 \end{pmatrix}$$

■ Generating Sparse Matrices

In sparse vectors or matrices, we typically have few nonzero elements compared with the number of zero elements. Using **SparseArray**, we can specify only the nonzero elements.

SparseArray[rules, dims, default] From **rules**, create a sparse array that has dimensions **dims** and takes unspecified elements to be **default**

SparseArray also accepts the shorter form **SparseArray[rules, dims]**, and then unspecified elements are taken to be 0. A still shorter form is **SparseArray[rules]**, and then the dimensions are exactly large enough to include all elements with explicitly specified positions.

The **rules** can be of the following forms:

$\{pos_1 \to val_1,\ pos_2 \to val_2,\ ...\}$

$\{pos_1,\ pos_2,\ ...\} \to \{val_1,\ val_2,\ ...\}$

$\{pos_1,\ pos_2,\ ...\} \to val$

In this way, we specify the positions and values of some elements; elements in other positions are set to **default**. The position specifications can contain patterns (see Chapter 16). Dimension **dims** is, for example, of the form $\{d_1\}$ for vectors and of the form $\{d_1, d_2\}$ for matrices.

As an example, we generate a 3×3 diagonal matrix with 5 as the diagonal element. The following is the easiest way:

```
s0 = DiagonalMatrix[{3, 3, 3}]
{{3, 0, 0}, {0, 3, 0}, {0, 0, 3}}
```

However, next we form the same matrix as a sparse matrix, in four different ways:

```
s1 = SparseArray[{{1, 1} → 5, {2, 2} → 5, {3, 3} → 5}]
SparseArray[<3>, {3, 3}]

s2 = SparseArray[{{1, 1}, {2, 2}, {3, 3}} → {5, 5, 5}];

s3 = SparseArray[{{1, 1}, {2, 2}, {3, 3}} → 5];

s4 = SparseArray[{{i_, i_} → 5}, {3, 3}];
```

The result is a **SparseArray** object. We only see the number of rules (three here) and the dimensions of the matrix ({3, 3} here). With **Normal**, we can see the sparse matrix in the usual list form and with **MatrixForm** in a two-dimensional form:

```
s1 // Normal    {{5, 0, 0}, {0, 5, 0}, {0, 0, 5}}
```

```
s1 // MatrixForm
```
$$\begin{pmatrix} 5 & 0 & 0 \\ 0 & 5 & 0 \\ 0 & 0 & 5 \end{pmatrix}$$

Normal[sparseArray] Create a list version of **sparseArray**

SparseArray[list] Create a sparse array version of **list**

ArrayRules[sparseArray] Give the rules of **sparseArray**

Normal converts a sparse array into a usual list. The inverse is done with **SparseArray**: It converts usual lists into sparse arrays. With **ArrayRules**, we get the list of rules that specify the elements of a sparse array. For example, we can transform the matrix **s0** calculated previously into a sparse matrix:

SparseArray[s0]   SparseArray["<"3">", {3, 3}]

ArrayRules[%]   {{1, 1} → 3, {2, 2} → 3, {3, 3} → 3, {_, _} → 0}

We can calculate with sparse matrices and vectors in a similar manner as we calculate with normal matrices and vectors. In general, matrix calculus such as **Eigenvalues**, **LinearSolve**, and **LinearProgramming** work with sparse matrices as they do with normal matrices.

■ Generating Band Matrices

Sparse matrices often consist of bands along the main diagonal and along diagonals above and beneath it. With **Band**, we can easily construct such sparse matrices.

Band[start] (※6)   Represents, in a sparse matrix, the sequence of positions on the diagonal band
  that starts at position **start** (**start** is of the form {*m, n*})
Band[start, end]   The band ends at position **end**
Band[start, end, step]   The band moves with **step**

Band[start] → a   All elements in the band **Band[start]** are equal to **a**
Band[start] → {a₁, a₂, … }   The elements in the band are equal to $a_1, a_2, \ldots$
Band[start] → {m₁, m₂, … }   The band contains a sequence of matrices $m_1, m_2, \ldots$

The starting and ending positions of the band can be anywhere in the matrix. Here are some examples of band matrices:

```
MatrixForm /@ {SparseArray[Band[{1, 1}] → 1, {4, 4}],
   SparseArray[Band[{1, 1}, {3, 3}] → 1, {4, 4}],
   SparseArray[Band[{1, 1}, {4, 4}, 2] → 1],
   SparseArray[Band[{2, 1}, {4, 4}, {0, 1}] → 1]}
```

$$\left\{ \begin{pmatrix} 1 & 0 & 0 & 0 \\ 0 & 1 & 0 & 0 \\ 0 & 0 & 1 & 0 \\ 0 & 0 & 0 & 1 \end{pmatrix}, \begin{pmatrix} 1 & 0 & 0 & 0 \\ 0 & 1 & 0 & 0 \\ 0 & 0 & 1 & 0 \\ 0 & 0 & 0 & 0 \end{pmatrix}, \begin{pmatrix} 1 & 0 & 0 & 0 \\ 0 & 0 & 0 & 0 \\ 0 & 0 & 1 & 0 \\ 0 & 0 & 0 & 0 \end{pmatrix}, \begin{pmatrix} 0 & 0 & 0 & 0 \\ 1 & 1 & 1 & 1 \\ 0 & 0 & 0 & 0 \\ 0 & 0 & 0 & 0 \end{pmatrix} \right\}$$

```
m1 = {{a, b}, {c, d}};
m2 = {{e, f}, {g, h}};
MatrixForm /@ {SparseArray[
   {Band[{1, 1}] → a, Band[{2, 1}] → b, Band[{1, 2}] → c, {1, 4} → d}, {4, 4}],
   SparseArray[Band[{2, 2}] → {a, b, c}, {4, 4}],
   SparseArray[Band[{2, 2}] → {{a, b}, {c, d}}, {4, 4}],
   SparseArray[Band[{1, 1}] → {m1, m2}, {5, 5}]}
```

$$\left\{ \begin{pmatrix} a & c & 0 & d \\ b & a & c & 0 \\ 0 & b & a & c \\ 0 & 0 & b & a \end{pmatrix}, \begin{pmatrix} 0 & 0 & 0 & 0 \\ 0 & a & 0 & 0 \\ 0 & 0 & b & 0 \\ 0 & 0 & 0 & c \end{pmatrix}, \begin{pmatrix} 0 & 0 & 0 & 0 \\ 0 & a & b & 0 \\ 0 & c & d & 0 \\ 0 & 0 & 0 & 0 \end{pmatrix}, \begin{pmatrix} a & b & 0 & 0 & 0 \\ c & d & 0 & 0 & 0 \\ 0 & 0 & e & f & 0 \\ 0 & 0 & g & h & 0 \\ 0 & 0 & 0 & 0 & 0 \end{pmatrix} \right\}$$

■ Plotting Matrices

We have two commands to plot matrices:

ArrayPlot[m]   Plot the matrix by showing zero elements in white, the element(s) with the maximum absolute value in black, and other elements in various gray levels

MatrixPlot[m] (✿6)   Plot the matrix by showing zero elements in white, the element(s) with the maximum positive value in orange, the element(s) with the maximum negative value in blue, and other elements in lighter orange and lighter blue

---

The following plots show the coloring schemes of the two commands:

(m = {Range[0, 5], Range[-5, 0]}) // MatrixForm

$$\begin{pmatrix} 0 & 1 & 2 & 3 & 4 & 5 \\ -5 & -4 & -3 & -2 & -1 & 0 \end{pmatrix}$$

{ArrayPlot[m], MatrixPlot[m]}

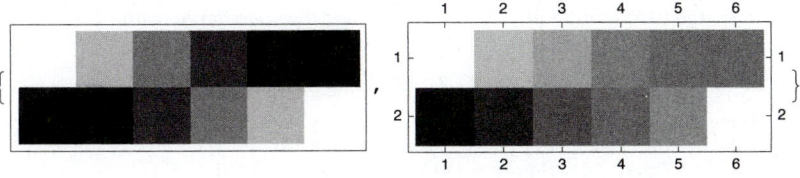

Both commands plot the rows of the matrix down the page: The first row is on the top.

The commands have all the options of **Graphics** and, in addition, the following options:

Options of **ArrayPlot** *and* **MatrixPlot**:

Mesh   Whether to draw a mesh; examples of values: **False, True, All, {5, 10}** (5 horizontal and 10 vertical lines)

MeshStyle   The style of the mesh; examples of values: **GrayLevel[GoldenRatio - 1], {{Blue, Thick}}**

ColorFunction   How the cells are colored; examples of values: **Automatic, Hue, (If[# == 0, White, Black] &)**

ColorFunctionScaling   Whether to scale the values of the elements into [0, 1]; possible values: **True, False**

ColorRules   Rules for determining colors from the values of elements; examples of values: **Automatic, {0 → White, _ → Black}, {1|-1 → Red, x_ /; x < -1 → Blue, x_ /; x > 1 → Green, _ → White}**

MaxPlotPoints   The maximum number of elements explicitly plotted; examples of values: ∞ (the default for **ArrayPlot**), **Automatic** (the default for **MatrixPlot**)

PixelConstrained   How to constrain cells to align with screen pixels; examples of values: **False, True, 1** (each cell is one pixel)

PlotRangeClipping   Whether to clip at the plot range; possible values: **True, False**

ClippingStyle   The color of cells whose values are clipped; examples of values: **None, Red, {Blue, Red}**

DataRange   The range of x and y values to assume; examples of values: **All, {{a, b}, {c, d}}** (the centers of successive cells should be at equally spaced positions between a and b in the x direction and between c and d in the y direction; the first item is centered at (a, d))

DataReversed   Whether to reverse the order of rows; possible values: **False, True**

In **ColorRules**, we can use patterns such as _ (meaning anything), **1|-1** (meaning 1 or -1), or **x_ /; x < -1** (meaning numbers that are < -1); for patterns, see Chapter 16.

For options other than those mentioned in the previous box, note the following.

The default value of **AspectRatio** is **Automatic**.

The default value of **Frame** is **Automatic** for **ArrayPlot** and **True** for **MatrixPlot**. For **ArrayPlot**, a frame is normally drawn, but if **Mesh** is **True**, the default is that a frame is not drawn.

The default value of **FrameTicks** is **None** for **ArrayPlot** and **All** for **MatrixPlot**. If **FrameTicks** is **Automatic**, then ticks are placed at round integers, but if **FrameTicks** is **All**, then ticks are also placed at the minimum and maximum row and column index.

The default value of **PlotRange** is **All**. For this option, similar values can be given as for **DensityPlot**; a value such as **s** plots only elements that are in [0, **s**]. For **ArrayPlot**, if an explicit value is given for **PlotRange**, then the minimum value is white and the maximum value black (e.g., the coloring is not based on absolute values).

```
m = {{0, 0, 1, 1, 0}, {1, 0, 1, 2, 1}, {2, 1, 2, 0, 1}, {2, 1, 1, 3, 2}};
```

```
GraphicsRow[{ArrayPlot[m, Mesh → All],
  ArrayPlot[m, ColorFunction → (If[# == 0, White, Black] &),
    Mesh → All, Frame → True, FrameTicks → {All, None, None, All}],
  ArrayPlot[m, ColorRules → {0 → White, _ → Black}, Mesh → All],
  ArrayPlot[SparseArray[{Band[{1, 1}] → 2, Band[{6, 1}] → 1, Band[{1, 6}] → 1,
    {40, 40}}]]}, ImageSize → 420]
```

■ Tensors

A tensor of rank *k* is a *k*-dimensional table of values; see tutorial/Tensors. Tensors of rank 0, 1, and 2 are scalars, vectors, and matrices, respectively. Tensors can be generated with **Table** or **Array**. With **ArrayDepth** we can obtain the rank of a tensor and with **Dimensions** a list of the dimensions.

21.2.2 Manipulating Matrices

■ Manipulating Rows

As matrices are lists, we can use all the list manipulation commands presented in Chapter 14. Elements and rows of matrices can be taken as follows:

---

m[[i, j]]   Take element (**i, j**)
m[[i, j]] = a   Set element (**i, j**) to **a**

m[[i]]   Take row **i**
m[[i]] = r   Set row **i** to vector **r**
m[[i ;; j]]   Take rows **i** through **j**
m[[{i, j, ...}]]   Take rows **i, j, ...**

Recall that ⟦ and ⟧ can be written as ESC [[ESC and ESC]] ESC; we can also use the character sequences [[and]].

Here are special commands for taking or dropping rows:

First[m], Rest[m] Take/drop the first row

Last[m], Most[m] Take/drop the last row

Take[m, i], Drop[m, i] Take/drop the first **i** rows

Take[m, -i], Drop[m, -i] Take/drop the last **i** rows

Take[m, {i}], Drop[m, {i}] Take/drop row **i**

Take[m, {i, j}], Drop[m, {i, j}] Take/drop rows **i**, ..., **j**

Take[m, {i, j, d}], Drop[m, {i, j, d}] Take/drop rows in steps of **d**

Next, we show commands for inserting, deleting, and replacing rows.

Prepend[m, r] or **Join[{r}, m]** Insert row **r** at the top of **m**

Append[m, r] or **Join[m, {r}]** Insert row **r** at the bottom of **m**

Insert[m, r, i] Insert row **r** between rows **i** - 1 and **i**

ReplacePart[m, r, i] Replace **i**th row with row **r**

Delete[m, i], Delete[m, -i] Delete the **i**th row, counting from the beginning/end

Delete[m, {{i}, {j}, … }] Delete rows **i**, **j**, ...

As an example, consider the matrix

```
MatrixForm[m = {{1, 2, 3}, {p, q, r}, {P, Q, R}}]
```

$$\begin{pmatrix} 1 & 2 & 3 \\ p & q & r \\ P & Q & R \end{pmatrix}$$

We arrange the rows in another order, take rows 2 to 3, and append a row at the bottom of the matrix:

```
m1 = m[[{2, 1, 3}]];
m2 = m[[2 ;; 3]];
m3 = Append[m, {α, β, γ}];
Map[MatrixForm, {m1, m2, m3}]
```

$$\left\{ \begin{pmatrix} p & q & r \\ 1 & 2 & 3 \\ P & Q & R \end{pmatrix}, \begin{pmatrix} p & q & r \\ P & Q & R \end{pmatrix}, \begin{pmatrix} 1 & 2 & 3 \\ p & q & r \\ P & Q & R \\ α & β & γ \end{pmatrix} \right\}$$

Map[f[#]&, m] or **f[#]& /@ m** Map each row with the pure function **f**

With **Map** we can easily do various manipulations for each row. As examples, we reverse the order of the elements of each row of **m**, pick the second element from each row (i.e., pick the second column), and calculate the sum of the elements of each row:

Reverse /@ m   {{3, 2, 1}, {r, q, p}, {R, Q, P}}

#[[2]] & /@ m   {2, q, Q}

Total /@ m   {6, p + q + r, P + Q + R}

■ Manipulating Columns and Diagonals

In the following, when we need the transpose of a matrix **m** or `Transpose[m]`, we use the short notation m^T. Here, ^{T} can be written as ⎡ESC⎤`tr`⎡ESC⎤.

`m[[All, i]]` or m^T`[[i]]` Take column **i**

`m[[All, i]] = c` Set the **i**th column to vector **c**

`m[[All, i ;; j]]` Take columns **i** through **j**

`m[[All, {i, j, … }]]` Take columns **i, j,** ...

`Take[m, All, i]` or `Take[`m^T`, i]`^{T} Take the first **i** columns

`Take[m, All, {i}]` or `Take[`m^T`, {i}]`^{T} Take column **i**

`Drop[m, None, i]` or `Drop[`m^T`, i]`^{T} Drop the first **i** columns

`Drop[m, None, {i}]` or `Drop[`m^T`, {i}]`^{T} Drop column **i**

`Prepend[`m^T`, c]`^{T} Insert column **c** at the left of **m**

`Append[`m^T`, c]`^{T} Insert column **c** at the right of **m**

As an example, consider the same **m** matrix discussed previously. We arrange the columns in another order, take columns 2 to 3, and append a column at the left of the matrix:

```
m4 = m[[All, {2, 1, 3}]];
m5 = m[[All, 2 ;; 3]];
m6 = Append[m , {4, s, S}] ;
MatrixForm /@ {m4, m5, m6}
```

$$\left\{ \begin{pmatrix} 2 & 1 & 3 \\ q & p & r \\ Q & P & R \end{pmatrix}, \begin{pmatrix} 2 & 3 \\ q & r \\ Q & R \end{pmatrix}, \begin{pmatrix} 1 & 2 & 3 & 4 \\ p & q & r & s \\ P & Q & R & S \end{pmatrix} \right\}$$

`Diagonal[m]` (❀6) Give the elements on the leading diagonal

`Diagonal[m, k]` Give the elements on the **k**th diagonal above the leading diagonal

`Diagonal[m, -k]` Give the elements on the **k**th diagonal below the leading diagonal

---

In `Diagonal`, the matrix need not be square:

```
{Diagonal[m6], Diagonal[m6, 1], Diagonal[m6, -1]}
```

```
{{1, q, R}, {2, r, S}, {p, Q}}
```

■ Taking Submatrices

`m[[{`i_1`, `i_2`, … }, {`j_1`, `j_2`, … }]]` Take the submatrix that has elements with the given row and column indices

`m[[`i_1`;; `i_2`, `j_1`;; `j_2`]]` or `Take[m, {`i_1`, `i_2`}, {`j_1`, `j_2`}]` Take the submatrix that has elements with row and column indices in the given ranges

`Drop[m, {`i_1`, `i_2`}, {`j_1`, `j_2`}]` Drop rows i_1 to i_2 and columns j_1 to j_2

`Partition[m, {i, j}]` Partition a matrix into blocks of size (**i**×**j**)

Consider matrix **m6** constructed previously:

```
m6 // MatrixForm
```

$$\begin{pmatrix} 1 & 2 & 3 & 4 \\ p & q & r & s \\ P & Q & R & S \end{pmatrix}$$

We do the following:

- We take the submatrix with row indices 1 and 3 and column indices 2 and 4.
- We take the submatrix with row indices 1, 2, and 3 and column indices 2, 3, and 4.
- We drop row 3 and columns 1 and 2.
- We partition **m6** to two matrices of the size (3, 2) (note that in partitioning, any leftover elements are dropped).

```
m7 = m6[[{1, 3}, {2, 4}]];
m8 = m6[[1 ;; 3, 2 ;; 4]];
m9 = Drop[m6, {3}, {1, 2}];
m10 = Partition[m6, {3, 2}];
MatrixForm /@ {m7, m8, m9, m10}
```

$$\left\{ \begin{pmatrix} 2 & 4 \\ Q & S \end{pmatrix}, \begin{pmatrix} 2 & 3 & 4 \\ q & r & s \\ Q & R & S \end{pmatrix}, \begin{pmatrix} 3 & 4 \\ r & s \end{pmatrix}, \left(\begin{pmatrix} 1 & 2 \\ p & q \\ P & Q \end{pmatrix} \begin{pmatrix} 3 & 4 \\ r & s \\ R & S \end{pmatrix} \right) \right\}$$

■ Combining and Extending Matrices

With **ArrayFlatten** and **Join** we can construct a single matrix from some blocks of matrices. In **ArrayFlatten**, the argument is a matrix of matrices.

ArrayFlatten[{{m}, {n}}] (❀6) or **Join[m, n]**   Create the matrix $\begin{pmatrix} m \\ n \end{pmatrix}$

ArrayFlatten[{{m, n}}] or **Join[m, n, 2]** or **Join[mᵀ, nᵀ]ᵀ**   Create the matrix (m   n)

ArrayFlatten[{{m, n}, {p, q}}]   Create the matrix $\begin{pmatrix} m & n \\ p & q \end{pmatrix}$

ArrayFlatten[{{m, n, … }, {p, q, … }, … }]   Create a matrix from matrices **m, n, p, q,** …

PadRight[m, {n₁, n₂}]   Extend **m** with zeros to an $n_1 \times n_2$ matrix
PadRight[m, {n₁, n₂}, a]   Extend **m** with replicates of matrix **a** to an $n_1 \times n_2$ matrix

---

```
m = {{1, 2}, {3, 4}}; n = {{a, b}, {c, d}};
o = {{0, 0}, {0, 0}}; p = {{S, T}, {U, V}};
m11 = Join[m, n];
m12 = Join[m, n, 2];
m13 = ArrayFlatten[{{m, n}, {o, p}}];
MatrixForm /@ {m11, m12, m13}
```

$$\left\{ \begin{pmatrix} 1 & 2 \\ 3 & 4 \\ a & b \\ c & d \end{pmatrix}, \begin{pmatrix} 1 & 2 & a & b \\ 3 & 4 & c & d \end{pmatrix}, \begin{pmatrix} 1 & 2 & a & b \\ 3 & 4 & c & d \\ 0 & 0 & S & T \\ 0 & 0 & U & V \end{pmatrix} \right\}$$

```
MatrixForm /@ {m, PadRight[m, {2, 4}], PadRight[m, {4, 2}], PadRight[m, {4, 4}]}
```

$$\left\{ \begin{pmatrix} 1 & 2 \\ 3 & 4 \end{pmatrix}, \begin{pmatrix} 1 & 2 & 0 & 0 \\ 3 & 4 & 0 & 0 \end{pmatrix}, \begin{pmatrix} 1 & 2 \\ 3 & 4 \\ 0 & 0 \\ 0 & 0 \end{pmatrix}, \begin{pmatrix} 1 & 2 & 0 & 0 \\ 3 & 4 & 0 & 0 \\ 0 & 0 & 0 & 0 \\ 0 & 0 & 0 & 0 \end{pmatrix} \right\}$$

■ **Reordering Rows and Columns**

`Transpose[m]`   Interchange rows and columns

`Reverse[m]`   Reverse the order of the rows

`Reverse[mᵀ]ᵀ` or `Reverse[m, 2]`   Reverse the order of the columns

`RotateLeft[m]`   Move the first row to be the last row

`RotateLeft[mᵀ]ᵀ`   Move the first column to be the last column

`RotateRight[m]`   Move the last row to be the first row

`RotateRight[mᵀ]ᵀ`   Move the last column to be the first column

21.2.3 Matrix Calculus

■ **Properties of Matrices**

`Length[m]`   The number of rows of matrix **m**

`Dimensions[m]`   The dimensions (numbers of rows and columns) of **m**

`MatrixQ[m]`   Test whether **m** is a matrix

`MatrixQ[v, test]`   Test whether **v** is a matrix with elements that satisfy `test`

`HermitianMatrixQ[m]` (✤6)   Test whether **m** is Hermitian (i.e., whether the conjugate transpose of
   the matrix is the same as the matrix; if the matrix is real, Hermitian means the same as symmetric)

`PositiveDefiniteMatrixQ[m]` (✤6)   Test whether **m** is positive definite

```
m = {{2, 7, 5}, {8, 1, 6}};

{Length[m], Dimensions[m]}        {2, {2, 3}}

{MatrixQ[{{4, 3}, {2, 7}}], MatrixQ[{4, 3}], MatrixQ[{{{4, 3}}}]}
{True, False, True}

MatrixQ[{{4, 3}, {2, 7}}, NumericQ]       True
```

We considered tests in more detail in Section 13.3.5, p. 431.

```
m = {{0.64, 0.61, 0.66}, {0.61, 0.76, 0.69}, {0.66, 0.69, 0.80}};

{HermitianMatrixQ[m], PositiveDefiniteMatrixQ[m]}        {True, True}
```

The positive definiteness can also be seen from the fact that all eigenvalues are positive:

```
Eigenvalues[m]        {2.04494, 0.0999083, 0.0551525}
```

■ **Arithmetic with Matrices**

Arithmetic with matrices is easy because *Mathematica* automatically does all operations element by element.

`a + m`   Add scalar **a** to each element of matrix **m**

`a m`   Multiply each element of **m** with scalar **a**

`1/m`   Calculate the reciprocal *of each element* of **m** (use `Inverse[m]` for a matrix inverse)

`m^a`   Calculate the ath power (**a** is a scalar) *of each element* of **m** (use `MatrixPower[m, a]` for a matrix
   power)

`a^m`   Calculate the powers of scalar **a** that are given in matrix **m**

```
n = {{a, b}, {c, d}};

MatrixForm /@ {3 + n, 5 n, n^-1, n^2, 2^n}
```

$$\left\{ \begin{pmatrix} 3+a & 3+b \\ 3+c & 3+d \end{pmatrix}, \begin{pmatrix} 5a & 5b \\ 5c & 5d \end{pmatrix}, \begin{pmatrix} \frac{1}{a} & \frac{1}{b} \\ \frac{1}{c} & \frac{1}{d} \end{pmatrix}, \begin{pmatrix} a^2 & b^2 \\ c^2 & d^2 \end{pmatrix}, \begin{pmatrix} 2^a & 2^b \\ 2^c & 2^d \end{pmatrix} \right\}$$

m + n	Add two matrices **m** and **n**
m n	Multiply *the corresponding elements* of **m** and **n** (use **m.n** for a matrix product)
m/n	Divide *the corresponding elements* of **m** and **n**
m^n	Calculate powers of *the corresponding elements* of **m** and **n**

```
m = {{1, 2}, {3, 4}};

MatrixForm /@ {m + n, m n, m / n, m^n}
```

$$\left\{ \begin{pmatrix} 1+a & 2+b \\ 3+c & 4+d \end{pmatrix}, \begin{pmatrix} a & 2b \\ 3c & 4d \end{pmatrix}, \begin{pmatrix} \frac{1}{a} & \frac{2}{b} \\ \frac{3}{c} & \frac{4}{d} \end{pmatrix}, \begin{pmatrix} 1 & 2^b \\ 3^c & 4^d \end{pmatrix} \right\}$$

As indicated previously, to get a true matrix inverse, power, or product, we cannot write **m^-1**, **m^a**, or **m n** but, rather, we have to write **Inverse[m]**, **MatrixPower[m, a]**, and **m.m**, respectively.

■ **Transpose and Trace**

Transpose[m]	Transpose
ConjugateTranspose[m]	Conjugate transpose
Tr[m]	Trace (the sum of diagonal elements)
Tr[m, List] or **Diagonal[m]** (✻6)	List of diagonal elements
Tr[m, Times]	Product of diagonal elements

Transpose[m] can also be written as **m$^\mathsf{T}$**, where $^\mathsf{T}$ can be written as ⎡ESC⎤**tr**⎡ESC⎤. Similarly, **ConjugateTranspose[m]** can also be written as **m$^\mathsf{t}$**, where $^\mathsf{t}$ can be written as ⎡ESC⎤**ct**⎡ESC⎤. **ConjugateTranspose[m]** is equivalent to **Conjugate[Transpose[m]]**.

```
m = {{2 + I, 5}, {-I, 5 - 3 I}};

MatrixForm /@ {m, m†}
```

$$\left\{ \begin{pmatrix} 2+i & 5 \\ -i & 5-3i \end{pmatrix}, \begin{pmatrix} 2-i & i \\ 5 & 5+3i \end{pmatrix} \right\}$$

For a triangular matrix, the product of diagonal elements is the same as the determinant:

```
m = {{a, 0, 0}, {b, c, 0}, {d, e, f}};

{Tr[m, Times], Det[m]}          {a c f, a c f}
```

■ **Sums, Minimums, and Maximums**

Total[list]	The sum(s) of the elements of **list** that are at level 1
Total[list, {n}]	The sum(s) of the elements of **list** that are at level **n**
Total[list, n]	The sum(s) of elements of **list** that are at levels 1 through **n**
Total[list, {m, n}]	The sum(s) of the elements of **list** that are at levels **m** through **n**

As special cases of these general commands for lists, we get the following commands for matrices:

> **Total[m, {1}]** or **Total[m]** The sum of the rows, giving the vector of column sums
> **Total[m, {2}]** or **Total /@ m** or **Total[mᵀ]** The sums of the elements of each row, giving the vector of row sums
> **Total[m, {1, 2}]** or **Total[m, 2]** or **Total[Flatten[m]]** The sum of all elements

Consider the following matrix:

```
(m = {{1, 2, 3}, {a, b, c}, {A, B, C}}) // MatrixForm
```

$$\begin{pmatrix} 1 & 2 & 3 \\ a & b & c \\ A & B & C \end{pmatrix}$$

Calculate the column sums, the row sums, and the overall sum:

> **Total[m]**   {1 + a + A, 2 + b + B, 3 + c + C}
>
> **Total[mᵀ]**   {6, a + b + c, A + B + C}
>
> **Total[m, 2]**   6 + a + A + b + B + c + C

For sums in which numerical errors cause problems, use **Total** with the option **Method → CompensatedSummation** to reduce the error.

Here are ways to calculate, with **Min**, various minimums for matrices; maximums can be calculated with **Max**:

> **Min /@ m** The smallest element of each row
> **Min /@ mᵀ** The smallest element of each column
> **Min[m]** The smallest element of the matrix

■ Differences

> **Differences[m]** (❀6) Differences of rows
> **Differences[m, i]** ith differences of rows
> **Differences[m, {0, j}]** jth differences of columns
> **Differences[m, {i, j}]** ith differences of rows and jth differences of columns

```
(m = {{1, 2, 3}, {a, b, c}, {A, B, C}}) // MatrixForm
```

$$\begin{pmatrix} 1 & 2 & 3 \\ a & b & c \\ A & B & C \end{pmatrix}$$

```
MatrixForm /@ {Differences[m], Differences[m, {0, 1}]}
```

$$\left\{ \begin{pmatrix} -1+a & -2+b & -3+c \\ -a+A & -b+B & -c+C \end{pmatrix}, \begin{pmatrix} 1 & 1 \\ -a+b & -b+c \\ -A+B & -B+C \end{pmatrix} \right\}$$

■ Determinant and Minors

> **Det[m]** Determinant of a square matrix
> **Minors[m]** Minors of a square matrix
> **Minors[m, k]** kth minors

Calculate the determinant of a Hilbert matrix by using exact arithmetic, usual decimal numbers, and high-precision decimal numbers:

```
m = HilbertMatrix[6];
```

```
{Det[m], Det[m // N], Det[N[m, 20]]}
```

$$\left\{ \frac{1}{186\,313\,420\,339\,200\,000}, \; 5.3673 \times 10^{-18}, \; 5.36729988736 \times 10^{-18} \right\}$$

According to Cramer's rule, the solution of a linear equation $A\,x = b$ is $x_i = |A(i/b)| \,/\, |A|$, where $|\;|$ means determinant and $A(i/b)$ is matrix A with column i replaced with b. The following program is from the Documentation Center.

```
cramersRule[a_, b_] := Module[{m, d = Det[a]},
  Table[m = a; m[[All, i]] = b; Det[m] / d, {i, Length[a]}]]
```

```
cramersRule[{{1, 2}, {2, -1}}, {7, 4}]     {3, 2}
```

```
LinearSolve[{{1, 2}, {2, -1}}, {7, 4}]     {3, 2}
```

■ Product, Power, and Exponential

> **m.n** Product of matrices
> **MatrixPower[m, a]** The ath power of a square matrix (**a** can also be negative)
> **MatrixExp[m]** Matrix exponential of a square matrix

Note that if we want the product of matrices **m** and **n** but write **m n**, the product is formed element by element, and this is probably not the product we want. The normal matrix product is formed with the use of the dot:

$$m = \begin{pmatrix} 1 & 2 \\ 3 & 4 \end{pmatrix}; \; n = \begin{pmatrix} a & b \\ c & d \end{pmatrix};$$

```
m.n // MatrixForm
```
$$\begin{pmatrix} a + 2\,c & b + 2\,d \\ 3\,a + 4\,c & 3\,b + 4\,d \end{pmatrix}$$

Multiplication with a vector, from left or right, is also formed with the dot:

```
m.{x, y}     {x + 2 y, 3 x + 4 y}
```

```
{x, y}.m     {x + 3 y, 2 x + 4 y}
```

The Kronecker product (or outer product) of two vectors must instead be calculated with a special command:

```
KroneckerProduct[{a , b, c}, {x, y, z}]
```
```
{{a x, a y, a z}, {b x, b y, b z}, {c x, c y, c z}}
```

Powers of matrices such as **m^3** are formed element by element, and this is again something we most likely do not want. If we want to calculate the true matrix power, we write **m.m.m** or **MatrixPower[m, 3]**:

```
m.m.m     {{37, 54}, {81, 118}}
```

Calculate the square root of **m**:

```
MatrixPower[m, 0.5]
```
```
{{0.553689 + 0.464394 i, 0.806961 - 0.212426 i},
 {1.21044 - 0.31864 i, 1.76413 + 0.145754 i}}
```
```
%.% // Chop     {{1., 2.}, {3., 4.}}
```

Next, we calculate a matrix exponential $e^{\mathbf{m}} = \sum\limits_{i=0}^{\infty} \frac{1}{i!}\,\mathbf{m}^i$:

```
MatrixExp[{{0, 1}, {-1, 0}} t]
```

```
{{Cos[t], Sin[t]}, {-Sin[t], Cos[t]}}
```

■ **Inverse and Pseudoinverse**

Inverse[m]	Inverse of a square matrix
PseudoInverse[m]	Pseudoinverse (of a possibly rectangular matrix)

Note that **m^-1** only inverts each element. Use **Inverse** to calculate the true matrix inverse:

Inverse[{{3, 2}, {4, 1}}] $\left\{\left\{-\frac{1}{5}, \frac{2}{5}\right\}, \left\{\frac{4}{5}, -\frac{3}{5}\right\}\right\}$

Matrix inversion is numerically an exacting and risky task. Avoid it whenever possible. Often, we do not really need an inverse because the problem can often be restated in a form in which the problem is to solve a linear system of equations.

Hilbert matrices are examples of matrices that behave badly, but no problems arise if we calculate with exact numbers:

h[n_] := HilbertMatrix[n]

(h[5].Inverse[h[5]])⟦2⟧   {0, 1, 0, 0, 0}

The result is as it should be. With decimal numbers, however, the result is not so good:

(N[h[5]].Inverse[N[h[5]]])⟦2⟧

$\left\{-3.13731 \times 10^{-13}, 1., 4.32606 \times 10^{-12}, -2.82196 \times 10^{-12}, 2.92681 \times 10^{-12}\right\}$

We can increase the precision of the calculations; here we use 20-digit precision, which gives a good result again:

(N[h[5], 22].Inverse[N[h[5], 20]])⟦2⟧

$\left\{0. \times 10^{-18}, 1.000000000000000, 0. \times 10^{-16}, 0. \times 10^{-16}, 0. \times 10^{-16}\right\}$

Even if the normal inverse does not exist, we can calculate a pseudoinverse:

c = {{3, 2}, {6, 4}};

Inverse[c]

Inverse::sing : Matrix {{3, 2}, {6, 4}} is singular. ≫
Inverse[{{3, 2}, {6, 4}}]

p = PseudoInverse[c] $\left\{\left\{\frac{3}{65}, \frac{6}{65}\right\}, \left\{\frac{2}{65}, \frac{4}{65}\right\}\right\}$

The pseudoinverse satisfies the following four identities:

$\left\{\text{c.p.c} == \text{c}, \text{p.c.p} == \text{p}, (\text{c.p})^{\top} == \text{c.p}, (\text{p.c})^{\top} == \text{p.c}\right\}$
{True, True, True, True}

The pseudoinverse is calculated with singular value decomposition. For a nonsingular square matrix, the pseudoinverse is the same as the usual inverse. For the option **Tolerance**, see **SingularValueList** in the next subsection. **PseudoInverse** is also considered in Section 22.1.2, p. 714.

■ Singular Values

SingularValueList[m] Singular values (of a possibly rectangular matrix)

SingularValueList[m, k] The **k** largest singular values

SingularValueList[{m, a}] Generalized singular values of **m** with respect to matrix **a**

SingularValueDecomposition[m] Singular value decomposition

Singular values are listed from largest to smallest:

```
m = {{3, 2}, {4, 1}};
```

```
SingularValueList[m]
```

$$\left\{ \sqrt{5\left(3 + 2\sqrt{2}\right)} , \sqrt{5\left(3 - 2\sqrt{2}\right)} \right\}$$

Singular values can also be calculated as follows:

```
Sqrt[Eigenvalues[m.m']]
```

$$\left\{ \sqrt{5\left(3 + 2\sqrt{2}\right)} , \sqrt{5\left(3 - 2\sqrt{2}\right)} \right\}$$

The singular values are sorted from the largest to the smallest (repeated singular values appear with their appropriate multiplicity). Only singular values considered to be nonzero are listed. We calculate the 12 singular values of the 12th Hilbert matrix and show their decimal values:

```
SingularValueList[HilbertMatrix[12]] // N
```

$\{1.79537, 0.380275, 0.0447385, 0.00372231, 0.000233089, 0.0000111634, 4.08238 \times 10^{-7},$
$1.12286 \times 10^{-8}, 2.25196 \times 10^{-10}, 3.11135 \times 10^{-12}, 2.64902 \times 10^{-14}, 1.04795 \times 10^{-16}\}$

If we calculate the singular values from the numerical Hilbert matrix, we get only 10 singular values:

```
SingularValueList[HilbertMatrix[12] // N]
```

$\{1.79537, 0.380275, 0.0447385, 0.00372231, 0.000233089,$
$0.0000111634, 4.08238 \times 10^{-7}, 1.12286 \times 10^{-8}, 2.25196 \times 10^{-10}, 3.11135 \times 10^{-12}\}$

The last two singular values are under a tolerance value and are considered to be zero. However, we can use an option to define a new tolerance value.

Tolerance An option for **SingularValueList** as well as for **SingularValueDecomposition** and **PseudoInverse**, to mention a few. The default value **Automatic** means that only singular values larger than 100×10^{-p}, where p is **Precision[m]**, are kept; in particular, for exact and symbolic matrices, all singular values are kept. Setting **Tolerance → t** keeps only singular values that are at least **t** times the largest singular value. Setting **Tolerance → 0** keeps all singular values.

By using a tolerance of 0, all singular values of the Hilbert matrix are calculated:

```
SingularValueList[HilbertMatrix[12] // N, Tolerance → 0]
```

$\{1.79537, 0.380275, 0.0447385, 0.00372231, 0.000233089, 0.0000111634, 4.08238 \times 10^{-7},$
$1.12286 \times 10^{-8}, 2.25196 \times 10^{-10}, 3.11135 \times 10^{-12}, 2.6492 \times 10^{-14}, 1.09518 \times 10^{-16}\}$

The 2-norm condition number of a matrix is the ratio of largest to smallest singular values:

```
conditionNumber[m_] :=
  With[{s = SingularValueList[m // N, Tolerance → 0]}, First[s] / Last[s]]
```

A large condition number indicates sensitivity to round-off errors. Hilbert matrices are well-known examples of matrices with large condition numbers:

```
Table[conditionNumber[HilbertMatrix[n]], {n, 1, 6}]
```

$$\{1., 19.2815, 524.057, 15\,513.7, 476\,607., 1.49511 \times 10^7\}$$

With a built-in command we can calculate the ∞-norm condition number:

```
Table[LinearAlgebra`MatrixConditionNumber[HilbertMatrix[n] // N], {n, 1, 6}]
```

$$\{1., 27., 748., 28\,375., 943\,656., 2.90703 \times 10^7\}$$

`SingularValueDecomposition` is considered in Section 21.2.4, p. 706.

■ Norms

`Norm[m]` The 2-norm

`Norm[m, p]` The p-norm; p is a number in $[1, \infty)$ or the value ∞ or `Frobenius`

The 2-norm is the largest of the singular values, the 1-norm is the largest of the absolute column sums, and the ∞-norm is the largest of the absolute row sums. The Frobenius norm is the square root of the sum of squared absolute values. For example,

```
m = {{3, 2}, {4, 1}};
```

```
Norm[m, #] & /@ {1, 2, Frobenius, ∞}
```

$$\left\{7, \sqrt{5\left(3 + 2\sqrt{2}\right)}, \sqrt{30}, 5\right\}$$

`% // N`   {7., 5.39835, 5.47723, 5.}

The basic matrix norms are easy to program:

```
matrixNorm[m_, p_] := Switch[p,
    1, Max[Total[Abs[m]]],
    2, SingularValueList[m][[1]],
    Frobenius, Sqrt[Total[Abs[m]^2, 2]],
    ∞, Max[Total[Abs[m], {2}]]]
```

```
matrixNorm[m, #] & /@ {1, 2, Frobenius, ∞}
```

$$\left\{7, \sqrt{5\left(3 + 2\sqrt{2}\right)}, \sqrt{30}, 5\right\}$$

■ Eigenvalues

`Eigenvalues[m]` Eigenvalues of a square matrix

`Eigenvalues[m, k]` The `k` eigenvalues that are largest in absolute value

`Eigenvalues[m, -k]` The `k` eigenvalues that are smallest in absolute value

`Eigenvalues[{m, a}]` The generalized eigenvalues of `m` with respect to matrix `a`

`Eigenvectors[m]` Eigenvectors

`Eigensystem[m]` Eigenvalues and eigenvectors

`CharacteristicPolynomial[m, x]` Characteristic polynomial

Here is an example of eigenvalues and eigenvectors:

```
m = {{3, 2}, {4, 1}};
```

```
Eigenvalues[m]      {5, -1}
```

```
{u, v} = Eigenvectors[m]      {{1, 1}, {-1, 2}}
```

Eigenvalues are given in the order of decreasing absolute value (repeated eigenvalues appear with their appropriate multiplicity). Eigenvalues and eigenvectors can also be calculated at the same time:

```
{{λ, μ}, {u, v}} = Eigensystem[m]      {{5, -1}, {{1, 1}, {-1, 2}}}
```

We can verify that the familiar conditions of eigenvalues and eigenvectors are satisfied:

```
{m.u == λ u, m.v == μ v}      {True, True}
```

Eigenvalues can also be easily calculated in the following ways:

```
Solve[Det[m - x IdentityMatrix[2]] == 0, x]      {{x → -1}, {x → 5}}
```

```
CharacteristicPolynomial[m, x]      -5 - 4 x + x²
```

```
Solve[% == 0, x]      {{x → -1}, {x → 5}}
```

If you do not want exact eigenvalues and eigenvectors, it is better to calculate with decimal numbers from the start; this can be done with `Eigenvalues[N[m]]`. If you want to start the calculations with k-digit precision, use `Eigenvalues[N[m, k]]`.

It may happen that a matrix has fewer eigenvectors than the number of rows; in this case, null vectors are added to get the same number of vectors as there are rows.

Roots of cubic and quadratic equations are often not written explicitly because they are often long expressions:

```
m = {{3, 2, 4}, {2, 2, 4}, {6, 4, 3}};
```

```
Eigenvalues[m]
```

$$\{\text{Root}\left[10 - 23 \text{\#1} - 8 \text{\#1}^2 + \text{\#1}^3 \&, 3\right],$$
$$\text{Root}\left[10 - 23 \text{\#1} - 8 \text{\#1}^2 + \text{\#1}^3 \&, 1\right], \text{Root}\left[10 - 23 \text{\#1} - 8 \text{\#1}^2 + \text{\#1}^3 \&, 2\right]\}$$

With some options, we can get explicit eigenvalues for (3×3) and (4×4) matrices.

> **Cubics, Quartics** Options of **Eigenvalues**, **Eigenvectors**, and **Eigensystem**. The default value
> **False** means that eigenvalues resulting as roots of cubic and quadratic equations are not written
> explicitly. Give value **True** to get explicit eigenvalues.

```
Eigenvalues[m, Cubics → True][[1]]
```

$$\frac{8}{3} + \frac{133}{3\left(1205 + 6 \text{ i }\sqrt{25\,017}\right)^{1/3}} + \frac{1}{3}\left(1205 + 6 \text{ i }\sqrt{25\,017}\right)^{1/3}$$

An eigenvalue may seem to be complex, although it actually is real:

```
% // N      10.1657+0. i
```

```
Im[%%] // FullSimplify      0
```

■ **Rank, Null Space, and Gaussian Elimination**

> `MatrixRank[m]` Rank
> `NullSpace[m]` Basis vectors of the null space
> `RowReduce[m]` Do Gaussian elimination to produce a reduced row echelon form

Consider the following matrix:

m = {{0, 1, 0, 0, 0}, {0, 0, 0, 1, 1}, {0, 1, 0, 1, 1}, {0, 0, 0, 0, 0}, {1, 1, 0, 0, 1}};

The matrix is singular. This can be seen in various ways. The determinant is zero and the inverse does not exist:

Det[m]   0

Inverse[m]

Inverse::sing :

  Matrix {{0, 1, 0, 0, 0}, {0, 0, 0, 1, 1}, {0, 1, 0, 1, 1}, {0, 0, 0, 0, 0}, {1, 1, 0, 0, 1}} is singular. ≫
  Inverse[
    {{0, 1, 0, 0, 0}, {0, 0, 0, 1, 1}, {0, 1, 0, 1, 1}, {0, 0, 0, 0, 0}, {1, 1, 0, 0, 1}}]

Also, we get only three nonzero singular values and only three nonzero eigenvalues, and the rank (the number of linearly independent rows or columns) of the matrix is only 3:

SingularValueList[m // N]   {2.53209, 1.3473, 0.879385}

Eigenvalues[m // N]

{1.83929, -0.419643 + 0.606291 i, -0.419643 - 0.606291 i, 0., 0.}

MatrixRank[m]   3

The null space of a matrix **m** consists of vectors **x** satisfying **m.x == 0**. **NullSpace[m]** gives the basis vectors of this space. For nonsingular matrices, the null space is empty. The previous matrix **m** has a null space spanned by the following two vectors:

NullSpace[m]   {{-1, 0, 0, -1, 1}, {0, 0, 1, 0, 0}}

The rank of a matrix equals the column dimension minus the dimension of the null space.

The reduced row echelon form of **m** has only three nonzero rows:

RowReduce[m] // MatrixForm

$$\begin{pmatrix} 1 & 0 & 0 & 0 & 1 \\ 0 & 1 & 0 & 0 & 0 \\ 0 & 0 & 0 & 1 & 1 \\ 0 & 0 & 0 & 0 & 0 \\ 0 & 0 & 0 & 0 & 0 \end{pmatrix}$$

RowReduce is also considered in Section 22.1.2, p. 713.

21.2.4 Decompositions

■ Decompositions into Triangular Matrices

LUDecomposition[m] PLU decomposition of a square matrix. For matrix M, the decomposition is $PM = LU$, where L is unit lower triangular (with 1's on the diagonal), U is upper triangular, and P is a permutation matrix. Output is $\{K, p, c\}$, where matrix K contains both L and U, p is a permutation vector (i.e., a list specifying rows used for pivoting), and c is the L^∞ condition number of m (however, for exact matrices with no decimal points, c is 1).

We use the following matrix as an example when we next compute various decompositions:

M = {{0.53, 0.88, 0.18}, {0.70, 0.44, 0.17}, {0.17, 0.56, 0.36}};

Calculate a PLU decomposition:

```
MatrixForm /@ ({LU, perm, cond} = LUDecomposition[M])
```

$$\left\{ \begin{pmatrix} 0.7 & 0.44 & 0.17 \\ 0.757143 & 0.546857 & 0.0512857 \\ 0.242857 & 0.828631 & 0.276217 \end{pmatrix}, \begin{pmatrix} 2 \\ 1 \\ 3 \end{pmatrix}, 12.7397 \right\}$$

In the following way, we can get explicit L and U matrices:

```
L = LU SparseArray[{i_, j_} /; i > j → 1, {3, 3}] + IdentityMatrix[3];
U = LU SparseArray[{i_, j_} /; i ≤ j → 1, {3, 3}];
```

```
MatrixForm /@ {L, U}
```

$$\left\{ \begin{pmatrix} 1 & 0 & 0 \\ 0.757143 & 1 & 0 \\ 0.242857 & 0.828631 & 1 \end{pmatrix}, \begin{pmatrix} 0.7 & 0.44 & 0.17 \\ 0 & 0.546857 & 0.0512857 \\ 0 & 0 & 0.276217 \end{pmatrix} \right\}$$

Now we can check that the decomposition is correct:

```
M[[perm]] - L.U // Chop        {{0, 0, 0}, {0, 0, 0}, {0, 0, 0}}
```

In the following way, we can get a function to solve linear equations of the form $M x = b$ for fixed M and any b:

```
luFunction = LinearSolve[M]
```

```
LinearSolveFunction[{3, 3}, <>]
```

For example, if $b = (4, 2, 7)$, then the solution of the equation $M x = b$ is

```
luFunction[{4, 2, 7}]        {-2.96588, 3.03303, 16.127}
```

LinearSolve uses the PLU decomposition. Indeed, some parts of **luFunction** are the same as the output from the PLU decomposition:

```
luFunction[[2, 3]] == {LUᵀ, perm, cond}        True
```

The logic of the LU decomposition for the system $M x = b$ is that when L and U have been obtained, the original system is replaced with two very easy triangular systems: First we solve $L y = b$ for y and then $U x = y$ for x. Therefore, $M x = (L U) x = L y = b$. This means that x is the solution. Triangular systems are solved very easily: y is solved by forward substitution, and then x is solved by backward substitution. We try these computations:

```
y = LinearSolve[L, {4, 2, 7}[[perm]]]        {2., 2.48571, 4.45455}
```

```
x = LinearSolve[U, y]        {-2.96588, 3.03303, 16.127}
```

From www.wolfram.com/products/mathematica/newin6/content/ExtendedArrayOperations , I found the following program for LU decomposition:

```
LUDecompose[A_] := Module[{m, n, L, U},
  {m, n} = Dimensions[A]; L = IdentityMatrix[n]; U = A;
  Do[L[[k ;; n, k]] = U[[k ;; n, k]] / U[[k, k]];
    U[[(k + 1) ;; n, k ;; n]] = U[[(k + 1) ;; n, k ;; n]] - L[[(k + 1) ;; n, {k}]].U[[{k}, k ;; n]],
    {k, n - 1}]; {L, U}]
```

```
LUDecompose[M]
```

$$\{\{\{1., 0, 0\}, \{1.32075, 1., 0\}, \{0.320755, -0.384535, 1\}\},$$
$$\{\{0.53, 0.88, 0.18\}, \{1.11022 \times 10^{-16}, -0.722264, -0.0677358\}, \{0., 0., 0.276217\}\}\}$$

CholeskyDecomposition[m] Cholesky decomposition of a Hermitian, positive definite matrix. For matrix M, the decomposition is $M = U^\dagger U$, where U is upper triangular. Output is U.

Here, $U^\dagger$ means conjugate transpose. A matrix M is Hermitian if the conjugate transpose of the matrix is the same as the matrix: $M^\dagger = M$; if M is real, Hermitian means the same as symmetric. To calculate a Cholesky decomposition, define the following matrix:

```
M2 = {{0.64, 0.61, 0.66}, {0.61, 0.76, 0.69}, {0.66, 0.69, 0.80}};
```

```
{HermitianMatrixQ[M2], PositiveDefiniteMatrixQ[M2]}
```

{True, True}

Then calculate and check the Cholesky decomposition:

```
(U = CholeskyDecomposition[M2]) // MatrixForm
```

$$\begin{pmatrix} 0.8 & 0.7625 & 0.825 \\ 0. & 0.422604 & 0.144195 \\ 0. & 0. & 0.313979 \end{pmatrix}$$

```
M2 - U†.U // Chop        {{0, 0, 0}, {0, 0, 0}, {0, 0, 0}}
```

> **HermiteDecomposition[m]** (❀6)   Hermite decomposition of a matrix with rational (real or complex) elements. For matrix M, the decomposition is $U M = R$, where U is unimodular (determinant is a unit) and R upper triangular. Output is $\{U, R\}$.

Calculate and check a Hermite decomposition:

```
M3 = {{3 / 5, 2 / 3}, {5 / 2, 1 / 4}};
```

```
MatrixForm /@ ({U, R} = HermiteDecomposition[M3])
```

$$\left\{ \begin{pmatrix} -4 & 1 \\ 25 & -6 \end{pmatrix}, \begin{pmatrix} \frac{1}{10} & -\frac{29}{12} \\ 0 & \frac{91}{6} \end{pmatrix} \right\}$$

```
{U.M3 == R, Det[U]}      {True, -1}
```

■ **Orthogonal Decompositions**

> **QRDecomposition[m]**   QR decomposition. For matrix M, the decomposition is $M = Q^\dagger R$, where Q is orthonormal and R is upper triangular. Output is $\{Q, R\}$.

Calculate and check a QR decomposition:

```
MatrixForm /@ ({Q, R} = QRDecomposition[M])
```

$$\left\{ \begin{pmatrix} -0.592632 & -0.782722 & -0.19009 \\ 0.523927 & -0.553844 & 0.647115 \\ -0.611791 & 0.283908 & 0.738315 \end{pmatrix}, \begin{pmatrix} -0.894315 & -0.972364 & -0.308169 \\ 0. & 0.579749 & 0.233115 \\ 0. & 0. & 0.203935 \end{pmatrix} \right\}$$

```
M - Q†.R // Chop       {{0, 0, 0}, {0, 0, 0}, {0, 0, 0}}
```

The rows of Q are orthonormal:

```
Q.Qᵀ // Chop       {{1., 0, 0}, {0, 1., 0}, {0, 0, 1.}}
```

> **SingularValueDecomposition[m]**   Singular value decomposition. For matrix M, the decomposition is $M = U W V^\dagger$, where U and V are orthonormal and W is diagonal with singular values as the diagonal elements. Output is $\{U, W, V\}$. The option **Tolerance** can be used; see **SingularValueList** in Section 21.2.3, p. 701.

Calculate and check a singular value decomposition:

```
MatrixForm /@ ({U, W, V} = SingularValueDecomposition[M])
```

$$\left\{\begin{pmatrix} -0.716157 & 0.156269 & 0.68022 \\ -0.54731 & -0.730522 & -0.408399 \\ -0.433096 & 0.664769 & -0.608695 \end{pmatrix},\right.$$

$$\left.\begin{pmatrix} 1.44299 & 0. & 0. \\ 0. & 0.394413 & 0. \\ 0. & 0. & 0.185784 \end{pmatrix}, \begin{pmatrix} -0.579564 & -0.800004 & -0.155239 \\ -0.771708 & 0.477566 & 0.419997 \\ -0.261863 & 0.363215 & -0.894149 \end{pmatrix}\right\}$$

```
M - U.W.V† // Chop       {{0, 0, 0}, {0, 0, 0}, {0, 0, 0}}
```

PseudoInverse uses singular value decomposition and returns $V\,W^{-1}\,U^\dagger$.

■ **Decompositions Related to Eigenvalue Problems**

JordanDecomposition[m]   Jordan decomposition of a square matrix. For matrix M, the decomposition is $M = S\,J\,S^{-1}$, where S is a similarity matrix and J is the Jordan canonical form of M. Output is $\{S, J\}$.

Calculate and check a Jordan decomposition:

```
MatrixForm /@ ({S, J} = JordanDecomposition[M])
```

$$\left\{\begin{pmatrix} -0.687844 & -0.61899 & -0.198684 \\ -0.584 & 0.672139 & -0.134772 \\ -0.431063 & -0.406301 & 0.970753 \end{pmatrix}, \begin{pmatrix} 1.38995 & 0 & 0 \\ 0 & -0.30741 & 0 \\ 0 & 0 & 0.24746 \end{pmatrix}\right\}$$

```
M - S.J.Inverse[S] // Chop       {{0, 0, 0}, {0, 0, 0}, {0, 0, 0}}
```

In this example, the Jordan decomposition is essentially the same as the result given by **Eigensystem**:

```
MatrixForm /@ Eigensystem[M]
```

$$\left\{\begin{pmatrix} 1.38995 \\ -0.30741 \\ 0.24746 \end{pmatrix}, \begin{pmatrix} -0.687844 & -0.584 & -0.431063 \\ -0.61899 & 0.672139 & -0.406301 \\ -0.198684 & -0.134772 & 0.970753 \end{pmatrix}\right\}$$

SchurDecomposition[m]   Schur decomposition of a square, numerical matrix with at least one entry that has a decimal point. For matrix M, the decomposition is $M = Q\,T\,Q^\dagger$, where Q is orthonormal and T is block upper triangular. Output is $\{Q, T\}$.

Calculate and check a Schur decomposition:

```
MatrixForm /@ ({Q, T} = SchurDecomposition[M])
```

$$\left\{\begin{pmatrix} -0.687844 & -0.486333 & -0.538843 \\ -0.584 & 0.811651 & 0.0129322 \\ -0.431063 & -0.32358 & 0.842307 \end{pmatrix}, \begin{pmatrix} 1.38995 & -0.361635 & 0.123309 \\ 0. & -0.30741 & -0.196508 \\ 0. & 0. & 0.24746 \end{pmatrix}\right\}$$

```
M - Q.T.Q† // Chop       {{0, 0, 0}, {0, 0, 0}, {0, 0, 0}}
```

HessenbergDecomposition[m]   Hessenberg decomposition of a square, numerical matrix with at least one entry that has a decimal point. For matrix M, the decomposition is $M = P\,H\,P^\dagger$, where P is a unitary and H a Hessenberg matrix. Output is $\{P, H\}$.

Calculate and check a Hessenberg decomposition:

```
MatrixForm /@ ({P, H} = HessenbergDecomposition[M])
```

$$\left\{ \begin{pmatrix} 1. & 0. & 0. \\ 0. & -0.971754 & -0.235997 \\ 0. & -0.235997 & 0.971754 \end{pmatrix}, \begin{pmatrix} 0.53 & -0.897623 & -0.032762 \\ -0.720347 & 0.602956 & -0.110996 \\ 0. & -0.500996 & 0.197044 \end{pmatrix} \right\}$$

```
M - P.H.P† // Chop      {{0, 0, 0}, {0, 0, 0}, {0, 0, 0}}
```

22

Equations

Introduction

> *Someone told me that each equation I included in the book would halve the sales.* —*Stephen Hawking*

Equations can be classified as *linear, polynomial, radical*, or *transcendental*. Polynomial equations consist of sums of integer powers of variables, whereas radical equations may contain rational powers. Transcendental equations contain transcendental functions such as $\sin(x)$ or $\log(x + y)$. The main command for linear, polynomial, and radical equations is **Solve**, and that for transcendental equations is **FindRoot**. However, other commands can also be used; these are explained next.

If we have linear equations in the form of a coefficient matrix and right-hand-side vector, then **LinearSolve** is easy to use.

For polynomial equations, **Solve** gives an answer for *generic* values of the possible parameters of the equations. If an exhaustive analysis of the solutions is wanted for *all* possible values of the parameters in polynomial equations, then **Reduce** can be used. If **Solve** cannot obtain exact solutions for polynomial equations, then **NSolve** can be used to calculate the solutions numerically.

For transcendental equations, we can sometimes apply **Solve** or **Reduce**, but usually we have to resort to the iterative methods provided by **FindRoot** (Newton's, Brent's, or the secant method).

Here is a summary of the commands for equations:

- *Linear equations:* `Solve`, `LinearSolve`
- *Polynomial and radical equations:* `Solve`, `Reduce`, `NSolve`
- *Transcendental equations:* `Solve`, `Reduce`, `FindRoot`

Exact symbolic solutions can be obtained with `Solve`, `LinearSolve`, and `Reduce`. Numerical methods are used with `NSolve` and `FindRoot`.

In this chapter, we also consider inequalities.

22.1 Linear Equations

22.1.1 Two Representations

Linear systems can be represented by either writing down the equations explicitly with variables or giving the left-hand-side coefficient matrix and the right-hand-side vector. Appropriate commands for these representations are `Solve` and `LinearSolve`, respectively.

■ **Giving the Equations**

`Solve[eqns]` Solve the equations `eqns` for the symbols therein
`Solve[eqns, vars]` Solve the equations `eqns` for the variables `vars`

Here, `eqns` is a list of equations, and `vars` is a list of variables. Remember that equations are defined with two equal signs (`==`), but *Mathematica* replaces them with the special symbol `=` (this symbol can also be directly written with ESC==ESC). First, we write the equations directly in the command:

$\quad$ `Solve[{x + 2 y == 7, 2 x - y == 4}]` $\quad$ `{{x → 3, y → 2}}`

We can also first define the equations and then solve them:

$\quad$ `eqns = {x + 2 y == 7, 2 x - y == 4};`

$\quad$ `sol = Solve[eqns]` $\quad$ `{{x → 3, y → 2}}`

The result of `Solve` is a list of transformation rules. A system of equations may have several solutions, and, accordingly, the general form of the result of `Solve` is {sol_1, sol_2, ... }, where each sol_i is a list of rules for the variables. In our example, the equations only have one solution, and so the result is {sol_1}.

For unique solutions, it may be convenient to get rid of the outermost curly braces. We can do this by picking the first and only element of `sol`:

$\quad$ `sol2 = sol[[1]]` $\quad$ `{x → 3, y → 2}`

We can check the solution:

$\quad$ `eqns /. sol2` $\quad$ `{True, True}`

We can also insert the solution into other expressions:

$\quad$ `x y /. sol2` $\quad$ `6`

If a list of values is wanted as the solution, we can write the following:

$\quad$ `{x, y} /. sol2` $\quad$ `{3, 2}`

If we want to assign the solution into `{x, y}`, we can write the following:

```
{x, y} = {x, y} /. sol2      {3, 2}
```

Now $x = 3$ and $y = 2$:

```
{x, y}      {3, 2}
```

You may want to reread Section 13.1.2, p. 416, in which we considered transformation rules. We now clear our assignments for x and y:

```
Clear[x, y]
```

If symbols appear as coefficients, then the variables have to be given:

```
Solve[{x + y == 2 a, x - y == 2 b}, {x, y}]      {{x → a + b, y → a - b}}
```

Here we have a system with no solutions:

```
Solve[{3 x + y == 9, 6 x + 2 y == 4}]      {}
```

The result is an empty list, which indicates that no solutions exist. Here we have infinitely many solutions:

```
Solve[{3 x + y == 9, 6 x + 2 y == 18}]
```

Solve::svars : Equations may not give solutions for all "solve" variables. ≫

$$\left\{\left\{x \to 3 - \frac{y}{3}\right\}\right\}$$

This means that y can be arbitrary and $x = 3 - y/3$.

For *sparse* systems, **Solve** uses special methods that are efficient for such systems (a linear system is sparse if the coefficient matrix contains many zeros). The special methods are available if the coefficients are real or complex machine numbers.

■ Giving the Coefficients

LinearSolve[a, b]	Solve the system **a.x == b** of linear equations

In the simplest case, **x** is the vector of unknowns, **a** is a square matrix, and **b** is a vector (generalizations are considered in Section 22.1.2, p. 712). For example, let **a** and **b** be as follows:

```
a = {{1, 2}, {2, -1}};
b = {7, 4};
```

The equations are then $x + 2 y = 7$ and $2 x - y = 4$. Solve the equations:

```
sol = LinearSolve[a, b]      {3, 2}
```

Thus, $x = 3$ and $y = 2$. We could also use **Solve**:

```
Solve[a.{x, y} == b]      {{x → 3, y → 2}}
```

To check the solution, write the following:

```
a.sol - b      {0, 0}
```

If the coefficient matrix is singular (i.e., has a zero determinant), the system usually has no solutions:

```
LinearSolve[{{3, 1}, {6, 2}}, {9, 4}]
```

LinearSolve::nosol : Linear equation encountered that has no solution. ≫

```
LinearSolve[{{3, 1}, {6, 2}}, {9, 4}]
```

Sometimes an infinite number of solutions exists:

    LinearSolve[{{3, 1}, {6, 2}}, {9, 18}]   {3, 0}

In this example, all solutions are of the form {3 − y/3, y}; **LinearSolve** gives one of the possible solutions (**Solve**, as we saw previously, gives all solutions).

Inverting the coefficient matrix is one possibility for solving linear equations, but this method is not recommended (inversion of a matrix is more demanding than solving linear equations):

    Inverse[a].b   {3, 2}

LinearSolve has a **Method** option. Settings for exact and symbolic matrices include **"CofactorExpansion"**, **"DivisionFreeRowReduction"**, and **"OneStepRowReduction"**. Settings for approximate numerical matrices include **"Cholesky"**, and those for sparse arrays include **"Multifrontal"** and **"Krylov"**. The default setting of **Automatic** switches among these methods depending on the matrix given.

■ **Transforming between the Two Representations**

> **{b, a} = CoefficientArrays[eqns, vars]** Find the coefficients **b** and **a** of the given linear equations; the coefficients correspond with the system **b + a.vars == 0**
>
> **eqns = Thread[a.vars == b]** Form explicit equations from the coefficients

Sometimes we would like to find the coefficient matrix and right-hand-side vector of explicitly written equations:

    eqns = {x + 2 y == 7, 2 x − y == 4};

    {b, a} = CoefficientArrays[eqns, {x, y}]

    {SparseArray[<2>, {2}], SparseArray[<4>, {2, 2}]}

The result is expressed in terms of sparse arrays. The normal arrays are as follows:

    % // Normal   {{−7, −4}, {{1, 2}, {2, −1}}}

We got the vector $b = \{−7, −4\}$ and the matrix $a = \{\{1, 2\}, \{2, −1\}\}$. This means that the equations are $b + a.x == 0$; that is, $a.x == −b$. This can be checked:

    Thread[a.{x, y} == −b]   {x + 2 y == 7, 2 x − y == 4}

22.1.2 Special Topics

■ **Eliminating Variables**

> **Solve[eqns, vars, elims]** Attempt to solve the equations **eqns** for the variables **vars**, eliminating the variables **elims**

Define two equations:

    eqns = {x − y == c, x + y == 2 c};

We can solve for x and y:

    Solve[eqns, {x, y}] $\left\{\left\{x \rightarrow \dfrac{3\,c}{2}, \ y \rightarrow \dfrac{c}{2}\right\}\right\}$

We can also ask to eliminate c:

```
Solve[eqns, {x, y}, {c}]
```

Solve::svars : Equations may not give solutions for all "solve" variables. ≫

$\{\{x \to 3\,y\}\}$

■ **Several Systems**

---

f = LinearSolve[a]   Give a function **f** for which **f[b]** solves the equation **a.x == b**

f[b1], f[b2], …   Solve the systems **a.x == b1, a.x == b2, …**

This is a handy way to solve several systems that have the same left-hand-side matrix but different right-hand-side vectors. In the following example, we have two right-hand sides, {1, 2, 3} and {−4, 5, 6}:

```
a = {{2, 1, 1}, {1, 1, 1}, {1, 0, 1}};
b1 = {1, 2, 3};
b2 = {-4, 5, 6};
```

First, ask for the solution as a function:

```
f = LinearSolve[a]      LinearSolveFunction[{3, 3}, <>]
```

The result is an object called **LinearSolveFunction**. We are then able to solve systems with various right-hand-side vectors:

```
{f[b1], f[b2]}      {{-1, -1, 4}, {-9, -1, 15}}
```

Thus, the corresponding solutions are {−1, −1, 4} and {−9, −1, 15}. The two solutions can also be calculated at the same time:

LinearSolve$\left[$a, {b1, b2}$^{\mathsf{T}}\right]^{\mathsf{T}}$ {{-1, -1, 4}, {-9, -1, 15}}

■ **Gaussian Elimination**

RowReduce[a]   Do Gaussian elimination to **a** to produce a reduced row echelon form

---

RowReduce can be used to solve linear systems. We solve the system **a.x == b1** we defined previously. First, we append the elements of the right-hand-side vector to the rows of the left-hand-side matrix:

m = Append$\left[$a^{T}, b1$\right]^{\mathsf{T}}$ {{2, 1, 1, 1}, {1, 1, 1, 2}, {1, 0, 1, 3}}

Then we do a Gaussian elimination:

```
r = RowReduce[m]      {{1, 0, 0, -1}, {0, 1, 0, -1}, {0, 0, 1, 4}}
```

The solution is the last column of this matrix:

r〚All, -1〛   {-1, -1, 4}

■ **Tridiagonal Systems**

Tridiagonal systems can be efficiently solved by first forming a sparse array from the sub-, main, and superdiagonals and then using **LinearSolve**. As an example, we solve a tridiagonal system in which the coefficient matrix is as follows:

```
{{4, 2, 0}, {1, 4, 2}, {0, 1, 4}} // MatrixForm
```

$$\begin{pmatrix} 4 & 2 & 0 \\ 1 & 4 & 2 \\ 0 & 1 & 4 \end{pmatrix}$$

The right-hand-side vector is {6, 4, 7}. First define the diagonals and the right-hand side and then form the tridiagonal matrix (for **Band**, see Section 21.2.1, p. 690):

```
sub = {1, 1}; main = {4, 4, 4}; super = {2, 2}; b = {6, 4, 7};

a = SparseArray[{Band[{2, 1}] → sub, Band[{1, 1}] → main, Band[{1, 2}] → super}]
SparseArray[<7>, {3, 3}]
```

We can check that the matrix is correct:

```
a // Normal      {{4, 2, 0}, {1, 4, 2}, {0, 1, 4}}
```

Then solve the system:

```
LinearSolve[a, b]      
```
$$\left\{ \frac{5}{3}, -\frac{1}{3}, \frac{11}{6} \right\}$$

■ Overdetermined Systems

In addition to square linear systems, *Mathematica* can handle rectangular linear systems. If there are more equations than variables, the system is *overdetermined,* and solutions usually do not exist. **Solve** and **LinearSolve** can be tried: They give a solution if it exists and otherwise give an empty list (**Solve**) or tell us that no solutions exist (**LinearSolve**).

If a solution of the equations $a.x = b$ does not exist, we can find a least-squares solution by minimizing the 2-norm $\| a.x - b \|$.

LeastSquares[a, b] (❀6) Give an approximating solution of **a.x == b**

Consider the following overdetermined system:

```
a = {{3, 1}, {2, 5}, {8, 1}};
b = {2, 7, 5};
```

An approximating solution is as follows:

```
apprsol = LeastSquares[a, b] // N      {0.450549, 1.20513}
```

The left-hand sides of the equations with these values for the variables are similar to the right-hand sides {2, 7, 5}:

```
a.apprsol      {2.55678, 6.92674, 4.80952}
```

Alternatively, we can use a pseudoinverse (or a generalized inverse or Moore–Penrose inverse):

```
PseudoInverse[a].b // N      {0.450549, 1.20513}
```

This can be compared with **Inverse[a].b** (see Section 22.1.1, p. 712).

■ Underdetermined Systems

If there are more variables than equations, the system is *underdetermined,* and an infinite number of solutions usually exist. For example,

```
a = {{4, 3, 1}, {3, 2, 5}};
b = {9, 4};
```

Solve gives all solutions:

```
Solve[a.{x, y, z} == b]
```

Solve::svars : Equations may not give solutions for all "solve" variables. ≫
$\{\{x \to -6 - 13 z, y \to 11 + 17 z\}\}$

LinearSolve gives one solution:

```
LinearSolve[a, b]     {-6, 11, 0}
```

LeastSquares or **PseudoInverse** also gives one solution (exact this time):

```
LeastSquares[a, b] // N    {1.50545, 1.18519, -0.577342}

PseudoInverse[a].b // N    {1.50545, 1.18519, -0.577342}
```

■ Jacobi's Method

There are a number of iterative solution methods for linear systems. Here, we implement Jacobi's method and the Gauss–Seidel method. Consider the following diagonally dominant system:

```
a = {{4, 1, -2}, {-1, 4, 3}, {2, 1, -3}};  b = {5, 2, 2};

n = 3;  y = Array[x, n]     {x[1], x[2], x[3]}

eqns = Thread[a.y == b]
{4 x[1] + x[2] - 2 x[3] == 5, -x[1] + 4 x[2] + 3 x[3] == 2, 2 x[1] + x[2] - 3 x[3] == 2}
```

The solution of this system is as follows:

```
sol = y /. Solve[eqns][[1]] // N
{1.30769, 0.538462, 0.384615}
```

To apply Jacobi's method, first we solve x_i from the ith equation:

```
newy = y /. Table[Solve[eqns[[i]], x[i]][[1, 1]], {i, n}]
```
$$\left\{\frac{1}{4}\ (5 - x[2] + 2\ x[3]),\ \frac{1}{4}\ (2 + x[1] - 3\ x[3]),\ \frac{1}{3}\ (-2 + 2\ x[1] + x[2])\right\}$$

We start from a point, for example, $(0, 0, ..., 0)$, and then iteratively calculate new values for $x_1, ..., x_n$ from **newy**. We do 30 iterations:

```
yi = {0, 0, 0};
Do[yi = newy /. Thread[y → yi] // N, {30}];
yi
{1.30769, 0.538462, 0.384615}
```

This is a good approximation to the solution:

```
sol - yi
```
$$\left\{2.96988 \times 10^{-8},\ 1.02619 \times 10^{-9},\ 2.91744 \times 10^{-8}\right\}$$

■ Gauss–Seidel Method

To accelerate convergence of Jacobi's method, the already calculated new values for $x_1, ..., x_i$ can be used in the calculation of the new values for x_{i+1}, $i = 1, ..., n - 1$. This is the Gauss–Seidel method. We again start from $(0, 0, 0)$ and do 30 iterations:

```
yi = {0, 0, 0};
Do[yi = Table[yi[[j]] = newy[[j]] /. Thread[y → yi], {j, n}] // N, {30}];
yi    {1.30769, 0.538462, 0.384615}
```

The point here is to do the update equation by equation, with **Table**, and to assign the new value **newy[[j]]** directly to **yi[[j]]** so that this value is used in the remaining updates when the substitution **y → yi** is made. The solution is much better than the solution offered by Jacobi's method:

```
sol - yi
```
$$\left\{8.90399 \times 10^{-14},\ -5.50671 \times 10^{-14},\ 4.10783 \times 10^{-14}\right\}$$

```
Remove["Global`*"]
```

22.2 Polynomial and Radical Equations

22.2.1 Polynomial Equations

■ Exact Solution

> `Solve[eqn]`   Solve the equation `eqn` for the symbol in the equation
> `Solve[eqn, x]`   Solve the equation `eqn` for the variable `x`

If the equation contains only one symbol, it need not be mentioned in the command. Here is the familiar second-order equation:

eqn = a x^2 + b x + c == 0;

sol = Solve[eqn, x]

$$\left\{\left\{x \to \frac{-b - \sqrt{b^2 - 4\,a\,c}}{2\,a}\right\},\ \left\{x \to \frac{-b + \sqrt{b^2 - 4\,a\,c}}{2\,a}\right\}\right\}$$

The solutions are given as transformation rules (see Section 13.1.2, p. 416). We can verify the solutions:

eqn /. sol // Simplify

{True, True}

For both solutions, the equation simplifies to 0 == 0, which is True.

If you want a list of values (rather than a list of rules), enter the following:

x /. sol

$$\left\{\frac{-b - \sqrt{b^2 - 4\,a\,c}}{2\,a},\ \frac{-b + \sqrt{b^2 - 4\,a\,c}}{2\,a}\right\}$$

You can also type the following:

x /. Solve[eqn, x]

$$\left\{\frac{-b - \sqrt{b^2 - 4\,a\,c}}{2\,a},\ \frac{-b + \sqrt{b^2 - 4\,a\,c}}{2\,a}\right\}$$

Note, however, that the transformation rules are handy in that they can easily be used to calculate the value of any expression:

b + 2 a x /. sol $\left\{-\sqrt{b^2 - 4\,a\,c},\ \sqrt{b^2 - 4\,a\,c}\right\}$

■ Special Questions

`Solve` is able to solve all polynomials up to order four. Polynomials of order five and greater can also sometimes be solved:

p = -6 + 23 x - 34 x^2 + 24 x^3 - 8 x^4 + x^5;

sol = Solve[p == 0]

{{x → 1}, {x → 1}, {x → 1}, {x → 2}, {x → 3}}

```
Plot[p, {x, 0.5, 3.1}, Epilog → Point[{x, 0} /. sol]]
```

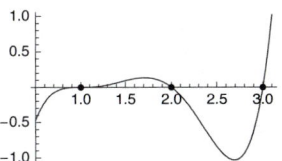

(Here, 1 is a zero of multiplicity 3.) However, it often happens that such high-order equations cannot be solved exactly:

```
sol = Solve[x^5 - x + 1 == 0]
```

$$\left\{\left\{x \to \text{Root}\left[1 - \#1 + \#1^5 \&, 1\right]\right\}, \left\{x \to \text{Root}\left[1 - \#1 + \#1^5 \&, 2\right]\right\},\right.$$
$$\left.\left\{x \to \text{Root}\left[1 - \#1 + \#1^5 \&, 3\right]\right\}, \left\{x \to \text{Root}\left[1 - \#1 + \#1^5 \&, 4\right]\right\}, \left\{x \to \text{Root}\left[1 - \#1 + \#1^5 \&, 5\right]\right\}\right\}$$

Solve gives a symbolic representation for the solution with the **Root** object. **Root[f, k]** represents the **k**th root of the equation **f == 0** (for **Root**, see Section 22.2.2, p. 720). We can, however, ask for numerical values for the roots:

```
sol // N
```

$$\{\{x \to -1.1673\}, \{x \to -0.181232 - 1.08395 \, i\}, \{x \to -0.181232 + 1.08395 \, i\},$$
$$\{x \to 0.764884 - 0.352472 \, i\}, \{x \to 0.764884 + 0.352472 \, i\}\}$$

If the solution contains powers of -1, we can transform them with **ComplexExpand** to expressions that contain trigonometric functions, which in turn sometimes automatically reduce to radicals (i.e., to arithmetic combinations of various roots):

```
sol = x /. Solve[x^3 + 1 == 0]
```
$$\left\{-1, \ (-1)^{1/3}, \ -(-1)^{2/3}\right\}$$

```
sol // ComplexExpand
```
$$\left\{-1, \ \frac{1}{2} + \frac{i\sqrt{3}}{2}, \ \frac{1}{2} - \frac{i\sqrt{3}}{2}\right\}$$

Solve uses explicit formulas up to degree four. For higher-order polynomials, **Solve** attempts to reduce polynomials using **Factor** and **Decompose**, and **Solve** recognizes cyclotomic and other special polynomials.

■ Numerical Solution

Solve[eqn]	Try exact solution
% // N	If this did not succeed, use numerical methods (**NSolve** is eventually used)
NSolve[eqn, x]	Solve **eqn** numerically for **x**
NSolve[eqn, x, n]	Solve **eqn** numerically for **x**, using **n**-digit precision in the calculations

NSolve is based on the Jenkins–Traub algorithm. For example,

```
p = x^5 - x + 1;
```

```
sol = NSolve[p == 0, x]
```

$$\{\{x \to -1.1673\}, \{x \to -0.181232 - 1.08395 \, i\}, \{x \to -0.181232 + 1.08395 \, i\},$$
$$\{x \to 0.764884 - 0.352472 \, i\}, \{x \to 0.764884 + 0.352472 \, i\}\}$$

The solution does not pass the test that we should have **p == 0** at the roots:

```
p == 0 /. sol     {False, False, False, False, False}
```

This is a normal situation with numerical solutions. If we instead ask for the value of the polynomial at the roots, we observe that the roots are very good:

`p /. sol`

$\{-2.22045 \times 10^{-15}, 0. - 2.22045 \times 10^{-16} \text{ i}, 0. + 2.22045 \times 10^{-16} \text{ i},$
$3.33067 \times 10^{-16} - 4.44089 \times 10^{-16} \text{ i}, 3.33067 \times 10^{-16} + 4.44089 \times 10^{-16} \text{ i}\}$

Here is a plot of the solutions in the complex plane:

`ListPlot[{Re[x], Im[x]} /. sol, PlotStyle → PointSize[Medium]]`

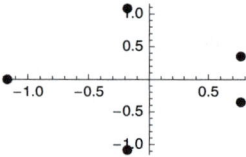

■ Several Polynomial Equations

`Solve[eqns, vars]`	Solve the equations for **vars**
`NSolve[eqns, vars]`	Solve the equations numerically for **vars**

For simultaneous equations we write a list of equations and a list of variables. When solving systems of polynomial equations, **Solve** actually constructs a Gröbner basis with **GroebnerBasis**.

In the following example, we get two solutions:

`Solve[{x^2 + y^2 == 5, x + y == 1}]` $\{\{x \to -1, y \to 2\}, \{x \to 2, y \to -1\}\}$

The following system has six solutions:

`eqns = {x^2 + y^3 - x y == 0, x + y + x^2 - 1 == 0};`

`sol = NSolve[eqns, {x, y}]`

$\{\{x \to 1.13665, y \to -1.42864\}, \{x \to -1.43152 + 0.695043 \text{ i}, y \to 0.865346 + 1.2949 \text{ i}\},$
$\{x \to -1.43152 - 0.695043 \text{ i}, y \to 0.865346 - 1.2949 \text{ i}\}, \{x \to -2.06867, y \to -1.21074\},$
$\{x \to 0.397534 - 0.0995355 \text{ i}, y \to 0.454339 + 0.178673 \text{ i}\},$
$\{x \to 0.397534 + 0.0995355 \text{ i}, y \to 0.454339 - 0.178673 \text{ i}\}\}$

Illustrate the real solutions:

`ContourPlot[eqns // Evaluate, {x, -2.5, 1.5}, {y, -1.7, 1.5}, Frame → False,`
`   Axes → True, AspectRatio → Automatic, Epilog → Point[{x, y} /. {sol[[1]], sol[[4]]}]]`

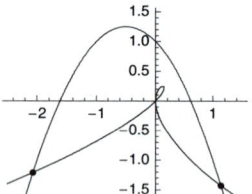

22.2.2 Special Topics

■ Eliminating Variables

`Solve[eqns, vars, elims]`	Solve **eqns** for **vars**, eliminating **elims**
`Eliminate[eqns, elims]`	Eliminate **elims** from **eqns**

In the following example, we ask to eliminate **y** and solve for **x**:

eqns = {x^2 + y^2 == a, x + y == b};

Solve[eqns, x, y]

$$\left\{\left\{x \to \frac{1}{2}\left(b - \sqrt{2\,a - b^2}\,\right)\right\}, \left\{x \to \frac{1}{2}\left(b + \sqrt{2\,a - b^2}\,\right)\right\}\right\}$$

Next, we ask for an elimination only. The result is an equation or several equations:

Eliminate[eqns, y] $a - 2\,x^2 == b^2 - 2\,b\,x$

Here is another example:

Eliminate[{x^2 + y^3 == x y, x + y + x^2 == 1}, x]

$-3\,y + y^2 + 4\,y^3 - y^4 + y^6 == -1$

■ Making Equations Valid for All Values of Some Variables

> **SolveAlways[eqns, vars]**   Give conditions for the parameters appearing in **eqns** that make the equations valid for all values of the variables **vars**

SolveAlways may be useful, for example, in the method of undetermined coefficients. In this method, we want to find conditions under which a trial expression is a solution of an equation. As an example, consider the following differential equation:

eqn = y'[t] + a y[t] + b == 0 $b + a\,y[t] + y'[t] == 0$

Define the following function:

z[t_] := c + d Exp[e t]

We examine whether this function could be a solution of the equation for some values of the parameters *c*, *d*, and *e*. First, we insert the function into the equation:

eqn /. y → z $b + d\,e\,e^{e\,t} + a\left(c + d\,e^{e\,t}\right) == 0$

Because **SolveAlways** does not handle transcendental equations, we form a series expansion:

Series[%, {t, 0, 2}]

$$(b + a\,c + a\,d + d\,e) + \left(a\,d\,e + d\,e^2\right)t + \frac{1}{2}\left(a\,d\,e^2 + d\,e^3\right)t^2 + O[t]^3 == 0$$

Then we find conditions under which this expression is identically zero for all *t*:

SolveAlways[%, t]

$\{\{b \to -a\,c, d \to 0\}, \{b \to -a\,c, e \to -a\}, \{b \to -a\,c - a\,d, e \to 0\}, \{b \to 0, e \to 0, a \to 0\}\}$

The second solution gives the solution $-\frac{b}{a} + d\,e^{-a\,t}$ for the differential equation. This solution is the same as that obtained with **DSolve**:

DSolve[eqn, y[t], t] $\left\{\left\{y[t] \to -\frac{b}{a} + e^{-a\,t}\,C[1]\right\}\right\}$

Another example is in Section 26.2.4, p. 848, where we solve a Fredholm integral equation.

■ **Unstable Systems**

> *An option for* **Solve**:
>
> **VerifySolutions** Whether solutions are verified: extraneous solutions rejected or inaccurate numerical solutions improved; possible values: **Automatic, True, False**

When solving equations with *exact* coefficients with **Solve**, the solution is automatically verified. The verification is important when solving radical equations, where extraneous solutions easily emerge (see Section 22.2.3, p. 724). For equations with *inexact* coefficients, the verification is not done automatically. For numerically unstable systems, verifying the convergence may improve the solution. As an example, consider the following system:

```
p1 = -49.8333 + 703.3295 x^2 + 1022.7811 x y + 895.0554 y^2;
p2 = -54.8990 + 791.4604 x^2 + 1150.9409 x y + 959.1353 y^2;
```

We use **Solve** and **NSolve** and calculate the values of the polynomials **p1** and **p2** at the first solution:

```
{p1, p2} /. Solve[{p1 == 0, p2 == 0}][[1]]
```
$\{-0.0016004, -0.00180094\}$

```
{p1, p2} /. NSolve[{p1 == 0, p2 == 0}, {x, y}][[1]]
```
$\{3.3932 \times 10^{-9}, 4.21171 \times 10^{-9}\}$

NSolve gave a better solution. To get more accurate solutions with **Solve**, we use the **VerifySolutions** option (note that **NSolve** does not have this option):

```
{p1, p2} /. Solve[{p1 == 0, p2 == 0}, VerifySolutions → True][[1]]
```
$\{1.42109 \times 10^{-14}, 1.42109 \times 10^{-14}\}$

■ **Root Objects**

> **Root[f, k]** The k th root of the polynomial equation **f[x] == 0** (**f** must be a pure function)
> **RootReduce[expr]** Attempts to reduce **expr** to a single **Root** object
> **RootSum[f, form]** The sum of **form[x]** for all **x** that satisfy the polynomial equation **f[x] == 0**
> **ToRadicals[expr]** Attempts to express a **Root** object in terms of radicals

If **Solve** cannot solve an equation, it represents the solution by the **Root** object. Consider the following equation and its solution:

```
sol = Solve[a - x + x^5 == 0, x]
```
$\left\{\left\{x \to \text{Root}\left[a - \#1 + \#1^5 \&, 1\right]\right\}, \left\{x \to \text{Root}\left[a - \#1 + \#1^5 \&, 2\right]\right\},\right.$
$\left.\left\{x \to \text{Root}\left[a - \#1 + \#1^5 \&, 3\right]\right\}, \left\{x \to \text{Root}\left[a - \#1 + \#1^5 \&, 4\right]\right\}, \left\{x \to \text{Root}\left[a - \#1 + \#1^5 \&, 5\right]\right\}\right\}$

Root objects are exact but implicit representations for the roots. They can be seen as representations of algebraic numbers. Several calculations are possible with root objects. For example, we can plot a root as a function of parameter or find a series expansion or calculate the product of the roots:

```
Plot[Evaluate[x /. sol[[1]]], {a, -2, 1}]
```

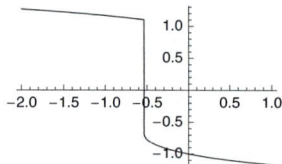

```
Series[x /. sol[[1]], {a, 0, 3}]
```

$$-1 - \frac{a}{4} + \frac{5\,a^2}{32} - \frac{5\,a^3}{32} + O[a]^4$$

```
Apply[Times, x /. sol /. a → 3] // FullSimplify    -3
```

RootSum can be used to calculate the sum of the values of a function at the solutions of a polynomial equation. We calculate the sum of roots and their inverses:

```
RootSum[a - # + #^5 &, # &]      0
```

```
RootSum[a - # + #^5 &, 1 / # &]    1/a
```

Some integrals are expressed in terms of **RootSum**:

```
Integrate[1 / (1 + x + x^3), {x, 0, ∞}]
```

$$-\text{RootSum}\!\left[1 + \#1 + \#1^3 \,\&,\ \frac{\text{Log}[-\#1]}{1 + 3\,\#1^2}\ \&\right]$$

```
% // N     0.921763+0. i
```

With **RootApproximant** (⌘6) we can find good approximations in terms of algebraic numbers:

```
Table[RootApproximant[π, i], {i, 1, 3}]
```

$$\left\{\frac{80\,143\,857}{25\,510\,582},\ \frac{198\,261 + \sqrt{105\,043\,517\,473}}{166\,274},\ \text{Root}\!\left[7622 + 2308\,\#1 - 2396\,\#1^2 + 283\,\#1^3\ \&,\ 2\right]\right\}$$

```
π - % // N
```

$$\left\{4.44089 \times 10^{-16},\ 0.,\ 2.66454 \times 10^{-15}\right\}$$

For algebraic number fields, see tutorial/AlgebraicNumberFields.

■ Detailed Solution

When solving equations, **Solve** produces solution candidates and then verifies which ones are correct. Note that **Solve** rejects only solution candidates that are incorrect for *all* values of parameters; candidates that are valid for at least some values of the parameters are accepted. Also note that the candidates that **Solve** accepts are *generic* solutions, which means that they are solutions that are valid for *general* values of the parameters; for special values of parameters, the solution may be different. For example, the solution for the general quadratic equation $a\,x^2 + b\,x + c = 0$ is valid only if a is not 0. If a is 0, the solution is $x = -c/b$. If b is also 0, then c must be 0 for the equation to be satisfied. Such a detailed solution can be obtained with **Reduce**.

Reduce[eqns, vars]   Give a detailed analysis of the solutions of given equations

Reduce[eqns, vars, dom]   Restrict all variables and parameters to belong to the given domain; examples of domains: **Complexes, Reals, Integers**

Options:

Cubics   Whether cubic equations are solved explicitly; possible values: **False, True**

Quartics   Whether quartic equations are solved explicitly; possible values: **False, True**

Backsubstitution   Whether values of later variables in the result are allowed to depend on earlier variables (**False**) or are given explicitly (**True**); possible values: **False, True**

Actually, **Reduce** can solve much more general conditions containing logical combinations of equations (**==**), inequations (**!=**), inequalities (**<**, **≤**, etc.), domain specifications (**x ∈ Reals**, etc.; see Section 13.2.1, p. 420), and universal (**∀** or **ForAll**) and existential (**∃** or **Exists**) quantifiers. For example,

Reduce[a x^2 + b x + c == 0, x]

$$\left(a \neq 0 \,\&\&\, \left(x == \frac{-b - \sqrt{b^2 - 4\,a\,c}}{2\,a} \,||\, x == \frac{-b + \sqrt{b^2 - 4\,a\,c}}{2\,a}\right)\right) \,||$$

$$\left(a == 0 \,\&\&\, b \neq 0 \,\&\&\, x == -\frac{c}{b}\right) \,||\, (c == 0 \,\&\&\, b == 0 \,\&\&\, a == 0)$$

Remember that **&&** means logical AND and **||** means logical OR. Later variables may be expressed in terms of earlier variables:

Reduce[{x + y == 1, x^2 - y == 2}, {x, y}]

$$\left(x == \frac{1}{2}\left(-1 - \sqrt{13}\right) \,||\, x == \frac{1}{2}\left(-1 + \sqrt{13}\right)\right) \,\&\&\, y == 1 - x$$

With the **Backsubstitution** option, however, we get explicit values:

Reduce[{x + y == 1, x^2 - y == 2}, Backsubstitution → True]

$$\left(y == \frac{1}{2}\left(3 - \sqrt{13}\right) \,\&\&\, x == \frac{1}{2}\left(-1 + \sqrt{13}\right)\right) \,||\, \left(y == \frac{1}{2}\left(3 + \sqrt{13}\right) \,\&\&\, x == \frac{1}{2}\left(-1 - \sqrt{13}\right)\right)$$

With **ToRules** we can get a list of rules:

{% // ToRules}

$$\left\{\left\{y \to \frac{1}{2}\left(3 - \sqrt{13}\right),\, x \to \frac{1}{2}\left(-1 + \sqrt{13}\right)\right\},\, \left\{y \to \frac{1}{2}\left(3 + \sqrt{13}\right),\, x \to \frac{1}{2}\left(-1 - \sqrt{13}\right)\right\}\right\}$$

Without a domain specification, the expression is reduced over complexes:

Reduce[x^2 + 2 y^2 == 9, {x, y}]

$$y == -\frac{\sqrt{9 - x^2}}{\sqrt{2}} \,||\, y == \frac{\sqrt{9 - x^2}}{\sqrt{2}}$$

Next, we reduce over reals:

Reduce[x^2 + 2 y^2 == 9, {x, y}, Reals]

$$-3 \leq x \leq 3 \,\&\&\, \left(y == -\frac{\sqrt{9 - x^2}}{\sqrt{2}} \,||\, y == \frac{\sqrt{9 - x^2}}{\sqrt{2}}\right)$$

If we ask for integer solutions, we are faced with a Diophantine equation:

Reduce[x^2 + 2 y^2 == 9, {x, y}, Integers]

(x == -3 && y == 0) || (x == -1 && y == -2) || (x == -1 && y == 2) ||
   (x == 1 && y == -2) || (x == 1 && y == 2) || (x == 3 && y == 0)

Now we find Pythagorean triples:

Table[{z, Reduce[x^2 + y^2 == z^2 && y > x > 0, {x, y}, Integers]}, {z, 13}]

{{1, False}, {2, False}, {3, False}, {4, False},
   {5, x == 3 && y == 4}, {6, False}, {7, False}, {8, False}, {9, False},
   {10, x == 6 && y == 8}, {11, False}, {12, False}, {13, x == 5 && y == 12}}

Select[%, #[[2]] =!= False &]

{{5, x == 3 && y == 4}, {10, x == 6 && y == 8}, {13, x == 5 && y == 12}}

In Sections 22.2.3, 22.2.4, 22.2.5, and 22.3.1 we use **Reduce** to solve radical equations, inequalities, quantified equations, and transcendental equations.

For more information about real, complex, and integer systems, see tutorial/RealReduce, tutorial/ComplexPolynomialSystems, and tutorial/DiophantineReduce.

■ Frobenius Equations

A Frobenius equation is a Diophantine equation of the form $a_1 x_1 + a_2 x_2 + \ldots + a_n x_n = b$, where the a_i are positive integers, b is an integer, and a solution $x_1, \ldots, x_n$ must consist of nonnegative integers. The Frobenius number of $a_1, \ldots, a_n$ is the largest integer b for which the corresponding Frobenius equation has no solutions.

`FrobeniusSolve[{a1, …, an}, b]` (❖6) Give all solutions of the Frobenius equation
`FrobeniusNumber[{a1, …, an}]` (❖6) Give the Frobenius number

Suppose we have many stamps of 5, 10, 20, and 50 cents. In how many ways can we pay a postage of 45 cents?

```
FrobeniusSolve[{5, 10, 20, 50}, 45]
```
```
{{1, 0, 2, 0}, {1, 2, 1, 0}, {1, 4, 0, 0}, {3, 1, 1, 0},
 {3, 3, 0, 0}, {5, 0, 1, 0}, {5, 2, 0, 0}, {7, 1, 0, 0}, {9, 0, 0, 0}}
```

Suppose we have many stamps of 5, 10, and 13 cents. We can pay all postages of at least 48 cents:

```
FrobeniusNumber[{5, 10, 13}]     47
```

■ Number of Roots; Isolating Intervals

`CountRoots[poly, x]` (❖6) Give the number of real roots
`CountRoots[poly, {x, a, b}]` Give the number of roots between **a** and **b**
`RootIntervals[poly]` (❖6) Give isolating intervals for real roots

In these commands, the polynomial can have rational coefficients.

The following polynomial has five real roots:

```
p = (x - 1/2)^3 (x^2 - 4) (x^2 + 1);
```

```
sol = Solve[p == 0]
```

$$\left\{ \{x \to -2\},\ \{x \to -i\},\ \{x \to i\},\ \left\{x \to \frac{1}{2}\right\},\ \left\{x \to \frac{1}{2}\right\},\ \left\{x \to \frac{1}{2}\right\},\ \{x \to 2\} \right\}$$

Count the number of real roots and find intervals for them:

```
CountRoots[p, x]     5
```

```
RootIntervals[p]     {{{-3, 0}, {0, 1}, {1, 3}}, {{1}, {1, 1, 1}, {1}}}
```

Thus, one root with multiplicity 1 is in $(-3, 0)$, one root with multiplicity 3 is in $(0, 1)$, and one root with multiplicity 1 is in $(1, 3)$.

22.2.3 Radical Equations

■ Solving Radical Equations

Solve, **NSolve**, **Eliminate**, and **Reduce** can also be used to solve radical equations, which are equations that contain rational powers. For example,

```
eqn = Sqrt[x + 1] + x^(1 / 3) == 2;
```

```
Solve[eqn]
```

$$\left\{\left\{x \to \frac{1}{3}\left(-2 - 161\left(\frac{2}{1703 + 459\sqrt{93}}\right)^{1/3} + \left(\frac{1}{2}\left(1703 + 459\sqrt{93}\right)\right)^{1/3}\right)\right\}\right\}$$

Reduce does not write an explicit solution for cubic or quartic equations if the solution is long:

```
Reduce[eqn]
```
$$x == \text{Root}\left[-27 + 55\,\#1 + 2\,\#1^2 + \#1^3 \,\&, \, 1\right]$$

With the **Cubics** and **Quartics** options, however, we get an explicit expression:

```
Reduce[eqn, Cubics → True]
```

$$x == \frac{1}{6}\left(-4 - 322\left(\frac{2}{1703 + 459\sqrt{93}}\right)^{1/3} + 2^{2/3}\left(1703 + 459\sqrt{93}\right)^{1/3}\right)$$

Another way is the use of **ToRadicals**:

```
ToRadicals[%%]
```

$$x == \frac{1}{3}\left(-2 - 161\left(\frac{2}{1703 + 459\sqrt{93}}\right)^{1/3} + \left(\frac{1}{2}\left(1703 + 459\sqrt{93}\right)\right)^{1/3}\right)$$

Here is another example:

```
f = Sqrt[3 x + 2] + Sqrt[2 x - 1] - 3 Sqrt[x - 1];
```

```
Solve[f == 0]
```
$$\left\{\left\{x \to \frac{3}{4}\left(-7 - \sqrt{73}\right)\right\}\right\}$$

■ Extraneous Solutions

When solving radical equations, extraneous solutions easily emerge (as the result of, for example, squaring). **Solve** automatically verifies solution candidates and rejects extraneous solutions if the equations contain exact coefficients. All solution candidates can be seen if the candidates are not verified. We continue with the preceding example:

```
sol = Solve[f == 0, VerifySolutions → False]
```

$$\left\{\left\{x \to \frac{3}{4}\left(-7 - \sqrt{73}\right)\right\}, \left\{x \to \frac{3}{4}\left(-7 + \sqrt{73}\right)\right\}\right\}$$

```
sol // N       {{x → -11.658}, {x → 1.158}}
```

In the preceding example, **Solve** rejected the second solution candidate. To graphically show the real root, we could plot f. However, because f is complex valued for $x < 1$, we plot $|f|$: f has a zero whenever $|f|$ has a zero. The figure shows that the first candidate is a solution but the second candidate is not:

```
Plot[Abs[f], {x, -20, 20}]
```

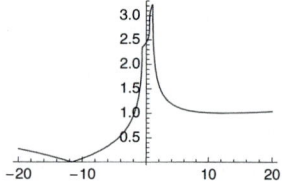

■ **Step-by-Step Solution**

Sometimes we would like to solve an equation by doing a series of transformations. Use **Thread** to apply a transformation to both sides of an equation:

Thread[f[expr1 == expr2], Equal]   Create the expression **f[expr1] == f[expr2]**

As an example, we again consider the familiar equation:

eqn = f == 0 $-3\sqrt{-1+x} + \sqrt{-1+2x} + \sqrt{2+3x} == 0$

First add $3\sqrt{x-1}$ to both sides and then square and expand both sides:

Thread[eqn + 3 Sqrt[x - 1], Equal] $\sqrt{-1+2x} + \sqrt{2+3x} == 3\sqrt{-1+x}$

Thread[%^2, Equal] $\left(\sqrt{-1+2x} + \sqrt{2+3x}\right)^2 == 9(-1+x)$

% // Expand $1 + 5x + 2\sqrt{-1+2x}\sqrt{2+3x} == -9 + 9x$

Subtract $1 + 5x$ from both sides, square both sides, move all terms into the left-hand side, and solve the resulting equation:

Thread[% - (1 + 5 x), Equal] $2\sqrt{-1+2x}\sqrt{2+3x} == -10 + 4x$

Thread[%^2, Equal] $4(-1+2x)(2+3x) == (-10+4x)^2$

Thread[% - (-10 + 4 x)^2, Equal] // Expand $-108 + 84x + 8x^2 == 0$

Solve[%] $\left\{\left\{x \to \frac{3}{4}\left(-7 - \sqrt{73}\right)\right\}, \left\{x \to \frac{3}{4}\left(-7 + \sqrt{73}\right)\right\}\right\}$

A verification shows that the second solution is extraneous:

f /. % // FullSimplify $\left\{0, \sqrt{-46 + 6\sqrt{73}}\right\}$

Similarly, we can, for example, take logarithms from both sides by **Thread[Log[eqn], Equal]** and add two equations by **Thread[eqn1 + eqn2, Equal]**.

22.2.4 Inequalities

■ **Solving Inequalities**

Reduce[expr, vars]   Give a detailed analysis of the solutions of given equations and inequalities
Reduce[expr, vars, dom]   Assume **dom** to be the domain of variables, parameters, and function values

---

In Section 22.2.2, p. 721, we used **Reduce** to get a detailed solution for equations. **Reduce** is also useful when solving inequalities and sets of equations and inequalities. In our first few examples, we do not have any parameters, and so the variable need not be declared:

Reduce[x^2 - 3 x - 2 > 0]

$x < \frac{1}{2}\left(3 - \sqrt{17}\right) \ || \ x > \frac{1}{2}\left(3 + \sqrt{17}\right)$

Remember that **&&** means logical AND and **||** means logical OR. All numbers should be exact. For problems with inexact numbers, **Reduce** solves a corresponding exact problem and numericizes the result:

> **Reduce[x^2 - 3 x - 2. > 0]**

Reduce::ratnz :

Reduce was unable to solve the system with inexact coefficients. The answer was obtained
by solving a corresponding exact system and numericizing the result. ≫
x < -0.561553 || x > 3.56155

Next, we solve a system of inequalities:

> **Reduce[{x^2 - 3 x - 2 > 0, Abs[x] < 1}]**

$$-1 < x < \frac{1}{2}\left(3 - \sqrt{17}\right)$$

Reduce gives a cylindrical algebraic decomposition of the region in question. **CylindricalDecomposition** is also available for the real domain.

■ **Specifying a Domain**

Reduce assumes that all quantities that appear algebraically in inequalities are real and that all other quantities are complex. With a third argument given to **Reduce**, we can restrict all variables, parameters, and function values to a given domain such as **Reals**, **Integers**, or **Complexes**. The domain **Reals** may be suitable in many problems with inequalities. In the following example, we get the condition that **a** and **b** should be real:

> **Reduce[{x + y == a, y < a b}, {x, y}]**

(a | b) ∈ Reals && Re[x] > a - a b && Im[x] == 0 && y == a - x

It is useful to declare that all variables are real:

> **Reduce[{x + y == a, y < a b}, {x, y}, Reals]**

x > a - a b && y == a - x

In Section 28.3.1, p. 950, we solve the following system of inequalities:

> **ineqs = Abs[2 - a] < 1 && Abs[b (a - 1) / a] < 1 && b ≥ 0;**

> **Reduce[ineqs, {a, b}, Reals]**

$$1 < a < 3 \&\& 0 \le b < \frac{a}{-1 + a}$$

We can plot the region satisfied by the inequalities with **RegionPlot** (see Section 5.2.5, p. 136):

> **RegionPlot[ineqs, {a, 0, 4}, {b, 0, 10}, AspectRatio → 1 / GoldenRatio]**

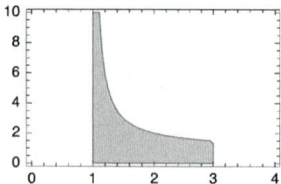

■ **Finding Instances**

FindInstance[expr, vars]   Find an instance of **vars** that makes **expr** be True

FindInstance[expr, vars, dom]   Assume domain **dom** for variables, parameters, and function values

FindInstance[expr, vars, dom, n]   Find **n** instances

A typical use of this command is to find a point that satisfies a set of inequalities (or, in terms of optimization, to find a feasible point). However, the expression can also contain equations and quantifiers. Inequalities in the following examples were considered previously with **Reduce**:

FindInstance[{x^2 - 3 x - 2 > 0, Abs[x] < 1}, {x}, Reals]

$$\left\{\left\{x \to -\frac{3}{4}\right\}\right\}$$

With the **RandomSeed** option we can get somewhat different points:

FindInstance[{x^2 - 3 x - 2 > 0, Abs[x] < 1}, {x}, Reals, 3, RandomSeed → 1]

$$\left\{\left\{x \to -\frac{655}{687}\right\}, \left\{x \to -\frac{184}{229}\right\}, \left\{x \to -\frac{143}{229}\right\}\right\}$$

SemialgebraicComponentInstances[ineqs, vars] (❀6)   Find at least one sample point in each connected component of the semialgebraic set defined by the inequalities **ineqs** in the variables **vars**

Find instances in which the following function is positive:

f = (x - 1) (x^2 - 7) (x^2 - 12);

s = SemialgebraicComponentInstances[f > 0, x]

$\{\{x \to -3\}, \{x \to 2\}, \{x \to 4\}\}$

Plot[f, {x, -3.8, 4.1}, Epilog → Point[{x, f} /. s], PlotRange → All]

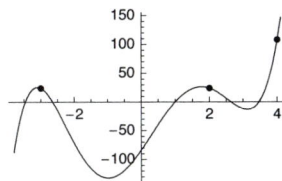

If we define a weak inequality, we also get the zeros:

s = SemialgebraicComponentInstances[f ≥ 0, x]

$$\left\{\left\{x \to -\frac{49}{16}\right\}, \{x \to 1\}, \left\{x \to \frac{15}{8}\right\}, \{x \to 4\}, \left\{x \to -2\sqrt{3}\right\}, \left\{x \to 2\sqrt{3}\right\}, \left\{x \to -\sqrt{7}\right\}, \left\{x \to \sqrt{7}\right\}\right\}$$

Plot[f, {x, -3.8, 4.1}, Epilog → Point[{x, f} /. %], PlotRange → All]

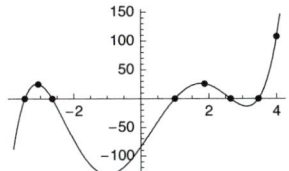

Find some points where the following function obtains negative values:

```
f = 10 x - x^3 + y^2;

s = SemialgebraicComponentInstances[f < 0, {x, y}]
```

$$\left\{\left\{x \to -\frac{3}{2}, \, y \to 0\right\}, \, \left\{x \to -\frac{3}{2}, \, y \to -\sqrt{6}\right\}, \, \left\{x \to -\frac{3}{2}, \, y \to \sqrt{6}\right\},\right.$$

$$\left.\{x \to 4, \, y \to 0\}, \, \left\{x \to 4, \, y \to -2\sqrt{3}\right\}, \, \left\{x \to 4, \, y \to 2\sqrt{3}\right\}\right\}$$

```
RegionPlot[f < 0, {x, -4, 6}, {y, -5, 5}, Epilog → Point[{x, y} /. s]]
```

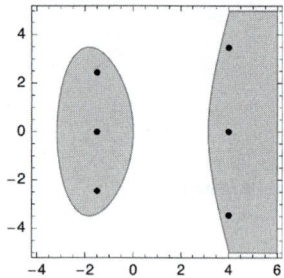

22.2.5 Quantifiers

In mathematics, we often use the quantifiers ∀ or "for all" and ∃ or "there exists." These can also be used in *Mathematica*.

ForAll[x, expr]   States that **expr** is True for all values of **x**

ForAll[x, cond, expr]   States that **expr** is True for all values of **x** satisfying the condition **cond**

ForAll[{x1, …, xn}, expr]   States that **expr** is True for all values of **x1, …, xn**

Exists[x, expr]   States that there exists a value of **x** for which **expr** is True

Exists[x, cond, expr]   States that there exists a value of **x** satisfying **cond** for which **expr** is True

Exists[{x1, …, xn}, expr]   States that there exist values of **x1, …, xn** for which **expr** is True

Implies[p, q]   States that **p** implies **q**

Resolve[expr]   Attempts to resolve **expr** into a form that eliminates **ForAll** and **Exists** quantifiers

Resolve[expr, dom]   Works over domain **dom**

Resolve[expr, vars, dom]   Solves for **vars**

In addition to **Resolve**, we can also use **Reduce** to eliminate quantifiers.

■ Example 1

Write the statement that, for three variables, harmonic mean ≤ geometric mean ≤ arithmetic mean:

```
ForAll[{x, y, z}, x > 0 && y > 0 && z > 0,
  3 / (1 / x + 1 / y + 1 / z) ≤ (x y z) ^ (1 / 3) ≤ (x + y + z) / 3]
```

$$\forall_{\{x,y,z\}, \, x>0\&\&y>0\&\&z>0} \; \frac{3}{\frac{1}{x}+\frac{1}{y}+\frac{1}{z}} \leq (x \, y \, z)^{1/3} \leq \frac{1}{3}(x+y+z)$$

Both **Reduce** and **Resolve** can be used to prove that this statement is true:

```
Reduce[%]    True
```

```
Resolve[%%]    True
```

■ Example 2

Write the statement that a quadratic expression is positive for all real x:

```
q = ForAll[x, {x, a, b, c} ∈ Reals, a x^2 + b x + c > 0]
```

$$\forall_{x, (x|a|b|c)\in Reals}\ c + b\,x + a\,x^2 > 0$$

Under what conditions is this statement true?

```
Reduce[q, {a, b, c}, Reals]
```

$$(a = 0\ \&\&\ b = 0\ \&\&\ c > 0)\ ||\ \left(a > 0\ \&\&\ c > \frac{b^2}{4\,a}\right)$$

Note that the result of **Reduce** may depend on the order in which the variables are declared; in this example, the simplest result is obtained by writing the variables as $\{a, b, c\}$. With **Resolve**, without declaring the variables, we get unsolved inequalities:

```
Resolve[q, Reals]
```

$$\left(a > 0\ \&\&\ -a\,b^2 + 4\,a^2\,c > 0\right)\ ||\ (a = 0\ \&\&\ b = 0\ \&\&\ c > 0)\ ||$$
$$\left(a \geq 0\ \&\&\ b = 0\ \&\&\ c > 0\ \&\&\ -a\,b^2 + 4\,a^2\,c > 0\right)$$

By declaring the variables, we get the same result as with **Reduce**:

```
Resolve[q, {a, b, c}, Reals]
```

$$(a = 0\ \&\&\ b = 0\ \&\&\ c > 0)\ ||\ \left(a > 0\ \&\&\ c > \frac{b^2}{4\,a}\right)$$

■ Example 3

Write the statement that there exists an x for which $a\,x^2 + b > 0$:

```
Exists[x, a x^2 + b > 0]        ∃ₓ b + a x² > 0
```

Under what conditions is this true?

```
Reduce[%, {a, b}, Reals]        (a ≤ 0 && b > 0) || a > 0
```

■ Example 4

Consider the following polynomial:

```
g[x_] := a + b x + c x^2 + x^3
```

Under what conditions are all roots equal?

```
q = ForAll[{x, y}, g[x] == 0 && g[y] == 0, x == y]
```

$$\forall_{\{x,y\},\,a+b\,x+c\,x^2+x^3=0\&\&a+b\,y+c\,y^2+y^3=0}\ x = y$$

```
Reduce[q, {c, a, b}]
```

$$\left(a = \frac{c^3}{27}\ \&\&\ b = \frac{c^2}{3}\right)\ ||\ (c = 0\ \&\&\ a = 0\ \&\&\ b = 0)$$

We can see that under these conditions all the roots are indeed the same:

```
Solve[(g[x] /. #) == 0, x] & /@ {ToRules[%]}
```

$$\left\{\left\{\left\{x \to -\frac{c}{3}\right\}, \left\{x \to -\frac{c}{3}\right\}, \left\{x \to -\frac{c}{3}\right\}\right\}, \{\{x \to 0\}, \{x \to 0\}, \{x \to 0\}\}\right\}$$

■ **Example 5**

Prove that if $y \le -x^2$, then $y \le x + 1$:

    ForAll[{x, y}, Implies[y ≤ -x^2, y ≤ x + 1]]

    $\forall_{\{x,y\}}$ Implies$\left[y \le -x^2, y \le 1 + x\right]$

    Reduce[%]   True

22.3 Transcendental Equations

22.3.1 Exact Solutions

If a transcendental equation or inequality is simple enough, **Solve** or **Reduce** may be able to get a solution; such equations are considered here. In most cases, we have to resort to iterative methods provided by **FindRoot**. These equations are considered in Sections 22.3.2 to 22.3.4.

> **Solve[eqns, vars]** Try to give some solutions of the given transcendental equations
>
> **Reduce[expr, vars, dom]** Try to give a complete solution of the given transcendental equations
>   and inequalities

■ **An Example**

A transcendental equation can be solved symbolically if the equation can be transformed into an equation in which a single transcendental function can be taken to be the variable and if the transformed equation can be solved for this transcendental function. The solution of the original equation is then obtained with the inverse function. As an example, we first try **Solve**:

    Solve[Sin[x] == Cos[x], x]

    Solve::ifun : Inverse functions are being used by Solve, so some
           solutions may not be found; use Reduce for complete solution information. ≫

    $\left\{\left\{x \to -\dfrac{3\pi}{4}\right\}, \left\{x \to \dfrac{\pi}{4}\right\}\right\}$

We obtained two solutions, with the warning that other solutions may exist. Indeed, the equation is valid whenever $x = \frac{\pi}{4} + n\pi$ and n is an integer. This is typical for solutions of transcendental equations given by **Solve**: We get some solutions, but other solutions may (and often do) exist. Then we try **Reduce**:

    Reduce[Sin[x] == Cos[x], x]

    $C[1] \in$ Integers && $\left(x == -2\,\text{ArcTan}\left[1 + \sqrt{2}\,\right] + 2\pi\,C[1] \,||\, x == -2\,\text{ArcTan}\left[1 - \sqrt{2}\,\right] + 2\pi\,C[1]\right)$

    % // FullSimplify

    $C[1] \in$ Integers && $(8\pi\,C[1] == 3\pi + 4x \,||\, \pi + 8\pi\,C[1] == 4x)$

Reduce was able to give all of the solutions: For an integer $C[1]$, solutions are $x = \frac{\pi}{4} + 2C[1]\pi$ and $x = -\frac{3\pi}{4} + 2C[1]\pi$, which is the same as $x = \frac{\pi}{4} + n\pi$.

By the way, **Solve** has the **InverseFunctions** option. The default value **Automatic** of this option means that inverse functions are used, and a warning message about possibly missing roots is printed. If we give the option the value **True**, then inverse functions are also used, but the warning is not printed. With the value **False**, inverse functions are not used:

> **Solve[Sin[x] == Cos[x], x, InverseFunctions → True]**

$$\left\{\left\{x \to -\frac{3\,\pi}{4}\right\},\ \left\{x \to \frac{\pi}{4}\right\}\right\}$$

> **Solve[Sin[x] == Cos[x], x, InverseFunctions → False]**

Solve::ifun2 : Cannot obtain a solution with the InverseFunctions –> False option setting. ≫

Solve[Sin[x] == Cos[x], x, InverseFunctions → False]

To save space, we now turn warnings about inverse functions off:

> **Off[Solve::"ifun"]**
> **Off[InverseFunction::"ifun"]**

■ More Examples

Here is a simple inequality:

> **Reduce[Sin[x] > 0, x, Reals]**

C[1] ∈ Integers && 2 π C[1] < x < π + 2 π C[1]

In the next example, we have hyperbolic functions:

> **Solve[Sinh[x] == Cosh[x] - 2, x] // FullSimplify**

{{x → -Log[2]}}

> **Reduce[Sinh[x] == Cosh[x] - 2, x]**

C[1] ∈ Integers && x == 2 i π C[1] - Log[2]

We can restrict variables and functions to be real:

> **Reduce[Sinh[x] == Cosh[x] - 2, x, Reals]**   x == -Log[2]

Sometimes the solution contains a product log function. The next example gives the definition of product log:

> **Solve[z == w Exp[w], w]**   {{w → ProductLog[z]}}

This shows that product log at z is the (principal) solution for w of $z = w\,e^w$. Here is another example:

> **f = a - x Exp[x^2];**

> **Solve[f == 0, x]**

$$\left\{\left\{x \to -\frac{\sqrt{\text{ProductLog}\left[2\,a^2\right]}}{\sqrt{2}}\right\},\ \left\{x \to \frac{\sqrt{\text{ProductLog}\left[2\,a^2\right]}}{\sqrt{2}}\right\}\right\}$$

From the result of **Reduce**, we see that the first solution is correct for $a \le 0$ and the second solution for $a \ge 0$.

```
Reduce[f == 0, x, Reals]
```

$$(a == 0 \, \&\& \, x == 0) \; || \; \left(a > 0 \, \&\& \, x == \frac{\sqrt{\text{ProductLog}\left[2\,a^2\right]}}{\sqrt{2}} \right) \; ||$$

$$\left(a \neq 0 \, \&\& \, a < 0 \, \&\& \, x == -\frac{\sqrt{\text{ProductLog}\left[2\,a^2\right]}}{\sqrt{2}} \right)$$

Now we turn the messages on:

```
On[Solve::"ifun"]
On[InverseFunction::"ifun"]
```

■ **Toward Numerical Methods**

The next equation is not solvable with **Solve** (or **Reduce**):

```
Solve[Sin[x] == x, x]
```

Solve::tdep : The equations appear to involve

  the variables to be solved for in an essentially non–algebraic way. ≫
```
Solve[Sin[x] == x, x]
```

However, this equation has the simple solution $x = 0$. If **Solve** or **Reduce** does not succeed, we can use numerical methods specially developed for transcendental equations. These methods include Newton's method, the secant method, and the bisection method. These methods are considered next.

22.3.2 Numerical Solutions

FindRoot is used to solve transcendental equations with iterative methods. The methods can be divided into two groups:

- a method that requires derivatives (Newton's method; this method needs one starting point);
- two methods not requiring derivatives (the secant method and Brent's method; these methods need two starting points).

 FindRoot decides the type of method to use from the number of starting points. Note that even when **FindRoot** starts with, for example, Newton's method, it may later move to other methods.

■ **Newton's Method**

The best-known method for solving a transcendental equation $f(x) = 0$ is Newton's method $x_{i+1} = x_i - f(x_i)/f'(x_i)$. It can be used in the following way:

FindRoot[eqn, {x, x0}] Find a solution for the equation starting from the point **x0**; use Newton's method

 If we write, in place of **{x, x0}**, a list **{x, x0, xmin, xmax}**, then iterations are stopped if the solution goes outside the interval (**xmin, xmax**). In place of an equation **expr1 == expr2**, we can also write a single expression **expr**, and then it is understood that the equation is **expr == 0**.

 Newton's method needs the derivative of the function. If the derivative or, more generally, the Jacobian cannot be calculated symbolically, a finite difference approximation is used.

 As an example, we solve the equation $e^{-x} - x^2 = 0$ by first defining and plotting the left-hand-side function:

```
f = Exp[-x] - x^2;

Plot[f, {x, -1, 2}]
```

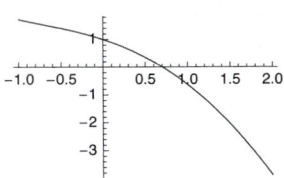

A root seems to be approximately 0.7. Apply Newton's method starting from zero:

```
sol = FindRoot[f, {x, 0}]     {x → 0.703467}
```

We could also have written an explicit equation:

```
sol = FindRoot[f == 0, {x, 0}]     {x → 0.703467}
```

The value of the function at the solution is, in fact, zero very accurately:

```
f /. sol     -1.4988 × 10^-15
```

If you only want the value of the zero (instead of the rule), then write the following:

```
x /. sol     0.703467
```

You can also write this:

```
sol = x /. FindRoot[f, {x, 0}]     0.703467
```

Complex zeros, too, can be searched by giving complex starting points:

```
FindRoot[f, {x, -1 + I}]     {x → -1.58805 + 1.54022 i}
```

■ The Secant Method and Brent's Method

Newton's method requires the derivative of the function, but we can also use methods that do not need the derivative. The secant method and Brent's method are two of the best-known methods of this kind. They can be used in the following way:

> `FindRoot[eqn, {x, x0, x1}]`   Find a solution starting from the points `x0` and `x1`; use the secant method or Brent's method

If we write, in place of `{x, x0, x1}`, a list `{x, x0, x1, xmin, xmax}`, then iterations are stopped if the solution goes outside the interval (`xmin, xmax`). *Mathematica* uses Brent's method if the values of the function at the points `x0` and `x1` are real and of opposite sign; otherwise, *Mathematica* uses the secant method.

The secant method applies the following recursion formula:

$$x_{n+1} = x_n - \frac{f(x_n)(x_n - x_{n-1})}{f(x_n) - f(x_{n-1})}.$$

Brent's method keeps the root bracketed (we always have one point where the function is positive and one point where the function is negative), and at each step a choice is made between an interpolated (secant) step and a bisection in such a way that convergence is guaranteed. For example,

```
FindRoot[Exp[-x] - x^2, {x, 0, 1}]     {x → 0.703467}
```

■ **Several Transcendental Equations**

> FindRoot[{eqn1, eqn2, ... }, {x, x0}, {y, y0}, ...]   Use Newton's method
> FindRoot[{eqn1, eqn2, ... }, {x, x0, x1}, {y, y0, y1}, ...]   Use the secant method

As an example, we solve the following pair of equations:

    eqns = {x^2 + y^2 - 1 == 0, Sin[x] - y == 0};

The situation can be visualized with contour plots:

    ContourPlot[eqns // Evaluate, {x, -2, 2}, {y, -1, 1},
      Frame → False, Axes → True, AspectRatio → Automatic]

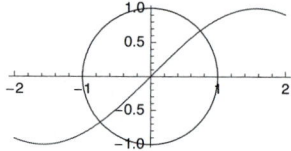

The system seems to have two solutions:

    FindRoot[eqns, {x, 1}, {y, 1}]
    {x → 0.739085, y → 0.673612}

    FindRoot[eqns, {x, -1}, {y, -1}]
    {x → -0.739085, y → -0.673612}

Simultaneous polynomial equations can also be formulated with matrices. Here, we find a normalized eigenvector and eigenvalue:

    a = {{1, -1, 1}, {-2, -0, 1}, {2, 1, 1}};

    FindRoot[{a.x == λ x, x.x == 1}, {{x, {1, 1, 1}}, {λ, 1}}]
    {x → {-0.222336, 0.738913, 0.636061}, λ → 1.4626}

The dimensions of the variables are taken from the dimensions of the starting points.

■ **Options**

As usual, the default values of options are mentioned first.

> *Options of* **FindRoot**:
>
> **WorkingPrecision**   Precision used in internal computations; examples of values:
>   **MachinePrecision, 20**
> **PrecisionGoal**   If the value of the option is **p**, the relative error of the root should be of the order
> 10^{-p}; examples of values: **Automatic** (usually means **8**), **10**
> **AccuracyGoal**   If the value of the option is **a**, the absolute error of the root and the absolute value of
>   the function at the root should be of the order 10^{-a}; examples of values: **Automatic** (usually means
>   **8**), **10**
> **Method**   Method used; possible values: **Automatic, "Newton", "Secant", "Brent"**
> **MaxIterations**   Maximum number of iterations used; default value: **100**
> **DampingFactor**   Damping factor; examples of values: **1, 2**
> **Jacobian**   Jacobian of the system in Newton's method; examples of values: **Automatic, "Symbolic",**
>   **"FiniteDifference"**

Compiled Whether the function should be compiled; possible values: **Automatic, True, False**

Evaluated Whether the function is evaluated; possible values: **True, False**

StepMonitor Command to be executed after each step of the iterative method; examples of values:

  None, Sow[x], ++n, AppendTo[iters, x]

EvaluationMonitor Command to be executed after each evaluation of the equation; examples of

  values: **None, Sow[x], ++n, AppendTo[points, x]**

The default is that iterations are stopped when the relative or absolute error of *the root* is less than 10^{-8} and the absolute value of *the function at the root* is less than 10^{-8}. The general default value of **PrecisionGoal** and **AccuracyGoal** is **WorkingPrecision**/2. For more information about these options, see Section 12.3.1, p. 409.

A slow convergence may be an indication of a multiple zero. If the zero is of multiplicity d, we get a quadratic convergence again if we give **DampingFactor** the value d. The iteration formula for Newton's method is now $x_{i+1} = x_i - d\, f(x_i) / f'(x_i)$. For equations in which the calculation of the Jacobian may cause problems, we can give **Jacobian** the value **FiniteDifference**, and then the Jacobian is approximated by numeric methods.

22.3.3 Special Topics

■ Looking at the Iterations: One Equation

Consider the following function:

```
f = Exp[-x] - x^2;
```

To see the iterations, write the following:

```
{zero, {points}} = Reap[FindRoot[f, {x, 1}, StepMonitor :> Sow[{x, f}]]]
```

$\{\{x \to 0.703467\}, \{\{\{0.733044, -0.0569084\}, \{0.703808, -0.000647392\},$
$\{0.703467, -8.7166 \times 10^{-8}\}, \{0.703467, -1.4988 \times 10^{-15}\}\}\}\}$

```
TableForm[points, TableHeadings -> {Range[Length[points]], {"x", "f(x)"}}]
```

	x	f(x)
1	0.733044	−0.0569084
2	0.703808	−0.000647392
3	0.703467	-8.7166×10^{-8}
4	0.703467	-1.4988×10^{-15}

■ Looking at the Iterations: Two Equations

Consider two equations:

```
eqns = {x^2 + y^2 - 1 == 0, Sin[x] - y == 0};
```

```
{zero, {points}} = Reap[FindRoot[eqns, {x, 1}, {y, -1}, StepMonitor :> Sow[{x, y}]]]
```

$\{\{x \to 0.739085, y \to 0.673612\},$
$\{\{\{1.29182, -0.658184\}, \{1.4143, -0.196652\}, \{1.20721, 0.955503\},$
$\{0.777205, 0.781706\}, \{0.743713, 0.677414\}, \{0.739096, 0.673627\},$
$\{0.739085, 0.673612\}, \{0.739085, 0.673612\}\}\}\}$

```
ContourPlot[eqns // Evaluate, {x, -1, 1.8}, {y, -1, 1}, Frame → False, Axes → True,
    AspectRatio → Automatic, Epilog → {Point[points], Line[points]}, ImageSize → 180]
```

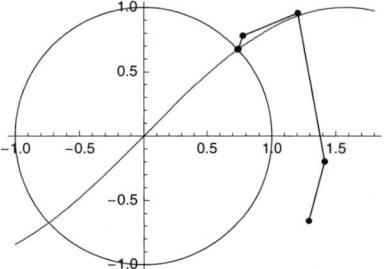

■ Zeros of Bessel and Airy Functions

We have special commands for the roots of the Bessel functions $J_n(x)$ (first kind) and $Y_n(x)$ (second kind) and of the Airy functions Ai(x) and Bi(x).

BesselJZero[n, k] (⌘6) The **k**th zero greater than 0 of $J_n(x)$
BesselYZero[n, k] (⌘6) The **k**th zero greater than 0 of $Y_n(x)$
AiryAiZero[k] (⌘6) The **k**th zero less than 0 of Ai(x)
AiryBiZero[k] (⌘6) The **k**th zero less than 0 of Bi(x)

In place of **k** we can also write a list of values; then we get the corresponding list of zeros. These commands also accept one additional argument, **x0**, and then we ask for the **k**th zero greater than **x0** for Bessel functions and for the **k**th zero smaller than **x0** for Airy functions.

The **NumericalMath`BesselZeros`** package defines more commands for zeros of various expressions that contain Bessel functions. It can be loaded from library.wolfram.com/infocenter/MathSource/6777/.

Zeros of Bessel functions are sometimes needed in the solutions of partial differential equations (see Section 27.2.4, p. 901).

Consider, for example, **BesselJ[2, x]**:

```
Plot[BesselJ[2, x], {x, 0, 20}]
```

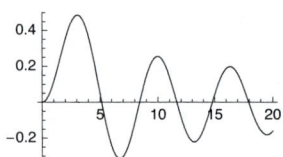

```
BesselJZero[2, Range[5]] // N
{5.13562, 8.41724, 11.6198, 14.796, 17.9598}
```

In addition, 0 is a zero.

■ Inverse Cubic Interpolation

In the **FunctionApproximations`** *package:*
InterpolateRoot[f, {x, a, b}] Find a zero for **f** near the starting points **a** and **b**

This package uses inverse cubic interpolation. The idea of this method is to take the last four points—for example a, b, c, and d—to calculate a cubic interpolation polynomial through the points $(f(a), a)$, $(f(b), b)$, $(f(c), c)$, and $(f(d), d)$ and to calculate the value of this polynomial at 0. This package is designed only for simple roots of a single function or an equation that is well behaved. The method can be advantageous in comparison to **FindRoot** in cases in which evaluating the function is extremely laborious, particularly for very high precision. The default value of **WorkingPrecision** is 40.

```
<< FunctionApproximations`
```

```
r = InterpolateRoot[Exp[-x] - x^2, {x, 0, -2.5}]
```

$\{x \to 0.7034674224983916520498186\}$

```
Exp[-x] - x^2 /. r     0. × 10^-27
```

22.3.4 Own Programs

■ Own Newton

In Chapter 18, we presented several implementations for Newton's method. Here, we present a version of the function **newton8** that was presented in Section 18.3.4, p. 579. We add a damping factor, which is used for multiple zeros and explained in Section 22.3.2, p. 735.

```
newtonSolve[f_, x_, x0_, d_: 1, n_: 20, opts___?OptionQ] :=
 With[{df = D[f, x]},
  FixedPointList[(x - d f / df) /. x -> # &, N[x0], n, opts]]
```

You may want to read Section 18.3.4 for explanations of this program. The iterations are stopped if two successive points are the same to 16-digit precision. The damping factor **d** has the default value 1. At most **n** iterations are done; **n** has the default value 20. We can write zero or more options (note the three underscores after **opts**). The option we can use is **SameTest** (see Section 18.3.4). (Note that if you want to use a different **n**, you must also enter a value for **d**, and if you want to write an option, you must also write values for **d** and **n**.) We try **newtonSolve** for the equation $f = 0$ that we considered previously:

```
f = Exp[-x] - x^2;
```

```
newtonSolve[f, x, 1]
```

$\{1., 0.733044, 0.703808, 0.703467, 0.703467, 0.703467, 0.703467\}$

Complex starting values can be given:

```
newtonSolve[f, x, -1.5 + 1.5 I]
```

$\{-1.5 + 1.5 \, i, -1.59551 + 1.54133 \, i, -1.58809 + 1.54021 \, i,$
$-1.58805 + 1.54022 \, i, -1.58805 + 1.54022 \, i, -1.58805 + 1.54022 \, i\}$

Consider the following function:

```
f2 = x^3 / 3 - x + 2 / 3;
```

```
Plot[f2, {x, -3, 3}]
```

It seems that a point at approximately 1 is a zero of multiplicity 2, so we use a damping factor of 2. We also use a custom stopping criterion. We stop the iterations after two successive approximations to the zero (represented by **#1** and **#2** in the **SameTest** option) differ by at most 10^{-6}:

```
newtonSolve[f2, x, 4, 2, 20, SameTest → (Abs[#1 - #2] < 10^-6 &)]
{4., 1.6, 1.04615, 1.00035, 1., 1.}
```

Then we stop after the value of the function at the last point is less than 10^{-14}:

```
newtonSolve[f2, x, 4, 2, 20, SameTest → (Abs[f2 /. x → #2] < 10^-14 &)]
{4., 1.6, 1.04615, 1.00035, 1.}
```

■ Illustrating Newton's Method

Newton's method for an equation $f(x) = 0$ is interpreted as follows. A tangent to $f(x)$ is drawn at the present point x_i. The next point x_{i+1} is where the tangent intersects the x axis. A new tangent is drawn at this point, and so we continue. To graphically show this process, we first form a set of points that consists of the iteration points at the x axis and at the function:

```
f = Exp[-x] - x^2;

it = newtonSolve[f, x, -2.5];

points = Flatten[{{#, 0}, {#, f /. x → #}} & /@ it, 1];
```

Then we show how Newton's method proceeds (the starting point is at the left: $x_0 = -2.5$):

```
Plot[f, {x, -2.7, 1.7}, Epilog → Line[points], ImageSize → 210]
```

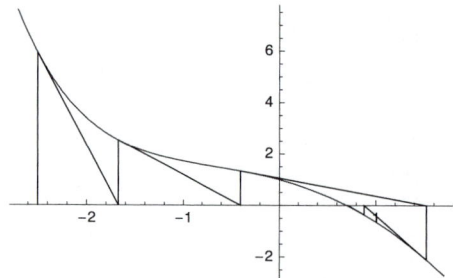

■ Own Newton for Several Equations

```
newtonSolveSystem[f_List, x_List, x0_List, eps_: 10^-6, n_: 20] :=
 With[{jac = D[f, {x}]},
  FixedPointList[# + LinearSolve[jac /. Thread[x → #], -f /. Thread[x → #]] &,
   N[x0], n, SameTest → (Norm[#1 - #2] < eps &)]]
```

This program is a generalization of **newtonSolve** for several equations. The stopping criterion is now the 2-norm (or the square root of the sum of the squares) of the difference between the last two iterations; we can give the **eps** for this criterion. The new point could be calculated from $x_{n+1} = x_n - J_n^{-1} f_n$, where J_n is the Jacobian of the system (**jac** in the program). However, we can avoid calculating the inverse of J_n by solving the linear system $J_n \delta_x = -f_n$ and then calculating $x_{n+1} = x_n + \delta_x$. This method is faster. For example,

```
f1 = x^2 + y^2 - 1; f2 = Sin[x] - y;
```

```
newtonSolveSystem[{f1, f2}, {x, y}, {1, -1}]
```

```
{{1., -1.}, {3.91816, 2.41816}, {2.75207, 0.130969},
 {1.48707, 1.54999}, {0.900059, 0.947405}, {0.766791, 0.700529},
 {0.739527, 0.674194}, {0.739085, 0.673612}, {0.739085, 0.673612}}
```

```
ContourPlot[{f1 == 0, f2 == 0}, {x, -1, 4}, {y, -1, 2.5}, Frame → False, Axes → True,
 AspectRatio → Automatic, Epilog → {Point[%], Line[%]}, ImageSize → 160]
```

■ Own Secant in Procedural Style

```
secantSolve[f_, x_, x0_, x1_, eps_: 10^-10, d_: 1, n_: 20] :=
 Module[{y0 = x0, y1 = x1, f0 = N[f /. x → x0], f1, newy, iters = {x0, x1}},
  Do[f1 = N[f /. x → y1];
   If[Abs[f1] < eps, Break[]];
   newy = y1 - d f1 (y1 - y0) / (f1 - f0);
   iters = {iters, newy};
   y0 = y1; y1 = newy; f0 = f1, {n}];
  Flatten[iters]]
```

This program is in the procedural style. If the value of the function becomes smaller than **eps** (default value is 10^{-10}), then iterations are stopped. Iterations are done at most **n** times (the default value is 20). The damping factor **d** has the default value 1.

The iterations are gathered to **iters**. In general, a module prints the result of the last command so that we have placed the simple command **Flatten[iters]** as the last command to get a list of all of the iterations. **Flatten** is needed because **iters = {iters, newy}** creates a nested list.

Note that we have used **y0** and **y1** to store the two successive iterations. We cannot use **x0** and **x1** because **x0** and **x1** are arguments of **secantSolve**, and the arguments of a function cannot be changed inside the function (see Section 18.2.4, p. 566). For example,

```
f = Exp[-x] - x^2;
```

```
secantSolve[f, x, 0.2, 0.5]
```
```
{0.2, 0.5, 0.753338, 0.699275, 0.703386, 0.703468, 0.703467}
```

■ Illustrating the Secant Method

The next point is where the secant through the last two points intersects the *x* axis.

```
it = secantSolve[f, x, -2.5, -2.0]; points = {};
```

```
Do[AppendTo[points, {it[[i]], 0}];
 AppendTo[points, {it[[i]], f /. x → it[[i]]}];
 AppendTo[points, {it[[i + 1]], f /. x → it[[i + 1]]}];
 AppendTo[points, {it[[i + 2]], 0}], {i, Length[it] - 2}]
```

```
Plot[f, {x, -2.7, 1.7}, Epilog → Line[points], ImageSize → 180]
```

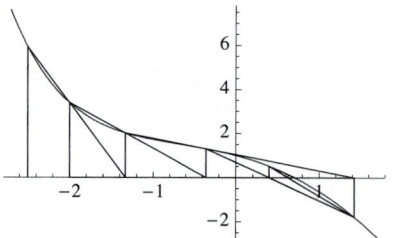

■ **Own Secant in Functional Style**

```
secantStep2[f_, x_, d_, {x0_, x1_, f0_, f1_}] :=
 With[{newx = x1 - d f1 (x1 - x0) / (f1 - f0)},
  {x1, newx, f1, f /. x → newx}]

secantSolve2[f_, x_, x0_, x1_, eps_: 10^-10, d_: 1, n_: 20] :=
 FixedPointList[secantStep2[f, x, d, #] &,
   {x0, x1, f /. x → x0, f /. x → x1} // N, n,
   SameTest → (Abs[#2[[4]]] < eps &)][[All, 2]]
```

This program is in the functional style. Now we have four values to be iterated, and we enclose them in a list {x0, x1, f0, f1} in **secantStep2** because **FixedPointList** requires one iteration variable (denoted by #). Note how easily one step can be done: Just calculate the new value (denoted by **newx**) and return the new values {x1, newx, f1, N[f /. x → newx]}.

The function **secantSolve2** does the needed iterations. The starting point is {x0, x1, f /. x → x0, f /. x → x1}. The stopping criterion is that the value of the function at the latest approximation is less than **eps**. The variable #2 represents the latest iteration that consists of the four values {x0, x1, f /. x → x0, f /. x → x1} so that #2[[4]] is the value of the function at the latest approximation.

The result of **FixedPointList** is an ($m \times 4$)-matrix if m iterations were done. By taking the second column, the values of the approximations to the root are obtained (however, the initial guess **x0** is lacking). For example,

```
secantSolve2[f, x, 0.2, 0.5]
```

```
{0.5, 0.753338, 0.699275, 0.703386, 0.703468, 0.703467}
```

Optimization

Introduction

> *Because the shape of the whole universe is most perfect and, in fact, designed*
> *by the wisest creator, nothing in the world will occur in which no maximum*
> *or minimum rule is somehow shining forth.—Leonhard Euler*

Mathematica contains tools for all basic types of optimization: We can solve linear and nonlinear, unconstrained and constrained, local and global, and continuous and discrete problems. The main minimization commands are **LinearProgramming**, **Minimize**, **NMinimize**, and **FindMinimum**; for maximization, we have **Maximize**, **NMaximize**, and **FindMaximum**. The problem may be that of choice: Given a problem, which of the commands is the most suitable one? The following list gives some guides for minimization:

- Linear optimization in matrix form: `LinearProgramming`
- Linear optimization in variable form: `Minimize`
- Nonlinear, global, exact optimization: `Minimize`
- Nonlinear, global, numerical optimization: `NMinimize`
- Nonlinear, local, numerical optimization: `FindMinimum`

All the methods apply to continuous and discrete and to unconstrained and constrained optimization. `LinearProgramming` and `Minimize` can find the *global* optimum for linear and polynomial problems, respectively. These commands give the *exact* solution if the problem is exact. `NMinimize` often also finds the global optimum. `FindMinimum` only searches for a local minimum near the starting point.

`Minimize` works mainly with polynomial optimization problems but can also solve many transcendental and piecewise problems, and it is also able to solve problems with symbolic parameters. This command is mainly suited to small exact problems. `NMinimize` is also suited to larger nonlinear problems. For large problems, `FindMinimum` may be suitable if we only need a local optimum, if the problem only has a single optimum, if a good starting point can be provided, or if the problem only has a small number of local optimums (they can perhaps be investigated separately by using suitable starting values).

`LinearProgramming`, `Minimize`, and `NMinimize` solve linear problems with the simplex or revised simplex method or with an interior point method; for linear integer problems, the branch-and-bound algorithm is used. For polynomial problems, `Minimize` uses cylindrical algebraic decomposition. `NMinimize` uses various derivative-free iterative methods: the method of Nelder and Mead, a genetic method, the method of simulated annealing, and a random search method. `FindMinimum` uses various iterative methods, most of which require the derivative: Mainly the quasi-Newton method is used, but also Newton's method, a conjugate gradient method, Brent's principal axis method (derivative-free), the Levenberg–Marquardt method, and an interior point method are available.

Note that even if we only want to get a local minimum, it is not necessary to use `FindMinimum`. Indeed, `Minimize` can be used by suitably restricting the area where the optimum is searched for. The latter command has the advantage that we can get an exact solution.

In addition to the built-in commands, we also consider classical optimization in which necessary conditions are used to find the optimum. The conditions are usually equations that contain derivatives of the object function or of a modified function if constraints are present. By solving the (nonlinear) equations, we get solution candidates for the problem. Often, we can use some sufficiency conditions to check whether the solution candidates are maximum or minimum points.

Furthermore, we also consider some special topics: the traveling salesman problem, dynamic programming, and calculus of variations.

For more information about optimization with *Mathematica*, see Bhatti (2000), Hastings (2006), and the tutorials tutorial/UnconstrainedOptimizationOverview and tutorial/ConstrainedOptimizationOverview. With `ExampleData` we have access to many test problems in linear optimization (see Section 9.3.4, p. 311). The **Optimization`UnconstrainedProblems`** package contains test problems in unconstrained optimization; see tutorial/UnconstrainedOptimizationTestProblems. In tutorial/ConstrainedOptimizationLocalNumerical we can find examples of solving minimax problems and using goal programming.

In the **Combinatorica`** package, there are several optimization functions for graph-theoretical problems: `Dijkstra`, `ShortestPath`, `MinimumSpanningTree`, `NetworkFlow`, etc. For integer programming, see Bulmer and Carter (1996); for genetic programming, see Nachbar (1995); and for dynamic programming, see Hastings (2006) and Wagner (1995). A multiplier method for constrained nonlinear problems can be found at library.wolfram.com/database/MathSource/795.

The following commercial products are available (see www.wolfram.com/products/fields):

- *Global Optimization* (global optimization for constrained and unconstrained nonlinear functions)
- *Industrial Optimization* (local optimization for linear, nonlinear, and queuing problems)
- *KNITRO for Mathematica* (large-scale nonlinear optimization)
- *MathOptimizer* (advanced modeling and optimization)
- *MathOptimizer Professional* (advanced global and local nonlinear optimization)
- *Operations Research* (constrained optimization with applications from operations research)

23.1 Global Optimization

23.1.1 Exact Global Optimization

■ Exact Global Minimums and Maximums

`Minimize[f, vars]`	Give the global minimum of **f** with respect to variables **vars**
`Minimize[{f, cons}, vars]`	Minimize subject to constraints **cons**
`Minimize[{f, cons}, vars, dom]`	Minimize over domain **dom**
`Maximize[f, vars]`	Give the global maximum of **f** with respect to variables **vars**
`Maximize[{f, cons}, vars]`	Maximize subject to constraints **cons**
`Maximize[{f, cons}, vars, dom]`	Maximize over domain **dom**

The function to be minimized or maximized and the equality and inequality constraints can be *algebraic expressions*. Typical examples are polynomial, rational, and radical expressions. Many transcendental problems are also solved. The functions can contain symbolic parameters.

The constraints are written as lists or as logical expressions containing, for example, the logical AND (`&&`). By specifying the domain to be `Integers`, we can solve integer problems (the default is to optimize over `Reals`). Within the constraints we can also restrict individual variables to be integer valued by writing, for example, `x ∈ Integers`.

`Minimize` and `Maximize` give an *exact* solution if the problem is exact—that is, does not contain decimal numbers. If the problem contains a decimal number, then actually `NMinimize` and `NMaximize` are used; these commands are considered in Section 23.1.2, p. 747.

`Minimize` and `Maximize` give the *global* optimum in the region in which the constraints hold. If the global optimum is not unique, the commands pick one of them. If we are interested in special local optimums, we have to add suitable constraints so that in the resulting feasible region, the local optimum we are searching for is also the global optimum.

`Minimize` and `Maximize` use *cylindrical algebraic decomposition* to solve the optimization problem. This method is not an iterative method; thus, it can give the exact solution. However, the method has double-exponential complexity in the number of variables, and this means that the method is not applicable to large nonlinear problems. For large problems or other problems not solved by the mentioned commands, we can resort to the numerical, iterative methods provided by `NMinimize` and `NMaximize`. For problems with equality constraints within a bounded box, `Minimize` and `Maximize` use the method of Lagrange multipliers. The application of `Minimize` and `Maximize` to solve linear problems is considered in Section 23.2.1, p. 753.

■ Example 1

Consider the following function:

```
f = 5 + 40 x^3 - 45 x^4 + 12 x^5;
```

```
Plot[f, {x, -0.7, 2.4}]
```

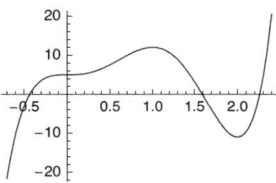

When using **Minimize**, we should remember that it calculates the *global* minimum. Thus, without any restrictions, the global minimum value of the function in our example is $-\infty$:

```
Minimize[f, x]
```

Minimize::natt : The minimum is not attained at any point satisfying the given constraints. ≫
$\{-\infty, \{x \to -\infty\}\}$

If we want to calculate the local minimum at approximately $x = 2$, we have to write suitable constraints to force the search to a region in which that local minimum is also the global minimum:

```
Minimize[{f, x > 0}, x]      {-11, {x → 2}}
```

Thus, at $x = 2$, we have a local minimum where the value of the function is -11. Next, we find a local maximum:

```
Maximize[{f, x < 3 / 2}, x]      {12, {x → 1}}
```

If we have a strict inequality constraint, the result may be that an optimum does not exist. In such a case, **Minimize** gives the point on the boundary:

```
Minimize[{f, x > 2}, x]
```

Minimize::wksol : Warning: There is no minimum in the
    region described by the constraints; returning a result on the boundary. ≫
$\{-11, \{x \to 2\}\}$

■ Example 2

The following function has a local minimum and a local maximum:

```
f = x^3 + y^3 + 2 x^2 + 4 y^2 + 6;
```

```
ContourPlot[f, {x, -2.5, 1}, {y, -3.5, 1}, Contours → 12]
```

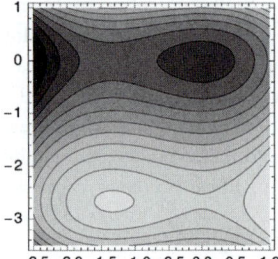

```
Minimize[{f, -1 < x < 1, -1 < y < 1}, {x, y}]
```

$\{6, \{x \to 0, y \to 0\}\}$

```
Maximize[{f, -2 < x < -1, -3 < y < -2}, {x, y}]
```

$\left\{\dfrac{50}{3}, \left\{x \to -\dfrac{4}{3}, y \to -\dfrac{8}{3}\right\}\right\}$

■ **Example 3**

The following function seems to have a minimum:

```
f = x^4 + 3 x^2 y + 5 y^2 + x + y;
```

```
ContourPlot[f, {x, -3, 3}, {y, -3, 3}, PlotRange → All]
```

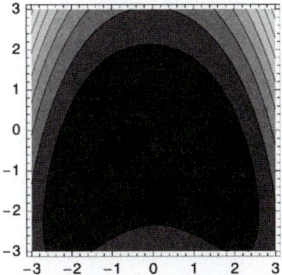

Here is the exact solution:

```
sol = Minimize[f, {x, y}]
```

$\{\text{Root}[6793 - 11\,208\,\#1 + 8976\,\#1^2 + 38\,720\,\#1^3\,\&,\,1],$
$\quad \{x \to \text{Root}[5 - 3\,\#1 + 11\,\#1^3\,\&,\,1], y \to \text{Root}[43 + 336\,\#1 + 2244\,\#1^2 + 4840\,\#1^3\,\&,\,1]\}\}$

The solution is expressed with the aid of **Root** objects (see Section 22.2.2, p. 720). We can ask for explicit expressions:

```
sol // ToRadicals
```

$\left\{\dfrac{1}{220}\left(-17 - 1653\left(\dfrac{2}{78\,373 - 1355\,\sqrt{2981}}\right)^{1/3} - 3\left(\dfrac{1}{2}\left(78\,373 - 1355\,\sqrt{2981}\right)\right)^{1/3}\right),\right.$

$\left\{x \to -\left(\dfrac{2}{11\left(55 - \sqrt{2981}\right)}\right)^{1/3} - \dfrac{\left(\frac{1}{2}\left(55 - \sqrt{2981}\right)\right)^{1/3}}{11^{2/3}},\right.$

$\left.\left. y \to \dfrac{1}{110}\left(-17 - 3\left(\dfrac{2}{273 - 5\,\sqrt{2981}}\right)^{1/3} - 3\left(\dfrac{1}{2}\left(273 - 5\,\sqrt{2981}\right)\right)^{1/3}\right)\right\}\right\}$

The numerical values are as follows:

```
sol // N
```

$\{-0.832579, \{x \to -0.886324, y \to -0.335671\}\}$

■ **Example 4**

The following problem has both equality and inequality constraints:

```
Minimize[{(x - y)^2 + 5 z, x + y + z == -5, y - 3 z == 1, x ≥ 0}, {x, y, z}]
```

$$\left\{\frac{19}{4}, \left\{x \to 0, y \to -\frac{7}{2}, z \to -\frac{3}{2}\right\}\right\}$$

Here is a constrained nonlinear integer problem:

```
Minimize[{x^2 + x y + z, x + x y ≥ 5, x ≥ 0, y ≥ 0, z ≥ 0}, {x, y, z}, Integers]
```

$$\{5, \{x \to 1, y \to 4, z \to 0\}\}$$

Here is a larger linear integer problem:

```
vars = Array[x, 15]; SeedRandom[7]; a = RandomInteger[{-10, 10}, {15, 15}];

Minimize[
  {Total[vars], Join[Thread[a.vars ≥ 1], Thread[vars ≥ 0], {vars ∈ Integers}]}, vars]
```

$$\{26, \{x[1] \to 8, x[2] \to 0, x[3] \to 3, x[4] \to 0, x[5] \to 0, x[6] \to 0, x[7] \to 1, x[8] \to 0,$$
$$x[9] \to 0, x[10] \to 0, x[11] \to 6, x[12] \to 0, x[13] \to 1, x[14] \to 5, x[15] \to 2\}\}$$

The following problem is somewhat difficult:

```
Minimize[{x^2 / 2 + (y^2 + z^2 + v^2) / 6, x ≥ 2, x + y ≥ 5, x + z ≥ 2, x + v ≥ 1, v ≥ 0},
  {x, y, z, v}] // Timing
```

$$\left\{24.6092, \left\{\frac{7}{2}, \{x \to 2, y \to 3, z \to 0, v \to 0\}\right\}\right\}$$

Note that all constraints except the constraint $x + v \geq 1$ are active on the optimum solution.

What is the minimum distance from the point $(a, 0)$ to the line $y = b x^2$?

```
Simplify[Minimize[{Sqrt[(x - a)^2 + y^2], y == b x}, {x, y}], a > 0 && b > 0]
```

$$\left\{\frac{a b}{\sqrt{1 + b^2}}, \left\{x \to \frac{a}{1 + b^2}, y \to \frac{a b}{1 + b^2}\right\}\right\}$$

Next, we optimize an algebraic expression:

```
Maximize[{Sqrt[3 x + 2] + Sqrt[2 x - 1] - 3 Sqrt[x - 1], 0 ≤ x ≤ 10}, x]
```

$$\left\{1 + \sqrt{5}, \{x \to 1\}\right\}$$

■ Example 5

In the next problem, the original solution contains **Root** objects but we succeed, with **FullSimplify** and **ToRadicals**, in getting quite simple expressions into the solution:

```
{val, point} =
  Minimize[{1 / (5 x y z) + 4 / x + 3 / z, 2 x z + x y ≤ 10, x ≥ 0, y ≥ 0, z ≥ 0}, {x, y, z}] //
    FullSimplify // ToRadicals
```

$$\left\{\frac{4}{5}\sqrt{\frac{1}{5}\left(76 + \sqrt{151}\right)}, \left\{x \to \frac{2}{3}\sqrt{\frac{1}{5}\left(76 - \sqrt{151}\right)}, y \to \frac{1}{\sqrt{10}}, z \to \frac{\sqrt{\frac{151}{10}}}{2}\right\}\right\}$$

```
% // N
```

$$\{3.36168, \{x \to 2.37976, y \to 0.316228, z \to 1.94294\}\}$$

The first inequality constraint is active at the optimum solution:

```
2 x z + x y /. point // FullSimplify        10
```

■ Example 6

Minimize the surface area $\pi r \sqrt{r^2 + h^2}$ of a cone given that its volume $\frac{1}{3}\pi r^2 h$ has a given value v:

```
sol =
  Simplify[Minimize[{π r Sqrt[r^2 + h^2], π r^2 h / 3 == v, r > 0, h > 0}, {r, h}], v > 0]
```

$$\left\{ \text{Root}\left[-2187\, \pi^2\, v^4 + 4\, \#1^6\ \&,\ 2 \right], \right.$$

$$\left\{ r \to \text{Root}\left[9\, v^2 - \text{Root}\left[-2187\, \pi^2\, v^4 + 4\, \#1^6\ \&,\ 2 \right]^2 \#1^2 + \pi^2\, \#1^6\ \&,\ 3 \right], \right.$$

$$\left. \left. h \to \frac{3\, v}{\pi\, \text{Root}\left[9\, v^2 - \text{Root}\left[-2187\, \pi^2\, v^4 + 4\, \#1^6\ \&,\ 2 \right]^2 \#1^2 + \pi^2\, \#1^6\ \&,\ 3 \right]^2} \right\}\right\}$$

By using **ToRadicals** several times together with **FullSimplify**, we succeed in getting the following explicit result:

```
FullSimplify[sol // ToRadicals // ToRadicals, v > 0] // ToRadicals
```

$$\left\{ \frac{3\, 3^{1/6}\, \pi^{1/3}\, (-v)^{2/3}}{2^{1/3}},\ \left\{ r \to -\frac{(-3)^{1/3}\, v^{1/3}}{2^{1/6}\, \pi^{1/3}},\ h \to -\left(-\frac{6}{\pi}\right)^{1/3} v^{1/3} \right\} \right\}$$

Calculate the third powers:

```
{%[[1]]^3, {r → %[[2, 1, 2]]^3, h → %[[2, 2, 2]]^3}}
```

$$\left\{ \frac{27}{2} \sqrt{3}\ \pi\, v^2,\ \left\{ r \to \frac{3\, v}{\sqrt{2}\ \pi},\ h \to \frac{6\, v}{\pi} \right\} \right\}$$

Thus, the optimum values are $r = \left(\dfrac{3}{\sqrt{2}\ \pi}\, v\right)^{\frac{1}{3}}$, $h = \left(\dfrac{6}{\pi}\, v\right)^{\frac{1}{3}}$ and the minimum surface area is $3\left(\dfrac{\sqrt{3}}{2}\, \pi\, v^2\right)^{\frac{1}{3}}$.

In Section 23.4.2, p. 769, and in Example 4 of Section 23.4.3, p. 776, we again consider this problem by using Lagrange's multipliers and Karush–Kuhn–Tucker conditions.

23.1.2 Numerical Global Optimization

■ Numerical Global Minimums and Maximums

`NMinimize[f, vars]`	Give the global minimum of **f** with respect to variables **vars**
`NMinimize[{f, cons}, vars]`	Minimize subject to constraints **cons**
`NMaximize[f, vars]`	Give the global maximum of **f** with respect to variables **vars**
`NMaximize[{f, cons}, vars]`	Maximize subject to constraints **cons**

These commands use iterative methods and *give a decimal approximation* to the optimum solution. Recall that **Minimize** and **Maximize** give the exact solution. Also, **NMinimize** and **NMaximize** *attempt to give the global optimum,* but this cannot be guaranteed. Recall that **Minimize** and **Maximize** give the global solution.

The function to be minimized and the equality and inequality constraints can be arbitrary expressions (e.g., nonlinear, noncontinuous, and nondifferentiable). Within the constraints we can restrict individual variables to be integer valued by writing, for example, **x ∈ Integers**. The functions cannot contain any symbolic parameters. The methods used are direct search methods and as such are derivative-free.

■ **Examples**

We solve some of the problems we solved previously with **Minimize**. In most cases, we get a result that is very close to the exact result.

```
NMinimize[{5 + 40 x^3 - 45 x^4 + 12 x^5, x > 0}, x]
```
$\{-11., \{x \to 2.\}\}$

```
NMinimize[{ x^3 + y^3 + 2 x^2 + 4 y^2 + 6, -1 < x < 1, -1 < y < 1}, {x, y}]
```
$\left\{6., \left\{x \to 1.20221 \times 10^{-15}, y \to -2.70515 \times 10^{-15}\right\}\right\}$

```
NMinimize[x^4 + 3 x^2 y + 5 y^2 + x + y, {x, y}]
```
$\{-0.832579, \{x \to -0.886324, y \to -0.335671\}\}$

```
NMinimize[{ (x - y)^2 + 5 z, x + y + z == -5, y - 3 z == 1, x ≥ 0}, {x, y, z}]
```
$\{4.75, \{x \to 0., y \to -3.5, z \to -1.5\}\}$

```
NMinimize[
  {x^2/2 + (y^2 + z^2 + v^2) / 6, x ≥ 2, x + y ≥ 5, x + z ≥ 2, x + v ≥ 1, v ≥ 0}, {x, y, z, v}]
```
$\left\{3.5, \left\{v \to 4.44159 \times 10^{-13}, x \to 2., y \to 3., z \to 5.45179 \times 10^{-9}\right\}\right\}$

```
NMinimize[{1 / (5 x y z) + 4 / x + 3 / z, 2 x z + x y ≤ 10, x ≥ 0, y ≥ 0, z ≥ 0}, {x, y, z}]
```
$\{3.36168, \{x \to 2.37977, y \to 0.316227, z \to 1.94293\}\}$

In Example 4 of Section 23.1.1, p. 745, we solved the following problem with **Minimize** and got the global optimum $x = 1$, $y = 4$, $z = 0$, with minimum value 5. Now we do not get the global optimum:

```
NMinimize[{x^2 + x y + z, x + x y ≥ 5, x ≥ 0, y ≥ 0, z ≥ 0, {x, y, z} ∈ Integers}, {x, y, z}]
```
$\{8., \{x \to 2, y \to 2, z \to 0\}\}$

Here is a larger linear integer problem:

```
vars = Array[x, 15]; SeedRandom[7]; a = RandomInteger[{-10, 10}, {15, 15}];
```

```
NMinimize[
  {Total[vars], Join[Thread[a.vars ≥ 1], Thread[vars ≥ 0], {vars ∈ Integers}]}, vars]
```
$\{26., \{x[1] \to 8, x[2] \to 0, x[3] \to 3, x[4] \to 0, x[5] \to 0, x[6] \to 0, x[7] \to 1, x[8] \to 0,$
$\quad x[9] \to 0, x[10] \to 0, x[11] \to 6, x[12] \to 0, x[13] \to 1, x[14] \to 5, x[15] \to 2\}\}$

■ **Initial Intervals**

NMinimize needs an initial interval in which to start the search for the optimum. The default is to use the interval $[-1, 1]$ for each variable. We can define initial intervals when defining the variables. For example, an initial interval $[0, 3]$ for x can be defined with **{x, 0, 3}**. If the constraints contain an interval such as **0 ≤ x ≤ 3**, then it is used as the initial interval (if not otherwise specified). With the initial interval we can force the search to a special region where we believe the optimum should lie. To ensure that we find the global optimum, we can solve the problem several times using different starting intervals.

In the following example, the intervals $[-2, 0]$ and $[-1, 0]$ are used to start the search for the optimum:

```
NMinimize[x^4 + 3 x^2 y + 5 y^2 + x + y, {{x, -2, 0}, {y, -1, 0}}]
```
$\{-0.832579, \{x \to -0.886324, y \to -0.335671\}\}$

23.1.3 Options for Numerical Global Optimization

NMinimize and **NMaximize** can resort to several methods. With regard to the options of the commands, some options are common to all methods, whereas others are specific to each method. First we consider the common options.

■ **Common Options**

Options of **NMinimize** *and* **NMaximize**:

WorkingPrecision Precision used in internal computations; examples of values:
  MachinePrecision, 20

PrecisionGoal If the value of the option is **p**, the relative error of the optimum point and of the
  value of the function (or a penalty function) at the optimum point should be of the order 10^{-p};
  examples of values: **Automatic** (usually means **8**), **10**

AccuracyGoal If the value of the option is **a**, the absolute error of the optimum point and of the
  value of the function (or a penalty function) at the optimum point should be of the order 10^{-a};
  examples of values: **Automatic** (usually means **8**), **10**

Method Method used; possible values: **Automatic**, **"NelderMead"**, **"DifferentialEvolution"**,
  "SimulatedAnnealing", **"RandomSearch"**

MaxIterations Maximum number of iterations used; examples of values: **100, 200**

StepMonitor Command to be executed after each step of the iterative method; examples of values:
  None, Sow[x], ++n, AppendTo[iters, x]

EvaluationMonitor Command to be executed after each evaluation of the function to be minimized; examples of values: **None, Sow[x], ++n, AppendTo[points, x]**

The default value of **PrecisionGoal** and **AccuracyGoal** is usually 8; their general default value is
WorkingPrecision/2. They both refer to both the minimum point and the minimum value of the
function. Thus, iteration is, by default, stopped when the estimated relative or absolute error of the
optimum point and of the optimum value is less than 10^{-8}. In more detail, if the precision goal is p and
accuracy goal a, then iterations are stopped if $\| x_k - x^* \| \le \max\left\{10^{-a},\ 10^{-p} \| x_k \|\right\}$ and $\| \nabla f(x_k) \| \le 10^{-a}$.

To ensure we get the global optimum, it may be advantageous to try several settings:

• Try some values of **MaxIterations** (the default is 100).
• Try some values of **Method** (the default is **Automatic**).
• Try some initial intervals (the default is $[-1, 1]$).
• Try some values of **"RandomSeed"** (the default is 0).

For example, we previously found that we obtained only a local minimum for the following problem:

```
problem = {x^2 + x y + z, x + x y ≥ 5, x ≥ 0, y ≥ 0, z ≥ 0, {x, y, z} ∈ Integers};
```

```
NMinimize[problem, {x, y, z}]
```

$\{8.,\ \{x \to 2,\ y \to 2,\ z \to 0\}\}$

It turns out that with the simulated annealing method we get the global optimum:

```
NMinimize[problem, {x, y, z}, Method → "SimulatedAnnealing"]
```

$\{5.,\ \{x \to 1,\ y \to 4,\ z \to 0\}\}$

Another solution is to give some initial intervals:

```
NMinimize[problem, {{x, 0, 10}, {y, 0, 10}, {z, 0, 10}}]
```

$\{5., \{x \to 1, y \to 4, z \to 0\}\}$

■ Looking at the iterations

Let us see all the points where the function is evaluated when using the various methods:

```
f = x^4 + 3 x^2 y + 5 y^2 + x + y;
methods =
  {"NelderMead", "DifferentialEvolution", "SimulatedAnnealing", "RandomSearch"};
solutions =
  Reap[NMinimize[f, {x, y}, EvaluationMonitor :> Sow[{x, y}], Method → #]] & /@ methods;

MapThread[ContourPlot[f, {x, -2, 2.5}, {y, -2, 2}, PlotRange → All, Contours → 30,
    ContourShading → False, PlotLabel → #1, Epilog → {Point[#2[[2, 1]]], Line[#2[[2, 1]]],
      Red, PointSize[Medium], Point[{x, y} /. #2[[1, 2]]]}] &, {methods, solutions}]
```

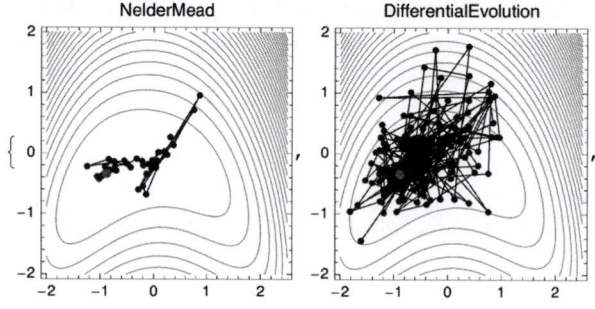

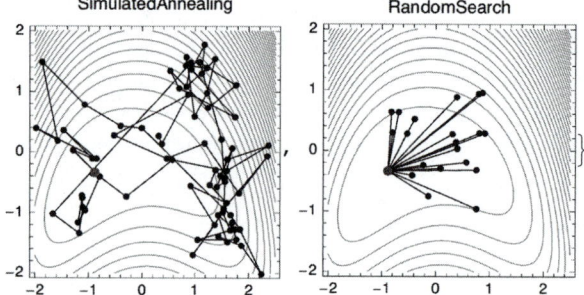

■ Methods

We can use four methods with **NMinimize** and **NMaximize**:

- **"NelderMead"** is the simplex method by J. A. Nelder and R. Mead. It is generally the fastest of the four methods, and it is well suited for problems with continuous variables. For a function of d variables, the algorithm maintains a set of $d + 1$ points forming the vertices of a polytope in d-dimensional space.

- **"DifferentialEvolution"** is a genetic method developed by K. Price and R. Storn. It may be the most robust method of the four, but it is also computationally demanding and often slower than other methods. This method is suggested, for example, for problems with integer variables. The algorithm maintains a population of m points, $m \gg d$. During each iteration a new population of m points is generated, based on the old population.

- **"SimulatedAnnealing"** starts from many points. For each starting point, a sequence of iterations is generated by moving from the current point to a random direction. If the move results in a better point, it is accepted; otherwise, the point is accepted with a certain probability. The best found point is chosen as the solution. The method can also be used for discrete problems.

- **"RandomSearch"** first generates a large number of points in the initial region and then uses each point as the starting point for a local optimizer to find a local optimum. The best local minimum is chosen as the solution. The default method for local optimization is **Automatic**, which means **FindMinimum**. This method requires that the objective function is locally continuous. The method is not well suited for discrete problems.

The default setting **Method → Automatic** tries to choose a good method, as follows:

- If the problem is linear (and does not have integer variables), use **"LinearProgramming"**.
- If any of the variables are integer valued, use **"DifferentialEvolution"**.
- Otherwise—that is, if the problem is nonlinear and continuous—use **"NelderMead"**, and if it does poorly, switch to **"DifferentialEvolution"**.

Each method has special options that we study next.

■ Common Method-Specific Options

The method-specific options are used inside the value of the **Method** option, as follows:

```
NMinimize[{f, cons}, vars, opts, Method → {"method", methodSpecOpts}]
```

Among the method-specific options, six options are not so method specific but are shared with all four methods (with one exception). These options are considered first.

> **"RandomSeed"**   Seed for random number generator; default value: **0**
> **"SearchPoints"**   Number of initial points (not for **"NelderMead"**); default value: **Automatic**, which
>   means, for d variables,
> $\min(10\,d, 50)$ for **"DifferentialEvolution"**,
> $\min(10\,d, 100)$ for **"RandomSearch"**, and
> $\min(2\,d, 50)$ for **"SimulatedAnnealing"**
>   (**NelderMead** uses $d + 1$ points)
> **"InitialPoints"**   Set of initial points; examples of values: **Automatic**, {{x1,y1}, {x2,y2}, … }
> **"PenaltyFunction"**   Function applied to constraints to penalize invalid points; default value:
>   **Automatic**
> **"Tolerance"**   Tolerance for accepting constraint violations; default value: **0.001**
> **"PostProcess"**   Whether and how to postprocess using local search methods; examples of values:
>   **Automatic**, **"FindMinimum"** (a penalty method), **"InteriorPoint"** (interior point method), **"KKT"**
>   (Karush–Kuhn–Tucker method), **True**, **False**

All of the methods use random numbers in choosing initial points and/or in each iteration. The default value of **"RandomSeed"** is zero. Thus, we get the same result if we solve a problem several times. However, with different seeds, we may get different results. In fact, it is advisable to solve a problem with several seeds and pick the best result. For example, here we use the differential evolution method with six different seeds:

$$\text{problem} = \left\{\frac{4}{x} + \frac{3}{z} + \frac{1}{5\,x\,y\,z}, \ x\,y + 2\,x\,z \le 10, \ x \ge 0, \ y \ge 0, \ z \ge 0\right\};$$

```
Table[{i, NMinimize[problem, {x, y, z},
     Method → {"DifferentialEvolution", "RandomSeed" → i}]}, {i, 0, 5}] // Quiet
{{0, {3.36168, {x → 2.37976, y → 0.316225, z → 1.94294}}},
 {1, {3.36168, {x → 2.37977, y → 0.316227, z → 1.94293}}},
 {2, {3.36991, {x → 2.51199, y → 0.332459, z → 1.82244}}},
 {3, {3.36168, {x → 2.37977, y → 0.316225, z → 1.94293}}},
 {4, {3.36168, {x → 2.37977, y → 0.316227, z → 1.94293}}},
 {5, {3.36838, {x → 2.49371, y → 0.28102, z → 1.8621}}}}
```

We got three different solutions; the value 3.36168 seems to be the global minimum.

All of the four methods start from a set of points. We can either give the number of initial points as the value of `"SearchPoints"` and let the algorithm choose the points at random from the initial intervals or give the points themselves as the value of `"InitialPoints"`. `"NelderMead"` and `"RandomSearch"` proceed deterministically after the initial points are chosen, but `"DifferentialEvolution"` and `"SimulatedAnnealing"` use random numbers in each iteration.

Near the solution of the problem, the current approximation of the solution is fine-tuned by postprocessing with a combination of KKT, interior point, and penalty methods. The postprocessing can be controlled via the `"PostProcess"` option.

■ **Special Method-Specific Options**

Some method-specific options are special for each method. For more about these options, see tutorial/ConstrainedOptimizationOverview. The options of each method can be seen as follows:

```
Options[NMinimize`NelderMead] // N
{ContractRatio → 0.5, ExpandRatio → 2., InitialPoints → Automatic,
 PenaltyFunction → Automatic, PostProcess → Automatic, RandomSeed → 0.,
 ReflectRatio → 1., ShrinkRatio → 0.5, Tolerance → 0.001}
```

Special options for `"NelderMead"`:

`"ContractRatio"`   Ratio used for contraction; default value: **1/2**

`"ExpandRatio"`   Ratio used for expansion; default value: **2**

`"ReflectRatio"`   Ratio used for reflection; default value: **1**

`"ShrinkRatio"`   Ratio used for shrinking; default value: **1/2**

Special options for `"DifferentialEvolution"`:

`"CrossProbability"`   Probability that a gene is taken from the parent; default value: **1/2**

`"ScalingFactor"`   Scale applied to the difference vector when creating a mate; default value: **3/5**

In integer problems, a larger value of `"ScalingFactor"` such as 1 may be tried in an effort to get better mobility with respect to the integer variables.

Special options for `"SimulatedAnnealing"`:

`"BoltzmannExponent"`   Exponent for the probability function; default value: **Automatic**

`"LevelIterations"`   Maximum number of iterations to stay at a given point; default value: **50**

`"PerturbationScale"`   Scale for the random jump; default value: **1**

Special options for `"RandomSearch"`:

`Method`   Which method to use for minimization; possible values: **Automatic**, `"InteriorPoint"`

23.2 Linear Optimization

23.2.1 Linear Problems by Variables

We can formulate a linear programming problem either by writing explicit expressions with variables or by giving only coefficient matrices and vectors. The corresponding commands are **Minimize** or **Maximize** and **LinearProgramming**, respectively. Linear optimization is a special case of global optimization considered in Section 23.1.

■ Formulation with Variables

Minimize[{f, cons}, vars]	Give the global minimum of **f** subject to constraints **cons**
Minimize[{f, cons}, vars, dom]	Minimize over domain **dom**
Maximize[{f, cons}, vars]	Give the global maximum of **f** subject to constraints **cons**
Maximize[{f, cons}, vars, dom]	Maximize over domain **dom**

By specifying the domain to be **Integers**, we can solve integer problems. Within the constraints we can also restrict individual variables to be integer valued by writing, for example, **x ∈ Integers**. Note that **Minimize** and **Maximize** give an *exact* solution if the problem is exact—that is, does not contain decimal numbers. If the problem contains a decimal number, then actually **NMinimize** and **NMaximize** are used.

Here is an example:

```
Maximize[{x + y, 2 x + y ≤ 2, x - y ≥ -1 / 2, x + 2 y ≤ 2}, {x, y}]
```

$$\left\{\frac{4}{3}, \left\{x \to \frac{2}{3}, y \to \frac{2}{3}\right\}\right\}$$

The problem can be visualized as follows:

```
Show[RegionPlot[2 x + y ≤ 2 && x - y ≥ -1 / 2 && x + 2 y ≤ 2,
  {x, 0, 1}, {y, 0, 0.9}, AspectRatio → Automatic],
  ContourPlot[x + y, {x, 0, 1}, {y, 0, 0.9}, ContourShading → False,
  Contours → Range[1 / 3, 5 / 3, 1 / 3], ContourStyle → Dashing[Small]],
  Graphics[{Red, PointSize[Medium], Point[{2 / 3, 2 / 3}]}]]]
```

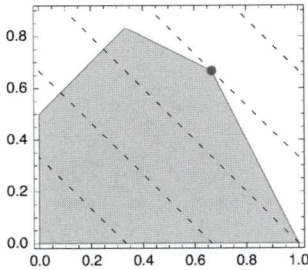

Here, the shaded region is the feasible region; the dashed lines are contours where the object function takes on the values 1/3, 2/3, 1, 4/3, and 5/3; and the red point is the optimum point. Indeed, the contour where the object function has the value 4/3 is the highest that still has a point in the feasible region; this point is the solution of the problem.

To solve an integer linear programming problem, *Mathematica* first solves the equational constraints, reducing the problem to one containing inequality constraints only. Then lattice reduction techniques are used to put the inequality system in a simpler form. Finally, the simplified optimization problem is solved by using a branch-and-bound method.

■ **Special Problems**

Now we solve an integer problem:

```
Maximize[{x + y, 2 x + y ≤ 2, x - y ≥ -1 / 2, x + 2 y ≤ 2, x ≥ 0, y ≥ 0}, {x, y}, Integers]
{1, {x → 1, y → 0}}
```

The problem can also contain symbolic parameters:

```
Maximize[{x + y, 2 x + y ≤ 2, x - y ≥ -1 / 2, x + 2 y ≤ a}, {x, y}]
```

$$\left\{ \left\{ \begin{array}{cc} \frac{3}{2} & a > \frac{5}{2} \\ \frac{2+a}{3} & \text{True} \end{array} \right., \left\{ x \to \begin{array}{cc} \frac{1}{2} & a > \frac{5}{2} \\ -\frac{2}{3}(-2-a)-a & \text{True} \end{array} \right., y \to \begin{array}{cc} 1 & a > \frac{5}{2} \\ \frac{2}{3}(-2-a)+a+\frac{2+a}{3} & \text{True} \end{array} \right\} \right\}$$

For a problem having many solutions, we get one of them:

```
Maximize[{2 x + y, 2 x + y ≤ 2, x - y ≥ -1 / 2, x + 2 y ≤ 2, x ≥ 0, y ≥ 0}, {x, y}]
{2, {x → 1, y → 0}}
```

```
Show[RegionPlot[2 x + y ≤ 2 && x - y ≥ -1 / 2 && x + 2 y ≤ 2, {x, 0, 1}, {y, 0, 0.9}],
  ContourPlot[2 x + y, {x, 0, 1}, {y, 0, 0.9}, ContourShading → False,
   Contours → Range[1 / 3, 8 / 3, 1 / 3], ContourStyle → Dashing[Small]],
  Epilog → {Red, PointSize[Medium], Point[{1, 0}]}, AspectRatio → Automatic]
```

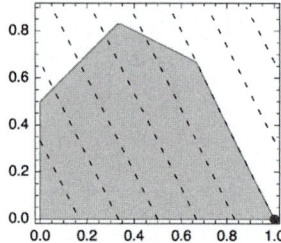

A strict inequality may cause the solution to be achieved only infinitesimally. In such a case, *Mathematica* chooses the closest point on the boundary. In general, avoid using strict inequalities.

```
Maximize[{x + y, 2 x + y < 2, x - y ≥ -1 / 2, x + 2 y ≤ 2}, {x, y}]
```

Maximize::wksol : Warning: There is no maximum in the
     region described by the constraints; returning a result on the boundary. ≫

$$\left\{ \frac{4}{3}, \left\{ x \to \frac{2}{3}, y \to \frac{2}{3} \right\} \right\}$$

■ **Example 1: A Transportation Problem**

Plants 1, 2, and 3 have a supply of a food, and cities 1, 2, 3, and 4 have demand for this food. The problem is to decide the amounts to be transported from the plants to the cities so as to minimize transportation costs. First, we draw a graph (see Section 8.5, p. 267):

```
p1 = "Plant 1"; p2 = "Plant 2"; p3 = "Plant 3";
c1 = "City 1"; c2 = "City 2"; c3 = "City 3"; c4 = "City 4";
edges = {p1 → c1, p1 → c2, p1 → c3, p1 → c4, p2 → c1,
   p2 → c2, p2 → c3, p2 → c4, p3 → c1, p3 → c2, p3 → c3, p3 → c4};
```

```
GraphPlot[edges, Method → {"LayeredDigraphDrawing", "Rotation" → -π / 2},
  DirectedEdges → True, VertexLabeling → True, AspectRatio → 1]
```

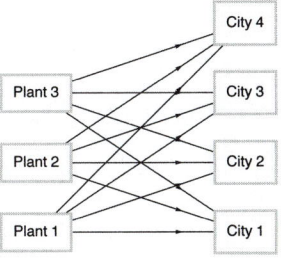

The plants, cities, supplies, demands, and transportation costs are as follows:

```
plants = {p1, p2, p3}; cities = {c1, c2, c3, c4};
supplies = {47, 36, 52}; demands = {38, 34, 29, 34};
costs = {{5, 7, 6, 10}, {9, 4, 6, 7}, {5, 8, 6, 6}};
```

For example, transportation of one unit from Plant 1 to City 1 costs $5. To create a tabular representation of the problem, we write a function (for **Grid**, see Section 15.2, p. 470):

```
tabulate[tt_] := Grid[Join[{Join[{""}, cities, {"Supply"}]},
  Join[{plants}ᵀ, tt, {supplies}ᵀ, 2],
  {Join[{"Demand"}, demands, {Total[demands]}]}],
  Dividers → {{2 → True, -2 → True}, {2 → True, -2 → True}},
  Alignment → {{Left, {Right}}}] // Text
```

Here is the problem:

```
tabulate[costs]
```

	City 1	City 2	City 3	City 4	Supply
Plant 1	5	7	6	10	47
Plant 2	9	4	6	7	36
Plant 3	5	8	6	6	52
Demand	38	34	29	34	135

Let $x_{i,j}$ be the amount transported from Plant i to City j:

```
vars = Table[x_{i,j}, {i, 3}, {j, 4}]
```

$$\{\{x_{1,1}, x_{1,2}, x_{1,3}, x_{1,4}\}, \{x_{2,1}, x_{2,2}, x_{2,3}, x_{2,4}\}, \{x_{3,1}, x_{3,2}, x_{3,3}, x_{3,4}\}\}$$

The total cost is as follows:

```
obj = {Total[Flatten[costs vars]]}
```

$$\{5 x_{1,1} + 7 x_{1,2} + 6 x_{1,3} + 10 x_{1,4} + 9 x_{2,1} + 4 x_{2,2} + 6 x_{2,3} + 7 x_{2,4} + 5 x_{3,1} + 8 x_{3,2} + 6 x_{3,3} + 6 x_{3,4}\}$$

Note that here we did not use the dot in matrix multiplication but, rather, the space, which does an element-by-element multiplication. Supply, demand, and nonnegativity constraints are as follows:

```
supplyConstr = Thread[(Total /@ vars) ≤ supplies]
```

$$\{x_{1,1} + x_{1,2} + x_{1,3} + x_{1,4} \le 47, x_{2,1} + x_{2,2} + x_{2,3} + x_{2,4} \le 36, x_{3,1} + x_{3,2} + x_{3,3} + x_{3,4} \le 52\}$$

```
demandConstr = Thread[Total[vars] ≥ demands]
```

$$\{x_{1,1} + x_{2,1} + x_{3,1} \ge 38, x_{1,2} + x_{2,2} + x_{3,2} \ge 34, x_{1,3} + x_{2,3} + x_{3,3} \ge 29, x_{1,4} + x_{2,4} + x_{3,4} \ge 34\}$$

```
nonneg = Thread[Flatten[vars] ≥ 0]
```

$\{x_{1,1} \geq 0, x_{1,2} \geq 0, x_{1,3} \geq 0, x_{1,4} \geq 0, x_{2,1} \geq 0,$
$\quad x_{2,2} \geq 0, x_{2,3} \geq 0, x_{2,4} \geq 0, x_{3,1} \geq 0, x_{3,2} \geq 0, x_{3,3} \geq 0, x_{3,4} \geq 0\}$

The problem to be solved is now as follows:

```
problem = Join[obj, supplyConstr, demandConstr, nonneg]
```

$\{5 x_{1,1} + 7 x_{1,2} + 6 x_{1,3} + 10 x_{1,4} + 9 x_{2,1} + 4 x_{2,2} + 6 x_{2,3} + 7 x_{2,4} + 5 x_{3,1} + 8 x_{3,2} + 6 x_{3,3} + 6 x_{3,4},$
$\quad x_{1,1} + x_{1,2} + x_{1,3} + x_{1,4} \leq 47, x_{2,1} + x_{2,2} + x_{2,3} + x_{2,4} \leq 36, x_{3,1} + x_{3,2} + x_{3,3} + x_{3,4} \leq 52,$
$\quad x_{1,1} + x_{2,1} + x_{3,1} \geq 38, x_{1,2} + x_{2,2} + x_{3,2} \geq 34, x_{1,3} + x_{2,3} + x_{3,3} \geq 29,$
$\quad x_{1,4} + x_{2,4} + x_{3,4} \geq 34, x_{1,1} \geq 0, x_{1,2} \geq 0, x_{1,3} \geq 0, x_{1,4} \geq 0, x_{2,1} \geq 0,$
$\quad x_{2,2} \geq 0, x_{2,3} \geq 0, x_{2,4} \geq 0, x_{3,1} \geq 0, x_{3,2} \geq 0, x_{3,3} \geq 0, x_{3,4} \geq 0\}$

Solve the problem:

```
{val, point} = Minimize[problem, Flatten[vars]]
```

$\{704, \{x_{1,1} \to 20, x_{1,2} \to 0, x_{1,3} \to 27, x_{1,4} \to 0, x_{2,1} \to 0,$
$\quad x_{2,2} \to 34, x_{2,3} \to 2, x_{2,4} \to 0, x_{3,1} \to 18, x_{3,2} \to 0, x_{3,3} \to 0, x_{3,4} \to 34\}\}$

Tabulate the optimal transportation amounts:

```
tabulate[vars /. point /. 0 → ""]
```

	City 1	City 2	City 3	City 4	Supply
Plant 1	20		27		47
Plant 2		34	2		36
Plant 3	18			34	52
Demand	38	34	29	34	135

■ Example 2: A Knapsack Problem

Let us solve a knapsack problem in which the benefits of four items are 14, 10, 15, 8, and 9 and the corresponding weights are 6, 8, 5, 6, and 4. What is the optimal collection of items when we want to maximize the total benefit subject to the constraint that the total weight has to be at most 18? Let x_i be 1 if the ith item is included and 0 otherwise. 0–1 variables can be formulated in *Mathematica* by constraining the variables in the interval [0, 1] and requiring that the variables are integers. Write the following:

```
benefits = {14, 10, 15, 8, 9}; weights = {6, 8, 5, 6, 4};
```

```
vars = Table[xᵢ, {i, 5}]
```

$\{x_1, x_2, x_3, x_4, x_5\}$

```
obj = benefits.vars
```

$14 x_1 + 10 x_2 + 15 x_3 + 8 x_4 + 9 x_5$

```
constr = weights.vars ≤ 18
```

$6 x_1 + 8 x_2 + 5 x_3 + 6 x_4 + 4 x_5 \leq 18$

```
intervals = Thread[0 ≤ vars ≤ 1]
```

$\{0 \leq x_1 \leq 1, 0 \leq x_2 \leq 1, 0 \leq x_3 \leq 1, 0 \leq x_4 \leq 1, 0 \leq x_5 \leq 1\}$

```
{val, point} = Maximize[{obj, constr, intervals}, vars, Integers]
```

$\{38, \{x_1 \to 1, x_2 \to 0, x_3 \to 1, x_4 \to 0, x_5 \to 1\}\}$

It is optimal to take the first, third, and fifth items; the total benefit is then 38. The total weight is as follows:

```
constr⟦1⟧ /. point        15
```

This means that 3 weight units are unused.

23.2.2 Linear Problems by Matrices

■ Formulation with Matrices

`LinearProgramming[c, m, b]` Minimize $c.x$ subject to $m.x \geq b$ and $x \geq 0$

`LinearProgramming[c, m, {{b₁, s₁}, {b₂, s₂}, … }]` The ith constraint is $m_i.x \leq b_i$, $m_i.x == b_i$, or $m_i.x \geq b_i$ according to whether s_i is $-1, 0$, or 1

`LinearProgramming[c, m, b, {l₁, l₂, … }]` Add the constraints $x_i \geq l_i$ (a single number can also be supplied for the lower bound; it is then applied for each variable)

`LinearProgramming[c, m, b, {{l₁, u₁}, {l₂, u₂}, … }]` Add the constraints $l_i \leq x_i \leq u_i$

`LinearProgramming[c, m, b, lu, dom]` Minimize over domain **dom**

`LinearProgramming[c, m, b, lu, {dom₁, dom₂, … }]` Assume $x_i \in dom_i$

If we have to maximize **c.x**, then we can minimize **-c.x**, instead. Lower and upper bounds can also be $-\infty$ and ∞. The fourth argument defining the lower and upper bounds can also be **Automatic** if we do not need special bounds but would like to define a domain. A domain can be **Reals** or **Integers**. All vectors and matrices can be defined with sparse arrays.

With the **Method** option we can set the method to be **"Simplex"**, **"RevisedSimplex"**, or **"InteriorPoint"**; the default value is **Automatic**. With the simplex methods, the algorithm moves from vertices to vertices of the polytope defined by the constraints. These methods use dense linear algebra, and they can also use exact or arbitrary-precision arithmetic. The interior point method operates from the interior of the polytope defined by the constraints. The method uses machine-precision sparse linear algebra. Thus, the interior point method is suited for large-scale problems.

As an example, we solve the same problem we solved with **Maximize**: Minimize $-x - y$ subject to $-2x - y \geq -2$, $x - y \geq -1/2$, and $-x - 2y \geq -2$ together with $x \geq 0$ and $y \geq 0$:

```
c = {-1, -1}; m = {{-2, -1}, {1, -1}, {-1, -2}};
b = {-2, -1 / 2, -2};
```

```
sol = LinearProgramming[c, m, b]          {2/3, 2/3}
```

Thus, $x = y = 2/3$. The value of the original objective function $x + y$ is as follows:

```
-c.sol          4/3
```

If we want to write the constraints as $2x + y \leq 2$, $x - y \geq -1/2$, and $x + 2y \leq 2$, we have to use the s_i numbers. A $\leq$ inequality is denoted with $s_i = -1$ and a $\geq$ inequality with $s_i = 1$:

```
c = {-1, -1}; m = {{2, 1}, {1, -1}, {1, 2}};
b = {2, -1 / 2, 2}; s = {-1, 1, -1};
bs = {b, s}ᵀ
```

$$\left\{ \{2, -1\}, \left\{-\frac{1}{2}, 1\right\}, \{2, -1\} \right\}$$

```
LinearProgramming[c, m, bs]          {2/3, 2/3}
```

In the following example, we add the bounds $\frac{1}{4} \leq x \leq \frac{1}{2}$ and $\frac{1}{5} \leq y \leq \frac{2}{5}$:

```
l = {1 / 4, 1 / 5}; u = {1 / 2, 2 / 5};
lu = {l, u}ᵀ
```

$$\left\{\left\{\frac{1}{4}, \frac{1}{2}\right\}, \left\{\frac{1}{5}, \frac{2}{5}\right\}\right\}$$

```
LinearProgramming[c, m, bs, lu]
```
$$\left\{\frac{1}{2}, \frac{2}{5}\right\}$$

Next, we solve an integer problem. For such problems, we get a message:

```
LinearProgramming[c, m, bs, {{0, 0}, {5, 5}}ᵀ, Integers]
```
LinearProgramming::lpip :

  Warning: integer linear programming will use a machine precision approximation of the inputs. ≫
  {1, 0}

To get rid of the message, put a decimal point in the problem:

```
LinearProgramming[c, m, bs, {{0., 0}, {5, 5}}ᵀ, Integers]
{1, 0}
```

■ Example 1: A Transportation Problem

Let us solve the transportation problem anew, now using vectors and matrices. The data were

```
supplies = {47, 36, 52}; demands = {38, 34, 29, 34};
costs = {{5, 7, 6, 10}, {9, 4, 6, 7}, {5, 8, 6, 6}};
```
Although variables will not be used in the formulation, they are as follows:

```
vars = Flatten[Table[xᵢ,ⱼ, {i, 3}, {j, 4}]]
```
$\{x_{1,1}, x_{1,2}, x_{1,3}, x_{1,4}, x_{2,1}, x_{2,2}, x_{2,3}, x_{2,4}, x_{3,1}, x_{3,2}, x_{3,3}, x_{3,4}\}$

The cost vector, the left-hand-side matrix, and the right-hand-side matrix are shown here:

```
c = Flatten[costs]
```
{5, 7, 6, 10, 9, 4, 6, 7, 5, 8, 6, 6}

```
m = {{1, 1, 1, 1, 0, 0, 0, 0, 0, 0, 0, 0},
     {0, 0, 0, 0, 1, 1, 1, 1, 0, 0, 0, 0},
     {0, 0, 0, 0, 0, 0, 0, 0, 1, 1, 1, 1},
     {1, 0, 0, 0, 1, 0, 0, 0, 1, 0, 0, 0},
     {0, 1, 0, 0, 0, 1, 0, 0, 0, 1, 0, 0},
     {0, 0, 1, 0, 0, 0, 1, 0, 0, 0, 1, 0},
     {0, 0, 0, 1, 0, 0, 0, 1, 0, 0, 0, 1}};
```

```
bs = {Join[supplies, demands], {-1, -1, -1, 1, 1, 1, 1}}ᵀ
```
{{47, -1}, {36, -1}, {52, -1}, {38, 1}, {34, 1}, {29, 1}, {34, 1}}

The solution is as follows:

```
sol = LinearProgramming[c, m, bs]
```
{20, 0, 27, 0, 0, 34, 2, 0, 18, 0, 0, 34}

We pick the variables that have a positive value:

```
Select[Thread[vars → sol], #[[2]] ≠ 0 &]
```
$\{x_{1,1} \to 20, x_{1,3} \to 27, x_{2,2} \to 34, x_{2,3} \to 2, x_{3,1} \to 18, x_{3,4} \to 34\}$

The minimum cost is as follows:

```
c.sol      704
```

■ Example 2: A Knapsack Problem

We solve the familiar knapsack problem by maximizing $14 x_1 + 10 x_2 + 15 x_3 + 8 x_4 + 9 x_5$ subject to $6 x_1 + 8 x_2 + 5 x_3 + 6 x_4 + 4 x_5 \leq 18$. In addition, variables are 0–1 variables so that we define the lower bound to be 0 and the upper bound to be 1 and optimize over integers:

```
benefits = {14, 10, 15, 8, 9}; weights = {6, 8, 5, 6, 4};

sol = LinearProgramming[-benefits, -{weights}, {-18.}, Table[{0, 1}, {5}], Integers]

{1, 0, 1, 0, 1}
```

The corresponding benefit and total weight are

```
{benefits.sol, weights.sol}    {38, 15}
```

23.3 Local Optimization

23.3.1 Numerical Local Optimization

■ Finding Local Minimums and Maximums

`FindMinimum[f, {x, x0}]`	Find a local minimum of **f** by starting from **x0**; use the gradient of **f**
`FindMinimum[f, {{x, x0}, {y, y0}, … }]`	Start from **x0, y0, …**
`FindMinimum[{f, cons}, {{x, x0}, {y, y0}, … }]`	Minimize subject to constraints **cons**
`FindMinimum[{f, cons}, {x, y, … }]`	Start from a point within the region defined by the constraints
`FindMaximum[f, {x, x0}]`	Start from **x0**; etc.

If we would like to avoid the use of the gradient, we can define two starting points for each variable:

```
FindMinimum[f, {x, x0, x1}]
FindMinimum[f, {{x, x0, x1}, {y, y0, y1}, … }]
```

In addition, if we write **{x, x0, xmin, xmax}** or **{x, x0, x1, xmin, xmax}**, then iterations are stopped if the solution is going outside the interval (**xmin, xmax**).

FindMinimum uses iterative methods to *approximate* a *local* minimum point (note that **Minimize**, introduced in Section 23.1.1, p. 743, finds an *exact global* minimum and **NMinimize**, introduced in Section 23.1.2, p. 747, finds an *approximate global* minimum). A domain constraint such as **x ∈ Integer** can only be used for linear problems.

■ Example 1

Consider the following function:

```
f = x Cos[x];

Plot[f, {x, 0, 8}]
```

One of the local minimums seems to be near 3.5:

```
{val, point} = FindMinimum[f, {x, 3}]
{-3.28837, {x → 3.42562}}
```

The corresponding minimum value is −3.28837. We can check whether the derivative is zero (a necessary condition) and the second derivative positive (a sufficient condition) at the point found:

```
{D[f, x], D[f, x, x]} /. point
{5.80336 × 10⁻¹², 3.84882}
```

$$\{5.80336 \times 10^{-12}, 3.84882\}$$

Next, we find a local point of maximum:

```
FindMaximum[f, {x, 6}]      {6.361, {x → 6.4373}}
```

A starting point is not needed, but then we cannot control which extremum point is sought:

```
FindMaximum[f, x]      {0.561096, {x → 0.860334}}
```

■ **Example 2**

Consider the following function:

```
f = x^4 + 3 x^2 y + 5 y^2 + x + y
```

$$x + x^4 + y + 3 x^2 y + 5 y^2$$

A local minimum point is as follows:

```
{val, point} = FindMinimum[f, {{x, 1}, {y, -2}}]
{-0.832579, {x → -0.886324, y → -0.335671}}
```

A starting point is not needed:

```
{val, point} = FindMinimum[f, {x, y}]
{-0.832579, {x → -0.886324, y → -0.335671}}
```

A necessary condition for a minimum or maximum is that the gradient is zero. A sufficient condition for a minimum is that the Hessian is positive definite. A sufficient condition for a maximum is that the Hessian is negative definite or that the Hessian multiplied by −1 is positive definite. To test the Hessian, we have **PositiveDefiniteMatrixQ**, but definiteness can also be concluded with the eigenvalues: A symmetric matrix is positive [negative] definite if and only if all eigenvalues are positive [negative].

For the previous example, check whether the gradient is zero and the Hessian positive definite:

```
D[f, {{x, y}}]      {1 + 4 x³ + 6 x y, 1 + 3 x² + 10 y}
% /. point      {-2.96061 × 10⁻¹¹, 1.79594 × 10⁻¹⁰}
D[f, {{x, y}, 2}]      {{12 x² + 6 y, 6 x}, {6 x, 10}}
PositiveDefiniteMatrixQ[% /. point]      True
```

We could also check that the eigenvalues are positive:

```
Eigenvalues[%% /. point]      {14.1794, 3.23339}
```

■ **Example 3**

The following function has a local minimum, a local maximum, and two saddle points, as can be seen from the contours:

```
f = x^3 + y^3 + 2 x^2 + 4 y^2 + 6
```
$6 + 2 x^2 + x^3 + 4 y^2 + y^3$

```
ContourPlot[f, {x, -2.5, 1}, {y, -3.5, 1}, AspectRatio → Automatic, Contours → 34]
```

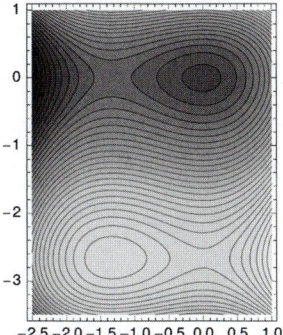

Calculate the Hessian:

```
hes = D[f, {{x, y}, 2}]     {{4 + 6 x, 0}, {0, 8 + 6 y}}
```

Here is a minimum point:

```
{val, point} = FindMinimum[f, {{x, 1}, {y, 1}}]
```
$\{6., \{x \to 2.07017 \times 10^{-9}, y \to 6.2673 \times 10^{-9}\}\}$

```
PositiveDefiniteMatrixQ[hes /. point]     True
```

Here is a maximum point:

```
{val, point} = FindMaximum[f, {{x, -1}, {y, -2}}]
```
$\{16.6667, \{x \to -1.33333, y \to -2.66667\}\}$

```
PositiveDefiniteMatrixQ[-hes /. point]      True
```

One of the saddle points is $(0, -8/3)$. If we start from this point, we get a warning. *Mathematica* finds that this point may be a saddle point:

```
{val, point} = FindMinimum[f, {{x, 0}, {y, -8 / 3}}]
```

FindMinimum::fmgz :

  Encountered a gradient that is effectively zero. The result returned may not be
      a minimum; it may be a maximum or a saddle point. ≫
$\{15.4815, \{x \to 0., y \to -2.66667\}\}$

Indeed, the Hessian is neither positive nor negative definite:

```
PositiveDefiniteMatrixQ[# /. point] & /@ {hes, -hes}
```
{False, False}

This can also be seen from the eigenvalues:

```
Eigenvalues[hes /. point]     {-8., 4.}
```

The other saddle point is $(-4/3, 0)$.

Even if the starting point is not a saddle point, it may happen that the search is stopped at a saddle point without a warning. However, calculating the eigenvalues of the Hessian reveals the nature of the point:

```
{val, point} = FindMinimum[f, {{x, -4/3}, {y, -2}}]
```

$$\{7.18519, \{x \to -1.33333, y \to 8.79606 \times 10^{-14}\}\}$$

```
Eigenvalues[hes /. point]     {8., -4.}
```

A slight modification of the starting point helps to avoid the saddle point:

```
FindMinimum[f, {{x, -1.33333}, {y, -2}}]
```

$$\{6., \{x \to 1.87457 \times 10^{-9}, y \to -5.59565 \times 10^{-10}\}\}$$

In general, it is recommended that no "special" points (e.g., $-4/3$) be chosen as starting points. A good starting point is near the minimum point but is otherwise rather random.

■ Example 4

Previously, we considered the following constrained problem:

```
problem = {x^2/2 + (y^2 + z^2 + v^2)/6, x ≥ 2, x + y ≥ 5, x + z ≥ 2, x + v ≥ 1, v ≥ 0};
```

Minimize gives the exact global solution:

```
Minimize[problem, {x, y, z, v}]
```

$$\left\{\frac{7}{2}, \{x \to 2, y \to 3, z \to 0, v \to 0\}\right\}$$

NMinimize gives an approximate global solution:

```
NMinimize[problem, {x, y, z, v}]
```

$$\{3.5, \{v \to 4.44159 \times 10^{-13}, x \to 2., y \to 3., z \to 5.45179 \times 10^{-9}\}\}$$

FindMinimum gives an approximate local minimum (in this case, the local minimum also happens to be the global minimum):

```
FindMinimum[problem, {x, y, z, v}]
```

$$\{3.5, \{x \to 2., y \to 3., z \to 0.0015887, v \to 0.00137299\}\}$$

To get a better solution, define a tighter accuracy goal:

```
FindMinimum[problem, {x, y, z, v}, AccuracyGoal → 10]
```

$$\{3.5, \{x \to 2., y \to 3., z \to 7.09418 \times 10^{-6}, v \to 6.95631 \times 10^{-6}\}\}$$

■ Example 5

The function to be minimized can be formulated with matrices and vectors. The dimension of the variable is then taken from the dimension of the initial value. For example,

```
A = {{4, 1, 1}, {1, 1, 1}, {1, 1, 3}}; b = {3, 1, 2}; c = 3;
```

```
FindMinimum[x.A.x + b.x + c, {x, {1, 1, 1}}]
```

$$\{2.29167, \{x \to \{-0.333333, 0.0833333, -0.25\}\}\}$$

23.3.2 Options for Numerical Local Optimization

■ Common Options

With the options we can, for example, choose the method from Newton, quasi-Newton, conjugate gradient, and others.

Options of **FindMinimum** *and* **FindMaximum**:

WorkingPrecision Precision used in internal computations; examples of values:
MachinePrecision, 20

PrecisionGoal If the value of the option is **p**, the relative error of the optimum point and of the value of the function at the optimum point should be of the order 10^{-p}; examples of values:
Automatic (usually means **8**), **10**

AccuracyGoal If the value of the option is **a**, the absolute error of the optimum point and of the value of the function at the optimum point should be of the order 10^{-a}; examples of values:
Automatic (usually means **8**), **10**

Method Method used; possible values: **Automatic**, **"Newton"**, **"QuasiNewton"**,
"ConjugateGradient", **"LevenbergMarquardt"**, **"PrincipalAxis"**, **"InteriorPoint"**

MaxIterations Maximum number of iterations used; examples of values: **Automatic, 500**

Gradient Gradient of the function; examples of values: **Automatic, Symbolic, FiniteDifference**

Compiled Whether the function should be compiled; possible values: **Automatic, True, False**

StepMonitor Command to be executed after each step of the iterative method; examples of values:
None, Sow[x], ++n, AppendTo[iters, x]

EvaluationMonitor Command to be executed after each evaluation of the function to be minimized; examples of values: **None, Sow[x], ++n, AppendTo[points, x]**

The default value of **PrecisionGoal** and **AccuracyGoal** is usually 8; their general default value is **WorkingPrecision**/2. They both refer to both the minimum point and the minimum value of the function. Therefore, iteration is, by default, stopped when the estimated relative or absolute error of the optimum point and of the optimum value is less than 10^{-8}. In more detail, if the precision goal is p and accuracy goal a, then iterations are stopped if $\| x_k - x^* \| \le \max\{10^{-a}, 10^{-p} \| x_k \|\}$ and $\| \nabla f(x_k) \| \le 10^{-a}$.

The default value **Automatic** of **Gradient** means that a symbolic gradient is calculated if possible; if not possible, a finite difference approximation is used. The value **Symbolic** works otherwise similarly but prints a warning if finite differences are used. If the option has the value **FiniteDifference**, then the gradient is approximated with finite differences. The value of **Gradient** can also be a mathematical expression containing the gradient.

■ Looking at the Iterations

First, we consider a function of one variable and show information about the steps:

```
f = x Cos[x]; df = D[f, x];
```

```
{sol, {points}} = Reap[FindMinimum[f, {x, 4}, StepMonitor :> Sow[{x, f, df}]]];
```

```
TableForm[points, TableSpacing -> {1, 2},
 TableHeadings -> {Range[Length[points]], {"x", "f", "f'"}}]
```

	x	f	f'
1	3.46189	−3.28582	0.140828
2	3.42795	−3.28836	0.00897361
3	3.42564	−3.28837	0.0000779862
4	3.42562	−3.28837	4.52149×10^{-8}
5	3.42562	−3.28837	2.27929×10^{-13}

Consider then a function of two variables and apply four different methods:

```
f = x^4 + 3 x^2 y + 5 y^2 + x + y;
methods = {"Newton", "QuasiNewton", "ConjugateGradient", "PrincipalAxis"};
```

```
solutions = Reap[FindMinimum[f, {{x, 1}, {y, -2}},
      Method → #, EvaluationMonitor :→ Sow[{x, y}]]] & /@ methods;
points = Prepend[#[[2, 1]], {1, -2}] & /@ solutions;
```

Plot all the points where the function was evaluated:

```
MapThread[ContourPlot[f, {x, -3.2, 2.1}, {y, -3.6, 1.6},
      Contours → 20, PlotRange → All, ContourShading → False,
      AspectRatio → Automatic, PlotLabel → #1, Epilog → {Point[#2], Line[#2],
      Red, PointSize[Medium], Point[{-0.886, -0.336}]}]] &, {methods, points}]
```

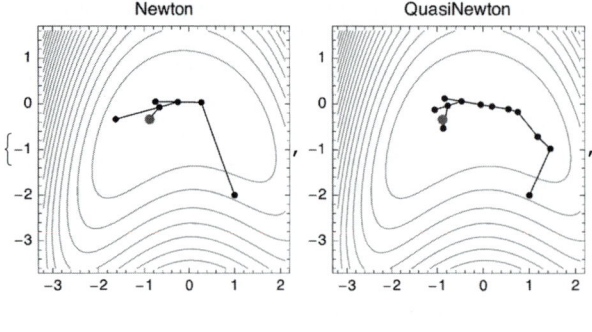

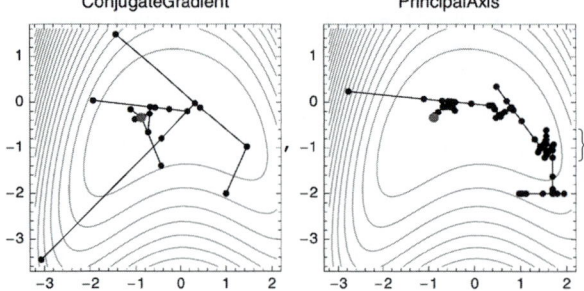

■ **Methods**

We can use several methods with **FindMinimum** and **FindMaximum**:

- **"Newton"** uses the exact Hessian (or a finite difference approximation to it).
- **"QuasiNewton"** uses the quasi-Newton Broyden–Fletcher–Goldfarb–Shanno approximation to the Hessian built up by updates based on past steps.
- **"ConjugateGradient"** is a nonlinear version (Polak–Ribiere) of the conjugate gradient method for solving linear systems; a model Hessian is never formed explicitly.
- **"LevenbergMarquardt"** (or **"GaussNewton"**) is a Gauss–Newton method for least-squares problems; the Hessian is approximated by $J^T J$, where J is the Jacobian of the residual function.
- **"PrincipalAxis"** by Brent works without using any derivatives, not even the gradient, by keeping values from past steps; it requires two starting conditions in each variable.
- **"InteriorPoint"** combines constraints and the objective function through the use of a barrier function and requires first and second derivatives of the objective and constraints.

The default setting **Method → Automatic** tries to choose a good method, as follows:

- If the objective is a sum of squares, `"LevenbergMarquardt"` is used.
- If two starting values are provided, `"PrincipalAxis"` is used.
- If the problem contains constraints, `"InteriorPoint"` is used (this is the only method available).
- Otherwise, `"QuasiNewton"` is used.

Thus, the quasi-Newton method is the main algorithm for unconstrained local optimization.

Each of the methods also has special options to fine-tune the algorithm; see tutorial/UnconstrainedOptimizationOverview. The options can be asked for as follows:

```
Options[FindMinimum`QuasiNewton]
```

{StepControl → LineSearch, StepMemory → Automatic}

23.3.3 Own Programs

■ Newton's Method

In Section 22.3.4, p. 737, we presented the following program to solve nonlinear equations $f(x) = 0$ using Newton's method:

```
newtonSolve[f_, x_, x0_, d_: 1, n_: 20, opts___?OptionQ] :=
 With[{df = D[f, x]}, FixedPointList[(x - d f / df) /. x → # &, N[x0], n, opts]]
```

Now we are interested in maximum and minimum points, and so we want to solve equations of the form $f'(x) = 0$. Thus, `newtonSolve` can also be used for optimization. The program stops if two successive points are the same to 16-digit precision. For example,

```
f = x Cos[x]; df = D[f, x];
```

```
Plot[f, {x, 0, 8}]
```

Here is a point of minimum:

```
newtonSolve[df, x, 4]
```

{4., 3.42503, 3.42562, 3.42562, 3.42562, 3.42562}

Here is a point of maximum:

```
newtonSolve[df, x, 1]
```

{1., 0.864536, 0.860339, 0.860334, 0.860334, 0.860334}

■ Davidon–Fletcher–Powell Method

Here, we write a program for the Davidon–Fletcher–Powell (DFP) method to minimize a function of several variables. First, we define functions for a substitution and for Newton's method (the optimal step size is calculated by Newton's method; we use a simplified version of the program `newtonSolve` we presented previously).

```
subst[f_, x_List, x0_List] := N[f /. Thread[x → x0]]
newton[f_, x_, x0_] := With[{df = D[f, x]}, FixedPoint[(x - f / df) /. x → # &, x0, 15]]
```

Then we write a program **dfpStep** for one step of the DFP method. Let $f(x)$ be the function to be minimized and $g(x)$ its gradient. Let x_i be the present point, $g_i = g(x_i)$, and H_i an approximation to the inverse of the Hessian at x_i. We calculate the next values x_{i+1}, g_{i+1}, and H_{i+1} as follows.

The direction of search is $d_i = -H_i g_i$ (normalized to have length 1). The optimal step size λ is calculated in the program **oneDim** by minimizing $h(\lambda) = f(x_i + \lambda d_i)$ with respect to λ. The minimization is done by first sampling $h(\lambda)$ at a set of values for λ. The best value λ_0 is chosen as the starting point for Newton's method to calculate the optimal value λ_i.

The updates are then calculated in the program **update**. The new step is $dx_i = \lambda_i d_i$, the new point is $x_{i+1} = x_i + dx_i$, and the new gradient is $g_{i+1} = g(x_{i+1})$. If we denote $dg_i = g_{i+1} - g_i$, the new approximation to the inverse of the Hessian is as follows:

$$H_{i+1} = H_i + \frac{dx_i \, dx_i^{\mathrm{T}}}{dx_i^{\mathrm{T}} \, dg_i} - \frac{H_i \, dg_i \, dg_i^{\mathrm{T}} \, H_i}{dg_i^{\mathrm{T}} \, H_i \, dg_i}.$$

```
dfpStep[f_, g_, x_, {xi_, gi_, Hi_}] := Module[{di, ndi, λi},
  di = -Hi.gi;                                  (*direction of search*)
  ndi = Norm[di];                               (*norm of the direction*)
  If[ndi == 0., Return[{xi, gi, Hi}], di = di / ndi]; (*normalize the dir.*)
  λi = oneDim[f, x, xi, di];                     (*λi is the optimal λ*)
  If[λi == 0., Return[{xi, gi, Hi}]];            (*test for stopping*)
  update[g, x, λi, di, xi, gi, Hi]]              (*return updated values*)
```

```
oneDim[f_, x_, xi_, di_] := Module[{λsample, h, hsample, λ0},
  λsample = Range[0., 10., 0.2];                 (*used to sample h*)
  h = subst[f, x, xi + λ di];                    (*step size λ is chosen optimally*)
  hsample = N[h /. λ → λsample];                 (*sample h at points λsample*)
  λ0 = λsample[[ Ordering[hsample, 1][[1]] ]];   (*best of λsample*)
  newton[D[h, λ], λ, λ0]]                         (*the optimal λ*)
```

```
update[g_, x_, λi_, di_, xi_, gi_, Hi_] := Module[{dx, xnew, gnew, dg, Hnew},
  dx = λi di;                                    (*the optimal step*)
  xnew = xi + dx;                                (*update x*)
  gnew = subst[g, x, xnew];                      (*update g*)
  dg = gnew - gi;                                (*needed next*)
  Hnew = Hi + 1 / (dx.dg) KroneckerProduct[dx, dx] -
    1 / (dg.Hi.dg) KroneckerProduct[Hi.dg, dg.Hi]; (*update H*)
  {xnew, gnew, Hnew}]                            (*return updated values*)
```

In **dfpStep**, **f** is the function to be minimized, **g** its gradient, and **{xi, gi, Hi}** the current state of the algorithm. The result of **dfpStep** is **{xnew, gnew, Hnew}**. Note that **Ordering[hsample, 1][[1]]**, which is used in **oneDim**, gives the position of the smallest element in **hsample**.

We then write the main program **dfpMinimize**. We apply the functional programming style in which **dfpStep** is iterated with **FixedPointList**:

```
dfpMinimize[f_, x_List, startx_List, ε1_: 10^-6, ε2_: 10^-6] :=
 Module[{g, x0, g0, H0},
  g = D[f, {x}];                              (*gradient of f*)
  x0 = N[startx];                             (*initial x*)
  g0 = subst[g, x, x0];                       (*initial g*)
  H0 = IdentityMatrix[Length[x]];             (*initial H*)
  iters = FixedPointList[dfpStep[f, g, x, #] &, {x0, g0, H0}, 100,
    SameTest → (Norm[#2[[2]]] < ε1 && Norm[#1[[1]] - #2[[1]]] < ε2 &)][[All, 1]];
                                              (*solve the problem*)
  {Length[iters] - 1, subst[f, x, Last[iters]], Last[iters]}]
                        (*number of iterations, optimal f, optimal x*)
```

The stopping criterion requires that the norm of the gradient is at most ε1 (default value is 10^{-6}) and the norm of the difference between the last two points is less than ε2 (default value is 10^{-6}). Remember that #1 is the next-to-last and #2 the last iteration, and that, for example, #2[[2]] is the second component of the last iteration—that is, the gradient gnew. With [[All, 1]] we pick the x values from the iterations. We have not defined iters (the whole list of points generated by dfpMinimize) as a local variable, so it is available outside the module. We can then, for example, plot the points.

■ **Example**

Consider the function that we have already used several times:

```
f = x^4 + 3 x^2 y + 5 y^2 + x + y;
```

```
dfpMinimize[f, {x, y}, {1, -2}]
```

```
{9, -0.832579, {-0.886324, -0.335671}}
```

We needed nine iterations, the minimum value of the function is −0.832579, and the minimum point is (−0.886324, −0.335671). The iterations are in the variable iters. The following program can be used to show the iterations:

```
showIterations[f_, x_, y_, iters_List, opts___] :=
 Module[{X, Y, e = 0.2}, {X, Y} = iters^T;
  ContourPlot[f, {x, Min[X] - e, Max[X] + e},
   {y, Min[Y] - e, Max[Y] + e}, ContourShading → False,
   PlotRange → All, opts, Epilog → {Point[iters], Line[iters]}]]
```

```
showIterations[f, x, y, iters, Contours → 30, AspectRatio → Automatic]
```

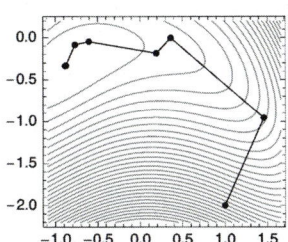

23.4 Classical Optimization

23.4.1 No Constraints

■ **Example 1: One Variable**

Let us examine the critical points of the following function:

```
f = 5 + 40 x^3 - 45 x^4 + 12 x^5;
```

The critical points are points where the derivative is zero. Among these are points of minimum and maximum. The critical points are as follows:

```
c = Solve[D[f, x] == 0]
```
$\{\{x \to 0\}, \{x \to 0\}, \{x \to 1\}, \{x \to 2\}\}$

If the function is a polynomial of high order, **Solve** may not be able to solve the equation, and you may want to use **NSolve** instead (or, if the function to be optimized is transcendental, you may need **FindRoot**). In general, the critical points may also be complex. However, we pick real and distinct points:

```
crit = c[[{1, 3, 4}]]
```
$\{\{x \to 0\}, \{x \to 1\}, \{x \to 2\}\}$

We check the second derivative:

```
D[f, x, x] /. crit
```
$\{0, -60, 240\}$

Thus, $x = 1$ is a maximum point and $x = 2$ a minimum point. At the point $x = 0$, the third derivative is nonzero:

```
D[f, x, x, x] /. crit[[1]]
```
240

Thus, this point is an inflection point. The critical points and their corresponding function values are as follows:

```
points = {x, f} /. crit
```
$\{\{0, 5\}, \{1, 12\}, \{2, -11\}\}$

Lastly, we plot the function and the critical points:

```
Plot[f, {x, -0.7, 2.4}, Epilog → {Red, Point[points]}]
```

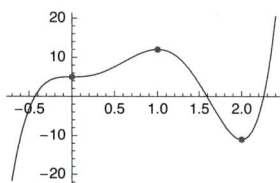

■ **Example 2: Several Variables**

Here is a function of two variables (see also Example 3 in Section 23.3.1, p. 760):

```
f = x^3 + y^3 + 2 x^2 + 4 y^2 + 6;
```

Calculate the gradient:

```
grad = D[f, {{x, y}}]
```
$\{4 x + 3 x^2, 8 y + 3 y^2\}$

Then calculate the critical points:

```
c = Solve[grad == 0]
```

$$\left\{\left\{x \to -\frac{4}{3}, \ y \to -\frac{8}{3}\right\}, \ \left\{x \to -\frac{4}{3}, \ y \to 0\right\}, \ \left\{x \to 0, \ y \to -\frac{8}{3}\right\}, \ \{x \to 0, \ y \to 0\}\right\}$$

These are all real and distinct, so we accept them all:

```
crit = c;
```

We plot the function and the points:

```
ContourPlot[f, {x, -2.5, 1}, {y, -3.5, 1}, AspectRatio → Automatic,
  Contours → 34, Epilog → {PointSize[Medium], Point[{x, y} /. crit]}]
```

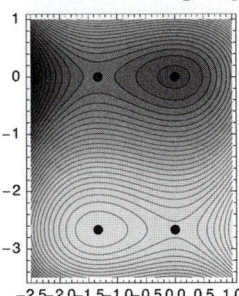

From this plot, we see that $(0, 0)$ is a minimum point, $\left(-\frac{4}{3}, -\frac{8}{3}\right)$ is a maximum point, and $\left(-\frac{4}{3}, 0\right)$ and $\left(0, -\frac{8}{3}\right)$ are saddle points. Values of the function at the critical points are as follows:

```
f /. crit // N     {16.6667, 7.18519, 15.4815, 6.}
```

A sufficient condition for a minimum [maximum] is that the Hessian is positive [negative] definite. Recall that a symmetric matrix is positive [negative] definite if and only if all eigenvalues are positive [negative]. Calculate the Hessian:

```
hess = D[f, {{x, y}, 2}]     {{4 + 6 x, 0}, {0, 8 + 6 y}}
```

Next, we write a table that contains the eigenvalues of the Hessian at each critical point:

```
Grid[Prepend[{x, y, f, Eigenvalues[hess]} /. crit, {"x", "y", "f", "eigenvalues"}],
  Spacings → 2, Dividers → {False, {2 → True}}]
```

x	y	f	eigenvalues
$-\frac{4}{3}$	$-\frac{8}{3}$	$\frac{50}{3}$	{-4, -8}
$-\frac{4}{3}$	0	$\frac{194}{27}$	{-4, 8}
0	$-\frac{8}{3}$	$\frac{418}{27}$	{4, -8}
0	0	6	{4, 8}

Thus, the first point is a maximum, the last point is a minimum, and the rest are saddle points.

23.4.2 Equality Constraints

■ Problem

Consider the following problem that we already solved in Example 6 of Section 23.1.1, p. 747. Given that the volume of a cone should be v, what should the height h and radius r of the cone be if we want to minimize the surface area of the cone? The surface area A and the volume V are as follows:

```
A = Pi r Sqrt[h^2 + r^2]        π r √(h² + r²)
```
$$\pi\, r \sqrt{h^2 + r^2}$$

```
V = Pi h r^2 / 3
```
$$\frac{1}{3}\, h\, \pi\, r^2$$

We want to minimize A with respect to h and r given that $V = v$. We solve the problem with three methods: graphically (for a numerical value of v), by the method of substitution, and by Lagrange's method.

■ **A Graphical Solution**

With the graphical method, we assume that $v = 150$. First, we plot contours of constant value of the surface area: We plot, on the (h, r)-surface, the contours where the surface area has the values 28, 58, 88, ..., 238:

```
p1 = ContourPlot[A, {h, 0, 10}, {r, 0, 7}, Contours → Range[28, 238, 30],
    ContourLabels → Automatic, ContourShading → False, FrameLabel → {"h", "r"},
    RotateLabel → False, PlotRangePadding → 0.6, AspectRatio → Automatic]
```

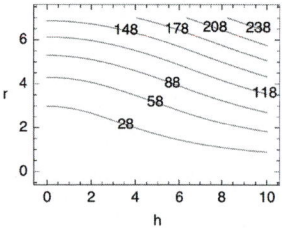

Then we plot the constraint $V - 150 = 0$:

```
p2 = ContourPlot[V - 150 == 0, {h, 0, 10},
    {r, 0, 7}, ContourShading → False, ContourStyle → Red,
    FrameLabel → {"h", "r"}, RotateLabel → False, AspectRatio → Automatic]
```

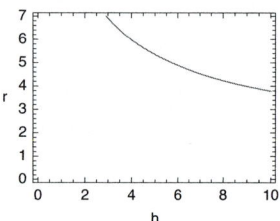

All points on this curve satisfy the constraint. Combine the plots:

```
Show[p1, p2, Epilog → Point[{6.6, 4.7}]]
```

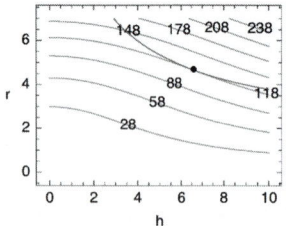

In the combined plot, we have added (by trial and error) a point that is approximately the solution of the problem. The point is approximately $(h, r) = (6.6, 4.7)$. Why is this point the solution? The point of solution is such that the constraint curve and one of the contours of constant value of the surface area have the same tangent. Such a contour seems to be the one that has the value 118; the smallest surface area is thus approximately 118.

The plot also shows how sensitive the solution is—how much the surface area increases if we move away from the optimum point but stay on the constraint curve. We see that the surface area does not increase much if h is in an interval of approximately (5.5, 7.5) and r is adjusted accordingly such that the volume has the value 150 (r changes from approximately 5.1 to approximately 4.4).

■ The Method of Substitution

Now we solve one variable from the constraint and substitute it into the object function, thereby reducing a 2D problem to a 1D one. Indeed, we solve h from the constraint and substitute it into the surface area:

h1 = Solve[V == v, h]〚1, 1〛 $\qquad h \rightarrow \dfrac{3 v}{\pi r^2}$

A1 = A /. h1 $\qquad \pi r \sqrt{r^2 + \dfrac{9 v^2}{\pi^2 r^4}}$

We find the optimum value of r:

r1 = Solve[D[A1, r] == 0, r]

$$\left\{\left\{r \rightarrow \dfrac{\left(-\frac{3}{\pi}\right)^{1/3} v^{1/3}}{2^{1/6}}\right\}, \left\{r \rightarrow -\dfrac{\left(\frac{3}{\pi}\right)^{1/3} v^{1/3}}{2^{1/6}}\right\}, \left\{r \rightarrow \dfrac{\left(\frac{3}{\pi}\right)^{1/3} v^{1/3}}{2^{1/6}}\right\},\right.$$

$$\left.\left\{r \rightarrow \dfrac{(-1)^{2/3} \left(\frac{3}{\pi}\right)^{1/3} v^{1/3}}{2^{1/6}}\right\}, \left\{r \rightarrow -\dfrac{(-3)^{1/3} v^{1/3}}{2^{1/6} \pi^{1/3}}\right\}, \left\{r \rightarrow -\dfrac{(-1)^{2/3} 3^{1/3} v^{1/3}}{2^{1/6} \pi^{1/3}}\right\}\right\}$$

Check which solutions are real and positive:

% // N

$$\left\{\left\{r \rightarrow (0.438654 + 0.759771 \,i)\, v^{1/3}\right\}, \left\{r \rightarrow -0.877308\, v^{1/3}\right\},\right.$$
$$\left\{r \rightarrow 0.877308\, v^{1/3}\right\}, \left\{r \rightarrow (-0.438654 + 0.759771 \,i)\, v^{1/3}\right\},$$
$$\left.\left\{r \rightarrow (-0.438654 - 0.759771 \,i)\, v^{1/3}\right\}, \left\{r \rightarrow (0.438654 - 0.759771 \,i)\, v^{1/3}\right\}\right\}$$

Only the third is real and positive. We choose this value as the optimum r and calculate the corresponding values of h and A:

{ropt = r1〚3, 1〛, hopt = h1 /. ropt, Aopt = A1 /. ropt /. hopt} // Simplify

$$\left\{r \rightarrow \dfrac{\left(\frac{3}{\pi}\right)^{1/3} v^{1/3}}{2^{1/6}}, \; h \rightarrow \left(\dfrac{6}{\pi}\right)^{1/3} v^{1/3}, \; \dfrac{3 \cdot 3^{1/6} \pi^{1/3} v^{2/3}}{2^{1/3}}\right\}$$

■ Lagrange's Method

Lastly, we form Lagrange's function, in which we have the object function and the left-hand side of the constraint $V - v = 0$ multiplied by a constant λ (Lagrange's multiplier):

L = A + λ (V - v) $\qquad \pi r \sqrt{h^2 + r^2} + \left(\dfrac{1}{3} h \pi r^2 - v\right) \lambda$

A necessary condition for the optimum solution is that partial derivatives of Lagrange's function with respect to h and r are zero and that the equality constraint is satisfied:

```
eqns = {D[L, h] == 0, D[L, r] == 0, V - v == 0}
```

$$\left\{ \frac{h \pi r}{\sqrt{h^2 + r^2}} + \frac{1}{3} \pi r^2 \lambda == 0, \quad \frac{\pi r^2}{\sqrt{h^2 + r^2}} + \pi \sqrt{h^2 + r^2} + \frac{2}{3} h \pi r \lambda == 0, \quad \frac{1}{3} h \pi r^2 - v == 0 \right\}$$

Solve these equations:

```
sol = Solve[eqns, {λ, h, r}];
```

Check which solutions are real and positive:

```
{h, r} /. sol // N
```

$$\left\{ \left\{ 1.2407\, v^{1/3},\ 0.877308\, v^{1/3} \right\}, \left\{ 1.2407\, v^{1/3},\ -0.877308\, v^{1/3} \right\}, \right.$$
$$\left\{ (-0.62035 + 1.07448\, i)\, v^{1/3},\ (0.438654 - 0.759771\, i)\, v^{1/3} \right\},$$
$$\left\{ (-0.62035 + 1.07448\, i)\, v^{1/3},\ (-0.438654 + 0.759771\, i)\, v^{1/3} \right\},$$
$$\left\{ (-0.62035 - 1.07448\, i)\, v^{1/3},\ (-0.438654 - 0.759771\, i)\, v^{1/3} \right\},$$
$$\left. \left\{ (-0.62035 - 1.07448\, i)\, v^{1/3},\ (0.438654 + 0.759771\, i)\, v^{1/3} \right\} \right\}$$

Of the six solutions, only the first solution has real and positive values for h and r, and so we choose this as the optimal solution:

```
λhropt = sol〚1〛
```

$$\left\{ \lambda \to - \frac{2^{2/3}\, 3^{1/6}\, \pi^{1/3}}{v^{1/3}},\ h \to \left(\frac{6}{\pi} \right)^{1/3} v^{1/3},\ r \to \frac{\left(\frac{3}{\pi} \right)^{1/3} v^{1/3}}{2^{1/6}} \right\}$$

Another way to state this is $h_{opt} = \sqrt[3]{6v/\pi}$ and $r_{opt} = \sqrt[3]{3v \big/ \left(\sqrt{2}\, \pi \right)}$. The smallest surface area is as follows:

```
Aopt = A /. λhropt // Simplify
```
$$\frac{3 \cdot 3^{1/6}\, \pi^{1/3}\, v^{2/3}}{2^{1/3}}$$

Another way to state this is $A_{opt} = 3 \sqrt[3]{\sqrt{3}\, \pi v^2/2}$.

We plot the optimal r and h and also the optimal surface area, all as functions of v:

```
{Plot[{r, h} /. λhropt, {v, 0, 200}, PlotRange → {{-5, 235}, All},
  Epilog → {Text[ropt, {220, 5}], Text[hopt, {220, 7.1}]}], Plot[Aopt, {v, 0, 200}]}
```

If the volume of the cone is, for example, 150, the optimal solution is as follows:

```
{A0, λhr0} = {Aopt, λhropt} /. v → 150.
```
```
{118.234, {λ → -0.525484, h → 6.59221, r → 4.66139}}
```

This optimal cone is displayed as follows:

```
line = h - h / r x /. λhr0
```

```
6.59221 - 1.41421 x
```

```
RevolutionPlot3D[line, {x, 0, r /. λhr0},
 BoxRatios → {1, 1, r / h /. λhr0}, ViewPoint → {1.6, -2.8, 1.0}]
```

23.4.3 Equality and Inequality Constraints

■ **Example 1**

We minimize $(x - y)^2 + 5z$ subject to $x + y + z + 5 = 0$, $y - 3z - 1 = 0$, and $-x \le 0$. First, we define the corresponding functions:

```
f = (x - y) ^ 2 + 5 z;
g1 = x + y + z + 5;
g2 = y - 3 z - 1;
h1 = -x;
```

Then we form Lagrange's function:

```
L = f + λ1 g1 + λ2 g2 + µ1 h1
```

$$(x - y)^2 + 5 z + (5 + x + y + z) \lambda1 + (-1 + y - 3 z) \lambda2 - x \mu1$$

A necessary condition for the optimum is that the derivatives of L with respect to x, y, and z are 0. Another necessary condition is that the equality constraints hold. A third necessary condition is the *complementary slackness condition*: $\mu_1 h_1 = 0$, which means that a given multiplier is 0 if the corresponding inequality constraint is satisfied with strict inequality ($h_1(x) < 0$), and a multiplier can be positive only if the corresponding inequality constraint is satisfied as an equality. We form these three types of equality conditions:

```
eqns = {D[L, x] == 0, D[L, y] == 0, D[L, z] == 0, g1 == 0, g2 == 0, µ1 h1 == 0}
```

$$\{2 (x - y) + \lambda1 - \mu1 == 0, -2 (x - y) + \lambda1 + \lambda2 == 0,$$
$$5 + \lambda1 - 3 \lambda2 == 0, 5 + x + y + z == 0, -1 + y - 3 z == 0, -x \mu1 == 0\}$$

Additional necessary conditions are that the original inequality conditions hold and that the Lagrange multipliers for the inequalities are nonnegative. We form these two types of inequality conditions:

```
ineqs = {h1 ≤ 0, µ1 ≥ 0}        {-x ≤ 0, µ1 ≥ 0}
```

The equality and inequality conditions together form the *Karush–Kuhn–Tucker conditions*. First, we solve the equality conditions:

```
sol = Solve[eqns]
```

$$\left\{\left\{x \to 0, \ y \to -\frac{7}{2}, \ z \to -\frac{3}{2}, \ \lambda1 \to 4, \ \lambda2 \to 3, \ \mu1 \to 11\right\},\right.$$

$$\left.\left\{y \to -\frac{211}{98}, \ z \to -\frac{103}{98}, \ x \to -\frac{88}{49}, \ \lambda1 \to -\frac{5}{7}, \ \lambda2 \to \frac{10}{7}, \ \mu1 \to 0\right\}\right\}$$

We then check which of the solutions also satisfy the inequality conditions:

```
ineqs /. sol     {{True, True}, {False, True}}
```

Only the first solution satisfies all inequality conditions. This can also be seen as follows (recall that **@@** is the same as **Apply** and **/@** the same as **Map**; see Section 14.2, p. 459):

```
(And @@ #) & /@ %     {True, False}
```

```
Position[%, True]     {{1}}
```

The candidate for a point of minimum and the corresponding minimum value are as follows:

```
cand = Extract[sol, %]
```

$$\left\{\left\{x \to 0, \ y \to -\frac{7}{2}, \ z \to -\frac{3}{2}, \ \lambda1 \to 4, \ \lambda2 \to 3, \ \mu1 \to 11\right\}\right\}$$

```
f /. %[[1]]     19
                --
                 4
```

■   **A General Program**

Consider the general problem of minimizing $f(x)$ (x is a vector) with respect to the equality constraints $g(x) = 0$ ($g(x)$ is a vector) and the inequality constraints $h(x) \le 0$ ($h(x)$ is a vector). The Lagrangian is $L = f(x) + \lambda^T g(x) + \mu^T h(x)$. The Karush–Kuhn–Tucker necessary conditions for a local minimum are as follows: A first set of conditions is that $\partial L / \partial x = 0$, $g(x) = 0$, and $\mu_i h_i(x) = 0$ for all i (the last condition is the complementary slackness condition); a second set of conditions is that $h(x) \le 0$ and $\mu \ge 0$.

The following program finds candidates for an optimum point—that is, points satisfying the necessary conditions. In the program, we proceed as we did in Example 1.

```
kktOptimize[f_, g_List, h_List, x_List, ε_: 10.^-12] :=
 Module[{λλ = Array[λ, Length[g]], μμ = Array[μ, Length[h]],
   xλμ, eqns, sol, realsol, nrealsol, ineqs, pos, cand},
  xλμ = Join[x, λλ, μμ];
  eqns = Thread[Join[D[f + λλ.g + μμ.h, {x}], g, μμ h] == 0];
  sol = Union[Solve[eqns, xλμ]];
  realsol = Select[sol, FreeQ[Chop[N[#], ε], Complex] &];
  If[realsol == {}, Return["No real solutions; try a larger ε"]];
  nrealsol = Thread[xλμ → #] & /@ Chop[N[xλμ /. realsol], ε];
  ineqs = Thread[Join[h, -μμ] ≤ ε];
  pos = Position[(And @@ #) & /@ (ineqs /. nrealsol), True];
  If[pos == {}, Return["Inequalities not satisfied; try a larger ε"]];
  cand = Union[Sort /@ Extract[realsol, pos]];
  {f /. cand, cand}ᵀ]
```

If the optimization problem does not contain some types of constraints, replace **g** and/or **h** with an empty list (**{}**).

In the program, we first form the set of variables **xλμ**, calculate the gradient of the Lagrangian, form the necessary equality conditions **eqns**, find the solution **sol** of the equations, select real solutions into **realsol**, and form a numerical version **nrealsol** of the set of real solutions (to be used to check the inequality conditions).

Then we form the set of necessary inequality conditions **ineqs** and check which solutions in **nrealsol** satisfy all the inequalities; the result is a list with the ith component **True** or **False** according to whether the ith solution in **nrealsol** satisfies all of the inequalities. The solutions of the equations that give **True** for all inequalities form the set **cand** of candidates for an optimum point. Lastly, we attach to each candidate the corresponding value of the object function.

■ Some Notes

In the program, we have used a small number ϵ for three points. When selecting the real solutions into **realsol**, we use ϵ to ignore (with **Chop**) small imaginary parts that may be the result of numerical inaccuracy (and not "true" imaginary parts). When forming the numerical solutions **nrealsol**, we use ϵ for the same reason. When forming the inequality conditions **ineqs**, we use ϵ to make the conditions slightly less tight: We replace the original conditions $h(x) \leq 0$ and $\mu \geq 0$ with $h(x) \leq \epsilon$ and $\mu \geq -\epsilon$; otherwise, numerical inaccuracy may cause the rejection of a feasible point when the inequalities are tested (see Example 7). (On the other hand, the less tight conditions probably do not cause the acceptance of solutions that do not satisfy the original tight inequality conditions.)

The default value 10^{-12} of ϵ has been shown to be suitable in many problems. Although in most problems ϵ could be exactly 0, in some problems this value is too small: Due to numerical inaccuracies, the inequality conditions do not hold exactly. When using **kktOptimize**, you can give a new value for ϵ by writing a fifth argument for the command.

■ Further Notes

The result of **kktOptimize** for problems that have only exact numbers may be a very long expression (even several pages) because **Solve** writes explicit expressions for the solutions of third- and fourth-order polynomials. Such exact results may be useless, so it is better to solve such problems with decimal numbers (simply insert a decimal point into the problem). A decimal solution is also calculated more quickly.

In the program, we do not simplify the candidates. If appropriate, you may apply **Simplify** or **FullSimplify** to the solution (note, however, that **Simplify** and especially **FullSimplify** may take a very long time for a long expression).

The multiplier method package mentioned in the introduction to this chapter contains a notebook of test problems for nonlinear constrained optimization. **kktOptimize** solved all of the test problems in no time. Some of these test problems are solved in the examples that follow. **Minimize** (see Section 23.1.1, p. 743) also solves all of the test problems.

The major advantage of **Minimize** is that it gives a definite result: the global minimum. **kktOptimize** only gives candidates for local optimum points (some of them may be local minimums, some local maximums, and some saddle points).

■ Example 2

First, we solve the problem of Example 1:

```
kktOptimize[(x - y) ^2 + 5 z, {x + y + z + 5, y - 3 z - 1}, {-x}, {x, y, z}]
```

$$\left\{\left\{\frac{19}{4}, \left\{x \to 0, y \to -\frac{7}{2}, z \to -\frac{3}{2}, \lambda[1] \to 4, \lambda[2] \to 3, \mu[1] \to 11\right\}\right\}\right\}$$

`Minimize` gives the same result:

```
Minimize[{(x - y)^2 + 5 z, x + y + z + 5 == 0, y - 3 z - 1 == 0, -x ≤ 0}, {x, y, z}]
```

$$\left\{\frac{19}{4}, \left\{x \to 0, y \to -\frac{7}{2}, z \to -\frac{3}{2}\right\}\right\}$$

■ Example 3

The following problem has only inequality constraints:

```
f = (x - 2)^2 + (y - 3)^2;
h1 = x + y - 4;
h2 = x - y - 2;
kktOptimize[f, {}, {h1, h2}, {x, y}]
```

$$\left\{\left\{\frac{1}{2}, \left\{x \to \frac{3}{2}, y \to \frac{5}{2}, \mu[1] \to 1, \mu[2] \to 0\right\}\right\}\right\}$$

A figure confirms that $\left(\frac{3}{2}, \frac{5}{2}\right)$ really is the minimum point of the constrained problem:

```
Show[RegionPlot[h1 ≤ 0 && h2 ≤ 0, {x, 0, 4}, {y, 0, 4}],
  ContourPlot[f, {x, 0, 4.1}, {y, 0, 4.1}, Contours → 20, ContourShading → False],
  Graphics[Point[{3 / 2, 5 / 2}]]]
```

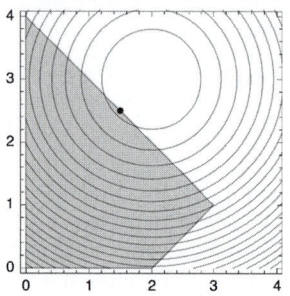

■ Example 4

Next, we solve the problem of Section 23.4.2, p. 769, which has an equality constraint:

```
kktOptimize[Pi r Sqrt[h^2 + r^2], {Pi h r^2 / 3 - v}, {}, {h, r}] // Simplify
```

$$\left\{\left\{-\frac{3 \cdot 3^{1/6} \pi^{1/3} v^{2/3}}{2^{1/3}}, \left\{h \to \left(\frac{6}{\pi}\right)^{1/3} v^{1/3}, r \to -\frac{\left(\frac{3}{\pi}\right)^{1/3} v^{1/3}}{2^{1/6}}, \lambda[1] \to \frac{2^{2/3} \cdot 3^{1/6} \pi^{1/3}}{v^{1/3}}\right\}\right\},\right.$$

$$\left.\left\{\frac{3 \cdot 3^{1/6} \pi^{1/3} v^{2/3}}{2^{1/3}}, \left\{h \to \left(\frac{6}{\pi}\right)^{1/3} v^{1/3}, r \to \frac{\left(\frac{3}{\pi}\right)^{1/3} v^{1/3}}{2^{1/6}}, \lambda[1] \to -\frac{2^{2/3} \cdot 3^{1/6} \pi^{1/3}}{v^{1/3}}\right\}\right\}\right\}$$

We obtained two candidates, but only the second one has a positive value for r, and so we choose this candidate as the solution.

■ Example 5

Now we solve the following familiar problem:

```
kktOptimize[x^3 + y^3 + 2 x^2 + 4 y^2 + 6, {}, {}, {x, y}]
```

$$\left\{\left\{\frac{50}{3}, \left\{x \to -\frac{4}{3}, y \to -\frac{8}{3}\right\}\right\}, \left\{\frac{194}{27}, \left\{x \to -\frac{4}{3}, y \to 0\right\}\right\},\right.$$

$$\left.\left\{\frac{418}{27}, \left\{x \to 0, y \to -\frac{8}{3}\right\}\right\}, \{6, \{x \to 0, y \to 0\}\}\right\}$$

The program gives us four candidates. These can be investigated as in Example 3 of Section 23.3.1, p. 760; the candidates contain local minimum and maximum points and saddle points.

■ Example 6

The following example was first encountered in Section 23.1.1, p. 745 (the computing time with **Minimize** was approximately 24 seconds):

```
kktOptimize[x^2 / 2 + (y^2 + z^2 + v^2) / 6, {},
  {2 - x, 5 - x - y, 2 - x - z, 1 - x - v, -v}, {x, y, z, v}] // Timing
```

$$\left\{0.042132,\right.$$

$$\left.\left\{\left\{\frac{7}{2}, \{v \to 0, x \to 2, y \to 3, z \to 0, \mu[1] \to 1, \mu[2] \to 1, \mu[3] \to 0, \mu[4] \to 0, \mu[5] \to 0\}\right\}\right\}\right\}$$

23.5 Special Topics

23.5.1 Traveling Salesman

■ Finding the Shortest Tour

FindShortestTour[points]   Try to solve the traveling salesman problem: Try to find the shortest tour through the points by visiting all the points once

Options:

DistanceFunction   The distance function to apply to pairs of points; examples of values:
   Automatic, EuclideanDistance, ManhattanDistance

Method   Method to use; examples of values: **Automatic, "OrZweig"** (default for 2D real points), **"OrOpt"** (default for non-2D or nonreal points), **"TwoOpt"** (performs exchanges of edge end points for improvement), **"CCA"** (Convex hull, Cheapest insertion, and Angle selection; intended for points in $\mathbb{R}^n$), **"Greedy"** (moves from one point to the nearest unvisited neighbor), **"GreedyCycle"** (a variant of the greedy algorithm with known upper bound), **"SimulatedAnnealing"** (uses simulated annealing to minimize the tour length)

For small numbers of points, the command usually finds the shortest tour. For larger numbers of points, the command will normally find a tour whose length is at least close to the minimum.

The default distance for points that are numbers is the Euclidean distance. In the following, we list all the built-in distance functions:

```
Names["*Distance"]
```

```
{BrayCurtisDistance, CanberraDistance, ChebyshevDistance,
 CorrelationDistance, CosineDistance, EditDistance, EuclideanDistance,
 HammingDistance, ManhattanDistance, SquaredEuclideanDistance}
```

In Section 23.5.2, p. 786, we solve small traveling salesman problems exactly with dynamic programming.

■ **Example 1**

Generate 100 points and try to find the shortest tour:

```
SeedRandom[2]; points = RandomReal[10, {100, 2}];
```

```
{length, tour} = FindShortestTour[points]
```

```
{78.6367, {1, 16, 39, 32, 59, 68, 6, 96, 75, 29, 80, 19, 85, 38, 78, 21, 65, 92, 97, 57,
   52, 53, 84, 17, 7, 70, 83, 89, 88, 26, 11, 47, 5, 95, 86, 79, 82, 30, 93, 48,
   9, 77, 94, 55, 71, 10, 12, 41, 27, 31, 2, 22, 58, 14, 99, 91, 42, 64, 36, 33,
   72, 50, 28, 69, 45, 40, 87, 60, 15, 25, 63, 34, 67, 13, 61, 20, 24, 4, 43, 46,
   51, 23, 54, 100, 62, 98, 90, 8, 74, 56, 18, 66, 73, 44, 81, 76, 3, 35, 37, 49}}
```

Thus, the tour starts from the first point, goes to the 16th point, and so on. The last point visited is the 49th point. The solution does not give the last step of going back to the starting point. The length of the tour (including the step of going back to the first point) is 78.6367. Here are the points and the corresponding shortest tour:

```
{ListPlot[points, AspectRatio → Automatic],
 ListLinePlot[points[[tour]], Mesh → All, AspectRatio → Automatic,
  Epilog → {Red, PointSize[Medium], Point[First[points]]}]}
```

We have shown the starting point with a red color. The plot of the tour does not contain the last step of going back to the starting point. Note that the optimal tour does not depend on the starting point.

To check the length of the tour, first order the **points**, with **points[[tour]]**, into the order given by **tour**. Then partition the points into a list of pairs, using offset 1 and adding the first point to the end of the list:

```
pairs = Partition[points[[tour]], 2, 1, 1];
```

This list of pairs has length 100 and the first point of the first pair and last point of the last pair are the same:

```
{pairs // Length, pairs // First, pairs // Last}
```

```
{100, {{7.2224, 1.09449}, {7.71263, 1.47503}},
  {{5.76362, 1.5754}, {7.2224, 1.09449}}}
```

The Euclidean length of the tour is as follows:

```
Total[EuclideanDistance[#[[1]], #[[2]]] & /@ pairs]
```

```
78.6367
```

■ **Example 2**

Next, we use the Chebyshev and Manhattan distances and also show the solution of a 3D problem:

```
{ListLinePlot[points[[FindShortestTour[points, DistanceFunction → #][[2]]]],
   Mesh → All, AspectRatio → Automatic,
   Epilog → {Red, PointSize[Medium], Point[First[points]]}] & /@
  {ChebyshevDistance, ManhattanDistance}, SeedRandom[0];
 With[{p = RandomReal[10, {20, 3}]}, Graphics3D[
   {Line[p[[FindShortestTour[p] // Last]]], Blue, PointSize[Medium], Point[p]}]]}
```

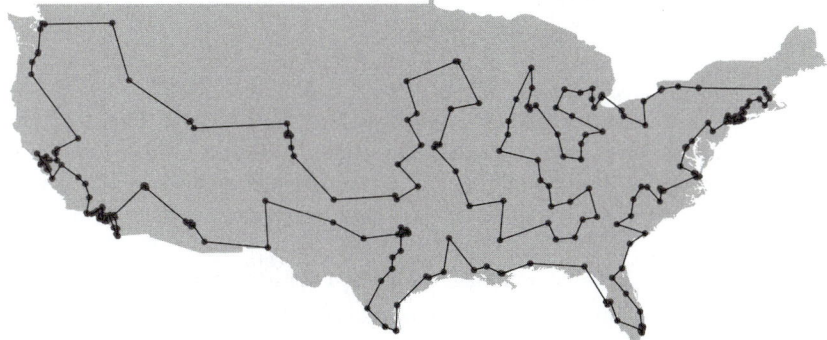

■ Example 3

Consider the large cities of United States but exclude Anchorage and Honolulu:

```
cities = CityData[{Large, "UnitedStates"}];

cities2 = DeleteCases[cities, c_ /; MemberQ[c, "Anchorage" | "Honolulu"]];

Length[cities2]      266
```

Create the coordinates of these cities. Then find and show the shortest tour:

```
points = Reverse[CityData[#, "Coordinates"]] & /@ cities2;

{length, tour} = FindShortestTour[points];

Graphics[{Green, CountryData["UnitedStates", "Polygon"], Red,
  Point[points[[tour]]], Blue, Line[points[[tour]]]}, ImageSize → 400]
```

■ Example 4

Thus far, we have specified the points with their coordinates. Now we specify the points with a distance matrix. Suppose we have five cities. Assume that the distances are as follows:

```
distances = SparseArray[{{1, 2} → 5, {2, 1} → 5, {1, 3} → 2, {3, 1} → 2, {3, 2} → 2,
    {2, 3} → 2, {2, 4} → 4, {4, 2} → 4, {3, 4} → 5, {4, 3} → 5, {4, 1} → 6,
    {1, 4} → 6, {4, 5} → 8, {5, 4} → 8, {1, 5} → 1, {5, 1} → 1}, {5, 5}, ∞];

distances // MatrixForm
```

$$\begin{pmatrix} \infty & 5 & 2 & 6 & 1 \\ 5 & \infty & 2 & 4 & \infty \\ 2 & 2 & \infty & 5 & \infty \\ 6 & 4 & 5 & \infty & 8 \\ 1 & \infty & \infty & 8 & \infty \end{pmatrix}$$

Find the shortest tour through the cities 1, 2, 3, 4, and 5 by using the distance matrix:

```
{length, tour} = FindShortestTour[Range[5], DistanceFunction → (distances[[#1, #2]] &)]

{17, {1, 3, 2, 4, 5}}
```

Use **GraphPlot** to draw the optimal tour with red, thick lines (in our example, the **"CircularEmbedding"** method works well):

```
GraphPlot[distances, Method → "CircularEmbedding", VertexLabeling → True,
    EdgeRenderingFunction → ({If[MemberQ[Partition[tour, 2, 1, 1], #2 | Reverse[#2]],
        Directive[Red, Thick], Black], Line[#1], Black,
        Text[distances[[#2[[1]], #2[[2]]]], Mean[#1], Background → White]} &)]
```

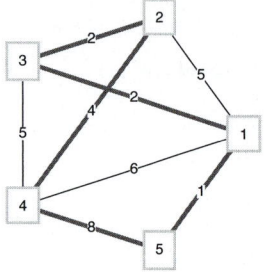

23.5.2 Dynamic Programming

■ Introduction

Dynamic programming (or dynamic optimization) is a versatile optimization method applicable to many different types of problem. The idea of this method is to break up the original problem into a series of smaller problems that can be solved easily. Typically, the solution procedure goes stagewise backward from the "end" of the problem toward the "beginning" by utilizing a *recursion formula*. The solution of the original problem is then found by going forward and utilizing the results of the backward computations.

Let $f_t(i)$ be the minimum cost incurred from stage t to the end of the problem, given that at stage t the state is i. The recursion formula of dynamic programming is often of the following form:

$$f_t(i) = \min[(\text{cost during stage } t) + f_{t+1}(\text{new state at stage } t + 1)],$$

where the minimum is over all feasible decisions when the state at stage t is i.

Suppose that we have T stages. Often, the calculation of $f_T(i)$ is very easy; this gives us the starting point of the recursion. With the recursion formula we then go backward and solve a series of simple optimization problems until we have found $f_1(a)$ for a given initial state a. The optimal value of the object function is $f_1(a)$. During the recursive calculation we write down the optimal decision for each state in each stage. This enables us to get the optimal values of the decision variables for all stages, by going forward.

Dynamic programming with pencil and paper easily becomes unmanageable for other than small problems. However, with *Mathematica* the situation is much more favorable. Indeed, *Mathematica* supports the recursive programming style (see Section 18.5, p. 596) so that writing down the recursion formulas of dynamic programming is straightforward. After we have coded the recursion formula, all we have to do is to ask for the solution! *Mathematica* does all the lengthy recursive computations behind the scenes. The optimal value of the object function is obtained simply by asking for the value of $f_1(a)$, whereas the optimal values of the decision variables can be seen with a small calculation.

As examples of dynamic programming, we next consider some of the examples in Winston (1994): a shortest path problem, a resource allocation problem, an inventory problem, and the traveling salesman problem.

Note that "dynamic programming" has two meanings. In programming, it refers to a technique in recursive programming in which rule bases containing the values of a function are updated dynamically—that is, during the computation. In optimization, dynamic programming refers to the recursive optimization method described previously. Note that in the optimization method of dynamic programming we use the programming technique of dynamic programming.

■ Shortest Path

Consider the following graph (Winston, 1994, p. 1005):

```
edges = {{1 → 2, 550}, {1 → 3, 900}, {1 → 4, 770}, {2 → 5, 680}, {2 → 6, 790},
    {2 → 7, 1050}, {3 → 5, 580}, {3 → 6, 760}, {3 → 7, 660}, {4 → 5, 510},
    {4 → 6, 700}, {4 → 7, 830}, {5 → 8, 610}, {5 → 9, 790}, {6 → 8, 540},
    {6 → 9, 940}, {7 → 8, 790}, {7 → 9, 270}, {8 → 10, 1030}, {9 → 10, 1390}};
vertices = {1 → {1, 1}, 2 → {2, 2}, 3 → {2, 1}, 4 → {2, 0},
    5 → {3, 2}, 6 → {3, 1}, 7 → {3, 0}, 8 → {4, 1.5}, 9 → {4, 0.5}, 10 → {5, 1}};
GraphPlot[edges, DirectedEdges → True,
    VertexCoordinateRules → vertices, VertexLabeling → True, ImageSize → 300]
```

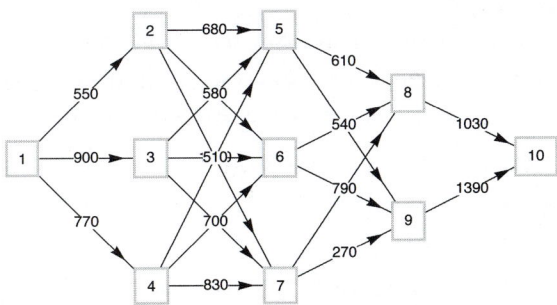

The numbers on the edges are their lengths c_{ij} (some of them overlap in the plot). We would like to find the shortest path from City 1 to City 10. Note that the network is acyclic and the vertices are numbered such that if we can go from i to j then $i < j$.

Write the lengths of the edges as a matrix:

```
T = 10;
c = SparseArray[{#[[1, 1]], #[[1, 2]]} → #[[2]] & /@ edges, {T, T}, ∞]
SparseArray[<20>, {10, 10}, ∞]
```

```
c // MatrixForm
```

$$
\begin{pmatrix}
\infty & 550 & 900 & 770 & \infty & \infty & \infty & \infty & \infty & \infty \\
\infty & \infty & \infty & \infty & 680 & 790 & 1050 & \infty & \infty & \infty \\
\infty & \infty & \infty & \infty & 580 & 760 & 660 & \infty & \infty & \infty \\
\infty & \infty & \infty & \infty & 510 & 700 & 830 & \infty & \infty & \infty \\
\infty & \infty & \infty & \infty & \infty & \infty & \infty & 610 & 790 & \infty \\
\infty & \infty & \infty & \infty & \infty & \infty & \infty & 540 & 940 & \infty \\
\infty & \infty & \infty & \infty & \infty & \infty & \infty & 790 & 270 & \infty \\
\infty & \infty & \infty & \infty & \infty & \infty & \infty & \infty & \infty & 1030 \\
\infty & \infty & \infty & \infty & \infty & \infty & \infty & \infty & \infty & 1390 \\
\infty & \infty & \infty & \infty & \infty & \infty & \infty & \infty & \infty & \infty
\end{pmatrix}
$$

Let f_t be the length of the shortest path from vertex t to the final vertex T (in our example, $T = 10$). It is clear that

$$f_t = \min_j (c_{tj} + f_j),$$

$t = T - 1, T - 2, \ldots, 1$. Let p_t be the optimal value of j for a given t. In addition, $p_T = T$ and $f_T = 0$. Here are recursive formulas to calculate f_t and p_t, $t = 1, \ldots, T$:

```
Clear[f, p]
p[T] = T;
f[T] = 0;
f[t_] := f[t] = Module[{v},
   v = Table[c[[t, j]] + f[j], {j, t + 1, T}];
   p[t] = Ordering[v, 1][[1]] + t;
   Min[v]]
```

Here, we use recursive programming: The function `f` calls itself. Indeed, the value of `f[t]` depends on `f[t + 1]`, `f[t + 2]` .... Note the use of the dynamic programming technique: `f[t_] := f[t] = …`, causing the list of values of `f` to be dynamically updated during the computation. For recursive programming, see Section 18.5, p. 596. We have written `Clear[f, p]` at the beginning of the code to get fresh values of `f` and `p` each time we use the code.

In calculating the value of f_t, we first create a table v for the values of $c_{tj} + f_j$ for all $j = t + 1, \ldots, T$. The value of f_t is the minimum of these values. The optimal j or p_t is the value of j that gives the minimum of $c_{tj} + f_j$. Recall that with `Ordering` we can get the position of the minimum value in a list:

```
Ordering[{40, 20, 50, 10, 30}, 1]      {4}
```

Now, the length of the shortest path from vertex 1 is f_1:

```
f[1]      2870
```

Here are the lengths of the shortest paths from all the vertices:

```
Table[{t, f[t]}, {t, 10}]
{{1, 2870}, {2, 2320}, {3, 2220}, {4, 2150},
  {5, 1640}, {6, 1570}, {7, 1660}, {8, 1030}, {9, 1390}, {10, 0}}
```

For a given vertex, what is the optimal next vertex? The optimal vertices are as follows:

```
Table[{t, p[t]}, {t, 10}]
```

$\{\{1, 2\}, \{2, 5\}, \{3, 5\}, \{4, 5\}, \{5, 8\}, \{6, 8\}, \{7, 9\}, \{8, 10\}, \{9, 10\}, \{10, 10\}\}$

This means that if we start from vertex 1, then we go to 2. From 2 we go to 5, then to 8, and lastly to 10. The shortest path can also be computed as follows:

```
y = 1; path = Join[{1}, Table[y = p[y], {t, 4}]]
```

$\{1, 2, 5, 8, 10\}$

Still another way is the following:

```
path = FixedPointList[p, 1] // Most
```

$\{1, 2, 5, 8, 10\}$

The shortest path is shown in the following figure:

```
GraphPlot[edges, DirectedEdges → True, VertexLabeling → True,
  VertexCoordinateRules → vertices, ImageSize → 300, EdgeRenderingFunction →
   ({If[MemberQ[Partition[path, 2, 1, 1], #2], Directive[Red, Thick], Black],
     Line[#1], Black, Text[c〚#2〚1〛, #2〚2〛〛, Mean[#1], Background → White]} &)]
```

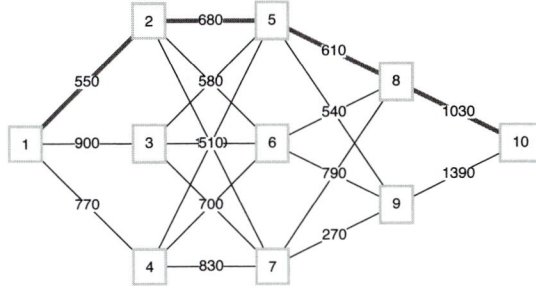

■ **Resource Allocation**

Suppose we have $\mathcal{D}$ thousand dollars to invest, and T investments are available (Winston, 1994, p. 1018). If d_j thousand dollars are invested in investment j, then a net present value (in thousands) of $r_j(d_j)$ is obtained (these functions are assumed to be known). The amount placed in each investment must be an exact multiple of $1000. How to allocate the $\$\mathcal{D}$ to maximize the net present value obtained from the investments? We should maximize $r_1(d_1) + \ldots + r_T(d_T)$ subject to the constraint that $d_1 + \ldots + d_T = \mathcal{D}$; the values of d_j should be nonnegative integers.

Assume that we have three possible investments. Define the $r_j(d_j)$ functions:

```
Remove["Global`*"]
T = 3;
r[1, d_] := Piecewise[{{7 d + 2, d > 0}, {0, d == 0}}]
r[2, d_] := Piecewise[{{3 d + 7, d > 0}, {0, d == 0}}]
r[3, d_] := Piecewise[{{4 d + 5, d > 0}, {0, d == 0}}]
```

Let $f_t(i)$ be the maximum net present value that can be obtained by investing i thousand dollars in investments $t, t + 1, \ldots, T$. It is clear that

$$f_t(i) = \max_d \left[r_t(d) + f_{t+1}(i - d) \right].$$

Let $p_t(i)$ be the optimal value of d. In addition, $f_T(i) = r_T(i)$; alternatively, we can define $f_{T+1}(i) = 0$. Here are recursive formulas to calculate $f_t(i)$ and $p_t(i)$, $t = 1, \ldots, T$:

```
Clear[f, p];
f[T + 1, i_] := 0
f[t_, i_] := f[t, i] = Module[{v},
    v = Table[r[t, d] + f[t + 1, i - d], {d, 0, i}];
    p[t, i] = Ordering[v, -1][[1]] - 1;
    Max[v]]
```

If we have $\mathcal{D}$ thousand dollars to invest, $\mathcal{D} = 1, \ldots, 6$, the maximum net present value that can be obtained can be seen from the following list:

```
𝒟𝒟 = 6; Table[{𝒟, f[1, 𝒟]}, {𝒟, 𝒟𝒟}]
```
```
{{1, 10}, {2, 19}, {3, 28}, {4, 35}, {5, 42}, {6, 49}}
```

For example, if we have $6000, the maximum net present value is 49. The optimal amounts of investments can be obtained by proceeding forward. First, ask what is the optimal amount for investment 1:

```
p[1, 6]      4
```

Thus, we should invest $4000 for investment 1. Now we have $2000 left, so the optimal amount for investment 2 is

```
p[2, 2]      1
```

Only $1000 is now left, so the optimal amount for investment 3 is

```
p[3, 1]      1
```

The optimal investments for various initial amounts of money $\mathcal{D}$ can also be calculated as follows:

```
Table[{Row[{"𝒟 = ", 𝒟}], Table[y = p[t, 𝒟]; 𝒟 = 𝒟 - y; y, {t, T}]}, {𝒟, 𝒟𝒟}] // Column
```
```
{𝒟 = 1, {0, 1, 0}}
{𝒟 = 2, {1, 1, 0}}
{𝒟 = 3, {1, 1, 1}}
{𝒟 = 4, {2, 1, 1}}
{𝒟 = 5, {3, 1, 1}}
{𝒟 = 6, {4, 1, 1}}
```

For example, if we have $6000, it is optimal to invest $4000 in investment 1 and $1000 in both investment 2 and investment 3.

Still another way to get the optimal solution is to prepare the following table:

```
Grid[Join[{Join[{"Money available"}, Range[0, 𝒟𝒟]]},
   Table[Join[{Row[{"Investment ", t}]}, Table[p[t, i], {i, 0, 𝒟𝒟}]], {t, T}]],
   Dividers → {2 → True, 2 → True}, Alignment → {Left, {Right}}, Spacings → 2,
   Background → {Automatic, Automatic, {{{2, 2}, {8, 8}} → LightGray,
      {{3, 3}, {4, 4}} → LightGray, {{4, 4}, {3, 3}} → LightGray}}]
```

Money available	0	1	2	3	4	5	6
Investment 1	0	0	1	1	2	3	4
Investment 2	0	1	1	1	1	1	1
Investment 3	0	1	2	3	4	5	6

From this table we can read the optimal amount of investments as a function of the amount of money available. For example, consider the situation in which the initial amount of money is 6 (this situation is shown with gray background in the table):

1. Investment: The optimal investment is 4 units.
2. Investment: Because we have 2 units of money left, the optimal investment is 1 unit.
3. Investment: Because we have 1 unit of money left, the optimal investment is 1 unit.

■ Inventory

Consider the following inventory problem (Winston, 1994, p. 1012). The planning period is T months. In month t, the demand for a product is d_t (they are known in advance) and the produced amount is x_t (to be determined optimally). The inventory at the end of month t is i_t (unknown quantities). We assume that the production in each month comes early enough to satisfy the demand for that month. Assume that in month t, the total cost is $c(x_t, i_t) = a + b x_t + h i_t$ if $x_t > 0$ and $c(x_t, i_t) = h i_t$ if $x_t = 0$. We have the restrictions that $0 \le x_t \le r$ and $0 \le i_t \le s$.

Set the following constants:

```
Remove["Global`*"]
a = 3; b = 1; h = 0.5; r = 5; s = 4; T = 4;
d = {1, 3, 2, 4};
```

Define the cost function:

```
c[x_, i_] = Piecewise[{{0, x == 0}, {a + b x, x > 0}}, ∞] + h i
```

$$0.5\,i + \begin{bmatrix} 0 & x == 0 \\ 3 + x & x > 0 \\ \infty & \text{True} \end{bmatrix}$$

Let $f_t(i)$ be the minimum cost of meeting demands for months $t, t+1, ..., T$ if i units are in inventory at the beginning of month t. It is clear that

$$f_t(i) = \min_{x_t}\left[c(x_t, i + x_t - d_t) + f_{t+1}(i + x_t - d_t)\right].$$

Let $p_t(i)$ be the optimal value of x_t. In addition, in the last month it is optimal to choose the produced amount such that the inventory is empty at the end of the month. Thus, $p_T(i) = d_T - i$ and $f_T(i) = c(d_T - i, 0)$. Here are recursive formulas to calculate $f_t(i)$ and $p_t(i)$, $t = 1, ..., T$:

```
Clear[f, p]
p[T, i_] := d[[T]] - i;
f[T, i_] := f[T, i] = c[d[[T]] - i, 0]
f[t_, i_] := f[t, i] = Module[{v, e = d[[t]]},
    v = Table[c[x, i + x - e] + f[t + 1, i + x - e], {x, e - i, Min[r, s + e - i]}];
    p[t, i] = Ordering[v, 1][[1]] + e - i - 1;
    Min[v]]
```

The minimum cost of production during the 4 months can be seen from the following list for various values of the initial inventory i:

```
Table[f[1, i], {i, 0, s}]
```
```
{20., 16., 15.5, 15., 13.5}
```

To get the optimal amounts produced, suppose the initial inventory is zero. Proceed forward by asking first what is the optimal amount produced in the first month:

```
p[1, 0]      1
```

Because the demand for this month is 1, the inventory is again zero at the beginning of the second month. Thus, the optimal production for this month is

```
p[2, 0]      5
```

Because the demand for the second month is 3, the inventory is 2 at the beginning of the third month. Thus, the optimal production for this month is

```
p[3, 2]    0
```

Because the demand for the third month is 2, the inventory is zero at the beginning of the fourth month. Thus, the optimal production for this month is

```
p[4, 0]    4
```

The optimal amounts of production for various values of initial inventory can also be calculated as follows:

```
Table[{Row[{"i = ", i}], Table[y = p[t, i]; i = i + y - d[[t]]; y, {t, T}]}, {i, 0, s}] //
  Column
{i = 0, {1, 5, 0, 4}}
{i = 1, {0, 5, 0, 4}}
{i = 2, {0, 4, 0, 4}}
{i = 3, {0, 3, 0, 4}}
{i = 4, {0, 0, 2, 4}}
```

For example, if the initial inventory is empty, it is optimal to produce 1, 5, 0, and 4 units in months 1, 2, 3, and 4, respectively.

Still another way to get the optimal solution is to prepare the following table:

```
Grid[Join[{Join[{"Init. inv."}, Range[0, s]]},
   Table[Join[{Row[{"Month ", t}]}, Table[p[t, i], {i, 0, s}]], {t, T}]],
   Dividers → {2 → True, 2 → True}, Alignment → {Left, {Right}}, Spacings → 2,
   Background → {Automatic, Automatic, {{{2, 3}, {2, 2}} → LightGray,
      {{4, 4}, {4, 4}} → LightGray, {{5, 5}, {2, 2}} → LightGray}}]
```

Init. inv.	0	1	2	3	4
Month 1	1	0	0	0	0
Month 2	5	4	3	0	0
Month 3	2	5	0	0	0
Month 4	4	3	2	1	0

From this table we can read the optimal amount of production as a function of the initial inventory of each month. For example, consider the situation in which the initial inventory is 0 (this situation is shown with gray background in the table):

1. Month: Because the initial inventory is 0, in the 1. month it is optimal to produce 1 unit.
2. Month: Because the demand for the 1. month was 1, the inventory at the beginning of the 2. month will again be 0, so in the 2. month it is optimal to produce 5 units.
3. Month: Because the demand for the 2. month was 3, the inventory at the beginning of the 3. month will be 2, so in the 3. month it is optimal to produce 0 units.
4. Month: Because the demand for the 3. month was 2, the inventory at the beginning of the 4. month will be 0, so in this month it is optimal to produce 4 units. Because the demand for the 4. month was 4, the inventory will be empty at the end of the 4. month.

■ **Traveling Salesman**

We have a set of T cities, numbered as 1, ..., T. A traveling salesman, currently in City 1, should visit all the other cities once and return to City 1. The salesman wants to minimize the total distance he must travel (Winston, 1994, p. 1039). In Section 23.5.1, p. 777, we solved traveling salesman problems with heuristic methods, giving the optimal solution with high probability. Now we try to solve small traveling salesman problems exactly.

Denote by c_{ij} the distance between cities i and j. As an example, let us generate 15 random points and calculate the distance matrix $\{c_{ij}\}$ for them:

```
Remove["Global`*"]
T = 15; SeedRandom[54];
points = RandomReal[10, {T, 2}];
c = Array[cc, {T, T}];
Table[c[[i, j]] = EuclideanDistance[points[[i]], points[[j]]], {i, T}, {j, T}];
```

Let $f_t(i, S)$ be the minimum distance that must be traveled to complete a tour if the $t - 1$ cities in the set S have been visited and city i was the last city visited. It is clear that

$$f_t(i, S) = \min_{j \in S^c}\left[c_{ij} + f_{t+1}\left(j, S \cup \{j\}\right)\right].$$

Here, S^c is the complement of S. Let $p_t(i, S)$ be the optimal value of j—that is, the city to which it is optimal to travel when the $t - 1$ cities in the set S have been visited and city i was the last city visited. In addition, $p_T(i, S) = 1$ and $f_T(i, S) = c_{i1}$. Here are recursive formulas to calculate $f_t(i, S)$ and $p_t(i, S)$, $t = 1, \ldots, T$:

```
Clear[f, p]; cities = Range[2, T];
p[T, i_, cities] := 1
f[T, i_, cities] := f[T, i, cities] = c[[i, 1]]
f[t_, i_, S_] := f[t, i, S] = Module[{unvisited, v},
   unvisited = Complement[cities, S];
   v = c[[i, #]] + f[t + 1, #, Union[S, {#}]] & /@ unvisited;
   p[t, i, S] = Extract[unvisited, Ordering[v, 1][[1]]];
   Min[v]]
```

The length of the shortest tour is obtained by asking what is the shortest length of a tour if 0 cities in the empty set {} have been visited and City 1 was the last city visited:

```
f[1, 1, {}] // Timing
{34.2202, 37.6344}
```

The computation time was approximately half a minute and the length of the optimal tour is 37.6344. To get the optimal tour, proceed forward step by step:

```
p[1, 1, {}]        7

p[2, 7, {7}]        13

p[3, 13, {7, 13}]        2

p[4, 2, {2, 7, 13}]        14

p[5, 14, {2, 7, 13, 14}]        10

p[6, 10, {2, 7, 10, 13, 14}]        5

p[7, 5, {2, 5, 7, 10, 13, 14}]        4

p[8, 4, {2, 4, 5, 7, 10, 13, 14}]        12

p[9, 12, {2, 4, 5, 7, 10, 12, 13, 14}]        15

p[10, 15, {2, 4, 5, 7, 10, 12, 13, 14, 15}]        3

p[11, 3, {2, 3, 4, 5, 7, 10, 12, 13, 14, 15}]        9

p[12, 9, {2, 3, 4, 5, 7, 9, 10, 12, 13, 14, 15}]        11

p[13, 11, {2, 3, 4, 5, 7, 9, 10, 11, 12, 13, 14, 15}]        6
```

```
p[14, 6, {2, 3, 4, 5, 6, 7, 9, 10, 11, 12, 13, 14, 15}]      8

p[15, 8, {2, 3, 4, 5, 6, 7, 8, 9, 10, 11, 12, 13, 14, 15}]      1
```

Thus, the optimal tour is {1, 7, 13, 2, 14, 10, 5, 4, 12, 15, 3, 9, 11, 6, 8, 1}. The optimal tour can also be calculated as follows:

```
i = 1; S = {}; tour1 = Join[{1}, Table[y = p[t, i, S]; i = y; S = Union[S, {y}]; y, {t, T}]]
{1, 7, 13, 2, 14, 10, 5, 4, 12, 15, 3, 9, 11, 6, 8, 1}
```

Note that more than 100,000 values were calculated for **f** and **p**:

```
{DownValues[f] // Length, DownValues[p] // Length}
{114 691, 114 676}
```

Although these figures might seem large, they are small compared to the amount of computations needed if we solved the problem by checking all the possible 14!/2 tours among the 15 cities:

```
14! / 2      43 589 145 600
```

Winston (1994, p. 1039) provides a smaller example in which a traveling salesman, now in New York, should visit Miami, Dallas, and Chicago and then return to New York:

```
T = 4;
c = {{0, 1334, 1559, 809},
     {1334, 0, 1343, 1397}, {1559, 1343, 0, 921}, {809, 1397, 921, 0}};
```

Evaluate the commands in the previous box. After that, ask for the length of the optimal tour and the optimal tour itself:

```
f[1, 1, {}]      4407

i = 1; S = {}; Join[{1}, Table[y = p[t, i, S]; i = y; S = Union[S, {y}]; y, {t, T}]]
{1, 2, 3, 4, 1}
```

■ A Comparison

The built-in **FindShortestTour** finds a solution candidate in a fraction of a second:

```
({length, tour2} = FindShortestTour[points]) // Timing
{0.743936, {38.6838, {1, 7, 13, 2, 14, 5, 10, 11, 4, 12, 15, 9, 3, 6, 8}}}
```

This time, **FindShortestTour** only gave a near-optimal solution; remember that this command only gives the optimal solution with a high probability. From the plots we can compare the solutions:

```
ListLinePlot[points[[#]], Mesh → All, AspectRatio → Automatic,
    Epilog → {Red, PointSize[Medium], Point[First[points]]}] & /@ {tour1, tour2}
```

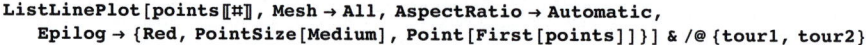

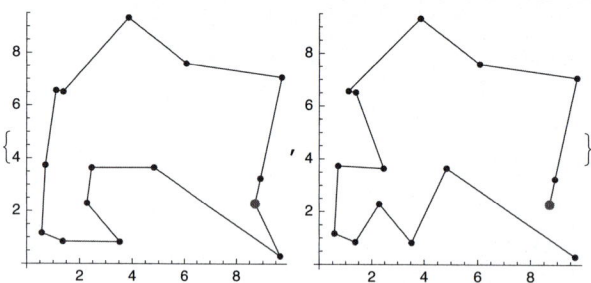

The first plot is the optimal solution (with length 37.6344) given by dynamic programming and the second plot the near-optimal solution (with length 38.6838) given by **FindShortestTour**.

Note that our problem, in which we used **SeedRandom[54]** in the generation of the points, is a very special problem. It is one of the rare cases in which **FindShortestTour** does not find the optimal solution. Indeed, when solving the problem with **SeedRandom[s]** when **s** gets on the values 0, 1, 2, ..., 100, we only found eight values of **s** where the solution given by **FindShortestTour** is suboptimal (these values of **s** are 3, 6, 26, 45, 54, 55, 76, and 85). Furthermore, the case in which **s** is 54 is the worst among the eight values: The difference between the lengths of the suboptimal and optimal tours is approximately 1.05. On average, this difference is only approximately 0.44 (among the eight cases). This means that in the rare cases **FindShortestTour** gives a suboptimal solution, it is very near to the optimal solution. Also, trying another method may give a better solution. In our example, the simulated annealing method gives a better (although not the optimal) solution:

```
({length, tour3} = FindShortestTour[points, Method → "SimulatedAnnealing"]) // Timing
{7.15697, {37.8455, {3, 15, 12, 9, 11, 4, 10, 5, 14, 2, 13, 7, 1, 8, 6}}}
```

23.5.3 Calculus of Variations

In a package we have commands that relate to the calculus of variations. We present here only one command from this package: **EulerEquations**. This command forms Euler's equation (or equations) for a problem in which we want to find the extremum for the integral $\int_a^b f(x, u(x), u'(x)) \, dx$. The command also accepts several independent variables $x, y, \ldots$ and several dependent variables $u(x, y, \ldots), v(x, y, \ldots)$. Here are some examples of the command:

In the **VariationalMethods`** *package*:

```
EulerEquations[f, u[x], x]
EulerEquations[f, u[x,y], {x,y}]
EulerEquations[f, {u[x,y],v[x,y]}, {x,y}]
```

Let us find the curve $y(x)$ of minimum length between $(0, 0)$ and (a, b) (the solution is, of course, the straight line that connects the points). The length of the curve is $\int_0^a \sqrt{1 + y'(x)^2} \, dx$. Euler's equation is as follows:

```
<< VariationalMethods`

eqn = EulerEquations[Sqrt[1 + y'[x]^2], y[x], x]
```

$$-\frac{y''[x]}{\left(1 + y'[x]^2\right)^{3/2}} == 0$$

The solution with the given boundary conditions is as follows:

```
DSolve[{eqn, y[0] == 0, y[a] == b}, y[x], x]
```

$$\left\{\left\{y[x] \to \frac{b\,x}{a}\right\}\right\}$$

Interpolation

Introduction

> *Life is the art of drawing sufficient conclusions from insufficient premises.—Samuel Butler*

With interpolation, we represent a set of points with a curve that passes exactly through all of the points. Interpolation can also be applied to functions by first sampling the function at some points. In this way, we can obtain for the data or function a suitable representation that may be sufficiently precise for practical purposes. However, if the observations contain errors (as they often do), then you may not want a function that passes exactly through the points but, rather, a simple function that represents the overall behavior of the observations. Approximation is then the correct technique to use (see Section 25.1).

■ Interpolation of Data

With *Mathematica,* we can do three kinds of interpolation of data. First, the usual *interpolating polynomial,* which is calculated with **`InterpolatingPolynomial`**, gives, for $n + 1$ data points, the unique polynomial of at most degree n that passes exactly through all of the points.

The object produced by **`ListInterpolation`** and **`Interpolation`** is called an *interpolating function.* It is a set of piecewise-calculated interpolating polynomials between successive points (the result is a continuous curve that has a discontinuous derivative). We can choose the degree of the piecewise polynomials. The commands also work for multidimensional data, producing, for example, an interpolating surface.

`SplineFit` calculates various *splines,* such as cubic splines that pass through all points and have a continuous first and second derivative.

■ Interpolation of Functions

For the interpolation of functions we have two commands. `RationalInterpolation` calculates, for a given function, a rational interpolating function (i.e., a quotient of two polynomials). We can give the interpolation points ourselves or let the command choose them; in the latter case, we obtain, in fact, a rational Chebyshev approximation (see Section 25.2.1, p. 826).

`FunctionInterpolation` forms, for mathematical expressions, interpolating functions that consist of a set of piecewise-calculated interpolating polynomials. For a complex expression, such an interpolating function may be a useful and efficient representation. Indeed, if an approximation for a function is required, then piecewise interpolation by `FunctionInterpolation` is a strong candidate, despite the fact that this command falls into the category of interpolation.

24.1 Usual Interpolation

24.1.1 Interpolating Polynomial

> `InterpolatingPolynomial[data, x]`   The interpolating polynomial through the points in `data`; the
>   result is a function of `x`
>
> The data points can be given in either of the following forms:
> `{f₁, f₂, … }`   Interpolate through the points `{1, f₁}, {2, f₂}, …`
> `{{x₁, f₁}, {x₂, f₂}, … }`   Interpolate through the points `{x₁, f₁}, {x₂, f₂}, …`

The command uses the Newton form of the interpolating polynomial and then writes the polynomial in the Horner form, suitable for numerical evaluation.

Recall that in Section 10.1.3, p. 329, we presented a manipulation to interactively study interpolating polynomials.

■ Simple Examples

We calculate the line that goes through the points $(1, f)$ and $(2, g)$:

```
InterpolatingPolynomial[{f, g}, x]
```

$$f + (-f + g) \ (-1 + x)$$

The next polynomial goes through the points (a, f) and (b, g):

```
InterpolatingPolynomial[{{a, f}, {b, g}}, x]
```

$$f + \frac{(-f + g) \ (-a + x)}{-a + b}$$

Here is the quadratic polynomial that goes through the points (a, f), (b, g), and (c, h):

```
InterpolatingPolynomial[{{a, f}, {b, g}, {c, h}}, x]
```

$$f + (-a + x) \left(\frac{-f + g}{-a + b} + \frac{\left(-\frac{-f+g}{-a+b} + \frac{-g+h}{-b+c} \right) \ (-b + x)}{-a + c} \right)$$

We check that this really goes through the given points:

```
% /. x → {a, b, c} // Simplify
```

$$\{f, g, h\}$$

■ A Numerical Example

Next, we consider numerical data:

```
data = {Range[14], {1, 2, 0, 2, 2, 2, 0, 0, 2, 3, 5, 4, 3, 1} // N}ᵀ
{{1, 1.}, {2, 2.}, {3, 0.}, {4, 2.}, {5, 2.}, {6, 2.}, {7, 0.},
 {8, 0.}, {9, 2.}, {10, 3.}, {11, 5.}, {12, 4.}, {13, 3.}, {14, 1.}}

p = ListPlot[data]
```

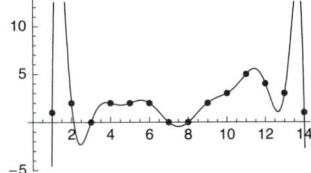

The interpolating polynomial is as follows:

```
int = InterpolatingPolynomial[data, x]
```

$$1. + \left(0. + \left(0.0238095 + \left(-0.0392857 + \right.\right.\right.$$
$$\left(-0.00423431 + \left(0.000800415 + \left(0.000513318 + \left(0.000222964 + \left(-0.0000382671 + \right.\right.\right.\right.\right.$$
$$\left(6.13777 \times 10^{-7} + \left(-2.14196 \times 10^{-6} + \left(-7.97492 \times 10^{-7} + \right.\right.\right.$$
$$\left(-7.7244 \times 10^{-7} + 2.50521 \times 10^{-8} \, (-6 + x)\right) \, (-10 + x)\right)$$
$$(-4 + x)\right) \, (-12 + x)\right) \, (-9 + x)\right) \, (-2 + x)\right) \, (-5 + x)\right)$$
$$(-13 + x)\right) \, (-3 + x)\right) \, (-11 + x)\right) \, (-7 + x)\right) \, (-1 + x)\right) \, (-14 + x)$$

Note that the interpolating polynomial should not be simplified or expanded because the unsimplified, nested form is the best for numerical computations. However, to show that the result in our example of 14 points is a polynomial of degree 13, we expand it:

```
int // Expand
```

$$-1674. + 4948.28 \, x - 6052.37 \, x^2 + 4139.98 \, x^3 - 1790.19 \, x^4 +$$
$$521.379 \, x^5 - 105.873 \, x^6 + 15.2428 \, x^7 - 1.56035 \, x^8 + 0.112324 \, x^9 -$$
$$0.00552547 \, x^{10} + 0.00017544 \, x^{11} - 3.20249 \times 10^{-6} \, x^{12} + 2.50521 \times 10^{-8} \, x^{13}$$

Here are some values of the polynomial:

```
int /. x → {1, 1.5, 2, 2.5, 3}
```

$$\left\{1., 17.9203, 2., -2.27272, 2.22045 \times 10^{-16}\right\}$$

We see that at the points 1, 2, 3, ... the polynomial really has the values 1, 2, 0, ..., but between these points the values may be far from the neighboring values. We plot the polynomial and also show the points:

```
Show[p, Plot[int, {x, 0.97, 14.03}]]
```

As can be seen, the polynomial goes through all of the points and is quite a good representation of the data in an interval of, for example, (3, 12). Outside of this interval—that is, near the end points—the polynomial behaves badly. Indeed, high-order interpolating polynomials should be used with caution. It may be better to use piecewise-interpolating polynomials (see Section 24.2), splines (see Section 24.3), or least-squares fits (see Section 25.1).

■ **Using Values of Derivatives**

`InterpolatingPolynomial` also accepts values of derivatives. For example, the data could be in the following forms:

```
{{x₁, f₁, df₁}, {x₂, f₂, df₂}, … }
{{x₁, f₁, df₁, ddf₁}, {x₂, f₂, df₂, ddf₂}, … }
```

Here, df_1 is the value of the derivative at point x_1, and ddf_1 is the value of the second derivative. Higher-order derivatives can also be given.

As an example, we calculate an interpolating polynomial for given values of the function and its first derivative at two points:

```
int = InterpolatingPolynomial[{{a, f, df}, {b, g, dg}}, x]
```

$$f + (-a + x)\left(df + (-a + x)\left(\frac{-df + \frac{-f+g}{-a+b}}{-a+b} + \frac{\left(\frac{dg - \frac{-f+g}{-a+b}}{-a+b} - \frac{-df + \frac{-f+g}{-a+b}}{-a+b}\right)(-b + x)}{-a+b}\right)\right)$$

The result is now a third-degree polynomial.

Some values of the function or its derivatives can also be `Automatic`:

```
int = InterpolatingPolynomial[{{a, f, df}, {b, Automatic, dg}}, x]
```

$$\frac{-a^2\, df + 2\, a\, b\, df - a^2\, dg + 2\, a\, f - 2\, b\, f}{2\,(a - b)} + x\left(\frac{-b\, df + a\, dg}{a - b} + \frac{(df - dg)\, x}{2\,(a - b)}\right)$$

In this example, the value of the function at $x = b$ was not specified. The resulting interpolating polynomial is a second-degree polynomial satisfying the other three conditions.

Next, we try to improve the interpolating polynomial of the numerical example by defining that the derivative of the polynomial at points 1, 2, 13, and 14 is 2, −3, −1, and −3, respectively (the values were found by trial and error):

```
data2 = {{1, 1, 2}, {2, 2, -3}, {3, 0}, {4, 2}, {5, 2}, {6, 2}, {7, 0},
    {8, 0}, {9, 2}, {10, 3}, {11, 5}, {12, 4}, {13, 3, -1}, {14, 1, -3}};
```

The result is now much better:

```
int2 = InterpolatingPolynomial[data2, x];
```

```
Show[p, Plot[int2, {x, 0.97, 14.03}]]
```

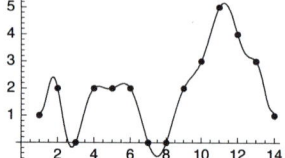

■ Multivariate Interpolating Polynomials

> InterpolatingPolynomial[data, {x, y, … }] (✺6) The interpolating polynomial through the
> points in **data**; the result is a function of **x, y**, …
>
> For example, the data are of the following form for a two-variate interpolating polynomial:
> {{{x₁, y₁}, f₁}, {{x₂, y₂}, f₂}, … } Interpolate through the points {x₁, y₁, f₁}, {x₂, y₂, f₂}, …

Gradients and higher derivatives can also be given. We construct an interpolating surface that goes through four points:

```
data3 = {{{0, 0}, 0}, {{3, 0}, 1}, {{0, 2}, 2}, {{3, 2}, 0}};
```

```
points = Partition[Flatten[data3], 3]
```

$$\{\{0, 0, 0\}, \{3, 0, 1\}, \{0, 2, 2\}, \{3, 2, 0\}\}$$

```
int3 = InterpolatingPolynomial[data3, {x, y}]
```

$$x \left(\frac{1}{3} - \frac{y}{2} \right) + y$$

```
Show[Plot3D[int3, {x, 0, 3}, {y, 0, 2}, BoxRatios → Automatic],
  Graphics3D[{Red, PointSize[Large], Point[points]}]]
```

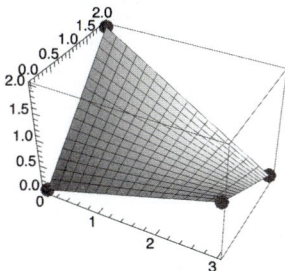

24.1.2 Own Programs

■ Lagrange Form

```
lagrangeInterpolation[xx_List, yy_List, x_] :=
  Sum[yy〚i〛 Apply[Times, (x - Drop[xx, {i}]) / (xx〚i〛 - Drop[xx, {i}])], {i, Length[xx]}]
```

Define some x values and their corresponding y values:

```
xx = {a, b, c}; yy = {f, g, h};
```

The interpolation polynomial in the Lagrange form is as follows:

```
lagrangeInterpolation[xx, yy, x]
```

$$\frac{h \, (-a + x) \, (-b + x)}{(-a + c) \, (-b + c)} + \frac{g \, (-a + x) \, (-c + x)}{(-a + b) \, (b - c)} + \frac{f \, (-b + x) \, (-c + x)}{(a - b) \, (a - c)}$$

To understand the program, note first that **Drop[xx, {i}]** deletes the ith element of **xx**. Note then that calculations with lists are done automatically element by element:

```
x - Drop[xx, {3}]      {-a + x, -b + x}
```

```
xx〚3〛 - Drop[xx, {3}]      {-a + c, -b + c}
```

Here is the quotient of these terms:

```
(x - Drop[xx, {3}]) / (xx[[3]] - Drop[xx, {3}])
```

$$\left\{ \frac{-a + x}{-a + c}, \frac{-b + x}{-b + c} \right\}$$

Note again that the division of the two lists was done automatically element by element. Then we multiply the elements of the last list:

```
Apply[Times, (x - Drop[xx, {3}]) / (xx[[3]] - Drop[xx, {3}])]
```

$$\frac{(-a + x) (-b + x)}{(-a + c) (-b + c)}$$

If we multiply this term by **yy[[3]]**, we get one term of the Lagrange interpolating polynomial. Summing all such terms gives the whole polynomial.

■ Newton Form

```
Remove[div]
div[z_List] := div[z] = (div[Rest[z]] - div[Most[z]]) / (Last[z] - First[z])

newtonInterpolation[xx_List, yy_List, x_] := With[{n = Length[xx]},
  Do[div[{xx[[i]]}] = yy[[i]], {i, n}];
  Sum[div[Take[xx, i]] Product[x - xx[[j]], {j, i - 1}], {i, n}]]
```

Here is an example (we use the same data as previously used):

```
newtonInterpolation[xx, yy, x]
```

$$f + \frac{(-f + g) (-a + x)}{-a + b} + \frac{\left(-\frac{-f+g}{-a+b} + \frac{-g+h}{-b+c} \right) (-a + x) (-b + x)}{-a + c}$$

This is of the form $\mathrm{div}(a) + \mathrm{div}(a, b)(x - a) + \mathrm{div}(a, b, c)(x - a)(x - b)$, where div denotes divided differences. Note that **InterpolatingPolynomial** writes the polynomial in the Horner form (see Section 24.1.1, p. 792).

The function **div** calculates the divided differences. We demonstrate how **div** works. First, note that to speed up the computations we have used dynamic programming: **div[z_List] := div[z] = ...** (see Section 18.5.1, p. 597). The starting values for **div** are calculated as follows:

```
Do[div[{xx[[i]]}] = yy[[i]], {i, 3}]
```

Now **div** is defined for **{a}**, **{b}**, and **{c}**:

```
{div[{a}], div[{b}], div[{c}]}      {f, g, h}
```

Then we can calculate the first- and second-order divided differences:

$$\{\mathtt{div[\{a, b\}], div[\{b, c\}]}\} \qquad \left\{ \frac{-f + g}{-a + b}, \frac{-g + h}{-b + c} \right\}$$

$$\mathtt{div[\{a, b, c\}]} \qquad \frac{-\frac{-f+g}{-a+b} + \frac{-g+h}{-b+c}}{-a + c}$$

We can show the divided differences in the form of a table:

```
TableForm[
  {{a, "", b, "", c}, {div[{a}], "", div[{b}], "", div[{c}]}, {"", div[{a, b}], "",
    div[{b, c}], ""}, {"", "", div[{a, b, c}], "", ""}}ᵀ, TableSpacing → {0, 3}]
```

a   f

$\qquad \dfrac{-f+g}{-a+b}$

$\qquad\qquad \dfrac{-\frac{-f+g}{-a+b}+\frac{-g+h}{-b+c}}{-a+c}$

b   g

$\qquad \dfrac{-g+h}{-b+c}$

c   h

24.2 Piecewise Interpolation

24.2.1 Two-Dimensional Data

As shown in the numerical example of Section 24.1.1, p. 793, if we have many points and thus an interpolating polynomial of a high order, the result may be bad; that is, the polynomial behaves badly, particularly near the end points. Often, it is better to proceed piecewise: Calculate low-order polynomials between successive points. This can be done with **ListInterpolation** or **Interpolation**.

These two commands differ in the way we specify the data. With 2D data, we have values f_i at some points x_i. With **ListInterpolation**, we specify separately the f_i values and the corresponding x_i values, whereas with **Interpolation** we form a list that consists of pairs $\{x_i, f_i\}$. Here is a summary of how to calculate an interpolating function that consists of polynomials between successive points (the default is that the polynomials are of third degree).

ListInterpolation[{f₁, …, fₘ}] x coordinates are assumed to be $\{1, …, m\}$

ListInterpolation[{f₁, …, fₘ}, {{xₘᵢₙ, xₘₐₓ}}] x coordinates are assumed to be evenly spaced
  between $[x_{min}, x_{max}]$

ListInterpolation[{f₁, …, fₘ}, {{x₁, …, xₘ}}] x coordinates are $\{x_1, …, x_m\}$

Interpolation[{f₁, …, fₘ}] x coordinates are assumed to be $\{1, …, m\}$

Interpolation[{{x₁, f₁}, …, {xₘ, fₘ}}] x coordinates are $x_1, …, x_m$

The values **fᵢ** can be real or complex (or even symbolic), whereas the values of **xᵢ** must be real (and numeric).

■ **Example 1: Piecewise Cubic Interpolation**

As an example, we consider the same data we used in Section 24.1.1, p. 793:

```
data = {1, 2, 0, 2, 2, 2, 0, 0, 2, 3, 5, 4, 3, 1};
```

We calculate a piecewise cubic interpolating function, assuming that the x coordinates are 1, …, 14:

```
int = ListInterpolation[data]
InterpolatingFunction[{{1, 14}}, <>]
```

The result of **ListInterpolation** (and **Interpolation**) is an object called **InterpolatingFunction**; it contains all of the information needed to handle the piecewise polynomial. Only the interval of definition of the piecewise polynomial is shown; all other information is hidden inside the marks <>.

We can calculate with the piecewise interpolating function in all possible ways. Just give a name to the function, such as `int`, and remember that its value, for example, at **a** is `int[a]` (similarly, the value of `Sin` at **a** is `Sin[a]`).

A typical use of an interpolating function:

`int = ListInterpolation[data]`   Calculate an interpolating function
`int[a]`   Calculate the value of the interpolating function at **a**

We calculate the value of the function at some points:

`{int[1], int[1.5], int[2]}`
`{1, 2.3125, 2}`

The function goes exactly through all data points and interpolates between them. If we ask for the value at a point outside of the interval of definition, extrapolation is used, and we get a warning:

`int[0.5]`

InterpolatingFunction::dmval : Input value {0.5} lies outside the
     range of data in the interpolating function. Extrapolation will be used. ≫
`-2.8125`

The interpolating function is displayed as follows:

`xf = {Range[14], data}ᵀ;`

`Plot[int[x], {x, 1, 14}, Epilog → Point[xf], AxesOrigin → {0, 0}]`

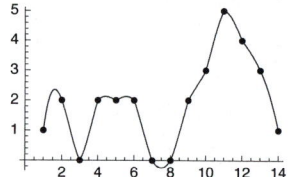

A cubic polynomial is calculated for each pair of points by using these points and their nearest neighbors. The resulting curve is smooth enough to make it an effective and useful way of summarizing and using large data sets.

Here is the derivative function:

`Show[Table[Plot[int'[x], {x, i, i + 1}], {i, 13}],`
`  PlotRange → {{0, 14}, Automatic}, AxesOrigin → {0, 0}]`

We see that the derivative is discontinuous at 3, 4, 5, ..., 12. Next, we integrate the function:

`Integrate[int[x], {x, 1, 14}] // N      26.75`

Previously, we assumed that the x coordinates are 1, ..., 14. If the x coordinates are, for example, 0, 1, ..., 13, then we have to write one of the following three commands:

```
int = ListInterpolation[data, {{0, 13}}]

int = ListInterpolation[data, {Range[0, 13]}]

int = Interpolation[{Range[0, 13], data}ᵀ]
```

When we later solve differential equations numerically, we will encounter **InterpolatingFunction** again: The result of **NDSolve** is an interpolating function (see Section 26.3.1, p. 849).

■ Options

Options of **ListInterpolation** *and* **Interpolation**:

InterpolationOrder   Degree of the piecewise interpolating polynomials; examples of values: **3**, **1**
PeriodicInterpolation   Whether a periodic interpolating function is formed; possible values:
  False, **True**

The default value of **InterpolationOrder** is 3 so that if you intend to use third-order polynomials, the option need not be written. Give the option the value 1 if you want a piecewise linear interpolation (see the next example). If we give **PeriodicInterpolation** the value **True**, then the interpolating function is considered a periodic function, with one period being the same as the range of the data (see Example 3).

■ Example 2: Piecewise Linear Interpolation

We calculate the piecewise linear interpolating function for the data given in Example 1:

```
int = ListInterpolation[data, InterpolationOrder → 1]
InterpolatingFunction[{{1, 14}}, <>]

Plot[int[x], {x, 1, 14}, Epilog → Point[xf], AxesOrigin → {0, 0}]
```

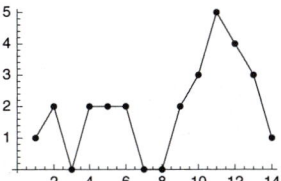

■ Example 3: Periodic Interpolation

Now we ask for a periodic interpolating function:

```
int = ListInterpolation[data, PeriodicInterpolation → True]
InterpolatingFunction[{{1, 14}}, <>]

Plot[int[x], {x, 1, 56}, AspectRatio → 0.2, ImageSize → 300]
```

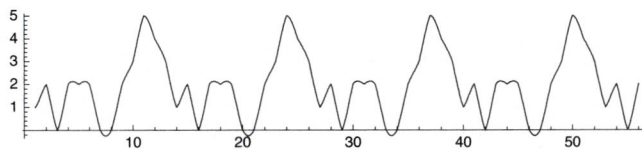

Note that for a periodic interpolating function, the data at the end points of the fundamental period must match: If the first and last data points are (x_1, f_1) and (x_n, f_n), we must have $f_1 = f_n$.

■ Using Values of Derivatives

We can also input the values of first and higher derivatives. For example, if we input the values of the first derivative, the commands are in the following forms:

```
ListInterpolation[{{f₁, df₁}, …, {fₘ, dfₘ}}]
Interpolation[{{{x₁}, f₁, df₁}, …, {{xₘ}, fₘ, dfₘ}}]
```

Here, df_i is the value of the first derivative at x_i. We try to improve the interpolating function of **data** by specifying that the derivative of the function at 1, 5, 6, and 7 is 1.3, 0, 0, and 0, respectively:

```
data2 = {{1, 1.3}, 2, 0, 2, {2, 0}, {2, 0}, {0, 0}, 0, 2, 3, 5, 4, 3, 1};

int2 = ListInterpolation[data2];

Plot[int2[x], {x, 1, 14}, Epilog → Point[xf], AxesOrigin → {0, 0}]
```

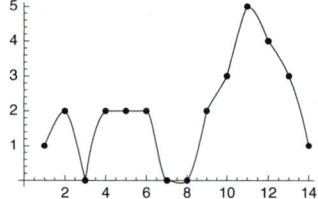

24.2.2 Higher-Dimensional Data

■ Regular 3D Data

ListInterpolation and **Interpolation** can also be used for higher-dimensional data to calculate piecewise interpolating surfaces. The choice of command depends on the form of the data.

For example, with 3D data, we have values f_i at some points (x_i, y_i). With **ListInterpolation**, we define separately the f_i values in a matrix form and the values of x_i and y_i in one of several easy ways, whereas **Interpolation** requires the points in the form $\{\{x_i, y_i\}, f_i\}$. Here is a summary of the ways in which a piecewise interpolating surface for 3D data can be calculated (the summary extends readily to higher-dimensional data):

```
ListInterpolation[data]   x and y coordinates are {1, …, m} and {1, …, n}
ListInterpolation[data, {{xₘᵢₙ, xₘₐₓ}, {yₘᵢₙ, yₘₐₓ}}]   x and y coordinates are evenly spaced
   between [xₘᵢₙ, xₘₐₓ] and [yₘᵢₙ, yₘₐₓ]
ListInterpolation[data, {{x₁, …, xₘ}, {y₁, …, yₙ}}]   x and y coordinates are {x₁, …, xₘ} and
   {y₁, …, yₙ}
```

data is of the matrix form:
$\{\{f_{11}, …, f_{1n}\}, …, \{f_{m1}, …, f_{mn}\}\}$ (each row corresponds to a fixed value of x)

```
Interpolation[data]
```

data is of the list form:
$\{\{\{x_1, y_1\}, f_1\}, …, \{\{x_k, y_k\}, f_k\}\}$

With both commands, the x and y coordinates must eventually form a *regular rectangular grid* on the (x, y) plane. Such a grid can be seen in the following figure:

```
xy = Outer[List, {0, 1, 4, 6, 7}, {0, 2, 5, 6}]

{{{0, 0}, {0, 2}, {0, 5}, {0, 6}},
  {{1, 0}, {1, 2}, {1, 5}, {1, 6}}, {{4, 0}, {4, 2}, {4, 5}, {4, 6}},
  {{6, 0}, {6, 2}, {6, 5}, {6, 6}}, {{7, 0}, {7, 2}, {7, 5}, {7, 6}}}

ListPlot[Flatten[xy, 1], PlotStyle → PointSize[Medium],
  Axes → None, AspectRatio → Automatic, PlotRange → All,
  Epilog → {Line[xy], Line[xyᵀ]}, ImageSize → 90]
```

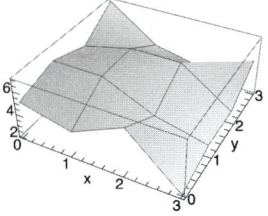

Note that unlike `ListInterpolation`, `Interpolation` does not accept the points in a *matrix* form: The points must be in a *flattened* list form in which the rows are not separated by curly braces; if your data is in a matrix form, use `Flatten[data, 1]` to remove the curly braces of the rows.

The options `InterpolationOrder` and `PeriodicInterpolation` can be used as they are for 2D data (see Section 24.2.1, p. 799). The default order is 3. The order can also be set separately for each independent variable (e.g., `InterpolationOrder → {2, 1}`). If there are not enough data for a requested order, the order is lowered automatically (with a warning). The periodicity can also be defined separately for each independent variable (e.g., `PeriodicInterpolation → {True, False}`).

■ An Example

Consider the following data:

```
data = {{5, 6, 5, 7}, {4, 6, 6, 4}, {6, 4, 6, 3}, {2, 3, 3, 5}};
```

Here, each row of f_i values corresponds to a fixed value of x. Before calculating an interpolating surface for these data, we plot the data (see Section 8.6.1, p. 275). If x and y are both in the interval $(0, 3)$, we can plot the surface as follows:

```
ListPlot3D[dataᵀ, DataRange → {{0, 3}, {0, 3}},
  Mesh → Full, AxesLabel → {"x", "y", None}]
```

Note that we have to transpose the data because the plotting commands interpret the data in such a way that each row corresponds to a fixed value of y, whereas with `ListInterpolate` each row corresponds to a fixed value of x. Then we calculate a piecewise third-order interpolating surface and plot it:

```
int = ListInterpolation[data, {{0, 3}, {0, 3}}]

InterpolatingFunction[{{0, 3}, {0, 3}}, <>]
```

```
Plot3D[int[x, y], {x, 0, 3}, {y, 0, 3}, AxesLabel → {"x", "y", ""}]
```

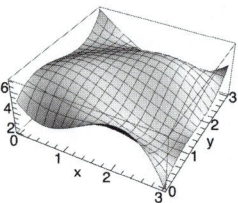

The surface goes exactly through all of the given points and interpolates between them. We calculate a value of the surface and integrate it in a region:

```
int[1.7, 2.1]     6.24158
```

```
Integrate[int[x, y], {x, 1, 2}, {y, 0, 1}] // N     4.76563
```

■ **Giving Derivatives**

```
ListInterpolation[{ …,{fᵢⱼ, {dfxᵢⱼ, dfyᵢⱼ}}, … }]

Interpolation[{ …, {{xᵢ, yᵢ}, fᵢ,{dfxᵢ, dfyᵢ}}, … }]
```

Derivatives are specified in **ListInterpolation** by replacing f_{ij} with a list $\{f_{ij}, \{dfx_{ij}, dfy_{ij}\}\}$ in which dfx_{ij} and dfy_{ij} are the derivatives with respect to x and y, respectively (if a derivative is lacking at a point, then give the value **Automatic** for the derivative).

■ **Irregular 3D Data**

We noted that for **ListInterpolation** and **Interpolation**, the 3D points must form a regular rectangular grid on the (x, y) plane (however, neither the x points nor the y points need be evenly spaced). If the points are spaced irregularly, there is no built-in command to calculate a piecewise interpolating surface as a mathematical function. However, with **ListPlot3D** we can show such a surface (see Section 8.6.1, p. 281):

```
SeedRandom[2]; data = Table[With[
    {x = RandomReal[{0, π}], y = RandomReal[{0, π}]}, {x, y, Sin[x] Sin[y]}], {10}];
gr = Graphics3D[{PointSize[Medium], Point[data]}];
{Show[ListPlot3D[data, Filling → Bottom, Mesh → None,
    ColorFunction → "SouthwestColors", InterpolationOrder → #], gr] & /@ {0, 1}}
```

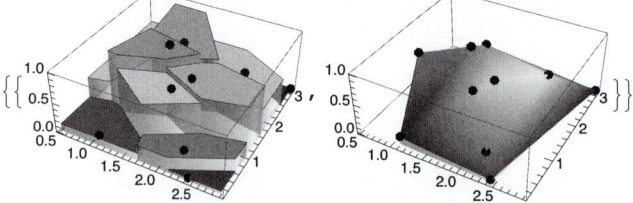

By defining the interpolation order to be 0, we get flat Voronoi regions (the first plot above). With a package we can show a triangular surface plot; a similar triangularization is obtained with **ListPlot3D** by defining **Mesh → All**:

```
<< ComputationalGeometry`
```

```
{Show[ListPlot3D[data, Filling → Bottom,
   Mesh → All, ColorFunction → "SouthwestColors", Ticks → None], gr],
 Show[TriangularSurfacePlot[data], gr]}
```

24.3 Splines

24.3.1 Cubic Splines

■ Introduction to Splines

We observed in Section 24.1.1, p. 793, that if one polynomial is required to pass through many points, the resulting polynomial may fluctuate in an undesirable manner. Piecewise interpolation is often better; such interpolating functions were considered in Section 24.2.1, p. 797. The resulting function does not have unnecessary fluctuations, but its derivative is not continuous, and so the function lacks this smoothness condition. *Splines* are piecewise interpolating functions that are smooth.

> *In the **Splines`** package:*
>
> **SplineFit[data, type]**   A spline of type **type** through the points in **data**

The points are given in the form **{{x₁, f₁}, {x₂, f₂}, … }**, and the type of spline can be **Cubic**, **Bezier**, or **CompositeBezier**. A *cubic spline* is made up of a set of cubic polynomials in such a way that the resulting function passes through each point, the first and second derivatives of the function are continuous, and the second derivative is zero at the end points. *Bezier splines* interpolate only the end points; other points "control" the curve. A *composite Bezier* spline interpolates the first, third, fifth, … points, while the other points control the curve.

■ Example 1

Consider again the numeric data that we used previously in this chapter:

```
data = {Range[14], {1, 2, 0, 2, 2, 2, 0, 0, 2, 3, 5, 4, 3, 1}}ᵀ
{{1, 1}, {2, 2}, {3, 0}, {4, 2}, {5, 2}, {6, 2}, {7, 0},
  {8, 0}, {9, 2}, {10, 3}, {11, 5}, {12, 4}, {13, 3}, {14, 1}}
```

Calculate the cubic spline:

```
<< Splines`
```

```
cub = SplineFit[data, Cubic]
SplineFunction[Cubic, {0., 13.}, <>]
```

The result is a **SplineFunction** object; it contains all of the information about the spline. Only the interval in which the spline is defined is shown. You may wonder about the interval {0., 13.} because our points were in the interval (1, 14). Generally, if you have n points, the interval of the spline is {0., $(n-1)$.}. Thus, the interval shown emerges simply by giving each observation an ordinal number, starting with 0. This may seem odd, but there are reasons for it, which will be discussed when we consider multiple-valued splines in Example 2.

If we want to calculate the value of the spline at a particular point, we have to reparameterize the point so that it complies with the numbering system used by **SplineFit**. For example, the point 3.5 is halfway between the third and fourth points, and so the appropriate argument is 2.5:

```
cub[2.5]     {3.5, 0.797698}
```

The result shows, besides the value of the spline (0.797698), the coordinate of the point in the normal x axis (3.5). Thus, **cub** is a parametric function. Accordingly, the spline can be plotted with **ParametricPlot**:

```
ParametricPlot[cub[t], {t, 0, 13},
  Epilog → Point[data], AspectRatio → 1 / GoldenRatio]
```

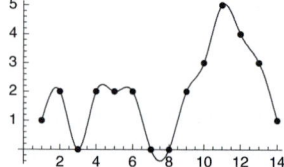

If we want to integrate the spline, we have to take the second component **cub[t]⟦2⟧** that contains the value of the spline. We integrate the spline numerically when x is in (1, 14), which also means that t is in (0, 13) (the warning we get is harmless):

```
NIntegrate[cub[t]⟦2⟧, {t, 0, 13}]
```

Part::partw : Part 2 of (SplineFunction[Cubic, {0., 13.}, <>])[t] does not exist. ≫
26.3721

Derivatives can also be calculated numerically. We calculate the derivative when x is 2.5—that is, when t is 1.5:

```
<< NumericalCalculus`
```

```
ND[cub[t], t, 1.5]⟦2⟧     -2.66597
```

■ **Example 2**

A spline can be drawn through any set of points in the (x, y) plane. Accordingly, the resulting curve may well be multiple-valued. Here is an example:

```
data2 = {{0, 1}, {1, 1}, {2, 1}, {2, 2}, {1, 2}, {1, 1}, {1, 0}, {1, -1}, {1, -2},
   {2, -2}, {2, -1}, {1, -1}, {0, -1}, {-1, -1}, {-2, -1}, {-2, -2}, {-1, -2},
   {-1, -1}, {-1, 0}, {-1, 1}, {-1, 2}, {-2, 2}, {-2, 1}, {-1, 1}, {0, 1}};
cub2 = SplineFit[data2, Cubic]
```

SplineFunction[Cubic, {0., 24.}, <>]

```
ParametricPlot[cub2[t], {t, 0, 24}, Epilog → Point[data2], ImageSize → 100]
```

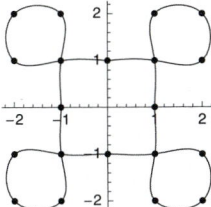

With multiple-valued splines, the reparameterization of the argument becomes clear. If, for example, we want the value of the spline at a point in the highest part of the top right loop, we must inform *Mathematica* that we want a value between the third and fourth points (when the counting begins from 0):

```
cub2[3.5]      {1.47835, 2.17933}
```

24.3.2 Bezier Splines

■ Ordinary Bezier Splines

Consider again the same data introduced previously, and calculate a Bezier spline:

```
<< Splines`
```

```
b = SplineFit[data, Bezier]
```

```
SplineFunction[Bezier, {0., 13.}, <>]
```

```
ParametricPlot[b[t], {t, 0, 13}, Epilog → Point[data],
  PlotRange → {All, {-0.3, 5.3}}, AxesOrigin → {0, 0}, AspectRatio → 1 / GoldenRatio]
```

As can be seen, a Bezier spline interpolates only the end points; other points control the curve.

■ Composite Bezier Splines

Now we calculate a composite Bezier spline:

```
cb = SplineFit[data, CompositeBezier]
```

```
SplineFunction[CompositeBezier, {0., 13.}, <>]
```

```
ParametricPlot[cb[t], {t, 0, 13}, Epilog → Point[data],
  PlotRange → All, AxesOrigin → {0, 0}, AspectRatio → 1 / GoldenRatio]
```

As can be seen, a composite Bezier spline interpolates the first, third, fifth, ... points, whereas the other points control the curve.

24.4 Interpolation of Functions

24.4.1 Usual Interpolation

■ Rational Interpolation

Thus far, we have considered the interpolation of given data. Another situation is the case in which we want to build an interpolating polynomial for a given function. One possibility is that the interpolation points are given (i.e., we cannot determine them ourselves), and another possibility is that we can choose the interpolation points. In the latter case, we can define the points so that the error of interpolation (between interpolation points) becomes smaller; the result is a Chebyshev approximation. With a package, we can calculate both polynomial and rational interpolating functions.

> *In the* **FunctionApproximations`** *package:*
>
> `RationalInterpolation[f, {x, m, n}, {`x_1`, `x_2`, ..., `x_{m+n+1}`}]`   Rational interpolating function of
>   degree (**m, n**) for **f** through the given points
> `RationalInterpolation[f, {x, m, n}, {x, a, b}]`   Rational interpolating function of degree (**m, n**)
>   for **f** in the interval (**a, b**) (i.e., rational Chebyshev approximation)

Here, **m** and **n** are the desired degrees of the numerator and the denominator. Giving **n** the value zero, we can calculate polynomial interpolating functions.

In Chebyshev approximation, the interpolation points are chosen in a special way: They are the zeros of the $(m + n + 1)$th-degree Chebyshev polynomial. It turns out that by choosing the point in this way, we get a good approximation to the function: The error is small throughout the interval (a, b). This means that the result is near the best approximation, which is the minimax approximation. Approximation of functions is considered in more detail in Section 25.2.

■ Example 1: Interpolation

Suppose we have to interpolate the following function (the cumulative distribution function of the standard normal distribution):

```
f = (1 + Erf[x / Sqrt[2]]) / 2;
```

The interpolation points are given as 0, 1/3, 2/3, ..., 2. We form the sixth-degree interpolating polynomial:

```
<< FunctionApproximations`

int = RationalInterpolation[f, {x, 6, 0}, Range[0, 2, 1 / 3]]
```

$0.5 + 0.398489\,x + 0.00309722\,x^2 -$
   $0.0739335\,x^3 + 0.00750692\,x^4 + 0.00801905\,x^5 - 0.0018339\,x^6$

Here is the absolute error:

```
Plot[f - int, {x, 0, 2}]
```

The error has relatively large values near the end points of the interval.

■ Example 2: Chebyshev Approximation

Now we calculate the sixth-degree Chebyshev approximation:

```
cheb = RationalInterpolation[f, {x, 6, 0}, {x, 0, 2}]
```

$0.500007 + 0.398604\, x + 0.00255252\, x^2 - $
$0.0732252\, x^3 + 0.00721336\, x^4 + 0.00800419\, x^5 - 0.00181145\, x^6$

```
Plot[f - cheb, {x, 0, 2}]
```

The points where the error is zero are chosen according to the zeros of the seventh-degree Chebyshev polynomial. Notice how small the error is over the whole interval. We could use the option **Bias** to fine-tune the points where the error is zero to produce an even more uniform error (see Section 25.2.2, p. 828).

24.4.2 Piecewise Interpolation

■ Piecewise Interpolation for Functions

FunctionInterpolation forms a piecewise interpolating function for a mathematical expression. The expression can contain built-in mathematical functions and possibly also interpolating functions. For example, if an expression is so complicated that working with it takes some time, we may consider forming an interpolating function for it because working with these latter functions is fast. Another example is an expression containing several interpolating functions. We may again consider forming a single interpolating function for the expression, thus making computations faster.

Of course, we could form an interpolating function manually by sampling the mathematical expression at some points and then using, for example, **ListInterpolation**. However, **FunctionInterpolation** does the job automatically, is adaptive, and offers some options for controlling the precision.

FunctionInterpolation[expr, {x, a, b}] Form an interpolating function for **expr** by sampling **expr** at sufficiently many points in (**a, b**)

The command generalizes for multivariate expressions. For example, if **expr** has two independent variables **x** and **y**, the command is of the following form:

```
FunctionInterpolation[expr, {x, a, b}, {y, c, d}]
```

The command has some options:

```
Options[FunctionInterpolation]
```

```
{InterpolationOrder → 3, InterpolationPrecision → Automatic,
 AccuracyGoal → Automatic, PrecisionGoal → Automatic,
 InterpolationPoints → 11, MaxRecursion → 6}
```

Of these, **InterpolationOrder** is the usual order of the polynomial pieces, whereas **InterpolationPrecision** is the precision of the values to be returned by the interpolating function generated. **InterpolationPoints** is the initial number of evenly spaced points (in each dimension) at which the expression is evaluated, and **MaxRecursion** is the maximum number of times a subinterval can be bisected (to achieve the desired precision). **PrecisionGoal** and **AccuracyGoal** are the standard options for controlling the precision and accuracy of the result.

■ Example 1: A Definite Integral

Suppose we want to treat the definite integral of $\sin(\sin(t^2))$ over $(0, x)$ as a function of x. First, we define this function:

```
g[x_?NumericQ] := NIntegrate[Sin[Sin[t^2]], {t, 0, x}]
```

We then form an interpolating function for this function:

```
int = FunctionInterpolation[g[x], {x, 0, π}]
```

```
InterpolatingFunction[{{0., 3.14159}}, <>]
```

Now we can, for example, plot the function:

```
Plot[int[x], {x, 0, π}]
```

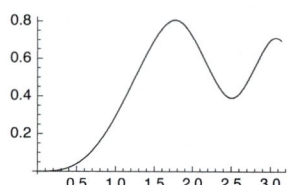

(The time taken to make this plot is a fraction of the time needed to make the plot from the original definite integral as **Plot[g[x], {x, 0, π}]**.) The error is small:

```
Plot[g[x] - int[x], {x, 0, π}, PlotRange → All, AspectRatio → 0.2, ImageSize → 400]
```

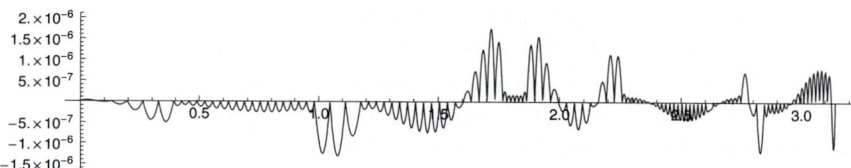

Below, we show the points at which the expression has been sampled. We see that the sampling is adaptive: More points are taken where the expression changes more rapidly.

```
p = {InputForm[int]〚1, 3, 1〛, InputForm[int]〚1, 4, 3〛}ᵀ;
```

```
Graphics[Line[{{#[[1]], 0}, #}] & /@ p, Axes → True,
  AspectRatio → 0.2, ImageSize → 400, ImagePadding → {{37, 3}, {3, 3}}]
```

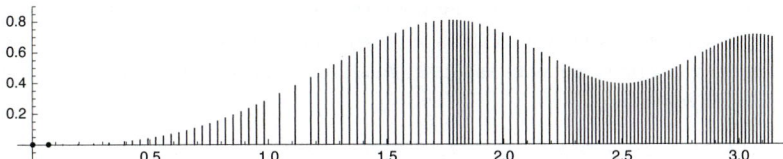

In Section 30.7.3, p. 1051, we use **FunctionInterpolation** in Bayesian statistics in the same way as in the example here.

■ Example 2: The Solution of a Nonlinear Equation

In Section 5.2.3, p. 134, we considered the following equation:

```
eqn = y Exp[r (1 - y)] == 2 - y;
```

The equation defined a function $y(r)$. We define that function:

```
yr[r_ ?NumericQ] := y /. FindRoot[eqn, {y, 0.1}]
```

Then we form an interpolating function for this function (we do not show here a series of warning messages) and plot it:

```
yy = FunctionInterpolation[yr[r], {r, 2, 2.8}] // Quiet
InterpolatingFunction[{{2., 2.8}}, <>]
```

```
Plot[yy[r], {r, 2, 2.8}]
```

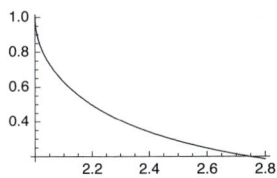

■ Example 3: A Function of an Interpolating Function

As a continuation of Example 2, we note that in Section 5.2.3 we also plotted the following expression as a function of **r**:

```
delta = Abs[(1 - r yy[r]) (1 - 2 r + r yy[r])];
```

We can first form an interpolating function for **delta** and then plot this function (we choose a large number of interpolation points to get a sufficiently sharp corner in the figure):

```
delta2 = FunctionInterpolation[delta, {r, 2, 2.8}, InterpolationPoints → 60]
InterpolatingFunction[{{2., 2.8}}, <>]
```

```
Plot[delta2[r], {r, 2, 2.8}, Epilog → Line[{{0, 1}, {2.8, 1}}]]
```

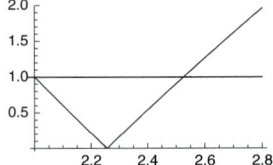

■ **Example 4: A Combination of Two Interpolating Functions**

In Section 26.4.3, p. 874, we present a module `linearBVP` that solves linear boundary value problems. The solution can be written as a linear combination of two interpolating functions (they are the results of solving two initial value problems with `NDSolve`). `FunctionInterpolation` can then be used to form a single interpolating function.

Approximation

Introduction

If the facts do not fit the theory, change the facts.—Albert Einstein

We will let Einstein rest (relatively speaking) with his strange advice, and we will do just the opposite. Indeed, we have facts in the form of data or functions, and we have theory in the form of approximating functions. We will try to find the approximations that fit the facts as well as possible. There are two basic areas in approximation: approximation of *data* by a function and approximation of a *function* by another function. Approximation of data is often done to get a good summary of the overall behavior of the data, whereas the reason to approximate a function may be to get an expression that requires less time to evaluate.

■ Approximation of Data

Approximation of data is useful if the data contain errors and we want to find a simple representation of the data by a function. We choose the *form* of the function by looking at the overall behavior of the data. The chosen function contains *parameters*, and for these parameters we try to find the best values according to a chosen criterion. The most widely used method is the *least-squares method*. There are two types of least-squares problems:

- *Linear* least squares: The parameters a, b, ... appear linearly in the function, which means that the function is of the form $a\,f(x) + b\,g(x) + ...$; for example, $a + b\,x + c\,x^2$ or $a\exp(-x) + b\sin(x)$.
- *Nonlinear* least squares: The parameters appear nonlinearly in the function, which means that the function is of the form $f(x, a, b, ...)$; for example, $\exp(a + b\,x)$ or $a\,x^b \exp(c\,x)$.

For linear least squares, we have **Fit**. Nonlinear least squares are done with **FindFit**. If we want statistical analysis of the fits, we can use **Regress** and **NonlinearRegress** from some packages; these commands are considered in Chapter 30, in which we also consider *local regression* and smoothing of data.

If the data do not contain observational errors, then interpolation or piecewise interpolation may be the appropriate technique for summarizing and using the data (see Chapter 24).

In Sections 26.4.5, p. 878, and 28.3.2, p. 954, we estimate differential and difference equation models from data.

■ **Approximation of Functions**

Approximation of a function is useful if we have a complicated function that is difficult or time-consuming to evaluate and handle and we want to find a simpler function that is close enough to the original function for practical purposes. We can distinguish two types of situations: approximation near a point and approximation in an interval.

For approximation near a point, we have, for example, Taylor polynomials. Another method is Padé approximation; in that case, the approximating function is a rational function.

For approximation in an interval, we can use interpolation and approximation techniques. In Chapter 24, we noted that piecewise interpolation by `FunctionInterpolation` is a strong candidate for an approximation of a complex function in an interval. We also noted that `RationalInterpolation` gives Chebyshev approximations; they are close to minimax approximations. The true minimax approximation is calculated by `MinimaxApproximation`; now the maximum error of the approximation over the whole interval is made as small as possible. The approximating function is a rational function, and the starting point for the iterative method is a Chebyshev approximation.

25.1 Approximation of Data

25.1.1 Linear Fitting

■ **Finding a Linear Fit**

`Fit` is designed for fitting problems in which the parameters appear linearly. The criterion used is the least-squares criterion.

`Fit[data, basis, var]`　Find the least-squares fit to `data` as a linear combination of functions `basis` of variable `var`

`data` can be given in either of the following forms:
`{f₁, f₂, … }`　Fit using the points `{1, f₁}`, `{2, f₂}`, …
`{{x₁, f₁}, {x₂, f₂}, … }`　Fit using the given points

Examples of `basis`:
`{1, x}`　The fitting function is of the form $a + b\,x$
`{1, x, x^2}`　The fitting function is of the form $a + b\,x + c\,x^2$
`{Exp[x], Cos[x]}`　The fitting function is of the form $a\exp(x) + b\cos(x)$

We also have `FindFit`, which suits both linear and nonlinear fitting.

`FindFit[data, funct, params, var]`　When `funct` is an expression of the variable `var` and contains the parameters `params`, find values for the parameters such that the function fits `data` in the best way (in the sense of least squares, by default)

We will consider **FindFit** mainly in Sections 25.1.3 and 25.1.4 in the context of nonlinear fitting. Note that with **FindFit** we can define our own norm function, and this enables us to find, for example, L_1-norm fits; an example is provided in Section 25.1.4, p. 822.

If you want statistical information about the fit, use **Regress** from the **LinearRegression`** package (see Section 30.5.1, p. 1030).

Recall that in Section 10.1.3, p. 329, we presented a manipulation to interactively study linear least-squares fitting. A modified manipulation is given in Section 25.1.4, p. 822.

■ A First Fit

In this example, we use simulated data:

```
xx = Range[0, 50]; SeedRandom[2];
rand = RandomReal[NormalDistribution[0, 1], 51];
data = {xx, 2 + xx - 0.004 xx^2 + 2 rand}ᵀ;
Short[data, 2]
{{0, 3.24673}, «49», {50, 42.7481}}
```

Recall that ᵀ means the transpose; write it as ⎚tr⎚. Note also that we utilized the fact that *Mathematica* automatically does all calculations with vectors element by element. The data look as follows:

```
pdata = ListLinePlot[data, AspectRatio → 0.4, Mesh → All, ImageSize → 200]
```

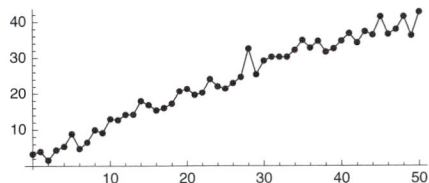

We try a linear fit for the data (with the simulated data in mind, we know that a quadratic fit would be better):

```
fit = Fit[data, {1, x}, x]
3.66138 + 0.78471 x
```

We could also use **FindFit**:

```
FindFit[data, a + b x, {a, b}, x]
{a → 3.66138, b → 0.78471}

fit = a + b x /. %
3.66138 + 0.78471 x
```

The fit looks as follows:

```
Show[pdata, Plot[fit, {x, 0, 50}]]
```

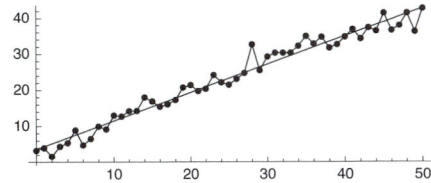

The fit seems quite good. Next, we do a simple graphical residual analysis to get information about the quality of the fit. But first we write, for later use, a program for fitting and showing the data and the fit:

```
showFit[data_, basis_, var_, opts___] := With[{fit = Fit[data, basis, var]},
  Print[Show[ListLinePlot[data, Mesh → All],
    Plot[fit, {var, Min[data[[All, 1]]], Max[data[[All, 1]]]}], opts]];
  fit]
```

```
fit = showFit[data, {1, x}, x, AspectRatio → 0.4, ImageSize → 200]
```

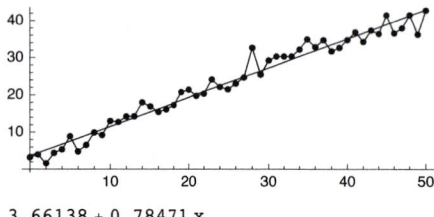

```
3.66138 + 0.78471 x
```

■ Graphical Residual Analysis

First, we extract the x and f values:

```
{xx, ff} = data⊤;
```

Now **xx** contains the x values and **ff** the f values. We then calculate the predicted values—that is, the values of the fit at the data points:

```
pred = fit /. x → xx;
```

We calculate the residuals:

```
resf = ff - pred;
```

The sum of the squared residuals is as follows:

```
resf.resf      239.262
```

(The parameters of **fit** were chosen by **Fit** such that the sum of the squared residuals is as small as possible. The minimum value is thus 239.262.) To plot the residuals, add the x values:

```
res = {xx, resf}⊤;
```

Here is a plot of the residuals:

```
pres = ListLinePlot[res, Mesh → All]
```

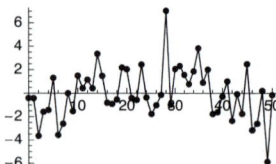

The residuals are quite random but not wholly random: A roughly quadratic pattern can be seen. We investigate the situation by fitting a cubic polynomial to the residuals:

```
resfit = Fit[res, {1, x, x^2, x^3}, x]
```

$$-1.65214 + 0.134729\,x + 0.00130549\,x^2 - 0.0000883095\,x^3$$

```
Show[pres, Plot[resfit, {x, 0, 50}]]
```

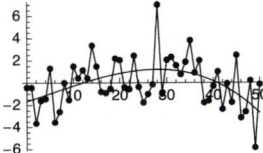

This plot confirms that the linear fit to the data is not adequate; the residuals contain some information.

For later use, we write a program for this kind of graphical residual analysis:

```
showResiduals[data_, fit_, var_, opts___] := Module[{xx, ff, resf, res, resfit},
  {xx, ff} = data^T;
  resf = ff - (fit /. var → xx);
  res = {xx, resf}^T;
  resfit = Fit[res, {1, var, var^2, var^3}, var];
  Show[ListLinePlot[res, Mesh → All],
   Plot[resfit, {var, Min[xx], Max[xx]}],
   PlotLabel → Row[{"Sum of squared residuals: ", resf.resf}], opts]]
```

```
showResiduals[data, fit, x]
```

Sum of squared residuals: 239.262

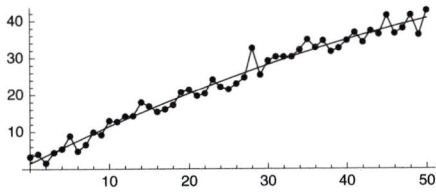

■ A Second Fit

Since the linear fit was not adequate, we next try a quadratic fit:

```
fit = showFit[data, {1, x, x^2}, x, AspectRatio → 0.4, ImageSize → 200]
```

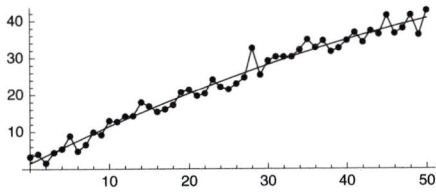

$$1.48998 + 1.0506\,x - 0.00531772\,x^2$$

The fit seems very good. We also show the residuals:

```
showResiduals[data, fit, x]
```

Sum of squared residuals: 185.163

It seems that the residuals do not contain significant information anymore.

25.1.2 More about Linear Fitting

■ **Multidimensional Data**

With `Fit`, the data can be multidimensional. If we have, for example, two independent variables x and y and a response variable f, we can find a fitting surface for the data. The following is a summary of 3D fitting, but the summary generalizes readily for higher-dimensional data.

`Fit[data, basis, vars]`   Find the least-squares fit to **data** as a linear combination of functions **basis** of variables **vars**

data is given in the following form:

`{{x₁, y₁, f₁}, {x₂, y₂, f₂}, … }`

Examples of **basis**:

`{1, x, y}`   The fitting function is of the form $a + bx + cy$

`{1, x, y, x y, x^2, y^2}`   The fitting function is of the form $a + bx + cy + dxy + ex^2 + fy^2$

Consider the following 3D data:

```
data = {{6, 4, 7.92}, {6, 5, 9.31}, {6, 6, 9.74},
    {7, 4, 11.24}, {7, 5, 12.09}, {7, 6, 12.62},
    {8, 4, 14.31}, {8, 5, 14.58}, {8, 6, 16.16}};
```

(In our example, the x and y arguments form a regular grid, but generally the points may be irregular.) We then fit a plane:

```
fit = Fit[data, {1, x, y}, {x, y}]
```
$-13.305 + 3.01333 x + 0.841667 y$

```
Plot3D[fit, {x, 6, 8}, {y, 4, 6}]
```

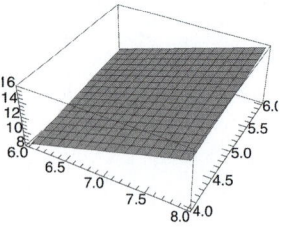

The sum of the squared residuals is as follows:

```
{xx, yy, ff} = data;
pred = fit /. {x → xx, y → yy};
(ff - pred).(ff - pred)
0.526717
```

■ Own Least Squares

The following program is based on the equation $X^T X a = X^T f$, where vector a contains the parameters to be estimated, vector f contains the f_i values, and the ith row of matrix X contains the values of the basis functions at x_i. If we solve the linear equations for a, we get the least-squares parameters.

```
dataLSQ[xx_List, ff_List, basis_List, t_] := With[{X = basis /. t → # & /@ xx},
    LinearSolve[X.X, X.ff].basis]
```

Here, **xx** and **ff** contain the x and f values of the data, **basis** contains the basis functions, and **t** is the variable of the basis functions. As an example, we find a quadratic fitting curve for the same data we considered in Section 25.1.1, p. 813:

```
xx = Range[0, 50]; SeedRandom[2];
rand = RandomReal[NormalDistribution[0, 1], 51];
ff = 2 + xx - 0.004 xx^2 + 2 rand;

dataLSQ[xx, ff, {1, x, x^2}, x]

1.48998 + 1.0506 x - 0.00531772 x^2
```

We could also use **LeastSquares** (**Fit** actually uses this command):

```
LeastSquares[{1, #, #^2} & /@ xx, ff]

{1.48998, 1.0506, -0.00531772}
```

■ Logarithmic Transform

If the data show an exponential growth, then one candidate for the fitting function is $f(x) = \exp(a + b\,x)$, but this is nonlinear in the parameters a and b, so **Fit** cannot be applied in the standard way. However, we can take logarithms of the values of the function: $\log(f(x)) = a + b\,x$. Thus, the logarithms of the data have a simple linear form to which we can apply **Fit**. After the fit $\hat{a} + \hat{b}\,x$ is found for $(x_i, \log(f_i))$, we do the inverse transform to find the fit $\exp(\hat{a} + \hat{b}\,x)$ for the original data.

Taking logarithms and then using linear least squares is a widely used procedure, but note that the resulting fit is not the best possible one. To find a true least-squares fit to the exponential model, **FindFit** should be used (see Section 25.1.3, p. 819).

First, we generate points that show an exponential growth:

```
xx = Range[0, 10, 0.2]; SeedRandom[0];
rand = RandomReal[NormalDistribution[0, 1], 51];
data = {xx, Exp[0.3 + 0.2 xx] + 0.5 rand};
```

Then we take logarithms of the f values:

```
{xx, ff} = data;
logdata = {xx, Log[ff]};
```

Now we fit a linear function to this data and make the inverse transform:

```
logfit = Fit[logdata, {1, x}, x]
```

$0.0902513 + 0.228637 \, x$

```
fit = Exp[logfit]
```

$e^{0.0902513 + 0.228637 \, x}$

The coefficients 0.09 and 0.23 are quite near the values 0.3 and 0.2 that we used in the simulation. We plot the fit and the data:

```
Show[ListLinePlot[data, Mesh → All],
  Plot[fit, {x, 0, 10}]]
```

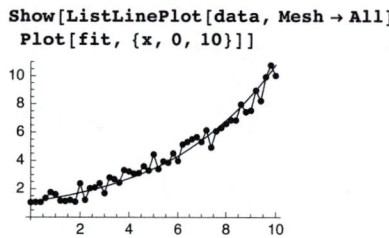

The fit seems to be good. We show the residuals (using the program **showResiduals** presented in Section 25.1.1, p. 815):

```
showResiduals[data, fit, x]
```

Sum of squared residuals: 11.8784

The residuals are quite near zero but not quite random. We will apply **FindFit** in Section 22.1.3, and then we get a slightly better fit.

25.1.3 Nonlinear Fitting

■ **Finding a Nonlinear Fit**

> **FindFit[data, funct, params, var]**   When **funct** is an expression of the variable **var** and contains the parameters **params**, find values for the parameters such that the function fits **data** in the best way (in the sense of least squares, by default)
>
> **data** can be given in either of the following forms:
> $\{f_1, f_2, \ldots\}$ Fit using the points $\{1, f_1\}, \{2, f_2\}, \ldots$
> $\{\{x_1, f_1\}, \{x_2, f_2\}, \ldots\}$ Fit using the given points
> Examples of **funct**: **Exp[a + b x], a/(1 + b Exp[-c x])**
>
> The parameter specification **params** is of the form $\{a, b, \ldots\}$ or of the form $\{\{a, a_0\}, \{b, b_0\}, \ldots\}$. In the former case, the starting value for each parameter is 1, and in the latter case, a_0, b_0, and so on are used as the starting values (all parameters need not have the same form of specification).

If you want statistics of the model, use **NonlinearRegress** (see Section 30.5.2, p. 1035).

By default, **FindFit** uses the Levenberg–Marquardt method to find the best values for the parameters of a function that is nonlinear in the parameters. The default criterion is to minimize the square root of the sum of the squares of the residuals; the result is a least-squares fit. The method is an iterative procedure, and initial guesses for the parameters can be provided. In general, **FindFit** only finds a locally optimal fit (**FindFit** effectively uses **FindMinimum** to minimize the norm of the residuals).

We can have constraints **cons** for the parameters. In place of **funct** in the previous box, just write {**funct, cons**}.

Multidimensional data can be entered in the same way as for **Fit**. For example, if we have two independent variables, the command is of the following form:

```
FindFit[{{x₁, y₁, f₁}, {x₂, y₂, f₂}, … }, funct, {a, b, … }, {x, y}]
```

■ Example 1: Exponential Growth

We consider the same data that were used when we introduced the logarithmic transform in Section 25.1.2, p. 817, and we fit the same model $\exp(a + bx)$, which we used there. We try the default starting value of 1 for both parameters:

```
xx = Range[0, 10, 0.2]; SeedRandom[0];
rand = RandomReal[NormalDistribution[0, 1], 51];
data = {xx, Exp[0.3 + 0.2 xx] + 0.5 rand}ᵀ;

f = Exp[a + b x];

ab = FindFit[data, f, {a, b}, x]
{a → 0.202349, b → 0.212479}

fit = f /. ab
```

$$e^{0.202349 + 0.212479\,x}$$

In Section 25.1.2 we obtained, by the logarithmic transform, the fit $e^{0.090\,x + 0.229\,x}$. Thus, the two fits differ somewhat. We plot the fit and show the residuals (using the program **showResiduals** we presented in Section 25.1.1, p. 815):

```
Show[ListLinePlot[data, Mesh → All],
  Plot[fit, {x, 0, 10}]]
```

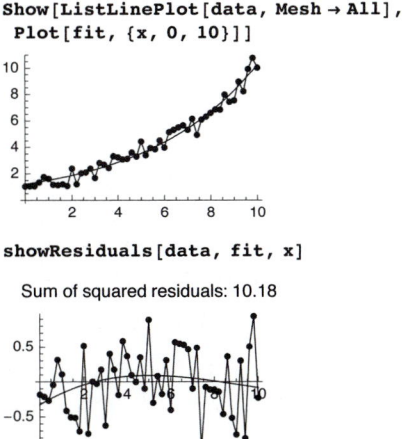

```
showResiduals[data, fit, x]
```

Sum of squared residuals: 10.18

The sum of the squared residuals, 10.18, is somewhat smaller than the value 11.88 obtained by the logarithmic transform.

■ Example 2: Logistic Growth

In an experiment, the growth of a yeast culture was measured at time instances 0, 1, 2, ..., 18 (hours). The measurements were as follows (Pearl, 1927):

```
yeast = {9.6, 18.3, 29.0, 47.2, 71.1, 119.1, 174.6, 257.3, 350.7,
    441.0, 513.3, 559.7, 594.8, 629.4, 640.8, 651.1, 655.9, 659.6, 661.8};
tt = Range[0, 18];
data2 = {tt, yeast}ᵀ;
p1 = ListPlot[data2]
```

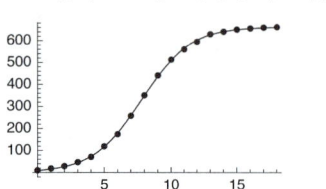

The growth seems to follow the logistic pattern $a / (1 + b\, e^{-cx})$. We try this model:

```
f = a / (1 + b Exp[-c t]);

abc = FindFit[data2, f, {a, b, c}, t]
```
$\{a \to 663.022, b \to 71.5763, c \to 0.546995\}$

```
fit = f /. abc
```
$$\frac{663.022}{1 + 71.5763\, e^{-0.546995\, t}}$$

The fit is very good:

```
Show[p1, Plot[fit, {t, 0, 18}]]
```

■ Example 3: A Modified Logistic Growth

Another parameterization of the logistic model is $M \Big/ \left[1 + \left(\frac{M}{y_0} - 1\right) \exp(-r M t)\right]$. This is the form of the solution of the logistic differential equation model $y'(t) = r\, y(t)\left[M - y(t)\right]$ with $y(0) = y_0$. The parameter M is the limiting value of $y(t)$ as time approaches infinity.

```
f2 = M / (1 + (M/y₀ - 1) Exp[-r M t]);
FindFit[data2, f2, {M, y₀, r}, t]
```
$\{M \to 393.039, y_0 \to 9.6, r \to 0.819405\}$

If we plotted this fit, we would see that the fit is very bad; **FindFit** reached a local optimum. For example, the value of M is approximately 393, whereas it should be approximately 660. To get a better fit, we give our own starting value for r:

```
Myr = FindFit[data2, f2, {M, y₀, {r, 0.1}}, t]
```
$\{M \to 663.022, y_0 \to 9.13552, r \to 0.000825002\}$

```
f2 /. Myr
```

$$\frac{663.022}{1 + 71.5763\, e^{-0.546995\, t}}$$

Now we obtained the same fit as in the preceding example. We could also use a constraint for **r**:

```
Myr = FindFit[data2, {f2, r < 0.1}, {M, y₀, r}, t, Method → "NMinimize"]
```
$\{M \to 663.051, r \to 0.000824771, y_0 \to 9.14387\}$

Here, we used the method **NMinimize**; the methods are listed in the next section.

25.1.4 More about Nonlinear Fitting

■ Finding the Global Optimum

As we saw in Example 3, if the starting values are not good enough, the result may not be the global optimum. Because of this, it is wise to try several starting values. We can also plot the criterion function by fixing all but two parameters. We fix y_0 to be 9.6 and form and plot the criterion (the sum of the squares of the residuals, which is called the χ^2 merit function) as a function of M and r:

```
res = yeast - (f2 /. {y₀ → 9.6, t → tt});
khi2 = res.res;
ContourPlot[khi2, {M, 300, 1000},
  {r, 0, 0.0015}, ContourShading → False, Contours → 30]
```

We see that the optimum values of M and r are near 650 and 0.0008, respectively.

We can also use **NMinimize** or **FindMinimum** (see Section 23.1.2, p. 747) to minimize the χ^2 merit function:

```
res = yeast - (f2 /. t → tt);
khi2 = res.res;
NMinimize[khi2, {M, y₀, {r, 0, 0.1}}]
```
$\{194.325, \{M \to 663.022, r \to 0.000825002, y_0 \to 9.13552\}\}$

```
FindMinimum[khi2, {M, y₀, {r, 0.1}}]
```
$\{194.325, \{M \to 663.022, y_0 \to 9.13552, r \to 0.000825002\}\}$

■ **Options**

Options of `FindFit`:

`WorkingPrecision` Precision used in internal computations; examples of values: `Automatic`, `20`

`PrecisionGoal` If the value of the option is `p`, the relative error of the optimum point and of the
value of the norm of the residuals should be of the order 10^{-p}; examples of values: `Automatic`
(usually means `8`), `10`

`AccuracyGoal` If the value of the option is `a`, the absolute error of the optimum point and of the
value of the norm of the residuals should be of the order 10^{-a}; examples of values: `Automatic`
(usually means `8`), `10`

`Method` Method used; possible values: `Automatic` (usually means `"LevenbergMarquardt"`),
`"LevenbergMarquardt"`, `"Gradient"`, `"ConjugateGradient"`, `"Newton"`, `"QuasiNewton"`,
`"NMinimize"`

`MaxIterations` Maximum number of iterations; examples of values: `Automatic` (usually means
`100`), `200`

`NormFunction` Norm of the residuals to be minimized; examples of values: `Norm` (means `(Norm[#,`
`2] &))`, `(Norm[#, 1] &)`

`Gradient` How the gradient is calculated; examples of values: `Automatic`, `"Symbolic"`,
`"FiniteDifference"`

`StepMonitor` Command to be executed after each step of the iterative method; examples of values:
`None`, `Sow[{a, b}]`, `n++`, `AppendTo[iters, {a, b}]`

`EvaluationMonitor` Command to be executed after each evaluation of the expression to be fitted;
examples of values: `None`, `Sow[{a, b}]`, `n++`, `AppendTo[points, {a, b}]`

The default is that iterations are stopped if the estimated relative or absolute error in the point
obtained and in the value of the norm of the residuals is less than 10^{-8}.

The norm of the residuals is minimized with `Method`. The default value `Automatic` of this option
means `LevenbergMarquardt` if the 2-norm is used (which is the default). The Levenberg-Marquardt
method initially uses the steepest descent method, but it shifts gradually to quadratic minimization.

The default norm is `Norm`, which is the 2-norm (the square root of the sum of the squares of the
residuals); we could also write this norm as `(Norm[#, 2] &)`. With the `NormFunction` option, we can
define other norms, such as the 1-norm or `(Norm[#, 1] &)` (the sum of the absolute values of the
residuals) or the ∞-norm or `(Norm[#, ∞] &)` (the maximum of the absolute values of the residuals; the
result is a minimax approximation).

■ **Using Norm Functions**

Outliers can cause problems in a least-squares fit. An outlier is an observation that has a value that
differs markedly from the general trend of the data. Because the least-squares fit is calculated by
minimizing the squared residuals, an outlier can have a considerable unwanted effect on the fit. An
excellent illustration of outliers is given in Shaw and Tigg (1994, pp. 315–319).

Let us consider an example given in the mentioned book. The data are otherwise regular but we have
two outliers:

```
lindata = Table[{x, 2 + x + 0.1 RandomReal[]}, {x, 0, 5, 0.5}];
lindata[[5]] = {2, 7};
lindata[[11]] = {5, 1};
```

We form three linear fits by using L_1-, L_2-, and L_∞-norms:

```
fits = a + b x /. FindFit[lindata, a + b x, {a, b}, x, NormFunction → (#)] & /@
   {Norm[#, 1] &, Norm[#, 2] &, Norm[#, ∞] &}
{2.04523 + 1.00085 x, 3.26496 + 0.399945 x, 4.74137 - 0.203543 x}
```

The corresponding plots are as follows:

```
MapThread[Plot[#1, {x, 0, 5}, PlotRange → {0, 7.3},
    PlotLabel → Row[{#2, "-norm"}], Epilog → Point[lindata]] &, {fits, {L₁, L₂, L∞}}]
```

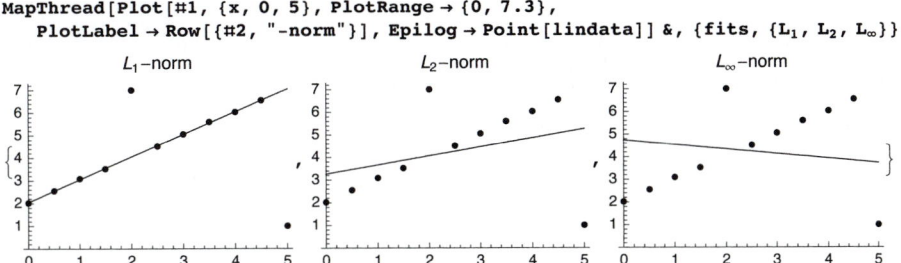

We see that the outliers have a clear bad effect to the usual least-squares or L_2 fit and still worse effect to the L_∞ fit (in the latter fit, the maximum error is as small as possible). However, the L_1 fit is very good: It is able to ignore the outliers.

With the following manipulation, we can study the effect of various norm functions when using polynomial fits. Note that the points can be moved with the mouse. New points can be added by holding down the [ALT] key (Windows) or ⌘ key (Macintosh) and then clicking on the plot. A point can be deleted by [ALT]- or ⌘-clicking on that point.

```
fitPlot[order_, p_, points_] :=
  With[{aa = Array[a, order + 1], xx = x^Range[0, order]}, Plot[
    Evaluate[aa.xx /. FindFit[points, aa.xx, aa, x, NormFunction → (Norm[#, p] &)]],
    {x, 0, 10}, PlotRange → {0, 10.3}, ImageSize → 230]]
Manipulate[
  fitPlot[order, p, points],
  {{order, 1}, 1, 10, 1, Appearance → "Labeled"},
  {p, {1, 2, ∞}}, {{points, {{0, 2}, {1, 2.5}, {2, 3}, {3, 3.5}, {4, 7},
     {5, 4.5}, {6, 5}, {7, 5.5}, {8, 6}, {9, 6.5}, {10, 1}}}, Locator,
   Appearance → •, LocatorAutoCreate → True}, SaveDefinitions → True]
```

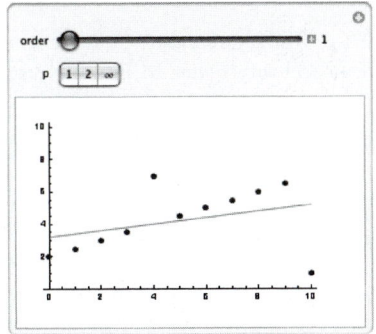

■ Showing the Steps

Let us again consider the exponential data we used in Example 1 of Section 25.1.3, p. 819:

```
xx = Range[0, 10, 0.2]; SeedRandom[0];
rand = RandomReal[NormalDistribution[0, 1], 51];
data = {xx, Exp[0.3 + 0.2 xx] + 0.5 rand}ᵀ;
```

Use **StepMonitor** to gather information about how the iterations proceed step by step:

```
f = Exp[a + b x]; {xx, ff} = dataᵀ; fxx = f /. x → xx;

{fit, {steps}} = Reap[FindFit[data, f, {a, b}, x,
    Method → "Gradient", StepMonitor :> Sow[{a, b, Norm[ff - fxx]}]]];
```

Prepare a table:

```
TableForm[steps, TableSpacing → {0.6, 2},
  TableHeadings → {None, {a, b, Norm[residual]}}] // TraditionalForm
```

a	b	$\|$residual$\|$
0.908741	0.124512	8.26372
0.364025	0.181292	4.78486
0.244722	0.207836	3.21582
0.202633	0.212418	3.19062
0.202351	0.212478	3.19061
0.202349	0.212479	3.19061

25.2 Approximation of Functions

25.2.1 Simple Methods

■ Introduction

When we next explain several methods for the approximation of functions, we will use the cumulative distribution function of the standard normal distribution as an example:

```
f = (1 + Erf[x / Sqrt[2]]) / 2;

Plot[f, {x, 0, 4}]
```

We apply different approximation methods either at point 1 or in the interval (0, 2). If we want an approximation over a wider interval, for example, (0, 4), it is probably better to find at least two approximations—one for (0, 2) and the other for (2, 4).

We will use the following module to show the absolute and relative errors of the approximation **appr** for **f** in the interval from **a** to **b**:

```
showError[f_, appr_, x_, a_, b_, opts___] :=
  Plot[#, {x, a, b}, PlotRange → All, opts] & /@ {f - appr, 1 - appr / f}
```

First, we consider some simple methods of approximation: Taylor polynomials and Padé, economized rational, Chebyshev, and least-squares approximation. In Section 25.2.2, we explain the minimax approximation. This method minimizes the maximum error in the interval considered to get an error that is evenly spread over the entire interval.

Note that in Section 24.4.2, p. 807, we discussed **FunctionInterpolation**. This command, although it uses interpolation, is very good for approximating functions because it is adaptive (i.e., gives special care to regions where the function changes rapidly), its precision can be controlled, and it is easy to calculate and use.

■ **Taylor Polynomials**

`Series[f, {x, a, m}]`   Taylor series of degree **m** for **f** about **a**

A Taylor polynomial (see Section 19.2.1, p. 624) gives an approximation at a point. For example,

`taylor6 = Normal[Series[f, {x, 1, 6}]] // N`

$0.841345 + 0.241971 \; (-1. + x) - 0.120985 \; (-1. + x)^2 +$
$0.0201642 \; (-1. + x)^4 - 0.00403285 \; (-1. + x)^5 - 0.00201642 \; (-1. + x)^6$

The absolute and relative errors show that the approximation is good in the interval (0.5, 1.5):

`showError[f, taylor6, x, 0, 2]`

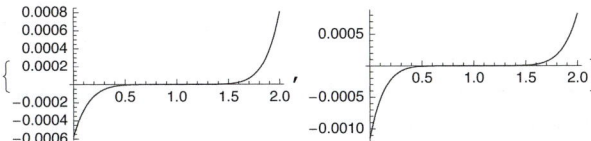

■ **Padé Approximation**

A Padé approximant gives a rational approximation for a function at a point.

`PadeApproximant[f, {x, a, {m, n}}]`   Padé approximant of degree (**m**, **n**) for **f** about **a**

Here, **m** and **n** are the degrees of the numerator and the denominator. The rational function $p(x)\big/q(x)$ is a Padé approximant of order (m, n) for $f(x)$ at $x = a$, if $p(x)$ and $q(x)$ are of order m and n, respectively, and the power series of $f(x)\,q(x) - p(x)$ about $x = a$ begins with the term $(x - a)^{m+n+1}$. Here is an example:

`pade23 = PadeApproximant[f, {x, 1, {2, 3}}] // N`

$$\frac{0.841345 + 0.274377 \; (-1. + x) - 0.0167285 \; (-1. + x)^2}{1. + 0.0385177 \; (-1. + x) + 0.112839 \; (-1. + x)^2 - 0.0269137 \; (-1. + x)^3}$$

`showError[f, pade23, x, 0, 2]`

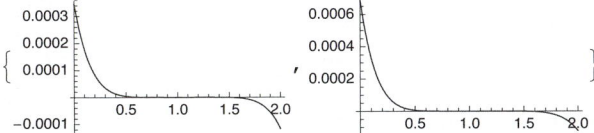

■ **Economized Rational Approximation**

In the **FunctionApproximations`** *package:*
`EconomizedRationalApproximation[f, {x, {a, b}, m, n}]`   Economized rational approximation of degree (**m**, **n**) for **f** in the interval (**a**, **b**)

This command first finds the Padé approximant about the midpoint of the interval (a, b) and then perturbs the approximant with Chebyshev polynomials to reduce the leading coefficient in the error. With this method, we get a somewhat better approximation than with Padé approximation:

```
<< FunctionApproximations`

econ23 = EconomizedRationalApproximation[f, {x, {0, 2}, 2, 3}] // N
```

$$\frac{0.834462 + 0.272397\,(-1. + x) - 0.0157965\,(-1. + x)^2}{0.991815 + 0.0385177\,(-1. + x) + 0.112839\,(-1. + x)^2 - 0.0269137\,(-1. + x)^3}$$

```
showError[f, econ23, x, 0, 2]
```

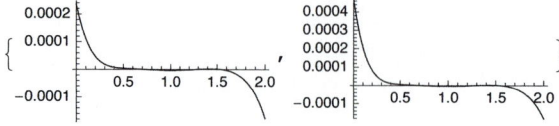

■ **Chebyshev Approximation**

> *In the* **FunctionApproximations`** *package:*
>
> **RationalInterpolation[f, {x, m, n}, {x, a, b}]**   Rational interpolating function of degree (**m, n**)
>   for **f** in the interval (**a, b**) (i.e., rational Chebyshev approximation)

In Section 24.4.1, p. 806, we considered rational Chebyshev approximation. We apply this method to our function by using degree (6, 0); that is, the approximating function is a sixth-degree polynomial:

```
cheb60 = RationalInterpolation[f, {x, 6, 0}, {x, 0, 2}]
```

$$0.500007 + 0.398604\,x + 0.00255252\,x^2 -$$
$$0.0732252\,x^3 + 0.00721336\,x^4 + 0.00800419\,x^5 - 0.00181145\,x^6$$

```
showError[f, cheb60, x, 0, 2]
```

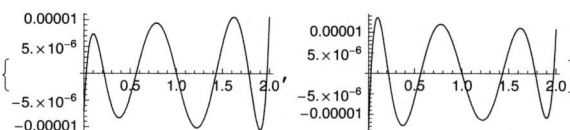

The approximation is very accurate, although the errors are not spread perfectly evenly over the interval.

With the option **Bias**, we can fine-tune the approximation. The value 0 means that the interpolation points are chosen symmetrically in the interval. A positive [negative] value causes the points to be shifted toward the right [left]. The value has to be between −1 and 1.

We try to make the relative error of the approximation **cheb60** more even. Using trial and error, we find that a bias of −0.015 is appropriate:

```
cheb60b = RationalInterpolation[f, {x, 6, 0}, {x, 0, 2}, Bias → -0.015]
```

$$0.500006 + 0.398636\,x + 0.00235795\,x^2 -$$
$$0.0727988\,x^3 + 0.00679414\,x^4 + 0.00819327\,x^5 - 0.00184333\,x^6$$

```
showError[f, cheb60b, x, 0, 2]
```

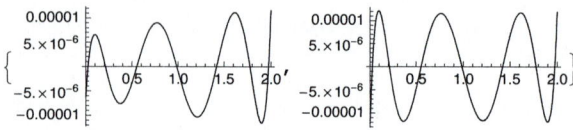

With **GeneralRationalInterpolation**, we can calculate Chebyshev approximations for parametrically defined functions.

■ Least-Squares Approximation

With the least-squares method, we minimize the integral of the square of the difference between the function and the approximation on the interval considered. If the approximating function is a linear combination of certain functions, the coefficients are obtained from a system of linear equations. Here is a module that calculates a least-squares approximation for **f** in the interval from **a** to **b**; the approximating function is a linear combination of the functions in the list **basis**.

```
functionLSQ[f_, x_, a_, b_, basis_] :=
 LinearSolve[NIntegrate[KroneckerProduct[basis, basis], {x, a, b}],
   NIntegrate[f basis, {x, a, b}]].basis
```

We calculate a sixth-degree least-squares approximation:

```
lsq6 = functionLSQ[f, x, 0, 2, x^Range[0, 6]]
```

$0.500019 + 0.398426\, x + 0.00319491\, x^2 -$
$0.0741471\, x^3 + 0.00780805\, x^4 + 0.00784075\, x^5 - 0.00179815\, x^6$

```
showError[f, lsq6, x, 0, 2]
```

The approximation is good, but at the end points the errors are considerably larger than at other points.

25.2.2 Minimax Approximation

■ Finding a Minimax Approximation

The goal of minimax approximation is to minimize the maximum error (absolute or relative) in an interval. This is clearly a very desirable goal: The error is then evenly low over the whole interval, which is in contrast with an error that is low over a subinterval but large elsewhere.

In the **FunctionApproximations`** *package:*

MiniMaxApproximation[f, {x, {a, b}, m, n}]   Rational minimax approximation of degree (**m**, **n**) for **f** in the interval (**a, b**)

Here, **m** and **n** are the degrees of the numerator and the denominator. Giving **n** the value zero, we can calculate polynomial minimax approximations.

The procedure starts with a rational interpolating function $r(x)$ using **RationalInterpolation**. This function is then iteratively modified according to Remez's algorithm: The interpolation points are adjusted to make the maximum relative error as small as possible.

The procedure uses the relative error $\left|1 - r(x)/f(x)\right|$ as the criterion. This means that $f(x)$ cannot have a zero in the interval. However, we can overcome this problem by dividing the zero out of the function (see the documentation of the package). Singularities must also first be eliminated. In addition, it is better to calculate several approximations for small intervals rather than one approximation for a long interval.

■ **Example**

Here is the (6, 0) degree rational minimax approximation (i.e., the sixth-degree polynomial minimax approximation) for our familiar function:

```
f = (1 + Erf[x / Sqrt[2]]) / 2;

<< FunctionApproximations`

{maxPoints, {miniMax60, maxError}} = MiniMaxApproximation[f, {x, {0, 2}, 6, 0}]
```

$\{\{0., 0.0946758, 0.364016, 0.761322, 1.20893, 1.61548, 1.89853, 2.\},$
$\{0.500006 + 0.398643 x + 0.00232708 x^2 - 0.0727457 x^3 +$
$0.00675247 x^4 + 0.00820846 x^5 - 0.00184541 x^6, -0.0000118083\}\}$

The result is of the following form: {(points where the maximum relative error occurs), {the minimax approximation, the maximum relative error}}. The relative error is perfectly even in the interval (0, 2):

```
showError[f, miniMax60, x, 0, 2]
```

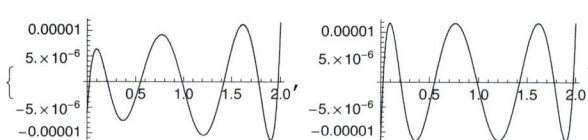

The Chebyshev approximations **cheb60** and **cheb60b** that we calculated in Section 25.2.1, p. 826, are close to the minimax approximation.

■ **Options**

> *Options of* **MiniMaxApproximation**:
>
> **WorkingPrecision** Precision used in internal computations; examples of values:
>   **MachinePrecision, 20**
> **Bias** Bias in the symmetry of the initial interpolation points; examples of values: **0, −0.1, 0.26**
> **Brake** Defines the braking properties of the algorithm; default value: **{5, 5}**
> **MaxIterations** Maximum number of iterations after braking has ended; default value: **20**
> **Derivatives** Specifies a function to use for the derivatives; default value: **Automatic**
> **PlotFlag** Whether to plot the relative error at each step; possible values: **False, True**
> **PrintFlag** Whether to print information about the relative error at each step; possible values:
>   **False, True**

The value of **Bias** is a number between −1 and 1. The default value 0 means that the initial interpolation points are chosen symmetrically in the interval. A positive [negative] value of **Bias** causes the points to shift toward the right [left].

Brake controls the braking of the iterations. If the change from one iteration to the next is too large, the procedure may go astray. Braking can prevent this. The default value of the option is {5, 5}. The first value in the list tells us how many iterations are to be affected by the braking, and the second value tells us how much braking is to be applied to the first iteration. The braking automatically decreases with the iterations.

The documentation of the package contains much more information about minimax approximation than is presented here. The package also defines **GeneralMinimaxApproximation**, which can be used to approximate parametrically defined functions.

Differential Equations

Introduction

> *God is not so cruel as to create situations described by nonlinear differential equations. —Edward Sexton*

Solving ordinary differential equations with *Mathematica* is straightforward: We have **DSolve** for symbolic solution and **NDSolve** for numerical solution. Both commands accept one or more equations, first- or higher-order equations, and linear and nonlinear equations, and they solve both initial and boundary value problems.

DSolve can solve linear, constant coefficient differential equations of any order. It can also solve many linear equations up to second order with nonconstant coefficients. In addition, it can solve almost all the nonlinear equations whose solutions are given in standard reference books such as Kamke.

We also consider solving differential equations with the Laplace transform, finding series solutions, and solving integral equations. We implement the Runge–Kutta method and some methods for boundary value problems. Some well-known nonlinear systems are considered, such as the logistic model, a predator–prey model, a competing species model, and the Lorenz model. We also consider estimating parameters of nonlinear differential equations.

For more about differential equations with *Mathematica*, see Abell and Braselton (1997) and Schwalbe and Wagon (1997). See also the advanced tutorials Differential Equation Solving with DSolve and Advanced Numerical Differential Equation Solving in *Mathematica*.

26.1 Symbolic Solutions

26.1.1 First-Order Equations

Here are some common commands for first-order differential equations:

`sol = y[t] /. DSolve[eqn, y[t], t]` Give the general solution
`sol = y[t] /. DSolve[{eqn, y[a] == α}, y[t], t]` Solve an initial value problem
`Plot[sol, {t, a, b}]` Plot the solution of an initial value problem

An example of a differential equation is `y'[t] == a y[t] + b t + c`. The dependent variable, which here is **y**, must contain the independent variable, here **t**, as the argument; this means that we cannot write the equation as `y' == a y + b t + c`. Note also that the equation must contain == (not =) and that the initial condition must also contain == (not =) (remember that *Mathematica* replaces == with ≟).

■ **Example 1: General Solution**

Consider the following logistic equation:

  `eqn = y'[t] == r y[t] (M - y[t])` $y'[t] == r (M - y[t]) y[t]$

The name of the equation is **eqn**. The solution is as follows:

$$\text{DSolve[eqn, y[t], t]} \qquad \left\{\left\{y[t] \rightarrow \frac{e^{M r t + M C[1]} M}{-1 + e^{M r t + M C[1]}}\right\}\right\}$$

The solution is in the form of a transformation rule (for more information about rules, see Section 13.1.2, p. 416). The arbitrary constant is `C[1]` (we can give it whatever value we want).

In general, the solution given by **DSolve** consists of a list of solutions:

  `{{solution 1}, {solution 2}, ...}`

Indeed, a given equation can have several solutions. Each solution is again a list that consists of as many elements as there are dependent variables. In our example, we have only one dependent variable, **y**, and we obtained only one solution. Thus, the solution is of the form `{{solution 1 for y}}`.

Often, it is convenient to ask for the value of `y[t]`:

$$\text{y[t] /. DSolve[eqn, y[t], t]} \qquad \left\{\frac{e^{M r t + M C[1]} M}{-1 + e^{M r t + M C[1]}}\right\}$$

If there is only one solution, we may also want to get rid of the curly braces by asking for the first component of the solution:

$$\texttt{y[t] /. DSolve[eqn, y[t], t]⟦1⟧} \qquad \frac{e^{Mrt+MC[1]}\, M}{-1 + e^{Mrt+MC[1]}}$$

We will use this method from now on.

Note that the solution **sol** is a *generic* solution, which is a solution that is valid for general values of the parameters **r** and **M**. For some particular values, the solution may be of a different form. For example, when **M** is zero, the solution is as follows:

$$\texttt{y[t] /. DSolve[eqn /. M → 0, y[t], t]⟦1⟧} \qquad \frac{1}{rt - C[1]}$$

■ Example 2: Initial Value Problem

Next, we solve an initial value problem:

```
y[t] /. DSolve[{eqn, y[0] == α}, y[t], t]⟦1⟧
```

Solve::ifun : Inverse functions are being used by Solve, so some
    solutions may not be found; use Reduce for complete solution information. ≫

$$\frac{e^{Mrt}\, M\, \alpha}{M - \alpha + e^{Mrt}\, \alpha}$$

For all transcendental equations, **Solve** gives the warning that because inverse functions are used, some solutions may not be found (see Section 22.3.1, p. 730). When solving differential equations, the solution is mostly unique, and in such cases the warning can be ignored. We turn the message off to save space:

```
Off[Solve::ifun]
```

Next, we give specific values for all constants:

```
sol = y[t] /. DSolve[{eqn /. {r → 1 / 10, M → 10}, y[0] == 1 / 4}, y[t], t]⟦1⟧
```

$$\frac{10\, e^{t}}{39 + e^{t}}$$

This solution can be plotted because it does not contain any parameters:

```
p1 = Plot[{10, sol}, {t, 0, 10}]
```

This is a logistic curve. We also plotted the asymptote 10 of this curve. We can easily calculate values of the solution:

```
sol /. t → 0.0     0.25
```

```
Table[{t, sol}, {t, 0., 4, 1}]
```
```
{{0., 0.25}, {1., 0.65158}, {2., 1.59284}, {3., 3.3994}, {4., 5.83325}}
```

Recall that in Section 10.1.2, p. 325, we presented a manipulation to study the form of the logistic curve for various values of $y(0)$, r, and M.

■ **Example 3: Direction Field**

We can learn the behavior of the solution of a differential equation by plotting a set of arrows that are tangent to the solution. The plot is called a direction field. It can be plotted with **VectorFieldPlot** from a package (see Section 5.3.2, p. 144). This command plots vectors **{exprx, expry}** for some values of **x** and **y**. In our example, we can choose **exprx** to be **1** and **expry** to be **y'[t]**:

```
<< VectorFieldPlots`

p2 = VectorFieldPlot[{1, 1 / 10 y (10 - y)},
    {t, 0, 10}, {y, 0, 10}, PlotPoints → 11, Axes → True];
```

We show both the direction field and one solution:

```
Show[p1, p2, ImageSize → 200]
```

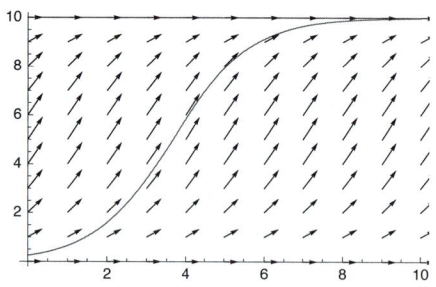

■ **Example 4: A Set of Trajectories**

Next, we plot a set of trajectories, which is a set of solutions to the equation with different starting points. To begin, we compute the solution with general values **a** and α for the starting time and starting value:

```
sol = y[t] /. DSolve[{eqn /. {r → 1 / 10, M → 10}, y[a] == α}, y[t], t][[1]]
```

$$\frac{10\,e^t\,\alpha}{10\,e^a - e^a\,\alpha + e^t\,\alpha}$$

We fix **a** to be 0 and give α various values:

```
solset = Table[sol /. a → 0, {α, 0.1, 15.1, 0.5}];

Plot[solset, {t, 0, 8}, PlotRange → All, PlotStyle → Black, ImageSize → 200]
```

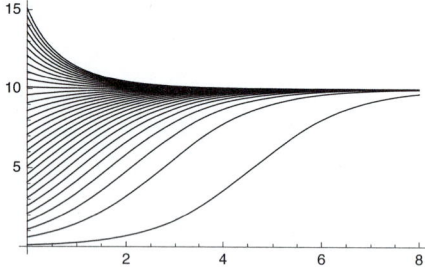

■ Example 5: Equilibrium Points

Recall from a course of differential equations that if the differential equation is $y' = f(y)$, then the equilibrium points y^* are the solutions of the equation $f(y^*) = 0$ and an equilibrium point is asymptotically stable if $f'(y^*) < 0$. In our model, the f function is as follows:

```
f = eqn[[2]]        r (M - y[t]) y[t]
```

Calculate the equilibrium points:

```
equi = Solve[f == 0, y[t]]        {{y[t] → 0}, {y[t] → M}}
```

Check the stability:

```
D[f, y[t]] /. equi        {M r, -M r}
```

Thus, if $M r > 0$, then 0 is an unstable and M an asymptotically stable equilibrium point. These properties can also be seen from the figure of the previous example.

■ Example 6: Equations Are Defined by ==

A common problem encountered when solving differential equations is the following:

```
eqn = {y'[t] == y[t] + t + 1, y[0] = 2};
```

```
DSolve[eqn, y[t], t]
```

DSolve::deqn :

   Equation or list of equations expected instead of 2 in the first argument $\{y'[t] == 1 + t + y[t], 2\}$. ≫
DSolve[{y'[t] == 1 + t + y[t], 2}, y[t], t]

DSolve tells you that it found, from the first argument, the element 2 and that this is not an equation. Indeed, we observe that the initial condition **y[0] = 2** is not a correct equation. It must be written as **y[0] == 2**. When we wrote **y[0] = 2**, we actually assigned the value **2** for **y[0]**, and this causes the error message. Before we solve the initial value problem, we must clear the value of **y[0]** and correct the initial condition:

```
y[0] =.
```

```
eqn = {y'[t] == y[t] + t + 1, y[0] == 2};
```

```
DSolve[eqn, y[t], t]        {{y[t] → -2 + 4 e^t - t}}
```

■ Example 7: Implicit Solutions

Sometimes the solution is given in an implicit form:

```
Off[InverseFunction::ifun]
sol = DSolve[ y'[t] - 2 t y[t]^2 - y[t]^3 == 0, y[t], t]
```
Solve::tdep : The equations appear to involve

    the variables to be solved for in an essentially non–algebraic way. ≫

$$\text{Solve}\left[\frac{t\, \text{AiryAi}\left[t^2 + \frac{1}{y[t]}\right] + \text{AiryAiPrime}\left[t^2 + \frac{1}{y[t]}\right]}{t\, \text{AiryBi}\left[t^2 + \frac{1}{y[t]}\right] + \text{AiryBiPrime}\left[t^2 + \frac{1}{y[t]}\right]} + C[1] == 0, y[t]\right]$$

```
On[InverseFunction::ifun]
```

If the equation inside the **Solve** command could be solved for **y[t]**, we would get the solution of the differential equation. A contour plot of the implicit function can be made as follows:

```
c1 = C[1] /. Solve[sol[[1]], C[1]] /. y[t] → y
```

$$\left\{ \frac{-t \, \text{AiryAi}\left[t^2 + \frac{1}{y}\right] - \text{AiryAiPrime}\left[t^2 + \frac{1}{y}\right]}{t \, \text{AiryBi}\left[t^2 + \frac{1}{y}\right] + \text{AiryBiPrime}\left[t^2 + \frac{1}{y}\right]} \right\}$$

```
ContourPlot[c1, {t, 0, 0.5}, {y, 0, 10}, FrameLabel → {t, y[t]}, RotateLabel → False]
```

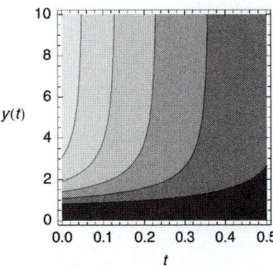

Each contour corresponds to a solution of the differential equation for a given value of the constant C[1].

■ Example 8: Several Solutions

Sometimes a problem has several general solutions:

```
sol = DSolve[y'[t] == 1 / y[t], y[t], t]
```

$$\left\{ \left\{ y[t] \to -\sqrt{2} \, \sqrt{t + C[1]} \right\}, \left\{ y[t] \to \sqrt{2} \, \sqrt{t + C[1]} \right\} \right\}$$

Given an initial value, a solution may not be obtained from all general solutions:

```
sol = DSolve[{y'[t] == 1 / y[t], y[0] == 1}, y[t], t]
```

DSolve::bvnul : For some branches of the general
    solution, the given boundary conditions lead to an empty solution. ≫

$$\left\{ \left\{ y[t] \to \sqrt{1 + 2 \, t} \right\} \right\}$$

■ Example 9: Piecewise Functions

The equation can contain piecewise functions:

```
sol = y[t] /. DSolve[y'[t] == r y[t] - h UnitStep[30 - t], y[t], t]
```

$$\left\{ e^{r \, t} \, C[1] - e^{r \, t} \, h \left(\begin{array}{cc} -\frac{e^{-r \, t}}{r} & t \le 30 \\ -\frac{e^{-30 \, r}}{r} & \text{True} \end{array} \right) \right\}$$

```
Plot[sol /. {r → 0.01, h → 2, C[1] → -100}, {t, 0, 100}]
```

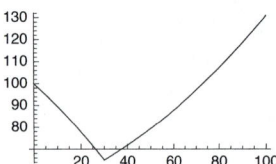

26.1.2 Second- and Higher-Order Equations

> `sol = y[t] /. DSolve[eqn, y[t], t]` General solution
>
> `sol = y[t] /. DSolve[{eqn, y[a] == α, y'[a] == β}, y[t], t]` Initial value problem
>
> `sol = y[t] /. DSolve[{eqn, y[a] == α, y[b] == β}, y[t], t]` Boundary value problem
>
> `Plot[sol, {t, a, b}]` Plot the solution of an initial or boundary value problem

These commands apply to second-order equations. They generalize directly to higher-order equations. The initial and boundary conditions mentioned are the simplest ones; the conditions can be more complex equations.

■ Example 1: Basic Techniques

We ask for a general solution of a second-order equation:

`eqn = y''[t] + y[t] == 1;`

`DSolve[eqn, y[t], t]`

$\{\{y[t] \to 1 + C[1] \, \text{Cos}[t] + C[2] \, \text{Sin}[t]\}\}$

The arbitrary constants are `C[1]` and `C[2]` (by the way, with the option **GeneratedParameters**, we can give the constants another name). Next, we give two initial conditions:

`sol = y[t] /. DSolve[{eqn, y[0] == 0, y'[0] == 1}, y[t], t][[1]]`

$1 - \text{Cos}[t] + \text{Sin}[t]$

`Plot[sol, {t, 0, 4 π}]`

Now we give two boundary conditions, one at $t = 0$ and the other at $t = 5$:

`y[t] /. DSolve[{eqn, y[0] == 0, y[5] == 7}, y[t], t][[1]]`

$1 - \text{Cos}[t] + \text{Cot}[5] \, \text{Sin}[t] + 6 \, \text{Csc}[5] \, \text{Sin}[t]$

The boundary conditions can be even more complex:

`y[t] /. DSolve[{eqn, y[0] == 2, y[5] + y'[5] == 1}, y[t], t][[1]] // FullSimplify`

$$1 + \text{Cos}[t] + \frac{(-\text{Cos}[5] + \text{Sin}[5]) \, \text{Sin}[t]}{\text{Cos}[5] + \text{Sin}[5]}$$

■ Example 2: Constant Coefficients

All linear second-order equations with constant coefficients can be solved. For example,

`eqn = y''[t] == a y'[t] + b y[t] + c;`

`DSolve[eqn, y[t], t]`

$$\left\{\left\{y[t] \to -\frac{c}{b} + e^{\frac{1}{2}\left(a - \sqrt{a^2 + 4b}\right)t} \, C[1] + e^{\frac{1}{2}\left(a + \sqrt{a^2 + 4b}\right)t} \, C[2]\right\}\right\}$$

Note that this is a generic solution. For special values of the parameters, the solution can be of another form. The following solution is of the preceding form:

```
DSolve[{eqn /. {a → 2, b → 3, c → 2}, y[0] == 2, y'[0] == 0}, y[t], t]
```

$$\left\{\left\{y[t] \rightarrow \frac{2}{3} \, e^{-t} \left(3 - e^{t} + e^{4 \, t}\right)\right\}\right\}$$

However, this solution is not:

```
DSolve[{eqn /. {a → 2, b → -1, c → 1}, y[0] == 2, y'[0] == 0}, y[t], t]
```

$$\left\{\left\{y[t] \rightarrow 1 + e^{t} - e^{t} \, t\right\}\right\}$$

Neither is this:

```
DSolve[{eqn /. {a → -1, b → -1, c → 1}, y[0] == 2, y'[0] == 0}, y[t], t]
```

$$\left\{\left\{y[t] \rightarrow \frac{1}{3} \, e^{-t/2} \left(3 \, e^{t/2} + 3 \cos\left[\frac{\sqrt{3} \, t}{2}\right] + \sqrt{3} \, \sin\left[\frac{\sqrt{3} \, t}{2}\right]\right)\right\}\right\}$$

■ Example 3: A Set of Trajectories

We continue the last example by varying `y[0]`:

```
sol = y[t] /.
    DSolve[{eqn /. {a → -1, b → -1, c → 1}, y[0] == α, y'[0] == 0}, y[t], t][[1]] // Simplify
```

$$\frac{1}{3} \, e^{-t/2} \left(3 \, e^{t/2} + 3 \, (-1 + \alpha) \cos\left[\frac{\sqrt{3} \, t}{2}\right] + \sqrt{3} \, (-1 + \alpha) \sin\left[\frac{\sqrt{3} \, t}{2}\right]\right)$$

```
solset = Table[sol, {α, 0, 5}];
```

```
Plot[solset, {t, 0, 10}, PlotRange → All, PlotStyle → Black]
```

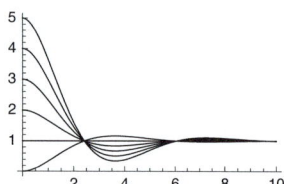

26.1.3 Simultaneous Equations

In simultaneous equations, we have several dependent variables. Here are some typical commands for two equations. They generalize easily to more equations and to different initial and boundary conditions.

`vars = {x[t], y[t]}`	Define the dependent variables
`eqns = {eqn1, eqn2}`	Define the differential equations
`inits = {x[a] == α, y[a] == β}`	Define the initial (or boundary) conditions
`sol = vars /. DSolve[eqns, vars, t]`	Give the general solution
`sol = vars /. DSolve[Join[eqns, inits], vars, t]`	Solve an initial value problem
`Plot[sol, {t, a, b}]`	Plot `x[t]` and `y[t]`
`ParametricPlot[sol, {t, a, b}]`	Plot a phase trajectory

■ Example 1: Basic Techniques

First, we ask for a general solution:

```
DSolve[{x'[t] == y[t], y'[t] == x[t]}, {x[t], y[t]}, t] // Simplify
```

$$\left\{\left\{x[t] \to \frac{1}{2} e^{-t} \left(\left(1 + e^{2t}\right) C[1] + \left(-1 + e^{2t}\right) C[2]\right), \right.\right.$$

$$\left.\left. y[t] \to \frac{1}{2} e^{-t} \left(\left(-1 + e^{2t}\right) C[1] + \left(1 + e^{2t}\right) C[2]\right)\right\}\right\}$$

We can also separately define the variables and the equations:

```
vars = {x[t], y[t]};
```

```
eqns = {x'[t] == y[t], y'[t] == x[t]};
```

```
DSolve[eqns, vars, t] // Simplify
```

$$\left\{\left\{x[t] \to \frac{1}{2} e^{-t} \left(\left(1 + e^{2t}\right) C[1] + \left(-1 + e^{2t}\right) C[2]\right), \right.\right.$$

$$\left.\left. y[t] \to \frac{1}{2} e^{-t} \left(\left(-1 + e^{2t}\right) C[1] + \left(1 + e^{2t}\right) C[2]\right)\right\}\right\}$$

We may want to directly ask for the values of **x[t]** and **y[t]**:

```
vars /. DSolve[eqns, vars, t][[1]] // Simplify
```

$$\left\{\frac{1}{2} e^{-t} \left(\left(1 + e^{2t}\right) C[1] + \left(-1 + e^{2t}\right) C[2]\right), \frac{1}{2} e^{-t} \left(\left(-1 + e^{2t}\right) C[1] + \left(1 + e^{2t}\right) C[2]\right)\right\}$$

By the way, the solution of this constant coefficient system can also be obtained using the matrix exponential:

```
MatrixExp[{{0, 1}, {1, 0}} t].{C[1], C[2]} // Simplify
```

$$\left\{\frac{1}{2} e^{-t} \left(\left(1 + e^{2t}\right) C[1] + \left(-1 + e^{2t}\right) C[2]\right), \frac{1}{2} e^{-t} \left(\left(-1 + e^{2t}\right) C[1] + \left(1 + e^{2t}\right) C[2]\right)\right\}$$

Next, we solve and plot an initial value problem:

```
inits = {x[0] == 1, y[0] == 0};
```

```
sol = vars /. DSolve[Join[eqns, inits], vars, t][[1]]
```

$$\left\{\frac{1}{2} e^{-t} \left(1 + e^{2t}\right), \frac{1}{2} e^{-t} \left(-1 + e^{2t}\right)\right\}$$

```
Plot[sol, {t, 0, 2}, PlotStyle → {{}, Dashing[{Tiny}]}]
```

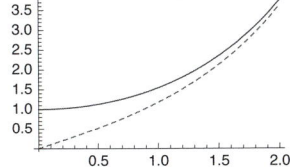

■ Example 2: Phase Trajectories

Consider the following linear system:

```
vars = {x[t], y[t]};
```

```
eqns = {x'[t] == x[t] - y[t], y'[t] == 3 x[t] - 2 y[t]};
```

```
inits = {x[0] == 10, y[0] == 0};
```

```
sol = vars /. DSolve[Join[eqns, inits], vars, t] [[1]]
```

$$\left\{ 10 \, e^{-t/2} \left(\cos\left[\frac{\sqrt{3} \ t}{2}\right] + \sqrt{3} \ \sin\left[\frac{\sqrt{3} \ t}{2}\right] \right), \ 20 \sqrt{3} \ e^{-t/2} \sin\left[\frac{\sqrt{3} \ t}{2}\right] \right\}$$

Plot the solution:

```
Plot[sol, {t, 0, 10}, PlotStyle → {{}, Dashing[{Tiny}]}]
```

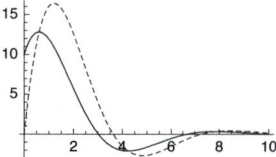

With **ParametricPlot**, we can plot phase trajectories for equations with two dependent variables. A plot of this kind describes how the point $(x(t), y(t))$ moves on the (x, y) plane:

```
ParametricPlot[sol, {t, 0, 30}, PlotRange → All, AxesLabel → {x, y}]
```

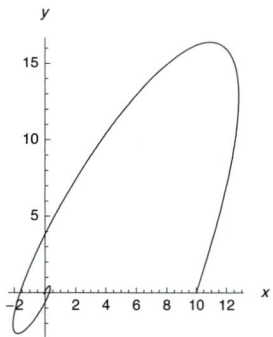

The curve approaches the point (0, 0) like a spiral (origin is a stable focus). Plotting individual points shows the speed of the point as it moves on a curve:

```
p = Table[sol, {t, 0, 20, 0.2}];
```

```
ListPlot[p, PlotRange → All, AxesLabel → {x, y}, AspectRatio → Automatic]
```

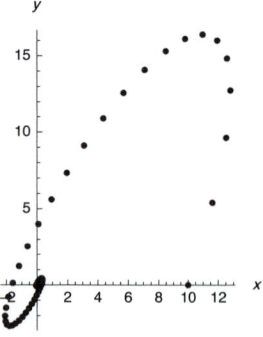

For a linear matrix system $y' = A \, y$, the origin is an equilibrium point. The nature of this point can be seen from the eigenvalues of the coefficient matrix A:

```
Eigenvalues[{{1, -1}, {3, -2}}]
```
$$\left\{(-1)^{2/3}, -(-1)^{1/3}\right\}$$

```
% // ComplexExpand
```
$$\left\{-\frac{1}{2} + \frac{i\sqrt{3}}{2}, -\frac{1}{2} - \frac{i\sqrt{3}}{2}\right\}$$

Because the eigenvalues are complex with a negative real part, origin is a stable focus.

■ Example 3: A Set of Trajectories

Plotting a direction field gives an impression about how the trajectories behave:

```
<< VectorFieldPlots`
```

```
VectorFieldPlot[{x - y, 3 x - 2 y}, {x, -20, 20},
  {y, -20, 20}, Axes → True, PlotPoints → 11, ImageSize → 140]
```

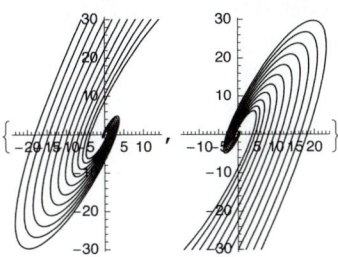

However, plotting several trajectories from different starting points gives a better description of the behavior of the system. We consider the system of Example 2 and give general initial values **x0** and **y0**:

```
inits = {x[0] == x0, y[0] == y0};
```

```
sol = vars /. DSolve[Join[eqns, inits], vars, t][[1]] // Simplify
```

$$\left\{\frac{1}{3} e^{-t/2}\left(3\,x0\,\cos\left[\frac{\sqrt{3}\,t}{2}\right] + \sqrt{3}\,(3\,x0 - 2\,y0)\,\sin\left[\frac{\sqrt{3}\,t}{2}\right]\right),\right.$$

$$\left. e^{-t/2}\left(y0\,\cos\left[\frac{\sqrt{3}\,t}{2}\right] + \sqrt{3}\,(2\,x0 - y0)\,\sin\left[\frac{\sqrt{3}\,t}{2}\right]\right)\right\}$$

When **y0** is 30, we let **x0** vary from −6 to 14 in steps of 2; when **y0** is −30, we let **x0** vary from −14 to 6 in steps of 2:

```
solset1 = Table[sol, {x0, -6, 14, 2}];
solset2 = Table[sol, {x0, -14, 6, 2}];
{p1 = ParametricPlot[Evaluate[solset1 /. y0 → 30],
   {t, 0, 15}, PlotStyle → Black, PlotRange → All, ImageSize → 70],
 p2 = ParametricPlot[Evaluate[solset2 /. y0 → -30], {t, 0, 15},
   PlotStyle → Black, PlotRange → All, ImageSize → 70]}
```

Combine the plots:

Show[p1, p2, ImageSize → 130]

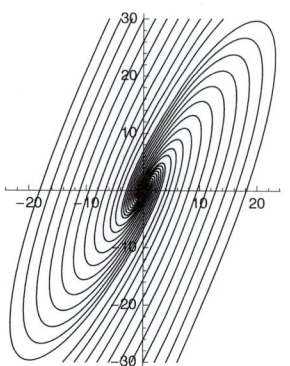

■ **Example 4: Three Equations**

Let $x(t)$, $y(t)$, and $z(t)$ stand for the amount of lead in blood, tissues, and bones, respectively; time is measured in days. Define $v = (x, y, z)^\text{T}$. Borrelli and Coleman (1998, p. 339) present the model $v' = A\,v + b$, in which the coefficients are as follows:

A = {{-0.0361, 0.0124, 0.000035}, {0.0111, -0.0286, 0}, {0.0039, 0, -0.000035}};
b = {49.3, 0, 0};

The equilibrium point is the v that satisfies $A\,v + b = 0$; that is, $A\,v = -b$:

LinearSolve[A, -b]   {1800.1, 698.639, 200 582.}

To solve the system of differential equations, define the variables and the equations:

vars = {x[t], y[t], z[t]};

eqns = Thread[D[vars, t] == A.vars + b]
{x′[t] == 49.3 − 0.0361 x[t] + 0.0124 y[t] + 0.000035 z[t],
  y′[t] == 0.0111 x[t] − 0.0286 y[t], z′[t] == 0.0039 x[t] − 0.000035 z[t]}
inits = {x[0] == 0, y[0] == 0, z[0] == 0};

Time is measured in days. Then solve the system:

(sol = vars /. DSolve[Join[eqns, inits], vars, t] // Expand // Chop) // Column

{1800.1 − 719.885 e^{−0.0446688 t} − 855.314 e^{−0.0200356 t} − 224.898 e^{−0.0000306322 t},
  698.639 + 497.283 e^{−0.0446688 t} − 1108.54 e^{−0.0200356 t} − 87.3791 e^{−0.0000306322 t},
  200 582. + 62.902 e^{−0.0446688 t} + 166.781 e^{−0.0200356 t} − 200 812. e^{−0.0000306322 t}}

(Here, we applied **Chop** to get rid of some negligible terms that originated from rounding errors.) The solution can be written as follows:

$$
\begin{pmatrix} x(t) \\ y(t) \\ z(t) \end{pmatrix} = \begin{pmatrix} 1800.1 \\ 698.639 \\ 200\,582. \end{pmatrix} + \begin{pmatrix} -719.885 & -855.314 & -224.898 \\ 497.283 & -1108.54 & -87.3791 \\ 62.902 & 166.781 & -200\,812. \end{pmatrix} \begin{pmatrix} e^{-0.0446688\,t} \\ e^{-0.0200356\,t} \\ e^{-0.0000306322\,t} \end{pmatrix}.
$$

The solution is displayed as follows:

```
Plot[sol, {t, 0, 400},
  PlotStyle → {Black, {Black, Dashing[{Tiny}]}, {Black, Thickness[Medium]}}]
```

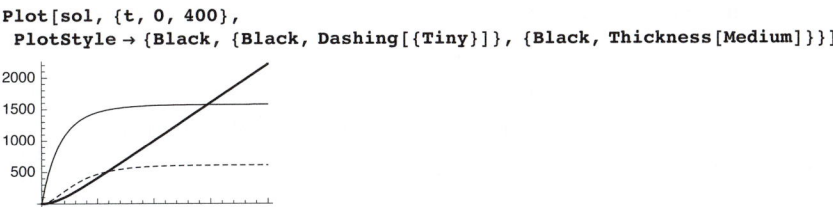

The amounts of lead in blood and tissues reach the equilibrium quite rapidly (in about a year), but it would take a very long time (several hundred years!) before the amount of lead in bones would reach the equilibrium:

```
Plot[sol[[1, 3]], {t, 0, 200 000}, PlotStyle → {Black, Thickness[Medium]},
  Ticks → {{100 000, 200 000}, Automatic}]
```

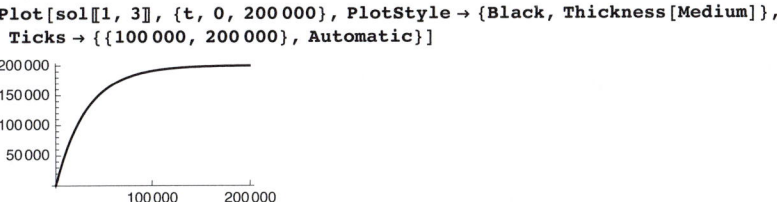

■ Example 5: Differential-Algebraic Equations

We can solve differential-algebraic equations. Here is an example:

```
DSolve[{x'[t] == 2 x[t] + y[t], x[t] + y[t] == 1, x[0] == 1}, {x[t], y[t]}, t]
```

$$\left\{\left\{x[t] \rightarrow -1 + 2\,e^t,\ y[t] \rightarrow -2\left(-1 + e^t\right)\right\}\right\}$$

26.2 More about Symbolic Solutions

26.2.1 Using the Laplace Transform

■ One Equation

We try to solve, using the Laplace transform (see Section 20.4.1, p. 670), the following problem:

```
eqn = y''[t] + y[t] == t;
```

```
inits = {y[0] → α, y'[0] → β};
```

First, take the Laplace transform of the equation:

```
LaplaceTransform[eqn, t, s]
```

$$\text{LaplaceTransform}[y[t], t, s] + s^2\,\text{LaplaceTransform}[y[t], t, s] - s\,y[0] - y'[0] == \frac{1}{s^2}$$

Insert the initial values:

```
lapeqn = % /. inits
```

$$-s\,\alpha - \beta + \text{LaplaceTransform}[y[t], t, s] + s^2\,\text{LaplaceTransform}[y[t], t, s] == \frac{1}{s^2}$$

From this equation, solve the transform:

```
lap = Solve[lapeqn, LaplaceTransform[y[t], t, s]]
```

$$\left\{\left\{\text{LaplaceTransform}[y[t], t, s] \to \frac{1 + s^3 \alpha + s^2 \beta}{s^2 (1 + s^2)}\right\}\right\}$$

Lastly, take the inverse transform, which is then the solution to the initial value problem:

```
sol = InverseLaplaceTransform[lap, s, t]
```

$$\{\{y[t] \to t + \alpha \text{Cos}[t] - \text{Sin}[t] + \beta \text{Sin}[t]\}\}$$

The same solution is obtained with **DSolve**:

```
DSolve[{eqn, y[0] == α, y'[0] == β}, y[t], t]
```

$$\{\{y[t] \to t + \alpha \text{Cos}[t] - \text{Sin}[t] + \beta \text{Sin}[t]\}\}$$

■ Simultaneous Equations

Now we solve the same initial value problem as was seen in Example 2 of Section 26.1.3, p. 837:

```
eqns = {y'[t] == y[t] - z[t], z'[t] == 3 y[t] - 2 z[t]};
```

```
inits = {y[0] → 10, z[0] → 0};
```

```
lapeqns = LaplaceTransform[eqns, t, s] /. inits
```

$$\{-10 + s \text{LaplaceTransform}[y[t], t, s] ==$$
$$\text{LaplaceTransform}[y[t], t, s] - \text{LaplaceTransform}[z[t], t, s],$$
$$s \text{LaplaceTransform}[z[t], t, s] ==$$
$$3 \text{LaplaceTransform}[y[t], t, s] - 2 \text{LaplaceTransform}[z[t], t, s]\}$$

```
lap = Solve[lapeqns, {LaplaceTransform[y[t], t, s], LaplaceTransform[z[t], t, s]}]
```

$$\left\{\left\{\text{LaplaceTransform}[y[t], t, s] \to \frac{10 (2 + s)}{1 + s + s^2}, \text{LaplaceTransform}[z[t], t, s] \to \frac{30}{1 + s + s^2}\right\}\right\}$$

```
sol = InverseLaplaceTransform[lap, s, t]
```

$$\left\{\left\{y[t] \to 10 \, e^{-t/2} \left(\text{Cos}\left[\frac{\sqrt{3}\,t}{2}\right] + \sqrt{3} \, \text{Sin}\left[\frac{\sqrt{3}\,t}{2}\right]\right), z[t] \to 20 \sqrt{3} \, e^{-t/2} \, \text{Sin}\left[\frac{\sqrt{3}\,t}{2}\right]\right\}\right\}$$

The solution is the same as the one obtained in Section 26.1.3.

■ The Step Function

Consider the following equation:

```
eqn = y'[t] == r y[t] - h UnitStep[30 - t];
```

It describes an exponential growth in which a harvesting of h units per time unit occurs for $0 \le t \le 30$. Apply the Laplace transform (this example was also solved with **DSolve** in Example 9 of Section 26.1.1, p. 834):

```
lapeqn = LaplaceTransform[eqn, t, s] /. y[0] → α
```

$$-\alpha + s \text{LaplaceTransform}[y[t], t, s] == -\frac{(1 - e^{-30 s}) h}{s} + r \text{LaplaceTransform}[y[t], t, s]$$

```
lap = Solve[lapeqn, LaplaceTransform[y[t], t, s]]
```

$$\left\{\left\{\text{LaplaceTransform}[y[t], t, s] \to \frac{e^{-30 s} (h - e^{30 s} h + e^{30 s} s \alpha)}{s (-r + s)}\right\}\right\}$$

```
lapsol = y[t] /. InverseLaplaceTransform[lap, s, t][[1]]
```

$$h - e^{r\,t}\,h + e^{r\,t}\,r\,\alpha + \left(-1 + e^{r\,(-30+t)}\right) h\,\text{HeavisideTheta}\,[-30 + t]$$
$$\overline{\phantom{h - e^{r\,t}\,h + e^{r\,t}\,r\,\alpha + \left(-1\right)}}$$
$$r$$

```
Plot[lapsol /. {α → 100, r → 0.01, h → 2}, {t, 0, 100}]
```

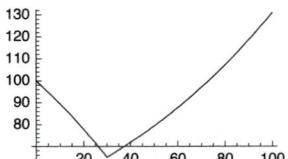

26.2.2 Series Solutions

■ Infinite Number of Terms

We have already considered series solutions in Section 19.2.3, p. 629. Now we consider the equation $y''(t) - 2\,t\,y'(t) - 2\,y(t) = 0$. Let us try to find the solution in the form $y(t) = \sum_{i=0}^{\infty} a_i\,t^i$ by inserting this into the equation. First, note the following:

$$t\,y'(t) = \sum_{i=1}^{\infty} i\,a_i\,t^i = \sum_{i=0}^{\infty} i\,a_i\,t^i, \quad y''(t) = \sum_{i=2}^{\infty} i(i-1)\,a_i\,t^{i-2} = \sum_{i=0}^{\infty} (i+2)\,(i+1)\,a_{i+2}\,t^i.$$

The equation then becomes $\sum_{i=0}^{\infty}\left[(i+2)\,(i+1)\,a_{i+2} - 2\,i\,a_i - 2\,a_i\right] t^i = 0$. Thus, the coefficients a_i satisfy the difference equation $(i+2)\,(i+1)\,a_{i+2} - 2\,(i+1)\,a_i = 0$; that is, $a_i = \frac{2}{i}\,a_{i-2}$. In addition, $y(0) = a_0$ and $y'(0) = a_1$.

Assume first that $y(0) = 1$ and $y'(0) = 0$ so that $a_0 = 1$ and $a_1 = 0$. It is easy to see that the solution of $a_i = \frac{2}{i}\,a_{i-2}$ is $a_i = 1 \big/ \left(\frac{i}{2}\right)!$ for i even and $a_i = 0$ for i odd. This can also be seen with **RSolve** (see Section 28.1.1, p. 924):

```
RSolve[{a[i] == 2 / i a[i - 2], a[0] == 1, a[1] == 0}, a[i], i]
```

$$\left\{\left\{a[i] \to \frac{1 + (-1)^i}{2\,\frac{i}{2}\,!}\right\}\right\}$$

Thus, we get the series solution $y(t) = \sum_{i\,\text{even}} t^i \big/ (i/2)!$. The value of the sum is as follows:

```
sol = Sum[t^i / (i / 2) !, {i, 0, ∞, 2}]        e^{t²}
```

Thus, $y(t) = e^{t^2}$ is the solution of the problem. **DSolve** gives the same solution:

```
DSolve[{y''[t] - 2 t y'[t] - 2 y[t] == 0, y[0] == 1, y'[0] == 0}, y[t], t]    {{y[t] → e^{t²}}}
```

Assume then that $y(0) = 0$ and $y'(0) = 1$ so that $a_0 = 0$ and $a_1 = 1$. From $a_i = \frac{2}{i}\,a_{i-2}$, it is easy to see that $a_{2\,i+1} = 2^i \big/ (2\,i + 1)!!$, $i = 0, 1, 2, \ldots$, and other a_i values are zero. **RSolve** gives the solution in another form:

```
RSolve[{a[i] == 2 / i a[i - 2], a[0] == 0, a[1] == 1}, a[i], i]
```

$$\left\{\left\{a[i] \to -\frac{\left(-1 + (-1)^i\right)\sqrt{\pi}}{4\,\frac{i}{2}\,!}\right\}\right\}$$

Therefore, we know the series expansion of $y(t)$, and *Mathematica* is able to calculate the infinite sum:

```
sol = Simplify[Sum[2^i / (2 i + 1) !! t^(2 i + 1), {i, 0, ∞}], t > 0]
```

$$\frac{1}{2} e^{t^2} \sqrt{\pi} \, \text{Erf}[t]$$

DSolve gives the same solution:

```
DSolve[{y''[t] - 2 t y'[t] - 2 y[t] == 0, y[0] == 0, y'[0] == 1}, y[t], t]
```

$$\left\{\left\{y[t] \rightarrow \frac{1}{2} e^{t^2} \sqrt{\pi} \, \text{Erf}[t]\right\}\right\}$$

The general solution of the equation is a linear combination of the two solutions:

```
DSolve[y''[t] - 2 t y'[t] - 2 y[t] == 0, y[t], t]
```

$$\left\{\left\{y[t] \rightarrow e^{t^2} C[2] + \frac{1}{2} e^{t^2} \sqrt{\pi} \, C[1] \, \text{Erf}[t]\right\}\right\}$$

If we are satisfied with a finite series, we can get an approximate solution, as is seen in the next two examples.

■ A Finite Number of Terms 1

We continue with the preceding example and calculate an approximate solution as a finite sum by using the formula $a_{2i+1} = 2^i/(2i+1)!!$:

```
apprsol = Sum[2^i / (2 i + 1) !! t^(2 i + 1), {i, 0, 5}]
```

$$t + \frac{2 t^3}{3} + \frac{4 t^5}{15} + \frac{8 t^7}{105} + \frac{16 t^9}{945} + \frac{32 t^{11}}{10395}$$

We can also directly use the difference equation $a_i = \frac{2}{i} a_{i-2}$ of the coefficients by first defining the equation:

```
a[0] = 0; a[1] = 1;
a[i_] := a[i] = 2 a[i - 2] / i
```

Then we calculate a finite sum:

```
apprsol = Sum[a[i] t^i, {i, 0, 12}]
```

$$t + \frac{2 t^3}{3} + \frac{4 t^5}{15} + \frac{8 t^7}{105} + \frac{16 t^9}{945} + \frac{32 t^{11}}{10395}$$

Before we continue, we have to remove the definition of the coefficients:

```
Remove[a]
```

■ A Finite Number of Terms 2

If we do not know the difference equation of the coefficients, we can proceed as follows. First, we define the differential equation and the initial conditions (it is now advantageous to use **&&** rather than a list, because later we will use **LogicalExpand**):

```
eqn = y''[t] - 2 t y'[t] - 2 y[t] == 0 && y[0] == 0 && y'[0] == 1;
```

Before we find an approximate series solution, we show the solution given by **DSolve**:

```
sol = y[t] /. DSolve[eqn, y[t], t][[1]]
```

$$\frac{1}{2} e^{t^2} \sqrt{\pi} \, \text{Erf}[t]$$

Then we try to find an approximate solution in the form of a 12th-degree power series:

```
y[t_] = Sum[a[i] t^i, {i, 0, 12}] + O[t]^13
```

$a[0] + a[1] t + a[2] t^2 + a[3] t^3 + a[4] t^4 + a[5] t^5 + a[6] t^6 +$
$a[7] t^7 + a[8] t^8 + a[9] t^9 + a[10] t^{10} + a[11] t^{11} + a[12] t^{12} + O[t]^{13}$

The equation is now as follows:

```
eqn
```

$(-2 a[0] + 2 a[2]) + (-4 a[1] + 6 a[3]) t + (-6 a[2] + 12 a[4]) t^2 + (-8 a[3] + 20 a[5]) t^3 +$
$(-10 a[4] + 30 a[6]) t^4 + (-12 a[5] + 42 a[7]) t^5 + (-14 a[6] + 56 a[8]) t^6 +$
$(-16 a[7] + 72 a[9]) t^7 + (-18 a[8] + 90 a[10]) t^8 + (-20 a[9] + 110 a[11]) t^9 +$
$(-22 a[10] + 132 a[12]) t^{10} + O[t]^{11} == 0 \&\& a[0] == 0 \&\& a[1] == 1$

We could find the conditions under which the equation is true:

```
cond = LogicalExpand[eqn]
```

$a[0] == 0 \&\& a[1] == 1 \&\& -2 a[0] + 2 a[2] == 0 \&\& -4 a[1] + 6 a[3] == 0 \&\&$
$-6 a[2] + 12 a[4] == 0 \&\& -8 a[3] + 20 a[5] == 0 \&\& -10 a[4] + 30 a[6] == 0 \&\&$
$-12 a[5] + 42 a[7] == 0 \&\& -14 a[6] + 56 a[8] == 0 \&\& -16 a[7] + 72 a[9] == 0 \&\&$
$-18 a[8] + 90 a[10] == 0 \&\& -20 a[9] + 110 a[11] == 0 \&\& -22 a[10] + 132 a[12] == 0$

Then we could solve the conditions with **Solve[cond]**. Actually, **Solve** can be applied directly to **eqn**:

```
aa = Solve[eqn]
```

$\left\{\left\{a[0] \to 0, a[1] \to 1, a[2] \to 0, a[3] \to \frac{2}{3}, a[4] \to 0, a[5] \to \frac{4}{15}, a[6] \to 0, \right.\right.$
$\left.\left. a[7] \to \frac{8}{105}, a[8] \to 0, a[9] \to \frac{16}{945}, a[10] \to 0, a[11] \to \frac{32}{10\,395}, a[12] \to 0\right\}\right\}$

The corresponding approximate series solution of the original equation is as follows:

```
apprsol = y[t] /. aa[[1]]
```

$t + \frac{2 t^3}{3} + \frac{4 t^5}{15} + \frac{8 t^7}{105} + \frac{16 t^9}{945} + \frac{32 t^{11}}{10\,395} + O[t]^{13}$

This is exactly the same as the series expansion of the exact solution:

```
Series[sol, {t, 0, 12}]
```

$t + \frac{2 t^3}{3} + \frac{4 t^5}{15} + \frac{8 t^7}{105} + \frac{16 t^9}{945} + \frac{32 t^{11}}{10\,395} + O[t]^{13}$

From a figure, we see that the approximate solution is quite good near the origin:

```
Plot[Evaluate[{sol, apprsol // Normal}], {t, 0, 2}, PlotStyle → {{}, Dashing[{Tiny}]}]
```

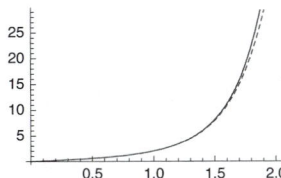

```
Remove["Global`*"]
```

26.2.3 Solution as a Pure Function

---

sol = y /. DSolve[eqn, y, t]   Give the solution in the form of a pure function

sol[a]   Calculate the value of the solution at **a**

eqn /. y → sol   Check the solution

Plot[sol[t], {t, a, b}]   Plot the solution

Thus far, we have used **DSolve** in the form **DSolve[eqn, y[t], t]**. However, **DSolve** also has another form, in which the dependent variable is not declared as **y[t]** but as **y**. Thus, we can also write **DSolve[eqn, y, t]**. The solution is then expressed as a function—specifically, a pure function (see Section 17.1.4, p. 520). As an example, we define an equation:

  eqn = {y'[t] == y[t] + t + 1, y[0] == 1};

We calculate the solution using both ways:

  DSolve[eqn, y[t], t] $\{\{y[t] \to -2 + 3\,e^t - t\}\}$

  DSolve[eqn, y, t] $\{\{y \to \text{Function}[\{t\}, -2 + 3\,e^t - t]\}\}$

It may seem as if the latter form of solution would not be as good as the former. In fact, the latter form has a clear advantage in that we can use the solution similarly as we can other functions. Next, we compare the methods in more detail.

First, we declare the dependent variable to be **y[t]**:

  sol = y[t] /. DSolve[eqn, y[t], t][[1]] $-2 + 3\,e^t - t$

A value calculation, differentiation, and checking of the correctness of the solution are done as follows:

  sol /. t → 2 $-4 + 3\,e^2$

  D[sol, t] $-1 + 3\,e^t$

  eqn /. {y[t] → sol, y'[t] → D[sol, t], y[0] → sol /. t → 0}

  {True, True}

When checking correctness, we inserted the solution into the equation and into the initial condition. The result of the checking shows that both the equation and the initial condition are satisfied.

Then we declare the dependent variable to be **y**:

  sol = y /. DSolve[eqn, y, t][[1]] $\text{Function}[\{t\}, -2 + 3\,e^t - t]$

Now we can use **sol** in the same way as any other function of *Mathematica*, such as **Sin** or **Exp**. We calculate a value and derivative and check the correctness of the solution as follows:

  sol[2] $-4 + 3\,e^2$

  sol'[t] $-1 + 3\,e^t$

  eqn /. y → sol   {True, True}

The checking in particular is very easy with this latter method.

In Example 7 of Section 26.1.1, p. 833, we encountered an implicit solution:

```
Off[InverseFunction::ifun]
sol = DSolve[ y'[t] - 2 t y[t]^2 - y[t]^3 == 0, y, t]
```
Solve::tdep : The equations appear to involve

the variables to be solved for in an essentially non–algebraic way. ≫

$$\text{Solve}\left[\frac{t\,\text{AiryAi}\left[t^2+\frac{1}{y[t]}\right]+\text{AiryAiPrime}\left[t^2+\frac{1}{y[t]}\right]}{t\,\text{AiryBi}\left[t^2+\frac{1}{y[t]}\right]+\text{AiryBiPrime}\left[t^2+\frac{1}{y[t]}\right]}+C[1]==0,\,y[t]\right]$$

To check this solution, do as follows:

```
Solve[D[sol[[1]], t], y'[t]]      {{y'[t] → y[t]² (2 t + y[t])}}
```

We see that the solution satisfies the original equation.

26.2.4 Integral and Integro-Differential Equations

■ Volterra Integral Equation

The following equation is a Volterra integral equation of the second kind:

```
eqn = y[t] == t^2 + Integrate[Sin[t - u] y[u], {u, 0, t}]
```

$$y[t] == t^2 + \int_0^t \text{Sin}[t-u]\, y[u]\, du$$

This can be solved using Laplace transform:

```
Solve[LaplaceTransform[eqn, t, s], LaplaceTransform[y[t], t, s]]
```

$$\left\{\left\{\text{LaplaceTransform}[y[t],\,t,\,s]\to\frac{2\,(1+s^2)}{s^5}\right\}\right\}$$

```
sol = y[t] /. InverseLaplaceTransform[%, s, t][[1]] // Simplify
```

$$t^2+\frac{t^4}{12}$$

We perform a check:

```
eqn /. {y[t] → sol, y[u] → (sol /. t → u)}      True

sol =.
```

■ Fredholm Integral Equation

The next equation is a Fredholm equation of the second kind:

```
eqn = y[t] == 2 + 3 t + 4 Integrate[(1 + t u^2 + t^2 u + t^3) y[u], {u, 0, 1}]
```

$$y[t] == 2 + 3 t + 4 \int_0^1 \left(1 + t^3 + t^2 u + t u^2\right) y[u]\, du$$

We guess (from the form of the equation) that the solution is of the following form:

```
sol[t_] := 2 + 3 t + 4 (a + b t + c t^2 + d t^3)
```

We then insert the guess into the equation:

```
eqn /. y → sol
```

$$2 + 3 t + 4 \left(a + b t + c t^2 + d t^3\right) ==$$

$$2 + 3 t + 4 \left(\frac{7}{2} + 4 a + 2 b + \frac{4 c}{3} + d + \frac{17 t}{12} + \frac{4 a t}{3} + b t + \frac{4 c t}{5} + \frac{2 d t}{3} + \right.$$

$$\left. 2 t^2 + 2 a t^2 + \frac{4 b t^2}{3} + c t^2 + \frac{4 d t^2}{5} + \frac{7 t^3}{2} + 4 a t^3 + 2 b t^3 + \frac{4 c t^3}{3} + d t^3\right)$$

We solve the coefficients that satisfy the equation identically (for **SolveAlways**, see Section 22.2.2, p. 719):

```
coeff = SolveAlways [%, t]
```

$$\left\{\left\{a \to -\frac{335}{636}, b \to -\frac{167}{424}, c \to -\frac{385}{848}, d \to -\frac{335}{636}\right\}\right\}$$

Thus, the solution is as follows:

```
sol2 = sol [t] /. coeff⟦1⟧
```

$$2 + 3 t + 4 \left(-\frac{335}{636} - \frac{167 t}{424} - \frac{385 t^2}{848} - \frac{335 t^3}{636}\right)$$

We check that the solution is correct:

```
eqn /. {y[t] → sol2, y[u] → (sol2 /. t → u)}       True
```

■ Integro-Differential Equation

Define an integro-differential equation:

```
eqn = y[t] + y ' [t] == 1 - Integrate[Exp[t - u] y[u], {u, a, t}]
```

$$y[t] + y'[t] == 1 - \int_a^t e^{t-u} y[u] \, du$$

Also define the initial condition $y(a) = c$. Differentiate the equation:

```
eqn2 = D[eqn, t]       y'[t] + y''[t] == - ∫_a^t e^{t-u} y[u] du - y[t]
```

Eliminate the integral from the two equations (for **Eliminate**, see Section 22.2.2, p. 718):

```
eqn3 = Eliminate[{eqn, eqn2}, Integrate[Exp[t - u] y[u], {u, a, t}]]       y''[t] == -1
```

We arrived at a differential equation. To get a second initial condition, substitute a for t in the original equation:

```
cond = eqn /. t → a       y[a] + y'[a] == 1
```

The solution of the initial value problem is as follows:

```
sol = y /. DSolve[{eqn3, y[a] == c, cond}, y, t]⟦1⟧
```

$$\text{Function}\left[\{t\}, \frac{1}{2}\left(-2 a - a^2 + 2 c + 2 a c + 2 t + 2 a t - 2 c t - t^2\right)\right]$$

```
Collect[sol[t], t, Factor]       \frac{1}{2}\left(-2 a - a^2 + 2 c + 2 a c\right) + (1 + a - c) t - \frac{t^2}{2}
```

This solves the original equation:

```
eqn /. y → sol // Simplify       True
```

26.3 Numerical Solutions

26.3.1 One Equation

Here are typical commands that are used when we numerically solve one differential equation with initial or boundary conditions.

```
sol = y[t] /. NDSolve[{eqn, conds}, y[t], {t, a, b}][[1]]   Solve the problem
Plot[sol, {t, a, b}]   Plot the solution
```

The arguments of **NDSolve** are otherwise the same as the ones of **DSolve** except that in place of **t**, we have to define the interval **{t, a, b}** where we ask for the numerical solution.

The conditions in **{eqn, conds}** are often initial conditions, but they can also be boundary conditions. The conditions need not be given at the end points of the solution interval. For example, if you ask for the solution in (0, 3), you can state all conditions at $t = 1$ or some conditions at $t = 1$ and other conditions at $t = 2$. Generally, if the highest-order derivative of the equation is n, a total of n conditions must be given so that the solution can be computed. In the simplest case, the conditions give values of $y(a)$, $y'(a)$, ..., $y^{(n-1)}(a)$.

NDSolve normally uses an adaptive Adams predictor–corrector method; the methods are considered in more detail in Section 26.4.1, p. 865. The basic principle of the solution method is that the solution is computed at a finite set of t points, and piecewise interpolation is used to calculate the values of the solution at other points. Note that options of **NDSolve** are considered in Section 26.4.1, p. 865.

■ **Example 1: A First-Order Initial Value Problem**

We solve a first-order nonlinear equation:

```
eqn = 1000 y[t] y'[t] == 160 - 3 y[t]^3;
sol = NDSolve[{eqn, y[0] == 0.00001}, y[t], {t, 0, 150}]
{{y[t] → InterpolatingFunction[{{0., 150.}}, <>][t]}}
```

The solution is an **InterpolatingFunction** object (see Section 24.2.1, p. 797). The object represents the solution as a collection of cubic interpolating polynomials. Only the interval of the solution is shown (**{0., 150.}** here); all other information is hidden behind the marks <>.

It may be preferable to ask directly for the value of **y[t]**:

```
sol = y[t] /. NDSolve[{eqn, y[0] == 0.00001}, y[t], {t, 0, 150}][[1]]
InterpolatingFunction[{{0., 150.}}, <>][t]
```

The solution looks like this:

```
Plot[sol, {t, 0, 150}, AxesOrigin → {0, 0}]
```

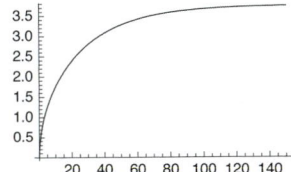

Next, we can ask for the value of the solution at a point and tabulate the solution at a set of points:

```
sol /. t → 100     3.6752

Table[{t, sol}, {t, 1, 4}]
{{1, 0.565302}, {2, 0.798468}, {3, 0.976357}, {4, 1.12527}}
```

We can plot the solution for several initial values:

```
solset = Table[
    y[t] /. NDSolve[{eqn, y[0] == y0}, y[t], {t, 0, 150}], {y0, 0.00001, 6.00001, 0.5}];
Plot[solset, {t, 0, 150}, PlotRange → All, PlotStyle → Black]
```

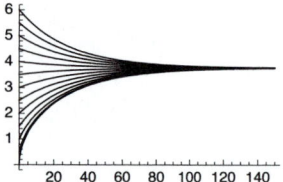

■ Example 2: A Second-Order Initial Value Problem

Consider the following second-order equation (the first Painlevé transcendent):

```
eqn = y''[t] == y[t]^2 - t;
```

Solve the equation and plot the solution and its derivative, both as functions of t and as a parametric plot:

```
sol = y[t] /. NDSolve[{eqn, y[0] == 0, y'[0] == -3.4}, y[t], {t, 0, 12}][[1]]
InterpolatingFunction[{{0., 12.}}, <>][t]

Plot[Evaluate[{sol, D[sol, t]}], {t, 0, 12}]
```

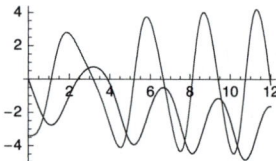

```
ParametricPlot[Evaluate[{sol, D[sol, t]}], {t, 0, 12}, ImageSize → 80]
```

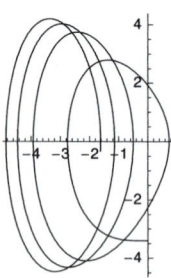

Solve the equation when $y(0) = 0$ and $y'(0)$ varies from -3.4 to 1 in steps of 0.25:

```
solset = Table[
    y[t] /. NDSolve[{eqn, y[0] == 0, y'[0] == d0}, y[t], {t, 0, 12}], {d0, -3.4, 1, 0.25}];
```

```
Plot[solset, {t, 0, 12}, PlotStyle → Black, ImageSize → 180]
```

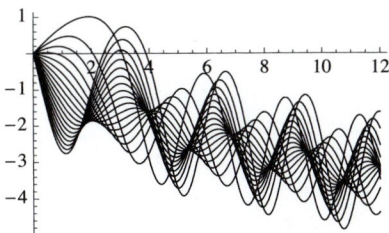

■ Example 3: A Second-Order Boundary Value Problem

Now we give boundary conditions:

```
eqn = y''[t] == y'[t] - t y[t];

sol = y[t] /. NDSolve[{eqn, y[0] == 2, y[4] == 2}, y[t], {t, 0, 4}];

Plot[sol, {t, 0, 4}]
```

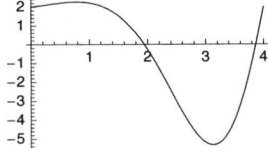

Boundary value problems are also considered in Section 26.4.3, p. 874.

■ Example 4: Several Solutions

More than one solution can be obtained:

```
sol = y[t] /. NDSolve[{y'[t]^2 == t, y[0] == 0}, y[t], {t, 0, π}]
```

```
{InterpolatingFunction[{{0., 3.14159}}, <>][t],
 InterpolatingFunction[{{0., 3.14159}}, <>][t]}
```

```
Plot[sol, {t, 0, π}]
```

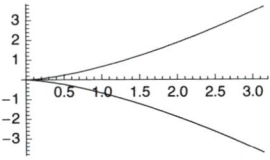

■ Example 5: Piecewise-Defined Equations

Consider the following equation:

```
eqn = y'[t] == 0.01 y[t] - 2 UnitStep[30 - t];
```

It describes an exponential growth in which a harvesting of *h* units per time unit occurs for $0 \le t \le 30$. In place of `2 UnitStep[30 - t]`, we could write `If[t ≤ 30, 2, 0]`. We have already solved the equation with `DSolve` (see Section 26.1.1, p. 834) and with the Laplace transform (see Section 26.2.1, p. 842), but now we solve the equation numerically:

```
sol = y[t] /. NDSolve[{eqn, y[0] == 100}, y[t], {t, 0, 100}];
```

```
Plot[sol, {t, 0, 100}]
```

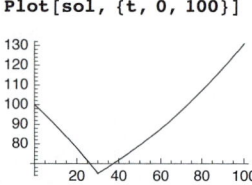

■ **Solution as a Pure Function**

`sol = y /. NDSolve[{eqn, conds}, y, {t, a, b}][[1]]`   Give the solution in the form of a pure function

`sol[a]`   Calculate the value of the solution at **a**

`Plot[sol[t], {t, a, b}]`   Plot the solution

Thus far, we have used **NDSolve** in the form **DSolve[{eqn, conds}, y[t], {t, a, b}]**. **NDSolve**, like **DSolve** (see Section 26.2.3, p. 846), also has another form in which the dependent variable is not declared as **y[t]** but as **y**. The solution is then expressed as a function, more exactly as a pure function. As an example, define an equation:

```
eqn = y'[t] == -t y[t] + Exp[-t];
```

Ask for the solution as a pure function:

```
sol = y /. NDSolve[{eqn, y[0] == 1}, y, {t, 0, 5}][[1]]
InterpolatingFunction[{{0., 5.}}, <>]
```

Now **sol** can be used like other functions. We calculate a value and plot the solution:

```
sol[1]      1.04613
```

```
Plot[sol[t], {t, 0, 5}]
```

■ **Delay Differential Equations**

For delay differential equations, see library.wolfram.com/database/MathSource/725. There, you can download a package defining **NDelayDSolve**.

26.3.2 Two Equations

`vars = {x[t], y[t]}`   Define the dependent variables

`eqns = {eqn1, eqn2}`   Define the differential equations

`inits = {x[a] == α, y[a] == β}`   Define the initial (or boundary) conditions

`sol = vars /. NDSolve[Join[eqns, inits], vars, {t, a, b}][[1]]`   Solve the problem

`Plot[sol, {t, a, b}]`   Plot **x[t]** and **y[t]**

`ParametricPlot[sol, {t, a, b}]`   Plot a phase trajectory

We present two examples of systems of two nonlinear differential equations. For more information about the examples, see Borrelli and Coleman (1998, pp. 276–298). For more information about studying systems of nonlinear equations with *Mathematica*, see Murrell (1994).

■ A Predator–Prey Model

Define the predator–prey model by Lotka and Volterra:

```
vars = {x[t], y[t]};

f = x[t] (p - q y[t]);
g = y[t] (-P + Q x[t]);
eqns = {x'[t] == f, y'[t] == g}
{x'[t] == x[t] (p - q y[t]), y'[t] == (-P + Q x[t]) y[t]}
```

The variables $x(t)$ and $y(t)$ are magnitudes of the prey and predator, respectively, at time t.

First, we calculate the equilibrium points. At these points, the derivatives are simultaneously zero:

```
equi = Solve[{f == 0, g == 0}, vars]
```

$$\left\{\{x[t] \to 0, y[t] \to 0\}, \left\{x[t] \to \frac{P}{Q}, y[t] \to \frac{p}{q}\right\}\right\}$$

To check the nature of these points, calculate the Jacobian of **f** and **g** (see Section 19.1.2, p. 619):

```
jac = D[{f, g}, {vars}]
{{p - q y[t], -q x[t]}, {Q y[t], -P + Q x[t]}}
```

The eigenvalues of the Jacobian at the equilibrium points are as follows:

```
eig = Eigenvalues[jac /. #] & /@ equi
```

$$\left\{\{p, -P\}, \left\{-i \sqrt{p} \sqrt{P}, i \sqrt{p} \sqrt{P}\right\}\right\}$$

Thus, if p and P are positive, the equilibrium point $(0, 0)$ is a saddle point and the point $\left(\frac{P}{Q}, \frac{p}{q}\right)$ a center or a focus.

Consider the following numerical values for the constants:

```
p = 3.0; q = 2;
P = 2.5; Q = 1;
```

The equilibrium points are now as follows:

```
equi
{{x[t] → 0, y[t] → 0}, {x[t] → 2.5, y[t] → 1.5}}
```

Calculate a solution of the system:

```
inits = {x[0] == 2.5, y[0] == 0.5};

sol = vars /. NDSolve[Join[eqns, inits], vars, {t, 0, 10}][[1]]
{InterpolatingFunction[{{0., 10.}}, <>][t],
  InterpolatingFunction[{{0., 10.}}, <>][t]}
```

We plot this solution in three ways. First, we plot $x(t)$ and $y(t)$ as functions of t:

```
Plot[sol, {t, 0, 10}, PlotStyle → {{Black}, {Black, Dashing[{Tiny}]}},
  AspectRatio → 0.2, ImageSize → 300]
```

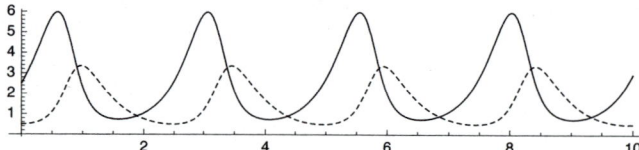

The sizes of the populations vary cyclically. Then we plot a phase trajectory in the (x, y) plane:

```
ParametricPlot[sol, {t, 0, 2.5},
  AxesLabel → {"prey", "predator"}, AxesOrigin → {0, 0}]
```

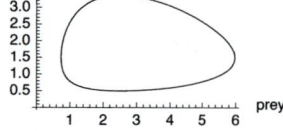

The trajectories seem to be closed curves. Lastly, we plot the phase trajectory as a collection of points:

```
ListPlot[Table[sol, {t, 0, 2.5, 0.1}], AxesLabel → {"prey", "predator"}]
```

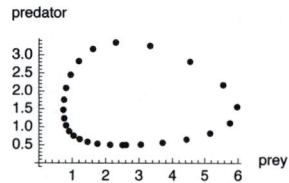

This figure gives an impression of the speed of motion in the plane.

A direction field gives an impression of the behavior of the model in an area:

```
<< VectorFieldPlots`
```

```
VectorFieldPlot[{f, g}, {x[t], 0, 12},
  {y[t], 0, 6}, Axes → True, PlotPoints → 11, ImageSize → 250]
```

A collection of phase trajectories is even more illustrative. We calculate solutions when $x(0) = 2.5$ and $y(0)$ gets a set of values:

```
solset = Table[vars /. NDSolve[Join[eqns, {x[0] == 2.5, y[0] == y0}], vars, {t, 0, 3.1}],
  {y0, 0.1, 1.5, 0.2}];
```

```
ParametricPlot[solset, {t, 0, 3.1}, PlotStyle → Black,
 PlotRange → All, AxesLabel → {"prey", "predator"}, ImageSize → 300,
 Epilog → {Arrowheads[0.02], Arrow[{{6.5, 3.2}, {6.0, 3.5}}]], Blue, Line[
   {{0, p / q}, {12, p / q}}], Line[{{P / Q, 0}, {P / Q, 6.5}}], Red, Point[{2.5, 1.5}]}]
```

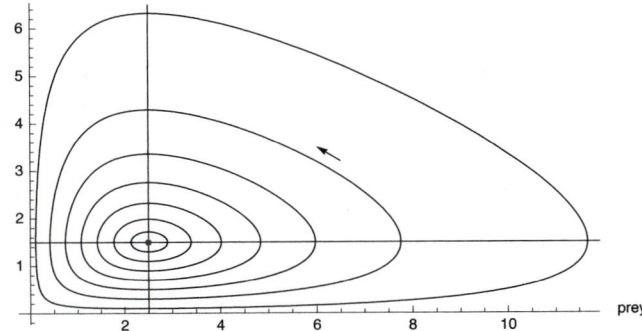

The red point is the center and the blue lines are the nullclines, which are the curves where $x'(t) = 0$ or $y'(t) = 0$. The direction of motion is counterclockwise.

In Sections 26.4.5, p. 879, and 26.4.6, p. 881, we study the predator–prey model by considering its estimation, showing a manipulation, and using a so-called equation trekker.

```
Remove["Global`*"]
```

■ Competing Species

Define the following model that describes competing species:

```
vars = {x[t], y[t]};

f = x[t] (p - q x[t] - r y[t]);
g = y[t] (P - Q x[t] - R y[t]);
eqns = {x'[t] == f, y'[t] == g};
```

The variables $x(t)$ and $y(t)$ are the population magnitudes at time t.

First, we calculate the nullclines where the derivatives $x'(t)$ and $y'(t)$ are zero:

```
h1 = y[t] /. Solve[f == 0, y[t]][[1]] /. x[t] → x
```

$$\frac{p - q x}{r}$$

```
h2 = y[t] /. Solve[g == 0, y[t]][[2]] /. x[t] → x
```

$$\frac{P - Q x}{R}$$

Then we calculate the equilibrium points:

```
equi = Solve[{f == 0, g == 0}, vars] // Simplify
```

$$\left\{ \{x[t] \to 0, \, y[t] \to 0\}, \, \left\{x[t] \to \frac{p}{q}, \, y[t] \to 0\right\}, \right.$$

$$\left\{x[t] \to \frac{P r - p R}{Q r - q R}, \, y[t] \to \frac{P q - p Q}{-Q r + q R}\right\}, \, \left\{y[t] \to \frac{P}{R}, \, x[t] \to 0\right\}\right\}$$

Calculate the Jacobian of **f** and **g**:

```
jac = D[{f, g}, {vars}]
```

$$\{\{p - 2 q x[t] - r y[t], -r x[t]\}, \{-Q y[t], P - Q x[t] - 2 R y[t]\}\}$$

The eigenvalues of the Jacobian at the equilibrium points are as follows:

```
eig = Eigenvalues[jac /. #] & /@ equi // Simplify
```

$$\left\{\{p, P\}, \left\{-p, P - \frac{pQ}{q}\right\}, \left\{-\frac{1}{2Qr - 2qR}\left(Pqr - pqR - PqR + pQR + \right.\right.\right.$$

$$\left.\sqrt{4 (Pq - pQ)(Pr - pR)(-Qr + qR) + (Pq(r - R) + p(-q + Q)R)^2}\right),$$

$$\frac{1}{2Qr - 2qR}\left(-Pqr + pqR + PqR - pQR + \right.$$

$$\left.\left.\left.\sqrt{4 (Pq - pQ)(Pr - pR)(-Qr + qR) + (Pq(r - R) + p(-q + Q)R)^2}\right)\right\}, \left\{-P, p - \frac{Pr}{R}\right\}\right\}$$

Define the following numerical values for the constants:

```
p = 2; q = 2 / 3; r = 2;
P = 2; Q = 4 / 3; R = 1;
```

The equilibrium points are now as follows:

```
equi
```

$$\left\{\{x[t] \to 0, y[t] \to 0\}, \{x[t] \to 3, y[t] \to 0\}, \left\{x[t] \to 1, y[t] \to \frac{2}{3}\right\}, \{y[t] \to 2, x[t] \to 0\}\right\}$$

Here are the eigenvalues:

```
eig
```
$$\left\{\{2, 2\}, \{-2, -2\}, \left\{-2, \frac{2}{3}\right\}, \{-2, -2\}\right\}$$

Thus, the equilibrium points are an unstable node, a stable node, a saddle point, and a stable node, respectively.

Calculate a solution of the system:

```
inits = {x[0] == 3.5, y[0] == 2};
```

```
sol = vars /. NDSolve[Join[eqns, inits], vars, {t, 0, 7}]⟦1⟧
```
```
{InterpolatingFunction[{{0., 7.}}, <>][t],
 InterpolatingFunction[{{0., 7.}}, <>][t]}
```

We plot this solution in three ways. First, we plot $x(t)$ and $y(t)$ as functions of t:

```
Plot[sol, {t, 0, 7}, PlotStyle → {{Black}, {Black, Dashing[{Tiny}]}}]
```

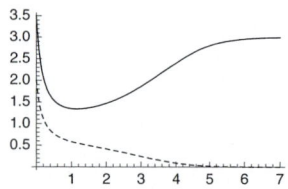

We see that the x population wins the competition and the y population vanishes; the limiting values are 3 and 0, respectively. Then we plot a phase trajectory in the (x, y) plane:

```
ParametricPlot[sol, {t, 0, 7}, PlotRange → All, AxesLabel → {x, y}]
```

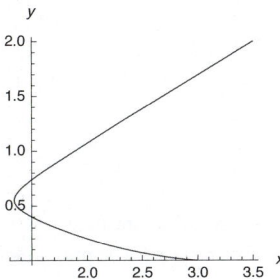

Lastly, we plot the phase trajectory as a collection of points to get an impression of the speed of motion in the plane:

```
ListPlot[Table[sol, {t, 0, 7, 0.1}], PlotRange → All,
  AxesLabel → {"x", "y"}, AspectRatio → Automatic]
```

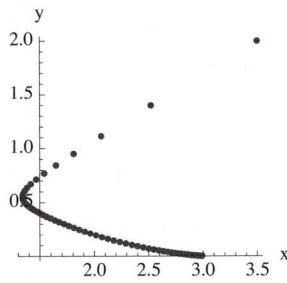

Next, we plot a direction field:

```
<< VectorFieldPlots`
```

```
VectorFieldPlot[{f, g}, {x[t], 0, 3.5},
  {y[t], 0, 2.5}, PlotPoints → 11, Axes → True, ImageSize → 250]
```

To plot a collection of phase trajectories, we first calculate solutions that go through the line $y = 1 - x$ at time $t = 4$ for some values of $x(4)$:

```
solset1 =
  Table[vars /. NDSolve[Join[eqns, {x[4] == x0, y[4] == 1 - x0}], vars, {t, 0, 20}],
    {x0, 0.05, 0.95, 0.1}];
```

```
p1 = ParametricPlot[solset1, {t, 0, 20}, PlotStyle → Black]
```

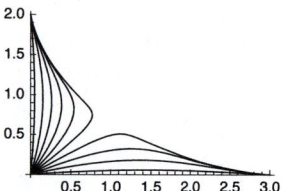

The time instance $t = 4$ was chosen by trial and error such that when the solution starts at $t = 0$, the starting points are close enough to the origin. Then we start from the line $y = 7 - x$ for some values of $x(0)$:

```
solset2 =
  Table[vars /. NDSolve[Join[eqns, {x[0] == x0, y[0] == 7 - x0}], vars, {t, 0, 10}],
    {x0, 0.1, 6.9, 0.25}];
p2 = ParametricPlot[solset2, {t, 0, 10}, PlotStyle → Black]
```

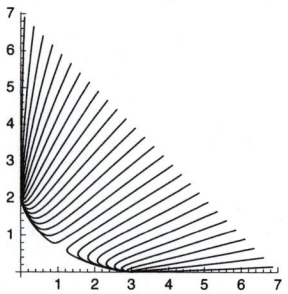

We also plot the nullclines:

```
p3 = Plot[{h1, h2}, {x, 0, 3.5}, PlotStyle → Blue]
```

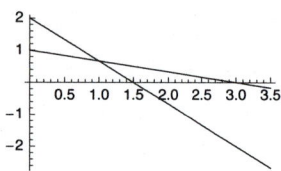

Then we show all of the plots in one figure:

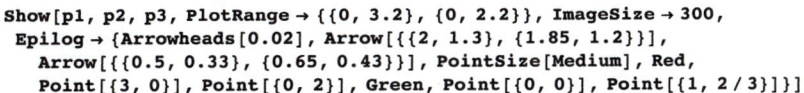

```
Show[p1, p2, p3, PlotRange → {{0, 3.2}, {0, 2.2}}, ImageSize → 300,
  Epilog → {Arrowheads[0.02], Arrow[{{2, 1.3}, {1.85, 1.2}}],
    Arrow[{{0.5, 0.33}, {0.65, 0.43}}]], PointSize[Medium], Red,
    Point[{3, 0}], Point[{0, 2}], Green, Point[{0, 0}], Point[{1, 2 / 3}]}}]
```

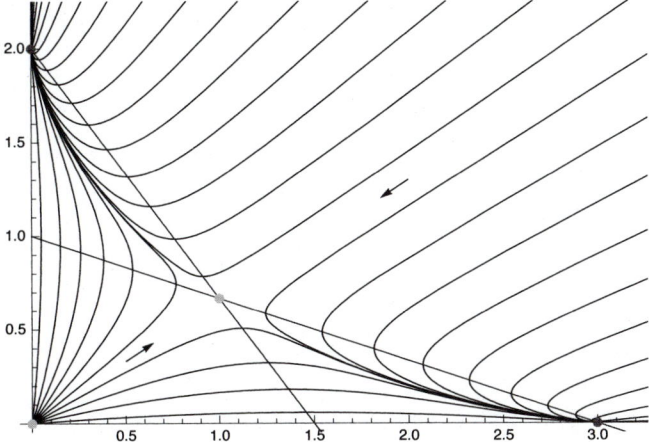

The red points are the stable equilibrium points, the green points are unstable equilibrium points, and the blue lines are the nullclines. The direction of motion is toward the stable equilibrium points $(3, 0)$ and $(0, 2)$. Note that for some other values of the parameters p, q, r, P, Q, and R, the behavior of the model may be wholly different.

Recall that in Section 10.1.3, p. 330, we presented a manipulation to study the trajectories of the model of competing species that we presented here.

■ A Matrix Equation

NDSolve accepts equations formulated with matrices. For example,

```
a = {{0.1, -0.5}, {0.3, -0.2}}; x0 = {15, 0};

sol = x[t] /. NDSolve[{x'[t] == a.x[t], x[0] == x0}, x[t], {t, 0, 80}]
{InterpolatingFunction[{{0., 80.}}, <>][t]}

ParametricPlot[sol, {t, 0, 80}, PlotRange → All]
```

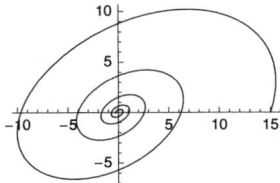

In this example, **x[t]** is a two-component vector. The dependent variable can also be a matrix.

■ A Differential-Algebraic System

NDSolve can solve differential-algebraic systems. Here is a simple example:

```
sol = {x[t], y[t]} /. NDSolve[
   {x'[t] == x[t]^2 + y[t] + 0.5, x[t] + y[t] == 1, x[0] == 0}, {x[t], y[t]}, {t, 0, 1.5}]
{{InterpolatingFunction[{{0., 1.5}}, <>][t],
   InterpolatingFunction[{{0., 1.5}}, <>][t]}}

Plot[sol, {t, 0, 1.5}]
```

```
Remove["Global`*"]
```

26.3.3 Three and More Equations

`vars = {x[t], y[t], z[t]}`	Define the dependent variables
`eqns = {eqn1, eqn2, eqn3}`	Define the differential equations
`inits = {x[a] == α, y[a] == β, z[a] == γ}`	Define the initial (or boundary) conditions
`sol = {sx, sy, sz} = vars /. DSolve[Join[eqns, inits], vars, {t, a, b}][[1]]`	Solve the problem
`Plot[sol, {t, a, b}]`	Plot x, y, and z in one figure
`Plot[#, {t, a, b}]& /@ sol`	Plot x, y, and z in three figures
`ParametricPlot[#, {t, a, b}]& /@ {{sx, sy}, {sx, sz}, {sy, sz}}]`	Plot three plane trajectories
`ParametricPlot3D[sol, {t, a, b}]`	Plot the space trajectory

■ The Lorenz Model

Define the following model:

```
vars = {x[t], y[t], z[t]};

f = p (y[t] - x[t]);
g = q x[t] - y[t] - x[t] z[t];
h = x[t] y[t] - r z[t];
eqns = {x'[t] == f, y'[t] == g, z'[t] == h};
```

This is the Lorenz model, which describes convective currents in atmosphere (see Borrelli and Coleman, 1998, pp. 500–514). The constants p, q, and r are positive. The variable x is the amplitude of the convective currents, y is the temperature difference between rising and falling currents, and z is the deviation from the normal temperatures. The equations result in approximating nonlinear partial differential equations of turbulent flow.

The equilibrium points are as follows:

```
equi = Solve[{f == 0, g == 0, h == 0}, vars]
```

$$\left\{\{y[t] \to 0, \ z[t] \to 0, \ x[t] \to 0\}, \ \left\{y[t] \to -\sqrt{-r + q\,r}, \ z[t] \to -1 + q, \ x[t] \to -\sqrt{-r + q\,r}\right\},\right.$$
$$\left.\left\{y[t] \to \sqrt{-r + q\,r}, \ z[t] \to -1 + q, \ x[t] \to \sqrt{-r + q\,r}\right\}\right\}$$

Define the following numerical values for the constants:

```
p = 3; q = 268 / 10; r = 1;
```

The equilibrium points are now as follows:

```
equi // N
{{y[t] → 0., z[t] → 0., x[t] → 0.}, {y[t] → -5.07937, z[t] → 25.8, x[t] → -5.07937},
  {y[t] → 5.07937, z[t] → 25.8, x[t] → 5.07937}}
```

■ **Sensitivity to Numerical Inaccuracies**

Calculating a numerical solution of the Lorenz model may cause trouble because the model is *chaotic*. The basic property of such models is extreme sensitivity of the solution to numerical inaccuracies and to the initial conditions. To demonstrate the effect of numerical inaccuracies, we first calculate one solution of the system with normal decimal numbers and ask for the value of the solution at $t = 31$:

```
inits = {x[0] == 0, y[0] == 1, z[0] == 0};

sol1 = vars /. NDSolve[Join[eqns, inits], vars, {t, 0, 35}][[1]];

sol1 /. t → 31
{-2.75379, -7.14803, 11.7918}
```

Then we calculate the solution with high-precision numbers by asking that the calculations are done to 20 decimals. We also increase the values of the precision and accuracy goals (see Section 26.4.1, p. 865):

```
sol2 = vars /. NDSolve[Join[eqns, inits], vars, {t, 0, 35},
    WorkingPrecision → 25, PrecisionGoal → 12, AccuracyGoal → 12][[1]];
sol2 /. t → 31.
{7.50215, 14.0227, 25.0762}
```

The values at $t = 31$ are wholly different in the two solutions. To see the differences more clearly, we plot the difference of the two solutions:

```
Plot[#, {t, 0, 35}, PlotRange → All] & /@ (sol1 - sol2)
```

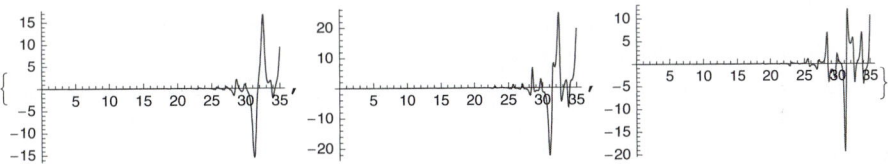

The figures show that from approximately $t = 25$ on, the solutions differ greatly.

We can trust the solution **sol2** more, but, actually, how accurate it is? We can solve the equations once more and use even tighter precision and accuracy conditions:

```
sol3 = {sx, sy, sz} =
    vars /. NDSolve[Join[eqns, inits], vars, {t, 0, 35}, WorkingPrecision → 25,
        PrecisionGoal → 16, AccuracyGoal → 16, MaxSteps → 13 000][[1]];
```

The difference of **sol2** and **sol3** is displayed as follows:

```
Plot[#, {t, 0, 35}, PlotRange → All] & /@ (sol2 - sol3)
```

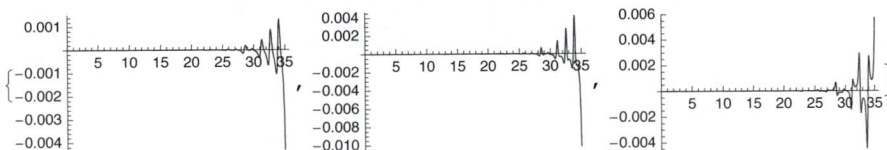

The solutions are practically the same up to approximately $t = 30$, so **sol2** may be quite good to this point. However, we use the more accurate **sol3** in the sequel.

To summarize, when calculating the solution of a chaotic model, it is important to use high precision in the calculations. Set **WorkingPrecision** to approximately 20 or higher. Also increase the precision and accuracy goals (their default value is 8). Remember to express all constants of the model with high accuracy. For example, do not set **q = 26.8** but, rather, **q = 268/10** or **q = 26.8`20** (the last definition sets the precision of 26.8 to 20; see Section 12.2.2, p. 406).

■ Sensitivity to Initial Conditions

To demonstrate the sensitivity of the Lorenz model to the initial conditions, we solve the equations by now setting $z(0) = 10^{-5}$ instead of $z(0) = 0$:

```
sol4 = vars /. NDSolve[Join[eqns, {x[0] == 0, y[0] == 1, z[0] == 10^-5}],
       vars, {t, 0, 30}, WorkingPrecision → 25, PrecisionGoal → 16,
       AccuracyGoal → 16, MaxSteps → 13 000][[1]];
```

We compare solutions **sol3**, where $z(0) = 0$, and **sol4** by plotting $x(t)$:

```
Plot[Evaluate[{sol3[[1]], sol4[[1]]}], {t, 0, 30},
  PlotStyle → {{Black, Dashing[{Tiny}]}, {Black}}, ImageSize → 160]
```

The solutions begin to differ from approximately $t = 25$ on, and from $t = 29$ on the solutions are wholly different. This sensitivity makes the prediction of a chaotic system impossible for a long time period because the initial conditions are hardly known exactly.

For a simple first-order equation sensitive for the initial condition, see Example 3 in Section 26.4.1, p. 868.

■ Plotting the Solution

First we plot the three curves in the same figure:

```
Plot[sol3, {t, 0, 30}, Ticks → {{30}, Automatic},
  PlotStyle → {{Black, Dashing[{Tiny}]}, {Black}, {Gray}},
  ImageSize → 420, AspectRatio → 0.2]
```

Then we plot three separate figures:

```
MapThread[Plot[#1, {t, 0, 30}, PlotStyle → #2, Ticks → {{30}, Automatic}] &,
  {sol3, {{Black, Dashing[{Tiny}]}, {Black}, {Gray}}}]
```

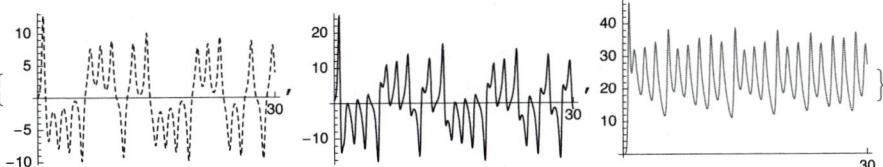

The components seem to evolve quite unpredictably and chaotically.

The pairwise plane trajectories show an interesting behavior:

```
ParametricPlot[#, {t, 0, 30}] & /@ {{sx, sy}, {sx, sz}, {sy, sz}}
```

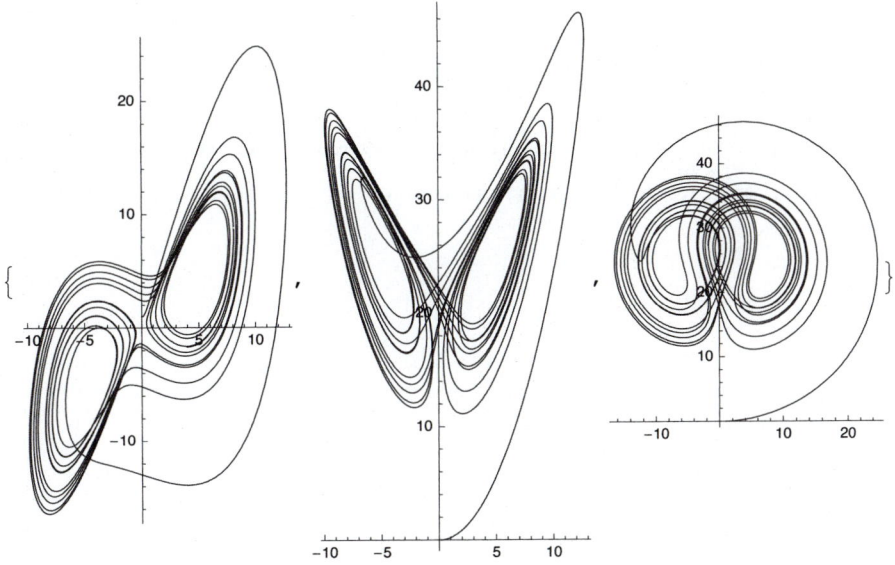

Lastly, we plot the space trajectory. To celebrate this extraordinarily fine figure, we produce a two-image stereogram, which involves two versions of the figure with slightly different viewpoints (see Section 5.3.3, p. 145). If you are able to superimpose the two figures with your eyes by focusing behind the paper, you will get an amazing stereographic view. (To prevent the figure from becoming too complex, we plotted only up to the value $t = 10.5$.)

```
GraphicsRow[ParametricPlot3D[sol3, {t, 0, 10.5}, ViewPoint → #] & /@
  {{1.3, -2.4, 0.8}, {1.4, -2.3, 0.8}}, ImageSize → 280, Spacings → -80]
```

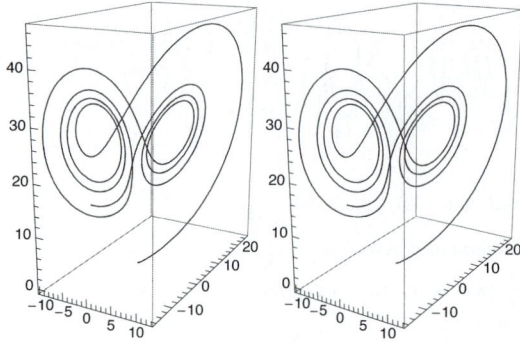

We can also plot points along the trajectory. A figure such as this gives an impression of the speed of the process:

```
GraphicsRow[
  Graphics3D[{PointSize[Tiny], Darker[Blue], Point[Table[sol3, {t, 0, 10.5, 0.03}]]}],
    ViewPoint → #] & /@
  {{1.3, -2.4, 0.8}, {1.4, -2.3, 0.8}}, ImageSize → 280, Spacings → -80]
```

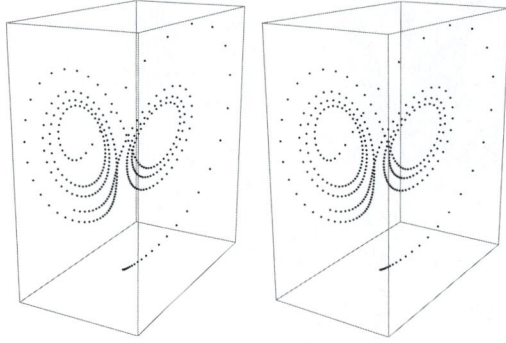

■ Larger Systems

In larger systems, we can consider the use of indexed dependent variables:

```
eqns = {y₁'[t] == -y₁[t], y₁[0] == 1,
    Table[{yᵢ'[t] == 1.5 yᵢ₋₁[t] - yᵢ[t], yᵢ[0] == 0}, {i, 2, 4}]};
vars = Table[yᵢ[t], {i, 4}];

sol = vars /. NDSolve[eqns, vars, {t, 0, 10}]
{{InterpolatingFunction[{{0., 10.}}, <>][t],
  InterpolatingFunction[{{0., 10.}}, <>][t],
  InterpolatingFunction[{{0., 10.}}, <>][t],
  InterpolatingFunction[{{0., 10.}}, <>][t]}}
```

```
Plot[sol, {t, 0, 10}]
```

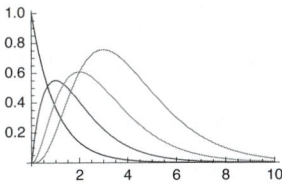

We can also ask for the solution of only one or some of the variables. Then the **DependentVariables** option has to be used to declare all the variables:

```
sol = y₄[t] /. NDSolve[eqns, y₄[t], {t, 0, 10}, DependentVariables → vars]
```

```
{InterpolatingFunction[{{0., 10.}}, <>][t]}
```

```
Plot[sol, {t, 0, 10}]
```

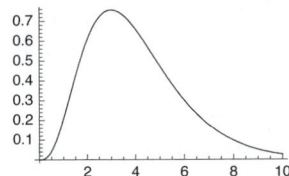

26.4 More about Numerical Solutions

26.4.1 Options

■ **Options of NDSolve**

As usual, for each option, the default value is mentioned first.

Options of **NDSolve**:

WorkingPrecision Precision used in internal computations; examples of values:
  MachinePrecision, 20

PrecisionGoal If the value of the option is **p**, the relative error of the solution at each point
  considered should be of the order 10^{-p}; examples of values: **Automatic** (usually means **8**), **10**

AccuracyGoal If the value of the option is **a**, the absolute error of the solution at each point
  considered should be of the order 10^{-a}; examples of values: **Automatic** (usually means **8**), **10**

Method Method to use; possible values: **Automatic** (means "**Adams**" for nonstiff and "**BDF**" for stiff
  problems), "**Adams**", "**BDF**", "**ExplicitRungeKutta**", "**ImplicitRungeKutta**",
  "**SymplecticPartitionedRungeKutta**"

StartingStepSize Initial step size used; examples of values: **Automatic, 0.01**

MaxStepSize Maximum size of each step; examples of values: **Automatic, 0.01**

MaxStepFraction Maximum fraction of the solution interval to cover in each step; examples of
  values: **1/10, 0.05**

MaxSteps Maximum number of steps to take; examples of values: **10000, 20000**

NormFunction Norm to use for error estimation in systems of equations; examples of values:
  Automatic (mostly means ∞), **1, 2, ∞**

DependentVariables List of all dependent variables; default value: **Automatic**

InterpolationOrder The continuity degree of the final output; examples of values: **Automatic**, **All** (the same as the underlying method used)

SolveDelayed Whether the derivatives are solved symbolically at the beginning (**False**) or at each step (**True**); possible values: **False**, **True**

Compiled Whether to compile the equations; possible values: **Automatic**, **True**, **False**

StepMonitor Command to be executed after each step of the method; examples of values: **None**, **Sow[{t, y[t]}]**, **n++**, **AppendTo[steps, {t, y[t]}]**

EvaluationMonitor Command to be executed after each evaluation of the equation; examples of values: **None**, **Sow[{t, y[t]}]**, **n++**, **AppendTo[points, t]**

The default value **MachinePrecision** of **WorkingPrecision** means that standard arithmetic is used in computations; a high value such as 20 should be given for numerically sensitive problems. The default value of **PrecisionGoal** is usually 8 (the general value is **WorkingPrecision**/2), so **NDSolve** tries to give you an answer with a relative error of order 10^{-8} (this is only a goal; the actual relative error may be smaller or larger). The default value of **AccuracyGoal** is also usually 8 (the general value is **WorkingPrecision**/2), so the goal is for the result to have an absolute error of the order 10^{-8}.

For a given value of the independent variable, **NDSolve** accepts the present value of the dependent value if the precision or accuracy goal is satisfied. If you increase **PrecisionGoal** or **AccuracyGoal**, you may also have to increase **WorkingPrecision**. See Section 12.3.1, p. 409, for more details about these options.

NDSolve uses an adaptive method in which the step size is varied as needed; this means that if the solution seems to change rapidly, the step size is reduced so that the solution can be followed more closely. For nonstiff problems, an Adams predictor–corrector method of order between 1 and 12 is used. For stiff problems, a backward difference formula method (i.e., Gear's method) of order between 1 and 5 is used. **NDSolve** detects the existence of stiffness automatically and chooses the correct method as well as a suitable order for it (the method and the order may be varied during the solution process). In a stiff system, the components vary at very different rates. An example could be the system $x'(t) = -100 x(t)$, $y'(t) = y(t) + x(t)$, $x(0) = 1$, $y(0) = 0$. Here, x decreases very rapidly, whereas y increases very slowly.

With the **Method** option, we can also select the method ourselves. **"Adams"** means the predictor–corrector Adams method with orders 1 through 12, and **"BDF"** means implicit backward differentiation formulas with orders 1 through 5. **"ExplicitRungeKutta"** uses adaptive embedded pairs of 2(1) through 9(8) Runge–Kutta methods, whereas **"ImplicitRungeKutta"** uses families of arbitrary-order implicit Runge–Kutta methods. **"SymplecticPartitionedRungeKutta"** applies interleaved Runge–Kutta methods for separable Hamiltonian systems.

The method can also be defined with so-called controller methods and submethods. Examples of controller methods are **"Shooting"** and **"EventLocator"**; see Sections 26.4.3 and 26.4.4. An example of submethods is **"ExplicitEuler"**; see Example 5. For more information about the controller methods and submethods, see the documentation.

If the interval of solution is long or the solution varies widely, you may receive a message that the maximum number of steps has been reached at a given point. You get the solution up to this point only. You can then increase the value of **MaxSteps** and solve anew. If the solution has a singular point, then **MaxSteps** prevents **NDSolve** from reducing the step size infinitely.

See Advanced Numerical Differential Equation Solving in *Mathematica* for advanced information about **NDSolve**.

■ Example 1: Precision and Accuracy

First, we solve an equation numerically:

```
eqn = y'[t] + t y[t] - Exp[-t] == 0;

numsol = NDSolve[{eqn, y[0] == 1}, y, {t, 0, 5}][[1]]
{y → InterpolatingFunction[{{0., 5.}}, <>]}

Plot[y[t] /. numsol, {t, 0, 5}]
```

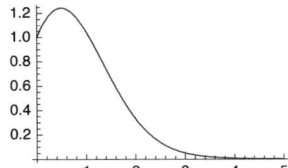

We can substitute the solution into the left-hand side of the equation. Because the right-hand side of the equation is zero, the result after substitution should be an expression that is very near to zero if the numerical solution is correct:

```
Plot[eqn[[1]] /. numsol, {t, 0, 5}, PlotRange → All]
```

The expression seems, indeed, to be very near to zero.

In this example, we can also calculate the symbolic solution, and this enables us to plot the absolute and relative errors:

```
numsol = y[t] /. NDSolve[{eqn, y[0] == 1}, y[t], {t, 0, 5}][[1]];

symbsol = y[t] /. DSolve[{eqn, y[0] == 1}, y[t], t][[1]]
```

$$\frac{1}{2} e^{-\frac{1}{2} - \frac{t^2}{2}} \left(2 \sqrt{e} + \sqrt{2\pi} \ \text{Erfi}\left[\frac{1}{\sqrt{2}} \right] + \sqrt{2\pi} \ \text{Erfi}\left[\frac{-1 + t}{\sqrt{2}} \right] \right)$$

```
{Plot[symbsol - numsol, {t, 0, 5}, PlotRange → All],
 Plot[1 - numsol / symbsol, {t, 0, 5}, PlotRange → All]}
```

The absolute error is at most approximately 3×10^{-7}, and the relative error is at most approximately 3×10^{-6}. The goal with **NDSolve** is that both of these errors should be less than 10^{-8} in the whole interval of solution. The goal is not strictly achieved, but, nevertheless, the errors are small enough for us to be well satisfied with the numerical solution.

■ **Example 2: Step Sizes**

In the following way, we can see all of the steps of the numerical solution:

```
{sol, {points}} = Reap[
    y[t] /. NDSolve[{eqn, y[0] == 1}, y[t], {t, 0, 5}, StepMonitor :> Sow[{t, y[t]}]]];
Graphics[{Line[{{#[[1]], 0}, #}] & /@ points}, Axes → True, ImageSize → 300]
```

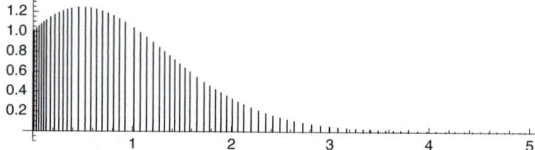

The step size varies during the solution process, which demonstrates the adaptivity property of **NDSolve**. With a package, we can also plot the used steps:

```
<< DifferentialEquations`NDSolveUtilities`

StepDataPlot[sol, ImageSize → 300, AspectRatio → 0.3]
```

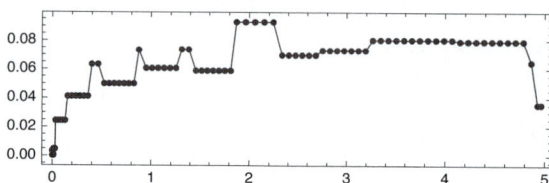

The plot shows how the step size is initially small and then grows, but at the end of the interval the step size is made smaller.

■ **Example 3: Increasing the Precision**

In Section 26.3.3, p. 860, we observed that the Lorenz model is very sensitive to numerical inaccuracies. Now we consider a very simple model that is also very sensitive; this example is taken from Wagon (2000, p. 302):

```
eqns[ε_] := {x'[t] == 2 x[t] + Cos[t], x[0] == -2 / 5 + ε};
```

We know the exact, symbolic solution:

```
sol[ε_] = x[t] /. DSolve[eqns[ε], x[t], t][[1]]
```

$$\frac{1}{5} \left(5 \, e^{2t} \, \epsilon - 2 \, Cos[t] + Sin[t]\right)$$

Note that the solution has the term $e^{2t} \epsilon$. Thus, for $\epsilon = 0$ we get a nicely oscillating solution but even a small ϵ will cause the term $e^{2t} \epsilon$ to, in due course, begin to dominate so that the solution is wholly different. This means that the solution is very sensitive for the initial value. Here are two solutions:

```
Plot[{sol[0], sol[10^-9]}, {t, 0, 12}]
```

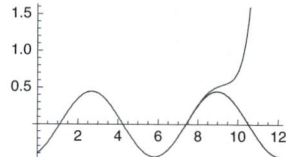

The sensitivity of the problem for the initial condition can also be seen when solving the problem numerically:

```
sol2 = x[t] /. NDSolve[eqns[0], x[t], {t, 0, 12}];
```

```
Plot[{sol[0], sol2}, {t, 0, 12}]
```

The numerical solution goes wrong from approximately $t = 7$ on. To get a better numerical solution, use either a higher working precision or tighter precision and accuracy goals:

```
sol3 = x[t] /. NDSolve[{x'[t] == 2 x[t] + Cos[t], x[0] == -2/5},
    x[t], {t, 0, 12}, WorkingPrecision → 30];
Plot[{sol[0], sol3}, {t, 0, 12}]
```

```
sol4 = x[t] /. NDSolve[{x'[t] == 2 x[t] + Cos[t], x[0] == -2/5},
    x[t], {t, 0, 12}, PrecisionGoal → 14, AccuracyGoal → 14];
Plot[{sol[0], sol4}, {t, 0, 12}]
```

■ **Example 4: A Singular Point**

If the solution has a singular point in the interval of solution, then to be able to follow the solution, the step size is reduced until it becomes effectively zero or until the maximum number of steps (10,000) is reached. In such situations, the solution process is stopped and a message is written. In the following example, we ask for the solution in the interval (0, 2) but get the solution only in the interval (0, 1) because there is a singularity at $t = 1$:

```
sol = y[t] /. NDSolve[{y'[t] == 1/(t-1), y[0] == 0}, y[t], {t, 0, 2}]
```

NDSolve::ndsz : At t == 0.999999999997489`,

step size is effectively zero; singularity or stiff system suspected. ≫
{InterpolatingFunction[{{0., 1.}}, <>][t]}

`Plot[sol, {t, 0, 1}]`

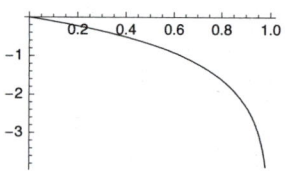

■ Example 5: Special Methods

We can ask to use a special method. Each method has its own options. For example, the explicit Runge–Kutta method has the following options:

```
Options[NDSolve`ExplicitRungeKutta] // N
```

```
{Coefficients → EmbeddedExplicitRungeKuttaCoefficients,
 DifferenceOrder → Automatic, EmbeddedDifferenceOrder → Automatic,
 StepSizeControlParameters → Automatic, StepSizeRatioBounds → {0.125, 4.},
 StepSizeSafetyFactors → Automatic, StiffnessTest → Automatic}
```

One of the options is **DifferenceOrder**. If we specify a difference order, then it is advantageous to also use the **InterpolationOrder** option to adjust the interpolation according to the order of the method:

```
eqn = y'[t] + t y[t] - Exp[-t] == 0;
```

```
sol1 = NDSolve[{eqn, y[0] == 1}, y, {t, 0, 2},
    Method → {"ExplicitRungeKutta", "DifferenceOrder" → 5}];
```

```
sol2 = NDSolve[{eqn, y[0] == 1}, y, {t, 0, 2},
    Method → {"ExplicitRungeKutta", "DifferenceOrder" → 5}, InterpolationOrder → All];
```

```
Plot[eqn[[1]] /. #, {t, 0, 2}, PlotRange → All] & /@ {sol1, sol2}
```

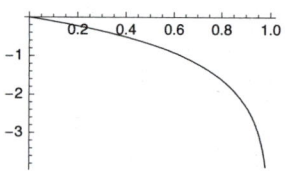

We substituted the solutions into the equation; the resulting expression should be very near to zero. In the first solution, we did not adjust the interpolation order, and the quality of the result is not as good as that in the second solution, where we did the adjustment.

The classical fixed step-size Euler method can be used as follows:

```
sol = y[t] /. NDSolve[{eqn, y[0] == 1}, y[t],
    {t, 0, 5}, Method → "ExplicitEuler", StartingStepSize → 0.1][[1]]
InterpolatingFunction[{{0., 5.}}, <>][t]
```

```
Plot[sol, {t, 0, 5}]
```

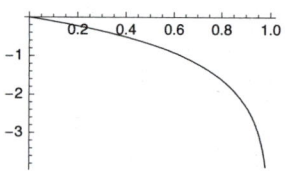

26.4.2 The Classical Runge–Kutta Method

■ A Coefficient Plug-in

The classical fourth-order Runge–Kutta method can be defined by giving its coefficients (these definitions are taken from tutorial/NDSolveExplicitRungeKutta):

```
crkamat = {{1 / 2}, {0, 1 / 2}, {0, 0, 1}};
crkbvec = {1 / 6, 1 / 3, 1 / 3, 1 / 6};
crkcvec = {1 / 2, 1 / 2, 1};
classicalRungeKuttaCoefficients[4, p_] := N[{crkamat, crkbvec, crkcvec}, p];
```

To use the classical Runge–Kutta method, define the method to be `"ExplicitRungeKutta"`, the difference order to be 4, and the coefficients to be as defined previously. Give a suitable starting step size (it will not be modified because the method uses a fixed step size).

```
sol = y /. NDSolve[{y'[t] == t - y[t]^2, y[0] == 3}, y,
     {t, 0, 6}, Method → {"ExplicitRungeKutta", "DifferenceOrder" → 4,
       "Coefficients" → classicalRungeKuttaCoefficients}, StartingStepSize → 0.2][[1]]
InterpolatingFunction[{{0., 6.}}, <>]
```

```
Plot[sol[t], {t, 0, 6}]
```

With a package we can extract the values of t and the corresponding approximate values of $y(t)$ (calculated with the classical Runge–Kutta method):

```
<< DifferentialEquations`InterpolatingFunctionAnatomy`
```

```
tt = InterpolatingFunctionCoordinates[sol][[1]]
```
```
{0., 0.2, 0.4, 0.6, 0.8, 1., 1.2, 1.4, 1.6, 1.8, 2., 2.2, 2.4, 2.6, 2.8,
 3., 3.2, 3.4, 3.6, 3.8, 4., 4.2, 4.4, 4.6, 4.8, 5., 5.2, 5.4, 5.6, 5.8, 6.}
```
```
yy = InterpolatingFunctionValuesOnGrid[sol]
```
```
{3., 1.88968, 1.41805, 1.18527, 1.07341, 1.03346, 1.03984, 1.07671,
 1.13307, 1.20094, 1.27459, 1.35007, 1.42488, 1.49761, 1.5676, 1.63464,
 1.69881, 1.76031, 1.81939, 1.87629, 1.93125, 1.98445, 2.03606,
 2.08624, 2.13511, 2.18277, 2.22932, 2.27483, 2.31938, 2.36302, 2.40582}
```
From the t values we can see that the method really used fixed-length steps.

The approximate solution at $t = 6$ is 2.40582. If we compare this value with the value given by **NDSolve**, we can see that the Runge–Kutta method gave an accurate solution:

```
y[t] /. NDSolve[{y'[t] == t - y[t]^2, y[0] == 3}, y[t], {t, 0, 6}][[1]] /. t → 6
2.40583
```

With the package NumericalDifferentialEquationAnalysis` we can study various Runge–Kutta methods.

■ A Method Plug-in

In *Mathematica*, we can easily write our own numerical solvers for differential equations in such a way that **NDSolve** can use them. To write the classical fourth-order Runge–Kutta method, we need the following three functions (they are taken from tutorial/NDSolvePlugIns):

```
CRK4[]["Step"[rhs_, t_, h_, y_, yp_]] := Module[{k0, k1, k2, k3},
  k0 = h yp;
  k1 = h rhs[t + h / 2, y + k0 / 2];
  k2 = h rhs[t + h / 2, y + k1 / 2];
  k3 = h rhs[t + h, y + k2];
  {h, (k0 + 2 k1 + 2 k2 + k3) / 6}]
CRK4[___]["DifferenceOrder"] := 4
CRK4[___]["StepMode"] := Fixed
```

Now we can use this fixed step-size method:

```
sol = y /. NDSolve[{y'[t] == t - y[t]^2, y[0] == 3},
    y, {t, 0, 6}, Method → CRK4, StartingStepSize → 0.2][[1]]
InterpolatingFunction[{{0., 6.}}, <>]

sol[6]     2.40582
```

■ An Own Program

Consider the following system of differential equations:

$$y_1'(t) = f_1(t, y_1, ..., y_m),$$

...

$$y_m'(t) = f_m(t, y_1, ..., y_m).$$

We form a function **rungeKuttaStep** to perform a single step of the Runge–Kutta method. We then use **NestList** to apply the step **n** times in **rungeKuttaSolve**.

```
rungeKuttaStep[f_List, ty_List, tyi_List, h_] := Module[{k1, k2, k3, k4},
  k1 = f /. Thread[ty → tyi];
  k2 = f /. Thread[ty → tyi + h / 2 Join[{1}, k1]];
  k3 = f /. Thread[ty → tyi + h / 2 Join[{1}, k2]];
  k4 = f /. Thread[ty → tyi + h Join[{1}, k3]];
  tyi + h Join[{1}, (k1 + 2 k2 + 2 k3 + k4) / 6]]

rungeKuttaSolve[f_List, ty_List, ty0_List, t1_, n_] :=
 With[{h = N[(t1 - ty0[[1]]) / n]},
  NestList[rungeKuttaStep[f, ty, #, h] &, N[ty0], n]]
```

Here, **f** is the list of right-hand-side expressions of the differential equations, **ty** is the list of the independent and dependent variables, **ty0** contains their initial values, **t1** is the end point of the interval of solution, and **n** is the number of steps to be performed.

Remember that the result of **Thread[ty → tyi]** is a list of the form {t → ti, y1 → y1i, ..., ym → ymi}. **Flatten** is needed to form unnested lists. **NestList** performs the iterations in which the result of one iteration is the starting point of the next iteration (see Section 18.3.3, p. 575, for details).

Recall that in Sections 18.1.1, p. 546, and 18.3.3, p. 576, we developed programs for Euler's method to solve differential equations. Because it is not sufficiently accurate, this method has mainly pedagogical value and is rarely used in practice.

■ **Example 1**

First, we solve the equation $y'(t) = t - y(t)^2$, $y(0) = 3$, in the interval $(0, 6)$ using 30 steps. We can write the dependent variable simply as **y**, but we could also write it more completely as **y[t]**.

```
sol = rungeKuttaSolve[{t - y^2}, {t, y}, {0, 3}, 6, 30];
```

```
sol // Last      {6., 2.40582}
```

```
ListLinePlot[sol, Mesh → All]
```

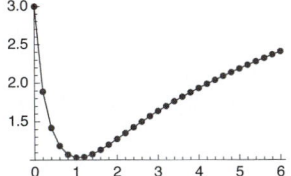

■ **Example 2**

Now we solve the system of two equations $y'(t) = t - z(t)^2$, $z'(t) = y(t)$, $y(0) = 1$, $z(0) = 2$:

```
sol = rungeKuttaSolve[{t - z^2, y}, {t, y, z}, {0, 1, 2}, 6, 40];
```

```
sol // Last      {6., 0.273551, 0.744243}
```

NDSolve gives quite similar numbers:

```
{y[t], z[t]} /. NDSolve[{y'[t] == t - z[t]^2, z'[t] == y[t], y[0] == 1, z[0] == 2},
   {y[t], z[t]}, {t, 0, 6}][[1]] /. t → 6
{0.273997, 0.74375}
```

To plot the solution, we first extract the three components of the solution:

```
{tt, yy, zz} = sol⊤;
```

We plot **yy** and **zz** as well as the trajectory:

```
ty = {tt, yy}⊤; tz = {tt, zz}⊤; yz = {yy, zz}⊤;
```

```
ListLinePlot[{ty, tz}, Mesh → All]
```

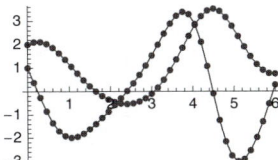

```
ListLinePlot[yz, Mesh → All]
```

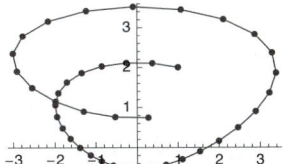

26.4.3 Boundary Value Problems

■ Linear Boundary Value Problems

Consider the following boundary value problem:

$$y''(t) = p(t)\, y'(t) + q(t)\, y(t) + r(t), \quad y(t_0) = y_0, \quad y(t_1) = y_1.$$

Here, p, q, and r do not depend on y. We solve this problem by solving two initial value problems:

$$z''(t) = p(t)\, z'(t) + q(t)\, z(t) + r(t), \quad z(t_0) = y_0, \quad z'(t_0) = 0,$$
$$v''(t) = p(t)\, v'(t) + q(t)\, v(t), \qquad v(t_0) = 0, \quad v'(t_0) = 1.$$

It is easy to show that the following function is the solution of the original boundary value problem:

$$y(t) = z(t) + v(t)\left(y_1 - z(t_1)\right)\big/ v(t_1).$$

This requires that $v(t_1) \neq 0$. If $v(t_1) = 0$ and $z(t_1) = y_1$, then $y(t) = z(t) + c\, v(t)$ is a solution for all constants c. If $v(t_1) = 0$ and $z(t_1) \neq y_1$, the problem has no solutions. Here is a module that implements this method (without the exceptional cases).

```
linearBVP[p_, q_, r_, t_, t0_, t1_, y0_, y1_] := Module[{eqn, sol1, sol2, sol},
  eqn = y''[t] == p y'[t] + q y[t] + r;
  sol1 = y[t] /. NDSolve[{eqn, y[t0] == y0, y'[t0] == 0}, y[t], {t, t0, t1}];
  sol2 = y[t] /. NDSolve[{eqn /. r -> 0, y[t0] == 0, y'[t0] == 1}, y[t], {t, t0, t1}];
  sol = sol1 + sol2 (y1 - sol1 /. t -> t1) / (sol2 /. t -> t1);
  FunctionInterpolation[sol, {t, t0, t1}][t]]
```

Here, `FunctionInterpolation` (see Section 24.4.2, p. 807) forms a single interpolating function from the solution `sol`, which contains the two interpolating functions `sol1` and `sol2`.

Let us solve the equation $y''(t) = y'(t) - t\, y(t)$ with the boundary conditions $y(0) = 2$ and $y(4) = 2$. Now $p = 1$, $q = -t$, and $r = 0$:

```
sol = linearBVP[1, -t, 0, t, 0, 4, 2, 2]
```

InterpolatingFunction[{{0., 4.}}, <>][t]

```
Plot[sol, {t, 0, 4}]
```

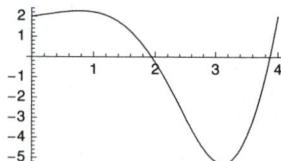

In Example 3 of Section 26.3.1, p. 851, we solved the same problem using **NDSolve**.

■ Nonlinear Boundary Value Problems

Here is a nonlinear boundary value problem:

```
eqn = y''[t] == y'[t] - y[t]^2 + t^2;
```

```
sol = y[t] /. NDSolve[{eqn, y[0] == 1, y[3] == 2}, y[t], {t, 0, 3}][[1]];
```

```
Plot[sol, {t, 0, 3}]
```

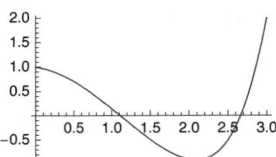

Is this the only solution? Let us define a function that calculates, for a given value of $y'(0)$, the value of $y(3)$:

```
y3[yp0_?NumericQ] := y[3] /. NDSolve[{eqn, y[0] == 1, y'[0] == yp0}, y, {t, 0, 3}][[1]]
```

Plot the function:

```
Plot[{2, y3[yp0]}, {yp0, -0.3, 1.7}, AxesLabel → {"y'(0)", "y(3)"}]
```

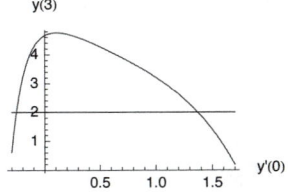

We can see that $y(3) = 2$ when $y'(0)$ is either approximately -0.3 or approximately 1.4. Thus, we have two solutions for the boundary value problem. Find the values of $y'(0)$ by solving the equation **y3[yp0] == 2**:

```
yp0a = yp0 /. FindRoot[y3[yp0] == 2, {yp0, -0.3, -0.2}]
-0.258224

yp0b = yp0 /. FindRoot[y3[yp0] == 2, {yp0, 1.3, 1.4}]
1.37153
```

Now we know the values of $y'(0)$ that give $y(3) = 2$. Thus, we have reduced the boundary value problem to two initial value problems:

```
sola = y[t] /. NDSolve[{eqn, y[0] == 1, y'[0] == yp0a}, y[t], {t, 0, 3}][[1]];
solb = y[t] /. NDSolve[{eqn, y[0] == 1, y'[0] == yp0b}, y[t], {t, 0, 3}][[1]];
Plot[{sola, solb}, {t, 0, 3}, Epilog → {Point[{0, 1}], Point[{3, 2}]}]
```

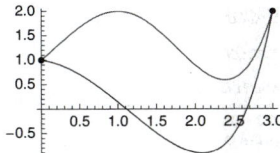

The method we used here is the shooting method. We "shoot" the solution at $t = 0$ with a guess for $y'(0)$ and try to iteratively improve the value $y'(0)$ until the end condition $y(3) = 2$ is satisfied. The iterations were done with **FindRoot**. With **NDSolve**, we can ask to use the shooting method, and we can also give starting initial conditions:

```
sol = y[t] /. NDSolve[{eqn, y[0] == 1, y[3] == 2}, y[t], {t, 0, 3}, Method → {"Shooting",
        "StartingInitialConditions" → {y[0] == 1, y'[0] == #}}][[1]] & /@ {-0.3, 1, 4};
Plot[sol, {t, 0, 3}]
```

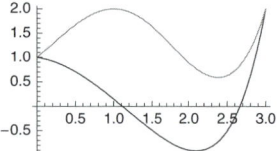

In addition to the **"Shooting"** method, we also have the **"Chasing"** method for boundary value problems.

The shooting can also be done experimentally with a manipulation (the following code is a modification of the code at demonstrations.wolfram.com/ShootingMethod by Bruce Miller):

```
Manipulate[Plot[y[s] /.
    NDSolve[ {y''[t] == y'[t] - y[t]^2 + t^2, y[0] == 1, y'[0] == yp0}, y, {t, 0, 3}][[1]],
    {s, 0, 3}, PlotRange → {-1.3, 4.1}, Prolog → {Red, PointSize[Large], Point[{3, 2}]},
    ImageSize → 200, MaxRecursion → ControlActive[0, 1]],
  {{yp0, 0, "y'(0)"}, -0.3, 1.8, Appearance → "Labeled"}]
```

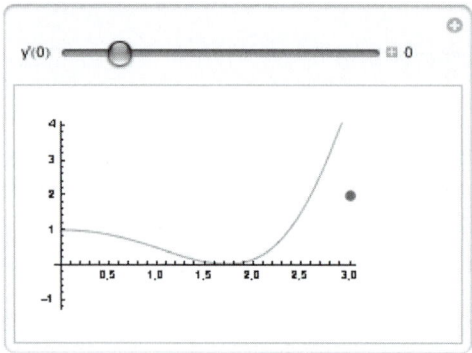

To speed up the redrawing, we have used the value 0 for **MaxRecursion** when the slider is dragged and the value 1 when the mouse is released.

26.4.4 Events

With the **"EventLocator"** method, we can solve a differential equation until a point where a given event happens. An event is a point where a given expression gets the zero value or where the value of a Boolean expression changes from True to False or vice versa. The event is defined with the **"Event"** option. With the **"EventAction"** option, we can tell what to do when an event occurs; the default is to stop the solution process. The points where events occur are, by default, calculated with **FindRoot** by using Brent's method.

■ **Example 1**

In the following example, we solve the problem until $y(t)$ becomes zero:

```
eqn = y''[t] == y'[t] - y[t]^2 + t^2;
```

```
sol = y[t] /. NDSolve[{eqn, y[0] == 1, y'[0] == 0.2},
    y[t], {t, 0, ∞}, Method → {"EventLocator", "Event" → y[t]}][[1]]
InterpolatingFunction[{{0., 4.30992}}, <>][t]
```

At $t = 4.30992$, $y(t)$ became zero. The value of t can be extracted as follows:

```
sol[[0, 1, 1, 2]]      4.30992
```

Plot the solution:

```
Plot[sol, {t, 0, sol[[0, 1, 1, 2]]}]
```

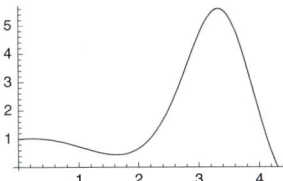

■ Example 2

In the next two examples, we solve the problem until $y(t)$ reaches the value 3 and until $y'(t)$ becomes zero:

```
sol1 = y[t] /. NDSolve[{eqn, y[0] == 1, y'[0] == 0.2},
    y[t], {t, 0, ∞}, Method → {"EventLocator", "Event" → y[t] - 3}][[1]];
sol2 = y[t] /. NDSolve[{eqn, y[0] == 1, y'[0] == 0.2},
    y[t], {t, 0, ∞}, Method → {"EventLocator", "Event" → y'[t]}][[1]];
{Plot[sol1, {t, 0, sol1[[0, 1, 1, 2]]}, PlotRange → All],
 Plot[sol2, {t, 0, sol2[[0, 1, 1, 2]]}]}
```

■ Example 3

We can specify an action when an event occurs (the default action is to stop the solution process). In the next example, we count the number of times the solution becomes zero:

```
problem = {eqn, y[0] == 1, y'[0] == -0.2};

n = 0; sol = y[t] /. NDSolve[problem, y[t], {t, 0, 4.5},
    Method → {"EventLocator", "Event" → y[t], "EventAction" :→ n++}];

n      3

Plot[sol, {t, 0, 4.5}]
```

Next, we form a list of the zeros of the solution:

```
zeros = {}; sol = y[t] /. NDSolve[problem, y[t], {t, 0, 5},
   Method → {"EventLocator", "Event" → y[t], "EventAction" :→ AppendTo[zeros, t]}];
zeros     {1.18674, 2.44972, 4.29136}
```

■ Example 4

If the `"Direction"` option has the value 1, only the events are considered where the value of the event expression changes from negative to positive; with the value −1 we can consider events where the value of the event expression changes from positive to negative. Next, we form a list of points where the solution changes from positive to negative:

```
zeros = {}; sol =
  y[t] /. NDSolve[problem, y[t], {t, 0, 5}, Method → {"EventLocator", "Event" → y[t],
     "EventAction" :→ AppendTo[zeros, t], "Direction" → -1}];
zeros     {1.18674, 4.29136}
```

26.4.5 Estimation of Differential Equations

How do we estimate unknown parameters that appear in a differential equation if we have some data? We can distinguish two situations. If the equation has a closed-form symbolic solution, then we use **FindFit** or **NonlinearRegress** to estimate the parameters of the solution (see Example 1). If a closed-form solution is not available, then we can use **FindFit** or **FindMinimum** together with a numerical solution of the equations to minimize a least-squares criterion (see Examples 2 and 3).

■ Example 1: Using the Symbolic Solution and FindFit

In Example 2 of Section 25.1.3, p. 820, we considered an experiment in which the growth of a yeast culture was measured at time instances 0, 1, 2, ..., 18 (Pearl, 1927). The measurements were as follows:

```
yeast = {9.6,  18.3, 29.0, 47.2, 71.1, 119.1, 174.6, 257.3, 350.7,
   441.0, 513.3, 559.7, 594.8, 629.4, 640.8, 651.1, 655.9, 659.6, 661.8};

data = {Range[0, 18], yeast}ᵀ;

p1 = ListPlot[data, Epilog → Line[{{0, 663}, {18, 663}}]]
```

The data show a logistic growth that can be described by the differential equation $y'(t) = r\, y(t)\,(M - y(t))$. This equation has a closed-form solution:

```
sol = y[t] /. DSolve[{y'[t] == r y[t] (M - y[t]), y[0] == y0}, y[t], t][[1]]
```

Solve::ifun : Inverse functions are being used by Solve, so some
    solutions may not be found; use Reduce for complete solution information. ≫

$$\frac{e^{M r t}\, M\, y0}{M - y0 + e^{M r t}\, y0}$$

Estimate the parameters of the solution:

```
fit = FindFit[data, sol, {{M, 600}, {y0, 10}, {r, 0.01}}, t]
```

$\{M \to 663.022, y0 \to 9.13552, r \to 0.000825002\}$

The fit is good:

```
Show[p1, Plot[sol /. fit, {t, 0, 18}]]
```

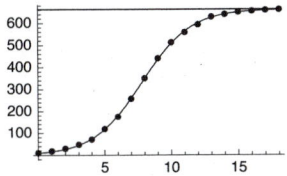

■ Example 2: Using the Numerical Solution and FindFit

Now we show how to solve the fitting problem of the preceding example if we do not have available the symbolic solution of the differential equation. First, define a model as the numerical solution of the differential equation:

```
model[r_?NumericQ, M_?NumericQ, y0_?NumericQ] :=
 model[r, M, y0] = y /. NDSolve[{y'[t] == r y[t] (M - y[t]), y[0] == y0}, y, {t, 0, 18}][[1]]
```

We used caching (`... := model[r, M, y0] = ...`) to speed up the fitting of the model. Then estimate the parameters of the model:

```
fit = FindFit[data, model[r, M, y0][t], {{M, 600}, {y0, 10}, {r, 0.01}}, t]
```

$\{M \to 663.022, y0 \to 9.13551, r \to 0.000825003\}$

```
Show[p1, Plot[model[r, M, y0][t] /. fit, {t, 0, 18}]]
```

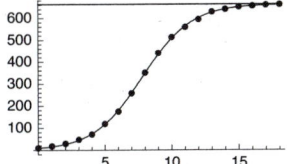

We obtained almost the same solution as in the preceding example.

```
Remove["Global`*"]
```

■ Example 3: Estimating the Predator–Prey Model

From 1909 through 1934, the yearly numbers of hare and lynx pelts sold to the Hudson Bay Trading Company were as follows:

```
hare = {25, 50, 55, 75, 70, 55, 30, 20, 15, 15,
    20, 35, 60, 80, 85, 60, 30, 20, 10, 5, 5, 10, 30, 80, 100, 80};
lynx = {2, 4, 10, 14, 19, 14, 8, 9, 2, 1, 1, 2, 4, 4, 8, 7, 9, 7, 4, 3, 2, 3, 3, 5, 7, 7};
```

(We considered an extended hare and lynx pelt data set in Section 8.1.3, p. 246.) The data are displayed as follows:

```
tt = Range[0, Length[hare] - 1];
th = {tt, hare}ᵀ; tl = {tt, lynx}ᵀ;
```

```
{p1 = ListLinePlot[{th, tl}, Mesh → All],
 p2 = ListLinePlot[{hare, lynx}ᵀ, Mesh → All]}
```

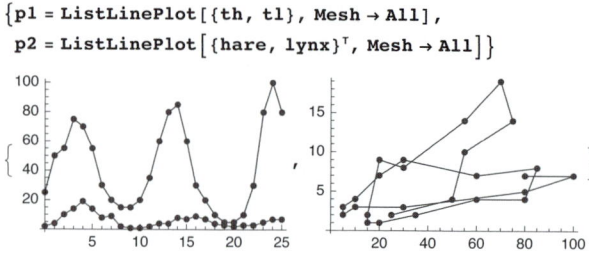

For later use, we combine the data into one list:

```
data = Flatten[{hare, lynx}ᵀ]
{25, 2, 50, 4, 55, 10, 75, 14, 70, 19, 55, 14, 30, 8,
 20, 9, 15, 2, 15, 1, 20, 1, 35, 2, 60, 4, 80, 4, 85, 8, 60, 7, 30,
 9, 20, 7, 10, 4, 5, 3, 5, 2, 10, 3, 30, 3, 80, 5, 100, 7, 80, 7}
```

The predator–prey model of Section 26.3.2, p. 853, would possibly suit these data. Let $x(t)$ and $y(t)$ be the amounts of hare and lynx pelts, respectively. The equations are $x' = x(p - q\,y)$ and $y' = y(-P + Q\,x)$:

```
vars = {x[t], y[t]}; pars = {p, q, P, Q};
eqns[{p_, q_, P_, Q_}] :=
 {x'[t] == x[t] (p - q y[t]),
  y'[t] == y[t] (-P + Q x[t]),
  x[0] == 25, y[0] == 2}
```

The initial conditions $x(0) = 25$ and $y(0) = 2$ are from the data. We also define the time interval and the time step of the data:

```
t0 = 0; t1 = 25; dt = 1;
```

To estimate the parameters of the model, define the following functions:

```
model[a_] := model[a] =
   vars /. NDSolve[eqns[a], vars, {t, t0, t1}, PrecisionGoal → 4, AccuracyGoal → 4][[1]]
pred[a_] := Flatten[Table[Evaluate[model[a]], {t, t0, t1, dt}]]
crit[a_] := EuclideanDistance[data, pred[a]]
```

Here, **model** gives the solution of the equations for given parameters, **pred** calculates predictions from the model (i.e., values of the solution at the same time instances as the data), and **crit** defines the expression to be minimized (i.e., the square root of the sum of the squares of the differences between the data and the model).

We will minimize **crit** using **FindMinimum**. To obtain reasonable starting values for the parameters, we can sample the criterion function at some random points. We guess that p, q, P, and Q may be in the intervals $(0.5, 1)$, $(0, 0.5)$, $(0, 0.5)$, and $(0, 0.1)$, respectively, and so we write the following:

```
SeedRandom[1];
s = Table[{RandomReal[{0.5, 1}], RandomReal[{0, 0.5}],
     RandomReal[{0, 0.5}], RandomReal[{0, 0.1}]}, {3000}];
(cs = crit[#] & /@ s;) // Timing
{18.0209, Null}
```

We ask for the best value and the corresponding values of the parameters:

```
{Min[cs], s[[Position[cs, Min[cs]][[1, 1]]]]}
{54.8889, {0.975995, 0.301868, 0.426301, 0.0101342}}
```

We choose starting values near these values (we need to set the value of a system option):

```
Developer`SetSystemOptions["EvaluateNumericalFunctionArgument" → False];

{minvalue, est} = FindMinimum[crit[{p, q, P, Q}],
   {p, 0.97, 0.98}, {q, 0.30, 0.31}, {P, 0.42, 0.43}, {Q, 0.010, 0.011}]
{47.1609, {p → 0.97326, q → 0.268641, P → 0.412891, Q → 0.0103831}}

Developer`SetSystemOptions["EvaluateNumericalFunctionArgument" → True];
```

The estimated equations are as follows:

```
eqns[pars /. est]

{x'[t] == x[t] (0.97326 - 0.268641 y[t]),
 y'[t] == (-0.412891 + 0.0103831 x[t]) y[t], x[0] == 25, y[0] == 2}
```

Here is the estimated model:

```
estModel = model[pars /. est]

{InterpolatingFunction[{{0., 25.}}, <>][t],
 InterpolatingFunction[{{0., 25.}}, <>][t]}
```

Then we show both the data and the model. The fit may be reasonable:

```
{Show[p1,
  Plot[estModel, {t, 0, 25}, PlotStyle → AbsoluteThickness[1.5]], ImageSize → 200],
  Show[p2, ParametricPlot[estModel, {t, 0, 10.5}, AxesOrigin → {0, 0}, PlotStyle →
    AbsoluteThickness[1.5], AspectRatio → 1 / GoldenRatio], ImageSize → 200]}
```

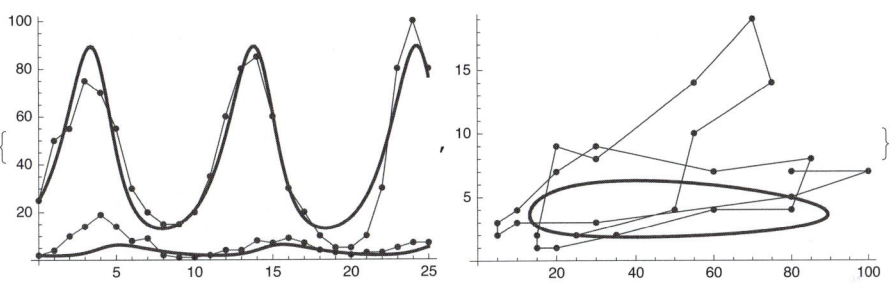

26.4.6 Manipulator and Equation Trekker

■ **Manipulator**

A manipulation is a very illustrative way to study the behavior of a system described by differential equations. Here is a manipulation to study the predator–prey model. We have set the initial values of the parameters to be the values that we obtained in Example 3 of the preceding section.

```
Manipulate[
 ParametricPlot[Evaluate[{x[t], y[t]} /. NDSolve[{x'[t] == x[t] (p - q y[t]),
     y'[t] == y[t] (-P + Q x[t]), x[0] == #[[1]], y[0] == #[[2]]},
     {x[t], y[t]}, {t, 0, 20}] & /@ start], {t, 0, 20},
  PlotRange → {{0, 123}, {0, 20.5}}, AspectRatio → 1 / GoldenRatio,
  ImageSize → 230],
 {{start, {{25, 1}, {25, 2}, {25, 3}}}, Locator,
  Appearance → •, LocatorAutoCreate → True},
 {{p, 0.97}, 0.5, 1.5, Appearance → "Labeled"},
 {{q, 0.27}, 0, 0.5, Appearance → "Labeled"},
 {{P, 0.41}, 0, 1, Appearance → "Labeled"},
 {{Q, 0.01}, 0, 0.05, Appearance → "Labeled"}]
```

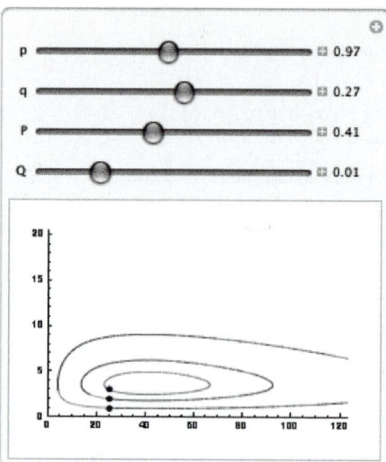

Here, the middle curve corresponds with initial conditions $x(0) = 25$ and $y(0) = 2$ of the hare–lynx example. The parameters can be adjusted with the sliders. The trajectories can be moved by dragging the points. New trajectories can be added by ⒜ᴸᵀ- or ⌘-clicking on the plotting area. A trajectory can be deleted by ⒜ᴸᵀ or ⌘ clicking on it.

Next, we consider a package that provides a similar functionality as **Manipulate** but has many positive features.

■ Equation Trekker

In the **EquationTrekker`** *package:*

EquationTrekker[{eqn1, eqn2}, {x, y}, {t, tmin, tmax}] Open a graphical interface for
specifying initial conditions and plotting the resulting numerical solution [$y(t)$ versus $x(t)$] of the
system of two first-order differential equations for t in (t_{min}, t_{max})

Options:

TrekParameters List of parameters and parameter ranges; examples of values: **{}**, **{p → 0.97, q →**
0.27}, **{p → {0.97, {0.5, 1.5}}, q → {0.27, {0, 0.5}}}**

TrekGenerator Method used to generate treks; possible values: **DifferentialEquationTrek,**
PoincareSection

PlotRange Range of values to include; default value: **{Automatic, {-1, 1}}**

ImageSize Absolute size of equation trekker window; default value: **{400, 400}**

As an example, we use the package for the same predator–prey model as discussed previously:

```
<< EquationTrekker`

EquationTrekker[{x'[t] == x[t] (p - q y[t]),
  y'[t] == y[t] (-P + Q x[t])}, {x, y}, {t, 0, 5},
  TrekParameters → {p → {0.97, {0.5, 1.5}}, q → {0.27, {0, 0.5}},
    P → {0.41, {0, 1}}, Q → {0.01, {0, 0.05}}}, PlotRange → {{0, 120}, {0, 20}}]
```

After executing this command, the following new window appears:

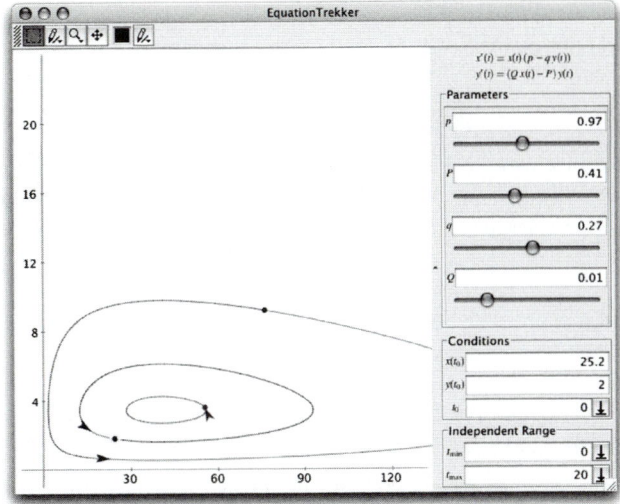

Initially, the window does not contain any curves or "treks" as they are also called. To create curves, CTRL-click on the plotting area. In Windows, we can also click with the right mouse button (without pressing the CTRL key). By each click, a curve appears on the window that starts, at $t = t_0$ (by default, $t_0 = t_{min}$), from the point clicked. The point is also shown in the plot and its coordinates can be read from the "Conditions" box at the lower right corner of the window. The curve ends with an arrowhead.

A curve can be moved by dragging on the point. A curve can be deleted by clicking on the corresponding point and then pressing the DEL key. The parameters can be changed with the sliders (or by typing and pressing the ENTER or RET key). The values of t_0, t_{min}, and t_{max} can be changed by typing a new value in the corresponding input field and then clicking the ⊥ button or pressing the ENTER or RET key; new curves use the changed values. A curve is only shown from $t = t_{min}$ to $t = t_{max}$.

The possible parameters can be defined in two ways. The first is like **p → value**. Then the initial value of **p** is the given value and with the slider we can adjust **p** in the range [0, 2 **value**] (or [−1, 1] if **value** is zero). The second is like **p → {value, {vmin, vmax}}**. Now the initial value of **p** is the given value and with the slider we can adjust **p** in the range [**vmin, vmax**].

The kernel is reserved for the trekker all the time when the trekker window is open. Thus, no other calculations can be made when the trekker window is open. When we close the trekker window, an output appears on the ordinary notebook, below the command that created the trekker window. The first component of the output is an **EquationTrekkerState** object containing a complete description of the trekker window that was closed. This object can be used to reopen the trekker window; for example, write **EquationTrekker[% // First]**. The second component of the output is the plot that was on the trekker window just before it was closed.

In the previous box, we showed the use of **EquationTrekker** when we have two first-order differential equations. We can also use the command for a singe second-order equation; then the plot shows $y'(t)$ versus $y(t)$. If we have a single first-order equation, then the plot shows $y(t)$ versus t.

By using the **TrekGenerator** option, we can also produce so-called *Poincare sections*; see the documentation of the package.

■ Controls for the Equation Trekker

At the upper left corner of the trekker window are the following buttons:

With the first four buttons, we can choose the drawing mode; the corresponding modes are the *select treks*, *create treks*, *zoom*, and *pan* mode. With the fifth button, we can change the color of a selected trek. The sixth button is used to change a selected trek from lines to points or vice versa. Details of the first four buttons follow.

By default, the *select treks* mode is chosen. Then we can select treks by clicking with the mouse and create treks by CTRL-clicking, as described previously.

 If the *create treks* mode is chosen, the role of clicking and CTRL-clicking is reversed: By clicking we can create treks and by CTRL-clicking we can select treks.

In this mode, we can also choose between lines and points: Drag the create treks button so that two menu items appear: *Line* and *Points*. Release the mouse button and then click on either menu item; new treks are then draw accordingly. By dragging the last or sixth button, we can change a chosen trek from line to points or vice versa.

With the *zoom* mode, we can zoom in and out. To zoom in, select the zoom button and drag over the plot: The rectangle shown will be the new plot range. Or, simply click on the plot: The plot will be zoomed such that the clicked point will be at the center of the plot. To zoom in, SHIFT-drag over the plot: The current plot region will be mapped to the rectangle shown. Or, simply SHIFT-click on the plot. After starting dragging or SHIFT-dragging, when also the CTRL-key is pressed, the aspect ratio of the plot can be changed. With a CTRL-click we can pan the plot.

When the zoom button is dragged, two menu items appear: *Zoom to Fit* and *Scale to Fit*. Release the mouse button and then click on either menu item. *Zoom to Fit* finds a plot range in which all of the trajectories fit, preserving aspect ratio. *Scale to Fit* finds the smallest plot range in which all of the trajectories fit.

By selecting the *pan* mode, we can move the plot. With a CTRL-click we can zoom the plot.

Partial Differential Equations

Introduction

> *A mathematician was given a test in which he had to produce steam starting with a block of ice which was stored in the refrigerator. He successfully described in great detail all the steps involved in the procedure, such as thawing the ice and boiling the water. Next he was asked to produce steam starting with the contents of a small pond. He replied: "Put a bucket of water from the pond into the refrigerator and apply the result of the previous problem."*

For partial differential equations (PDEs), *Mathematica* has the same two commands, **DSolve** and **NDSolve**, as for ordinary differential equations (ODEs).

One class of PDEs for which **DSolve** is suitable is quasi-linear first-order equations, for which we can find general solutions (containing unspecified functions) and also solutions that satisfy initial or boundary conditions. For linear second-order equations (containing only second-order derivatives) **DSolve** is able to find general solutions. For some nonlinear equations we can find, with **DSolve**, a so-called complete integral (containing unspecified parameters).

Hence, the set of PDEs for which a symbolic, closed-form solution can be found is restricted. Many problems can only be solved if initial and boundary conditions are present, and even then the solution cannot usually be expressed in terms of standard functions but only as infinite series. The Laplace transform can also be tried, and if *Mathematica* is not able to find the inverse transform, we can consult a table of Laplace transforms.

The basic method for PDEs with initial and boundary conditions is the method of separation of variables, which leads to series solutions. We present several examples of how to handle such solutions. The idea in the use of this kind of series is to truncate it and use the resulting finite sum as an approximate solution.

With the series solutions obtained by the method of separation of variables or by using the Laplace transform, we can obtain accurate numerical results (with up to, for example, six-digit precision). Thus, if you can find the series solution for your problem in a book or can derive such a solution by yourself, use it; otherwise, resort to numerical methods.

NDSolve uses the method of lines. The number of independent variables is not restricted, and the problem can also have more than one dependent variable. In the case of two independent variables (typically one space and one time variable), the region of solution is a rectangle that has initial and boundary conditions on up to three sides; typical examples of such problems are 1D *parabolic* and *hyperbolic* problems. Note that *elliptic* problems cannot be solved with **NDSolve**. For elliptic problems, we present a finite difference method in Section 27.3.4.

Using numerical methods, we can quickly get a low- or medium-precision solution (with up to, for example, three-digit precision). However, obtaining a high-precision result may either require a lot of computing time and memory or simply be impossible.

For more information about PDEs with *Mathematica*, see Abell and Braselton (1997), Kythe, Puri, and Schäferkotter (1996), or Ganzha and Vorozhtsov (1996). See also the advanced tutorials

- Differential Equation Solving with DSolve and
- Advanced Numerical Differential Equation Solving in *Mathematica*.

27.1 Symbolic Solutions

27.1.1 Using Symbolic Solver

■ **Introduction**

DSolve is able to solve linear and quasi-linear first-order equations. A first-order PDE for an unknown function $u(x, y)$ is *linear* if it is of the form

$$a(x, y) u_x + b(x, y) u_y + c(x, y) u = d(x, y),$$

where the function u and its first-order partial derivatives with respect to the independent variables x and y appear linearly. A first-order PDE is *quasi-linear* if it is of the form

$$a(x, y, u) u_x + b(x, y, u) u_y = c(x, y, u),$$

where the first-order partial derivatives of u appear linearly but u may appear nonlinearly in the functions a, b, and c. Here are typical commands used for first-order PDEs:

DSolve[eqn, u[x, y], {x, y}]	Give the general solution of the PDE
DSolve[{eqn, bound}, u[x, y], {x, y}]	Give the solution of the boundary value problem

If the solution is requested with the command **DSolve[eqn, u, {x, y}]**, the solution is given as a pure function. A solution of this kind is useful if we want to calculate with it (e.g., if we want to check the solution).

DSolve is also able to give general solutions of linear, constant coefficient, second-order partial differential equations of the form

$$a\, u_{xx} + b\, u_{xy} + c\, u_{yy} = 0.$$

Note that a, b, and c are constants and the equation only has second-order terms. An equation of the previous form is classified to be an elliptic, a parabolic, or a hyperbolic equation according to whether $b^2 - 4\,a\,c$ is negative, zero, or positive.

In addition, **DSolve** is able to find solutions of some nonlinear PDEs as so-called complete integrals.

Next, we consider some examples. The first four examples consider first-order PDEs, whereas the last two examples are devoted to linear second-order PDEs and nonlinear PDEs, respectively.

■ Example 1: Linear Constant Coefficient Equations

First, we consider an equation with constant coefficients:

```
eqn = a D[u[x, y], x] + b D[u[x, y], y] + c u[x, y] == d
```

$$c\, u[x, y] + b\, u^{(0,1)}[x, y] + a\, u^{(1,0)}[x, y] == d$$

The equation can also be written as follows:

```
eqn = a ∂ₓ u[x, y] + b ∂ᵧ u[x, y] + c u[x, y] == d
```

$$c\, u[x, y] + b\, u^{(0,1)}[x, y] + a\, u^{(1,0)}[x, y] == d$$

Here we used the **BasicMathInput** palette to write the partial derivatives $\partial_\square\ \square$. We use this shorter notation from now on. The general solution is as follows:

```
sol = u[x, y] /. DSolve[eqn, u[x, y], {x, y}]〚1〛 // Simplify
```

$$\frac{d}{c} + e^{-\frac{c\,x}{a}}\, C[1]\left[-\frac{b\,x}{a} + y\right]$$

The solution contains an arbitrary function $C[1]$ with the argument $-\frac{b\,x}{a} + y$ (we could simplify the arbitrary function to $C[1][a\,y - b\,x]$). To check the solution, we ask for it as a pure function and then substitute it into the equation:

```
DSolve[eqn, u, {x, y}]
```

$$\left\{\left\{u \to \text{Function}\left[\{x, y\},\ \frac{e^{-\frac{c\,x}{a}}\left(d\, e^{\frac{c\,x}{a}} + c\, C[1]\left[\frac{-b\,x+a\,y}{a}\right]\right)}{c}\right]\right\}\right\}$$

```
eqn /. % // Simplify        {True}
```

We can set the arbitrary function to be, for example, the sin function:

```
sol2 = sol /. C[1][e_] → Sin[e]
```
$$\frac{d}{c} - e^{-\frac{c\,x}{a}}\, \text{Sin}\left[\frac{b\,x}{a} - y\right]$$

```
Plot3D[sol2 /. {a → 1, b → 1, c → 1, d → 1}, {x, 0, π}, {y, 0, 2 π}]
```

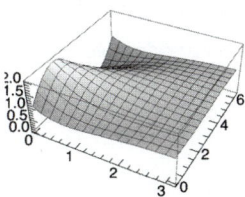

Next, we solve a boundary value problem:

```
sol3 = u[x, y] /. DSolve[{eqn, u[0, y] == Cos[y]}, u[x, y], {x, y}][[1]]
```

$$\frac{e^{-\frac{cx}{a}}\left(-d + d\, e^{\frac{cx}{a}} + c\, Cos\left[\frac{-bx+ay}{a}\right]\right)}{c}$$

```
Plot3D[sol3 /. {a → 1, b → 1, c → 1, d → 1}, {x, 0, π}, {y, 0, 2 π}]
```

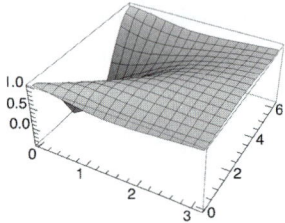

The following equation is a so-called transport equation:

```
eqn = ∂x u[x, y] + c ∂y u[x, y] == 0;

sol = u[x, y] /. DSolve[eqn, u[x, y], {x, y}][[1]] // Simplify
C[1][-c x + y]
```

This solution is constant along all lines of the form $y = cx + d$.

Now we solve an equation with three independent variables:

```
eqn = a ∂x u[x, y, z] + b ∂y u[x, y, z] + c ∂z u[x, y, z] + d u[x, y, z] == e;

DSolve[eqn, u[x, y, z], {x, y, z}] // Simplify
```

$$\left\{\left\{u[x,y,z] \to \frac{e}{d} + e^{-\frac{dx}{a}} C[1]\left[-\frac{bx}{a}+y, -\frac{cx}{a}+z\right]\right\}\right\}$$

■ Example 2: Linear Variable-Coefficient Equations

Here is a variable-coefficient linear equation:

```
eqn = x^2 ∂x u[x, y] - x y ∂y u[x, y] + y u[x, y] == 0;

DSolve[eqn, u[x, y], {x, y}]
```

$$\left\{\left\{u[x,y] \to e^{\frac{y}{2x}} C[1][x y]\right\}\right\}$$

Consider then a boundary value problem:

```
DSolve[{eqn, u[x, 1] == Sin[x]}, u[x, y], {x, y}]
```

$$\left\{\left\{u[x,y] \to e^{-\frac{1}{2xy}+\frac{y}{2x}} Sin[x y]\right\}\right\}$$

■ Example 3: A Birth–Death Process

Let $X(t)$ be the size of a population at time t, and suppose that the population develops according to a birth–death process with birth rate λ and death rate μ (suppose $\lambda \neq \mu$). It can be shown that the probability-generating function $p(s, t)$ of $X(t)$ satisfies the following linear first-order PDE:

```
eqn = (1 - s) (μ - λ s) ∂s p[s, t] - ∂t p[s, t] == 0;
```

If there are k individuals in the population at time 0, we know that $p(s, 0) = s^k$. Solve the initial value problem:

```
sol = p[s, t] /. DSolve[{eqn, p[s, 0] == s^k}, p[s, t], {s, t}][[1]] // FullSimplify
```

$$\left(1 + \frac{-\lambda + \mu}{\lambda + \frac{e^{t\,(-\lambda+\mu)}\,(-s\,\lambda+\mu)}{-1+s}}\right)^k$$

We can now calculate, for example, the expectation of $X(t)$:

```
Limit[D[sol, s], s → 1]        e^{t\,(\lambda-\mu)} k
```

We can see that when t approaches infinity, the expected value of $X(t)$ goes to 0 if $\lambda < \mu$ and to infinity if $\lambda > \mu$. The probability of 0 individuals at time t (meaning extinction) is as follows:

```
sol /. s → 0        (1 + \frac{-\lambda + \mu}{\lambda - e^{t\,(-\lambda+\mu)}\,\mu})^k
```

$$\left(1 + \frac{-\lambda + \mu}{\lambda - e^{t\,(-\lambda+\mu)}\,\mu}\right)^k$$

From this we can show that when t approaches infinity, the probability of 0 individuals goes to 1 if $\lambda < \mu$ and to $(\mu/\lambda)^k$ if $\lambda > \mu$. By also solving the case $\lambda = \mu$, we could show that the probability of 0 individuals goes to 1 even in this case, although the mean number of individuals is the constant k for all t.

■ **Example 4: Quasi-linear Equations**

Here is a quasi-linear equation:

```
eqn = ∂_x u[x, y] + ∂_y u[x, y] == x y u[x, y]^2;
sol = DSolve[eqn, u[x, y], {x, y}]
```

$$\left\{\left\{u[x, y] \to \frac{6}{x^3 - 3 x^2 y - 6 C[1][-x + y]}\right\}\right\}$$

Solve an initial value problem:

```
sol = DSolve[{eqn, u[x, 0] == x}, u[x, y], {x, y}]
```

$$\left\{\left\{u[x, y] \to -\frac{6\,(x - y)}{-6 + 3 x^2 y^2 - 4 x y^3 + y^4}\right\}\right\}$$

The following equation is the so-called Burgers' equation:

```
eqn = ∂_x u[x, y] + u[x, y] ∂_y u[x, y] == 0;

sol = DSolve[eqn, u[x, y], {x, y}]
Solve[C[1][u[x, y], y - x u[x, y]] == 0, u[x, y]]
```

The solution is in an implicit form. To check the solution, first pick the equation:

```
e = sol[[1]]        C[1][u[x, y], y - x u[x, y]] == 0
```

Differentiate this equation with respect to x and y and solve for u_x and u_y:

```
ux = Solve[D[e, x], ∂_x u[x, y]][[1, 1]];
uy = Solve[D[e, y], ∂_y u[x, y]][[1, 1]];
```

Substitute these expressions into the original PDE:

```
eqn /. {ux, uy}        True
```

■ **Example 5: Linear Second-Order Equations**

The wave equation is the most well-known PDE of hyperbolic type (use the $\partial_{\square,\square}\square$ button in the **Basic-MathInput** palette to write the second-order partial derivatives):

    eqn = ∂ₜ,ₜ u[x, t] - c² ∂ₓ,ₓ u[x, t] == 0;

We could also write

    eqn = D[u[x, t], t, t] - c² D[u[x, t], x, x] == 0;

Ask for the general solution:

    Simplify[DSolve[eqn, u[x, t], {x, t}], c > 0]

$$\left\{\left\{u[x, t] \to C[1]\left[t - \frac{x}{c}\right] + C[2]\left[t + \frac{x}{c}\right]\right\}\right\}$$

Consider the initial value problem $u_{tt} - c^2 u_{xx} = F(x, t)$, $u(x, 0) = f(x)$, $u_t(x, 0) = g(x)$, $-\infty < x < \infty$, $t \geq 0$. The solution of this problem (the so-called d'Alambert's solution) is

$$u(x, t) = \frac{1}{2}\left[f(x + c\,t) + f(x - c\,t)\right] + \frac{1}{2c}\int_{x-ct}^{x+ct} g(\xi)\,d\xi + \frac{1}{2c}\int_0^t\left[\int_{x-c(t-\tau)}^{x+c(t-\tau)} F(\xi, \tau)\,d\xi\right]d\tau.$$

    dAlambert[c_, F_, f_, g_, x_, t_] := 1/2 ((f /. x → x + c t) + (f /. x → x - c t)) +

      1/(2 c) ∫ₓ₋꜀ₜˣ⁺ᶜᵗ (g /. x → ξ) dξ + 1/(2 c) ∫₀ᵗ (∫ₓ₋꜀ (ₜ₋τ)ˣ⁺ᶜ ⁽ᵗ⁻τ⁾ (F /. {x → ξ, t → τ}) dξ) dτ

    sol = dAlambert[c, 1, Sin[ω x], 0, x, t] // Simplify

$$\frac{t^2}{2} + \text{Cos}[c\,t\,\omega]\,\text{Sin}[x\,\omega]$$

Laplace's equation is the most well-known PDE of elliptic type:

    eqn = ∂ₓ,ₓ u[x, y] + ∂_{y,y} u[x, y] == 0;

    DSolve[eqn, u[x, y], {x, y}]

    {{u[x, y] → C[1][i x + y] + C[2][-i x + y]}}

Here is an example of a parabolic equation:

    eqn = a² ∂ₓ,ₓ u[x, y] + 2 a c ∂ₓ,_y u[x, y] + c² ∂_{y,y} u[x, y] == 0;

    DSolve[eqn, u[x, y], {x, y}]

$$\left\{\left\{u[x, y] \to C[1]\left[-\frac{c\,x}{a} + y\right] + x\,C[2]\left[-\frac{c\,x}{a} + y\right]\right\}\right\}$$

■ **Example 6: Nonlinear Equations**

The general solution of linear and quasi-linear equations contains arbitrary functions. For most nonlinear PDEs, general solutions cannot be obtained. However, for some nonlinear equations we can get a special solution that contains arbitrary parameters. Such a solution is called a *complete integral*. Here is an example:

    eqn = ∂ₓ u[x, y] ∂_y u[x, y] == c;

```
sol = DSolve[eqn, u, {x, y}][[1]]
```

DSolve::nlpde :

Solution requested to nonlinear partial differential equation. Trying to build a special solution. ≫

$$\left\{u \to \text{Function}\left[\{x, y\}, C[1] + \frac{c\,x}{C[2]} + y\,C[2]\right]\right\}$$

Here, $C[1]$ and $C[2]$ are arbitrary parameters. The solution satisfies the equation:

```
eqn /. sol      True
```

Here is the so-called eikonal equation:

```
eqn = (∂x u[x, y])² + (∂y u[x, y])² == 1;
sol = DSolve[eqn, u[x, y], {x, y}]
```

DSolve::nlpde :

Solution requested to nonlinear partial differential equation. Trying to build a special solution. ≫

$$\left\{\left\{u[x, y] \to C[1] + y\,C[2] - x\,\sqrt{1 - C[2]^2}\right\}, \left\{u[x, y] \to C[1] + y\,C[2] + x\,\sqrt{1 - C[2]^2}\right\}\right\}$$

Consider the Korteweg–deVries equation:

```
eqn = ∂y u[x, y] + ∂x,x,x u[x, y] + 6 u[x, y] ∂x u[x, y] == 0;

sol = u[x, y] /. DSolve[eqn, u[x, y], {x, y}][[1]]
```

DSolve::nlpde :

Solution requested to nonlinear partial differential equation. Trying to build a special solution. ≫

$$-\frac{-8\,C[1]^3 + C[2] + 12\,C[1]^3\,\text{Tanh}[x\,C[1] + y\,C[2] + C[3]]^2}{6\,C[1]}$$

```
Plot3D[sol /. {C[1] → 1, C[2] → -4, C[3] → 1/4},
  {x, 0, 5}, {y, -0.5, 2}, PlotRange → All]
```

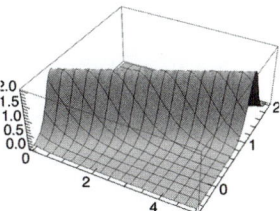

27.1.2 Using Laplace Transform

In using the Laplace transform to solve PDEs, the critical point is to find the inverse transform. It is probable that *Mathematica* cannot find it. In particular, *Mathematica* cannot find inverse transforms that are in the form of an infinite sum.

■ Example 1: A Wave Problem

Consider the following wave equation:

$$u_{tt} - c^2\,u_{xx} = 0, \quad 0 < x < 1, \quad t > 0,$$

$$u(x, 0) = d\,\sin(2\,\pi\,x), \quad u_t(x, 0) = 0, \quad u(0, t) = u(1, t) = 0.$$

First, take the Laplace transform of the equation:

```
eqn = ∂_{t,t} u[x, t] - c² ∂_{x,x} u[x, t] == 0;

lapeqn1 = LaplaceTransform[eqn, t, s]
```

$$s^2 \text{ LaplaceTransform}[u[x, t], t, s] -$$
$$c^2 \text{ LaplaceTransform}[u^{(2,0)}[x, t], t, s] - s\,u[x, 0] - u^{(0,1)}[x, 0] == 0$$

Simplify the notation, and take the initial conditions into account:

```
lapeqn2 = lapeqn1 /. {LaplaceTransform[u[x, t], t, s] → U[x],
   LaplaceTransform[∂_{x,x} u[x, t], t, s] → U''[x],
   u[x, 0] → d Sin[2 π x], Derivative[0, 1][u][x, 0] → 0}
```
$$-d\,s\,\text{Sin}[2\,\pi\,x] + s^2\,U[x] - c^2\,U''[x] == 0$$

Solve the transformed equation by using the boundary conditions:

```
lap = U[x] /. DSolve[{lapeqn2, U[0] == 0, U[1] == 0}, U[x], x] ⟦1⟧
```

$$\frac{d\,s\,\text{Sin}[2\,\pi\,x]}{4\,c^2\,\pi^2 + s^2}$$

Mathematica succeeds in inverting this:

```
sol = InverseLaplaceTransform[lap, s, t]
```
$$d\,\text{Cos}[2\,c\,\pi\,t]\,\text{Sin}[2\,\pi\,x]$$

Plot the solution:

```
Plot3D[sol /. {c → 1, d → 1}, {x, 0, 1}, {t, 0, 2},
  AxesLabel → {"x", "t", ""}, Ticks → {{0, 1}, {0, 1, 2}, {-1, 0, 1}}]
```

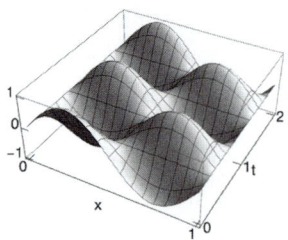

■ Example 2: A Heat Problem

Consider the following heat problem:

$$u_t - c\,u_{xx} = 0, \quad 0 < x < a, \quad t > 0,$$

$$u(x, 0) = u_0, \quad u_x(0, t) = 0, \quad u(a, t) = u_1.$$

Proceed as in Example 1:

```
eqn = ∂_t u[x, t] - c ∂_{x,x} u[x, t] == 0;

lapeqn1 = LaplaceTransform[eqn, t, s]
```

$$s\,\text{LaplaceTransform}[u[x, t], t, s] - c\,\text{LaplaceTransform}[u^{(2,0)}[x, t], t, s] - u[x, 0] == 0$$

```
lapeqn2 = lapeqn1 /. {LaplaceTransform[u[x, t], t, s] → U[x],
   LaplaceTransform[ ∂_{x,x} u[x, t], t, s] → U''[x], u[x, 0] → u0}
```
$$-u0 + s\,U[x] - c\,U''[x] == 0$$

The solution of the transformed equation is as follows:

```
lap =
  U[x] /. DSolve[{lapeqn2, U'[0] == 0, U[a] == u1/s}, U[x], x][[1]] // ExpToTrig // Simplify
```

$$\frac{u0 - (u0 - u1)\, \text{Cosh}\left[\frac{\sqrt{s}\, x}{\sqrt{c}}\right]\, \text{Sech}\left[\frac{a\sqrt{s}}{\sqrt{c}}\right]}{s}$$

Mathematica cannot invert this, but we can consult a table of Laplace transforms. For example, formula 33.153 in Spiegel (1999) is appropriate, and we get the solution in the form of an infinite series:

$$u_0 + (u_1 - u_0)\left(1 + \frac{4}{\pi}\sum_{n=1}^{\infty}\frac{(-1)^n}{2n-1}\exp\left(-\frac{c(2n-1)^2\pi^2 t}{4a^2}\right)\cos\left(\frac{(2n-1)\pi x}{2a}\right)\right).$$

Let us assume that $u_0 = 0$ and $u_1 = c = a = 1$. We take 60 terms from the sum and plot the resulting approximate solution:

```
uappr = 1 + 4/π Σ[n=1,60] (-1)^n/(2 n - 1) Exp[- (2 n - 1)^2 π^2 t / 4] Cos[(2 n - 1) π x / 2];

Plot3D[uappr, {x, 0, 1}, {t, 0, 2}, ViewPoint → {-2, -1.5, 0.8},
  AxesLabel → {"x", "t", ""}, Ticks → {{0, 1}, {0, 1, 2}, {0, 1}}, PlotRange → All]
```

27.2 Series Solutions

27.2.1 1D Parabolic Problems

■ **Series Solution**

Consider the following heat problem:

$$u_t - c\,u_{xx} = F(x, t),\quad 0 < x < a,\ \ t > 0,$$

$$u(x, 0) = f(x),\quad u(0, t) = u(a, t) = 0.$$

This model can be interpreted as follows. The value of $u(x, t)$ is the temperature of a bar at position x and time t. The bar is assumed to be slender and homogeneous and of uniform cross section. The bar lies along the x axis with ends at $x = 0$ and $x = a$. The lateral surface of the bar is insulated. The value of $F(x, t)$ is the amount of heat per unit volume per unit time generated at the point x at time t. The constant c depends on the properties of the bar (c is the thermal conductivity divided by the product of the specific heat and the material density). The ends of the bar are kept at the constant temperature 0. The initial temperature of the bar at x is $f(x)$. The goal is to determine the temperature of the bar for $t > 0$.

Separation of variables is a well-known method for solving PDEs. The solution is usually in the form of an infinite sum. The solution of this problem, using the method of separation of variables, is as follows (Dennemeyer, 1968, p. 309):

$$u(x, t) = \sum_{n=1}^{\infty} \left[A_n \exp\left(-c\, v_n^2\, t\right) + H_n(t)\right] \sin(v_n\, x), \quad v_n = \frac{n\,\pi}{a}, \quad A_n = \frac{2}{a} \int_0^a f(x) \sin(v_n\, x)\, dx,$$

$$F_n(t) = \frac{2}{a} \int_0^a F(x, t) \sin(v_n\, x)\, dx, \quad H_n(t) = \int_0^t F_n(\tau) \exp\left[-c\, v_n^2(t - \tau)\right] d\tau.$$

■ **Calculating the General Term**

Consider the following example:

```
a = 1; c = 1; F = 0; f = x (1 - x);
```

First, calculate the coefficients:

```
vₙ = n π / a;
```

```
Aₙ = Simplify[2/a ∫₀ᵃ f Sin[vₙ x] dx, n ∈ Integers]
```

$$-\frac{4\left(-1 + (-1)^n\right)}{n^3\, \pi^3}$$

```
Fₙ = 2/a ∫₀ᵃ F Sin[vₙ x] dx        0
```

```
Hₙ = ∫₀ᵗ (Fₙ /. t → τ) Exp[-c vₙ² (t - τ)] dτ        0
```

Here is the nth term of the series solution:

```
term = (Aₙ Exp[-c vₙ² t] + Hₙ) Sin[vₙ x]
```

$$-\frac{4\left(-1 + (-1)^n\right) e^{-n^2 \pi^2 t} \sin[n\,\pi\,x]}{n^3\, \pi^3}$$

If exact integration with **Integrate** does not succeed, we can use **NIntegrate**. Exact integration is handy because we have to do only one integration for a general n, and from the result we obtain all of the coefficients we are going to use. If we have to use numerical integration, we must separately integrate each coefficient we need.

■ **Choosing an Approximation**

How many terms from the infinite series should we choose so that the results would be precise enough? We investigate the solution when $x = 0.5$. By making experiments with series of different lengths, we would find that to obtain the correct value 0.25 for $u(0.5, 0)$ to six decimal places, we need 63 terms; here are the corresponding values of $u(0.5, t)$ for $t = 0, 0.1, 0.2, 0.3, 0.4$, and 0.5:

```
Table[Sum[term /. x → 0.5, {n, 63}], {t, 0, 0.5, 0.1}]
```

```
{0.25, 0.0961619, 0.0358408, 0.0133581, 0.00497868, 0.00185559}
```

If we are satisfied with five correct decimals for $u(0.5, 0)$, then 31 terms suffice. For four-digit precision, 13 terms suffice.

We choose to take 15 terms and form an approximate solution (note that the terms corresponding with even n values are zero):

```
uappr1[x_, t_] = Sum[term, {n, 15}]
```

$$\frac{8\, e^{-\pi^2\, t}\, \text{Sin}[\pi\, x]}{\pi^3} + \frac{8\, e^{-9\,\pi^2\, t}\, \text{Sin}[3\,\pi\, x]}{27\,\pi^3} + \frac{8\, e^{-25\,\pi^2\, t}\, \text{Sin}[5\,\pi\, x]}{125\,\pi^3} + \frac{8\, e^{-49\,\pi^2\, t}\, \text{Sin}[7\,\pi\, x]}{343\,\pi^3} +$$

$$\frac{8\, e^{-81\,\pi^2\, t}\, \text{Sin}[9\,\pi\, x]}{729\,\pi^3} + \frac{8\, e^{-121\,\pi^2\, t}\, \text{Sin}[11\,\pi\, x]}{1331\,\pi^3} + \frac{8\, e^{-169\,\pi^2\, t}\, \text{Sin}[13\,\pi\, x]}{2197\,\pi^3} + \frac{8\, e^{-225\,\pi^2\, t}\, \text{Sin}[15\,\pi\, x]}{3375\,\pi^3}$$

▪ Using the Solution

Calculate some numerical values:

```
Table[uappr1[0.5, t], {t, 0, 0.5, 0.1}]
```

```
{0.249969, 0.0961619, 0.0358408, 0.0133581, 0.00497868, 0.00185559}
```

Note that the exact value of $u(0.5, 0)$ is 0.25 according to the initial condition, so the approximate value 0.249969 is quite accurate. Here is a plot of the approximate solution:

```
Plot3D[uappr1[x, t], {t, 0, 0.5}, {x, 0, 1}, PlotRange → All,
  AxesLabel → {"t", "x", ""}, Ticks → {{0, 0.5}, {0, 1}, {0, 0.2}}]
```

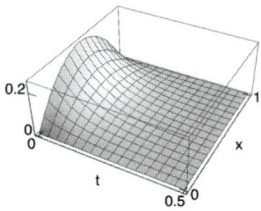

To manipulate the time evolution of the solution, do as follows:

```
Manipulate[Plot[Evaluate[uappr1[x, t]], {x, 0, 1}, PlotRange → {0, 0.26},
  Ticks → {{1}, {0.1, 0.2}}, ImageSize → 200], {t, 0, 0.5}, SaveDefinitions → True]
```

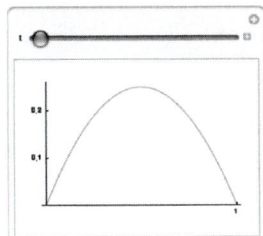

Here we used **Evaluate** to speed up the computation. In a printed material, it may be also useful to see a collection of plots:

```
GraphicsArray[Partition[Table[Plot[uappr1[x, t], {x, 0, 1}, PlotRange → {0, 0.26},
  Ticks → {{1}, {0.1, 0.2}}], {t, 0, 0.45, 0.05}], 5], ImageSize → 420]
```

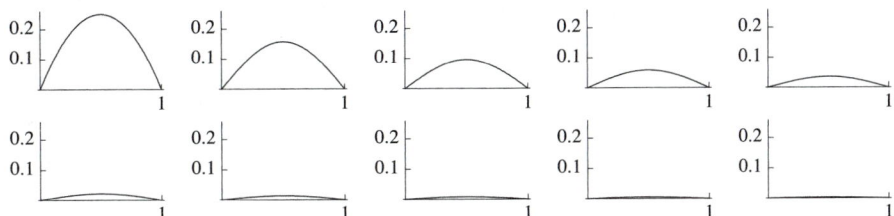

27.2.2 1D Hyperbolic Problems

■ Series Solution

Consider the following wave problem:

$$u_{tt} - c^2 u_{xx} = F(x, t), \quad 0 < x < a, \quad t > 0,$$

$$u(x, 0) = f(x), \quad u_t(x, 0) = g(x), \quad u(0, t) = u(a, t) = 0.$$

One interpretation of the model is as follows. The value of $u(x, t)$ is the transverse displacement of a homogeneous thin string at time t of the point on the string with the abscissa of x. The string is assumed to be perfectly flexible and subject to uniform tension. In addition to the tension, an external force transverse $F(x, t)$ (force/unit mass) is acting on the string; one example is gravity. The constant c^2 depends on the properties of the string (c^2 is the tension divided by the linear density of the string). The ends of the string are fastened at $x = 0$ and $x = a$. The string is pulled aside according to the function $f(x)$ and then released at the speed $g(x)$. The problem is to determine the subsequent motion of the string.

Using the method of separation of variables (Dennemeyer, 1968, pp. 170–175), we obtain the following solution:

$$u(x, t) = \sum_{n=1}^{\infty} [A_n \cos(c\, v_n\, t) + B_n \sin(c\, v_n\, t) + H_n(t)] \sin(v_n\, x), \quad v_n = \frac{n\pi}{a},$$

$$A_n = \frac{2}{a} \int_0^a f(x) \sin(v_n\, x)\, dx, \quad B_n = \frac{2}{n\pi c} \int_0^a g(x) \sin(v_n\, x)\, dx,$$

$$F_n(t) = \frac{2}{a} \int_0^a F(x, t) \sin(v_n\, x)\, dx, \quad H_n(t) = \frac{1}{c\, v_n} \int_0^t F_n(\tau) \sin[c\, v_n(t - \tau)]\, d\tau.$$

■ Calculating the General Term

Consider the following example:

```
a = 1; c = 1; F = -9.80665; f = 10 x² (1 - x)²; g = 0;
```

(9.80665 is acceleration due to gravity.) First we calculate the coefficients:

```
vₙ = n π / a;
```

$$A_n = \text{Simplify}\left[\frac{2}{a} \int_0^a f\, \text{Sin}[v_n\, x]\, dx,\ n \in \text{Integers}\right]$$

$$\frac{40\,(-1 + (-1)^n)\,(-12 + n^2\,\pi^2)}{n^5\,\pi^5}$$

$$B_n = \frac{2}{n\,\pi\,c} \int_0^a g\, \text{Sin}[v_n\, x]\, dx \qquad 0$$

$$F_n = \text{Simplify}\left[\frac{2}{a} \int_0^a F\, \text{Sin}[v_n\, x]\, dx,\ n \in \text{Integers}\right]$$

$$\frac{-6.24311 + 6.24311\,(-1)^n}{n}$$

$$H_n = \frac{1}{c\,v_n} \int_0^t (F_n \,/.\, t \to \tau) \, Sin[c\,v_n\,(t-\tau)] \, d\tau$$

$$\frac{-1.98724 + 1.98724\,(-1)^n + \left(1.98724 - 1.98724\,(-1)^n\right)\,Cos[n\,\pi\,t]}{n^3\,\pi}$$

Here is the nth term of the series solution:

```
term = (Aₙ Cos[c vₙ t] + Bₙ Sin[c vₙ t] + Hₙ) Sin[vₙ x]
```

$$\left(\frac{40\,\left(-1 + (-1)^n\right)\,\left(-12 + n^2\,\pi^2\right)\,Cos[n\,\pi\,t]}{n^5\,\pi^5} \right. +$$

$$\left. \frac{-1.98724 + 1.98724\,(-1)^n + \left(1.98724 - 1.98724\,(-1)^n\right)\,Cos[n\,\pi\,t]}{n^3\,\pi} \right) Sin[n\,\pi\,x]$$

■ Choosing an Approximation

Approximately 150 terms are needed for six-digit precision; here are the corresponding values of $u(0.5, t)$ for $t = 0, 0.4, ..., 2.4$:

```
Table[Sum[term /. x → 0.5, {n, 150}], {t, 0, 2.4, 0.4}]
```

```
{0.625, -0.703533, -2.69653, -2.69653, -0.703533, 0.625, -0.703533}
```

If we want five-digit precision, approximately 65 terms are needed, and approximately 30 terms suffice for four-digit precision.

We choose 30 terms from the series and form an approximation to the solution:

```
Style[uappr2[x_, t_] = Sum[term, {n, 30}] // N // Chop, 8] // Text
```

(0.55693 Cos[3.14159 t] + 0.31831 (−3.97449 + 3.97449 Cos[3.14159 t])) Sin[3.14159 x] +
  (−0.0826504 Cos[9.42478 t] + 0.0117893 (−3.97449 + 3.97449 Cos[9.42478 t])) Sin[9.42478 x] +
  (−0.0196371 Cos[15.708 t] + 0.00254648 (−3.97449 + 3.97449 Cos[15.708 t])) Sin[15.708 x] +
  (−0.00733557 Cos[21.9911 t] + 0.000928017 (−3.97449 + 3.97449 Cos[21.9911 t])) Sin[21.9911 x] +
  (−0.00348614 Cos[28.2743 t] + 0.000436639 (−3.97449 + 3.97449 Cos[28.2743 t])) Sin[28.2743 x] +
  (−0.00191901 Cos[34.5575 t] + 0.000239151 (−3.97449 + 3.97449 Cos[34.5575 t])) Sin[34.5575 x] +
  (−0.00116594 Cos[40.8407 t] + 0.000144884 (−3.97449 + 3.97449 Cos[40.8407 t])) Sin[40.8407 x] +
  (−0.00076035 Cos[47.1239 t] + 0.000094314 (−3.97449 + 3.97449 Cos[47.1239 t])) Sin[47.1239 x] +
  (−0.000522953 Cos[53.4071 t] + 0.0000647893 (−3.97449 + 3.97449 Cos[53.4071 t])) Sin[53.4071 x] +
  (−0.000374899 Cos[59.6903 t] + 0.0000464076 (−3.97449 + 3.97449 Cos[59.6903 t])) Sin[59.6903 x] +
  (−0.000277833 Cos[65.9734 t] + 0.000034371 (−3.97449 + 3.97449 Cos[65.9734 t])) Sin[65.9734 x] +
  (−0.000211572 Cos[72.2566 t] + 0.0000261617 (−3.97449 + 3.97449 Cos[72.2566 t])) Sin[72.2566 x] +
  (−0.000164807 Cos[78.5398 t] + 0.0000203718 (−3.97449 + 3.97449 Cos[78.5398 t])) Sin[78.5398 x] +
  (−0.000130865 Cos[84.823 t] + 0.0000161718 (−3.97449 + 3.97449 Cos[84.823 t])) Sin[84.823 x] +
  (−0.000105637 Cos[91.1062 t] + 0.0000130514 (−3.97449 + 3.97449 Cos[91.1062 t])) Sin[91.1062 x]

■ Using the Solution

First, we calculate some numerical values:

```
Table[uappr2[0.5, t], {t, 0, 2.4, 0.4}]
```

```
{0.624953, -0.703626, -2.69652, -2.69652, -0.703626, 0.624953, -0.703626}
```

The exact value $u(0.5, 0)$ is 0.625 according to the initial condition $u(x, 0) = 10\,x^2(1-x)^2$. We plot the movement of the center point $u(0.5, t)$ when t is in the interval (0, 10):

```
Plot[Evaluate[uappr2[0.5, t]], {t, 0, 10}, AspectRatio → 0.3]
```

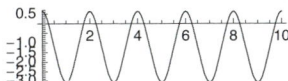

Here is a plot of the solution for t in (0, 4):

```
Plot3D[Evaluate[uappr2[x, t]], {t, 0, 4}, {x, 0, 1},
  AxesLabel → {"t", "x", ""}, Ticks → {{0, 1, 2, 3, 4}, {0, 1}, {-3, 0}}]
```

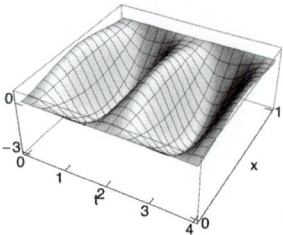

A manipulation is as follows:

```
Manipulate[Plot[Evaluate[uappr2[x, t]], {x, 0, 1}, PlotRange → {-3.3, 0.7},
  Ticks → {{1}, {-3}}, ImageSize → 200], {t, 0, 1.95}, SaveDefinitions → True]
```

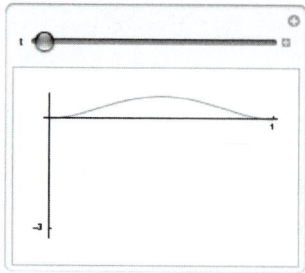

Here is a collection of plots:

```
GraphicsArray[
  Partition[Table[Plot[Evaluate[uappr2[x, t]], {x, 0, 1}, PlotRange → {-3.3, 0.7},
    Ticks → None], {t, 0, 2, 0.1}], 7], ImageSize → 420]
```

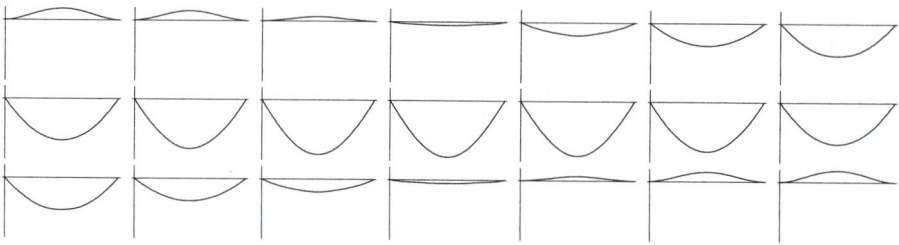

27.2.3 2D Hyperbolic Problems in Cartesian Coordinates

■ **Series Solution**

Consider the following wave problem:

$$u_{tt} - c^2(u_{xx} + u_{yy}) = F(x, y, t), \quad 0 < x < a, \quad 0 < y < b, \quad t > 0,$$

$$u(x, y, 0) = f(x, y), \quad u_t(x, y, 0) = g(x, y),$$

$$u(0, y, t) = u(a, y, t) = u(x, 0, t) = u(x, b, t) = 0.$$

Using the method of separation of variables (Dennemeyer, 1968, pp. 191–194, 263–264), we obtain the following solution:

$$u(x, y, t) = \sum_{m=1}^{\infty} \sum_{n=1}^{\infty} [A_{mn} \cos(\omega_{mn} t) + B_{mn} \sin(\omega_{mn} t) + H_{mn}(t)] \phi_{mn}(x, y),$$

$$\phi_{mn}(x, y) = \sin(v_m x) \sin(w_n y), \quad v_m = \frac{m\pi}{a}, \quad w_n = \frac{n\pi}{b}, \quad \lambda_{mn} = v_m^2 + w_n^2, \quad \omega_{mn} = c\sqrt{\lambda_{mn}},$$

$$A_{mn} = \frac{4}{ab} \int_0^a \int_0^b f(x, y) \phi_{mn}(x, y) \, dx \, dy, \quad B_{mn} = \frac{4}{ab\,\omega_{mn}} \int_0^a \int_0^b g(x, y) \phi_{mn}(x, y) \, dx \, dy,$$

$$F_{mn}(t) = \frac{4}{ab} \int_0^a \int_0^b F(x, y, t) \phi_{mn}(x, y) \, dx \, dy, \quad H_{mn}(t) = \frac{1}{\omega_{mn}} \int_0^t F_{mn}(\tau) \sin[\omega_{mn}(t - \tau)] \, d\tau.$$

■ Calculating the General Term

Consider the following example:

```
a = 1; b = 1; c = 1; F = 0; f = 10 x y (1 - x) (1 - y); g = 0;
```

First, we calculate the coefficients:

```
vₘ = m π / a ; wₙ = n π / b ; λₘₙ = vₘ² + wₙ² ; ωₘₙ = c √λₘₙ ;
```

$$v_m = \frac{m\pi}{a}; \quad w_n = \frac{n\pi}{b}; \quad \lambda_{mn} = v_m^2 + w_n^2; \quad \omega_{mn} = c\sqrt{\lambda_{mn}};$$

$$\phi_{mn} = \mathrm{Sin}[v_m\,x]\,\mathrm{Sin}[w_n\,y];$$

$$A_{mn} = \mathrm{Simplify}\!\left[\frac{4}{ab}\int_0^a \left(\int_0^b f\,\phi_{mn}\,dy\right) dx, \{m, n\} \in \mathrm{Integers}\right]$$

$$\frac{160\left(-1 + (-1)^m\right)\left(-1 + (-1)^n\right)}{m^3\,n^3\,\pi^6}$$

$$B_{mn} = \frac{4}{ab\,\omega_{mn}}\int_0^a \left(\int_0^b g\,\phi_{mn}\,dy\right) dx \qquad 0$$

$$F_{mn} = \frac{4}{ab}\int_0^a \left(\int_0^b F\,\phi_{mn}\,dy\right) dx \qquad 0$$

$$H_{mn} = \frac{1}{\omega_{mn}}\int_0^t (F_{mn}\,/.\,t \to \tau)\,\mathrm{Sin}[\omega_{mn}\,(t - \tau)]\,d\tau \qquad 0$$

Here is the (m, n)th term of the series solution:

$$\mathrm{term} = (A_{mn}\,\mathrm{Cos}[\omega_{mn}\,t] + B_{mn}\,\mathrm{Sin}[\omega_{mn}\,t] + H_{mn})\,\phi_{mn} \,//\, \mathrm{Simplify}$$

$$\frac{160\left(-1 + (-1)^m\right)\left(-1 + (-1)^n\right)\mathrm{Cos}\!\left[\sqrt{m^2 + n^2}\,\pi\,t\right]\mathrm{Sin}[m\,\pi\,x]\,\mathrm{Sin}[n\,\pi\,y]}{m^3\,n^3\,\pi^6}$$

■ Choosing an Approximation

The values of $u(0.5, 0.5, t)$ for $t = 0$, 0.3, 0.6, 0.9, 1.2, and 1.5 are as follows if we use 55 as the upper bound for both m and n:

```
Table[Evaluate[Sum[term /. {x → 0.5, y → 0.5}, {m, 55}, {n, 55}]], {t, 0, 1.5, 0.3}]
{0.624996, 0.202005, -0.650998, -0.394842, 0.355062, 0.659583}
```

These values for the upper bounds seem to suffice for five-digit precision (for six-digit precision, it seems that the upper bound for m and n has to be approximately 130). For four- and three-digit precision, 25 and 15, respectively, suffice as the upper bound.

We choose 25 as the upper bound for m and n (this gives us a total of 625 terms, many of which, however, are zero) and form an approximation to the solution:

```
uappr3[x_, y_, t_] = Sum[term, {m, 25}, {n, 25}] // N;
```

```
Short[uappr3[x, y, t], 5]
```

0.665703 Cos[4.44288 t] Sin[3.14159 x] Sin[3.14159 y] +

$\ll$167$\gg$ + 2.72672 × 10^{-9} Cos[111.072 t] Sin[78.5398 x] Sin[78.5398 y]

■ Using the Solution

First, we calculate the solution at the point (0.5, 0.5) for some values of t:

```
Table[uappr3[0.5, 0.5, t], {t, 0, 1.5, 0.3}]
```

{0.625036, 0.202054, -0.651003, -0.394831, 0.355097, 0.659562}

The exact value when t is 0 is 0.625 according to the initial condition. Then we plot the movement of the point $u(0.5, 0.5, t)$ when t is in the interval (0, 20):

```
Plot[Evaluate[uappr3[0.5, 0.5, t]], {t, 0, 20}, AspectRatio → 0.2, ImageSize → 300]
```

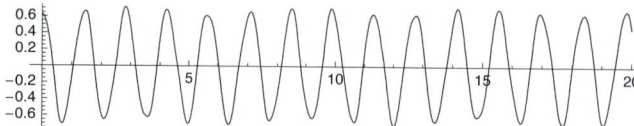

Here is a manipulation to show the surface profile for t in (0, 1.4) when y is 0.5:

```
Manipulate[Plot[Evaluate[uappr3[x, 0.5, t]], {x, 0, 1}, PlotRange → 0.71,
    Ticks → None, ImageSize → 200], {t, 0, 1.4}, SaveDefinitions → True]
```

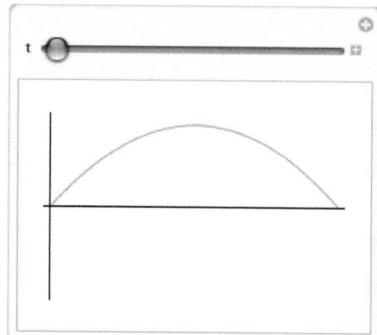

Here we only show a collection of plots:

```
GraphicsArray[Partition[
  Table[Plot[Evaluate[uappr3[x, 0.5, t]], {x, 0, 1}, PlotRange → 0.71, Ticks → None],
   {t, 0, 1.4, 0.1}], 5], ImageSize → 420]
```

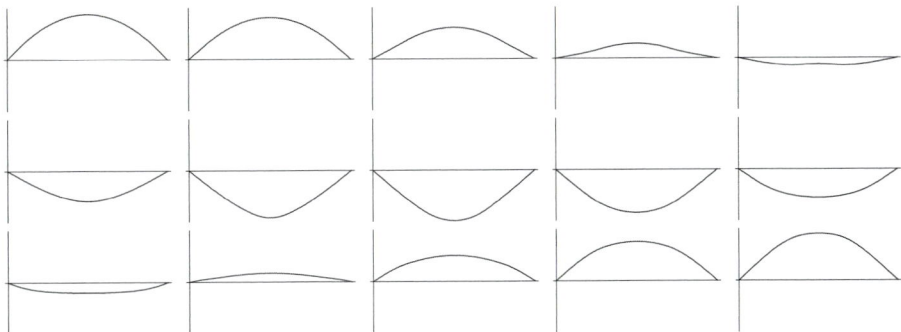

Next, we show the whole surface:

```
Manipulate[Plot3D[Evaluate[uappr3[x, y, t]], {x, 0, 1}, {y, 0, 1}, PlotRange → 0.71,
  Boxed → False, Axes → False, BoxRatios → {1, 1, 1}, ImageSize → 200,
  PlotRegion → {{0, 1}, {-0.4, 1.3}}], {t, 0, 1.4}, SaveDefinitions → True]
```

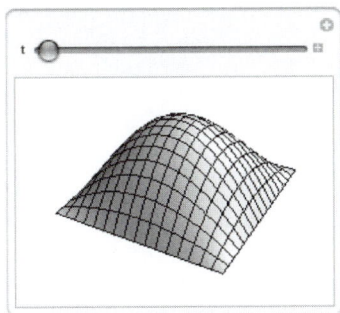

Here we only show a collection of plots:

```
Plot3D[Evaluate[uappr3[x, y, #]], {x, 0, 1}, {y, 0, 1},
  PlotRange → 0.71, Boxed → False, Axes → False, BoxRatios → {1, 1, 1},
  PlotRegion → {{0, 1}, {-0.4, 1.3}}, ImageSize → 100] & /@ {0, 0.2, 0.4, 0.6}
```

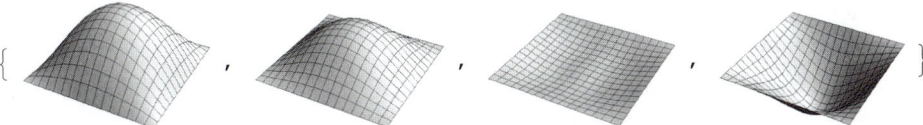

27.2.4 2D Hyperbolic Problems in Polar Coordinates

■ Series Solution

Consider the following circularly symmetric wave problem in polar coordinates:

$$u_{tt} - c^2(u_{rr} + u_r/r) = F(r), \quad 0 < r < a, \quad t > 0,$$

$$u(r, 0) = f(r), \quad u_t(r, 0) = g(r), \quad u(a, t) = 0.$$

By the method of separation of variables (Dennemeyer, 1968, pp. 201–202), we obtain the following solution:

$$u(r, t) = \sum_{n=1}^{\infty} [A_n \cos(v_n t) + B_n \sin(v_n t) + H_n(t)] J_0(w_n),$$

$$v_n = \frac{c \xi_n}{a}, \quad w_n = \frac{r \xi_n}{a}, \quad u_n = \frac{2}{a^2 J_1(\xi_n)^2},$$

$$A_n = u_n \int_0^a r f(r) J_0(w_n) \, dr, \quad B_n = \frac{u_n}{v_n} \int_0^a r g(r) J_0(w_n) \, dr,$$

$$G_n = u_n \int_0^a r F(r) J_0(w_n) \, dr, \quad H_n(t) = \frac{2 G_n}{v_n^2} \sin^2(v_n t / 2).$$

Here, J_0 and J_1 are Bessel functions of the first kind of order zero and one, and ξ_n is the nth zero of J_0.

■ Calculating the General Term

Consider the following example:

```
a = 1; c = 1; F = 0; f = 1 - r²; g = 0;
```

First, we calculate the coefficients:

```
ξn = BesselJZero[0, n];
```

$$v_n = \frac{c \, \xi_n}{a}; \quad w_n = \frac{r \, \xi_n}{a}; \quad u_n = \frac{2}{a^2 \, \text{BesselJ}[1, \, \xi_n]^2};$$

$$A_n = u_n \int_0^a r \, f \, \text{BesselJ}[0, \, w_n] \, dr$$

$$\frac{4 \, \text{BesselJ}[2, \, \text{BesselJZero}[0, n]]}{\text{BesselJ}[1, \, \text{BesselJZero}[0, n]]^2 \, \text{BesselJZero}[0, n]^2}$$

$$B_n = \frac{u_n}{v_n} \int_0^a r \, g \, \text{BesselJ}[0, \, w_n] \, dr \quad 0$$

$$G_n = u_n \int_0^a r \, F \, \text{BesselJ}[0, \, w_n] \, dr \quad 0$$

$$H_n = \frac{2 \, G_n}{v_n^2} \, \text{Sin}\left[\frac{v_n \, t}{2}\right]^2 \quad 0$$

Here is the nth term of the series solution:

```
term = (Aₙ Cos[vₙ t] + Bₙ Sin[vₙ t] + Hₙ) BesselJ[0, wₙ]
(4 BesselJ[0, r BesselJZero[0, n]]
   BesselJ[2, BesselJZero[0, n]] Cos[t BesselJZero[0, n]]) /
 (BesselJ[1, BesselJZero[0, n]]² BesselJZero[0, n]²)
```

```
% // TraditionalForm
```

$$\frac{4 J_0(r \, j_{0,n}) J_2(j_{0,n}) \cos(t \, j_{0,n})}{J_1(j_{0,n})^2 \, j_{0,n}^2}$$

■ Choosing an Approximation

Here are values of $u(0, t)$ for t = 0, 0.2, 0.4, ..., 1, using 100 terms:

```
Table[Evaluate[Sum[term /. r → 0, {n, 100}] // N], {t, 0, 1, 0.2}]
{0.999997, 0.919997, 0.679997, 0.279996, -0.280007, -0.999731}
```

It turns out that the use of 100 terms from the series solution does not quite suffice for five-digit precision. Approximately 50 terms suffice for four-digit precision, and 30 terms are sufficient for three-digit precision. We take 30 terms:

```
Style[uappr4[r_, t_] = Sum[term, {n, 30}] // N, 8] // Text
```

1.10802 BesselJ[0., 2.40483 r] Cos[2.40483 t] − 0.139778 BesselJ[0., 5.52008 r] Cos[5.52008 t] +
0.0454765 BesselJ[0., 8.65373 r] Cos[8.65373 t] − 0.0209909 BesselJ[0., 11.7915 r] Cos[11.7915 t] +
0.0116362 BesselJ[0., 14.9309 r] Cos[14.9309 t] − 0.00722118 BesselJ[0., 18.0711 r] Cos[18.0711 t] +
0.00483787 BesselJ[0., 21.2116 r] Cos[21.2116 t] − 0.00342568 BesselJ[0., 24.3525 r] Cos[24.3525 t] +
0.00252953 BesselJ[0., 27.4935 r] Cos[27.4935 t] − 0.00193015 BesselJ[0., 30.6346 r] Cos[30.6346 t] +
0.00151221 BesselJ[0., 33.7758 r] Cos[33.7758 t] − 0.00121077 BesselJ[0., 36.9171 r] Cos[36.9171 t] +
0.000987185 BesselJ[0., 40.0584 r] Cos[40.0584 t] − 0.000817394 BesselJ[0., 43.1998 r] Cos[43.1998 t] +
0.000685835 BesselJ[0., 46.3412 r] Cos[46.3412 t] − 0.000582113 BesselJ[0., 49.4826 r] Cos[49.4826 t] +
0.000499091 BesselJ[0., 52.6241 r] Cos[52.6241 t] − 0.000431745 BesselJ[0., 55.7655 r] Cos[55.7655 t] +
0.000376465 BesselJ[0., 58.907 r] Cos[58.907 t] − 0.000330609 BesselJ[0., 62.0485 r] Cos[62.0485 t] +
0.000292208 BesselJ[0., 65.19 r] Cos[65.19 t] − 0.000259772 BesselJ[0., 68.3315 r] Cos[68.3315 t] +
0.000232161 BesselJ[0., 71.473 r] Cos[71.473 t] − 0.000208491 BesselJ[0., 74.6145 r] Cos[74.6145 t] +
0.000188066 BesselJ[0., 77.756 r] Cos[77.756 t] − 0.000170336 BesselJ[0., 80.8976 r] Cos[80.8976 t] +
0.000154861 BesselJ[0., 84.0391 r] Cos[84.0391 t] − 0.000141285 BesselJ[0., 87.1806 r] Cos[87.1806 t] +
0.000129319 BesselJ[0., 90.3222 r] Cos[90.3222 t] − 0.000118724 BesselJ[0., 93.4637 r] Cos[93.4637 t]

■ Using the Solution

Here are some values of the solution:

```
Table[uappr4[0, t], {t, 0, 1, 0.2}]
{0.999943, 0.919941, 0.679934, 0.279917, -0.280132, -0.998375}
```

The exact value of $u(0, 0)$ is 1 according to the initial condition. Here is the movement of the center point when t is in the interval $(0, 30)$:

```
Plot[Evaluate[uappr4[0, t]], {t, 0, 30}, AspectRatio → 0.2, ImageSize → 280]
```

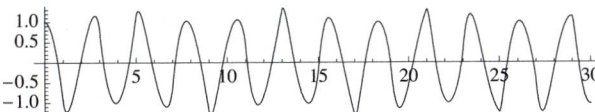

Next, we show the form of the surface for r in $(-1, 1)$ for various values of t:

```
Manipulate[Plot[Evaluate[uappr4[r, t]], {r, -1, 1}, PlotRange → 1.25,
  Ticks → None, ImageSize → 200, MaxRecursion → ControlActive[0, 2]],
  {t, 0, 2.5}, SaveDefinitions → True]
```

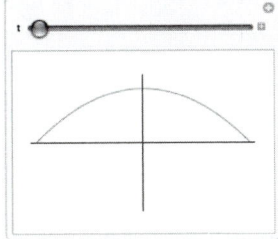

Here we only show a collection of plots:

```
GraphicsArray[Partition[
    Table[Plot[Evaluate[uappr4[r, t]], {r, -1, 1}, PlotRange → 1.25, Ticks → None],
    {t, 0, 1.1, 0.1}], 6], ImageSize → 420]
```

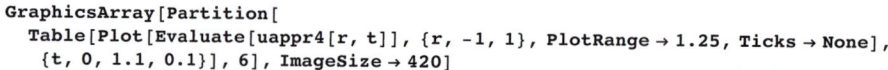

Next, we show the whole surface. To speed up the manipulation, we only take the first 10 terms of the series solution and ask not to refine the plot (**MaxRecursion → 0**):

```
Manipulate[ParametricPlot3D[Evaluate[{r Cos[θ], r Sin[θ], Take[uappr4[r, t], 10]}],
    {r, 0, 1}, {θ, 0, 2 π}, PlotRange → 1.25, Boxed → False, Axes → False,
    ImageSize → 200, MaxRecursion → 0, PlotRegion → {{0, 1}, {-0.6, 1.4}}],
    {t, 0, 2.5}, SaveDefinitions → True]
```

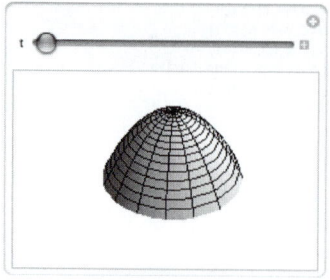

Here we only show a collection of plots:

```
ParametricPlot3D[Evaluate[{r Cos[θ], r Sin[θ], Take[uappr4[r, #], 10]}],
    {r, 0, 1}, {θ, 0, 2 π}, PlotRange → 1.25, Boxed → False, Axes → False,
    PlotRegion → {{-0.3, 1.3}, {-0.6, 1.4}}, ImageSize → 100] & /@ {0, 0.35, 0.7, 1.05}
```

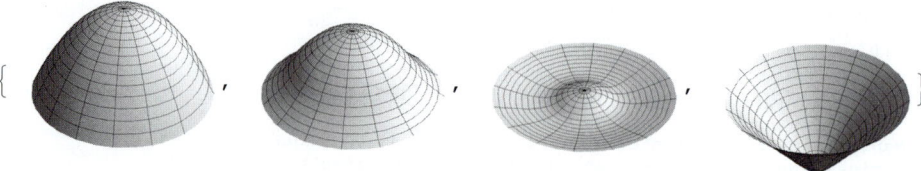

27.2.5 2D Elliptic Problems

■ Series Solution

Consider the following elliptic problem:

$$u_{xx} + u_{yy} = 0, \quad 0 < x < a, \quad 0 < y < b,$$

$$u(0, y) = u(a, y) = u(x, b) = 0, \quad u(x, 0) = f(x).$$

One interpretation of the model is as follows. A thin rectangular homogeneous thermally conducting plate lies in the (x, y) plane and occupies the rectangle $0 \leq x \leq a$, $0 \leq y \leq b$. The value of $u(x, y)$ is the steady-state temperature of the plate at point (x, y). The faces of the plate are insulated, and no internal sources or sinks of heat are present. The edge $y = 0$ is kept at temperature $f(x)$, whereas the remaining edges are kept at zero temperature.

The series solution by the method of separation of variables is as follows (Dennemeyer, 1968, pp. 147–148; the solution is derived later):

$$u(x, y) = \sum_{n=1}^{\infty} A_n \sin(v_n x) \sinh(v_n(b - y)), \quad v_n = \frac{n\pi}{a}, \quad A_n = \frac{2}{a \sinh(v_n b)} \int_0^a f(x) \sin(v_n x)\, dx.$$

■ Calculating the General Term

Consider the following example:

```
a = 1; b = 1; f = 4 x (1 - x);
```

Calculate the coefficient and the nth term of the series:

```
v_n = n π / a;
```

```
A_n = Simplify[ 2/(a Sinh[v_n b]) ∫_0^a f Sin[v_n x] dx, n ∈ Integers]
```

$$-\frac{16 \left(-1 + (-1)^n\right) \mathrm{Csch}[n\pi]}{n^3 \pi^3}$$

```
term = A_n Sin[v_n x] Sinh[v_n (b - y)]
```

$$-\frac{16 \left(-1 + (-1)^n\right) \mathrm{Csch}[n\pi] \mathrm{Sin}[n\pi x] \mathrm{Sinh}[n\pi (1 - y)]}{n^3 \pi^3}$$

■ Choosing an Approximation

To get six-digit precision, approximately 65 suffices for the upper bound of n. Here are the corresponding values of $u(0.5, y)$ for $y = 0, 0.1, ..., 0.5$:

```
Table[Sum[term, {n, 65}] /. x → 0.5, {y, 0, 0.5, 0.1}]
```

```
{1., 0.739132, 0.542517, 0.395755, 0.28663, 0.205315}
```

For five- and four-digit precision, approximately 20 and 10 terms suffice, respectively.

We choose 10 terms from the series and form an approximate solution:

```
uappr5[x_, y_] = Sum[term, {n, 1, 10}] // N
```

```
0.0893647 Sin[3.14159 x] Sinh[3.14159 (1. - 1. y)] +
  6.16932 × 10^-6 Sin[9.42478 x] Sinh[9.42478 (1. - 1. y)] +
  2.48851 × 10^-9 Sin[15.708 x] Sinh[15.708 (1. - 1. y)] +
  1.69356 × 10^-12 Sin[21.9911 x] Sinh[21.9911 (1. - 1. y)] +
  1.48804 × 10^-15 Sin[28.2743 x] Sinh[28.2743 (1. - 1. y)]
```

■ Using the Solution

Some numerical values are as follows:

```
Table[uappr5[0.5, y], {y, 0, 0.5, 0.1}]
```

```
{1.00049, 0.73915, 0.542518, 0.395755, 0.28663, 0.205315}
```

The exact value of $u(0.5, 0)$ is 1 (see the boundary condition). Next, we plot the solution:

```
Plot3D[Evaluate[uappr5[x, y]], {x, 0, 1}, {y, 0, 1}, ViewPoint → {2.0, -2.4, 0.9},
  AxesLabel → {"x", "y", ""}, Ticks → {{0, 1}, {0, 1}, {0, 1}}]
```

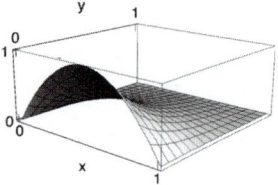

■ Derivation of the Separable Solution

As an example of the method of separation of variables, we solve the elliptic problem considered at the beginning of this section. We try to find the solution in the form $u(x, y) = X(x) Y(y)$ (i.e., in a form in which the variables x and y are separated). For the homogeneous boundary conditions to be satisfied, we set $X(0) = X(a) = 0$ and $Y(b) = 0$. For the PDE to be satisfied, we must have $X'' Y + X Y'' = 0$. This can also be written as $X''/X = -Y''/Y$. Because this must hold for all x and y, both X''/X and $-Y''/Y$ must be the same constant; let us denote it $-\lambda$. We get two ODEs, $X''/X = -\lambda$ and $Y''/Y = \lambda$. The general solution of the first equation is as follows:

```
Remove["Global`*"]

solx = DSolve[{X''[x] + λ X[x] == 0}, X, x]〚1〛
```

$$\left\{X \to \text{Function}\left[\{x\}, \text{C[1]} \cos\left[x \sqrt{\lambda}\,\right] + \text{C[2]} \sin\left[x \sqrt{\lambda}\,\right]\right]\right\}$$

From the boundary conditions for X, we get the following conditions:

```
{X[0] == 0, X[a] == 0} /. solx
```

$$\left\{\text{C[1]} == 0, \text{C[1]} \cos\left[a \sqrt{\lambda}\,\right] + \text{C[2]} \sin\left[a \sqrt{\lambda}\,\right] == 0\right\}$$

Thus, C[1] is 0. To get a nontrivial solution, we do not choose C[2] to be 0 but rather $\sqrt{\lambda}$ to be $n\pi/a$; we denote $v_n = n\pi/a$. Thus, we get $X(x) = d_1 \sin(v_n x)$. For Y, we get the following solution:

```
soly = DSolve[{Y''[y] - λ Y[y] == 0, Y[b] == 0}, Y[y], y]〚1〛 // ExpToTrig // FullSimplify
```

$$\left\{Y[y] \to 2 e^{-b \sqrt{\lambda}}\, \text{C[2]} \sinh\left[(b - y) \sqrt{\lambda}\,\right]\right\}$$

Thus, we can write $Y(y) = d_2 \sinh[v_n(b - y)]$. We denote $u_n(x, y) = A_n \sin(v_n x) \sinh[v_n(b - y)]$ and take an infinite sum of these terms to get the superposition $u(x, y) = \sum_{n=1}^{\infty} u_n(x, y)$. This satisfies all other conditions but still not the condition $u(x, 0) = f(x)$. We form the equation $u(x, 0) = f(x)$, multiply this equation by $\sin(v_m x)$, and then integrate both sides for x from 0 to a. From the resulting infinite sum on the left-hand side, only the mth term is nonzero, and it is $a/2$. This can be seen as follows:

```
Simplify[∫₀ᵃ Sin[mπx/a] Sin[nπx/a] dx, {m, n} ∈ Integers]        0
```

```
Simplify[∫₀ᵃ Sin[mπx/a]² dx, m ∈ Integers]        a/2
```

We then know that $A_m \sinh(v_m b) a/2 = \int_0^a f(x) \sin(v_m x)\, dx$. From this equation, we can solve A_m, and so we get the solution mentioned previously.

27.2.6 3D Elliptic Problems

■ Series Solution

Consider the following elliptic problem:

$$u_{xx} + u_{yy} + u_{zz} = 0, \quad 0 < x < a, \quad 0 < y < b, \quad 0 < z < c,$$

$$u(0, y, z) = u(a, y, z) = u(x, 0, z) = u(x, b, z) = u(x, y, c) = 0, \quad u(x, y, 0) = f(x, y).$$

Here, we find the steady-state temperature in a solid, the bottom of which (at $z = 0$) is kept at a temperature $f(x, y)$ and the other sides at a temperature 0. Using the method of separation of variables, we obtain the following solution (Dennemeyer, 1968, pp. 150–151):

$$u(x, y, z) = \sum_{m=1}^{\infty} \sum_{n=1}^{\infty} A_{mn} \sin(v_m x) \sin(w_n y) \sinh(\omega_{mn}(c - z)), \quad v_m = \frac{m\pi}{a}, \quad w_n = \frac{n\pi}{b},$$

$$A_n = \frac{4}{a b \sinh(\omega_{mn} c)} \int_0^a \int_0^b f(x, y) \sin(v_m x) \sin(w_n y) \, dx \, dy, \quad \omega_{mn} = \sqrt{v_m^2 + w_n^2}.$$

■ Calculating the General Term

Consider the following example:

```
a = b = c = 1; f = 20 x y (1 - x) (1 - y);
```

The highest temperature, 1.25, is at the center of the bottom. Calculate the general term:

```
       m π        n π
vm = ─── ; wn = ─── ; ωmn = √vm² + wn² ;
        a          b
```

$$A_{mn} = Simplify\left[\frac{4}{a b \, Sinh[\omega_{mn} c]} \int_0^a \left(\int_0^b f \, Sin[v_m x] \, Sin[w_n y] \, dy\right) dx, \{m, n\} \in Integers\right]$$

$$\frac{320 \left(-1 + (-1)^m\right) \left(-1 + (-1)^n\right) Csch\left[\sqrt{m^2 + n^2} \; \pi\right]}{m^3 \, n^3 \, \pi^6}$$

```
term = Amn Sin[vm x] Sin[wn y] Sinh[ωmn (c - z)]
```

$$\frac{1}{m^3 \, n^3 \, \pi^6} 320 \left(-1 + (-1)^m\right) \left(-1 + (-1)^n\right)$$

$$Csch\left[\sqrt{m^2 + n^2} \; \pi\right] Sin[m \pi x] \, Sin[n \pi y] \, Sinh\left[\sqrt{m^2 \pi^2 + n^2 \pi^2} \; (1 - z)\right]$$

■ Choosing an Approximation

To get an answer at six-digit precision, the upper bound for m and n must be approximately 65:

```
Table[Evaluate[Sum[term, {m, 65}, {n, 65}] /. {x → 0.5, y → 0.5}], {z, 0, 1, 0.2}]
{1.25, 0.534528, 0.222296, 0.0897094, 0.0316101, 0.}
```

For five- and four-digit precision, the upper bound should be approximately 35 and 15, respectively.

We take 15 as the upper value for both m and n:

```
uappr5[x_, y_, z_] = Sum[term, {m, 15}, {n, 15}] // N;
```

```
Short[uappr5[x, y, z], 6]
```

$0.0313243\ \text{Sin}[3.14159\ x]\ \text{Sin}[3.14159\ y]\ \text{Sinh}[4.44288\ (1.-1.\ z)] + \ll 62\gg +$
$2.66686 \times 10^{-36}\ \text{Sin}[47.1239\ x]\ \text{Sin}[47.1239\ y]\ \text{Sinh}[66.6432\ (1.-1.\ z)]$

■ **Using the Solution**

The steady-state temperature at some points along a vertical line inside the solid is as follows:

```
Table[uappr5[0.5, 0.5, z], {z, 0, 1, 0.2}]
```

$\{1.24969,\ 0.534528,\ 0.222296,\ 0.0897094,\ 0.0316101,\ 0.\}$

(The exact value at the bottom of the solid is 1.25.) We plot the temperature along this line:

```
Plot[uappr5[0.5, 0.5, z], {z, 0, 1}, AxesLabel → {"z", "temp"}]
```

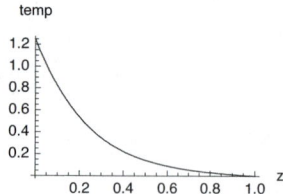

We also show a contour plot at the height $z = 0.02$ (we plot the contours only in the region $0 < x < 0.5$, $0 < y < 0.5$):

```
ContourPlot[Evaluate[uappr5[x, y, 0.02]], {x, 0, 0.5}, {y, 0, 0.5},
  Contours → Range[0.1, 1.1, 0.2], FrameLabel → {"x", "y"}, RotateLabel → False]
```

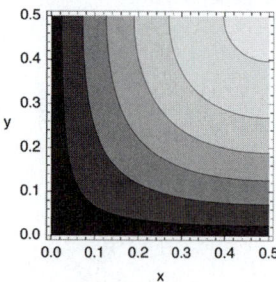

Then we plot surfaces of constant value (again in the region $0 < x < 0.5$, $0 < y < 0.5$) (for **ContourPlot3D**, see Section 5.4.2, p. 149):

```
ContourPlot3D[Evaluate[uappr5[x, y, z]], {x, 0, 0.5}, {y, 0, 0.5}, {z, 0, 0.6},
  Contours → Range[0.1, 1.1, 0.2], Axes → True, AxesLabel → {"x", "y", "z"},
  AxesEdge → {{-1, -1}, {1, -1}, {1, 1}}, ViewPoint → {3.1, -1.3, -0.2},
  ImageSize → 200, Ticks → {{0, 0.2, 0.4}, Range[0, 0.5, 0.1], Range[0, 0.6, 0.1]}}]
```

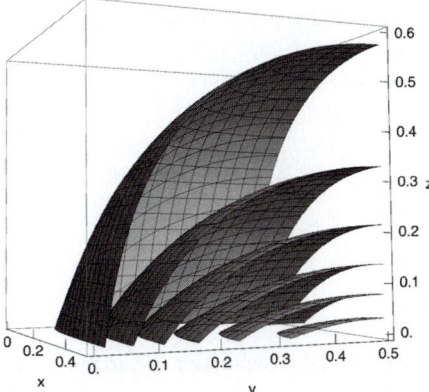

Here we can see the surfaces where the temperature is 0.1 (the highest surface), 0.3, 0.5, 0.7, 0.9, and 1.1 (the lowest surface).

27.3 Numerical Solutions

27.3.1 Parabolic and Hyperbolic Problems

■ Method of Lines

NDSolve uses the method of lines and is typically suitable for solving problems of parabolic or hyperbolic type. The problem may also consist of several equations with several dependent variables. The command is not suitable for elliptic problems; for 2D elliptic problems, we present a finite difference method in Section 27.3.4, p. 921.

In problems with one space and one time variable, initial and boundary conditions can be given on three sides in a rectangular region of the space–time plane. The boundary conditions may contain derivatives, and they may be time dependent. The problem can include periodic boundary conditions such as $u(-1, t) = u(1, t)$.

Here are typical commands for problems with one equation of parabolic or hyperbolic type:

```
sol = u[x, t] /. NDSolve[eqns, u[x, t], {x, a, b}, {t, c, d}][[1]]  Solve the problem
Plot3D[sol, {x, a, b}, {t, c, d}]  Plot the solution
```

Next, we show several examples of using **NDSolve**. In Section 27.3.2, we will demonstrate the working of the method of lines. After that, it may be easier to understand the options of **NDSolve**; the options are considered in Section 27.3.3.

For advanced information about **NDSolve**, see tutorial/NDSolveOverview.

■ Example 1: A 1D Heat Problem

We solve the same problem that we solved in Section 27.2.1, p. 894:

```
eqns = {∂ₜ u[x, t] - ∂ₓ,ₓu[x, t] == 0, u[x, 0] == x (1 - x), u[0, t] == 0, u[1, t] == 0};

sol = u[x, t] /. NDSolve[eqns, u[x, t], {x, 0, 1}, {t, 0, 0.5}][[1]]
InterpolatingFunction[{{0., 1.}, {0., 0.5}}, <>][x, t]
```

The result of **NDSolve** is a 2D interpolating function (see Section 24.2.2, p. 800). Here is the solution:

```
Plot3D[sol, {t, 0, 0.5}, {x, 0, 1}, PlotRange → All,
  AxesLabel → {"t", "x", ""}, Ticks → {{0, 0.5}, {0, 1}, {0.1, 0.2}}]
```

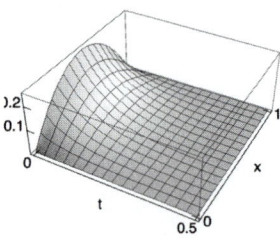

In Section 27.2.1, we obtained the following values by the method of separation of variables, using 63 terms:

```
{0.25, 0.0961619, 0.0358408, 0.0133581, 0.00497868, 0.00185559}
```

These numbers were of approximately six-digit precision. Here are the corresponding values from the solution given by **NDSolve**:

```
Table[sol /. x → 0.5, {t, 0, 0.5, 0.1}]
{0.25, 0.0961717, 0.0359048, 0.0133206, 0.00494963, 0.00182372}
```

The numbers are of approximately two-digit precision. To get a more accurate result, we can set the goals for precision and accuracy:

```
sol2 = u[x, t] /. NDSolve[eqns, u[x, t],
    {x, 0, 1}, {t, 0, 0.5}, PrecisionGoal → 6, AccuracyGoal → 6][[1]];
```

Now we get numbers of approximately four-digit precision:

```
Table[sol2 /. x → 0.5, {t, 0, 0.5, 0.1}]
{0.25, 0.0961619, 0.0358402, 0.0133585, 0.00497952, 0.00185651}
```

■ Example 2: A 1D Wave Problem

We solve the same problem that we solved in Section 27.2.2, p. 896:

```
eqns = {∂ₜ,ₜu[x, t] - ∂ₓ,ₓu[x, t] == -9.80665`20, u[x, 0] == 10 x² (1 - x)²,
    Derivative[0, 1][u][x, 0] == 0, u[0, t] == 0, u[1, t] == 0};
```

Note that **Derivative** is handy for specifying both the orders of the derivative and the point at which it is calculated (see Section 19.1.1, p. 618). We could also have written (**D[u[x, t], t] /. t → 0) == 0**. The solution is as follows:

```
sol = u[x, t] /. NDSolve[eqns, u[x, t], {x, 0, 1}, {t, 0, 4}, PrecisionGoal → 3][[1]]
InterpolatingFunction[{{0., 1.}, {0., 4.}}, <>][x, t]
```

```
Plot3D[sol, {t, 0, 4}, {x, 0, 1}, AxesLabel → {"t", "x", ""},
  Ticks → {{0, 1, 2, 3, 4}, {0, 1}, {-3, 0}}]
```

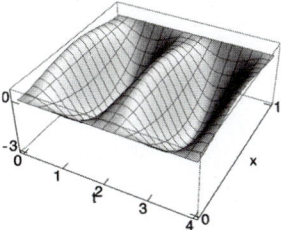

In Section 27.2.2, we obtained the following values via the method of separation of variables using 150 terms:

```
{0.625, -0.703533, -2.69653, -2.69653, -0.703533, 0.625, -0.703533}
```

These numbers were of approximately six-digit precision. Here are the corresponding values from the solution given by **NDSolve**:

```
Table[sol /. x → 0.5, {t, 0, 2.4, 0.4}]
```
```
{0.625, -0.703436, -2.69782, -2.69656, -0.704925, 0.624696, -0.702448}
```

The numbers have approximately three digits of precision.

■ Example 3: A 2D Wave Problem

We solve the same problem that we solved in Section 27.2.3, p. 899:

```
eqns = {∂t,t u[x, y, t] - (∂x,x u[x, y, t] + ∂y,y u[x, y, t]) == 0,
  u[x, y, 0] == 10 x y (1 - x) (1 - y), Derivative[0, 0, 1][u][x, y, 0] == 0,
  u[0, y, t] == u[1, y, t] == u[x, 0, t] == u[x, 1, t] == 0};
sol = u[x, y, t] /. NDSolve[eqns, u[x, y, t], {x, 0, 1}, {y, 0, 1}, {t, 0, 1.4}][[1]]
```
```
InterpolatingFunction[{{0., 1.}, {0., 1.}, {0., 1.4}}, <>][x, y, t]
```

Plot the surface at times $t = 0, 0.2, 0.4$, and 0.6:

```
Plot3D[Evaluate[sol /. t → #], {x, 0, 1}, {y, 0, 1},
  PlotRange → 0.71, Boxed → False, Axes → False, BoxRatios → {1, 1, 1},
  PlotRegion → {{0, 1}, {-0.4, 1.3}}, ImageSize → 100] & /@ {0, 0.2, 0.4, 0.6}
```

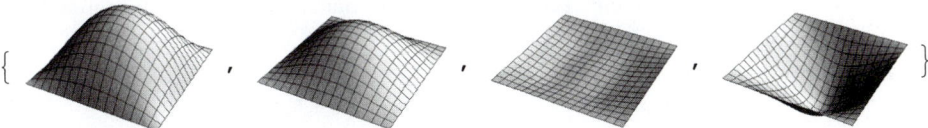

■ Example 4: A System of Equations

In the following example, we have two dependent variables, u and v:

```
eqns = {∂t u[x, t] - ∂x,x u[x, t] == v[x, t], ∂t,t v[x, t] - ∂x,x u[x, t] == 0,
  u[x, 0] == x (1 - x), u[0, t] == 0, u[1, t] == 0, v[x, 0] == 10 x² (1 - x)²,
  Derivative[0, 1][v][x, 0] == 0, v[0, t] == 0, v[1, t] == 0};
sol = {u[x, t], v[x, t]} /. NDSolve[eqns, {u[x, t], v[x, t]}, {x, 0, 1}, {t, 0, 4}][[1]]
```
```
{InterpolatingFunction[{{0., 1.}, {0., 4.}}, <>][x, t],
  InterpolatingFunction[{{0., 1.}, {0., 4.}}, <>][x, t]}
```

```
Plot3D[#, {t, 0, 4}, {x, 0, 1}, PlotRange → All, AxesLabel → {"t", "x", ""}] & /@ sol
```

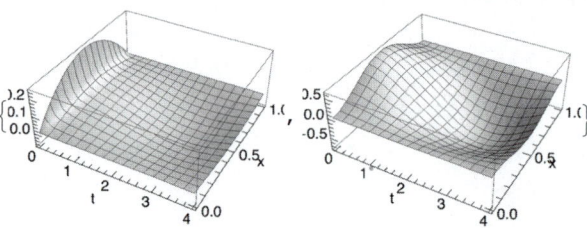

27.3.2 Method of Lines

■ **Method of Lines**

When using the method of lines, a typical situation is as follows:

```
Graphics[{Table[Line[{{0, i}, {10, i}}], {i, 5}],
  Thickness[Medium], Line[{{10, 0}, {0, 0}, {0, 6}, {10, 6}}],
  Text[u[x, 0] == f[x], {-0.3, 3}, {1, 0}], Text[u[0, t] == b₀[t], {5, -0.3}, {0, 1}],
  Text[u[a, t] == b₁[t], {5, 6.2}, {0, -1}]}, ImageSize → 180]
```

$u(a, t) = b_1(t)$

$u(x, 0) = f(x)$

$u(0, t) = b_0(t)$

The x axis goes from bottom to top and the t axis from left to right. At $t = 0$, we have initial conditions such as $u(x, 0) = f(x)$. At $x = 0$ and $x = a$, we have boundary conditions such as $u(0, t) = b_0(t)$ and $u(a, t) = b_1(t)$ [periodic boundary conditions of the form $u(0, t) = u(a, t)$ can also be given].

Divide the interval $[0, a]$ into n subintervals of length $h = a/n$ with the points $x_i = ih$, $i = 0, 1, ..., n$. Let $u_i(t)$ be the solution along the line $x = x_i$; that is, $u_i(t) = u(x_i, t)$. To derive differential equations for the $u_i(t)$ functions, approximate the spatial derivatives u_x and u_{xx} by finite differences. Here are examples of such approximations:

$$u_x(x, t) \simeq \frac{u_{i+1}(t) - u_{i-1}(t)}{2h}, \quad u_{xx}(x, t) \simeq \frac{u_{i+1}(t) - 2u_i(t) + u_{i-1}(t)}{h^2}, \quad i = 1, ..., n-1.$$

Thus, if the PDE is $F(x, t, u, u_x, u_{xx}, u_t, u_{tt}) = 0$, we obtain a system of $n - 1$ simultaneous ODEs:

$$F\left(x_i, t, u_i(t), \frac{u_{i+1}(t) - u_{i-1}(t)}{2h}, \frac{u_{i+1}(t) - 2u_i(t) + u_{i-1}(t)}{h^2}, u_i'(t), u_i''(t)\right) = 0, \quad i = 1, ..., n-1.$$

The solutions along the lines $x = 0$ and $x = a$ are known from the boundary conditions, and from the initial conditions we get initial conditions for the ODEs. The simultaneous system can then be solved using **NDSolve**.

■ **Example: The Wave Equation**

Let us see how the method of lines proceeds by solving the following familiar wave problem:

$$u_{tt} - c^2 u_{xx} = F(x, t), \quad 0 < x < a, \quad t > 0,$$

$$u(x, 0) = f(x), \quad u_t(x, 0) = g(x), \quad u(0, t) = u(a, t) = 0.$$

Define

```
a = 1; c = 1; F = -9.80665; f = 10 x² (1 - x)²; g = 0;
n = 10; h = a / n; vars = Table[uᵢ[t], {i, 0, n}];
```

Create the set of space-discretized differential equations from $u_{tt} - c^2 u_{xx} = F(x, t)$:

$$\text{eqns = Table}\left[u_i{}''[t] - c^2 \frac{u_{i-1}[t] - 2 u_i[t] + u_{i+1}[t]}{h^2} == F \; / . \; x \to i\,h, \; \{i, 1, n-1\}\right]$$

```
{-100 (u₀[t] - 2 u₁[t] + u₂[t]) + u₁″[t] == -9.80665,
 -100 (u₁[t] - 2 u₂[t] + u₃[t]) + u₂″[t] == -9.80665,
 -100 (u₂[t] - 2 u₃[t] + u₄[t]) + u₃″[t] == -9.80665,
 -100 (u₃[t] - 2 u₄[t] + u₅[t]) + u₄″[t] == -9.80665,
 -100 (u₄[t] - 2 u₅[t] + u₆[t]) + u₅″[t] == -9.80665,
 -100 (u₅[t] - 2 u₆[t] + u₇[t]) + u₆″[t] == -9.80665,
 -100 (u₆[t] - 2 u₇[t] + u₈[t]) + u₇″[t] == -9.80665,
 -100 (u₇[t] - 2 u₈[t] + u₉[t]) + u₈″[t] == -9.80665,
 -100 (u₈[t] - 2 u₉[t] + u₁₀[t]) + u₉″[t] == -9.80665}
```

Form the initial conditions from $u(x, 0) = f(x)$ and $u_t(x, 0) = g(x)$:

```
inits1 = Table[uᵢ[0] == f /. x → i h, {i, 1, n - 1}]
```

$$\left\{u_1[0] == \frac{81}{1000}, \; u_2[0] == \frac{32}{125}, \; u_3[0] == \frac{441}{1000}, \; u_4[0] == \frac{72}{125}, \right.$$

$$\left. u_5[0] == \frac{5}{8}, \; u_6[0] == \frac{72}{125}, \; u_7[0] == \frac{441}{1000}, \; u_8[0] == \frac{32}{125}, \; u_9[0] == \frac{81}{1000}\right\}$$

```
inits2 = Table[Derivative[1][uᵢ][0] == g /. x → i h, {i, 1, n - 1}]
```

```
{u₁′[0] == 0, u₂′[0] == 0, u₃′[0] == 0, u₄′[0] == 0,
 u₅′[0] == 0, u₆′[0] == 0, u₇′[0] == 0, u₈′[0] == 0, u₉′[0] == 0}
```

Add the boundary conditions $u_0(t) = 0$ and $u_n(t) = 0$. In this way, we get a differential-algebraic system. Solve the system when t is, for example, in the interval $(0, 4)$:

```
sol = NDSolve[{eqns, inits1, inits2, u₀[t] == 0, uₙ[t] == 0}, vars, {t, 0, 4}]〚1〛
```

```
{u₀[t] → InterpolatingFunction[{{0., 4.}}, <>][t],
 u₁[t] → InterpolatingFunction[{{0., 4.}}, <>][t],
 u₂[t] → InterpolatingFunction[{{0., 4.}}, <>][t],
 u₃[t] → InterpolatingFunction[{{0., 4.}}, <>][t],
 u₄[t] → InterpolatingFunction[{{0., 4.}}, <>][t],
 u₅[t] → InterpolatingFunction[{{0., 4.}}, <>][t],
 u₆[t] → InterpolatingFunction[{{0., 4.}}, <>][t],
 u₇[t] → InterpolatingFunction[{{0., 4.}}, <>][t],
 u₈[t] → InterpolatingFunction[{{0., 4.}}, <>][t],
 u₉[t] → InterpolatingFunction[{{0., 4.}}, <>][t],
 u₁₀[t] → InterpolatingFunction[{{0., 4.}}, <>][t]}
```

Thus, now we know approximate values of the solution along the lines $x = x_i$, $i = 0, 1, \ldots, n$. Plot the solution along these lines:

```
ParametricPlot3D[Evaluate[Table[{t, i h, u_i[t] /. sol}, {i, 0, n}]],
 {t, 0, 4}, AxesLabel → {t, x, ""}, BoxRatios → {1, 1, 0.4},
 Ticks → {{0, 1, 2, 3, 4}, {0, 1}, {-3, 0}}]
```

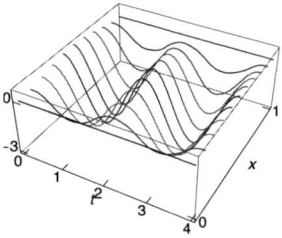

To get a surface plot, tabulate the solution along the lines:

```
ListPlot3D[Table[vars /. sol, {t, 0, 4, 0.1}]^T,
 DataRange → {{0, 4}, {0, 1}}, Ticks → {{0, 1, 2, 3, 4}, {0, 1}, {-3, 0}}]
```

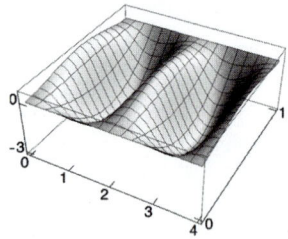

■ Approximating Derivatives with Finite Differences

Mathematica has a command to calculate various finite difference approximations of derivatives.

`NDSolve`FiniteDifferenceDerivative[m, grid, values, DifferenceOrder → n]` Calculate an nth-order finite difference approximation for the values of the mth-order derivative of a function that takes on **values** on **grid**; the default value of **n** is 4

As an example, let the values of a function $f(t)$ at the points ih, $i = 0, 1, \ldots, 5$, be as follows:

```
h =.; f =.; grid = h Range[0, 5]
{0, h, 2 h, 3 h, 4 h, 5 h}
```

```
values = f /@ grid
{f[0], f[h], f[2 h], f[3 h], f[4 h], f[5 h]}
```

Here are the second-order approximations for the first-order derivative:

```
NDSolve`FiniteDifferenceDerivative[1, grid, values, DifferenceOrder → 2] //
 Simplify // TraditionalForm
```
$$\left\{-\frac{3 f(0) - 4 f(h) + f(2 h)}{2 h}, \frac{f(2 h) - f(0)}{2 h}, \frac{f(3 h) - f(h)}{2 h}, \frac{f(4 h) - f(2 h)}{2 h}, \frac{f(5 h) - f(3 h)}{2 h}, \frac{f(3 h) - 4 f(4 h) + 3 f(5 h)}{2 h}\right\}$$

At points $x = h, 2h, 3h$, and $4h$ we got the familiar difference formula, but note that we also got second-order approximations at $x = 0$ and $x = 5h$. Next we calculate the fourth-order approximations; this is the default, so we do not need the option:

```
NDSolve`FiniteDifferenceDerivative[1, grid, values] // Simplify // TraditionalForm
```

$$\left\{ -\frac{25\,f(0) - 48\,f(h) + 36\,f(2\,h) - 16\,f(3\,h) + 3\,f(4\,h)}{12\,h}, \quad \frac{-3\,f(0) - 10\,f(h) + 18\,f(2\,h) - 6\,f(3\,h) + f(4\,h)}{12\,h}, \right.$$

$$\frac{f(0) - 8\,f(h) + 8\,f(3\,h) - f(4\,h)}{12\,h}, \quad \frac{f(h) - 8\,f(2\,h) + 8\,f(4\,h) - f(5\,h)}{12\,h},$$

$$\left. -\frac{f(h) - 6\,f(2\,h) + 18\,f(3\,h) - 10\,f(4\,h) - 3\,f(5\,h)}{12\,h}, \quad \frac{3\,f(h) - 16\,f(2\,h) + 36\,f(3\,h) - 48\,f(4\,h) + 25\,f(5\,h)}{12\,h} \right\}$$

At $x = 2\,h$ and $3\,h$ we got the usual fourth-order approximation, but note that we also got fourth-order approximations at $x = 0, h, 4\,h$, and $5\,h$.

Here are formulas for second- and fourth-order approximations of the second-order derivative:

```
NDSolve`FiniteDifferenceDerivative[2, grid, values, DifferenceOrder → 2] //
   Simplify // TraditionalForm
```

$$\left\{ \frac{2\,f(0) - 5\,f(h) + 4\,f(2\,h) - f(3\,h)}{h^2}, \quad \frac{f(0) - 2\,f(h) + f(2\,h)}{h^2}, \quad \frac{f(h) - 2\,f(2\,h) + f(3\,h)}{h^2}, \right.$$

$$\left. \frac{f(2\,h) - 2\,f(3\,h) + f(4\,h)}{h^2}, \quad \frac{f(3\,h) - 2\,f(4\,h) + f(5\,h)}{h^2}, \quad -\frac{f(2\,h) - 4\,f(3\,h) + 5\,f(4\,h) - 2\,f(5\,h)}{h^2} \right\}$$

```
NDSolve`FiniteDifferenceDerivative[2, grid, values] // Simplify // TraditionalForm
```

$$\left\{ \frac{45\,f(0) - 154\,f(h) + 214\,f(2\,h) - 156\,f(3\,h) + 61\,f(4\,h) - 10\,f(5\,h)}{12\,h^2}, \right.$$

$$\frac{10\,f(0) - 15\,f(h) - 4\,f(2\,h) + 14\,f(3\,h) - 6\,f(4\,h) + f(5\,h)}{12\,h^2}, \quad -\frac{f(0) - 16\,f(h) + 30\,f(2\,h) - 16\,f(3\,h) + f(4\,h)}{12\,h^2},$$

$$-\frac{f(h) - 16\,f(2\,h) + 30\,f(3\,h) - 16\,f(4\,h) + f(5\,h)}{12\,h^2}, \quad \frac{f(0) - 6\,f(h) + 14\,f(2\,h) - 4\,f(3\,h) - 15\,f(4\,h) + 10\,f(5\,h)}{12\,h^2},$$

$$\left. \frac{-10\,f(0) + 61\,f(h) - 156\,f(2\,h) + 214\,f(3\,h) - 154\,f(4\,h) + 45\,f(5\,h)}{12\,h^2} \right\}$$

If the grid and the function values are numerical, we get numerical approximations of derivatives. As an example, we calculate fourth-order approximations of the second-order derivative of $\sin(x)$ at some given points:

```
grid = Range[0, π, π / 10]; values = Table[Sin[x], {x, 0., π, π / 10}];
```

```
fdd = NDSolve`FiniteDifferenceDerivative[2, grid, values]
```

```
{0.0044102, -0.309449, -0.587722, -0.80893, -0.950954,
 -0.999893, -0.950954, -0.80893, -0.587722, -0.309449, 0.0044102}
```

The approximations are quite near the true values:

```
(-values) - fdd
```

```
{-0.0044102, 0.00043234, -0.0000630593, -0.0000867937, -0.000102032, -0.000107283,
 -0.000102032, -0.0000867937, -0.0000630593, 0.00043234, -0.0044102}
```

■ Example: Using Higher-Order Approximations

Let us now again solve the wave problem we considered previously:

$$u_{tt} - c^2\,u_{xx} = F(x, t), \quad 0 < x < a, \quad t > 0,$$

$$u(x, 0) = f(x), \quad u_t(x, 0) = g(x), \quad u(0, t) = u(a, t) = 0.$$

Define, as previously,

```
a = 1; c = 1; F = -9.80665; f = 10 x² (1 - x)²; g = 0;
n = 10; h = a / n; vars = Table[uᵢ[t], {i, 0, n}];
```

Create a set of space-discretized differential equations from $u_{tt} - c^2 u_{xx} = F(x, t)$, now by using fourth-order finite differences for spatial derivatives:

```
(eqns = Thread[D[vars, t, t] -
      c^2 NDSolve`FiniteDifferenceDerivative[2, Range[0., a, a / n], vars] ==
      Table[F /. x → i h, {i, 0, n}]]) // TraditionalForm
```

$\{-375. u_0(t) + 1283.33 u_1(t) - 1783.33 u_2(t) + 1300. u_3(t) - 508.333 u_4(t) + 83.3333 u_5(t) + u_0''(t) = -9.80665,$

$-83.3333 u_0(t) + 125. u_1(t) + 33.3333 u_2(t) - 116.667 u_3(t) + 50. u_4(t) - 8.33333 u_5(t) + u_1''(t) = -9.80665,$

$8.33333 u_0(t) - 133.333 u_1(t) + 250. u_2(t) - 133.333 u_3(t) + 8.33333 u_4(t) + u_2''(t) = -9.80665,$

$8.33333 u_1(t) - 133.333 u_2(t) + 250. u_3(t) - 133.333 u_4(t) + 8.33333 u_5(t) + u_3''(t) = -9.80665,$

$8.33333 u_2(t) - 133.333 u_3(t) + 250. u_4(t) - 133.333 u_5(t) + 8.33333 u_6(t) + u_4''(t) = -9.80665,$

$8.33333 u_3(t) - 133.333 u_4(t) + 250. u_5(t) - 133.333 u_6(t) + 8.33333 u_7(t) + u_5''(t) = -9.80665,$

$8.33333 u_4(t) - 133.333 u_5(t) + 250. u_6(t) - 133.333 u_7(t) + 8.33333 u_8(t) + u_6''(t) = -9.80665,$

$8.33333 u_5(t) - 133.333 u_6(t) + 250. u_7(t) - 133.333 u_8(t) + 8.33333 u_9(t) + u_7''(t) = -9.80665,$

$8.33333 u_6(t) - 133.333 u_7(t) + 250. u_8(t) - 133.333 u_9(t) + 8.33333 u_{10}(t) + u_8''(t) = -9.80665,$

$-8.33333 u_5(t) + 50. u_6(t) - 116.667 u_7(t) + 33.3333 u_8(t) + 125. u_9(t) - 83.3333 u_{10}(t) + u_9''(t) = -9.80665,$

$83.3333 u_5(t) - 508.333 u_6(t) + 1300. u_7(t) - 1783.33 u_8(t) + 1283.33 u_9(t) - 375. u_{10}(t) + u_{10}''(t) = -9.80665\}$

However, from the boundary conditions we know that $u_0(t) = 0$ and $u_{10}(t) = 0$:

```
eqns⟦1⟧ = u₀[t] == 0; eqns⟦11⟧ = uₙ[t] == 0;
```

Form the initial conditions from $u(x, 0) = f(x)$ and $u_t(x, 0) = g(x)$, as previously:

```
inits1 = Table[uᵢ[0] == f /. x → i h, {i, 1, n - 1}];
```

```
inits2 = Table[Derivative[1][uᵢ][0] == g /. x → i h, {i, 1, n - 1}];
```

Solve the differential-algebraic system when t is, for example, in the interval (0, 4):

```
sol = NDSolve[{eqns, inits1, inits2}, vars, {t, 0, 4}]⟦1⟧;
```

Plot the solution:

```
ListPlot3D[Table[vars /. sol, {t, 0, 4, 0.1}]ᵀ,
  DataRange → {{0, 4}, {0, 1}}, Ticks → {{0, 1, 2, 3, 4}, {0, 1}, {-3, 0}}]
```

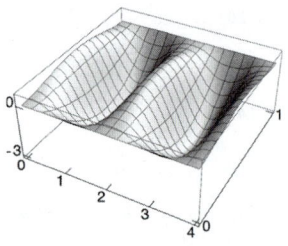

27.3.3 Options

■ **Options of NDSolve**

The options of **NDSolve** were mentioned in Section 26.4.1, p. 865, in the context of ODEs. The same options can also be used for PDEs. The meanings of some options, however, are somewhat new. In particular, some options accept a several-component list as a value; this reflects the number of independent variables. Which value belongs to which variable is inferred from the order of the independent variables in **NDSolve**.

Options of **NDSolve** *when solving PDEs:*

WorkingPrecision Precision used in internal computations; examples of values: **MachinePrecision, 20**

PrecisionGoal If the value of the option is **p**, the relative error of the solution at each point considered should be of the order 10^{-p}; examples of values: **Automatic** (usually means **4**), **6**, **{6, 8}**

AccuracyGoal If the value of the option is **a**, the absolute error of the solution at each point considered should be of the order 10^{-a}; examples of values: **Automatic** (usually means **4**), **6**, **{6, 8}**

Method Method to use; possible values: **Automatic** (means **"Adams"** for nonstiff and **"BDF"** for stiff problems), **"Adams"**, **"BDF"**, **"ExplicitRungeKutta"**, **"ImplicitRungeKutta"**, **"SymplecticPartitionedRungeKutta"**, **"MethodOfLines"**

StartingStepSize Initial step size used; examples of values: **Automatic, 0.01, {0.01, 0.02}**

MaxStepSize Maximum size of each step; examples of values: **Automatic, 0.01, {0.01, 0.03}**

MaxStepFraction Maximum fraction of the solution interval to cover in each step; examples of values: **1/10, 0.05**

MaxSteps Maximum number of steps to take; examples of values: **10000, 20000, {1000, 5000}**

NormFunction Norm to use for error estimation in systems of equations; default value: **Automatic** (mostly means ∞), **1, 2, ∞**

DependentVariables List of all dependent variables; default value: **Automatic**

InterpolationOrder The continuity degree of the final output; examples of values: **Automatic, All** (the same as the underlying method used)

SolveDelayed Whether the derivatives are solved symbolically at the beginning (**False**) or at each step (**True**); possible values: **False, True**

Compiled Whether to compile the equations; possible values: **Automatic, True, False**

StepMonitor Command to be executed after each step of the method; examples of values: **None,** **Sow[ListPlot[u[x, t], PlotRange → {-3, 1}]**

EvaluationMonitor Command to be executed after each evaluation of the equation; examples of values: **None, Sow[ListPlot[u[x, t], PlotRange → {-3, 1}]**

The **Method** option now determines the method used to solve the space-discretized system. To use some special options of the method of lines, we have to use **"MethodOfLines"**.

With **StartingStepSize, MaxStepSize**, and **MaxSteps**, we can specify properties of the steps separately for each independent variable. If one value is given, it is used for all variables.

To determine a suitable *spatial* grid (the lines), **NDSolve** uses spatial error estimates on the initial conditions. A suitable *temporal* grid is determined by the adaptive ODE methods.

■ **Using the StepMonitor Option**

Consider again the familiar wave problem:

$$\text{eqns} = \{\partial_{t,t}u[x, t] - \partial_{x,x}u[x, t] == -9.80665`20, u[x, 0] == 10\,x^2\,(1-x)^2,$$
$$\text{Derivative}[0, 1][u][x, 0] == 0, u[0, t] == 0, u[1, t] == 0\};$$

To see the state of the computation for various values of t, use the **StepMonitor** option. The default is to treat the value of the solution for a given t as an interpolating function so that we can, for example, plot the solution for a given t:

```
{sol, {plots}} =
  Reap[NDSolve[eqns, u, {x, 0, 1}, {t, 0, 4}, PrecisionGoal → 3, StepMonitor :>
    Sow[Plot[u[x, t], {x, 0, 1}, PlotRange → {-3.2, 0.7}, ImageSize → 200]]]];
```

Now the variable **plots** contains the plots of the solution for all the values of t that **NDSolve** used during the solution. With **Show[plots]** we can show all the curves in one plot. To get an animation of the plots, do as follows:

```
ListAnimate[plots, AnimationRunning → False]
```

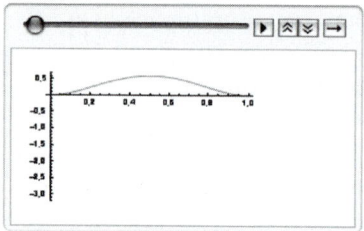

■ **Special Options of the Method of Lines**

The method of lines has some special options. They can be asked as follows:

```
Options[NDSolve`MethodOfLines]
```

```
{DifferentiateBoundaryConditions → True, DiscretizedMonitorVariables → False,
 ExpandEquationsSymbolically → False, Method → Automatic,
 SpatialDiscretization → TensorProductGrid, TemporalVariable → Automatic}
```

Here are three of the special options:

Some special options of the method of lines:

"SpatialDiscretization"   What method to use for spatial discretization; possible value: **"TensorProductGrid"**

"DifferentiateBoundaryConditions"   Whether to differentiate the boundary conditions with respect to the temporal variable; possible values: **True, False**

"DiscretizedMonitorVariables"   Whether to interpret, for a given t, the dependent variable (given, for example, in monitors such as **StepMonitor**) as an interpolating function (**False**) or as a list of values (**True**); possible values: **False, True**

Currently, the only method implemented for **"SpatialDiscretization"** is the **"TensorProductGrid"** method. It uses discretization methods for each spatial dimension and then uses an outer tensor product of these grids to derive a grid for multiple spatial dimensions on rectangular regions. In the next subsection, we consider the options of the **"TensorProductGrid"** method.

To handle the boundary conditions, we have two methods; the method can be selected with the **"DifferentiateBoundaryConditions"** option. In the first method—this is the default—the boundary conditions are differentiated with respect to the temporal variable and the resulting differential equations are added to the set of ODEs. In the second method, the boundary conditions are either used as such [in simple cases such as $u(0, t) = b_0(t)$] or discretized (if they contain spatial derivatives); this leads to algebraic equations so that the result is a differential-algebraic system.

If **"DiscretizedMonitorVariables"** has the default value **False**, we can use, in **StepMonitor**, **Plot** to plot the solution for the various values of t; see a previous example. If the value of the option is **True**, then we can, in **StepMonitor**, use **ListPlot** to plot the points.

The special options of the method of lines are written as suboptions of the **Method** option as follows: **Method → {"MethodOfLines", specOpts}**; see examples in the next subsection.

■ Special Options of the Tensor Product Grid

The **"TensorProductGrid"** method has the following special options:

> **Options[NDSolve`MethodOfLines`TensorProductGrid]**
>
> {AccuracyGoal → Automatic, Coordinates → Automatic,
> DifferenceOrder → Automatic, MaxPoints → Automatic, MaxStepSize → Automatic,
> MinPoints → Automatic, MinStepSize → Automatic, PrecisionGoal → Automatic,
> StartingPoints → Automatic, StartingStepSize → Automatic}

Two of these options are explained here. These options are written as suboptions of the **"SpatialDiscretization"** option; see the examples that follow.

Some special options of **"TensorProductGrid"**:

"DifferenceOrder"   The order of finite difference approximation to use for spatial discretization; examples of values: **Automatic** (usually means **4**), **2**, **4**, **6**, **"Pseudospectral"**

"MinPoints"   The minimum number of points to be used for each dimension in the grid; examples of values: **Automatic**, **300**

As an example of the use of these options, consider the familiar wave problem:

> **eqns = {∂$_{t,t}$u[x, t] − ∂$_{x,x}$u[x, t] == −9.80665`20, u[x, 0] == 10 x^2 (1 − x)^2,**
>   **Derivative[0, 1][u][x, 0] == 0, u[0, t] == 0, u[1, t] == 0};**
> **sol = u /. NDSolve[eqns, u, {x, 0, 1}, {t, 0, 4}][[1]]**

NDSolve::eerr :
  Warning: Scaled local spatial error estimate of 17.55676955558677` at t = 4.` in the
      direction of independent variable x is much greater than prescribed error
      tolerance. Grid spacing with 25 points may be too large to achieve the desired
      accuracy or precision. A singularity may have formed or you may want to specify
      a smaller grid spacing using the MaxStepSize or MinPoints method options. ≫
InterpolatingFunction[{{0., 1.}, {0., 4.}}, <>]

```
Plot3D[sol[x, t], {t, 0, 4}, {x, 0, 1},
  AxesLabel → {"t", "x", ""}, Ticks → {{0, 1, 2, 3, 4}, {0, 1}, {-3, 0}}]
```

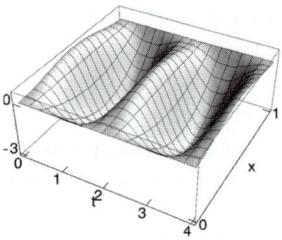

Although the solution seems to be good, we got a warning that the spatial error at the end of the time interval is not small enough. Recall that to determine a suitable spatial grid (the lines), **NDSolve** uses spatial error estimates on the initial conditions. In addition to this *a priori* error estimation, an *a posteriori* error estimation at the end of the time interval is also carried out. In our example, this *a posteriori* checking revealed that the spatial error exceeds the tolerance. We can ask for the number of points used in each dimension:

```
<< DifferentialEquations`InterpolatingFunctionAnatomy`
```

```
Length /@ InterpolatingFunctionCoordinates[sol]
```
{25, 66}

Thus, from the *x* interval (0, 1) **NDSolve** used 25 points—that is, the method of lines used 25 lines (this is the default)—whereas from the *t* interval (0, 4) **NDSolve** used 66 points—that is, the ODEs were solved by using 66 time points. The warning message states that 25 points may be too few, giving rise to a too large grid spacing. The reason may be a singularity. If this is not the case, to resolve the problem we could try the **MaxStepSize** or **MinPoints** option.

Let us try the **MinPoints** option and require the use of at least 300 spatial points:

```
sol = u /. NDSolve[eqns, u, {x, 0, 1}, {t, 0, 4}, Method → {"MethodOfLines",
    "SpatialDiscretization" → {"TensorProductGrid", "MinPoints" → 300}}] [[1]]
InterpolatingFunction[{{0., 1.}, {0., 4.}}, <>]
```

The problem disappeared. We can see that now 301 spatial points and 265 temporal points were used:

```
Length /@ InterpolatingFunctionCoordinates[sol]
```
{301, 265}

Note that to use the **"MinPoints"** option, we have to also use the **Method** option, although the **"MethodOfLines"** is the only method available, and we have to specify to use the **"TensorProductGrid"** method, although this is the only spatial discretization method available.

Another possibility to resolve the problem of a too large local spatial error may be to use a higher-order difference approximation for the spatial derivatives (the default is to use fourth-order approximation). This works in our example:

```
sol = u /. NDSolve[eqns, u, {x, 0, 1}, {t, 0, 4}, Method → {"MethodOfLines",
    "SpatialDiscretization" → {"TensorProductGrid", "DifferenceOrder" → 8}}] [[1]]
InterpolatingFunction[{{0., 1.}, {0., 4.}}, <>]
```

Using **"DifferenceOrder" → "Pseudospectral"** also resolves the problem. Still another possibility to resolve the problem is to lower the precision requirement from the default value 4:

```
u /. NDSolve[eqns, u, {x, 0, 1}, {t, 0, 4}, PrecisionGoal → 3] [[1]]
InterpolatingFunction[{{0., 1.}, {0., 4.}}, <>]
```

27.3.4 2D Elliptic Problems

■ A Finite Difference Method

Consider the following elliptic problem:

$$u_{xx} + u_{yy} = F(x, y), \quad x_0 < x < x_1, \quad y_0 < y < y_1,$$

$$u(x, y_1) = a_1(x),$$

$$u(x_0, y) = b_0(y), \qquad u(x_1, y) = b_1(y),$$

$$u(x, y_0) = a_0(x).$$

To derive a finite difference method for this problem, we divide the interval (x_0, x_1) into n_x subintervals of length h_x, the interval (y_0, y_1) into n_y subintervals of length h_y, and then obtain a mesh of $(n_x + 1)(n_y + 1)$ points. Denote the value of u at a mesh point with $u_{i,j}$. Then approximate the partial derivatives with finite differences at the mesh points. The following formula uses the familiar finite difference approximation of a second-order derivative:

$$\frac{1}{h_x^2} \left(u_{i+1,j} - 2 u_{i,j} + u_{i-1,j} \right) + \frac{1}{h_y^2} \left(u_{i,j+1} - 2 u_{i,j} + u_{i,j-1} \right) = F_{i,j}.$$

In this way, we obtain a system of $(n_x - 1)(n_y - 1)$ linear equations. The boundary conditions can be inserted into these equations or used as further equations. The solution of the linear system is an approximate solution of the original problem.

■ Example

As an example, we solve the following elliptic problem:

$$u_{xx} + u_{yy} = 0, \quad 0 < x < 1, \quad 0 < y < 1,$$

$$u(x, 0) = 4 x(1 - x), \quad u(x, 1) = u(0, y) = u(1, y) = 0.$$

This is the same problem for which we considered the series solution in Section 27.2.5, p. 904. We use 30 subintervals in both directions:

```
Remove["Global`*"]
```

```
F = 0; a₀ = 4 x (1 - x); a₁ = 0; b₀ = 0; b₁ = 0;
x₀ = 0; x₁ = 1; y₀ = 0; y₁ = 1; nₓ = 30; n_y = 30;
```

Calculate the mesh points and define the variables:

```
hₓ = N[(x₁ - x₀) / nₓ]; h_y = N[(y₁ - y₀) / n_y];
Do[ξᵢ = x₀ + i hₓ, {i, 0, nₓ}];
Do[ηⱼ = y₀ + j h_y, {j, 0, n_y}];
vars = Table[uᵢ,ⱼ, {i, 0, nₓ}, {j, 0, n_y}];
```

Apply the boundary conditions:

```
bound1 = Table[{uᵢ,₀ == a₀ /. x → ξᵢ, uᵢ,n_y == a₁ /. x → ξᵢ}, {i, 0, nₓ}];
bound2 = Table[{u₀,ⱼ == b₀ /. y → ηⱼ, unₓ,ⱼ == b₁ /. y → ηⱼ}, {j, n_y - 1}];
```

Use the finite difference approximations of the partial derivatives and write down the corresponding linear equations:

$$\text{eqns = Table}\left[\frac{u_{i+1,j} - 2\,u_{i,j} + u_{i-1,j}}{h_x{}^2} + \frac{u_{i,j+1} - 2\,u_{i,j} + u_{i,j-1}}{h_y{}^2} == F\ /.\ \{x \to \xi_i,\ y \to \eta_j\},\right.$$

$$\left.\{i,\ n_x - 1\},\ \{j,\ n_y - 1\}\right];$$

We are ready to solve the problem:

```
sol = vars /. Solve[Flatten[{eqns, bound1, bound2}], Flatten[vars]]〚1〛;
```

A neat way to represent the solution is to form an interpolating function from the solution with **ListInterpolation** (see Section 24.2.2, p. 800):

```
uappr = ListInterpolation[sol, {{x₀, x₁}, {y₀, y₁}}]
InterpolatingFunction[{{0., 1.}, {0., 1.}}, <>]
```

```
Plot3D[uappr[x, y], {x, 0, 1}, {y, 0, 1}, ViewPoint → {2.0, -2.4, 0.9},
 AxesLabel → {"x", "y", ""}, Ticks → {{0, 1}, {0, 1}, {0, 1}}]
```

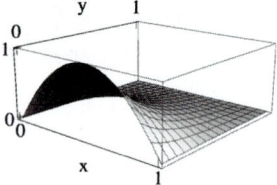

Tabulate some values:

```
Table[uappr[0.5, y], {y, 0, 0.5, 0.1}]
{1., 0.739267, 0.542751, 0.396036, 0.286914, 0.205572}
```

To investigate the precision of the numerical solution, we show the most accurate values of $u(0.5, y)$, $y = 0, 0.1, ..., 0.5$, that we obtained by the series solution in Section 27.2.5:

```
{1., 0.739132, 0.542517, 0.395755, 0.28663, 0.205315}
```

The values we obtained using 30 and 30 subintervals (961 variables and equations) seem to be of approximately three-digit precision. If we use 60 and 60 subintervals (3721 variables and equations), we get nearly four-digit precision.

28

Difference Equations

Introduction

> *Don't let pessimistic statistics about the future worry you. Remember in 1850*
> *it was predicted that if the traffic kept increasing at the same rate, the entire*
> *surface of the earth would be covered in six feet of horse manure by 1970.*

For solving difference or recurrence equations, we have **RSolve**. It can solve all constant coefficient linear equations, many variable coefficient linear equations, and also quite a few nonlinear equations. Note, however, that nonlinear difference equations are much more difficult to solve than nonlinear differential equations. For example, a solution for the logistic difference equation is only known for two positive values of the parameter of the model. We can study nonlinear equations by other means, and we will encounter interesting features such as bifurcation, cycles, and chaos.

Here, we also consider some other discrete systems, namely fractals, Lindenmayer systems, and cellular automata.

A good short introduction to linear difference equations is in Spiegel (1971). More comprehensive treatments of difference equations can be found in Sandefur (1990), Kelley and Peterson (2001), and Martelli (1999). For difference equations with *Mathematica*, see Kulenovic and Merino (2002).

28.1 Solving Difference Equations

28.1.1 One Linear Equation

■ **Solving a Difference Equation**

RSolve solves difference equations or recurrence equations. It is used in the same way that **DSolve** is used for differential equations. The following box shows typical examples of the use of **RSolve**:

`RSolve[eqn, y[n], n]`	Give the general solution
`RSolve[{eqn, inits}, y[n], n]`	Solve an initial value problem

An example of a difference equation is $F_n = F_{n-1} + F_{n-2}$, with initial conditions $F_1 = 1$ and $F_2 = 1$; this equation defines the Fibonacci numbers. In *Mathematica,* the equation is written as `F[n] == F[n-1] + F[n-2]` and the initial conditions as `F[1] == 1` and `F[2] == 1` (remember that all equations have to be defined with `==`).

RSolve can solve all linear constant coefficient difference equations and systems of such equations (the solution is sought with matrix powers). **RSolve** can also solve many linear variable coefficient difference equations in which the coefficients are polynomial or rational. Furthermore, **RSolve** can solve many nonlinear difference equations, and it can solve q-difference equations—with terms such as $y(q\,x)$ and $y(q^2\,x)$—and some partial difference equations.

■ **Constant Coefficient Equations**

We ask for a general solution to a first-order constant coefficient difference equation:

```
eqn = y[n + 1] == 4 / 5 y[n] + 1 / 5;
```

```
RSolve[eqn, y[n], n]
```

$$\left\{\left\{y[n] \to 1 - \left(\frac{4}{5}\right)^n + \left(\frac{5}{4}\right)^{1-n} C[1]\right\}\right\}$$

Here, `C[1]` is an arbitrary constant. Now we give an initial value:

```
RSolve[{eqn, y[0] == 0}, y[n], n]
```

$$\left\{\left\{y[n] \to 1 - \left(\frac{4}{5}\right)^n\right\}\right\}$$

The indices can be written in various forms in the equation. The equation of the preceding example can also be written as follows:

```
eqn2 = y[n] == 4 / 5 y[n - 1] + 1 / 5;
```

```
RSolve[eqn2, y[n], n]
```

$$\left\{\left\{y[n] \to 1 - \left(\frac{4}{5}\right)^n + \left(\frac{5}{4}\right)^{1-n} C[1]\right\}\right\}$$

As with **DSolve**, the solution can also be requested as a pure function:

```
RSolve[eqn, y, n]
```

$$\left\{\left\{y \to \text{Function}\left[\{n\}, 1 - \left(\frac{4}{5}\right)^n + \left(\frac{5}{4}\right)^{1-n} C[1]\right]\right\}\right\}$$

This solution has the advantage that it can easily be inserted into the equation to check the correctness of the solution:

```
eqn /. % // Simplify      {True}
```

■ **Calculating Values**

If we calculate with the solution, it may be useful to ask for the value of y_n (as was the case with **DSolve**). We continue with the preceding example:

```
sol = y[n] /. RSolve[{eqn, y[0] == 2}, y[n], n][[1]] // FullSimplify
```

$$1 + \left(\frac{4}{5}\right)^n$$

Once we have the solution, values y_n can be calculated and plotted:

```
vals = Table[{n, sol}, {n, 0, 20}];

ListLinePlot[vals, Mesh → All]
```

Note, however, that if you calculate a large number of values, it is often more efficient to calculate values directly from the recursive relation. One way to implement this is the following:

```
y = 2.; Prepend[Table[y = 4 / 5 y + 1 / 5, {6}], 2.]
{2., 1.8, 1.64, 1.512, 1.4096, 1.32768, 1.26214}

y =.
```

Another way is to define a recursive function (see Section 18.5.1, p. 596):

```
z[0] = 2.;
z[n_] := z[n] = 4 / 5 z[n - 1] + 1 / 5
```

Then use it:

```
Table[z[n], {n, 0, 6}]
{2., 1.8, 1.64, 1.512, 1.4096, 1.32768, 1.26214}

Remove[z]
```

However, the most compact and fastest way to calculate the values of a difference equation is the use of **NestList** (see Section 18.3.3, p. 575):

```
NestList[4 / 5 # + 1 / 5 &, 2., 6]
{2., 1.8, 1.64, 1.512, 1.4096, 1.32768, 1.26214}
```

If the values of n are needed, you can get them in one of the following ways:

```
{Range[0, 6], NestList[4 / 5 # + 1 / 5 &, 2., 6]}ᵀ
{{0, 2.}, {1, 1.8}, {2, 1.64}, {3, 1.512}, {4, 1.4096}, {5, 1.32768}, {6, 1.26214}}

NestList[{#[[1]] + 1, 4 / 5 #[[2]] + 1 / 5} &, {0, 2.}, 6]
{{0, 2.}, {1, 1.8}, {2, 1.64}, {3, 1.512}, {4, 1.4096}, {5, 1.32768}, {6, 1.26214}}
```

■ **A Set of Trajectories**

A direction field may be informative; it shows the direction of movement at several points. First, write the equation in the form $y_{n+1} - y_n = f(n, y_n)$, and then plot a direction field of $(1, f)$. In our example, the equation $y_{n+1} = \frac{4}{5} y_n + \frac{1}{5}$ can be written as $y_{n+1} - y_n = -\frac{1}{5} y_n + \frac{1}{5}$:

```
<< VectorFieldPlots`

VectorFieldPlot[{1, -1/5 y + 1/5}, {n, 0, 20}, {y, 0, 1.5},
  PlotPoints → 11, Axes → True, AspectRatio → 1/GoldenRatio, ImageSize → 250]
```

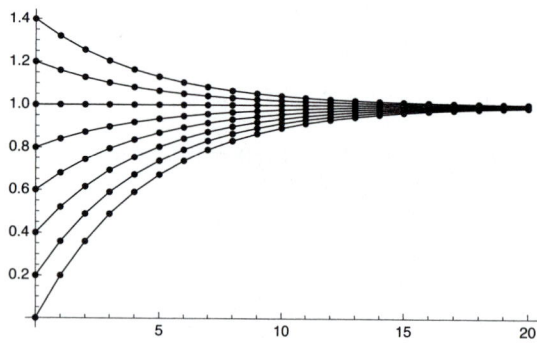

We then plot a set of solutions starting from various points. The solution of the equation with a general starting value **y0** is as follows:

```
sol = y[n] /. RSolve[{eqn, y[0] == y0}, y[n], n][[1]]
```

$$1 - \left(\frac{4}{5}\right)^n + \left(\frac{4}{5}\right)^n y0$$

Compute a set of solutions using various starting points:

```
solset = Table[Table[{n, sol}, {n, 0, 20}], {y0, 0, 1.4, 0.2}];
```

Show the solutions:

```
ListLinePlot[solset, Mesh → All, PlotRange → All, PlotStyle → Black, ImageSize → 250]
```

■ **Variable Coefficient Equations**

We try some variable coefficient equations:

```
RSolve[{y[n + 1] == (n + 1) y[n], y[0] == 1}, y[n], n]
```

$\{\{y[n] \to \text{Gamma}[1 + n]\}\}$

```
RSolve[{y[n + 1] == a y[n] + b n, y[0] == c}, y[n], n] // FullSimplify
```

$$\left\{\left\{y[n] \to \frac{a^n \left(b + (-1 + a)^2 c - \left(\frac{1}{a}\right)^n b (1 + (-1 + a) n)\right)}{(-1 + a)^2}\right\}\right\}$$

```
RSolve[{y[n + 1] == y[n] / (n + 1) + 1, y[0] == 1}, y[n], n] // FullSimplify
```

$$\left\{\left\{y[n] \to \frac{1 + \text{Subfactorial}[-2] - (-1)^n \text{Gamma}[2 + n] \text{Subfactorial}[-2 - n]}{\text{Gamma}[1 + n]}\right\}\right\}$$

■ Fibonacci Numbers

The first few Fibonacci numbers are as follows:

```
F[1] = 1; F[2] = 1;
F[n_] := F[n] = F[n - 1] + F[n - 2]
Table[F[n], {n, 15}]
```

$\{1, 1, 2, 3, 5, 8, 13, 21, 34, 55, 89, 144, 233, 377, 610\}$

```
Remove[F]
```

We can see that the solution of the equations for $F(n)$ is really the nth Fibonacci number:

```
RSolve[{F[n] == F[n - 1] + F[n - 2], F[1] == F[2] == 1}, F[n], n]
```

$\{\{F[n] \to \text{Fibonacci}[n]\}\}$

■ Chebyshev Polynomials

Consider the following second-order equation:

```
eqn = T[n + 2] - 2 x T[n + 1] + T[n] == 0;
```

With initial conditions $T_0(x) = 1$ and $T_1(x) = x$, it defines Chebyshev polynomials. The solution is as follows:

```
sol = T[n] /. RSolve[{eqn, T[0] == 1, T[1] == x}, T[n], n][[1]]
```

$$\frac{1}{2}\left(\left(x - \sqrt{-1 + x^2}\right)^n + \left(x + \sqrt{-1 + x^2}\right)^n\right)$$

This is not of the familiar form $T_n(x) = \cos(n \arccos(x))$, but for given values of n we can verify that the expressions agree:

```
Table[sol == Cos[n ArcCos[x]], {n, 0, 5}] // Simplify
```

$\{\text{True, True, True, True, True, True}\}$

Note that *Mathematica* has the built-in `ChebyshevT[n,x]`.

■ Partial Difference Equations

With `RSolve` we can also solve partial difference equations:

```
RSolve[y[m + 1, n] + a y[m, n + 1] == b, y[m, n], {m, n}]
```

$$\left\{\left\{y[m, n] \to \frac{\left(1 - \left(-\frac{1}{a}\right)^m\right) (-a)^{-1+m} a b}{1 + a} + (-a)^{-1+m} C[1][m + n]\right\}\right\}$$

Here, C[1] in an undetermined function.

As another example, consider a game played by persons A and B (see Spiegel, 1971, pp. 186, 211). Person A [B] needs k [m] points in order to win the game. The probability of A [B] getting one point is p [$q = 1 - p$]. What is the probability that A wins? Let $u_{k,m}$ be this probability. By considering separately the events that {A wins the first point and then wins the game} and {A loses the first point and then wins the game}, we get at once $u_{k,m} = p\,u_{k-1,m} + q\,u_{k,m-1}$. This is a partial difference equation. We have the boundary conditions $u_{k,0} = 0$ for $k \geq 0$ and $u_{0,m} = 1$ for $m > 0$.

To solve the partial difference equation, define the generating function $G_k(t) = \sum_{m=0}^{\infty} u_{k,m}\,t^m$ for $k \geq 0$. We try to find a closed-form expression for $G_k(t)$. Then we expand it as a series and read the coefficient of t^m; this coefficient is $u_{k,m}$. Assume $k \geq 1$, multiply the difference equation by t^m, and sum from 1 to ∞:

$$\sum_{m=1}^{\infty} u_{k,m}\,t^m = p \sum_{m=1}^{\infty} u_{k-1,m}\,t^m + q \sum_{m=1}^{\infty} u_{k,m-1}\,t^m.$$

This can be written as $G_k(t) - u_{k,0} = p\left[G_{k-1}(t) - u_{k-1,0}\right] + q\,t\,G_k(t)$, or, using the boundary conditions, $G_k(t) = p\,G_{k-1}(t) + q\,t\,G_k(t)$. So we arrived at a difference equation for $G_k(t)$. From the boundary conditions we also get the initial condition $G_0(t) = \sum_{m=1}^{\infty} t^m = \frac{t}{1-t}$. The solution of the difference equation is as follows:

```
sol = G[k] /. RSolve[{G[k] == p G[k - 1] + (1 - p) t G[k], G[0] == t / (1 - t)}, G[k], k][[1]]
```

$$-\frac{t\left(\dfrac{p}{1+(-1+p)\,t}\right)^k}{-1+t}$$

Find the coefficient of t^m in the series expansion of $G_k(t)$; this is $u_{k,m}$:

```
u[k_, m_, p_] = SeriesCoefficient[sol, {t, 0, m}] // Simplify
```

$$1 - \frac{(1-p)^m\,p^k\,\text{Gamma}[k+m]\,\text{Hypergeometric2F1}[1, k+m, 1+m, 1-p]}{\text{Gamma}[k]\,\text{Gamma}[1+m]}$$

Plot this probability as a function of p when $k = 20$ and $m = 10$:

```
Plot[u[20, 10, p], {p, 0, 1}]
```

We see that if person A needs 20 points to win and person B only needs 10 points, it is nearly impossible for A to win if his or her probability of getting 1 point is less than approximately 0.5; however, if this probability is at least 0.8, then A wins almost always.

■ Q-Difference Equations

In a q-difference equation, the values of the function in the difference equation are not given at successive integers such as in $y(n + 1) = a\,y(n) + b$. Instead, we can have equations such as $y(2\,n) = a\,y(n) + b\,n$ or, more generally, $y(q\,n) = a\,y(n) + b$. Here is an example:

```
RSolve[y[q n] == 2 y[n] + n, y[n], n]
```

$$\left\{\left\{y[n] \rightarrow -\frac{2^{\frac{\text{Log}[n]}{\text{Log}[q]}} - n}{-2 + q} + 2^{-1 + \frac{\text{Log}[n]}{\text{Log}[q]}} C[1]\right\}\right\}$$

An equivalent way to write the equation is as follows:

```
RSolve[y[n] == 2 y[n / q] + n / q, y[n], n]
```

$$\left\{\left\{y[n] \rightarrow -\frac{2^{\frac{\text{Log}[n]}{\text{Log}[q]}} - n}{-2 + q} + 2^{-1 + \frac{\text{Log}[n]}{\text{Log}[q]}} C[1]\right\}\right\}$$

Next, we solve an initial value problem:

```
sol = y[n] /. RSolve[{y[2 n] == 2 y[n] + n, y[1] == 1}, y[n], n][[1]] // Simplify
```

$$\frac{n \, \text{Log}[4 \, n]}{\text{Log}[4]}$$

28.1.2 Two Linear Equations

RSolve[eqns, {x[n], y[n]}, n]	Solve two difference equations

■ A Constant Coefficient System

Consider the following system:

```
eqns = {x[n + 1] == (7 x[n] + 4 y[n]) / 10,
        y[n + 1] == (-3 x[n] + 6 y[n]) / 10};
```

Here is the coefficient matrix:

```
A = 1 / 10 {{7, 4}, {-3, 6}};
```

Its eigenvalues and their absolute values are as follows:

```
Eigenvalues[A]
```
$$\left\{\frac{1}{20}\left(13 + i\sqrt{47}\right), \frac{1}{20}\left(13 - i\sqrt{47}\right)\right\}$$

```
Abs[%] // N      {0.734847, 0.734847}
```

Because the common absolute value is smaller than 1, the trajectories are spirals approaching the origin (Kelley and Peterson, 2001, p. 148).

Solve the system with general starting values α and β:

```
vars = {x[n], y[n]};
inits = {x[0] == α, y[0] == β};
sol[α_, β_] =
    vars /. RSolve[Join[eqns, inits], vars, n][[1]] // ComplexExpand // FullSimplify
```

$$\left\{\frac{1}{47} 2^{-n/2} 3^{3n/2} 5^{-n}\left(47 \, \alpha \, \text{Cos}\left[n \, \text{ArcCot}\left[\frac{13}{\sqrt{47}}\right]\right] + \sqrt{47} \, (\alpha + 8\,\beta) \, \text{Sin}\left[n \, \text{ArcCot}\left[\frac{13}{\sqrt{47}}\right]\right]\right),\right.$$

$$\left.\frac{1}{47} 2^{-n/2} 3^{3n/2} 5^{-n}\left(47 \, \beta \, \text{Cos}\left[n \, \text{ArcCot}\left[\frac{13}{\sqrt{47}}\right]\right] - \sqrt{47} \, (6\,\alpha + \beta) \, \text{Sin}\left[n \, \text{ArcCot}\left[\frac{13}{\sqrt{47}}\right]\right]\right)\right\}$$

Calculate a trajectory of 20 steps from the starting point (60, 60) by using the solution:

```
xy = Table[sol[60., 60.], {n, 0, 20}];
```

The trajectory can also be calculated directly from the difference equations:

```
xy = NestList[A.# &, {60., 60.}, 20];
```

Plot the trajectory in the (x, y) plane:

```
ListLinePlot[xy, Mesh → All, PlotRange → All, AspectRatio → Automatic]
```

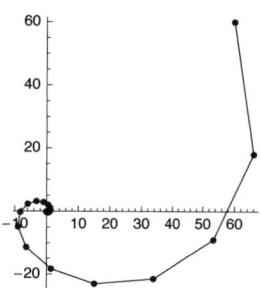

To plot the components as functions of n, write the following:

```
{xx, yy} = xyᵀ; tt = Range[0, 20];
xt = {tt, xx}ᵀ; yt = {tt, yy}ᵀ;
ListLinePlot[{xt, yt}, Mesh → All, PlotStyle → {Black, Gray}, PlotRange → All]
```

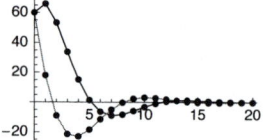

■ A Set of Trajectories

We continue the preceding example by plotting a direction field. To this end, we write the equations in the form $x_{n+1} - x_n = f(n, x_n, y_n)$, $y_{n+1} - y_n = g(n, x_n, y_n)$ and then plot a direction field of (f, g):

```
<< VectorFieldPlots`

VectorFieldPlot[{-3 x + 4 y, -3 x - 4 y} / 10, {x, -60, 60},
  {y, -60, 60}, PlotPoints → 11, Axes → True, ImageSize → 160]
```

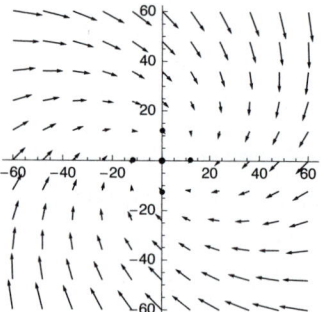

Then we calculate some trajectories from a set of starting points:

```
solset = NestList[Function[{z}, A.z], #, 20] & /@ {{60, 60}, {30, 60},
    {-60, 60}, {-60, 30}, {-60, -60}, {-30, -60}, {60, -60}, {60, -30}};
```

The trajectories are displayed as follows:

```
ListLinePlot[solset, Mesh → All, AspectRatio → Automatic,
  PlotRange → All, PlotStyle → Black, ImageSize → 160]
```

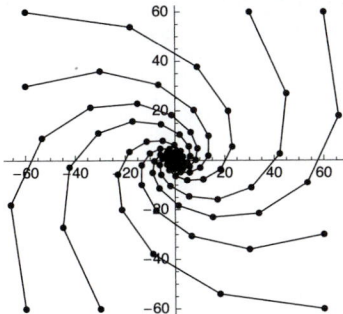

28.1.3 Some Techniques

■ **Using Generating Functions**

One method of solving difference equations is the use of generating functions. Consider, for example, the equation $y_{n+1} = a\,y_n + b$. Define the generating function $G(t) = \sum_{m=0}^{\infty} y_n\,t^n$. Multiply both sides of the equation by t^n and sum from 0 to ∞:

$$\sum_{n=0}^{\infty} y_{n+1}\,t^n = a \sum_{n=0}^{\infty} y_n\,t^n + b \sum_{n=0}^{\infty} t^n.$$

Because the left-hand side can be written as $\frac{1}{t} \sum_{n=1}^{\infty} y_n\,t^n$, we get the equation

$$\frac{1}{t}\big(G(t) - y_0\big) = a\,G(t) + b\,\frac{1}{1-t}.$$

The solution of this equation for $G(t)$ is as follows:

```
eqn = - (G - y0) == a G + b ----- ;
      t                      1 - t

Gsol = G /. Solve[eqn, G]〚1〛
```

$$\frac{b\,t + y0 - t\,y0}{(-1 + t)\,(-1 + a\,t)}$$

The coefficient of the term t^n in the series expansion of this expression is y_n, which is also the solution of the difference equation:

```
SeriesCoefficient[Gsol, {t, 0, n}]
```

$$\frac{(-1 + a^n)\,b + (-1 + a)\,a^n\,y0}{-1 + a}$$

(We considered **SeriesCoefficient** in Section 19.2.2, p. 627.) This expression is in agreement with the solution given by **RSolve**:

```
eqn = {y[n + 1] == a y[n] + b, y[0] == y0};

RSolve[eqn, y[n], n] // Simplify
```

$$\left\{\left\{y[n] \to \frac{(-1 + a^n)\,b + (-1 + a)\,a^n\,y0}{-1 + a}\right\}\right\}$$

■ Using the *Z*-Transform

Difference equations can also be solved with the *Z*-transform (see Section 20.4.2, p. 672), in a way that is similar to how differential equations can be solved with the Laplace transform. To solve the difference equation of the Fibonacci numbers, we rewrite the equation and the initial values as follows:

```
eqn = F[n + 2] == F[n + 1] + F[n];
inits = {F[0] → 0, F[1] → 1};
```

Take the *Z*-transform of the equation:

```
ZTransform[eqn, n, z]
```

$-z^2\, F[0] - z\, F[1] + z^2\, \text{ZTransform}[F[n],\, n,\, z] ==$
$\quad - z\, F[0] + \text{ZTransform}[F[n],\, n,\, z] + z\, \text{ZTransform}[F[n],\, n,\, z]$

Use the initial conditions:

```
zeqn = % /. inits
```

$-z + z^2\, \text{ZTransform}[F[n],\, n,\, z] == \text{ZTransform}[F[n],\, n,\, z] + z\, \text{ZTransform}[F[n],\, n,\, z]$

Solve the *Z*-transform:

```
Solve[zeqn, ZTransform[F[n], n, z]]
```

$$\left\{\left\{\text{ZTransform}[F[n],\, n,\, z] \to \frac{z}{-1 - z + z^2}\right\}\right\}$$

Find the inverse *Z*-transform:

```
sol = F[n] /. InverseZTransform[%, z, n][[1]]
```

$$\frac{-\left(\frac{1}{2}\left(1 - \sqrt{5}\right)\right)^n + \left(\frac{1}{2}\left(1 + \sqrt{5}\right)\right)^n}{\sqrt{5}}$$

■ Cobweb Plot

The *cobweb* is an interesting way to illustrate how the values computed from a difference equation proceed. Consider the equation $y_{n+1} = -0.8\, y_n + 0.2$, $y_0 = 0.6$. Calculate and plot some values:

```
vals = {Range[0, 20], NestList[-0.8 # + 0.2 &, 0.6, 20]}ᵀ;
```

```
ListLinePlot[vals, Mesh → All]
```

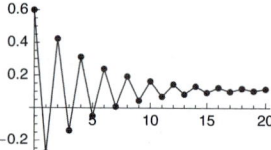

We then introduce the program **cobwebPlot**:

```
cobwebPlot[f_, y_, y0_, n_, a_, b_, opts___] := Module[{x = y0},
  Plot[{y, f}, {y, a, b}, Epilog :>
    Line[Join[{{y0, 0}}, Flatten[Table[{{x, x = f /. y → x}, {x, x}}, {n}], 1]]], opts]]
```

Here, **f** is the right-hand-side function of the difference equation $y_{n+1} = f(y_n)$; in the previous example, **f** is **-0.8 y + 0.2**. In addition, **y0** is the starting value, **n** is the number of values to be calculated (in addition to **y0**), (**a, b**) is the interval in which the figure is plotted, and **opts** is a set of options of **Plot**. For our example, the figure is displayed as follows:

```
cobwebPlot[-0.8 y + 0.2, y, 0.6, 20, -0.4, 0.7,
  PlotRange → {-0.30, 0.43}, ImageSize → 200, AspectRatio → Automatic]
```

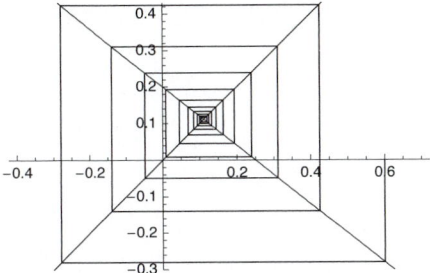

In the figure, the functions y and $f(y)$ are plotted. The starting point for the cobweb is (0.6, 0). From this point on, follow the broken line. Each time you meet a vertical line, the point where this line (or its extension) intersects the x axis represents the next value of the solution sequence. The horizontal lines lead you to the consecutive vertical lines. The figure shows the convergence of the sequence to a point.

28.1.4 Nonlinear Equations

■ **Example 1**

Several nonlinear difference equations can also be solved. An example is an equation of the form $\text{expr}_1 = \text{expr}_2$ in which the two expressions contain products, quotients, and powers (not sums). For example,

```
sol = RSolve[{y[n + 2] y[n] == y[n + 1], y[0] == α, y[1] == β}, y[n], n]
```

$$\left\{\left\{y[n] \to e^{\frac{1}{3}\left(-\sqrt{3}\, \text{Log}[\alpha] + 2\sqrt{3}\, \text{Log}[\beta]\right) \text{Sin}\left[\frac{n\pi}{3}\right]} \alpha^{\text{Cos}\left[\frac{n\pi}{3}\right]}\right\}\right\}$$

The equation could have been reduced, with a logarithmic transform, to a linear one. The solution has period six:

```
Table[y[n] /. sol[[1]] /. {α → 1, β → 2}, {n, 0, 17}]
```

$$\left\{1, 2, 2, 1, \frac{1}{2}, \frac{1}{2}, 1, 2, 2, 1, \frac{1}{2}, \frac{1}{2}, 1, 2, 2, 1, \frac{1}{2}, \frac{1}{2}\right\}$$

Here is another example:

```
sol = RSolve[{y[n + 2] == y[n + 1] y[n], y[0] == α, y[1] == β}, y[n], n]
```

$$\left\{\left\{y[n] \to e^{\text{Fibonacci}[n]\left(-\text{Log}\left[-\sqrt{\alpha}\right] + \text{Log}[\beta]\right)} \left(-\sqrt{\alpha}\right)^{\text{LucasL}[n]}\right\},\right.$$

$$\left.\left\{y[n] \to e^{\frac{1}{2}\text{Fibonacci}[n]\left(-\text{Log}[\alpha] + 2\text{Log}[\beta]\right)} \alpha^{\frac{\text{LucasL}[n]}{2}}\right\}\right\}$$

■ **Example 2**

This example is a homogeneous constant coefficient Riccati equation:

```
sol = RSolve[{y[n + 1] y[n] + p y[n + 1] + q y[n] == 0, y[0] == α}, y[n], n]
```

$$\left\{\left\{y[n] \to \frac{\left(-\frac{p}{q}\right)^{-n}(p+q)\,\alpha}{p+q+\alpha-\left(-\frac{q}{p}\right)^{n}\alpha}\right\}\right\}$$

The equation could have been reduced, with the transformation $y_n = \frac{1}{z_n}$, to a linear one.

■ Example 3

Now we have an equation that contains a convolution:

```
eqns = {y[n + 1] == Sum[y[i] y[n - i], {i, 0, n}], y[0] == α}
```

$$\left\{y[1+n] == \sum_{i=0}^{n} y[i]\,y[-i+n],\ y[0] == \alpha\right\}$$

```
sol = RSolve[eqns, y[n], n]
```

$$\left\{\left\{y[n] \to \frac{\alpha^{1+n}\,\text{Binomial}[2\,n,\,n]}{1+n}\right\}\right\}$$

■ Example 4

If we solve the equation $f(x) = 0$ with Newton's method $x_{i+1} = x_i - f(x_i)/f'(x_i)$, we get a difference equation. For example, to calculate the square root of a, define $f(x) = x^2 - a$. We can solve the resulting difference equation:

```
sol = x[n] /. RSolve[{x[n + 1] == x[n] - (x[n]^2 - a) / (2 x[n]), x[0] == x0}, x[n], n][[1]]
```

$$\sqrt{a}\ \text{Coth}\left[2^n\,\text{ArcCoth}\left[\frac{x0}{\sqrt{a}}\right]\right]$$

Here are the first few values calculated by Newton's method, when $a = 10$ and $x_0 = 5$:

```
Table[sol /. {a → 10., x0 → 5}, {n, 1, 6}] // Chop
{3.5, 3.17857, 3.16232, 3.16228, 3.16228, 3.16228}
```

■ Example 5: The Logistic Equation

The most famous nonlinear difference equation is the logistic equation:

```
eqns = {y[n + 1] == a y[n] (1 - y[n]), y[0] == α};
```

The solution of this equation is unknown, except for two positive values of the parameter a, $a = 2$ and $a = 4$ (see Kelley and Peterson, 2001, p. 173). In addition, *Mathematica* knows the solution for one negative value of a, $a = -2$:

```
Off[Solve::ifun]
```

```
RSolve[eqns /. a → 2, y[n], n]
```
$$\left\{\left\{y[n] \to \frac{1}{2}\left(1 - (1 - 2\,\alpha)^{2^n}\right)\right\}\right\}$$

```
RSolve[eqns /. a → 4, y[n], n]
```
$$\left\{\left\{y[n] \to \frac{1}{2}\left(1 - \text{Cos}\left[2^n\,\text{ArcCos}[1 - 2\,\alpha]\right]\right)\right\}\right\}$$

```
RSolve[eqns /. a → -2, y[n], n]
```
$$\left\{\left\{y[n] \to \frac{1}{2}\left(1 + 2\,\text{Cos}\left[2^n\,\text{ArcCos}\left[\frac{1}{2}(-1 + 2\,\alpha)\right]\right]\right)\right\}\right\}$$

A realization of the solution for $a = 2$ is as follows:

```
vals = NestList[2 # (1 - #) &, 0.001, 15];
```

```
ListLinePlot[{Range[0, 15], vals}ᵀ, Mesh → All]
```

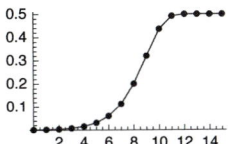

The solution for $a = 4$ behaves completely differently. To calculate the values accurately, we start with 70 digits of precision:

```
vals = NestList[4 # (1 - #) &, 0.001`70, 100];
```

```
ListLinePlot[{Range[0, 100], vals}ᵀ, Mesh → All, AspectRatio → 0.2, ImageSize → 340]
```

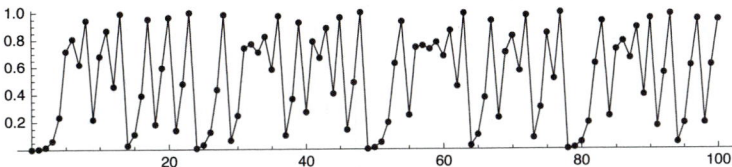

The values seem to develop rather chaotically. For $a = -2$, the solution is also very unpredictable:

```
vals = NestList[-2 # (1 - #) &, 0.001`70, 100];
```

```
ListLinePlot[{Range[0, 100], vals}ᵀ, Mesh → All, AspectRatio → 0.2, ImageSize → 340]
```

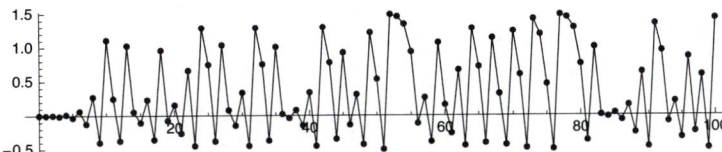

Next, we study the logistic equation in more detail.

```
Remove["Global`*"]
```

28.2 The Logistic Equation

28.2.1 Trajectories

■ The Logistic Model

Most nonlinear difference equations cannot be solved to a closed-form expression. We have to resort to other means of investigating them. As an example of nonlinear equations, we consider in this section the famous logistic model $y_{n+1} = a\, y_n (1 - y_n)$. The methods presented for this model may also be used for other nonlinear models.

A logistic equation is often written as follows:

```
eqn = z[n + 1] - z[n] == k z[n] (K - z[n])
-z[n] + z[1 + n] == k (K - z[n]) z[n]
```

This can be reduced to the standard form by making the change of variable $z_n = \frac{1 + kK}{k}\, y_n$:

```
eqn /. z[n_] → (1 + k K) y[n] / k // FullSimplify
```

$$\frac{(1 + k\,K)\,((1 + k\,K)\,(-1 + y[n])\,y[n] + y[1 + n]))}{k} == 0$$

Now denote $1 + k\,K = a$.

■ Sensitivity to Numerical Inaccuracies

For some values of the parameter a, the logistic model $y_{n+1} = a\,y_n\,(1 - y_n)$ is very sensitive to numerical inaccuracies. To see this, we calculate 100 values from the model with $a = 4$, first by using normal decimal numbers and then by using high-precision numbers. In the latter case, we start with numbers that have a precision of 65 digits:

```
vals1 = NestList[4 # (1 - #) &, 0.01, 100];
vals2 = NestList[4 # (1 - #) &, 0.01`65, 100];
```

Look at the first values of both sequences and check the precision of these numbers:

```
First /@ {vals1, vals2}
```

```
{0.01, 0.0100000000000000000000000000000000000000000000000000000000000000000}
```

```
Precision /@ %      {MachinePrecision, 65.}
```

Look also at the last values of both sequences:

```
Last /@ {vals1, vals2}      {0.224142, 0.598853}
```

```
Precision /@ %      {MachinePrecision, 6.57132}
```

The last values differ quite a lot. Plot both sequences:

```
ListLinePlot[{vals1, vals2}, Mesh → All,
  PlotStyle → {{}, Thickness[Medium]}, AspectRatio → 0.18, ImageSize → 400]
```

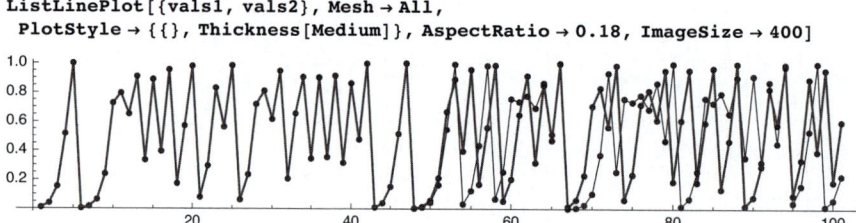

Values corresponding to **vals2** are thick. From approximately iteration 50 on, the values differ greatly.

In calculating **vals2**, we started with numbers having 65 digits of precision. During the calculation, many digits were lost so that the last value 0.598853 only has a precision of approximately 6.6. Look at some elements of **vals2**:

```
vals2[[Range[11, 101, 10]]] // Column
```

```
0.79515396263983587159021493124186948784676458997382822834 5435
0.07992833029428810569388179674398376762532984133074014 0139
0.945561800274275776014142302425504943869633496313
0.47500024203452384603544426557955466840 8401
0.2113501716692765412892965634787079 14
0.647005629669639008172470697581
0.830552409049308180739491
0.609007649692576357
0.290999858817
0.598853
```

These values clearly show the loss of significant digits during the iteration. Approximately six digits are lost during each 10 iterations.

High-precision numbers were considered in Section 12.2.2, p. 406. Recall that when we calculate with such numbers, *Mathematica* takes care that all the digits in the result are correct. Thus, we know that all the digits of **vals2** are correct. This means that the values in **vals1** are incorrect from approximately iteration 50 on. This demonstrates the sensitivity to numerical inaccuracies of the logistic model for some values of the parameter a. Thus, if we calculate long sequences from the logistic model, it is important to use a high enough precision during the calculation.

From the plot of **vals2**, we see that the series behaves quite chaotically. It is known that *chaotic* models are very sensitive to numerical inaccuracies. It can be shown that the logistic model is chaotic for a from approximately 3.57 to 4, although inside this interval there are also some small nonchaotic intervals.

■ Sensitivity to Initial Values

Chaotic models are also very sensitive to the initial value. To show this, compute, with $a = 4$, 50 iterations using starting points $0.02 + 10^{-i}$, $i = 1, ..., 25$. Then plot the 20th value of each of the 25 series. Also plot the 50th value of each of the 25 series:

```
vals = Table[NestList[4 # (1 - #) &, 0.02`50 + 10^-i, 50], {i, 25}];

{ListPlot[vals[[All, 20]], PlotRange → {-0.05, 1.05}, PlotLabel → "20th value"],
 ListPlot[vals[[All, 50]], PlotRange → {-0.05, 1.05}, PlotLabel → "50th value"]}
```

From the first plot, we see that even if the starting point differs from 0.02 by 10^{-7} or more (see the first seven points in the plot), the value of y_{20} significantly differs from the value that results when starting from 0.02. From the second plot, we see that if the starting point differs from 0.02 by 10^{-16} or more, the value of y_{50} differs significantly from the value that results when starting from 0.02.

■ Trajectories

Now that we know about the numerical sensitivity of the logistic model, we can study its behavior in more detail.

From a direction field, we get an impression of the solution. Write the equation in the form $y_{n+1} - y_n = a\,y_n(1 - y_n) - y_n$, and assume that $a = 1.5$:

```
<< VectorFieldPlots`
```

```
VectorFieldPlot[{1, 1.5 y (1 - y) - y}, {n, 0, 20}, {y, 0, 0.5},
  PlotPoints → 11, Axes → True, AspectRatio → 1 / GoldenRatio, ImageSize → 250]
```

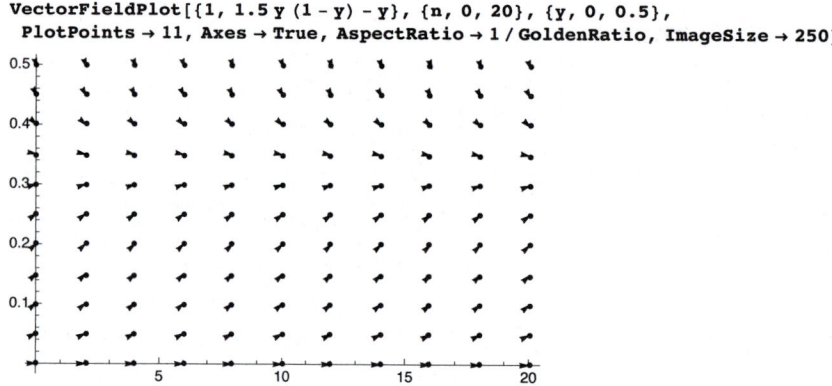

To plot sets of trajectories, we write the following program:

```
logisticPlot[a_, n_, y01_, y02_, dy0_, opts___] :=
ListLinePlot[
  Table[{Range[0, n], NestList[a # (1 - #) &, y0, n]}ᵀ, {y0, y01, y02 + 10^-5, dy0}],
  Mesh → All, PlotStyle → Black, PlotRange → All, opts]
```

In the program, we first calculate a solution set by starting from various points and iterating the equation **n** times. The starting points are chosen between **y01** and **y02** in steps of **dy0**. When $a = 1.5$, we get the following trajectories:

```
logisticPlot[1.5, 30, 0.001, 0.41, 0.025, AspectRatio → 0.25, ImageSize → 400]
```

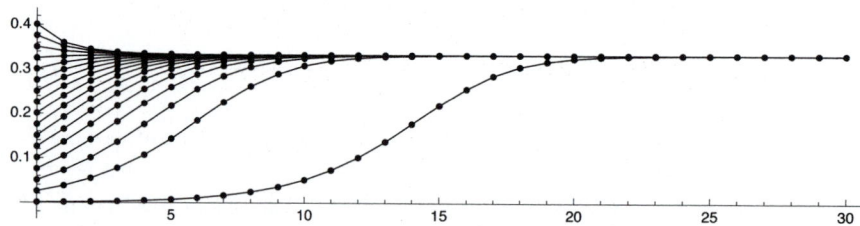

All trajectories in this figure seem to approach a certain value. When $a = 3.3$, the trajectories seem to approach a cycle of two points:

```
logisticPlot[3.3, 30, 0.01, 0.31, 0.05, AspectRatio → 0.25, ImageSize → 400]
```

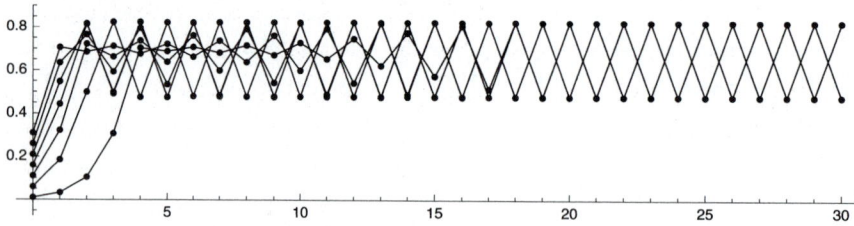

For $a = 3.52$, we get a cycle of four points:

```
logisticPlot[3.52, 30, 0.01, 0.31, 0.05, AspectRatio → 0.25, ImageSize → 400]
```

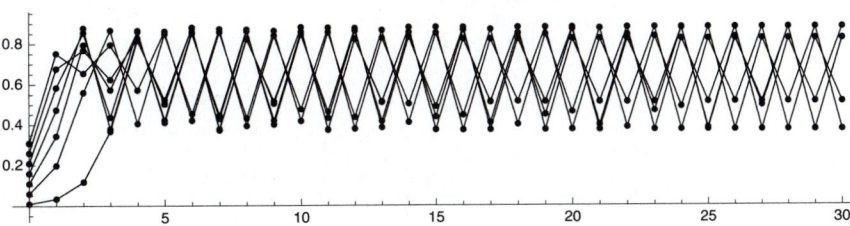

When $a = 3.7$, the trajectories appear to be chaotic:

```
logisticPlot[3.7`100, 30, 0.01`100, 0.31`100,
  0.05`100, AspectRatio → 0.25, ImageSize → 400]
```

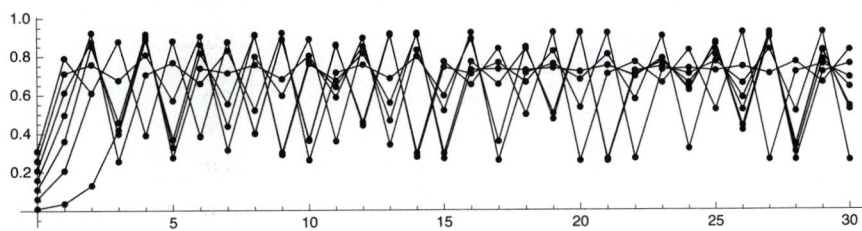

To calculate the values accurately, we started with 100 digits of precision. This precision suffices to calculate 160 values accurately.

■ **Cobwebs**

We previously plotted a cobweb in Section 28.1.3, p. 932. We repeat the program here.

```
cobwebPlot[f_, y_, y0_, n_, a_, b_, opts___] := Module[{x = y0},
  Plot[{y, f}, {y, a, b}, Epilog :→
    Line[Join[{{y0, 0}}, Flatten[Table[{{x, x = f /. y → x}, {x, x}}, {n}], 1]]], opts]]
```

Now we investigate the logistic model by using the same values of a as used previously. Again, to calculate the values accurately for $a = 3.7$, we start with 50 digits of precision:

```
GraphicsArray[{{cobwebPlot[1.5 y (1 - y), y, 0.03, 20, 0, 0.4],
    cobwebPlot[3.3 y (1 - y), y, 0.03, 20, 0, 0.9]},
   {cobwebPlot[3.52 y (1 - y), y, 0.08, 30, 0, 0.95],
    cobwebPlot[3.7`50 y (1 - y), y, 0.02`50, 40, 0, 1]}}, ImageSize → 400]
```

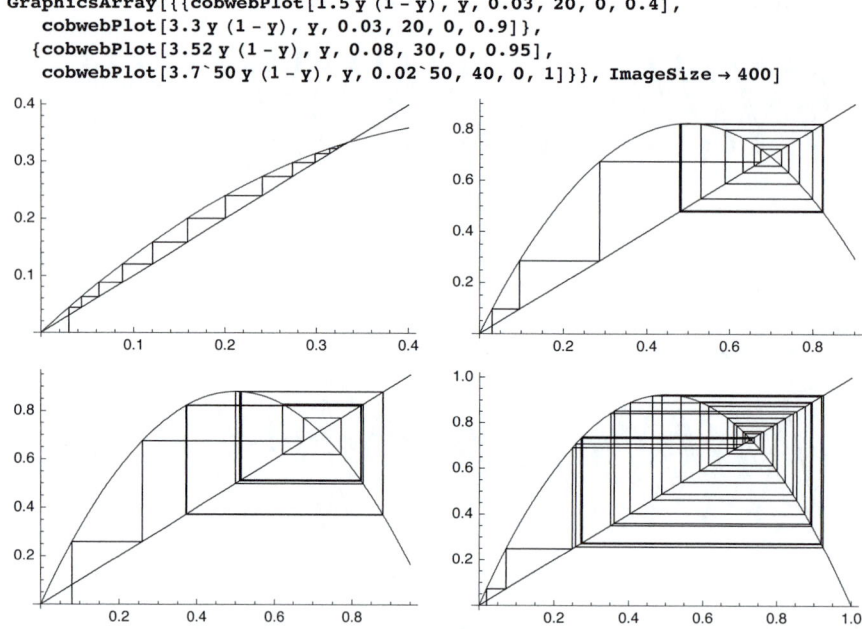

From these figures we can see the same things as we did with the trajectories: convergence to a point, a two-point cycle, a four-point cycle, and chaos.

With the following manipulation we can interactively study the behavior of the cobweb. Here, *a* is the parameter of the logistic model, y_0 is the starting point, *iters* is the numbers of iterations done, and *drop* is the number of initial iterations dropped from the iterations (to better see the limiting behavior like cycles).

```
Manipulate[Module[{x = SetPrecision[y₀, 70], a1 = SetPrecision[a, 70]},
  Plot[{y, a y (1 - y)}, {y, 0, 1}, Epilog ⧴ Line[Drop[Join[{{y₀, 0}},
        Flatten[Table[{{x, x = a1 y (1 - y) /. y → x}, {x, x}}, {iters}], 1]], 2 drop]],
    AspectRatio → Automatic, ImageSize → 250]],
  {{a, 3.8}, 0, 4, Appearance → "Labeled"}, {{y₀, 0.1}, 0, 1, Appearance → "Labeled"},
  {{iters, 50}, 0, 100, 1, Appearance → "Labeled"},
  {drop, 0, iters, 1, Appearance → "Labeled"}]
```

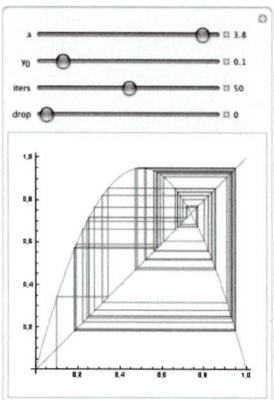

28.2.2 Bifurcation Diagrams

■ Limit Values

The *bifurcation diagram* or *final-state diagram* shows the long-run values calculated from the difference equation when a parameter of the model gets a range of values. The trajectories calculated in Section 28.2.1 showed that for different values of the parameter a, the logistic model behaves very differently. We say that a bifurcation occurs for a certain value of the parameter if the behavior of the final values undergoes a qualitative change at this point.

To prepare a bifurcation diagram, first choose a set of values of the parameter, then directly calculate a long sequence from the difference equation for these values, and lastly plot the limit values. Therefore, first we write a function to calculate limit values (Dickau, 1997):

```
limits = Compile[{a}, {a, #} & /@ Union[Drop[NestList[a # (1 - #) &, 0.5, 1000], 301]]];
```

The function `limits` calculates the limit values for a single value of a. `NestList` does most of the work: We iterate the recursion function 1000 times starting from 0.5. Of these values, the first 301—considered to be transient—are dropped. Of the remaining 700 values we drop duplicates with `Union`; lastly, with `Map`, we attach the value of a to each limit value of y. We have compiled the function `limits` to speed up this process (compiling was explained in Section 17.2.3, p. 528). (If your computer is not fast and does not have much RAM, you may prefer to replace 1000 with a smaller value such as 600.)

Try the function for $a = 1.3$:

```
limits[1.3]        {{1.3, 0.230769}}
```

The approximate limiting value is 0.230769. Try $a = 3.3$:

```
limits[3.3]
{{3.3, 0.479427}, {3.3, 0.479427}, {3.3, 0.479427},
 {3.3, 0.823603}, {3.3, 0.823603}, {3.3, 0.823603}}
```

Here we see the two points forming a cycle. Each point appears three times: `Union` has found that all 16 digits are not the same. Consider $a = 3.7$:

```
limits[3.7] // Length      700
```

All 700 values are different. Indeed, for this value of a, the system behaves chaotically.

Normal decimal numbers are used in `limit`. If we remember the numerical problems of a chaotic logistic model, we may wonder whether the limiting points are correct at all. Contrary to expectation, the bifurcation diagram obtained with `limit` correctly displays all of the essential things. We could write a function such as the following:

```
limits2[a_] := {a, #} & /@ Union[Drop[NestList[a # (1 - #) &, 0.5`400, 1000], 301]];
```

High-precision numbers are used here (the starting value 0.5 has 400 digits of precision!). However, the bifurcation diagrams obtained with `limit` and `limit2` differ very little: Only close examination of the chaotic region reveals some slight differences. Thus, we can be satisfied with `limit`, particularly because it is many times faster than `limit2` and uses much less memory.

■ Bifurcation Diagram

Now we form a function to calculate the limit values for a range of values for a:

```
bifurcation[a0_, a1_, n_] := Flatten[Table[limits[a], {a, a0, a1, (a1 - a0) / n}], 1]
```

The function **bifurcation** constructs, for $n + 1$ values of a between a_0 and a_1, the corresponding set of limit values of y for the logistic model. As an example, we take 801 values of a in the interval $(0, 4)$ and plot the corresponding limit points of the recursion formula (if your computer has limited resources, replace 800 with a smaller value such as 400). Note that the next two figures and similar figures later are of low quality because *Mathematica* 6 is not able to print small enough points.

```
points = bifurcation[0, 4, 800];
```

```
ListPlot[points, PlotStyle → PointSize[0.001],
  AxesOrigin → {0, -0.02}, AxesLabel → {a, y}, ImageSize → 400]
```

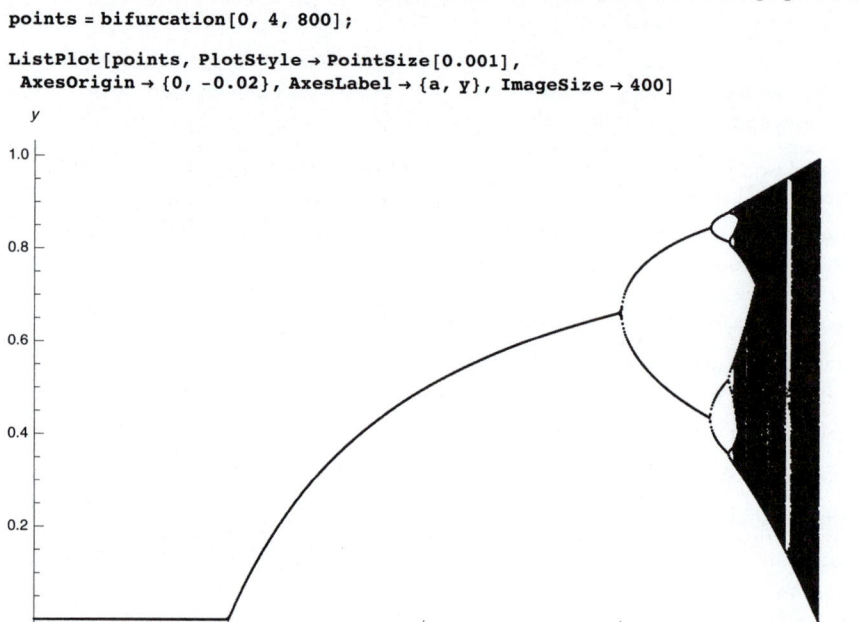

The figure shows that for $0 < a < 1$, the limit value is 0, and for $1 < a < 3$, it is another value (depending on a). Then we have cycles of 2, 4, 8, ... points, and for approximately $3.6 < a < 4$, the model behaves chaotically. For $a > 4$, the limit value is $-\infty$. Next, we take a closer look at the interesting interval $(3.5, 4)$:

```
points = bifurcation[3.5, 4, 800];
```

```
ListPlot[points, PlotStyle → PointSize[0.001],
  AxesOrigin → {3.5, 0}, ImageSize → 400]
```

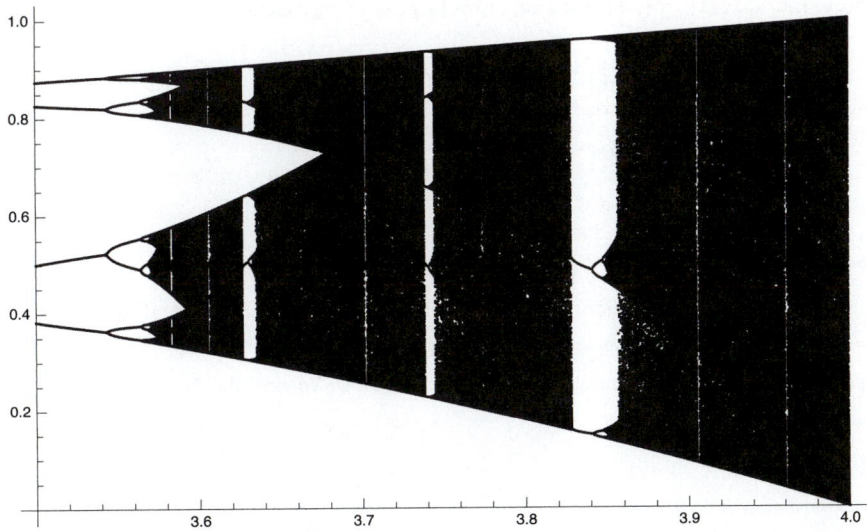

We see that after the chaotic region begins at approximately 3.57, here and there we suddenly again have small intervals of periodic behavior.

28.2.3 Equilibrium and Periodic Points

■ **Equilibrium Points**

The bifurcation diagram shows limit values, but we also want to know their mathematical expressions. Thus, now we calculate equilibrium points and points that make up periods of various lengths. With them, we can plot a better bifurcation diagram for $a < 3.57$.

Consider the difference equation $y_{n+1} = f(y_n)$. If $f(y_n) = y_n$, then $y_{n+1} = y_n$ and the state remains the same. Such a y_n is an *equilibrium point*. For the logistic equation, the equilibrium points are as follows:

```
f[y_] := a y (1 - y)
```

```
soll = Solve[f[y] == y, y]     {{y → 0}, {y → -1 + a / a}}
```

An equilibrium point y^* is asymptotically stable if $\left| f'(y^*) \right| < 1$. For the logistic model, we have the following:

```
f'[y] /. soll // Simplify     {a, 2 - a}
```

Assume that $a > 0$. Points $y^* = 0$ and $y^* = 1 - \frac{1}{a}$ are asymptotically stable if $|a| < 1$ and $|2 - a| < 1$, respectively—that is, if $0 < a < 1$ and $1 < a < 3$, respectively. We plot these equilibrium points in their regions of stability:

`{p1 = Plot[y /. sol1[[1]], {a, 0, 1}], p2 = Plot[y /. sol1[[2]], {a, 1, 3}]}`

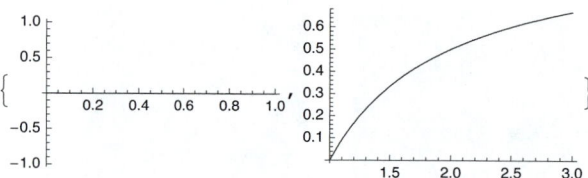

■ 2-Periodic Points

Consider again the difference equation $y_{n+1} = f(y_n)$, and calculate $y_{n+2} = f(f(y_n))$. If $f(f(y_n)) = y_n$, we have $y_{n+2} = y_n$, which means that the same point appears by doing two iterations. Such a point y_n is called a *2-periodic point,* and the points y_n and $f(y_n)$ form a *cycle.* To find the 2-periodic points of the logistic model, first calculate $f(f(y))$:

`f2[y_] = f[f[y]]` $a^2 (1 - y) y (1 - a (1 - y) y)$

Then solve the equation $f(f(y)) = y$:

`sol2 = Solve[f2[y] == y, y] // FullSimplify`

$$\left\{ \{y \to 0\}, \left\{y \to \frac{-1 + a}{a}\right\}, \left\{y \to \frac{1 + a - \sqrt{(-3 + a)(1 + a)}}{2a}\right\}, \left\{y \to \frac{1 + a + \sqrt{(-3 + a)(1 + a)}}{2a}\right\} \right\}$$

The first two points are the equilibrium points (they are, of course, also 2-periodic points). The last two points are genuine 2-periodic points; they exist if $a > 3$. We pick these points:

`sol2a = Take[sol2, {3, 4}];`

A 2-periodic point is asymptotically stable if the absolute value of the derivative of $f(f(y))$ is less than 1 at the 2-periodic points. The derivatives are as follows:

`df2 = f2'[y] /. sol2a // Simplify`

$\{4 + 2a - a^2, 4 + 2a - a^2\}$

Thus, the 2-periodic points are asymptotically stable if $\left| 4 + 2a - a^2 \right| < 1$. We solve this inequality, also taking into account the requirement $a > 3$:

`Reduce[Abs[df2[[1]]] < 1 && a > 3, a]` $3 < a < 1 + \sqrt{6}$

`% // N` $3. < a < 3.44949$

For example, when $a = 3.3$, the equilibrium and 2-periodic points are as follows:

`y2 = sol2 /. a → 3.3`

$\{\{y \to 0\}, \{y \to 0.69697\}, \{y \to 0.479427\}, \{y \to 0.823603\}\}$

From a cobweb plot (see Section 28.2.1, p. 939), we can see the 2-periodic points and how the sequence $y_{n+2} = f(f(y_n))$ approaches one of the 2-periodic points:

```
Show[cobwebPlot[f2[y] /. a → 3.3, y, 0.09, 40, 0, 1],
  Graphics[{Red, PointSize[Medium], Point[{y, y}] /. y2[[{3, 4}]]}], ImageSize → 200]
```

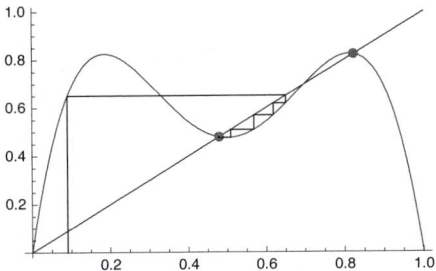

We plot the points that form the cycle:

```
p3 = Plot[y /. sol2a, {a, 3, 1 + Sqrt[6]}]
```

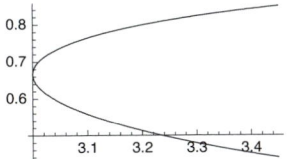

■ 4-Periodic Points

To calculate 4-periodic points, define the 4-times nested function:

```
f4[y_] = f[f[f[f[y]]]]
```

$a^4 (1 - y) y (1 - a (1 - y) y) \left(1 - a^2 (1 - y) y (1 - a (1 - y) y)\right)$
$\left(1 - a^3 (1 - y) y (1 - a (1 - y) y) \left(1 - a^2 (1 - y) y (1 - a (1 - y) y)\right)\right)$

We can solve the condition of 4-periodic points:

```
sol4 = Solve[f4[y] == y, y];
```

However, the solution is not (and cannot be) expressed as closed-form formulas but, rather, as **Root**-objects. Here is one of the roots:

```
Style[sol4[[5]], 7]
```

$\{y \to \text{Root}[1 + a^2 + a^2 (-1 - a - a^2 - a^3) \#1 + a^3 (2 + a + 4 a^2 + a^3 + 2 a^4) \#1^2 + a^3 (-1 - 5 a^2 - 4 a^3 - 5 a^4 - 4 a^5 - a^6) \#1^3 +$
$a^5 (2 + 6 a + 4 a^2 + 14 a^3 + 5 a^4 + 3 a^5) \#1^4 + a^6 (-4 - a - 18 a^2 - 12 a^3 - 12 a^4 - 3 a^5) \#1^5 +$
$a^6 (1 + 10 a^2 + 17 a^3 + 18 a^4 + 15 a^5 + a^6) \#1^6 + a^8 (-2 - 14 a - 12 a^2 - 30 a^3 - 6 a^4) \#1^7 +$
$a^9 (6 + 3 a + 30 a^2 + 15 a^3) \#1^8 + a^9 (-1 - 15 a^2 - 20 a^3) \#1^9 + a^{11} (3 + 15 a) \#1^{10} - 6 a^{12} \#1^{11} + a^{12} \#1^{12} \&, 1]\}$

For a given value of *a*, we can ask for the solution:

```
sol4 /. a → 3.52
```

$\{\{y \to 0\}, \{y \to 0.715909\}, \{y \to 0.424275\}, \{y \to 0.859816\}, \{y \to 0.373084\},$
$\{y \to 0.512076\}, \{y \to 0.823301\}, \{y \to 0.879487\}, \{y \to 0.0489016 - 0.0232193 i\},$
$\{y \to 0.0489016 + 0.0232193 i\}, \{y \to 0.165614 - 0.0737384 i\},$
$\{y \to 0.165614 + 0.0737384 i\}, \{y \to 0.505554 - 0.173586 i\}, \{y \to 0.505554 + 0.173586 i\},$
$\{y \to 0.985956 - 0.00678703 i\}, \{y \to 0.985956 + 0.00678703 i\}\}$

Here are the equilibrium and 2-periodic points:

```
sol2 /. a → 3.52
```

$\{\{y \to 0\}, \{y \to 0.715909\}, \{y \to 0.424275\}, \{y \to 0.859816\}\}$

Thus, it follows that the 4-periodic points are the next four points: 0.373084, 0.512076, 0.823301, and 0.879487. We select these solutions from **sol4**:

```
sol4a = Take[sol4, {5, 8}];
```

To find the values of *a* for which the 4-periodic points are asymptotically stable, we have to find the values of *a* for which the absolute value of the derivative of $f(f(f(f(y))))$ at the 4-periodic points is less than 1. At $a = 1 + \sqrt{6}$, the derivatives are 1:

```
df4 = f4'[y];
```

```
(df4 /. a → 1. + Sqrt[6]) /. (sol4a /. a → 1. + Sqrt[6])
{1., 1., 1., 1.}
```

At *a* = 3.5, the derivatives are −0.0305:

```
(df4 /. a → 3.5) /. (sol4a /. a → 3.5)
{-0.0305, -0.0305, -0.0305, -0.0305}
```

At *a* = 3.544, the derivatives are almost −1:

```
(df4 /. a → 3.544) /. (sol4a /. a → 3.544)
{-0.997943, -0.997943, -0.997943, -0.997943}
```

To find the value for which the derivatives are exactly −1, we solve an equation:

```
a4 = a /. FindRoot[(df4 /. sol4a〚1〛) == -1, {a, 3.544, 3.545}]
3.54409
```

Thus, the 4-periodic points are asymptotically stable for $1 + \sqrt{6} < a < 3.54409$. We plot the 4-periodic points in this interval:

```
p4 = Plot[Evaluate[y /. sol4a], {a, 1 + Sqrt[6], a4}]
```

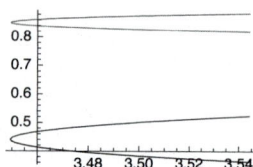

■ 8-Periodic Points

To find 8-periodic points, the method we have used does no more work; the algebra is too demanding. We proceed in a different way. Define first the 8-times nested function and its derivative:

```
f8[a_, y_] = Nest[f, y, 8];
```

```
df8[a_?NumericQ, y_] = D[f8[a, y], y];
```

Define then a function that gives, for a given *a*, the 8-periodic point near y_0:

```
period8[a_?NumericQ, y0_] := y /. FindRoot[f8[a, y] == y, {y, y0}]〚1〛
```

For example, for *a* = 3.55, the 8-periodic point that is near 0.35 is as follows:

```
period8[3.55, 0.35]      0.3548
```

To find the last value of *a* for which the 8-periodic points are stable, we find the value of *a* for which the derivative of **f8** at an 8-periodic point is −1:

```
a8 = a /. FindRoot[df8[a, period8[a, 0.35]] == -1, {a, 3.55, 3.56}]
3.56441
```

So, the 8-periodic points are stable in $3.54409 < a < 3.56441$.

From the second bifurcation diagram in Section 28.2.2, p. 942, we can see that 8-periodic points are near the points 0.35, 0.37, 0.50, 0.55, 0.81, 0.83, 0.88, and 0.89. To plot the 8-periodic points as functions of a, we produce eight figures that correspond with these eight starting values (the plotting takes some time):

```
GraphicsArray[Partition[p5 = Plot[period8[a, #], {a, a4, a8}] & /@
    {0.35, 0.37, 0.50, 0.55, 0.81, 0.83, 0.88, 0.89}, 4], ImageSize → 400]
```

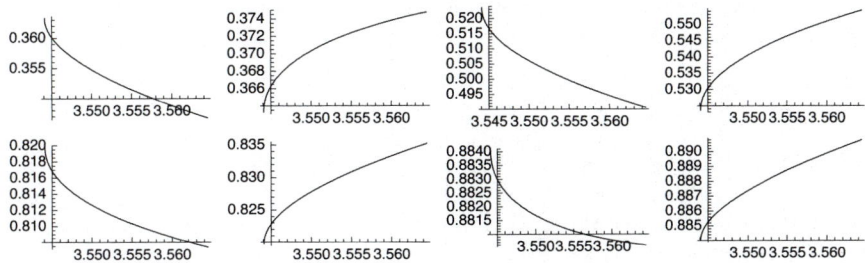

■ A Bifurcation Diagram

Now we can combine all of the plots to produce a better bifurcation diagram for $0 < a < 3.56441$:

```
Show[p1, p2, p3, p4, p5, PlotRange → All, AxesOrigin → {0, -0.02}, ImageSize → 250]
```

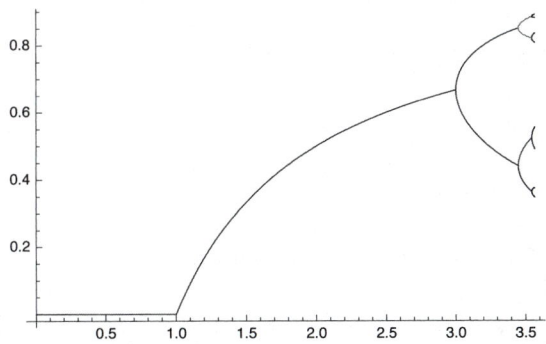

28.2.4 Lyapunov Exponents

■ Calculating the Lyapunov Exponent

The *Lyapunov exponent* gives still another view of a difference equation. It shows stability and instability and measures the sensitive dependence on initial conditions. It is defined, for a given a, by $\lambda = \lim_{n \to \infty} \frac{1}{n} \sum_{i=0}^{n-1} \log \left| f'(y_i) \right|$, where $\{y_i\}$ is the sequence calculated from the iteration formula $y_{i+1} = f(y_i)$. In numerical calculations, we approximate the limit by calculating only a finite sequence. Because the derivative of the right-hand-side function of the logistic model is $a(1 - 2y)$, we can write the following function to calculate the Lyapunov exponent for a given a:

```
lyapunovExponent = Compile[{a, {n, _Integer}},
   With[{p = NestList[a # (1 - #) &, 0.6, n - 1]}, Total[Log[Abs[a (1 - 2 p)]]] / n]];
```

Note that because **p** is a list, **a (1 - 2 p)** is the list of values $\{f'(y_0), \ldots, f'(y_{n-1})\}$.

■ **Numerical Questions**

When calculating the Lyapunov exponent, we again face the numerical problems if the system is chaotic. To examine the effect of loss of precision for an a for which the system is chaotic, we define another function in which we use high-precision numbers:

```
lyapunovExponent2[a_, n_] :=
   With[{p = NestList[a # (1 - #) &, 0.6`1150, n - 1]}, Total[Log[Abs[a (1 - 2 p)]] / n]]
```

We calculate the same exponent with both functions by using 2000 calculated values:

```
lyapunovExponent[3.7, 2000]        0.357374

lyapunovExponent2[3.7`1150, 2000]        0.3457317224630509
```

The latter result has no rounding errors, and all decimals are correct (the result, of course, contains a truncation error due to the finiteness of the sum). The difference of the results is approximately 0.01; this cannot be regarded as very small, but at least if we are interested in plotting the exponent, using normal decimal numbers may give acceptable results.

To examine the convergence of the Lyapunov exponent, we calculate estimates of the exponent for $a = 3.3$ and $a = 3.7$ using sequences of length 1000, 2000, ..., 100,000:

```
ListPlot[Table[lyapunovExponent[#, n], {n, 1000, 100 000, 1000}], PlotRange → All,
   ImageSize → 180, Ticks → {{{50, 50 000}, {100, 100 000}}, Automatic}] & /@ {3.3, 3.7}
```

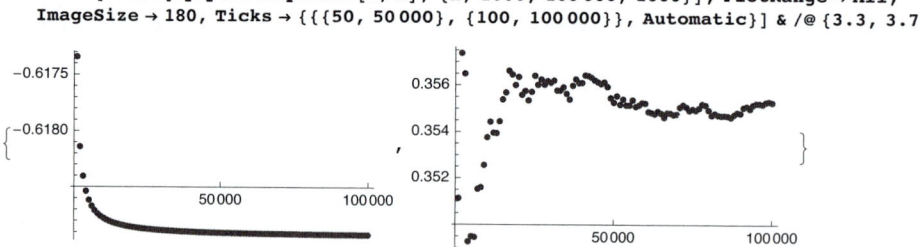

It seems that for an a for which the system is not chaotic, the estimate of the exponent converges rapidly, and that for plotting purposes, where approximately two correct decimals suffice, a few thousand values give satisfactory results. For an a for which the system is chaotic, the estimate does not converge as rapidly, but for plotting purposes, perhaps 20,000 values suffice. Next, when we plot the Lyapunov exponent, we use 30,000 values.

■ **Plotting the Lyapunov Exponent**

We plot, for the logistic model, the Lyapunov exponent as a function of a when a is in (0, 4):

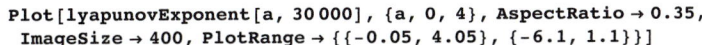

```
Plot[lyapunovExponent[a, 30 000], {a, 0, 4}, AspectRatio → 0.35,
  ImageSize → 400, PlotRange → {{-0.05, 4.05}, {-6.1, 1.1}}]
```

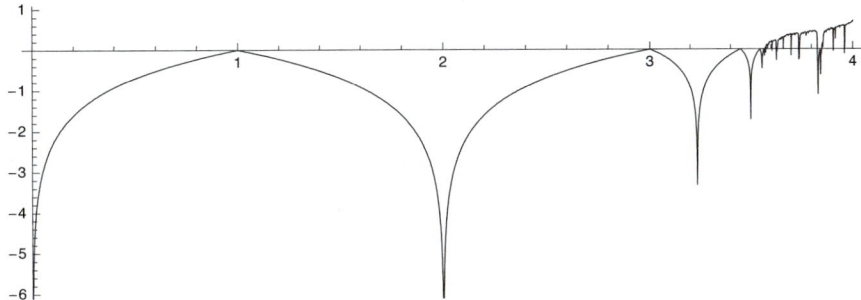

For values of *a* for which the system has a stable equilibrium point or a stable cycle, the Lyapunov exponent λ is negative. For these *a* values, the trajectories are not sensitive to initial conditions. At bifurcation points, we have $\lambda = 0$. For values of *a* for which the system is chaotic, the exponent is positive. For these *a* values, the trajectories are sensitive to initial conditions.

As noted previously, for $\lambda > 0$, the numeric behavior of the sequence is problematic, and the plot is not accurate. However, the plot does describe the overall behavior of the Lyapunov exponent. As with the bifurcation diagram, we take a closer look at the exponent when $3.5 < a < 4$ (with my somewhat slow computer, the preparation of this plot took approximately 7 minutes):

```
Plot[lyapunovExponent[a, 30 000], {a, 3.5, 4}, PlotPoints → 200, AspectRatio → 0.35,
  PlotRange → {-1.15, 0.85}, AxesOrigin → {3.5, 0}, ImageSize → 400]
```

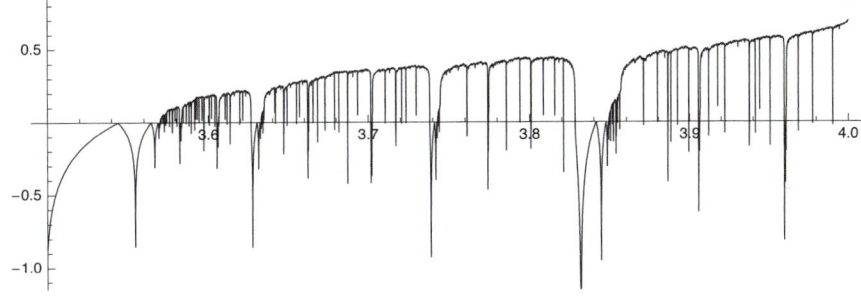

For more about studying the logistic model, function iteration, and chaos with *Mathematica*, see Wagon (2000) and Maeder (1995b); from www.mathematica-journal.com/issue/v5i2, you can load the programs of the article of Maeder (1995b). Knapp and Sofroniou (1997) use external C programs with *MathLink* to perform the heavy calculations needed to produce the bifurcation diagrams and plots of the Lyapunov exponent. Kaplan and Glass (1995) provide an introduction to nonlinear time series analysis.

28.3 More about Discrete Systems

28.3.1 A Predator–Prey Model

■ **A Predator–Prey Model**

In Section 26.3.2, p. 853, we analyzed a predator–prey model governed by the differential equations $x'(t) = x(t)\,(p - q\,y(t))$ and $y'(t) = y(t)\,(-P + Q\,x(t))$. By approximating the derivatives by divided differences, we could introduce the discrete-time model $x_{n+1} - x_n = x_n(p - q\,y_n)$, $y_{n+1} - y_n = y_n(-P + Q\,x_n)$. However, it turns out that this model does not have realistic properties. Indeed, predators and prey fluctuate in a growing manner, and the trajectories are outgoing spirals.

Other discrete-time predator–prey models have been introduced, one of which is the following: $x_{n+1} = a\,x_n(1 - x_n - y_n)$, $y_{n+1} = b\,x_n\,y_n$ (see Kelley and Peterson [2001, p. 184] or Martelli [1999, p. 246]). We assume that $a > 0$ and $b > 0$ and begin to analyze the model.

■ **Equilibrium Points**

Define the right-hand-side expressions:

```
f = a x (1 - x - y);
g = b x y;
```

Calculate the equilibrium points:

```
equi = Solve[{x == f, y == g}, {x, y}]
```

$$\left\{ \{x \to 0,\, y \to 0\},\, \left\{x \to 1 - \frac{1}{a},\, y \to 0\right\},\, \left\{y \to 1 - \frac{1}{a} - \frac{1}{b},\, x \to \frac{1}{b}\right\} \right\}$$

We denote these points with e_1, e_2, and e_3. The Jacobian is as follows:

```
jac = D[{f, g}, {{x, y}}]
```

$$\{\{-a\,x + a\,(1 - x - y),\, -a\,x\},\, \{b\,y,\, b\,x\}\}$$

Calculate the eigenvalues of the Jacobian at e_1:

```
eig1 = Eigenvalues[jac /. equi[[1]]]        {0, a}
```

An equilibrium point is asymptotically stable if the spectral radius (the largest of the absolute values of the eigenvalues) is less than 1. Thus, e_1 is asymptotically stable if $0 < a < 1$.

Calculate the eigenvalues of the Jacobian at e_2:

```
eig2 = Eigenvalues[jac /. equi[[2]]] // Simplify
```

$$\left\{2 - a,\, \frac{(-1 + a)\,b}{a}\right\}$$

The requirement that the absolute values of the eigenvalues are less than 1 is in simplified form as follows:

```
Reduce[Abs[eig2[[1]]] < 1 && Abs[eig2[[2]]] < 1 && b > 0, {a, b}, Reals]
```

$$1 < a < 3 \ \&\& \ 0 < b < \frac{a}{-1 + a}$$

Thus, the point e_2 is asymptotically stable if $1 < a < 3$ and $0 < b < \frac{a}{a-1}$. In this region, the x coordinate of e_2 or $\frac{a-1}{a}$ is positive (the y coordinate is 0).

Consider the point e_3:

```
eig3 = Eigenvalues[jac /. equi[[3]]]
```

$$\left\{ \frac{-a + 2\,b - \sqrt{a^2 + 4\,a\,b + 4\,b^2 - 4\,a\,b^2}}{2\,b}, \quad \frac{-a + 2\,b + \sqrt{a^2 + 4\,a\,b + 4\,b^2 - 4\,a\,b^2}}{2\,b} \right\}$$

Compute the conditions under which the absolute values of these are less than 1:

```
Reduce[Abs[eig3[[1]]] < 1 && Abs[eig3[[2]]] < 1 && b > 0, {a, b}, Reals]
```

$$\left(1 < a \leq 3 \,\&\&\, \frac{a}{-1 + a} < b \leq \frac{a}{2\,(-1 + a)} + \frac{1}{2}\sqrt{\frac{a^3}{(-1 + a)^2}} \right) \,||\,$$

$$\left(3 < a < 9 \,\&\&\, \frac{3\,a}{3 + a} < b \leq \frac{a}{2\,(-1 + a)} + \frac{1}{2}\sqrt{\frac{a^3}{(-1 + a)^2}} \right)$$

Here, the upper bound of b can be simplified as follows:

```
FullSimplify[ a/(2 (-1 + a)) + 1/2 Sqrt[a^3/(-1 + a)^2] , a > 1]
```

$$\frac{a}{2\left(-1 + \sqrt{a}\right)}$$

Thus, the requirement is that either $\{1 < a < 3$ and $\frac{a}{a-1} < b < \frac{a}{2\left(\sqrt{a}-1\right)}\}$ or $\{3 < a < 9$ and $\frac{3a}{3+a} < b < \frac{a}{2\left(\sqrt{a}-1\right)}\}$.

In summary, the situation is as follows:

```
p1 = Plot[{a / (a - 1), a / (2 (Sqrt[a] - 1))}, {a, 1, 3}];
p2 = Plot[{3 a / (3 + a), a / (2 (Sqrt[a] - 1))}, {a, 3, 9}];
Show[p1, p2, Ticks → {{1, 3, 9}, {1, 2, 3}}, PlotRange → {0, 3.6},
  AxesLabel → {a, b}, AxesOrigin → {0, 0}, ImageSize → 220,
  Epilog → {Line[{{1, 0}, {1, 3.6}}], Line[{{3, 0}, {3, 3 / 2}}],
    Text["e₁ stable", {0.57, 1.5}, {0, 0}, {0, 1}],
    Text["e₂ stable", {2.1, 0.8}], Text["e₃ stable", {3.15, 1.77}]}]
```

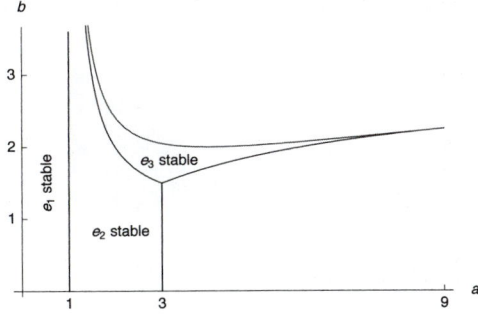

■ **Bifurcations**

Let us investigate how the trajectories behave for various values of a and b. First, we define the programs **limits** and **bifurcation** anew so that they compute both x and y values:

```
limits = Compile[{a, b},
    Drop[NestList[{a #[[1]] (1 - #[[1]] - #[[2]]), b #[[1]] #[[2]]} &, {0.4, 0.4}, 600], 301]ᵀ];
limits2[a_, b_] := With[{lims = limits[a, b]},
    {{a, #} & /@ Union[lims[[1]]], {a, #} & /@ Union[lims[[2]]]}]
bifurcation[a0_, a1_, n_, b_] :=
  Flatten[#, 1] & /@ (Table[limits2[a, b], {a, a0, a1, (a1 - a0) / n}]ᵀ)
```

For example, when $b = 1.6$, we expect, from the previous figure, that when a grows from 0, the trajectory approaches first e_1, then e_2, then e_3, and then after that it will behave in some other way. The bifurcation diagrams for x and y confirm this:

```
{xx, yy} = bifurcation[0., 4., 400, 1.6];
```

```
{ListPlot[xx, PlotRange → All, AxesOrigin → {0, -0.02},
   PlotStyle → AbsolutePointSize[0.2], AxesLabel → {a, x}, ImageSize → 200],
  ListPlot[yy, PlotRange → All, AxesOrigin → {0, -0.002},
   PlotStyle → AbsolutePointSize[0.2], AxesLabel → {a, y}, ImageSize → 200]}
```

In the case in which $b = 3.1$, we have interesting behavior near $a = 4$:

```
{xx, yy} = bifurcation[0., 4.2, 200, 3.1];
```

```
{ListPlot[xx, PlotRange → All, AxesOrigin → {0, -0.02},
   PlotStyle → AbsolutePointSize[0.2], AxesLabel → {a, x}, ImageSize → 200],
  ListPlot[yy, PlotRange → All, AxesOrigin → {0, -0.02},
   PlotStyle → AbsolutePointSize[0.2], AxesLabel → {a, y}, ImageSize → 200]}
```

■ Trajectories

```
predatorPreyPlot[a_, b_, n_, x0_, y0_, opts___] :=
 ListLinePlot[NestList[{a #[[1]] (1 - #[[1]] - #[[2]]), b #[[1]] #[[2]]} &, {x0, y0}, n],
  Mesh → All, PlotRange → All, opts]
```

With this program, we can plot trajectories of the predator–prey model. For example, when $a = 3$ and $b = 1.6$, the equilibrium points are as follows:

```
equi /. {a → 3., b → 1.6}
```

```
{{x → 0, y → 0}, {x → 0.666667, y → 0}, {y → 0.0416667, x → 0.625}}
```

The trajectory approaches the last point, e_3:

```
predatorPreyPlot[3, 1.6, 50, 0.6, 0.02, AxesLabel → {x, y}]
```

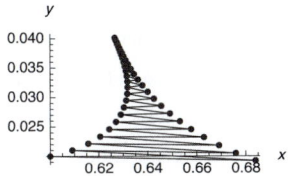

We then consider some cases in which $b = 3.1$. First, here are plots for $a = 1.4$ and $a = 2.5$:

```
{predatorPreyPlot[1.4, 3.1, 30, 0.3, 0.1], predatorPreyPlot[2.5, 3.1, 60, 0.4, 0.25]}
```

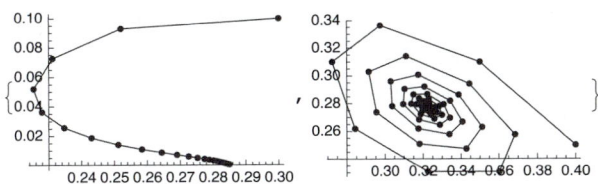

In the first plot, the trajectory approaches e_2, and in the second plot, the trajectories approach e_3 as a spiral. Next, we use the value $a = 3$ (b still has the value 3.1):

```
{predatorPreyPlot[3, 3.1, 200, 0.32, 0.33],
 ListPlot[
  Take[NestList[{3 #[[1]] (1 - #[[1]] - #[[2]]), 3.1 #[[1]] #[[2]]} &, {0.32, 0.33}, 600], -400],
  PlotStyle → PointSize[Tiny]]}
```

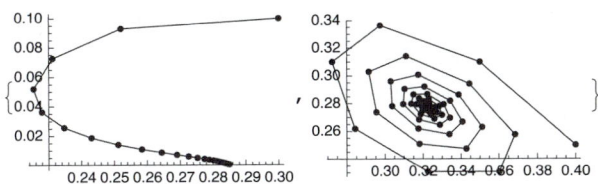

In the first plot, the trajectories approach a cycle whose center is e_3. The cycle can be seen more clearly from the second plot, where we have dropped some early points and plotted only the points, not the connecting lines. The limiting points seem to be on a closed curve. A more detailed study would reveal that the points evolve around the cycle with an approximate period of 6. Lastly, we use the values $a = 3.1$ and $a = 4.2$:

```
{predatorPreyPlot[3.1, 3.1, 100, 0.32, 0.34],
 predatorPreyPlot[4.2`150, 3.2`150, 200, 0.32`150, 0.43`150]}
```

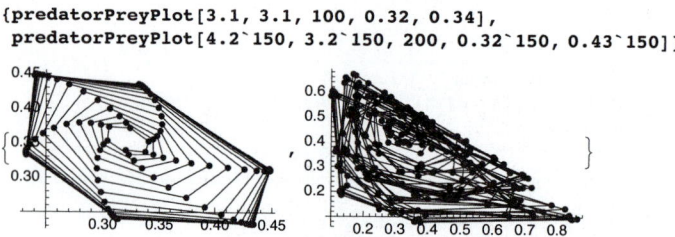

In the first plot, we have a 6-cycle, whereas the second plot shows chaotic behavior.

28.3.2 Estimation of Difference Equations

■ Return Plot

A useful way to explore a data set is to plot a *return plot*. If we have data $y_1, \ldots, y_n$, then the return plot simply contains points at (y_i, y_{i+1}), $i = 1, \ldots, n-1$. If we assume that the data are generated by a difference equation model $y_{i+1} = f(y_i)$, then the return plot contains the points $(y_i, f(y_i))$, which means that the points are on the curve $f(y)$. For example, if the model is $y_{i+1} = 3.7 \, y_i(1 - y_i)$ (a logistic model), then the points in the return plot are on the curve $3.7 \, y(1 - y)$.

Let us compare three series. The first series contains white noise (i.e., random numbers from a normal distribution; the distribution has mean 1 and standard deviation 0.5). The second series is obtained from the linear difference equation $y_{n+1} = 1.1 \, y_n$, $y_0 = 1$, and the third is obtained from the logistic difference equation $y_{n+1} = 3.7 \, y_n(1 - y_n)$, $y_0 = 0.02$. The series are as follows:

```
ss = Array[s, 3]; s[1] = RandomReal[NormalDistribution[1, 0.5], {50}];
s[2] = NestList[1.1 # &, 1, 50];
s[3] = NestList[3.7 # (1 - #) &, 0.02, 50];
ListLinePlot[#, Mesh → All] & /@ ss
```

For each sequence, we produce the return plot:

```
pairs = Table[{Drop[s[i], -1], Drop[s[i], 1]}ᵀ, {i, 3}];
```

```
ListPlot[#, AspectRatio → Automatic] & /@ pairs
```

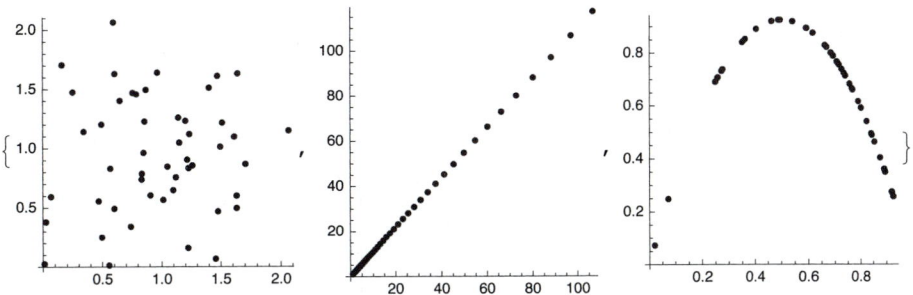

The points in the first figure do not have a clear form, but they are concentrated around the point (1, 1). The points in the second figure are on the line $1.1\,y$, whereas the points in the third figure are on the curve $3.7\,y(1-y)$.

If we have data and we are searching a difference equation model for it, a return plot may give us information about a suitable difference equation model. Indeed, one possibility to estimate a first-order difference equation is to fit a function to the paired observations. If a fit is $f(y)$, then an approximate difference equation model is $y_{n+1} = f(y_n)$. This model may be good enough, or it can be used as a starting point for nonlinear fitting to get an even better model. Two examples follow.

■ Drug in the Blood

A drug dosage of 1 mg was injected into the blood, and the amount of the drug still in the blood was measured after 1, 2, …, 12 days. The measurements and the return plot are as follows (the data are not real):

```
drug = {1, 0.739, 0.537, 0.394, 0.298,
    0.211, 0.161, 0.112, 0.088, 0.060, 0.048, 0.032, 0.023};

pairs = {Drop[drug, -1], Drop[drug, 1]}ᵀ;

{ListLinePlot[{Range[0, 12], drug}ᵀ, Mesh → All],
 ListPlot[pairs, AspectRatio → Automatic]}
```

Fit a linear function to the pairs:

```
f = Fit[pairs, {y}, y]     0.735232 y
```

A possible difference equation model is thus $y_{n+1} = 0.735232\,y_n$.

To validate the model, calculate predictions from the model:

```
g = Function[{y}, Evaluate[f]]     Function[{y}, 0.735232 y]
```

```
pred = NestList[g, 1, 12]
```

{1, 0.735232, 0.540565, 0.397441, 0.292211, 0.214843, 0.157959,
  0.116137, 0.0853872, 0.0627794, 0.0461574, 0.0339364, 0.0249511}

The differences between the data and the predictions are negligible:

```
ListPlot[drug - pred]
```

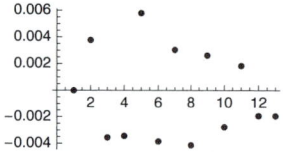

We can be satisfied with the model.

In this example, a slightly better model can be obtained using nonlinear fitting. A linear difference equation can be solved, and then it is straightforward to use **FindFit**. In the drug example, the solution of the difference equation is as follows:

```
sol = y[n] /. RSolve[{y[n + 1] == a y[n], y[0] == 1}, y[n], n][[1]]      aⁿ
```

a^n

We estimate the parameter a (for nonlinear fitting, see Section 25.1.3, p. 818):

```
FindFit[{Range[0, 12], drug}ᵀ, sol, a, n]      {a → 0.734947}
```

Thus, a slightly better model would be $y_{n+1} = 0.734947\, y_n$.

■ **Yeast Culture: First Estimation**

Consider the example of yeast culture (Pearl, 1927) that we discussed in Section 26.4.5, p. 878. The measurements and the return plot are as follows:

```
yeast = {9.6, 18.3, 29.0, 47.2, 71.1, 119.1, 174.6, 257.3, 350.7,
    441.0, 513.3, 559.7, 594.8, 629.4, 640.8, 651.1, 655.9, 659.6, 661.8};
```

```
pairs = {Drop[yeast, -1], Drop[yeast, 1]}ᵀ;
```

```
{p1 = ListLinePlot[yeast, Mesh → All], ListPlot[pairs, AspectRatio → Automatic]}
```

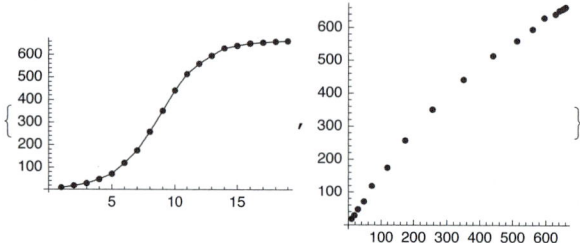

A quadratic fit may be adequate for the pairs:

```
f = Fit[pairs, {y, y^2}, y]      1.55769 y - 0.000852666 y²
```

Thus, a possible difference equation model would be $y_{n+1} = 1.55769\, y_n - 0.000852666\, y_n^2$. We can also write this equation in the form $y_{n+1} = y_n + k\, y_n(K - y_n)$:

```
SolveAlways[f == y + k y (K - y), {y}]
```

{{K → 654.049, k → 0.000852666}}

Thus, we have the logistic model $y_{n+1} = y_n + 0.000852666\, y_n(654.049 - y_n)$.

To validate the model, calculate predictions from it:

```
pred = NestList[Function[{y}, Evaluate[f]], 9.6, 18];
```

Plot the data and the predictions:

```
Show[p1, ListPlot[pred]]
```

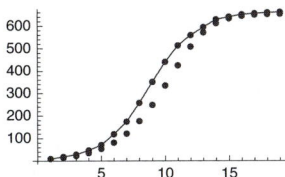

The unconnected points of the predictions differ too much from the connected data points, and so we have to conclude that our model is not adequate.

```
pred =.
```

■ Yeast Culture: Second Estimation

We use the preliminary model obtained previously as a starting point to nonlinear optimization to find a better logistic model $y_{n+1} = y_n + k\, y_n(K - y_n)$. First, define the right-hand side of the equation:

```
eqn = # + k # (K - #) &;
```

Then define the following functions:

```
pred[eqn_, data_] := NestList[eqn, data[[1]], Length[data] - 1]
crit[eqn_, data_] := Total[(data - pred[eqn, data])^2]
```

Here, **pred** calculates predictions from the equation, and **crit** defines a least-squares criterion to be minimized. Now use **FindMinimum**:

```
est = FindMinimum[crit[eqn, yeast], {k, 0.0008, 0.0009}, {K, 654, 655}]
```

```
{3756.96, {k → 0.000999256, K → 643.887}}
```

This gives the model $y_{n+1} = y_n + 0.000999256\, y_n(643.887 - y_n)$.

To validate the model, calculate predictions from it:

```
Show[p1, ListPlot[pred[eqn /. est[[2]], yeast]]]
```

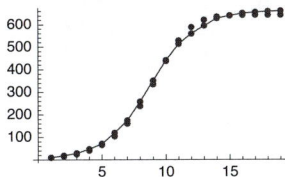

The model seems to be acceptable. (Remember from Section 26.4.5, p. 878, that a continuous logistic model was very good.)

28.3.3 Fractals and More

■ **Fractals**

Among fractal images, Mandelbrot figures are the best known. They are based on the nonlinear difference equation $z_{i+1} = z_i^2 + c$ for different complex numbers c. The values of z_i may tend toward infinity, but the numbers c for which the values do not tend toward infinity constitute the Mandelbrot set. Here is a program for investigating whether a point belongs to the Mandelbrot set:

```
mandelbrot = Compile[{{c, _Complex}},
   Module[{z = 0 + 0. I, n = 0}, While[Abs[z] < 2 && n < 50, z = z^2 + c; n++]; n]];
```

We have compiled the function `mandelbrot` to speed up the execution (compiling was explained in Section 17.2.3, p. 528). We have given the complex number $0 + 0.\,i$ as the starting value for z so that the compiler understands that z is a complex variable. The program returns, for a given number c, the number of iterations (n) needed for the absolute value of z to exceed 2; at most, 50 iterations are done. The points c for which the full 50 iterations can be done are considered to belong to the Mandelbrot set. For example, the point $0.2 + i$ does not belong to the Mandelbrot set, but the point $0.2 + 0.2\,i$ does:

```
mandelbrot[0.2 + I]     4
```

```
mandelbrot[0.2 + 0.2 I]      50
```

Giving c complex values $x + i\,y$ for many x and y, we get the corresponding numbers of iterations n. These numbers can then be plotted with `DensityPlot` to get a Mandelbrot image. We give two versions of a Mandelbrot image, having different colorings:

```
DensityPlot[mandelbrot[x + I y], {x, -2, 1}, {y, -1.5, 1.5}, ColorFunction → #,
   Mesh → False, PlotPoints → 200, ImageSize → 200] & /@ {Automatic, Hue}
```

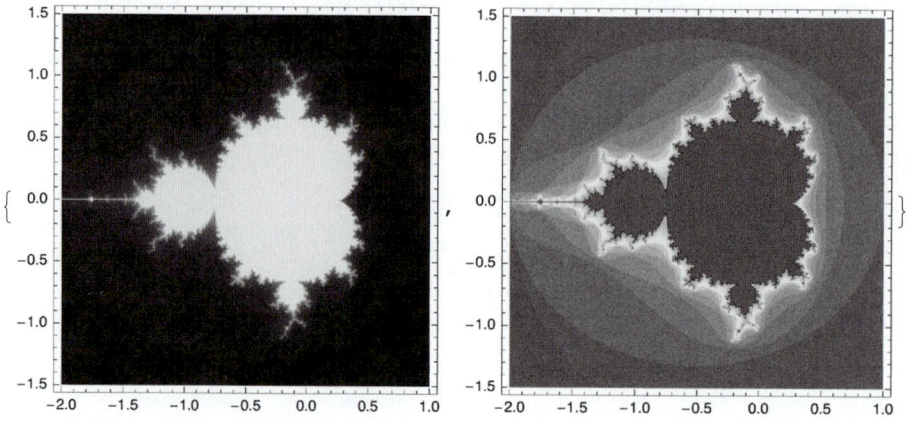

We could also use `FixedPointList`:

```
mandelbrot2 = Compile[{{c, _Complex}},
   Length[FixedPointList[#^2 + c &, 0., 50, SameTest → (Abs[#2] > 2.0 &)]]];
```

However, this program is somewhat slower than the program that uses `While`.

■ Lindenmayer Systems

L-systems or Lindenmayer systems generate, from an initial object or *axiom*, a sequence of lines and branches. The system is described with the aid of a few *characters*:

F go from the current point with the current step to a new point
- turn left by a given angle
+ turn right by a given angle
[save the current point and step
] recall the last saved point and step
X do nothing

The initial object consists of a sequence of these characters, such as the single character F. The development of the system is determined by one or more replacement *rules*; for example, F → F [- F] F [+ F] F. The rules are applied recursively a given number of times.

The following implementation of L-systems is adapted from Trott (2004a, pp. 366–392); see also Wagon (2000, pp. 167–185). Unfortunately, we do not have the space to explain the workings of the programs.

```
LSystemPoints[axiom_, rules_, iterations_, α_] :=
 Module[{point = {0, 0}, step = {0, 1}, memory = {}, toleft, toright, f, characters},
  toleft = {{Cos[-α], Sin[-α]}, {-Sin[-α], Cos[-α]}};
  toright = {{Cos[α], Sin[α]}, {-Sin[α], Cos[α]}};
  f[x_] := Which[
    x == "F", point = point + step,
    x == "-", step = toleft.step; Null,
    x == "+", step = toright.step; Null,
    x == "[", memory = Append[memory, {point, step}]; Null,
    x == "]", {point, step} = Last[memory];
             memory = Most[memory];
             Apply[Sequence, {"split", point}],
    x == "X", Null];
  characters = Flatten[Nest[(# /. rules) &, axiom, iterations]];
  Select[Prepend[f[#] & /@ characters, {0, 0}], # =!= Null &]]
```

```
LSystemLines[ls_] := Module[{a, b, c},
  a = {# - 1, # + 1} & /@ Position[ls, "split"];
  b = Partition[Flatten[{1, a, Length[ls]}], 2];
  c = Take[ls, #] & /@ b;
  Line /@ c]
```

```
LSystemPlot[axiom_, rules_, iterations_, α_] :=
  With[{ls = LSystemPoints[axiom, rules, iterations, α]},
   Graphics[LSystemLines[ls], AspectRatio → Automatic]];
```

With L-systems, we can get many figures resembling plants. Here are some examples:

```
GraphicsRow[
 {LSystemPlot[{"F"}, {"F" → Characters["FF+[+F-F-F]-[-F+F+F]"]}, 4, Pi / 8.],
  LSystemPlot[{"X"},
   {"F" → {"F", "F"}, "X" → Characters["F[-X][+X]FX"]}, 6, Pi / 8.],
  LSystemPlot[{"X"}, {"F" → {"F", "F"}, "X" → Characters["F+[[X]-X]-F[-FX]+X"]},
   5, Pi / 8.]}, ImageSize → 420]
```

■ Cellular Automata and Turing Machines

Cellular automata (see Wolfram, 2002) are based on recursive rules: The next step depends on the previous step.

> **CellularAutomaton[rule, init, n]**   Generate a list that represents the evolution of the cellular automaton rule **rule** from initial condition **init** for **n** steps

To generate three steps of cellular automaton rule 30, starting from a single 1 surrounded by 0's, write

```
CellularAutomaton[30, {{1}, 0}, 3] // MatrixForm
```

$$\begin{pmatrix} 0 & 0 & 0 & 1 & 0 & 0 & 0 \\ 0 & 0 & 1 & 1 & 1 & 0 & 0 \\ 0 & 1 & 1 & 0 & 0 & 1 & 0 \\ 1 & 1 & 0 & 1 & 1 & 1 & 1 \end{pmatrix}$$

Now we generate 40 steps:

```
ArrayPlot[CellularAutomaton[30, {{1}, 0}, 40]]
```

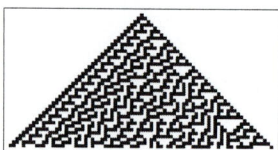

For Turing machines, see **TuringMachine** in the Documentation Center.

Probability

Introduction

> *The generation of random numbers is too important to be left to chance.* —Robert R. Coveyou

Mathematica has a wealth of information about most standard probability distributions. For each distribution, we can ask for cumulative distribution function, probability density function, quantiles, mean, variance, random numbers, and so on. As examples of discrete, continuous, and multivariate distributions, we consider in more detail the binomial, multinomial, normal, and bivariate normal distributions. Recall that we have considered the calculation of the distribution of sums of random variables in Section 18.5.1, p. 602, by writing recursive programs for convolutions and in Section 19.2.2, p. 628, by using probability-generating functions.

One very useful property of *Mathematica* is the ease of random number generation; this makes it very convenient to perform simulations. We will give several examples of simulation, particularly when we consider stochastic processes. A variety of processes are simulated, from a simple random walk to coin tossing, gambler's ruin, Brownian motion, discrete-time Markov process, Poisson process, birth–death process, and M/M/1 queue. From the simulations, we can see how realizations of the processes are displayed.

29.1 Random Numbers and Sampling

29.1.1 Uniform Random Numbers

■ Generating Uniform Random Numbers

With *Mathematica*, we can generate integer, real, and complex pseudorandom numbers. They are, respectively, from distributions that are uniformly distributed over a set of consecutive integers, over a real interval, and over a rectangle in the complex plane. The commands are as follows:

`RandomInteger[]` (⌘6) A uniform integer random number from {0, 1}

`RandomInteger[b]` A uniform integer random number from {0, 1, ..., b}

`RandomInteger[{a, b}]` A uniform integer random number from {a, $a + 1$, ..., b}

`RandomReal[]` (⌘6) A uniform real random number from [0, 1]

`RandomReal[b]` A uniform real random number from [0, b]

`RandomReal[{a, b}]` A uniform real random number from [a, b]

`RandomComplex[]` (⌘6) A uniform complex random number from the unit square

`RandomComplex[b]` A uniform complex random number from the rectangle defined by 0 and **b**

`RandomComplex[{a, b}]` A uniform complex random number from the rectangle defined by **a** and **b**

In addition, we can add a second argument to tell how many random numbers we would like to get. As an example, we show how to get lists or matrices of real random numbers; even higher-dimensional arrays of random numbers can be asked for. Integer and complex random numbers can be obtained similarly.

`RandomReal[b, n]` A list of **n** uniform random numbers from [0, b]

`RandomReal[{a, b}, n]` A list of **n** uniform random numbers from [a, b]

`RandomReal[b, {m, n}]` A matrix of (**m**×**n**) uniform random numbers from [0, b]

`RandomReal[{a, b}, {m, n}]` A matrix of (**m**×**n**) uniform random numbers from [a, b]

Recall from Section 12.1.1 that we also have `RandomPrime`.

By default, a cellular-automata-based random number generator called `"ExtendedCA"` is used. According to the documentation, this generator produces an extremely high level of randomness. See tutorial/RandomNumberGeneration for details about generating uniform random numbers.

■ Integer Random Numbers

Let DU(a, b) be the integer-valued uniform distribution among the integers a, $a + 1$, ..., b. The probability function of this distribution is $P(X = k) = 1/(b - a + 1)$ when $k = a$, $a + 1$, ..., b. First, we generate 20 random numbers from DU(0, 1)—that is, random numbers being 0 or 1 with probability $1/2$:

```
RandomInteger[1, 20]

{1, 0, 0, 1, 1, 1, 1, 0, 1, 0, 1, 1, 1, 0, 1, 0, 0, 1, 0, 0}
```

This sequence can be interpreted as being a realization of 20 coin tosses (0 could mean a head and 1 a tail). Next, we generate 20 random numbers from DU(1, 6)—that is, $P(X = k) = 1/6$, $k = 1, 2, ..., 6$. An interpretation of the simulation is that we throw a die 20 times:

```
RandomInteger[{1, 6}, 20]
```

{6, 4, 6, 5, 5, 1, 2, 6, 2, 3, 3, 5, 3, 5, 6, 6, 5, 4, 3, 6}

■ Real Random Numbers

A real uniform distribution over [a, b] is sometimes denoted as U(a, b). The probability density function of this distribution is $1/(b-a)$ for $a \le x \le b$ and 0 otherwise. Here are random numbers from the distribution U(0, 1):

```
RandomReal[1, 6]
```

{0.701228, 0.143366, 0.731902, 0.410061, 0.858202, 0.693984}

We also generate random numbers from the distributions U(1, 10) and U(−1, 1):

```
RandomReal[{1, 10}, 6]
```

{5.8545, 9.33834, 6.70059, 8.04156, 4.49274, 3.37093}

```
RandomReal[{-1, 1}, 6]
```

{−0.129994, 0.76516, 0.566518, −0.519139, 0.4803, −0.197397}

Next, we ask for real random numbers from U(0, 1) with a precision of 20:

```
RandomReal[1, 3, WorkingPrecision → 20]
```

{0.24965256350185449509, 0.40653546880408769238, 0.89580367807447865401}

■ Complex Random Numbers

Uniform complex random numbers are uniformly distributed in a rectangle in the complex plane. Here are uniform complex random numbers from the unit square, which means that the real and imaginary parts are distributed as U(0, 1):

```
RandomComplex[1 + I, 3]
```

{0.705152 + 0.398677 i, 0.444455 + 0.895985 i, 0.372455 + 0.0818215 i}

Next, we generate complex random numbers with real parts distributed as U(5, 6) and imaginary parts as U(3, 4):

```
RandomComplex[{5 + 3 i, 6 + 4 i}, 3]
```

{5.09451 + 3.18198 i, 5.65826 + 3.55053 i, 5.21453 + 3.22406 i}

The **WorkingPrecision** option can also be used.

■ Controlling the Random Numbers

SeedRandom[ ] Reseed the random number generator with the time of day and certain attributes of the current *Mathematica* session

SeedRandom[n] Reseed the random number generator with the integer *n*

SeedRandom[Method → "meth"] Use the given method to generate the random numbers; possible methods: **"ExtendedCA"** (extended cellular automaton generator; the default), **"Congruential"** (linear congruential generator; for low-quality randomness), **"Legacy"** (default generators from before *Mathematica* 6.0), **"MersenneTwister"** (Mersenne twister shift register generator), **"MKL"** (Intel MKL generator; in Intel-based systems), **"Rule30CA"** (Wolfram Rule 30 generator)

SeedRandom[n, Method → "meth"] Reseed the random number generator with the integer *n* and use the given method to generate the random numbers

Each time we generate a sequence of random numbers, we get a different sequence because *Mathematica* uses a different seed for the numbers each time (the seed is based on the time of day measured in small fractions of a second and certain attributes of the current *Mathematica* session):

```
RandomReal[1, 6]
```
{0.0409815, 0.560409, 0.920074, 0.952521, 0.449596, 0.902222}

```
RandomReal[1, 6]
```
{0.684339, 0.310975, 0.191823, 0.499712, 0.353737, 0.688333}

If we want the same sequence several times, we can set the seed with **SeedRandom[n]**:

```
SeedRandom[15]; RandomReal[1, 6]
```
{0.958907, 0.71308, 0.17847, 0.427434, 0.298721, 0.15273}

```
SeedRandom[15]; RandomReal[1, 6]
```
{0.958907, 0.71308, 0.17847, 0.427434, 0.298721, 0.15273}

BlockRandom[expr] (⌘6) Save the states of all random number generators, then evaluate **expr**, and lastly restore the states of all random number generators

---

With **BlockRandom**, we can do subsidiary random number generations in such a way that these random numbers do not affect the forthcoming random numbers. To see the effect of this command, do the following two computations:

```
SeedRandom[15]; {a = RandomReal[], b = RandomReal[]}
```
{0.958907, 0.71308}

```
SeedRandom[15];
{a = RandomReal[], BlockRandom[{RandomReal[], RandomReal[]}], b = RandomReal[]}
```
{0.958907, {0.71308, 0.17847}, 0.71308}

In both cases, the values of **a** and **b** are the same, despite the fact that in the second calculation we made some subsidiary generations.

■ Testing the Equality of Two Expressions

Random numbers can be used—in addition to simulation—in the testing of the equality of two complex expressions. As an example, consider the following integral:

```
math = Integrate[Sin[a x^n], {x, 0, ∞}, Assumptions → n > 1 && a > 0]
```

$$a^{-1/n} \, \text{Gamma}\left[1 + \frac{1}{n}\right] \, \text{Sin}\left[\frac{\pi}{2\,n}\right]$$

Spiegel (1999), formula 18.51, gives the following value for the integral:

```
spiegel = 1 / (n a^(1 / n)) Gamma[1 / n] Sin[π / (2 n)];
```

Mathematica is able to show that the two expressions are equal:

```
FullSimplify[math - spiegel]      0
```

Another way to test the equality of the two expressions is to insert random numbers in place of *a* and *n* and check whether the difference of the expressions is practically zero:

```
math - spiegel /. {a → RandomReal[], n → RandomInteger[{1, 10}]}
```
0.

By repeating this command several times, we could observe that the result is always approximately zero, and this confirms the equality of the two expressions.

29.1.2 Sampling

■ Sampling with Replacement

`RandomChoice[list, n]` (⌘6) Take, with replacement, **n** random elements from **list** (Random͵
  `Choice[list]` takes one element)

`RandomChoice[list, {m, n}]` Generate **m** samples of size **n**

`RandomChoice[weights → list, n]` Give each element a weight (e.g., probability)

`RandomChoice[weights → list, {m, n}]` Give each element a weight

If **list** contains k elements, then each time an element is chosen from **list**, each element has the probability of $1/k$ of becoming chosen into the sample. Thus, **RandomChoice** performs sampling with replacement so that an element of the given list can be in the sample several times.

For example, we throw a die 10 times both with **RandomInteger** and with **RandomChoice**:

```
RandomInteger[{1, 6}, 10]
```
{4, 1, 1, 2, 5, 4, 1, 6, 2, 5}

```
RandomChoice[Range[6], 10]
```
{2, 4, 5, 3, 1, 5, 4, 4, 2, 5}

Next, we throw a die 3 times and repeat this experiment 6 times:

```
RandomChoice[Range[6], {6, 3}]
```
{{4, 4, 2}, {2, 5, 5}, {5, 4, 1}, {4, 4, 3}, {2, 2, 1}, {6, 2, 1}}

Now we throw a biased coin for which the probability of heads (or 0) is 0.6 and that of tails (or 1) 0.4:

```
RandomChoice[{0.6, 0.4} → {0, 1}, 10]
```
{0, 1, 0, 0, 1, 0, 0, 0, 0, 0}

The weights are automatically normalized so that we can also write the following:

```
RandomChoice[{6, 4} → {0, 1}, 10]
```
{0, 1, 1, 0, 0, 0, 0, 1, 0, 0}

The weights make it possible to sample with replacement from any discrete distribution with finite domain.

Choose characters randomly:

```
RandomChoice[CharacterRange["a", "z"], 10]
```
{b, d, u, o, f, n, y, y, q, h}

Choose randomly one of Olli, Heikki, and Ulla:

```
RandomChoice[{"Olli", "Heikki", "Ulla"}]     Olli
```

The sample is, with high probability, each time different. If we want to get the same sample several times, we can use **SeedRandom** similarly as with the uniform random numbers.

■ Sampling without Replacement

`RandomSample[list, n]` (⌘6) Take, without replacement, **n** elements from **list**

`RandomSample[weights → list, n]` Give each element a weight (e.g., probability)

This command performs sampling without replacement so that an element of the given list cannot be in the sample more than once.

In the popular lotto game in Finland, seven numbers are randomly chosen from numbers 1, 2, ..., 39. Now we do such a lottery:

```
RandomSample[Range[39], 7]        {27, 8, 2, 13, 11, 24, 21}
```

■ **Random Permutations**

`RandomSample[list]`   Give a random permutation of the elements of **list**

`RandomSample[weights → list]`   Give a random permutation by giving each element a weight

---

Here is a random permutation:

```
RandomSample[Range[0, 9]]        {3, 6, 1, 8, 9, 2, 5, 7, 4, 0}
```

Divide 20 elements randomly into four groups:

```
Partition[RandomSample[Range[20]], 5]
{{1, 7, 15, 20, 13}, {12, 17, 14, 2, 11}, {8, 9, 10, 3, 5}, {4, 18, 6, 16, 19}}
```

29.2 Discrete Probability Distributions

29.2.1 Probability Distributions

Mathematica defines many discrete and continuous probability distributions. We now begin to explore these distributions. As an introduction, we briefly study the binomial distribution.

■ **Example: The Binomial Distribution**

Let X be the number of sixes when a die is tossed six times; X has the binomial distribution with parameters 6 and $\frac{1}{6}$:

```
dist = BinomialDistribution[6, 1 / 6]
```

$$\text{BinomialDistribution}\left[6, \frac{1}{6}\right]$$

Ask for the mean, variance, and standard deviation:

```
{Mean[dist], Variance[dist], StandardDeviation[dist]}
```

$$\left\{1, \frac{5}{6}, \sqrt{\frac{5}{6}}\right\}$$

The probability density function (PDF) and cumulative distribution function (CDF) at k are as follows:

```
{PDF[dist, k], CDF[dist, k]}
```

$$\left\{\frac{5^{6-k} \text{Binomial}[6, k]}{46\,656}, \text{BetaRegularized}\left[\frac{5}{6}, 6 - \text{Floor}[k], 1 + \text{Floor}[k]\right]\right\}$$

Calculate the numerical values of the density and distribution functions:

```
t1 = Table[{k, PDF[dist, k]}, {k, 0, 6}] // N
{{0., 0.334898}, {1., 0.401878}, {2., 0.200939}, {3., 0.0535837},
 {4., 0.00803755}, {5., 0.000643004}, {6., 0.0000214335}}
```

```
t2 = Table[{k, CDF[dist, k]}, {k, 0, 6}] // N
```
{{0., 0.334898}, {1., 0.736776}, {2., 0.937714},
 {3., 0.991298}, {4., 0.999336}, {5., 0.999979}, {6., 1.}}

Plot the functions:

```
ListPlot[#, PlotRange → {-0.05, 1.05}] & /@ {t1, t2}
```

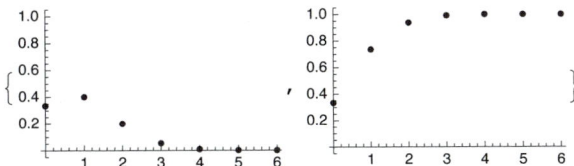

■ Properties of Distributions

For each univariate distribution (both discrete and continuous), we can ask for the following information. In place of **dist**, simply write a particular distribution, such as **BinomialDistribution[6, 1/6]**.

Mean[dist], Variance[dist], StandardDeviation[dist],
Skewness[dist], Kurtosis[dist]
PDF[dist, x]   Value of the probability density function at **x**
CDF[dist, x]   Value of the cumulative distribution function at **x**
Quantile[dist, q]   The **q** quantile
InverseCDF[dist, q]   The inverse of the CDF
ExpectedValue[f, dist]   Expected value of a pure function **f**
ExpectedValue[f, dist, x]   Expected value of a function **f** of **x**
CharacteristicFunction[dist, t]   Value of the characteristic function at **t**

Let X be a random variable. The mean μ and the variance σ^2 are the expected values of X and $(X - \mu)^2$. The standard deviation σ is the square root of the variance.

Skewness is the expected value of $(X - \mu)^3 / \sigma^3$. A positive [negative] skewness indicates a distribution with a long right [left] tail; symmetric distributions have a skewness of 0.

Kurtosis is the expected value of $(X - \mu)^4 / \sigma^4$. Kurtosis measures the concentration of the distribution around the peak and in the tails versus the concentration in the flanks. The kurtosis of all normal distributions is 3. A kurtosis larger than 3 indicates a distribution that is more peaked and has heavier tails than a normal distribution with the same variance, whereas a kurtosis smaller than 3 indicates a distribution that is flatter.

The CDF is $F(x) = P(X \le x)$. The PDF in *Mathematica* means, for continuous random variables, the derivative of the CDF and, for discrete variables, the function often called the probability function or probability mass function. In mathematical terms, the PDF in *Mathematica* is $f(x) = F'(x)$ if X is continuous and $f(x) = P(X = x)$ if X is discrete.

For probability distributions, **Quantile** and **InverseCDF** are the same (**Quantile** can also be used for data). For continuous distributions, the q quantile is the value x such that $F(x) = q$. For discrete distributions, the q quantile is the smallest integer x such that $F(x) \ge q$. For most discrete distributions, a symbolic, closed-form expression is not available for the quantile function; however, numerical values of quantiles can always be calculated.

In `ExpectedValue`, the option `Assumptions` can be used.

The characteristic function is the expected value of e^{itx} for real t.

In addition, we have an undocumented command for the domain of a distribution:

```
DistributionDomain[BinomialDistribution[n, p]]
Range[0, n]
```

■ Random Numbers from Distributions

> *Random numbers from discrete distributions:*
> `RandomInteger[dist]`   A random number from the distribution
> `RandomInteger[dist, n]`   A list of **n** random numbers
> `RandomInteger[dist, {m, n}]`   A matrix of (**m**×**n**) random numbers
>
> *Random numbers from continuous distributions:*
> `RandomReal[dist]`   A random number from the distribution
> `RandomReal[dist, n]`   A list of **n** random numbers
> `RandomReal[dist, {m, n}]`   A matrix of (**m**×**n**) random numbers

Recall from a course of probability that we can get random numbers from a distribution by calculating the inverse CDF (or the quantile function) at points that are random numbers from the continuous uniform distribution of (0, 1). Here, we calculate random numbers from an exponential distribution both by using `RandomReal` and by using the quantile function:

```
RandomReal[ExponentialDistribution[0.5], 6]
{2.24403, 1.78748, 0.265188, 0.912953, 0.833566, 1.24243}
```

```
Quantile[ExponentialDistribution[0.5], RandomReal[1, 6]]
{2.61057, 0.121715, 3.80131, 1.26564, 1.43354, 3.50496}
```

For details about generating random numbers from various distributions, see the document tutorial/RandomNumberGeneration.

■ Tabulating Distributions

When we later list univariate distributions, we use the following function, which shows the name of the distribution and the PDF, CDF, mean, and variance:

```
listDistributions[distributions_List] :=
  {Apply[StringDrop[ToString[#[[0]]], -12], #], PDF[#, x],
    CDF[#, x], Mean[#], Variance[#]} & /@ distributions
```

All names of distributions end with `Distribution`. The previous function shortens the names by dropping the "`Distribution`," to make room in a table. The table is produced with the following function:

```
tabulateDistributions[list_, opts___] := TraditionalForm[
  Grid[Join[{Style[#, Bold] & /@ {"Distribution", "PDF", "CDF", "Mean", "Variance"}},
    list], Alignment → Center, Dividers → {False, {2 → True, -1 → True}}, opts]]
```

29.2.2 Univariate Discrete Distributions

■ Discrete Distributions with a Finite Domain

Mathematica has the following univariate discrete probability distributions with a finite domain:

```
discr1 = {DiscreteUniformDistribution[{m, n}],
    BernoulliDistribution[p], BinomialDistribution[n, p],
    BetaBinomialDistribution[α, β, n], HypergeometricDistribution[n, K, N]};
```

They could be tabulated with the programs we presented in Section 29.2.1:

```
tabulateDistributions[listDistributions[discr1], Spacings → {1, 0.3}]
```

However, we do not show the result here. Instead, we present a table in which we have somewhat simplified and rearranged the expressions:

Distribution	PDF	CDF	Mean	Variance
DiscreteUniform[{m, n}]	$\frac{1}{n-m+1}$	$\begin{cases} \frac{\lfloor x-m\rfloor+1}{n-m+1} & m \le x < n \\ 1 & x \ge n \end{cases}$	$\frac{m+n}{2}$	$\frac{(n-m)(n-m+2)}{12}$
Bernoulli[p]	$\begin{cases} q & x=0 \\ p & x=1 \end{cases}$	$\begin{cases} q & 0 \le x < 1 \\ 1 & x \ge 1 \end{cases}$	p	$p\,q$
Binomial[n, p]	$\binom{n}{x} p^x q^{n-x}$	$I_q(n-\lfloor x\rfloor, \lfloor x\rfloor+1)$	$n\,p$	$n\,p\,q$
BetaBinomial[α, β, n]	$\binom{n}{x} \frac{B(x+\alpha,n-x+\beta)}{B(\alpha,\beta)}$	*	$\frac{n\,\alpha}{\alpha+\beta}$	$\frac{n\,\alpha\,\beta\,(n+\alpha+\beta)}{(\alpha+\beta)^2\,(\alpha+\beta+1)}$
Hypergeometric[n, K, N]	$\frac{\binom{K}{x}\binom{N-K}{n-x}}{\binom{N}{n}}$	*	$n\,\frac{K}{N}$	$n\,\frac{K}{N}\left(1-\frac{K}{N}\right)\frac{N-n}{N-1}$

As we stated in Section 29.2.1, the names of the distributions are shortened by dropping "**Distribu‍tion**". Also, $1-p$ is replaced with q. For some distributions, the CDF takes a lot of space and is not shown. Here, $\lfloor x\rfloor$ is **Floor[x]**, B is the **Beta**, and I_q is the **BetaRegularized** function.

The domains (or ranges of values) of these discrete distributions are as follows:

- Discrete uniform distribution: $\{m, m+1, ..., n\}$
- Bernoulli distribution: $\{0, 1\}$
- Binomial and beta binomial distribution: $\{0, 1, ..., n\}$
- Hypergeometric distribution: $\{\max(0, K+n-N), ..., \min(K, n)\}$

Many discrete distributions can be interpreted in terms of some experiments.

In the discrete uniform distribution, the experiment has n equally probable outcomes $m, m+1, ..., n$. An example is die tossing with $m = 1$ and $n = 6$.

With the Bernoulli distribution, we have two outcomes, 1 (success) and 0 (failure), which occur with probabilities p and $1-p$, respectively. An example is coin tossing with $p = \frac{1}{2}$.

The binomial distribution is the distribution of the successes in n trials, with each trial being a success or failure with probabilities p and $1 - p$. An example is $p = \frac{1}{6}$, corresponding with tossing a die n times and counting the 6's.

The beta binomial distribution emerges when we consider the parameter p in the binomial distribution to be a random variable having the Beta(α, β) distribution. A similar interpretation holds for the beta negative binomial distribution.

The hypergeometric distribution may be interpreted as follows: From an urn containing N balls—K of them being black and $N - K$ being white—we pick, without replacement, n balls and count the number of black balls.

■ Discrete Distributions with an Infinite Domain

Mathematica has the following univariate discrete probability distributions with an infinite domain:

```
discr2 = {GeometricDistribution[p], NegativeBinomialDistribution[n, p],
    BetaNegativeBinomialDistribution[α, β, n], PoissonDistribution[μ],
    LogSeriesDistribution[θ], ZipfDistribution[ρ]};
```

They could be tabulated with the programs we presented in Section 29.2.1:

```
tabulateDistributions[listDistributions[discr2], Spacings → {1, 0.3}]
```

However, we present here a table in which we have somewhat simplified and rearranged the expressions:

Distribution	PDF	CDF	Mean	Variance
Geometric$[p]$	$p\,q^x$	$1 - q^{\lfloor x \rfloor + 1}$	$\dfrac{q}{p}$	$\dfrac{q}{p^2}$
NegativeBinomial$[n, p]$	$\dbinom{n+x-1}{n-1} p^n\,q^x$	$I_p(n, \lfloor x \rfloor + 1)$	$\dfrac{nq}{p}$	$\dfrac{nq}{p^2}$
BetaNegativeBinomial$[\alpha, \beta, n]$	$\dbinom{n+x-1}{n-1} \dfrac{B(n+\alpha, x+\beta)}{B(\alpha, \beta)}$	*	$\dfrac{n\beta}{\alpha-1}$	$\dfrac{n\beta(n+\alpha-1)(\alpha+\beta-1)}{(\alpha-1)^2(\alpha-2)}$
Poisson$[\mu]$	$e^{-\mu}\dfrac{\mu^x}{x!}$	$Q(\lfloor x \rfloor + 1, \mu)$	μ	μ
LogSeries$[\theta]$	$\dfrac{-1}{\log(1-\theta)}\dfrac{\theta^x}{x}$	*	$\dfrac{-\theta}{(1-\theta)\log(1-\theta)}$	$\dfrac{-\theta\left[\theta+\log(1-\theta)\right]}{(1-\theta)^2\log^2(1-\theta)}$
Zipf$[\rho]$	$\dfrac{1}{\zeta(\rho+1)}\dfrac{1}{x^{\rho+1}}$	$\dfrac{H_{\lfloor x \rfloor}^{(\rho+1)}}{\zeta(\rho+1)}$	$\dfrac{\zeta(\rho)}{\zeta(\rho+1)},\ \rho > 1$	$\dfrac{\zeta(\rho-1)}{\zeta(\rho+1)} - \left(\dfrac{\zeta(\rho)}{\zeta(\rho+1)}\right)^2,\ \rho > 2$

Here, $1 - p$ is replaced with q. For some distributions, the CDF takes a lot of space and is not shown. Here, $\lfloor x \rfloor$ is **Floor[k]**, B is the **Beta**, ζ is the **Zeta**, I_p is the **BetaRegularized**, Q is the **GammaRegularized**, and H is the **HarmonicNumber** function.

The domains (or ranges of values) of the geometric, negative binomial, beta negative binomial, and Poisson distribution are {0, 1, 2, ...}, whereas the domain of the logarithmic series and Zipf distribution is {1, 2, ...}.

In the geometric distribution, we repeat an experiment and count the number of failures before the first success. In the negative binomial distribution, we count the number of failures before the nth success.

As an example of the discrete distributions, we next consider in more detail the binomial distribution.

29.2.3 The Binomial Distribution

■ **PDF**

Toss a die 120 times and count the occurrences of sixes:

```
dist = BinomialDistribution[120, 1 / 6.];
```

Ask for some properties:

```
{Mean[dist], Variance[dist], Skewness[dist], Kurtosis[dist]}
```

```
{20., 16.6667, 0.163299, 3.01}
```

Probabilities can be plotted as points and also as bars. To plot the bar chart, we add, in **t2**, a third element of 1 that defines the width of a bar:

```
t1 = Table[{i, PDF[dist, i]}, {i, 0, 40}];
t2 = Table[{i, PDF[dist, i], 1}, {i, 0, 40}];
<< BarCharts`
```

```
{ListPlot[t1, PlotRange → All, Filling → Axis, ImageSize → 200],
 GeneralizedBarChart[t2, ImageSize → 200] }
```

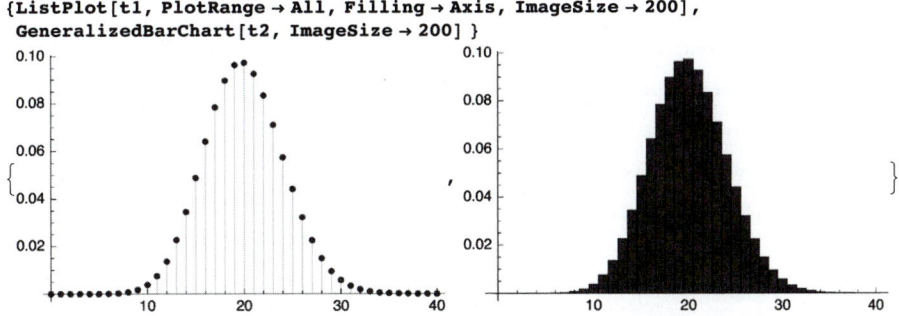

In Section 11.2.3, p. 384, we presented a dynamic interface to study the probabilities of binomial distributions.

■ **CDF and Quantiles**

The value of CDF at a point x is the probability that we will obtain at most x sixes. The value of the quantile function at q is the smallest integer x such that $F(x) \geq q$.

```
{Plot[CDF[dist, x], {x, -2, 40}],
 Plot[Quantile[dist, q], {q, 0, 1}, PlotRange → {0, 40}]}
```

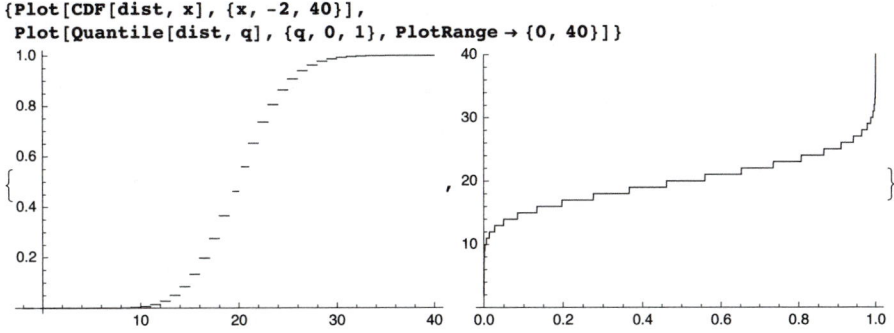

What is the probability that we will obtain at most 30 sixes?

```
CDF[dist, 30]      0.992859
```

What is the probability that we obtain 10 to 30 sixes? The probability is most easily calculated with the CDF, but we can also use the PDF:

    CDF[dist, 30] - CDF[dist, 9]   0.990158

    Sum[PDF[dist, k], {k, 10, 30}]   0.990158

Ask for some quantiles:

    Quantile[dist, #] & /@ {0.5, 0.9, 0.95, 0.99, 0.999}
    {20, 25, 27, 30, 33}

For example, the 0.5 quantile (or the median) is 20. This means that 20 is the smallest integer x satisfying $F(x) \geq 0.5$. Indeed, $F(19) < 0.5$ but $F(20) \geq 0.5$:

    {CDF[dist, 19], CDF[dist, 20]}   {0.462038, 0.559339}

Because the median is 20, we know that when we do repeatedly the experiment of tossing a die 120 times, in approximately half of the experiments the number of sixes is at most 20 and in approximately half of the cases it is more than 20.

■ Expectations

Calculate the first and second moments of a general binomial distribution:

    dist = BinomialDistribution[n, p];

    m1 = Simplify[ExpectedValue[k, dist, k], 0 < p < 1]   n p

    m2 = Simplify[ExpectedValue[k^2, dist, k], 0 < p < 1]   n p (1 + (-1 + n) p)

The variance can then be calculated in two ways:

    m2 - m1^2 // Simplify   -n (-1 + p) p

    Simplify[ExpectedValue[(k - m1)^2, dist, k], 0 < p < 1]   -n (-1 + p) p

Of course, the easiest way to calculate the mean and variance is as follows:

    {Mean[dist], Variance[dist]}   {n p, n (1 - p) p}

■ Characteristic Function

Ask for the characteristic function of the general binomial distribution:

    char = CharacteristicFunction[dist, t] $\left(1 - p + e^{i t} p\right)^n$

Then calculate some raw moments:

    Table[Limit[D[char, {t, k}], t → 0] / I^k, {k, 4}] // FullSimplify
    {n p, n p (1 + (-1 + n) p), n p (1 + (-1 + n) p (3 + (-2 + n) p)),
     n p (1 + (-1 + n) p (7 + (-2 + n) p (6 + (-3 + n) p)))}

From the characteristic function we can get the moment-generating function:

    mom = char /. t → -I t $\left(1 - p + e^{t} p\right)^n$

We can also get the probability-generating function:

    prob = char /. t → -I Log[z] $(1 - p + p z)^n$

The coefficients in the series expansion of the probability-generating function are the probabilities of the distribution:

```
CoefficientList[prob /. {n → 6, p → 1. / 6}, z]
```

{0.334898, 0.401878, 0.200939, 0.0535837, 0.00803755, 0.000643004, 0.0000214335}

From the probability-generating function, we can also calculate the mean and the variance:

```
m1 = D[prob, z] /. z → 1        n p
```

```
(D[prob, z, z] /. z → 1) + m1 - m1^2 // Simplify        -n (-1 + p) p
```

■ Random Numbers

Toss a die six times and count the occurrence of sixes. Simulate this experiment 20 times:

```
dist = BinomialDistribution[6, 1 / 6];
```

```
SeedRandom[1]; RandomInteger[dist, 20]
```

{2, 0, 2, 0, 0, 0, 1, 0, 1, 1, 0, 2, 1, 0, 3, 2, 2, 1, 0, 0}

We got 2 sixes in the first try of six tosses, 0 sixes in the second try, and so on.

Next, we do 100,000 tries of tossing a die six times:

```
SeedRandom[1]; t1 = RandomInteger[dist, 10^5];
```

The frequencies of the results 1, 2, …, 6 are as follows:

```
freq = Tally[t1] // Sort
```

{{0, 33 516}, {1, 40 327}, {2, 19 979}, {3, 5290}, {4, 821}, {5, 65}, {6, 2}}

We succeeded in getting two results in which all six throws were sixes. Look at the following:

```
Position[t1, 6]        {{13 127}, {99 323}}
```

Thus, the two remarkable results occurred on the 13,127th and 99,323rd tries. Calculate relative frequencies by dividing the frequencies by 100,000:

```
freq〚All, 2〛 / 100 000.
```

{0.33516, 0.40327, 0.19979, 0.0529, 0.00821, 0.00065, 0.00002}

These numbers are quite close to the exact probabilities:

```
Table[PDF[dist, i], {i, 0, 6}] // N
```

{0.334898, 0.401878, 0.200939, 0.0535837, 0.00803755, 0.000643004, 0.0000214335}

29.2.4 Multivariate Discrete Distributions

■ Properties of Multivariate Distributions

The **MultivariateStatistics`** package defines several multivariate distributions and their properties. The following information exists for multivariate discrete and continuous distributions:

In the **MultivariateStatistics`** *package:*

Mean, Variance, StandardDeviation, Skewness, Kurtosis,

PDF, CDF, ExpectedValue, CharacteristicFunction,

RandomInteger (for discrete distributions), **RandomReal** (for continuous distributions)

These are otherwise the same commands mentioned for univariate distributions in Section 29.2.1, p. 967, but **Quantile** and **InverseCDF** are lacking. In place of **Quantile**, we can use, for the multinormal and multivariate t distributions, **EllipsoidQuantile**. In addition, for multivariate distributions, we have the following:

In the **MultivariateStatistics`** *package:*

Covariance[dist]   Covariance matrix
Correlation[dist]   Correlation matrix
MultivariateSkewness[dist]   Multivariate coefficient of skewness
MultivariateKurtosis[dist]   Multivariate coefficient of kurtosis

---

■ **Multivariate Discrete Distributions**

In the **MultivariateStatistics`** *package:*

MultinomialDistribution[n, p]   (**p** is a vector)
NegativeMultinomialDistribution[n, p]   (**p** is a vector)
MultiPoissonDistribution[μ_0, μ]   (μ is a vector)

Here is an example of the PDF of a multinomial distribution:

```
<< MultivariateStatistics`
```

```
PDF[MultinomialDistribution[n, {p₁, p₂, p₃}], {k₁, k₂, k₃}]
```

$$\text{If}\left[k_1 + k_2 + k_3 == n, \text{Multinomial}[k_1, k_2, k_3]\, p_1^{k_1}\, p_2^{k_2}\, p_3^{k_3}, 0\right]$$

This can be interpreted as follows. Repeat an experiment independently n times. Each experiment can have three results, 1, 2, and 3, with probabilities p_1, p_2, and p_3, respectively. The previous probability is the probability of getting k_1 times result 1, k_2 times result 2, and k_3 times result 3.

Here is an example of the PDF of a negative multinomial distribution:

```
PDF[NegativeMultinomialDistribution[n, {p₁, p₂, p₃}], {k₁, k₂, k₃}]
```

$$\text{Multinomial}[-1 + n, k_1, k_2, k_3]\, p_1^{k_1}\, p_2^{k_2}\, (1 - p_1 - p_2 - p_3)^n\, p_3^{k_3}$$

This can be interpreted as follows. Repeat an experiment independently. Each experiment can be a success or one of three modes of failure. The failures have probabilities p_1, p_2, and p_3, respectively. The experiments are repeated until we have gotten n successes. The previous probability is the probability of getting k_1 times failure 1, k_2 times failure 2, and k_3 times failure 3.

Here is an example of the PDF of a multiple Poisson distribution:

```
PDF[MultiPoissonDistribution[μ₀, {μ₁, μ₂}], {k₁, k₂}]
```

$$\frac{e^{-\mu_0 - \mu_1 - \mu_2}\, \text{HypergeometricU}\left[-k_1,\, 1 - k_1 + k_2,\, -\frac{\mu_1 \mu_2}{\mu_0}\right]\, \mu_1^{k_1}\, \mu_2^{k_2}\, \left(-\frac{\mu_1 \mu_2}{\mu_0}\right)^{-k_1}}{\text{Gamma}[1 + k_1]\, \text{Gamma}[1 + k_2]}$$

Next, we consider the multinomial distribution in more detail.

■ **The Multinomial Distribution**

We toss a die three times. The numbers of the results 1, 2, ..., 6 form a multinomial distribution:

```
<< MultivariateStatistics`
```

```
dist = MultinomialDistribution[3, Table[1 / 6, {6}]]
```

$$\text{MultinomialDistribution}\left[3, \left\{\frac{1}{6}, \frac{1}{6}, \frac{1}{6}, \frac{1}{6}, \frac{1}{6}, \frac{1}{6}\right\}\right]$$

The PDF is as follows:

```
PDF[dist, Table[kᵢ, {i, 6}]]
```

$$\text{If}\left[k_1 + k_2 + k_3 + k_4 + k_5 + k_6 == 3, \ 6^{-k_1-k_2-k_3-k_4-k_5-k_6} \text{Multinomial}[k_1, k_2, k_3, k_4, k_5, k_6], 0\right]$$

For example, the probability of getting 0, 1, 0, 0, 2, 0 times the results 1, 2, 3, 4, 5, 6, respectively, is

```
PDF[dist, {0, 1, 0, 0, 2, 0}]
```
$$\frac{1}{72}$$

All possible different results are

```
Style[dom = DistributionDomain[dist], 6]
```

{{3, 0, 0, 0, 0, 0}, {0, 3, 0, 0, 0, 0}, {0, 0, 3, 0, 0, 0}, {0, 0, 0, 3, 0, 0}, {0, 0, 0, 0, 3, 0}, {0, 0, 0, 0, 0, 3}, {2, 1, 0, 0, 0, 0},
{2, 0, 1, 0, 0, 0}, {2, 0, 0, 1, 0, 0}, {2, 0, 0, 0, 1, 0}, {2, 0, 0, 0, 0, 1}, {1, 2, 0, 0, 0, 0}, {1, 0, 2, 0, 0, 0}, {1, 0, 0, 2, 0, 0},
{1, 0, 0, 0, 2, 0}, {1, 0, 0, 0, 0, 2}, {0, 2, 1, 0, 0, 0}, {0, 2, 0, 1, 0, 0}, {0, 2, 0, 0, 1, 0}, {0, 2, 0, 0, 0, 1},
{0, 1, 2, 0, 0, 0}, {0, 1, 0, 2, 0, 0}, {0, 1, 0, 0, 2, 0}, {0, 1, 0, 0, 0, 2}, {0, 0, 2, 1, 0, 0}, {0, 0, 2, 0, 1, 0}, {0, 0, 2, 0, 0, 1},
{0, 0, 1, 2, 0, 0}, {0, 0, 1, 0, 2, 0}, {0, 0, 1, 0, 0, 2}, {0, 0, 0, 2, 1, 0}, {0, 0, 0, 2, 0, 1}, {0, 0, 0, 1, 2, 0}, {0, 0, 0, 0, 2, 1},
{0, 0, 0, 1, 0, 2}, {0, 0, 0, 0, 1, 2}, {1, 1, 1, 0, 0, 0}, {1, 1, 0, 1, 0, 0}, {1, 1, 0, 0, 1, 0}, {1, 1, 0, 0, 0, 1},
{1, 0, 1, 1, 0, 0}, {1, 0, 1, 0, 1, 0}, {1, 0, 1, 0, 0, 1}, {1, 0, 0, 1, 1, 0}, {1, 0, 0, 1, 0, 1}, {1, 0, 0, 0, 1, 1}, {1, 0, 0, 1, 1, 0},
{1, 0, 0, 1, 0, 1}, {1, 0, 0, 0, 1, 1}, {0, 1, 1, 1, 0, 0}, {0, 1, 1, 0, 1, 0}, {0, 1, 1, 0, 0, 1}, {0, 1, 0, 1, 1, 0},
{0, 1, 0, 1, 0, 1}, {0, 1, 0, 0, 1, 1}, {0, 0, 1, 1, 1, 0}, {0, 0, 1, 1, 0, 1}, {0, 0, 1, 0, 1, 1}, {0, 0, 0, 1, 1, 1}}

Here are their probabilities:

```
Style[pr = PDF[dist, #] & /@ dom, 7]
```

$$\left\{\frac{1}{216}, \frac{1}{216}, \frac{1}{216}, \frac{1}{216}, \frac{1}{216}, \frac{1}{216}, \frac{1}{72}, \frac{1}{72}, \frac{1}{72}, \frac{1}{72}, \frac{1}{72}, \frac{1}{72}, \frac{1}{72}, \frac{1}{72}, \frac{1}{72}, \frac{1}{72},\right.$$

$$\frac{1}{72}, \frac{1}{72}, \frac{1}{72}, \frac{1}{72}, \frac{1}{72}, \frac{1}{72}, \frac{1}{72}, \frac{1}{72}, \frac{1}{72}, \frac{1}{72}, \frac{1}{72}, \frac{1}{72}, \frac{1}{72}, \frac{1}{72}, \frac{1}{72}, \frac{1}{72}, \frac{1}{72}, \frac{1}{72}, \frac{1}{72}, \frac{1}{72},$$

$$\left.\frac{1}{36}, \frac{1}{36}, \frac{1}{36}, \frac{1}{36}, \frac{1}{36}, \frac{1}{36}, \frac{1}{36}, \frac{1}{36}, \frac{1}{36}, \frac{1}{36}, \frac{1}{36}, \frac{1}{36}, \frac{1}{36}, \frac{1}{36}, \frac{1}{36}, \frac{1}{36}, \frac{1}{36}, \frac{1}{36}, \frac{1}{36}, \frac{1}{36}\right\}$$

```
pr // Total          1
```

The mean and variance vectors are as follows:

```
{Mean[dist], Variance[dist]}
```

$$\left\{\left\{\frac{1}{2}, \frac{1}{2}, \frac{1}{2}, \frac{1}{2}, \frac{1}{2}, \frac{1}{2}\right\}, \left\{\frac{5}{12}, \frac{5}{12}, \frac{5}{12}, \frac{5}{12}, \frac{5}{12}, \frac{5}{12}\right\}\right\}$$

We repeat four times the experiment of tossing a die three times. Here are the results:

```
RandomInteger[dist, {4}]
```

{{1, 0, 0, 0, 1, 1}, {0, 0, 2, 1, 0, 0}, {1, 0, 0, 1, 1, 0}, {2, 0, 0, 0, 0, 1}}

The covariance and correlation matrices are as follows:

```
MatrixForm /@ {Covariance[dist], Correlation[dist]}
```

$$\left\{\begin{pmatrix} \frac{5}{12} & -\frac{1}{12} & -\frac{1}{12} & -\frac{1}{12} & -\frac{1}{12} & -\frac{1}{12} \\ -\frac{1}{12} & \frac{5}{12} & -\frac{1}{12} & -\frac{1}{12} & -\frac{1}{12} & -\frac{1}{12} \\ -\frac{1}{12} & -\frac{1}{12} & \frac{5}{12} & -\frac{1}{12} & -\frac{1}{12} & -\frac{1}{12} \\ -\frac{1}{12} & -\frac{1}{12} & -\frac{1}{12} & \frac{5}{12} & -\frac{1}{12} & -\frac{1}{12} \\ -\frac{1}{12} & -\frac{1}{12} & -\frac{1}{12} & -\frac{1}{12} & \frac{5}{12} & -\frac{1}{12} \\ -\frac{1}{12} & -\frac{1}{12} & -\frac{1}{12} & -\frac{1}{12} & -\frac{1}{12} & \frac{5}{12} \end{pmatrix}, \begin{pmatrix} 1 & -\frac{1}{5} & -\frac{1}{5} & -\frac{1}{5} & -\frac{1}{5} & -\frac{1}{5} \\ -\frac{1}{5} & 1 & -\frac{1}{5} & -\frac{1}{5} & -\frac{1}{5} & -\frac{1}{5} \\ -\frac{1}{5} & -\frac{1}{5} & 1 & -\frac{1}{5} & -\frac{1}{5} & -\frac{1}{5} \\ -\frac{1}{5} & -\frac{1}{5} & -\frac{1}{5} & 1 & -\frac{1}{5} & -\frac{1}{5} \\ -\frac{1}{5} & -\frac{1}{5} & -\frac{1}{5} & -\frac{1}{5} & 1 & -\frac{1}{5} \\ -\frac{1}{5} & -\frac{1}{5} & -\frac{1}{5} & -\frac{1}{5} & -\frac{1}{5} & 1 \end{pmatrix}\right\}$$

Here are the multivariate coefficients of skewness and kurtosis:

```
{MultivariateSkewness[dist], MultivariateKurtosis[dist]}
```

$$\left\{\frac{20}{3}, \frac{95}{3}\right\}$$

29.3 Continuous Probability Distributions

29.3.1 Univariate Continuous Distributions

■ Properties of Continuous Distributions

For univariate continuous distributions, we can use the same commands **Mean**, **Variance**, and so on that were mentioned in Section 29.2.1, p. 967:

Mean[dist], Variance[dist], StandardDeviation[dist],
Skewness[dist], Kurtosis[dist]

PDF[dist, x]　Value of the probability density function at **x**

CDF[dist, x]　Value of the cumulative distribution function at **x**

Quantile[dist, q]　The **q** quantile

InverseCDF[dist, q]　The inverse of the CDF

ExpectedValue[f, dist]　Expected value of a pure function **f**

ExpectedValue[f, dist, x]　Expected value of a function **f** of **x**

CharacteristicFunction[dist, t]　Value of the characteristic function at **t**

RandomReal[dist]　A random number from the distribution
RandomReal[dist, n]　A list of **n** random numbers
RandomReal[dist, {m, n}]　A matrix of (**m×n**) random numbers

---

We now introduce the continuous distributions of *Mathematica*. We have classified the distributions as follows:

- Distributions with a finite domain
- Distributions with a half-infinite domain
- Distributions with an infinite domain
- Statistical distributions

■ Continuous Distributions with a Finite Domain

Mathematica has the following continuous distributions with a finite domain:

```
cont1 = {UniformDistribution[{a, b}], TriangularDistribution[{a, b}],
    TriangularDistribution[{a, b}, c], BetaDistribution[α, β]};
```

They could be tabulated with the program we presented in Section 29.1.3:

```
tabulateDistributions[listDistributions[cont1], Spacings → {1, 0.3}]
```

However, we present here a table in which we have somewhat simplified the expressions:

Distribution	PDF	CDF	Mean	Variance
Uniform[{a, b}]	$\begin{cases} \frac{1}{b-a} & a \le x \le b \end{cases}$	$\begin{cases} \frac{x-a}{b-a} & a \le x \le b \\ 1 & x > b \end{cases}$	$\frac{a+b}{2}$	$\frac{(b-a)^2}{12}$
Triangular[{a, b}]	$\begin{cases} \frac{4(x-a)}{(b-a)^2} & a \le x \le \frac{a+b}{2} \\ \frac{4(b-x)}{(b-a)^2} & \frac{a+b}{2} < x \le b \end{cases}$	$\begin{cases} 2\left(\frac{x-a}{b-a}\right)^2 & a \le x \le \frac{a+b}{2} \\ 1 - 2\left(\frac{b-x}{b-a}\right)^2 & \frac{a+b}{2} < x \le b \\ 1 & x > b \end{cases}$	$\frac{a+b}{2}$	$\frac{(b-a)^2}{24}$
Triangular[{a, b}, c]	$\begin{cases} \frac{2(x-a)}{(b-a)(c-a)} & a \le x \le c \\ \frac{2(b-x)}{(b-a)(b-c)} & c < x \le b \end{cases}$	$\begin{cases} \frac{(x-a)^2}{(b-a)(c-a)} & a \le x \le c \\ 1 - \frac{(b-x)^2}{(b-a)(b-c)} & c < x \le b \\ 1 & x > b \end{cases}$	$\frac{a+b+c}{3}$	$\frac{(a-b)^2+(a-c)^2+(b-c)^2}{36}$
Beta[α, β]	$\frac{1}{\mathrm{B}(\alpha,\beta)} x^{\alpha-1}(1-x)^{\beta-1}$	$I_x(\alpha, \beta)$	$\frac{\alpha}{\alpha+\beta}$	$\frac{\alpha\beta}{(\alpha+\beta)^2(\alpha+\beta+1)}$

Note again that here the names of the distributions are shortened by dropping "**Distribution**" from the names. Recall that B is the **Beta** function and I_x the **BetaRegularized** function. The domain of the uniform and both triangular distributions is the interval $[a, b]$, and the domain of the beta distribution is $[0, 1]$.

If $c = (a + b)/2$, then the triangular distribution with parameters $\{a, b\}$ and c (having the peak at $x = c$) reduces to the symmetric triangular distribution with parameter $\{a, b\}$.

■ Continuous Distributions with a Half-Infinite Domain

Mathematica has the following continuous distributions with a half-infinite domain:

```
cont2 = {ExponentialDistribution[λ], RayleighDistribution[σ],
    WeibullDistribution[α, λ], ParetoDistribution[k, α], GammaDistribution[α, λ],
    HalfNormalDistribution[θ], LogNormalDistribution[μ, σ],
    MaxwellDistribution[σ], InverseGaussianDistribution[μ, λ]};
```

They could be tabulated with

```
tabulateDistributions[listDistributions[cont2], Spacings → {1, 0.3}]
```

but we present here a table in which we have somewhat simplified and rearranged the expressions:

Distribution	PDF	CDF	Mean	Variance
Exponential[λ]	$\lambda e^{-x\lambda}$	$1 - e^{-\lambda x}$	$\frac{1}{\lambda}$	$\frac{1}{\lambda^2}$
Rayleigh[σ]	$\frac{x}{\sigma^2} e^{-\frac{1}{2}\left(\frac{x}{\sigma}\right)^2}$	$1 - e^{-\frac{1}{2}\left(\frac{x}{\sigma}\right)^2}$	$\sqrt{\frac{\pi}{2}}\,\sigma$	$\left(2 - \frac{\pi}{2}\right)\sigma^2$
Weibull[α, λ]	$\frac{\alpha}{\lambda}\left(\frac{x}{\lambda}\right)^{\alpha-1} e^{-\left(\frac{x}{\lambda}\right)^\alpha}$	$1 - e^{-\left(\frac{x}{\lambda}\right)^\alpha}$	$\lambda\,\Gamma\!\left(1 + \frac{1}{\alpha}\right)$	$\lambda^2\left[\Gamma\!\left(1 + \frac{2}{\alpha}\right) - \Gamma\!\left(1 + \frac{1}{\alpha}\right)^2\right]$
Pareto[k, α]	$\frac{\alpha}{k}\left(\frac{k}{x}\right)^{\alpha+1}$	$1 - \left(\frac{k}{x}\right)^\alpha,\ x > k$	$\frac{\alpha k}{\alpha-1},\ \alpha > 1$	$\frac{\alpha k^2}{(\alpha-1)^2(\alpha-2)},\ \alpha > 2$
Gamma[α, λ]	$\frac{1}{\lambda\,\Gamma(\alpha)}\left(\frac{x}{\lambda}\right)^{\alpha-1} e^{-\frac{x}{\lambda}}$	$Q\!\left(\alpha, 0, \frac{x}{\lambda}\right)$	$\alpha\lambda$	$\alpha\lambda^2$
HalfNormal[θ]	$\frac{2\theta}{\pi} e^{-\left(\frac{\theta x}{\sqrt{\pi}}\right)^2}$	$\mathrm{erf}\!\left(\frac{\theta x}{\sqrt{\pi}}\right)$	$\frac{1}{\theta}$	$\frac{\pi-2}{2\theta^2}$

Distribution	PDF	CDF	Mean	Variance
LogNormal$[\mu, \sigma]$	$\dfrac{1}{\sqrt{2\pi}\,\sigma x}e^{-\left(\frac{\log(x)-\mu}{\sqrt{2}\,\sigma}\right)^2}$	$\dfrac{1}{2} + \dfrac{1}{2}\,\mathrm{erf}\left(\dfrac{\log(x)-\mu}{\sqrt{2}\,\sigma}\right)$	$e^{\mu+\frac{1}{2}\sigma^2}$	$e^{2\mu+\sigma^2}\left(e^{\sigma^2}-1\right)$
Maxwell$[\sigma]$	$\dfrac{\sqrt{2}\,x^2}{\sqrt{\pi}\,\sigma^3}e^{-\left(\frac{x}{\sqrt{2}\,\sigma}\right)^2}$	$*$	$2\dfrac{\sqrt{2}}{\sqrt{\pi}}\,\sigma$	$\left(3-\dfrac{8}{\pi}\right)\sigma^2$
InverseGaussian$[\mu, \lambda]$	$\dfrac{\sqrt{\lambda}}{\sqrt{2\pi x}\,x}e^{-\frac{\lambda}{2x}\left(\frac{x-\mu}{\mu}\right)^2}$	$*$	μ	$\dfrac{\mu^3}{\lambda}$

Remember that Q is the `GammaRegularized` function. The domain of the Pareto distribution is (k, ∞), whereas the domain of all the other distributions is $(0, \infty)$.

The lognormal distribution is the distribution followed by the exponential of a normally distributed random variable.

■ Continuous Distributions with an Infinite Domain

Here are continuous distributions for which the domain is the interval $(-\infty, \infty)$:

```
cont3 = {NormalDistribution[μ, σ], LaplaceDistribution[μ, β],
   LogisticDistribution[μ, β], CauchyDistribution[a, b],
   ExtremeValueDistribution[α, β], GumbelDistribution[α, β]};
```

They could be tabulated with

```
tabulateDistributions[listDistributions[cont3], Spacings → {1, 0.3}]
```

but we present here a table in which we have somewhat simplified and rearranged the expressions:

Distribution	PDF	CDF	Mean	Variance
Normal$[\mu, \sigma]$	$\dfrac{1}{\sqrt{2\pi}\,\sigma}e^{-\left(\frac{x-\mu}{\sqrt{2}\,\sigma}\right)^2}$	$\dfrac{1}{2} + \dfrac{1}{2}\,\mathrm{erf}\left(\dfrac{x-\mu}{\sqrt{2}\,\sigma}\right)$	μ	σ^2
Laplace$[\mu, \beta]$	$\dfrac{1}{2\beta}e^{-\frac{|x-\mu|}{\beta}}$	$\begin{cases}\dfrac{1}{2}e^{\frac{x-\mu}{\beta}} & x \le \mu \\[2mm] 1-\dfrac{1}{2}e^{\frac{\mu-x}{\beta}} & x > \mu\end{cases}$	μ	$2\beta^2$
Logistic$[\mu, \beta]$	$\dfrac{1}{\beta}e^{-\frac{x-\mu}{\beta}}\left(1+e^{-\frac{x-\mu}{\beta}}\right)^{-2}$	$\left(1+e^{-\frac{x-\mu}{\beta}}\right)^{-1}$	μ	$\dfrac{\pi^2}{3}\beta^2$
Cauchy$[a, b]$	$\dfrac{1}{\pi b}\left[1+\left(\dfrac{x-a}{b}\right)^2\right]^{-1}$	$\dfrac{1}{2}+\dfrac{1}{\pi}\tan^{-1}\left(\dfrac{x-a}{b}\right)$	$-$	$-$
ExtremeValue$[\alpha, \beta]$	$\dfrac{1}{\beta}e^{-\frac{x-\alpha}{\beta}-e^{-\frac{x-\alpha}{\beta}}}$	$e^{-e^{-\frac{x-\alpha}{\beta}}}$	$\alpha+\gamma\beta$	$\dfrac{\pi^2}{6}\beta^2$
Gumbel$[\alpha, \beta]$	$\dfrac{1}{\beta}e^{\frac{x-\alpha}{\beta}-e^{\frac{x-\alpha}{\beta}}}$	$1-e^{-e^{\frac{x-\alpha}{\beta}}}$	$\alpha-\gamma\beta$	$\dfrac{\pi^2}{6}\beta^2$

Note that the second parameter in the normal distribution is the standard deviation and not the variance. The standard normal distribution with mean 0 and standard deviation 1 can be represented as `NormalDistribution[]`.

The mean and variance do not exist for the Cauchy distribution.

The extreme value distribution is the limiting distribution for the largest values in large samples drawn from a variety of distributions, including the normal distribution. The limiting distribution for the smallest values in such samples is the Gumbel distribution. The γ in the mean of the extreme value and Gumbel distribution is `EulerGamma`.

■ Statistical Distributions

Here are well-known statistical distributions:

```
cont4a = {ChiSquareDistribution[n], ChiDistribution[n],
    StudentTDistribution[n], FRatioDistribution[m, n]};
```

They could be tabulated with

```
tabulateDistributions[listDistributions[cont4a], Spacings → {1, 0.3}]
```

but we present here a table in which we have somewhat simplified and rearranged the expressions:

Distribution	PDF	CDF	Mean	Variance
ChiSquare[n]	$\dfrac{1}{2\Gamma(\frac{n}{2})}\left(\frac{x}{2}\right)^{\frac{n}{2}-1} e^{-\frac{x}{2}}$	$Q(\frac{n}{2}, 0, \frac{x}{2})$	n	$2n$
Chi[n]	$\dfrac{\sqrt{2}}{\Gamma(\frac{n}{2})}\left(\frac{x^2}{2}\right)^{\frac{n-1}{2}} e^{-\frac{x^2}{2}}$	$Q(\frac{n}{2}, 0, \frac{x^2}{2})$	$\sqrt{2}\,\dfrac{\Gamma\left(\frac{n+1}{2}\right)}{\Gamma\left(\frac{n}{2}\right)}$	$n - 2\dfrac{\Gamma\left(\frac{n+1}{2}\right)^2}{\Gamma\left(\frac{n}{2}\right)^2}$
StudentT[n]	$\dfrac{1}{B\left(\frac{n}{2}, \frac{1}{2}\right)}\sqrt{\dfrac{n^n}{(n+x^2)^{n+1}}}$	$\frac{1}{2} + \frac{\text{sgn}(x)}{2} I_{\left(\frac{n}{n+x^2}, 1\right)}\left(\frac{n}{2}, \frac{1}{2}\right)$	$0, \quad n > 1$	$\dfrac{n}{n-2}, \quad n > 2$
FRatio[m, n]	$\dfrac{1}{B\left(\frac{m}{2}, \frac{n}{2}\right)}\sqrt{\dfrac{m^m n^n x^{m-2}}{(n+mx)^{m+n}}}$	$I_{\frac{mx}{n+mx}}\left(\frac{m}{2}, \frac{n}{2}\right)$	$\dfrac{n}{n-2}, \quad n > 2$	$\dfrac{2n^2(m+n-2)}{m(n-2)^2(n-4)}, \quad n > 4$

Here, B is the `Beta` function, Q is the `GammaRegularized`, and I_p is the `BetaRegularized` function. The domain of the χ^2 and F-ratio distributions is $(0, \infty)$, and that of the Student t distribution is $(-\infty, \infty)$.

The $\chi^2(n)$ distribution (with degrees of freedom n) is followed by the sum of the squares of n independent $N(0, 1)$ random variables. The $\chi(n)$ distribution is the distribution of the square root of a $\chi^2(n)$ random variable. If X has the $N(0, 1)$ and Z the $\chi^2(n)$ distribution and X and Z are independent, then $X \big/ \sqrt{Z/n}$ has a $t(n)$ distribution. The F-ratio distribution is followed by the ratio of two independent χ^2 variables divided by their respective degrees of freedom.

The limit of a Student $t(n)$ distribution when $n \to \infty$ is the standard normal distribution:

```
Limit[PDF[StudentTDistribution[n], x], n → ∞]
```

$$\frac{e^{-\frac{x^2}{2}}}{\sqrt{2\pi}}$$

In addition, we have the following noncentral statistical distributions that are derived from normal distributions with nonzero means:

```
cont4b = {NoncentralChiSquareDistribution[n, λ],
    NoncentralStudentTDistribution[n, λ], NoncentralFRatioDistribution[m, n, λ]};
```

As an example of the continuous distributions, we next consider in more detail the normal distribution.

29.3.2 The Normal Distribution

■ **Probabilities**

Note that the parameters of the normal distribution in *Mathematica* are the mean and the standard deviation of the distribution (in some books, the second parameter is the variance). Here are plots of the PDF, CDF, and quantile function of a normal distribution with mean 2 and standard deviation 1.5:

```
dist = NormalDistribution[2, 1.5];

{Plot[PDF[dist, x], {x, -3, 7}], Plot[CDF[dist, x], {x, -3, 7}],
 Plot[Quantile[dist, q], {q, 0, 1}, PlotRange → {-3, 7}]}
```

What is the probability of getting a value in the interval (−1, 3)? The most convenient way to answer this question is to use the distribution function, but we can also integrate the density function over the interval:

```
CDF[dist, 3] - CDF[dist, -1]        0.724757

Integrate[PDF[dist, x], {x, -1, 3}]        0.724757
```

What is a value of x such that we obtain, with probability p, at most the value x? A quantile provides the solution to this problem:

```
Quantile[dist, #] & /@ {0.5, 0.9, 0.95, 0.99, 0.999}
{2., 3.92233, 4.46728, 5.48952, 6.63535}
```

We can check that, for example, the 0.95-quantile is correct:

```
CDF[dist, % // Last]        0.999
```

The quantile function of a general random distribution is as follows:

```
Quantile[NormalDistribution[μ, σ], q]
```

$$\mu + \sqrt{2}\ \sigma\ \text{InverseErf}[-1 + 2\,q]$$

■ **Tabulating Probabilities**

For situations in which we do not have access to *Mathematica*, we can prepare a table of probabilities (for **Grid**, see Section 15.2.1, p. 470):

```
tt = Table[NumberForm[CDF[NormalDistribution[0, 1], y + x], 4],
    {y, 0.0, 3.9, 0.1}, {x, 0, 0.09, 0.01}];
rowLabels = Style[#, Bold] & /@ Range[0.0, 3.9, 0.1];
colLabels = Style[#, Bold] & /@ Range[0, 9];

tt2 = ArrayFlatten[{{{{""}}, {colLabels}, {{""}}},
    {{rowLabels}ᵀ, tt, {rowLabels}ᵀ},
    {{{""}}, {colLabels}, {{""}}}}];

Text[Style[Grid[tt2, Spacings → {1, 0}, Alignment → Decimal, Dividers →
    {{2 → True, 12 → True}, {2 → True, 12 → True, 22 → True, 32 → True, 42 → True}}], 6]]
```

	0	1	2	3	4	5	6	7	8	9	
0.	0.5	0.504	0.508	0.512	0.516	0.5199	0.5239	0.5279	0.5319	0.5359	**0.**
0.1	0.5398	0.5438	0.5478	0.5517	0.5557	0.5596	0.5636	0.5675	0.5714	0.5753	**0.1**
0.2	0.5793	0.5832	0.5871	0.591	0.5948	0.5987	0.6026	0.6064	0.6103	0.6141	**0.2**
0.3	0.6179	0.6217	0.6255	0.6293	0.6331	0.6368	0.6406	0.6443	0.648	0.6517	**0.3**
0.4	0.6554	0.6591	0.6628	0.6664	0.67	0.6736	0.6772	0.6808	0.6844	0.6879	**0.4**
0.5	0.6915	0.695	0.6985	0.7019	0.7054	0.7088	0.7123	0.7157	0.719	0.7224	**0.5**
0.6	0.7257	0.7291	0.7324	0.7357	0.7389	0.7422	0.7454	0.7486	0.7517	0.7549	**0.6**
0.7	0.758	0.7611	0.7642	0.7673	0.7704	0.7734	0.7764	0.7794	0.7823	0.7852	**0.7**
0.8	0.7881	0.791	0.7939	0.7967	0.7995	0.8023	0.8051	0.8078	0.8106	0.8133	**0.8**
0.9	0.8159	0.8186	0.8212	0.8238	0.8264	0.8289	0.8315	0.834	0.8365	0.8389	**0.9**
1.	0.8413	0.8438	0.8461	0.8485	0.8508	0.8531	0.8554	0.8577	0.8599	0.8621	**1.**
1.1	0.8643	0.8665	0.8686	0.8708	0.8729	0.8749	0.877	0.879	0.881	0.883	**1.1**
1.2	0.8849	0.8869	0.8888	0.8907	0.8925	0.8944	0.8962	0.898	0.8997	0.9015	**1.2**
1.3	0.9032	0.9049	0.9066	0.9082	0.9099	0.9115	0.9131	0.9147	0.9162	0.9177	**1.3**
1.4	0.9192	0.9207	0.9222	0.9236	0.9251	0.9265	0.9279	0.9292	0.9306	0.9319	**1.4**
1.5	0.9332	0.9345	0.9357	0.937	0.9382	0.9394	0.9406	0.9418	0.9429	0.9441	**1.5**
1.6	0.9452	0.9463	0.9474	0.9484	0.9495	0.9505	0.9515	0.9525	0.9535	0.9545	**1.6**
1.7	0.9554	0.9564	0.9573	0.9582	0.9591	0.9599	0.9608	0.9616	0.9625	0.9633	**1.7**
1.8	0.9641	0.9649	0.9656	0.9664	0.9671	0.9678	0.9686	0.9693	0.9699	0.9706	**1.8**
1.9	0.9713	0.9719	0.9726	0.9732	0.9738	0.9744	0.975	0.9756	0.9761	0.9767	**1.9**
2.	0.9772	0.9778	0.9783	0.9788	0.9793	0.9798	0.9803	0.9808	0.9812	0.9817	**2.**
2.1	0.9821	0.9826	0.983	0.9834	0.9838	0.9842	0.9846	0.985	0.9854	0.9857	**2.1**
2.2	0.9861	0.9864	0.9868	0.9871	0.9875	0.9878	0.9881	0.9884	0.9887	0.989	**2.2**
2.3	0.9893	0.9896	0.9898	0.9901	0.9904	0.9906	0.9909	0.9911	0.9913	0.9916	**2.3**
2.4	0.9918	0.992	0.9922	0.9925	0.9927	0.9929	0.9931	0.9932	0.9934	0.9936	**2.4**
2.5	0.9938	0.994	0.9941	0.9943	0.9945	0.9946	0.9948	0.9949	0.9951	0.9952	**2.5**
2.6	0.9953	0.9955	0.9956	0.9957	0.9959	0.996	0.9961	0.9962	0.9963	0.9964	**2.6**
2.7	0.9965	0.9966	0.9967	0.9968	0.9969	0.997	0.9971	0.9972	0.9973	0.9974	**2.7**
2.8	0.9974	0.9975	0.9976	0.9977	0.9977	0.9978	0.9979	0.9979	0.998	0.9981	**2.8**
2.9	0.9981	0.9982	0.9982	0.9983	0.9984	0.9984	0.9985	0.9985	0.9986	0.9986	**2.9**
3.	0.9987	0.9987	0.9987	0.9988	0.9988	0.9989	0.9989	0.9989	0.999	0.999	**3.**
3.1	0.999	0.9991	0.9991	0.9991	0.9992	0.9992	0.9992	0.9992	0.9993	0.9993	**3.1**
3.2	0.9993	0.9993	0.9994	0.9994	0.9994	0.9994	0.9994	0.9995	0.9995	0.9995	**3.2**
3.3	0.9995	0.9995	0.9995	0.9996	0.9996	0.9996	0.9996	0.9996	0.9996	0.9997	**3.3**
3.4	0.9997	0.9997	0.9997	0.9997	0.9997	0.9997	0.9997	0.9997	0.9997	0.9998	**3.4**
3.5	0.9998	0.9998	0.9998	0.9998	0.9998	0.9998	0.9998	0.9998	0.9998	0.9998	**3.5**
3.6	0.9998	0.9998	0.9999	0.9999	0.9999	0.9999	0.9999	0.9999	0.9999	0.9999	**3.6**
3.7	0.9999	0.9999	0.9999	0.9999	0.9999	0.9999	0.9999	0.9999	0.9999	0.9999	**3.7**
3.8	0.9999	0.9999	0.9999	0.9999	0.9999	0.9999	0.9999	0.9999	0.9999	0.9999	**3.8**
3.9	1.	1.	1.	1.	1.	1.	1.	1.	1.	1.	**3.9**
	0	**1**	**2**	**3**	**4**	**5**	**6**	**7**	**8**	**9**	

■ Confidence Intervals

What is an interval $(\mu - a\,\sigma, \mu + a\,\sigma)$ such that a normal random variable with mean μ and standard deviation σ is in that interval with probability p? If $\Phi(x)$ is the CDF of the normal distribution, the constant satisfies the equation $\Phi(\mu + a\,\sigma) - \Phi(\mu - a\,\sigma) = p$. By symmetry, $\Phi(\mu - a\,\sigma) = 1 - \Phi(\mu + a\,\sigma)$, so we get the equation $\Phi(\mu + a\,\sigma) = (1 + p)/2$. Thus, $\mu + a\,\sigma$ is the $(1 + p)/2$ quantile. This quantile is as follows:

```
dist = NormalDistribution[μ, σ];
```

```
Quantile[dist, (1 + p) / 2]      μ + √2 σ InverseErf[p]
```

Thus, we get a as the solution of an equation:

```
Solve[μ + a σ == %, a]     {{a → √2 InverseErf[p]}}
```

Calculate the value of a for some values of p:

```
% /. p → {0.5, 0.9, 0.95, 0.99, 0.999}
{{a → {0.67449, 1.64485, 1.95996, 2.57583, 3.29053}}}
```

Check, for example, the last value of a:

```
CDF[dist, μ + 3.29053 σ] - CDF[dist, μ - 3.29053 σ]      0.999
```

■ Expectations and Characteristic Function

Calculate the first and second moments of a general normal distribution:

```
m1 = ExpectedValue[x, dist, x]      μ
```

```
m2 = ExpectedValue[x^2, dist, x]      μ² + σ²
```

The variance can then be calculated in two ways:

```
m2 - m1 ^ 2      σ²
```

```
ExpectedValue[(x - m1) ^ 2, dist, x]      σ²
```

Of course, the easiest way to calculate the mean and variance is as follows:

```
{Mean[dist], Variance[dist]}      {μ, σ²}
```

Ask for the characteristic function of the general normal distribution:

```
char = CharacteristicFunction[dist, t]      e^{i t μ - \frac{t² σ²}{2}}
```

Calculate some raw moments:

```
Table[Limit[D[char, {t, k}], t → 0] / I^k, {k, 4}] // FullSimplify
```

$$\{μ, μ² + σ², μ³ + 3 μ σ², μ⁴ + 6 μ² σ² + 3 σ⁴\}$$

■ Normal Approximation

A binomial distribution with parameters n and p can be approximated with a normal distribution with mean np and variance npq, if n is large. Let us toss a die 100 times and count sixes. Here are the probabilities of getting 0, 1, ..., 30 sixes (the probability of getting more than 30 sixes is practically zero):

```
dist = BinomialDistribution[100, 1 / 6];
```

```
t1 = Table[{i, N[PDF[dist, i]], 1}, {i, 0, 30}];
```

The 1 as the third element of the sublists is the width of the bars in a bar chart to be plotted next. We also plot the PDF of the approximating normal distribution:

```
apprN = NormalDistribution[100 × \frac{1}{6}, Sqrt[100 × \frac{1}{6} × \frac{5}{6}]];
```

```
<< BarCharts`
```

```
Show[GeneralizedBarChart[t1, BarStyle → LightGray],
  Plot[PDF[apprN, x], {x, 0, 30}], ImageSize → 200]
```

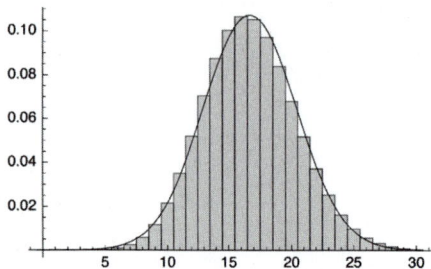

As is seen, the binomial distribution is close to the normal distribution.

■ Random Numbers

Now we generate 100 observations from a normal distribution with mean 2 and standard deviation 1.5. We plot both the original observations and the sorted observations. Most observations can be seen to be close to the mean 2:

```
dist = NormalDistribution[2, 1.5];
```

```
SeedRandom[1]; t1 = RandomReal[dist, 100];

{ListPlot[t1], ListPlot[Sort[t1], PlotStyle → PointSize[Tiny]]}
```

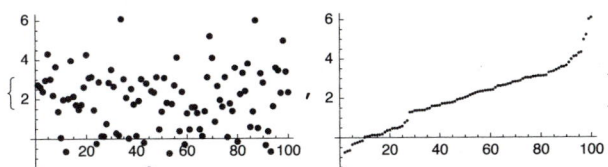

By the way, at support.wolfram.com/mathematica/graphics/decorations/probabilityplot.html [Note: This document no longer exists], you can find a program that plots sorted data on a *probability graph paper* where the *y* axis is scaled according to a given distribution (normal distribution is the default). Normally distributed data are near to a straight line on such a paper:

```
NormalProbabilityPlot[t1, ImageSize → 200, PlotStyle → PointSize[Small]]
```

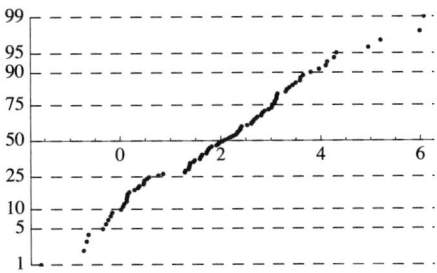

We also plot a histogram:

```
<< Histograms`

Histogram[t1, Ticks → {Range[-2, 7, 1], Automatic}]
```

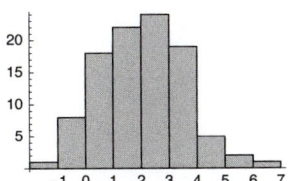

■ Truncated Normal Distribution

A truncated normal distribution is otherwise like the normal distribution but the domain is truncated to nonnegative reals. Thus, to get the PDF of the truncated normal distribution, divide the PDF of the normal distribution by the probability that the normally distributed variable is nonnegative:

```
PDFTrN =
PDF[NormalDistribution[μ, σ], x] / (1 - CDF[NormalDistribution[μ, σ], 0]) // Simplify
```

$$\frac{e^{-\frac{(x-\mu)^2}{2\sigma^2}}\sqrt{\frac{2}{\pi}}}{\sigma\left(1 + \text{Erf}\left[\frac{\mu}{\sqrt{2}\,\sigma}\right]\right)}$$

Calculate the CDF of the truncated normal distribution:

```
CDFTrN = Integrate[PDFTrN /. x → t, {t, 0, x}]
```

$$1 + \frac{\mathrm{Erfc}\left[\frac{x-\mu}{\sqrt{2}\ \sigma}\right]}{-2 + \mathrm{Erfc}\left[\frac{\mu}{\sqrt{2}\ \sigma}\right]}$$

To calculate the quantile function, solve an equation:

```
QuantileTrN = x /. Solve[CDFTrN == p, x][[1]] // Simplify
```

Solve::ifun : Inverse functions are being used by Solve, so some

   solutions may not be found; use Reduce for complete solution information. ≫

$$\mu + \sqrt{2}\ \sigma\ \mathrm{InverseErfc}\left[\langle -1 + p\rangle\ \left(-2 + \mathrm{Erfc}\left[\frac{\mu}{\sqrt{2}\ \sigma}\right]\right)\right]$$

To calculate random numbers from the truncated normal distribution, evaluate the quantile function at uniform random numbers:

```
ListPlot[QuantileTrN /. {μ → 10, σ → 6, p → RandomReal[1, 1000]},
  PlotRange → All, PlotStyle → PointSize[Small]]
```

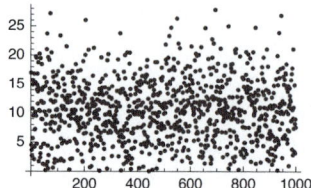

29.3.3 Multivariate Continuous Distributions

■ **Multivariate Continuous Distributions**

In the **MultivariateStatistics`** *package:*

MultinormalDistribution[μ, Σ] μ = mean vector, Σ = covariance matrix

MultivariateTDistribution[R, m] R = correlation matrix, m = dof

WishartDistribution[Σ, m] Σ = scale matrix, m = dof

HotellingTSquareDistribution[p, m] p = dimensionality parameter, m = dof

QuadraticFormDistribution[{A, b, c}, {μ, Σ}]   Distribution of $z^{\mathrm{T}} A z + b^{\mathrm{T}} z + c$, where z has the multinormal distribution with parameters μ and Σ

Here, dof is degrees of freedom. The multinormal and the multivariate t distribution are vector-valued distributions, the Wishart distribution is a matrix-valued distribution, and the Hotelling T^2 and the quadratic form distribution are scalar-valued distributions.

For the multivariate continuous distributions, we can apply the same **Mean, Variance, PDF, CDF, RandomReal, Covariance, Correlation,** etc. that were mentioned in Section 29.2.4, p. 973. Only for the Hotelling T^2 distribution, we can use **Quantile**. For the multinormal and multivariate t distributions, we have, in place of **Quantile, EllipsoidQuantile**; its inverse is **EllipsoidProbability**.

> *In the* **MultivariateStatistics`** *package:*
>
> **EllipsoidQuantile[dist, q]**   Ellipsoid containing $100\,q\%$ of the probability
>
> **EllipsoidProbability[dist, ellipsoid]**   Probability of the ellipsoid

■ The Bivariate Normal Distribution

Consider a bivariate normal distribution with means 1 and 2, variances 2 and 1, and covariance $1/2$:

```
<< MultivariateStatistics`

dist = MultinormalDistribution[{1, 2}, {{2, 1 / 2}, {1 / 2, 1}}];
```

Ask for some information:

```
{Mean[dist], Variance[dist]}      {{1, 2}, {2, 1}}

MatrixForm /@ {Covariance[dist], Correlation[dist]}
```

$$\left\{ \begin{pmatrix} 2 & \frac{1}{2} \\ \frac{1}{2} & 1 \end{pmatrix}, \begin{pmatrix} 1 & \frac{1}{2\sqrt{2}} \\ \frac{1}{2\sqrt{2}} & 1 \end{pmatrix} \right\}$$

Ask for the PDF:

```
f = PDF[dist, {x, y}]
```

$$\frac{e^{\frac{1}{2}\left(-(-1+x)\left(\frac{4}{7}(-1+x)-\frac{2}{7}(-2+y)\right)-\left(-\frac{2}{7}(-1+x)+\frac{8}{7}(-2+y)\right)(-2+y)\right)}}{\sqrt{7}\,\pi}$$

Plot the PDF both as a surface and as contours:

```
{Plot3D[f, {x, -2.5, 4.5}, {y, -0.5, 4.5}, BoxRatios → {7, 5, 3}],
 ContourPlot[f, {x, -2.5, 4.5}, {y, -0.5, 4.5}, AspectRatio → Automatic]}
```

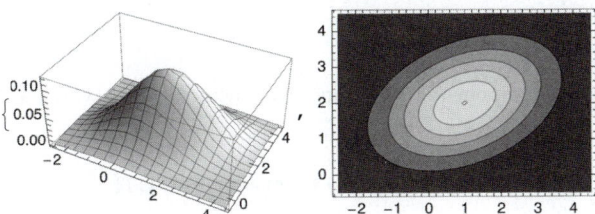

We plot the CDF (this plotting takes some time):

```
Plot3D[Evaluate[CDF[dist, {x, y}] // N],
  {x, -2, 4}, {y, 0, 4}, BoxRatios → {7, 5, 3}] // Timing
```

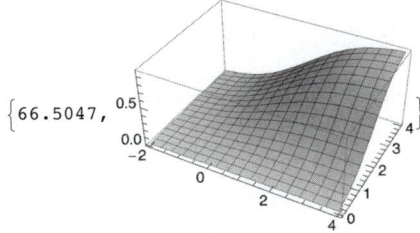

With the CDF, we can calculate the probabilities of the form $P(X \le x, Y \le y)$. For example, here is the probability that $X \le 2$ and $Y \le 3$:

```
CDF[dist, {2., 3.}]      0.669716
```

An ellipsoid centered on the mean that encompasses 99% of the probability is given as follows:

```
e99 = EllipsoidQuantile[dist, 0.99]

Ellipsoid[{1, 2}, {4.50868, 2.70237}, {{0.92388, 0.382683}, {-0.382683, 0.92388}}]

p1 = Graphics[{e99, Point[{1, 2}]}, Axes → True]
```

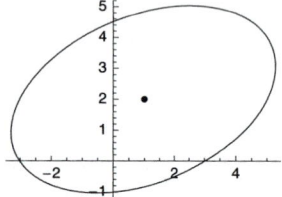

The ellipsoid really does contain 99% of the probability:

```
EllipsoidProbability[dist, e99]      0.99
```

We plot the ellipsoids that encompass $100\,p\%$ of the probability, for $p = 0.04, 0.09, 0.14, 0.19, ..., 0.94, 0.99$:

```
Graphics[Table[EllipsoidQuantile[dist, p], {p, 0.04, 0.99, 0.05}], Axes → True]
```

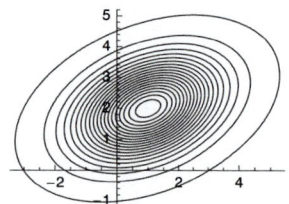

Lastly, we generate 1000 random two-component vectors and plot them together with the 99% ellipsoid:

```
SeedRandom[1];
Show[p1, ListPlot[RandomReal[dist, 1000], PlotStyle → PointSize[Small]]]
```

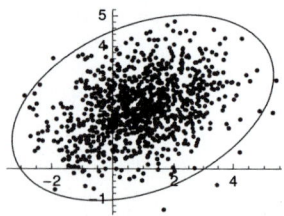

Of the points, 7 are outside the 99% region. Theoretically, the region should contain approximately 990 points out of 1000.

29.4 Stochastic Processes

29.4.1 Random Walks and Brownian Motion

■ **1D Random Walk**

In a simple random walk, we start from the point 0 and at each step we move one unit either left or right. With **RandomChoice** we can generate the steps:

```
SeedRandom[1]; steps = RandomChoice[{-1, 1}, 20]
```
```
{1, 1, -1, 1, -1, -1, -1, 1, -1, 1, -1, -1, -1, -1, -1, -1, -1, 1, 1, 1}
```

Use **Accumulate** to sum the steps up and thus to get the realization of the random walk. In addition, we have to add the starting point 0 to the walk:

```
walk = Join[{0}, Accumulate[steps]]
```
```
{0, 1, 2, 1, 2, 1, 0, -1, 0, -1, 0, -1, -2, -3, -4, -5, -6, -7, -6, -5, -4}
```

To plot the random walk, it is useful to show the time on the x axis and the movement on the y axis:

```
ListLinePlot[{Range[0, 20], walk}ᵀ, Mesh → All]
```

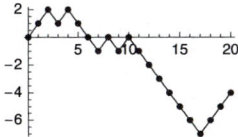

So we arrive at the following program for a simple random walk:

```
randomWalk[n_, opts___] :=
 ListLinePlot[{Range[0, n], Join[{0}, Accumulate[RandomChoice[{-1, 1}, n]]]}ᵀ, opts]
```

Next, we simulate 2000 steps:

```
SeedRandom[1]; randomWalk[2000, AspectRatio → 0.2, ImageSize → 400]
```

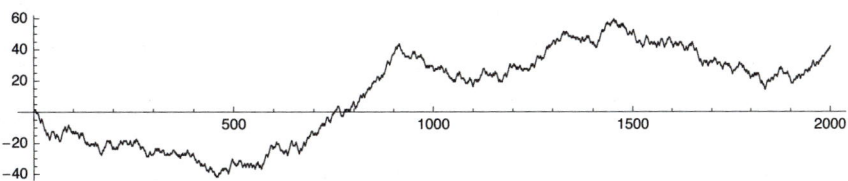

Next, we do 20 simulations of 2000 steps:

```
SeedRandom[1]; Show[Table[randomWalk[2000, PlotStyle → Hue[RandomReal[]]], {20}],
 PlotRange → All, AspectRatio → 0.2, ImageSize → 400]
```

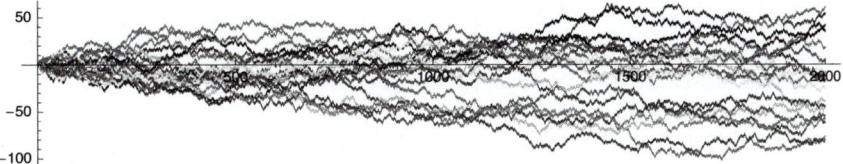

The following code (taken from the documentation of *Mathematica*) produces a dynamic plot that shows a random walk going forever (to stop the walk, delete the plot):

```
DynamicModule[{n = 199, data = Accumulate[RandomChoice[{-1, 1}, 200]]},
  Dynamic[ListLinePlot[n = n + 1; {Range[n - 199, n], data = Append[Rest[data],
    Last[data] + RandomChoice[{-1, 1}]]}ᵀ, AxesOrigin → {n - 199, 0}]]]
```

■ Coin Tossing

Here we simulate coin tossing and investigate how the relative frequency of heads evolves as we toss the coin increasingly more times. It is expected that the relative frequency approaches the value 0.5 if the coin is unbiased.

```
coinTossing[n_, opts___] := ListLinePlot[
  Accumulate[RandomChoice[{0, 1}, n]] / Range[n], AxesOrigin → {0, 0.5}, opts]
```

Here, we first generate a list of heads and tails; let us say that 1 means heads and 0 tails. Cumulative sums are then calculated, and relative frequencies of heads are obtained from the cumulative sums. Note that by dividing the cumulative sums with **Range[n]**, we divide an *n* list by another *n* list; the division is automatically done element by element so that the result is a list of relative frequencies. As an example, we toss a coin 5000 times and so calculate a sequence of empirical estimates for the probability of a head:

```
SeedRandom[5]; coinTossing[5000, AspectRatio → 0.2, ImageSize → 400, Filling → 0.5]
```

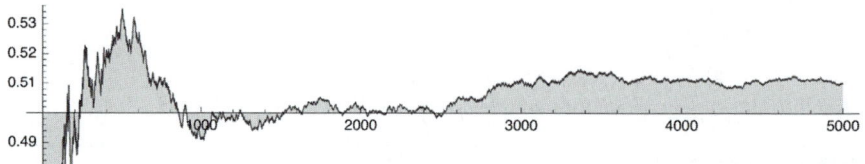

As we see, the convergence to 0.5 may not be especially fast. It is better to just trust that the probability is 0.5.

■ Gambler's Ruin

Suppose you have $*k* and your friend has $(*K* − *k*). You play the following game. You toss a coin. If the result is heads, your friend gives you $1, but if the result is tails, you have to give your friend $1. The game stops when either of the players has lost the last dollar. Let us simulate this game.

Consider the following example:

```
NestList[# + (-1)^RandomInteger[] &, 0, 10]
{0, 1, 2, 1, 0, -1, 0, 1, 0, 1, 0}
```

Note that **RandomInteger[]** is 0 or 1, with each having the probability of 1/2. We see that **(-1)^RandomInteger[]** is then 1 or −1. The previous command thus adds to the current state # a step of 1 or −1. The result is a random walk. To stop the walk when the walk reaches 0 or *K*, we use **FixedPointList** with a stopping test:

```
gamblersRuin[k_, K_, opts___] :=
  With[{gr = FixedPointList[# + (-1)^RandomInteger[] &, k,
      1000, SameTest → (#2 == 0 || #2 == K &)]},
    ListLinePlot[{Range[0, Length[gr] - 1], gr}ᵀ, opts]]
```

Recall that in `SameTest`, `#2` refers to the latest value. In the previous program, we use at most 1000 steps. As an example, suppose both you and your friend have $10 initially:

```
SeedRandom[11]; gamblersRuin[10, 20, Mesh → All]
```

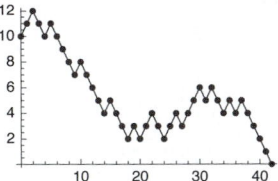

Sorry, you lost. Next, we show 20 paths:

```
SeedRandom[4];
Show[Table[gamblersRuin[10, 20, PlotStyle → Hue[RandomReal[]]], {20}],
 PlotRange → All, AxesOrigin → {0, 0}, AspectRatio → 0.25, ImageSize → 400]
```

You won 13 times and lost 7 times. One game lasted for approximately 270 iterations.

Now we simulate the game 10,000 times and only show the frequencies of the duration of the game:

```
SeedRandom[1]; tt = Table[FixedPointList[# + (-1) ^ RandomInteger[] &,
 10, 1000, SameTest → (#2 == 0 || #2 == 20 &)], {10 000}];
ListPlot[Sort[Tally[(Length /@ tt) - 1]],
 Filling → Axis, AspectRatio → 0.3, ImageSize → 400]
```

We see that most games last at most for approximately 200 iterations, but in this simulation one game lasted for more than 800 iterations.

■ 2D and 3D Random Walk

In two dimensions, we can consider several different random walks. In the following plots, the last position is shown as a red point:

```
{SeedRandom[2]; With[{rw = Accumulate[RandomChoice[{-1, 1}, {2000, 2}]]},
  Graphics[{Line[rw], Red, PointSize[Large], Point[Last[rw]]},
   Axes → True, ImageSize → 200]],
 SeedRandom[4]; With[{rw = Accumulate[
     RandomChoice[{{1, 0}, {0, 1}, {-1, 0}, {0, -1}}, 2000]]},
  Graphics[{Line[rw], Red, PointSize[Large], Point[Last[rw]]},
   Axes → True, ImageSize → 200]]}
```

```
{SeedRandom[3]; With[{rw = Accumulate[RandomChoice[{-1, 0, 1}, {2000, 2}]]},
  Graphics[{Line[rw], Red, PointSize[Large], Point[Last[rw]]},
   Axes → True, ImageSize → 200]],
 SeedRandom[2]; With[{rw = Accumulate[{Cos[#], Sin[#]} & /@ RandomReal[2 π, 2000]]},
  Graphics[{Line[rw], Red, PointSize[Large], Point[Last[rw]]},
   Axes → True, ImageSize → 200]]}
```

In the last figure, the random walk goes, from the current point, one unit in a random direction.

The following code produces a dynamic plot that shows a 2D random walk going forever (to stop the walk, delete the plot):

```
DynamicModule[{data = {{0, 0}}}, Dynamic[
  Graphics[{new = Last[data] + RandomChoice[{{1, 0}, {0, 1}, {-1, 0}, {0, -1}}];
    Line[data = Append[data, new]], Red, PointSize[Large],
    Point[new]}, Axes → True, AxesOrigin → {0, 0}]]]
```

Similarly, we can simulate a random walk in three dimensions:

```
SeedRandom[10]; With[{rw = Accumulate[RandomChoice[{-1, 1}, {500, 3}]]},
  Graphics3D[{Line[rw], Red, PointSize[Large], Point[Last[rw]]}, ImageSize → 160]]
```

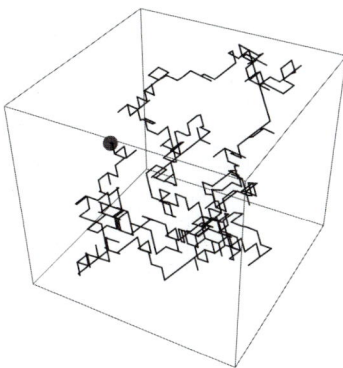

■ **Brownian Motion**

A Brownian motion or a Wiener process is a continuous-time continuous-state process. Realizations of a Brownian motion are continuous, but they are nowhere differentiable. An approximating discretization has to be done for simulation purposes. The following approximation is from Cox and Miller (1965, p. 205):

```
brownianMotion[μ_, σ_, τ_, n_, opts___] :=
  With[{δ = σ Sqrt[τ], p = 0.5 (1 + μ Sqrt[τ] / σ)},
    ListLinePlot[{Range[0, n τ, τ], NestList[# + If[Random[] ≤ p, δ, -δ] &, 0, n]}ᵀ,
      PlotLabel → Row[{"δ = ", δ, ", p = ", p}], opts]]
```

This program simulates a Brownian motion where the state $X(t)$ at time t has the normal distribution with mean μt and variance $\sigma^2 t$. The approximate process moves n times a small step τ. At each step, the next value of the process is the previous value plus either δ or $-\delta$, with probabilities p and $1 - p$, respectively. Here, $\delta = \sigma \sqrt{\tau}$ and $p = 0.5 \left(1 + \frac{\mu}{\sigma} \sqrt{\tau}\right)$. The time step τ should be small so that δ is considerably larger than τ. The probability p should not differ much from 0.5. The Brownian motion is, mathematically, the result of the discretization if τ approaches 0.

First, we show a path where the drift μ is 0:

```
SeedRandom[2]; brownianMotion[0, 0.5, 0.01, 2000, AspectRatio → 0.25, ImageSize → 400]
```

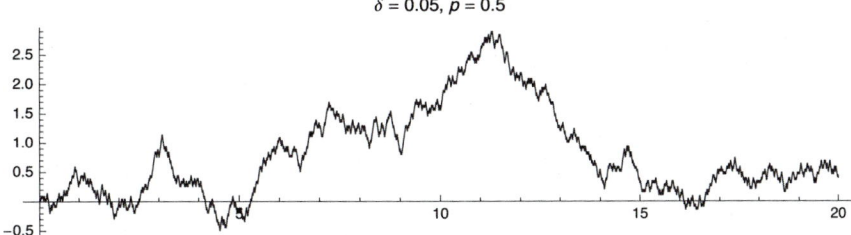

Next, we draw 20 paths with a drift of 0.3:

```
SeedRandom[2]; Show[
  Table[brownianMotion[0.3, 0.5, 0.01, 2000, PlotStyle → Hue[RandomReal[]]], {20}],
  PlotRange → All, AspectRatio → 0.25, ImageSize → 340]
```

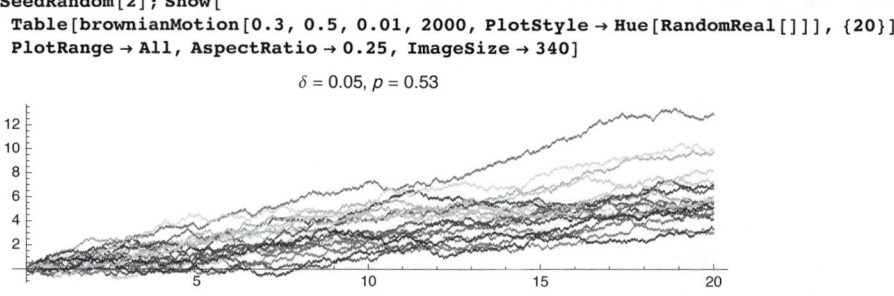

$\delta = 0.05,\ p = 0.53$

29.4.2 Discrete-Time Markov Chains

■ **Precipitation Data**

There is a precipitation monitor at the Snoqualmie Falls in western Washington. It defines a day as wet if the precipitation is as least 0.01 inches. Such days are denoted by 1. Other days are denoted by 0; let us call these days dry. The file **precipitation** (available on the CD-ROM of this book) contains codes for all days of January for the 36 years from 1948 to 1983. The data are reprinted with permission from Guttorp (1995, p. 17, Figure 2.1); copyright CRC Press, Boca Raton, Florida.

I have saved the file in the **MNata** folder of my **Documents** folder. We read the file (see Section 4.2.1, p. 100) and prepare a table for the data:

```
prec = Import["/Users/heikki/Documents/MNData/precipitation", "Table"];
```

```
prec2 = ArrayFlatten[{{{{""}}, {Range[31]}}, {{Range[1948, 1983]}ᵀ, prec}}];
```

```
Style[Grid[prec2, Spacings → {0.2, 0.25},
  ItemSize → {{3.5, {1.4}}}, Dividers → {2 → True, 2 → True}], 6]
```

	1	2	3	4	5	6	7	8	9	10	11	12	13	14	15	16	17	18	19	20	21	22	23	24	25	26	27	28	29	30	31
1948	1	1	1	1	1	1	1	1	1	1	1	1	0	0	0	0	0	0	0	1	0	1	1	1	1	0	0	0	1	1	1
1949	1	1	0	0	1	1	1	1	0	0	0	0	0	1	0	0	1	0	1	0	1	0	0	0	0	1	1	0	0	1	0
1950	1	1	1	1	1	1	1	1	1	1	1	1	0	1	1	1	1	1	1	1	1	1	1	1	1	1	1	0	0	0	0
1951	1	1	1	1	1	0	1	1	0	1	1	1	1	1	1	1	1	1	1	1	1	1	1	1	1	1	0	0	0	0	0
1952	0	0	1	1	1	1	1	1	1	1	1	1	1	0	1	0	0	0	1	1	1	1	1	1	1	1	1	0	1	1	1
1953	1	1	1	1	0	1	1	1	1	1	1	1	1	1	1	1	1	1	1	1	1	1	1	1	1	1	1	1	1	1	1
1954	1	1	1	1	1	1	1	1	1	0	0	0	1	1	1	1	1	1	1	1	1	1	1	1	1	1	1	0	0	0	1
1955	1	1	0	1	1	0	0	0	0	0	0	1	1	1	1	1	1	0	0	1	0	1	1	1	1	0	0	1	0	1	1
1956	1	1	1	1	1	1	1	1	1	1	1	0	1	1	1	1	1	1	1	1	1	1	1	0	0	0	0	1	1	0	0
1957	1	1	1	1	0	0	1	1	0	1	1	0	1	1	1	0	0	1	0	1	1	1	1	0	0	0	0	0	0	1	1
1958	0	1	1	0	0	0	0	1	1	1	1	1	1	1	1	1	1	1	0	1	1	1	1	1	1	0	1	1	1	1	1
1959	1	1	0	0	1	1	1	1	1	1	1	1	1	1	1	1	1	1	1	1	1	1	1	0	1	1	1	1	1	1	1
1960	0	1	0	0	1	1	1	0	1	1	1	0	0	1	0	0	0	0	0	1	1	1	1	1	1	1	1	1	1	1	0
1961	1	0	0	1	1	1	1	1	1	1	1	1	1	1	1	1	1	0	0	0	0	1	1	0	0	0	1	1	1	1	1
1962	0	1	1	0	1	1	1	0	0	0	1	1	1	1	1	1	1	0	0	0	0	0	1	1	1	1	0	0	0	1	1
1963	1	1	1	0	1	1	1	1	1	1	0	0	1	1	0	1	0	0	0	0	0	0	1	0	0	0	0	0	0	1	1
1964	1	1	1	1	1	1	1	1	1	1	0	0	1	1	1	1	1	1	1	1	1	0	1	0	1	1	1	1	1	1	1
1965	1	1	1	1	1	1	1	1	0	0	0	0	1	0	0	1	1	1	1	1	1	1	1	1	1	1	1	1	1	1	0
1966	1	1	1	1	1	1	1	0	1	1	1	1	0	1	1	0	0	0	0	1	0	1	0	1	1	0	1	1	1	1	1
1967	1	1	1	1	1	1	0	0	1	1	1	1	1	1	1	1	1	1	1	1	1	1	0	1	1	1	1	0	0	0	0
1968	1	0	1	1	0	1	0	1	1	1	0	1	1	1	1	1	1	1	1	1	1	0	0	1	1	1	0	0	0	1	1
1969	1	0	1	1	1	1	1	1	1	1	1	1	1	1	1	1	1	1	1	0	1	0	0	1	0	1	0	1	1	1	1
1970	0	0	1	0	0	0	0	1	1	1	1	1	1	1	1	1	1	1	1	1	1	1	1	1	1	1	1	0	0	0	1
1971	0	0	0	0	0	0	1	1	1	1	1	1	1	1	1	1	1	1	1	1	1	1	1	1	1	1	0	0	1	1	0
1972	1	1	0	1	1	1	1	1	1	1	1	1	1	1	0	1	1	1	1	1	1	1	1	1	0	0	0	0	0	0	0
1973	1	1	0	1	1	0	0	0	0	1	1	1	1	1	1	1	1	1	1	0	0	1	1	1	0	0	0	1	1	0	0
1974	0	0	0	0	0	0	0	0	1	1	1	1	1	1	1	1	1	1	1	1	1	1	1	1	1	1	1	1	1	1	1
1975	1	1	1	1	1	1	1	1	1	1	1	1	1	1	1	1	1	1	1	0	0	1	1	1	1	1	0	0	0	0	0
1976	0	0	1	1	1	1	1	1	1	1	1	1	1	1	1	0	0	0	1	1	0	0	1	0	0	0	1	0	0	0	0
1977	1	1	1	0	0	0	0	0	0	1	1	1	1	1	1	1	1	1	1	1	0	0	0	0	0	0	0	0	0	1	1
1978	0	1	1	1	1	1	1	1	1	1	1	1	0	1	0	1	0	0	0	1	1	0	1	1	1	0	1	1	1	0	1
1979	0	0	0	0	0	0	0	1	1	1	1	0	1	1	0	1	1	0	0	1	1	0	1	1	0	0	0	0	0	0	0
1980	1	1	0	1	1	0	1	1	1	1	1	1	1	1	1	0	0	0	1	0	0	0	1	0	0	0	0	0	0	0	1
1981	0	0	0	0	1	1	0	1	1	0	0	0	0	0	0	1	1	0	1	1	1	1	1	1	1	1	1	1	1	1	0
1982	1	1	1	1	0	0	0	0	0	1	1	1	1	1	1	1	1	0	0	1	1	1	1	1	1	1	1	1	1	1	1
1983	1	1	1	1	1	1	1	1	1	1	0	1	1	1	1	1	1	1	1	1	0	1	1	1	1	1	1	0	1	0	0

We see that the first 12 days of January 1948 were wet, the next 7 days dry, and so on. Here is a plot of the data. Each black square denotes a wet day:

```
ArrayPlot[prec, Mesh → True, Frame → True,
  FrameTicks → {{Range[3, 35, 5], Range[1950, 1980, 5]}ᵀ, None, None, Range[5, 30, 5]},
  ImageSize → 180]
```

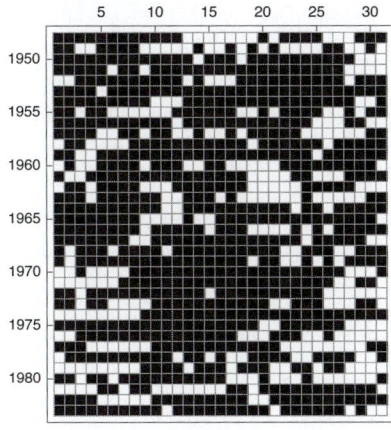

■ **Runs**

With **Split**, we can find runs, which are sequences of like values. For example, here are the seven runs of January 1948:

```
Split[prec[[1]]]
```

{{1, 1, 1, 1, 1, 1, 1, 1, 1, 1, 1, 1},
 {0, 0, 0, 0, 0, 0, 0}, {1}, {0}, {1, 1, 1, 1}, {0, 0, 0}, {1, 1, 1}}

Next, we calculate the numbers of runs for all 36 Januaries:

```
Length /@ (Split /@ prec)
```

{7, 16, 6, 6, 8, 3, 5, 13, 6, 13, 8, 5, 11, 7, 9, 11, 7,
 6, 11, 6, 11, 11, 6, 5, 6, 10, 2, 4, 7, 5, 12, 11, 11, 9, 5, 8}

Calculate the mean number of runs per month:

```
Total[%] / 36.       7.97222
```

What would the expected number of runs be if the weather were to change randomly from day to day? If the probability of a wet day is p and there are n days, the expected number of runs is as follows:

```
expRuns[n_, p_] := 1 + 2 (n - 1) p (1 - p)
```

We have a total of 1116 days, out of which 791 are wet days:

```
{ndays, nwet} = {Length[Flatten[prec]], Total[Flatten[prec]]}
```

{1116, 791}

We estimate the probability of a wet day:

```
probDry = nwet / ndays // N       0.708781
```

The expected number of runs in a month for random weather is thus approximately the following:

```
expRuns[31, probDry]       13.3846
```

The actual mean number was only approximately eight. This is an indication that the weather does not change randomly. The weather on a certain day has a tendency to remain stable for a few days. Next, we estimate the transition probabilities.

■ **Estimating the Transition Probabilities**

The data for January 1948 are as follows:

```
d = prec[[1]]
{1, 1, 1, 1, 1, 1, 1, 1, 1, 1, 1, 1, 0, 0, 0, 0, 0, 0, 0, 1, 0, 1, 1, 1, 1, 0, 0, 0, 1, 1, 1}
```

With **ReplaceList**, we can easily obtain information about all kinds of transitions. For example, we form a list consisting of as many ones as there are pairs {0, 0} in the list:

```
ReplaceList[d, {___, 0, 0, ___} → 1]     {1, 1, 1, 1, 1, 1, 1, 1}
```

The three underscores (___) mean zero or more elements (see Section 16.1.3, p. 500). A pair {0, 0} means a transition from a dry day to another dry day. The sum of the ones is the number of such pairs of days:

```
Total[ReplaceList[d, {___, 0, 0, ___} → 1]]     8
```

Similarly, we can find the numbers of other kinds of transitions. We write a function for doing this:

```
transitions[d_List] := Total[ReplaceList[d, # → 1]] & /@
  {{___, 0, 0, ___}, {___, 0, 1, ___}, {___, 1, 0, ___}, {___, 1, 1, ___}}
```

For example,

```
transitions[d]     {8, 3, 3, 16}
```

To get the numbers of transitions for all years, we use **Map**:

```
transitions[#] & /@ prec
{{8, 3, 3, 16}, {10, 7, 8, 5}, {2, 2, 3, 23}, {3, 2, 3, 22}, {3, 4, 3, 20}, {0, 1, 1, 28},
 {3, 2, 2, 23}, {7, 6, 6, 11}, {3, 2, 3, 22}, {8, 6, 6, 10}, {3, 4, 3, 20},
 {1, 2, 2, 25}, {6, 5, 5, 14}, {6, 3, 3, 18}, {8, 4, 4, 14}, {10, 5, 5, 10},
 {1, 3, 3, 23}, {4, 2, 3, 21}, {3, 5, 5, 17}, {2, 2, 3, 23}, {2, 5, 5, 18},
 {1, 5, 5, 19}, {6, 3, 2, 19}, {6, 2, 2, 20}, {5, 2, 3, 20}, {6, 4, 5, 15},
 {7, 1, 0, 22}, {4, 1, 2, 23}, {8, 3, 3, 16}, {12, 2, 2, 14}, {3, 6, 5, 16},
 {11, 5, 5, 9}, {9, 5, 5, 11}, {9, 4, 4, 13}, {5, 2, 2, 21}, {1, 3, 4, 22}}
```

With **Transpose**, we can form separate lists for the four kinds of transitions:

```
{t00, t01, t10, t11} = %ᵀ;
```

For example,

```
t00
{8, 10, 2, 3, 3, 0, 3, 7, 3, 8, 3, 1, 6, 6, 8, 10,
 1, 4, 3, 2, 2, 1, 6, 6, 5, 6, 7, 4, 8, 12, 3, 11, 9, 9, 5, 1}
```

We calculate the sum of each kind of transition:

```
Total[#] & /@ %%     {186, 123, 128, 643}
```

Thus, we know that there were 186 transitions $0 \to 0$ (from a dry day to another dry day), 123 transitions $0 \to 1$, 128 transitions $1 \to 0$, and 643 transitions $1 \to 1$. Overall, there were $186 + 123 = 309$ transitions from a dry day and $128 + 643 = 771$ transitions from a wet day. It is thus natural to estimate the probability of $0 \to 0$ as $186/309 = 0.602$, of $0 \to 1$ as $123/309 = 0.398$, of $1 \to 0$ as $128/771 = 0.166$, and of $1 \to 1$ as $643/771 = 0.834$. (The same estimates could also be obtained by the method of maximum likelihood.) Thus, we arrive at the following transition matrix:

```
P = {{0.602, 0.398}, {0.166, 0.834}};
```

■ Predicting

Suppose Monday is dry; the initial distribution is thus {1, 0}. We predict the weather for Tuesday, Wednesday, Thursday, and Friday:

```
TableForm[Table[{1, 0}.MatrixPower[P, i], {i, 0, 4}], TableHeadings →
   {{"Monday", "Tuesday", "Wednesday", "Thursday", "Friday"}, {"P(dry)", "P(wet)"}}]
```

	P(dry)	P(wet)
Monday	1.	0.
Tuesday	0.602	0.398
Wednesday	0.428472	0.571528
Thursday	0.352814	0.647186
Friday	0.319827	0.680173

The predictions converge to a limit, which is achieved to six-digit precision in 18 steps:

```
{1, 0}.MatrixPower[P, 18]     {0.294326, 0.705674}
```

■ Stationary Distribution

The stationary distribution is obtained with the aid of the following linear equations:

```
eqns = Thread[{r1, r2}.P == {r1, r2}]
```

$\{0.602 \, r1 + 0.166 \, r2 == r1, \; 0.398 \, r1 + 0.834 \, r2 == r2\}$

One of these equations can be dropped, and the equation **r1 + r2 == 1** has to be taken into account:

```
Solve[Prepend[Most[eqns], r1 + r2 == 1]]
```

$\{\{r1 \to 0.294326, \; r2 \to 0.705674\}\}$

This solution is the same as the limit we obtained previously. A general program to calculate the stationary distribution is as follows:

```
stationaryDistribution[P_?MatrixQ, s0_, s1_] :=
 Module[{st = Array[v, {s1 - s0 + 1}], eqn},
  eqn = Prepend[Most[Thread[st.P == st]], Total[st] == 1];
  {Range[s0, s1], st /. Solve[eqn, st][[1]]}ᵀ]
```

Here, **P** is the transition matrix, and **s0** and **s1** are the minimum and the maximum state, respectively. For example,

```
stationaryDistribution[P, 0, 1]     {{0, 0.294326}, {1, 0.705674}}
```

■ Simulating the Process

By using the following functions, we can simulate a discrete-time Markov chain:

```
dtMarkovChainStep[dist_?VectorQ] :=
 With[{r = RandomReal[]}, Position[dist, x_ /; x ≥ r, {1}, 1][[1, 1]]]

dtMarkovChainPath[x0_, P_?MatrixQ, n_] :=
 {Range[0, n], NestList[dtMarkovChainStep[(Accumulate /@ P)[[#]]] &, x0 + 1, n] - 1}ᵀ

dtMarkovChain[x0_, P_?MatrixQ, n_, opts___] :=
 ListLinePlot[dtMarkovChainPath[x0, P, n],
  AxesOrigin → {0, -0.2}, PlotRange → {-0.2, Length[P] - 0.8}, opts]
```

The program `dtMarkovChainStep` calculates the next state of the process. The states are numbered 1, 2, ... The input for this function is a cumulative distribution `dist` for the next state. For example, if `dist` is {0.3, 0.8, 1}, the next state is 1, 2, or 3 with probabilities 0.3, 0.5, and 0.2, respectively. With the help of `Position`, the program searches for the positions of the cumulative distribution in which the cumulative probability is at least a given uniform random number. The `{1}` in `Position` means that the search is done at the first level of `dist`. This specification is unnecessary, but because we use a fourth argument in `Position`, we have to write something as the third argument. The fourth argument `1` specifies that we, in fact, only need the first position that satisfies $x \geq r$. The result of this `Position` could be, for example, {{2}}. Finally, taking the part ⟦1, 1⟧ gives the position as a number, for example, 2.

The program `dtMarkovChainPath` generates `n` steps for a chain with initial state `x0` and transition matrix `P`. The function first calculates the cumulative distribution `Accumulate/@P` for the rows of the transition matrix. In `NestList`, the current state is denoted by `#`. From this state, we go on to the next state according to a cumulative distribution, which is the `#`th row of the cumulative transition matrix. Thus, the next state is given by `dtMarkovChainStep[(Accumulate/@P)⟦#⟧]`. `NestList` does this iteration `n` times. The result is a list of states, and we subtract 1 from all of the states because we want them to be numbered 0, 1, 2, ... Lastly, we add the ordinal numbers of the steps to the states.

For example, if the initial day is dry, the initial state is 0. Here are 10 simulated steps:

```
SeedRandom[1]; dtMarkovChainPath[0, P, 10]

{{0, 0}, {1, 1}, {2, 0}, {3, 1}, {4, 1}, {5, 1}, {6, 0}, {7, 0}, {8, 0}, {9, 0}, {10, 1}}
```

We see that the next day is wet, followed by a dry day and so on. Next, we simulate 100 steps, assuming that the starting day is dry:

```
SeedRandom[1]; dtMarkovChain[0, P, 100, Mesh → All,
  AspectRatio → 0.1, Ticks → {Automatic, {0, 1}}, ImageSize → 420]
```

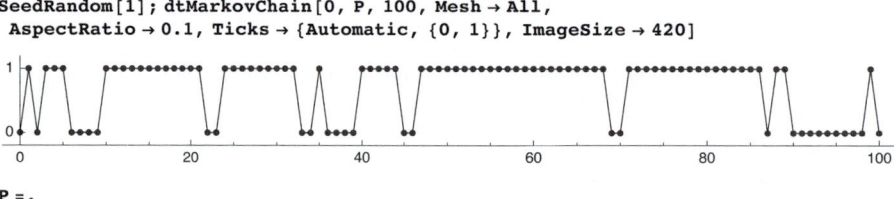

```
P = .
```

■ An Example of Diffusion

We have `n` black and `n` white balls. They are put into urns `A` and `B` so that there are `n` balls in both urns. Then we take, at times 1, 2, …, one ball at random from both urns, and then they are put back into the urns: The ball taken from urn `A` is put into urn `B`, and the ball taken from urn `B` is put into urn `A`. This is a simple model of diffusion.

We define the state of the system to be the number of black balls in urn `A`. The transition probabilities are $p_{i,i-1} = \left(\frac{i}{n}\right)^2$, $p_{i,i} = \frac{2i(n-i)}{n^2}$, and $p_{i,i+1} = \left(\frac{n-i}{n}\right)^2$. The transition matrix is as follows:

```
P[n_] := Table[Which[j == i - 1, (i/n)^2, j == i,

    2 i (n - i)
    ───────────, j == i + 1, ((n - i)/n)^2, True, 0], {i, 0, n}, {j, 0, n}]
      n^2
```

As an example, suppose we have 50 black and 50 white balls. Calculate and plot the stationary distribution:

```
st = stationaryDistribution[P[50] // N, 0, 50];
```

```
ListPlot[st, PlotRange → All]
```

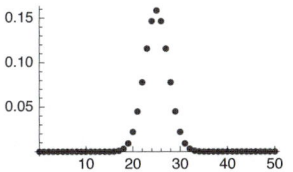

We see that, with high probability, there are, in the long run, approximately 18 to 32 black balls in urn *A*. Calculate the exact probability:

```
Sum[st〚i, 2〛, {i, 19, 33}]     0.997476
```

We simulate 200 steps of the diffusion, assuming that there are initially 0 black balls in urn *A*:

```
SeedRandom[1]; dtMarkovChain[0, P[50] // N, 200,
  AspectRatio -> 0.2, ImageSize → 420, Epilog → Line[{{0, 25}, {200, 25}}]]]
```

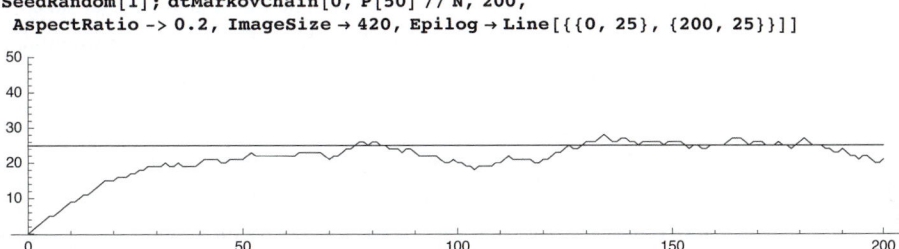

29.4.3 Continuous-Time Markov Chains

■ The Poisson Process

In a Poisson process, some events happen as time goes on. For example, calls arrive at a database, customers arrive at a service point, or particles arrive at a particle detector. When we look at the realization of a Poisson process, we should observe that the events are on the time axes randomly and uniformly. The state of the process at time *t* is the number of events that have occurred up to that time. A Poisson process is a special case of a continuous-time discrete-state Markov chain.

Let λ be the expected number of events in a time unit. Then the number of events in a time interval of length *t* has a Poisson distribution with mean λt, and the interarrival times have an exponential distribution with parameter λ (i.e., with mean $1/\lambda$); the interarrival times are independent.

A Poisson process can be simulated by generating the interarrival times from an exponential distribution:

```
poissonProcess[λ_, n_, opts___] :=
 With[{pp = Join[{0}, Accumulate[RandomReal[ExponentialDistribution[λ], n]]]},
  Graphics[Table[Line[{{pp〚i〛, i - 1}, {pp〚i + 1〛, i - 1}}], {i, n}], Axes → True, opts]]
```

Here, we first calculate the cumulative sums of the interarrival times, resulting in a list **pp** of the instants of the events. The program draws, at levels 0, 1, 2, and so on, lines that have lengths that are the interarrival times.

Let us assume that, on average, four calls arrive at a database per minute. The following is a simulated sequence of 50 arrivals: 50 events occurred in approximately 12 minutes.

```
SeedRandom[4]; poissonProcess[4, 50, AspectRatio → 0.3, ImageSize → 420]
```

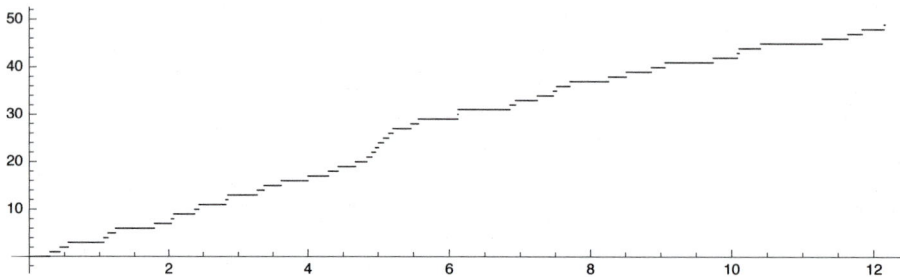

■ General Continuous-Time Markov Chains

Consider a continuous-time discrete-state process. Let T_{ij} be the time the system stays at state i if the system then goes to state j. Assume that the random variable T_{ij} has an exponential distribution with parameter q_{ij} (or with mean $1/q_{ij}$) and that these random variables are independent of each other and also independent of the history of the process before arriving at state i (define $q_{ii} = -\sum_{j \neq i} q_{ij}$). Collect the parameters q_{ij} into a matrix Q, which is called the rate matrix or the generator of the process. It can be shown that the process is a continuous-time Markov chain. In addition, the system stays at state i for an exponential time with parameter $q_i = \sum_{k \neq i} q_{ik}$ and then goes to state j with probability $p_{ij} = q_{ij}/q_i$ (if $q_i = 0$, then define $p_{ii} = 1$) (see Kulkarni, 1995, p. 245)

The following programs simulate **n** steps for a continuous-time Markov chain. The initial state is the scalar **x0**, and the generator matrix is **Q**.

```
ctMarkovChainPath[x0_, Q_?MatrixQ, n_] :=
  Module[{R = Q, m = Length[Q], q, P, cumP, t = 0, r, x = x0, tt = {0}, xx = {x0}},
    Do[R[[i, i]] = 0, {i, m}];
    q = Total[#] & /@ R;
    P = Table[If[q[[i]] ≠ 0, R[[i]] / q[[i]], Table[If[j ≠ i, 0, 1], {j, m}]], {i, m}];
    cumP = Accumulate /@ P;
    Do[If[q[[x + 1]] ≠ 0, t = t + RandomReal[ExponentialDistribution[q[[x + 1]]]],
      Print["Absorption at ", x]; Break[]];
      r = RandomReal[];
      x = Position[cumP[[x + 1]], z_ /; z ≥ r, {1}, 1][[1, 1]] - 1;
      tt = {tt, t}; xx = {xx, x}, {n}];
    {Flatten[tt], Flatten[xx]}]

ctMarkovChain[x0_, Q_?MatrixQ, n_, opts___] := Module[{tt, xx},
    {tt, xx} = ctMarkovChainPath[x0, Q, n];
    Graphics[Table[Line[{{tt[[i]], xx[[i]]}, {tt[[i + 1]], xx[[i]]}}], {i, Length[tt] - 1}],
      Axes → True, AxesOrigin → {0, -0.2}, PlotRange → {-0.2, Max[xx] + 0.1}, opts]]
```

The states are numbered 0, 1, and so on. The program first sets the diagonal elements of Q to 0 and then calculates the q_i and p_{ij} numbers into vector **q** and matrix **P**. The rows of **cumP** are the cumulative sums of elements of the rows of **P** (they are needed to generate the next state). The vector **q** is used to generate the exponential visiting times. The instances of events are gathered in the list **tt**, and the states are gathered in the list **xx**.

As examples of continuous-time Markov chains, we consider the Poisson process, a birth–death process, and the M/M/1 queue.

■ The Poisson Process Revisited

In a Poisson process, the state goes from i to $i+1$ when the next event happens so that $T_{i,i+1}$ is an exponential random variable with parameter λ. Thus, if we generate at most m events, the generator can be written as follows:

```
Q[λ_, m_] := Table[Which[j == i, -λ, j == i + 1, λ, True, 0], {i, 0, m}, {j, 0, m}]
```

Here is a small example of the generator:

```
Q[λ, 4] // Grid
```

$$
\begin{array}{ccccc}
-\lambda & \lambda & 0 & 0 & 0 \\
0 & -\lambda & \lambda & 0 & 0 \\
0 & 0 & -\lambda & \lambda & 0 \\
0 & 0 & 0 & -\lambda & \lambda \\
0 & 0 & 0 & 0 & -\lambda
\end{array}
$$

Here is a simulation:

```
SeedRandom[4]; ctMarkovChain[0, Q[4, 50], 50, AspectRatio -> 0.3, ImageSize → 420]
```

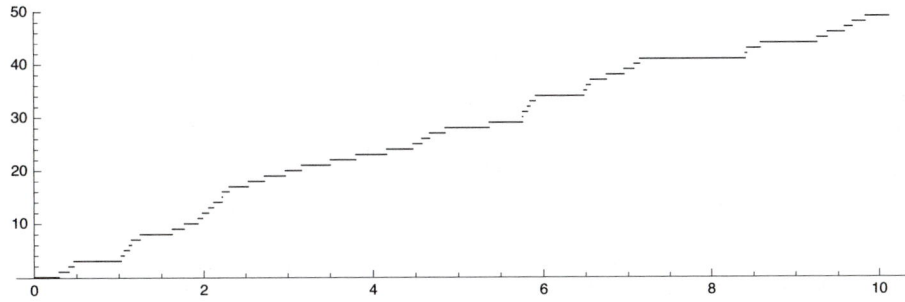

■ A Birth–Death Process

Consider a population that initially consists of x_0 individuals. The length of the life of each individual has the exponential distribution with mean $1/\mu$. Each individual produces, during his or her entire lifetime, descendants in such a way that the time between successive births has the exponential distribution with mean $1/\lambda$. Such a process is a simple birth–death process with a state that is the size of the population. This process was also considered in Section 27.1.1, p. 888.

To derive the generator matrix, note that the state goes from i to $i+1$ if one of the i individuals gives birth to a child so that $T_{i,i+1}$ is the minimum of i exponential random variables with parameter λ; that is, $T_{i,i+1}$ has an exponential distribution with parameter $i\lambda$. Similarly, $T_{i,i-1}$ is exponentially distributed with parameter $i\mu$. Thus, if we suspect that the population will not exceed the value m, the generator is as follows:

```
Q[λ_, μ_, m_] := Table[
   Which[j == i - 1, i μ, j == i, -i (λ + μ), j == i + 1, i λ, True, 0], {i, 0, m}, {j, 0, m}]
```

Here is a small example:

```
Q[λ, μ, 4] // Grid
```

0	0	0	0	0
μ	−λ − μ	λ	0	0
0	2 μ	−2 (λ + μ)	2 λ	0
0	0	3 μ	−3 (λ + μ)	3 λ
0	0	0	4 μ	−4 (λ + μ)

As an example, let the average lifetime be 2 time units and the average time between births, for a given individual, 3 time units. Then $1/\mu = 2$ and $1/\lambda = 3$ so that $\mu = \frac{1}{2}$ and $\lambda = \frac{1}{3}$. Suppose we initially have 20 individuals. We generate six steps:

```
SeedRandom[5]; ctMarkovChainPath[20, Q[1 / 3, 1 / 2, 50], 6]
```

```
{{0, 0.428564, 0.429228, 0.525535, 0.663282, 0.743467, 0.865291},
 {20, 19, 20, 21, 22, 23, 22}}
```

Thus, at time 0 we have 20 individuals, at time 0.4285 one of these individuals dies, at time 0.4292 an individual gives birth to a child, and so on. Here is a longer simulation in which the population dies out at approximately $t = 9$:

```
SeedRandom[1];
ctMarkovChain[20, Q[1 / 3, 1 / 2, 50], 200, AspectRatio → 0.3, ImageSize → 420]
Absorption at 0
```

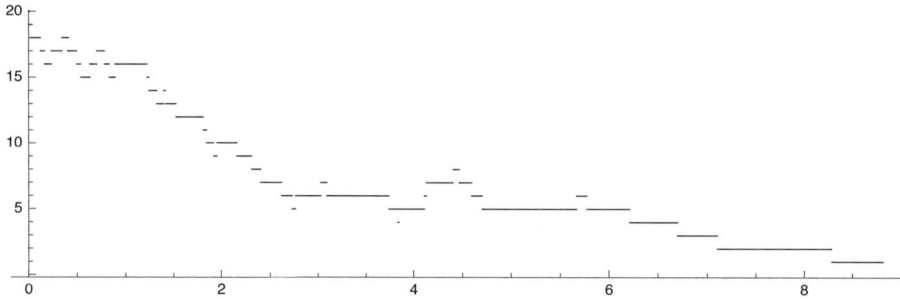

The population died out at approximately $t = 9$.

■ The M/M/1 Queue

In an M/M/1 queuing model, customers arrive at a service place, and a single person does the serving. Customers are served in the order of their arrival. There is room for all arriving customers to queue if the server is busy, and all customers wait until they get service (i.e., customers cannot leave the place). Customers arrive as a Poisson process with a mean of λ customers per time unit. The service time for each customer has an exponential distribution with mean $1/\mu$. The state of the system is the number of customers in the service place (customers in the system consist of the one receiving service and the others standing in the queue).

The state goes from i to $i + 1$ if a new customer arrives so that $T_{i,i+1}$ has an exponential distribution with parameter λ. Similarly, the state goes from i to $i - 1$ if a customer is served so that $T_{i,i-1}$ has an exponential distribution with parameter μ. Thus, if we suspect that the population will not exceed the value m, the generator is as follows:

```
Q[λ_, μ_, m_] := Table[Which[i == 0 && j == 0, -λ, j == i - 1,
    μ, j == i, -λ - μ, j == i + 1, λ, True, 0], {i, 0, m}, {j, 0, m}]
```

Here is a small example:

```
Q[λ, μ, 4] // Grid
```

$$
\begin{array}{ccccc}
-\lambda & \lambda & 0 & 0 & 0 \\
\mu & -\lambda-\mu & \lambda & 0 & 0 \\
0 & \mu & -\lambda-\mu & \lambda & 0 \\
0 & 0 & \mu & -\lambda-\mu & \lambda \\
0 & 0 & 0 & \mu & -\lambda-\mu
\end{array}
$$

We will simulate a queuing system in which customers arrive at the service point at the mean rate of 4.0 arrivals per hour (one customer every 15 minutes) and in which the server has a mean service rate of 4.55 customers per hour. This means that $\lambda = 4$ and $\mu = 4.55$. One customer is then served in an average of $1/4.55$ hour = 13.2 minutes. We simulate 100 events (arrivals and departures), when there are initially 0 customers:

```
SeedRandom[6];
ctMarkovChain[0, Q[4., 4.55, 50], 100, AspectRatio → 0.3, ImageSize → 420]
```

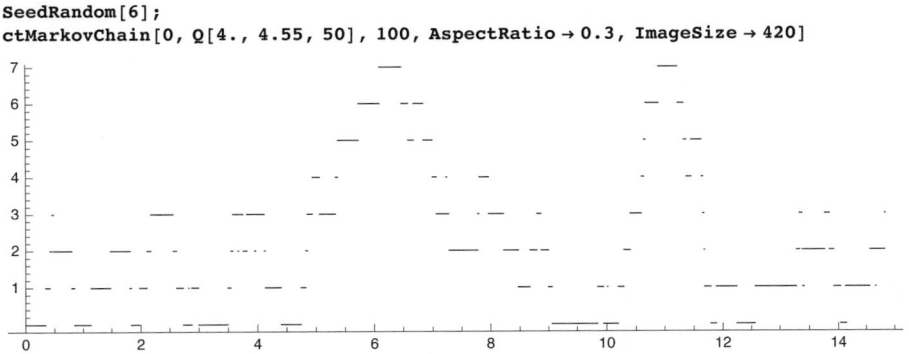

In the simulation, 100 events took approximately 15 hours to occur. The highest number of customers in the service place was 7. By simulating more events we could observe that the length of the queue can grow to large values so that it seems that one server is too few for this system: The queue is too long for too large a portion of the time.

With the following module, we can calculate various average values if the system has reached a steady state:

```
steadyStateAverages[λ_, μ_] := Module[{ρ = λ / μ, L, W, Wq, Lq},
  L = ρ / (1 - ρ); W = L / λ; Wq = W - 1 / μ; Lq = λ Wq;
  Print["L  = ", L, " (steady-state mean number of customers)\n",
   "Lq = ", Lq, " (steady-state mean length of the queue)\n",
   "W  = ", W, " (steady-state mean time in the system)\n",
   "Wq = ", Wq, " (steady-state mean time of queueing)\n",
   "ρ  = ", ρ, " (steady-state server utilization)"]]
```

For our example, the averages are as follows:

```
steadyStateAverages[4.0, 4.55]
```

```
L  = 7.27273 (steady-state mean number of customers)
Lq = 6.39361 (steady-state mean length of the queue)
W  = 1.81818 (steady-state mean time in the system)
Wq = 1.5984 (steady-state mean time of queueing)
ρ  = 0.879121 (steady-state server utilization)
```

Thus, if our system is in steady state, the mean number of customers in the system in a long time interval should be near 7.3. Approximately 6.4 customers are queuing, and they each spend approximately 1.8 hours in the system (this time consists of queuing time and service time). Queuing takes approximately 1.6 hours. The server is busy a fraction of 0.88 of the time. In steady state, the number of customers in the system has the geometric distribution with parameter ρ: The probability of n customers being in the system is $(1 - \rho) \rho^n$, $n = 0, 1, 2, \ldots$.

Statistics

Introduction

> *A statistician was about to undergo a serious operation and asked the surgeon what his chances of survival were. "Your chances are excellent," said the surgeon, "Nine people out of ten die from this operation, and the last nine patients I've operated on have died."*

With *Mathematica*, we can do all kinds of basic statistical analyses, from descriptive statistics to maximum likelihood, frequencies, confidence intervals, hypothesis testing, analysis of variance (ANOVA), and linear and nonlinear regression. Additional topics include finding clusters of data, smoothing data, local regression analysis, and Bayesian statistics. Regarding the last topic, the power of *Mathematica* for integration, interpolation, and random number generation helps with the solving of statistical problems that are related to Bayesian models. Two of the methods we consider are Gibbs sampling and Markov chain Monte Carlo (MCMC).

Regarding fitting and regression, note that we have four commands. For linear models, we have `Fit` (built-in) and `Regress` (in the **LinearRegression`** package); the latter also gives statistical information about the fit. For nonlinear models, we have `FindFit` (built-in) and `NonlinearRegress` (in the **Nonlinear-Regression`** package); the latter also gives statistical information about the fit. `Fit` and `FindFit` were considered in Sections 25.1.1, p. 812, and 25.1.3, p. 818.

Estimation of differential and difference equation models was considered in Sections 26.4.5, p. 878, and 28.3.2, p. 954.

Probability distributions were considered in Chapter 29. We have, for example, the normal, Student *t*, chi-square, and *F*-ratio distributions. The **MultivariateStatistics`** package contains the multinormal distribution and other related distributions.

Note that the plotting of data was considered in Chapter 8. Some of the plots are especially useful in statistical reasoning. These plots include pairwise scatter plots, quantile–quantile plots, histograms, stem-and-leaf plots, dot and multiway dot plots, and box-and-whisker plots. Recall also probability graph paper plots from Section 29.3.2, p. 983.

Application packages related to statistics include Experimental Data Analyst, Statistical Inference Package, and Time Series. For more information about statistics with *Mathematica*, see Abell, Braselton, and Rafter (1999) and Rose and Smith (2002).

30.1 Descriptive Statistics

30.1.1 Descriptive Statistics

■ Location Statistics

For the sake of brevity, we have mostly not shown the argument of the commands. The argument, if not shown, is a list of observations.

`Mean` $\left(m = \frac{1}{n}\sum x_i\right)$, `TrimmedMean[data, f]`, `TrimmedMean[data, {f1, f2}]`

`Commonest`, `Median`, `Quantile[list, q]`, `Quartiles`

`GeometricMean` $\left(\left(\prod x_i\right)^{\frac{1}{n}}\right)$, `HarmonicMean` $\left(\frac{n}{\sum 1/x_i}\right)$, `RootMeanSquare` $\left(\sqrt{\frac{1}{n}\sum x_i^2}\right)$

The mean *m* is, for a random sample, an unbiased estimate of the population mean μ. A trimmed mean is the mean of remaining entries, when a fraction *f* is removed from each end of the sorted list of data.

The commonest observation is the value with the highest frequency; this element is often called the mode. **Commonest** gives a list because there can potentially be several elements, each having the highest frequency. The median is the central value—that is, the observation in the center of the sorted observations (or the average of the two most central observations if there is an even number of observations).

The q quantile gives a value such that $100q\%$ of the observations are at most this value ($0 < q < 1$). For a list with n elements, the q quantile is computed as **Sort[list][[Ceiling[n q]]]** (note that the median and the 0.5 quantile are not necessarily the same). In addition, a third argument in **Quantile** can be used to define the method used to calculate the quantile in more detail (see the documentation). The quartiles are the 0.25, 0.5, and 0.75 (interpolated) quantiles.

Here are examples:

```
d = {1, 1, 2, 2, 3, 4, 4, 4, 5, 5};

{Mean[d], Commonest[d], Median[d],
   Quantile[d, 0.5], Quantile[d, 0.9], Quartiles[d]} // N
{3.1, {4.}, 3.5, 3., 5., {2., 3.5, 4.}}
```

Recall that in Section 8.4.2, p. 264, we considered box-and-whisker plots. Such a plot is simply a way to show the quartiles and the minimum and maximum of the data.

■ **Dispersion and Shape Statistics**

> **Variance** $\left(s^2 = \frac{1}{n-1} \sum (x_i - m)^2\right)$, **StandardDeviation** (s)
>
> **MeanDeviation** $\left(\frac{1}{n} \sum |x_i - m|\right)$, **MedianDeviation**,
>
> **QuartileDeviation**, **InterquartileRange**

The sample variance s^2 is, for a random sample, an unbiased estimate of the population variance σ^2.

The median deviation is the median of the $|x_i - \text{median}|$ values. The quartile deviation is the difference between the first and third quartiles, and interquartile range is half the quartile deviation.

Some other dispersion statistics can be calculated with the aid of the built-in commands. For example, the maximum likelihood estimate of the variance is $\frac{1}{n} \sum (x_i - m)^2$. It can be calculated from $\frac{n-1}{n} s^2$. The theoretical variance of the sample mean m is σ^2/n, and an unbiased estimate of this is the variance of sample mean $s_m^2 = s^2/n$; standard error of sample mean is s_m. Coefficient of variation is s/m.

> **CentralMoment[data, r]** $\left(m_r = \frac{1}{n} \sum (x_i - m)^r\right)$
>
> **Skewness** $\left(m_3/m_2^{3/2}\right)$, **Kurtosis** $\left(m_4/m_2^2\right)$, **QuartileSkewness**

Skewness describes the amount of asymmetry. Kurtosis measures the concentration of data around the peak and in the tails versus the concentration in the flanks.

■ **Method of Moments**

Descriptive statistics can be used in estimating parameters of distributions with the method of moments. As an example, generate some random data from an extreme value distribution:

```
dist = ExtremeValueDistribution[α, β];

SeedRandom[1]; data = RandomReal[dist /. {α → 5, β → 3}, 100];
```

ListPlot[data]

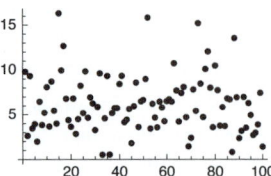

To calculate estimates of α and β with the method of moments, calculate the first two moments of the distribution and of the data:

{μ1 = Mean[dist], μ2 = Variance[dist] + Mean[dist] ^2}

$$\left\{\alpha + \text{EulerGamma } \beta,\ \frac{\pi^2\ \beta^2}{6} + (\alpha + \text{EulerGamma } \beta)^2\right\}$$

{m1 = Mean[data], m2 = Mean[data^2]}

{6.14438, 47.7309}

Equate the corresponding moments of distribution and data:

eqns = {μ1 == m1, μ2 == m2}

$$\left\{\alpha + \text{EulerGamma } \beta == 6.14438,\ \frac{\pi^2\ \beta^2}{6} + (\alpha + \text{EulerGamma } \beta)^2 == 47.7309\right\}$$

We solve these equations with respect to α and β. We can first show the situation with a contour plot:

ContourPlot[{μ1 == m1, μ2 == m2}, {α, 0, 9}, {β, -6, 6}, Frame → False,
  Axes → True, AxesLabel → {α, β}, AspectRatio → 1 / GoldenRatio]

The curves seem to have two solutions. In one of the solutions, α is approximately 5 and β approximately 2.5. The other solution can be rejected because β is negative. Solve the equations:

Solve[{μ1 == m1, μ2 == m2}, {α, β}]

{{α → 4.72279, β → 2.46284}, {α → 7.56597, β → -2.46284}}

The first solution gives the estimates of α and β.

■ **Method of Maximum Likelihood**

At the same time, we can also show how to estimate parameters with the method of maximum likelihood. First, form the log-likelihood function:

logL = Total[Log[PDF[dist, data]]];

Here are the first and last terms of the sum of 100 terms:

{First@logL, Last@logL}

$$\left\{\text{Log}\left[\frac{-e^{\frac{-16.3047+\alpha}{\beta}} + \frac{-16.3047+\alpha}{\beta}}{\beta}\right],\ \text{Log}\left[\frac{-e^{\frac{-0.530298+\alpha}{\beta}} + \frac{-0.530298+\alpha}{\beta}}{\beta}\right]\right\}$$

We maximize the log-likelihood function with respect to α and β. To get good starting values, prepare a contour plot:

```
ContourPlot[logL // Evaluate, {α, 0, 10}, {β, 1, 6}, Contours → Range[-351, -251, 10]]
```

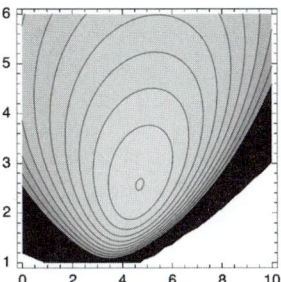

We can see that at the maximum, α is approximately 4.5 and β approximately 2.5. Find the maximum either with **FindMaximum** or with **NMaximize**:

```
FindMaximum[logL, {{α, 4.5}, {β, 2.5}}]
```
$\{-250.816, \{\alpha \to 4.69347, \beta \to 2.56823\}\}$

```
NMaximize[logL, {{α, 4, 5}, {β, 2, 3}}]
```
$\{-250.816, \{\alpha \to 4.69347, \beta \to 2.56823\}\}$

■ Autocorrelation

Autocorrelation is important in time series analysis. Let ρ_k be the autocorrelation at lag k. An estimate of ρ_k is $r_k = \sum_{t=1}^{n-k} (x_t - m)(x_{t+k} - m) \big/ \sum_{t=1}^{n} (x_t - m)^2$. The following program calculates all autocorrelations up to lag k:

```
autocorrelation[list_, k_] := With[{diff = list - Mean[list]},
   Table[{i, Drop[diff, -i].Drop[diff, i] / Total[diff^2]}, {i, 0, k}]]
```

As an example, we consider the same data as we did in Section 8.2.1, p. 249 (the data file **environmental** is on the CD-ROM that comes with this book):

```
data =
   Rest[Import["/Users/heikki/Documents/MNData/visdata/environmental", "Table"]];
```
The file contains 111 observations of ozone, radiation, temperature, and wind. Extract the components of the data:

```
{no, ozone, radiation, temperature, wind} = data⊤;
```
Consider the temperature:

```
ListLinePlot[temperature, Mesh → All, AspectRatio → 0.15, ImageSize → 400]
```

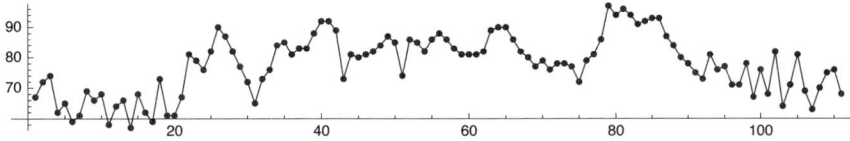

The estimated autocorrelation function is as follows:

```
ac = autocorrelation[temperature // N, 25]
```
{{0, 1.}, {1, 0.777313}, {2, 0.694894}, {3, 0.632474}, {4, 0.546984},
 {5, 0.487101}, {6, 0.39759}, {7, 0.401579}, {8, 0.345652}, {9, 0.248692},
 {10, 0.241805}, {11, 0.199781}, {12, 0.220929}, {13, 0.208489},
 {14, 0.205918}, {15, 0.193053}, {16, 0.148181}, {17, 0.0833282},
 {18, 0.0378245}, {19, 0.0199929}, {20, -0.029604}, {21, -0.0628506},
 {22, -0.0468014}, {23, -0.0179397}, {24, -0.0147237}, {25, -0.0562379}}

```
Graphics[{AbsoluteThickness[1.2], Line[{{#[[1]], 0}, #}] & /@ ac},
 Axes → True, AspectRatio → 1 / GoldenRatio]
```

■ Multivariate Descriptive Statistics

Assume that the data are in the form of a matrix that contains as many rows as there are observations and as many columns as there are variables. The rows are treated as independent identically distributed multivariate observations. All the commands that we have considered in this section for univariate data, except for **Commonest**, can also be used for multivariate data. The statistics are calculated for each column of the data separately. As an example, consider the same data for which we previously calculated the autocorrelation function:

```
{no, ozone, radiation, temperature, wind} =
  Rest[Import["/Users/heikki/Documents/MNData/visdata/environmental", "Table"]]ᵀ;
```
Prepare a matrix from the values of ozone, radiation, temperature, and wind:

```
data = {ozone, radiation, temperature, wind}ᵀ;
```
Calculate some descriptive statistics:

```
Mean[data] // N      {42.0991, 184.802, 77.7928, 9.93874}
```
```
Variance[data] // N      {1107.29, 8308.74, 90.8203, 12.668}
```

Commonest has to be used with **Map**:

```
Commonest /@ (dataᵀ)      {{23}, {238}, {81}, {11.5, 10.3}}
```

Recall that in Section 8.2.1, p. 249, we considered scatter plots and scatter plot matrices. These plots are useful when studying relationships among a number of dependent variables. In Section 8.2.2, p. 252, we considered quantile–quantile plots for comparing the distributions of data sets.

■ Dispersion and Association Statistics

Covariance[list1, list2]	Covariance coefficient between two lists
Covariance[matrix]	Covariance matrix for a matrix
Covariance[matrix1, matrix2]	Covariance matrix for two matrices
Correlation[list1, list2]	Correlation coefficient between two lists
Correlation[matrix]	Correlation matrix for a matrix
Correlation[matrix1, matrix2]	Correlation matrix for two matrices

The formula $\text{cov}(X, Y) = \frac{1}{n-1}\sum(x_i - m_x)(y_i - m_y)$ gives the unbiased covariance of two variables. The correlation is $\text{cov}(X, Y)\big/\sqrt{\text{var}(X)\,\text{var}(Y)}$. The (i, j)th element of the covariance [correlation] matrix of a matrix is the covariance [correlation] between the ith and jth columns of the matrix. Calculate a single covariance and a matrix of covariances:

```
Covariance[ozone, wind]     -72.5957
```

```
Covariance[data] // MatrixForm
```

$$\begin{pmatrix} 1107.29 & 1056.58 & 221.521 & -72.5957 \\ 1056.58 & 8308.74 & 255.468 & -41.3213 \\ 221.521 & 255.468 & 90.8203 & -16.8628 \\ -72.5957 & -41.3213 & -16.8628 & 12.668 \end{pmatrix}$$

Similarly, we can calculate correlations:

```
Correlation[ozone, wind]     -0.612951
```

```
Correlation[data] // MatrixForm
```

$$\begin{pmatrix} 1. & 0.348342 & 0.698541 & -0.612951 \\ 0.348342 & 1. & 0.294088 & -0.127366 \\ 0.698541 & 0.294088 & 1. & -0.497146 \\ -0.612951 & -0.127366 & -0.497146 & 1. \end{pmatrix}$$

The (i, j)th element of the covariance [correlation] matrix of two matrices is the covariance [correlation] of the ith column of the first matrix and the jth column of the second matrix.

The **MultivariateStatistics`** package defines more descriptive statistics for multivariate data.

30.1.2 Exploratory Data Analysis

■ Finding Clusters of Data

FindClusters[data] (⌘6) Partition **data** into lists of similar elements
FindClusters[data, n] Partition **data** into exactly **n** lists of similar elements

Options:
Method The clustering method to use; possible values: **Automatic**, **"Optimize"**, **"Agglomerate"**
DistanceFunction The distance or dissimilarity measure to use; examples of values: **Automatic**,
  EuclideanDistance, **ManhattanDistance**, **ChebyshevDistance**
RandomSeed Starting value for the random number generation; examples of values: **Automatic, 1**

Generate some random data:

```
SeedRandom[5]; data = RandomInteger[{0, 9}, 20]
```

```
{0, 0, 3, 3, 4, 5, 1, 9, 1, 0, 5, 1, 8, 0, 5, 9, 5, 3, 7, 3}
```

Find clusters:

```
FindClusters[data]
```

```
{{0, 0, 0, 0}, {3, 3, 4, 3, 3}, {5, 5, 5, 5}, {1, 1, 1}, {9, 8, 9, 7}}
```

We got five clusters. The order of the elements in the data may have an effect on the clusters found:

```
FindClusters[data // Sort]
```

{{0, 0, 0, 0, 1, 1, 1}, {3, 3, 3, 3, 4, 5, 5, 5, 5}, {7, 8, 9, 9}}

We can also ask for a given number of clusters:

```
FindClusters[data, 3]
```

{{0, 0, 1, 1, 0, 1, 0}, {3, 3, 4, 5, 5, 5, 5, 3, 3}, {9, 8, 9, 7}}

As another example, generate a set of number pairs by joining random numbers from four two-variate normal distributions:

```
<< MultivariateStatistics`
```

```
SeedRandom[1];
data = Flatten[RandomReal[MultinormalDistribution[#, {{1, 0}, {0, 1}}], 20] & /@
    {{3, 3}, {5, 5}, {8, 2}, {9, 6}}, 1];
```

Plot the default clusters (in this example, we get two clusters) and clusters obtained by creating four clusters:

```
{ListPlot[FindClusters[data]],
 ListPlot[FindClusters[data, 4]]}
```

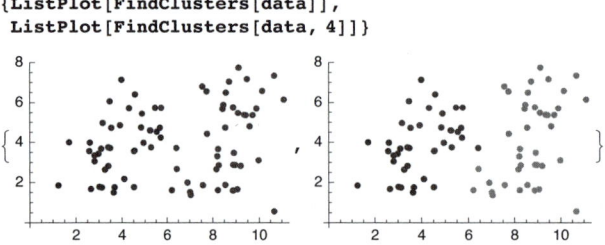

See also the **HierarchicalClustering`** package.

■ Picking the Nearest Points

Nearest[{a, b, …}, x] (✿6)   Find the elements from **a**, **b**, …, to which **x** is nearest

Nearest[{a, b, …}, x, n]   Give the **n** nearest elements to **x**

Nearest[{a, b, …}, x, {n, r}]   Give up to the **n** nearest elements to **x** within a radius **r**

Nearest[{a, b, …}, x, {∞, r}]   Give all elements nearest to **x** within a radius of **r**

An option:

DistanceFunction   The distance measure to use; examples of values: **Automatic**,
   EuclideanDistance, **ManhattanDistance**, **ChebyshevDistance**

Generate some random data:

```
SeedRandom[5]; data = RandomInteger[{0, 9}, 20]
```

{0, 0, 3, 3, 4, 5, 1, 9, 1, 0, 5, 1, 8, 0, 5, 9, 5, 3, 7, 3}

Pick the elements that are nearest to 2:

```
Nearest[data, 2]     {3, 3, 1, 1, 1, 3, 3}
```

Pick all elements that are nearest to 2 within a radius of 2:

```
Nearest[data, 2, {∞, 2}]     {3, 3, 1, 1, 1, 3, 3, 0, 0, 4, 0, 0}
```

Generate random 2D data:

```
data = RandomReal[{-1, 1}, {3000, 2}];
```

Pick the 1000 nearest points by using various distance functions:

```
ListPlot[{data, Nearest[data, {0, 0}, 1000, DistanceFunction → #]},
    PlotStyle → {{PointSize[Small], Gray}, {PointSize[Small], Black}},
    AspectRatio → Automatic] & /@
  {EuclideanDistance, ManhattanDistance, ChebyshevDistance}
```

Note that **Nearest** has more advanced uses. The following example is taken from www.wolfram.com/products/mathematica/newin6:

```
SeedRandom[1]; data = Table[RandomReal[{-4, 4}, 2] → i, {i, 30}];
```

```
nf = Nearest[data, DistanceFunction → (Norm[#1 - #2, ∞] &)]
```

```
NearestFunction[{30, 2}, <>]
```

```
ContourPlot[First[nf[{x, y}]], {x, -5, 5}, {y, -5, 5}, Contours → Range[1 / 2, 30],
    ColorFunction → "Pastel", Epilog → {Red, PointSize[Medium], Point[First /@ data]},
    PlotPoints → 50, ImageSize → 150]
```

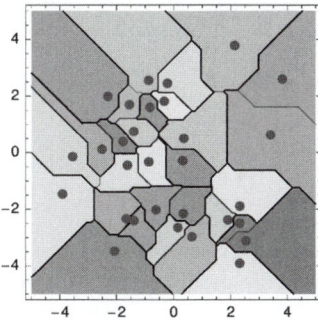

30.2 Frequencies

30.2.1 Frequencies of Discrete Data

■ **Introduction**

We consider calculating frequencies separately for data arising from discrete variables and for data arising from continuous variables. In the discrete case, we often would like to count the frequencies for all different elements, and **Tally** is then the suitable command. In the continuous case, we would like to group the elements into bins of intervals, and then we can use **BinCounts**. However, before we consider these commands, we recall some basic data manipulation techniques and the plotting of histograms.

■ **Data Manipulation**

Manipulating rows and columns is frequently needed when working with data. This topic was considered in Section 21.2.2, p. 692; here, we recall some basic techniques.

---

data[[n]] Give row number **n**

data[[All, n]] Give column number **n**

xy = Transpose[{x, y}] Pair the corresponding elements of **x** and **y**

{x, y} = Transpose[xy] Extract the first and second components of **xy**

In place of **Transpose[{x, y}]**, we can write **{x, y}$^\top$**; here, $^\top$ can be written as ⎡ESC⎤tr⎡ESC⎤. Suppose we have separate lists for the independent and dependent variables:

x = {1, 2, 3, 4, 5}; y = {14, 12, 15, 16, 13};

We want to pair the corresponding elements of **x** and **y**:

xy = {x, y}$^\top$ {{1, 14}, {2, 12}, {3, 15}, {4, 16}, {5, 13}}

Then we extract the **x** and **y** values:

{x, y} = xy$^\top$ {{1, 2, 3, 4, 5}, {14, 12, 15, 16, 13}}

Now the value of **x** is {1, 2, 3, 4, 5, 6} and the value of **y** is {14, 12, 15, 16, 13}.

■ **Histograms**

With **Histogram** we can plot frequencies as histograms; **Histogram** was introduced in Section 8.3.2, p. 258. The data can be either raw data or frequencies. In the former case, the command first calculates the frequencies.

In the **Histograms`** *package:*

Histogram[{x$_1$, x$_2$, … }] Plot the frequencies of the given raw data

Histogram[{f$_1$, …, f$_n$}, FrequencyData → True, HistogramCategories → cats] Plot the given frequencies

Options:

HistogramCategories How the data are categorized—that is, for which intervals the frequencies are calculated; possible values: **Automatic** (use an internal algorithm), a positive integer **n** (use exactly **n** categories of equal width, if **ApproximateIntervals → False**, and about **n** categories, if **ApproximateIntervals → True**), or a list of cutoff values **{b$_0$, b$_1$, …, b$_n$}** (calculate the frequencies in the intervals [b$_0$, b$_1$), …, [b$_{n-1}$, b$_n$))

ApproximateIntervals Whether interval boundaries should be approximated by simple numbers; possible values: **Automatic** (usually means **True**), **True**, **False**

HistogramScale Whether to scale the heights of the bars; examples of values: **Automatic** (means **False** for categories with equal widths and **True** for categories with unequal widths), **False** (no scaling: plot frequencies as such), **True** (scale by dividing the heights by the widths of the bars to get a frequency density), **1** (scale to get the sum of the areas of the bars equal to 1 so that the histogram approximates the PDF of the data; other constants can also be used)

HistogramRange Range of data to be included in the histogram; examples of values: **Automatic** (means that all data are included), **{0, 10}**

BarOrientation, BarStyle, BarEdgeStyle, BarEdges (see Section 8.3.1, p. 254)

Histogram also has the options of **Graphics**. For **Histogram**, the default value of **AspectRatio** is **1/GoldenRatio** and that of **Axes** is **True**. Examples of the use of the package are presented next.

■ Frequencies of Integer Data

Tally[list] (✿6)   Calculate the frequencies of the distinct elements of **list**

Toss a die 20 times and calculate the frequencies:

```
SeedRandom[7]; a = RandomInteger[{1, 6}, 20]
```

```
{6, 1, 3, 3, 6, 4, 2, 1, 4, 5, 3, 4, 3, 2, 4, 2, 2, 6, 4, 3}
```

```
Tally[a]
```

```
{{6, 3}, {1, 2}, {3, 5}, {4, 5}, {2, 4}, {5, 1}}
```

Thus, 6, 1, 3, 4, 2, and 5 occurred 3, 2, 5, 5, 4, and 1 times, respectively. As we see, we get the elements in the order they appear in the list. To get the elements in standard order, sort the list of the frequencies:

```
fr = Tally[a // Sort]
```

```
{{1, 2}, {2, 4}, {3, 5}, {4, 5}, {5, 1}, {6, 3}}
```

Plot the frequencies in one of the following ways:

```
<< Histograms`
```

```
{ListPlot[fr, Filling → Axis, AxesOrigin → {0, 0}],
 Histogram[a, HistogramCategories → Range[0.5, 6.5]],
 Histogram[fr[[All, 2]], FrequencyData → True, HistogramCategories → Range[0.5, 6.5]]}
```

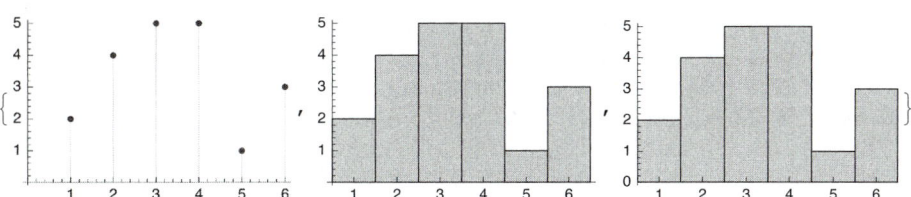

In the first and third plots, we used the frequencies **fr**, whereas in the second plot we used the original data **a**. To get the correct *x* ticks to the histograms, we used the **HistogramCategories** option.

■ Showing Zero Frequencies

Toss again a die 20 times and calculate the frequencies:

```
SeedRandom[27]; a = RandomInteger[{1, 6}, 20]
```

```
{5, 2, 2, 1, 1, 2, 2, 6, 1, 4, 1, 6, 4, 6, 6, 4, 5, 1, 4, 5}
```

```
fr = Tally[a // Sort]
```

```
{{1, 5}, {2, 4}, {4, 4}, {5, 3}, {6, 4}}
```

As we see, the result 3 did not appear at all. Note that **Tally** only reports nonzero frequencies. To get frequencies that also contain possible zero frequencies, use **Count**:

```
fr2 = {#, Count[a, #]} & /@ Range[6]
```

```
{{1, 5}, {2, 4}, {3, 0}, {4, 4}, {5, 3}, {6, 4}}
```

Plot the frequencies in one of the following ways:

```
{ListPlot[fr, Filling → Axis, AxesOrigin → {0, 0}],
 Histogram[a, HistogramCategories → Range[0.5, 6.5]],
 Histogram[fr2[All, 2],
  FrequencyData → True, HistogramCategories → Range[0.5, 6.5]]}
```

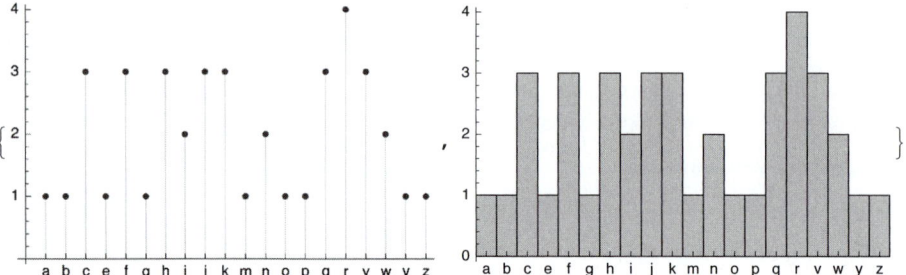

■ Frequencies of Nonnumerical Data

As an example of calculating frequencies of nonnumerical data, consider random characters:

```
chars = CharacterRange["a", "z"];
SeedRandom[4]; a = RandomChoice[chars, 40]
{j, f, q, w, r, n, k, f, m, q, w, r, v, j, z, i, e, n,
 i, c, y, a, k, c, c, g, v, p, b, r, v, h, h, f, k, r, j, h, o, q}
```

Calculate the frequencies:

```
fr = Tally[a // Sort]
{{a, 1}, {b, 1}, {c, 3}, {e, 1}, {f, 3}, {g, 1}, {h, 3}, {i, 2}, {j, 3}, {k, 3},
 {m, 1}, {n, 2}, {o, 1}, {p, 1}, {q, 3}, {r, 4}, {v, 3}, {w, 2}, {y, 1}, {z, 1}}
```

Then plot the frequencies:

```
{ListPlot[fr[All, 2], Filling → Axis, AxesOrigin → {0, 0},
  ImageSize → 200, Ticks → {{Range[Length[fr]], fr[All, 1]}ᵀ, Automatic}],
 Histogram[fr[All, 2], FrequencyData → True, ImageSize → 200,
  Ticks → {{Range[Length[fr]] - 0.5, fr[All, 1]}ᵀ, Automatic}]}
```

If we would also like to get the zero frequencies, do as follows:

```
fr2 = {#, Count[a, #]} & /@ chars
{{a, 1}, {b, 1}, {c, 3}, {d, 0}, {e, 1}, {f, 3}, {g, 1}, {h, 3},
 {i, 2}, {j, 3}, {k, 3}, {l, 0}, {m, 1}, {n, 2}, {o, 1}, {p, 1}, {q, 3},
 {r, 4}, {s, 0}, {t, 0}, {u, 0}, {v, 3}, {w, 2}, {x, 0}, {y, 1}, {z, 1}}
```

Then plot the frequencies:

```
{ListPlot[fr2〚All, 2〛, Filling → Axis, AxesOrigin → {0, 0},
   ImageSize → 200, Ticks → {{Range[26], chars}ᵀ, Automatic}],
 Histogram[fr2〚All, 2〛, FrequencyData → True, ImageSize → 200,
   Ticks → {{Range[26] - 0.5, chars}ᵀ, Automatic}]}
```

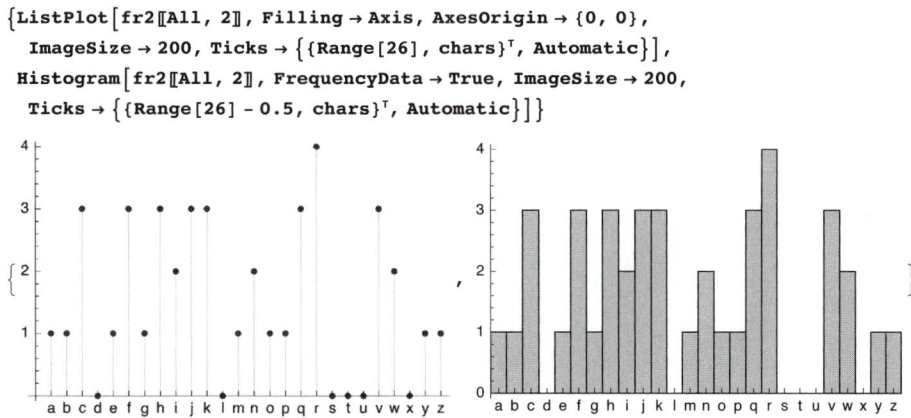

30.2.2 Frequencies of Continuous Data

■ Frequencies of 1D Real Data

For a 1D data set, to count the number of elements that lie in successive bins, use the following commands:

BinCounts[list] (❀6)   Use successive integer bins (equivalent to **BinCounts[list, 1]**)

BinCounts[list, d]   Use bins of width **d** (the first bin starts at **Ceiling[Min[list] − d, d]** and the last bin ends at **Floor[Max[list] + d, d]**)

BinCounts[list, {min, max}]   Use bins of width 1 from **min** to **max**

BinCounts[list, {min, max, d}]   Use bins of width **d** from **min** to **max**

BinCounts[list, {{b₀, b₁, …, bₙ}}]   Use bins $[b_0, b_1)$, $[b_1, b_2)$, … (b_0 can be $-\infty$ and b_n $+\infty$)

BinLists[list] (❀6), etc.   To get the individual elements in the bins

---

As an example, consider the list $a = \{1, 1.4, 2, 3\}$.

- In **BinCounts[a]**, the bins are $[0, 1)$ (!), $[1, 2)$, $[2, 3)$, and $[3, 4)$.
- In **BinCounts[a, 0.5]**, the bins are $[0.5, 1)$ (!), $[1, 1.5)$, $[1.5, 2)$, $[2, 2.5)$, $[2.5, 3)$, and $[3, 3.5)$.
- In **BinCounts[a, 2]**, the bins are $[0, 2)$ and $[2, 4)$.
- In **BinCounts[a, {1, 4}]**, the bins are $[1, 2)$, $[2, 3)$, and $[3, 4)$.

Generate 20 random numbers from the interval $(0, 10)$:

```
SeedRandom[4]; a = RandomReal[10, 20] // Sort
```

```
{1.06783, 1.29313, 1.86905, 2.15046, 2.17094, 2.94215,
 3.11376, 3.16852, 3.61809, 3.70108, 3.7091, 5.66004, 5.74075,
 6.27155, 6.54409, 6.67946, 7.89113, 7.91636, 8.72012, 8.97782}
```

Calculate the frequencies of numbers in integer bins:

```
BinCounts[a]     {3, 3, 5, 0, 2, 3, 2, 2}
```

Here are the individual elements in each bin:

```
BinLists[a]
```

```
{{1.06783, 1.29313, 1.86905}, {2.15046, 2.17094, 2.94215},
 {3.11376, 3.16852, 3.61809, 3.70108, 3.7091}, {}, {5.66004, 5.74075},
 {6.27155, 6.54409, 6.67946}, {7.89113, 7.91636}, {8.72012, 8.97782}}
```

We got but eight frequencies. Indeed, there were no numbers in the intervals [0, 1) and [9, 10). To get frequencies for all intervals of interest—that is, also possible zero frequencies for some first and last intervals—it is safe to define the whole interval:

```
fr = BinCounts[a, {0, 10}]      {0, 3, 3, 5, 0, 2, 3, 2, 2, 0}
```

Draw a histogram in one of the following ways:

```
<< Histograms`
```

```
{Histogram[a],
 Histogram[a, HistogramCategories → Range[0, 10],
  HistogramRange → {0, 10}, Ticks → {Range[0, 10], Automatic}],
 Histogram[fr, FrequencyData → True, HistogramCategories → Range[0, 10],
  Ticks → {Range[0, 10], Automatic}]}
```

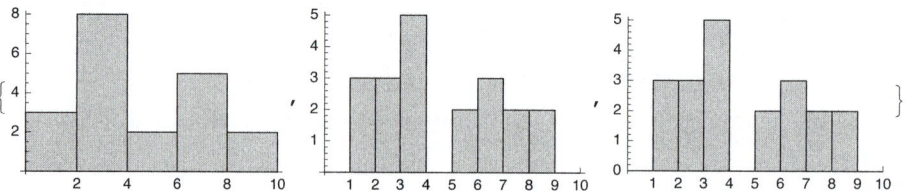

In the first and second plots, we used the original data, whereas in the third plot we used the frequencies. In the first plot, we used automatically generated histogram categories, whereas in the second plot we defined our own categories; in the second plot we also defined a suitable histogram range to get the empty bars at the beginning and end.

Next, we ask for frequencies of numbers in the intervals [0, 2), ..., [8, 10):

```
BinCounts[a, {0, 10, 2}]      {3, 8, 2, 5, 2}
```

Now we ask for frequencies of numbers in the intervals [0, 3), [3, 7), and [7, 10):

```
BinCounts[a, {{0, 3, 7, 10}}]      {6, 10, 4}
```

■ **Example: Normal Data**

Generate a set of 1000 observations from the standard normal distribution:

```
SeedRandom[1]; data = RandomReal[NormalDistribution[0, 1], 1000];
```

Calculate the number of these observations falling in the intervals [−4, −3.75), [−3.75, −3.5), ..., [3.75, 4):

```
fr = BinCounts[data, {-4, 4, 0.25}]
```

```
{0, 0, 0, 0, 0, 6, 8, 9, 14, 30, 44, 57, 64, 85, 98,
 106, 102, 98, 74, 51, 58, 43, 22, 16, 6, 4, 3, 0, 2, 0, 0, 0}
```

Check that all observations have been taken into account:

```
Total[fr]      1000
```

Plot the frequencies. First, we use the original data:

```
{Histogram[data],
 Histogram[data, HistogramCategories → Range[-4, 4, 0.25]],
 Histogram[data, HistogramCategories → Range[-4, 4, 0.25], HistogramRange → {-4, 4}]}
```

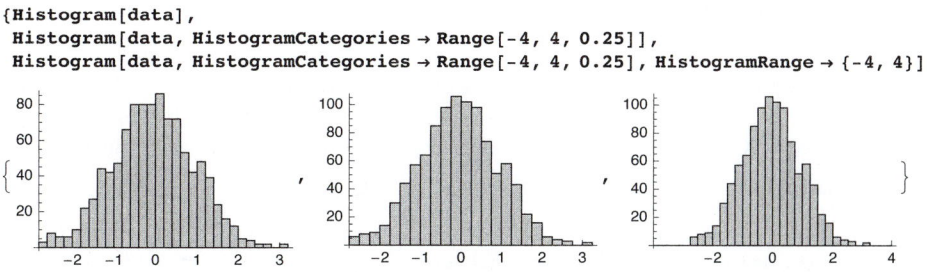

In the first plot, we have automatically generated histogram categories (they have width 0.2), whereas in the second and third plots we use our own categories (that have width 0.25). In the third plot, we have also defined our own histogram range to get a somewhat wider domain. Next, we use the calculated frequencies:

```
Histogram[fr, FrequencyData → True,
 HistogramCategories → Range[-4, 4, 0.25], Ticks → {Range[-4, 4], Automatic}]
```

The cumulative relative frequencies are as follows:

```
cumfr = Accumulate[fr] / 1000.
```

```
{0, 0, 0, 0, 0, 0.006, 0.014, 0.023, 0.037, 0.067, 0.111, 0.168,
 0.232, 0.317, 0.415, 0.521, 0.623, 0.721, 0.795, 0.846, 0.904,
 0.947, 0.969, 0.985, 0.991, 0.995, 0.998, 0.998, 1., 1., 1., 1.}
```

Next, we first plot the relative frequencies and show them together with the probability density function of the normal distribution; to get the histogram scaled so that the total area of the bars is 1, we use the **HistogramScale** option. Then we plot the cumulative relative frequencies and the cumulative distribution function of the normal distribution:

```
{Show[Histogram[data, HistogramScale → 1, HistogramRange → {-4, 4},
   HistogramCategories → Range[-4, 4, 0.25], ImageSize → 200],
  Plot[PDF[NormalDistribution[0, 1], x], {x, -4, 4}]],
 Show[Histogram[cumfr, FrequencyData → True,
   HistogramCategories → Range[-4, 4, 0.25], ImageSize → 200],
  Plot[CDF[NormalDistribution[0, 1], x], {x, -4, 4}]]}
```

■ **Frequencies of 2D Real Data**

`BinCounts[data, xbins, ybins]`	Prepare a 2D frequency table from 2D data

Next, we generate 100 lists, with each list consisting of two random numbers from the interval (0, 10):

```
SeedRandom[1]; b = RandomReal[10, {100, 2}];
```

To state it another way, we generated 100 observations, with each observation consisting of measurements of two variables. Calculate a 2D table of frequencies where 5 bins are used for the first variable and 10 bins for the second variable:

```
MatrixForm[fb = BinCounts[b, {0, 10, 2}, {0, 10, 1}]]
```

$$\begin{pmatrix} 5 & 0 & 3 & 2 & 1 & 3 & 1 & 0 & 0 & 1 \\ 3 & 6 & 1 & 0 & 3 & 4 & 1 & 3 & 5 & 1 \\ 2 & 4 & 3 & 2 & 1 & 4 & 1 & 2 & 1 & 0 \\ 3 & 2 & 4 & 3 & 2 & 2 & 0 & 0 & 2 & 4 \\ 2 & 1 & 0 & 0 & 1 & 4 & 3 & 3 & 1 & 0 \end{pmatrix}$$

This means that there are 5 observations where the first variable is in [0, 2) and the second variable in [0, 1). There are 0 observations where the first variable is in [0, 2) and the second variable in [1, 2), etc. Plot the frequencies:

```
ArrayPlot[fb, FrameTicks → {{Range[0.5, 5.5], Range[0, 10, 2]}ᵀ,
    False, False, {Range[0.5, 10.5], Range[0, 10]}ᵀ},
  FrameLabel → {x, None, None, y}, RotateLabel → False, ImageSize → 150]
```

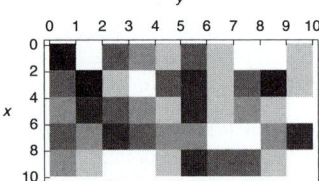

We can also plot a 3D histogram (for another example, see Section 8.6.1, p. 280):

```
<< Histograms`
```

```
Histogram3D[b, HistogramCategories → {Range[0, 10, 2], Range[0, 10, 1]},
  BoxRatios → {5, 10, 5}, ImageSize → 150]
```

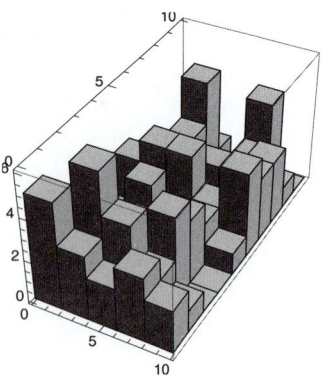

Calculate the row sums of the frequencies:

f1 = Total /@ fb   {16, 27, 20, 22, 15}

The row sums are the frequencies of the first variable:

{b1, b2} = b$^\mathsf{T}$;

BinCounts[b1, {0, 10, 2}]   {16, 27, 20, 22, 15}

Calculate the column sums:

f2 = Total[fb]   {15, 13, 11, 7, 8, 17, 6, 8, 9, 6}

The column sums are the frequencies of the second variable:

BinCounts[b2, {0, 10, 1}]   {15, 13, 11, 7, 8, 17, 6, 8, 9, 6}

The total number of observations is 100:

ft = Total[%]   100

Thus, we arrive at the following frequency table:

ff = ArrayFlatten$\big[\{\{$fb, {f1}$^\mathsf{T}\}$, {{f2}, {{ft}}}$\}\big]$

{{5, 0, 3, 2, 1, 3, 1, 0, 0, 1, 16}, {3, 6, 1, 0, 3, 4, 1, 3, 5, 1, 27},
 {2, 4, 3, 2, 1, 4, 1, 2, 1, 0, 20}, {3, 2, 4, 3, 2, 2, 0, 0, 2, 4, 22},
 {2, 1, 0, 0, 1, 4, 3, 3, 1, 0, 15}, {15, 13, 11, 7, 8, 17, 6, 8, 9, 6, 100}}

Grid[ff, Alignment → Right, ItemSize → {{{1}, 1.8}}, Dividers → {-2 → True, -2 → True}]

5	0	3	2	1	3	1	0	0	1	16
3	6	1	0	3	4	1	3	5	1	27
2	4	3	2	1	4	1	2	1	0	20
3	2	4	3	2	2	0	0	2	4	22
2	1	0	0	1	4	3	3	1	0	15
15	13	11	7	8	17	6	8	9	6	100

We can also add column and row labels:

clb = Table[Row[{"[", i, ",", i + 1, ")"}], {i, 0, 9}]

{[0,1), [1,2), [2,3), [3,4), [4,5), [5,6), [6,7), [7,8), [8,9), [9,10)}

rlb = Table[Row[{"[", i, ",", i + 2, ")"}], {i, 0, 8, 2}]

{[0,2), [2,4), [4,6), [6,8), [8,10)}

ff = ArrayFlatten$\big[$
**   {{{{""}}, {clb}, {{""}}}, {{rlb}$^\mathsf{T}$, fb, {f1}$^\mathsf{T}$}, {{{""}}, {f2}, {{ft}}}}$\big]$;**

Grid[ff, Alignment → {{Left, {Right}}}, ItemSize → 2.8,
**   Dividers → {{2 → True, -2 → True}, {2 → True, -2 → True}}] // Text**

	[0,1)	[1,2)	[2,3)	[3,4)	[4,5)	[5,6)	[6,7)	[7,8)	[8,9)	[9,10)	
[0,2)	5	0	3	2	1	3	1	0	0	1	16
[2,4)	3	6	1	0	3	4	1	3	5	1	27
[4,6)	2	4	3	2	1	4	1	2	1	0	20
[6,8)	3	2	4	3	2	2	0	0	2	4	22
[8,10)	2	1	0	0	1	4	3	3	1	0	15
	15	13	11	7	8	17	6	8	9	6	100

■ **Frequencies of Multidimensional Real Data**

`BinCounts[data, xbins, ybins, …]`   Prepare a multidimensional frequency table

In the next example, we have three variables:

`SeedRandom[1]; b = RandomReal[10, {100, 3}];`

`MatrixForm[BinCounts[b, {0, 10, 2}, {0, 10, 1}, {0, 10, 5}], TableDepth → 3]`

$$\begin{pmatrix} \begin{pmatrix} 0 \\ 2 \end{pmatrix} & \begin{pmatrix} 0 \\ 1 \end{pmatrix} & \begin{pmatrix} 1 \\ 1 \end{pmatrix} & \begin{pmatrix} 0 \\ 3 \end{pmatrix} & \begin{pmatrix} 1 \\ 2 \end{pmatrix} & \begin{pmatrix} 2 \\ 2 \end{pmatrix} & \begin{pmatrix} 1 \\ 1 \end{pmatrix} & \begin{pmatrix} 1 \\ 4 \end{pmatrix} & \begin{pmatrix} 1 \\ 0 \end{pmatrix} & \begin{pmatrix} 0 \\ 1 \end{pmatrix} \\ \begin{pmatrix} 1 \\ 0 \end{pmatrix} & \begin{pmatrix} 1 \\ 3 \end{pmatrix} & \begin{pmatrix} 0 \\ 2 \end{pmatrix} & \begin{pmatrix} 1 \\ 0 \end{pmatrix} & \begin{pmatrix} 5 \\ 0 \end{pmatrix} & \begin{pmatrix} 0 \\ 0 \end{pmatrix} & \begin{pmatrix} 0 \\ 1 \end{pmatrix} & \begin{pmatrix} 1 \\ 0 \end{pmatrix} & \begin{pmatrix} 1 \\ 3 \end{pmatrix} & \begin{pmatrix} 0 \\ 0 \end{pmatrix} \\ \begin{pmatrix} 0 \\ 0 \end{pmatrix} & \begin{pmatrix} 1 \\ 1 \end{pmatrix} & \begin{pmatrix} 1 \\ 2 \end{pmatrix} & \begin{pmatrix} 3 \\ 1 \end{pmatrix} & \begin{pmatrix} 0 \\ 1 \end{pmatrix} & \begin{pmatrix} 3 \\ 2 \end{pmatrix} & \begin{pmatrix} 0 \\ 0 \end{pmatrix} & \begin{pmatrix} 1 \\ 2 \end{pmatrix} & \begin{pmatrix} 2 \\ 0 \end{pmatrix} & \begin{pmatrix} 0 \\ 0 \end{pmatrix} \\ \begin{pmatrix} 2 \\ 0 \end{pmatrix} & \begin{pmatrix} 1 \\ 2 \end{pmatrix} & \begin{pmatrix} 0 \\ 2 \end{pmatrix} & \begin{pmatrix} 2 \\ 2 \end{pmatrix} & \begin{pmatrix} 2 \\ 1 \end{pmatrix} & \begin{pmatrix} 0 \\ 1 \end{pmatrix} & \begin{pmatrix} 0 \\ 0 \end{pmatrix} & \begin{pmatrix} 1 \\ 0 \end{pmatrix} & \begin{pmatrix} 0 \\ 1 \end{pmatrix} & \begin{pmatrix} 0 \\ 0 \end{pmatrix} \\ \begin{pmatrix} 2 \\ 1 \end{pmatrix} & \begin{pmatrix} 1 \\ 2 \end{pmatrix} & \begin{pmatrix} 0 \\ 1 \end{pmatrix} & \begin{pmatrix} 1 \\ 0 \end{pmatrix} & \begin{pmatrix} 1 \\ 0 \end{pmatrix} & \begin{pmatrix} 1 \\ 1 \end{pmatrix} & \begin{pmatrix} 0 \\ 0 \end{pmatrix} & \begin{pmatrix} 2 \\ 3 \end{pmatrix} & \begin{pmatrix} 1 \\ 1 \end{pmatrix} & \begin{pmatrix} 0 \\ 2 \end{pmatrix} \end{pmatrix}$$

30.3   Confidence Intervals

30.3.1   Confidence Intervals for a Mean

■ **Introduction**

With the **HypothesisTesting`** package, we can compute confidence intervals for a mean, for the difference of two means, for a variance, and for the ratio of two variances. Recall that a confidence interval, such as for the population mean, gives an interval within which the population mean lies with a given probability—for example, 0.95. We assume that the observations follow a normal distribution. We will also present a confidence interval for the probability of success of independent trials.

We use the following terminology and notation: μ is the population mean, m is the sample **Mean**, σ^2 is the population variance, s^2 is the sample **Variance**, σ is the population standard deviation, s is the sample **StandardDeviation**, $\sigma_m = \sigma / \sqrt{n}$ is the standard deviation of sample mean, $s_m = s / \sqrt{n}$ is the standard error of sample mean, and $s_m^2 = s^2 / n$ is the variance of sample mean.

■ **Confidence Intervals for a Mean**

`MeanCI[data]`   `StudentTCI[m, `s_m`, n-1]`

For the confidence interval of the population mean, we have two main commands. **MeanCI** uses the original data, whereas **StudentTCI** uses only computed values for the sample mean m, standard error of sample mean s_m, and degrees of freedom $n - 1$ (n is the size of the sample). These two commands use the Student t distribution to calculate the confidence interval.

`MeanCI[data, KnownVariance → `σ^2`]`   `NormalCI[m, `σ_m`]`

Sometimes the population variance σ^2 is known. With **MeanCI**, we can add the option **KnownVariance**, whereas **NormalCI** is used if we input the sample mean and the standard deviation of sample mean. These commands use the normal distribution to calculate the confidence interval.

The confidence level is 0.95 by default, but the level can be set with the option **ConfidenceLevel** (this holds true not only for the commands mentioned here but also for all commands used to calculate confidence intervals).

■ An Example

To illustrate these commands, we first generate data from a normal distribution with a mean of 50 and a standard deviation of 3:

```
SeedRandom[1];
data = RandomReal[NormalDistribution[50, 3], 100];

ListPlot[data]
```

Thus, in this demonstration example, we know that the population mean is 50 and the standard deviation is 3. Now we proceed as if we did not know these values, and we calculate a 95% confidence interval for the population mean:

```
<< HypothesisTesting`

MeanCI[data]      {49.3673, 50.5747}
```

Therefore, we know that with a probability of 0.95, the population mean is within this interval (from the simulated data, we have the knowledge that the population mean really is in this interval). We could also first calculate the sample mean and the standard error of sample mean:

```
{m, sm} = {Mean[data], StandardDeviation[data] / Sqrt[100]}

{49.971, 0.304251}
```

We then use **StudentTCI**:

```
StudentTCI[m, sm, 99]      {49.3673, 50.5747}
```

For intervals other than 95%, we add the confidence level as an option:

```
MeanCI[data, ConfidenceLevel → 0.99]      {49.1719, 50.7701}
```

If we know that the population standard deviation is 3 or the variance is 9, the 95% confidence interval is as follows:

```
MeanCI[data, KnownVariance → 9]      {49.383, 50.559}
```

We could also use **NormalCI**:

```
NormalCI[m, Sqrt[9 / 100]]      {49.383, 50.559}
```

With **NormalCI**, we can easily calculate the well-known confidence intervals for the normal distribution:

```
NormalCI[0, 1, ConfidenceLevel → #] & /@ {0.95, 0.99, 0.999}

{{-1.95996, 1.95996}, {-2.57583, 2.57583}, {-3.29053, 3.29053}}
```

■ **The Meaning of a Confidence Interval**

If we have a 95% confidence interval, we know that there is a 5% probability that the population mean is not in the interval. Thus, if we take 100 samples and calculate the corresponding 95% confidence intervals, we can expect that approximately 5% of the intervals will not contain the population mean. To illustrate this, we generate 100 samples of 100 observations:

```
SeedRandom[1];
samples = RandomReal[NormalDistribution[50, 3], {100, 100}];
```

Calculate for each sample the confidence interval:

```
cis = MeanCI[#] & /@ samples;
```

Investigate how many of these intervals contain the population mean 50:

```
Length[Select[cis, #[[1]] < 50 < #[[2]] &]]      94
```

From the 100 samples, 94 generated a 95% confidence interval that actually contained the true population mean 50. Here are all of the 100 confidence intervals:

```
showConfidenceIntervals[cis_, mu_, n_, opts___] := Graphics[
   {Line[{{1, mu}, {n, mu}}], Table[Line[{{i, cis[[i, 1]]}, {i, cis[[i, 2]]}}], {i, n}]},
   Axes → {False, True}, AspectRatio → 0.2, PlotRange → All, opts];
```

```
showConfidenceIntervals[cis, 50, 100, ImageSize → 370]
```

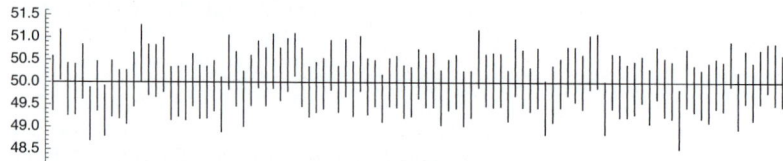

Here are the confidence intervals in ascending order according to the mean of the samples (the means are shown with points):

```
me = {Range[100], Sort[Mean /@ samples]}ᵀ;
```

```
showConfidenceIntervals[Sort[cis, Mean[#1] < Mean[#2] &],
  50, 100, Epilog → Point[me], ImageSize → 370]
```

30.3.2 Other Confidence Intervals

■ **Confidence Intervals for the Difference between Two Means**

```
MeanDifferenceCI[data1, data2]
```

If the populations are known to have equal variances, we can add the option **EqualVariances → True**. If we know this common value, we can simply add the option **KnownVariance → σ^2**. If the known variances are different, we can add the option **KnownVariance → $\{\sigma_1^2, \sigma_2^2\}$**.

■ **Confidence Intervals for a Variance**

```
VarianceCI[data]
ChiSquareCI[s², n-1]
```

The first command uses the original data, whereas the other uses only the sample variance s^2 and degrees of freedom. The default confidence level is 0.95, but a different level can be set with **ConfidenceLevel**. These two commands use the chi-square distribution.

As an example, we use the same simulated data that were calculated in the preceding section:

```
SeedRandom[1];
data = RandomReal[NormalDistribution[50, 3], 100];
<< HypothesisTesting`

VarianceCI[data]      {7.13607, 12.492}
```

Thus, with a probability of 0.95, the population variance is within this interval (the true variance 9 really is in this interval). We could also first calculate the sample variance and then use the other command:

```
var = Variance[data]      9.25685

ChiSquareCI[var, 99]      {7.13607, 12.492}
```

■ **Confidence Intervals for the Ratio of Two Variances**

```
VarianceRatioCI[data1, data2]
FRatioCI[s₁²/s₂², n₁-1, n₂-1]
```

The first command uses the two data sets, whereas the other uses only the ratio of the sample variances s_1^2/s_2^2 and degrees of freedom (n_1 and n_2 are the sizes of the samples from the two populations). These two commands use the F-ratio distribution.

■ **Confidence Intervals for a Probability**

Suppose we have made n independent trials, of which k have succeeded. The estimate of the probability of success is k/n. The following module gives an approximate $100\,\alpha\,\%$ confidence interval for the true probability of success (Johnson, Kotz, and Kemp, 1992, p. 130). The probability of the true p being within the computed interval is at least α.

```
probabilityCI[succ_, total_, α_] := Module[
   {n1 = 2 succ, n2 = 2 (total - succ + 1), n3 = 2 (succ + 1), n4 = 2 (total - succ), q1, q2},
   q1 = Quantile[FRatioDistribution[n1, n2], (1 - α) / 2];
   q2 = Quantile[FRatioDistribution[n3, n4], (1 + α) / 2];
   {n1 q1 / (n2 + n1 q1), n3 q2 / (n4 + n3 q2)}]
```

Generate a sequence of successes and failures by assuming that each trial succeeds with a probability of 0.3:

```
SeedRandom[1];
data2 = RandomInteger[BernoulliDistribution[0.3], 100]
{1, 0, 1, 0, 0, 0, 0, 0, 0, 1, 0, 1, 0, 0, 1, 1, 1, 0, 0, 0, 0, 0, 0, 1,
 0, 1, 0, 0, 0, 0, 0, 1, 0, 0, 1, 0, 0, 0, 0, 1, 1, 0, 0, 0, 0, 0, 1, 0, 0,
 0, 1, 1, 0, 0, 0, 0, 0, 0, 0, 0, 0, 0, 1, 0, 0, 0, 1, 0, 0, 0, 0, 0, 1, 1,
 0, 1, 1, 0, 0, 1, 0, 0, 0, 0, 0, 0, 0, 1, 0, 0, 0, 0, 0, 0, 0, 0, 0, 0, 0}
```

Calculate the number of successes:

```
k = Total[data2]     24
```

Then calculate an approximate 95% confidence interval:

```
probabilityCI[k, 100, 0.95]     {0.160225, 0.335735}
```

30.4 Hypothesis Testing

30.4.1 Tests for a Mean

■ **Tests for a Mean**

With the **HypothesisTesting`** package, we can test the mean, the difference of two means, the variance, and the ratio of two variances. Recall that when testing, for example, the population mean, we want to infer, on the basis of a sample from the population, whether the population mean is a certain value μ or whether it is different from that value. We assume that the data follow a normal distribution. We also present a test for the probability of success of independent trials and a test for goodness of fit. The material in this section is closely analogous to that of Section 30.3. First, we consider the testing of the mean.

```
MeanTest[data, μ]
StudentTPValue[ (m−μ) / s_m, n−1]
```

To test whether the population mean could be μ, use one of the previous commands. **MeanTest** uses the original data and the hypothetical value μ of the population mean, whereas **StudentTPValue** uses only the value of the test statistic $t = (m - \mu)/s_m$ and degrees of freedom $n - 1$. The two commands use the Student t distribution to perform the testing.

The result of the testing is a p value. This is the probability that if the hypothetical mean value μ is true, the test statistic t (treated as a random variable) has a value at least as extreme as its computed value. If the p value is sufficiently small—smaller than, for example, 0.05 (a significance level)—then the hypothetical mean value μ can be rejected: The observations do not give sufficient support for this mean value.

```
MeanTest[data, μ, KnownVariance → σ²]
NormalPValue[ (m−μ) / σ_m]
```

The commands (that use the normal distribution) can be used if the population variance is known. In **MeanTest** we can add the option **KnownVariance**, whereas **NormalPValue** is used if we input the value of the test statistic.

■ **Options**

Options for hypothesis testing:

SignificanceLevel Significance level of the test; examples of values: **None, 0.05, 0.01**
TwoSided Whether to perform a two-sided test; possible values: **False, True**
FullReport Whether to include additional information; possible values: **False, True**

These options can be used for all hypothesis test commands that end with **Test**. We can add a significance level (e.g., **SignificanceLevel → 0.05**), in which case the result of the test contains the conclusion whether the hypothesis is accepted or rejected at this significance level.

The default is a one-sided test. This means that we test whether, for example, the population mean is μ against the alternative that the population mean is greater than μ or against the alternative that the population mean is smaller than μ. If we want to test that the population mean is μ against the alternative that the population mean is different from μ, then we add the option **TwoSided → True**. This option can also be used with the commands ending with **PValue**.

If we add the option **FullReport → True**, we get, in addition to the p value, the sample mean, the value of the test statistic, and the distribution used in calculating the p value.

■ An Example

To illustrate the testing of a mean, we use the same simulated data we used in Section 30.3:

```
SeedRandom[1];
data = RandomReal[NormalDistribution[50, 3], 100];
{m, sm} = {Mean[data], StandardDeviation[data] / Sqrt[100]}
{49.971, 0.304251}
```

The sample mean is close to 50. We test whether the population mean could be $\mu = 50$ against the alternative that the population mean is smaller than 50:

```
<< HypothesisTesting`
```

```
MeanTest[data, 50]      OneSidedPValue → 0.462117
```

This probability is not small (not smaller than, for example, 0.05), and so we cannot reject the hypothesis that the population mean is 50 (from the simulated data, we have the knowledge that the population mean really is 50). We could also first calculate the value of the test statistic t from the values of the sample mean m and the standard error of sample mean s_m:

```
t = (m - 50) / sm      -0.0953436
```

```
StudentTPValue[t, 99]      OneSidedPValue → 0.462117
```

We can add a significance level as an option:

```
MeanTest[data, 50, SignificanceLevel → 0.05]
{OneSidedPValue → 0.462117,
  Fail to reject null hypothesis at significance level → 0.05}
```

To test whether the population mean is 50 against the alternative that the population mean is other than 50, we write the following:

```
MeanTest[data, 50, TwoSided → True]      TwoSidedPValue → 0.924235
```

This probability is again not small, and so we cannot reject the hypothesis that the population mean is 50. Then we ask for a full report:

```
MeanTest[data, 50, FullReport → True]
```

$$\left\{ \text{FullReport} \to \frac{\text{Mean} \quad \text{TestStat} \quad \text{Distribution}}{49.971 \quad -0.0953436 \quad \text{StudentTDistribution[99]}} ,\right.$$

$$\left. \text{OneSidedPValue} \to 0.462117 \right\}$$

If we know that the population standard deviation is 3 or the variance 9, we get the following p value:

```
MeanTest[data, 50, KnownVariance → 9]      OneSidedPValue → 0.461485
```

We could also use `NormalCI`:

```
NormalPValue[(m - 50) / Sqrt[9 / 100]]    OneSidedPValue → 0.461485
```

■ Type I Error

If we accept a correct hypothesis or reject a wrong one, we make a correct decision. However, if we reject a correct hypothesis or accept a wrong one, we make a wrong decision.

If we perform several tests and always use significance level 0.05 (i.e., we always reject the hypotheses if the p value is smaller than 0.05), then we know that, in approximately 5% of the tests, we reject a correct hypothesis. This is a *type I error*. To illustrate this error, we generate 100 samples of 100 observations from a normal distribution with mean 50:

```
SeedRandom[1];
samples = RandomReal[NormalDistribution[50, 3], {100, 100}];
```

Perform for each sample the test that tells us whether the population mean is 50 against the alternative that the mean is other than 50:

```
pvalues1 =
  TwoSidedPValue /. Partition[MeanTest[#, 50, TwoSided → True] & /@ samples, 1];
```

Investigate the number of tests in which we draw the correct conclusion that the population mean is 50 when we use the significance level 0.05:

```
Length[Select[pvalues1, # ≥ 0.05 &]]      94
```

From the 100 samples, in 94 we draw the true conclusion that the population mean is 50. To plot the p values, we write the following function:

```
showPValues[pvalues_, α_, n_, opts___] :=
  Graphics[{Point[{Range[n], pvalues}ᵀ], Line[{{1, α}, {n, α}}]}, Axes → True,
    AxesOrigin → {0, -0.05}, AspectRatio → 0.2, PlotRange → All, opts]
```

Here are the p values for all 100 samples:

```
showPValues[pvalues1, 0.05, 100, ImageSize → 400]
```

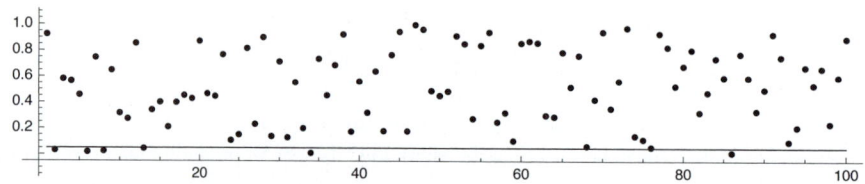

We see that six p values are below the significance level 0.05. In these cases, we made the type I error: We rejected the correct hypothesis that the population mean is 50.

■ Type II Error

A *type II error* is made if the hypothesis is not true but we still accept it. To illustrate this error, we use the 100 samples of 100 observations generated in the preceding example and test whether the population mean is 51 against the alternative that the mean is not 51. We know that in this case, the hypothesis is wrong and should be rejected:

```
pvalues2 =
  TwoSidedPValue /. Partition[MeanTest[#, 51, TwoSided → True] & /@ samples, 1];
Length[Select[pvalues2, # < 0.05 &]]      91
```

We rightly rejected the wrong hypothesis 91 times. Here are all of the *p* values:

```
showPValues[pvalues2, 0.05, 100, ImageSize → 400]
```

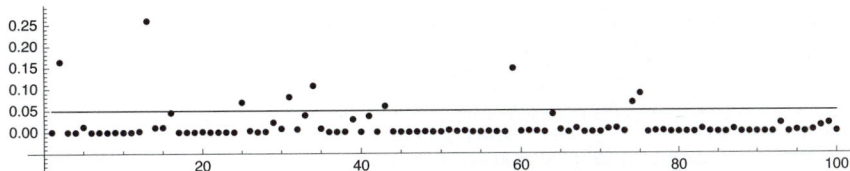

We see that nine points are above the significance level 0.05. In these cases, we made the type II error: We accepted the wrong hypothesis, which was that the population mean is 51.

30.4.2 Other Tests

■ Testing the Difference between Two Means

```
MeanDifferenceTest[data1, data2, d]
```

Here, we test whether the difference between two population means could be *d*. In addition to the options **EqualVariances → True** and **KnownVariance → σ^2** or **KnownVariance → $\{\sigma_1^2, \sigma_2^2\}$**, we can also use the three options mentioned in Section 30.4.1, p. 1024: **SignificanceLevel → α, TwoSided → True**, and **FullReport → True**.

■ Testing a Variance

```
VarianceTest[data, σ²]
ChiSquarePValue[(n-1) s²/σ², n-1]
```

To test whether the population variance could be σ^2, use one of these commands. As an example, we use the simulated data that were calculated in Section 30.4.1 and test the hypothesis that the population variance is 9 against the alternative that the variance is other than 9:

```
SeedRandom[1];
data = RandomReal[NormalDistribution[50, 3], 100];
<< HypothesisTesting`

VarianceTest[data, 9, TwoSided → True]
TwoSidedPValue → 0.80553
```

We could also use the test statistic:

```
ChiSquarePValue[99 Variance[data] / 9, 99, TwoSided → True]
TwoSidedPValue → 0.80553
```

This probability is not small, so we cannot reject the hypothesis that the population variance is 9 (the true variance is 9 in this example).

■ Testing the Ratio of Two Variances

```
VarianceRatioTest[data1, data2, r]
FRatioPValue[ (s₁²/s₂²)/r, n₁-1, n₂-1]
```

Use these commands to test whether the ratio of two population variances could be r.

■ Testing a Probability

Suppose we have made n independent trials of which k have succeeded. The estimate of the probability of success is k/n. The following module finds the p value to test the probability of success (Allen, 1990, p. 508).

```
probabilityTest[succ_, total_, p0_, type_] := Module[{p1, p2},
  p1 = 1 - CDF[BinomialDistribution[total, p0], succ - 1];
  p2 = CDF[BinomialDistribution[total, p0], succ];
  Which[type === g, p1, type === l,
   p2, type === d, If[succ > total p0, 2 p1, 2 p2], True, Null]]
```

Here, **succ** is the number of successes, **total** is the number of trials, **p0** is the hypothetical value of the probability of success, and **type** is the type of the alternative hypothesis. The type is **g**, **l**, or **d**, depending on whether the alternative hypothesis claims that the probability is greater than, less than, or different from **p0**. As an example, we generate a sequence of successes and failures. Each trial succeeds with a probability of 0.3:

```
SeedRandom[1];
data2 = RandomInteger[BernoulliDistribution[0.3], 100];
```

Calculate the number of successes:

```
k = Apply[Plus, data2]      24
```

We test whether the true probability of success is 0.3 against the alternative that the probability is less than 0.3:

```
probabilityTest[k, 100, 0.3, l]      0.11357
```

This probability is not small, so we cannot reject the hypothesis that the probability of success is 0.3 (from the simulated observations, we know that the true probability is 0.3).

■ Goodness-of-Fit Test

The chi-square distribution can be used to test whether a given set of observations may have arisen from a certain distribution. Note that plotting the data on probability graph paper may help in determining whether the data follow a given distribution (see Section 29.3.2, p. 983).

When investigating days of absence at a firm during a period of 50 days, a statistician obtained the following results (these are not real data):

```
obs = {{8, 0}, {12, 1}, {14, 2}, {8, 3}, {3, 4}, {4, 5}, {1, 6}};
```

This means that there were 8 days with no absences, 12 days with one absence, and so on. Do these observations follow a Poisson distribution? To find out, first separate the observed frequencies and the values:

```
{obsfreq, obsval} = obsᵀ
{{8, 12, 14, 8, 3, 4, 1}, {0, 1, 2, 3, 4, 5, 6}}
```

Check that the number of days is 50:

```
n = Total[obsfreq]      50
```

Then calculate the mean number of absences per day:

```
lambda = Total[obsfreq obsval] / n // N      2.04
```

Thus, in the mean, there were 2.04 absences per day. Then we calculate the first six Poisson probabilities with this parameter and multiply them by the number of days to obtain the expected frequencies:

```
expfreq = n Table[PDF[PoissonDistribution[lambda], i], {i, 0, 5}]
```
{6.50144, 13.2629, 13.5282, 9.19917, 4.69158, 1.91416}

To these frequencies we add the expected frequency of at least six absences:

```
AppendTo[expfreq, n (1 - CDF[PoissonDistribution[lambda], 5])]
```
{6.50144, 13.2629, 13.5282, 9.19917, 4.69158, 1.91416, 0.902544}

All of the expected frequencies should be at least 5 for the test to be sufficiently accurate, so we combine the last three classes to form one class:

```
expfreq = Append[Take[expfreq, 4], Total[Take[expfreq, -3]]]
```
{6.50144, 13.2629, 13.5282, 9.19917, 7.50828}

The sum of the expected frequencies is 50, as it should be:

```
Total[%]       50.
```

Similarly, we combine the last three observed frequencies:

```
obsfreq = Append[Take[obsfreq, 4], Total[Take[obsfreq, -3]]]
```
{8, 12, 14, 8, 8}

Now we calculate the chi-square statistic:

```
chi2 = Total[(obsfreq - expfreq)^2 / expfreq]       0.670651
```

Lastly, we calculate the p value, which shows the probability of getting a chi-square value at least as extreme as `chi2`. The parameter of the chi-square distribution is the number of classes minus the number of estimated parameters (in this example, we estimated one parameter, namely the parameter of the Poisson distribution) minus 1:

```
1 - CDF[ChiSquareDistribution[5 - 1 - 1], chi2]       0.880084
```

This probability is not small, so we cannot reject the hypothesis that the observations follow a Poisson distribution.

Note that to calculate the p value, we cannot here use the built-in command:

```
ChiSquarePValue[chi2, 5 - 1 - 1]       OneSidedPValue → 0.119916
```

This is the probability of getting a chi-square value *at most* `chi2`.

30.4.3 Analysis of Variance (ANOVA)

*In the **ANOVA`** package:*

`ANOVA[data]` Perform a one-way ANOVA

`ANOVA[data, model, vars]` Perform a general ANOVA

With the `PostTests` option, we can tell what tests we want to apply to find significant differences. Possibilities are `Bonferroni`, `Duncan`, `StudentNewmanKeuls`, `Tukey`, and `Dunnett`. `SignificanceLevel` has the default value 0.05.

ANOVA is a way to investigate whether several populations—having normal distributions with equal variances—have equal means.

Consider the following example (Rohatgi, 1984, p. 811). When a farmer investigated four fertilizers for soybeans, he received the following yields for plots of equal size:

```
fertilizer[1] = {47, 42, 43, 46, 44, 42};
fertilizer[2] = {51, 58, 62, 49, 53, 51, 50, 59};
fertilizer[3] = {37, 39, 41, 38, 39, 37, 42, 36, 40};
fertilizer[4] = {42, 43, 42, 45, 47, 50, 48};
```

We define the number of fertilizers and the number of plots for each fertilizer:

```
k = 4; n =.; n[1] = 6; n[2] = 8; n[3] = 9; n[4] = 7;
```

To do the one-way ANOVA, write the data as follows:

```
data = Flatten[Table[{Table[i, {n[i]}], fertilizer[i]}ᵀ, {i, k}], 1]
{{1, 47}, {1, 42}, {1, 43}, {1, 46}, {1, 44}, {1, 42}, {2, 51},
 {2, 58}, {2, 62}, {2, 49}, {2, 53}, {2, 51}, {2, 50}, {2, 59},
 {3, 37}, {3, 39}, {3, 41}, {3, 38}, {3, 39}, {3, 37}, {3, 42}, {3, 36},
 {3, 40}, {4, 42}, {4, 43}, {4, 42}, {4, 45}, {4, 47}, {4, 50}, {4, 48}}
```

The results of ANOVA are as follows:

```
<< ANOVA`
```

```
ANOVA[data, PostTests → {Bonferroni, Tukey}]
```

		DF	SumOfSq	MeanSq	FRatio	PValue
ANOVA →	Model	3	1015.51	338.503	31.6746	7.77406×10^{-9}
	Error	26	277.859	10.6869		
	Total	29	1293.37			

	All	45.4333
	Model[1]	44.
CellMeans →	Model[2]	54.125
	Model[3]	38.7778
	Model[4]	45.2857

PostTests → {Model → { Bonferroni   {{1, 2}, {1, 3}, {2, 3}, {2, 4}, {3, 4}}
                        Tukey   {{1, 2}, {1, 3}, {2, 3}, {2, 4}, {3, 4}} }}

The *p* value is very small, which indicates that there are significant differences between the fertilizers. Both the Tukey and the Bonferroni test arrived at the conclusion that fertilizers 1 and 2, 1 and 3, 2 and 3, 2 and 4, and 3 and 4 differ significantly. The tests did not find a significant difference between fertilizers 1 and 4.

For more information about ANOVA, see ANOVA/tutorial/ANOVA.

30.5 Regression

30.5.1 Linear Regression

■ **Linear Regression**

The command `Fit[data, basis, vars]`, which was considered in Section 25.1.1, p. 812, calculates linear least-squares fits to data: It finds the linear combination of the functions in the basis that gives the least squared error.

The command **Regress** in the **LinearRegression`** package does the same but, in addition, can print statistical information about the fit. **Regress** is used in the same way as **Fit**. For parameters that appear nonlinearly, use **NonlinearRegress** (see Section 30.5.2, p. 1035). For local regression, see Section 30.5.3, p. 1038.

In the **LinearRegression`** *package:*

Regress[data, basis, vars] Fit **data** by a linear combination of functions of **vars** in **basis**

Options:

IncludeConstant Whether a constant term is automatically included in the model; possible values:
True, False

Weights List of weights for each data point or a pure function of the response; default value:
Automatic

ConfidenceLevel Used for confidence intervals; examples of values: **0.95, 0.99**

RegressionReport Statistics to be included in output; default value: **SummaryReport**

BasisNames Names of basis elements for table headings; default value: **Automatic**

Method Method used to compute singular values; default value: **Automatic**

Tolerance Numerical tolerance to use in computing singular values; default value: **Automatic**

Data are normally given in the form **{{x1, f1}, {x2, f2}, … }**. An example of the basis is **{1, x, x^2}**.

The default value of **IncludeConstant** is **True**, which means that the constant term is automatically included even if it is not mentioned in the list of basis functions (**Fit** only uses the functions in the list of basis functions). You have to write **IncludeConstant → False** if you do not want the constant term.

Next, we explain the use of the option **RegressionReport**.

■ Obtainable Information

RegressionReport controls the amount of information that is printed. The default value **SummaryReport** means the list **{ParameterTable, RSquared, AdjustedRSquared, EstimatedVariance, ANOVATable}**. The command **RegressionReportValues[Regress]** gives all possible items. Here are the items classified into groups (many of these items are explained in the following examples):

- To get the fit and information about the estimated parameters: **BestFit, BestFitParameters, ParameterTable, ParameterCITable, ParameterConfidenceRegion, CovarianceMatrix, CorrelationMatrix**

- To analyze variances: **ANOVATable, EstimatedVariance, CoefficientOfVariation, RSquared, AdjustedRSquared**

- To analyze predictions: **FitResiduals, PredictedResponse, SinglePredictionCITable, MeanPredictionCITable**

- To detect correlated errors: **DurbinWatsonD**

- To evaluate basis functions and detect collinearity: **PartialSumOfSquares, SequentialSumOfSquares, VarianceInflation, EigenstructureTable**

- To detect outliers: **HatDiagonal, JackknifedVariance, StandardizedResiduals, StudentizedResiduals, CookD, PredictedResponseDelta, BestFitParametersDelta, CovarianceMatrixDetRatio**

- To get the catcher matrix: **CatcherMatrix**

■ **An Example: Basic Information**

Consider the example from Section 25.1.1, p. 813:

```
xx = Range[0, 50]; SeedRandom[2];
rand = RandomReal[NormalDistribution[0, 1], 51];
data = {xx, 2 + xx - 0.004 xx^2 + 2 rand}ᵀ;
p1 = ListLinePlot[data, Mesh → All]
```

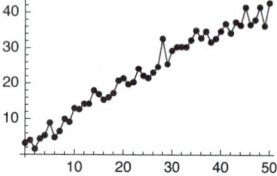

We could use **Fit** to calculate the least-squares fit. Try a second-order polynomial:

```
Fit[data, {1, x, x^2}, x]
```

$1.48998 + 1.0506 \, x - 0.00531772 \, x^2$

We can also use **Regress** and get more information about the fit:

```
<< LinearRegression`
```

```
Regress[data, {1, x, x^2}, x]
```

$$\left\{ \text{ParameterTable} \rightarrow \begin{array}{c|cccc} & \text{Estimate} & \text{SE} & \text{TStat} & \text{PValue} \\ \hline 1 & 1.48998 & 0.793743 & 1.87716 & 0.066581 \\ x & 1.0506 & 0.0734167 & 14.31 & 0. \\ x^2 & -0.00531772 & 0.00141999 & -3.7449 & 0.000483345 \end{array} \right. ,$$

RSquared → 0.973712, AdjustedRSquared → 0.972616, EstimatedVariance → 3.85756,

$$\text{ANOVATable} \rightarrow \left. \begin{array}{c|ccccc} & \text{DF} & \text{SumOfSq} & \text{MeanSq} & \text{FRatio} & \text{PValue} \\ \text{Model} & 2 & 6858.35 & 3429.18 & 888.95 & 0. \\ \text{Error} & 48 & 185.163 & 3.85756 & & \\ \text{Total} & 50 & 7043.51 & & & \end{array} \right\}$$

ParameterTable contains the estimates of the parameters and information to test whether a specific parameter is zero against the alternative that the parameter is not zero. The test can be done with the t statistic and the corresponding p value. A small p value (e.g., not larger than 0.05) indicates that the observations do not support the hypothesis that the parameter is zero. In our example, all p values are small, which tells us that all coefficients are statistically significantly different from zero.

RSquared—the square of the multiple correlation coefficient—is also called the coefficient of determination and is in the interval [0, 1]. It tells us how much the full, fitted model improves a reduced model that only contains a constant term. **EstimatedVariance** is the estimated error variance or the residual mean square.

With **ANOVATable**, we can test the null hypothesis that the data could be described by a model containing only the constant term. A large F-ratio and a small p value indicate that we can reject the null hypothesis.

■ **More Information**

Next, we ask for more information:

```
result = Regress[data, {1, x, x^2}, x,
    RegressionReport → {BestFit, BestFitParameters, ParameterCITable, DurbinWatsonD}]
```

$\{$BestFit $\to 1.48998 + 1.0506\,x - 0.00531772\,x^2,$

BestFitParameters $\to \{1.48998, 1.0506, -0.00531772\},$

ParameterCITable $\to$

	Estimate	SE	CI
1	1.48998	0.793743	$\{-0.105945, 3.08591\}$
x	1.0506	0.0734167	$\{0.902981, 1.19821\}$
x^2	-0.00531772	0.00141999	$\{-0.0081728, -0.00246264\}$

DurbinWatsonD $\to 2.29017\}$

BestFit gives the fit in the same form as **Fit**, whereas **BestFitParameters** gives only the parameters. **ParameterCITable** contains 95% confidence intervals for the parameters.

The Durbin–Watson d statistic is between 0 and 4. Values near 2 mean uncorrelated errors (this is the assumption in regression analysis). Values that are less than 2 mean positive correlation, and values greater than 2 indicate negative correlation. In our model, the d statistic is near 2, so the errors are not correlated.

We show the data and the fit:

```
Show[p1, Plot[BestFit /. result, {x, 0, 50}], ImageSize → 200]
```

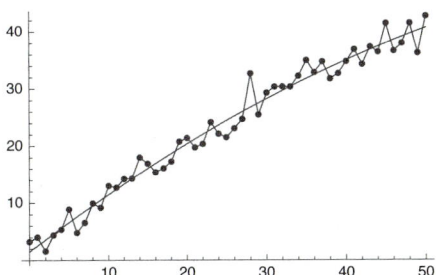

Next, we continue the analysis of our example by studying the residuals, the confidence region of the data, the confidence region of the fitted curve, and the confidence regions of the parameters.

■ **Residuals**

The residuals are as follows:

```
res = FitResiduals /. Regress[data, {1, x, x^2}, x, RegressionReport → FitResiduals];
```

```
ListLinePlot[{xx, res}ᵀ, Mesh → All]
```

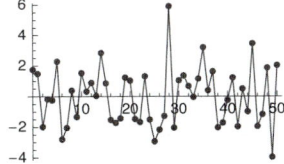

■ **Confidence Region of the Data**

With `SinglePredictionCITable`, we can ask for the confidence interval for *a single observed response* at each of the values of the independent variables. In this way, we get a region that is likely to contain all possible observations. First, we extract the components of the result: the observed values, the predicted values, the standard errors of the predicted response, and the confidence intervals:

```
{observed, predicted, se, ci} =
   (SinglePredictionCITable /. Regress[data, {1, x, x^2},
       x, RegressionReport → SinglePredictionCITable])[[1]]ᵀ;
```

Then we plot the data, the predicted values, and the lower and upper values of the 95% confidence intervals:

```
pred = {xx, predicted}ᵀ;
lowerCI = {xx, First /@ ci}ᵀ;
upperCI = {xx, Last /@ ci}ᵀ;

Graphics[{Point[data], Line[pred], Gray, Line[lowerCI], Line[upperCI]},
  Axes → True, AspectRatio → 1 / GoldenRatio, ImageSize → 200]
```

■ **Confidence Region of the Curve**

With `MeanPredictionCITable`, we can ask for the confidence interval for the *mean response* at each of the values of the independent variables. In this way, we get a region that is likely to contain the regression curve. We do as we did previously:

```
{observed, predicted, se, ci} = (MeanPredictionCITable /.
      Regress[data, {1, x, x^2}, x, RegressionReport → MeanPredictionCITable])[[1]]ᵀ;

lowerCI = {xx, First /@ ci}ᵀ;
upperCI = {xx, Last /@ ci}ᵀ;

Graphics[{Point[data], Line[pred], Gray, Line[lowerCI], Line[upperCI]},
  Axes → True, AspectRatio → 1 / GoldenRatio, ImageSize → 200]
```

■ Confidence Regions of the Parameters

The correlation matrix gives the correlations between the parameters:

Regress[data, {1, x, x^2}, x, RegressionReport → CorrelationMatrix]

$$\left\{\text{CorrelationMatrix} \rightarrow \begin{pmatrix} 1. & -0.856214 & 0.7305 \\ -0.856214 & 1. & -0.967074 \\ 0.7305 & -0.967074 & 1. \end{pmatrix}\right\}$$

For example, the coefficients of 1 and x have a negative correlation -0.86, whereas 1 and x^2 have a positive correlation 0.73. The sign of the correlation can also be seen from the confidence regions of the parameters. We plot the 95% confidence region of the coefficients of 1 and x:

cr = ParameterConfidenceRegion[{1, x}] /. Regress[data,
   {1, x, x^2}, x, RegressionReport → {ParameterConfidenceRegion[{1, x}]}]
Ellipsoid[{1.48998, 1.0506}, {2.01141, 0.0955139},
   {{0.996865, -0.0791262}, {0.0791262, 0.996865}}]

Graphics[cr, Axes → True, AspectRatio → 1 / GoldenRatio]

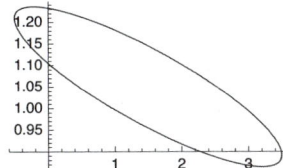

■ Designed Regress

> *In the* **LinearRegression`** *package:*
>
> **DesignedRegress[designmatrix, response]** Fit the model represented by **designmatrix** given the vector **response** of response data
>
> **DesignMatrix[data, basis, vars]** Give the design matrix

30.5.2 Nonlinear Regression

■ Nonlinear Regression

Recall from Section 25.1.3, p. 818, that we have **FindFit** for nonlinear fitting. However, if you want statistical information about the fit, then use **NonlinearRegress**.

> *In the* **NonlinearRegression`** *package:*
>
> **NonlinearRegress[data, funct, params, var]** When **funct** is an expression of the variable **var** and contains the parameters **params**, find values for the parameters such that the function fits **data** in the best way (in the sense of least squares)
>
> *Options:*
>
> **WorkingPrecision** Precision used in internal computations; examples of values: **MachinePrecision, 20**
>
> **PrecisionGoal** If the value of the option is **p**, the relative error of the χ^2 merit function should be of the order 10^{-p}; examples of values: **Automatic** (usually means **8**), **10**
>
> **AccuracyGoal** If the value of the option is **a**, the absolute error of the χ^2 merit function should be of the order 10^{-a}; examples of values: **Automatic** (usually means **8**), **10**

Method Method used; possible values: **Automatic** (usually means **"LevenbergMarquardt"**),
   "LevenbergMarquardt", **"Gradient"**, **"ConjugateGradient"**, **"Newton"**, **"QuasiNewton"**,
   "NMinimize"

MaxIterations Maximum number of iterations; examples of values: **100**, **200**

Weights List of weights for each data point or a pure function; default value: **Automatic**

Tolerance Numerical tolerance for certain matrix operations; default value: **Automatic**

Gradient How the gradient is calculated; examples of values: **Automatic**, **"Symbolic"**,
   "FiniteDifference"

ConfidenceLevel Used for confidence intervals; examples of values: **0.95**, **0.99**

RegressionReport Statistics to be included in output; default value: **SummaryReport**

The method of finding the best parameters is based on the so-called χ^2 merit function, which is the sum of the squares of the residuals.

The parameter specification **params** is of the form **{aspec, bspec, cspec, … }**. Each specification is of one of the following forms:

- **a** Start from **1.0** for parameter **a**
- **{a, a0}** Start from **a0**
- **{a, a0, a1}** Start from **a0** and **a1** (this form must be used if symbolic derivatives of χ^2 with respect to the parameters cannot be found)
- **{a, a0, amin, amax}** Start from **a0**; stop iteration if it goes outside of [**amin**, **amax**]
- **{a, a0, a1, amin, amax}** Start from **a0** and **a1**; stop iteration if it goes outside of [**amin**, **amax**]

We can have constraints **cons** for the parameters. In place of **funct** in the previous box, just write **{funct, cons}**.

■ Obtainable Information

The **RegressionReport** option controls the amount of information printed. The default value **SummaryReport** means the list **{BestFitParameters, ParameterCITable, EstimatedVariance, ANOVATable, AsymptoticCorrelationMatrix, FitCurvatureTable}**. All possible items can be seen by giving the command **RegressionReportValues[NonlinearRegress]**. Many of the items are the same for **NonlinearRegress** as they are for **Regress**, but the former also has five unique items, marked with (nlr) here:

- To get the fit and information about the estimated parameters: **BestFit**, **BestFitParameters**, **ParameterTable**, **ParameterCITable**, **ParameterConfidenceRegion**, **AsymptoticCovarianceMatrix** (nlr), **AsymptoticCorrelationMatrix** (nlr)

- To analyze variances: **ANOVATable**, **EstimatedVariance**

- To analyze predictions: **FitResiduals**, **PredictedResponse**, **SinglePredictionCITable**, **MeanPredictionCITable**

- To detect outliers: **HatDiagonal**, **StandardizedResiduals**

- To get other information: **FitCurvatureTable** (nlr), **ParameterBias** (nlr), **StartingParameters** (nlr)

■ An Example

We consider the same model of exponential growth as in Section 25.1.3, p. 819:

```
xx = Range[0, 10, 0.2]; SeedRandom[0];
rand = RandomReal[NormalDistribution[0, 1], 51];
data = {xx, Exp[0.3 + 0.2 xx] + 0.5 rand}ᵀ;
```

Calculate an exponential fit:

```
<< NonlinearRegression`
```

```
result = NonlinearRegress[data, Exp[a + b x], {a, b}, x]
```

$\Big\{$BestFitParameters → {a → 0.202349, b → 0.212479},

		Estimate	Asymptotic SE	CI
ParameterCITable →	a	0.202349	0.0511552	{0.0995487, 0.305149},
	b	0.212479	0.00628525	{0.199848, 0.225109}

EstimatedVariance → 0.207755,

		DF	SumOfSq	MeanSq
	Model	2	1272.17	636.084
ANOVATable →	Error	49	10.18	0.207755,
	Uncorrected Total	51	1282.35	
	Corrected Total	50	354.964	

AsymptoticCorrelationMatrix → $\begin{pmatrix} 1. & -0.968294 \\ -0.968294 & 1. \end{pmatrix}$,

		Curvature
	Max Intrinsic	0.0236575
FitCurvatureTable →	Max Parameter-Effects	0.0385863
	95. % Confidence Region	0.560193

For information about curvature, see NonlinearRegression/tutorial/NonlinearRegression. Plot the data and the fit:

```
Show[ListPlot[data], Plot[Exp[a + b x] /. (BestFitParameters /. result), {x, 0, 10}]]
```

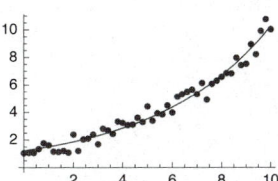

The residuals are as follows:

```
res = FitResiduals /.
    NonlinearRegress[data, Exp[a + b x], {a, b}, x, RegressionReport → FitResiduals];
```

```
ListLinePlot[{xx, res}ᵀ, Mesh → All]
```

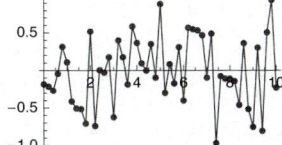

30.5.3 Local Regression

■ **Introduction**

Sometimes the form of the data is so complex or obscure that it does not easily suggest a form for the approximating function—that is, a parametric family of functions. In such situations, a *local regression* may be suitable (local regression or locally weighted regression falls in the category of *nonparametric regression*).

In local regression, we choose a set of points from the range of the independent variable and fit a set of low-order polynomials, with each polynomial describing the behavior of the data only near one of the chosen points (this is achieved by appropriately weighing the observations). Each polynomial is evaluated at the corresponding point, and so we obtain smoothed values. When these points are connected, the result is a local regression curve. Each part of the curve describes the average behavior of the data near that part.

In the following, we apply the local regression method described in Cleveland (1993, pp. 91–101); the method is also called *loess*. According to Cleveland, the method has some desirable statistical properties, is easy to compute (but computing intensive), and is easy to use.

■ **Explaining Ozone by Wind**

To illustrate the method, we read the **environmental** data that come on the CD-ROM of this book (we already considered this data set in Section 8.2.1, p. 249; the same data are also analyzed in Cleveland (1994, pp. 172–175):

```
env =
    Rest[Import["/Users/heikki/Documents/MNData/visdata/environmental", "Table"]];
```

The data have 111 observations, each of which contains the number of the observation and the value of ozone, radiation, temperature, and wind. First, we separate the components:

```
{no, ozone, radiation, temperature, wind} = env⊤;
```

We consider only wind and ozone:

```
data = {wind, ozone}⊤;
```

```
pdata = ListPlot[data, AspectRatio → 1, PlotRange → All, AxesOrigin → {0, 0}]
```

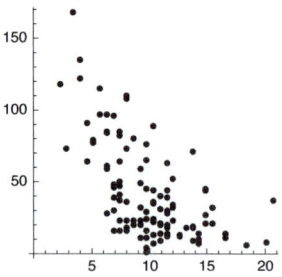

A descending pattern is clear, but otherwise the form of the data is somewhat obscure. A third-order polynomial fit is quite good:

```
fit = Fit[data, {1, x, x^2, x^3}, x]
```

$$201.663 - 31.8806\,x + 1.84631\,x^2 - 0.0350707\,x^3$$

```
Show[pdata, Plot[fit, {x, 2, 21}]]
```

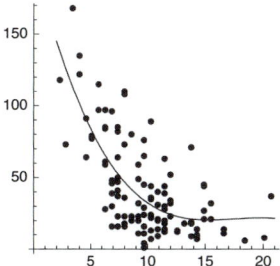

■ Local Regression

In the following box, we have a program to calculate a local regression curve:

```
<< LinearRegression`
T = Compile[{u}, If[Abs[u] < 1, (1 - Abs[u] ^ 3) ^ 3, 0]];
ww = Compile[{{i, _Integer}, {x, _Real}, {xx, _Real, 1}, {q, _Integer}},
    T[Abs[xx〚i〛 - x] / Sort[Abs[xx - x]]〚q〛]];
localRegress[data_, localpols_, α_, λ_] := Module[{xx, ff, a, b, n, x, q, xwei, y},
    {xx, ff} = dataᵀ; a = Min[xx]; b = Max[xx]; n = Length[xx];
    x = Range[a, b, (b - a) / (localpols - 1)]; q = Floor[α n];
    xwei = {#, Table[ww[i, #, xx, q], {i, n}]} & /@ x;
    Interpolation[
    {#〚1〛, BestFit /. Regress[data, y ^ Range[0, λ], y, Weights → #〚2〛 + 10 ^ -15,
        RegressionReport → BestFit] /. y → #〚1〛} & /@ xwei]]
```

Now we try to explain the program (you may want to first move to Using the Program below and then come back here later). The package **LinearRegression`** has to be loaded because we use **Regress** from that package; we cannot use **Fit** because **Fit** does not have an option to weigh the data. The function **T** is the key to use to weigh the data; it is $\left(1 - |u|^3\right)^3$ for $|u| < 1$ and 0 otherwise. It looks like this:

```
Plot[T[u], {u, -1, 1}]
```

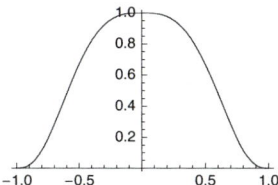

Most weight is given if the argument is near zero, and the weight is zero outside $(-1, 1)$. The function **ww** then computes the weights for the data. The wind varies between the following values:

```
{Min[wind], Max[wind]}     {2.3, 20.7}
```

Consider, for example, the wind observations with ordering numbers 1, 5, 9, 3, and 7:

```
wind[{1, 5, 9, 3, 7}]     {7.4, 8.6, 9.2, 12.6, 20.1}
```

We plot the weights of these observations as functions of the point where the local low-order polynomial will be fitted:

```
GraphicsRow[
  Plot[ww[#, x, wind, 88], {x, 2.3, 20.7}, PlotRange → {{0, 21}, {-0.05, 1}}, Epilog →
    {AbsolutePointSize[4], Point[{wind[[#]], 0}]}, Ticks → {Range[5, 20, 5], {1}}] & /@
  {1, 5, 9, 3, 7}, ImageSize → 420, Spacings → -40]
```

For example, the first observation with wind value 7.4 receives most weight when a low-order polynomial is fitted at 7.4. For polynomials fitted at points far from 7.4, the first observation receives less weight.

Assume that we have n data points. Let $\Delta_i(x) = |x_i - x|$ be the distance between x_i and x, and let $\Delta_{(i)}(x)$ be the ith smallest of these distances. Let $\alpha \leq 1$ be given, and let q be the product αn truncated to an integer. The function **ww** is $T(\Delta_i(x)/\Delta_{(q)}(x))$. If α is near 1, the smoothing of the data is strong. Lower values of α smooth less.

The program **localRegress** calculates the weight for each data point **xx[[i]]** and for each point **x[[k]]** where a local polynomial is calculated (this is quite a computing-intensive task, and to speed up the computations we have compiled the functions **T** and **ww**). Local polynomials are calculated a total of **localpols** times. For **localRegress**, we input also α and λ; λ is the degree of the local polynomials (either 1 or 2). The calculated points are connected by calculating a piecewise third-order interpolating function through the points (see Section 24.2.1, p. 797).

■ Using the Program

We compute a local regression curve by computing 20 first-order polynomials ($\lambda = 1$) and using the value 0.9 for α:

```
fit = localRegress[data, 20, 0.9, 1]
InterpolatingFunction[{{2.3, 20.7}}, <>]
```

The result is an interpolating function. Plot it and show the curve together with the data:

```
Show[pdata, Plot[fit[x], {x, 2.3, 20.7}]]
```

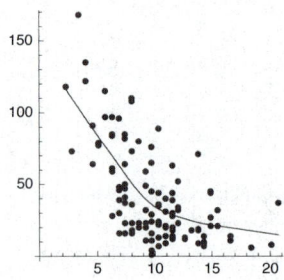

We use the following program to compute the residuals and a local regression curve for them (by calculating 20 local first-order polynomials with $\alpha = 0.8$).

```
showLocalResiduals[data_, fit_, opts___] := Module[{xx, ff, resf, res, resfit},
  {xx, ff} = dataᵀ; resf = ff - (fit[#] & /@ xx); res = {xx, resf}ᵀ;
  Print["Sum of squared residuals: ", resf.resf];
  resfit = localRegress[res, 20, 0.8, 1];
  Show[ListPlot[res, PlotRange → All],
   Plot[resfit[x], {x, Min[xx], Max[xx]}], opts]]
```

```
showLocalResiduals[data, fit, AxesOrigin → {0, 0}]
```
Sum of squared residuals: 59 829.9

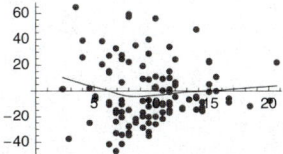

The local fit of the residuals should be near zero. We see that the local fit in the figure does not quite satisfy this property. Accordingly, we calculate a new curve—now using a lower value for α—to adapt the curve more closely to the data. Try the value $\alpha = 0.6$:

```
fit = localRegress[data, 20, 0.6, 1];
```

```
{Show[pdata, Plot[fit[x], {x, 2.3, 20.7}]],
 showLocalResiduals[data, fit, AxesOrigin → {0, 0}]}
```
Sum of squared residuals: 58 054.1

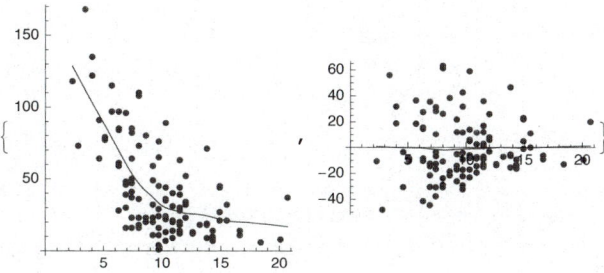

The fit seems good, and the local fit to the residuals is now practically zero.

Another example of local regression is provided in Section 30.6.2, p. 1044, where we consider smoothing.

30.6 Smoothing

30.6.1 Smoothing with a Kernel

■ Using a Kernel

Suppose we have data $\{x_1, …, x_n\}$ and we want to smooth the data with a *kernel* $\{k_1, …, k_m\}$ $(m < n)$ by forming the sums $\sum_{j=1}^{m} k_j x_{i+j}$, $i = 0, 1, …, n - m$. For example, if $n = 5$ and $m = 3$, the result is the following smoothed values:

```
{k₁, k₂, k₃}.# & /@ {{x₁, x₂, x₃}, {x₂, x₃, x₄}, {x₃, x₄, x₅}}
{k₁ x₁ + k₂ x₂ + k₃ x₃, k₁ x₂ + k₂ x₃ + k₃ x₄, k₁ x₃ + k₂ x₄ + k₃ x₅}
```

We can also use **ListCorrelate**:

ListCorrelate[kernel, list]	Form the correlation of **kernel** with **list**
ListConvolve[kernel, list]	Form the convolution of **kernel** with **list**

We can verify that we get with **ListCorrelate** the same result that we obtained previously:

```
ListCorrelate[{k₁, k₂, k₃}, {x₁, x₂, x₃, x₄, x₅}]
{k₁ x₁ + k₂ x₂ + k₃ x₃, k₁ x₂ + k₂ x₃ + k₃ x₄, k₁ x₃ + k₂ x₄ + k₃ x₅}
```

On the other hand, **ListConvolve** gives the following result:

```
ListConvolve[{k₁, k₂, k₃}, {x₁, x₂, x₃, x₄, x₅}]
{k₃ x₁ + k₂ x₂ + k₁ x₃, k₃ x₂ + k₂ x₃ + k₁ x₄, k₃ x₃ + k₂ x₄ + k₁ x₅}
```

Thus, **ListConvolve[kernel, list]** means **ListCorrelate[Reverse[kernel], list]**. Note that both commands have more general forms and that they also apply to multidimensional kernels and data.

■ Examples of Kernels

Moving averages are obtained by giving a constant kernel:

```
ListCorrelate[{1, 1, 1} / 3, {x₁, x₂, x₃, x₄, x₅}] // Simplify
```
$$\left\{ \frac{1}{3} (x_1 + x_2 + x_3), \frac{1}{3} (x_2 + x_3 + x_4), \frac{1}{3} (x_3 + x_4 + x_5) \right\}$$

The kernel {−1, 1} gives successive differences:

```
ListCorrelate[{-1, 1}, {x₁, x₂, x₃, x₄, x₅}]
{-x₁ + x₂, -x₂ + x₃, -x₃ + x₄, -x₄ + x₅}
```

A Gaussian kernel is of the following form:

```
gaussianKernel[denom_, max_] :=
 With[{t = Table[Exp[-n^2 / denom] // N, {n, -max, max}]}, t / Total[t]]
```

For example,

```
gk = gaussianKernel[3, 3]
{0.0162712, 0.0861476, 0.234173, 0.326815, 0.234173, 0.0861476, 0.0162712}
```

```
ListPlot[gk, AxesOrigin → {0, 0}]
```

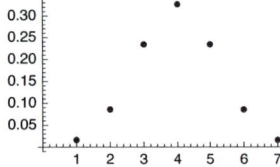

In the following example, we use the Gaussian kernel.

■ An Example of Smoothing

We try to smooth noisy data. Our signal is as follows:

```
s = Sin[5 × 2 Pi t] - 0.8 Cos[9 × 2 Pi t];

signalPlot = Plot[s, {t, 0, 1}, AspectRatio → 0.2, PlotStyle → Gray, ImageSize → 300]
```

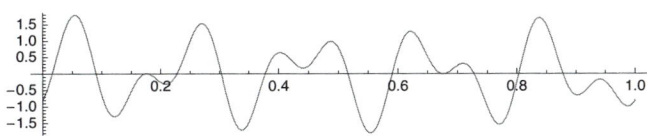

To generate noisy observations from this signal, we first sample the signal:

```
tt = Range[0, 1, 0.01];
xx = s /. t → tt;
```

Then we generate noise from a normal distribution with mean of 0 and standard deviation of 0.4:

```
SeedRandom[1];
noise = RandomReal[NormalDistribution[0, 0.4], 101];
```

Add the noise to the signal and plot the resulting data:

```
xdata = xx + noise;
data = {tt, xdata}ᵀ;
dataPlot = ListPlot[data, AspectRatio → 0.2, ImageSize → 300]
```

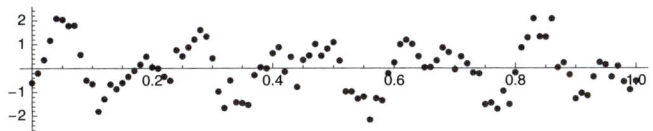

We smooth the data using the Gaussian kernel we calculated previously:

```
smooth = ListCorrelate[gaussianKernel[3, 3], xdata];
```

When plotting the smoothed values, note that data are lost at both ends:

```
smoothPlot = ListLinePlot[{Range[0.03, 0.97, 0.01], smooth}ᵀ,
    PlotStyle → {Black, Thickness[Medium]}];
```

We can compare the smoothed values with the data and with the signal:

```
Show[dataPlot, smoothPlot]
```

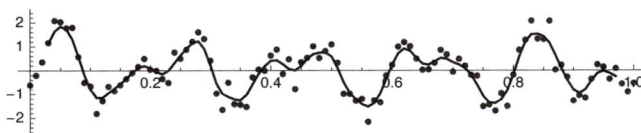

```
Show[signalPlot, smoothPlot]
```

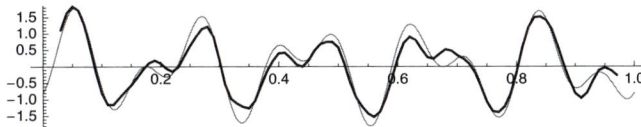

30.6.2 Other Methods of Smoothing

■ Moving Averages

> **MovingAverage[data, r]** (✎6) Simple *r*-term moving average
>
> **MovingAverage[data, {w$_0$, …, w$_r$}]** (✎6) Use the given weights
>
> **MovingMedian[data, r]** (✎6) Moving median using spans of *r* elements
>
> **ExponentialMovingAverage[data, a]** (✎6) Exponential smoothing with smoothing constant *a*

A simple *r*-term moving average smoother calculates the average of all *r* successive terms. A moving median smoother of span *r* calculates the median of all *r* successive terms. **MovingAverage** and **MovingMedian** give a list of length **Length[data]** - **r** + 1. Note that

- **MovingAverage[data, r]** is equivalent to **ListCorrelate[Table[1/r, {r}], data]**.
- **MovingAverage[data, wts]** is equivalent to **ListCorrelate[wts/Total[wts], data]**.

If $\{x_1, x_2, \ldots\}$ is the original set of data, then an exponential moving average smoother calculates the values $y_{t+1} = y_t + a(x_{t+1} - y_t)$, $t = 0, 1, \ldots$, $y_0 = x_1$, for a smoothing constant a, $0 < a < 1$. The smaller a is, the stronger the smoothing. The list of smoothed values has the same length as the data.

The smoothing commands work for both univariate and multivariate data. The data only contain the dependent variable(s), not the independent variable (such as time).

As an example, try exponential moving average:

```
es = ExponentialMovingAverage[xdata, 0.6];
```

```
smoothPlot = ListLinePlot[{tt, es}ᵀ, PlotStyle → {Black, Thickness[Medium]}];
Show[dataPlot, smoothPlot]
```

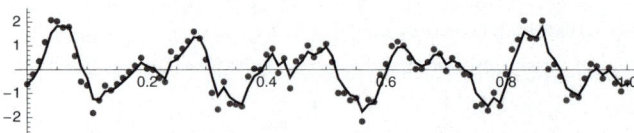

```
Show[signalPlot, smoothPlot]
```

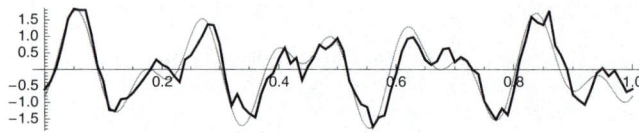

■ Local Regression

In Section 30.5.3, p. 1038, we presented a program to calculate a local regression curve. The method finds a smooth curve through the points by fitting a sequence of low-order polynomials. We try this method with our data:

```
fit = localRegress[data, 30, 0.05, 1]
```

```
InterpolatingFunction[{{0., 1.}}, <>]
```

Plot the result and show the curve together with the data and also with the signal:

```
smoothPlot = Plot[fit[x], {x, 0, 1}, PlotStyle → {Black, Thickness[Medium]}];
```

```
Show[dataPlot, smoothPlot]
```

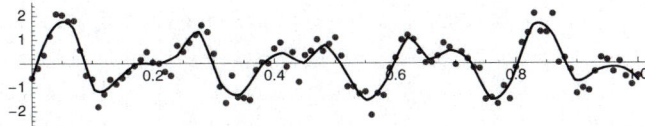

```
Show[signalPlot, smoothPlot]
```

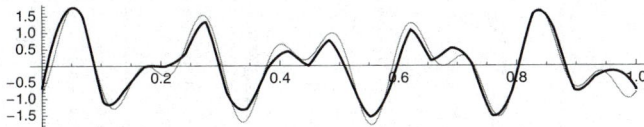

■ Discrete Fourier Transform

The discrete Fourier transform (see Section 20.4.4, p. 675) can be used to smooth or filter data. First, we calculate the Fourier transform of the y values and plot the absolute values of the transform:

```
fou = Fourier[xdata];
```

```
ListPlot[{Range[0, 100], Abs[fou]}ᵀ,
  PlotRange → All, AspectRatio → 0.2, ImageSize → 300]
```

We see two peaks at the frequencies 5 and 9 (and the corresponding symmetric peaks on the right-hand side), which correspond with the frequencies 5 and 9 of the signal. All other frequencies can be considered to be caused by the noise.

We try to filter the data by simply replacing with zeros all of the values of the Fourier transform except for the four peaks. We do this by replacing with zeros all frequencies whose absolute value is less than, in this case, 2:

```
filtFou = Chop[fou, 2]
{0, 0, 0, 0, 0, 4.84479 i, 0, 0, 0, -3.97842, 0, 0, 0, 0, 0, 0, 0, 0, 0, 0, 0, 0, 0, 0,
 0, 0, 0, 0, 0, 0, 0, 0, 0, 0, 0, 0, 0, 0, 0, 0, 0, 0, 0, 0, 0, 0, 0, 0, 0, 0,
 0, 0, 0, 0, 0, 0, 0, 0, 0, 0, 0, 0, 0, 0, 0, 0, 0, 0, 0, 0, 0, 0, 0, 0, 0, 0,
 0, 0, 0, 0, 0, 0, 0, 0, 0, 0, 0, -3.97842, 0, 0, 0, -4.84479 i, 0, 0, 0, 0}
```

Then we find the inverse transform:

```
filtData = {tt, Chop[InverseFourier[filtFou]]}ᵀ;
```

Next, we compare the filtered values with the data and the signal. The fit is very good:

```
smoothPlot = ListLinePlot[filtData, PlotStyle → {Black, Thickness[Medium]}];
```

`Show[dataPlot, smoothPlot]`

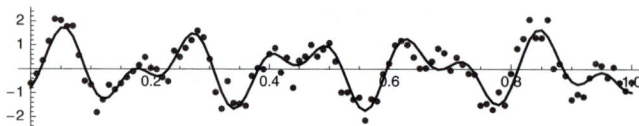

`Show[signalPlot, smoothPlot]`

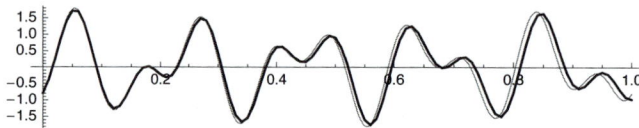

30.7 Bayesian Statistics

30.7.1 Introduction

■ Posterior Joint Density

Suppose we have data $Y = (y_1, y_2, \ldots)$ and we want to describe the data with a model that contains unknown parameters $\theta = (\alpha, \beta, \gamma, \ldots)$. We have some prior information about the parameters in the form of a probability density function $f(\theta)$, which is called the *prior (joint) density*. In addition, we know the conditional density $f(y_i \mid \theta)$.

We want to derive statistical information about θ based on the data Y. The solution is sought in the form of the density $f(\theta \mid Y)$, which is called the *posterior (joint) density*. From this density we can then calculate the mean, variance, and so on of θ. According to Bayes' theorem, $f(\theta \mid Y) = f(\theta) f(Y \mid \theta) / f(Y)$, or, in terms of proportionality, $f(\theta \mid Y) \propto f(\theta) f(Y \mid \theta)$. Assuming independent data (conditionally on θ), we have $f(Y \mid \theta) = \prod f(y_i \mid \theta)$, which is often called the *likelihood function*. Thus, we arrive at the formula $f(\theta \mid Y) \propto f(\theta) \prod f(y_i \mid \theta)$, which means that *posterior* $\propto$ *prior* $\times$ *likelihood*.

■ Example

Green (2001, p. 5) considers the following model. Let the data $Y = (y_1, \ldots, y_n)$ come from a normal distribution: $(y_i \mid \mu, \sigma) \sim N(\mu, \sigma)$, where $\mu \sim N\left(\xi, \frac{1}{\kappa}\right)$, $\frac{1}{\sigma^2} \sim \Gamma\left(\alpha, \frac{1}{\beta}\right)$; assume that μ and σ^{-2} are independent and that ξ, κ, α, and β are known. The parameters of interest are μ and σ^{-2}. The posterior joint density of these parameters is as follows:

$$f(\mu, \sigma^{-2} \mid Y) \propto f(\mu, \sigma^{-2}) f(Y \mid \mu, \sigma^{-2}) = f(\mu) f(\sigma^{-2}) \prod_{i=1}^{n} f(y_i \mid \mu, \sigma).$$

To write down this expression, we denote the argument of the density of σ^{-2} by **σ2i** (**i** for inverse) so that the prior joint density is as follows:

```
prior = PDF[NormalDistribution[ξ, 1 / κ], μ]
    PDF[GammaDistribution[α, 1 / β], σ2i] // PowerExpand
```

$$\frac{e^{-\frac{1}{2} \kappa^2 (\mu-\xi)^2 - \beta \, \sigma2i} \, \beta^\alpha \, \kappa \, \sigma2i^{-1+\alpha}}{\sqrt{2 \pi} \; \text{Gamma}[\alpha]}$$

To form the likelihood function $f(Y \mid \mu, \sigma^{-2})$, we look at the density function $f(y_i \mid \mu, \sigma)$:

```
PDF[NormalDistribution[μ, σ], yi]
```
$$\frac{e^{-\frac{(yi-\mu)^2}{2\sigma^2}}}{\sqrt{2 \pi} \; \sigma}$$

If we denote $\sum y_i$ with **sum** and $\sum y_i^2$ with **sum2**, we see that the likelihood function is as follows:

```
likelihood = (σ2i / 2π)^(n/2) Exp[- 1/2 σ2i (sum2 - 2 μ sum + n μ²)]
```

$$e^{-\frac{1}{2}(\text{sum2}-2\,\text{sum}\,\mu+n\,\mu^2)\,\sigma2i} \, (2 \pi)^{-n/2} \, \sigma2i^{n/2}$$

The posterior joint density $f(\mu, \sigma^{-2} \mid Y)$ is now proportional to the following:

```
prior likelihood
```

$$\frac{e^{-\frac{1}{2} \kappa^2 (\mu-\xi)^2 - \beta \, \sigma2i - \frac{1}{2}(\text{sum2}-2\,\text{sum}\,\mu+n\,\mu^2)\,\sigma2i} \, (2 \pi)^{-\frac{1}{2}-\frac{n}{2}} \, \beta^\alpha \, \kappa \, \sigma2i^{-1+\frac{n}{2}+\alpha}}{\text{Gamma}[\alpha]}$$

We can simplify this further by picking up only the terms that contain μ or σ^{-2}:

```
pμσ2i = Select[%, ! FreeQ[#, μ] || ! FreeQ[#, σ2i] &]
```

$$e^{-\frac{1}{2} \kappa^2 (\mu-\xi)^2 - \beta \, \sigma2i - \frac{1}{2}(\text{sum2}-2\,\text{sum}\,\mu+n\,\mu^2)\,\sigma2i} \, \sigma2i^{-1+\frac{n}{2}+\alpha}$$

Thus, the posterior joint density is proportional to this expression.

■ Posterior Marginal Densities

Posterior *marginal* densities $f(\alpha \mid Y)$, $f(\beta \mid Y)$ and so on can be obtained by integrating out all other parameters from the posterior joint density. For example, $f(\alpha \mid Y) = \int_{-\infty}^{\infty} d\beta \int_{-\infty}^{\infty} d\gamma \dots f(\theta \mid Y)$. We consider four methods to calculate the posterior marginal densities: using integration, using interpolation, Gibbs sampling, and Markov chain Monte Carlo.

Using integration (see Section 30.7.2, p. 1049). The integrals needed to calculate the marginal densities are often tedious or difficult (or both) to handle with pencil and paper, but mathematical programs such as *Mathematica* should be tried. A program may be able to calculate difficult integrals, thereby reducing the need to resort to approximative methods. However, even *Mathematica* may not be able to calculate some of the integrals you need. Consider then the interpolation method.

Using interpolation (see Section 30.7.3, p. 1051). This method relies on *Mathematica*'s ability to form a representation of a complicated expression by interpolation. Indeed, **FunctionInterpolation** samples the expression at many points and forms an interpolating function that passes through these points (see Section 24.4.2, p. 807). With this method, we can get numerical but accurate representations of the posterior marginal densities. If this method fails, try Gibbs sampling.

Gibbs sampling (see Section 30.7.4, p. 1053). This method uses the *full conditional distributions* of the parameters (the conditional distributions of each parameter given the other parameters and the data) and requires that random numbers from these distributions be generated. The posterior marginal densities are approximated by using a random sample from the full conditional distributions. What if we do not have a random number generator for these distributions? Consider the next method.

Markov chain Monte Carlo (see Section 30.7.5, p. 1057). This method is often shortened as "MCMC." The method has two significant advantages over Gibbs sampling. First, MCMC only requires the expressions of the full conditional densities up to proportionality. Second, random numbers are not required from these distributions. The posterior marginal densities are approximated by using random numbers from a suitably selected distribution for which we have a random number generator.

Note that the advanced methods of integration and interpolation of *Mathematica* may somewhat reduce the need to resort to Gibbs sampling or MCMC.

■ Data

In Sections 30.7.2 through 30.7.5, we continue with the previous example. The starting point is always **pμσ2i**, which is the posterior joint density (up to proportionality) of the parameters μ and σ^{-2}. The data to be used are generated from $N(15, 2)$:

```
SeedRandom[1]; data = RandomReal[NormalDistribution[15, 2], 20]
```
```
{15.9714, 15.8183, 15.5238, 16.2713, 18.0886, 16.3536,
 15.2614, 17.1973, 14.1665, 12.4085, 14.9703, 11.4765, 15.0413,
 17.6238, 15.1955, 14.6172, 14.3047, 14.6287, 15.8166, 18.0261}
```

Calculate their sum and sum of squares:

```
{datasum = Total[data], datasum2 = Total[data^2]}
```
```
{308.761, 4819.95}
```

(The symbolic values of these sums were denoted by **sum** and **sum2** in the example.) Use the following values of constants (which lead to [improper] noninformative priors):

```
vals = {ξ → 0, κ → 0, α → 0, β → 0, n → 20, sum → datasum, sum2 → datasum2};
```

Non-Bayesian estimates of μ and σ are as follows:

```
{Mean[data], StandardDeviation[data]}
```
```
{15.4381, 1.67452}
```

■ Posterior Joint Density of μ and σ^{-2}

Given the data and the constants, we get the following expression for the posterior joint density of μ and σ^{-2} in our example:

```
pμσ2i /. vals        e^{-\frac{1}{2}(4819.95-617.523 μ+20 μ²) σ2i} σ2i^9
```

Although the scaling constant of this density is lacking, we can still plot its contours of constant value because the forms of the contours do not depend on scaling:

```
ContourPlot[pμσ2i /. vals, {μ, 14.5, 16.4},
  {σ2i, 0.15, 0.63}, FrameLabel → {"μ", "σ⁻²"}, RotateLabel → False]
```

30.7.2 Using Integration

■ Calculating the Posterior Densities

We continue the example of Section 30.7.1, p. 1046, and try to calculate the posterior marginal densities by direct integration. The marginal density of μ, which is $f(\mu \,|\, Y)$, is proportional to the following:

```
Integrate[pμσ2i, {σ2i, 0, ∞}, Assumptions → {α, β, μ, κ, ξ, σ2i, n, sum, sum2} ∈ Reals]
```

$$\text{If}\Big[n > -2\,\alpha \;\&\&\; 2\,\text{sum}\,\mu < \text{sum2} + 2\,\beta + n\,\mu^2 ,$$

$$2^{\frac{n}{2}+\alpha}\, e^{-\frac{1}{2}\kappa^2\,(\mu-\xi)^2}\, (\text{sum2} + 2\,\beta + \mu\,(-2\,\text{sum} + n\,\mu))^{-\frac{n}{2}-\alpha}\, \text{Gamma}\Big[\frac{n}{2} + \alpha\Big],$$

$$\text{Integrate}\Big[e^{-\frac{1}{2}\kappa^2\,(\mu-\xi)^2 - \frac{1}{2}\,(\text{sum2}+2\,\beta-2\,\text{sum}\,\mu+n\,\mu^2)\,\sigma2i}\, \sigma2i^{-1+\frac{n}{2}+\alpha},\ \{\sigma2i, 0, \infty\},\ \text{Assumptions} \to$$

$$(\alpha \mid \beta \mid \mu \mid \kappa \mid \xi \mid \sigma2i \mid n \mid \text{sum} \mid \text{sum2}) \in \text{Reals}\ \&\&\ !\ \big(n > -2\,\alpha\ \&\&\ 2\,\text{sum}\,\mu < \text{sum2} + 2\,\beta + n\,\mu^2\big)\Big]\Big]$$

To simplify the expression, we make the assumptions given in the result and, at the same time, select only the terms that contain μ:

```
pμ = Select[Integrate[pμσ2i, {σ2i, 0, ∞},
  Assumptions → n > -2 α && 2 sum μ < sum2 + 2 β + n μ^2], ! FreeQ[#, μ] &]
```

$$e^{-\frac{1}{2}\kappa^2\,(\mu-\xi)^2}\, (\text{sum2} + 2\,\beta + \mu\,(-2\,\text{sum} + n\,\mu))^{-\frac{n}{2}-\alpha}$$

In the same way, we get that the marginal density of σ^{-2}, which is $f(\sigma^{-2}\,|\,Y)$, is proportional to the following:

```
pσ2i =
  Select[Integrate[pμσ2i, {μ, -∞, ∞}, Assumptions → κ^2 + n σ2i > 0], ! FreeQ[#, σ2i] &]
```

$$\frac{e^{\frac{1}{2}\left(-\kappa^2\,\xi^2 - \text{sum2}\,\sigma2i - 2\,\beta\,\sigma2i + \frac{(\kappa^2\,\xi + \text{sum}\,\sigma2i)^2}{\kappa^2 + n\,\sigma2i}\right)}\, \sigma2i^{-1+\frac{n}{2}+\alpha}}{\sqrt{\kappa^2 + n\,\sigma2i}}$$

Next, we study these densities in more detail.

■ Posterior Density of μ

Calculate the scaling constant of $f(\mu \,|\, Y)$:

```
cμ = 1 / NIntegrate[pμ /. vals, {μ, 0, 30}]     1.93716 × 10¹⁷
```

The posterior density of μ is then the following:

fμ = cμ pμ /. vals

$$\frac{1.93716 \times 10^{17}}{(4819.95 + \mu \, (-617.523 + 20 \, \mu))^{10}}$$

Plot the density:

plotfμ = Plot[fμ, {μ, 13, 17}, PlotStyle → Black]

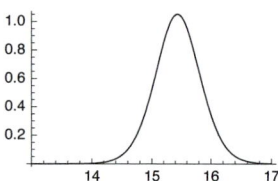

From here on, we do not show some warning messages about valid limits of integration:

Off[NIntegrate::nlim]

Calculate the mean, median, and mode:

NIntegrate[μ fμ, {μ, 10, 20}]   15.4381

FindRoot[NIntegrate[fμ, {μ, 10, medianμ}] == 0.5, {medianμ, 15, 16}]

{medianμ → 15.4381}

FindMaximum[fμ /. μ → modeμ, {modeμ, 15}][[2]]

{modeμ → 15.4381}

Calculate a 95% confidence interval:

q /. {FindRoot[NIntegrate[fμ, {μ, 10, q}] == 0.025, {q, 14, 15}],
  FindRoot[NIntegrate[fμ, {μ, q, 20}] == 0.025, {q, 16, 17}]}
{14.6544, 16.2218}

■ Posterior Densities of σ^{-2}, σ^2, and σ

Calculate the scaling constant of $f(\sigma^{-2} \mid Y)$:

cσ2i = 1 / NIntegrate[pσ2i /. vals, {σ2i, 0, ∞}]   1.30675 × 10^9

The posterior density of σ^{-2} is then the following:

fσ2i = cσ2i pσ2i /. vals

$2.92199 \times 10^8 \, e^{-26.6382 \, \sigma 2i} \, \sigma 2i^{17/2}$

(This is a gamma distribution.) Plot the density:

Plot[fσ2i, {σ2i, 0, 0.8}]

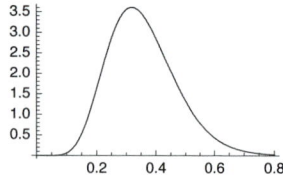

Actually, we are interested in the posterior density of σ^2 or σ. Because the cumulative distribution function of σ^2 is $F_{\sigma^2}(s) = P(\sigma^2 \le s) = P\left(\frac{1}{\sigma^2} \ge \frac{1}{s}\right) = 1 - F_{\sigma^{-2}}\left(\frac{1}{s}\right)$, the density function of σ^2 is $f_{\sigma^2}(s) = \frac{1}{s^2} f_{\sigma^{-2}}\left(\frac{1}{s}\right)$:

```
fσ2 = (fσ2i /. σ2i → 1/σ2) 1/σ2²;

Plot[fσ2, {σ2, 0, 8}]
```

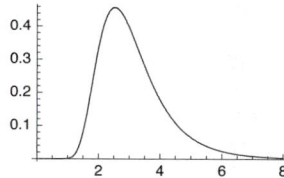

Calculate the posterior density of σ:

```
fσ = (fσ2 /. σ2 → σ²) 2 σ;

plotfσ = Plot[fσ, {σ, 1, 3}, PlotStyle → Black]
```

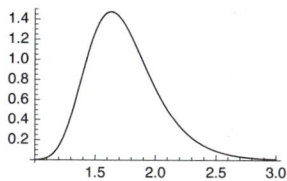

Its mean, median, and mode are as follows:

```
NIntegrate[σ fσ, {σ, 0, 10}]      1.74445

FindRoot[NIntegrate[fσ, {σ, 0, medianσ}] == 0.5, {medianσ, 1, 2}]
{medianσ → 1.70449}

FindMaximum[fσ /. σ → modeσ, {modeσ, 1.5}][[2]]
{modeσ → 1.63212}
```

Calculate a 95% confidence interval:

```
q /. {FindRoot[NIntegrate[fσ, {σ, 0, q}] == 0.025, {q, 1, 1.5}],
   FindRoot[NIntegrate[fσ, {σ, q, 10}] == 0.025, {q, 2, 2.5}]}
{1.27346, 2.44576}
```

30.7.3 Using Interpolation

■ **Introduction**

In Section 30.7.2, we succeeded in obtaining the posterior marginal densities of μ and σ^{-2} by integrating the posterior joint density. If the integration fails, one possibility is to use interpolation to get a close approximation of the marginal densities. In our example, we define anew the posterior joint density that we calculated in Section 39.7.1:

```
p[μ_, σ2i_] = pμσ2i /. vals
```

$$e^{-\frac{1}{2}\left(4819.95-617.523\,\mu+20\,\mu^2\right)\sigma2i}\; \sigma2i^9$$

■ Posterior Density of μ

Suppose that (contrary to reality) *Mathematica* is not able to integrate the posterior joint density with respect to μ or σ^{-2}. We move to numerical integration. To calculate the posterior density of μ, we first define the numerical integral as a function of μ:

```
pμa[μ_ ? NumberQ] := NIntegrate[10^8 p[μ, σ2i], {σ2i, 0.0001, 1}]
```

To help the numerical calculations, we multiplied the function by 10^8 and replaced the upper bound ∞ with 1. Now we can find a representation of the posterior density of μ as an interpolating function (see Section 24.4.2, p. 807):

```
pμ = FunctionInterpolation[pμa[μ], {μ, 10, 20}]
InterpolatingFunction[{{10., 20.}}, <>]
```

Calculate the scaling constant:

```
cμ = 1 / NIntegrate[pμ[μ], {μ, 10, 20}]        5.21308
```

The density and its mean are as follows:

```
fμ = cμ pμ[μ];

Plot[fμ, {μ, 13, 17}]
```

```
NIntegrate[μ fμ, {μ, 10, 20}]        15.4382
```

■ Posterior Densities of σ^{-2} and σ

Similarly, we can calculate the posterior density of σ^{-2}:

```
pσ2ia[σ2i_ ? NumberQ] := NIntegrate[10^8 p[μ, σ2i], {μ, 10, 20}]

pσ2i = FunctionInterpolation[pσ2ia[σ2i], {σ2i, 0.0001, 4}]
InterpolatingFunction[{{0.0001, 4.}}, <>]
```

Calculate the scaling constant:

```
cσ2i = 1 / NIntegrate[pσ2i[σ2i], {σ2i, 0.0001, 4}]        5.21319
```

The density is then as follows:

```
fσ2i = cσ2i pσ2i[σ2i];
```

```
Plot[fσ2i, {σ2i, 0.0001, 0.8}]
```

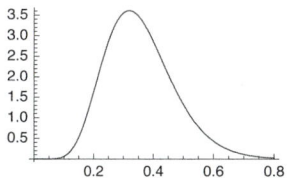

The posterior density of σ and its mean are as follows:

```
fσ = (fσ2i /. σ2i → (1 / σ^2)) 2 / σ^3;
```

```
Plot[fσ, {σ, 1, 3}]
```

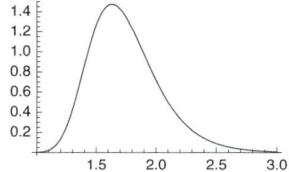

```
NIntegrate[σ fσ, {σ, 0.1, 10}]      1.74446
```

30.7.4 Gibbs Sampling

■ Full Conditional Distributions

Gibbs sampling is based on the *full conditional distributions*. They are the conditional distributions of each parameter given the other parameters and the data. Thus, if the parameters are α, β, γ, ... and we have data Y, then the full conditional distributions are the distributions of $(\alpha \mid \beta, \gamma, \delta, ..., Y)$, $(\beta \mid \alpha, \gamma, \delta, ..., Y)$, $(\gamma \mid \alpha, \beta, \delta, ..., Y)$, ...

Because $f(\alpha \mid \beta, \gamma, ..., Y) = f(\alpha, \beta, \gamma, ... \mid Y) / f(\beta, \gamma, ... \mid Y)$ where the denominator does not depend on α, then $f(\alpha \mid \beta, \gamma, ..., Y) \propto f(\alpha, \beta, \gamma, ... \mid Y)$, which means that $f(\alpha \mid \beta, \gamma, ..., Y)$ is proportional to the terms of the posterior joint density that contain α.

■ Full Conditional Distribution of μ

In Section 30.7.1, p. 1047, $f(\mu, \sigma^{-2} \mid Y)$ was proportional to

```
pμσ2i
```

$$e^{-\frac{1}{2}\kappa^2 (\mu-\xi)^2 - \beta \sigma2i - \frac{1}{2}\left(\text{sum2} - 2\,\text{sum}\,\mu + n\,\mu^2\right)\sigma2i}\; \sigma2i^{-1+\frac{n}{2}+\alpha}$$

Now we want to find full conditional densities $f(\mu \mid \sigma^{-2}, Y)$ and $f(\sigma^{-2} \mid \mu, Y)$. First, $f(\mu \mid \sigma^{-2}, Y)$ is proportional to the terms of **pμσ2i** that contain μ, which makes it proportional to the following:

```
pμ = Select[pμσ2i, ! FreeQ[#, μ] &]
```

$$e^{-\frac{1}{2}\kappa^2 (\mu-\xi)^2 - \frac{1}{2}\left(\text{sum2} - 2\,\text{sum}\,\mu + n\,\mu^2\right)\sigma2i}$$

The exponent is a quadratic polynomial of μ:

```
Collect[pμ[[2]], μ, Simplify]
```

$$\frac{1}{2}\mu^2 \left(-\kappa^2 - n\,\sigma2i\right) + \mu \left(\kappa^2\,\xi + \text{sum}\,\sigma2i\right) + \frac{1}{2}\left(-\kappa^2\,\xi^2 - \text{sum2}\,\sigma2i - 2\,\beta\,\sigma2i\right)$$

Pick the coefficients of μ^2 and μ:

    `{a = Coefficient[%, μ^2], b = Coefficient[%, μ]}`

$$\left\{\frac{1}{2}\left(-\kappa^2 - \text{n } \sigma 2\text{i}\right), \kappa^2 \,\xi + \text{sum } \sigma 2\text{i}\right\}$$

More generally, consider a probability density function $f(x)$ that is proportional to an expression of the form e^{ax^2+bx+c} with $a < 0$. Because $ax^2 + bx + c = \frac{1}{2}\left(x + \frac{b}{2a}\right)^2 / \left(\frac{1}{2a}\right) + d$ for a constant d that does not depend on x, the random variable has a normal distribution with mean $-\frac{b}{2a}$ and variance $-\frac{1}{2a}$. Accordingly, the full conditional distribution of μ is the following normal distribution:

    `fμ = NormalDistribution[` $\frac{-b}{2\,a}$ `,` $\sqrt{\frac{-1}{2\,a}}$ `] // Simplify`

    `NormalDistribution[` $\dfrac{\kappa^2\,\xi + \text{sum }\sigma 2\text{i}}{\kappa^2 + \text{n }\sigma 2\text{i}}$ `,` $\sqrt{\dfrac{1}{\kappa^2 + \text{n }\sigma 2\text{i}}}$ `]`

■ **Full Conditional Distribution of σ^{-2}**

Recall $f(\mu, \sigma^{-2} \mid Y)$:

    `pμσ2i` $e^{-\frac{1}{2}\kappa^2\,(\mu-\xi)^2 - \beta\,\sigma 2\text{i} - \frac{1}{2}\left(\text{sum2} - 2\,\text{sum}\,\mu + \text{n}\,\mu^2\right)\,\sigma 2\text{i}}\,\sigma 2\text{i}^{-1+\frac{n}{2}+\alpha}$

The density function $f(\sigma^{-2} \mid \mu, Y)$ is proportional to the terms of this expression that contain σ^{-2}. The density function of a general gamma distribution is proportional to the following:

    `Select[PDF[GammaDistribution[γ, λ], x], ! FreeQ[#, x] &]` $e^{-\frac{x}{\lambda}}\,x^{-1+\gamma}$

We now see that the full conditional distribution of σ^{-2} is a gamma distribution with the following parameters:

    `{λ = -1 / Coefficient[pμσ2i⟦1, 2⟧, σ2i], γ = 1 + Exponent[pμσ2i, σ2i]} // Simplify`

$$\left\{\frac{2}{\text{sum2} + 2\,\beta + \mu\,(-2\,\text{sum} + \text{n}\,\mu)}, \frac{n}{2} + \alpha\right\}$$

Thus, the full conditional distribution of σ^{-2} is as follows:

    `fσ2i = GammaDistribution[γ, λ] // Simplify`

    `GammaDistribution[` $\dfrac{n}{2} + \alpha$ `,` $\dfrac{2}{\text{sum2} + 2\,\beta + \mu\,(-2\,\text{sum} + \text{n}\,\mu)}$ `]`

■ **Gibbs Sampling**

With Gibbs sampling, we need to generate random numbers (or sample or draw) from all of the full conditional distributions. This is easy if we can infer that each of these distributions is one of the well-known distributions (as in our example) and we have a random number generator for it.

    Gibbs sampling is a method to sample the posterior joint distribution. The sample is then used to infer the form of the marginal densities. First, select initial values $\alpha_0, \beta_0, \ldots$ for the parameters. In the ith step, draw from each full conditional distribution in turn, using the values of the previous step and the values already drawn in the current step. Thus, draw from $f(\alpha \mid \beta_{i-1}, \gamma_{i-1}, \delta_{i-1}, \ldots, Y)$, $f(\beta \mid \alpha_i, \gamma_{i-1}, \delta_{i-1}, \ldots, Y)$, $f(\gamma \mid \alpha_i, \beta_i, \delta_{i-1}, \ldots, Y)$, $\ldots$ in turn until you have a random number from all full conditional distributions. Then go to the next step.

The values of the parameters from steps 0, 1, 2, and so on in Gibbs sampling form a Markov chain. It can be shown that the chain converges to the posterior joint distribution and that the iterative sampling scheme draws—in the limit—a value from this distribution. In practice, a sample from the posterior joint distribution can be obtained by deleting some early draws and then considering the remaining draws to be—approximately—draws from the posterior joint distribution. For Gibbs sampling, see Gamerman (1997) or Green (2001).

■ **An Example**

In our example, we first insert the numerical values of the constants into the full conditional densities:

```
gμ = fμ /. vals // PowerExpand
```

$$\text{NormalDistribution}\left[15.4381, \frac{1}{2\sqrt{5}\ \sqrt{\sigma 2i}}\right]$$

```
go2i = fo2i /. vals
```

$$\text{GammaDistribution}\left[10, \frac{2}{4819.95 + \mu\,(-617.523 + 20\,\mu)}\right]$$

We initialize Gibbs sampling with the sample values:

```
rμ = Mean[data]; ro2i = 1 / Variance[data];
```

Then we do 11,000 iterations of Gibbs sampling and drop the first 1000 points:

```
SeedRandom[8];
gibbs = Drop[Table[{rμ = RandomReal[gμ /. σ2i → ro2i],
    ro2i = RandomReal[go2i /. μ → rμ]}, {11 000}], 1000];
```

The points are as follows:

```
ListPlot[gibbs, AxesOrigin → {13, 0}, PlotStyle → PointSize[Tiny], ImageSize → 200]
```

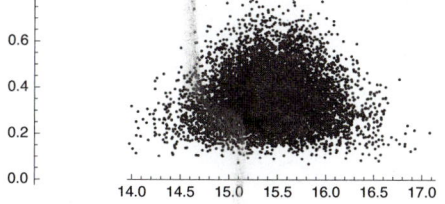

■ **An Illustration**

To illustrate the method, we plot the first 10 points (the starting point is large in the plot):

```
{gibbsμ, gibbso2i} = gibbs⊤;
μμ = Flatten[Partition[Take[gibbsμ, 11], 2, 1]];
σσ =
  Most[Prepend[Flatten[Partition[Take[gibbso2i, 11], 2, 1]], First[gibbso2i]]];
μμσσ = {μμ, σσ}⊤;
```

```
Graphics[{Point[Take[μμσσ, {1, 20, 2}]], Line[Most[μμσσ]], Red,
   AbsolutePointSize[5], Point[First[μμσσ]]}, Axes → True, AxesLabel → {"μ", "σ⁻²"},
   AxesOrigin → {14.7, 0.1}, AspectRatio → 1 / GoldenRatio, ImageSize → 200]
```

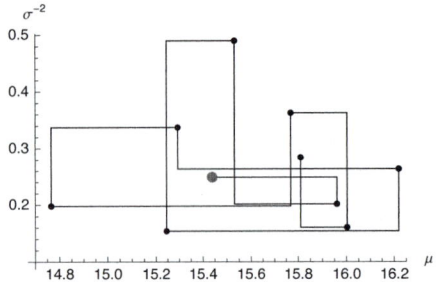

■ **Posterior Histograms of μ and σ**

The draws for μ and σ are as follows:

```
gibbsσ = 1 / Sqrt[gibbsσ2i];
```

```
ppp = ListPlot[{gibbsμ, gibbsσ}ᵀ, AxesOrigin → {13, 0.7},
   PlotStyle → PointSize[Tiny], PlotRange → All, ImageSize → 200]
```

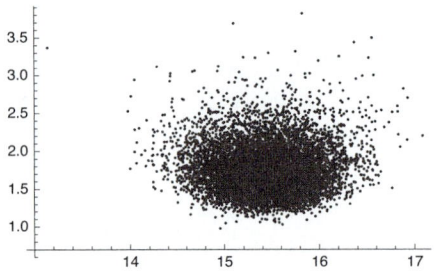

The histograms for μ and σ are as follows:

```
<< Histograms`
```

```
{hμ =
   Histogram[gibbsμ, HistogramScale → 1, HistogramCategories → Range[13, 18, 0.1]],
 hσ = Histogram[gibbsσ, HistogramScale → 1, HistogramCategories → Range[0, 5, 0.1]]}
```

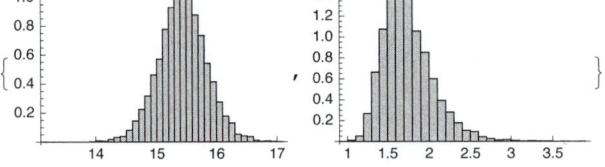

If we compare the histograms with the exact densities that we calculated in Section 30.7.2, p. 1049, we can see a close agreement, which confirms that Gibbs sampling is a working method:

```
{Show[hμ, plotfμ], Show[hσ, plotfσ]}
```

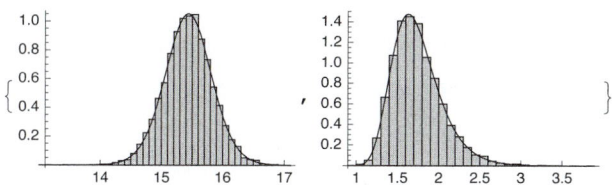

The means and variances of the samples are

    `{Mean[gibbsμ], Variance[gibbsμ]}`   `{15.4336, 0.155218}`

    `{Mean[gibbsσ], Variance[gibbsσ]}`   `{1.74527, 0.0918638}`

For another example of Gibbs sampling, see tutorial/RandomNumberGeneration.

30.7.5 Markov Chain Monte Carlo

■ The MCMC Method

The major advantage of MCMC is that we do not need to *draw* from the full conditional distributions as in Gibbs sampling; we only need to *evaluate* the full conditional densities at given points, and even the scaling constants are not needed. For MCMC, see Gilks, Richardson, and Spiegelhalter (1996), Gamerman (1997), or Green (2001).

MCMC has some variants. The basic method is called the *Metropolis method,* and a generalization is called the *Metropolis–Hastings method.* Actually, Gibbs sampling is also a variant of MCMC. We describe here the *random walk Metropolis method.*

Suppose the parameters of interest are α, β, γ, …, and let $\theta = (\alpha, \beta, \gamma, \ldots)$, which is a *d*-vector. We have a posterior joint distribution $f(\theta \mid Y)$ from which we want to draw in order to estimate the posterior marginal distributions of the parameters [in the literature, $f(\theta \mid Y)$ is often denoted by $\pi(\theta)$ and called the target distribution]. Let the successive draws be denoted as θ_1, θ_2, …. If we already have draws θ_1, …, θ_i, then the next draw θ_{i+1} is obtained as follows.

Choose an arbitrary (really!) *d*-variate distribution that is symmetric around zero (a *d*-variate normal distribution with mean zero, for example), and draw a random number w_i from this density. Calculate $\phi_i = \theta_i + w_i$, which is called the *proposal.* The proposal is accepted with probability $\min\{1, f(\phi_i \mid Y) / f(\theta_i \mid Y)\}$ [this can be implemented by drawing a random number that is uniform on $(0, 1)$]. If the proposal is accepted, we set $\theta_{i+1} = \phi_i$, but if the proposal is rejected, we set $\theta_{i+1} = \theta_i$.

It can be shown that this Metropolis method generates a Markov chain that converges to the posterior joint distribution, and the iterative sampling scheme draws—in the limit—a value from this distribution. In practice, a sample from the posterior joint distribution can be obtained by deleting some early draws and then considering the remaining draws to be—approximately—draws from the posterior joint distribution.

■ Programming the MCMC Method

We continue studying the model from Section 30.7.1, p. 1046. The posterior joint density $p(\mu, \sigma^{-2} \mid Y)$ was shown to be proportional to the following:

```
p[μ_, σ2i_] = pμσ2i /. vals
```

$$e^{-\frac{1}{2}\left(4819.95-617.523\,\mu+20\,\mu^2\right)\,\sigma2i}\,\sigma2i^9$$

We calculate the ratio needed in the acceptance probability of the Metropolis method:

```
ratio[{μ1_, σ2i1_}, {μ2_, σ2i2_}] = p[μ2, σ2i2] / p[μ1, σ2i1] // Simplify
```

$$\frac{e^{\left(2409.98-308.761\,\mu1+10.\,\mu1^2\right)\,\sigma2i1+\left(-2409.98+308.761\,\mu2-10.\,\mu2^2\right)\,\sigma2i2}\,\sigma2i2^9}{\sigma2i1^9}$$

Let $\theta = (\mu, \sigma^{-2})$. We generate the proposals with a two-variate normal distribution that has independent components with means of zero and known standard deviations:

```
proposal[θ_, std_] := θ + (RandomReal[NormalDistribution[0, #]] & /@ std)
```

The proposals may need to satisfy some restrictions. For example, a draw for σ^{-2} has to be positive, so in our example we write the following test:

```
test[ϕ_] := If[ϕ[[2]] > 0, True, False]
```

The Metropolis method can then be written as follows:

```
metropolisStep[θ_, std_] := With[{ϕ = proposal[θ, std], r = RandomReal[]},
  If[test[ϕ], If[r ≤ ratio[θ, ϕ], ϕ, θ], θ]]
metropolis[initialState_, std_, steps_] :=
  NestList[metropolisStep[#, std] &, initialState, steps]
```

Here, `metropolisStep` calculates one step of the method, and `metropolis` repeats the step a total of `steps` times. In this way, we get a sequence of random vectors.

■ **Adjusting the Standard Deviations**

The working of MCMC is greatly affected by the standard deviations of the distribution that gives the proposals. In our example, we first try the standard deviations 2 (used to draw μ) and 0.5 (used to draw σ^{-2}). We start from the sample estimates and generate 500 draws:

```
m = Mean[data]; s2i = 1 / Variance[data];
```

```
SeedRandom[2]; {metroμ, metroσ2i} = metropolis[{m, s2i}, {2, 0.5}, 500]ᵀ;
```

The draws for μ are as follows:

```
ListPlot[metroμ, PlotStyle → PointSize[Tiny], AspectRatio → 0.2, ImageSize → 400]
```

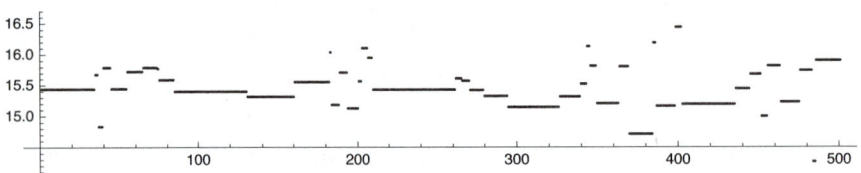

We see that the chain stays for long amounts of time in the same state before going to a new state. Calculate the different values of the chain:

```
Length[Union[metroμ]]    45
```

The small number of different values is an indication that the standard deviations are too large: A large standard deviation allows for large values for the proposal ϕ_i, and then the ratio $p(\phi_i \mid Y)/p(\theta_i \mid Y)$ is often small, which results in small acceptance probabilities and frequent rejection of the proposal.

We try smaller standard deviations 0.5 and 0.1:

```
SeedRandom[2];
{metroμ, metroσ2i} = metropolis[{m, s2i}, {0.5, 0.1}, 500]ᵀ;
ListPlot[metroμ, PlotStyle → PointSize[Tiny], AspectRatio → 0.2, ImageSize → 400]
```

```
Length[Union[metroμ]]     246
```

Now the values vary more, and approximately half of the proposals are accepted. We then use even smaller standard deviations 0.1 and 0.02:

```
SeedRandom[2];
{metroμ, metroσ2i} = metropolis[{m, s2i}, {0.1, 0.02}, 500]ᵀ;
ListPlot[metroμ, PlotStyle → PointSize[Tiny], ImageSize → 400, AspectRatio → 0.2]
```

```
Length[Union[metroμ]]     450
```

Now approximately 90% of proposals are accepted, but the sequence has a high autocorrelation, and it moves too slowly to various parts of the domain of the posterior distribution of μ, which means that a long sequence will be needed to get a satisfactory estimate of the posterior density.

In conclusion, experimenting is often needed to adjust the standard deviations of the proposal distribution. The standard deviations should be small enough so that a sufficient number of proposals will be accepted but large enough so that the autocorrelation is not too large.

For a manipulation to demonstrate the effect of standard deviations, see demonstrations.wolfram.com/MarkovChainMonteCarloSimulationUsingTheMetropolisAlgorithm   by Philip Gregory.

Next, we use the standard deviations 0.5 and 0.1 and study MCMC in more detail.

■ Using the Results

We calculate 21,000 steps and reject the first 1000, which are considered to be transient:

```
SeedRandom[2];
metro = Drop[metropolis[{m, s2i}, {0.5, 0.1}, 21 000], 1000];
```

Calculate the number of different points:

```
Length[Union[metro]]     10 208
```

The points are displayed as follows:

```
ListPlot[metro, PlotStyle → PointSize[Tiny], PlotRange → All,
 AxesLabel → {"μ", "σ⁻²"}, AxesOrigin → {13.5, 0}, ImageSize → 200]
```

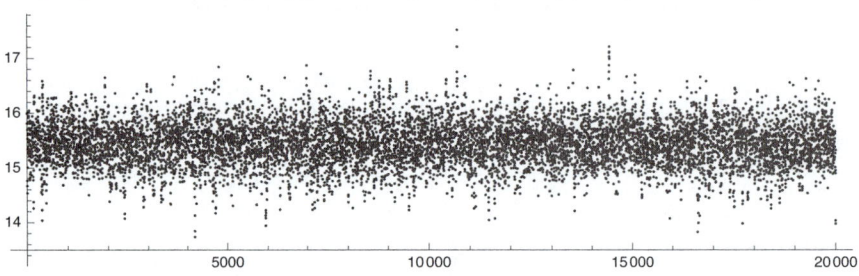

The draws for μ and σ^{-2} are as follows:

```
{metroμ, metroσ2i} = metroᵀ;
```

```
ListPlot[metroμ, AspectRatio → 0.3, PlotRange → All,
 PlotStyle → PointSize[Tiny], AxesOrigin → {0, 13.5}, ImageSize → 400]
```

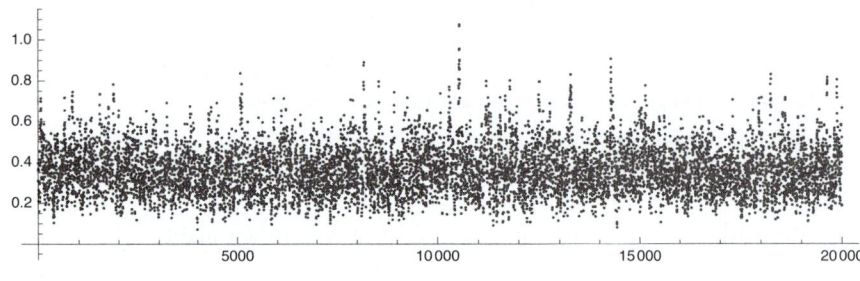

```
ListPlot[metroσ2i, AspectRatio → 0.3, PlotRange → All,
 PlotStyle → PointSize[Tiny], AxesOrigin → {0, 0}, ImageSize → 400]
```

```
{Mean[metroμ], Mean[metroσ2i]}     {15.4427, 0.360577}
```

Show the draws for μ and σ:

```
metroσ = 1 / Sqrt[metroσ2i];
```

```
ListPlot[{metroμ, metroσ}ᵀ, PlotStyle → PointSize[Tiny], PlotRange → All,
  AxesLabel → {"μ", "σ"}, AxesOrigin → {13.5, 0.8}, ImageSize → 200]
```

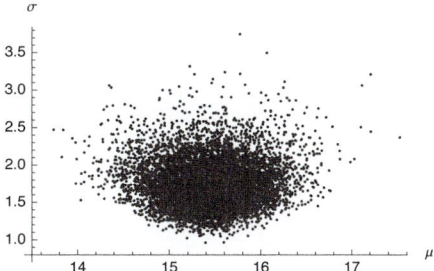

The histograms for μ and σ are as follows:

```
<< Histograms`
```

```
{hμ =
  Histogram[metroμ, HistogramScale → 1, HistogramCategories → Range[13, 18, 0.1]],
  hσ = Histogram[metroσ, HistogramScale → 1, HistogramCategories → Range[0, 5, 0.1]]}
```

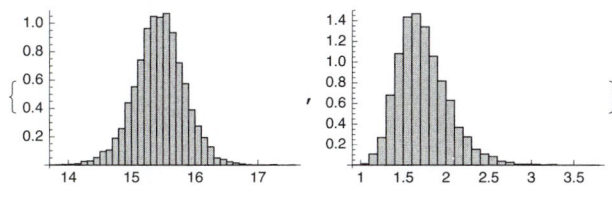

If we compare the histograms with the exact densities that we calculated in Section 30.7.2, p. 1049, we can see a close agreement, which confirms that MCMC is a working method:

```
{Show[hμ, plotfμ], Show[hσ, plotfσ]}
```

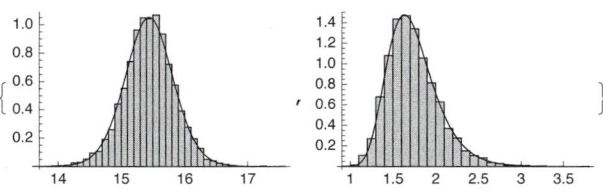

References

The references can also be found from the Index.

Abell, M. L., and J. P. Braselton (1997). *Differential Equations with Mathematica,* 2nd ed. Academic Press, Boston.

Abell, M. L., J. P. Braselton, and J. A. Rafter (1999). *Statistics with Mathematica.* Academic Press, Boston.

Allen, A. O. (1990). *Probability, Statistics, and Queueing Theory: With Computer Science Applications,* 2nd ed. Academic Press, Boston.

Bhatti, M. A. (2000). *Practical Optimization Methods with Mathematica Applications.* Springer, New York.

Borrelli, R. L., and C. S. Coleman (1998). *Differential Equations: A Modeling Perspective.* Wiley, New York.

Bulmer, M., and M. Carter (1996). Integer programming with Mathematica. *Mathematica Journal* 6(3), 28–36.

Burghes, D. N., and M. S. Borrie (1981). *Modelling with Differential Equations.* Horwood, Chichester, UK.

Cheng, A. H.-D., P. Sidauruk, and Y. Abousleiman (1994). Approximate inversion of the Laplace transform. *Mathematica Journal* 4(2), 76–82.

Cleveland, W. S. (1993). *Visualizing Data.* Hobart Press, Summit, NJ.

Cleveland, W. S. (1994). *The Elements of Graphing Data.* Hobart Press, Summit, NJ.

Cox, D. R., and H. D. Miller (1965). *The Theory of Stochastic Processes.* Methuen, London.

Dennemeyer, R. (1968). *Introduction to Partial Differential Equations and Boundary Value Problems.* McGraw-Hill, New York.

Dickau, R. M. (1997). Compilation of iterative and list operations. *Mathematica Journal* 7(1), 14–15.

Gamerman, D. (1997): *Markov Chain Monte Carlo: Stochastic Simulation for Bayesian Inference.* Chapman & Hall, London.

Ganzha, V. G., and E. V. Vorozhtsov (1996). *Numerical Solutions for Partial Differential Equations: Problem Solving Using Mathematica.* CRC Press, Boca Raton, FL.

Gilks, W. R., S. Richardson, and D. J. Spiegelhalter (Eds.) (1996). *Markov Chain Monte Carlo in Practice.* Chapman & Hall, London.

Giordano, F. R., M. D. Weir, and W. P. Fox (1997). *A First Course in Mathematical Modeling,* 2nd ed. Brooks/Cole, Pacific Grove, CA.

Gray, J. W. (1997). *Mastering Mathematica: Programming Methods and Applications,* 2nd ed. Academic Press, Boston.

Green, P. J. (2001): A primer on Markov chain Monte Carlo. In O. E. Barndorff-Nielsen, D. R. Cox, and C. Klüpperberg (Eds.), *Complex Stochastic Systems.* Chapman & Hall, Boca Raton, FL.

Guttorp, P. (1995). *Stochastic Modeling of Scientific Data.* Chapman & Hall, London.

Hastings, K. J. (2006). *Introduction to the Mathematics of Operations Research with Mathematica,* 2nd ed. Chapman & Hall/CRC Press, Boca Raton, FL.

Johnson, N. L., S. Kotz, and A. W. Kemp (1992). *Univariate Discrete Distributions,* 2nd ed. Wiley, New York.

Kaplan, D., and L. Glass (1995). *Understanding Nonlinear Dynamics.* Springer, New York.

Kelley, W. G., and A. C. Peterson (2001). *Difference Equations: An Introduction with Applications,* 2nd ed. Academic Press, San Diego.

Knapp, R., and M. Sofroniou (1997). Difference equations and chaos in Mathematica: Symbolic and numerical mathematics at work. *Dr. Dobb's Journal* 22(11), 84–90, 95–99. (A version of the article can be found as item number 0209-012 in *MathSource*.)

Kulenovic, M. R. S., and O. Merino (2002). *Discrete Dynamical Systems and Difference Equations with Mathematica.* Chapman & Hall/CRC Press, Boca Raton, FL.

Kulkarni, V. G. (1995). *Modeling and Analysis of Stochastic Systems.* Chapman & Hall, London.

Kythe, P. K., P. Puri, and M. R. Schäferkotter (1996). *Partial Differential Equations and Mathematica.* CRC Press, Boca Raton, FL.

MacHale, D. (1993). *Comic Sections: The Book of Mathematical Jokes, Humour, Wit and Wisdom.* Boole Press, Dublin.

Maeder, R. E. (1994). *The Mathematica Programmer.* Academic Press, Boston.

Maeder, R. E. (1995a). Single-image stereograms. *Mathematica Journal* 5(1), 50–61.

Maeder, R. E. (1995b). Function iteration and chaos. *Mathematica Journal* 5(2), 28–40.

Maeder, R. E. (1997). *Programming in Mathematica,* 3rd ed. Addison-Wesley, Reading, MA.

Martelli, M. (1999). *Introduction to Discrete Dynamical Systems and Chaos.* Wiley, New York.

Mesterton-Gibbons, M. (1989). *A Concrete Approach to Mathematical Modelling.* Addison-Wesley, Redwood City, CA.

Murrell, H. (1994). Planar phase plots and bifurcation animations. *Mathematica Journal* 4(3), 76–81.

Nachbar, R. B. (1995). Genetic programming. *Mathematica Journal* 5(3), 36–47.

Pearl, R. (1927). The growth of populations. *Quarterly Review of Biology* 2(4), 532–548.

Rohatgi, V. K. (1984). *Statistical Inference.* Wiley, New York.

Rose, C., and M. D. Smith (2002). *Mathematical Statistics with Mathematica.* Springer, New York.

Rowland, E. S. (2008). A natural prime-generating recurrence. *Journal of Integer Sequences* 11, Article 08.2.8, 1–13.

Ruskeepää, H. (2007). Mathematica. In L. Hogben (Ed.), *Handbook of Linear Algebra.* Chapman & Hall/CRC Press, Boca Raton, FL.

Ruskeepää, H. (2008a): Mathematica: Introductory examples related to Ramanujan, Part 1. *Mathematics Newsletter* 18(2), 33–57.

Ruskeepää, H. (2008b): Mathematica: Introductory examples related to Ramanujan, Part 2. *Mathematics Newsletter* 18(3), 69–91.

Sandefur, J. T. (1990). *Discrete Dynamical Systems: Theory and Applications.* Clarendon, Oxford.

Schwalbe, D., and S. Wagon (1997). *VisualDSolve: Visualizing Differential Equations with Mathematica.* Springer/TELOS, New York.

Shaw, W. T., and J. Tigg (1994). *Applied Mathematica: Getting Started, Getting It Done.* Addison-Wesley, Reading, MA.

Skeel, R. D., and J. B. Keiper (2001). *Elementary Numerical Computing with Mathematica.* Stipes, Champaign, IL.

Smith, C., and N. Blachman (1995). *The Mathematica Graphics Guidebook.* Addison-Wesley, Reading, MA.

Spiegel, M. R. (1971). *Schaum's Outline of Theory and Problems of Calculus of Finite Differences and Difference Equations.* McGraw-Hill, New York.

Spiegel, M. R. (1999). *Mathematical Handbook of Formulas and Tables,* 2nd ed. McGraw-Hill, New York.

Szabo, F. (2000). *Linear Algebra: An Introduction Using Mathematica.* Academic Press, San Diego.

Szabo, F. (2001). *Student Solutions Manual for Linear Algebra: An Introduction Using Mathematica.* Academic Press, San Diego.

Trott, M. (2004a). *The Mathematica Guidebook for Graphics.* Springer, New York.

Trott, M. (2004b). *The Mathematica Guidebook for Programming.* Springer, New York.

Trott, M. (2006a). *The Mathematica Guidebook for Numerics.* Springer, New York.

Trott, M. (2006b). *The Mathematica Guidebook for Symbolics.* Springer, New York.

Wagner, D. B. (1995). Dynamic programming. *Mathematica Journal* 5(4), 42–51.

Wagner, D. B. (1996). *Power Programming with Mathematica: The Kernel.* McGraw-Hill, New York.

Wagon, S. (2000). *Mathematica in Action,* 2nd ed. Springer/TELOS, New York.

Wellin, P., R. Gaylord, and S. Kamin (2005). *An Introduction to Programming with Mathematica.* Cambridge University Press, Cambridge, UK.

Wickham-Jones, T. (1994). *Mathematica Graphics.* Springer, New York.

Winston, W. L. (1994). *Operations Research: Applications and Algorithms,* 3rd ed. Duxbury Press, Belmont, CA.

Wolfram, S. (2002). *A New Kind of Science.* Wolfram Media, Champaign, IL.

Wrede, R. C., and Spiegel, M. R. (2002). *Schaum's Outline of Theory and Problems of Advanced Calculus.* McGraw-Hill, New York.

Index

Mathematica names are written in this style: **Factorial**; menu commands are shown like this: **Abort Evaluation**. Names that begin with a lowercase letter are either names of programs, such as **bifurca⌐ tion**, or names of data sets, such as **barley**. Names such as **FunctionApproximations`** are packages. The adjectives one-, two-, three-, and four-dimensional are written as 1D, 2D, 3D, and 4D, respectively. *Mathematica* names beginning with **$** are listed at the end of the index.

Note that some of the names listed here are in packages. Such names are denoted by a bullet after the name, such as **AdjustedRSquared•**. To use such a command, you first have to load the correct package, which is explained in Section 4.1.1.

Note also that if a cell in a *Mathematica* document is too long to fit at the bottom of a page, *Mathematica* may divide the cell into two parts, and the parts are printed on consecutive pages. However, index entries are associated with whole cells, and *Mathematica* has adopted the convention that the page number of an index entry will correspond with the page where the cell *ends*. Thus, the page numbers of the index entries that are associated with a divided cell and are located at the bottom of the first page are actually one less than their page number in the index.

About the CD-ROM

The CD-ROM of *Mathematica Navigator* contains

- **the entire book**, easily installable into the Help Browser;
- material that describes the new properties of **Mathematica 7**, also easily installable into the Help Browser; and
- **all the data sets** discussed in the book and several other data sets.

The CD-ROM also contains instructions for installation and usage. When you have installed the book into the Help Browser of *Mathematica*,

- the entire book is accessible from within *Mathematica* so that you can easily read sections of the book, experiment with the examples, interact with manipulations and animations, see the figures in color, and copy material from the book;
- the index entries contain hyperlinks to the correct positions in the book so that you can easily find information about various topics; and
- references in the text to other parts of the book are also hyperlinks that directly lead you to the appropriate points of the book.

The CD-ROM can be read using Windows, Macintosh, and Linux computers.

Here is a short installation instruction for the book:

- Drag the **MathematicaNavigator3** and **MathematicaNavigator3NewIn7** folders from the CD-ROM into the **Applications** folder found in `$UserBaseDirectory`.